CLIMATES OF THE STATES

National Oceanic and Atmospheric Administration
Narrative Summaries, Tables, and Maps
for Each State

with

Current Tables of Normals, 1941-1970
Means and Extremes to 1975
Overview of State Climatologist Programs

New Material by
James A. Ruffner

Volume 2
Nebraska-Wyoming
Puerto Rico and U.S. Virgin Islands
Hurricane Data
The State Climatologist Program 1954-1973
Appendixes

Gale Research Company
BOOK TOWER • DETROIT, MICHIGAN 48226

Bibliographic Note

Materials on the individual states were originally issued in parts by the United States Weather Bureau in the series, *Climates of the States, Climatography of the United States No. 60: parts 1-52* (part 50 vacant), 1959-60. The series was partially revised and partially reprinted by the Environmental Science Services Administration, 1967-1972. The parts were compiled for this edition in a revised standard format and augmented to include the most recent data for normals, means, and extremes from *Local Climatological Data* issued by the National Oceanic and Atmospheric Administration, 1975.

Library of Congress Cataloging in Publication Data

Main entry under title:

Climates of the States, with current tables of
    normals 1941-1970 and means and extremes to
    1975.

    Based on Climatology of the United States,
no. 60, issued in 1959-60 , by the U. S. Weather
Bureau and data from the National Oceanic and
Atmospheric Administration.
    1. United States--Climate--Tables. I. Ruffner,
James A. II. Bair, Frank E.
QC983.C55   551.6′9′730212   76-11672
ISBN 0-8103-1042-2

TABLE OF CONTENTS

CLIMATES OF THE STATES

NEBRASKA

(Normals, Means and Extremes tables revised 1973 and 1975. Basic report revised December 1959.)

Climate of Nebraska

W. R. Stevens, Weather Bureau State Climatologist

Nebraska, one of the Great Plains States, is located in the north-central portion of the United States. The area of the State is 76,653 square miles, of which about 600 are water. On the eastern boundary, along the Missouri River, the elevation rised from less than 900 feet in the southeast to 1,200 feet in the northeast. The elevation also increases westward to about 3,000 feet in the southwest and 5,000 feet in the northwest. The landscape changes from level or gently rolling prairie in the east, to rounded sandhills in the north-central part, and thence westward to high plains.

All of Nebraska is drained by the Missouri River System. The direction of flow is mostly west to east, but in the southeastern section the flow is from northwest to southeast. The Missouri River forms the eastern boundary of the State and a part of the northern boundary. The major tributary is the Platte River, with its two main branches which rise in the high elevations of Colorado. Other important tributaries are the Niobrara River in the north and the Republican and Big Blue Rivers in the south.

Greatest volume of flow occurs during May, June, and July, the months of heaviest rainfall. Although the heaviest snowfall occurs in February and March, it usually does not accumulate to any considerable depth and so the resultant runoff does not materially affect river stages.

The climate is typical of the interior of large continents in middle latitudes; that is, rather light rainfall, low humidity, hot summers, cold winters, great variations in temperature and rainfall from year to year, and frequent changes in weather from day to day. The rapid changes in weather are brought about by invasion of large masses of air of different characteristics, such as warm, moist air from the Gulf of Mexico; hot, dry air from the Southwest; cool, dry air from the north Pacific Ocean; and cold, dry air from northwestern Canada.

The Rocky Mountains to the west have a profound influence on the climate of Nebraska. Air crossing the mountains from the west loses much of its moisture on the windward side and becomes warmer and drier as it lescends on the eastern slopes; therefore, no significant amount of moisture which falls as rain or snow reaches the State from the Pacific Ocean. The moisture supply for precipitation comes from the Gulf of Mexico. The remoteness from the source of supply is one of the reasons for the wide variation in rainfall from year to year. Moist air from the Gulf is often deflected eastward before it reaches Nebraska. Downslope winds from the Rocky Mountains occasionally cause large, rapid changes to higher temperatures, particularly during the winter.

Although hot nights in summer occur rather frequently in the east, they are almost unknown in the higher elevations of the western, less humid, part of the State where rapid cooling after sunset generally occurs.

The mean annual temperature varies from about 53°F. along the eastern half of the southern border to about 45°F. in the northwest corner.

Maximum temperatures above 100°F. have occurred throughout the State in the months of June, July, August, and September. Temperatures of 110°F. or higher have been recorded over most of the State, except in parts of the northwest, the highest being 118°F. at Geneva, Hartington, and Minden. Minimum temperatures of zero or below occur on an average about 10 days a year in the southeast and 25 days in the northwest. Minima below -40°F. have been recorded a few times at northern and western stations, the lowest being -47°F. on February 12, 1899, at Camp Clarke, near Bridgeport. Although the winter climate is classed as cold, there are frequent periods of mild, pleasant weather.

The average date of the last freeze (32°F.) in spring ranges from about April 25 in the extreme southeast to about May 21 in a small area in the northwest portion, while the first in the fall varies from about October 6 in the southeast to about September 20 in the extreme northwest. Hence the average length of the growing season (freeze free season) ranges from 164 days in the southeast to 122 in the northwest. There is considerable variation, however, in the length of the growing season from year to year. Stations in the southeast show a difference of about 50 days between the shortest and longest growing seasons, while the difference is as much as 100 days at places in the north-central portion of the State.

The average annual precipitation in the eastern third of the State is about 27 inches; in the central third, about 22 inches; and in the western third, about 18 inches. The amount decreases rather uniformly from 33 inches in the southeast corner to about 14 inches in a small area near the western border. On the average nearly 80 percent of the yearly total falls in the 6 months from April to September. During July and August, rainfall normally diminishes slowly in the east portion. In the west it decreases more rapidly, so that the August average is only a little over one-half the June average in many localities. Consequently, relatively little corn is grown on unirrigated land in the western portion.

Since the economy of the State is largely agricultural and precipitation is often insufficient for good crop yields in some sections, extensive areas are now irrigated. In 1953 more than 1,200,000 acres were under irrigation, an increase of 25 percent in 5 years. Although the largest irrigation projects are in the western half of the State, nearly every county has some irrigated land.

Excessive rates of rainfall for short periods occur frequently in summer thundershowers. Since they damage crops and erode the soil, these excessively heavy showers are agriculturally unfavorable. In some seasons thundershowers are numerous and well distributed, but sometimes they are scattered and infrequent. The result is great variability in the monthly amounts of rainfall in different years and also in the annual amounts from year to year. In dry years, periods of 15 to 20 days without appreciable rain may occur in June, July, and August; and under such conditions, hot, dry winds often cause serious and extensive damage to crops.

The precipitation records show successions of wet and dry epochs as follows:

1876-1892, Wet period with one very dry year.

1893-1901, Dry period with one rather wet year.

1902-1909, Wet period with one rather dry year.

1910-1920, Tendency irregular; most years dry, but 1915 was the wettest year of record.

1921-1940, Long dry period with only one wet year; especially dry after 1930.

1941-1951, Wet period with one dry year.

1952-1956, Moderately dry period.

Floods may be expected once or twice in most years in smaller streams in the eastern third of the State, but less frequently over the west and central portions, and are generally caused by short duration, high intensity rainfall. Severe flooding occurs infrequently on the Missouri and is usually caused by rapid melting of heavy snowpacks in the upper portion of the basin, attended by moderate to heavy rains. Notable flood years in Nebraska are 1881, 1935, 1939, 1943, 1944, 1947, 1950 and 1952.

The average seasonal snowfall is approximately 29 inches. Snowfall usually increases during the late winter and reaches a maximum in March over most of the State. The higher regions in the west portion frequently have heavy snows in April, and occasionally in May.

Sunshine for the year averages about 65 percent of the possible amount, ranging from about 55 percent in December to nearly 80 percent in July.

There are frequent changes in wind direction at all seasons of the year, but the prevailing direction is from the south or southeast from May to September, and from the northwest or north during the remainder of the year, except that westerly winds predominate in the southwest portion during the autumn and winter months. The average is about 9 miles per hour.

A few tornadoes occur within the State nearly every year; the average is about 10 per year based on the 1916-58 period. The number of fatalities averages less than one for each storm. Although tornadoes are usually very small, both in width and in length of path, there is almost total destruction where the whirling funnel cloud touches the ground.

The number of hailstorms averages between 20 and 25 per year, occurring mostly in June, July, and August. A few each year cause severe agricultural losses.

Industry is limited mostly to processing farm products or production of materials used on the farm. The largest industry is centered around the stockyards, slaughter houses, and packing houses. Other important industries are grain milling, beet sugar refining, cement production, food processing and canning, oil and gas production, manufacture of farm implements, wire fencing, electric motors, small internal-combustion engines, communications equipment, chemicals, animal medicines, biologicals and serums, stock feed.

The principal crops in descending order of acreage are corn, hay, winter wheat, and oats. Receipts from farm marketing show cattle far in the lead, followed by hogs, wheat, corn, and dairy products. The total number of cattle on farms is about 4 3/4 million, hogs about 2 million, and sheep about 800 thousand. Sugar beets, dry edible beans, and potatoes are important crops in irrigated areas, where most of the yields are very high compared with unirrigated areas. For example, some areas produce 300 to 400 bushels of potatoes to the acre.

The main recreational activities are fishing and the hunting of small game, such as pheasants, quail, and migratory water fowl.

REFERENCES

(1) Weather Bureau Technical Paper No. 16-Maximum 24-Hour Precipitation in the United States. Washington, D. C. 1952.

(2) Weather Bureau Technical Paper No. 25-Rainfall Intensity-Duration-Frequency Curves. For selected stations in the United States, Alaska, Hawaiian Islands, and Puerto Rico.

BIBLIOGRAPHY

(A) Climatic Summary of the United States (Bulletin W) 1930 edition, Sections 38 and 39. U. S. Weather Bureau

(B) Climatic Summary of the United States, Nebraska - Supplement for 1931 through 1952 (Bulletin W Supplement). U. S. Weather Bureau

(C) Climatological Data - Nebraska. U. S. Weather Bureau

(D) Climatological Data National Summary. U. S. Weather Bureau

(E) Hourly Precipitation Data - Nebraska. U. S. Weather Bureau

(F) Local Climatological Data, U. S. Weather Bureau, for Grand Island, Lincoln, Norfolk, North Platte, Omaha, Scottsbluff, and Valentine, Nebraska

FREEZE DATA

STATION	Freeze threshold temperature	Mean date of last Spring occurrence	Mean date of first Fall occurrence	Mean No. of days between dates	Years of record Spring	No. of occurrences in Spring	Years of record Fall	No. of occurrences in Fall
ALBION	32	05-06	10-01	148	31	31	31	31
	28	04-27	10-09	165	31	31	31	31
	24	04-13	10-20	190	31	31	31	31
	20	04-01	10-30	212	31	31	31	31
	16	03-24	11-09	230	31	31	31	31
ALLIANCE	32	05-13	09-28	138	26	26	27	27
	28	04-27	10-10	166	26	26	27	27
	24	04-21	10-19	181	26	26	27	27
	20	04-12	10-26	197	26	26	27	27
	16	04-02	11-06	217	26	26	27	27
BEATRICE NO 1	32	04-26	10-13	170	30	30	30	30
	28	04-10	10-24	197	30	30	30	30
	24	03-30	11-02	218	30	30	30	30
	20	03-23	11-12	234	30	30	30	30
	16	03-15	11-23	254	30	30	30	30
BEAVER CITY	32	05-05	10-07	155	31	31	31	31
	28	04-22	10-19	180	31	31	31	31
	24	04-10	10-30	203	31	31	31	31
	20	03-31	11-06	220	31	31	31	31
	16	03-22	11-14	237	31	31	31	31
BRIDGEPORT	32	05-16	09-24	130	31	31	31	31
	28	05-02	10-02	153	31	31	31	31
	24	04-23	10-17	177	31	31	31	31
	20	04-13	10-25	195	31	31	31	31
	16	04-01	10-31	213	31	31	31	31
BROKEN BOW	32	05-11	09-26	138	31	31	31	31
	28	04-30	10-04	158	31	31	31	31
	24	04-19	10-19	182	31	31	31	31
	20	04-04	10-28	208	31	31	31	31
	16	03-28	11-05	222	31	31	31	31
CHADRON CAA AP	32	05-10	10-01	144	31	31	31	31
	28	04-29	10-10	164	31	31	31	31
	24	04-17	10-20	186	31	31	31	31
	20	04-11	10-25	198	31	31	31	31
	16	04-03	11-05	216	31	31	31	31
CLAY CENTER	32	04-30	10-11	163	31	31	31	31
	28	04-18	10-21	185	31	31	31	31
	24	04-03	11-01	212	31	31	31	31
	20	03-28	11-05	222	31	31	31	31
	16	03-22	11-16	239	30	30	31	31
CRETE	32	04-24	10-14	173	31	31	31	31
	28	04-09	10-27	201	31	31	31	31
	24	03-31	11-04	219	31	31	31	31
	20	03-25	11-11	232	31	31	31	31
	16	03-19	11-23	249	31	31	31	31
CULBERTSON	32	05-08	10-02	147	31	31	31	31
	28	04-26	10-12	169	31	31	31	31
	24	04-12	10-27	199	31	31	31	31
	20	04-01	11-03	216	31	31	31	31
	16	03-24	11-07	228	31	31	31	31
DAVID CITY	32	04-27	10-12	169	31	31	29	29
	28	04-18	10-20	185	31	31	29	29
	24	04-02	10-30	211	30	30	29	29
	20	03-26	11-06	225	30	30	29	29
	16	03-21	11-16	240	30	30	29	29
FAIRBURY	32	04-25	10-15	174	31	31	31	31
	28	04-11	10-26	198	31	31	31	31
	24	03-31	11-04	217	31	31	31	31
	20	03-24	11-13	234	31	31	31	31
	16	03-16	11-27	256	31	31	31	30
FRANKLIN	32	05-02	10-06	157	31	31	31	31
	28	04-18	10-19	183	31	31	31	31
	24	04-03	10-30	210	31	31	31	31
	20	03-28	11-04	221	31	31	31	31
	16	03-21	11-14	237	31	31	31	31
FREMONT	32	04-30	10-10	163	31	31	31	31
	28	04-17	10-18	185	31	31	31	31
	24	04-04	10-28	207	31	31	31	31
	20	03-26	11-06	225	31	31	31	31
	16	03-20	11-16	240	31	31	31	31
GENEVA	32	04-29	10-11	165	31	31	31	31
	28	04-18	10-24	188	31	31	31	31
	24	04-04	11-01	211	31	31	31	31
	20	03-28	11-11	229	31	31	31	31
	16	03-20	11-22	247	31	31	31	31
GORDON	32	05-25	09-16	114	30	30	30	30
	28	05-12	09-26	136	30	30	30	30
	24	05-02	10-09	160	30	30	30	30
	20	04-21	10-16	178	29	29	30	30
	16	04-11	10-24	196	29	29	30	30
GOTHENBURG	32	05-09	10-02	146	31	31	31	31
	28	04-26	10-12	169	31	31	31	31
	24	04-13	10-23	193	31	31	31	31
	20	04-01	11-03	217	31	31	31	31
	16	03-25	11-07	228	31	31	31	31
HARRISON	32	05-22	09-21	122	31	31	30	30
	28	05-09	09-29	143	31	31	30	30
	24	04-30	10-09	162	31	31	30	30
	20	04-21	10-21	183	31	31	30	30
	16	04-10	10-27	200	31	31	30	30
HARTINGTON	32	05-03	10-07	158	31	31	31	31
	28	04-21	10-19	180	31	31	31	31
	24	04-08	10-27	202	31	31	31	31
	20	04-01	11-02	215	31	31	31	31
	16	03-25	11-12	232	31	31	31	31
HAY SPRINGS	32	05-18	09-25	131	31	31	31	31
	28	05-08	10-05	150	31	31	31	31
	24	04-25	10-15	173	31	31	31	31
	20	04-17	10-23	189	31	31	31	31
	16	04-06	10-31	208	31	31	31	31
HEBRON	32	04-28	10-11	167	31	31	31	31
	28	04-13	10-23	193	31	31	31	31
	24	04-02	11-01	213	30	30	31	31
	20	03-27	11-10	227	30	30	31	31
	16	03-20	11-22	248	30	30	31	30
HOLDREGE	32	04-29	10-14	169	31	31	31	31
	28	04-16	10-23	191	31	31	31	31
	24	04-04	11-04	214	31	31	31	31
	20	03-28	11-08	225	31	31	31	31
	16	03-22	11-16	239	31	31	31	31
HYANNIS	32	05-15	09-23	131	23	23	24	24
	28	05-05	10-06	154	23	23	24	24
	24	04-22	10-16	177	23	23	24	24
	20	04-11	10-24	196	23	23	24	24
	16	04-01	11-06	219	23	23	24	24
IMPERIAL	32	05-09	10-05	150	31	31	31	31
	28	04-24	10-16	175	31	31	31	31
	24	04-14	10-24	193	31	31	31	31
	20	04-05	11-02	211	31	31	31	31
	16	03-30	11-08	223	31	31	31	31
KIMBALL	32	05-14	09-27	135	31	31	31	31
	28	05-03	10-06	156	31	31	31	31
	24	04-22	10-17	178	30	30	31	31
	20	04-13	10-24	194	30	30	31	31
	16	04-03	11-02	213	30	30	31	31
LINCOLN AGRO FARM	32	04-20	10-17	180	31	31	31	31
	28	04-07	10-27	204	31	31	31	31
	24	03-29	11-05	221	31	31	31	31
	20	03-21	11-15	239	31	31	31	31
	16	03-14	11-27	259	31	31	31	30

FREEZE DATA

STATION	Freeze threshold temperature	Mean date of last Spring occurrence	Mean date of first Fall occurrence	Mean No. of days between dates	Years of record Spring	No. of occurrences in Spring	Years of record Fall	No. of occurrences in Fall	STATION	Freeze threshold temperature	Mean date of last Spring occurrence	Mean date of first Fall occurrence	Mean No. of days between dates	Years of record Spring	No. of occurrences in Spring	Years of record Fall	No. of occurrences in Fall
LOUP CITY	32	05-08	10-01	146	30	30	29	29	PAWNEE CITY	32	04-26	10-12	170	31	31	31	31
	28	04-28	10-11	166	30	30	29	29		28	04-11	10-23	195	31	31	31	31
	24	04-13	10-22	192	30	30	29	29		24	04-01	10-31	213	31	31	31	31
	20	04-01	11-01	214	30	30	29	29		20	03-24	11-12	233	30	30	31	31
	16	03-25	11-08	227	29	29	29	29		16	03-15	11-22	252	30	30	31	31
MADRID	32	05-10	10-04	147	31	31	31	31	RAVENNA	32	05-09	09-29	142	31	31	30	30
	28	04-28	10-13	168	31	31	21	31		28	04-28	10-08	163	31	31	30	30
	24	04-14	10-24	193	31	31	30	30		24	04-10	10-22	195	31	31	30	30
	20	04-05	10-31	210	31	31	30	30		20	04-01	11-02	215	31	31	30	30
	16	03-29	11-07	223	31	31	30	30		16	03-27	11-08	226	31	31	30	30
MERRIMAN	32	05-18	09-21	126	31	31	30	30	RED CLOUD	32	05-01	10-04	156	31	31	31	31
	28	05-05	10-01	148	31	31	30	30		28	04-20	10-16	179	31	31	31	31
	24	04-25	10-10	168	31	31	30	30		24	04-04	10-29	208	31	31	31	31
	20	04-15	10-23	191	31	31	30	30		20	03-30	11-05	220	31	31	31	31
	16	04-04	11-02	212	31	31	30	30		16	03-22	11-14	237	31	31	31	31
MINDEN	32	05-04	10-09	159	30	30	31	31	SAINT PAUL	32	05-04	10-03	151	31	31	31	31
	28	04-18	10-20	185	30	30	31	31		28	04-22	10-13	174	31	31	31	31
	24	04-06	10-31	207	30	30	31	31		24	04-09	10-25	199	31	31	31	31
	20	03-30	11-06	221	29	29	31	31		20	03-28	11-03	219	31	31	31	31
	16	03-24	11-13	234	29	29	31	31		16	03-24	11-11	232	31	31	31	31
NEWPORT	32	05-08	09-30	145	31	31	31	31	SCOTTSBLUFF WB AP	32	05-14	09-26	135	31	31	31	31
	28	04-27	10-10	165	31	31	31	31		28	05-01	10-06	158	31	31	31	31
	24	04-18	10-19	185	31	31	31	31		24	04-21	10-17	178	31	31	31	31
	20	04-09	10-29	202	31	31	31	31		20	04-13	10-25	195	31	31	31	31
	16	03-30	11-08	223	31	31	31	31		16	04-02	11-02	213	31	31	31	31
NIOBRARA SANTEE	32	05-04	10-02	152	30	30	28	28	STANTON	32	05-04	10-04	153	31	31	31	31
	28	04-24	10-10	169	29	29	28	28		28	04-26	10-13	170	31	31	31	31
	24	04-13	10-23	193	29	29	28	28		24	04-12	10-24	196	31	31	31	31
	20	04-02	11-01	212	28	28	28	28		20	04-01	11-01	214	31	31	31	31
	16	03-25	11-09	228	28	28	28	28		16	03-24	11-06	227	31	31	31	31
NORFOLK	32	05-04	10-03	152	31	31	31	31	TECUMSEH	32	04-30	10-08	161	31	31	31	31
	28	04-23	10-10	171	31	31	31	31		28	04-16	10-17	184	31	31	31	31
	24	04-10	10-24	197	31	31	31	31		24	04-04	10-31	210	31	31	30	30
	20	04-01	11-02	215	31	31	31	31		20	03-27	11-07	225	31	31	30	30
	16	03-25	11-07	227	31	31	31	31		16	03-18	11-19	246	31	31	30	30
NORTH LOUP	32	05-07	10-01	147	31	31	31	31	TEKAMAH	32	05-02	10-13	164	30	30	31	31
	28	04-26	10-11	168	31	31	31	31		28	04-15	10-20	188	30	30	31	31
	24	04-12	10-22	193	31	31	31	31		24	04-02	10-29	210	30	30	31	31
	20	04-03	11-01	212	31	31	31	31		20	03-27	11-06	223	30	30	31	31
	16	03-25	11-07	227	31	31	31	31		16	03-20	11-17	242	30	30	31	31
NORTH PLATTE 1 E	32	04-30	10-07	160	31	31	31	31	VALENTINE LAKES GAME RE	32	05-07	09-30	146	14	14	14	14
	28	04-21	10-18	181	31	31	31	31		28	04-24	10-09	168	14	14	14	14
	24	04-05	10-30	208	31	31	31	31		24	04-14	10-20	190	14	14	14	14
	20	03-29	11-04	220	31	31	31	31		20	04-12	10-28	199	14	14	14	14
	16	03-22	11-08	232	31	31	31	31		16	04-02	11-05	217	14	14	14	14
OAKDALE	32	05-05	10-01	149	31	31	31	31	WAKEFIELD	32	05-07	10-02	148	31	31	31	31
	28	04-28	10-10	165	31	31	31	31		28	04-25	10-12	169	31	31	31	31
	24	04-16	10-24	192	31	31	31	31		24	04-13	10-23	193	31	31	30	30
	20	04-02	10-30	211	31	31	31	31		20	04-01	11-01	214	31	31	30	30
	16	03-27	11-05	224	31	31	31	31		16	03-26	11-07	226	31	31	30	30
OSCEOLA	32	05-01	10-07	159	27	27	28	28	WEEPING WATER	32	05-02	10-08	159	30	30	30	30
	28	04-21	10-19	181	27	27	28	28		28	04-17	10-15	181	30	30	30	30
	24	04-04	10-31	210	27	27	28	28		24	04-07	10-25	201	30	30	30	30
	20	03-28	11-06	223	27	27	28	28		20	03-29	11-01	216	30	30	30	30
	16	03-21	11-12	236	27	27	28	28		16	03-21	11-13	237	30	30	30	30
OSHKOSH	32	05-07	10-03	150	31	31	31	31	WESTPOINT	32	05-02	10-08	159	30	30	31	31
	28	04-26	10-14	171	31	31	31	31		28	04-19	10-17	181	30	30	31	31
	24	04-17	10-23	189	31	31	31	31		24	04-07	10-27	203	30	30	31	31
	20	04-06	10-31	209	31	31	31	31		20	03-30	11-05	220	30	30	31	31
	16	03-29	11-06	223	31	31	31	31		16	03-21	11-13	236	30	30	31	31
									YORK	32	05-02	10-09	160	31	31	31	31
										28	04-17	10-21	187	31	31	31	31
										24	04-04	10-30	209	31	31	31	31
										20	03-28	11-06	223	31	31	31	31
										16	03-21	11-15	238	31	31	31	31

Data in the above table are based on the period 1921-1950, or that portion of this period for which data are available.

Means have been adjusted to take into account years of non-occurrence.

A freeze is a numerical substitute for the former term "killing frost" and is the occurrence of a minimum temperature at or below the threshold temperature of 32°, 28°, etc.

Freeze data tabulations in greater detail are available and can be reproduced at cost.

*MEAN TEMPERATURE AND PRECIPITATION

STATION	JANUARY Temperature	JANUARY Precipitation	FEBRUARY Temperature	FEBRUARY Precipitation	MARCH Temperature	MARCH Precipitation	APRIL Temperature	APRIL Precipitation	MAY Temperature	MAY Precipitation	JUNE Temperature	JUNE Precipitation	JULY Temperature	JULY Precipitation	AUGUST Temperature	AUGUST Precipitation	SEPTEMBER Temperature	SEPTEMBER Precipitation	OCTOBER Temperature	OCTOBER Precipitation	NOVEMBER Temperature	NOVEMBER Precipitation	DECEMBER Temperature	DECEMBER Precipitation	ANNUAL Temperature	ANNUAL Precipitation
EAST CENTRAL DIVISION																										
ASHLAND 3 NE		.76		.97		1.32		2.54		3.33		4.91		3.02		3.82		3.33		1.58		1.20		.87		27.65
AURORA		.61		.81		1.10		2.28		3.62		4.07		2.57		2.87		2.40		1.10		.88		.70		23.01
BLAIR	22.5	.89	26.2	1.09	36.6	1.60	51.5	2.50	62.1	3.27	72.3	4.26	78.0	2.85	75.3	3.19	66.5	2.63	55.5	1.62	38.7	1.26	27.2	.91	51.0	26.07
BRADSHAW		.86		.75		1.09		2.58		3.43		4.56		2.69		2.89		2.24		1.29		.88		.70		23.96
CENTRAL CITY	24.2	.62	28.1	.77	37.2	1.24	51.0	2.08	61.3	3.82	72.1	3.85	78.7	3.06	76.5	2.67	67.0	2.34	55.0	1.02	38.6	.91	28.2	.72	51.5	23.20
COLUMBUS	23.9		27.5		37.6		51.5		62.1		72.5		78.6		76.3		67.4		55.5		39.0		28.0		51.7	
DAVID CITY	23.1	.97	27.1	1.05	36.6	1.50	50.7	2.34	61.5	3.67	72.0	4.90	78.3	2.56	76.0	3.06	67.0	2.63	55.2	1.30	38.3	1.15	27.5	.80	51.1	25.93
FREMONT	23.4	1.11	27.4	1.20	37.7	1.74	51.9	2.49	62.6	3.37	72.7	4.45	78.6	3.02	76.3	3.50	67.6	2.65	56.0	1.53	39.2	1.09	27.9	.93	51.8	27.08
GENOA	23.8	.64	27.6	.82	37.5	1.12	51.3	2.11	61.9	3.52	72.1	4.51	78.3	2.69	76.1	2.57	67.1	2.25	55.2	1.06	38.7	.87	27.7	.78	51.4	22.94
LINCOLN AGRO FARM	25.4	.84	29.7	.98	39.4	1.52	52.9	2.52	63.3	3.38	74.0	4.48	80.0	2.65	77.7	3.78	69.1	2.76	57.4	1.61	40.8	1.35	29.8	.92	53.3	26.79
LINCOLN COL VIEW		.86		1.00		1.69		2.63		3.29		4.37		2.78		3.50		2.78		1.57		1.34		.94		26.75
LINCOLN WB CITY	25.0	.82	29.3	.92	39.4	1.47	52.6	2.29	62.5	3.10	72.8	4.10	79.2	3.10	76.9	3.08	68.1	2.92	56.4	1.66	40.3	1.41	29.1	.86	52.7	25.73
MC COOL JUNCTION		.82		.88		1.25		2.41		3.61		4.66		3.08		3.40		2.51		1.25		1.19		.84		25.90
OMAHA WB AP	23.0	.85	27.2	.87	38.2	1.32	52.2	2.15	62.7	2.75	72.6	4.51	78.5	3.34	75.7	3.12	67.0	3.16	55.5	1.73	38.9	1.29	27.4	.81	51.6	25.90
OMAHA N OMAHA AP	20.6	.85	24.7	.87	35.9	1.32	50.0	2.15	60.6	2.75	70.5	4.51	76.4	3.34	73.8	3.12	65.2	3.16	53.7	1.73	37.0	1.29	25.3	.81	49.5	25.90
OSCEOLA	23.6	.81	27.5	1.00	37.5	1.22	51.1	2.33	61.9	3.64	72.3	3.96	79.1	2.81	76.5	2.67	67.7	2.40	55.4	1.21	39.1	1.02	28.0	.82	51.6	23.89
SCHUYLER		.87		1.01		1.47		2.63		3.54		4.57		2.84		3.46		2.45		1.18		.98		.88		25.88
SEWARD	24.5	.86	28.7	.92	38.1	1.45	51.8	2.43	61.8	3.39	72.4	4.81	78.6	3.25	75.9	3.54	67.3	2.48	55.7	1.56	39.0	1.20	28.6	.96	51.9	26.85
UTICA		.68		.87		1.11		2.25		3.57		4.65		3.12		3.40		2.35		1.31		1.01		.79		25.11
WAHOO		.99		1.06		1.50		2.64		3.29		4.77		2.75		3.70		2.63		1.62		1.18		.91		27.04
WEEPING WATER	22.9	.86	27.5	.99	37.6	1.74	51.6	2.67	62.0	3.64	72.5	5.14	78.0	2.77	75.2	4.42	66.7	2.89	55.1	1.62	38.6	1.40	27.6	.96	51.3	29.10
YORK	25.1	.79	29.0	.91	38.3	1.23	51.9	2.43	62.0	3.79	72.8	4.87	79.1	3.31	76.9	3.19	67.7	2.42	55.9	1.32	39.6	1.05	29.0	.87	52.3	26.18
DIVISION	23.8	.83	27.8	.96	37.5	1.40	51.5	2.41	62.0	3.55	72.4	4.57	78.6	2.93	76.2	3.30	67.3	2.62	55.6	1.40	39.0	1.12	28.1	.85	51.7	25.94
SOUTHWEST DIVISION																										
BENKELMAN	29.5	.40	33.7	.35	40.4	.98	52.0	1.77	61.5	2.72	71.8	2.59	78.9	2.07	76.9	2.25	67.4	1.52	55.0	.68	40.4	.47	32.0	.39	53.3	16.19
CULBERTSON	27.4	.44	32.2	.52	39.3	1.21	51.5	1.94	61.4	3.20	71.8	3.19	78.9	2.61	76.4	2.50	66.7	1.68	54.0	.87	38.6	.68	29.7	.45	52.3	19.29
CURTIS	26.8	.45	31.1	.47	39.0	1.26	51.1	1.95	60.8	3.38	71.0	3.47	77.9	2.21	75.8	2.38	65.8	1.58	53.7	.92	38.1	.55	29.2	.51	51.7	19.13
HAYES CENTER	27.7	.60	31.1	.63	37.8	1.42	49.8	2.31	59.9	3.41	70.2	3.30	77.9	2.55	75.9	2.11	66.5	1.94	54.6	.94	39.1	.70	30.5	.49	51.8	20.40
IMPERIAL	27.3	.42	31.1	.46	37.5	1.06	49.4	1.94	59.1	3.38	69.6	3.36	77.4	2.19	75.2	2.23	65.5	1.79	53.2	.80	38.4	.54	29.9	.45	51.1	18.62
MADRID	26.6	.51	30.7	.50	37.5	1.38	49.5	2.29	59.6	3.01	70.3	3.17	77.9	2.44	75.6	2.20	65.9	1.84	53.1	.73	37.9	.60	29.1	.54	51.1	19.21
MC COOK	28.4	.50	32.7	.59	40.2	1.35	52.3	2.06	61.6	3.12	71.7	3.17	78.6	2.80	76.3	2.30	66.8	1.70	54.9	.87	39.9	.76	31.0	.53	52.9	19.75
NORTH PLATTE WB AP	24.0	.39	28.7	.34	36.4	.91	48.7	2.06	58.9	2.67	68.8	2.92	75.8	2.40	[illegible]	2.11	63.8	1.75	51.7	1.04	36.8	.55	27.2	.40	49.5	17.54
OGALLALA	26.5	.44	30.4	.52	37.3	1.05	49.6	2.08	59.4	2.91	69.4	2.71	76.9	2.34	74.4	2.32	64.7	1.76	52.4	.73	37.7	.52	29.1	.50	50.7	17.88
STRATTON		.48		.44		1.23		2.25		3.29		3.11		2.61		2.10		1.62		.87		.60		.45		19.05
WAUNETA		.50		.61		1.27		1.99		3.18		3.10		2.12		2.46		1.79		.70		.65		.49		18.86
DIVISION	27.4	.46	31.5	.50	38.5	1.17	50.5	1.99	60.4	3.05	70.5	3.04	77.9	2.38	75.7	2.24	66.2	1.73	53.9	.81	38.8	.59	30.2	.47	51.8	18.43
SOUTH CENTRAL DIV																										
ALMA	26.9	.41	31.4	.58	39.9	.95	52.9	2.27	62.7	3.21	73.5	3.66	80.0	2.86	77.9	2.48	68.5	2.19	56.0	1.03	39.8	.77	29.9	.49	53.3	20.90
BEAVER CITY	28.2	.55	32.3	.60	40.6	1.11	53.1	2.09	62.5	3.64	73.1	3.34	79.9	2.93	77.7	2.37	68.2	1.72	56.0	.96	40.3	.78	30.8	.52	53.6	20.61
ELWOOD 9 SSW		.40		.45		1.03		2.10		3.67		3.97		2.43		2.36		1.99		1.04		.71		.46		20.61
FRANKLIN	27.4	.48	31.4	.68	40.2	1.13	52.7	2.24	62.3	3.05	72.6	3.87	79.2	2.79	77.5	2.80	67.5	2.22	55.9	1.11	40.2	.83	30.8	.62	53.1	21.82
GUIDE ROCK		.46		.69		1.16		2.17		3.38		4.52		2.98		2.85		2.44		1.22		.91		.67		23.45
HASTINGS	26.2	.59	30.1	.84	38.7	1.27	51.5	2.39	61.7	3.60	72.4	3.93	79.1	2.48	77.0	3.00	67.9	2.31	56.2	1.13	40.0	.98	29.9	.72	52.6	23.24
HOLDREGE	27.0	.56	31.0	.71	39.6	1.15	52.3	2.31	62.1	3.98	72.9	4.05	79.9	2.48	77.8	2.31	68.3	2.38	55.9	1.11	39.7	.85	30.0	.62	53.0	22.51
MINDEN	26.1	.49	30.2	.64	38.7	1.11	51.3	2.17	61.4	3.91	72.0	4.02	79.1	2.65	77.1	2.54	67.7	2.31	55.8	.97	39.4	.85	29.5	.54	52.4	22.20
RED CLOUD	26.2	.58	30.7	.87	39.8	1.25	52.8	2.04	62.6	3.33	73.5	4.00	79.8	2.98	77.8	2.48	68.2	2.53	55.7	1.21	39.5	.95	29.6	.76	53.0	22.98
UPLAND		.66		.73		1.04		2.29		3.31		4.32		2.66		2.42		2.40		1.08		.89		.62		22.42
DIVISION	26.7	.52	31.1	.68	39.4	1.14	52.2	2.18	62.0	3.56	72.6	3.84	79.2	2.77	77.4	2.53	67.9	2.22	55.8	1.07	39.7	.86	29.8	.60	52.8	21.97
SOUTHEAST DIVISION																										
AUBURN		1.03		1.21		2.22		2.96		4.52		5.65		3.23		4.56		3.96		2.22		1.56		1.19		34.31
BEATRICE NO 1	24.7	.70	29.3	.96	38.3	1.56	52.0	2.33	62.2	3.79	73.2	4.96	79.5	3.09	77.2	4.04	67.9	3.25	55.6	1.64	39.3	1.20	28.8	.85	52.3	28.37
BRUNING		.73		.85		1.23		2.55		4.02		4.70		2.48		3.46		2.58		1.31		1.23		.97		26.11
CLAY CENTER	25.2	.59	29.8	.82	38.7	1.10	51.7	2.19	61.4	3.44	72.5	3.87	78.9	2.32	76.9	2.86	68.2	2.35	55.9	1.23	39.6	.97	29.2	.78	52.3	22.52
CRETE	24.9	.76	29.3	1.03	39.0	1.47	52.4	2.60	62.6	3.81	73.0	4.87	79.4	2.75	77.1	3.65	68.2	2.87	56.6	1.53	39.7	1.31	29.0	.91	52.6	27.56
FAIRBURY	26.3	.70	30.9	1.02	40.3	1.50	53.4	2.34	63.3	4.10	74.2	4.94	80.4	3.15	78.3	4.09	69.1	2.98	57.5	1.53	40.9	1.27	30.3	.92	53.7	28.54
FAIRMONT	23.3	.66	27.7	.85	36.9	1.13	50.5	2.48	61.2	3.46	72.3	4.63	78.9	2.90	76.4	3.88	67.3	2.60	55.3	1.36	38.5	1.16	28.1	.82	51.4	25.93
FALLS CITY	26.9	1.16	31.0	1.18	40.3	2.23	53.1	3.16	62.8	4.85	73.2	5.45	79.0	2.96	76.6	4.53	68.4	3.75	57.4	2.18	41.4	1.74	30.9	1.32	53.4	34.51
GENEVA	24.6	.81	28.7	1.12	37.8	1.40	50.9	2.64	61.2	3.64	72.2	4.50	78.8	2.67	76.4	3.82	67.4	2.46	55.7	1.43	39.0	1.25	28.7	.95	51.8	26.69
HEBRON	25.1	.64	29.5	.92	38.5	1.28	51.8	2.51	61.8	4.11	73.2	5.20	79.4	2.79	77.4	3.25	67.8	2.97	55.9	1.43	39.5	1.17	29.3	.89	52.4	27.16
NEBRASKA CITY 1 WNW	25.7	.83	30.0	1.00	39.5	1.84	53.2	2.55	63.0	3.86	72.9	5.57	78.5	3.21	76.1	3.88	68.3	3.12	57.6	2.10	41.1	1.32	30.1	.89	53.0	30.17
NELSON		.64		.91		1.28		2.40		3.49		4.58		2.68		2.88		2.60		1.22		1.06		.84		24.58
PAWNEE CITY	27.0	.87	31.4	1.04	40.8	1.89	53.7	2.60	63.5	4.32	74.1	5.39	80.0	2.79	77.9	4.39	69.1	3.68	57.9	2.11	41.5	1.34	30.9	1.06	54.0	31.48
SUPERIOR		.58		.80		1.24		2.16		3.43		4.60		2.90		2.85		2.64		1.23		1.09		.76		24.28
SYRACUSE	25.3	.82	29.7	.95	39.5	1.70	53.0	2.71	63.0	3.85	73.6	4.96	79.4	2.87	77.2	3.77	68.5	3.01	56.9	1.78	40.4	1.43	29.5	.90	53.0	28.75
TABLE ROCK 5 N		.83		.96		1.83		2.54		3.78		5.00		2.98		4.20		3.64		1.89		1.24		.93		29.82
TECUMSEH	25.4	.99	29.8	1.11	39.3	1.90	52.4	2.49	62.0	4.33	72.7	5.38	78.4	3.26	76.0	4.09	67.3	3.67	55.9	1.90	40.1	1.41	29.4	1.02	52.4	31.55
WESTERN		.64		.90		1.15		2.29		3.61		4.44		2.95		3.83		2.82		1.39		1.16		.83		26.01
DIVISION	25.5	.80	29.9	1.01	39.1	1.61	52.4	2.57	62.4	3.97	73.1	5.00	79.2	2.90	77.0	3.84	68.1	3.09	56.5	1.69	40.2	1.30	29.6	.95	52.8	28.73

* **Averages for period 1931-1955, except for stations marked WB which are "normals" based on period 1921-1950. Divisional means may not be the arithmetical average of individual stations published, since additional data from shorter period stations are used to obtain better areal representation.**

*MEAN TEMPERATURE AND PRECIPITATION

STATION	JANUARY		FEBRUARY		MARCH		APRIL		MAY		JUNE		JULY		AUGUST		SEPTEMBER		OCTOBER		NOVEMBER		DECEMBER		ANNUAL	
	Temperature	Precipitation	Temperature	Precipitation	Temperature	Precipitation	Temperature	Precipitation	Temperature	Precipitation	Temperature	Precipitation	Temperature	Precipitation	Temperature	Precipitation	Temperature	Precipitation	Temperature	Precipitation	Temperature	Precipitation	Temperature	Precipitation	Temperature	Precipitation
PANHANDLE DIVISION																										
ALLIANCE	25.3	.44	28.7	.43	34.4	.91	46.1	1.95	56.2	2.79	66.0	2.81	74.2	1.72	72.1	1.61	62.1	1.34	50.0	.78	35.6	.47	28.2	.38	48.2	15.63
BOX BUTTE EXP FARM	23.7	.41	31.2	.44	32.9	.81	44.6	1.85	54.5	2.87	64.4	2.71	73.4	1.51	71.4	1.49	61.2	1.30	48.9	.77	34.3	.50	26.4	.37	47.2	15.03
BRIDGEPORT	26.2	.38	30.2	.32	36.9	.78	48.2	1.49	57.5	2.52	66.9	3.13	74.6	1.95	72.7	1.40	63.1	1.40	51.2	.63	36.7	.42	28.9	.36	49.4	14.78
CHADRON CAA AP	24.5	.48	28.1	.42	34.5	1.00	47.0	2.01	56.8	2.90	66.6	2.84	76.0	1.51	73.7	1.31	63.3	1.34	51.1	.84	36.5	.47	28.1	.43	48.9	15.55
DALTON		.39		.43		.96		1.90		2.79		3.34		1.79		1.80		1.55		.82		.62		.42		16.81
FORT ROBINSON	24.7	.49	27.7	.60	33.7	1.21	45.8	2.30	55.4	3.30	65.1	3.20	74.0	1.71	72.0	1.61	61.7	1.53	50.0	.93	35.7	.62	28.2	.43	47.8	17.93
GORDON	22.6	.52	26.4	.42	32.8	1.13	46.0	1.73	55.8	2.46	65.5	2.95	73.7	2.17	71.3	1.82	61.3	1.45	49.4	1.01	34.3	.54	25.9	.40	47.1	16.60
HARRISBURG 10 NW		.38		.41		.84		1.84		2.60		2.65		1.46		1.29		1.28		.72		.54		.46		14.47
HARRISON	23.1	.62	26.0	.70	31.5	1.41	43.4	2.35	52.9	3.24	62.5	2.98	72.0	1.73	70.0	1.37	60.0	1.45	48.4	.91	34.0	.69	26.2	.59	45.8	18.04
HAY SPRINGS	23.4	.91	26.9	.77	32.8	1.43	45.4	2.17	55.1	2.99	64.9	3.15	73.7	2.10	71.7	1.68	61.7	1.57	49.5	.96	34.8	.77	26.4	.70	47.2	19.20
KIMBALL	27.0	.42	30.0	.49	35.4	1.10	45.7	1.91	55.2	2.78	65.4	2.80	72.8	2.40	70.8	1.90	61.6	1.31	50.0	.68	36.7	.64	29.7	.54	48.4	16.97
LODGEPOLE	27.6	.37	30.8	.44	36.8	.98	48.3	2.07	57.8	2.90	67.8	3.09	76.1	2.15	73.8	2.20	64.2	1.55	52.0	.80	37.6	.57	29.9	.51	50.2	17.63
LYMAN		.37		.46		.88		2.04		2.56		2.63		1.17		1.03		1.07		.83		.52		.46		14.02
MITCHELL 5 E	25.0	.26	28.4	.24	34.8	.62	46.3	1.50	56.0	2.21	65.5	2.99	73.9	1.30	71.4	1.21	61.6	1.18	50.1	.67	35.9	.38	28.2	.32	48.1	12.88
OSHKOSH	26.0	.39	30.1	.37	37.1	.91	49.1	1.91	58.9	2.56	68.3	2.83	76.2	2.07	73.6	1.88	64.1	1.50	52.2	.80	37.1	.54	28.8	.40	50.1	16.16
OSHKOSH 8 SW		.36		.49		1.16		2.37		2.95		2.73		2.04		1.89		1.58		.82		.64		.52		17.55
SCOTTSBLUFF WB AP	23.5	.33	28.2	.41	34.9	.88	46.1	1.99	56.2	2.64	66.5	2.78	74.6	1.45	72.4	1.22	62.2	1.29	50.5	1.01	36.1	.57	27.0	.43	48.2	15.00
SIDNEY	26.0	.32	29.5	.32	36.0	.85	46.8	1.90	56.6	2.74	66.4	2.99	74.8	1.67	72.4	1.85	62.7	1.32	50.7	.76	36.6	.48	28.8	.45	48.9	15.65
DIVISION	25.1	.46	28.6	.46	34.6	1.01	46.5	1.93	56.1	2.80	65.8	2.97	74.2	1.93	72.1	1.64	62.2	1.42	50.3	.82	35.8	.53	28.0	.43	48.3	16.40
NORTH CENTRAL DIV																										
AINSWORTH	23.2	.49	26.4	.69	34.2	1.24	48.1	2.05	59.0	3.24	68.9	3.95	77.1	2.06	74.6	2.44	64.6	1.88	52.7	1.10	37.0	.50	27.1	.53	49.4	20.17
ARTHUR	24.1	.33	27.8	.37	35.0	.82	47.9	2.03	57.2	2.88	66.7	3.13	74.8	2.24	72.4	2.09	63.0	1.59	50.7	.86	35.7	.42	27.3	.39	48.6	17.15
BREWSTER		.51		.59		1.18		2.05		3.02		3.44		2.13		2.49		2.12		1.14		.48		.52		19.67
BURWELL		.34		.45		1.01		1.76		3.48		3.54		2.10		1.94		1.91		1.04		.56		.43		18.56
BUTTE	21.5	.61	24.6	.86	33.5	1.48	48.4	2.16	59.2	3.38	69.7	4.07	77.4	2.55	75.2	2.94	65.2	1.90	52.9	1.12	36.8	.69	26.0	.64	49.2	22.40
CHAMBERS	22.0	.50	25.4	.59	34.2	1.20	48.2	1.88	58.9	3.51	68.7	3.91	75.7	2.05	73.7	2.34	64.0	1.94	52.6	1.36	36.2	.58	25.8	.53	48.8	20.39
ERICSON 6 WNW		.59		.67		1.24		2.14		3.56		3.77		2.62		2.24		1.97		1.05		.76		.61		21.22
EWING	21.2	.51	24.6	.64	34.3	1.19	48.7	1.97	59.6	3.46	69.8	4.15	76.4	2.71	74.1	2.94	64.3	1.97	52.2	1.30	35.7	.72	25.1	.59	48.8	22.15
HALSEY 2 W	22.8	.47	26.3	.59	34.4	1.22	47.9	2.20	58.7	3.06	68.8	3.24	75.8	2.14	73.5	2.59	63.6	1.87	51.2	1.00	35.6	.57	26.6	.51	48.8	19.46
HYANNIS	25.0	.43	28.2	.54	34.7	1.14	47.2	2.19	56.8	2.57	66.8	2.77	74.2	2.24	72.4	2.03	62.8	1.49	51.0	.76	36.3	.52	27.9	.41	48.6	17.09
KOSHOPAH 7 NE		.54		.65		1.32		2.16		3.58		3.93		2.32		2.30		1.84		1.15		.54		.62		20.95
MERRIMAN	22.6		25.9		33.0		46.7		57.2		66.8		75.2		72.4		62.2		50.3		35.1		26.4		47.8	
NEWPORT	21.5	.67	24.6	.91	33.4	1.42	47.8	2.22	59.3	3.37	69.3	3.90	76.7	2.14	74.1	2.55	64.2	1.89	51.9	1.23	35.9	.55	25.7	.64	48.7	21.49
PURDUM	24.3	.59	27.4	.74	35.1	1.56	48.3	2.24	58.8	3.00	68.7	3.37	76.5	2.27	74.1	2.62	64.6	1.82	52.3	1.06	36.7	.56	27.8	.56	49.6	20.39
STAPLETON 5 SSE	25.2	.49	28.6	.56	35.9	1.17	48.5	2.17	58.8	3.00	68.9	3.42	76.0	1.96	73.8	2.08	64.2	2.04	52.2	.99	37.1	.57	28.1	.52	49.8	18.97
VALENTINE WB AP	20.0	.56	25.3	.46	33.3	1.25	46.5	1.97	57.0	2.66	66.8	3.16	74.6	2.70	72.2	2.00	61.6	1.20	49.5	1.17	34.4	.64	25.2	.46	47.2	18.23
DIVISION	22.8	.52	26.2	.65	34.0	1.24	48.0	2.08	58.4	3.12	68.4	3.63	75.9	2.24	73.7	2.37	64.0	1.78	52.0	1.08	36.2	.59	26.7	.52	48.9	19.82
NORTHEAST DIVISION																										
ALBION	22.4	.72	26.0	.81	35.7	1.41	49.7	2.23	60.5	3.35	71.0	4.03	77.9	2.42	75.5	2.87	66.1	1.92	54.2	1.06	37.6	.81	26.9	.83	50.3	22.46
ELGIN 9 WSW ARDEN		.80		.88		1.75		2.19		3.39		4.18		2.53		2.46		1.87		1.14		.91		.92		23.02
HARTINGTON	20.3	.95	23.9	1.17	34.1	1.76	49.3	2.34	60.7	3.27	70.5	4.13	77.2	2.64	74.9	2.89	65.6	2.43	53.7	1.26	36.5	1.06	25.1	.80	49.3	24.70
MADISON	22.2	.91	25.9	1.02	35.8	1.54	50.1	2.50	61.2	3.66	71.5	4.62	77.8	3.26	75.2	2.85	66.0	2.28	53.9	1.24	37.6	1.08	26.6	.89	50.3	25.85
NORFOLK		.79		.82		1.55		2.24		3.69		4.58		2.82		2.82		2.46		1.25		.97		.84		24.83
NORFOLK WB AP	19.3	.79	23.6	.79	34.9	1.55	48.3	2.28	59.1	3.63	69.3	4.78	76.2	3.03	73.9	2.87	64.5	2.77	52.1	1.36	34.9	1.06	23.7	.78	48.3	25.69
OAKDALE	19.9	.75	23.6	.81	33.6	1.44	48.4	2.18	59.9	3.38	70.1	4.19	76.5	2.41	73.7	2.99	63.6	2.08	51.6	1.32	35.2	.91	24.5	.76	48.4	23.22
OSMOND		.79		.93		1.64		2.43		3.75		4.02		2.54		2.68		2.43		1.28		1.02		.80		24.31
STANTON	21.7	.83	25.3	.89	35.7	1.52	50.2	2.23	60.8	3.24	71.0	4.34	77.3	3.01	74.9	2.79	65.7	2.48	53.8	1.20	37.3	.98	26.2	.74	50.0	24.25
TEKAMAH	22.4	.93	26.1	1.27	36.5	1.82	51.2	2.70	62.1	3.78	72.5	4.35	78.2	3.39	75.4	3.65	66.8	2.82	55.3	1.59	38.3	1.17	27.1	.90	51.0	28.37
WAKEFIELD	21.1	.73	24.8	.93	35.4	1.43	49.8	2.12	61.1	3.29	71.5	4.23	77.6	2.61	74.8	3.25	65.5	2.42	53.6	1.04	36.9	.94	25.4	.79	49.8	23.78
WALTHILL	20.2	.87	23.9	.97	34.5	1.41	49.1	2.28	60.8	3.33	70.8	4.40	76.5	3.01	73.8	2.74	64.9	2.35	53.1	1.39	36.4	.99	25.2	.73	49.1	24.47
WESTPOINT	21.8	.98	25.5	1.10	36.1	1.49	50.9	2.54	62.1	3.41	72.6	4.55	78.7	3.00	76.1	3.15	66.9	2.72	54.8	1.42	37.9	1.05	26.7	.86	50.8	26.27
DIVISION	21.3	.84	25.0	.96	35.1	1.55	49.8	2.34	61.0	3.46	71.2	4.32	77.4	2.79	74.9	2.96	65.6	2.41	53.7	1.29	36.9	.98	25.8	.79	49.8	24.69
CENTRAL DIVISION																										
ARCADIA		.42		.51		1.10		2.03		3.21		3.70		2.59		2.38		1.86		1.08		.60		.55		20.03
BROKEN BOW 2 W	23.8	.60	27.4	.62	35.4	1.10	48.2	2.07	58.6	3.24	69.0	4.07	75.9	2.66	73.8	2.30	63.6	2.10	51.7	.97	36.4	.62	27.2	.56	49.3	20.91
GOTHENBURG	25.9	.44	29.9	.52	38.0	1.11	50.8	1.99	60.6	3.23	70.5	4.06	77.6	1.99	75.2	2.43	65.5	1.88	53.6	.90	38.0	.58	28.7	.46	51.2	19.59
GRAND ISLAND WB AP	23.0	.58	27.7	.66	37.6	1.28	50.9	2.22	61.1	3.87	71.6	3.66	78.9	2.63	76.2	2.39	66.8	2.57	54.0	1.31	37.6	.98	27.0	.55	51.0	22.70
KEARNEY	24.5	.51	28.5	.61	36.6	1.11	49.9	2.41	60.5	3.91	71.3	4.04	78.2	2.57	76.0	2.05	66.0	2.42	54.0	1.17	37.9	.82	28.1	.58	51.0	22.20
LEXINGTON 7 ESE	25.4		30.3		38.1		50.6		60.2		70.8		77.2		75.0		65.5		53.8		38.6		29.2		51.2	
LOUP CITY	23.9	.45	27.9	.46	37.3	1.08	50.3	1.95	60.5	3.13	70.7	3.93	77.1	2.63	75.2	2.35	65.4	1.99	53.3	1.03	37.3	.67	27.4	.57	50.5	20.24
MASON CITY 1 NNW		.63		.73		1.21		1.86		3.19		3.51		2.56		2.10		1.94		1.28		.68		.59		20.28
NORTH LOUP	22.6	.54	26.6	.59	35.7	1.11	49.2	1.99	60.0	3.20	70.7	3.79	77.2	2.67	75.1	2.42	65.2	1.85	53.0	1.03	37.0	.76	26.6	.63	49.9	20.58
ORD		.62		.75		1.42		2.24		3.26		3.74		2.36		2.18		1.98		1.03		.75		.75		21.08
RAVENNA	25.0	.47	29.0	.57	38.1	1.09	51.1	2.11	61.2	3.48	71.8	4.36	78.3	2.64	76.5	2.26	66.6	2.16	54.4	1.08	38.3	.74	28.3	.51	51.6	21.47
SAINT PAUL	24.6	.49	28.4	.66	37.9	1.11	51.6	2.20	61.8	3.64	72.2	4.67	78.7	2.96	76.6	2.34	67.0	2.13	54.8	1.09	38.5	.90	28.0	.70	51.7	22.79
DIVISION	24.6	.49	28.5	.58	37.1	1.12	50.2	2.08	60.4	3.41	70.9	4.07	77.5	2.62	75.4	2.32	65.6	2.06	53.6	1.03	37.8	.69	27.9	.56	50.8	21.03

* Averages for period 1931-1955, except for stations marked WB which are "normals" based on period 1921-1950. Divisional means may not be the arithmetical average of individual stations published, since additional data from shorter period stations are used to obtain better areal represe.tation.

CONFIDENCE LIMITS

In the absence of trend or record changes, the chances are 9 out of 10 that the true mean will lie in the interval formed by adding and subtracting the values in the following table from the means for any station in the State. Because of the wider variation in mean precipitation, the corresponding monthly means and annual mean must be substituted for "p" in the precipitation table below to obtain mean precipitation confidence limits.

2.5	$.22\sqrt{p}$	2.2	$.19\sqrt{p}$	1.6	$.25\sqrt{p}$	1.1	$.39\sqrt{p}$	1.3	$.32\sqrt{p}$	1.2	$.28\sqrt{p}$	1.1	$.38\sqrt{p}$	1.1	$.35\sqrt{p}$	1.2	$.44\sqrt{p}$	1.4	$.36\sqrt{p}$	1.1	$.33\sqrt{p}$	1.5	$.22\sqrt{p}$	.6	$.32\sqrt{p}$

COMPARATIVE DATA

Data in the following table are the mean temperature and average precipitation for Crete, Nebraska, for the period 1906-1930 and are included in this publication for comparative purposes:

24.3	.48	29.6	.99	39.8	1.10	51.7	2.31	61.4	3.74	71.3	4.32	77.2	3.59	75.6	3.40	67.6	3.33	54.5	2.03	40.4	1.39	27.0	.90	51.7	27.58

NORMALS, MEANS AND EXTREMES
(Table Revised 1972. Base Period for Climatological Normals: 1931-1960)

Station: LINCOLN, NEBRASKA MUNICIPAL AIRPORT Standard time used: CENTRAL Latitude: 40°51'N Longitude: 96°45'W Elevation (ground): 1180 feet

Month	Temperature Normal Daily maximum	Daily minimum	Monthly	Extremes Record highest	Year	Record lowest	Year	Normal heating degree days (Base 65°)	Precipitation Normal total	Maximum monthly	Year	Minimum monthly	Year	Maximum in 24 hrs.	Year	Snow, Ice pellets Mean total	Maximum monthly	Year	Maximum in 24 hrs.	Year
	(b)	(b)	(b)	7		7		(b)	(b)	16		16		16		16	16		16	
J	34.0	16.2	25.1	60	1971	-18	1966	1237	0.92	1.48	1960	0.08	1970	1.15	1971	6.3	15.1	1971	12.4	1971
F	38.0	19.2	28.7	75	1967	-14	1967	1016	1.09	1.09	1965	0.11	1968	1.06	1965	6.9	26.1	1965	9.7	1965
M	48.0	28.2	38.1	87	1968+	2	1965	834	1.73	4.56	1959	0.11	1958	1.75	1959	6.9	15.8	1959	3.7	1959
A	63.4	39.2	51.8	90	1968	22	1968	402	2.48	4.58	1965	0.96	1966	1.76	1969	0.2	5.4	1969	3.0	1969
M	73.4	51.0	62.5	100	1967	30	1967	171	3.48	8.91	1966	1.26	1959	2.90	1969	0.1	3.0	1967	2.0	1967
J	84.4	61.8	73.1	102	1971	44	1967	30	4.50	12.93	1967	1.14	1958	5.38	1963	0.0	0.0		0.0	
J	92.3	67.9	80.1	105	1956	46	1971	0	3.38	11.40	1958	1.66	1959	4.86	1957	0.0	0.0		0.0	
A	89.5	66.1	77.7	100	1970	48	1967	6	3.38	9.45	1971	0.45	1971	2.37	1956	0.0	0.0		0.0	
S	80.2	55.9	68.2	100	1965	32	1965	75	2.87	6.78	1958	0.77	1958	3.58	1956	0.4	6.6	1970	6.6	1970
O	68.2	44.3	56.7	91	1969	13	1969	301	1.58	4.36	1957	0.10	1961	2.69	1961	2.5	12.6	1957	9.0	1957
N	50.7	30.3	40.8	81	1970	-13	1968	726	1.26	3.41	1971	0.12	1969	2.07	1969	5.0	15.8	1968	9.0	1968
D	39.5	21.6	30.6	64	1967			1066	0.90	1.89	1968	0.08	1960+	1.28	1968	5.0				
YR	63.6	41.9	52.8	105 JUL. 1966		-18 JAN. 1966		5864	27.43	12.93 JUN. 1967		0.05 OCT. 1958		5.38 JUN. 1963		28.9	26.1 FEB. 1965		19.0 FEB. 1965	

Wind: Mean speed 10.6; Prevailing direction S; Fastest mile speed 64; direction SW; August 1956.

Pct. of possible sunshine YR 64.7; Mean sky cover YR 6.1.

Means and extremes above are from existing and comparable exposures. Annual extremes have been exceeded at other sites in the locality as follows: Highest temperature 115 in July 1936; lowest temperature -29 in January 1892; maximum monthly precipitation 14.21 in August 1910; minimum monthly precipitation 0.00 in December 1899; maximum precipitation in 24 hours 8.38 in August 1910; fastest mile of wind 66 from NW in June 1946.

NORMALS, MEANS AND EXTREMES
(Table Revised 1975. Base Period for Climatological Normals: 1941-1970)

Station: LINCOLN, NEBRASKA Elev. 1189 feet m.s.l. Average station pressure 972.6 mb.

Month	Temperatures °F Normal Daily maximum	Daily minimum	Monthly	Extremes Record highest	Year	Record lowest	Year	Normal Degree days Base 65°F Heating	Cooling	Precipitation in inches Normal	Maximum monthly	Year	Minimum monthly	Year	Maximum in 24 hrs.	Year	Snow, Ice pellets Water equivalent Maximum monthly	Year	Minimum monthly	Year	Maximum in 24 hrs.	Year	Maximum monthly	Year	Maximum in 24 hrs.	Year
	(a)			3		3				3	3		3		3		3		3		3		3		3	
J	32.8	11.7	22.2	55	1974	-33	1974	1327	0	0.62	1.12	1972	0.22	1973	0.40	1974	1.12	1972	0.22	1973	0.40	1974	11.0	1973	4.5	1973
F	36.3	17.4	27.9	84	1972	-7	1972	1039	0	0.90	0.62	1973	0.08	1974	0.24	1972	0.62	1973	0.08	1974	0.24	1972	5.1	1973	3.1	1973
M	47.0	26.0	36.5	86	1972	5	1974	884	0	1.51	6.65	1973	0.49	1973	1.61	1973	1.64	1974	0.49	1973	1.61	1973	1.6	1974	1.5	1974
A	63.4	39.2	51.0	91	1972	15	1973	419	8	2.51	4.23	1973	0.59	1973	2.34	1973	2.31	1972	0.59	1973	2.34	1973	0.0		0.0	
M	73.1	50.6	62.0	93	1973	32	1974	166	73	3.99	5.86	1974	0.64	1972	2.71	1973	2.72	1973	0.64	1972	2.71	1973	0.0		0.0	
J	83.1	60.6	72.0	105	1973	43	1973	22	232	4.99	2.15	1972	0.77	1972	0.89	1973	2.15	1972	0.77	1972	0.89	1973	0.0		0.0	
J	88.9	65.7	77.3	106	1974	42	1972	0	386	3.32	4.48	1973	0.46	1973	1.80	1973	4.48	1973	0.46	1973	1.80	1973	0.0		0.0	
A	87.5	63.2	75.6	104	1973	45	1974	0	333	3.27	4.52	1974	0.75	1974	2.05	1974	4.52	1974	0.75	1974	2.05	1974	0.0		0.0	
S	78.7	53.6	65.6	94	1972	33	1972	83	101	2.92	7.52	1973	0.53	1972	2.31	1972	7.52	1973	0.53	1972	2.31	1972	T	1972	T	1972
O	65.7	41.9	53.6	86	1974	14	1974	329	15	1.67	4.23	1973	0.49	1972	1.72	1972	8.6	1972					8.6	1972	7.7	1972
N	50.3	27.8	39.0	80	1974	-3	1973	780	0	0.87	3.58	1972	1.08	1972	1.72	1973	3.58	1972	1.08	1972	1.72	1973	19.8	1973	10.4	1973
D	37.7	16.9	27.3	64	1973	-20	1973	1169	0	0.73	2.15	1973	0.69	1973	0.89	1973	2.15	1973	0.69	1973	0.89	1973	19.8	1973	10.4	1973
YR	62.2	39.7	51.0	106 JUL 1974		-33 JAN 1974		6218	1148	26.66	7.52 SEP 1973		0.08 FEB 1974		2.71 MAY 1973		7.52 SEP 1973		0.08 FEB 1974		2.71 DEC 1973		19.8 DEC 1973		10.4 DEC 1973

Wind: Mean speed 10.1; Fastest mile m.p.h. 54; direction N; MAR 1972.

Means and extremes above are from existing and comparable exposures. Annual extremes have been exceeded at other sites in the locality as follows: Highest temperature 115 in July 1936; lowest temperature -29 in August 1910; maximum monthly precipitation 14.21 in August 1910; minimum monthly precipitation 0.00 in December 1899; maximum precipitation in 24 hours 8.38 in August 1910; maximum monthly snowfall 26.1 in February 1965; maximum snowfall in 24 hours 19.0 in February 1965; fastest mile of wind 66 from NW in June 1946.

REFERENCE NOTES APPLYING TO TABLES APPEAR ON THE PAGE FOLLOWING LAST TABLE.
(Caution: Letters and symbols may have different meanings in 1941-1970 tables than in earlier tables. See notes.)

NORMALS, MEANS AND EXTREMES
(Table Revised 1973. Base Period for Climatological Normals: 1931-1960)

Station: GRAND ISLAND AIRPARK GRAND ISLAND, NEBRASKA Latitude: 40° 58' N Longitude: 98° 19' W Standard time used: CENTRAL Elevation (ground): 1841 feet

Temperature (°F)

Month	Normal Daily maximum	Normal Daily minimum	Normal Monthly	Extremes Record highest	Year	Extremes Record lowest	Year	Normal heating degree days (Base 65°)
	(b)	(b)	(b)	12		12		(b)
J	34.1	11.0	22.6	71	1963	-28	1963	1314
F	37.0	15.2	26.1	77	1972	-17	1965+	1089
M	47.1	24.3	35.7	88	1967	-5	1965+	908
A	62.0	37.1	49.6	94	1964	17	1972+	462
M	73.0	48.1	60.6	101	1967	23	1967	211
J	83.4	58.5	71.0	107	1970	39	1964	45
J	90.7	63.7	77.2	105	1966+	42	1971	0
A	88.5	62.0	75.3	106	1964	42	1964	6
S	79.0	51.3	65.2	101	1971+	28	1965	108
O	68.2	38.6	53.4	92	1963	16	1972	381
N	50.1	24.3	37.2	75	1965	-5	1964	834
D	38.2	16.1	27.2	76	1964	-17	1967	1172
YR	62.6	37.5	50.1	107	JUN. 1970	-28	JAN. 1963	6530

Precipitation (inches)

Month	Normal total	Max monthly	Year	Min monthly	Year	Max in 24 hrs	Year	Snow, Ice pellets Mean total	Max monthly	Year	Max in 24 hrs	Year
	(b)	34		34		34		34	34		34	
J	0.63	1.65	1960	T	1961	1.38	1947	5.7	16.1	1960	8.3	1960
F	0.74	3.39	1971	0.07	1967	2.21	1971	6.2	21.5	1969	10.6	1969
M	1.27	4.35	1959	0.01	1967	2.06	1959	5.7	15.0	1944	7.4	1944
A	2.28	6.25	1944	0.35	1946	3.30	1964	1.8	7.2	1957	5.4	1957
M	3.85	7.65	1951	0.43	1964	2.95	1961	0.3	4.5	1947	4.5	1947
J	3.79	13.96	1967	0.71	1967	4.54	1967	0.0	0.0		0.0	
J	2.51	9.00	1950	0.63	1970	5.41	1950	0.0	0.0		0.0	
A	2.88	6.40	1954	0.50	1940	2.68	1964	0.0	0.0		0.0	
S	2.15	4.92	1965	0.12	1939	3.17	1945	T	T	1961	T	1961
O	0.96	4.62	1946	T	1958	2.75	1968	0.2	4.4	1969	2.4	1969
N	0.76	2.40	1972	T	1939	1.52	1972	2.6	11.5	1972	8.6	1972
D	0.56	2.17	1968	0.02	1943	1.20	1968	6.0	21.8	1968	12.0	1968
YR	21.85	13.96	JUN. 1967	0.00	OCT. 1958	5.41	JUL. 1950	28.5	21.8	DEC. 1968	12.0	DEC. 1968

Relative humidity, Wind, Sky cover, Mean number of days

Month	Rel. hum. Hour 00	06	12	18	Mean speed	Fastest mile Speed	Direction	Year	Mean sky cover sunrise to sunset	Days Clear	Partly cloudy	Cloudy	Precip .01+	Snow 1.0+	Thunderstorms	Heavy fog	Max 90°+	Max 32°-	Min 32°-	Min 0°-
	11	11	11	11	23	10	10		23	34	34	34	34	34	34	34	11	11	11	11
J	75	76	66	64	11.7	48	35	1969	6.2	9	8	14	5	2	0	2	0	15	31	9
F	74	76	60	61	11.8	40	34	1971	6.4	7	7	14	6	2	*	1	0	10	27	4
M	69	74	45	45	13.0	48	34	1971+	6.4	8	8	14	7	1	1	1	0	6	25	1
A	71	80	47	49	14.0	47	NW	1968	6.2	8	9	14	9	*	4	1	0	0	8	0
M	74	82	52	48	12.2	57	S	1972+	5.2	8	10	10	11	0	8	1	3	0	1	0
J								1964	4.4	10	10	10	10	0	11	1	8	0	0	0
J	76	84	53	49	10.7	57	32	1972	4.4	13	11	7	7	0	9	1	13	0	0	0
A	85	87	57	57	11.3	46	32	1964	4.4	13	11	7	8	0	8	1	12	0	0	0
S	81	87	57	57	11.3	40	25	1970	4.6	13	10	7	7	*	5	1	6	0	*	0
O	76	81	53	53	11.3	41	31	1966	5.3	13	8	9	5	1	2	1	*	0	7	0
N	76	81	62	57	11.9	44	NNW	1964	5.9	9	7	13	4	2	*	1	0	2	23	*
D	76	79	68	64	11.6	46	27	1970+	6.0	10	7	14	4	2	*	2	0	13	30	4
YR	74	80	55	55	12.0	57	19	JUL. 1972+	5.5	122	102	141	88	9	50	18	39	46	151	17

Ø For period May 1961 through the current year.
Means and extremes above are from existing and comparable exposures. Annual extremes have been exceeded at other sites in the locality as follows:
Highest temperature 117 in July 1936; maximum monthly snowfall 26.0 in January 1932.
Highest temperature 117 in February 1899; lowest temperature -34 in February 1932.

NORMALS, MEANS AND EXTREMES
(Table Revised 1975. Base Period for Climatological Normals: 1941-1970)

Temperature (°F)

Month	Normal Daily maximum	Normal Daily minimum	Normal Monthly	Extremes Record highest	Year	Extremes Record lowest	Year	Normal Degree days Base 65° Heating	Cooling
	(a)			14		14			
J	33.3	11.2	22.3	71	1963	-28	1963	1324	0
F	38.7	16.7	27.7	77	1972	-17	1965	1044	0
M	46.7	24.3	35.5	88	1967	-5	1965	915	8
A	62.4	37.6	49.9	94	1964	17	1972	461	51
M	72.4	48.7	60.7	101	1967	23	1967	184	206
J	82.4	58.9	70.7	107	1970	39	1964	35	356
J	88.8	63.9	76.3	105	1966+	42	1971	6	315
A	87.4	62.5	75.0	106	1964	42	1964	5	89
S	78.7	51.6	65.4	101	1971+	28	1965	107	11
O	67.3	40.1	53.7	92	1963	16	1963	362	0
N	49.8	26.0	37.9	75	1965	-5	1964	804	0
D	37.8	16.2	27.0	76	1964	-20	1973	1178	0
YR	62.1	38.1	50.1	107	JUN. 1970	-28	JAN. 1963	6425	1036

Precipitation (inches)

Month	Normal	Max monthly	Year	Min monthly	Year	Water equiv. Max 24 hrs	Year	Snow, Ice pellets Max monthly	Year	Max 24 hrs	Year
	36	36		36		36		36		36	
J	0.52	1.65	1960	T	1961	1.38	1947	16.1	1960	8.3	1960
F	0.76	3.39	1971	0.07	1967	2.21	1971	21.5	1969	10.6	1969
M	1.18	5.57	1959	0.01	1967	2.06	1959	15.0	1944	7.4	1944
A	2.47	7.65	1944	0.35	1946	3.30	1964	7.2	1957	5.4	1957
M	3.78	7.65	1951	0.43	1964	2.95	1961	4.5	1947	4.5	1947
J	4.40	13.96	1967	0.71	1967	4.54	1967	0.0		0.0	
J	3.00	9.00	1950	0.63	1970	5.41	1950	0.0		0.0	
A	2.54	6.40	1954	0.50	1940	2.68	1964	0.0		0.0	
S	2.51	4.92	1965	0.12	1965	3.17	1945	T	1951	T	1951
O	1.08	4.42	1946	0.00	1958	2.75	1968	4.4	1959	2.4	1959
N	0.61	2.40	1972	T	1973	1.90	1972	11.5	1972	8.6	1972
D	0.56	2.17	1968	0.02	1943	1.20	1968	26.0	1973	12.0	1968
YR	23.41	13.96	JUN. 1967	0.00	OCT. 1958	5.41	JUL. 1950	26.0	DEC. 1973	12.0	DEC. 1968

Relative humidity, Wind, Sky cover, Mean number of days, Station pressure

Month	Rel. hum. Hour 00	06	12	18	Mean speed m.p.h.	Fastest mile m.p.h.	Direction	Year	Mean sky cover	Days Clear	Partly cloudy	Cloudy	Precip .01+	Snow 1.0+	Thunderstorms	Heavy fog	Max 90°+	Max 32°-	Min 32°-	Min 0°-	Avg station pressure mb.
	13	13	13	13	25	12	14		25	36	36	36	36	36	36	36	13	13	13	13	2
J	73	75	64	66	11.7	48	35	1969	6.2	9	8	14	5	2	0	2	0	15	30	9	950.9
F	82	74	60	61	11.9	40	34	1971	6.4	7	8	14	6	2	*	2	0	10	27	2	950.9
M	80	70	52	56	13.6	55	NNW	1971	6.5	8	8	14	7	1	1	1	0	5	23	*	945.9
A	84	78	50	52	14.2	47	NNW	1968	6.1	8	9	14	9	*	4	1	0	1	8	0	945.8
M	87	81	47	48	12.2	57	S	1972	6.0	8	10	11	11	0	8	1	3	0	1	0	945.7
J								1964	4.4	10	10	10	10	0	11	1	8	0	0	0	946.4
J	73	82	51	47	10.7	57	32	1972	4.4	13	11	7	7	0	9	1	14	0	0	0	949.2
A	80	84	56	52	11.2	46	32	1964	4.6	13	11	7	8	0	8	1	12	0	0	0	949.4
S	80	87	53	53	11.2	40	25	1970	4.7	14	10	8	7	*	5	1	6	0	*	0	951.1
O	76	81	53	53	11.3	41	31	1966	4.6	14	8	8	5	1	2	1	*	0	6	0	951.6
N	76	81	64	57	11.9	44	NNW	1964	5.9	9	8	13	4	2	*	2	0	3	23	*	950.1
D	79	76	68	64	11.6	46	NNW	1970	6.0	9	8	14	4	2	*	2	0	14	30	4	949.4
YR	74	80	55	55	12.0	57	19	S JUL. 1972	5.5	121	103	141	88	9	50	18	40	45	149	16	948.9

Elev. 1856 feet m.s.l.

Means and extremes above are from existing and comparable exposures. Annual extremes have been exceeded at other sites in the locality as follows:
Highest temperature 117 in July 1936; lowest temperature -34 in February 1899.

REFERENCE NOTES APPLYING TO TABLES APPEAR ON THE PAGE FOLLOWING LAST TABLE.
(Caution: Letters and symbols may have different meanings in 1941-1970 tables than in earlier tables. See notes.)

NORMALS, MEANS AND EXTREMES

(Table Revised 1973. Base Period for Climatological Normals: 1931-1960)

Station: NORFOLK, NEBRASKA KARL STEFAN MEMORIAL AIRPORT Standard time used: CENTRAL Latitude: 41° 59' N Longitude: 97° 26' W Elevation (ground): 1544 feet

Month	Temperature Normal Daily maximum	Temperature Normal Daily minimum	Temperature Normal Monthly	Extremes Record highest	Year	Extremes Record lowest	Year	Normal heating degree days (Base 65°)	Precipitation Normal total	Maximum monthly	Year	Minimum monthly	Year	Maximum in 24 hrs.	Year	Snow, Ice pellets Mean total	Maximum monthly	Year	Maximum in 24 hrs.	Year	Relative humidity Hour 00 (Local time)	Hour 06	Hour 12	Hour 18	Wind Mean speed	Prevailing direction	Fastest mile Speed	Direction	Year	Pct. of possible sunshine	Mean sky cover sunrise to sunset	Sunrise to sunset Clear	Partly cloudy	Cloudy	Precipitation .01 inch or more	Snow, Ice pellets 1.0 inch or more	Thunderstorms	Heavy fog	Temperatures Max. 90° and above ‡	Max. 32° and below	Min. 32° and below	Min. 0° and below	Average daily solar radiation - langleys
(a)	(b)	(b)	(b)	27		27		(b)	(b)	27		27		27		27	27		27			27	27	27							27	27	27	27	27	27	19	19	27	27	27	27	
J	30.1	8.6	19.4	62	1964+	-26	1966+	1414	0.78	2.33	1949	0.10	1970+	1.28	1949	5.4	16.2	1960	8.3	1960		76	63	68							6.2	9	7	15	5	2	0	1	0	17	31	10	
F	33.2	12.6	22.9	73	1946	-26	1962	1179	0.78	3.18	1971	0.04	1949	2.41	1971	5.8	19.1	1969	8.3	1969		79	64	68							6.3	8	6	14	5	2	*	2	0	13	28	5	
M	43.1	23.5	33.3	87	1968	-20	1960	983	1.59	3.19	1959	0.06	1967	1.52	1959	7.1	20.8	1960	9.7	1966		81	61	61							6.7	7	8	16	8	2	1	1	0	8	25	1	
A	59.7	37.0	48.4	94	1965	13	1957	498	2.18	4.12	1953	0.23	1969	1.63	1960	1.2	4.3	1957	4.2	1960		78	49	47							6.3	8	8	14	9	*	4	1	*	*	11	0	
M	71.6	48.8	60.2	103	1967	25	1967+	233	3.66	7.50	1959	1.01	1948	2.35	1950	0.1	2.7	1947	2.7	1947		80	51	50							6.2	7	10	14	11	*	9	1	1	0	1	0	
J	82.1	59.6	70.9	106	1946	38	1964	48	4.26	12.22	1967	2.17	1970	4.23	1969	0.0	0.0		0.0			83	53	51							5.5	10	9	11	10	0	10	1	7	0	0	0	
J	89.2	64.7	77.0	113	1954	42	1971	9	3.02	9.11	1950	0.33	1954	3.46	1952	0.0	0.0		0.0			84	53	50							4.7	12	12	7	9	0	9	1	11	0	0	0	
A	87.2	63.1	75.2	107	1947	40	1967+	0	2.61	5.53	1951	0.53	1971	2.57	1966	0.0	0.0		0.0			85	54	52							4.6	13	10	8	8	0	10	1	10	0	0	0	
S	78.0	52.0	65.0	101	1971+	27	1956	111	2.25	8.13	1970	0.30	1956	3.88	1970	T	0.7	1961	0.7	1961		84	52	53							4.9	13	7	10	8	0	3	1	3	0	1	0	
O	66.2	39.6	52.9	95	1963	13	1972	397	1.18	4.57	1968	T	1958+	2.79	1968	0.2	1.6	1966	1.6	1966		80	49	54							4.8	13	8	10	5	*	2	1	*	0	7	0	
N	47.2	24.5	35.9	82	1945	-15	1964	873	0.91	2.50	1946	0.04	1966	1.29	1946	2.4	9.7	1948	7.4	1946		79	57	63							6.2	8	8	14	4	1	1	1	0	3	24	1	
D	35.4	15.0	25.2	71	1962	-22	1967	1234	0.73	1.75	1968	0.08	1958+	1.28	1953	6.1	19.1	1968	10.5	1968		78	64	69							6.3	9	7	15	5	2	*	1	0	14	30	4	
YR	60.3	37.4	48.9	113	JUL. 1954	-26	JAN. 1966+	6979	23.95	12.22	JUN. 1967	T	OCT. 1958+	4.23	JUN. 1969	28.3	20.8	MAR. 1960	10.5	DEC. 1968		80	56	57							5.7	117	100	148	87	9	50	11	33	55	158	21	

Means and extremes above are from existing and comparable exposures. Annual extremes have been exceeded at other sites in the locality as follows: Highest temperature 116 in July 1936; lowest temperature -39 in January 1912; maximum monthly precipitation 12.28 in June 1924; maximum precipitation in 24 hours 7.78 in June 1940.

NORMALS, MEANS AND EXTREMES

(Table Revised 1975. Base Period for Climatological Normals: 1941-1970)

Month	Temperatures °F Normal Daily maximum	Normal Daily minimum	Normal Monthly	Extremes Record highest	Year	Extremes Record lowest	Year	Normal Degree days Base 65 °F Heating	Cooling	Precipitation in inches Water equivalent Normal	Maximum monthly	Year	Minimum monthly	Year	Maximum in 24 hrs.	Year	Snow, Ice pellets Maximum monthly	Year	Maximum in 24 hrs.	Year	Relative humidity pct. Hour 00 (Local time)	Hour 06	Hour 12	Hour 18	Wind Mean speed m.p.h.	Prevailing direction	Fastest mile Speed m.p.h.	Direction	Year	Pct. of possible sunshine	Mean sky cover, tenths, sunrise to sunset	Sunrise to sunset Clear	Partly cloudy	Cloudy	Precipitation .01 inch or more	Snow, Ice pellets 1.0 inch or more	Thunderstorms	Heavy fog, visibility ¼ mile or less	Temperatures °F Max. (b) 90° and above	Max. 32° and below	Min. 32° and below	Min. 0° and below	Average station pressure mb. Elev. 1551 feet m.s.l.
(a)				29		29					29		29		29		29		29			29	29	29							29	29	29	29	29	29	19	19	29	29	29	29	
J	29.6	8.1	18.9	62	1964	-27	1974	1429	0	0.62	2.33	1949	0.10	1970	1.28	1949	16.2	1960	8.3	1960		76	64	68							6.2	9	7	15	5	2	0	1	0	17	31	10	
F	34.4	13.4	23.9	73	1946	-26	1962	1151	0	0.78	3.18	1971	0.04	1949	2.41	1971	19.1	1969	8.3	1969		79	65	68							6.3	8	6	14	5	2	*	2	0	12	27	5	
M	42.9	22.7	32.8	87	1968	-20	1960	998	0	1.37	5.14	1973	0.06	1967	1.52	1959	20.8	1960	9.7	1966		81	61	61							6.7	7	8	16	8	2	1	1	0	7	25	1	
A	60.3	36.6	48.5	94	1965	13	1957	500	5	2.15	4.12	1953	0.23	1969	1.63	1960	4.3	1957	4.2	1960		78	49	47							6.3	8	8	14	8	*	4	1	*	*	11	0	
M	71.4	48.5	60.0	103	1967	25	1967	203	48	3.69	7.50	1959	1.01	1948	4.12	1973	2.7	1947	2.7	1947		80	51	50							6.2	7	10	14	11	*	9	1	1	0	1	0	
J	80.9	58.8	69.9	106	1946	38	1964	37	184	4.88	12.22	1967	1.33	1973	5.51	1974	0.0		0.0			82	52	50							5.4	10	10	10	10	0	10	1	7	0	0	0	
J	87.0	63.9	75.5	113	1954	42	1971	6	331	3.18	9.11	1950	0.33	1954	3.46	1952	0.0		0.0			83	53	50							4.7	12	12	7	9	0	9	1	12	0	0	0	
A	85.3	62.2	73.8	107	1947	40	1967	11	283	2.66	5.53	1951	0.53	1971	2.57	1966	0.0		0.0			85	54	52							4.6	13	10	8	8	0	10	1	10	0	0	0	
S	75.2	51.1	63.2	101	1971	27	1956	123	69	2.41	8.13	1970	0.30	1956	3.88	1970	0.7	1961	0.7	1961		84	52	53							4.9	12	8	10	8	0	3	1	3	0	1	0	
O	65.5	39.4	52.5	95	1963	13	1972	397	10	1.33	4.57	1968	T	1958	2.79	1968	1.6	1966	1.6	1966		80	50	54							4.9	13	7	11	5	*	2	1	*	0	7	0	
N	47.6	24.9	36.3	82	1945	-15	1964	861	0	0.62	2.50	1946	0.04	1966	1.49	1973	9.7	1948	7.4	1946		79	58	63							6.2	8	8	14	5	1	1	1	0	3	24	*	
D	34.6	13.8	24.2	71	1962	-22	1967	1265	0	0.63	1.75	1968	0.08	1958	1.28	1953	19.1	1968	10.5	1968		78	64	70							6.3	8	8	15	5	2	*	1	0	14	30	4	
YR	59.6	36.9	48.3	113	JUL 1954	-27	JAN 1974	6981	930	24.32	12.22	JUN 1967	T	OCT 1958	5.51	JUN 1974	20.8	MAR 1960	10.5	DEC 1968		80	56	57							5.7	115	102	148	87	9	50	11	33	54	157	20	

Means and extremes above are from existing and comparable exposures. Annual extremes have been exceeded at other sites in the locality as follows: Highest temperature 116 in July 1936; lowest temperature -39 in January 1912; maximum monthly precipitation 12.28 in June 1924; maximum precipitation in 24 hours 7.78 in June 1940.

REFERENCE NOTES APPLYING TO TABLES APPEAR ON THE PAGE FOLLOWING LAST TABLE.

(Caution: Letters and symbols may have different meanings in 1941-1970 tables than in earlier tables. See notes.)

NORMALS, MEANS AND EXTREMES
(Table Revised 1973. Base Period for Climatological Normals: 1931-1960)

Station: NORTH PLATTE, NEBRASKA   LEE BIRD FIELD   Standard time used: CENTRAL   Latitude: 41° 08' N   Longitude: 100° 41' W   Elevation (ground): 2775 feet

Temperature °F / Precipitation

Month	Normal Daily max	Normal Daily min	Normal Monthly	Record highest	Year	Record lowest	Year	Normal heating degree days (Base 65°)	Precip. Normal total	Snow mean total	Mean sky cover	Pct. possible sunshine
J	37.3	10.7	24.0	68	1968	-20	1968	1271	0.43	5.3	6.3	59
F	40.8	14.9	27.9	76	1972	-11	1966	1039	0.52	5.5	6.5	62
M	47.8	22.1	35.0	86	1965	-6	1965	930	0.98	8.1	6.4	60
A	61.4	34.5	47.7	96	1973	13	1973	519	2.01	3.0	6.2	62
M	71.1	45.8	58.5	96	1967	29	1967	248	2.95	0.3	6.4	61
J	81.9	56.2	69.1	100	1969+	41	1971	57	3.25	0.0	5.2	74
J	90.0	62.2	76.1	102	1969	45	1971	0	2.52	0.0	4.6	73
A	88.5	60.0	74.5	103	1971	41	1964	6	2.13	0.0	4.6	69
S	79.3	48.1	63.7	100	1971	23	1964	123	1.67	T	4.6	69
O	67.4	34.6	51.0	91	1969	-7	1969	440	0.91	1.8	4.6	60
N	50.2	20.8	35.5	77	1965+	-17	1964	885	0.50	2.9	5.8	60
D	40.5	14.2	27.4	68	1964	-25	1967	1166	0.40	4.6	5.8	65
YR	63.0	35.4	49.2	103 AUG. 1970		-25 DEC. 1967		6684	18.27	31.5	5.6	65

Ø For period September 1964 through the current year.

Means and extremes above are from existing and comparable exposures. Annual extremes have been exceeded at other sites in the locality as follows:
Highest temperature 112 in July 1954; lowest temperature -35 in February 1899 and earlier; maximum monthly precipitation 10.47 in June 1951; maximum precipitation in 24 hours 6.23 in September 1942; maximum snowfall 27.8 in March 1912; maximum snowfall in 24 hours 13.0 in March 1949 and earlier.

NORMALS, MEANS AND EXTREMES
(Table Revised 1975. Base Period for Climatological Normals: 1941-1970)

Elev. 2787 feet m.s.l.   Average station pressure 917.0 mb.

Temperature °F / Degree days / Precipitation

Month	Normal Daily max	Normal Daily min	Normal Monthly	Record highest	Year	Record lowest	Year	Heating degree days Base 65°F	Cooling degree days Base 65°F	Precip. Normal
J	36.6	10.1	23.4	68	1967	-23	1967	1290	0	0.45
F	42.4	15.2	28.1	76	1972	-11	1966	1033	0	0.52
M	47.4	21.3	34.3	86	1968	-6	1968	952	0	0.99
A	61.4	34.7	47.8	96	1965	13	1965	522	6	1.93
M	71.4	45.2	58.3	96	1967	21	1967	238	30	3.26
J	80.6	55.3	68.0	100	1974	38	1969	65	155	3.77
J	87.6	61.0	74.3	105	1973	41	1971	7	295	2.98
A	86.7	59.3	73.0	103	1970	38	1974	8	256	2.07
S	77.0	47.6	62.3	100	1971	22	1974	141	60	2.01
O	65.0	35.0	50.0	91	1968	-7	1969	439	0	0.99
N	47.6	21.8	36.2	77	1965	-11	1964	864	0	0.52
D	39.8	13.8	26.8	74	1964	-25	1967	1184	0	0.41
YR	62.2	35.0	48.6	105 JUL 1973		-25 DEC 1967		6743	807	19.90

Means and extremes above are from existing and comparable exposures. Annual extremes have been exceeded at other sites in the locality as follows:
Highest temperature 112 in July 1954; lowest temperature -35 in February 1899 and earlier; maximum monthly precipitation 10.47 in June 1951; maximum precipitation in 24 hours 6.23 in September 1942; maximum snowfall 27.8 in March 1912; maximum snowfall in 24 hours 13.0 in March 1949 and earlier.

REFERENCE NOTES APPLYING TO TABLES APPEAR ON THE PAGE FOLLOWING LAST TABLE.
(Caution: Letters and symbols may have different meanings in 1941-1970 tables than in earlier tables. See notes.)

NORMALS, MEANS AND EXTREMES
(Table Revised 1973. Base Period for Climatological Normals: 1931-1960)

Station: OMAHA, NEBRASKA   EPPLEY AIRFIELD   Standard time used: CENTRAL   Latitude: 41° 18' N   Longitude: 95° 54' W   Elevation (ground): 977 feet

Month	Normal Daily maximum	Normal Daily minimum	Normal Monthly	Record highest	Year	Record lowest	Year	Normal heating degree days (Base 65°)	Precip. Normal total	Max. monthly	Year	Min. monthly	Year	Max. in 24 hrs.	Year	Snow Mean total	Max. monthly	Year	Max. in 24 hrs.	Year
(a)	(b)	(b)	(b)	9	9	9		(b)	(b)	37		37		37		37	37		37	
J	31.4	12.9	22.3	64	1964	-17	1966	1324	0.82	3.70	1949	0.05	1943	1.52	1967	7.9	25.7	1936	13.1	1949
F	36.0	17.0	26.5	78	1972	-17	1971	1078	0.95	2.97	1954	0.10	1968	2.24	1965	7.5	25.4	1965	18.3	1965
M	46.4	27.3	36.9	89	1968+	18	1965+	871	1.45	3.59	1959	0.12	1958	2.96	1959	8.6	27.2	1948	13.0	1948
A	62.4	40.9	51.7	93	1967	31	1967	405	2.45	6.45	1951	0.23	1936	2.56	1938	0.8	8.0	1962	8.6	1938
M	73.9	51.7	63.0	97	1967	40	1969	164	3.48	10.33	1959	0.56	1964	3.58	1962	0.1	2.0	1967	2.0	1967
J	83.8	62.1	73.1	103	1972	51	1972	30	4.53	10.81	1947	1.03	1972	3.48	1945	0.0	0.0		0.0	
J	89.7	67.3	78.5	102	1970	44	1972	0	3.37	9.60	1958	0.52	1936	3.37	1958	0.0	0.0		0.0	
A	86.6	65.5	76.2	107	1954	43	1967	6	3.98	9.12	1959	0.73	1941	3.40	1959	0.0	0.0		0.0	
S	78.5	55.3	66.9	102	1955	31	1953+	90	2.63	13.75	1965	0.41	1953+	6.47	1961	T	T	1961	T	1961
O	67.5	43.8	55.7	95	1953	13	1972	326	1.73	4.99	1961	0.04	1952	3.13	1948	0.3	7.2	1941	7.2	1941
N	48.9	28.8	38.9	80	1964	-9	1964	783	1.26	3.30	1955+		1955+	2.53	1957	2.4	12.0	1957	8.7	1957
D	37.2	19.1	28.2	67	1954	-13	1968+	1141	0.80	3.30	1941		1943	1.79	1941	5.9	19.9	1969	10.2	1969
YR	61.9	41.0	51.5	107	JUL 1954	-17	FEB 1971+	6218	27.56	13.75	SEP. 1965	T	OCT. 1952+	6.47	SEP. 1965	32.0	27.2	MAR. 1948	18.3	FEB. 1965

Month	RH Hour 00	06	12	18	Wind Mean speed	Prevailing direction	Fastest mile Speed	Direction	Year	Pct. of possible sunshine	Mean sky cover sunrise to sunset	Clear	Partly cloudy	Cloudy	Precip. .01 inch or more	Snow, Ice pellets 1.0 inch or more	Thunderstorms	Heavy fog	Max. 90° and above	Max. 32° and below	Min. 32° and below	Min. 0° and below	Avg. daily solar radiation - langleys
	9	9	9	9	9	15	37	37		37	37	37	37	37	37	37	37	37	9	9	9	9	9
J	73	75	64	66	11.2	NNW	57	NW	1938	55	6.1	9	8	14	7	7	*	2	0	15	30	7	
F	75	76	61	60	11.5	NNW	57	NW	1947	55	6.1	8	8	13	7	7	*	1	0	11	26	3	
M	80	77	53	51	12.7	NNW	73	NW	1950	55	6.4	8	8	16	9	5	2	1	0	5	21	0	
A	78	73	50	54	13.2	NNW	65	NW	1937	59	6.3	7	11	14	12	*	4	1	*	0	5	0	
M	80	75	53	50	11.5	SSE	73	N	1936	62	6.3	8	11	14	11	0	8	1	3	0	0	0	
J	78	78	58	57	9.1	SSE	109	N	1936	76	4.8	11	13	7	8	0	8	1	8	0	0	0	
J	79	83	59	58	8.6	SSE	66	E	1944	71	4.4	13	13	6	8	0	5	1	12	0	0	0	
A	82	81	61	58	9.7	SSE	62	NW	1966	67	4.7	13	11	7	8	0	5	1	8	0	0	0	
S	81	80	55	57	9.7	SSE	62	NW	1951	67	4.7	13	10	8	6	0	3	1	2	*	0	0	
O	76	80	55	57	11.2	SSE	56	SSE	1938	65	4.9	14	10	6	6	*	1	1	*	0	1	0	
N	77	79	66	70	10.8	SSE	52	SSE		48	6.4	8	9	13	6	1	1	2	0	1	10	1	
D																							
YR	75	79	58	58	10.9		109	N	JUL 1936	62	5.7	112	107	146	99	9	49	15	33	40	139	13	

$ For period August 1963 through the current year.
Means and extremes above are from existing and comparable exposures. Annual extremes have been exceeded at other sites in the locality as follows:
Highest temperature 114 in July 1936; lowest temperature -32 in January 1884; maximum precipitation in 24 hours 7.03 in August 1903; maximum monthly snowfall 29.2 in March 1912.

NORMALS, MEANS AND EXTREMES
(Table Revised 1975. Base Period for Climatological Normals: 1941-1970)

Average station pressure: 982 feet m.s.l.   Elev. 982 feet m.s.l.

Month	Normal Daily maximum	Normal Daily minimum	Normal Monthly	Highest	Year	Lowest	Year	Normal Degree days Base 65°F Heating	Cooling	Precip. Normal	Water equivalent Max. monthly	Year	Min. monthly	Year	Max. in 24 hrs.	Year	Snow, Ice pellets Max. monthly	Year	Max. in 24 hrs.	Year
(a)				11	11	11				39	39		39		39		39		39	
J	32.7	12.4	22.6	64	1964	-22	1974	1314	0	0.76	3.70	1949	0.05	1943	1.52	1967	25.7	1936	13.1	1949
F	38.5	17.4	28.0	78	1972	-17	1971	1036	0	0.98	2.97	1968	0.10	1968	2.24	1954	25.4	1965	18.3	1965
M	47.7	26.1	37.1	89	1968	-17	1971	865	0	1.59	5.96	1973	0.12	1958	2.96	1959	27.2	1948	13.0	1948
A	64.4	40.1	52.3	93	1965	18	1965	391	9	2.97	6.45	1951	0.23	1936	2.56	1938	8.0	1964	8.6	1938
M	74.4	52.3	63.3	97	1967	31	1967	148	86	4.11	10.33	1951	0.56	1964	3.58	1962	2.0	1967	2.0	1967
J	83.1	61.3	72.2	103	1974	40	1969	20	236	4.94	10.81	1947	1.03	1972	3.48	1942	0.0		0.0	
J	88.6	65.8	77.2	110	1974	44	1972	0	378	3.71	9.60	1958	0.52	1936	3.37	1958	0.0		0.0	
A	87.2	64.0	75.6	107	1971	43	1967	6	334	3.98	9.12	1959	0.73	1941	3.40	1959	0.0		0.0	
S	78.6	54.0	66.3	100	1971	31	1965	88	110	3.27	13.75	1965	0.41	1961	6.47	1961	T	1961	T	1961
O	67.1	40.0	54.0	95	1964	13	1973	301	19	1.93	4.99	1952	0.04	1952	3.13	1948	7.2	1941	7.2	1941
N	50.9	29.1	40.0	80	1964	-9	1964	750	0	1.11	4.05	1948		1948	2.53	1943	12.0	1957	8.7	1957
D	37.8	18.1	28.0	67	1968	-13	1968	1147	0	0.84	3.30	1941		1943	1.79	1941	19.9	1969	10.2	1969
YR	62.8	40.2	51.5	110	JUL 1974	-22	JAN 1974	6049	1173	30.18	13.75	SEP 1965	T	OCT 1952	6.47	SEP 1965	27.2	MAR 1948	18.3	FEB 1965

Month	RH Hour 00	06	12	18	Wind Mean speed m.p.h.	Prevailing direction	Fastest mile Speed m.p.h.	Direction	Year	Pct. of possible sunshine	Mean sky cover, tenths, sunrise to sunset	Clear	Partly cloudy	Cloudy	Precip. .01 inch or more	Snow, Ice pellets 1.0 inch or more	Thunderstorms	Heavy fog, visibility ¼ mile or less	Max. 90° and above	Max. 32° and below	Min. 32° and below	Min. 0° and below	Average station pressure mb.
	11	11	11	11	11	15	39	39		39	39	39	39	39	39	39	39	39	11	11	11	11	2
J	74	76	65	68	11.1	NNW	57	NW	1938	55	6.1	9	7	14	7	7	*	2	0	15	30	8	983.1
F	73	76	61	61	11.5	NNW	57	NW	1947	55	6.1	8	7	13	7	7	*	2	0	11	27	3	982.6
M	77	75	54	53	12.7	NNW	73	NW	1950	55	6.7	8	8	16	9	5	1	1	0	6	21	0	977.2
A	68	68	50	50	13.2	NNW	65	NW	1937	59	6.3	8	10	14	11	*	4	1	*	0	6	0	976.6
M	75	75	54	53	11.5	SSE	72	N	1936	62	6.3	8	11	11	11	0	8	1	3	0	0	0	976.6
J	77	82	57	56	9.1	SSE	109	N	1936	76	4.8	12	13	7	10	0	8	1	13	0	0	0	976.5
J	80	87	59	58	8.9	SSE	66	E	1944	71	4.8	13	13	7	8	0	8	1	8	0	0	0	979.7
A	82	86	62	62	9.7	SSE	62	NW	1966	67	4.7	13	13	9	8	0	5	1	8	0	0	0	980.2
S	77	85	56	58	9.7	SSE	56	NW	1951	67	4.7	13	10	6	7	0	3	1	2	*	0	0	980.7
O	78	81	56	60	10.1	SSE	56	SSE	1938	65	4.7	14	10	11	6	*	1	2	*	0	5	0	982.8
N	80	80	68	71	10.9	SSE	52	SSE		48	6.4	8	7	16	6	1	1	2	0	2	11	2	981.4
D																					11	29		981.1
YR	75	80	59	58	10.9		109	N	JUL 1936	62	5.7	114	104	147	99	9	48	15	34	40	137	12	979.9

Means and extremes above are from existing and comparable exposures. Annual extremes have been exceeded at other sites in the locality as follows:
Highest temperature 114 in July 1936; lowest temperature -32 in January 1884; maximum precipitation in 24 hours 7.03 in August 1903; maximum monthly snowfall 29.2 in March 1912.

REFERENCE NOTES APPLYING TO TABLES APPEAR ON THE PAGE FOLLOWING LAST TABLE.
(Caution: Letters and symbols may have different meanings in 1941-1970 tables than in earlier tables. See notes.)

NORMALS, MEANS AND EXTREMES

(Table Revised 1973. Base Period for Climatological Normals: 1931-1960)

Station: SCOTTS BLUFF, NEBRASKA   SCOTTS BLUFF COUNTY AIRPORT   Standard time used: MOUNTAIN   Latitude: 41° 52' N   Longitude: 103° 36' W   Elevation (ground): 3957 feet

Means and extremes above are from existing and comparable exposures. Annual extremes have been exceeded at other sites in the locality as follows:
Highest temperature 110 in July 1939; lowest temperature -45 in February 1899; minimum monthly precipitation 0.00 in November 1939; maximum monthly snowfall 29.7 in April 1927; maximum snowfall 17.6 in 24 hours in April 1935.

NORMALS, MEANS AND EXTREMES

(Table Revised 1975. Base Period for Climatological Normals: 1941-1970)

Means and extremes above are from existing and comparable exposures. Annual extremes have been exceeded at other sites in the locality as follows:
Highest temperature 110 in July 1939; lowest temperature -45 in February 1899; minimum monthly precipitation 0.00 in November 1939; maximum monthly snowfall 29.7 in April 1927; maximum snowfall 17.6 in 24 hours in April 1935.

REFERENCE NOTES APPLYING TO TABLES APPEAR ON THE PAGE FOLLOWING LAST TABLE.

(Caution: Letters and symbols may have different meanings in 1941-1970 tables than in earlier tables. See notes.)

NORMALS, MEANS AND EXTREMES
(Table Revised 1973. Base Period for Climatological Normals: 1931-1960)

Station: VALENTINE, NEBRASKA   MILLER FIELD
Standard time used: CENTRAL   Latitude: 42° 52' N   Longitude: 100° 33' W   Elevation (ground): 2587 feet

Month	Normal Daily maximum	Normal Daily minimum	Normal Monthly	Extremes Record highest	Year	Extremes Record lowest	Year	Normal heating degree days (Base 65°)	Precip Normal total
J	33.3	6.7	20.0	68	1957	-29	1963+	1395	0.40
F	36.7		23.0	75	1962	-27	1960	1176	0.56
M	43.7	18.9	31.2	83	1952	-27	1960	1046	0.91
A	59.1	32.3	45.7	93	1962	6	1968+	579	1.98
M	70.1		57.1	98	1959+	19	1967	288	2.67
J	81.2	53.8	67.5	106	1968	30	1969	84	3.11
J	90.6	60.2	75.4	109	1964	38	1971	9	2.32
A	87.7	57.9	72.8	108	1965+	37	1965	12	2.29
S	77.1	45.3	61.2	102	1964+	24	1972+	165	1.25
O	63.8	34.1	49.0	95	1963	13	1971+	493	0.88
N	47.9	19.3	33.6	82	1959	-13	1959	942	0.49
D	38.3	11.9	25.1	69	1956	-27	1967	1237	0.34
YR	61.0	32.7	46.9	109 JUL 1964		-29 JAN 1963+		7425	17.30

Means and extremes above are from existing and comparable exposures. Annual extremes have been exceeded at other sites in the locality as follows: Highest temperature 110 in July 1934; lowest temperature -38 in January 1894; minimum monthly precipitation 0.00 in October 1895; maximum precipitation in 24 hours 4.21 in May 1920; maximum monthly snowfall 33.7 in November 1919; maximum snowfall in 24 hours 16.0 in March 1933.

NORMALS, MEANS AND EXTREMES
(Table Revised 1975. Base Period for Climatological Normals: 1941-1970)

Elev. 2598 feet m.s.l.

Month	Normal Daily maximum	Normal Daily minimum	Normal Monthly	Extremes Record highest	Year	Extremes Record lowest	Year	Degree days Heating	Degree days Cooling	Normal precipitation
J	33.3	7.5	20.4	70	1974	-29	1963	1383	0	0.31
F	37.2	11.8	24.5	75	1962	-27	1962	1134	0	0.53
M	43.5	19.5	31.5	83	1972	-27	1960	1048	0	0.76
A	59.0	32.5	45.8	93	1962	6	1968	576	0	1.77
M	70.1	43.2	56.9	98	1969	19	1967	273	22	2.80
J	79.7	54.1	66.9	107	1974	30	1969	73	130	3.60
J	88.4	59.8	74.1	109	1964	38	1971	8	291	2.50
A	87.0	58.0	72.5	108	1965	37	1965	10	242	2.48
S	76.2	46.5	61.4	102	1963	24	1971	154	45	1.42
O	65.4	34.6	50.0	95	1963	-13	1971	470	0	0.45
N	48.0	21.2	34.6	82	1959	-21	1959	912	0	0.30
D	36.9	11.9	24.4	69	1956	-27	1967	1259	0	0.30
YR	60.4	33.4	46.9	109 JUL 1964		-29 JAN 1963		7300	736	17.80

Means and extremes above are from existing and comparable exposures. Annual extremes have been exceeded at other sites in the locality as follows: Highest temperature 110 in July 1934; lowest temperature -38 in January 1894; minimum monthly precipitation 0.00 in October 1895; maximum precipitation in 24 hours 4.21 in May 1920; maximum monthly snowfall 33.7 in November 1919; maximum snowfall in 24 hours 16.0 in March 1933.

REFERENCE NOTES APPLYING TO TABLES APPEAR ON THE PAGE FOLLOWING LAST TABLE.
(Caution: Letters and symbols may have different meanings in 1941-1970 tables than in earlier tables. See notes.)

Reference notes applying to Normals, Means, and Extremes tables for 1931–1960 base period.

(a) Length of record, years, based on January data. Other months may be for more or fewer years if there have been breaks in the record.
Climatological standard normals (1931-1960).
(b)
• Less than one half.
+ Also on earlier dates, months, or years.
T Trace, an amount too small to measure.
Below zero temperatures are preceded by a minus sign.
The prevailing direction for wind in the Normals, Means, and Extremes table is from records through 1964.
‡ ≥ 70° at Alaskan stations.

Unless otherwise indicated, dimensional units used in this bulletin are: temperature in degrees F.; precipitation, including snowfall, in inches; wind movement in miles per hour; and relative humidity in percent. Heating degree day totals are the sums of negative departure of average daily temperatures from 65° F. Cooling degree day totals are the sums of positive departures of average daily temperatures from 65° F. Sleet was included in snowfall totals beginning with July 1948. The term "Ice pellets" includes solid grains of ice (sleet) and particles consisting of snow pellets encased in a thin layer of ice. Heavy fog reduces visibility to 1/4 mile or less.

Sky cover is expressed in a range of 0 for no clouds or obscuring phenomena to 10 for complete sky cover. The number of clear days is based on average cloudiness 0-3, partly cloudy days 4-7, and cloudy days 8-10 tenths.

Solar radiation data are the averages of direct and diffuse radiation on a horizontal surface. The langley denotes one gram calorie per square centimeter.

& Figures instead of letters in a direction column indicate direction in tens of degrees from true North; i.e., 09-East, 18-South, 27-West, 36-North, and 00-Calm. Resultant wind is the vector sum of wind directions and speeds divided by the number of observations. If figures appear in the direction column under "Fastest mile" the corresponding speeds are fastest observed 1-minute values.
# To 8 compass points only.   ¢ Through 1964
% Through 1964. The station did not operate 24 hours daily. Fog and thunderstorm data may be incomplete.
$ Data from City Office through 1964 and from the Airport Station thereafter.

⊬ The National Weather Service considers the accuracy of solar radiation data questionable; therefore, publication is suspended pending determination of corrected values.

Reference notes applying to Normals, Means, and Extremes tables for 1941–1970 base period.

(a) Length of record, years, through the current year unless otherwise noted, based on January data.
(b) 70° and above at Alaskan stations.
* Less than one half.
T Trace.

NORMALS - Based on record for the 1941-1970 period.
DATE OF AN EXTREME - The most recent in cases of multiple occurrence.
PREVAILING WIND DIRECTION - Record through 1964.
WIND DIRECTION - Numerals indicate tens of degrees clockwise from true north. 00 indicates calm.
FASTEST MILE WIND - Speed is fastest observed 1-minute value when the direction is in tens of degrees.

% Through 1964. The station did not operate 24 hours daily. Fog and thunderstorm data may be incomplete.
¢ Through 1964.

Mean Maximum Temperature (°F.), January

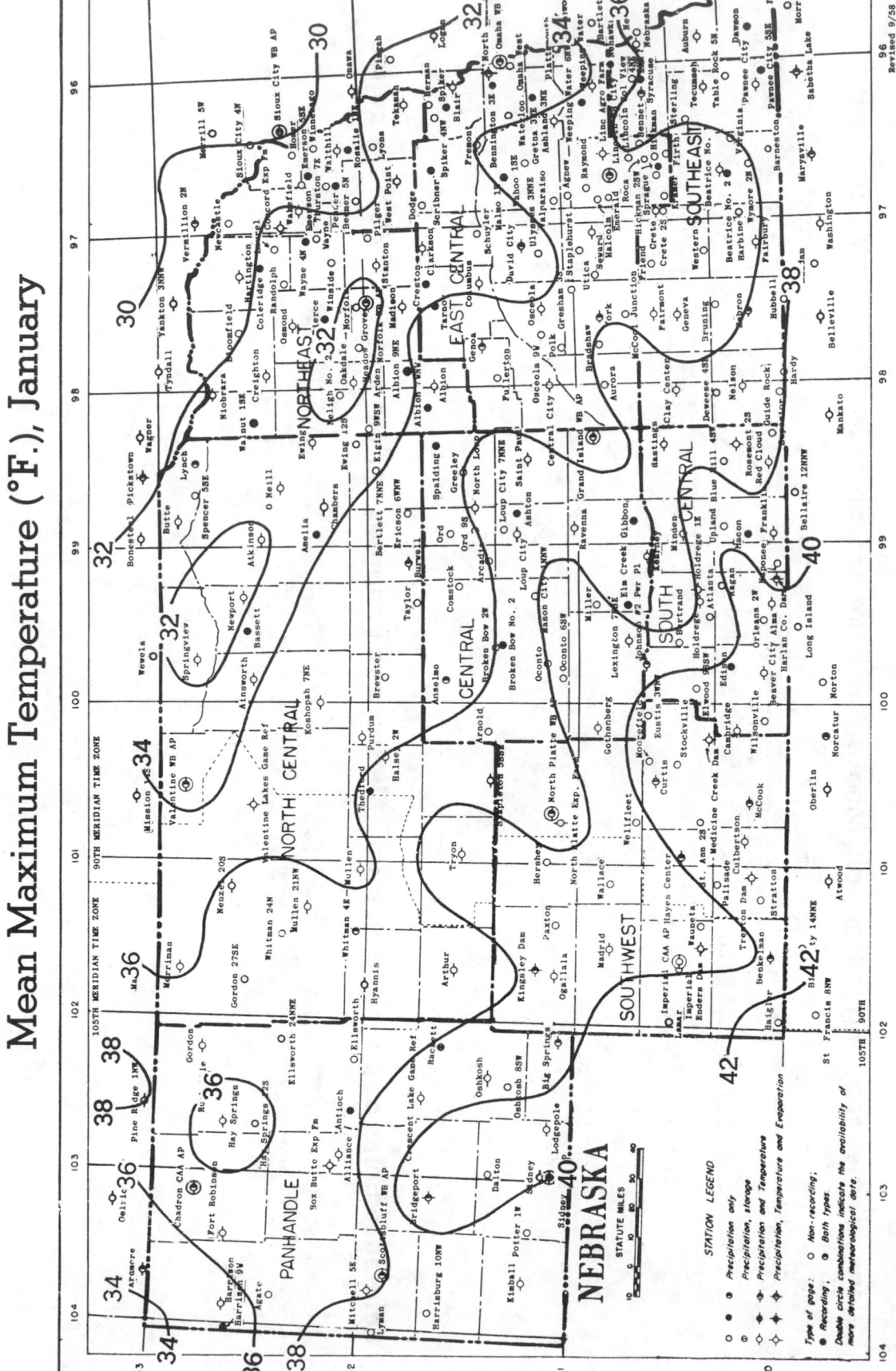

Based on period 1931-52

Isolines are drawn through points of approximately equal value. Caution should be used in interpolating on these maps.

The following precipitation stations are concentrated in such a small area that space does not permit plotting them on the map. Please refer to Station Index for location.

Bennet 4SW	Hallam 3W	Hickman 2WSW	Martell 2NNW	Princeton 3N	Roca 2SE
Crete 3ESE	Hallam 2NNE	Hickman 3W	Martell 5W	Princeton 2N	Roca 6NE
Crete 7NE	Hickman 1N	Holland	Panama 2NW	Princeton 2NW	Sprague 1ESE
				Roca 1NE	

Mean Minimum Temperature (°F.), January

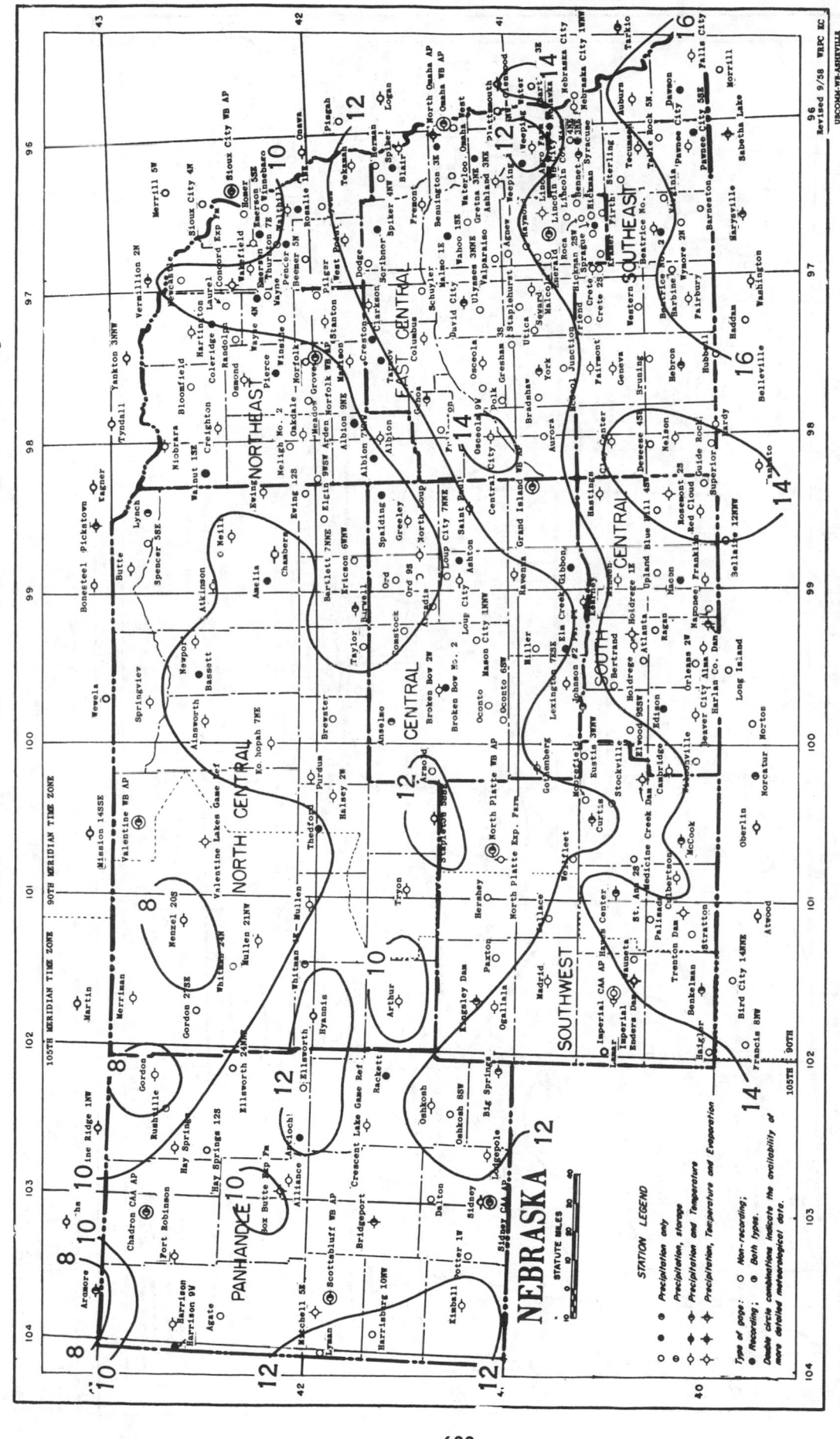

Based on period 1931-52

Isolines are drawn through points of approximately equal value. Caution should be used in interpolating on these maps.

The following precipitation stations are concentrated in such a small area that space does not permit plotting them on the map. Please refer to Station Index for location.

Bennet 4SW	Hallam 3W	Hickman 2WSW	Princeton 3N	Roca 2SE
Crete 3ESE	Hallam 2NNE	Martell 2NW	Princeton 2N	Roca 6SW
Crete 7NE	Hickman 1W	Martell 5W	Princeton 2NW	Sprague 1ESE
	Hickman 3W	Panama 2NW		Roca 1NE

Mean Maximum Temperature (°F.), July

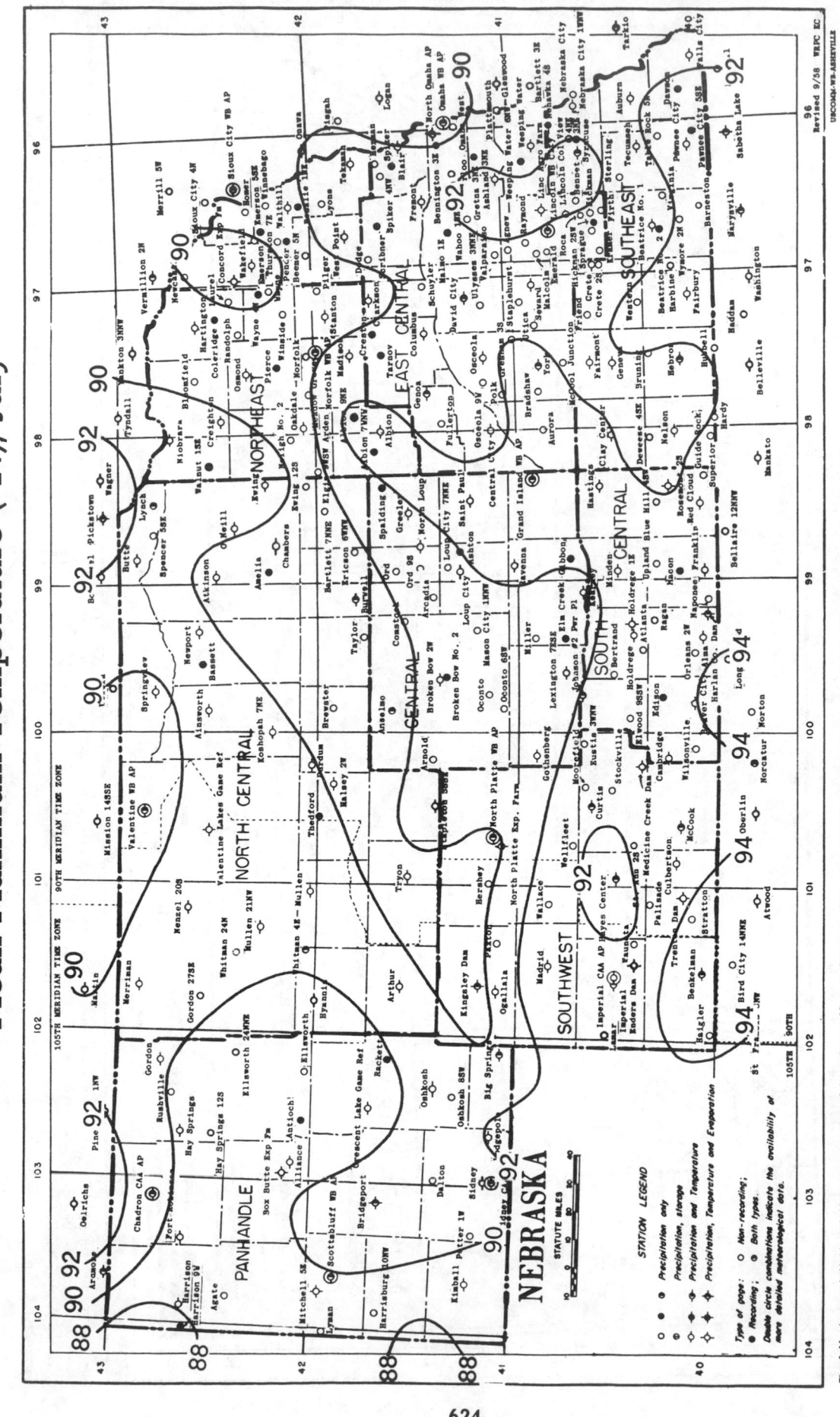

Based on period 1931-52

The following precipitation stations are concentrated in such a small area that space does not permit plotting them on the map. Please refer to Station Index for location.

Bennet 4SW	Hallam 3W	Hickman 2WSW	Martell 2NNW	Princeton 3N	Princeton 3N	Roca 2SE
Crete 3ESE	Hallam 2WNE	Hickman 4W	Martell 5W	Princeton 2N	Roca 2S	Roca 6NE
Crete 7NE	Hickman 1N	Holland	Panama 2NW	Princeton 2NW	Roca 1NE	Sprague 1ESE

Isolines are drawn through points of approximately equal value. Caution should be used in interpolating on these maps.

UNCONG-WB-ASHVILLE

624

Mean Minimum Temperature (°F.), July

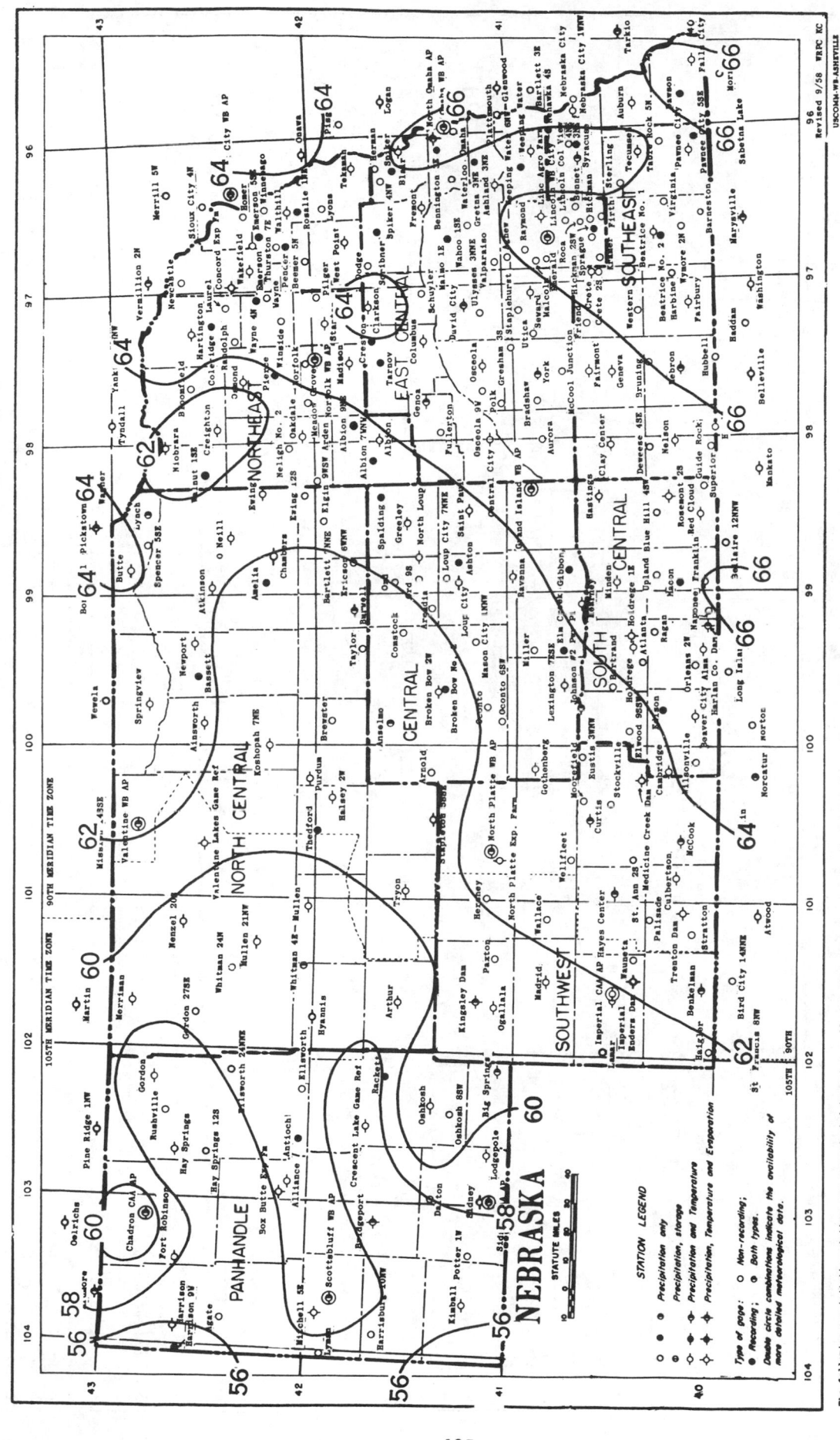

Based on period 1931-52

The following precipitation stations are concentrated in such a small area that space does not permit plotting them on the map. Please refer to Station Index for location.

Bennet 4SW	Hallam 3W	Hickman 2WSW	Princeton	Martell 2NNW	Roca 2SE	Princeton 3N
Crete 3ESE	Hallam 2NNE	Hickman 3W	Princeton 2E	Martell 5W	Roca 2S	Princeton 2N
Crete 7NE	Hickman 1N	Holland	Princeton 2N	Panama 2NW	Roca 1NE	Princeton 2NW
						Sprague 1ESE

Isolines are drawn through points of approximately equal value. Caution should be used in interpolating on these maps.

Mean Annual Precipitation, Inches

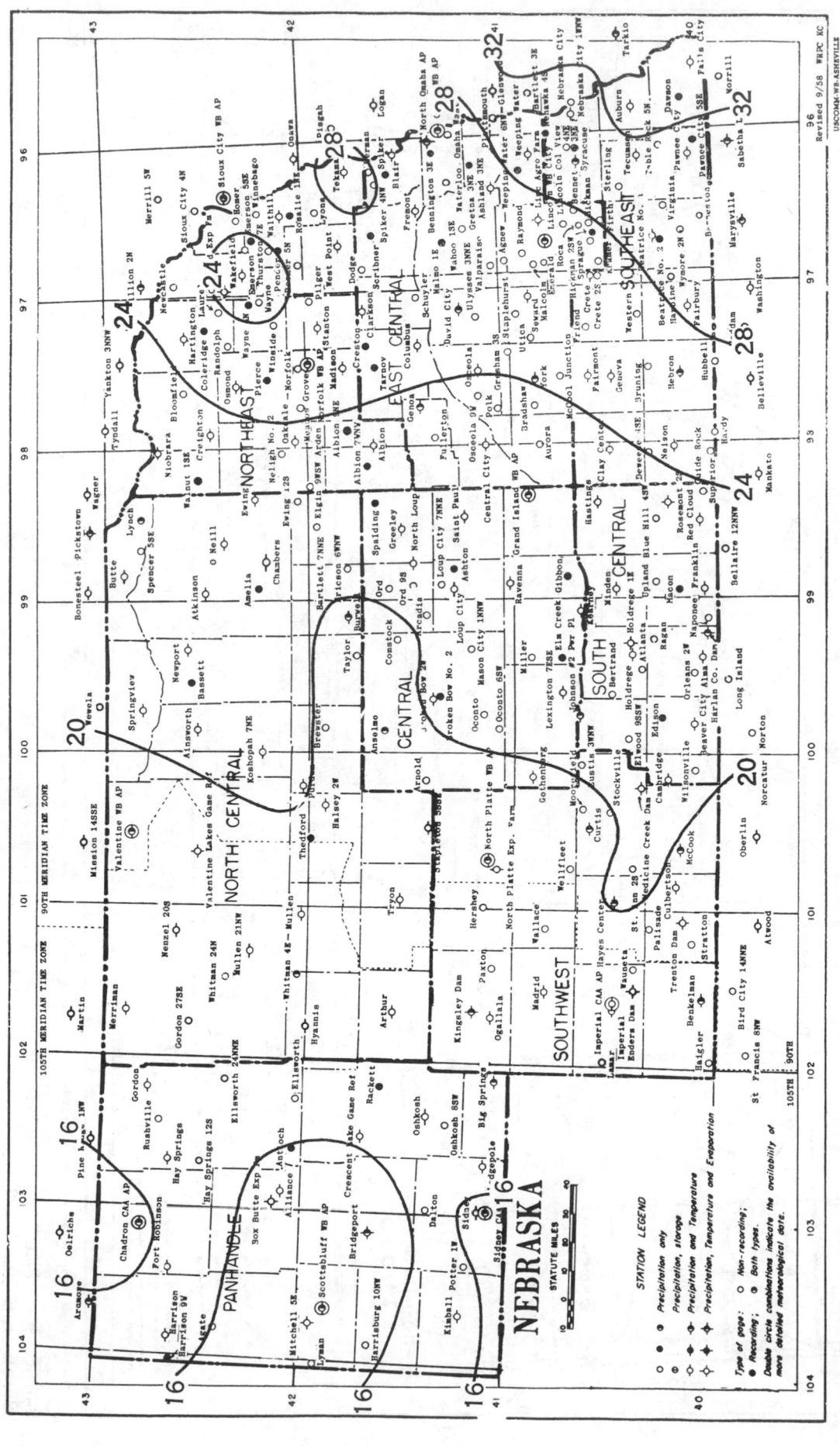

Based on period 1931-55

Isolines are drawn through points of approximately equal value. Caution should be used in interpolating on these maps.

USCOMM-WB-ASHEVILLE
Revised 9/58 WRPC KC

626

CLIMATES OF THE STATES

NEVADA

(Normals, Means and Extremes tables revised 1973 and 1975. Basic report revised February 1960.)

Climate of Nevada

Merle Brown, Weather Bureau State Climatologist

Nevada is primarily a plateau area. The eastern part has an average elevation of between 5,000 and 6,000 feet above sea level; the western portion between, 3,800 and 5,000 feet, the lower limit being in the vicinity of Pyramid Lake and Carson Sink; and the southern part generally between 2,000 and 3,000 feet. From the lower elevations of the west portion there is a fairly rapid rise westward to the summits of the eastern ranges of the Sierra Nevada. The southwestern part slopes down toward Death Valley, California, and the southern portion toward the channel of the Colorado River, the elevation of which is less than 1,000 feet above sea level. The extreme northeastern part slopes northerly, draining into the Snake River and thence into the Columbia.

On the Nevada plateau there are many mountain ranges, most of them 50 to 100 miles long, running generally north and south. The only east-west range is in the northeast, it forms the southern limit of the Columbia River Basin.

With the exception of this small drainage area and another limited region in the southeast which drains into the Colorado River, the State lies within the confines of the Great Basin, and the waters of its streams disappear into sinks or flow into lakes with no outlets.

Nevada lies just east and to the leeward of the Sierra Nevada Range, a massive mountain barrier which has a marked influence on the climate of the State. One of the greatest contrasts in precipitation found within a short distance in the United States occurs between the western, or California, slopes of the Sierras and the valleys just to the east of this range. The prevailing winds are from the west, and as moist air associated with storms from the Pacific Ocean ascends the western slopes of the Sierras, a large portion of the original moisture falls as precipitation. As the air descends the eastern slope, it is warmed by compression, so that very little precipitation occurs. The effects of this mountain barrier are felt not only in the extreme western part, but generally throughout the State, with the result that the lowlands of Nevada are largely desert or semidesert.

With its varied and rugged topography--its mountain ranges, narrow valleys and low, sage-covered deserts, ranging in elevation from about 1,500 to more than 10,000 feet--Nevada presents wide local variations of temperature and rainfall. The most striking climatic features are bright sunshine, small annual precipitation in the valleys and deserts, heavy snowfall in the higher mountains, dryness and purity of air, and phenomenally large daily ranges of temperature.

The mean annual temperatures vary from the middle 40's in the northeastern part to around 50°F. in the west, and to the middle 60's in the south.

In the northeastern portion summers are short and hot, winters long and cold. In the west the summers are also short and hot, but the winters are only moderately cold; while in the south the summers are long and hot and the winters short and mild. Prolonged periods of extremely cold weather are rare, due primarily to the mountains east and north of the State which act as a barrier to the intensely cold continental arctic air masses.

In Nevada there is relatively strong insolation of heat during the day and rapid nighttime cooling, because of the clear air, resulting in wide daily ranges in temperature. Even after the hottest days, the nights are usually cool. At Reno the average range between the highest and the lowest daily temperatures is 29°F. in January, increasing month by month to 45°F. in July.

In summer temperatures above 100°F. occur rather frequently in the extreme southern portion and occasionally over the remainder of the State. However, the humidity is normally low so that corresponding temperatures are less disagreeable in Nevada than in more humid climates. During the warmer season of the year, air conditioning is used in a large percentage of the commercial establishments and family dwellings over the southern portion. Due to the extreme dryness of the air, evaporative coolers operate very efficiently in Nevada's climate. The highest temperature of record is 122°F. observed June 23, 1954 at Overton and August 12 and 18, 1914 at Leeland; the lowest, -50°F. on January 8, 1937 at San Jacinto.

Nevada's precipitation mostly occurs during the winter season and on the average is less than in any other State. Precipitation is lightest over the lower parts of the western plateau, a series of long valleys extending from the state border opposite Death Valley in California northward to the Idaho line. Over the more southerly of those valleys the average annual precipitation is less than 5 inches. From this low average it ranges upward to 18 inches in Lamoille Canyon on the western side of the Ruby Mountains of northeast Nevada, and up to about 28 inches at Marlette Lake high in the most easterly range of the Sierras. Variations in precipitation are due mainly to differences in elevation and exposure to precipitation-bearing winds.

The average annual number of days with measurable precipitation varies considerably. Reno averages 47, Las Vegas 25, Ely 68, Elko 76, and Winnemucca 71 days.

Snowfall is usually heavy in the mountains, particularly in the north. This is conductive to considerable winter sports activity, including skiing and hunting. When winter and spring snowfall is light, as sometimes happens, the result is a shortage of water for irrigation. Long dry spells in summer, which occur rather frequently, are injurious to ranges and pastures, but they are of little moment to irrigated crops which depend almost entirely on stored waters.

Mountain snowfall forms the main source of water for streamflow. In years when winter and spring snowfall is light, the result is a shortage of water for irrigation. Melting of the mountain snowpack in the spring usually causes some flooding in northern and extreme western streams during the period April to June, but damaging floods of this type are infrequent. Rain floods usually occur from November to March in which snowmelt is also a factor. Heavy summer thunderstorms cause flooding in local streams, but they occur usually over sparsely settled mountainous areas and, there-

fore, are seldom destructive. Extensive flooding from melting of a heavy snowpack occurred in northern Nevada from April to June 1952. Heavy rains and melting snow produced damaging floods in western sections in November 1950 and in December 1937.

The State has a generous supply of sunshine, the average percentage of the possible amount at northern and central locations being generally between 65 and 75 percent and at southern locations above 80 percent.

The low humidity and abundant sunshine produce rapid evaporation. Annual amounts in the extreme southern portion of the State, as measured in evaporation pans, average over 100 inches. In northern and central sections amounts average roughly half as much.

Winds are generally light. Storms with high winds rarely occur, and still more rarely cause appreciable damage, except locally along the east slope of the Sierras. The prevailing wind direction is west, although at a few stations, because of local topography, it is south or southwest.

Dust or sandstorms occur occasionally, particularly over the southern part during the spring months when storms are moving through the region more frequently than at other seasons of the year.

Thunderstorms are infrequent, the average annual number being 12 at Winnemucca, 13 at Reno and Las Vegas, 22 at Elko, and 30 at Ely. Summer thunderstorms develop occasionally into heavy local downpours of rain. These storms, locally termed cloudbursts, may bring to a locality as much rain in a few hours as would normally fall in several months. Tornadoes are extremely rare.

Because of the generally arid climate, less than one percent of the land in Nevada is under cultivation. The cultivated area lies mostly in the valleys of the Walker, Carson, and Truckee Rivers, where irrigation is maintained by impounding the waters from melting snows in the Sierra Nevada; in the vicinity of Lovelock, in the lower Humboldt River Valley; and in the valleys of the Muddy and Virgin Rivers, in the southeastern part of the State. A small additional area in pastures and wild hay is watered by flooding when snows melt in the spring. Livestock raising is one of the principal activities of the State. Hay is by far the most important agricultural crop, although small quantities of grains, fruits, vegetables, and cotton are grown.

Over the northern and central portions of the State, freezes continue until late in spring and begin early in autumn. The shortest freeze-free season is in the extreme northeast, and the longest in the extreme south, the range being from less than 100 days at several stations in the northeast to around 140 in the west, and to over 225 in the extreme south.

Mining is the other basic industry of Nevada. The State ranks high in the amount and value of minerals it produces each year, principally manganese, tungsten, mercury, copper, silver, gold, lead, and zinc.

Many tourists travel to Nevada each year to vacation at Lake Tahoe, the "sky-high" lake which straddles the Nevada-California line in the rugged Sierras; to visit the numerous ghost towns; to fish in the lakes or cool mountain streams of the State; and to see the spectacular manmade landmark of the Far West, Hoover Dam, which is located on the State line of Nevada and Arizona.

REFERENCES

Weather Bureau Technical Paper No. 16 - Maximum 24-Hour Precipitation in the United States. Washington, D. C. 1952.

Weather Bureau Technical Paper No. 25 - Rainfall Intensity-Duration-Frequency Curves. For selected stations in the United States, Alaska, Hawaiian Islands and Puerto Rico.

BIBLIOGRAPHY

(A) Climatic Summary of the United States (Bulletin W) 1930 edition, Section 19. U. S. Weather Bureau

(B) Climatic Summary of the United States, Nevada - Supplement for 1931 through 1952 (Bulletin W Supplement). U. S. Weather Bureau

(C) Climatological Data - Nevada. U. S. Weather Bureau

(D) Climatological Data National Summary. U. S. Weather Bureau

(E) Hourly Precipitation Data - Nevada. U. S. Weather Bureau

(F) Local Climatological Data, U.S. Weather Bureau, for Elko, Ely, Las Vegas, Reno, Winnemucca, Nevada.

FREEZE DATA

STATION	Freeze threshold temperature	Mean date of last Spring occurrence	Mean date of first Fall occurrence	Mean No. of days between dates	Years of record Spring	No. of occurrences in Spring	Years of record Fall	No. of occurrences in Fall
ADAVEN	32	06-01	09-28	119	30	30	30	30
	28	05-21	10-09	141	30	30	30	30
	24	04-27	10-21	177	30	30	30	30
	20	04-12	11-02	203	30	30	30	30
	16	03-26	11-14	233	30	30	30	30
ALAMO	32	05-01	10-14	166	25	25	27	27
	28	04-14	10-25	194	25	25	27	27
	24	03-28	11-09	226	25	25	27	27
	20	03-08	11-23	260	25	23	26	25
	16	02-18	12-02	288	25	22	26	21
AUSTIN	32	06-06	09-23	108	28	28	28	28
	28	05-24	10-03	131	28	28	28	28
	24	04-30	10-16	168	28	28	28	28
	20	04-17	10-26	193	28	28	28	28
	16	04-02	11-10	221	28	28	28	28
BEATTY	32	04-16	10-28	195	25	25	26	26
	28	03-27	11-10	227	25	25	26	26
	24	03-15	11-22	253	24	24	26	25
	20	02-24	12-07	286	23	23	26	22
	16	01-29	12-21	326	23	19	26	11
BEOWAWE	32	06-01	09-12	103	26	26	25	25
	28	05-12	09-21	132	26	26	25	25
	24	04-22	10-06	168	23	23	23	23
	20	04-10	10-15	188	24	24	23	23
	16	03-19	10-26	222	23	23	23	23
CARSON CITY	32	05-28	09-20	116	28	28	28	28
	28	05-11	09-24	136	28	28	28	28
	24	04-18	10-11	176	27	27	28	28
	20	03-31	10-25	208	27	27	28	28
	16	03-09	11-07	243	27	27	27	27
ELKO WB AP	32	06-06	09-03	89	30	30	30	30
	28	05-22	09-18	119	30	30	30	30
	24	05-05	09-29	147	30	30	30	30
	20	04-14	10-16	185	30	30	30	30
	16	04-02	10-25	207	30	30	30	30
FALLON EXP STA	32	05-18	09-24	130	30	30	30	30
	28	04-26	10-08	165	30	30	30	30
	24	04-11	10-19	191	30	30	30	30
	20	04-02	11-02	214	30	30	30	30
	16	03-18	11-09	236	30	30	30	30
GERLACH	32	05-12	10-05	146	30	30	30	30
	28	04-24	10-15	174	30	30	30	30
	24	04-06	10-30	206	30	30	30	30
	20	03-18	11-07	234	29	29	30	30
	16	03-03	11-21	263	28	27	30	28
GOLCONDA	32	05-26	09-20	117	28	28	28	28
	28	05-08	09-29	144	27	27	28	28
	24	04-20	10-12	176	26	26	28	28
	20	04-08	10-21	195	27	27	28	28
	16	03-15	11-08	238	27	27	28	28
GOLDFIELD	32	05-16	10-07	144	30	30	29	29
	28	05-02	10-19	169	30	30	29	29
	24	04-14	10-30	200	30	30	29	29
	20	03-28	11-12	229	30	30	28	28
	16	03-11	11-21	254	30	30	28	27
HAWTHORNE	32	04-27	10-19	175	14	14	14	14
	28	04-06	10-29	206	14	14	14	14
	24	03-23	11-09	231	14	14	14	14
	20	03-07	11-25	262	14	14	14	13
	16	02-22	12-09	290	13	13	13	11
IMLAY	32	05-21	09-25	126	27	27	26	26
	28	05-07	10-06	152	27	27	26	26
	24	04-17	10-19	185	26	26	27	27
	20	04-06	10-31	208	26	26	27	27
	16	03-17	11-11	239	26	26	27	27
INDIAN SPRINGS	32	04-06	10-22	200	12	12	12	12
	28	03-25	11-01	221	12	12	12	12
	24	03-03	11-13	255	12	12	12	12
	20	02-19	11-23	278	11	11	12	12
	16	01-31	12-13	316	11	10	12	9
LAMOILLE P H	32	06-06	09-16	102	30	30	30	30
	28	05-17	10-02	138	30	30	30	30
	24	04-29	10-18	172	30	30	30	30
	20	04-13	11-03	204	30	30	30	30
	16	04-04	11-11	222	30	30	30	20
LAS VEGAS WB AP	32	03-13	11-13	245	26	26	28	28
	28	02-17	11-23	279	27	27	29	29
	24	01-28	12-12	318	28	25	29	21
	20	01-12	12-26	348	27	13	29	9
	16	01-04	*	*	28	5	29	3
LOVELOCK	32	05-13	09-25	135	29	29	30	30
	28	04-30	10-09	162	29	29	30	30
	24	04-16	10-21	188	29	29	30	30
	20	04-01	11-02	215	29	29	30	30
	16	03-13	11-11	243	28	28	30	30
MALA VISTA RCH	32	06-20	08-11	52	12	12	12	12
	28	05-31	09-05	96	12	12	11	11
	24	05-17	09-16	122	11	11	10	10
	20	04-30	09-30	153	11	11	11	11
	16	04-13	10-20	190	11	11	11	11
MC GILL	32	05-24	09-27	126	30	30	30	30
	28	05-05	10-12	160	30	30	30	30
	24	04-17	10-22	188	30	30	30	30
	20	04-03	11-04	215	30	30	30	30
	16	03-21	11-11	236	30	30	30	30
MINA	32	04-30	10-06	159	29	29	29	29
	28	04-16	10-20	187	29	29	29	29
	24	04-02	10-31	212	28	28	29	29
	20	03-18	11-09	236	27	27	28	28
	16	03-04	11-21	262	27	27	28	28
MINDEN	32	05-31	09-14	107	27	27	28	28
	28	05-10	09-27	140	27	27	28	28
	24	04-22	10-08	169	28	28	28	28
	20	04-09	10-26	200	28	28	27	27
	16	03-18	11-08	235	28	27	27	27
MONTELLO	32	05-30	09-12	104	28	28	30	30
	28	05-16	09-23	130	28	28	30	30
	24	04-29	10-05	159	28	28	29	29
	20	04-13	10-17	187	28	28	27	27
	16	04-01	10-29	210	27	27	25	25
OROVADA	32	06-04	09-18	106	26	26	24	24
	28	05-16	09-27	134	26	26	24	24
	24	04-25	10-12	169	26	26	23	23
	20	04-08	10-21	196	26	26	24	24
	16	03-24	11-01	222	26	26	24	24
OVERTON	32	03-09	11-11	247	11	11	13	13
	28	02-20	11-28	281	11	11	13	13
	24	01-24	12-16	326	11	9	13	10
	20	01-12	12-29	351	11	5	13	2
	16	01-04	*	*	11	2	13	0
OWYHEE	32	06-05	09-08	95	15	15	12	12
	28	05-20	09-25	128	15	15	12	12
	24	05-04	10-06	155	15	15	12	12
	20	04-13	10-24	193	15	15	12	12
	16	04-01	11-04	218	15	15	12	12
PIOCHE	32	05-25	10-18	145	12	12	12	12
	28	04-23	10-23	183	12	12	12	12
	24	04-05	11-05	214	12	12	12	12
	20	03-21	11-14	239	12	12	12	12
	16	03-05	11-28	268	12	12	12	11
RENO WB AP	32	05-14	10-02	141	30	30	30	30
	28	04-22	10-13	175	30	30	30	30
	24	04-05	10-26	204	30	30	30	30
	20	03-18	11-12	239	30	30	30	30
	16	02-21	11-23	276	30	28	30	30
RUBY LAKE	32	06-03	09-10	99	11	11	11	11
	28	05-15	09-27	134	11	11	11	11
	24	04-24	10-05	163	11	11	11	11
	20	04-15	10-20	188	11	11	11	11
	16	03-24	11-09	230	11	11	11	11

FREEZE DATA

STATION	Freeze threshold temperature	Mean date of last Spring occurrence	Mean date of first Fall occurrence	Mean No. of days between dates	Years of record Spring	No. of occurrences in Spring	Years of record Fall	No. of occurrences in Fall
SAND PASS	32	05-07	10-03	148	29	29	28	28
	28	04-22	10-18	178	29	29	28	28
	24	04-07	10-29	206	29	29	28	28
	20	03-22	11-05	228	29	29	28	28
	16	03-04	11-19	260	29	28	28	27
SCHURZ	32	05-15	09-21	130	29	29	29	29
	28	04-27	10-06	161	29	29	27	27
	24	04-09	10-23	197	29	29	27	27
	20	03-30	11-04	220	29	29	26	26
	16	03-14	11-10	241	29	29	26	26
SEARCHLIGHT	32	03-12	11-23	257	28	28	27	26
	28	02-15	12-08	296	27	25	27	19
	24	01-21	12-21	334	26	15	27	14
	20	01-12	12-24	346	26	11	27	6
	16	01-03	⊕	⊕	26	4	27	3
SHELDON	32	06-25	08-05	40	16	16	17	17
	28	06-17	08-30	74	17	17	17	17
	24	06-02	09-20	110	17	17	18	18
	20	05-14	09-29	138	17	17	18	18
	16	04-21	10-23	185	17	17	18	18
SULPHUR	32	05-21	09-28	130	25	25	25	25
	28	04-28	10-11	166	24	24	25	25
	24	04-11	10-21	193	24	24	25	25
	20	04-01	10-28	210	24	24	25	25
	16	03-16	11-09	238	24	24	25	25
THORNE	32	04-23	10-14	174	29	29	29	29
	28	04-08	10-30	205	29	29	28	28
	24	03-25	11-07	227	29	29	28	28
	20	03-11	11-18	252	29	29	28	28
	16	02-18	11-30	285	29	27	27	23
TONOPAH	32	05-14	10-13	152	29	29	30	30
	28	04-23	10-25	185	30	30	30	30
	24	04-15	11-05	204	30	30	30	30
	20	03-28	11-13	231	30	30	30	30
	16	03-03	11-28	271	29	27	30	25
WINNEMUCCA WB AP	32	05-18	09-21	125	30	30	30	30
	28	05-02	09-29	150	30	30	30	30
	24	04-12	10-16	188	30	30	30	30
	20	03-31	10-23	206	30	30	30	30
	16	03-19	11-08	234	30	30	30	30
YERINGTON	32	05-30	09-09	102	30	30	30	30
	28	05-14	09-26	135	30	30	30	30
	24	04-29	10-08	162	28	28	30	30
	20	04-14	10-21	190	28	28	30	30
	16	04-02	11-04	217	28	28	30	30

Data in the above table are based on the period 1921-1950, or that portion of this period for which data are available.

⊕ When the frequency of occurrence in either spring or fall is one year in ten, or less, mean dates are not given.

Means have been adjusted to take into account years of non-occurrence.

A freeze is a numerical substitute for the former term "killing frost" and is the occurrence of a minimum temperature at or below the threshold temperature of 32°, 28°, etc.

Freeze data tabulations in greater detail are available and can be reproduced at cost.

*MEAN TEMPERATURE AND PRECIPITATION

STATION	JAN Temp	JAN Precip	FEB Temp	FEB Precip	MAR Temp	MAR Precip	APR Temp	APR Precip	MAY Temp	MAY Precip	JUN Temp	JUN Precip	JUL Temp	JUL Precip	AUG Temp	AUG Precip	SEP Temp	SEP Precip	OCT Temp	OCT Precip	NOV Temp	NOV Precip	DEC Temp	DEC Precip	ANN Temp	ANN Precip
NORTHWESTERN																										
CARSON CITY	31.7	2.21	35.9	1.70	41.2	1.23	48.1	.56	55.1	.46	61.9	.38	69.6	.18	67.8	.09	61.2	.29	51.4	.70	40.2	1.24	34.2	2.46	49.9	11.50
EMPIRE	30.2	.82	35.4	.79	42.0	.52	51.0	.44	58.7	.68	66.0	.56	76.1	.17	74.0	.12	65.8	.15	54.2	.50	40.5	.56	33.5	.85	52.3	6.16
FALLON EXP STATION	29.9	.57	35.5	.66	42.1	.55	50.4	.51	57.5	.61	64.2	.42	72.4	.17	68.9	.12	62.0	.20	51.7	.50	39.4	.35	32.6	.68	50.6	5.34
GOLCONDA		.66		.60		.70		.53		.60		.63		.16		.08		.22		.55		.69		.80		6.20
IMLAY		.72		.66		.60		.61		.83		.67		.19		.10		.21		.69		.58		.78		6.64
LAHONTAN DAM	31.8	.47	36.7	.59	43.6	.34	52.4	.37	60.3	.44	67.9	.38	77.9	.11	76.2	.11	67.0	.24	55.6	.40	42.3	.30	34.7	.56	53.9	4.31
LOVELOCK	30.0	.82	35.6	.71	42.5	.54	51.1	.53	59.0	.46	66.1	.62	75.1	.13	72.5	.14	64.7	.20	53.2	.54	40.2	.42	33.2	.65	51.9	5.76
MINDEN	30.8	1.36	35.7	1.26	41.0	.91	48.4	.56	55.3	.43	62.0	.48	69.5	.32	67.7	.21	60.8	.17	51.2	.56	40.0	.83	33.9	1.87	49.7	8.96
OROVADA		1.14		1.14		1.04		1.25		1.51		1.21		.26		.11		.40		1.05		.96		1.21		11.28
PARADISE VALLEY 1 NW		1.37		1.00		.70		.66		.85		.82		.22		.14		.31		.66		.90		1.15		8.78
RENO WB AP	31.2	1.04	36.3	1.05	40.6	.70	47.7	.46	55.3	.48	61.5	.42	69.6	.23	67.4	.23	60.5	.22	50.7	.55	40.2	.64	33.2	.94	49.5	6.96
SAND PASS	29.7	1.01	35.4	.80	42.4	.53	50.5	.47	57.7	.50	64.7	.49	73.1	.18	71.0	.07	63.5	.19	52.9	.45	39.8	.61	33.1	1.23	51.2	6.53
SMITH		.88		.96		.57		.49		.49		.54		.34		.26		.15		.49		.75		1.32		7.24
WINNEMUCCA WB AP	27.8	.96	34.5	1.01	39.4	.86	46.8	.83	55.9	.84	64.0	.79	74.2	.31	69.7	.18	59.9	.34	48.6	.79	37.6	.84	30.0	1.00	49.1	8.75
YERINGTON	30.9	.59	35.9	.56	41.7	.41	49.6	.37	56.3	.51	62.6	.50	70.3	.23	68.6	.16	61.3	.22	51.3	.49	39.4	.45	32.7	.83	50.1	5.32
DIVISION	29.9	1.13	34.8	.98	39.6	.83	48.1	.62	55.3	.75	62.3	.58	71.4	.21	69.2	.14	61.9	.31	51.2	.65	39.6	.81	32.6	1.42	49.7	8.43
NORTHEASTERN																										
ARTHUR 5 NW		2.06		1.70		1.67		1.41		1.58		1.24		.57		.43		.55		1.11		1.47		1.65		15.44
AUSTIN	27.9	1.21	30.5	1.22	35.4	1.50	43.7	1.55	51.3	1.33	59.2	.85	70.0	.57	68.5	.46	60.7	.40	49.5	1.11	37.6	.91	31.7	1.24	47.2	12.35
ELKO WB AP	21.9	1.07	28.4	.95	36.1	.69	44.6	.93	53.0	.95	60.3	.70	70.2	.37	67.6	.29	57.8	.39	47.4	.81	34.5	.93	26.9	1.05	45.7	9.13
ELY WB AP	23.0	.94	28.1	.90	35.3	1.29	43.7	1.20	51.7	1.18	59.6	.50	68.4	.55	66.3	.89	57.8	.68	46.9	.82	35.2	.69	26.9	.88	45.2	10.52
KIMBERLY	24.3	1.55	27.0	1.58	32.5	1.61	42.6	1.35	50.5	.94	59.1	.67	69.1	.90	66.9	.88	58.6	.71	47.6	.90	35.0	.87	28.0	1.58	45.1	13.54
LAMOILLE PH	25.5	1.46	28.3	1.56	34.8	2.01	43.4	2.67	51.2	2.17	58.5	1.48	69.1	.74	67.4	.50	58.1	.76	48.4	1.48	36.0	1.61	28.7	1.72	45.8	18.16
MC GILL	26.0	.70	29.6	.64	35.6	.77	44.9	1.02	53.3	.80	61.8	.65	71.4	.76	69.6	.74	61.1	.52	49.4	.79	36.9	.60	29.9	.70	47.5	8.69
MONTELLO	23.3	.57	28.5	.38	37.2	.26	46.9	.59	55.1	.76	62.9	.65	72.4	.63	69.5	.46	59.9	.37	49.0	.44	34.9	.61	27.0	.60	47.2	6.32
OWYHEE		1.38		1.19		1.25		1.35		1.64		1.22		.36		.25		.40		1.05		1.06		1.38		12.53
DIVISION	24.1	1.13	28.3	.88	34.4	1.03	43.8	1.15	51.3	1.12	58.7	.88	68.4	.51	66.2	.41	57.7	.44	47.4	.76	35.1	.94	27.4	1.21	45.2	10.46
SOUTH CENTRAL																										
ADAVEN	28.6	1.54	31.1	1.58	37.0	1.50	46.0	1.13	53.7	.76	61.8	.55	69.9	1.02	67.9	1.15	61.4	.55	50.1	.96	38.9	.87	31.9	1.39	48.2	13.00
CALIENTE	30.0	.86	35.7	.80	43.6	.92	52.5	.74	60.5	.51	68.2	.38	76.0	.84	74.0	1.06	66.1	.53	54.1	.85	41.7	.63	33.4	.99	53.0	9.11
MINA	31.8	.34	36.6	.26	43.3	.31	52.0	.42	60.5	.40	68.9	.27	78.3	.24	75.7	.25	66.4	.14	54.3	.39	41.6	.21	34.3	.34	53.6	3.57
SCHURZ	32.5	.56	37.5	.57	43.5	.44	51.3	.46	58.7	.63	65.5	.43	73.4	.38	71.7	.19	63.8	.20	53.7	.44	41.5	.40	34.5	.69	52.3	5.39
DIVISION	29.5	.57	34.3	.55	39.7	.62	49.0	.64	56.6	.37	64.4	.24	73.4	.66	71.3	.53	63.8	.38	52.7	.48	40.5	.48	33.5	.71	50.7	6.23
EXTREME SOUTHERN																										
ALAMO		.75		.71		.74		.58		.40		.18		.80		.83		.33		.47		.38		.69		6.86
BEATTY		.66		.72		.54		.43		.20		.08		.22		.26		.18		.30		.38		.62		4.59
BOULDER CITY	45.3	.74	49.5	.60	56.4	.55	65.6	.38	73.9	.13	82.2	.06	89.1	.62	87.0	.86	81.2	.52	68.9	.24	55.5	.28	47.6	.61	66.9	5.39
LAS VEGAS WB AP	44.2	.74	50.4	.58	56.5	.35	65.6	.24	74.1	.16	83.6	.13	90.5	.46	88.4	.53	80.7	.34	67.4	.32	53.9	.22	46.8	.58	66.8	4.35
SEARCHLIGHT		.96		.86		.85		.36		.15		.10		1.10		1.07		.77		.41		.41		.95		7.99
DIVISION	42.7	.67	47.2	.51	53.5	.49	62.9	.28	70.7	.14	78.6	.09	86.3	.46	84.2	.46	77.3	.36	65.5	.26	52.4	.32	44.6	.52	63.8	4.56

* Averages for period 1931 – 1955, except for stations marked WB which are "normals" based on period 1921 – 1950. Divisional means may not be the arithmetical average of individual stations published, since additional data from shorter period stations are used to obtain better areal representation.

CONFIDENCE LIMITS

In the absence of trend or record changes, the chances are 9 out of 10 that the true mean will lie in the interval formed by adding and subtracting the values in the following table from the means for any station in the State. Because of the wider variation in mean precipitation, the corresponding monthly means and annual mean must be substituted for "p" in the precipitation table below to obtain mean precipitation confidence limits.

2.2	$.27\sqrt{p}$	1.9	$.28\sqrt{p}$	1.2	$.24\sqrt{p}$	1.1	$.23\sqrt{p}$	1.1	$.26\sqrt{p}$	1.1	$.24\sqrt{p}$	.6	$.22\sqrt{p}$	.7	$.27\sqrt{p}$	1.0	$.29\sqrt{p}$	1.0	$.28\sqrt{p}$	1.2	$.26\sqrt{p}$	1.5	$.27\sqrt{p}$	.5	$.26\sqrt{p}$

COMPARATIVE DATA

Data in the following table are the mean temperature and average precipitation for Fallon, Nevada for the period 1906-1930 and are included in this publication for comparative purposes:

30.0	.60	36.7	.45	43.3	.42	50.2	.51	57.3	.50	65.5	.32	73.5	.15	71.0	.27	61.0	.33	50.4	.37	39.9	.33	31.4	.53	50.8	4.81

NORMALS, MEANS AND EXTREMES
(Table Revised 1973. Base Period for Climatological Normals: 1931-1960)

Station: ELKO, NEVADA — MUNICIPAL AIRPORT
Standard time used: PACIFIC Latitude: 40° 50' N Longitude: 115° 47' W Elevation (ground): 5050 feet

Temperature

Month	Normal Daily maximum	Normal Daily minimum	Normal Monthly	Record highest	Year	Record lowest	Year
(a)	(b)	(b)	(b)	8		8	
J	35.1	10.0	22.6	62	1970	-18	1971
F	40.4	15.6	28.0	64	1968	-13	1966
M	49.0	22.2	35.6	77	1966	-4	1971
A	60.1	28.5	44.3	86	1969+	10	1970
M	69.2	34.8	52.0	91	1968	19	1965
J	78.8	41.1	60.0	98	1968	28	1966
J	90.8	48.3	69.6	102	1967	34	1966
A	88.8	44.9	66.9	102	1967	29	1965
S	79.5	36.2	57.9	94	1967	12	1971
O	65.8	28.0	46.9	86	1965	10	1971
N	49.3	20.6	34.2	74	1965	-9	1964
D	39.1	13.7	26.4	57	1964	-28	1972
YR	62.2	28.5	45.4	102 AUG. 1967+		-28 DEC. 1972	

Precipitation

Month	Normal total (b)	Max monthly	Year	Min monthly	Year	Max in 24 hrs	Year
J	1.16	3.35	1956	0.04	1961	1.27	1951
F	0.89	2.93	1932	0.08	1967	0.89	1936
M	0.83	1.82	1957	0.18	1956	0.97	1967
A	0.82	2.17	1963	0.10	1949	1.10	1963
M	0.96	4.09	1971	T	1954	1.73	1971
J	0.71	2.61	1963	0.01	1951	1.85	1968
J	0.40	2.35	1950	0.00	1950	1.04	1950
A	0.30	4.41	1970	0.00	1959	4.13	1970
S	0.75	1.74	1951+	T	1951+	1.19	1959
O	0.88	2.76	1938	T	1958+	1.31	1939
N	1.01	2.77	1942	T	1942	1.31	1950
D		3.30	1963	0.12	1962	1.62	1955
YR	9.05	4.61 AUG. 1970		0.00 JUL. 1963		4.13 AUG. 1970	

Normal heating degree days (Base 65°F) (b): 1314, 1036, 911, 621, 409, 192, 9, 34, 225, 561, 924, 1197 | YR 7433

Snow, Ice pellets

Month	Mean total	Max monthly	Year	Max in 24 hrs	Year
J	10.2	27.4	1950	16.7	1951
F	8.0	26.1	1932	9.1	1949
M	5.9	23.2	1967	13.8	1967
A	2.5	14.3	1963	8.5	1963
M	1.1	11.3	1971	8.6	1971
J	0.7	T	1971+	T	1971+
J	0.0	0.0	1950+	T	1950+
A	0.0	0.0	1949+	T	1949+
S	0.8	1.9	1948	1.9	1948
O	3.4	5.4	1971	5.2	1963
N	4.3	16.6	1944	9.0	1944
D	8.6	31.2	1955	9.2	1955
YR	39.7	31.2 DEC. 1955		16.7 JAN. 1951	

Relative humidity (Local time)

Month	Hour 04	Hour 10	Hour 16	Hour 22
J	76	70	59	74
F	77	64	47	72
M	75	50	35	64
A	68	36	25	53
M	71	37	29	55
J	(—)	—	—	—
J	56	25	19	43
A	60	29	23	46
S	62	32	23	47
O	65	40	27	53
N	73	61	49	68
D	74	71	65	74
YR	69	47	37	58

Wind

Month	Mean speed	Prevailing direction	Fastest mile Speed	Direction	Year
J	5.5	SW	40	23	1951
F	5.8	SW	39	27	1963
M	6.7	SW	41	29	1952
A	7.1	SW	48	25	1956
M	6.6	SW	55	34	1955
J		SW	45	23	1953
J	6.2	SW	36	16	1955
A	6.0	SW	35	16	1966+
S	5.5	SW	58	27	1959
O	5.2	SW	58	25	1953
N	5.0	SW	40	27	1957
D	5.1	SW	50	27	1952
YR	6.0	SW	58	27	SEP. 1959

Mean number of days

Month	Mean sky cover sunrise to sunset (tenths)	Clear	Partly cloudy	Cloudy	Precip ≥ .01 in	Snow ≥ 1.0 in	Thunderstorms	Heavy fog	Max 90° and above	Max 32° and below	Min 32° and below	Min 0° and below	Avg daily solar radiation (langleys)
J	6.9	7	7	17	9	3	*	8	0	8	28	5	
F	6.5	6	8	14	8	2	*	4	0	4	28	2	
M	6.5	7	8	16	7	1	*	1	0	1	28	*	
A	6.5	7	10	13	6	0	1	1	0	0	23	0	
M	6.1	8	13	8	6	0	4	*	0	0	9	0	
J	4.4	13	13	4	4	0	4	*	2	0	1	0	
J	3.2	18	9	4	3	0	5	*	20	0	*	0	
A	3.3	17	9	5	3	0	5	*	17	0	*	0	
S	3.2	18	8	5	3	0	5	*	7	0	2	*	
O	4.3	14	6	8	5	1	2	*	0	0	7	0	
N	6.1	9	6	14	7	4	*	1	0	0	22	2	
D	6.7	7	7	17	9	4	*	1	0	8	23	3	
YR	5.3	133	97	135	78	15	22	6	44	29	196	14	

Ø For period November 1964 through the current year.

Means and extremes above are from existing and comparable exposures. Annual extremes have been exceeded at other sites in the locality as follows: Highest temperature 107 in July 1937; lowest temperature -43 in January 1890; maximum monthly precipitation 6.00 in January 1903; maximum monthly snowfall 48.5 in January 1916.

NORMALS, MEANS AND EXTREMES
(Table Revised 1975. Base Period for Climatological Normals: 1941-1970)

Temperatures °F

Month	Normal Daily maximum	Normal Daily minimum	Normal Monthly	Record highest	Year	Record lowest	Year
(a)				10		10	
J	36.0	10.4	23.2	62	1970	-18	1971
F	41.6	16.8	29.2	64	1968	-13	1966
M	48.4	22.0	35.2	77	1966	-4	1971
A	58.8	28.1	43.5	86	1969	10	1970
M	68.5	35.2	51.9	91	1968	19	1965
J	77.5	41.6	59.6	98	1968	28	1966
J	90.4	48.6	69.5	102	1967	34	1966
A	88.2	46.4	67.0	102	1967	29	1965
S	78.8	36.4	57.6	94	1967	12	1971
O	65.8	26.6	36.4	86	1965	-9	1971
N	49.0	20.6	34.8	74	1965	-9	1964
D	38.2	13.5	25.9	57	1964	-28	1972
YR	61.8	28.9	45.4	102 AUG. 1967		-28 DEC. 1972	

Normal Degree days Base 65°F

Month	Heating	Cooling
J	1296	0
F	1002	0
M	930	0
A	645	0
M	406	0
J	190	28
J	27	166
A	60	122
S	248	0
O	561	0
N	906	0
D	1212	0
YR	7483	342

Precipitation in inches

Month	Water equivalent Normal	Max monthly	Year	Min monthly	Year	Max in 24 hrs	Year
J	1.16	3.35	1956	0.08	1961	1.27	1951
F	0.77	2.93	1932	0.08	1967	0.89	1936
M	0.83	2.17	1956	0.18	1956	0.97	1967
A	0.82	2.17	1963	0.10	1949	1.10	1963
M	1.01	4.09	1974	T	1974	1.73	1971
J	1.01	2.61	1963	T	1963	1.85	1968
J	0.41	2.35	1950	0.00	1963	1.04	1950
A	0.61	4.61	1970	0.00	1969	4.13	1970
S	0.34	1.74	1959	T	1951	1.19	1959
O	0.66	2.76	1938	T	1958	1.31	1939
N	1.01	2.77	1942	T	1942	1.31	1950
D	1.13	3.30	1964	0.12	1950	1.62	1955
YR	9.78	4.61 AUG. 1970		0.00 JUL 1963		4.13 AUG. 1970	

Snow, Ice pellets

Month	Max monthly	Year	Max in 24 hrs	Year
J	27.4	1950	16.7	1951
F	26.1	1932	9.1	1949
M	23.2	1967	13.8	1967
A	14.3	1963	8.5	1963
M	11.3	1971	8.6	1971
J	T	1971	T	1971
J	0.0	1950	T	1950
A	0.0	1949	T	1949
S	1.9	1948	1.9	1948
O	5.4	1971	5.2	1963
N	16.6	1944	9.0	1944
D	31.2	1955	9.2	1955
YR	31.2 DEC. 1955		16.7 JAN. 1951	

Relative humidity pct. (Local time)

Month	Hour 04	Hour 10	Hour 15	Hour 22
J	74	69	58	73
F	75	64	48	71
M	73	51	36	63
A	68	43	30	61
M	66	34	28	50
J	67	33	26	45
J	55	25	19	37
A	56	28	21	40
S	64	32	22	47
O	64	40	27	53
N	73	62	46	67
D	70	62	53	73
YR	67	46	36	56

Wind

Month	Mean speed m.p.h.	Prevailing direction	Fastest mile Speed m.p.h.	Direction	Year
J	5.5	SW	40	23	1951
F	5.8	SW	39	27	1963
M	6.7	SW	41	29	1952
A	7.2	SW	48	25	1956
M	6.7	SW	55	34	1955
J		SW	45	23	1953
J	6.2	SW	36	16	1955
A	6.0	SW	35	16	1966
S	5.8	SW	58	29	1959
O	5.1	SW	35	25	1953
N	5.1	SW	40	27	1957
D	5.2	SW	50	27	1952
YR	6.0	SW	58	27	SEP. 1959

Mean number of days

Month	Mean sky cover sunrise to sunset (tenths)	Clear	Partly cloudy	Cloudy	Precip ≥ .01 in	Snow, Ice pellets ≥ 1.0 in	Thunderstorms	Heavy fog, visibility ¼ mile or less	Max 90° and above	Max 32° and below	Min 32° and below	Min 0° and below	Average station pressure mb. Elev. 5077 feet m.s.l.
	25	38	38	44	26	26	26		10	10	10	10	2
J	6.8	7	7	17	9	3	*	1	0	10	28	5	844.4
F	6.5	6	8	14	9	2	*	1	0	4	26	1	845.8
M	6.5	7	7	16	8	2	1	*	0	1	26	*	840.5
A	6.5	8	11	12	7	1	3	*	0	0	23	0	843.4
M	6.3	8	11	6	6	*	4	*	0	0	9	0	843.3
J	4.0	13	9	6	4	0	5	*	20	0	1	0	842.4
J	3.2	18	9	4	3	0	5	*	16	0	*	0	845.6
A	3.1	19	8	4	3	0	5	*	0	0	*	0	845.1
S	3.1	18	8	4	3	0	2	*	0	0	1	*	846.4
O	4.3	14	9	8	5	1	1	*	0	0	7	0	846.8
N	6.2	9	6	15	7	4	*	1	0	1	21	*	846.0
D	6.7	7	7	17	9	4	*	1	0	13	29	6	847.0
YR	5.3	134	96	135	78	15	21	6	44	28	193	13	844.8

Means and extremes above are from existing and comparable exposures. Annual extremes have been exceeded at other sites in the locality as follows: Highest temperature 107 in July 1937; lowest temperature -43 in January 1890; maximum monthly precipitation 6.00 in January 1903; maximum monthly snowfall 48.5 in January 1916.

REFERENCE NOTES APPLYING TO TABLES APPEAR ON THE PAGE FOLLOWING LAST TABLE.
(Caution: Letters and symbols may have different meanings in 1941-1970 tables than in earlier tables. See notes.)

NORMALS, MEANS AND EXTREMES
(Table Revised 1973. Base Period for Climatological Normals: 1931-1960)

Station: ELY, NEVADA   YELLAND FIELD   Standard time used: PACIFIC   Latitude: 39° 17' N   Longitude: 114° 51' W   Elevation (ground): 6253 feet

Temperature (°F) and Heating Degree Days

Month	Normal Daily maximum	Normal Daily minimum	Normal Monthly	Record highest	Year	Record lowest	Year	Normal heating degree days (Base 65°)
J	36.8	8.7	22.8	68	1951	-27	1949	1308
F	39.2	13.3	26.6	66	1963	-25	1949	1075
M	47.2	19.7	33.5	73	1966	-13	1952	977
A	57.6	27.5	42.6	78	1962+	-5	1963	672
M	67.6	33.8	50.3	87	1974	7	1950	456
J	77.2	40.0	58.6	99	1954	19	1950	225
J	86.8	48.1	67.5	97	1960	30	1968+	28
A	85.1	46.1	65.6	96	1972	24	1960	43
S	76.9	38.1	57.5	93	+50	15	1968	234
O	63.0	28.7	45.9	84	1967	-3	1971	592
N	49.1	18.2	33.7	67	1956	-15	1964	939
D	40.8	12.7	26.8	67	1958	-28	1972	1184
YR	60.6	28.0	44.3	99	JUN. 1954	-28	DEC. 1972	7733

Precipitation

Month	Normal total	Maximum monthly	Year	Minimum monthly	Year	Maximum in 24 hrs.	Year
J	0.78	1.92	1952	T	1948	0.95	1952
F	0.70	2.19	1969	0.01	1972	1.54	1969
M	0.85	2.40	1952	0.07	1972+	0.86	1952
A	0.95	2.77	1964	0.16	1966+	1.04	1947
M	0.85	3.53	1963	T	1948	1.42	1955
J	0.50		1963	T	1946+	1.50	1963
J	0.65	1.81	1970	T	1948+	1.22	1952
A	0.56	1.23	1960	T	1962+	0.91	1957
S	0.73	2.26	1945	T	1953+	1.25	1963
O	0.59	1.76	1961	0.00	1941	1.09	1968
N	0.68	1.82	1960	T	1959	1.29	1960
D		2.11	1966	T	1962	1.12	1966
YR	8.33	3.53	JUN. 1963	0.00	OCT. 1952	1.54	FEB. 1969

Snow, Ice pellets

Month	Mean total	Maximum monthly	Year	Maximum in 24 hrs.	Year
J	8.5	24.8	1967	13.1	1967+
F	8.6	19.9	1959	10.4	1959
M	8.6	24.8	1958	10.6	1952
A	6.4	24.8	1963	10.7	1972+
M	1.8	10.8	1964	7.2	1964
J	0.3	5.6	1939	5.6	1939
J	0.0	0.0		0.0	
A	0.0	0.0		0.0	
S	0.1	2.2	1971	2.3	1971
O	1.8	9.1	1954	7.3	1954
N	4.6	15.3	1946	10.4	1946
D	7.9	22.3	1968	12.7	1968
YR	46.7	24.8	JAN. 1967+	13.1	JAN. 1967+

Relative humidity pct. (Local time) — Wind

Month	Hour 04	Hour 10	Hour 16	Hour 22	Mean speed	Prevailing direction	Fastest mile Speed	Direction	Year
J	71	59	51	71	10.7	S	66	SE	1952
F	75	58	47	72	10.7	S	56	SE	1954
M	74	54	46	57	11.1	S	59	S	1951
A	68	40	34	44	11.4	S	59	S	1948
M	60	34	30		11.0	S	74	SW	1952
J	60	33	25		10.7	S	63		
J	57	23	21	38	10.4	S	50	S	1957
A	56	26	22	41	10.4	S	57	S	1954
S	65	37	24	43	10.5	S	57	S	1953
O	72	52	46	68	10.4	S	65	S	1950
N	72	55	55	71	10.2	S	51	S	1954
D					10.3	S	61	SE	1952
YR	65	41	36	56	10.6	S	74	S	MAY 1948

Sunshine / Sky cover / Mean number of days

Month	Pct. of possible sunshine	Mean sky cover (sunrise to sunset)	Clear	Partly cloudy	Cloudy	Precipitation .01 in. or more	Snow, Ice pellets 1.0 in. or more	Thunderstorms	Heavy fog	Max 90° and above	Max 32° and below	Min 32° and below	Min 0° and below
J	65	6.2	9	7	15	6	3	*	*	0	8	31	7
F	66	6.3	7	7	14	7	2	*	*	0	5	28	4
M	67	6.1	9	8	14	8	3	*	*	0	3	30	1
A	67	6.1	8	10	12	8	3	1	*	0	*	24	*
M					7	7	1	4	*	0	0	13	0
J	78	4.4				5	*	5	*	2	0	*	0
J	80	3.9	15	10	5	5	0	8	0	9	0	1	0
A	81	3.8	15	13	5	5	0	8	0	5	0	8	0
S	83	4.2	15	8	8	5	*	3	*	1	0	22	*
O	75		11	8	12	5	1	1	*	0	*	28	1
N	66	5.8	9	8	12	6	3	*	*	0	3	30	5
D	64	6.1			14					0	7		
YR	73	5.2	133	108	124	72	15	32	2	16	27	218	19

NORMALS, MEANS AND EXTREMES
(Table Revised 1975. Base Period for Climatological Normals: 1941-1970)

Temperature (°F) and Degree Days (Base 65°F)

Month	Normal Daily maximum	Normal Daily minimum	Normal Monthly	Record highest	Year	Record lowest	Year	Heating	Cooling
J	38.0	9.2	23.6	68	1951	-27	1949	1283	0
F	41.4	14.3	27.9	66	1963	-25	1949	1039	0
M	46.4	18.2	32.8	73	1966	-13	1952	998	0
A	56.4	24.1	41.3	78	1962+	-7	1963	711	0
M	66.3	33.0	50.0	87	1974	11	1950	470	5
J	75.7	39.7	57.7	99	1954	19	1950	241	22
J	86.3	48.0	67.2	97	1960	30	1968	23	92
A	84.2	46.8	65.5	96	1972	24	1960	62	97
S	76.1	37.3	56.7	93	+50	15	1968	265	16
O	63.0	28.4	46.0	84	1967	-3	1971	589	0
N	49.1	18.9	34.0	67	1956	-15	1964	930	0
D	41.1	12.0	26.2	67	1958	-28	1972	1203	0
YR	60.4	27.8	44.1	99	JUN 1954	-28	DEC 1972	7814	212

Precipitation in inches — Water equivalent

Month	Normal	Maximum monthly	Year	Minimum monthly	Year	Maximum in 24 hrs.	Year
J	0.64	1.92	1952	T	1948	0.95	1952
F	0.60	2.06	1969	0.01	1972	1.54	1969
M	0.85	2.40	1952	0.07	1972+	0.86	1952
A	1.00	2.77	1964	0.16	1966+	1.04	1947
M	0.93	3.05	1967	T	1948	1.42	1955
J	0.93	3.53	1963	T	1974	1.50	1963
J	0.61	1.81	1948	T	1948+	1.22	1952
A	0.56	2.06	1970	T	1962	0.91	1957
S	0.61	1.76	1953	T	1953	1.25	1963
O	0.60	1.76	1961	0.00	1941	1.09	1968
N	0.66	1.82	1960	T	1959	1.29	1960
D	0.71	2.11	1966	T	1962	1.12	1966
YR	8.70	3.53	JUN 1963	0.00	OCT 1952	1.54	FEB 1969

Snow, Ice pellets

Month	Maximum monthly	Year	Maximum in 24 hrs.	Year
J	24.8	1967	13.1	1967
F	19.9	1959	10.4	1959
M	24.8	1958	10.7	1952
A	24.5	1963	10.7	1972+
M	10.8	1964	7.2	1964
J	5.6	1939	5.6	1939
J	0.0		0.0	
A	0.0		0.0	
S	2.2	1971	2.3	1971
O	9.1	1954	7.3	1954
N	15.3	1946	10.4	1946
D	22.3	1968	12.7	1968
YR	24.8	JAN 1967	13.1	JAN 1967

Relative humidity pct. (Local time) — Wind

Month	Hour 04	Hour 10	Hour 16	Hour 22	Mean speed m.p.h.	Prevailing direction	Fastest mile Speed m.p.h.	Direction	Year
J	71	60	55	70	10.6	S	66	SE	1952
F	75	59	51	72	10.6	S	56	SE	1954
M	74	51	41	57	11.1	S	59	S	1951
A	68	40	34	44	11.4	S	59	S	1948
M	64	34	29	43	11.0	S	74	SW	1952
J	58	23	21	38	10.7	S	63		
J	51	23	21	38	10.4	S	50	S	1957
A	55	26	22	41	10.4	S	57	S	1954
S	65	37	31	43	10.5	S	65	S	1950
O	72	52	46	68	10.4	S	65	S	1950
N	73	55	58	71	10.2	S	51	SE	1954
D					10.3	S	61		1952
YR	65	41	36	56	10.6	S	74	S	MAY 1948

Sunshine / Sky cover / Mean number of days

Month	Pct. of possible sunshine	Mean sky cover, tenths (sunrise to sunset)	Clear	Partly cloudy	Cloudy	Precipitation .01 in. or more	Snow, Ice pellets 1.0 in. or more	Thunderstorms	Heavy fog, visibility ¼ mile or less	Max 90° and above	Max 32° and below	Min 32° and below	Min 0° and below	Average station pressure mb.
J	65	6.2	9	7	15	7	2	*	*	0	8	31	8	807.2
F	66	6.3	7	7	14	8	3	*	*	0	5	28	4	808.9
M	71	6.1	9	8	14	8	3	*	*	0	3	30	1	803.8
A	68	6.1	8	9	13	7	2	1	*	0	*	25	*	806.7
M	75	5.9	14	12	7	5	1	4	*	*	0	13	0	807.3
J	79	4.4	14	13	5	5	*	5	*	5	0	1	0	809.2
J	80	3.9	14	14	5	4	0	8	0	9	0	*	0	811.2
A	83	3.8	14	13	5	4	0	8	0	5	0	8	0	810.7
S	83	4.2	15	8	8	5	*	3	*	1	0	22	*	810.9
O	75		15	8	12	5	1	1	*	0	*	27	1	810.7
N	66	5.8	8	12	12	6	3	*	*	0	3	30	5	808.4
D	63	6.1	8	8	15	6	3	*	*	0	7			809.7
YR	73	5.2	131	110	124	71	15	32	2	16	26	218	19	808.7

Elev. 6262 feet m.s.l.

REFERENCE NOTES APPLYING TO TABLES APPEAR ON THE PAGE FOLLOWING LAST TABLE.
(Caution: Letters and symbols may have different meanings in 1941-1970 tables than in earlier tables. See notes.)

NORMALS, MEANS AND EXTREMES

(Table Revised 1973. Base Period for Climatological Normals: 1931-1960)

Station: LAS VEGAS, NEVADA — MCCARRAN INTERNATIONAL AIRPORT Standard time used: PACIFIC Latitude: 36° 05' N Longitude: 115° 10' W Elevation (ground): 2162 feet

Month	Temperature Normal Daily maximum	Daily minimum	Monthly	Record highest	Year	Record lowest	Year	Normal heating degree days (Base 65°)
J	54.2	32.0	43.1	75	1963	8	1963	688
F	59.4	36.9	47.8	82	1972	19	1972	487
M	67.6	42.0	54.8	91	1966	23	1971+	335
A	77.6	51.0	64.5	96	1962	32	1967	111
M	87.9	59.0	73.8	106	1969+	40	1964	16
J	97.2	68.5	82.9	115	1970	51	1971+	0
J	103.5	75.8	89.7	116	1972	62	1970	0
A	101.0	73.4	87.2	116	1969	56	1968	0
S	94.3	65.6	80.0	115	1971	46	1965	6
O	80.1	53.1	66.6	99	1963	26	1971	78
N	65.0	39.9	52.5	83	1967	28	1968	387
D	56.2	33.9	45.1	71	1966	15	1968	617
YR	78.7	52.6	65.7	116 JUL 1972		8 JAN 1963		2709

Precipitation

Month	Normal total	Maximum monthly	Year	Minimum monthly	Year	Maximum in 24 hrs.	Year
J	0.53	2.41	1949	T	1971+	1.01	1949
F	0.44	0.96	1969	0.00	1967	0.63	1966+
M	0.35	1.50	1952	0.00	1972+	1.14	1963+
A	0.23	2.44	1965	0.00	1970+	0.80	1970+
M	0.08	0.96	1969	0.00	1970+	0.75	1969
J	0.04	0.82	1967	0.00	1958+		1967
J	0.50	1.64	1956	0.00	1963	1.32	1963
A	0.48	2.59	1957	0.00	1956	2.59	1957
S	0.34	1.58	1963	0.00	1971+	1.07	1971+
O	0.20	1.12	1972	0.00	1967+	0.65	1967
N	0.31	2.22	1965	0.00	1956	1.78	1956
D	0.40	1.38	1959	0.00	1963+	0.83	1959
YR	3.90	2.59 AUG 1957		0.00 MAR 1972+		2.59 AUG 1957	

Snow, ice pellets

Month	Mean total	Maximum monthly	Year	Maximum in 24 hrs.	Year
J	0.8	16.7	1949	5.0	1949
F	T	T	1966+	T	1966+
M	T	T	1963+	T	1963+
A	T	T	1970+	T	1970+
M	0.0	0.0		0.0	
J	0.0	0.0		0.0	
J	0.0	0.0		0.0	
A	0.0	0.0		0.0	
S	0.0	0.0		0.0	
O	0.0	0.0		0.0	
N	0.2	4.0	1964	4.0	1964
D	0.1	2.0	1967	2.0	1967
YR	1.1	16.7 JAN 1949		5.0 JAN 1949	

Relative humidity, wind, sunshine, mean sky cover and mean-number-of-days sections follow in additional columns.

Ø For period August 1964 through the current year.
Means and extremes above are from existing and comparable exposures. Annual extremes have been exceeded at other sites in the locality as follows:
Highest temperature 117 in July 1942; maximum monthly precipitation 3.39 in September 1939.

NORMALS, MEANS AND EXTREMES

(Table Revised 1975. Base Period for Climatological Normals: 1941-1970)

Month	Temperatures °F Normal Daily maximum	Daily minimum	Monthly	Record highest	Year	Record lowest	Year	Normal Degree days Base 65°F Heating	Cooling
J	55.7	32.6	44.2	75	1963	8	1963	645	0
F	61.3	36.9	49.1	82	1972	19	1972	451	8
M	67.8	41.7	54.8	91	1966	23	1971	324	0
A	77.5	50.0	63.8	96	1962	32	1967	126	90
M	87.5	59.0	73.3	106	1964	40	1967	10	268
J	97.2	67.4	82.3	115	1970	51	1971	0	519
J	103.9	75.3	89.6	116	1973	62	1970	0	763
A	101.5	73.3	87.4	116	1969	56	1968	0	694
S	94.6	65.4	80.1	107	1971	46	1965	0	453
O	81.0	53.1	67.1	99	1963	26	1971	74	139
N	65.7	40.8	53.3	85	1973	27	1973	357	6
D	56.7	33.7	45.2	72	1968	15	1968	614	0
YR	79.2	52.4	65.8	116 JUL 1973		8 JAN 1963		2601	2946

Precipitation in inches

Month	Water equivalent Normal	Maximum monthly	Year	Minimum monthly	Year	Maximum in 24 hrs.	Year
J	0.45	2.41	1949	T	1971	1.01	1952
F	0.45	1.64	1973	0.00	1967	0.63	1958
M	0.33	1.58	1973	0.00	1972	1.14	1963
A	0.22	2.44	1965	0.00	1970	0.97	1965
M	0.10	0.96	1969	0.00	1970	0.80	1969
J	0.09	0.82	1967	0.00	1974	0.75	1967
J	0.44	1.64	1956	0.00	1956	1.32	1956
A	0.49	2.59	1957	0.00	1957	2.59	1957
S	0.27	1.58	1963	0.00	1963	1.07	1963
O	0.21	1.12	1972	0.00	1972	0.45	1972
N	0.43	2.22	1965	0.00	1965	1.78	1956
D	0.37	1.38	1959	0.00	1959	0.83	1959
YR	3.76	2.59 AUG 1957		0.00 AUG 1957		2.59 AUG 1957	

Snow, ice pellets

Month	Maximum monthly	Year	Maximum in 24 hrs.	Year
J	16.7	1949	9.0	1974
F	T	1966	T	1966
M	T	1973	T	1973
A	T	1970	T	1970
M	0.0		0.0	
J	0.0		0.0	
J	0.0		0.0	
A	0.0		0.0	
S	T		T	
O	0.0		0.0	
N	4.0	1964	4.0	1964
D	2.0	1967	2.0	1967
YR	16.7 JAN 1949		9.0 JAN 1974	

Average station pressure: 937.6 mb Elev. 2180 feet m.s.l.

Means and extremes above are from existing and comparable exposures. Annual extremes have been exceeded at other sites in the locality as follows:
Highest temperature 117 in July 1942; maximum monthly precipitation 3.39 in September 1939.

REFERENCE NOTES APPLYING TO TABLES APPEAR ON THE PAGE FOLLOWING LAST TABLE.

(Caution: Letters and symbols may have different meanings in 1941-1970 tables than in earlier tables. See notes.)

NORMALS, MEANS AND EXTREMES

(Table Revised 1973. Base Period for Climatological Normals: 1931-1960)

Station: RENO, NEVADA   INTERNATIONAL AIRPORT   Standard time used: PACIFIC   Latitude: 39° 30' N   Longitude: 119° 47' W   Elevation (ground): 4404 feet

Month	Temperature °F — Normal Daily maximum	Daily minimum	Monthly	Extremes Record highest	Year	Record lowest	Year	Precipitation Normal total	Normal heating degree days (Base 65°)	Precipitation Normal total	Maximum monthly	Year	Minimum monthly	Year	Maximum in 24 hrs.	Year	Snow, ice pellets Mean total	Maximum monthly	Year	Maximum in 24 hrs.	Year	Relative humidity Hour 04	Hour 10	Hour 16	Hour 22	Wind Mean speed	Prevailing direction	Fastest mile Speed	Direction	Year	Mean sky cover	Pct. of possible sunshine	Clear	Partly cloudy	Cloudy	Precipitation .01 inch or more	Snow, ice pellets 1.0 inch or more	Thunderstorms	Heavy fog	Temperatures Max. 90° and above	32° and below	Min. 32° and below	0° and below	Avg daily solar radiation
(a)	(b)		(b)			9		(b)	(b)		31		31		31		30	31		31		9	9	9	9	30	15	13	13	13	30	30	30	30	30	30	30	30	30	9	9	9	9	9
J	44.6	16.2	30.4	70	1967	-11	1971	1.19	1073	1.19	4.13	1969	T	1966	2.37	1943	6.8	20.0	1956	12.0	1956	72	65	49	68	6.0	S	80	SW	1968	6.2	66	15	8	7	7	5	0	2	0	3	27	1
F	49.8	21.4	35.6	74	1967+	0	1968	1.02	823	1.02	3.69	1962	T	1967	1.55	1962	4.8	23.5	1969	13.9	1969	70	55	36	60	6.1	S	54	SW	1960	6.1	68	13	9	6	5	4	0	*	1	2	25	1
M	57.3	25.6	41.5	83	1966	0	1971	0.68	729	0.68	2.02	1952	0.03	1972	1.21	1943	5.5	29.0	1952	16.9	1952	64	44	30	51	7.5	WNW	80	SW	1968	5.8	75	13	9	9	6	2	1	*	2	*	27	*
A	65.4	30.5	48.0	88	1966	15	1970	0.54	510	0.54	2.04	1958	T	1962+	1.64	1958	1.1	14.1	1958	7.3	1958	63	38	28	45	8.6	WNW	48	SE	1972	5.5	79	11	10	9	4	3	*	*	1	0	22	*
M	71.5	36.3	53.9	95	1970+	18	1964	0.52	357	0.52	2.89	1963	T	1970+	1.29	1963	T	14.1	1964	9.0	1964	65	34	26	45	7.6	WNW	39	W	1968	4.9	79	11	11	9	3	2	2	0	*	1	7	0
J	80.3	39.8	60.1	100	1972	29	1969	0.37	189	0.37	1.31	1965	0.00	1959+	0.79	1969	T	0.2	1970	0.2	1970	70	36	26	45	7.1	WNW	42	NW	1968	3.6	83	17	7	6	2	0	3	*	1	0	1	0
J	89.6	45.9	67.7	103	1972+	33	1965	0.27	43	0.27	1.06	1971	0.00	1951+	0.80	1951+	0.0	0.0		0.0		67	29	19	38	6.5	WNW	44	SW	1962	2.1	92	23	6	2	1	0	4	*	7	0	0	0
A	88.3	42.7	65.5	103	1972+	29	1968	0.17	87	0.17	1.65	1965	0.00	1957+	0.97	1957+	0.0	0.0		0.0		69	35	22	40	6.1	WNW	43	SW	1968	2.1	93	23	5	3	2	0	3	*	5	0	0	0
S	81.2	36.4	58.8	96	1971	20	1965	0.23	204	0.23	2.48	1950	0.00	1964+	0.80	1964+	T	T		T		71	35	21	38	5.4	WNW	42	W	1970+	2.4	91	22	5	3	2	0	1	*	3	0	1	0
O	69.0	29.4	49.2	91	1965	8	1971	0.47	490	0.47	2.14	1945	T	1962	1.55	1962	0.4	5.1	1971	3.7	1961	74	41	27	56	5.3	WNW	50	S	1962	4.0	82	16	7	8	3	1	0	*	6	0	21	0
N	55.7	20.8	38.3	76	1969	8	1964	0.57	801	0.57	2.04	1946	0.00	1959	1.23	1964	1.7	8.7	1961	6.7	1961	74	55	43	65	5.5	NNW	52	SW	1968	5.3	70	13	8	9	5	2	0	*	8	0	24	0
D	47.3	16.4	31.9	70	1969	-16	1972	1.08	1026	1.08	5.23	1955	0.01	1947	2.16	1955	5.1	25.6	1971	14.9	1971	73	65	55	70	5.1	S	68	SW	1968	6.3	62	10	8	13	6	3	0	1	3	6	28	2
YR	66.7	30.1	48.4	103	AUG. 1972+	-16	DEC. 1972	7.15	6332	7.15	5.25	DEC. 1955	0.00	SEP. 1964+	2.37	JAN. 1943	26.2	29.0	MAR. 1952	16.9	MAR. 1952	69	44	32	53	6.3	WNW	80	SW	MAR. 1968+	4.5	80	166	89	110	49	8	14	8	53	10	188	3

Means and extremes above are from existing and comparable exposures. Annual extremes have been exceeded at other sites in the locality as follows:
Highest temperature 106 in July 1931; lowest temperature -19 in January 1890; maximum monthly precipitation 6.76 in January 1916; maximum precipitation in 24 hours 2.71 in January 1903; maximum monthly snowfall 65.7 in January 1916; maximum snowfall in 24 hours 22.5 in January 1916.

NORMALS, MEANS AND EXTREMES

(Table Revised 1975. Base Period for Climatological Normals: 1941-1970)

Elevation: 4400 feet m.s.l.   Average station pressure mb.

Month	Temperatures °F Normal Daily maximum	Daily minimum	Monthly	Extremes Record highest	Year	Record lowest	Year	Normal Degree days Base 65°F Heating	Cooling	Precipitation in inches Normal	Water equivalent Maximum monthly	Year	Minimum monthly	Year	Maximum in 24 hrs.	Year	Snow, ice pellets Maximum monthly	Year	Maximum in 24 hrs.	Year	Relative humidity pct. Hour 04	Hour 10	Hour 16	Hour 22	Wind Mean speed m.p.h.	Prevailing direction	Fastest mile Speed m.p.h.	Direction	Year	Mean sky cover, tenths, sunrise to sunset	Pct. of possible sunshine	Clear	Partly cloudy	Cloudy	Precipitation .01 inch or more	Snow, ice pellets 1.0 inch or more	Thunderstorms	Heavy fog, visibility ¼ mile or less	Temperatures Max. 90° and above	32° and below	Min. 32° and below	0° and below
(a)					11		11				33		33		33		33		33		11	11	11	11	32	15	15	15	15	32	32	32	32	32	32	32	32	32	11	11	11	11
J	45.4	18.3	31.9	70	1967	-11	1971	1026	0	1.21	4.13	1969	T	1966	2.37	1943	20.0	1956	12.0	1956	73	66	51	69	6.0	S	80	SW	1968	6.2	66	15	7	9	7	5	0	2	0	4	27	1
F	51.1	23.0	37.1	74	1967	0	1968	781	0	0.86	3.69	1962	T	1967	1.55	1962	23.5	1969	13.9	1969	70	56	37	60	6.1	S	60	SW	1974	6.1	68	13	9	8	6	5	*	*	1	2	26	1
M	56.0	24.6	40.3	83	1966	0	1971	766	0	0.70	2.04	1972	0.03	1972	1.21	1943	29.0	1952	16.9	1952	64	50	31	53	7.5	WNW	80	SW	1968	5.5	74	11	9	11	6	4	1	*	2	1	26	*
A	64.0	29.6	46.8	88	1966	15	1970	546	0	0.47	2.04	1958	T	1962	1.64	1958	7.5	1958	7.3	1958	64	37	27	48	8.0	WNW	48	SE	1973	5.5	80	12	10	8	4	3	1	*	1	0	22	*
M	72.4	37.0	54.6	95	1970+	18	1964	328	0	0.66	2.89	1963	T	1970+	1.29	1963	14.1	1964	9.0	1964	65	33	25	46	7.0	WNW	44	NW	1974	5.0	80	17	10	4	5	2	2	0	*	1	7	0
J	80.4	42.5	61.5	100	1972	29	1974	145	40	0.40	1.31	1965	0.00	1959	0.79	1969	0.2	1970	0.2	1970	69	34	24	43	7.2	WNW	46	NW	1974	3.6	84	17	6	7	3	0	3	*	7	0	1	0
J	91.1	47.4	69.3	103	1972	33	1974	17	150	0.26	1.06	1971	0.00	1951	0.80	1951	0.0		0.0		67	29	19	38	6.5	WNW	44	SW	1962	2.1	92	23	6	2	1	0	4	*	22	0	0	0
A	89.0	44.8	66.9	103	1972	29	1968	50	109	0.22	1.65	1965	0.00	1957	0.97	1957	0.0		0.0		69	31	20	40	5.5	WNW	43	SW	1968	2.1	93	22	6	3	2	0	3	*	19	0	0	0
S	81.8	38.6	60.2	96	1971	20	1965	168	24	0.23	1.02	1974	0.00	1950	0.80	1950	T	1971	T	1971	71	32	21	46	5.5	WNW	42	W	1970	2.4	84	21	8	3	2	0	1	*	4	0	21	0
O	70.0	30.5	50.3	91	1965	8	1971	456	0	0.42	2.14	1945	0.00	1966	1.55	1962	5.1	1971	3.7	1961	72	41	27	56	5.3	WNW	50	S	1962	4.0	83	16	7	8	3	1	0	*	7	0	21	0
N	56.3	23.9	40.1	76	1969	8	1964	747	0	0.68	2.04	1946	0.00	1959	1.23	1964	8.7	1961	6.7	1961	73	58	43	66	5.2	S	52	SW	1967	5.3	70	13	7	10	5	2	0	*	13	0	23	0
D	46.4	19.6	33.0	70	1969	-16	1972	992	0	1.09	5.25	1955	0.01	1947	2.16	1955	25.6	1971	14.9	1971	74	66	54	70	5.1	SW	68	SW	1968	6.4	63	8	6	16	6	3	0	1	16	5	28	2
YR	67.0	31.7	49.4	103	AUG. 1972	-16	DEC. 1972	6022	329	7.20	5.25	DEC. 1955	0.00	SEP. 1974	2.37	JAN. 1943	29.0	MAR. 1952	16.9	MAR. 1952	69	44	32	53	6.3	WNW	80	SW	MAR. 1968	4.5	80	166	88	111	49	13	9	2	54	10	188	3

Means and extremes above are from existing and comparable exposures. Annual extremes have been exceeded at other sites in the locality as follows:
Highest temperature 106 in July 1931; lowest temperature -19 in January 1890; maximum monthly precipitation 6.76 in January 1916; maximum precipitation in 24 hours 2.71 in January 1903; maximum monthly snowfall 65.7 in January 1916; maximum snowfall in 24 hours 22.5 in January 1916.

REFERENCE NOTES APPLYING TO TABLES APPEAR ON THE PAGE FOLLOWING LAST TABLE.
(Caution: Letters and symbols may have different meanings in 1941-1970 tables than in earlier tables. See notes.)

NORMALS, MEANS AND EXTREMES
(Table Revised 1973. Base Period for Climatological Normals: 1931-1960)

Station: WINNEMUCCA, NEVADA MUNICIPAL AIRPORT Standard time used: PACIFIC Latitude: 40° 54' N Longitude: 117° 48' W Elevation (ground): 4301 feet Elevation: 4301 feet

Month	Temperature Normal Daily maximum (b)	Daily minimum (b)	Monthly (b)	Extremes Record highest	Year	Record lowest	Year	Normal heating degree days (Base 65°) (b)	Precipitation Normal total (b)	Max monthly	Year	Min monthly	Year	Max in 24 hrs.	Year	Snow, Ice pellets Mean total	Max monthly	Year	Max in 24 hrs.	Year
J	39.7	14.7	27.2	68	1971	-24	1963	1172	1.05	2.70	1956	0.04	1966	0.67	1956	5.7	16.5	1950	6.1	1957
F	44.9	21.4	32.3	69	1971	-9	1955	916	0.94	2.17	1962	0.08	1967	0.72	1960	3.9	13.8	1969	9.9	1959
M	51.3	23.0	38.0	81	1972	-3	1971	837	0.81	1.66	1952	0.06	1959+	1.01	1963	4.6	23.4	1952	8.7	1952
A	60.2	29.0	45.9	84	1966	6	1972+	573	0.73	1.88	1967	0.04	1972+	1.01	1958	2.9	12.0	1964	7.2	1971
M	71.7	35.6	53.7	96	1954	12	1953	363	0.94	2.82	1957	0.08	1969	0.97	1957	0.7	5.4	1954	4.3	1954
J	80.3	43.2	61.5	104	1954	23	1954	153	0.76	2.86	1958	T	1960	1.79	1958	0.0	T		T	
J	92.0	49.2	71.0	106	1971	29	1955	34	0.27	0.92	1952	0.00	1950	0.63	1950	0.0	0.0		0.0	
A	89.7	45.4	67.6	106	1972	28	1960	34	0.15	1.26	1958	0.00	1969	0.75	1958	0.0	0.0		0.0	
S	80.7	36.8	58.6	103	1950	12	1958	210	0.34	1.26	1950	T	1963	0.79	1951	0.0	0.2	1971	0.2	1971
O	67.3	28.2	47.7	89	1964+	-5	1971	536	0.81	2.19	1951	T	1952	1.64	1951	0.4	3.5	1969	3.5	1969
N	52.3	19.2	35.8	75	1965	-5	1961	876	0.80	2.54	1955	T	1952	1.58	1955	2.0	8.6	1955	5.7	1955
D	43.3	16.2	29.8	64	1958	-34	1972	1091	0.94	2.54	1955	0.26	1962	0.95	1969	5.5	17.5	1971	8.4	1971
YR	64.8	30.0	47.4	106 AUG. 1972/1971		-34 DEC. 1972		6761	8.63	2.86 JUN. 1958		0.00 AUG. 1969+		1.79 JUN. 1958		25.7	23.4 MAR. 1952		9.9 FEB. 1959	

Means and extremes above are from existing and comparable exposures. Annual extremes have been exceeded at other sites in the locality as follows: Highest temperature 108 in July 1931; lowest temperature -36 in January 1937; maximum monthly precipitation 5.23 in March 1884; maximum monthly snowfall 33.0 in January 1890; maximum snowfall in 24 hours 18.2 in January 1890; fastest mile of wind 69 from the East in December 1941.

NORMALS, MEANS AND EXTREMES
(Table Revised 1975. Base Period for Climatological Normals: 1941-1970)

Station: WINNEMUCCA, NEVADA Average station pressure: 869.1 mb. Elev. 4314 feet m.s.l.

Month	Temperatures °F Normal Daily maximum (a)	Daily minimum	Monthly	Extremes Record highest	Year	Record lowest	Year	Normal Degree days Base 65° F Heating	Cooling	Precipitation in inches Water equivalent Normal	Max monthly	Year	Min monthly	Year	Max in 24 hrs.	Year	Snow, Ice pellets Max monthly	Year	Max in 24 hrs.	Year
J	40.6	15.7	28.2	68	1971	-24	1963	1141	0	0.97	2.70	1956	0.04	1966	0.67	1956	16.5	1950	6.1	1957
F	46.7	21.4	34.1	69	1971	-9	1955	865	0	0.81	2.17	1962	0.08	1967	0.72	1960	13.8	1969	9.9	1959
M	51.9	25.1	38.5	81	1972	-3	1971	849	0	0.71	1.66	1952	0.06	1959	1.01	1963	23.4	1952	8.7	1952
A	61.0	29.1	45.1	84	1966	6	1972	597	11	0.73	1.86	1967	0.01	1974	1.01	1958	12.0	1964	7.2	1971
M	70.3	37.3	53.1	96	1954	11	1954	359	50	0.91	2.82	1957	0.97	1960	0.97	1957	5.4	1954	4.3	1954
J	78.8	44.5	61.7	105	1954	23	1954	149	192	1.01	2.86	1958	1.79	1958	1.79	1958	T		T	
J	91.4	50.9	71.0	106	1971	29	1955	6		0.23	0.92	1952	0.00	1950	0.63	1950	0.0		0.0	
A	88.7	46.5	67.8	106	1971	28	1960	42	129	0.26	1.26	1958	0.00	1969	0.75	1958	0.0		0.0	
S	79.9	38.5	59.2	103	1950	12	1958	199	23	0.28	1.26	1950	1.64	1952	0.79	1951	0.2	1971	0.2	1971
O	67.3	28.3	48.3	89	1964	-5	1970	518	0	0.65	2.19	1951	T	1952	1.64	1951	3.5	1969	3.5	1969
N	52.3	22.3	37.3	75	1965	-5	1961	831	0	0.97	2.66	1955	T	1959	1.58	1955	8.6	1955	5.7	1955
D	42.9	17.8	30.4	64	1958	-34	1972	1073	0	0.94	2.54	1955	0.26	1962	0.95	1969	17.5	1971	8.4	1971
YR	64.3	31.4	47.9	106 AUG. 1972		-34 DEC. 1972		6629	407	8.47	2.86 JUN. 1958		0.00 SEP. 1958		1.79 JUN. 1958		23.4 MAR. 1952		9.9 FEB. 1959	

Means and extremes above are from existing and comparable exposures. Annual extremes have been exceeded at other sites in the locality as follows: Highest temperature 108 in July 1931; lowest temperature -36 in January 1937; maximum monthly precipitation 5.23 in March 1884; maximum monthly snowfall 33.0 in January 1890; maximum snowfall in 24 hours 18.2 in January 1890; fastest mile of wind 69 from the East in December 1941.

REFERENCE NOTES APPLYING TO TABLES APPEAR ON THE PAGE FOLLOWING LAST TABLE.
(Caution: Letters and symbols may have different meanings in 1941-1970 tables than in earlier tables. See notes.)

637

Reference notes applying to Normals, Means, and Extremes tables for 1931–1960 base period.

(a) Length of record, years, based on January data. Other months may be for more or fewer years if there have been breaks in the record.
Climatological standard normals (1931-1960).
(b) Less than one half.
* Also on earlier dates, months, or years.
+ Trace, an amount too small to measure.
- Below zero temperatures are preceded by a minus sign.
The prevailing direction for wind in the Normals, Means, and Extremes table is from records through 1963.
‡ ≥70° at Alaskan stations.

Unless otherwise indicated, dimensional units used in this bulletin are: temperature in degrees F.; precipitation, including snowfall, in inches; wind movement in miles per hour; and relative humidity in percent. Heating degree day totals are the sums of negative departures of average daily temperatures from 65° F. Cooling degree day totals are the sums of positive departures of average daily temperatures from 65° F. Sleet was included in snowfall totals beginning with July 1948. The term "Ice pellets" includes solid grains of ice (sleet) and particles consisting of snow pellets encased in a thin layer of ice. Heavy fog reduces visibility to 1/4 mile or less.

Sky cover is expressed in a range of 0 for no clouds or obscuring phenomena to 10 for complete sky cover. The number of clear days is based on average cloudiness 0-3, partly cloudy days 4-7, and cloudy days 8-10 tenths.

Solar radiation data are the averages of direct and diffuse radiation on a horizontal surface. The langley denotes one gram calorie per square centimeter.

& Figures instead of letters in a direction column indicate direction in tens of degrees from true North; i.e., 09 - East, 18 - South, 27 - West, 36 - North, and 00 - Calm. Resultant wind is the vector sum of wind directions and speeds divided by the number of observations. If figures appear in the direction column under "Fastest mile" the corresponding speeds are fastest observed 1-minute values.

To 8 compass points only.

** The National Weather Service considers the accuracy of solar radiation data questionable; therefore, publication is suspended pending determination of corrected values.

Reference notes applying to Normals, Means, and Extremes tables for 1941–1970 base period.

(a) Length of record, years, through the current year unless otherwise noted, based on January data.
(b) 70° and above at Alaskan stations.
* Less than one half.
T Trace.

NORMALS - Based on record for the 1941-1970 period.
DATE OF AN EXTREME - The most recent in cases of multiple occurrence.
PREVAILING WIND DIRECTION - Record through 1963.
WIND DIRECTION - Numerals indicate tens of degrees clockwise from true north. 00 indicates calm.
FASTEST MILE WIND - Speed is fastest observed 1-minute value when the direction is in tens of degrees.

Mean Maximum Temperature (°F.), January

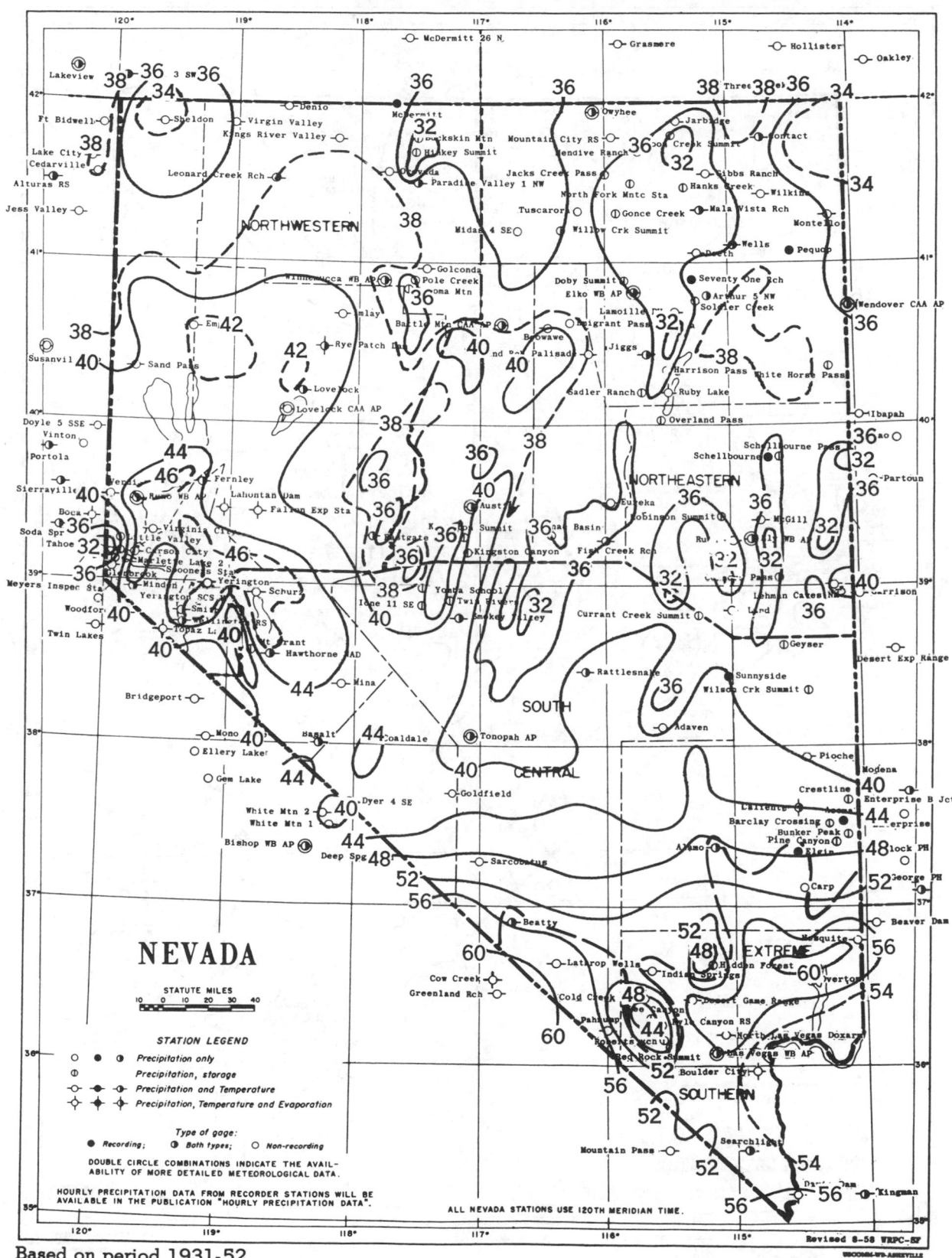

Based on period 1931-52

Isolines are drawn through points of approximately equal value. Caution should be used in interpolating on these maps, particularly in mountainous areas.

Mean Minimum Temperature (°F.), January

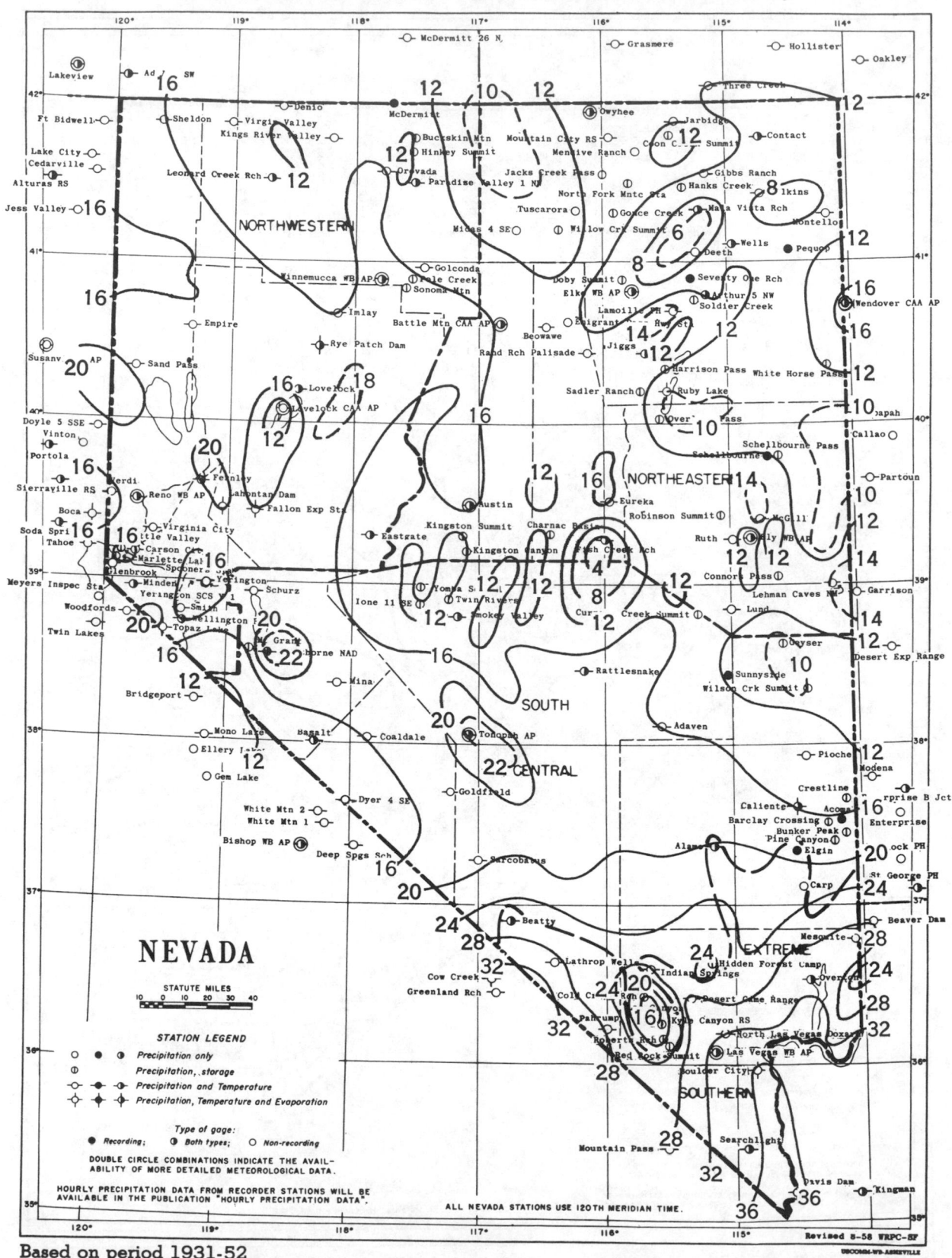

NEVADA

STATUTE MILES
10 0 10 20 30 40

STATION LEGEND

○ ● ◑ *Precipitation only*

◐ *Precipitation, storage*

○– ●– ◑– *Precipitation and Temperature*

◇– ◆– ◈– *Precipitation, Temperature and Evaporation*

Type of gage:
● *Recording;* ◑ *Both types;* ○ *Non-recording*

DOUBLE CIRCLE COMBINATIONS INDICATE THE AVAIL-
ABILITY OF MORE DETAILED METEOROLOGICAL DATA.

HOURLY PRECIPITATION DATA FROM RECORDER STATIONS WILL BE
AVAILABLE IN THE PUBLICATION "HOURLY PRECIPITATION DATA".

ALL NEVADA STATIONS USE 120TH MERIDIAN TIME.

Revised 8-58 WRPC-SF

USCOMM-WB-ASHEVILLE

Based on period 1931-52

Isolines are drawn through points of approximately equal value. Caution should be used
in interpolating on these maps, particularly in mountainous areas.

Mean Maximum Temperature (°F.), July

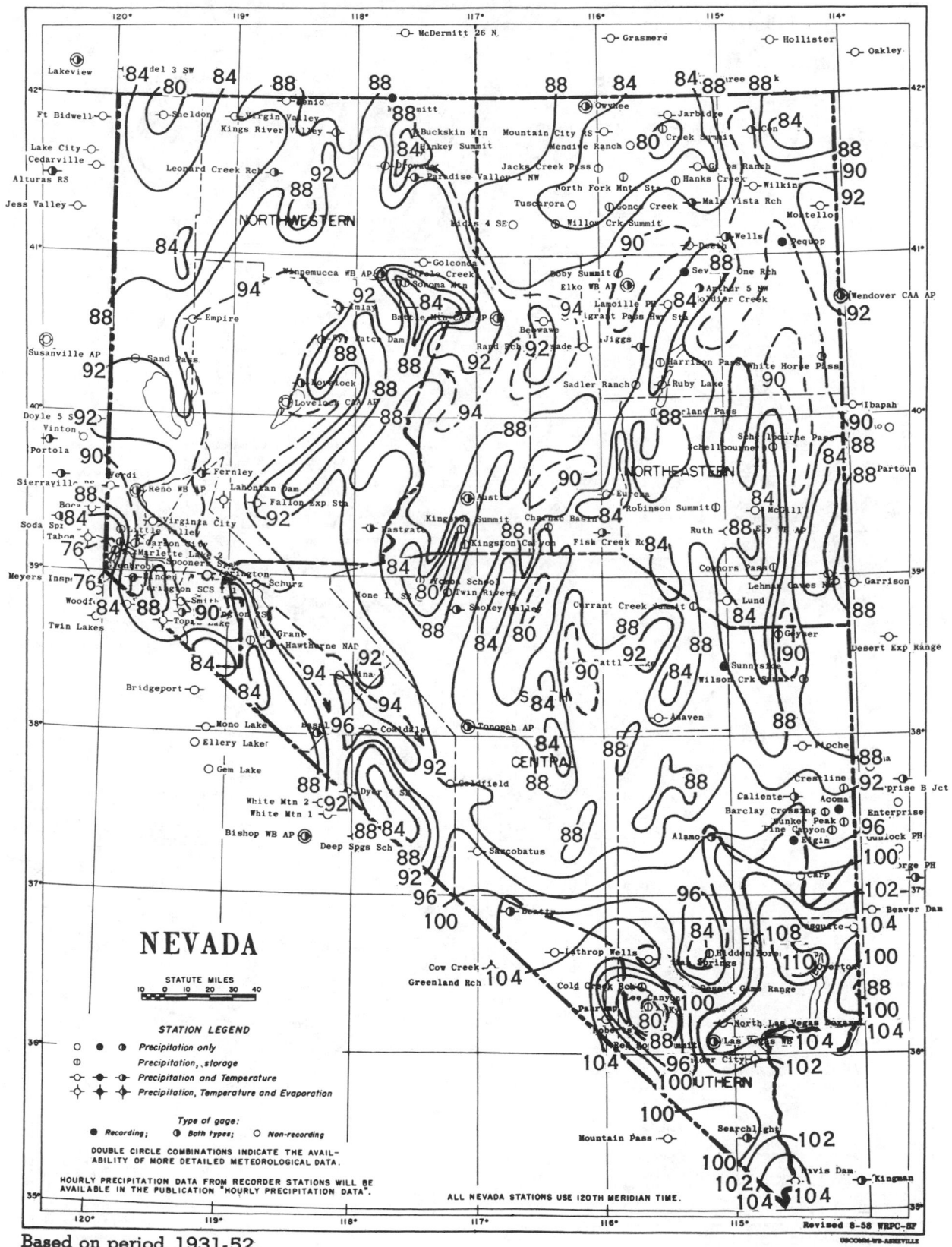

NEVADA

STATUTE MILES

10 0 10 20 30 40

STATION LEGEND

○	●	◑	Precipitation only
⊙			Precipitation, storage
-○-	-●-	-◑-	Precipitation and Temperature
-◇-	-◆-	-◈-	Precipitation, Temperature and Evaporation

Type of gage:

● Recording; ◑ Both types; ○ Non-recording

DOUBLE CIRCLE COMBINATIONS INDICATE THE AVAIL-
ABILITY OF MORE DETAILED METEOROLOGICAL DATA.

HOURLY PRECIPITATION DATA FROM RECORDER STATIONS WILL BE
AVAILABLE IN THE PUBLICATION "HOURLY PRECIPITATION DATA".

ALL NEVADA STATIONS USE 120TH MERIDIAN TIME.

Revised 8-58 WRPC-5F

USCOMM-WB-ASHEVILLE

Based on period 1931-52

Isolines are drawn through points of approximately equal value. Caution should be used
in interpolating on these maps, particularly in mountainous areas.

Mean Minimum Temperature (°F.), July

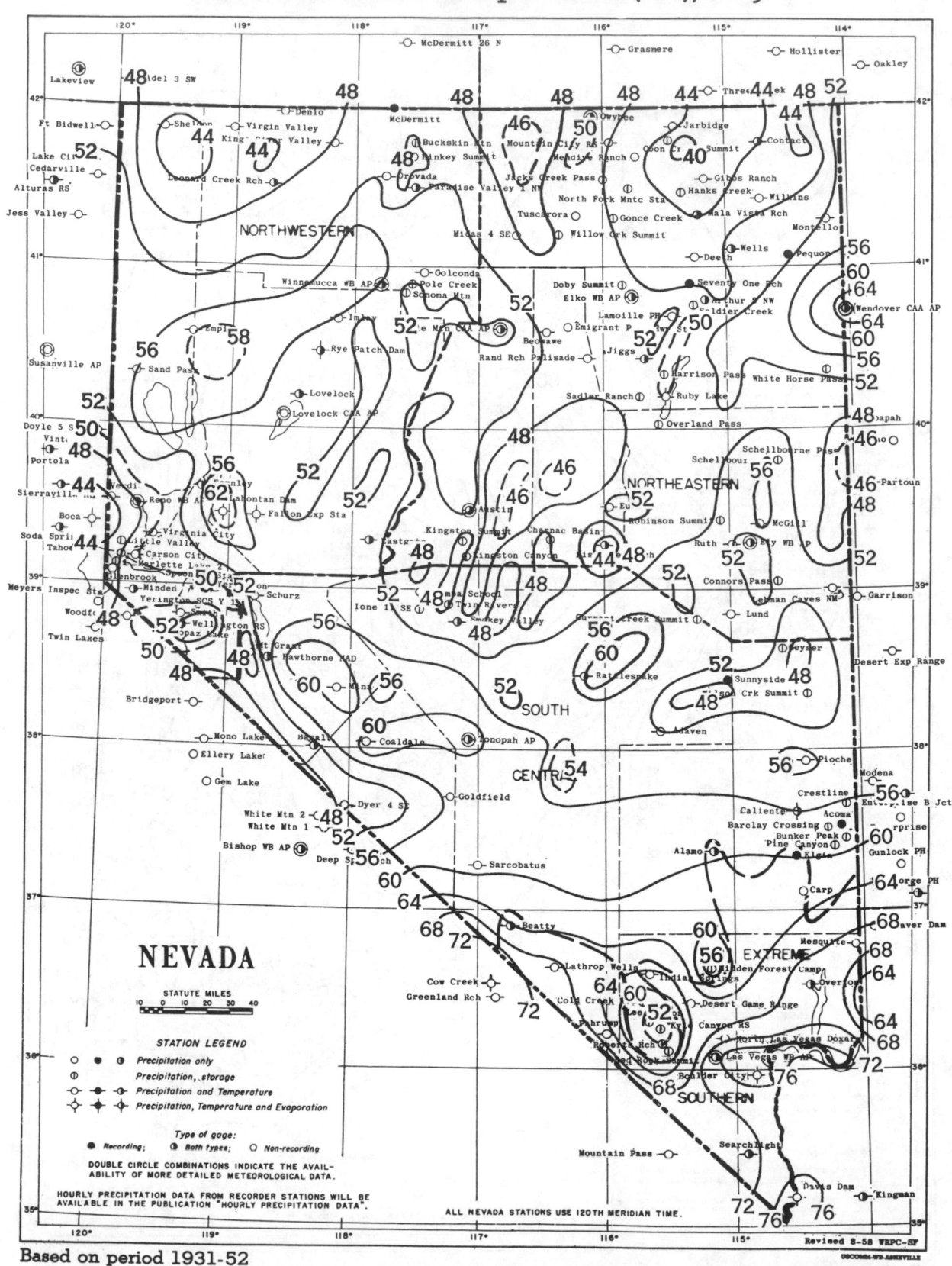

NEVADA

STATUTE MILES

10 0 10 20 30 40

STATION LEGEND

o • ◐ *Precipitation only*
ⓘ *Precipitation, storage*
-o- -•- -◐- *Precipitation and Temperature*
-⊕- -⬥- -⊕- *Precipitation, Temperature and Evaporation*

● Recording; ◐ Both types; o Non-recording

Type of gage:

DOUBLE CIRCLE COMBINATIONS INDICATE THE AVAIL-
ABILITY OF MORE DETAILED METEOROLOGICAL DATA.

HOURLY PRECIPITATION DATA FROM RECORDER STATIONS WILL BE
AVAILABLE IN THE PUBLICATION "HOURLY PRECIPITATION DATA".

ALL NEVADA STATIONS USE 120TH MERIDIAN TIME.

Revised 8-58 WRPC-SF

USCOMM-WB-ASHEVILLE

Based on period 1931-52

Isolines are drawn through points of approximately equal value.   Caution should be used
in interpolating on these maps, particularly in mountainous areas.

642

Mean Annual Precipitation, Inches

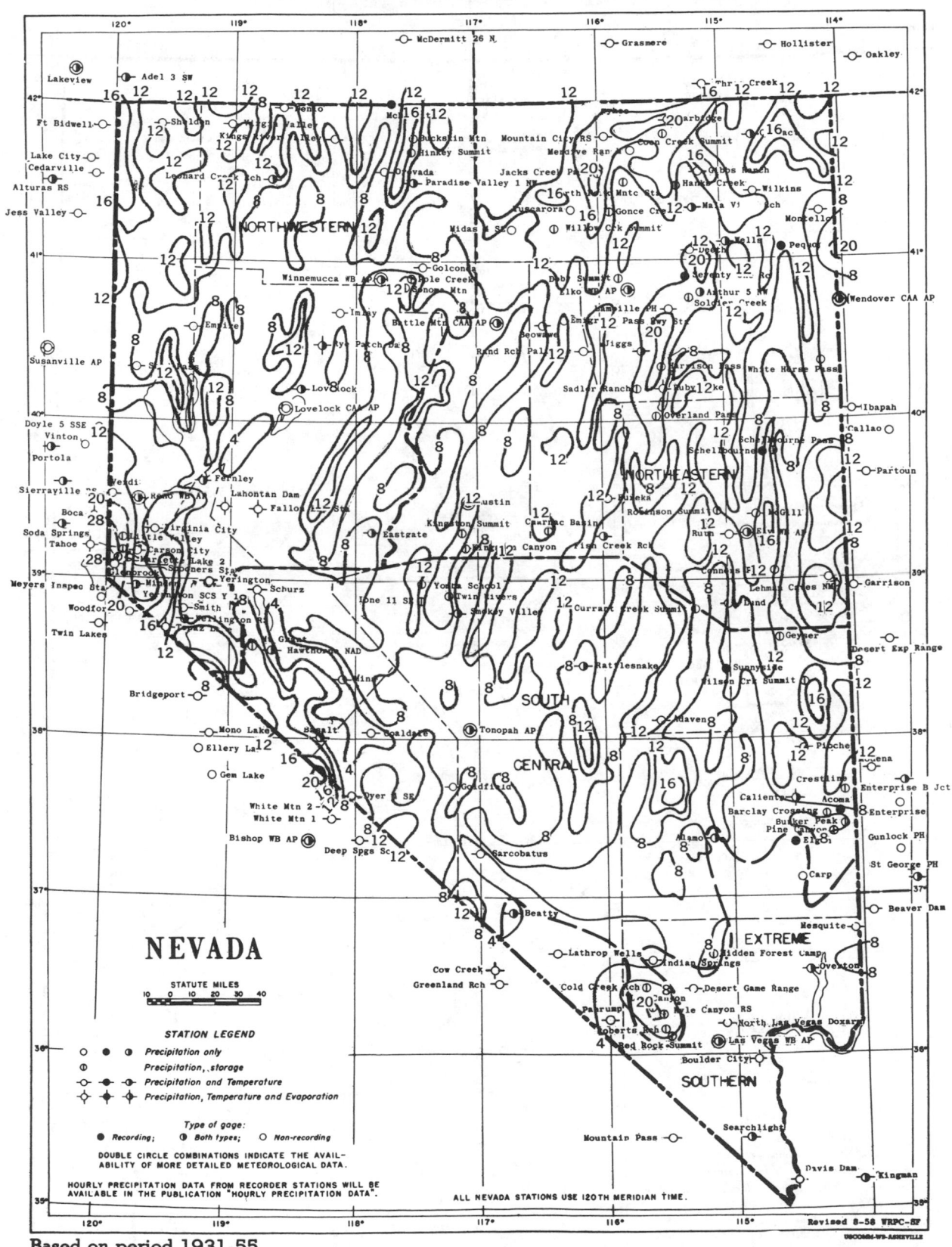

NEVADA

STATUTE MILES
10 0 10 20 30 40

STATION LEGEND

○ ● ◐ Precipitation only
◑ Precipitation, storage
○– ●– ◐– Precipitation and Temperature
✧ ✦ ✧ Precipitation, Temperature and Evaporation

Type of gage:

● Recording;   ◐ Both types;   ○ Non-recording

DOUBLE CIRCLE COMBINATIONS INDICATE THE AVAIL-
ABILITY OF MORE DETAILED METEOROLOGICAL DATA.

HOURLY PRECIPITATION DATA FROM RECORDER STATIONS WILL BE
AVAILABLE IN THE PUBLICATION "HOURLY PRECIPITATION DATA".

ALL NEVADA STATIONS USE 120TH MERIDIAN TIME.

Revised 8-58 WRPC-SF

USCOMM-WB-ASHEVILLE

Based on period 1931-55

Isolines are drawn through points of approximately equal value.   Caution should be used
in interpolating on these maps, particularly in mountainous areas.

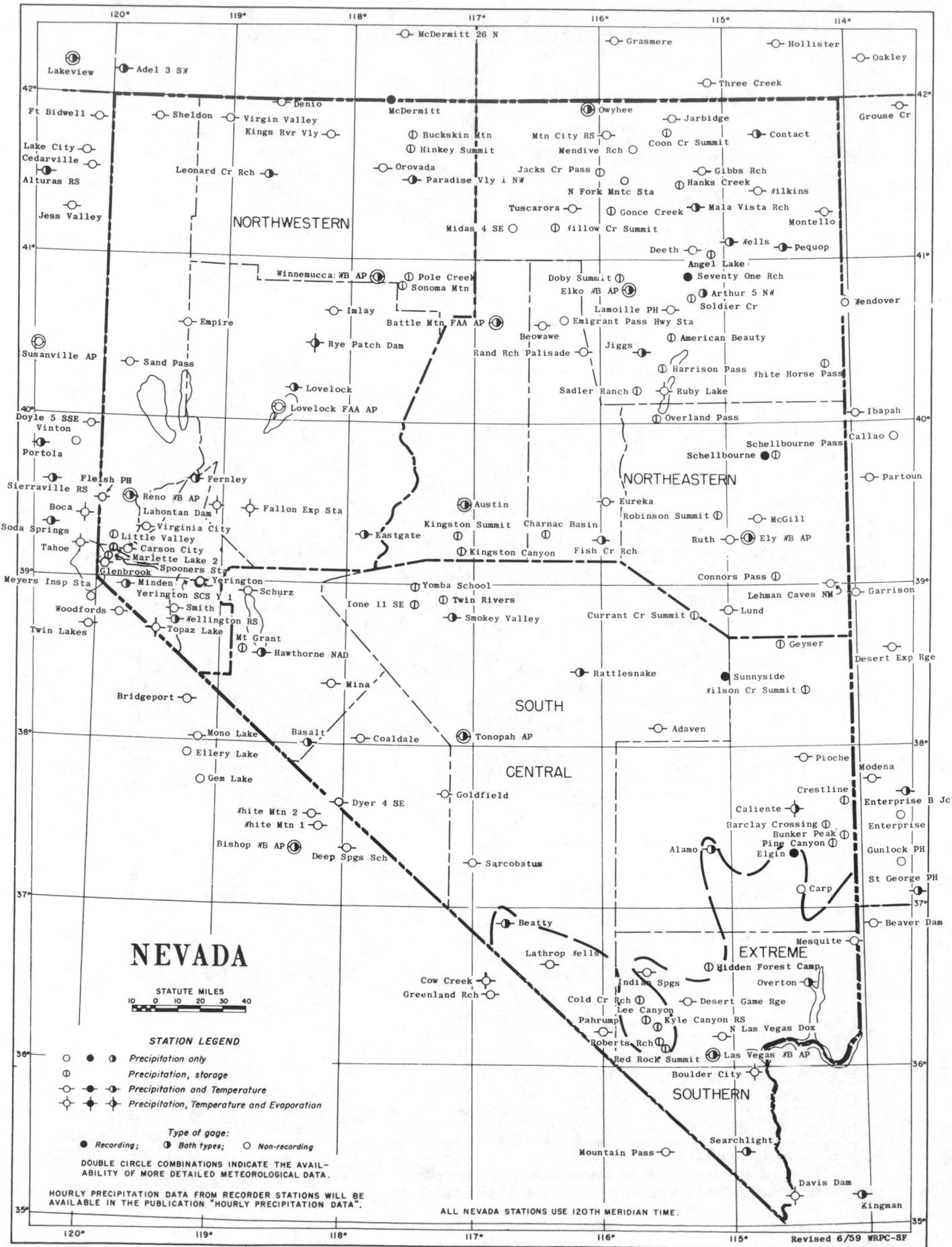

NEVADA

STATUTE MILES
10 0 10 20 30 40

STATION LEGEND

○ ● ◑ Precipitation only
◐ Precipitation, storage
○- ●- ◑- Precipitation and Temperature
○- ●- ◑- Precipitation, Temperature and Evaporation

Type of gage:
● Recording; ◑ Both types; ○ Non-recording

DOUBLE CIRCLE COMBINATIONS INDICATE THE AVAIL-
ABILITY OF MORE DETAILED METEOROLOGICAL DATA.

HOURLY PRECIPITATION DATA FROM RECORDER STATIONS WILL BE
AVAILABLE IN THE PUBLICATION "HOURLY PRECIPITATION DATA".

ALL NEVADA STATIONS USE 120TH MERIDIAN TIME.

Revised 6/59 WRPC-SF

USCOMM-WB-ASHEVILLE

644

CLIMATES OF THE STATES

NEW HAMPSHIRE

(Normals, Means and Extremes tables revised 1973 and 1975. Basic report revised November 1959.)

Climate of New Hampshire

Robert E. Lautzenheiser, Weather Bureau State Climatologist

PHYSICAL DESCRIPTION: -- New Hampshire occupies 9,304 square miles, nearly one-seventh of New England's total area. From below the 43d parallel of latitude it extends nearly 200 miles northward to beyond the 45th parallel. At its southern border, New Hampshire extends westward from the Atlantic coastline for nearly 100 miles. It narrows to less than 20 miles in width at its northern tip. The eastern border lies near 71°W. longitude. Its western border is the Connecticut River, except in the extreme north.

The terrain is hilly to mountainous. Elevations of less than 500 feet above sea level are found only in the coastal area of the southeast, the Merrimac River Valley, and the central and southern portions of the Connecticut River Valley. Elsewhere the general elevation is from 500 to 1,500 feet, excepting up to near 2,500 feet in the extreme north. Numerous hills and mountains extend to heights of 2,000 to 4,000 feet above sea level over most of the State except in the southeast. There most hills are not more than 500 to 1,000 feet in height. Many White Mountain peaks rise above 4,000 feet. The elevation of eight peaks in the Presidential Range exceed 1 mile. Mt. Washington reaches 6,288 feet above sea level. This is the highest mountain in the northeastern United States. The extreme climate on top of Mt. Washington makes this location valuable for cold weather research and testing. (A complete weather observatory is maintained there). However, these extreme conditions are not representative of the more temperate climate of the State in general. Therefore, Mt. Washington data are omitted from extremes and average weather statistics for New Hampshire. Some flatland is found near the coast and in the river valleys, with about 10 percent of the total State area being classified as farmland. The glacier of the great Ice Age accounts for much of the topography, including many of the numerous lakes. About 1,300 lakes and ponds add to the State's attractions. The largest is Lake Winnepesaukee, which covers an area of 71 square miles in the central part of the State. Inland waters cover about 280 square miles. The Atlantic coast is 18 miles in length and has several fine beaches.

The two principal rivers in the State are the Connecticut and the Merrimack Rivers, both of which flow in a southerly direction. The larger of the two, the Connecticut, rises in extreme northern New Hampshire and forms the border between it and Vermont. Other rivers include the Androsscoggin and Saco Rivers which rise in the east slopes of the White Mountains and flow eastward into Maine.

Approximately 85 percent of New Hampshire is forested. Considerable area, especially in the north, is sparsely settled. The mountains, hills, lakes, streams, and forests combine to make New Hampshire a state of scenic beauty.

GENERAL CLIMATIC FEATURES: -- Characteristics of

New Hampshire climate are: (1) Changeableness of the weather, (2) large range of temperature, both daily and annual, (3) great differences between the same seasons in different years, (4) equable distribution of precipitation, and (5) considerable diversity from place to place. The regional climatic influences are modified in New Hampshire by varying distances from the ocean, elevations, and types of terrain. The State has been divided into two climatological divisions (Northern and Southern), which take into account the main features of these modifying factors. To take all local factors into consideration would require an impractical number of such areal divisions.

New Hampshire lies in the "prevailing westerlies", the belt of generally eastward air movement which encircles the globe in middle latitudes. Embedded in this circulation are extensive masses of air originating in higher or lower latitudes and interacting to produce low-pressure storm systems. Relative to most other sections of the country, a large number of such storms pass over or near New Hampshire. The majority of air masses affecting this State belong to three types: (1) Cold, dry air pouring down from subarctic North America, (2) warm, moist air streaming up on a long overland journey from the Gulf of Mexico and eastward, and (3) cool, damp air moving in from the North Atlantic. Because the atmospheric flow is usually offshore, New Hampshire is more influenced by the first two types than it is by the third. In other words, the adjacent ocean constitutes an important modifying factor, particularly on the immediate coast, but does not dominate the climate.

The procession of contrasting air masses and the relatively frequent passage of storms bring about approximately twice-weekly alternation from fair to cloudy or stormy conditions, often attended by abrupt changes in temperature, moisture, sunshine, wind direction and speed. There is no regular or persistent rhythm to this sequence, and it is interrupted by periods during which the weather patterns continue the same for several days, infrequently for several weeks. New Hampshire weather, however, is cited for variety rather than monotony. Changeability is also one of its features on a longer time-scale. That is, the same month or season will exhibit varying characteristics over the years, sometimes in close alternation, sometimes arranged in similar groups for successive years. A "normal" month, season, or year is indeed the exception rather than the rule.

The basic climate, as outlined above, obviously does not result from the predominance of any single controlling weather regime, but is rather the integrated effect of a variety of weather patterns. Hence, "weather averages" in New Hampshire usually are not sufficient for important planning purposes without further climatological analysis.

The Northern Division contains approximately one-third of the State, including the northern and west-central areas. Its southern border is roughly parallel to the coast, except where it bends northward near the Connecticut River. This Division represents that area least affected by the ocean influences and most affected by higher elevations as well as by its more northerly latitude. The Southern Division comprises the remaining area. Its lower elevation and latitude tend to cause higher temperatures, though this is modified seasonally by ocean influences. A strip near the coast could be a third division, but its small size hardly merits delineation.

TEMPERATURE: -- The annual temperature averages near 41°F. in the Northern Division and near 46°F. in the Southern. Within the Northern Division it ranges from about 38°F. in the extreme north to about 44°F. in the extreme south. Averages vary within the divisions also from causes other than latitude. Elevation, slope, and other environmental effects, including urbanization, each has an effect. As an extreme example of the effect of altitude is Mt. Washington, whose summit has an annual average of 27°F., compared to averages of 40° to 42°F. at other stations in the general area. The highest temperature of record is 106°F. observed July 4, 1911 at Nashua; the lowest, -46°F., January 28, 1925, at Pittsburg.

Summer temperatures are delightfully comfortable for the most part. They are reasonably uniform over the State, excepting topographical extremes. Hot days with maxima of 90°F. or higher average from only a few per year in the extreme north to 5 to 15 per year over most of the rest of the State. The frequency varies from place to place and from year to year. They range, in frequency of occurrence, from only a few in cool summers to as many as 30 to 40 in the Southern Division in the warmest summers. The diurnal range may reach 40°F. or more during cool, dry weather in valleys and lowlands. Freezing temperatures may be a threat even in the warmer months in a few of the more susceptible areas.

Average temperatures vary from place to place more in the winter than in summer. Days with subzero readings are relatively few along the immediate coast but are common inland. They average from 25 to 50 in number per year in most of the Northern Division and from 10 to 25 in the Southern Division.

The growing season for vegetation subject to injury from freezing temperature averages from 90 to 120 days in the Northern Division. In the Southern Division the average is 120 to 140 days except up to 160 to 180 days in the extreme southeast, a coastal effect. Local topography causes exceptions to the above averages. Swampy areas, particularly, may have a shorter season. The average date of the last freezing temperature in spring ranges from early in June at the colder locations to late in April at a few southern stations. For most of the State the growing season begins in May, and usually ends in the latter part of September.

PRECIPITATION: -- New Hampshire is fortunate in having its precipitation rather evenly distributed through the year. Low pressure, or frontal, storm systems are the principal year-round moisture producers. This activity ebbs somewhat in summer, but thunderstorms are of increased activity at this time, tending to make up the difference. Though brief and often of small extent, the thunderstorms produce the heaviest local rainfall intensities, and sometimes cause minor washouts of roads and soils. Rains of 1 to 2 inches in 1 hour can be expected at least once in a 10-year period.

Variations in monthly precipitation totals are extreme, ranging from no measurable amount to 10 inches or more. A large majority of monthly totals falls in the range of from 50 to 200 percent of normal. As prolonged droughts are infrequent, irrigation water is available during the fairly common shorter dry spells of summer. Similarly, widespread floods are infrequent. However, hurricane Edna, in September 1954, brought the second occurrence that year of heavy flooding and washing rains in southern New Hampshire. Other floods of note were in 1785, 1826, 1852, 1870, 1895, 1896, 1927, the outstandingly disastrous flood of March 1936, and the flood of September 1938. Floods occur most often in the spring when they are caused by a combination of rain and melting snow. At other times of the year high flows and major flooding from rainfall alone occur less frequently. The mean annual runoff in the streams ranges from 14 inches in the north-central Connecticut River Valley to 50 inches in the White Mountains area.

Total annual precipitation averages near 44 inches in the Northern Division and 41 inches in the Southern. The distribution is quite uniform over the Southern Division, ranging from about 37 to 46 inches. The mountainous character of much of central and northern New Hampshire, and the generally higher elevations there, account for the greater annual totals and variability from place to place. As an extreme example, Bethlehem, elevation 1,470 feet, has an annual average of only about one-half that of Mt. Washington (70 inches), where the gage elevation is 6,262 feet above sea level. These stations are only 20 miles apart.

Considerable rain or wet snow falls along the coast in winter, while farther inland snow is more generally the rule. Occasionally freezing rain occurs, coating exposed surfaces with troublesome ice. This problem is less frequent in northern New Hampshire. Most areas can expect at least one occurrence of glaze in the season.

Measurable amounts of precipitation fall on an average of 1 day in 3. Frequency is higher at higher elevations and in extreme northern New Hampshire, up to 140 to 150 days per year. At the Mt. Washington station measurable amounts occur on more than one-half the days. As much as 6 inches of rain in 24 hours is rare in New Hampshire. Most stations have never recorded that much in a single day. However, Warren, N. H., recorded 6.31 inches in 6 hours, and the 24-hour maximum is 8.00 inches at New Durham.

SNOWFALL: -- Average annual amounts of snowfall in the Southern Division increase from around 50 inches near the coast to 60 to 80 inches inland. Totals vary greatly in the Northern Division. Along the Connecticut River in the southern portion, totals average near 60 inches but increase to over 100 inches at the higher elevations of the northern and western portions. The summit of Mt. Washington receives nearly 185 inches. As an example of great variation in a short distance, Bethlehem, only about 20 miles to the west, receives only about 70 inches per year.

The number of days with 1 inch or more of snowfall varies from near 20 per season over much of the Southern Division up to 30 to 40 in the Northern Division and even to 50 or more at the highest elevations. Most winters will have several snowstorms of 5 inches or more. Storms of this magnitude temporarily disrupt transportation.

On November 22-23, 1943, a single storm dropped 56 inches of snow at Randolph, N. H., and over 50 inches at other nearby stations. This was the heaviest snowstorm of record in New Hampshire. However, snowstorms of as much as 20 inches or more are unusual in any part of the State. Heaviest 24-hour falls of record at many stations do not exceed 20 inches.

Snowfall is highly variable from year to year or for the same month in different years, as well as from place to place. Totals for the least snowy seasons range from one-fourth to one-half of the greatest seasonal amounts. In 24 years of record at Mt. Washington, for example, seasonal snowfall totals ranged from a maximum of 317 inches to a minimum of 135 inches. At Concord, in 64 seasons, the totals ranged from 29 to 103 inches. Month to month variations are much greater. Concord's maximum monthly total is 59.0 inches in February, 1893, but only 1.4 inches for the same month in 1941.

Snow cover is continuous through the whole winter season as a rule. Most frequent exceptions are found along the immediate coast and sometimes in extreme southern New Hampshire. Snow cover reaches its maximum depth, on the average, during the latter half of February in the Southern Division. In the Northern Division, the greatest depth comes in early March, excepting the higher elevations where the date is deferred to the middle of March. Some stations have a tendency for a secondary maximum in January or even at the end of December. Water stored in the snow makes an important contribution to a continuous water supply. The spring melting is usually too gradual to produce serious flooding.

OTHER CLIMATIC FEATURES: -- Sunshine averages over 50 percent of the possible amount in the Southern Division. The percentage is near 50 in the lower elevations of the Northern Division. Higher elevations and peaks are cloudier, especially in winter, reducing the percentage to less than 50 percent generally. Mt. Washington reports an average of only 33 percent.

Heavy fog occurrence varies remarkably with location and topography, but not enough data are available to describe this in detail. Persistent fogs are sometimes experienced along the coast and on the higher elevations inland. Duration of fogs diminishes inland over flat and valley locations. But the shorter duration heavy ground fogs of early morning occur frequently at susceptible places in these areas. The number of days with fog probably varies from about 20 to 90 per year over the State, except that it is much higher on the highest mountain peaks.

WINDS AND STORMS: -- The prevailing wind, on a yearly basis, comes from a westerly direction. It is predominantly from the northwest in winter and from the southwest in summer. Topography has a strong influence on prevailing direction. Points in major river valleys, for example, may have prevailing directions paralleling the valley. Along the coast in spring and summer the sea breeze is important. These onshore winds, from the cool ocean, may come inland for 10 miles or so; infrequently, they may reach as far as 30 miles into the interior. They tend to retard spring growth, but are pleasingly cool in summer.

Coastal storms or "northeasters" can be a serious weather hazard in southeastern New Hampshire, decreasing in importance northward. They generate very strong winds and heavy rain or snow. They can produce abnormally high wind-driven tides. These can heavily damage coastal installations and beaches. Some of the heaviest snowstorms result from these storms. Occasionally in summer or fall storms of tropical origin affect New Hampshire. These may often be similar (except for snow) to the northeasters described above. Only a very few retain near or full hurricane force and cause widespread damage. Damage is usually confined to the effects of high tides and heavy rainfall. Hurricane Carol traversed the whole State from south to north on August 31, 1954, with near hurricane force winds in the southern portion. Storms of tropical origin affect or threaten New Hampshire about once in 2 to 3 years, on the average. Two such storms in the same year would not be expected more than once in 10 years.

Tornadoes are not common phenomena, yet, on a per unit area basis, New Hampshire may rank fairly high among the states. Many years may have one or more. The Sunapee tornado of 1821 has been described as perhaps the worst ever experienced in New England. Fortunately, most tornadoes are very small, affecting a very localized area. Due to the extent of forested or sparsely settled areas, a large percentage of tornadoes in New Hampshire are neither seen, recorded, nor do appreciable damage. They may occur even in the northern portion of the State. One such notable occurrence was at Berlin in 1929. About 80 percent of torna-

647

does occur between May 15 and September 15. About 78 percent strike between 2 and 7 p.m. The peak months are June and July and the peak hour of occurrence 4 to 5 p.m. The chance of a tornado striking any given spot is extremely small.

Thunder and hailstorms have a similar frequency maximum from midspring to early fall. Thunderstorms occur on 15 to 30 days per year. The most severe are attended by hail. Hailstorms can severely injure or even ruin field crops, break glass, dent automobiles, and damage other vulnerable exposed objects. However, this danger is minimized because the size of an area struck by hail is usually small. Glaze (ice) storms of winter can produce perilous conditions for travel. These are usually of brief duration. A few widespread and prolonged ice storms have occurred. Besides affecting travel and transport, they break trees and limbs, utility lines and poles. In designing structures such as steel towers, possible ice load should be considered. An ice load also magnifies the wind stress by increasing the area exposed to the wind.

CLIMATE AND ECONOMY: -- Activities in New Hampshire are profoundly influenced by climate. Tree growth is especially favored. Covering more than four-fifths of the area, forests constitute a major scenic attraction. The spectacular coloration of foliage in the autumn is of special interest, drawing countless visitors. Forests also provide material for forest product industries. These include lumbering, paper-making, wood products manufacturing, and related industries. The ample supply of rainfall provides not only for the growth of trees but also the huge amounts of water required in the making of paper and other manufactures.

Favored industries include textiles and leather goods. A great diversity of other interests take advantage of the abundant water supply. Approximately two-thirds of the State's electrical power requirements are met by water power through a well developed hydroelectric system.

Climate is a significant factor in the State's agriculture. It favors the production of high value specialized crops. New Hampshire, therefore, ranks well in the Nation in cash receipts per acre from farm marketing. The principal farm specialties are poultry raising, dairying, tree fruits, and truck gardening. Broiler production is a large factor in the poultry industry. Apples are the most prolific of the tree fruits, with quality production an important commercial pursuit. Top quality maple syrup is produced in commercial quantity. Considerable acreage is devoted to pasture, hay, oats, and, in the southern portion, corn. Potatoes are also grown. The best soils are found largely in the river valleys, such as those of the Connecticut and Merrimac Rivers.

Climate is particularly important to a major industry, the tourist and vacation trade. Abundant game and teeming lakes and streams draw sportsmen from far and near. Much of the vacation trade is in the summer and fall when pleasant temperatures prevail at coastal and lake resorts. Skiing, with related winter sports, is developing into a very important winter attraction, made possible by the abundant snowfall.

In summary, the climate of New Hampshire contributes greatly to its industrial, agricultural, and vacation activities. The climate is a rich, natural asset, invigorating to persons in average health, and is favorable to further economic development.

SELECTED REFERENCES

General:

1. National Planning Association: The Economic State of New England (1954).

2. U. S. Dept. of Agriculture: Atlas of American Agriculture (1936)

3. --- : Climate and Man (Yearbook of Agriculture for 1941), Part 5, Climatic data, with special reference to agriculture in the United States.*

4. --- : Soil (Yearbook of Agriculture for 1957).

5. --- : Climatological Data New England (issued monthly and annually, Jan. 1888 ---; pub. under various other titles previous to Jan. 1921).*

Specialized:

1. Brooks, C. F.: "New England Snowfall", Monthly Weather Review, Vol. 45 (1917).

2. --- : "The Rainfall of New England. General Statement", Journ. N. Eng. Water Works Assoc., Vol. 44 (1930).

3. Brown, Rodger A.: "Twisters in New England", unpublished manuscript, Antioch College, 1957.*

4. Church, P. E.: "A Geographical Study of New England Temperatures", Geogr. Review, Vol. 26 (1936).

5. Eustis, R. S.: "Winds over New England in relation to topography", Bull. Amer. Met. Soc., Vol. 23 (1942).

6. Galway, Joseph G.: "A Statistical Study of New England Snowfall", unpublished manuscript of U. S. Weather Bureau (1954).*

7. Goodnough, X. H.: "Rainfall in New England", Journ. N. Eng. Water Works Assoc., Vols. 29 (1915), 35 (1921) and 40 (1926).*

8. Palmer, Robert S.: "Agricultural Drought in New England". Technical Bulletin 97, Agricultural Experiment Station, U. of New Hampshire, Durham, N. H. (1958).

9. Perley, S.: Historic Storms of New England (1891).*

10. Stone, R. G.: "Distribution of snow depths over New York and New England", Trans. Amer. Geophy. Union (1940).

11. --- : "The average length of the season with snow cover of various depths in New England", Trans. Amer. Geophy. Union (1944).

12. Upton, W.: "Characteristics of the New England Climate", Annals Harvard Astron. Obser. (1890).

13. U. S. Weather Bureau: Tabulations of frequencies of various climatic elements for various selected stations. Available on microfilm at library of Weather Bureau State Climatologist, 1900 Post Office Bldg., Boston 9, Mass.

14. Weber, J. H.: "The Rainfall of New England. Historical Statement. Annual Rainfall. Seasonal Rainfall. Mean Monthly Rainfall of Southern New England. Maximum and Minimum Rainfall of Southern New England." Journ. N. Eng. Water Works Assn., Vol. 44 (1930).

15. White, C. V.: "Rainfall in New England", Journ. N. Eng. Water Works Assn., Vols. 56 (1942) and 57 (1943).*

*References marked with an asterisk are useful sources of data; the others are principally studies of the important climatic elements.

16. Weather Bureau Technical Paper No. 15 - Maximum Station Precipitation for 1, 2, 3, 6, 12, and 24 Hours.

17. Weather Bureau Technical Paper No. 16 - Maximum 24-Hour Precipitation in the United States. Washington, D. C. 1952.

18. Weather Bureau Technical Paper No. 25 - Rainfall Intensity-Duration-Frequency Curves. For selected Stations in the United States, Alaska, Hawaiian Islands, and Puerto Rico.

BIBLIOGRAPHY

(A) Climatic Summary of the United States (Bulletin W) 1930 edition, Section 84 (New Hampshire and Vermont). U. S. Weather Bureau

(B) Climatic Summary of the United States, New England - Supplement for 1931 through 1952 (Bulletin W Supplement). U. S. Weather Bureau

(C) Climatological Data - New Hampshire. U. S. Weather Bureau

(D) Climatological Data National Summary. U. S. Weather Bureau

(E) Hourly Precipitation Data - New England. U. S. Weather Bureau

(F) Local Climatological Data, U. S. Weather Bureau, for Concord and Mt. Washington, New Hampshire.

TROPICAL CYCLONE DATA HAVING IMPORTANCE FOR THIS STATE IS INCLUDED IN STATISTICS AND CHARTS ON PAGES 1161 THROUGH 1164 .

FREEZE DATA

STATION	Freeze threshold temperature	Mean date of last Spring occurrence	Mean date of first Fall occurrence	Mean No. of days between dates	Years of record Spring	No. of occurrences in Spring	Years of record Fall	No. of occurrences in Fall	STATION	Freeze threshold temperature	Mean date of last Spring occurrence	Mean date of first Fall occurrence	Mean No. of days between dates	Years of record Spring	No. of occurrences in Spring	Years of record Fall	No. of occurrences in Fall
BERLIN	32	05-29	09 15	109	30	30	30	30	KEENE	32	05-25	09 19	117	30	30	30	30
	28	05-11	09 28	140	30	30	30	30		28	05-12	10 02	143	30	30	30	30
	24	04-25	10 13	171	30	30	30	30		24	04-28	10 13	168	30	30	30	30
	20	04-15	10 27	195	30	30	30	30		20	04-15	10 26	194	30	30	30	30
	16	04-05	11 09	217	30	30	30	30		16	03-31	11 11	224	30	30	30	30
BETHLEHEM	32	05-19	09 25	129	29	29	30	30	MANCHESTER	32	04-29	10 13	167	22	22	22	22
	28	05-06	10 06	153	29	29	30	30		28	04-16	10 28	195	22	22	22	22
	24	04-22	10 20	181	29	29	30	30		24	04-02	11 10	222	22	22	22	22
	20	04-15	11 01	201	29	29	30	30		20	03-20	11 23	248	22	22	22	22
	16	04-03	11 10	220	29	29	30	30		16	03-15	12 02	262	22	22	22	22
CONCORD	32	05-11	09 30	142	30	30	30	30	NASHUA 3 N	32	05-22	09 23	124	25	25	26	26
	28	04-27	10 12	168	30	30	30	30		28	05-08	10 06	151	25	25	26	26
	24	04-13	10 22	191	30	30	30	30		24	04-21	10 20	182	25	25	26	26
	20	03-31	11 05	219	30	30	30	30		20	04-05	10 31	209	25	25	26	26
	16	03-23	11 20	242	30	30	30	30		16	03-22	11 15	238	25	25	26	26
FIRST CONN LAKE (P.O.Pittsburg)	32	06-09	09 09	93	30	30	30	30	PINKHAM NOTCH	32	05-23	09 22	121	21	21	21	21
	28	05-23	09 22	123	30	30	30	30		28	05-09	09 30	144	21	21	21	21
	24	05-06	10 06	153	30	30	30	30		24	04-28	10 19	174	21	21	21	21
	20	04-25	10 21	179	30	30	30	30		20	04-20	10 28	192	21	21	21	21
	16	04-19	10 30	195	30	30	30	30		16	04-05	11 08	217	21	21	21	21
FRANKLIN 1 NW	32	05-21	09 25	127	30	30	30	30	PLYMOUTH	32	05-26	09 17	114	30	30	30	30
	28	05-13	10 05	146	30	30	30	30		28	05-13	09 30	140	30	30	30	30
	24	04-27	10 16	172	30	30	30	30		24	04-28	10 14	170	30	30	30	30
	20	04-15	10 27	195	30	30	30	30		20	04-16	10 25	192	30	30	30	30
	16	03-31	11 12	227	30	30	30	30		16	03-29	11 08	224	30	30	30	30
GLENCLIFF	32	05-17	09 27	133	30	30	30	30	WOLFEBORO FALLS	32	05-19	09 28	132	15	15	16	16
	28	05-01	10 09	161	30	30	30	30		28	05-06	10 07	154	15	15	16	16
	24	04-23	10 27	187	30	30	30	30		24	04-27	10 21	177	15	15	16	16
	20	04-12	11 06	208	30	30	30	30		20	04-14	11 06	206	15	15	16	16
	16	04-01	11 13	226	30	30	30	30		16	04-02	11 18	231	15	15	16	16

Data in the above table are based on the period 1921-1950, or that portion of this period for which data are available.

Means have been adjusted to take into account years of non-occurrence.

A freeze is a numerical substitute for the former term "killing frost" and is the occurrence of a minimum temperature at or below the threshold temperature of 32°, 28°, etc.

Freeze data tabulations in greater detail are available and can be reproduced at cost.

*MEAN TEMPERATURE AND PRECIPITATION

STATION	JANUARY Temperature	JANUARY Precipitation	FEBRUARY Temperature	FEBRUARY Precipitation	MARCH Temperature	MARCH Precipitation	APRIL Temperature	APRIL Precipitation	MAY Temperature	MAY Precipitation	JUNE Temperature	JUNE Precipitation	JULY Temperature	JULY Precipitation	AUGUST Temperature	AUGUST Precipitation	SEPTEMBER Temperature	SEPTEMBER Precipitation	OCTOBER Temperature	OCTOBER Precipitation	NOVEMBER Temperature	NOVEMBER Precipitation	DECEMBER Temperature	DECEMBER Precipitation	ANNUAL Temperature	ANNUAL Precipitation
NEW HAMPSHIRE																										
NORTHERN																										
BERLIN	16.0	2.77	17.1	2.35	26.8	3.31	40.2	3.12	52.4	3.21	61.7	4.07	66.6	3.60	64.2	3.19	56.1	3.96	46.0	3.01	33.9	3.56	20.1	2.98	41.8	39.13
BETHLEHEM	17.2	1.73	18.5	1.49	27.9	2.27	41.0	2.67	54.1	3.47	67.9	4.23	67.6	4.08	65.7	3.56	57.6	4.04	47.1	3.13	34.1	3.10	20.4	2.25	42.8	35.97
ERROL		2.86		2.46		3.06		3.21		3.50		4.12		3.80		3.26		3.72		2.98		3.59		2.93		39.49
FIRST CONN LAKE	11.4	3.12	11.9	2.82	21.4	3.20	35.5	3.45	48.8	3.99	58.5	4.79	63.4	4.97	61.2	3.71	53.4	4.37	43.1	3.77	30.7	3.90	16.0	3.42	37.9	45.51
MILAN 7 N		2.80		2.24		2.94		3.09		3.22		4.13		3.55		3.14		3.40		2.84		3.20		2.75		37.30
MONROE 5 NNE	15.0	2.29	15.8	1.99	26.7	2.37	40.6	3.00	53.5	3.31	63.1	3.73	67.9	3.54	65.7	3.30	57.2	3.57	46.5	3.02	34.0	3.09	19.5	2.44	42.1	35.65
MOUNT WASHINGTON	5.4	5.10	5.5	4.75	12.0	5.55	22.0	5.87	35.3	5.32	44.4	6.26	49.3	6.36	47.7	6.15	41.1	6.56	31.7	5.90	19.9	6.58	8.9	5.85	27.0	70.25
PINKHAM NOTCH	16.8	4.43	17.5	3.98	26.1	5.82	37.7	4.93	50.2	5.02	59.0	5.08	63.7	4.97	61.6	4.69	54.3	5.03	45.0	4.95	32.6	5.80	19.9	4.70	40.4	58.97
YORK POND		3.16		2.74		3.80		3.56		3.90		4.66		3.95		3.46		4.39		3.34		3.80		3.26		44.02
DIVISION	16.0	2.88	17.0	2.54	26.3	3.43	39.2	3.50	51.8	3.94	61.0	4.45	65.9	4.25	63.7	3.80	55.9	4.21	45.8	3.57	33.3	4.01	19.8	3.28	41.3	43.86
SOUTHERN																										
CONCORD WB AIRPORT	20.1	2.91	21.2	2.30	31.8	3.04	43.0	3.08	54.8	3.04	64.1	3.62	69.0	3.57	66.5	3.10	58.8	3.39	48.0	2.80	36.7	3.57	24.0	2.81	44.8	37.23
DURHAM	24.4	3.53	25.4	2.97	34.3	4.16	45.0	3.76	56.0	3.40	65.1	3.39	70.5	3.37	68.7	3.36	61.1	3.52	50.7	3.04	39.5	4.18	27.5	3.59	47.4	42.27
FITZWILLIAM 3 SW		3.14		2.54		3.65		3.69		3.94		4.36		4.23		3.97		3.99		3.44		3.86		3.11		43.92
FRANKLIN 1 NW	20.8	3.20	21.9	2.69	31.4	3.69	43.6	3.49	55.7	4.16	65.2	3.76	70.3	3.46	68.0	3.15	60.1	3.79	49.0	2.86	37.1	4.00	24.3	3.45	45.6	41.69
HANOVER	19.2	2.75	20.7	2.34	30.7	2.78	43.3	3.19	55.3	3.37	64.6	3.44	69.4	4.30	67.2	3.13	59.2	3.23	48.4	2.78	36.1	3.36	22.7	2.70	44.7	37.37
KEENE	23.0	3.27	24.2	2.52	33.3	3.33	44.9	3.58	56.3	3.72	65.0	3.94	69.8	3.75	67.7	3.32	60.2	3.77	49.9	2.71	38.3	3.63	26.0	3.03	46.6	40.57
LINCOLN		3.60		2.89		3.63		3.69		3.77		4.07		4.28		4.03		4.59		3.64		4.14		3.76		46.09
MANCHESTER	24.1	3.70	25.0	2.79	33.5	4.00	44.6	3.77	56.2	3.83	65.2	3.73	70.4	3.48	67.8	3.58	60.0	3.74	50.0	2.92	39.2	4.18	27.7	3.54	47.0	43.26
NASHUA 3 N	23.9	3.44	24.9	2.71	33.9	3.78	45.1	3.57	56.4	3.63	65.4	3.89	70.5	3.33	68.2	3.52	60.4	3.75	50.3	3.29	39.0	3.91	27.2	3.31	47.1	42.13
NEWPORT		2.94		2.44		3.26		3.55		3.62		3.70		3.68		3.24		3.61		2.64		3.54		2.86		39.08
WEST LEBANON		2.62		2.20		2.68		3.00		3.22		3.35		3.98		3.02		3.22		2.66		3.17		2.51		35.63
WOLFEBORO FALLS		3.68		3.10		4.01		3.56		3.75		3.48		3.42		2.96		3.57		2.79		4.24		3.85		42.41
DIVISION	21.7	3.30	22.7	2.66	31.9	3.57	43.7	3.55	55.4	3.67	64.6	3.72	69.6	3.70	67.3	3.33	59.5	3.76	49.2	2.91	37.6	3.85	25.1	3.31	45.7	41.33

*   Averages for period 1931-1955, except for stations marked WB which are "normals" based on period 1921-1950. Divisional means may not be the arithmetical average of individual stations published, since additional data from shorter period stations are used to obtain better areal representation.

CONFIDENCE LIMITS

In the absence of trend or record changes, the chances are 9 out of 10 that the true mean will lie in the interval formed by adding and subtracting the values in the following table from the means for any station in the State:

1.7	.38	1.4	.24	1.5	.60	1.1	.50	.9	.48	.7	.69	.6	.59	.8	.50	.8	.79	.9	.57	1.0	.68	1.3	.49	.4	2.27

COMPARATIVE DATA

Data in the following table are the mean temperature and average precipitation for Franklin, New Hampshire for the period 1906-1930 and are included in this publication for comparative purposes:

20.0	2.82	20.0	2.87	30.9	3.23	43.0	3.28	54.6	3.16	63.7	3.65	69.4	3.87	66.4	3.60	59.2	3.57	48.5	2.90	36.2	3.42	24.3	2.97	44.7	39.34

NORMALS, MEANS AND EXTREMES
(Table Revised 1973. Base Period for Climatological Normals: 1931-1960)

Station: CONCORD, NEW HAMPSHIRE   MUNICIPAL AIRPORT   Standard time used: EASTERN   Latitude: 43° 12' N   Longitude: 71° 30' W   Elevation (ground): 342 feet

Ø For period April 1965 through current year.
Means and extremes above are from existing and comparable exposures. Annual extremes have been exceeded at other sites in the locality as follows:
Lowest temperature -37 in February 1943; maximum monthly precipitation 10.97 in September 1888; minimum monthly precipitation T in March 1915; maximum precipitation in 24 hours 5.97 in September 1932; maximum monthly snowfall 59.0 in February 1893.

NORMALS, MEANS AND EXTREMES
(Table Revised 1975. Base Period for Climatological Normals: 1941-1970)

Elev. 346 feet m.s.l.

Ø For period April 1965 through current year.
Means and extremes above are from existing and comparable exposures. Annual extremes have been exceeded at other sites in the locality as follows:
Lowest temperature -37 in February 1943; maximum monthly precipitation 10.97 in September 1888; minimum monthly precipitation T in March 1915; maximum precipitation in 24 hours 5.97 in September 1932; maximum monthly snowfall 59.0 in February 1893.

REFERENCE NOTES APPLYING TO TABLES APPEAR ON THE PAGE FOLLOWING LAST TABLE.
(Caution: Letters and symbols may have different meanings in 1941-1970 tables than in earlier tables. See notes.)

NORMALS, MEANS AND EXTREMES
(Table Revised 1973. Base Period for Climatological Normals: 1931-1960)

Station: MOUNT WASHINGTON OBSERVATORY GORHAM, NEW HAMPSHIRE   Standard time used: EASTERN   Latitude: 44°16'N   Longitude: 71°18'W   Elevation (ground): 6262 feet

Month	Temperature Normal — Daily maximum	Daily minimum	Monthly	Extremes Record highest	Year	Record lowest	Year	Normal heating degree days (Base 65°)	Precipitation Normal total	Max monthly	Year	Min monthly	Year	Max in 24 hrs	Year	Snow, Ice pellets Mean total	Max monthly	Year	Max in 24 hrs	Year
(a)	(b)	(b)	(b)	40		40		(b)	(b)	40		40		40		40	40		40	
J	14.7	-2.2	6.3	44	1950	-47	1934	1820	5.44	18.23	1958	1.80	1951	3.40	1958	32.3	92.5	1958	20.2	1958
F	13.9	-2.8	5.6	42	1937	-46	1943	1663	5.21	25.56	1969	2.64	1969	10.38	1962	40.2	172.8	1969	49.3	1962
M	19.0	4.4	11.8	48	1947	-38	1950	1652	5.74	15.57	1936	3.99	1946	3.99	1936	37.2	98.0	1970	27.4	1970
A	29.6	16.3	23.0	60	1941	-20	1954	1260	5.89	10.69	1961	2.19	1959	4.49	1959	25.1	74.4	1972+	23.1	1972
M	41.3	28.7	35.0	64	1956	-2	1966	930	5.84	13.41	1960	1.78	1960	4.60	1960	10.8	52.2	1967	22.2	1967
J	51.1	38.6	44.9	71	1933	8	1945	603	6.50	11.02	1968	3.30	1944	3.80	1951	1.2	8.1	1959	4.4	1959
J	54.6	43.3	49.1	71	1953	25	1969	493	6.71	15.53	1969	2.69	1955	7.37	1969	T	1.1	1957	1.1	1957
A	53.2	42.1	47.7	71	1947	20	1965	536	6.65	13.14	1955	2.77	1947	5.20	1955	0.1	2.5	1965	2.5	1965
S	46.8	35.2	41.0	67	1960+	11	1942	720	7.00	14.07	1948	2.74	1948	4.80	1949	11.3	7.8	1949	5.9	1949
O	37.0	24.8	30.9	59	1938	-5	1939	1057	6.19	13.49	1959	0.75	1947	7.03	1959	11.8	34.1	1969	17.0	1969
N	26.8	13.8	20.3	51	1956	-20	1958	1341	6.61	17.57	1963	2.31	1937	6.07	1963	28.8	86.6	1968	25.0	1968
D	16.5	1.0	8.8	45	1967+	-46	1933	1742	6.31	17.23	1969	1.49	1955	8.64	1969	38.6	103.7	1968	37.5	1968
YR	33.7	20.3	27.0	71	JUL 1953+	-47	JAN 1934	13817	74.09	25.56	FEB 1969	0.75	OCT 1947	10.38	FEB 1970	227.2	172.8	FEB 1969	49.3	FEB 1970

Month	Relative humidity (Local time) 01	07	13	19	Wind Mean speed	Prevailing direction	Peak gust Speed	Direction	Year	Mean sky cover sunrise to sunset	Pct. of possible sunshine
	22	22	22	19	38	26	40	40	40	34	34
J	81	82	83	83	46.1	W	170	W	1943	7.6	33
F	82	81	84	83	45.0	W	166	E	1972	7.6	33
M	86	83	83	83	42.0	W	180	SE	1942	7.6	34
A	86	86	86	86	36.5	W	231	SE	1934	7.5	35
M	90	89	84	87	30.5	W	164	S	1945	7.6	32
J	93	86	85	90	27.5	W	136	NW	1949	8.0	31
J	93	90	85	90	24.9	W	110	W	1933	8.1	31
A	85	85	88	85	25.2	W	142	ENE	1954	7.8	31
S	85	85	84	84	28.5	W	157	SE	1938	7.3	40
O	85	85	84	84	31.5	W	161	SE	1943	6.9	46
N	85	84	86	85	38.2	W	160	SE	1950	7.9	29
D	85	84	86	85	44.3	W	175	SE	1942	7.7	31
YR	86	85	84	86	35.2	W	231	SE	APR 1934	7.6	34

Mean number of days — Sunrise to sunset (Clear, Partly cloudy, Cloudy); Precipitation .01 inch or more; Snow, Ice pellets 1.0 inch or more; Thunderstorms; Heavy fog; Temperatures Max (90° and above, 32° and below), Min (32° and below, 0° and below); Average daily solar radiation - langleys

Month	Clear	Partly cloudy	Cloudy	Precip .01+	Snow 1.0+	Thunderstorms	Heavy fog	Max 90°+	Max 32°−	Min 32°−	Min 0°−	Solar radiation
	40	40	40	40	40	40	40	40	40	40	40	40
J	6	6	20	18	10	*	25	0	29	31	17	
F	4	5	19	17	10	*	25	0	27	28	16	
M	5	5	21	18	10	*	27	0	27	31	11	
A	7	4	19	17	7	1	24	0	18	20	1	
M	7	4	20	16	3	2	26	0	7	6	*	
J	2	8	20	17	1	4	27	0	1	1	0	
J	3	7	22	17	*	5	27	0	*	3	0	
A	5	8	16	16	*	4	25	0	0	12	0	
S	7	6	18	15	1	1	27	0	2	23	*	
O	9	5	14	14	4	*	25	0	10	28	4	
N	4	5	21	19	8	*	27	0	21	31	4	
D	5	5	21	20	11	*	27	0	27		14	
YR	51	75	239	206	64	16	309	0	168	243	65	

NORMALS, MEANS AND EXTREMES
(Table Revised 1975. Base Period for Climatological Normals: 1941-1970)

Month	Temperatures °F Normal — Daily maximum	Daily minimum	Monthly	Extremes Record highest	Year	Record lowest	Year	Normal Degree days Base 65°F Heating	Cooling	Precipitation in inches Water equivalent Normal	Max monthly	Year	Min monthly	Year	Max in 24 hrs	Year	Snow, Ice pellets Max monthly	Year	Max in 24 hrs	Year
(a)				42		42		42		42	42		42		42		42		42	
J	14.1	-2.8	5.7	44	1950	-47	1934	1838	0	5.12	18.23	1958	1.80	1951	3.40	1958	92.5	1958	20.2	1958
F	13.5	-3.5	5.0	42	1937	-46	1943	1680	0	6.51	25.56	1969	2.64	1969	10.38	1962	172.8	1969	49.3	1962
M	19.1	4.4	11.8	48	1947	-38	1950	1649	0	5.60	15.57	1936	2.15	1946	3.99	1936	98.0	1970	27.4	1970
A	29.3	16.0	22.7	60	1941	-20	1954	1269	0	5.46	14.62	1959	2.19	1959	4.49	1959	74.4	1972	23.1	1974
M	40.7	28.0	34.4	64	1966	-2	1966	949	0	5.84	13.41	1960	1.78	1960	4.60	1960	52.2	1967	22.2	1967
J	50.9	38.5	44.7	71	1933	8	1945	609	0	6.50	16.00	1973	3.30	1944	6.50	1973	8.1	1959	4.4	1959
J	54.6	43.4	48.8	71	1953	25	1969	502	0	6.77	15.53	1969	2.69	1955	7.37	1969	1.1	1957	1.1	1957
A	52.5	41.4	47.0	71	1947	20	1965	558	0	7.19	13.14	1955	2.77	1947	5.20	1955	2.5	1965	2.5	1965
S	46.9	35.4	41.2	67	1960	11	1942	714	0	6.36	14.07	1948	2.74	1948	4.80	1949	7.8	1949	5.9	1949
O	37.5	24.8	31.5	59	1939	-5	1939	1039	0	6.12	13.49	1959	0.75	1947	7.03	1959	34.1	1969	17.0	1969
N	27.1	14.0	20.6	51	1956	-20	1958	1332	0	7.67	17.57	1963	2.31	1939	6.07	1968	86.6	1968	25.0	1968
D	16.6	1.2	8.9	45	1967	-46	1933	1739	0	7.03	17.95	1973	1.49	1955	8.64	1969	103.7	1968	37.5	1968
YR	33.6	20.1	26.9	71	JUL 1953+	-47	JAN 1934	13878	0	76.17	25.56	FEB 1969	0.75	OCT 1947	10.38	FEB 1970	172.8	FEB 1969	49.3	FEB 1970

Month	Relative humidity pct. (Local time) 01	07	13	19	Wind Mean speed m.p.h.	Prevailing direction	Peak gust Speed m.p.h.	Direction	Year	Mean sky cover tenths, sunrise to sunset	Pct. of possible sunshine
	24	24	24	24	40	26	42	42	42	36	36
J	83	83	83	83	46.4	W	170	W	1943	7.6	33
F	82	81	84	83	44.8	W	166	E	1972	7.6	33
M	85	84	84	84	41.9	W	180	SE	1942	7.7	34
A	85	86	85	85	36.6	W	231	SE	1934	7.6	35
M	89	89	84	85	30.3	W	164	S	1945	7.6	32
J	91	86	85	89	27.4	W	136	NW	1949	8.1	31
J	93	91	86	90	24.9	W	110	W	1933	8.1	31
A	87	87	88	91	25.1	W	142	ENE	1954	7.9	36
S	84	84	83	86	28.5	W	157	SE	1943	7.3	41
O	82	83	83	83	33.5	W	161	SE	1943	6.9	46
N	85	84	85	84	38.4	W	160	SE	1942	7.9	30
D	85	84	84	84	44.1	W	175	SE	1942	7.8	31
YR	86	86	84	85	35.2	W	231	SE	APR 1934	7.7	33

Mean number of days

Month	Clear	Partly cloudy	Cloudy	Precip .01+	Snow, Ice pellets 1.0+	Thunderstorms	Heavy fog, visibility ¼ mile or less	Max 90°+	Max 32°−	Min 32°−	Min 0°−	Average station pressure mb.	Elev. feet m.s.l.
	42	42	42	42	42	42	42	(b) 42	42	42	42		
J	5	6	20	18	10	*	26	0	29	31	17		
F	5	5	18	17	10	*	25	0	27	28	16		
M	5	5	21	19	10	*	27	0	27	31	11		
A	4	7	19	18	8	1	24	0	18	20	2		
M	4	7	21	16	3	2	26	0	7	7	*		
J	2	7	21	17	1	4	27	0	1	1	0		
J	3	8	22	17	*	5	27	0	*	3	0		
A	5	9	19	15	*	3	26	0	0	12	0		
S	7	6	18	14	1	1	25	0	2	23	*		
O	9	6	18	14	4	1	25	0	10	28	4		
N	5	5	21	20	8	*	27	0	21	31	4		
D	5	5	21	20	11	*	27	0	27		14		
YR	51	73	241	206	66	16	310	0	168	243	65		

REFERENCE NOTES APPLYING TO TABLES APPEAR ON THE PAGE FOLLOWING LAST TABLE.
(Caution: Letters and symbols may have different meanings in 1941-1970 tables than in earlier tables. See notes.)

Reference notes applying to Normals, Means, and Extremes tables for 1931–1960 base period.

(a) Length of record, years, based on January data. Other months may be for more or fewer years if there have been breaks in the record.
Climatological standard normals (1931-1960).

(b) Less than one half.

* Also on earlier dates, months, or years.

T Trace, an amount too small to measure.
Below zero temperatures are preceded by a minus sign. The prevailing direction for wind in the Normals, Means, and Extremes table is from records through 1963.

‡ >70° at Alaskan stations.

Unless otherwise indicated, dimensional units used in this bulletin are: temperature in degrees F.; precipitation, including snowfall, in inches; wind movement in miles per hour; and relative humidity in percent. Heating degree day totals are the sums of negative departures of average daily temperatures from 65° F. Cooling degree day totals are the sums of positive departures of average daily temperatures from 65° F. Sleet was included in snowfall totals beginning with July 1948. The term "ice pellets" includes solid grains of ice (sleet) and particles consisting of snow pellets encased in a thin layer of ice. Heavy fog reduces visibility to 1/4 mile or less.

Sky cover is expressed in a range of 0 for no clouds or obscuring phenomena to 10 for complete sky cover. The number of clear days is based on average cloudiness 0-3, partly cloudy days 4-7, and cloudy days 8-10 tenths.

Solar radiation data are the averages of direct and diffuse radiation on a horizontal surface. The langley denotes one gram calorie per square centimeter.

& Figures instead of letters in a direction column indicate direction in tens of degrees from true North; i.e., 09-East, 18-South, 27-West, 36-North, and 00-Calm. Resultant wind is the vector sum of wind directions and speeds divided by the number of observations. If figures appear in the direction column under "Fastest mile" the corresponding speeds are fastest observed 1-minute values.

To 8 compass points only.

Precipitation gages are shielded and tilted so that the orifice is parallel to the slope.

Reference notes applying to Normals, Means, and Extremes tables for 1941–1970 base period.

(a) Length of record, years, through the current year unless otherwise noted, based on January data.

(b) 70° and above at Alaskan stations.

* Less than one half.

T Trace.

NORMALS - Based on record for the 1941-1970 period.
DATE OF AN EXTREME - The most recent in cases of multiple occurrence.
PREVAILING WIND DIRECTION - Record through 1963.
WIND DIRECTION - Numerals indicate tens of degrees clockwise from true north. 00 indicates calm.
FASTEST MILE WIND - Speed is fastest observed 1-minute value when the direction is in tens of degrees.

Precipitation gages are shielded and tilted so that the orifice is parallel to the slope.

Mean Annual Precipitation, Inches

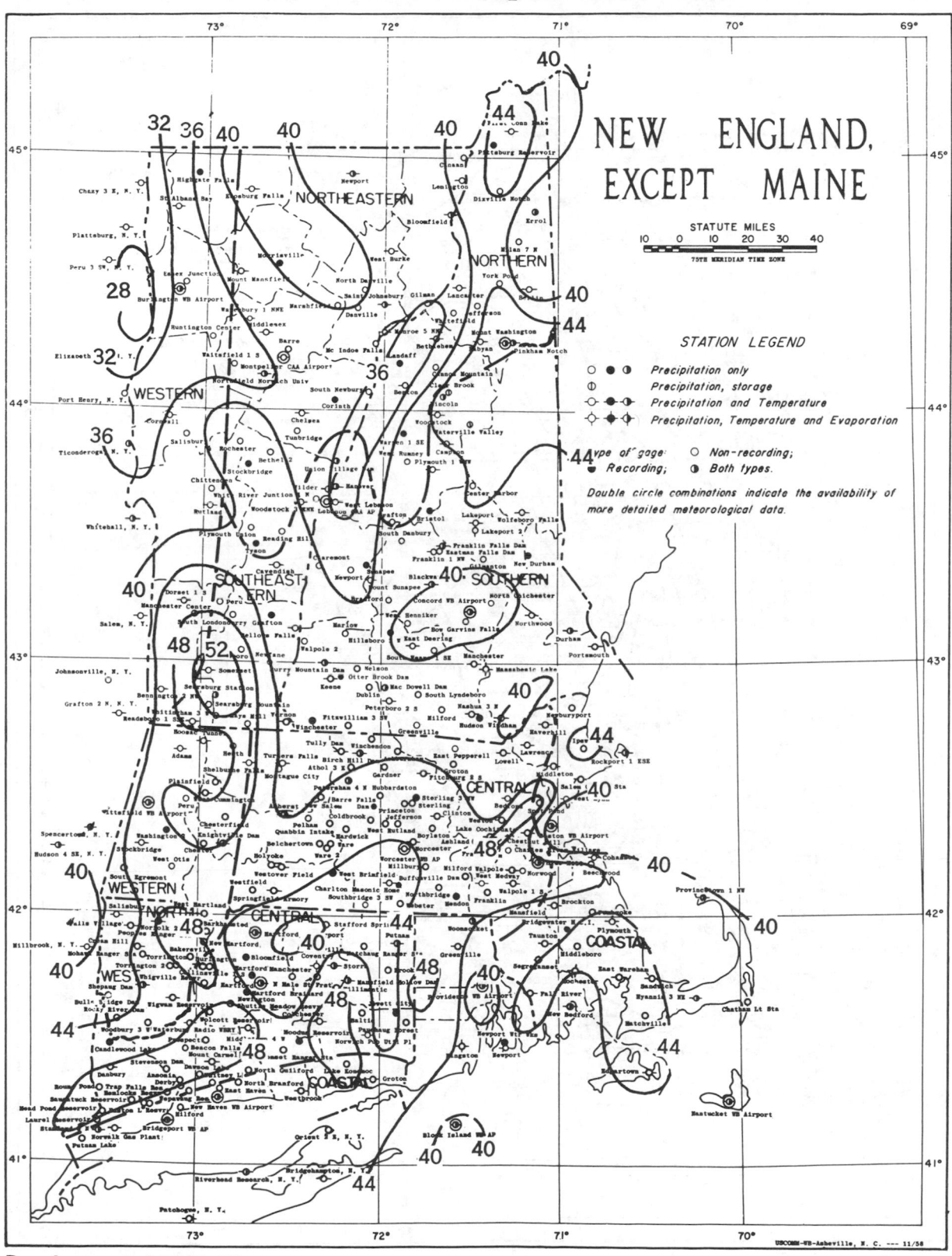

NEW ENGLAND, EXCEPT MAINE

STATUTE MILES

75TH MERIDIAN TIME ZONE

STATION LEGEND

Precipitation only

Precipitation, storage

Precipitation and Temperature

Precipitation, Temperature and Evaporation

Type of gage: Non-recording; Recording; Both types.

Double circle combinations indicate the availability of more detailed meteorological data.

Based on period 1931-55

Isolines are drawn through points of approximately equal value. Caution should be used in interpolating on these maps, particularly in mountainous areas.

Mean Maximum Temperature (°F.), January

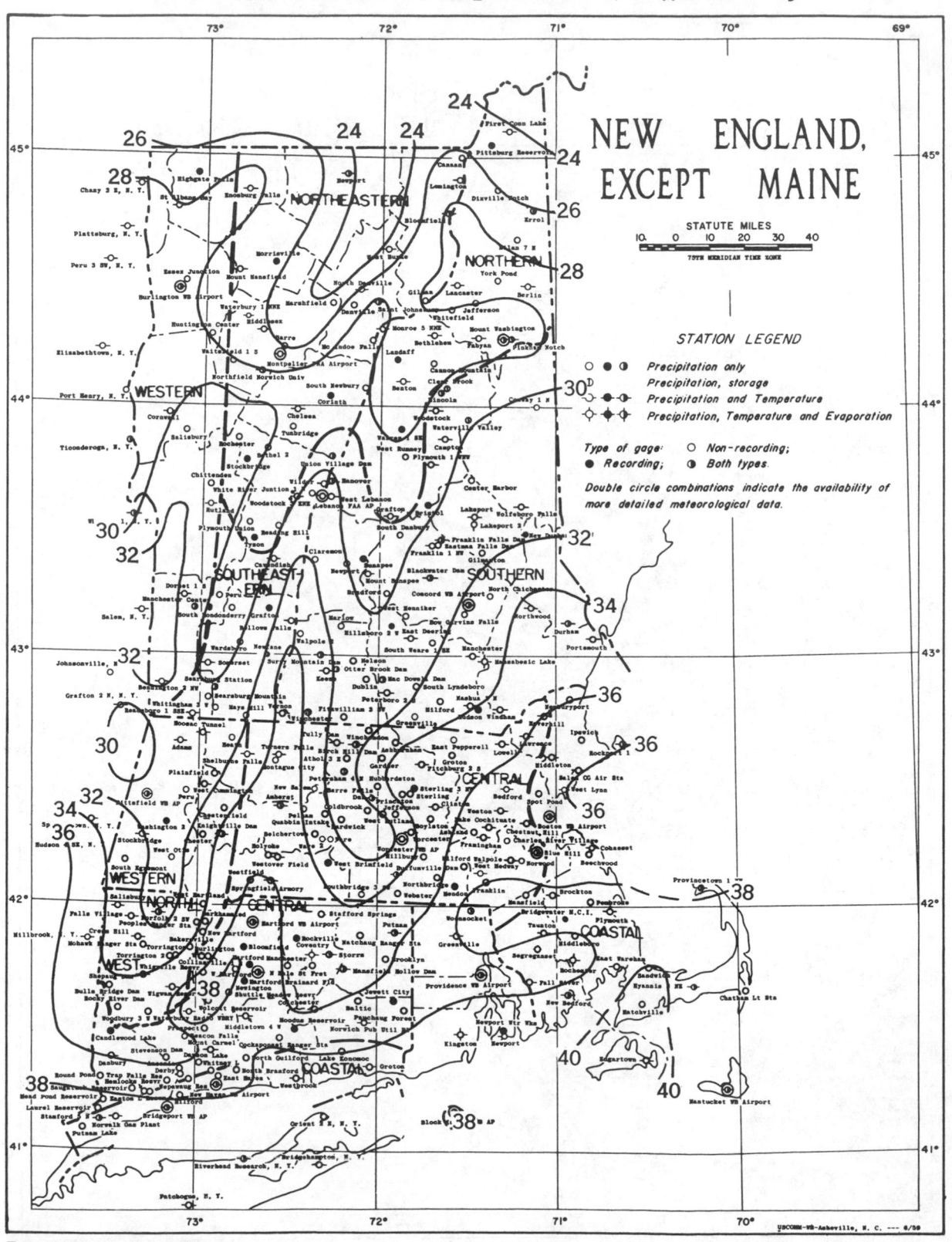

NEW ENGLAND, EXCEPT MAINE

STATUTE MILES

75TH MERIDIAN TIME ZONE

STATION LEGEND

○ ● ◐ *Precipitation only*
◑ *Precipitation, storage*
◈ *Precipitation and Temperature*
✦ *Precipitation, Temperature and Evaporation*

Type of gage: ○ *Non-recording;*
● *Recording;* ◑ *Both types.*

Double circle combinations indicate the availability of more detailed meteorological data.

Based on period 1931-52

Isolines are drawn through points of approximately equal value. Caution should be used in interpolating on these maps, particularly in mountainous areas.

656

Mean Minimum Temperature (°F.), January

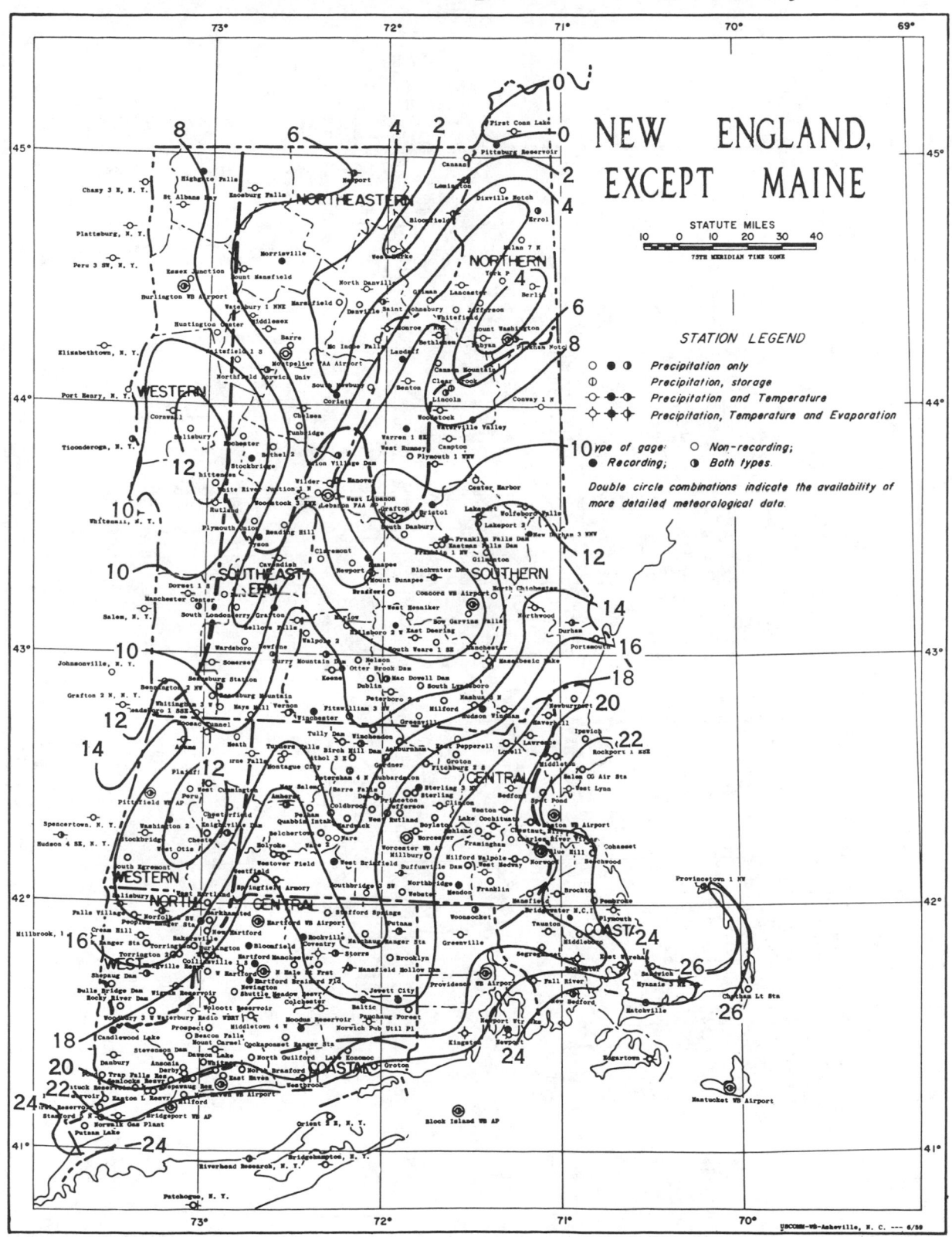

Based on period 1931-52

Isolines are drawn through points of approximately equal value. Caution should be used in interpolating on these maps, particularly in mountainous areas.

Mean Maximum Temperature (°F.), July

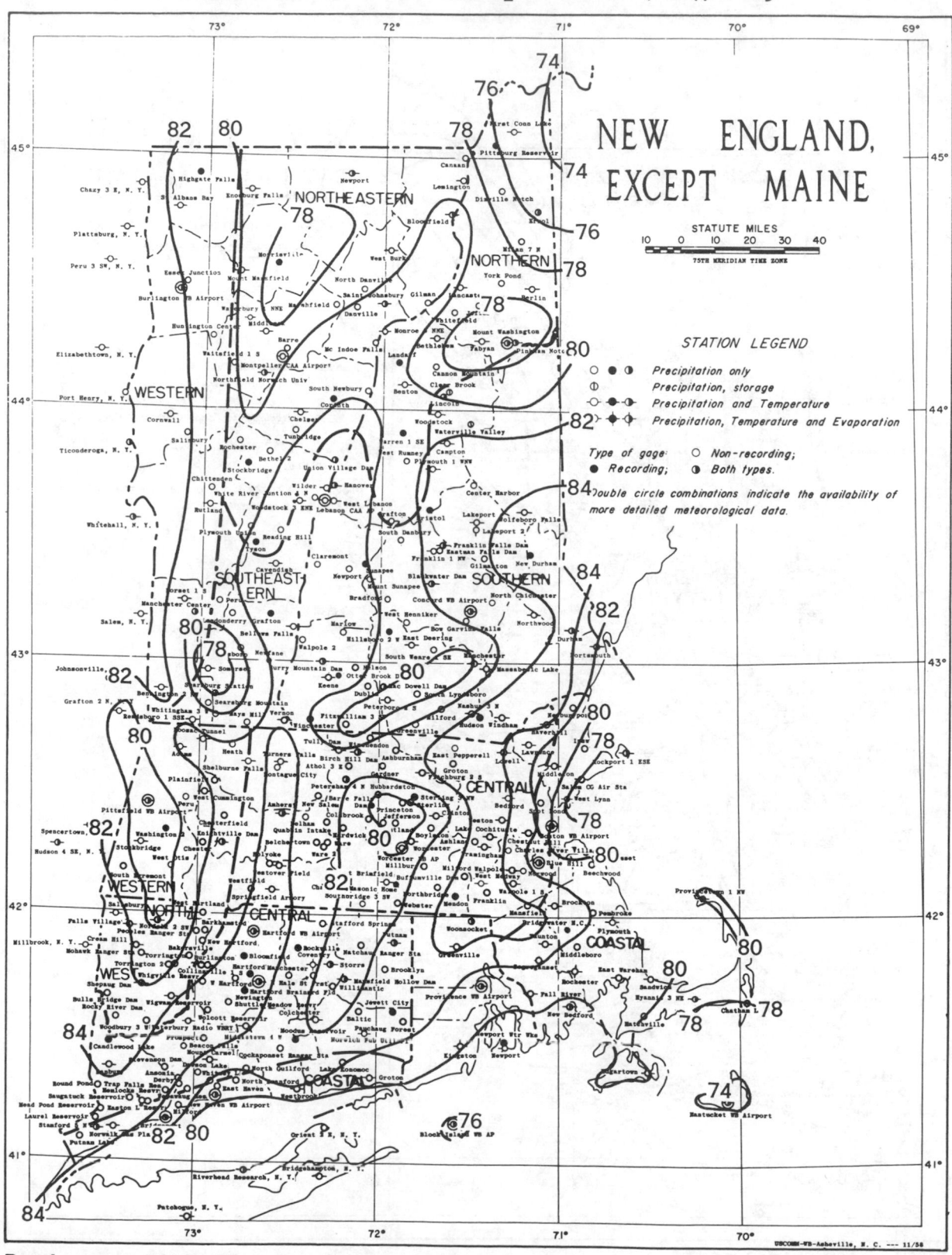

NEW ENGLAND, EXCEPT MAINE

STATUTE MILES

75TH MERIDIAN TIME ZONE

STATION LEGEND

○ ● ◐	*Precipitation only*
◑	*Precipitation, storage*
◒●◓	*Precipitation and Temperature*
◇◆◈	*Precipitation, Temperature and Evaporation*

Type of gage: ○ *Non-recording;*
Recording; ● *Both types.* ◐

Double circle combinations indicate the availability of more detailed meteorological data.

Based on period 1931-52

Isolines are drawn through points of approximately equal value. Caution should be used in interpolating on these maps, particularly in mountainous areas.

Mean Minimum Temperature (°F.), July

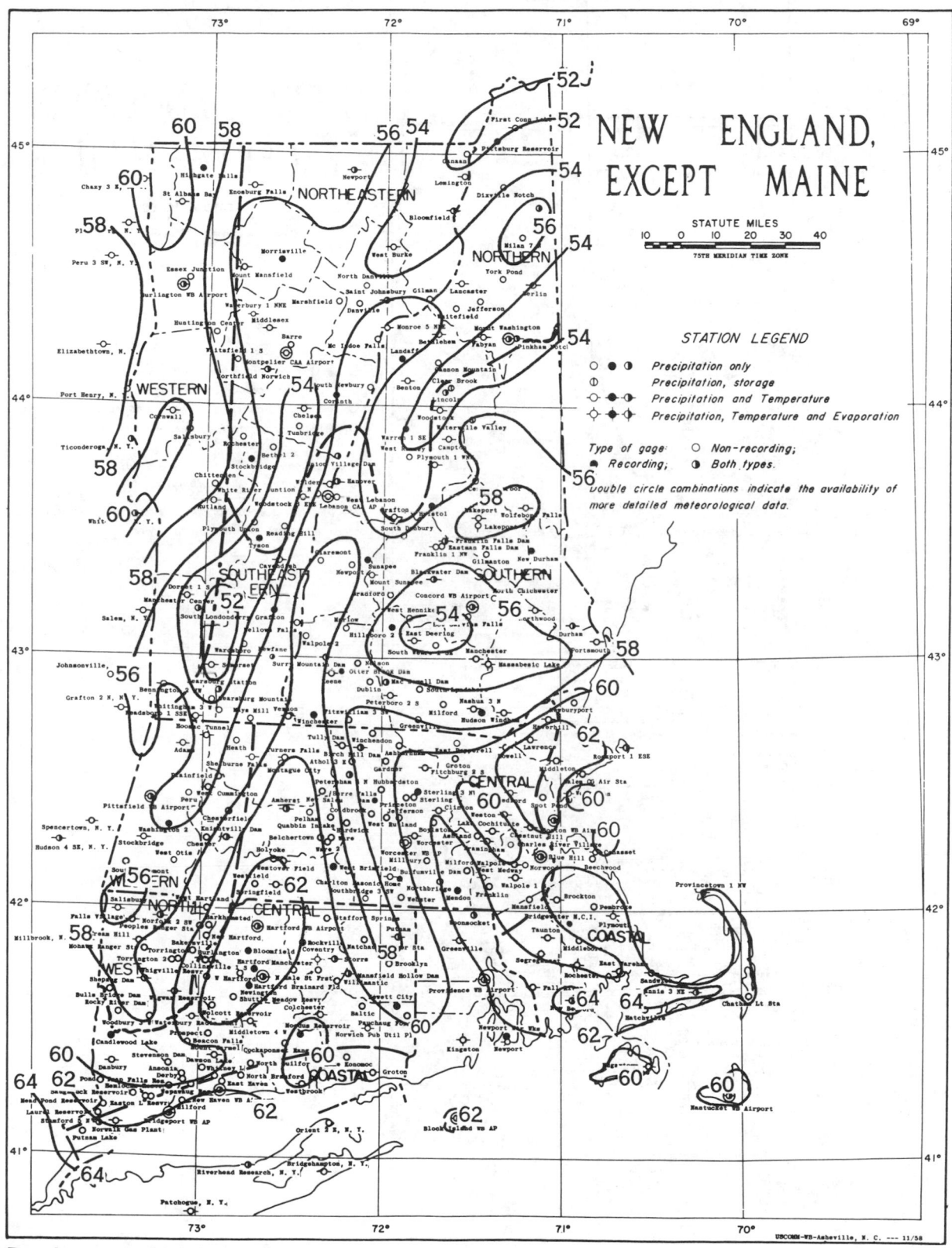

Based on period 1931-52

Isolines are drawn through points of approximately equal value. Caution should be used in interpolating on these maps, particularly in mountainous areas.

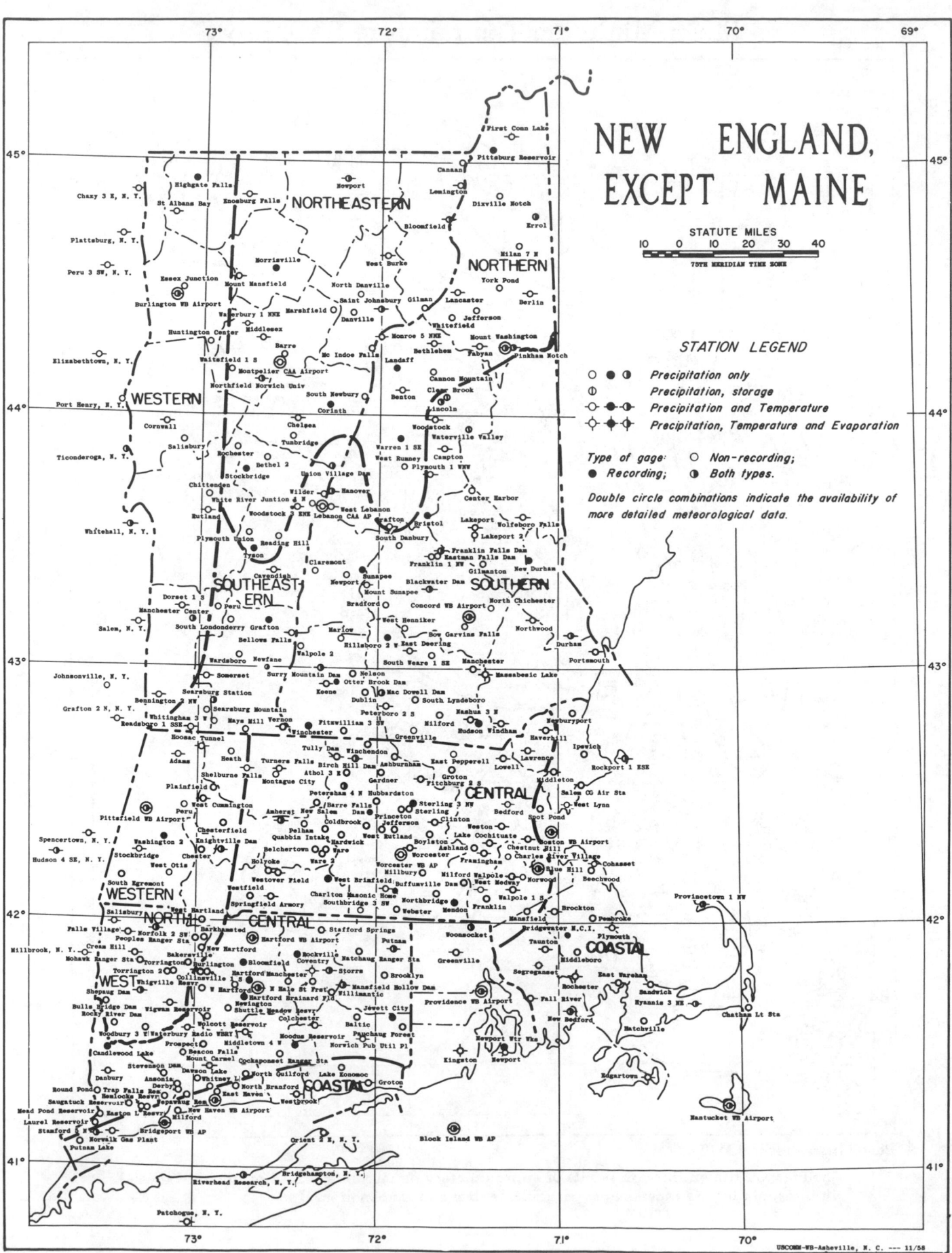

NEW ENGLAND, EXCEPT MAINE

STATUTE MILES

10 0 10 20 30 40

75TH MERIDIAN TIME ZONE

STATION LEGEND

○ ● ◑ *Precipitation only*
◑ *Precipitation, storage*
⦶ ⬤ ⦷ *Precipitation and Temperature*
✦ ✦ ✦ *Precipitation, Temperature and Evaporation*

Type of gage: ○ *Non-recording;*
● *Recording;* ◑ *Both types.*

*Double circle combinations indicate the availability of
more detailed meteorological data.*

CLIMATES OF THE STATES

NEW JERSEY

(Normals, Means and Extremes tables revised 1967 and 1975. Basic report Revised March 1967.)

Climate of New Jersey

Donald V. Dunlap, ESSA State Climatologist

New Jersey, though one of the smaller states, has a varied topography. In the northwestern part a section comprising about one-fifth of the area of the State is known as the Highlands and Kittatinny Valley. This region is traversed by several low mountain ridges extending northeasterly across the State with valleys and rolling hills between. The highest of these ranges is the Kittatinny, which rises from the banks of the Delaware River at the famous Delaware Water Gap. To the eastward the region is studded with numerous lakes, some of the largest of which are Lakes Hopatcong, Mohawk, and Greenwood. Elevations up to 1,800 feet above sea level are found in the Kittatinny Mountains near the New York State line.

South and east of the Highlands is a region of about equal area known as the Red Sandstone Plain, or the Piedmont of New Jersey. It is generally hilly in its northwestern part, becoming rolling and then flat toward the south and southeast. At its northeastern corner are the Palisades, cliffs which rise abruptly from the Hudson River to heights of 200 to 500 feet. The seacoast section extends from Sandy Hook to Cape May, or about 125 miles. This area is characterized by long stretches of sandy beaches, now occupied largely by summer resorts. Tidewater marshes become numerous toward the south.

In the southern interior a region known as the Pines is covered with scrubby forests of pine and some oak. The land is low and some of it is swampy. Here are found the large cranberry bogs of New Jersey. In fact, most of the State that lies south of a line connecting Jersey City and Trenton is low and flat with few elevations higher than 100 feet above mean sea level, these being mainly in Monmouth County.

About 30 percent of the area of New Jersey drains into the Delaware River and Delaware Bay, which form the western boundary. Nearly half of Sussex County, in the northwest, drains northward through the Wallkill River into the Hudson River of New York. The remainder of the State drains directly into the Atlantic Ocean through the Passaic, Hackensack, and Raritan Rivers in the north, and a number of small rivers and streams in the south.

Over the southern interior the soil changes from sandy near the coast to clay and marl in the western part. However, there is no steady transition, the change being effected mostly by alternating stretches of the different soils and combinations of them. In the most productive sections in the southwestern part, light to medium sandy loams predominate. Immense quantities of garden truck for commercial canning, especially tomatoes, are grown in Cumberland, Salem, Gloucester, Camden, and western Burlington Counties.

The extreme length of the State is 166 miles and its greatest width only about 65. The difference in climate is quite marked between the southern tip at Cape May and the northern extremity in the Kittatinny Mountains. The former locality is almost surrounded by water and is fairly well removed from the influence of the frequent storms that cross the Great Lakes region and move out the St. Lawrence Valley. The northern extremity is well within the zone of influence of these storms and, in addition, lies at elevations varying from 800 to 1,800 feet. The influence of these high elevations on the temperature is considerable. The differences between these two localities are particularly marked in the winter, Cape May having a normal January temperature about the same as that of southwestern Virginia, while that of Layton, in the extreme northwest, is similar to that of the northern area of Ohio. Since the prevailing winds are mostly offshore, the ocean influence does not have full effect.

Temperature differences between the northern and southern parts of the State are greatest in winter and least in summer. Nearly every station has registered readings of 100° F. or higher at some time, and all of them have records of zero or below. The highest temperature of record is 110° F. observed July 10, 1936 at Runyon; the lowest, -34° F., January 5, 1904, at River Vale.

In the northern highland area, the average date of last freeze (32° F.) in spring is about May 2, and that of the first in Fall, October 12. On the seacoast corresponding dates are April 6 and November 9, while in the central and southern interior the dates are April 23 and October 19. Freeze-free days in the northern highlands average 163, with 217 along the seacoast and 179 in the central and southern interior.

Northern New Jersey is near enough to the paths of the storms which cross the Great Lakes region and pass down the St. Lawrence Valley to receive part of its precipitation from that source. However, the heaviest general rains are produced by coastal storms of tropical origin. The centers of these storms usually pass some distance offshore, with heaviest rainfall and strongest wind near the coast. On several occasions tropical storms have moved inland along the south Atlantic coast, and then moved northward either through or to the west of New Jersey. Noteworthy storms of this type in recent years include hurricanes Able in 1952, Hazel in 1954, and Connie and Diane in 1955.

The damage by high tides to coastal installations during the passage of a tropical storm is often severe, whether the storm passes offshore or inland.

The average annual precipitation ranges from about 40 inches along the southeast coast to 51 inches in north-central parts of the State. In other sections the annual averages are mostly between 43 and 47 inches. Rainfall is well distributed during the warm months. Heavy 24-hour falls of 7 or 8 inches are occasionally recorded.

Brief periods of drought during the growing season are not uncommon, but prolonged droughts are relatively rare, occurring on the average once in 15 years.

Flooding in New Jersey is usually caused by heavy general rains, at times associated with storms of tropical origin. Local flooding results from ice gorging.

Important flooding occurred along the Raritan River in 1934, 1935, 1938, 1940, and 1948. The Passaic River flooded seriously in 1903, 1936, and 1952, and the Delaware River in 1903, 1936, and 1955.

The season during which measurable quantities of snow are likely to fall extends from about October 15 to April 20 in the Highlands, and from about November 15 to March 15 in the vicinity of Cape May. Average seasonal amounts range from about 13 inches at Cape May to nearly 50 inches in the Highlands. Snowfalls of 10 or more inches in a single storm are occasional occurrences.

The number of days a month with measurable precipitation averages 8 for each of the fall months, September, October, and November, and 9 to 12 for the other months of the year; the average yearly number is 120. Midday relative humidity averages 68 percent along the seacoast and 57 percent or less at inland locations.

Normally, sunshine varies from slightly over one-half of the possible amount in the northern counties to about 60 percent in the south. The prevailing wind is from the northwest from October to April, inclusive, and from the southwest for the other months of the year.

Thirty-four tornadoes were reported in New Jersey for the 50-year period ending with 1965. Two deaths and damage estimated at 2.5 million dollars resulted from these tornadoes. Damage from hail in the 22-year period ending with 1965 totaled about $500,000, with half of that amount resulting from a single storm which passed through Hunterdon and Warren Counties on June 10, 1956.

Damage from windstorms other than tornadoes amounted to about 1.5 million during the period 1938-1965. The most destructive of these storms was that of November 25, 1950, with the September 1944 hurricane second in the amount of damage, followed by hurricane Hazel in October 1954 in third place.

The most damaging storm on record was the wind storm accompanied by a tidal surge March 6-8, 1962, with the loss of 21 lives and damage which was estimated at 80 million dollars. Tidal damage covered the entire coastline of New Jersey, including Delaware and Raritan Bays.

The leading farm products are eggs, milk, vegetables, and poultry. The farm economy of the State is dependent upon an adequate water supply which must be met by irrigation in many areas. During periods of drought, as in the summers of 1962-1966, there is insufficient water for irrigation usage. The storage of even a small percentage of the State's runoff water would meet the requirements of industrial and agricultural users. Precipitation is both plentiful and reliable, thus guaranteeing an adequate water supply for industrial uses.

The resort industry along the seacoast serves the New York City and Philadelphia populace, as well as New Jerseyites. The mean daily maximum temperature for the summer months of June, July, and August at Atlantic City is 77.7° F., giving evidence of the seabreeze effect along the immediate coast of New Jersey. Numerous lakes in the Highlands also provide summer resort facilities, with a moderate climate during the summer months.

The invigorating climate of the north and central portions of the State, with marked changes in weather, generally neither extreme nor severe, provides an excellent setting for industrial and commercial interests, as evidenced by the concentration of population in the northeastern counties.

REFERENCES

(1)   Biel, E. R. and A. V. Havens, "Science and Land", 1948-49 Annual Report, Rutgers University, N. J. Agricultural Experiment Station, Meteorology Section.

(2)   Blood, Richard D. W., "A Study of Certain Basic and Applied Wind Problems of the Climate of New Jersey", M. S. Thesis, Rutgers Univ., Sept. 1953.

(3)   Cantlon, J. E. , "Vegetation and Microclimates on North and South Slopes of Cushetunk Mountain, New Jersey", Ph.D. Thesis, Rutgers Univ., 1950.

(4)   Havens, A. V., "A Preliminary Dynamic Climatology for New Jersey and Its Application to Human Activities", M. S. Thesis, Rutgers Univ., 1948.

(5)   Janifer, Clarence S., Jr., "A Study of Certain Thermal Features of the Climate of New Jersey", M. S. Thesis, Rutgers Univ., May 1952.

(6)   Marlatt, Wm. E., "A Comparison of Evapotranspiration Computed from Climatic Data with Field and Lysimeter Measurements"; M. S. Thesis, Rutgers, May 1958.

(7)   McGuire, James K., and Wayne C. Palmer, "The 1957 Drought in the Eastern United States", Monthly Weather Review, Vol. 85, No. 9, Sept. 1957.

(8)   Rutgers University Press, "The Economy of New Jersey", a survey published June 20, 1958, including a chapter on "The Climate of New Jersey" by Dr. E. R. Biel.

(9)   Weather Bureau Technical Paper No. 15 - Maximum Station Precipitation for 1,2,3, 6,12, and 24 Hours.

(10) Weather Bureau Technical Paper No. 16 - Maximum 24-Hour Precipitation in the United States. Washington, D. C. 1952.

(11) Weather Bureau Technical Paper No. 25 - Rainfall Intensity - Duration - Frequency Curves. For selected stations in the United States, Alaska, Hawaiian Islands and Puerto Rico.

(12) Weather Bureau Technical Paper No. 29 - Rainfall Intensity - Frequency Regime. Washington, D. C.

BIBLIOGRAPHY

(A)   Climatic Summary of the United States (Bulletin W) 1930 edition, Section 90. U. S. Weather Bureau

(B)   Climatic Summary of the United States, New Jersey - Supplement for 1931 through 1952 (Bulletin W Supplement). U. S. Weather Bureau

(C)   Climatic Summary of the United States, New Jersey - Supplement for 1951 through 1960 (Bulletin W Supplement). U. S. Weather Bureau

(D)   Climatological Data - New Jersey. U. S. Weather Bureau

(E)   Climatological Data National Summary. U. S. Weather Bureau

(F)   Hourly Precipitation Data - New Jersey. U. S. Weather Bureau

(G)   Local Climatological Data, U. S. Weather Bureau, for Atlantic City, Newark and Trenton, New Jersey.

FREEZE DATA

STATION	Freeze threshold temperature	Mean date of last Spring occurrence	Mean date of first Fall occurrence	Mean No. of days between dates	Years of record Spring	No. of occurrences in Spring	Years of record Fall	No. of occurrences in Fall
ATLANTIC CITY	32	03-31	11-11	225	10	10	10	10
	28	03-18	11-29	256	10	10	10	10
	24	03-04	12-04	275	10	10	10	10
	20	02-15	12-14	302	10	10	10	9
	16	02-09	12-18	312	10	9	10	9
BELLEPLAIN ST FOREST	32	04-27	10-14	170	29	29	29	29
	28	04-13	10-24	194	29	29	29	29
	24	03-31	11-04	218	29	29	28	28
	20	03-19	11-20	246	29	29	28	28
	16	03-08	12-03	270	29	29	28	28
BELVIDERE	32	04-26	10-13	170	30	30	30	30
	28	04-12	10-27	198	30	30	30	30
	24	03-23	11-14	236	30	30	30	30
	20	03-14	11-24	255	30	30	30	30
	16	03-07	12-05	273	30	30	30	30
BOONTON 1 SE	32	04-28	10-09	164	30	30	30	30
	28	04-13	10-22	192	30	30	30	30
	24	03-26	11-07	226	30	30	30	30
	20	03-15	11-23	253	30	30	30	30
	16	03-07	12-06	274	30	30	29	29
BURLINGTON	32	04-12	10-30	201	30	30	30	30
	28	03-27	11-12	230	29	29	30	30
	24	03-17	11-27	255	30	30	30	30
	20	03-08	12-07	274	30	30	30	30
	16	02-27	12-16	292	30	30	30	26
CANOE BROOK	32	05-04	10-10	159	10	10	10	10
	28	04-17	10-19	185	10	10	10	10
	24	04-06	11-06	214	10	10	10	10
	20	03-21	11-11	235	10	10	10	10
	16	03-15	11-25	255	10	10	10	10
CAPE MAY 3 W	32	03-31	11-17	231	21	21	20	20
	28	03-19	11-30	256	21	21	19	19
	24	03-05	12-09	279	20	20	19	19
	20	02-21	12-16	298	21	21	19	16
	16	02-05	12-24	322	21	17	18	10
CHARLOTTEBURG	32	05-17	09-25	131	30	30	30	30
	28	04-30	10-09	162	30	30	30	30
	24	04-15	10-25	193	30	30	30	30
	20	03-29	11-06	222	30	30	30	30
	16	03-19	11-23	249	30	30	30	30
CLAYTON	32	04-12	10-29	200	10	10	10	10
	28	03-25	11-07	227	10	10	10	10
	24	03-20	11-14	239	10	10	10	10
	20	03-04	11-28	269	10	10	10	10
	16	02-11	12-15	307	10	10	10	9
ELIZABETH	32	04-23	10-19	179	30	30	30	30
	28	04-04	11-02	212	30	30	30	30
	24	03-22	11-17	240	30	30	30	30
	20	03-11	11-29	263	30	30	30	30
	16	03-01	12-09	283	30	30	30	29
FLEMINGTON 1 NE	32	04-29	10-13	167	30	30	30	30
	28	04-16	10-23	190	30	30	30	30
	24	03-30	11-09	224	30	30	30	30
	20	03-15	11-21	251	30	30	30	30
	16	03-09	12-06	272	30	30	30	29
FREEHOLD	32	04-23	10-18	178	29	29	30	30
	28	04-07	11-02	209	29	29	30	30
	24	03-24	11-17	238	29	29	30	30
	20	03-16	11-29	258	29	29	30	30
	16	03-05	12-07	277	29	29	30	29
HAMMONTON 2 NNE	32	04-21	10-16	178	28	28	29	29
	28	04-06	10-31	208	28	28	29	29
	24	03-21	11-17	241	27	27	28	28
	20	03-10	11-28	263	27	27	28	28
	16	03-02	12-07	280	26	26	28	27
HIGHTSTOWN 1 N	32	04-22	10-12	173	10	10	10	10
	28	04-06	11-03	211	10	10	10	10
	24	03-20	11-16	241	10	10	10	10
	20	03-08	12-06	273	10	10	10	10
	16	02-24	12-13	292	10	10	10	10
INDIAN MILLS 2 W	32	05-01	10-08	160	29	29	30	30
	28	04-19	10-21	185	29	29	30	30
	24	04-02	10-31	212	30	30	30	30
	20	03-20	11-17	242	30	30	30	30
	16	03-09	12-01	267	30	30	30	30
JERSEY CITY	32	04-06	11-07	215	29	29	29	29
	28	03-24	11-21	242	28	28	29	29
	24	03-16	11-28	257	28	28	28	28
	20	03-08	12-06	273	28	28	28	28
	16	02-26	12-14	291	28	28	28	25
LAMBERTVILLE	32	04-23	10-15	175	29	29	29	29
	28	04-10	11-02	206	29	29	28	28
	24	03-23	11-15	237	28	28	27	27
	20	03-13	11-24	256	28	28	27	27
	16	03-02	12-08	281	29	28	26	25
LAURELTON 1 E	32	04-28	10-14	169	29	29	29	29
	28	04-10	10-28	201	28	28	28	28
	24	03-26	11-11	230	27	27	28	28
	20	03-15	11-27	257	27	27	28	28
	16	03-03	12-05	277	27	26	29	29
LAYTON 3 NW	32	05-21	09-27	129	30	30	30	30
	28	05-09	10-09	153	30	30	30	30
	24	04-26	10-17	174	30	30	30	30
	20	04-12	10-31	202	30	30	30	30
	16	03-24	11-13	234	30	30	30	30
LITTLE FALLS	32	04-23	10-14	174	30	30	30	30
	28	04-06	10-27	204	30	30	30	30
	24	03-25	11-11	231	30	30	30	30
	20	03-12	11-28	261	30	30	30	30
	16	03-04	12-09	280	30	30	30	30
LONG BRANCH 2 N	32	04-13	10-29	199	30	30	30	30
	28	03-27	11-14	232	30	30	30	30
	24	03-18	11-26	253	30	30	30	30
	20	03-06	12-06	275	30	30	30	30
	16	02-26	12-13	290	30	30	30	29
LONG VALLEY	32	05-08	10-01	146	28	28	29	29
	28	04-23	10-14	174	27	27	28	28
	24	04-04	10-29	208	27	27	28	28
	20	03-21	11-11	235	26	26	28	28
	16	03-14	11-28	259	26	26	27	27
MOORESTOWN	32	04-16	10-19	186	30	30	30	30
	28	03-30	11-04	219	30	30	30	30
	24	03-17	11-19	247	30	30	29	29
	20	03-08	12-05	272	30	30	29	29
	16	02-26	12-14	291	30	30	29	26
MORRIS PLAINS 1 W	32	05-09	10-09	153	15	15	16	16
	28	04-14	10-21	190	15	15	16	16
	24	03-31	11-08	222	15	15	16	16
	20	03-20	11-18	243	15	15	16	16
	16	03-09	12-06	272	15	15	16	16
NEWARK WB AIRPORT	32	04-03	11-08	219	10	10	10	10
	28	03-23	11-21	243	10	10	10	10
	24	03-18	12-01	258	10	10	10	10
	20	03-03	12-12	284	10	10	10	10
	16	02-15	12-16	304	10	10	10	10
NEW BRUNSWICK EXP STA	32	04-18	10-19	184	30	30	30	30
	28	04-04	11-03	213	30	30	30	30
	24	03-21	11-18	242	30	30	30	30
	20	03-12	12-02	265	30	30	30	30
	16	03-05	12-11	281	30	30	30	29

STATION	Freeze threshold temperature	Mean date of last Spring occurrence	Mean date of first Fall occurrence	Mean No. of days between dates	Years of record Spring	No. of occurrences in Spring	Years of record Fall	No. of occurrences in Fall
NEWTON	32	05-04	10-05	154	30	30	29	29
	28	04-20	10-20	183	30	30	29	29
	24	04-06	11-03	211	30	30	30	30
	20	03-23	11-13	235	30	30	30	30
	16	03-15	11-30	260	30	30	30	30
PATERSON	32	04-13	10-27	197	29	29	28	28
	28	03-30	11-12	227	28	28	28	28
	24	03-20	11-22	247	27	27	28	28
	20	03-12	12-02	265	27	27	28	28
	16	03-05	12-10	280	27	27	27	26
PEMBERTON 3 E	32	04-24	10-15	174	30	30	30	30
	28	04-08	10-29	204	30	30	30	30
	24	03-23	11-15	237	30	30	30	30
	20	03-13	11-26	258	30	30	30	30
	16	03-04	12-09	280	30	30	30	29
PHILLIPSBURG	32	04-24	10-16	175	30	30	30	30
	28	04-09	10-31	205	30	30	30	30
	24	03-21	11-15	239	30	30	30	30
	20	03-15	11-28	258	30	30	29	29
	16	03-05	12-08	278	30	30	29	28
PLAINFIELD	32	04-21	10-17	179	30	30	30	30
	28	04-04	11-03	213	30	30	30	30
	24	03-23	11-15	237	30	30	30	30
	20	03-14	11-30	261	30	30	30	30
	16	03-03	12-10	282	30	30	30	30
PLEASANTVILLE 1 N	32	04-27	10-15	171	28	28	27	27
	28	04-14	10-29	198	28	28	27	27
	24	03-31	11-10	224	27	27	27	27
	20	03-17	11-25	253	27	27	27	27
	16	03-06	12-04	273	27	27	27	27
RIDGEFIELD	32	04-16	10-21	188	30	30	29	29
	28	03-31	11-07	221	30	30	29	29
	24	03-20	11-24	249	30	30	29	29
	20	03-13	12-02	264	30	30	29	29
	16	03-03	12-12	284	30	30	29	28

STATION	Freeze threshold temperature	Mean date of last Spring occurrence	Mean date of first Fall occurrence	Mean No. of days between dates	Years of record Spring	No. of occurrences in Spring	Years of record Fall	No. of occurrences in Fall
SANDY HOOK LB STA	32	03-29	11-21	237	28	28	27	27
	28	03-22	11-29	252	29	29	27	26
	24	03-12	12-09	272	29	29	27	26
	20	03-04	12-13	284	29	29	27	23
	16	02-17	12-21	307	29	27	27	19
SHILOH	32	04-15	10-25	193	29	29	27	27
	28	03-30	11-09	224	28	28	28	28
	24	03-19	11-25	251	29	29	28	28
	20	03-06	12-04	273	28	28	28	28
	16	02-26	12-14	291	27	27	28	25
SOMERVILLE	32	04-27	10-13	169	28	28	28	28
	28	04-15	10-28	196	28	28	29	29
	24	03-30	11-07	222	28	28	29	29
	20	03-16	11-22	251	28	28	30	30
	16	03-10	12-05	270	29	29	30	30
SUSSEX 1 SE	32	05-06	10-05	152	28	28	30	30
	28	04-23	10-17	177	28	28	30	30
	24	04-01	10-30	212	28	28	30	30
	20	03-21	11-15	239	28	28	29	29
	16	03-12	11-29	262	29	29	29	29
TRENTON WB CITY	32	04-04	11-08	218	30	30	30	30
	28	03-26	11-21	240	30	30	30	30
	24	03-14	11-30	261	30	30	30	30
	20	03-06	12-09	278	30	30	30	30
	16	02-23	12-15	295	30	30	30	26
TUCKERTON	32	04-17	10-20	186	23	23	26	26
	28	03-31	11-06	220	23	23	25	25
	24	03-18	11-24	251	23	23	25	25
	20	03-11	11-30	264	23	23	26	26
	16	03-03	12-12	284	22	22	24	21

Data in the above table are based on the period 1931-1960, or that portion of this period for which data are available.

⊕ When the frequency of occurrence in either spring or fall is one year in ten, or less, mean dates are not given.

Means have been adjusted to take into account years of non-occurrence.

A freeze is a numerical substitute for the former term "killing frost" and is the occurrence of a minimum temperature at or below the threshold temperature of 32°, 28°, etc.

Freeze data tabulations in greater detail are available and can be reproduced at cost.

TROPICAL CYCLONE DATA HAVING IMPORTANCE FOR THIS STATE IS INCLUDED IN STATISTICS AND CHARTS ON PAGES 1161 THROUGH 1164 .

Taken from "Climatography of the United States No. 81-4, Decennial Census of U. S. Climate"

TEMPERATURE (°F)   PRECIPITATION (In.)

STATIONS (By Divisions)	Jan.	Feb.	Mar.	Apr.	May	June	July	Aug.	Sept.	Oct.	Nov.	Dec.	Annual	Jan.	Feb.	Mar.	Apr.	May	June	July	Aug.	Sept.	Oct.	Nov.	Dec.	Annual
NORTHERN																										
BELVIDERE	29.9	31.3	39.3	51.1	62.2	70.6	75.1	72.9	66.0	54.9	42.8	31.9	52.3	3.63	3.12	4.08	4.14	4.03	4.44	5.30	4.99	3.87	3.42	3.86	3.66	48.54
BOONTON 1 SE	28.9	29.5	37.3	49.0	59.1	67.9	72.8	70.8	63.5	53.3	42.5	31.5	50.5	3.24	2.73	3.86	3.86	4.16	3.89	4.24	4.93	4.35	3.40	4.02	3.42	46.10
CANOE BROOK	29.0	29.6	37.6	48.7	58.9	68.0	73.0	71.1	64.2	53.7	43.0	31.2	50.7	3.66	2.96	4.30	4.02	4.15	4.03	4.55	5.23	4.32	3.41	4.10	3.65	48.38
CHARLOTTEBURG	28.6	29.1	36.6	48.0	58.4	66.5	70.9	69.2	62.4	52.3	42.0	30.8	49.6	3.67	3.17	4.47	4.21	4.15	4.24	4.73	4.96	4.41	3.37	4.19	3.96	49.35
CHATHAM	*	*	*	*	*	*	*	*	*	*	*	*	*	3.72	3.17	4.42	4.12	4.17	4.01	4.58	5.25	4.39	3.37			
ELIZABETH	32.8	33.3	40.9	51.9	62.2	70.8	75.8	74.1	67.2	56.7	45.7	34.7	53.8	3.97	3.28	4.49	3.95	4.06	3.76	4.51	4.97	4.34	3.53	3.88	3.60	48.34
FLEMINGTON 1 NE	31.5	32.3	40.1	51.2	61.8	70.6	75.4	73.4	66.6	55.8	44.3	33.3	53.0	3.32	2.78	3.99	3.80	4.01	3.80	4.52	5.02	3.59	3.32	3.74	3.42	45.31
JERSEY CITY	32.5	32.7	39.8	50.5	61.2	70.1	75.3	73.8	66.9	56.7	45.7	34.8	53.3	3.37	2.90	4.03	3.59	3.75	3.45	4.17	4.62	3.81	3.18	3.42	3.29	43.58
LAMBERTVILLE	31.7	32.5	39.8	51.1	61.6	70.1	74.8	73.0	66.2	55.9	44.4	33.7	52.9	3.34	2.72	4.06	3.64	4.08	3.84	4.51	5.14	3.29	3.00	3.71	3.38	44.71
LAYTON 3 NW	26.9	27.8	35.9	47.7	58.3	67.2	71.6	69.7	62.3	51.7	40.2	28.8	49.0	2.87	2.49	3.81	3.64	4.08	3.84	4.09	4.90	4.42	4.08	3.64	2.92	44.13
LITTLE FALLS	31.1	31.7	39.3	50.7	61.2	69.7	74.8	72.8	65.8	55.4	44.6	33.6	52.6	3.65	3.07	4.30	4.26	4.33	4.25	4.54	5.09	4.55	3.70	4.28	3.90	49.92
LONG VALLEY	*	*	*	*	*	*	*	*	*	*	*	*	*	3.33	2.80	4.28	3.97	4.17	4.13	5.20	5.06	3.92	3.57	4.25	3.79	48.90
NEWARK WB AIRPORT	33.3	33.7	41.5	52.3	62.5	72.3	77.3	75.4	68.3	57.6	45.9	35.3	54.6	3.44	2.81	4.09	3.51	3.65	3.44	3.67	4.43	3.76	3.11	3.37	3.22	42.38
NEW MILFORD	*	*	*	*	*	*	*	*	*	*	*	*	*	3.44	2.81	3.99	3.84	3.85	3.76	4.38	4.45	3.75	3.19	3.66	3.46	44.58
NEWTON	27.5	28.1	36.4	48.3	59.0	67.4	72.2	70.1	63.0	52.5	41.2	29.7	49.6	3.04	2.57	3.45	3.83	3.95	4.30	4.87	4.54	3.99	3.42	3.63	3.20	44.79
PATERSON	31.6	32.1	39.5	51.1	61.9	70.7	75.7	73.6	66.4	56.1	45.1	34.1	53.2	3.60	3.15	4.45	4.07	4.30	4.21	4.63	4.87	4.21	3.47	4.04	3.83	48.83
PHILLIPSBURG	30.5	31.4	39.2	50.7	61.4	70.0	74.3	72.4	65.5	54.5	43.4	32.7	52.2	3.68	3.06	4.09	4.07	4.15	4.26	5.29	5.04	4.00	3.42	3.67	3.64	48.37
PLAINFIELD	31.9	32.5	39.9	50.9	61.2	69.9	74.9	73.1	66.2	55.9	44.8	34.0	52.9	3.61	3.02	4.31	3.87	4.14	4.33	4.61	5.19	4.33	3.46	3.83	3.59	48.29
SOMERVILLE	31.3	32.0	39.5	50.6	61.1	69.8	74.6	72.8	66.0	55.3	43.9	33.1	52.5	3.27	2.75	3.97	3.65	3.81	3.88	4.74	5.01	3.77	3.35	3.66	3.33	45.19
SUSSEX 1 SE	27.7	28.9	37.0	49.1	59.9	68.5	72.9	70.8	63.5	53.2	42.0	30.2	50.3	3.37	2.72	3.40	3.95	3.77	4.09	4.90	4.42	4.08	3.40	3.81	3.36	45.78
WOODCLIFF LAKE	*	*	*	*	*	*	*	*	*	*	*	*	*	3.57	2.89	4.13	3.98	3.95	3.52	4.44	4.71	3.93	3.28	3.87	3.59	45.86
DIVISION	30.2	30.9	38.6	50.0	60.5	69.2	74.0	72.1	65.0	54.5	43.5	32.4	51.7	3.47	2.92	4.10	3.91	4.07	4.01	4.65	4.91	4.06	3.42	3.88	3.56	46.96
SOUTHERN																										
ATLANTIC CITY WB AP	34.8	34.7	41.1	51.0	61.3	70.0	75.1	73.7	67.2	57.2	46.7	36.6	54.1	3.56	3.13	3.91	3.41	3.51	2.83	3.72	4.90	3.31	3.20	3.66	3.22	42.36
BELLEPLAIN ST FOREST														3.47	3.18	4.42	3.47	3.83	3.39	4.84	5.75	3.86	3.69	4.03	3.65	47.58
BURLINGTON	34.2	34.9	42.2	53.3	64.2	72.8	77.2	75.0	68.8	58.0	46.6	36.0	55.3	3.27	2.90	4.34	3.44	3.76	3.65	3.65	3.93	3.35	3.35	3.96	3.06	44.16
FREEHOLD	32.7	33.0	40.0	50.8	61.6	70.2	74.9	73.1	66.5	56.1	45.5	34.7	53.3	3.58	3.14	4.24	3.58	4.03	3.68	4.24	4.76	3.78	3.39	3.96	3.73	47.68
HAMMONTON 2 NNE	34.2	34.5	41.9	52.3	62.7	71.6	76.1	74.2	67.7	57.2	46.7	37.1	54.7	3.57	3.25	4.26	3.60	3.92	3.95	4.56	5.64	3.77	3.69	3.74	3.73	47.73
HIGHTSTOWN 1 N	32.4	32.9	40.2	50.9	61.6	69.9	74.8	72.9	66.2	55.9	45.2	34.3	53.1	3.22	2.73	3.82	3.38	3.72	3.83	4.46	4.52	3.99	3.32	3.39	3.03	43.41
INDIAN MILLS 2 W	33.3	33.8	40.7	51.4	62.0	70.2	74.7	72.9	66.3	55.9	45.2	34.8	53.4	3.62	3.11	4.28	3.42	3.88	3.90	4.27	5.46	3.60	3.41	3.71	3.23	45.89
LAURELTON 1 E	33.2	33.3	40.2	50.5	61.1	69.6	74.4	72.4	65.8	55.5	45.2	35.0	53.0	3.76	3.36	4.34	3.76	3.87	3.40	4.48	4.85	3.75	4.00	4.16	3.32	47.05
MOORESTOWN	33.0	33.4	40.7	51.4	61.7	70.2	75.1	73.1	66.4	55.9	45.0	34.8	53.4	3.12	2.73	3.81	3.44	4.07	3.56	4.17	4.75	3.75	3.06	3.61	2.91	42.98
NEW BRUNSWICK EXP STA	32.3	32.8	40.2	51.1	61.6	70.1	75.0	73.2	66.5	56.1	45.3	34.2	53.2	3.34	2.77	3.75	3.48	3.75	3.63	4.53	4.70	4.06	3.16	3.64	3.17	43.98
PEMBERTON 3 E	33.9	34.3	41.6	52.3	62.7	71.0	75.5	73.8	67.4	56.8	46.3	35.6	54.3	3.30	2.88	3.76	3.70	4.65	5.05	3.77	3.27	3.51	3.08			44.27
SHILOH	34.3	34.9	42.2	52.7	63.0	71.7	76.4	74.6	68.0	57.3	46.3	36.1	54.8	3.48	2.89	3.76	3.11	3.96	3.41	4.29	4.95	3.94	3.16	3.73	3.06	43.74
TRENTON WB CITY	33.1	33.4	40.7	51.7	62.3	71.0	76.0	73.9	67.1	56.8	45.8	35.2	53.9	3.10	2.59	3.84	3.21	3.62	3.60	4.18	4.77	3.50	2.84	3.16	2.87	41.28
DIVISION	33.4	33.7	40.9	51.4	62.0	70.7	75.4	73.6	67.0	56.5	45.7	35.3	53.8	3.45	2.99	4.08	3.45	3.78	3.62	4.34	5.06	3.79	3.35	3.74	3.26	44.91
COASTAL																										
ATLANTIC CITY	36.0	35.7	41.1	49.9	59.5	69.0	74.2	73.7	68.4	58.7	48.3	38.3	54.4	3.75	3.38	4.01	3.38	3.16	3.04	3.86	5.14	3.40	3.39	3.79	3.48	43.78
LONG BRANCH 2 N	33.0	33.0	39.8	49.6	59.8	69.1	74.4	72.8	66.2	56.2	45.7	35.3	52.9	3.78	3.54	4.37	3.63	3.49	3.41	4.25	5.25	3.92	3.71	3.81	3.63	46.79
DIVISION	34.5	34.3	40.5	50.0	60.0	69.4	74.6	73.6	67.8	58.0	47.4	37.0	53.9	3.52	3.12	3.92	3.32	3.26	3.07	3.88	4.94	3.67	3.30	3.59	3.34	42.93

* Normals for the period 1931-1960. Divisional normals may not be the arithmetical average of individual stations published, since additional data for shorter period stations are used to obtain better areal representation.

TEMPERATURE   PRECIPITATION

Jan.	Feb.	Mar.	Apr.	May	June	July	Aug.	Sept.	Oct.	Nov.	Dec.	Annual	Jan.	Feb.	Mar.	Apr.	May	June	July	Aug.	Sept.	Oct.	Nov.	Dec.	Annual

CONFIDENCE LIMITS

In the absence of trend or record changes, the chances are 9 out of 10 that the true mean will lie in the interval formed by adding and subtracting the values in the following table from the means for any station in the State. Because of the wider variation in mean precipitation, the corresponding monthly means and annual mean must be substituted for "p" in the precipitation table below to obtain mean precipitation confidence limits.

| 1.4 | 1.2 | 1.3 | .8 | .7 | .6 | .5 | .6 | .7 | .8 | .8 | .9 | .4 | $.25\sqrt{p}$ | $.20\sqrt{p}$ | $.25\sqrt{p}$ | $.25\sqrt{p}$ | $.29\sqrt{p}$ | $.30\sqrt{p}$ | $.37\sqrt{p}$ | $.37\sqrt{p}$ | $.47\sqrt{p}$ | $.37\sqrt{p}$ | $.34\sqrt{p}$ | $.25\sqrt{p}$ | $.32\sqrt{p}$ |

COMPARATIVE DATA

Data in the following table are the mean temperature and average precipitation for New Brunswick, New Jersey, for the period 1906-1930 and are included in this publication for comparative purposes:

| 30.7 | 30.7 | 40.2 | 50.0 | 60.2 | 68.8 | 73.7 | 71.6 | 66.1 | 55.1 | 43.7 | 32.9 | 52.0 | 3.33 | 3.30 | 3.27 | 3.70 | 3.83 | 3.77 | 5.16 | 5.17 | 3.20 | 3.54 | 2.63 | 3.53 | 44.43 |

NATIONAL AVIATION FACILITIES EXPERIMENTAL CENTER
LATITUDE 39° 27' N
LONGITUDE 74° 35' W
ELEVATION (ground) 64 Feet

NORMALS, MEANS AND EXTREMES
(Table Revised 1967. Base Period for Climatological Normals: 1931-1960)

Month	Temperature Normal Daily maximum	Daily minimum	Monthly	Extremes Record highest	Year	Record lowest	Year	Normal degree days	Precipitation Normal total	Max monthly $	Year	Min monthly $	Year	Max in 24 hrs	Year	Snow/Sleet Mean total	Max monthly	Year	Max in 24 hrs	Year	RH 1:00 A.M.	7:00 A.M.	1:00 P.M.	7:00 P.M.	Wind Mean hourly speed	Prevailing direction	Mean sky cover	Pct. possible sunshine	Clear	Partly cloudy	Cloudy	Precip .01"+	Snow 1.0"+	Thunderstorms	Heavy fog	Max 90+	Max 32−	Min 32−	Min 0−
(a)	(b)	(b)	(b)	2	2	2	2	(b)	(b)	23		23		23		22	22		22																				
J	42.9	26.6	34.8	65	1966+	−8	1965	936	3.56	7.71	1948	0.26	1955	2.86	1944	5.5	15.9	1961	14.4	1961	73	74	58	69	13.0	WNW	6.4	58	9	9	13	11	2	*	3	0	9	28	2
F	43.3	26.1	34.7	69	1965	0	1966	848	3.13	5.98	1958	1.46	1946	2.59	1966	3.9	14.5	1958	10.0	1958	78	80	57	70	12.2	WNW	5.9	52	8	6	15	10	2	*	5	0	7	21	1
M	49.7	32.4	41.1	67	1965	13	1965	741	3.91	6.80	1953	0.62	1945	2.22	1962	3.0	13.9	1960	9.5	1960	76	78	52	64	12.6	WNW	6.0	56	7	8	14	11	1	1	4	0	1	8	0
A	60.3	41.7	51.0	85	1965	24	1965	420	3.41	7.95	1952	1.24	1957	3.37	1965	0.3	3.2	1965	3.2	1965	74	76	52	66	12.8		6.5	56	6	10	14	10	*	2	3	0	0	0	0
M	71.0	51.5	61.3	93	1965	25	1966	133	3.51	11.51	1948	0.40	1964	4.15	1948	0.0	0.0		0.0		77	74	53	66	10.7		6.2	59	6	10	12	10	0	3	4	*	0	0	0
J	79.2	60.7	70.0	96	1966	39	1966	15	2.83	5.73	1951	0.10	1954	2.91	1952	0.0	0.0		0.0		81	77	51	62	9.5		5.9	62	8	10	12	9	0	4	3	1	0	0	0
J	83.8	66.3	75.1	104	1966	46	1965	0	3.72	13.09	1959	1.30	1957	6.46	1959	0.0	0.0		0.0		85	79	51	70	9.5		5.9	65	8	9	14	10	0	6	4	7	0	0	0
A	82.2	65.1	73.7	95	1966	40	1965	0	4.90	11.02	1948	0.34	1943	6.40	1966	0.0	0.0		0.0		87	82	51	74	9.5		6.2	62	8	8	12	9	0	6	4	4	0	0	0
S	76.0	58.4	67.2	104	1966	35	1965	39	3.20	6.21	1966	0.46	1953	3.98	1954	0.0	0.0		0.0		87	86	55	74	9.7	ENE	5.6	65	10	7	10	7	0	2	4	2	0	0	0
O	66.5	47.8	57.2	78	1966	26	1965	251	3.66	7.50	1943	0.15	1963	2.95	1943	T	0.2	1953	0.2	1953	83	86	51	71	10.5		4.8	65	14	7	10	7	*	1	4	0	0	1	0
N	55.5	37.9	46.7	76	1965	11	1964	549	3.66	8.60	1944	0.72	1946	2.93	1953	0.7	5.2	1953	3.2	1953	79	83	57	75	11.0		5.4	58	10	7	13	9	*	1	5	0	0	15	0
D	45.1	28.1	36.6	72	1964	6	1964	880	3.22	6.57	1948	0.62	1955	2.75	1951	2.8	8.6	1960	7.5	1960	80	83	57	75	11.2	WNW	6.0	48	10	7	14	9	1	*	5	0	3	23	3
YR	63.0	45.2	54.1	104 JUL 1966		−8 JAN 1965		4812	42.36	13.09 JUL 1959		0.10 JUN 1954		6.46 JUL 1959		16.1	15.9 JAN 1961		14.4 JAN 1961		81	80	54	70	11.0	S	5.9	58	105	104	156	112	4	24	44	22	19	120	3

Ø For period November 1964 through the current year.
Means and extremes in the above table are from existing or comparable location(s). Annual extremes have been exceeded at other locations as follows: Lowest temperature −9 in February 1934; maximum monthly precipitation 14.87 in August 1882; minimum monthly precipitation .01 in September 1941; maximum precipitation in 24 hours 9.21 in October 1903; maximum snowfall 27.9 in February 1899; maximum snowfall in 24 hours 18.0 in February 1902. % Based on U.S. Naval Air Station and Weather Bureau Airport Station records. $ Beginning with August 1943.

The prevailing direction for wind in the Normals, Means, and Extremes table is from records through 1963.

NORMALS, MEANS AND EXTREMES
(Table Revised 1975. Base Period for Climatological Normals: 1941-1970)

Month	Temp Normal Daily max	Daily min	Monthly	Extremes Record highest	Year	Record lowest	Year	Normal Degree days base 65°F Heating	Cooling	Precip Water equivalent Normal	Max monthly	Year	Min monthly	Year	Max in 24 hrs	Year	Snow/Ice pellets Max monthly	Year	Max in 24 hrs	Year	RH (Local time) 01	07	13	19	Wind Mean speed m.p.h.	Prevailing direction	Fastest mile m.p.h.	Direction	Year	Mean sky cover	Pct possible sunshine	Clear	Partly cloudy	Cloudy	Precip .01"+	Snow/Ice pellets 1.0"+	Thunderstorms	Heavy fog	Max 90+	Max 32−	Min 32−	Min 0−	Avg station pressure mb.
(a)	10			10		10				31	31		31		31		30		30		10	10	10	10	16	15	15	15		14	16	16	16	16	31	30	16	16	10	10	10	10	2
J	41.4	24.0	32.7	78	1967	−6	1965	1001	0	3.56	7.71	1948	0.26	1955	2.86	1944	15.9	1961	14.4	1961	72	74	56	67	11.8	WNW	47	29	1971	6.2	50	9	7	15	11	2	*	4	0	7	25	1	1017.6
F	42.4	24.9	33.9	78	1968	−7	1971	1071	0	3.17	5.98	1958	0.26	1945	2.59	1966	35.2	1967	13.1	1967	74	75	55	66	12.3	WNW	49	23	1967	6.0	49	9	6	13	10	1	*	4	0	5	23	0	1015.6
M	50.7	31.5	41.1	84	1968	12	1967	741	0	4.31	6.80	1953	1.00	1962	2.27	1962	17.6	1969	11.5	1969	75	77	52	63	12.5	WNW	46	24	1973	6.3	53	9	7	15	11	2	1	4	0	1	8−6	0	1013.9
A	62.4	41.7	52.1	90	1969	25	1968	399	0	3.47	7.95	1952	1.04	1970	3.37	1957	0.0		0.0		76	78	53	63	10.9	WNW	35	31	1973	6.3	54	8	10	12	11	*	2	5	*	0	0	0	1012.5
M	72.4	51.6	62.0	99	1969	37	1966	131	25	3.54	11.51	1948	0.40	1957	4.15	1948	0.0		0.0		78	79	55	69	9.8		37	29	1969	6.4	59	8	12	11	11	0	4	5	1	0	0	0	1011.7
J	80.8	60.7	70.8	106	1969	36	1970	15	168	3.38	6.36	1970	0.10	1954	2.91	1952	0.0		0.0		84	84	58	73	9.3		34	27	1968	6.4	58	7	12	11	9	0	5	4	2	0	0	0	1013.2
J	84.7	65.4	75.1	104	1966	46	1959	0	313	4.36	13.09	1959	1.30	1957	6.46	1959	0.0		0.0		88	87	57	72	9.1		37	26	1970	6.4	59	7	13	11	10	0	6	5	2	0	0	0	1013.4
A	83.0	63.8	73.4	97	1967	40	1966	0	260	4.90	11.02	1948	0.34	1959	6.40	1966	0.0		0.0		90	87	58	76	8.4		35	12	1971	6.0	59	8	13	10	10	0	5	5	2	0	0	0	1015.6
S	77.3	56.8	67.1	93	1973	32	1963	35	98	2.99	7.50	1966	0.41	1970	3.98	1954	T	1972	T	1972	87	87	57	77	9.4	ENE	41	29	1961	5.2	58	10	11	9	8	0	2	5	1	0	0	0	1016.6
O	67.5	46.0	56.8	81	1974	23	1969	262	0	4.21	9.65	1972	0.72	1963	3.93	1953	7.8	1967	7.8	1967	86	87	51	74	9.8		40	27	1960	5.1	61	12	9	10	8	0	1	5	0	*	13	0	1018.3
N	55.9	36.1	46.0	72	1966	10	1972	570	0	4.01	8.60	1944	0.62	1946	2.75	1951	8.6	1960	7.5	1960	85	82	59	72	11.5		55	30	1960	5.6	51	8	8	14	9	1	1	4	0	4	21	0	1015.1
D	44.2	26.0	35.1	72	1968	0	1968	927	0																																	
YR	63.6	43.8	53.7	106 JUN 1969		−8 JAN 1965		4946	869	45.46	13.09 JUL 1959		0.10 JUN 1954		6.46 JUL 1959		35.2 FEB 1967		14.4 JAN 1961		81	81	56	71	10.6	S	60	32	SEP 1960	6.1	54	98	105	162	112	4	26	48	17	15	110	1	1014.8

Means and extremes above are from existing and comparable exposures. Annual extremes have been exceeded at other sites in the locality as follows: Lowest temperature −9 in February 1934; maximum monthly precipitation 14.87 in August 1882; minimum monthly precipitation .01 in September 1941; maximum precipitation in 24 hours 9.21 in October 1903; maximum snowfall in 24 hours 18.0 in February 1902.

REFERENCE NOTES APPLYING TO TABLES APPEAR ON THE PAGE FOLLOWING LAST TABLE.
(Caution: Letters and symbols may have different meanings in 1941-1970 tables than in earlier tables. See notes.)

NEWARK, NEW JERSEY
NEWARK AIRPORT

LATITUDE	40° 42' N
LONGITUDE	74° 10' W
ELEVATION (ground)	7 Feet

NORMALS, MEANS AND EXTREMES
(Table Revised 1967. Base Period for Climatological Normals: 1931-1960)

Month	Daily maximum	Daily minimum	Monthly	Record highest	Year	Record lowest	Year	Normal degree days	Precip Normal total	Max monthly	Year	Min monthly	Year	Max 24 hrs	Year	Snow total	Snow max monthly	Year	Snow max 24 hrs	Year
J	39.5	25.0	32.3	74	1950	0	1957+	1014	3.33	5.57	1953	0.81	1955	1.78	1962	7.6	22.2	1961	13.7	1961
F	40.7	24.7	32.7	76	1949	-7	1943	904	2.80	4.47	1956	1.89	1959	2.45	1961	7.5	22.3	1961	20.0	1961
M	48.8	32.1	40.5	89	1945	7	1943	760	4.09	6.29	1954	1.12	1966	2.37	1962	5.3	26.0	1956	17.6	1956
A	60.9	41.7	51.3	91	1960+	23	1954	411	3.51	6.41	1958	0.90	1963	2.01	1958	0.6	4.1	1957	4.1	1957
M	72.1	51.9	62.0	98	1962	33	1947	127	3.65	4.86	1966	0.52	1964	2.36	1966	T	T	1956	T	1956
J	81.3	61.2	71.2	102	1952+	43	1945	9	3.44	3.89	1966	0.49	1966	1.52	1959	0.0	0.0		0.0	
J	86.1	66.5	76.3	105	1949	52	1945+	0	3.67	7.95	1961	0.89	1966	3.15	1961	0.0	0.0		0.0	
A	83.8	64.9	74.4	105	1948	51	1942	0	4.43	11.84	1955	0.50	1964	4.47	1955	0.0	0.0		0.0	
S	77.0	57.6	67.3	105	1953	35	1947	39	3.76	7.86	1966	1.30	1964	4.71	1966	0.0	0.0		0.0	
O	66.2	47.0	56.6	85	1949	28	1948	276	3.11	6.70	1955	0.21	1963	2.65	1966	T	0.3	1952	0.3	1952
N	53.5	37.3	45.4	72	1950	15	1955	588	3.37	5.68	1963	1.48	1965	2.09	1963	0.4	2.9	1945	2.9	1945
D	42.0	27.4	34.7	72	1946	-1	1942	939	3.22	5.74	1955	0.27	1955	1.83	1947	8.3	29.1	1947	26.0	1947
YR	62.7	44.8	53.7	105	SEP. 1953	-7	FEB. 1943	5067	42.38	11.84	AUG. 1955	0.21	OCT. 1963	4.71	SEP. 1966	29.7	29.1	DEC. 1947	26.0	DEC. 1947

Relative humidity / Wind / Mean number of days columns (EST hours 1:00 A.M., 7:00 A.M., 1:00 P.M., 7:00 P.M.):

Month	RH 1AM	RH 7AM	RH 1PM	RH 7PM	Prevailing dir	Mean hourly speed	Mean sky cover	Clear	Partly cloudy	Cloudy	Precip .01+	Snow 1.0+	Thunderstorms	Heavy fog	Temp max 90+	Temp max 32-	Temp min 32-	Temp min 0-
J	69	72	58	64	NE	11.4	6.4	8	7	16	11	2	*	3	0	7	25	*
F	68	70	56	60	NW	11.5	6.3	7	7	13	10	2	*	2	0	5	22	*
M	68	66	52	59	NW	12.1	6.3	8	9	14	12	1	1	2	0	1	14	0
A	70	69	49	60	NW	11.2	6.5	7	9	14	12	*	2	1	*	0	4	0
M	75	71	51	61	WNW	9.4	5.9	7	11	13	11	0	4	2	1	0	0	0
J	77	73	52	62	SW		6.0	7	12	11	10	0	5	1	4	0	0	0
J	78	74	51	63	SW	8.9	6.0	7	13	11	9	0	6	1	6	0	0	0
A	80	78	53	66	SW	8.7	5.1	10	12	9	10	0	6	2	4	0	0	0
S	77	80	53	67	SW	9.0	5.2	12	11	7	8	0	4	3	2	0	*	0
O	78	76	55	65	SW	9.4	5.6	12	11	8	8	*	1	2	*	0	*	*
N	73	76	56	65	SW	10.1	6.0	9	8	13	10	*	*	2	0	0	6	*
D	71	73	58	65	SW	10.6	6.2	8	8	14	11	2	*	2	0	6	22	*
YR	74	73	53	63	SW	10.2	6.1	99	114	152	122	8	25	22	25	18	94	*

(b) Climatological standard normals (1931-1960) Revised December 1966.

Means and extremes in the above table are from existing or comparable location(s). Annual extremes have been exceeded at other locations as follows:
Lowest temperature -14 in February 1934; maximum monthly precipitation 22.48 in August 1843; minimum monthly precipitation 0.07 in June 1949.

NORMALS, MEANS AND EXTREMES
(Table Revised 1975. Base Period for Climatological Normals: 1941-1970)

Month	Daily maximum	Daily minimum	Monthly	Record highest	Year	Record lowest	Year	Heating deg days	Cooling deg days	Precip Normal	Max monthly	Year	Min monthly	Year	Max 24 hrs	Year	Snow total	Snow max monthly	Year	Snow max 24 hrs	Year
J	38.5	24.3	31.4	69	1974	1	1968	1042	0	2.91	5.12	1964	0.81	1955	1.78	1962	7.6	22.2	1961	13.7	1961
F	40.2	24.9	32.6	66	1974	4	1967	907	0	2.95	4.47	1956	1.22	1968	2.45	1961	7.5	25.4	1961	17.0	1961
M	48.4	32.4	40.4	81	1968	4	1967	756	0	3.93	6.29	1954	1.12	1966	2.58	1956	5.3	26.0	1956	17.6	1956
A	61.2	41.9	51.6	91	1974	26	1974	399	0	3.44	6.41	1958	0.50	1963	4.11	1958	0.6	4.1	1957	4.1	1957
M	71.1	52.1	61.6	96	1966	35	1966	143	47	3.60	6.28	1968	0.52	1964	2.41	1968	T	T	1956	T	1956
J	81.0	61.6	71.4	101	1966	44	1972	5	197	2.99	6.94	1972	0.49	1966	2.13	1973	0.0	0.0		0.0	
J	85.6	67.2	76.4	105	1966	59	1973	0	353	4.03	7.95	1961	0.89	1966	3.40	1971	0.0	0.0		0.0	
A	83.7	65.5	74.6	100	1973	50	1965	0	298	4.27	11.84	1955	0.50	1964	7.84	1971	0.0	0.0		0.0	
S	77.0	58.6	67.8	95	1973	41	1967	34	118	3.44	7.86	1966	1.03	1972	5.47	1973	0.0	0.0		0.0	
O	66.9	48.1	57.5	87	1967	24	1969	243	11	2.82	6.70	1955	0.21	1963	3.04	1973	T	0.3	1952	0.3	1952
N	54.2	38.2	46.2	81	1974	20	1967	564	0	3.61	5.68	1972	0.80	1974	3.78	1972	0.4	3.1	1967	3.1	1967
D	41.5	27.4	34.5	68	1966	8	1968	946	0	3.46	7.24	1973	0.27	1955	2.14	1973	8.3	29.1	1947	26.0	1947
YR	62.5	45.2	53.9	105	JUL 1966	1	JAN 1968	5039	1024	41.45	11.84	AUG 1955	0.21	OCT 1963	7.84	AUG 1971	29.7	29.1	DEC 1947	26.0	DEC 1947

Relative humidity / Wind / Mean number of days (Local time hours 01, 07, 13, 19):

Month	RH 01	RH 07	RH 13	RH 19	Mean speed mph	Prevailing dir	Fastest mph	Dir	Year	Mean sky cover	Clear	Partly	Cloudy	Precip .01+	Snow 1.0+	Thunderstorms	Heavy fog	Temp max 90+	Temp max 32-	Temp min 32-	Temp min 0-	Avg sta pressure mb
J	70	73	58	63	11.2	NE	45	25	1951	6.4	8	8	15	11	2	*	3	0	3	23	0	1018.5
F	71	71	55	60	11.1	NW	46	23	1965	6.3	7	8	13	10	2	*	2	0	2	22	0	1016.5
M	68	66	50	57	12.1	NW	43	27	1950	6.4	7	9	14	11	1	1	2	0	1	13	0	1015.1
A	65	65	47	53	11.3	WNW	50	32	1951	6.4	7	11	14	11	*	2	1	*	0	3	0	1013.2
M	73	71	54	61	9.4	SW	55	07	1963	6.1	7	12	12	12	0	4	2	1	0	0	0	1012.6
J	72	71	51	59	8.8	SW	45	18	1950	6.2	7	12	12	10	0	5	1	4	0	0	0	1014.3
J	75	76	52	60	8.7	SW	46	05	1955	5.6	8	16	11	9	0	6	1	6	0	0	0	1016.5
A	76	77	52	64	9.0	SW	45	05	1960	5.3	10	11	11	8	0	5	2	4	0	0	0	1017.0
S	77	77	52	64	9.4	SW	48	11	1954	5.3	11	11	8	7	0	2	2	2	0	*	0	1019.5
O	71	76	62	67	10.2	SW	42	09	1962	5.3	11	8	12	7	*	1	2	*	0	1	0	1019.5
N	68	71	56	65	10.8	SW	55	32	1962	6.4	8	8	14	10	*	*	2	0	0	6	0	1016.3
D	75	73	62	67	10.8	SW	42	09		6.1	7	7	15	11	2	*	2	0	3	18	0	
YR	72	72	54	61	10.2	SW	82	09	NOV 1950	6.1	96	114	155	122	7	25	20	21	18	84	0	1015.8

Ø For period June 1965 through current year.
Means and extremes above are from existing and comparable exposures. Annual extremes have been exceeded at other sites in the locality as follows:
Lowest temperature -14 in February 1934; maximum monthly precipitation 22.48 in August 1843; minimum monthly precipitation 0.07 in June 1949.

REFERENCE NOTES APPLYING TO TABLES APPEAR ON THE PAGE FOLLOWING LAST TABLE.
(Caution: Letters and symbols may have different meanings in 1941-1970 tables than in earlier tables. See notes.)

FEDERAL BUILDING
LATITUDE 40° 13' N
LONGITUDE 74° 46' W
ELEVATION (ground) 56 Feet

NORMALS, MEANS AND EXTREMES

(Table Revised 1967. Base Period for Climatological Normals: 1931-1960)

Month	Normal Daily maximum (b)	Normal Daily minimum	Normal Monthly (b)	Record highest	Yr	Record lowest	Yr	Normal total (b)	Normal degree days (b)	Max monthly ppt	Yr	Min monthly ppt	Yr	Max 24 hrs ppt	Yr	Snow mean total	Snow max monthly	Yr	Snow max 24 hrs	Yr	Mean hourly wind	Prevailing dir	Fastest mile speed	Dir	Yr	Pct poss sun	Mean sky cover
	(b)	(b)	(b)	34	34	34	34	(b)	(b)	34	34	34	34	34	34	34	34	34	34	34	32	32	34	34	34	34	23
J	40.0	26.2	33.1	72	1950	-3	1936	3.10	989	6.00	1936	0.52	1955	2.03	1936	5.8	16.1	1961	10.1	1961	9.8	NW	48	N	1958	51	6.3
F	40.9	25.9	33.4	73	1949	-14	1934	2.59	885	5.56	1939	1.17	1966	2.45	1966	6.7	23.1	1934	23.1	1958	10.2	NW	49	NW	1960+	55	6.1
M	48.8	32.5	40.7	86	1945	8	1943	3.84	753	7.53	1953	1.17	1966	2.55	1953	4.4	21.5	1958	14.3	1958	10.7	NW	43	NW	1950	56	6.1
A	61.3	42.0	51.7	91	1941	24	1954	3.21	399	5.93	1952	0.83	1963	2.46	1952	0.4	4.2	1956	4.2	1956	10.4	S	43	S	1957	58	6.3
M	72.3	52.3	62.3	96	1962	34	1947	3.62	121	8.03	1964	0.25	1964	2.68	1966	T	T	1963	T	1963	9.0	S	37	NW	1953+	63	6.3
J	80.7	61.3	71.0	100	1952+	43	1938	3.60	12	9.00	1938	0.06	1966	4.79	1938	0.0	0.0		0.0		8.4	S	36	SE	1955	65	5.9
J	85.2	66.7	76.0	106	1936	53	1963+	4.18	0	10.19	1941	0.37	1944	4.85	1964	0.0	0.0		0.0		7.8	S	46	SW	1945+	67	5.8
A	82.8	65.0	73.9	100	1955	48	1947	4.77	0	14.10	1955	0.47	1964	4.76	1964	0.0	0.0		0.0		7.6	S	41	NW	1947	65	5.8
S	76.2	57.9	67.1	100	1953	36	1947	3.50	57	10.49	1934	0.05	1941	4.01	1960	0.0	0.0		0.0		7.9	S	56	NW	1960	62	5.6
O	65.9	47.7	56.8	94	1941	27	1940+	2.84	264	6.77	1951	0.19	1963	3.46	1966	0.1	1.6	1962	1.6	1962	8.3	NW	60	NW	1951	62	5.6
N	53.6	38.0	45.8	83	1950	14	1938	3.16	576	6.97	1936	0.75	1951	2.37	1963	1.0	13.0	1938	7.7	1938	9.2	NW	64	NE	1950	54	6.0
D	42.2	28.2	35.2	72	1966	-1	1942+	2.87	924	6.08	1948	0.19	1955	2.67	1948	4.9	21.5	1960	16.6	1960	9.3	NW	48	NW	1962	50	6.1
YR	62.5	45.3	53.9	106	JUL. 1936	-14	FEB. 1934	41.28	4980	14.10	AUG. 1955	0.05	OCT. 1963	4.85	JUL. 1964	23.3	23.1	FEB. 1934	16.6	DEC. 1960	9.0	S	64	E	NOV. 1950	60	5.9

Mean number of days (1931-1960):

Month	Clear	Partly cloudy	Cloudy	Ppt .01"+	Snow/Sleet 1.0"+	Thunderstorms	Heavy fog	Max 90°+	Max 32°-	Min 32°-	Min 0°-
	34	34	34	34	34	32		34	34	34	34
J	8	8	15	12	2	*		0	8	24	*
F	8	8	12	11	1	*		0	6	21	*
M	9	8	14	12	2	1		0	1	14	0
A	7	11	12	12	*	1		*	0	2	0
M	7	11	11	11	0	2		1	0	0	0
J	7	12	11	10	0	5		4	0	0	0
J	8	12	11	10	0	6		7	0	0	0
A	8	12	11	10	0	5		4	0	0	0
S	10	9	11	8	0	3		1	0	0	0
O	12	9	10	8	*	1		*	0	1	0
N	9	9	12	10	*	*		0	*	7	0
D	8	9	14	12	1	*		0	6	21	*
YR	101	116	148	121	7	33		18	20	90	*

Means and extremes in the above table are from the existing or comparable location(s). Annual extremes have been exceeded at other locations as follows: Maximum monthly precipitation 15.22 in July 1880; maximum precipitation in 24 hours 5.42 in October 1903; maximum monthly snowfall 34.0 in February 1899; fastest mile wind 73 in July 1914.

NORMALS, MEANS AND EXTREMES

(Table Revised 1975. Base Period for Climatological Normals: 1941-1970)

Month	Normal Daily max (a)	Normal Daily min	Normal Monthly	Record highest	Yr	Record lowest	Yr	Heating DD	Cooling DD	Normal ppt	Max monthly	Yr	Min monthly	Yr	Max 24 hrs	Yr	Snow mean total	Snow max monthly	Yr	Snow max 24 hrs	Yr	Mean wind	Prevailing dir	Fastest speed	Dir	Yr	Pct poss sun	Mean sky cover
J	38.8	25.3	32.1	72	1950	-3	1936	1020	0	2.76	6.00	1936	0.52	1955	2.03	1936	5.8	16.1	1961	10.1	1961	9.8	NW	48	N	1958	51	6.2
F	40.6	26.1	33.4	73	1949	-14	1934	885	0	2.70	5.56	1939	1.15	1968	2.53	1973	6.7	24.3	1973	13.0	1967	10.2	NW	51	S	1974	55	6.1
M	49.2	33.1	41.2	86	1945	8	1943	738	0	3.81	7.53	1953	1.17	1966	2.55	1953	4.4	21.5	1958	14.3	1958	10.7	NW	49	NW	1968	56	6.1
A	61.8	42.5	52.2	91	1941	24	1972	384	45	3.15	6.61	1973	0.83	1963	2.46	1952	0.4	4.4	1971	4.4	1971	10.4	S	43	NW	1957	60	6.3
M	72.0	52.2	62.1	96	1962	34	1972	135	194	3.40	8.03	1964	0.25	1964	3.11	1968	T	T	1963	T	1963	9.0	S	37	NW	1953	63	6.0
J	80.9	61.6	71.3	100	1952	43	1972	5		3.21	9.00	1938	0.06	1949	4.79	1938	0.0	0.0		0.0		8.4	S	36	SE	1955	63	6.0
J	84.9	66.8	75.9	106	1936	53	1963	0	338	4.74	10.19	1941	0.37	1944	4.85	1964	0.0	0.0		0.0		7.8	S	46	SW	1945	65	5.5
A	82.8	65.0	73.9	100	1955	48	1940	0	276	4.77	14.10	1955	0.47	1955	7.55	1971	0.0	0.0		0.0		7.5	NE	43	NE	1971	65	5.7
S	76.2	58.1	67.2	100	1953	36	1947	52	105	2.53	10.49	1934	0.05	1941	3.46	1966	0.0	0.0		0.0		8.1	S	56	NW	1960	61	5.5
O	66.2	48.2	57.2	94	1941	27	1940	252	10	2.13	6.77	1972	0.19	1963	2.90	1972	0.1	2.5	1972	2.5	1972	8.3	NW	60	NW	1951	62	6.2
N	53.9	38.7	46.3	83	1950	14	1938	561	0	3.25	6.97	1936	0.75	1936	2.90	1938	1.0	13.0	1938	7.7	1938	9.2	NW	64	NE	1950	53	6.2
D	41.5	28.3	34.9	72	1966	-1	1942	933	0	3.28	6.08	1973	0.19	1973	2.67	1948	4.9	21.5	1960	16.6	1960	9.3	NW	52	NE	1974	48	6.2
YR	62.4	45.5	54.0	106	JUL 1936	-14	FEB 1934	4952	968	40.17	14.10	AUG 1955	0.05	OCT 1963	7.55	AUG 1971	24.3	24.3	FEB 1934	16.6	DEC 1960	9.0	S	64	E	NOV 1950	59	6.0

Mean number of days (1941-1970):

Month	Clear	Partly cloudy	Cloudy	Ppt .01"+	Snow/Ice 1.0"+	Thunderstorms	Heavy fog	Max 90°+	Max 32°-	Min 32°-	Min 0°-
	42	42	42	42	42	32		42	42	42	42
J	8	9	14	11	2	*		0	8	24	*
F	8	9	12	11	1	*		0	6	22	*
M	9	8	14	12	2	1		0	1	14	0
A	7	10	13	11	1	1		*	0	2	0
M	7	11	12	11	0	2		1	0	0	0
J	7	11	11	10	0	5		4	0	0	0
J	8	12	11	10	0	6		7	0	0	0
A	8	11	11	10	0	5		4	0	0	0
S	10	9	11	8	0	3		1	0	0	0
O	12	9	10	7	*	1		*	0	1	0
N	9	9	13	10	*	*		0	*	7	0
D	8	9	14	11	1	*		0	5	20	*
YR	100	117	148	121	7	33		18	20	88	*

Average station pressure 1001.9 mb. Elev. 190 feet m.s.l.

Means and extremes above are from existing and comparable exposures. Annual extremes have been exceeded at other sites in the locality as follows: Maximum monthly precipitation 15.22 in July 1880; maximum monthly snowfall 34.0 in February 1899; fastest mile wind 73 in July 1914.

REFERENCE NOTES APPLYING TO TABLES APPEAR ON THE PAGE FOLLOWING LAST TABLE.

(Caution: Letters and symbols may have different meanings in 1941-1970 tables than in earlier tables. See notes.)

Reference notes applying to Normals, Means, and Extremes tables for 1931–1960 base period.

(a) Length of record, years.
(b) Climatological standard normals (1931-1960).
* Less than one half.
+ Also on earlier dates, months or years.
T Trace, an amount too small to measure.

Below-zero temperatures are preceded by a minus sign.

The prevailing direction for wind in the Normals, Means, and Extremes table is from records through 1963.

\# To 8 compass points only.

Unless otherwise indicated, dimensional units used in this bulletin are: temperature in degrees F.; precipitation, including snowfall, in inches; wind movement in miles per hour; and relative humidity in percent. Degree day totals are the sums of the negative departures of average daily temperatures from 65°F. Sleet was included in snowfall totals beginning with July 1948. Heavy fog reduces visibility to 1/4 mile or less.

Sky cover is expressed in a range of 0 for no clouds or obscuring phenomena to 10 for complete sky cover. The number of clear days is based on average cloudiness 0-3; partly cloudy days 4-7; and cloudy days 8-10 tenths.

& Figures instead of letters in a direction column indicate direction in tens of degrees from true North; i.e., 09 - East, 18 - South, 27 - West, 36 - North, and 00 - Calm. Resultant wind is the vector sum of wind directions and speeds divided by the number of observations. If figures appear in the direction column under "Fastest mile" the corresponding speeds are fastest observed 1-minute values.

¢ Temperature extremes and relative humidity means in the Normals, Means, and Extremes table are for comparable locations through 1964. Summaries for the present location of temperature sensors will be published when more data are accumulated.

Reference notes applying to Normals, Means, and Extremes tables for 1941–1970 base period.

(a) Length of record, years, through the current year unless otherwise noted, based on January data.
(b) 70° and above at Alaskan stations.
* Less than one half.
T Trace.

NORMALS - Based on record for the 1941-1970 period.
DATE OF AN EXTREME - The most recent in cases of multiple occurrence.
PREVAILING WIND DIRECTION - Record through 1964.
WIND DIRECTION - Numerals indicate tens of degrees clockwise from true north. 00 indicates calm.
FASTEST MILE WIND - Speed is fastest observed 1-minute value when the direction is in tens of degrees.

% Through 1964. The station did not operate 24 hours daily and thunderstorm data may be incomplete.

Ø Through 1964.

½ Based on U.S. Naval Air Station and Weather Bureau Airport Station records.

Mean Annual Precipitation, Inches

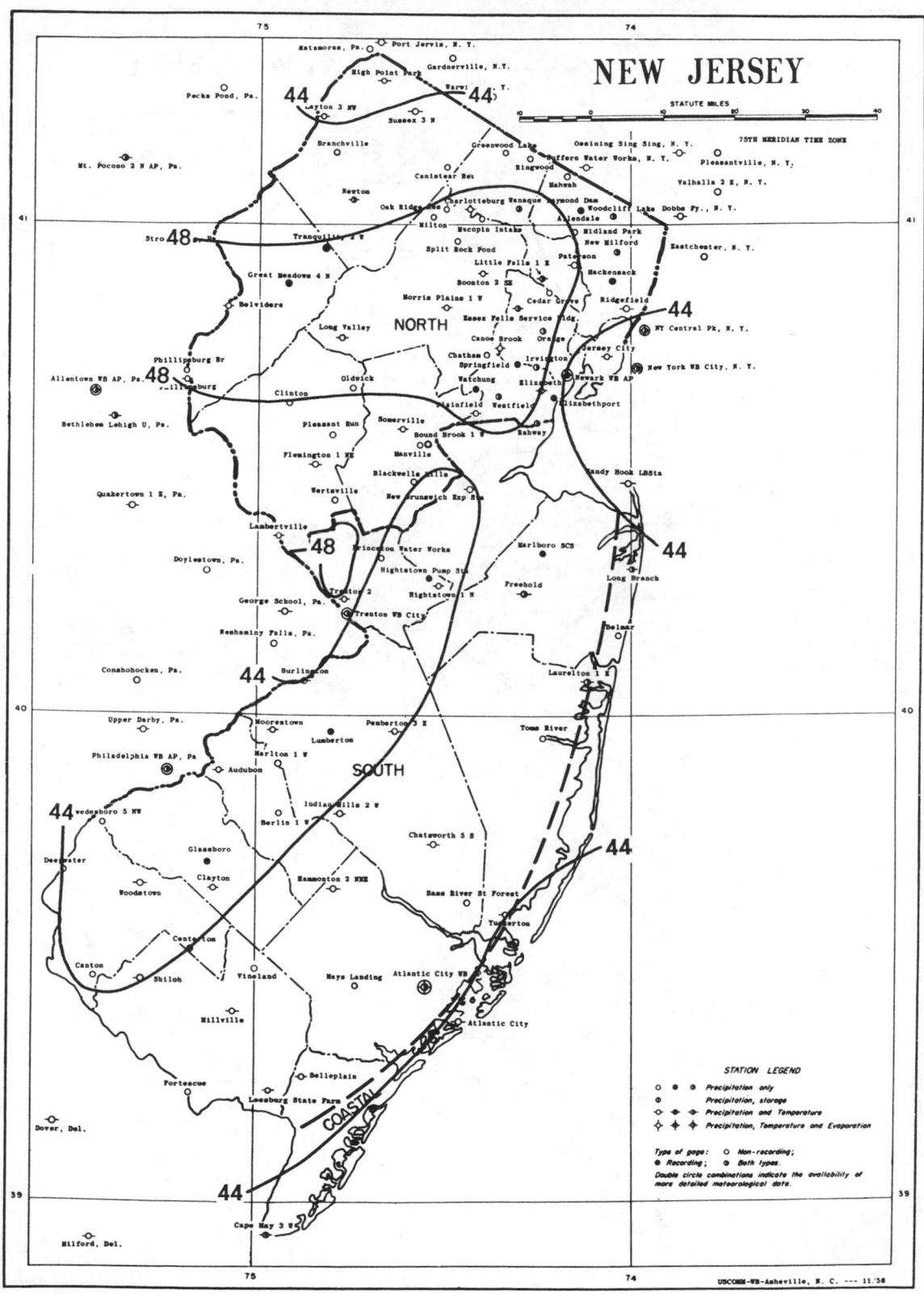

NEW JERSEY

STATUTE MILES

75TH MERIDIAN TIME ZONE

NORTH

SOUTH

COASTAL

STATION LEGEND

○ ◐ ● Precipitation only
◉ Precipitation, storage
○– ◐– ●– Precipitation and Temperature
◇ ◈ ◆ Precipitation, Temperature and Evaporation

Type of gage: ○ Non-recording;
● Recording; ◐ Both types.
Double circle combinations indicate the availability of
more detailed meteorological data.

USCOMM-WB-Asheville, N. C. --- 11/58

Based on period 1931-55

Isolines are drawn through points of approximately equal value. Caution should be used
in interpolating on these maps.

Mean Maximum Temperature (°F.), January

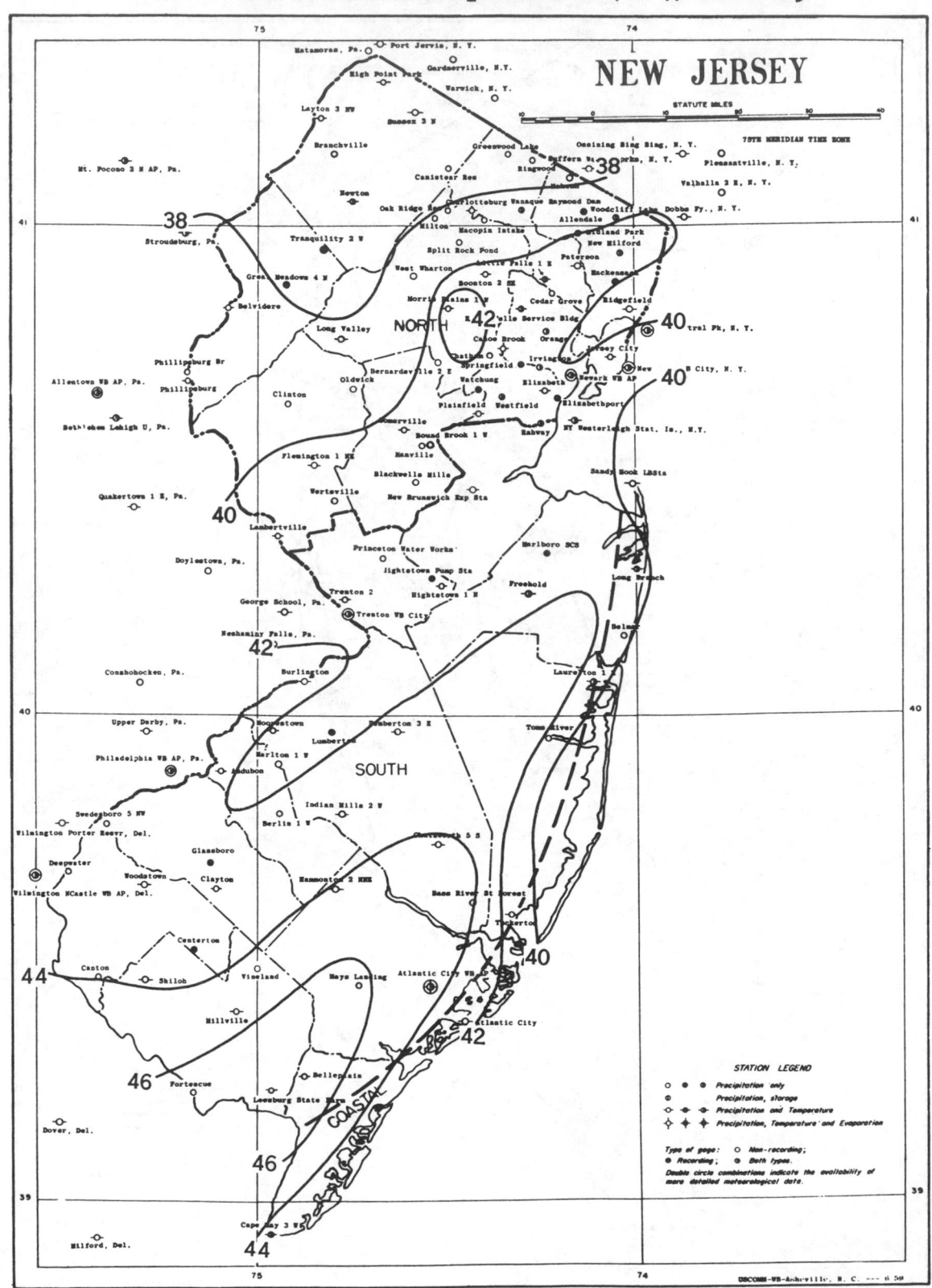

Based on period 1931-52

Isolines are drawn through points of approximately equal value. Caution should be used in interpolating on these maps.

Mean Minimum Temperature (°F.), January

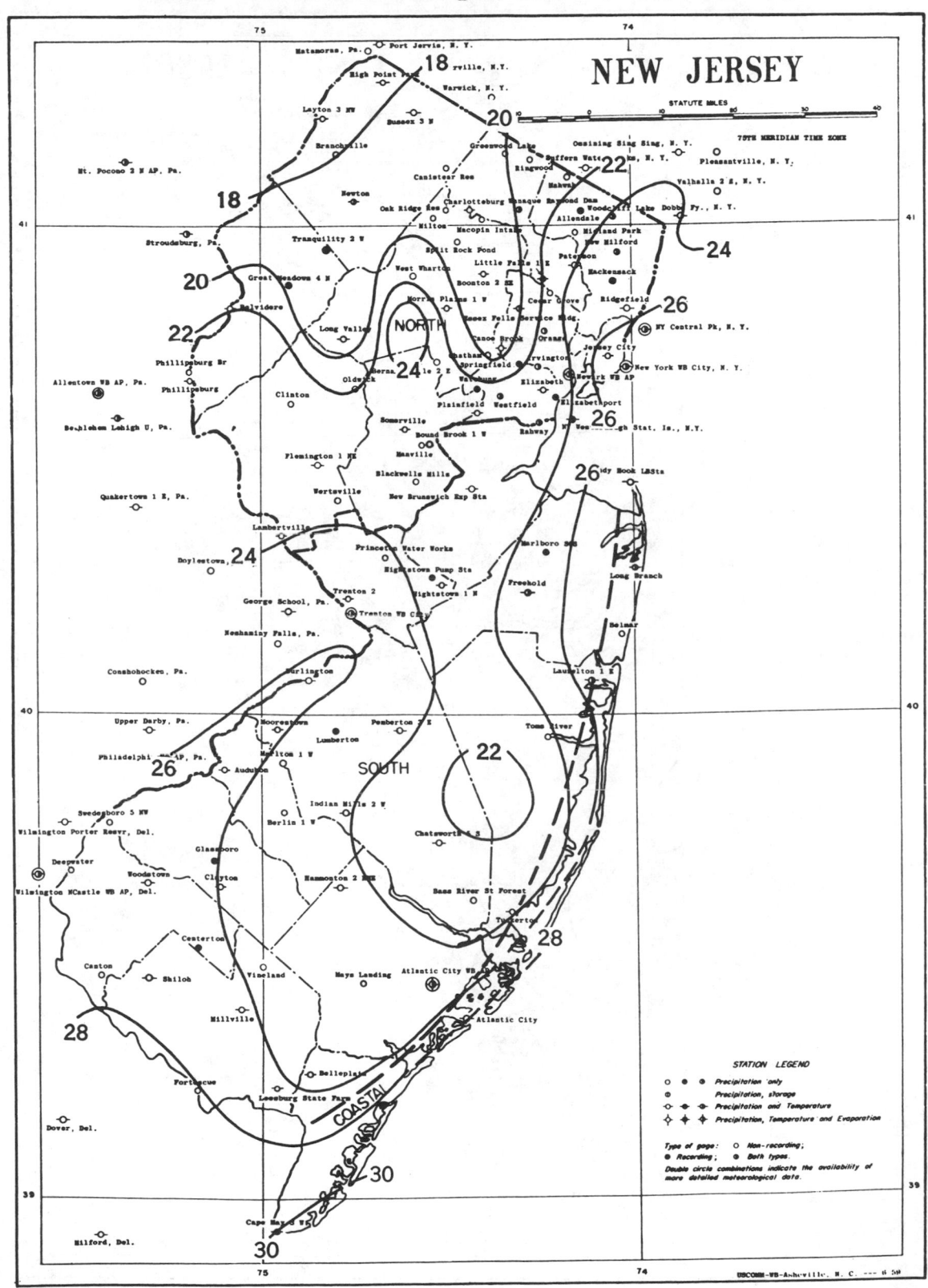

NEW JERSEY

Based on period 1931-52

Isolines are drawn through points of approximately equal value.   Caution should be used
in interpolating on these maps,

Mean Maximum Temperature (°F.), July

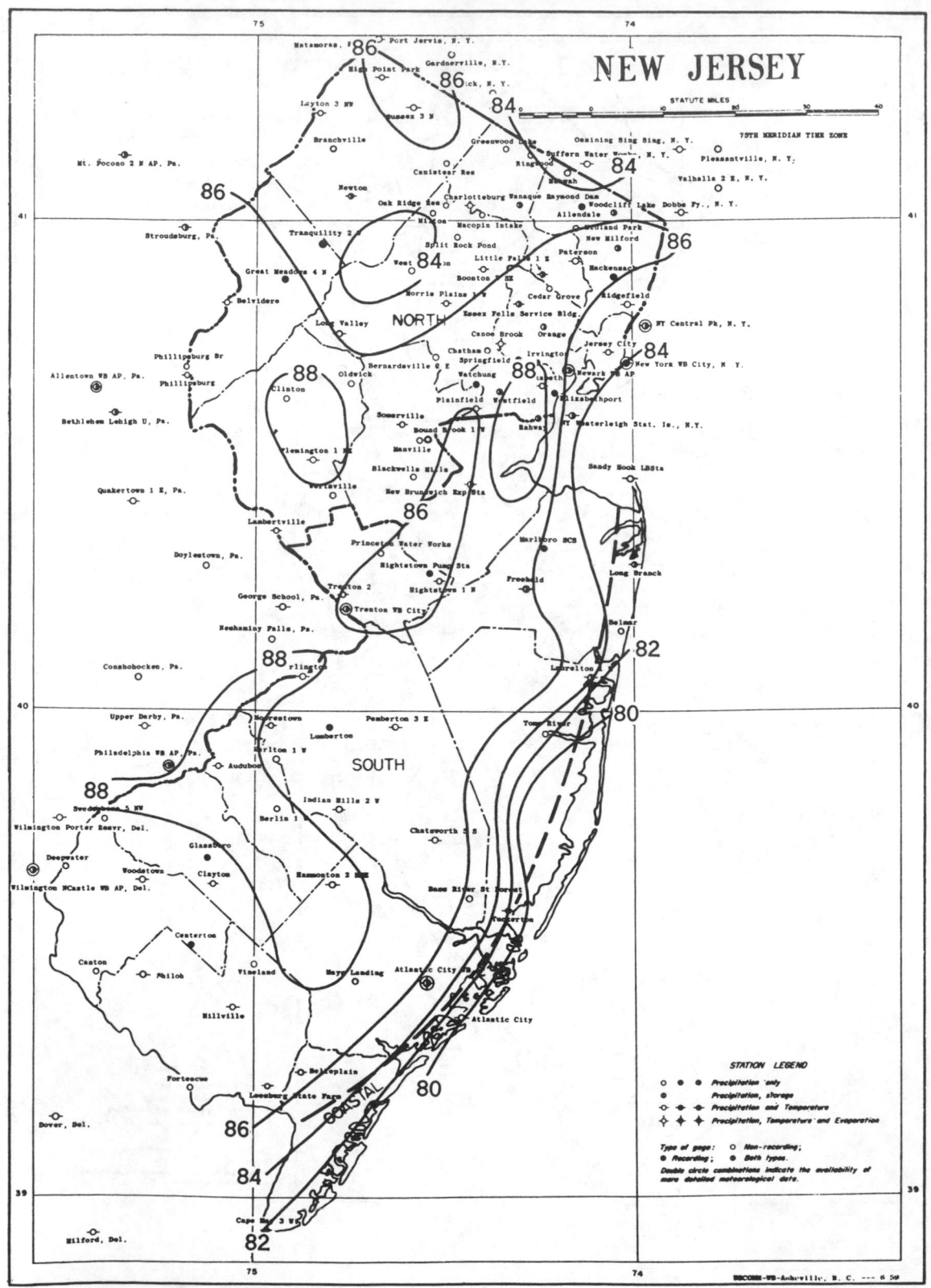

Based on period 1931-52

Isolines are drawn through points of approximately equal value.   Caution should be used in interpolating on these maps.

Mean Minimum Temperature (°F.), July

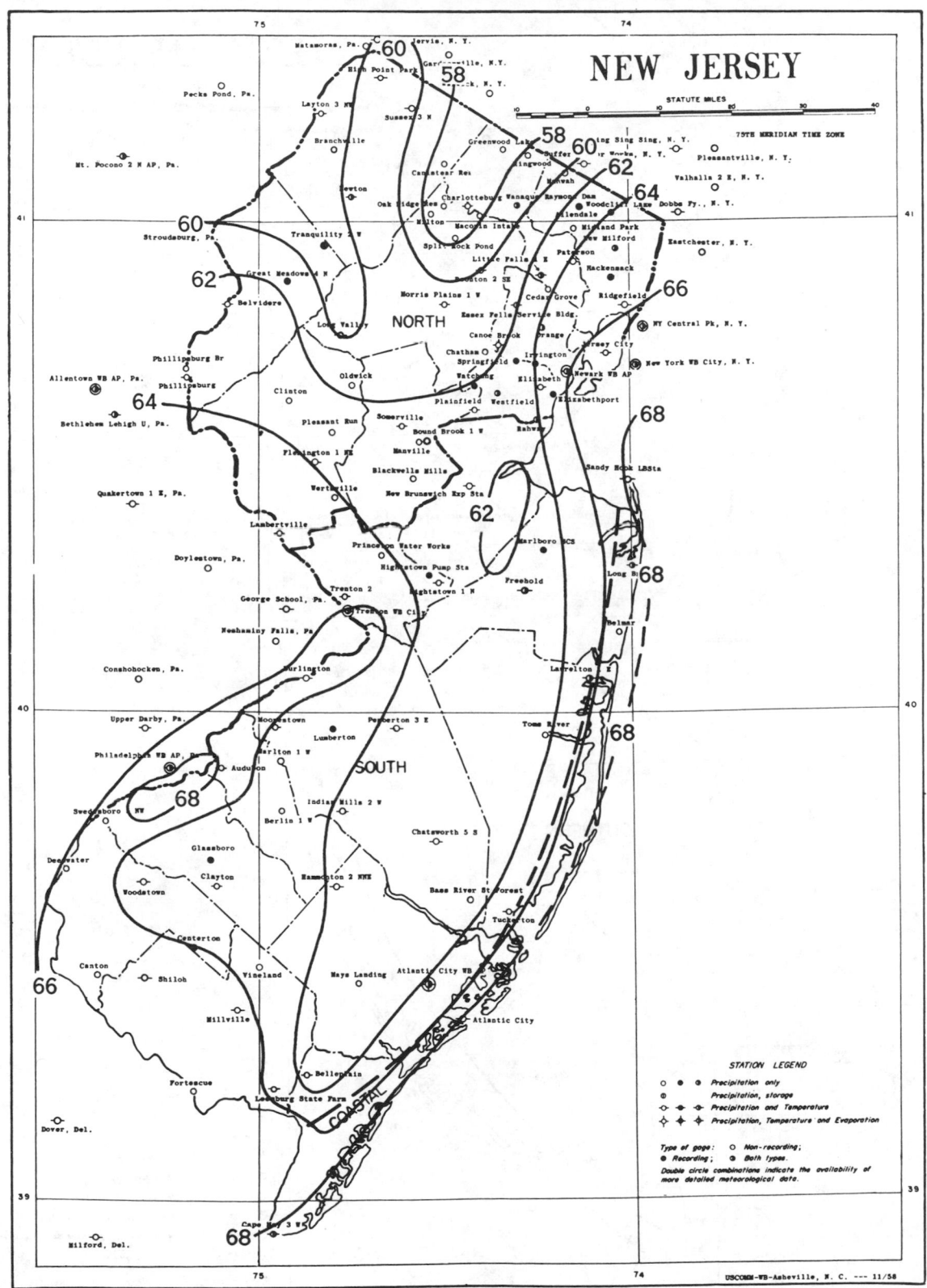

Based on period 1931-52

Isolines are drawn through points of approximately equal value. Caution should be used in interpolating on these maps.

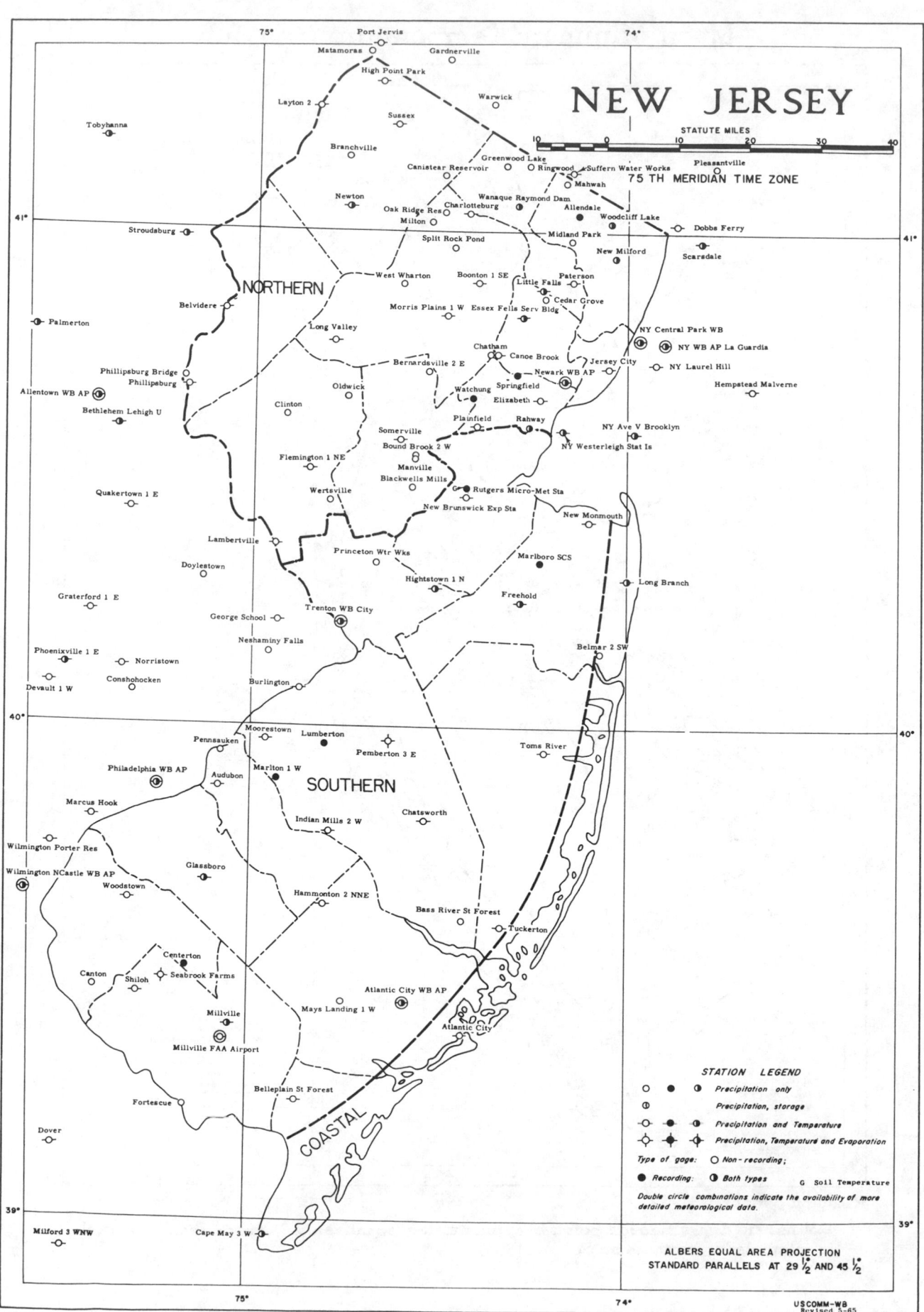

NEW JERSEY

STATUTE MILES

75 TH MERIDIAN TIME ZONE

NORTHERN

SOUTHERN

COASTAL

STATION LEGEND

○ ● ◑ *Precipitation only*

◑ *Precipitation, storage*

◐ ● ◑ *Precipitation and Temperature*

◇ ◆ ◈ *Precipitation, Temperature and Evaporation*

Type of gage: ○ *Non-recording;*

● *Recording;* ◑ *Both types*   G *Soil Temperature*

Double circle combinations indicate the availability of more detailed meteorological data.

ALBERS EQUAL AREA PROJECTION
STANDARD PARALLELS AT 29 ½° AND 45 ½°

USCOMM-WB
Revised 5-65

Port Jervis
Matamoras
Gardnerville
High Point Park
Warwick
Layton 2
Sussex
Tobyhanna
Branchville
Greenwood Lake
Pleasantville
Canistear Reservoir
Ringwood
Suffern Water Works
Newton
Mahwah
Wanaque Raymond Dam
Oak Ridge Res
Charlotteburg
Allendale
Milton
Woodcliff Lake
Stroudsburg
Split Rock Pond
Midland Park
Dobbs Ferry
New Milford
Scarsdale
West Wharton
Boonton 1 SE
Paterson
Little Falls
Belvidere
Morris Plains 1 W
Essex Fells Serv Bldg
Cedar Grove
Palmerton
Long Valley
NY Central Park WB
Chatham
NY WB AP La Guardia
Bernardsville 2 E
Canoe Brook
Phillipsburg Bridge
Oldwick
Newark WB AP
Jersey City
NY Laurel Hill
Phillipsburg
Watchung
Springfield
Allentown WB AP
Clinton
Elizabeth
Hempstead Malverne
Bethlehem Lehigh U
Plainfield
Rahway
Somerville
NY Ave V Brooklyn
Quakertown 1 E
Flemington 1 NE
Bound Brook 2 W
NY Westerleigh Stat Is
Manville
Wertsville
Blackwells Mills
G Rutgers Micro-Met Sta
New Brunswick Exp Sta
New Monmouth
Lambertville
Princeton Wtr Wks
Doylestown
Marlboro SCS
Hightstown 1 N
Long Branch
Graterford 1 E
Freehold
George School
Trenton WB City
Phoenixville 1 E
Neshaminy Falls
Norristown
Belmar 2 SW
Conshohocken
Devault 1 W
Burlington
Moorestown
Pennsauken
Lumberton
Pemberton 3 E
Toms River
Philadelphia WB AP
Marlton 1 W
Audubon
SOUTHERN
Marcus Hook
Indian Mills 2 W
Chatsworth
Wilmington Porter Res
Glassboro
Wilmington NCastle WB AP
Woodstown
Hammonton 2 NNE
Bass River St Forest
Centerton
Tuckerton
Canton
Seabrook Farms
Shiloh
Atlantic City WB AP
Millville
Mays Landing 1 W
Atlantic City
Millville FAA Airport
Belleplain St Forest
Fortescue
Dover
Milford 3 WNW
Cape May 3 W

676

CLIMATES OF THE STATES

NEW MEXICO

(Normals, Means and Extremes tables revised 1972 and 1975. Basic report revised May 1972.)

Frank E. Houghton, Climatologist for New Mexico

INTRODUCTION. New Mexico, fifth largest State in the Union, with a total area of 121,666 square miles, is in the southwestern part of the country. The State, approximately 350 miles square, lies mostly between latitudes 32° and 37° N. and longitudes 103° and 109° W. The State's topography consists mainly of high plateaus or mesas, with numerous mountain ranges, canyons, valleys, and normally dry arroyos. Average elevation is about 5,700 feet above sea level. The lowest point is upstream from the Red Bluff Reservoir at 2,817 feet where the Pecos River flows into Texas. The highest point is Wheeler Peak at 13,161 feet above sea level. The principal sources of moisture for the scant rains and snows that fall on the State are the Pacific Ocean, 500 miles to the west, and the Gulf of Mexico, 500 miles to the southeast. New Mexico has a mild, arid or semiarid, continental climate characterized by light precipitation totals, abundant sunshine, low relative humidities, and a relatively large annual and diurnal temperature range. The highest mountains have climate characteristics common to the Rocky Mountains.

PHYSICAL FEATURES. New Mexico is divided into three major areas by mountain ranges and highlands, oriented in a general north-south direction, which merge in the north. The Northern Mountains and Central Highlands, between longitudes 105° and 106° W., are the western boundary of the Northeastern and Southeastern Plains which slope gradually eastward and southeastward.

The northern part of these eastern plains lies within the Arkansas River Basin and is drained mostly by the Canadian River, which flows southward then eastward into Oklahoma to its confluence with the Arkansas, and the Cimarron River in the extreme northeastern corner. The Pecos River rises in the Sangre de Cristo Mountains and flows southward through the Southeastern Plains into Texas, and then southeastward to join the Rio Grande. West of the mountain ranges that form the Continental Divide, whose height decreases to a markedly lower elevation in southern New Mexico, rivers drain into the Gulf of California through the Colorado River system. Principal tributaries flowing westward into the Colorado River are the San Juan River in the north, the Gila River in the south, and the San Francisco tributary of the Gila and other headwater streams of the Little Colorado River in the west-central area. The largest closed basins in the west are the Plains of St. Augustine in Catron County and the Rio Mimbres Basin in Grant and Luna Counties. Between the Northern Mountains and the Central Highland system and the Continental Divide system is the Rio Grande Valley which widens toward the south. The Rio Grande rises in the San Juan Mountains of southern Colorado, flows southward through New Mexico, then southeastward along the Texas-Mexico border into the Gulf of Mexico. The closed Tularosa Basin in southern New Mexico is in an intermountain area east of the Central Valley.

Location and topography play major roles in determining the climate of New Mexico, particularly true for any specific locality. Both the ruggedness of the terrain and its direction of slope are important. The eastern plains open to the Great Plains of Texas and Oklahoma and to their northward extension into central Canada. At times during winter months, cold continental air masses move southward out of central Canada and invade this area, producing blizzard and cold-wave conditions. These air masses occasionally cross the Central Highlands, which greatly modify and warm the air masses before they reach the Rio Grande Valley.

PRECIPITATION. Average annual precipitation ranges from less than 10 inches over much of the southern desert and the Rio Grande and San Juan Valleys to more than 20 inches at higher elevations in the State. A wide variation in annual totals is characteristic of arid and semiarid climates as illustrated by annual extremes of 2.95 and 33.94 inches at Carlsbad during a period of record of more than 71 years.

Summer rains fall almost entirely during brief, but frequently intense, thunderstorms. The general southeasterly circulation from the Gulf of Mexico brings moisture for these storms into the State, and strong surface heating combined with orographic lifting as the air moves over higher terrain causes convective air currents and condensation. July and August are the rainiest months over most of the State, with from 30 to 40 percent of the year's total moisture falling at that time. The San Juan Valley area is least affected by this summer circulation, receiving about 25 percent of its annual rainfall during July and August. During the warmest 6 months of the year, May through October, total precipitation averages from 60 percent of the annual total in the Northwestern Plateau to 80 percent of the annual total in the eastern plains.

Winter precipitation is caused mainly by frontal activity associated with the general movement of Pacific Ocean storms across the country from west to east. As these storms move inland, much of the moisture is precipitated over the coastal and inland mountain ranges of California, Nevada, Arizona, and Utah. Much of the remaining moisture falls on the western slope of the Continental Divide and over northern and high central mountain ranges. Winter is the driest season in New Mexico except for the portion west of the Continental Divide. This dryness is most noticeable in the Central Valley and on eastern slopes of the mountains.

Much of the winter precipitation falls as snow in the mountain areas, but it may occur as either rain or snow in the valleys. Average annual snowfall ranges from about 3 inches at the Southern Desert and Southeastern Plains stations to well over 100 inches at Northern Mountain stations. It may exceed 300 inches in the highest mountains of the north.

FLOODS. General floods are seldom widespread in New Mexico. Heavy summer thunderstorms may bring several inches of rain to small areas in a short time. Because of the rough terrain and sparse vegetation in many areas, runoffs from these storms frequently cause local flash floods. Normally dry arroyos may overflow their banks for several hours, halting traffic where water crosses highways; damaging bridges, culverts, and roadways; and if in an urban area, possibly causing considerable property damage. Snowmelt during April to June, especially in combination with a warm rain, and heavy general rains during August to October may occasionally cause flooding of the larger rivers. Although streams in New Mexico have risen substantially during several floods, the overflows cannot be termed disastrous because comparatively little real property damage has resulted in this lightly industrialized and sparsely populated State. During spring snowmelt, main rivers may exceed flood stage and cause some damage to property along their banks.

Years in which there have been high flood discharges in major New Mexico river basins since 1903 are: Rio Grande--1904, 1905, 1929, 1935, and 1941; Pecos--1904, 1905, 1915, 1916, 1937, 1941, 1942, and 1966; Canadian--1904, 1913, 1937, and 1965; San Juan--1909, 1911, 1927, 1929, and 1942; and Gila--1941 and 1965.

TEMPERATURE. Mean annual temperatures range from 64° in the extreme southeast to 40° or lower in high mountains and valleys of the north; elevation is a greater factor in determining the temperature of any specific locality than its latitude. This factor is shown by only a 3° difference in mean temperature between stations at similar elevations, one in the extreme northeast and the other in the extreme southwest; however, at two stations only 15 miles apart, but differing in elevation by 4,700 feet, the mean annual temperatures are 61° and 45°-- a difference of 16° or a little more than 3° decrease in temperature for each 1,000-foot increase in elevation.

During the summer months, individual daytime temperatures quite often exceed 100° at elevations below 5,000 feet; but the average monthly maximum temperatures during July, the warmest month, range from slightly above 90° at lower elevations to the upper 70s at high elevations. Warmest days quite often occur in June before the thunderstorm season sets in; during July and August, afternoon convective storms tend to shut off afternoon solar insolation, lowering temperatures before they reach their potential daily high. The highest temperatures of record in New Mexico are 116° at Orogrande on July 14, 1934, and at Artesia on June 29, 1918. A preponderance of clear skies and low relative humidities permit rapid cooling by radiation from the earth after sundown; consequently, nights are usually comfortable in summer. The average range between daily high and low temperatures is from 25° to 35°.

In January, the coldest month, average daytime temperatures range from the middle 50s in the southern and central valleys to the middle 30s in the higher elevations of the north. Minimum temperatures below freezing are common in all sections of the State during the winter, but subzero temperatures are rare except in the mountains. The lowest temperature recorded at regular observing stations in the State was -50° at Gavilan on February 1, 1951. An unofficial low temperature of -57° at Ciniza on January 13, 1963, was widely reported by the press.

The freeze-free season ranges from more than 200 days in the southern valleys to less than 80 days in the northern mountains where some high mountain valleys have freezes in summer months.

SEVERE STORMS. On rare occasions, a tropical hurricane may cause heavy rain in eastern and central New Mexico as it moves inland from the western part of the Gulf of Mexico, but there is no record of serious wind damage from these storms. Also on rare occasions, a tropical storm moving inland from the Gulf of California area may cause heavy rain to fall in southwestern New Mexico.

Tornadoes are occasionally reported in New Mexico, most frequently during afternoon and early evening hours from May through August. There has been an average of three tornadoes a year, but damage has been light because most occur over open, sparsely populated country. The tornado causing the most loss of life and injuries occurred in 1930 at Wagon Mound with 3 deaths, 19 injuries, and property loss of $150,000. Greater property damage, $450,000, but fewer casualties--1 death and 8 injuries--resulted from a destructive tornado at Maxwell in 1964.

Thunderstorms are relatively frequent in summer, averaging in numbers from 40 in the south to more than 70 in the northeast, the latter area having the second greatest thunderstorm frequency in the country. Occasionally, these heavy thunderstorms are accompanied by hail, with the greatest hail frequency occurring near and to the east of Los Alamos. When hail falls over an agricultural area, considerable local crop damage may result.

SUNSHINE. Plentiful sunshine occurs in New Mexico, with from 75 to 80 percent of the possible sunshine being received. In winter, this prevalence is particularly noticeable with from 70 to 75 percent of the possible sunshine being received. It is not uncommon for as much as 90 percent of the possible sunshine to occur in November and in some of the spring months.

The average number of hours of annual sunshine ranges from near 3,700 in the southwest to 2,800 in the north-central portions.

RELATIVE HUMIDITY. Average relative humidities are lower in the valleys but higher in the mountains because of the lower mountain temperatures. Relative humidity ranges from an average of near 65 percent about sunrise to near 30 percent in midafternoon; however, afternoon humidities in warmer months are often less than 20 percent and occasionally may go as low as 4 percent. The prevalent low relative humidities during periods of extreme temperatures ease the effect of summer and winter temperatures on comfort.

WIND. Wind speeds over the State are usually moderate, although relatively strong winds often accompany occasional frontal activity during late winter and spring months and sometimes occur just in advance of thunderstorms. Frontal winds may exceed 30 m.p.h. for several hours and reach peak speeds of more than 50 m.p.h. Spring is the windy season. Blowing dust and serious soil erosion of unprotected fields may be a problem during dry spells. Winds are generally stronger in the eastern plains than in other parts of the State. Winds generally predominate from the southeast in summer and from the west in winter, but local surface wind directions will vary greatly because of local topography and mountain and valley breezes.

EVAPORATION. Potential evaporation in New Mexico is much greater than average annual precipitation. Evaporation from a Class A pan ranges from near 56 inches in the north-central mountains to more than 110 inches in southeastern valleys. During the warm months, May through October, evaporation ranges from near 41 inches in the north-central to 73 inches in the southeast portions of the State.

DROUGHT. Periods of recent extreme meteorological drought, as defined by a Palmer drought index of -4.0 or lower, have been noted in the mid-1930's in the Northeastern Plains and Central Highlands, in 1947 in the Central Highlands, in the 1950's throughout the State, in 1963-64 in the Northern Mountains, in 1964 in the Southeastern Plains, and in 1967 in the Northern Mountains. The longest general drought since 1930 was in the 1950's.

RECREATION AND HEALTH. Large primitive areas and many campgrounds are in the more than 8 million acres of forestland. There are many National Monuments and State Parks and one National Park--Carlsbad Caverns. Hunting and fishing areas are available in most sections of the State, and several reservoirs have facilities for boating. Snows in mountain areas permit skiing during winter months. These features, combined with generally mild, dry, sunny climate, make New Mexico a mecca for outdoor recreation. Many people seeking a mild and dry climate for health reasons might find the State a desirable place to settle.

INDUSTRY. Principal industries of New Mexico are agriculture, mining, lumbering, gas and oil production, and recreation. Of these, the influence of climate upon agriculture and recreation is of major importance. Less than 4 percent of the State's area is under cultivation, and about one-third of this area is irrigated. Farming on this latter portion is intensive. More than one-half of the area of the State is pastureland; about 28 percent is woodland. The remainder is generally classified as wasteland and urban. Most irrigated land is in the southern valleys, although some is found in the middle Rio Grande Valley, the Canadian Valley in the northeast, the San Juan Valley in the northwest, and in east-central counties. These irrigated lands draw on stored surface water as well as underground water supplies for irrigation. Most dryland farming is in the eastern plains, but short-season dryland summer crops are grown in some small areas in the Central Highlands. Dryland crops are divided primarily between winter grains, which require favorable moisture conditions from early fall throughout winter and spring, and short-season row and feed crops, which depend mainly on summer showers to produce a yield. Stored surface water for irrigation, used principally for cotton, truck and feed crops, and fruit, depends on adequate winter snows in the mountains of both the northern part of the State and in southern Colorado for its initial source. Livestock raising is the most extensive agricultural pursuit. Sufficient moisture usually falls, providing for the growth of good range forage. Because of the mild climate, livestock can live on the open range throughout the year, grazing in the higher mountain ranges during the summer and in the lower valleys and plains during the winter.

BIBLIOGRAPHY

ESSA, Environmental Data Service, Climatic Atlas of the United States, Washington, D. C., June 1968, 80 pp.

Gifford, R. O., Ashcroft, G. L., and Magnuson, M. D., Probability of Selected Precipitation Amounts in the Western Region of the United States, Agricultural Experiment Station, University of Nevada, Reno, Nev., 1967, 255 pp.

Houghton, Frank E., "Carlsbad Caverns, New Mexico, Climatic Summaries of Resort Areas," Climatography of the United States No. 21-29-2, ESSA, Environmental Data Service, Silver Spring, Md., Oct. 1967, 4 pp.

_____, "Cloudcroft, New Mexico, Climatic Summaries of Resort Areas," Climatography of the United States No. 21-29-3, ESSA, Environmental Data Service, Silver Spring, Md., Feb. 1970, 4 pp.

_____, "Red River, New Mexico, Climatic Summaries of Resort Areas," Climatography of the United States No. 21-29-1, ESSA, Environmental Data Service, Silver Spring, Md., 1966, 4 pp.

NOAA, Environmental Data Service, "Albuquerque, Clayton, and Roswell, New Mex.," Local Climatological Data, Asheville, N. C., monthly plus annual summary.

_____, Climatological Data--National Summary, Asheville, N. C., monthly plus annual summary.

_____, Climatological Data--New Mexico, Asheville, N. C., monthly plus annual summary.

_____, Hourly Precipitation Data--New Mexico, Asheville, N. C., monthly plus annual summary.

_____, Storage-gage Precipitation Data for Western United States, Asheville, N. C., annually.

_____, Storm Data, Asheville, N. C., monthly.

New Mexico Secretary of State, New Mexico Blue Book, 1969-1970, published by Ernestine D. Evans, Santa Fe, N. Mex., 1970, 136 pp.

Stout, Glenn E., and Changnon, Stanley A., Jr., "Climatography of Hail in the Central United States," CHIAA Research Report No. 38, Crop Hail Insurance Actuarial Association, Chicago, Ill., 1968, 49 pp.

U. S. Weather Bureau, Climatic Summary of the United States (Bulletin W), Sections 27, 28, and 29, Washington, D. C., 1930, 87 pp.

_____, "Climatic Summary of the United States--Supplement for 1931 through 1952, New Mexico" (Bulletin W Supplement), Climatography of the United States No. 11-25, Washington, D. C., 1956, 96 pp.

_____, "Climatic Summary of the United States--Supplement for 1951 through 1961, New Mexico" (Bulletin W Supplement), Climatography of the United States No. 86-25, Washington, D. C., 1965, 98 pp.

_____, "Monthly Averages for State Climatic Divisions, 1931-1960, New Mexico," Climatography of the United States No. 85-25, Washington, D. C., 1963, 3 pp.

_____, "Rainfall Frequency Atlas of the United States for Durations from 30 Minutes to 24 Hours and Return Periods from 1 to 100 Years," Technical Paper No. 40, Washington, D. C., May 1961, 115 pp.

_____, "Summary of Hourly Observations, Albuquerque, New Mexico, 1951-1960," Climatography of the United States No. 82-29, Washington, D. C., 1963, 15 pp.

_____, "Two- to Ten-day Precipitation for Return Periods of 2 to 100 Years in the Contiguous United States," Technical Paper No. 49, Washington, D. C., 1964, 29 pp.

FREEZE DATA

STATION	Freeze threshold temperature	Mean date of last Spring occurrence	Mean date of first Fall occurrence	Mean No. of days between dates	Years of record Spring	No. of occurrences in Spring	Years of record Fall	No. of occurrences in Fall
AGRICULTURAL COLLEGE	32	04-15	10-29	197	30	30	30	30
	28	04-02	11-05	217	30	30	30	30
	24	03-20	11-14	240	30	30	30	30
	20	02-27	11-23	268	30	30	30	30
	16	02-09	12-06	300	30	28	30	23
ALAMOGORDO	32	04-06	11-09	217	27	27	26	26
	28	03-22	11-16	239	27	27	26	26
	24	02-27	11-27	273	27	27	26	25
	20	02-11	12-09	301	27	25	26	23
	16	01-25	12-22	331	27	20	26	10
ALBUQUERQUE	32	04-16	10-29	196	30	30	30	30
	28	04-02	11-06	218	30	30	30	30
	24	03-22	11-15	238	30	30	30	30
	20	03-05	11-26	266	30	30	30	29
	16	02-07	12-08	304	30	28	30	26
AZTEC RUINS NAT MON	32	05-16	10-10	147	28	28	30	30
	28	05-02	10-20	171	28	28	30	30
	24	04-21	10-28	190	28	28	30	30
	20	04-09	11-05	210	28	28	30	30
	16	03-28	11-13	230	28	28	30	30
BELL RANCH	32	04-23	10-20	180	29	29	29	29
	28	04-09	10-31	204	29	29	29	29
	24	03-31	11-08	222	29	29	29	29
	20	03-20	11-13	238	29	29	29	29
	16	03-08	11-23	260	29	29	29	29
BLUEWATER 3 WSW	32	06-05	09-21	108	25	25	26	26
	28	05-13	09-29	139	24	24	26	26
	24	05-03	10-09	159	23	23	26	26
	20	04-20	10-23	186	23	23	26	26
	16	04-10	10-30	203	23	23	25	26
CARLSBAD FAA AP	32	04-02	11-07	220	30	30	29	29
	28	03-22	11-14	237	30	30	29	29
	24	03-08	11-26	262	30	30	29	29
	20	02-20	12-09	292	30	28	30	29
	16	02-02	12-20	321	29	24	30	17
CLAYTON WSO AP	32	05-02	10-15	166	29	29	30	30
	28	04-18	10-27	192	29	29	30	30
	24	04-06	11-05	213	29	29	30	30
	20	03-30	11-12	227	29	29	30	30
	16	03-21	11-19	243	29	29	30	30
CLOVIS	32	04-16	10-31	198	30	30	30	30
	28	04-04	11-08	219	29	29	30	30
	24	03-27	11-15	233	29	29	30	30
	20	03-16	11-26	255	29	29	30	29
	16	03-06	12-12	281	29	29	30	23
COLUMBUS	32	03-29	11-09	224	29	29	30	29
	28	03-16	11-19	249	29	29	30	29
	24	02-24	11-30	279	29	29	30	27
	20	02-08	12-11	306	29	26	30	21
	16	01-18	12-24	340	29	19	29	9
CORONA	32	05-06	10-14	161	28	28	28	28
	28	04-19	10-23	187	28	28	28	28
	24	04-11	11-01	204	28	28	27	27
	20	03-30	11-14	229	27	27	27	27
	16	03-17	11-22	250	27	27	28	28
DEMING	32	04-16	11-02	200	29	29	29	29
	28	04-03	11-07	218	29	29	29	29
	24	03-18	11-17	244	29	29	29	29
	20	02-26	12-03	280	29	28	29	26
	16	02-02	12-15	317	28	24	29	17
DES MOINES	32	05-15	10-02	140	26	26	28	28
	28	05-02	10-11	162	25	25	28	28
	24	04-20	10-24	187	24	24	28	28
	20	04-12	10-30	202	24	24	28	28
	16	04-04	11-07	217	24	24	26	26
ELEPHANT BUTTE DAM	32	03-31	11-13	227	30	30	30	30
	28	03-13	11-21	254	30	30	30	30
	24	02-15	12-05	293	30	30	30	28
	20	01-29	12-17	323	30	27	30	23
	16	01-15	12-26	346	30	16	30	9
ESTANCIA	32	05-18	10-02	137	28	28	27	27
	28	05-04	10-11	160	28	28	27	27
	24	04-24	10-20	179	28	28	27	27
	20	04-11	10-30	202	28	28	27	27
	16	03-27	11-09	228	28	28	27	27
FT BAYARD	32	04-25	10-27	185	29	29	28	28
	28	04-11	11-06	209	29	29	28	28
	24	03-25	11-19	239	29	29	28	28
	20	03-07	11-26	264	29	29	28	28
	16	02-09	12-13	307	27	26	28	21
FT STANTON	32	04-30	10-12	165	28	28	27	27
	28	04-21	10-21	184	28	28	27	27
	24	04-06	11-01	209	28	28	26	26
	20	03-28	11-11	228	28	28	26	26
	16	03-15	11-16	246	28	28	26	26
FT SUMNER	32	04-16	10-26	193	26	26	28	28
	28	04-02	11-05	217	26	26	27	27
	24	03-25	11-13	233	26	26	27	27
	20	03-16	11-21	250	26	26	26	26
	16	03-04	12-02	273	26	26	26	25
HACHITA 1 N	32	04-06	11-10	217	30	30	28	28
	28	03-20	11-19	244	30	30	28	27
	24	02-28	11-28	272	30	28	28	23
	20	02-07	12-07	303	30	25	28	23
	16	01-18	12-20	336	30	19	28	10
HOPE	32	04-16	10-30	197	22	22	20	20
	28	04-01	11-12	225	21	21	20	20
	24	03-23	11-18	240	20	20	20	20
	20	03-10	12-01	267	20	20	20	19
	16	02-26	12-17	288	20	20	19	14
JEMEZ SPRINGS	32	05-03	10-19	168	29	29	29	29
	28	04-22	10-28	189	29	29	29	29
	24	04-06	11-07	215	29	29	29	29
	20	03-27	11-17	235	29	29	29	29
	16	03-12	11-24	257	29	29	29	29
LAS VEGAS 2 NW	32	05-10	10-06	149	26	26	26	26
	28	04-29	10-17	172	25	25	26	26
	24	04-18	10-29	195	25	25	26	26
	20	04-09	11-03	208	25	25	25	25
	16	03-31	11-11	225	25	25	25	25
LOVINGTON 2 WNW	32	04-14	10-27	196	23	23	25	25
	28	04-05	11-04	213	23	23	24	24
	24	03-24	11-12	233	22	22	24	24
	20	03-12	11-21	253	22	22	24	24
	16	02-21	12-05	287	22	21	24	22
PASAMONTE	32	05-05	10-05	154	23	23	21	21
	28	04-25	10-19	177	21	21	21	21
	24	04-15	10-30	198	21	21	20	20
	20	04-04	11-05	215	22	22	20	20
	16	03-29	11-10	226	22	22	20	20
QUEMADO RS	32	06-09	10-02	115	14	14	16	16
	28	05-20	10-09	143	14	14	16	16
	24	05-07	10-22	168	15	15	16	16
	20	04-25	10-29	187	14	14	16	16
	16	04-16	11-06	204	14	14	15	15
RATON	32	05-20	09-26	129	30	30	30	30
	28	05-06	10-06	153	30	30	30	30
	24	04-24	10-18	177	30	30	30	30
	20	04-16	10-26	193	30	30	30	30
	16	04-07	11-07	214	30	30	30	30
ROSWELL WSO AP	32	04-09	11-02	208	28	28	29	29
	28	03-28	11-11	228	28	28	29	29
	24	03-16	11-17	246	28	28	29	29
	20	03-04	11-26	267	28	28	29	29
	16	02-13	12-10	300	28	25	29	22
SAN FIDEL	32	05-08	10-08	154	27	27	26	26
	28	04-28	10-21	176	27	27	26	26
	24	04-18	10-31	196	25	25	26	26
	20	03-29	11-06	223	25	25	26	26
	16	03-13	11-14	246	25	25	26	26

STATION	Freeze threshold temperature	Mean date of last Spring occurrence	Mean date of first Fall occurrence	Mean No. of days between dates	Years of record Spring	No. of occurrences in Spring	Years of record Fall	No. of occurrences in Fall
SANTA FE	32	05-03	10-15	165	26	26	28	28
	28	04-20	10-24	186	27	27	28	28
	24	04-07	11-05	211	27	27	28	28
	20	03-29	11-15	230	27	27	28	28
	16	03-17	11-23	251	27	27	28	28
SANTA ROSA	32	04-13	10-28	199	29	29	28	28
	28	03-30	11-06	221	29	29	28	28
	24	03-21	11-12	236	29	29	28	28
	20	03-09	11-23	259	29	29	28	28
	16	02-21	12-10	292	29	27	28	24
SHIPROCK	32	05-02	10-11	163	20	20	21	21
	28	04-22	10-22	183	20	20	21	21
	24	04-05	10-29	208	19	19	21	21
	20	03-25	11-08	228	19	19	21	21
	16	03-04	11-22	263	19	18	21	21
SOCORRO	32	04-17	10-23	189	29	29	29	29
	28	04-04	11-05	215	29	29	29	29
	24	03-24	11-10	231	29	29	29	29
	20	03-08	11-20	257	29	29	28	27
	16	02-13	12-02	292	27	25	28	24
SPRINGER	32	05-08	10-05	150	29	29	29	29
	28	04-25	10-15	173	29	29	29	29
	24	04-15	10-22	190	29	29	29	29
	20	04-08	10-29	203	28	28	29	29
	16	03-30	11-07	223	28	28	29	29
TAOS	32	05-22	09-30	130	26	26	26	26
	28	05-07	10-09	155	26	26	26	26
	24	04-25	10-18	177	26	26	26	26
	20	04-13	10-30	200	26	26	26	26
	16	04-01	11-08	221	26	26	26	26
TUCUMCARI	32	04-18	10-26	190	30	30	30	30
	28	04-03	11-06	217	30	30	30	30
	24	03-28	11-14	231	30	30	30	30
	20	03-15	11-19	248	30	30	30	30
	16	03-05	12-02	272	30	28	30	29

Data in the above table are based on the period 1921-1950, or that portion of this period for which data are available.

Means have been adjusted to take into account years of non-occurrence.

A freeze is a numerical substitute for the former term "killing frost" and is the occurrence of a minimum temperature at or below the threshold temperature of 32°, 28°, etc.

Freeze data tabulations in greater detail are available and can be reproduced at cost.

*NORMALS BY CLIMATOLOGICAL DIVISIONS

Taken from "Climatography of the United States No. 81-4, Decennial Census of U. S. Climate"

STATIONS (By Divisions)	TEMPERATURE (°F) JAN	FEB	MAR	APR	MAY	JUNE	JULY	AUG	SEPT	OCT	NOV	DEC	ANN	PRECIPITATION (In.) JAN	FEB	MAR	APR	MAY	JUNE	JULY	AUG	SEPT	OCT	NOV	DEC	ANN
NORTHWESTERN PLATEAU																											
AZTEC RUINS NAT MON	29.0	34.4	41.3	50.2	58.6	67.2	73.8	71.9	64.7	53.6	39.2	31.1	51.3	.74	.76	.79	.60	.67	.44	.94	1.32	1.12	1.09	.49	.86	9.82
BLOOMFIELD 3 SE	·	·	·	·	·	·	·	·	·	·	·	·	·	.52	.67	.59	.51	.63	.33	.81	1.32	.87	.93	.44	.60	8.22
ZUNI FAA AIRPORT	29.4	34.1	40.1	48.3	56.2	65.3	70.8	69.2	63.1	52.2	39.1	32.0	50.0	.79	.83	.77	.61	.45	.46	1.68	1.81	1.09	1.03	.62	.77	10.91
DIVISION	27.8	32.6	39.2	48.2	56.8	66.1	71.9	70.0	63.2	52.0	38.0	30.1	49.7	.76	.81	.79	.64	.64	.47	1.42	1.78	1.13	1.01	.60	.79	10.84
NORTHERN MOUNTAINS																											
ABBOTT 2 SE	·	·	·	·	·	·	·	·	·	·	·	·	·	.32	.23	.42	.77	2.19	1.63	2.51	3.23	1.42	1.29	.43	.25	14.69
BANDELIER NAT MON	·	·	·	·	·	·	·	·	·	·	·	·	·	.79	.78	.94	.86	1.36	.86	2.36	2.60	1.62	1.41	.80	.88	15.26
BLACK LAKE	·	·	·	·	·	·	·	·	·	·	·	·	·	.62	.55	.96	1.15	1.89	1.50	2.84	3.28	1.49	1.35	.62	.43	16.68
CERRO 4 NE	·	·	·	·	·	·	·	·	·	·	·	·	·	.64	.60	.58	.97	1.34	.87	1.70	1.92	1.43	1.19	.61	.61	12.46
CHACON	·	·	·	·	·	·	·	·	·	·	·	·	·	1.11	.78	1.07	1.22	1.82	1.46	2.97	3.73	1.75	1.29	.90	.91	19.01
CHAMA	·	·	·	·	·	·	·	·	·	·	·	·	·	1.80	1.67	1.51	1.37	1.40	.92	1.58	2.32	1.86	1.57	.96	1.63	18.59
CIMARRON	32.2	35.4	40.5	48.8	57.0	66.2	69.8	68.4	62.3	52.4	40.1	34.6	50.6	.42	.45	.67	1.20	2.21	1.41	2.22	2.53	1.36	1.24	.60	.34	14.65
DES MOINES	·	·	·	·	·	·	·	·	·	·	·	·	·	.50	.47	.88	1.46	2.55	1.79	3.07	3.36	1.64	1.15	.65	.39	17.91
EAGLE NEST	·	·	·	·	·	·	·	·	·	·	·	·	·	.73	.78	.92	1.06	1.62	1.08	2.48	2.55	1.27	1.14	.75	.59	14.97
EL RITO	·	·	·	·	·	·	·	·	·	·	·	·	·	.74	.67	.77	.75	1.16	.69	1.60	1.99	1.33	1.12	.55	.62	11.99
JEMEZ SPRINGS	32.9	36.4	42.1	50.6	58.2	66.8	70.8	69.3	63.9	53.7	41.8	35.0	51.8	.88	.95	1.05	1.01	1.32	1.33	2.46	3.16	2.02	1.75	.87	.88	17.68
LAS VEGAS 2 NW	32.3	35.8	40.5	48.6	56.5	66.0	69.0	68.1	61.8	52.3	40.3	34.5	50.5	.65	.47	.78	.91	1.82	1.76	2.58	2.74	2.16	1.09	.54	.44	15.94
LOS ALAMOS	·	·	·	·	·	·	·	·	·	·	·	·	·	.78	.64	.88	.92	1.35	1.38	2.73	3.92	1.89	1.65	.70	.83	17.67
PECOS RANGER STATION	·	·	·	·	·	·	·	·	·	·	·	·	·	.68	.68	.98	.80	1.26	1.19	2.79	3.07	1.73	1.15	.49	.68	15.50
PENASCO RANGER STATION	·	·	·	·	·	·	·	·	·	·	·	·	·	1.04	1.02	1.20	1.14	1.37	.99	1.98	2.10	1.34	1.32	.87	.82	15.19
RED RIVER	19.4	22.2	28.2	37.6	45.1	53.5	58.1	57.3	51.5	42.1	28.9	21.7	38.8	1.07	1.10	1.35	1.60	1.80	1.24	2.56	3.07	1.49	1.47	.98	.93	18.66
SANTA FE	29.9	33.7	39.2	48.1	56.7	66.5	70.5	68.7	62.9	52.0	39.0	32.1	49.9	.69	.69	.79	.83	1.37	1.14	2.19	2.27	1.44	1.05	.60	.70	13.76
TAOS	·	·	·	·	·	·	·	·	·	·	·	·	·	.82	.72	.79	.92	1.20	.76	1.58	1.77	1.11	1.13	.71	.57	12.08
TIERRA AMARILLA 4 NNW	·	·	·	·	·	·	·	·	·	·	·	·	·	1.41	1.06	1.04	1.15	1.10	.91	1.85	2.33	1.64	1.19	.88	1.24	15.80
TRES PIEDRAS	·	·	·	·	·	·	·	·	·	·	·	·	·	.85	.86	.76	.85	1.11	.69	1.93	2.32	1.36	1.05	.59	.80	13.17
VALMORA	30.7	34.0	38.7	47.2	55.3	64.4	68.7	67.4	61.4	51.1	38.6	33.3	49.2	.45	.31	.59	.89	1.93	1.74	2.79	2.89	1.82	1.30	.55	.41	15.67
WOLF CANYON	·	·	·	·	·	·	·	·	·	·	·	·	·	1.64	1.77	1.90	1.50	1.45	1.04	2.91	3.28	2.14	1.62	1.16	1.51	21.92
DIVISION	26.7	30.2	36.0	44.8	53.0	62.1	66.4	65.1	58.9	48.6	36.0	29.2	46.6	.81	.80	.95	1.12	1.65	1.24	2.39	2.76	1.64	1.28	.70	.73	16.07
NORTHEASTERN PLAINS																											
AMISTAD 1 SSW	·	·	·	·	·	·	·	·	·	·	·	·	·	.47	.43	.66	1.25	2.49	1.54	2.46	2.39	1.57	1.04	.44	.60	15.34
BELL RANCH	36.6	40.9	47.1	56.4	64.8	74.9	78.3	77.2	70.4	58.5	44.9	38.0	57.3	.33	.33	.63	1.05	1.81	1.60	2.47	2.78	1.68	1.03	.38	.43	14.52
CLAYTON WSO	33.2	36.0	40.9	50.7	59.8	70.2	74.5	73.3	66.0	55.3	41.7	36.0	53.1	.35	.37	.63	1.19	2.74	1.48	2.33	2.09	1.65	1.00	.33	.35	14.51
CLOVIS	37.4	41.3	47.2	56.7	65.7	75.4	78.5	77.4	70.2	59.3	46.1	39.4	57.9	.55	.41	.49	.92	2.51	2.30	2.64	2.62	2.10	1.89	.46	.59	17.48
ELIDA	·	·	·	·	·	·	·	·	·	·	·	·	·	.41	.35	.50	.77	1.86	1.63	2.64	2.38	2.16	1.47	.39	.53	15.09
MELROSE	·	·	·	·	·	·	·	·	·	·	·	·	·	.43	.36	.46	.77	2.06	1.63	2.64	2.53	1.73	1.63	.36	.53	15.13
MOSQUERO	33.6	37.0	41.9	50.8	59.4	69.1	73.0	71.3	64.6	54.6	42.2	35.4	52.7	.48	.41	.57	1.09	2.36	1.82	2.85	2.94	1.89	.98	.38	.53	16.25
NEWKIRK	·	·	·	·	·	·	·	·	·	·	·	·	·	.43	.48	.62	.88	1.74	1.64	2.13	2.30	1.69	.92	.39	.59	13.81
PASAMONTE	·	·	·	·	·	·	·	·	·	·	·	·	·	.37	.31	.49	1.02	2.12	1.80	2.49	2.54	1.59	1.02	.27	.31	14.33
PORTALES	37.3	41.8	48.2	57.2	65.8	75.1	77.5	76.5	69.5	58.8	45.7	39.1	57.7	.46	.36	.54	.77	2.42	2.38	2.91	2.66	1.91	1.50	.45	.58	16.94
ROY	·	·	·	·	·	·	·	·	·	·	·	·	·	.41	.43	.63	.91	2.50	1.62	2.32	2.22	1.72	1.40	.43	.53	15.12
SAN JON	37.6	42.0	48.1	57.3	65.9	76.1	79.2	78.0	71.3	59.9	46.3	39.5	58.4	.41	.38	.51	.89	2.28	1.67	2.81	2.62	1.33	1.17	.46	.65	15.18
SPRINGER	30.0	34.9	40.9	49.9	58.7	67.7	71.5	70.1	63.3	52.6	39.0	31.3	50.8	.46	.33	.59	1.11	1.96	1.68	2.39	3.00	1.76	1.35	.61	.35	15.59
STEAD	·	·	·	·	·	·	·	·	·	·	·	·	·	.34	.35	.53	1.37	2.77	1.57	2.69	2.33	1.69	1.01	.38	.40	15.43
TUCUMCARI 3 NE	37.7	41.3	47.5	57.0	65.7	75.8	79.1	77.8	71.0	60.0	46.5	39.4	58.2	.42	.43	.59	.89	2.27	1.61	2.75	2.50	1.35	1.13	.49	.64	15.07
DIVISION	35.3	39.1	45.0	54.3	62.9	72.6	76.1	74.9	68.0	57.2	44.0	37.4	55.6	.41	.37	.55	.99	2.29	1.80	2.70	2.53	1.69	1.25	.41	.48	15.47
SOUTHWESTERN MOUNTAIN																											
AUGUSTINE	·	·	·	·	·	·	·	·	·	·	·	·	·	.44	.52	.53	.47	.45	.64	2.06	2.24	1.44	.90	.27	.55	10.51
FORT BAYARD	37.7	41.0	45.5	53.0	60.7	70.2	72.7	70.9	66.4	56.7	45.4	40.1	55.0	.84	.97	.62	.41	.42	.77	2.79	3.23	1.97	1.01	.57	.81	14.41
LUNA RANGER STATION	·	·	·	·	·	·	·	·	·	·	·	·	·	1.04	.80	.76	.62	.48	.73	2.59	2.90	1.71	1.36	.59	.95	14.53
MIMBRES RANGER STATION	·	·	·	·	·	·	·	·	·	·	·	·	·	1.16	1.15	.87	.54	.45	.93	2.97	3.40	1.95	1.33	.67	1.05	16.47
PINOS ALTOS	·	·	·	·	·	·	·	·	·	·	·	·	·	1.33	1.70	1.18	.70	.46	1.06	3.27	4.17	2.04	1.41	.69	1.29	19.30
QUEMADO RANGER STATION	·	·	·	·	·	·	·	·	·	·	·	·	·	.73	.54	.62	.53	.47	.65	1.83	2.18	1.30	.66	.39	.50	10.40
RESERVE RANGER STATION	·	·	·	·	·	·	·	·	·	·	·	·	·	1.11	.85	1.12	.67	.42	.72	2.10	2.69	1.61	1.30	.61	.95	14.15
WINSTON	·	·	·	·	·	·	·	·	·	·	·	·	·	.55	.60	.41	.45	.52	.82	2.25	2.75	1.95	.97	.30	.57	12.14
DIVISION	31.5	35.3	40.6	48.4	56.4	65.6	69.5	67.5	61.9	51.8	39.7	33.5	50.1	.73	.67	.66	.54	.50	.75	2.27	2.71	1.61	.98	.46	.71	12.59

*NORMALS BY CLIMATOLOGICAL DIVISIONS

Taken from "Climatography of the United States No. 81-4, Decennial Census of U. S. Climate"

TEMPERATURE (°F)   PRECIPITATION (In.)

STATIONS (By Divisions)	JAN	FEB	MAR	APR	MAY	JUNE	JULY	AUG	SEPT	OCT	NOV	DEC	ANN	JAN	FEB	MAR	APR	MAY	JUNE	JULY	AUG	SEPT	OCT	NOV	DEC	ANN
CENTRAL VALLEY																										
ALBUQUERQUE WSO	35.0	39.9	45.8	55.7	65.1	74.9	78.5	76.2	70.0	58.0	43.6	37.0	56.6	.41	.38	.48	.47	.75	.57	1.20	1.33	.95	.75	.38	.46	8.13
BOSQUE DEL APACHE	•	•	•	•	•	•	•	•	•	•	•	•	•	.39	.38	.28	.26	.42	.76	1.10	1.46	1.25	.91	.19	.39	7.79
CARRIZOZO	37.0	40.9	46.8	55.0	63.6	72.9	75.4	73.8	68.0	57.2	44.0	38.2	56.1	.73	.75	.77	.67	.92	1.20	2.32	2.25	1.95	.99	.53	.77	13.85
ELEPHANT BUTTE DAM	41.3	46.3	52.3	60.8	69.1	78.3	80.3	78.5	73.0	62.7	49.5	42.5	61.2	.37	.45	.31	.33	.34	.68	1.50	1.95	1.22	.74	.20	.46	8.55
SOCORRO	37.5	43.0	49.6	58.5	66.5	75.8	79.0	77.1	70.5	59.2	45.5	38.3	58.4	.43	.44	.33	.43	.70	.66	1.34	1.48	1.32	.88	.23	.51	8.75
DIVISION	36.2	41.0	47.4	56.2	64.6	73.8	77.4	75.7	69.0	57.9	44.3	37.3	56.7	.41	.44	.41	.45	.61	.72	1.46	1.61	1.20	.81	.32	.47	8.91
CENTRAL HIGHLANDS																										
ANCHO	•	•	•	•	•	•	•	•	•	•	•	•	•	.95	.74	.93	.84	1.03	1.14	2.06	2.14	1.75	1.03	.59	.99	14.19
CLOUDCROFT LODGE	•	•	•	•	•	•	•	•	•	•	•	•	•	1.80	1.70	1.63	.85	1.09	1.86	5.79	4.73	2.37	1.61	.85	1.46	25.74
CORONA	33.9	37.0	42.4	50.2	58.4	67.4	69.9	68.6	63.2	53.7	42.0	35.5	51.9	.84	.80	.82	1.04	1.37	1.27	2.59	2.78	2.01	1.09	.56	.70	15.87
GRAN QUIVIRA NAT MON	•	•	•	•	•	•	•	•	•	•	•	•	•	.67	.65	.69	.61	.91	1.18	2.37	2.98	1.61	1.00	.49	.72	13.88
MAYHILL RANGER STATION	•	•	•	•	•	•	•	•	•	•	•	•	•	.75	.73	.79	.59	1.23	1.90	3.53	3.81	2.94	1.45	.38	.80	18.90
MC INTOSH 4 NW	30.2	34.7	40.5	49.0	57.4	66.6	70.1	68.5	62.2	51.6	39.0	32.4	50.2	.48	.44	.53	.72	1.10	1.03	2.17	2.53	1.46	1.11	.39	.50	12.46
MESCALERO	•	•	•	•	•	•	•	•	•	•	•	•	•	1.05	1.03	.91	.65	.87	1.42	3.59	3.71	2.13	1.27	.67	1.07	18.37
MOUNTAIN PARK	•	•	•	•	•	•	•	•	•	•	•	•	•	1.16	1.05	.97	.57	.73	1.28	3.27	3.26	1.89	1.41		.96	17.16
MOUNTAINAIR	•	•	•	•	•	•	•	•	•	•	•	•	•	.81	.58	.67	.67	.93	1.01	2.69	2.43	1.39	1.15	.55	.88	13.76
PROGRESSO	•	•	•	•	•	•	•	•	•	•	•	•	•	.55	.55	.58	.65	.88	1.13	2.84	2.51	1.56	.87	.44	.80	13.36
DIVISION	31.7	35.3	40.5	48.6	56.7	65.6	68.5	67.1	61.5	51.7	39.8	33.8	50.1	.90	.84	.93	.79	1.10	1.25	2.94	2.95	1.86	1.20	.61	.92	16.29
SOUTHEASTERN PLAINS																										
ARTESIA 6 S	40.8	44.9	51.8	60.9	69.4	78.4	80.0	79.4	72.7	62.1	48.8	41.8	60.9	.47	.35	.50	.54	1.47	1.32	1.78	1.43	1.62	1.13	.29	.43	11.33
CARLSBAD	•	•	•	•	•	•	•	•	•	•	•	•	•	.45	.37	.46	.54	1.76	1.33	1.56	1.60	1.94	1.62	.35	.47	12.45
CARLSBAD CAVERNS	45.6	48.8	53.7	62.9	70.5	78.2	78.9	78.4	72.9	64.4	53.6	47.4	62.9	.52	.43	.45	.76	1.51	1.55	2.01	1.92	2.51	1.71	.37	.55	14.29
CROSSROADS 2 NE	•	•	•	•	•	•	•	•	•	•	•	•	•	.44	.42	.47	.77	2.13	1.86	2.39	2.59	2.39	1.76	.45	.52	16.19
ELK 3 E	38.0	40.5	45.4	52.5	59.9	67.8	69.6	68.7	63.5	54.9	44.6	39.7	53.8	.61	.50	.61	.59	1.33	1.64	2.53	3.00	2.61	1.36	.42	.64	15.84
FLYING H	•	•	•	•	•	•	•	•	•	•	•	•	•	.50	.48	.55	.56	1.30	1.31	2.15	2.34	2.52	1.37	.44	.52	14.04
FORT SUMNER	•	•	•	•	•	•	•	•	•	•	•	•	•	.48	.38	.56	.87	1.64	1.12	2.53	2.30	1.80	1.28	.34	.44	13.74
LAKE AVALON	•	•	•	•	•	•	•	•	•	•	•	•	•	.40	.33	.37	.47	1.92	1.28	1.47	1.41	1.80	1.54	.34	.40	11.73
ROSWELL WSO	37.9	42.1	49.5	59.0	68.0	77.1	78.6	76.6	69.7	59.0	45.9	39.0	58.5	.48	.42	.50	.73	1.28	1.05	1.77	1.62	1.82	1.07	.34	.54	11.62
SANTA ROSA	38.5	42.3	47.9	56.8	65.2	74.4	77.3	76.1	69.2	58.5	46.4	40.5	57.8	.43	.42	.63	.72	1.77	1.36	2.47	2.47	1.63	1.17	.34	.51	13.92
DIVISION	40.2	44.1	50.1	58.7	66.9	76.0	77.9	77.1	70.6	60.6	48.0	41.8	59.3	.47	.39	.49	.66	1.64	1.33	2.13	2.02	1.97	1.34	.36	.50	13.30
SOUTHERN DESERT																										
ALAMOGORDO	•	•	•	•	•	•	•	•	•	•	•	•	•	.73	.55	.39	.34	.48	.72	1.61	1.66	1.46	.88	.40	.58	9.80
ANIMAS	•	•	•	•	•	•	•	•	•	•	•	•	•	.61	.62	.53	.22	.23	.49	1.79	2.39	1.25	.85	.41	.60	9.99
DEMING	40.7	45.3	51.0	59.2	67.6	77.4	80.5	78.7	72.8	62.0	49.0	42.2	60.5	.46	.59	.32	.30	.25	.50	1.62	1.70	1.46	.80	.30	.54	8.84
GAGE 4 ESE	•	•	•	•	•	•	•	•	•	•	•	•	•	.58	.75	.42	.26	.20	.41	1.43	1.97	1.49	.82	.38	.66	9.37
HACHITA 1 N	41.1	45.3	50.8	58.7	66.7	76.1	78.6	76.7	71.5	61.0	48.8	42.4	59.8	.58	.67	.45	.21	.17	.40	1.76	2.33	1.10	.86	.36	.65	9.54
HATCH 2 W	•	•	•	•	•	•	•	•	•	•	•	•	•	.45	.40	.28	.37	.29	.53	1.69	1.81	1.37	.98	.23	.54	8.94
HILLSBORO	•	•	•	•	•	•	•	•	•	•	•	•	•	.62	.58	.46	.41	.42	.71	1.89	2.00	1.87	1.06	.30	.66	10.98
JORNADA EXP RANGE	39.3	43.8	49.7	57.9	66.0	75.7	79.3	77.5	71.2	60.0	46.5	39.9	58.9	.54	.42	.27	.24	.29	.53	1.60	1.71	1.44	.91	.36	.55	8.86
LATHAM RANCH	•	•	•	•	•	•	•	•	•	•	•	•	•	.79	.91	.66	.48	.37	.86	2.66	2.44	1.96	1.06	.47	.79	13.45
LORDSBURG 4 SE	•	•	•	•	•	•	•	•	•	•	•	•	•	.79	.90	.61	.33	.13	.45	1.51	2.21	1.29	.75	.49	.64	10.10
OROGRANDE	42.3	46.9	53.5	61.7	70.8	80.7	82.4	80.5	74.3	63.7	49.3	43.0	62.4	.50	.39	.28	.33	.43	.84	1.52	1.57	1.26	.92	.30	.45	8.79
STATE UNIVERSITY	40.9	45.4	51.1	59.3	67.3	76.6	79.5	77.8	71.5	61.0	47.7	42.0	60.0	.47	.51	.30	.17	.31	.53	1.29	1.68	1.22	.75	.30	.48	8.01
DIVISION	41.2	45.4	51.2	59.3	67.4	76.8	79.6	77.8	72.2	61.7	48.6	42.4	60.3	.63	.61	.44	.31	.30	.56	1.77	1.95	1.42	.90	.38	.64	9.91

* Normals for the period 1931-1960. Divisional normals may not be the arithmetical average of individual stations published, since additional data for shorter period stations are used to obtain better areal representation.

COMPARATIVE DATA

Data in the following table are the mean temperature and average precipitation for New Mexico State University, New Mexico, for the period 1906-1930 and are included in this publication for comparative purposes.

| 41.9 | 46.2 | 50.0 | 59.1 | 66.7 | 76.5 | 79.3 | 77.2 | 71.5 | 60.5 | 48.8 | 41.1 | 59.9 | .28 | .31 | .43 | .24 | .36 | .43 | 1.49 | 1.65 | .93 | .71 | .55 | .55 | 7.93 |

CONFIDENCE - LIMITS

In absence of trend or record changes, the chances are 9 out of 10 that the true mean will lie in the interval formed by adding and subtracting the values in the following table from the means for any station in the State. Because of the wider variation in mean precipitation, the corresponding monthly means and annual mean must be substituted for "p" in the precipitation table below to obtain mean precipitation confidence limits.

| 1.1 | 1.1 | .8 | .9 | .7 | .6 | .5 | .5 | .5 | .7 | .9 | 1.2 | .4 | $.22\sqrt{p}$ | $.19\sqrt{p}$ | $.21\sqrt{p}$ | $.27\sqrt{p}$ | $.35\sqrt{p}$ | $.31\sqrt{p}$ | $.29\sqrt{p}$ | $.26\sqrt{p}$ | $.39\sqrt{p}$ | $.26\sqrt{p}$ | $.22\sqrt{p}$ | $.21\sqrt{p}$ | $.27\sqrt{p}$ |

NORMALS, MEANS AND EXTREMES

(Table Revised 1972. Base Period for Climatological Normals: 1931-1960)

Station: ALBUQUERQUE, NEW MEXICO SUNPORT-KIRTLAND AFB   Standard time used: MOUNTAIN   Latitude: 35° 03' N   Longitude: 106° 37' W   Elevation (ground): 5311 feet

Month	Normal Daily max	Normal Daily min	Normal Monthly	Ext. Record highest	Year	Record lowest	Year	Normal heating degree days (Base 65°)	Precip Normal total	Max monthly	Year	Min monthly	Year	Max 24 hrs	Year	Snow Mean total	Max monthly	Year	Max 24 hrs	Year	RH 05	RH 11	RH 17	RH 23	Wind Mean speed	Prevail dir	Fastest mile speed	dir	Year	Pct sunshine	Mean sky cover	Clear	Ptly cldy	Cldy	Precip .01+	Snow 1.0+	T-storms	Hvy fog	Max 90+	Max 32−	Min 32−	Min 0−	Solar rad. (ly)
(yrs)	(b)	(b)	(b)	12		12		(b)	32	32		32		32		32	32		32		11	11	11	11	32	15	32	32		32	32	32	32	32	32	32	32	32	32	11	11	11	11	23
J	46.4	23.5	35.0	69	1971	−17	1971	930	0.41	1.17	1941	T	1970	0.87	1962	1.9	6.0	1951	4.6	1962	66	47	36	57	7.8	N	61	E	1949	73	4.8	13	8	10	3	3	*	1	0	3	29	1	301
F	52.2	27.5	39.9	72	1963	9	1964	703	0.38	1.42	1948	0.04	1959	0.48	1957	1.7	8.2	1946	4.2	1964	62	42	31	51	8.2	N	68	NW	1946	74	5.0	12	7	9	4	4	*	1	0	1	23	*	384
M	58.8	32.7	45.8	85	1971	9	1948	595	0.48	1.71	1973	T	1971	0.77	1968	1.7	7.3	1958	3.0	1958	53	32	24	43	10.0	NE	80	NW	1943	76	5.0	12	9	10	4	3	1	1	0	*	17	0	504
A	69.1	42.2	55.7	89	1965	23	1973	288	0.47	1.97	1942	T	1967	1.14	1969	0.3	4.0	1949	T	1951	42	25	17	30	10.4	SE	72	W	1942	77	4.6	15	10	8	4	*	1	*	*	0	4	0	606
M	78.3	51.9	65.1	94	1969+	28	1967	81	0.57	3.07	1941	T	1945+	1.64	1951	T	T	1951	T	1951	42	23	16	29	10.4	SE	72	W	1950	83	4.1	18	10	6	4	0	2	*	3	0	0	0	694
J	88.6	61.1	74.9	102	1969+	42	1952	0	0.57	1.71	1969	T	1944	1.64	1969	0.0	0.0		0.0		44	24	17	30	9.9	S	82	SE	1946	83	3.2	18	8	4	4	0	5	0	16	0	0	0	729
J	91.2	65.6	78.4	104	1971	54	1964	0	1.20	3.33	1968	0.14	1958	1.77	1958	0.0	0.0		0.0		61	36	28	48	9.0	SE	68	E	1945	76	4.5	14	13	4	9	0	12	0	24	0	0	0	675
A	88.0	64.3	76.2	99	1969	52	1968	0	1.33	3.30	1967	T	1962	1.22	1961	0.0	0.0		0.0		65	40	31	52	8.1	SE	61	SE	1951	81	4.1	16	11	4	10	0	12	0	16	0	0	0	620
S	83.4	58.0	70.7	97	1971	37	1971+	12	0.95	1.99	1941	T	1967	1.92	1957	0.5	0.5	1971+	0.5	1971	57	36	28	46	8.5	SE	66	NW	1959	81	3.4	18	7	5	5	*	3	*	3	0	2	0	537
O	70.7	45.3	58.0	87	1963	25	1973	229	0.75	2.88	1960	0.00	1952	1.80	1969	1.2	9.3	1940	5.5	1940	63	43	35	53	7.7	N	57	NW	1948+	78	3.4	18	8	5	4	*	1	1	0	*	14	0	435
N	56.1	31.6	43.9	72	1971+	13	1968	642	0.38	1.45	1940	0.00	1949	0.76	1940	2.8	9.3	1957	5.5	1957	69	52	44	60	7.5	N	57	SE	1943	78	3.9	14	8	8	4	1	*	1	0	*	28	*	323
D	46.3	25.6	36.0	68	1966	4	1968	868	0.46	1.85	1958	0.00	1959	1.35	1958	2.8	14.7	1959	14.2	1959	69	52	44	60	7.5	N	57	SE	1943	71	4.6	14	8	9	4	1	*	1	0	1	28	1	270
YR	69.2	44.1	56.6	104	JUL 1971	−17	JAN 1971	4348	8.13	3.33	JUL 1968	0.00	DEC 1959	1.92	SEP 1955	9.7	14.7	DEC 1959	14.2	DEC 1959	57	37	28	46	8.9	SE	90	SE	DEC 1943	77	4.2	175	110	80	58	4	43	5	61	6	120	1	507

Means and extremes above are from existing and comparable exposures. Annual extremes have been exceeded at other sites in the locality as follows: Maximum monthly precipitation 8.15 in June 1852 (measured by Medical Officers of Army at Army Post near plaza).

To 8 compass points only.

NORMALS, MEANS AND EXTREMES

(Table Revised 1975. Base Period for Climatological Normals: 1941-1970)

Elev. 5314 feet m.s.l.   Average station pressure mb. 838.1 (YR)

Month	Normal Daily max	Normal Daily min	Normal Monthly	Ext. Record highest	Year	Record lowest	Year	Normal degree days Base 65° Heating	Cooling	Precip Normal	Max monthly	Year	Min monthly	Year	Max 24 hrs	Year	Snow Max monthly	Year	Max 24 hrs	Year	RH 05	RH 11	RH 17	RH 23	Wind Mean speed mph	Prevail dir	Fastest mile mph	dir	Year	Pct sunshine	Mean sky cover tenths	Clear	Ptly cldy	Cldy	Precip .01+	Snow 1.0+	T-storms	Hvy fog ≤¼ mi	Max 90+	Max 32−	Min 32−	Min 0−
(yrs)	(a)			15		15				35	35		35		35		35		35		14	14	14	14	35	15	35	35		35	35	35	35	35	35	35	35	35	14	14	14	14
J	46.9	23.5	35.2	69	1971	−17	1971	924	0	0.30	1.17	1941	T	1970	0.87	1962	9.5	1973	5.1	1973	68	48	37	58	7.9	N	61	E	1949	73	4.8	13	8	10	4	3	*	1	0	3	29	1
F	52.6	27.4	40.0	75	1972	−7	1964	700	0	0.39	1.42	1948	T	1959	0.48	1957	8.2	1964	4.2	1964	65	42	32	54	8.8	N	68	NW	1946	74	5.0	12	7	9	4	3	*	1	0	1	24	*
M	59.2	31.5	45.4	85	1971	9	1966	595	0	0.47	2.18	1973	0.04	1966	1.11	1973	13.1	1973	10.7	1973	54	32	23	41	10.0	SE	80	NW	1943	74	5.0	13	9	9	4	3	1	1	0	*	17	0
A	70.1	41.4	55.8	89	1965	22	1973	282	6	0.48	1.97	1942	T	1972	1.66	1972	6.6	1973	6.6	1973	44	25	16	30	10.5	SE	72	W	1942	77	4.5	15	10	5	4	*	2	*	*	0	5	0
M	79.1	50.9	65.0	95	1974	28	1967	58	67	0.53	3.07	1941	T	1969	1.14	1951	0.0		0.0		43	23	16	30	10.5	SE	72	W	1950	80	4.1	18	10	3	4	0	4	*	5	0	*	0
J	89.5	59.7	74.6	102	1974	42	1964	0	291	0.50	1.71	1969	T	1964	1.64	1951	0.0		0.0		44	23	17	32	10.0	S	82	SE	1946	83	3.2	18	9	3	4	0	5	0	16	0	0	0
J	92.2	65.4	78.8	104	1971	54	1964	0	425	1.39	3.33	1968	0.14	1958	1.77	1958	0.0		0.0		61	35	28	48	9.1	SE	68	E	1945	76	4.5	14	12	5	9	0	12	0	23	0	0	0
A	88.7	63.4	76.1	99	1971	52	1968	0	360	1.34	3.30	1967	T	1962	1.22	1961	0.0		0.0		65	39	30	52	8.5	SE	61	SE	1951	81	4.1	14	12	5	10	0	12	0	16	0	0	0
S	83.4	56.7	70.1	95	1957	37	1971	7	160	0.77	1.99	1957	T	1957	1.80	1969	0.5	1971	0.5	1971	59	38	30	54	8.3	SE	66	NW	1959	81	3.5	16	10	5	5	*	3	*	3	0	2	0
O	71.7	44.5	58.1	87	1952	25	1973	218	17	0.79	3.08	1952	0.00	1949	1.80	1970	9.3	1945	5.5	1945	61	43	35	54	7.9	N	57	NW	1959	79	3.5	16	9	6	5	*	1	1	0	*	15	0
N	57.1	31.8	44.5	73	1940	13	1959	615	0	0.29	1.45	1940	0.00	1949	0.76	1940	9.3	1957	7.6	1958	69	51	43	61	7.6	N	57	SE	1943	77	3.9	15	8	7	3	1	*	1	0	*	29	*
D	47.5	24.9	36.2	68	1966	3	1958	893	0	0.52	1.85	1959	0.00	1956	1.35	1958	14.7	1959	14.2	1959					7.6	N	90	SE	1943	72	4.6	14	8	9	4	1	*	1	0	1	29	1
YR	70.0	43.5	56.8	105	JUN 1974	−17	JAN 1971	4292	1316	7.77	3.33	JUL 1968	0.00	DEC 1956	1.92	SEP 1955	14.7	DEC 1959	14.2	DEC 1959	58	37	28	47	8.9	SE	90	SE	DEC 1943	77	4.2	175	108	82	59	4	43	5	62	6	120	1

Means and extremes above are from existing and comparable exposures. Annual extremes have been exceeded at other sites in the locality as follows: Maximum monthly precipitation 8.15 in June 1852 (measured by Medical Officers of Army at Army Post near plaza).

REFERENCE NOTES APPLYING TO TABLES APPEAR ON THE PAGE FOLLOWING LAST TABLE.

(Caution: Letters and symbols may have different meanings in 1941-1970 tables than in earlier tables. See notes.)

NORMALS, MEANS AND EXTREMES
(Table Revised 1972. Base Period for Climatological Normals: 1931-1960)

Station: CLAYTON, NEW MEXICO   MUNICIPAL AIRPARK   Standard time used: MOUNTAIN   Latitude: 36° 27' N   Longitude: 103° 09' W   Elevation (ground): 4969 feet

Month	Temp Normal Daily max	Temp Normal Daily min	Temp Normal Monthly	Extremes Record highest	Year	Extremes Record lowest	Year	Normal heating degree days (Base 65°)	Precip Normal total	Precip Max monthly	Year	Precip Min monthly	Year	Precip Max in 24 hrs	Year	Snow Mean total	Snow Max monthly	Year	Snow Max in 24 hrs	Year	Mean sky cover sunrise to sunset
J	46.2	20.2	33.2	78	1956	-21	1959	986	0.35	1.06	1960	T	1970	0.71	1960	3.0	11.2	1960	7.0	1945	5.1
F	49.0	22.9	36.0	81	1963	-17	1951	812	0.37	1.07	1948	T	1950	0.75	1948	3.3	8.1	1953	7.5	1953	5.2
M	55.0	26.7	40.9	85	1971+	-11	1948	747	0.40	1.07	1957	0.01	1966	0.90	1959	5.0	16.0	1958	8.6	1958	5.2
A	64.6	36.7	50.7	91	1965	17	1945	429	1.19	4.67	1944	0.09	1945	2.30	1944	1.7	10.9	1955	10.9	1955	5.4
M	73.2	46.4	59.8	95	1964	23	1967	183	2.74	4.67	1949	0.35	1955	1.61	1944	T	0.6	1969+	0.0	1969+	5.2
J	84.1	56.2	70.2	104	1968	37	1964	21	1.48	4.51	1950	0.19	1955	2.48	1950	0.0	0.0		0.0		4.2
J	88.4	60.6	74.5	102	1964+	45	1958	0	2.33	7.77	1946	0.66	1950	2.59	1950	0.0	0.0		0.0		4.7
A	86.5	60.0	73.3	102	1964	45	1964	0	2.09	5.60	1962	0.54	1963	4.41	1963	0.0	0.0		0.0		4.7
S	79.5	52.5	66.0	99	1948+	28	1945	66	1.65	5.22	1960	T	1960	3.48	1960	0.1	1.1	1971	1.1	1971	3.7
O	69.3	40.9	55.3	90	1973	17	1969	310	1.00	3.84	1952	T	1965	3.90	1961	0.5	5.2	1970	2.6	1960	4.4
N	55.4	27.7	41.7	81	1971	-1	1951	699	0.33	1.58	1966+	T	1961	1.58	1961	2.5	14.8	1961	12.7	1961	4.3
D	49.3	22.6	36.0	83	1955	-4	1949	899	0.35	1.10	1957	0.00	1960	0.71	1958	2.9	9.0	1947	8.1	1958	4.6
YR	66.7	39.5	53.1	104	JUN 1968	-21	JAN 1959	5158	14.51	7.77	JUL 1950	0.00	DEC 1957	4.67	MAY 1954	19.0	16.0	MAR 1958	12.7	NOV 1961	4.6

Relative humidity pct. (Local time) — Hour 05 / 11 / 17 / 23:

Month	05	11	17	23
J	63	47		
F	66	43		
M	66	40		
A	64	38		
M	64	34		
J	72	32		
J	77	42		
A	79	42		
S	73	41		
O	64	35		
N	63	41		
D	68	40	41	

Mean number of days (Clear / Partly cloudy / Cloudy; Precip .01"+; Snow 1.0"+; Thunderstorms; Heavy fog; Max 90°+; Max 32°-; Min 32°-; Min 0°-):

Month	Clear	Partly	Cloudy	Precip .01"+	Snow 1.0"+	Thunderstorms	Heavy fog	Max 90°+	Max 32°-	Min 32°-	Min 0°-
J	12	7	12	3	1	*	1	0	4	29	2
F	11	7	10	4	1	*	1	0	3	25	*
M	11	9	11	5	1	1	2	0	2	24	*
A	12	9	10	5	*	2	1	*	*	7	0
M	11	9	10	8	0	8	1	1	0	1	0
J	13	11	6	8	0	9	1	9	0	0	0
J	13	11	7	11	0	14	1	14	0	0	0
A	15	10	6	11	0	14	1	10	0	0	0
S	18	6	6	5	0	5	*	3	0	*	0
O	15	7	9	4	*	1	1	*	*	5	*
N	15	6	9	3	1	*	1	0	1	21	*
D	13	7	10	3	1	*	1	0	3	28	3
YR	163	100	102	67	7	54	11	37	15	143	3

Ø Extremes are for period April 1944 through the current year. Means and extremes above are from existing and comparable exposures. Annual extremes have been exceeded at other sites in the locality as follows: Highest temperature 105 in July 1934 and earlier; maximum monthly precipitation 10.51 in July 1899; maximum precipitation in 24 hours 6.20 in April 1914; maximum monthly snowfall 21.0 in December 1918.

% Through 1964, the station did not operate 24 hours daily. Fog and thunderstorm data may be incomplete.

NORMALS, MEANS AND EXTREMES
(Table Revised 1975. Base Period for Climatological Normals: 1941-1970)

Month	Temp Normal Daily max	Temp Normal Daily min	Temp Normal Monthly	Extremes Record highest	Year	Extremes Record lowest	Year	Degree days Base 65°F Heating	Cooling	Precip Normal	Precip Max monthly	Year	Precip Min monthly	Year	Precip Max in 24 hrs	Year	Water equiv Max monthly	Year	Water equiv Min monthly	Year	Snow Max monthly	Year	Snow Max in 24 hrs	Year	Mean sky cover sunrise to sunset
J	47.5	18.6	33.1	78	1956	-21	1959	989	0	0.28	1.06	1970	T	1946	0.71	1945	11.2	1960	7.0	1945	5.1				
F	50.3	21.4	36.1	81	1963	-17	1951	809	0	0.36	1.07	1950	T	1962	0.75	1953	8.1	1953	7.5	1953	5.3				
M	55.6	25.9	40.8	85	1971	-11	1948	431	0	0.67	2.50	1957	0.01	1957	0.90	1958	16.0	1958	10.0	1958	5.2				
A	65.6	36.0	50.8	91	1965	17	1945	172	5	1.27	4.67	1944	0.08	1944	1.64	1944	10.9	1955	10.0	1955	5.1				
M	74.4	45.6	60.0	95	1967	23	1967	38	17	1.64	6.77	1974	0.19	1974	2.48	1974	0.6	1969	0.0	1969	4.8				
J	83.4	54.9	69.2	104	1968	37	1964	0	164	1.75	4.51	1955		1955	2.48	1955	0.0		0.0		4.3				
J	87.3	59.9	73.6	102	1964	45	1958	0	271	2.76	7.77	1950	0.66	1950	2.59	1950	0.0		0.0		3.8				
A	85.7	59.3	72.5	102	1964	45	1963	0	234	2.67	5.60	1963	0.54	1963	4.41	1963	0.0		0.0		3.8				
S	78.9	50.1	65.0	99	1948	28	1960	73	73	1.77	5.22	1960	T	1960	3.48	1965	1.1	1971	1.1	1971	3.6				
O	69.3	40.2	54.7	90	1973	17	1969	324	0	1.11	3.91	1965	T	1965	3.90	1966	5.2	1970	1.58	1960	4.4				
N	56.9	27.7	42.3	81	1971	-1	1951	681	0	0.33	1.84	1966	T	1960	1.58	1947	14.8	1961	12.7	1961	4.6				
D	49.3	20.9	35.1	83	1955	-6	1947	927	0	0.30	1.10	1957	0.00	1957	0.71		9.0	1947	8.1	1958	4.6				
YR	67.0	38.4	52.7	104	JUN 1968	-21	JAN 1959	5212	772	15.91	7.77	JUL 1950	0.00	DEC 1957	4.67	MAY 1954	16.0	MAR 1958	12.7	NOV 1961	4.6				

Relative humidity pct. (Local time) — Hour 05 / 11 / 17 / 23:

Month	05	11	17	23
J	63	42	47	
F	65	42	43	
M	65	40	37	
A	65	34	36	
M	65	34	35	
J	72	38	35	
J	77	42	42	
A	79	42	42	
S	74	35	40	
O	65	40	47	
N	65	40	52	
D	69	40	41	

Mean number of days:

Month	Clear	Partly	Cloudy	Precip .01"+	Snow 1.0"+	Thunderstorms	Heavy fog, visibility ½ mile or less	Max 90°+	Max 32°-	Min 32°-	Min 0°-
J	12	6	12	3	1	*	1	0	3	29	2
F	11	7	10	4	1	*	1	0	3	25	*
M	11	9	11	5	1	1	2	0	2	24	*
A	12	9	10	5	*	2	1	*	1	10	0
M	11	9	10	8	0	8	1	1	0	1	0
J	13	11	6	8	0	10	1	9	0	0	0
J	12	12	7	11	0	14	1	13	0	0	0
A	15	10	6	11	0	11	*	9	0	0	0
S	17	6	7	5	0	5	1	3	0	*	0
O	13	7	4	4	*	1	1	*	1	5	*
N	13	6	4	3	1	*	1	0	1	21	*
D	14	7	10	3	1	*	1	0	3	28	3
YR	162	98	105	68	7	54	11	36	15	143	3

Means and extremes above are from existing and comparable exposures. Annual extremes have been exceeded at other sites in the locality as follows: Highest temperature 105 in July 1934 and earlier; maximum monthly precipitation 10.51 in July 1899; maximum precipitation in 24 hours 6.20 in April 1914; maximum monthly snowfall 21.0 in December 1918.

REFERENCE NOTES APPLYING TO TABLES APPEAR ON THE PAGE FOLLOWING LAST TABLE.
(Caution: Letters and symbols may have different meanings in 1941-1970 tables than in earlier tables. See notes.)

NORMALS, MEANS AND EXTREMES

(Table Revised 1972. Base Period for Climatological Normals: 1931-1960)

Station: ROSWELL, NEW MEXICO   MUNICIPAL AIRPORT   Standard time used: MOUNTAIN   Latitude: 33° 24' N   Longitude: 104° 32' W   Elevation (ground): 3612 feet

Ø June 1950 through August 1960 and February 1969 to date.   % September 1950 through August 1960 and January 1969 to date.

Means and extremes above are from existing and comparable exposures. Annual extremes have been exceeded at other sites in the locality as follows:
Highest temperature 110 in July 1958; lowest temperature -29 in February 1905; maximum monthly precipitation 9.56 in August 1916; maximum
precipitation in 24 hours 5.65 in November 1901; maximum snowfall 23.3 in February 1905.

To 8 compass points only.   ¢ Through 1968.   †† Normals applicable to hygrothermometer site on field.

NORMALS, MEANS AND EXTREMES

(Table Revised 1975. Base Period for Climatological Normals: 1941-1970)

Station: ROSWELL, NEW MEXICO   ROSWELL INDUSTRIAL AIR CENTER   Standard time used: MOUNTAIN   Latitude: 33° 18' N   Longitude: 104° 32' W   Elevation (ground): 3649 feet

Means and extremes above are from existing and comparable exposures. Annual extremes have been exceeded at other sites in the locality as follows:
Highest temperature 110 in July 1958; lowest temperature -29 in February 1905; maximum monthly precipitation 9.56 in August 1916; maximum precipitation in 24 hours 5.65 in November 1901; minimum monthly precipitation 0.00 in February 1972; maximum monthly snowfall 23.3 in February 1905;
maximum snowfall in 24 hours 15.3 in December 1960; fastest mile wind 75 from West in April 1953.

REFERENCE NOTES APPLYING TO TABLES APPEAR ON THE PAGE FOLLOWING LAST TABLE.
(Caution: Letters and symbols may have different meanings in 1941-1970 tables than in earlier tables. See notes.)

Reference notes applying to Normals, Means, and Extremes tables for 1931–1960 base period.

(a) Length of record, years, based on January data. Other months may be for more or fewer years if there have been breaks in the record. Climatological standard normals (1931-1960).

(b) Less than one half.

* Also on earlier dates, months, or years.

T Trace, an amount too small to measure.

Below zero temperatures are preceded by a minus sign.

The prevailing direction for wind in the Normals, Means, and Extremes table is from records through 1963.

‡ ≥ 70° at Alaskan stations.

Unless otherwise indicated, dimensional units used in this bulletin are: temperature in degrees F.; precipitation, including snowfall, in inches; wind movement in miles per hour; and relative humidity in percent. Heating degree day totals are the sums of negative departures of average daily temperatures from 65° F. Cooling degree day totals are the sums of positive departures of average daily temperatures from 65° F. Sleet was included in snowfall totals beginning with July 1948. The term "Ice pellets" includes solid grains of ice (sleet) and particles consisting of snow pellets encased in a thin layer of ice. Heavy fog reduces visibility to 1/4 mile or less.

Sky cover is expressed in a range of 0 for no clouds or obscuring phenomena to 10 for complete sky cover. The number of clear days is based on average cloudiness 0-3, partly cloudy days 4-7, and cloudy days 8-10 tenths.

Solar radiation data are the averages of direct and diffuse radiation on a horizontal surface. The langley denotes one gram calorie per square centimeter.

& Figures instead of letters in a direction column indicate direction in tens of degrees from true North; i.e., 09 - East, 18 - South, 27 - West, 36 - North, and 00 - Calm. Resultant wind is the vector sum of wind directions and speeds divided by the number of observations. If figures appear in the direction column under "Fastest mile" the corresponding speeds are fastest observed 1-minute values.

Reference notes applying to Normals, Means, and Extremes tables for 1941–1970 base period.

(a) Length of record, years, through the current year unless otherwise noted, based on January data.

(b) 70° and above at Alaskan stations.

* Less than one half.

T Trace.

NORMALS - Based on record for the 1941-1970 period.

DATE OF AN EXTREME - The most recent in cases of multiple occurrence.

PREVAILING WIND DIRECTION - Record through 1963.

WIND DIRECTION - Numerals indicate tens of degrees clockwise from true north. 00 indicates calm.

FASTEST MILE WIND - Speed is fastest observed 1-minute value when the direction is in tens of degrees.

% Through 1964. The station did not operate 24 hours daily. Fog and thunderstorm data may be incomplete.

MEAN MAXIMUM TEMPERATURE (°F.), JANUARY

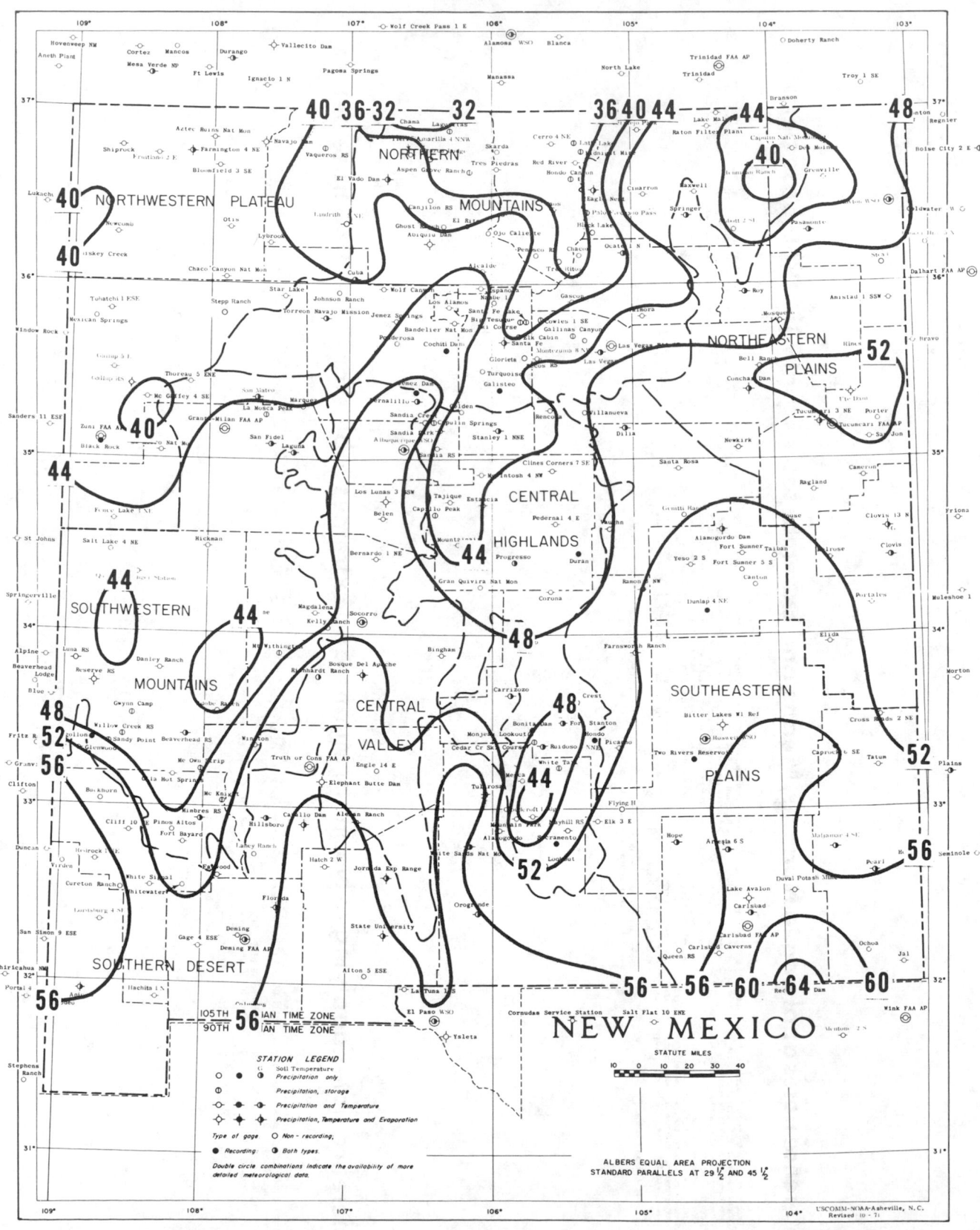

Data are based on the period 1931-52. Isolines are drawn through points of approximately equal value. Caution should be used in interpolating on these maps, particularly in mountainous areas.

MEAN MINIMUM TEMPERATURE (°F.), JANUARY

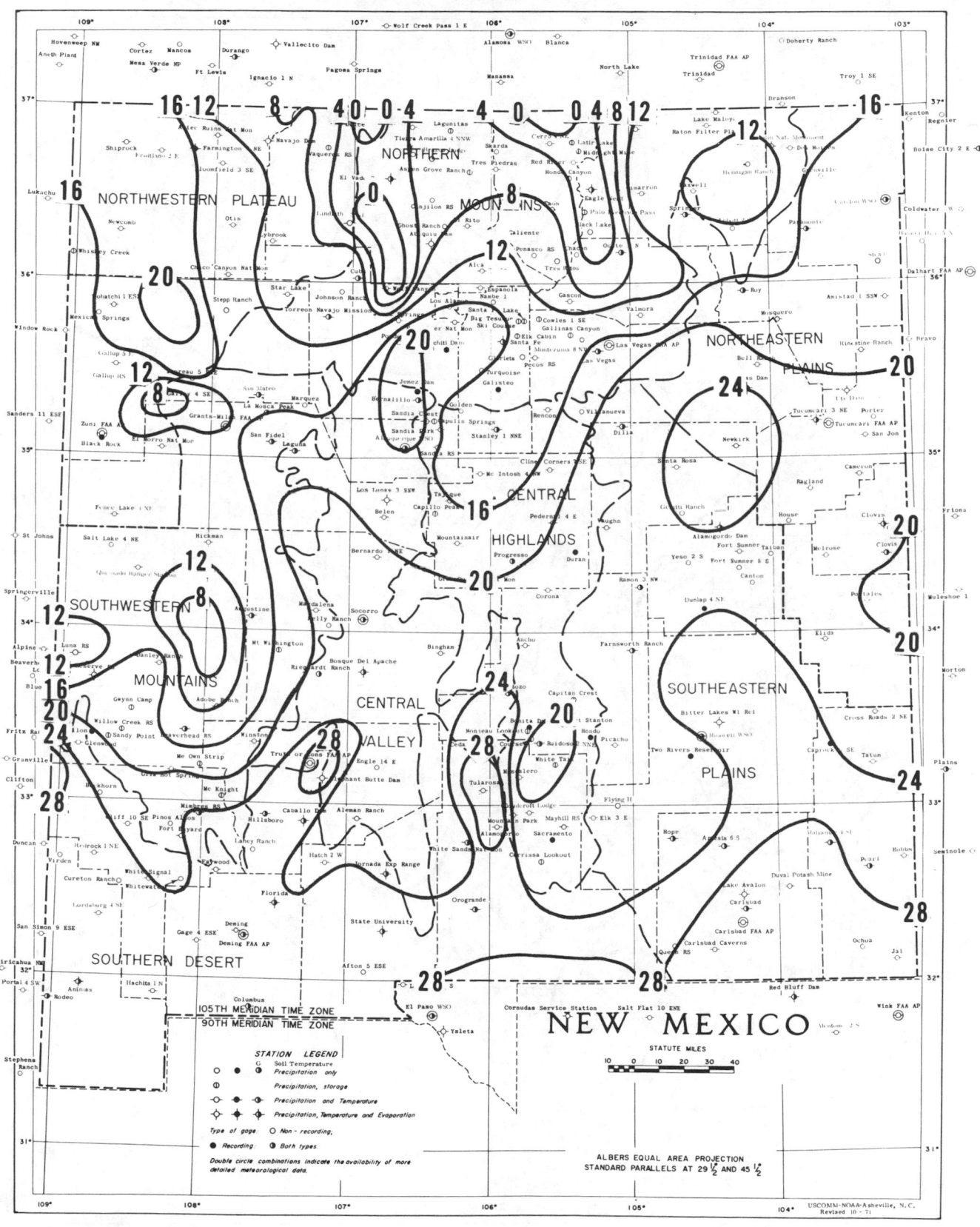

Data are based on the period 1931-52. Isolines are drawn through points of approximately equal value. Caution should be used in interpolating on these maps, particularly in mountainous areas.

MEAN MAXIMUM TEMPERATURE (°F.), JULY

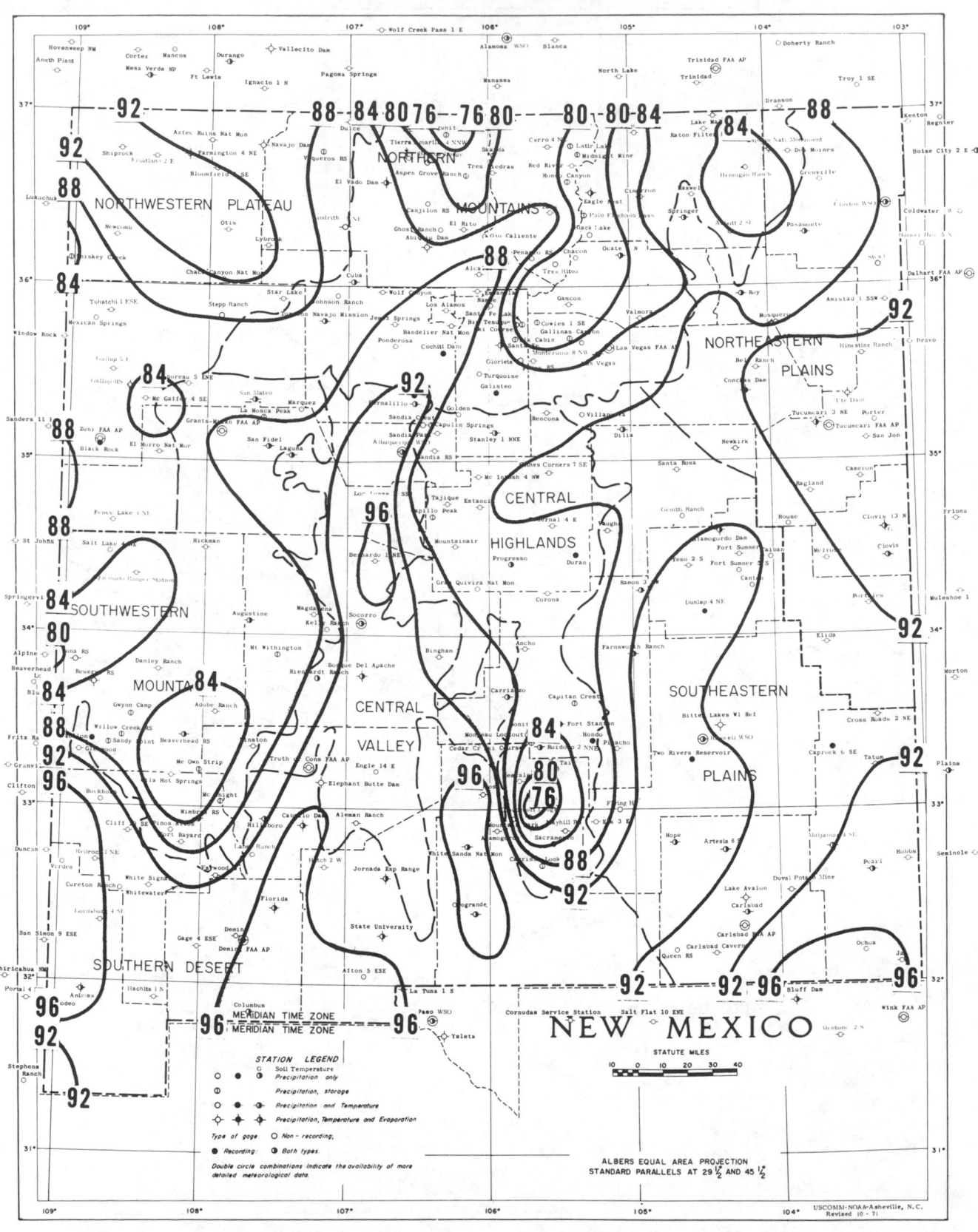

Data are based on the period 1931-52. Isolines are drawn through points of approximately equal value. Caution should be used in interpolating on these maps, particularly in mountainous areas.

MEAN MINIMUM TEMPERATURE (°F.), JULY

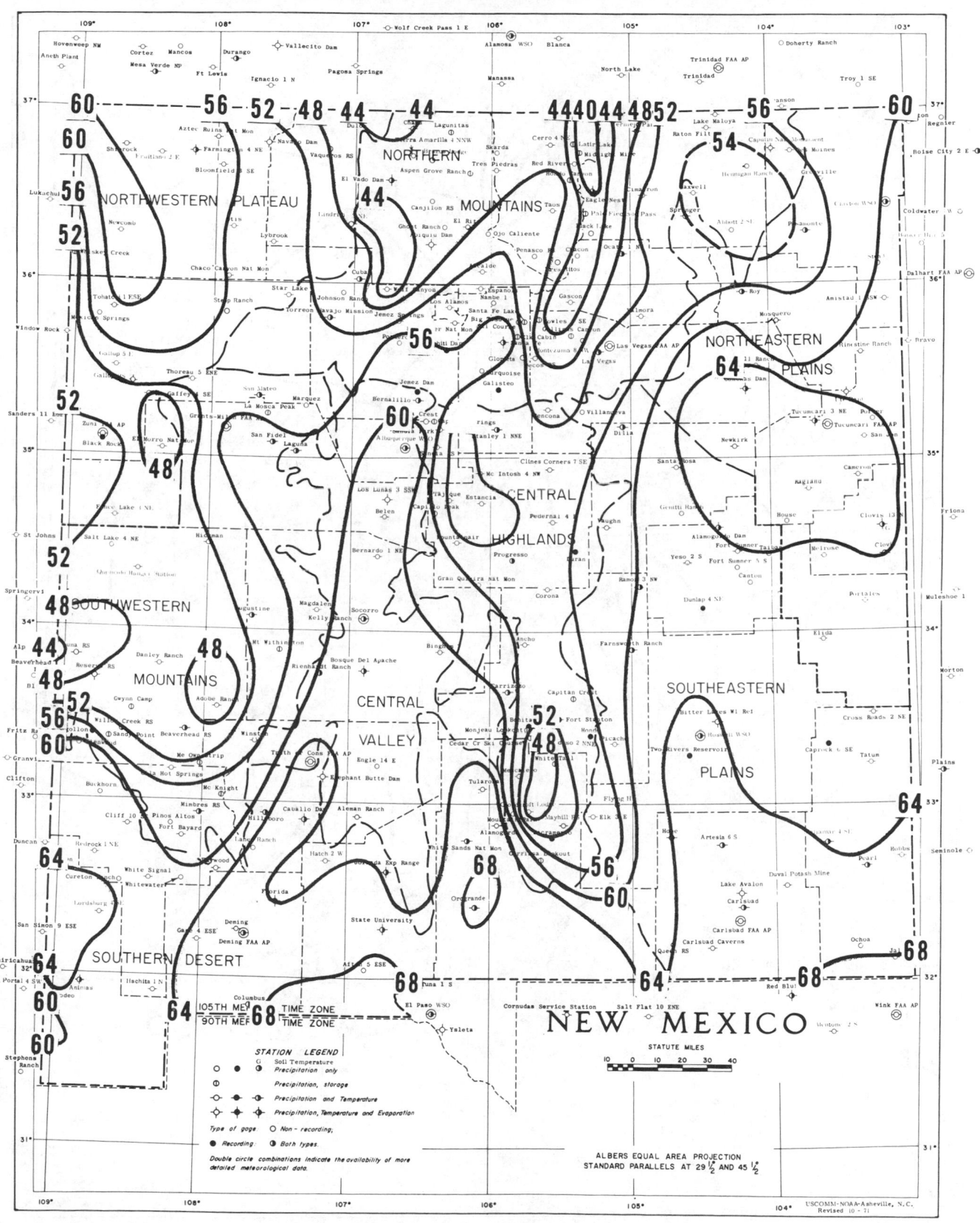

Data are based on the period 1931-52. Isolines are drawn through points of approximately equal value. Caution should be used in interpolating on these maps, particularly in mountainous areas.

MEAN ANNUAL PRECIPITATION, INCHES

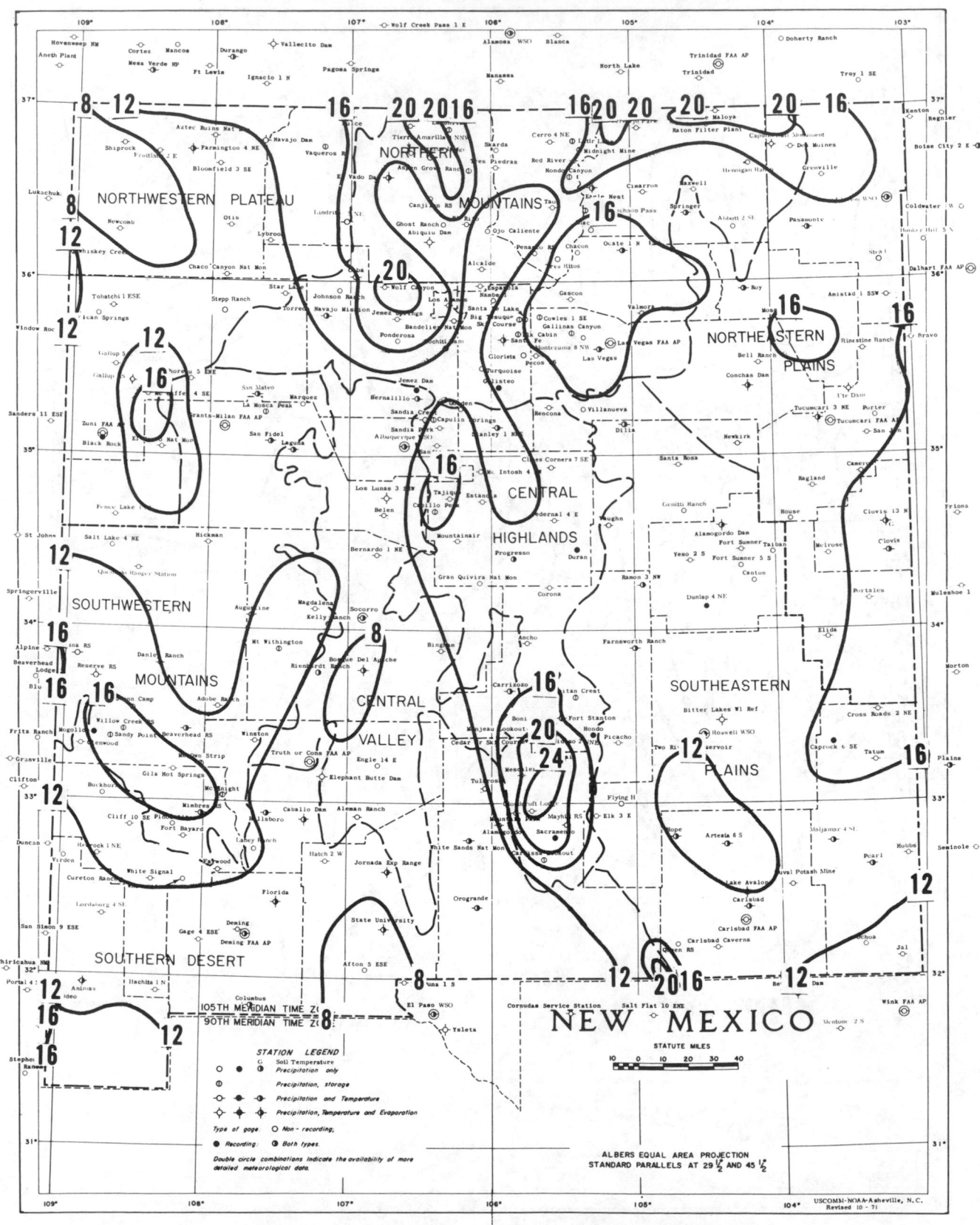

Data are based on the period 1931-55. Isolines are drawn through points of approximately equal value. Caution should be used in interpolating on these maps, particularly in mountainous areas.

NEW MEXICO

STATUTE MILES

CLIMATES OF THE STATES

NEW YORK

(Normals, Means and Extremes tables revised 1972 and 1975. Basic report revised June 1972.)

Dr. A. Boyd Pack, Climatologist for New York

PHYSICAL DESCRIPTION. New York State contains 49,576 square miles, inclusive of 1,637 square miles of inland water, but exclusive of the boundary-water areas of Long Island Sound, New York Harbor, Lake Ontario, and Lake Erie. The major portion of the State lies generally between latitudes 42° and 45° N. and between longitudes 73° 30' and 79° 45' W. However, in the extreme southeast, a triangular portion extends southward to about latitude 40° 30' N., while Long Island lies eastward to about longitude 72° W.

The principal highland regions of the State are the Adirondacks in the northeast and the Appalachian Plateau (Southern Plateau) in the south. The latter Plateau is subdivided by the deep channel of Seneca Lake, which extends from the Lake Plain of Lake Ontario southward to the Chemung River Valley, into the Western and Eastern Plateaus. The former extends from the eastern Finger Lakes across the hills of southwestern New York to the narrow Lake Plain bordering Lake Erie; the latter extends from the eastern Finger Lakes to the Hudson River Valley and includes the Catskill Mountains.

A minor highland region occurs in southeastern New York where the Hudson River has cut a Valley between the Palisades on the west, near the New Jersey border, and the Taconic Mountains on the east, along the Connecticut and Massachusetts border. Just west of the Adirondacks and the upper Black River Valley in Lewis County is another minor highland known as Tug Hill.

Much of the eastern border of the State consists of a long, narrow lowland region which is occupied by Lake Champlain, Lake George, and the middle and lower portions of the Hudson Valley. Another lowland region, the Great Lakes Plain, on the northern and western boundaries of the State adjoins the St. Lawrence River, Lake Ontario, and Lake Erie. This latter region is widest south of the eastern end of Lake Ontario, but does narrow to a width of less than 5 miles in the western portion of the State. A third lowland region, which contains Lake Oneida and a deep valley cut by the Mohawk River, connects the Hudson Valley and the Great Lakes Plain. Long Island, which is a part of the Atlantic Coastal Plain, comprises the fourth lowland region of the State.

Approximately 40 percent of New York State has an elevation of more than 1,000 feet above sea level. In northwestern Essex County, confined to an area of 500 or 600 square miles, are a number of peaks with an elevation of between 4,000 to 5,000 feet. The highest point, Mount Marcy, reaches a height of 5,344 feet above sea level. Nearby Mount MacIntyre ranges to a height of 5,112 feet. With the exception of the Blue Ridge of North Carolina and the White Mountains of New Hampshire, these are the loftiest mountains in eastern North America.

The Appalachian Plateau merges variously

into the Great Lakes Plain of western New York with gradual- to steep-sloping terrain. This Plateau is penetrated by the valleys of the Finger Lakes which, resembling the appearance of outstretched fingers on the hand, extend southward from the Great Lakes Plain. The major Finger Lakes going from west to east, are Canandaigua, Keuka, Seneca, Cayuga, and Skaneateles. Other prominent lakes in the State include Lake George in the central part of the eastern boundary, Lake Oneida in central New York between Syracuse and Rome, and Chautauqua Lake in the extreme southwest. Sacandaga and Pepacton Reservoirs are sizeable manmade bodies of water in the eastern portion of the State. Innumerable smaller lakes and ponds dot the landscape, with more than 1,500 in the Adirondack region alone.

Rivers of New York State may be divided into those that are tributary to the Great Lakes and St. Lawrence River and those that flow in a general southward direction. The first group includes rivers such as the Genesee, Oswego, Black, Oswegatchie, Grass, Raquette, Saranac, and Ausable. The Chemung, Susquehanna, Delaware, and Hudson River systems which are part of the Atlantic slope drainage and the Allegheny River which is part of the Ohio Basin drainage comprise the second group.

GENERAL CLIMATIC FEATURES. The climate of New York State is broadly representative of the humid continental type which prevails in the Northeastern United States, but its diversity is not usually encountered within an area of comparable size. The geographical position of the State and the usual course of air masses, governed by the large-scale patterns of atmospheric circulation, provide general climatic controls. Differences in latitude, character of the topography, and proximity to large bodies of water have pronounced effects on the climate.

The planetary atmospheric circulation brings a great variety of air masses to New York State. Masses of cold, dry air frequently arrive from the northern interior of the continent. Prevailing winds from the south and southwest transport warm, humid air which has been conditioned by the Gulf of Mexico and adjacent subtropical waters. These two air masses provide the dominant continental characteristics of the climate. The third great air mass flows inland from the North Atlantic Ocean and produces cool, cloudy, and damp weather conditions. This maritime influence is important to New York's climatic regime, especially in the southeastern portion of the State, but it is secondary to that of the more prevalent air mass flow from the continent.

Nearly all storm and frontal systems moving eastward across the continent pass through or in close proximity to New York State. Storm systems often move northward along the Atlantic coast and have an important influence on the weather and climate of Long Island and the lower Hudson Valley. Frequently, areas deep in the interior of the State feel the effects of such coastal storms.

Lengthy periods of either abnormally cold or warm weather result from the movement of great high pressure (anticyclonic) systems into and through the Eastern United States. Cold winter temperatures prevail over New York whenever Arctic air masses, under high barometric pressure, flow southward from central Canada or from Hudson Bay. High pressure systems often move just off the Atlantic coast, become more or less stagnant for several days, and then a persistent air flow from the southwest or south affects the State. This circulation brings the very warm, often humid weather of the summer season and the mild, more pleasant temperatures during the fall, winter, and spring seasons.

TEMPERATURE. Many atmospheric and physiographic controls on the climate result in a considerable variation of temperature conditions over New York State. The average annual mean temperature ranges from about 40° in the Adirondacks to near 55° in the New York City area. In January, the average mean temperature is approximately 16° in the Adirondacks and St. Lawrence Valley, but increases to about 26° along Lake Erie and in the lower Hudson Valley and to 31° on Long Island. The highest temperature of record in New York State is 107°, observed at Lewiston, Elmira, Poughkeepsie, and New York City. The record coldest temperature is -52° at Stillwater Reservoir (northern Herkimer County). Some 30 communities have recorded temperatures of -40° or colder, most of them occurring in the northern one-half of the State and the remainder in the Western Plateau Division and in localities just south of the Mohawk Valley.

The winters are long and cold in the Plateau Divisions of the State. In the majority of winter seasons, a temperature of -25° or lower can be expected in the northern highlands (Northern Plateau) and -15° or colder in the southwestern and east-central highlands (Southern Plateau). The Adirondack region records from 35 to 45 days with below zero temperatures in normal to severe winters, with a somewhat fewer number of such days occurring near Lake Champlain and the St. Lawrence River. In the Southern Plateau and in the upper Hudson Valley Division, below zero minimums are observed on about 15 days in most winters and on more than 25 days in notably cold seasons.

Winter temperatures are moderated considerably in the Great Lakes Plain of western New York. The moderating influence of Lakes Erie and Ontario is comparable to that produced by the Atlantic Ocean in the southern portion of the Hudson Valley. In both regions, the coldest temperature in most winters will range between 0° and -10°. Long Island and New York City experience below zero minimums in 2 or 3 winters out of 10, with the low temperature generally near -5°.

The summer climate is cool in the Adirondacks, Catskills, and higher elevations of the Southern Plateau. The New York City area and lower portions of the Hudson Valley have rather warm summers by comparison, with some periods of high, uncomfortable humidity. The remainder of New York State enjoys pleasantly warm summers, marred by only occasional, brief intervals of sultry conditions. Summer daytime temperatures usually range from the upper 70s to mid-80s over much of the State, producing an atmospheric environment favorable to many athletic, recreational, and other outdoor activities.

Temperatures of 90° or higher occur from late May to mid-September in all but the normally cooler portions of the State. The New York City area and most of the Hudson Valley record an average of from 18 to 25 days with such temperatures during the warm season, but in the Northern and Southern Plateaus the normal quota does not exceed 2 or 3 days. While temperatures of 100° are rare, many long-term weather stations, especially in the southern one-half of the State, have recorded maximums in the 100° to 105° range on one or more occasions. Minimum, or nighttime, temperatures drop to the 40s and upper 30s with some frequency during the summer season in the interior portions of the Plateau Divisions. It is not uncommon for temperatures to approach the freezing level in the Adirondacks and Southern Plateau during June and the latter one-half of August, but rarely in July.

The moderating effect of Lakes Erie and Ontario on temperatures assumes practical importance during the spring and fall seasons. The lake waters warm slowly in the spring, the effect of which is to reduce the warming of the atmosphere over adjacent land areas. Plant growth is thereby retarded, allowing a great variety of freeze-sensitive crops, especially tree and vine fruits, to reach critical early stages of development when the risk of freeze injury is minimized or greatly reduced. In the fall season, the lake waters cool more slowly than the land areas and thus serve as a heat source. The cooling of the atmosphere at night is moderated or reduced, the occurrence of freezing temperatures is delayed, and the growing season is lengthened for freeze-sensitive crops and vegetation.

The average length of the freeze-free season in New York State varies from 100 to 120 days in the Adirondacks, Catskills, and higher elevations of the Western Plateau Division to 180 to 200 days on Long Island. The important fruit and truck crop areas in the Great Lakes Plain enjoy a frost-free growing season of from 150 to 180 days in duration. A freeze-free season of similar length also prevails in the Hudson Valley from Albany southward to Westchester and Orange Counties, another zone of valuable crop production. The Southern Plateau, St. Lawrence Valley, and Lake Champlain regions have an average duration of 120 to 150 days between the last spring and first fall freezes.

PRECIPITATION. Moisture for precipitation in New York State is transported primarily from the Gulf of Mexico and Atlantic Ocean through circulation patterns and storm systems of the atmosphere. Distribution of precipitation within the State is greatly influenced by topography and proximity to the Great Lakes or Atlantic Ocean. Average annual amounts in excess of 50 inches occur in the western Adirondacks, Tug Hill area, and the Catskills, while slightly less than that amount is noted in the higher elevations of the Western Plateau southeast of Lake Erie. Areas of least rainfall, with average accumulations of about 30 inches, occur near Lake Ontario in the extreme western counties, in the lower half of the Genesee River Valley, and in the vicinity of Lake Champlain.

New York State has a fairly uniform distribution of precipitation during the year. There are no distinctly dry or wet seasons which are regularly repeated on an annual basis. Minimum precipitation occurs in the winter season, with an average monthly accumulation ranging from about 3.5 inches on Long Island to 2.2 inches in the Finger Lakes and Lake Champlain regions. Maximum amounts are noted in the summer season throughout the State except along the Great Lakes where slight peaks of similar magnitude occur in both the spring and fall seasons. Average monthly amounts in the summer vary from 3.0 inches in the lowlands south of Lake Ontario (Great Lakes Division) to 4.0 inches in the Eastern Plateau, Hudson Valley, and Coastal Divisions. New York's precipitation tends to be distributed most uniformly over the year in counties along the coast and the Great Lakes.

Variations in precipitation amounts from month to month or for the same month in different years can be wide for any individual area. Usually such variations range from near 1 inch to about 6 inches; in extreme cases, the variation is from less than 1 inch to 10 inches or more. Almost any calendar month has the potential of having the lightest, or heaviest, monthly accumulation of precipitation within a calendar year at a given location. The greatest monthly precipitation of record in New York State was a total of 25.27 inches at West Shokan (Ulster County) in October 1955. On the other hand, wide areas of the State measured less than 0.3 inch of rain in October 1963. Within relatively short distances, precipitation in the same month may be strikingly different. An extreme example occurred in August 1971 with a total of 16.7 inches falling at New York City's Borough of Richmond (Staten Island), but only 2.9 inches at Riverhead, about

90 miles away in eastern Long Island.

The amount and distribution of precipitation are normally sufficient for the maintenance of the State's water resources for municipal and industrial supplies, transportation, and recreation. Rainfall is usually adequate during the growing season for economic crops, lawns, gardens, shrubs, forests, and woodlands. Severe droughts are rare, but deficiencies of precipitation may occur from time to time which cause at least temporary concern over declining water supplies and moisture stress in crops and other vegetation. In some years, a pronounced shortage of precipitation during the spring or fall months results in a considerable fire hazard in the State's woodlands.

SNOWFALL. The climate of New York State is marked by abundant snowfall. With the exception of the Coastal Division, the State receives an average seasonal amount of 40 inches or more. The average snowfall is greater than 70 inches over some 60 percent of New York's area. The moderating influence of the Atlantic Ocean reduces the snow accumulation to 25 to 35 inches in the New York City area and on Long Island. About one-third of the winter season precipitation in the Coastal Division occurs from storms which also yield at least 1 inch of snow. The great bulk of the winter precipitation in upstate New York comes as snow.

Topography, elevation, and proximity to large bodies of water result in a great variation of snowfall in the State's interior, even within relatively short distances. Maximum seasonal snowfall, averaging more than 175 inches, occurs on the western and southwestern slopes of the Adirondacks and Tug Hill. A secondary maximum of 150 to 180 inches prevails in the southwestern highlands, some 10 to 30 miles inland from Lake Erie. Three separate areas of the Eastern Plateau record heavy snow accumulations, averaging from 100 to 120 inches: (1) the uplands of southeastern Onondaga County and adjoining counties; (2) the Cherry Valley section of northern Otsego and southern Herkimer Counties; and (3) the Catskill highlands in Ulster, Delaware, and Sullivan Counties. Minimum seasonal snowfall of 40 to 50 inches occurs upstate in (1) Niagara County, near the south shore of Lake Ontario, (2) the Chemung and mid-Genesee River Valleys of western New York, and (3) near the Hudson River in Orange, Rockland, and Westchester Counties upstream to the southern portion of Albany County.

In northern New York, the Adirondack region has an average seasonal snowfall in excess of 90 inches, but amounts decrease to 60 to 70 inches in the lowlands of the St. Lawrence Valley and to about 60 inches in the vicinity of Lake Champlain.

Snow produced in the lee of Lakes Erie and Ontario is a prominent and very important aspect of New York's climate. As cold air crosses the unfrozen lake waters, it is warmed in the lower layers, picks up moisture, and reaches the land in an unstable condition. Precipitation in the form of snow is released as the airstream moves inland and over the gradually sloping higher terrain. Heavy snow squalls frequently occur, generating from 1 to 2 feet of snow and occasionally 4 feet or more. Snowfall produced by this "lake-effect" usually extends into the Mohawk Valley and often inland as far as the southern Finger Lakes and nearby southern tier of counties. Counties to the lee of Lake Erie are subject to heavy lake-effect snows in November and December, but as the lake surface gradually freezes by midwinter, these snows become less frequent. Areas near Lake Ontario, especially those to the southeast and east, are exposed to severe snow squalls well into February because the Lake generally retains considerable open water throughout the winter months.

In the heavy snowbelts near Lake Erie and Ontario as well as in the plateau regions of eastern and northern New York, monthly snowfall amounts in excess of 24 inches are experienced in most winters; accumulations of more than 50 inches within 2 consecutive months are not uncommon. Monthly accumulations of between 3 to 10 inches usually occur in New York City and Long Island during the winter season, but occasionally the amounts may exceed 20 inches as a result of recurring coastal storms (northeasters).

A durable snow cover generally begins to develop in the Adirondacks and northern lowlands by late November and remains on the ground until various times in April, depending upon late winter snowfall and early spring temperatures. The Southern Plateau, Great Lakes Plain in southern portions of western upstate New York, and the Hudson Valley experience a continuous snow cover from about mid-December to mid-March, with maximum depths usually occurring in February. Bare ground may occur briefly in the lower elevations of these regions during some winters. From late December or early January through February, the Atlantic coastal region of the State experiences alternating periods of measurable snow cover and bare ground.

FLOODS. Although major floods are relatively infrequent, appreciable damage usually occurs every year in one or more localities of New York State. Floods that arise from a variety of causes have been recorded in all seasons. The greatest potential and frequency for floods occur in the early spring when substantial rains combine with rapid snowmelting to produce a heavy runoff. Since the turn of the century, several historic floods from this cause have occurred in the major river basins of southern and eastern New York. In northern New York, the normally colder early spring temperatures are conducive to a slower rate of snowmelt. In combination with other factors, major spring floods have been less frequent along streams draining into the St. Lawrence River. Ice jams sometimes contribute to serious flooding in very localized areas.

Damaging floods are caused at other times of the year by prolonged periods of heavy rainfall. Examples in recent years were those in southwestern New York in September 1967, in the lower Hudson Valley in May 1968, and in the Catskills in July 1969. In combination with heavy showers and thundershowers, the rugged terrain of the Adirondacks and Southern Plateau is conducive to occasional severe flash floods on smaller streams. The metropolitan New York City area and other heavily urbanized areas of the State are becoming increasingly subject to severe flooding of highways, streets, and low-lying ground. Replacement of the natural soil cover with cement, asphalt, and other impervious materials encourages such floods from rains of not more than moderately heavy intensity, that formerly were easily absorbed.

The shores of Long Island, especially those facing the Atlantic Ocean, are subject to tidal flooding during storm surges. Winds generated by hurricanes and great coastal storms may drive tidal waters well inland, causing extensive property damage and beach erosion. The great storm of November 1950, hurricane Carol in August 1954, and the historic Atlantic storm of March 1962 are some examples of severe, but infrequent, occurrences of this type of flooding.

WINDS AND STORMS. The prevailing wind is generally from the west in New York State. A southwest component becomes evident in winds during the warmer months while a northwest component is characteristic of the colder one-half of the year. Occasionally, well-developed storm systems moving across the continent or along the Atlantic coast are accompanied by very strong winds which cause considerable property damage over wide areas of the State. A unique effect of strong cyclonic winds from the southwest is the rise of water to abnormally high levels at the northeastern end of Lake Erie.

Thunderstorms occur on an average of about 30 days in a year throughout the State. Destructive winds and lightning strikes in local areas are common with the more vigorous warm-season thunderstorms. Locally, hail occurs with more severe thunderstorms, but extensive, crippling losses to property and crops are rare.

Tornadoes are not common. About three or four of these storms strike limited, localized areas of New York State in most years. The paths of destruction, mostly in rural, semi-rural, or wooded areas, are usually short and narrow. Tornadoes occur generally between late May and late August.

Storms of freezing rain occur on one or more occasions during the winter season and often affect a wide area of the State in any one incident. While such storms are usually limited to a thin but dangerous coating of ice on highways, sidewalks, and exposed surfaces, crippling destruction of utility lines, transmission towers, and trees over an extensive portion of the State may result on rare occasions. Such a destructive ice storm affected east-central and southeastern New York in December 1964.

Hurricanes and tropical storms periodically cause serious and heavy losses in the vicinity of Long Island and southeastern upstate New York. Only one such storm in recent years (October 1954) has brought serious damage to the interior portion of the State.

The greatest storm hazard in terms of area and number of people affected is heavy snow. Coastal northeaster storms occur with some frequency in most winters. Snow yields of from 12 to 24 inches or more from such storms have fallen over the southeastern one-quarter of the State, including Long Island, and will often extend into western and northern interior New York. Snow squalls along the Great Lakes have been previously cited. These may persist over a period of 1 week or more, bringing snow amounts in excess of 40 inches to local areas that lie to the eastern lee of Lakes Erie and Ontario. During heavy snow squalls, surface visibility is reduced to zero. Blizzard conditions of heavy snow, high winds, and rapidly falling temperature occur occasionally, but are much less characteristic of New York's climate than in the plains of Midwestern United States.

OTHER CLIMATIC ELEMENTS. The climate of the State features much cloudy weather during the months of November, December, and January in upstate New York, especially those regions that adjoin the Great Lakes and Finger Lakes and include the southern tier of counties. From June through September, however, about 60 to 70 percent of the possible sunshine hours is received. In the Atlantic coastal region, the sunshine hours increases from 50 percent of possible in the winter to about 65 percent of possible in the summer.

The Atlantic Coastal Plain and lower Hudson Valley experience conditions of high temperature and high humidity with some frequency and duration during the summer. By comparison, such conditions occur less frequently in the broad interior of New York State where they are usually shortened by the arrival of cooler, drier air masses from the northwest.

The occurrence of heavy dense fog is variable over the State. The valleys and ridges of the Southern Plateau are most subject to periods of fog, with occurrences averaging about 50 days in a year. In the Great Lakes Plain and northern valleys, the frequency decreases to only 10 to 20 days annually. In those portions of the State with greater maritime influence on the climate, the frequency of dense fog in a year ranges from about 35 days on the south shore of Long Island to 25 days in the Hudson Valley.

CLIMATE AND THE ECONOMY. New York State's diversified economy, involving agriculture, industry, commerce, and recreation, is greatly influenced by the climate. Human activities, whether in labor pursuits or recreation, are stimulated by an invigorating winter climate and a generally comfortable atmospheric environment during summer.

The general climate as well as regional variations in climate throughout New York State support diversified agriculture. Dairying is the largest, most widespread enterprise. Precipitation and temperature conditions favor the growth of alfalfa and grasses for hay and of corn for silage throughout rural New York, except where limitations are imposed by soils and topography. Corn for grain is produced on some 850,000 acres, mostly in the Great Lakes Plain, Southern Plateau, and Hudson Valley; climatic conditions couple with technology to realize an average statewide yield of 70 to 80 bushels per acre. The amount and distribution of rainfall, warm (rather than hot) daytime temperatures, and frequent cool nights in western and central New York are important environmental factors that aid in the growing of 450,000 acres of small grains. Dry beans, snap beans, and sugar beets are additional valuable crops which thrive well in New York's climate.

A nationally important production area of apples and other tree fruits is found along Lake Ontario, largely the result of favorable climatic conditions induced by the nearby Lake. The climate over the Great Lakes Plain is also benevolent for a wide variety of vegetable crops. New York is a leading producer of grapes, with suitable weather conditions for viticulture existing in the western Great Lakes counties and on the sloping terrain along the Finger Lakes where good air drainage and moderating influence of lake waters produce a suitable temperature regime. The lower Hudson Valley has a climate which also supports important acreage of tree fruits and truck crops.

The warmer climate of eastern Long Island permits a significant production of potatoes for the early season market. Late-season potato varieties are grown in the cooler climate of the Southern Plateau and of northeastern New York. The uplands northwest of the Catskill Mountains have a cool climate very suitable for cauliflower production.

The sugar maple tree (Acer saccharum Marsh.) finds a climate optimum for growth in New York State. Thus, the production of syrup and other maple products constitutes a valuable segment of the agricultural and forestry economy.

Ample precipitation, dependable runoff, and adequate ground water supplies contribute to vast water resources in the Empire State. These water resources have supported the growth of many large metropolitan areas, the establishment of diverse industries, and the development of waterways and impoundments for transportation, power, recreation, and municipal supplies.

Though rigorous and sometimes severe, New York's winter climate is an asset to the economy. Abundant snowfall has made possible the development of skiing and snowmobiling into very important activities for winter sports and recreation. The climate at other times of the year is a prominent factor in attracting tourists and vacationers to the State.

In summary, the climate contributes greatly to the agricultural, industrial, commercial, and recreational economy. It has been an unquestionable asset to the historical development of New York State and to its economic expansion of recent decades. Undoubtedly, the climate will continue its important role in the remainder of this century and beyond.

TROPICAL CYCLONE DATA HAVING IMPORTANCE FOR THIS STATE IS INCLUDED IN STATISTICS AND CHARTS ON PAGES 1161 THROUGH 1164.

BIBLIOGRAPHY

Dethier, Bernard E., "Precipitation in New York State," Cornell University Agricultural Experiment Station, Bulletin 1009, New York State College of Agriculture, Ithaca, N. Y., July 1966, 78 pp.

Dethier, B. E., and McGuire, J. K., "The Climate of the Northeast, Probability of Selected Weekly Precipitation Amounts in the Northeast Region of the U. S.," Cornell University Agricultural Experiment Station, Agronomy Mimeo 61-4, Ithaca, N. Y., Nov. 1961, 302 pp.

Dethier, B. E., and Pack, A. B., "The Climate of Northern New York," Division of Meteorology, Department of Agronomy, Cornell University, Agronomy Mimeo 67-14, Ithaca, N. Y., Dec. 1967, 17 pp.

Dethier, B. E., and Vittum, M. T., "Growing Degree Days for New York State," Cornell University Agricultural Experiment Station, Bulletin 1017, Ithaca, N. Y., Nov. 1967, 50 pp.

Dickerson, W. H., "The Climate of the Northeast. Heating Degree Days," West Virginia Agricultural Experiment Station, Bulletin 483T, Morgantown, W. Va., June 1963, 26 pp.

Dunlap, D. C., "The Climate of the Northeast, Probabilities of Extreme Snowfalls and Snow Depths," New Jersey Agricultural Experiment Station, Bulletin 821, Rutgers University, The State University of New Jersey, New Brunswick, N. J., 1970, 15 pp.

Fieldhouse, Donald J., and Palmer, Wayne C., "The Climate of the Northeast. Meteorological and Agricultural Drought," University of Delaware Agricultural Experiment Station, Bulletin 353, Newark, Del., Feb. 1965, 71 pp.

Frederick, Ralph H., Johnson, Ernest C., and MacDonald, H. A., "Spring and Fall Freezing Temperatures in New York State," New York State College of Agriculture, Cornell Miscellaneous Bulletin 33, Ithaca, N. Y., Aug. 1959, 15 pp.

Havens, A. V., and McGuire, J. K., "The Climate of the Northeast. Spring and Fall Low-Temperature Probabilities," New Jersey Agricultural Experiment Station, Bulletin 801, Rutgers, The State University, New Brunswick, N. J., June 1961, 32 pp.

Johnson, Ernest C., "Climate of New York, Climates of the States," Climatography of the United States No. 60-30, U. S. Weather Bureau, Washington, D. C., Feb. 1960, 20 pp.

Miller, J. F., and Frederick, R. H., "The Precipitation Regime of Long Island, New York," U. S. Department of the Interior, U. S. Geological Survey Professional Paper 627-A, Washington, D. C., 1969, A1-A21 pp.

Mordoff, Richard Alan, "The Climate of New York State," Cornell University Extension, Bulletin 764, Ithaca, N. Y., Dec. 1949, 72 pp.

NOAA, Environmental Data Service, "Albany, Binghamton, Buffalo, New York City (Central Park, J. F. Kennedy International Airport, and La Guardia Airport), Rochester, and Syracuse, N. Y.," Local Climatological Data, Asheville, N. C., monthly plus annual summary.

_____, Climatological Data--National Summary, Asheville, N. C., monthly plus annual summary.

_____, Climatological Data--New York, Asheville, N. C., monthly plus annual summary.

_____, Hourly Precipitation Data--New York, Asheville, N. C., monthly plus annual summary.

Pack, A. Boyd, "Average Seasonal Snowfall in New York State (map)." What's Cropping Up in Agronomy, Vol. XIV, No. 1, New York State College of Agriculture, Cornell University, Ithaca, N. Y., Jan. 1970, 3 pp.

_____, "The Water Content of Snowstorms in New York State: Variations Among Different Physiographic Regions," in Proceedings of The Twenty-Sixth Annual Eastern Snow Conference, Feb. 6-7, 1969, Portland, Maine, Vol. 14, pp. 46-54.

Pack, A. Boyd and Dethier, B. E., "The Climate of Western New York," Division of Meteorology, Department of Agronomy, Cornell University, Agronomy Mimeo 69-12. Ithaca, N. Y., 1969, 19 pp.

U.S. Weather Bureau, Climatic Summary of the United States (Bulletin W), Sections 80, 81, 82, and 83, Washington, D.C., 1930, 113 pp.

_____, "Climatic Summary of the United States--Supplement for 1931 through 1952, New York" (Bulletin W Supplement), Climatography of the United States No. 11-26, Washington, D.C., 1953, 66 pp.

_____, "Climatic Summary of the United States--Supplement for 1951 through 1960, New York" (Bulletin W Supplement), Climatography of the United States No. 86-26, Washington, D.C., 1964, 111 pp.

_____, "Decadal Census of Weather Stations, New York," Key to Meteorological Records Documentation No. 6.11, Washington, D.C., 1963, 8 pp.

_____, "Maximum Station Precipitation for 1, 2, 3, 6, 12, and 24 Hours, Part X: New York," Technical Paper No. 15, Washington, D.C., Dec. 1954, 113 pp.

_____, "Monthly Averages for State Climate Divisions, 1931-1960, New York," Climatography of the United States No. 85-26, Washington, D.C., 1963, 4 pp.

_____, "Monthly Normals of Temperature, Precipitation, and Heating Degree Days, New York," Climatography of the United States No. 81-26, Washington, D.C., 1962, 2 pp.

_____, "Rainfall Frequency Atlas of the United States for Durations from 30 Minutes to 24 Hours and Return Periods from 1 to 100 Years," Technical Paper No. 40, Washington, D.C., May 1961, 115 pp.

_____, "Summary of Hourly Observations, Albany, Binghamton, Buffalo, New York City (J.F. Kennedy International Airport and La Guardia Airport), Rochester, and Syracuse, New York, 1951-1960," Climatography of the United States No. 82-30, Washington, D.C., 1962 and 1963, 105 pp.

STATION	Freeze threshold temperature	Mean date of last Spring occurrence	Mean date of first Fall occurrence	Mean No. of days between dates	Years of record Spring	No. of occurrences in Spring	Years of record Fall	No. of occurrences in Fall
ADDISON	32	05-21	09-30	131	23	23	20	20
	28	05-06	10-10	158	23	23	20	20
	24	04-22	10-24	185	22	22	20	20
	20	04-08	11-05	211	22	22	20	20
	16	03-25	11-22	242	22	22	20	20
ALBANY WSO AP	32	04-27	10-13	169	29	29	29	29
	28	04-14	10-26	195	29	29	29	29
	24	03-31	11-09	224	29	29	29	29
	20	03-19	11-22	248	29	29	29	29
	16	03-13	11-29	261	29	29	29	29
ALBION 3 ENE	32	05-01	10-15	166	12	12	12	12
	28	04-22	11-03	196	12	12	12	12
	24	04-02	11-18	230	12	12	12	12
	20	03-25	11-28	248	12	12	12	12
	16	03-20	12-08	263	12	12	12	12
ALEXANDRIA BAY	32	05-02	10-10	160	14	14	14	14
	28	04-18	10-27	192	14	14	14	14
	24	04-08	11-11	217	14	14	14	14
	20	03-31	11-20	234	14	14	14	14
	16	03-22	11-28	251	14	14	14	14
ALFRED	32	05-27	09-19	115	29	29	29	29
	28	05-14	10-05	143	29	29	29	29
	24	04-27	10-14	171	29	29	29	29
	20	04-15	11-01	201	29	29	28	28
	16	04-03	11-16	227	29	29	27	27
ALLEGANY STATE PARK	32	06-07	09-05	90	26	26	26	26
	28	05-23	09-29	130	26	26	26	26
	24	05-08	10-15	160	26	26	26	26
	20	04-20	11-03	197	26	26	26	26
	16	04-07	11-16	223	26	26	26	26
ANGELICA	32	06-01	09-12	103	30	30	27	27
	28	05-18	09-29	134	30	30	26	26
	24	05-02	10-12	163	30	30	26	26
	20	04-20	10-26	188	30	30	26	26
	16	04-04	11-10	220	30	30	26	26
AUBURN WTR WKS	32	05-02	10-21	172	28	28	27	27
	28	04-18	11-08	204	28	28	26	26
	24	04-05	11-19	228	28	28	26	26
	20	03-24	11-28	249	27	27	26	26
	16	03-16	12-03	262	26	26	26	25
BAINBRIDGE	32	05-15	09-26	134	14	14	14	14
	28	05-07	10-11	157	14	14	14	14
	24	04-21	10-23	185	14	14	13	13
	20	04-03	11-06	217	14	14	13	13
	16	03-27	11-23	241	14	14	13	13
BEDFORD HILLS	32	04-29	10-13	167	24	24	23	23
	28	04-15	10-26	194	24	24	23	23
	24	03-30	11-08	223	23	23	21	21
	20	03-19	11-23	249	22	22	21	21
	16	03-12	12-04	267	22	22	21	21
BINGHAMTON	32	05-04	10-06	154	29	29	28	28
	28	04-19	10-20	184	29	29	28	28
	24	04-04	11-07	217	29	29	28	28
	20	03-23	11-21	243	29	29	28	28
	16	03-14	12-02	263	29	29	28	28
BRIDGEHAMPTON	32	04-20	11-02	196	20	20	21	21
	28	04-05	11-17	225	20	20	21	21
	24	03-19	11-29	256	19	19	21	21
	20	03-10	12-05	270	19	19	21	21
	16	03-06	12-12	282	19	19	21	19
BROCKPORT 2 NW	32	05-07	10-20	166	16	16	14	14
	28	04-21	11-03	196	16	16	14	14
	24	04-07	11-11	218	16	16	14	14
	20	03-24	11-26	247	16	16	14	14
	16	03-11	11-30	264	16	16	13	13
BUFFALO WSO AP	32	04-30	10-25	179	29	29	29	29
	28	04-20	11-09	202	29	29	29	29
	24	04-03	11-20	231	29	29	29	29
	20	03-23	11-30	251	29	29	29	28
	16	03-15	12-07	267	29	29	29	28
CANTON 4 SE	32	05-09	09-26	140	29	29	29	29
	28	04-28	10-10	166	29	29	29	29
	24	04-15	10-24	192	29	29	29	29
	20	04-03	11-08	219	29	29	29	29
	16	03-26	11-19	238	29	29	29	29
CARMEL 1 SW	32	05-03	10-10	160	29	29	28	28
	28	04-22	10-22	183	29	29	28	28
	24	04-09	11-04	209	29	29	28	28
	20	03-24	11-18	240	29	29	28	28
	16	03-16	11-30	259	29	29	28	28
CHASM FALLS	32	05-25	09-14	112	17	17	17	17
	28	05-10	10-02	145	17	17	17	17
	24	04-30	10-11	164	17	17	17	17
	20	04-16	10-27	193	17	17	17	17
	16	03-31	11-14	228	17	17	17	17
CHAZY	32	05-15	09-26	134	26	26	24	24
	28	04-30	10-08	161	26	26	23	23
	24	04-17	10-27	192	26	26	22	22
	20	04-06	11-08	216	26	26	19	19
	16	03-24	11-22	243	26	26	19	19
COOPERSTOWN	32	05-22	09-22	123	24	24	22	22
	28	05-12	10-05	145	24	24	22	22
	24	04-26	10-18	176	23	23	22	22
	20	04-11	11-03	207	23	23	21	21
	16	04-01	11-16	229	22	22	20	20
CORTLAND	32	05-13	09-30	140	29	29	29	29
	28	04-25	10-19	177	29	29	29	29
	24	04-12	11-07	209	29	29	29	29
	20	03-29	11-21	237	29	29	29	29
	16	03-20	11-28	253	29	29	29	29
DANNEMORA	32	05-13	10-02	142	29	29	29	29
	28	05-01	10-12	164	29	29	28	28
	24	04-23	10-26	185	29	29	28	28
	20	04-11	11-07	210	29	29	27	27
	16	04-04	11-12	223	29	29	27	27
DANSVILLE	32	05-10	10-07	150	27	27	27	27
	28	04-24	10-27	187	27	27	27	27
	24	04-13	11-11	212	26	26	27	27
	20	03-29	11-25	242	25	25	27	27
	16	03-16	11-30	260	25	25	27	26
DELHI 2 SW	32	05-29	09-23	117	26	26	24	24
	28	05-15	10-04	143	26	26	24	24
	24	04-30	10-16	169	26	26	24	24
	20	04-17	10-30	196	26	26	24	24
	16	03-29	11-14	229	26	26	24	24
ELMIRA	32	05-06	10-09	156	30	30	30	30
	28	04-19	10-25	188	30	30	30	30
	24	04-08	11-06	212	30	30	30	30
	20	03-22	11-24	247	30	30	30	30
	16	03-13	12-04	266	30	30	30	29
FARMINGDALE 2 NE	32	04-30	10-19	173	11	11	11	11
	28	04-16	10-27	194	11	11	11	11
	24	03-25	11-14	234	9	9	11	11
	20	03-15	12-01	261	9	9	10	10
	16	03-07	12-08	276	9	9	9	9
FRANKLINVILLE	32	05-31	09-13	105	14	14	13	13
	28	05-18	09-29	134	12	12	13	13
	24	05-06	10-14	161	11	11	13	13
	20	04-17	10-31	197	11	11	12	12
	16	04-05	11-11	219	11	11	12	12
FREDONIA	32	05-07	10-24	171	30	30	30	30
	28	04-20	11-04	198	30	30	30	30
	24	04-05	11-21	230	30	30	30	30
	20	03-24	12-01	253	30	30	30	30
	16	03-14	12-07	268	30	30	30	29

FREEZE DATA

STATION	Freeze threshold temperature	Mean date of last Spring occurrence	Mean date of first Fall occurrence	Mean No. of days between dates	Years of record Spring	No. of occurrences in Spring	Years of record Fall	No. of occurrences in Fall
GENEVA	32	05-06	10-11	158	30	30	29	29
	28	04-20	10-27	190	30	30	29	29
	24	04-08	11-12	218	30	30	29	29
	20	03-26	11-25	244	30	30	29	29
	16	03-13	12-03	264	30	30	29	29
GLENHAM	32	04-29	10-07	161	19	19	19	19
	28	04-18	10-24	189	19	19	19	19
	24	03-30	11-05	220	19	19	19	19
	20	03-22	11-21	244	19	19	19	19
	16	03-14	12-01	261	19	19	19	19
GLOVERSVILLE	32	05-14	09-30	139	23	23	23	23
	28	04-30	10-10	163	23	23	23	23
	24	04-18	10-26	192	23	23	23	23
	20	04-08	11-06	213	23	23	23	23
	16	03-27	11-19	237	23	23	22	22
GOUVERNEUR	32	05-19	09-18	122	13	13	13	13
	28	05-05	09-26	144	13	13	13	13
	24	04-23	10-10	170	13	13	12	12
	20	04-09	10-26	200	13	13	12	12
	16	03-31	11-13	227	13	13	12	12
HEMLOCK	32	05-08	10-12	158	30	30	30	30
	28	04-25	10-26	185	30	30	30	30
	24	04-11	11-08	210	30	30	30	30
	20	03-31	11-22	236	30	30	30	30
	16	03-20	12-01	255	30	30	30	29
INDIAN LAKE 2 SW	32	06-11	08-27	78	27	27	27	27
	28	05-27	09-20	117	28	28	27	27
	24	05-14	10-04	143	28	28	27	27
	20	04-30	10-16	169	28	28	26	26
	16	04-15	10-26	194	28	28	26	26
ITHACA CORNELL UNIV	32	05-12	10-04	145	29	29	29	29
	28	04-23	10-21	181	29	29	29	29
	24	04-12	11-05	207	29	29	29	29
	20	03-29	11-21	237	29	29	29	29
	16	03-18	11-29	255	29	29	29	29
JAMESTOWN	32	05-16	10-07	144	26	26	24	24
	28	04-25	10-23	181	25	25	22	22
	24	04-10	11-11	215	25	25	22	22
	20	03-31	11-23	237	24	24	20	20
	16	03-17	12-06	264	24	24	19	18
LAKE PLACID CLUB	32	06-04	09-11	99	29	29	29	29
	28	05-18	09-22	127	28	28	29	29
	24	05-03	10-09	159	28	28	29	29
	20	04-22	10-21	182	28	28	29	29
	16	04-12	11-02	204	28	28	28	28
LAWRENCEVILLE	32	05-14	09-22	130	18	18	18	18
	28	05-09	10-09	153	16	16	18	18
	24	04-26	10-17	174	16	16	18	18
	20	04-13	11-01	202	16	16	17	17
	16	03-31	11-14	228	16	16	17	17
LEWISTON 1 N	32	05-07	10-11	157	15	15	16	16
	28	04-14	10-29	197	15	15	16	16
	24	04-05	11-17	226	15	15	16	16
	20	03-20	11-25	250	15	14	16	16
	16	03-10	12-07	271	15	14	16	16
LOCKPORT 2 NE	32	05-09	10-13	157	27	27	27	27
	28	04-23	10-31	191	27	27	27	27
	24	04-10	11-14	218	27	27	27	27
	20	03-28	11-23	240	26	26	26	26
	16	03-16	12-03	262	25	24	26	26
LOWVILLE	32	05-18	09-21	126	29	29	30	30
	28	05-05	10-04	152	29	29	30	30
	24	04-22	10-20	180	29	29	30	30
	20	04-08	11-03	209	29	29	30	30
	16	03-29	11-13	229	29	29	30	30
MINEOLA	32	04-07	11-16	223	13	13	12	12
	28	03-26	11-29	249	13	13	12	12
	24	03-15	12-06	267	13	13	12	12
	20	03-09	12-14	280	13	13	12	11
	16	03-05	12-17	288	13	13	12	11
MOHONK LAKE	32	04-29	10-22	177	27	27	27	27
	28	04-16	11-04	201	27	27	27	27
	24	04-05	11-16	225	27	27	27	27
	20	03-25	11-23	243	27	27	26	26
	16	03-17	12-01	259	27	27	26	26
MORRISVILLE	32	05-21	09-20	122	26	26	25	25
	28	05-10	10-03	146	25	25	24	24
	24	04-26	10-19	176	25	25	24	24
	20	04-12	10-30	201	25	25	23	23
	16	03-30	11-14	229	25	25	23	23
NEWCOMB 4 WNW	32	05-27	09-16	112	10	10	11	11
	28	05-12	09-27	138	10	10	10	10
	24	04-29	10-10	164	10	10	10	10
	20	04-17	10-29	195	10	10	10	10
	16	04-04	11-09	219	9	9	10	10
NY CENTRAL PK WSO	32	04-07	11-12	219	29	29	29	29
	28	03-24	11-24	244	29	29	29	29
	24	03-15	12-02	262	29	29	29	29
	20	03-10	12-09	273	29	29	29	27
	16	03-03	12-16	289	29	29	29	23
NORWICH 1 NE	32	05-19	09-24	127	28	28	30	30
	28	05-07	10-06	152	28	28	30	30
	24	04-24	10-19	182	27	27	30	30
	20	04-10	10-30	203	27	27	30	30
	16	03-26	11-18	238	27	27	30	30
OGDENSBURG 3 NE	32	05-08	10-08	153	27	27	27	27
	28	04-25	10-24	182	27	27	27	27
	24	04-13	11-02	203	26	26	27	27
	20	03-31	11-15	229	24	24	27	27
	16	03-25	11-26	247	24	24	26	26
ONEONTA	32	05-19	09-28	133	27	27	27	27
	28	05-01	10-11	163	27	27	27	27
	24	04-20	10-25	188	27	27	27	27
	20	04-01	11-09	221	27	27	27	27
	16	03-25	11-21	241	27	27	27	27
OSWEGO EAST	32	04-24	10-24	184	29	29	29	29
	28	04-10	11-09	214	28	28	29	29
	24	03-28	11-20	237	28	28	29	29
	20	03-19	11-28	254	28	28	29	29
	16	03-13	12-04	266	28	28	29	29
PATCHOGUE 2 N	32	04-28	10-12	167	12	12	13	13
	28	04-11	10-27	199	12	12	13	13
	24	03-31	11-15	228	12	12	13	13
	20	03-22	11-24	247	12	12	13	13
	16	03-10	12-08	273	12	12	13	13
PENN YAN 2 SW	32	05-13	10-05	146	26	26	22	22
	28	04-25	10-18	176	25	25	21	21
	24	04-12	11-03	205	24	24	21	21
	20	03-30	11-15	231	23	23	21	21
	16	03-19	11-28	254	23	23	21	21
PORT JERVIS	32	05-10	10-04	146	29	29	29	29
	28	04-24	10-17	176	29	29	29	29
	24	04-09	10-28	202	29	29	28	28
	20	03-26	11-10	229	29	29	28	28
	16	03-16	11-27	256	29	29	28	28

FREEZE DATA

STATION	Freeze threshold temperature	Mean date of last Spring occurrence	Mean date of first Fall occurrence	Mean No. of days between dates	Years of record Spring	No. of occurrences in Spring	Years of record Fall	No. of occurrences in Fall
RIVERHEAD RESEARCH	32	04-15	11-08	208	13	13	13	13
	28	04-02	11-22	233	13	13	13	13
	24	03-16	12-05	265	13	13	13	13
	20	03-10	12-10	275	13	13	13	13
	16	03-02	12-18	292	13	13	13	11
ROCHESTER WSO AP	32	04-28	10-21	176	29	29	27	27
	28	04-16	11-05	203	29	29	27	27
	24	03-31	11-19	233	29	29	27	27
	20	03-20	11-28	253	29	29	27	27
	16	03-12	12-05	267	29	29	27	26
ROXBURY	32	05-28	09-18	113	27	27	25	25
	28	05-12	09-30	141	26	26	25	25
	24	04-28	10-10	165	24	24	24	24
	20	04-20	10-26	189	24	24	24	24
	16	04-04	11-04	214	24	24	24	24
SALISBURY	32	05-29	09-18	113	30	30	30	30
	28	05-12	09-29	140	30	30	30	30
	24	04-25	10-13	171	30	30	30	30
	20	04-15	10-31	199	30	30	30	30
	16	04-05	11-08	217	30	30	30	30
SCARSDALE	32	04-21	10-20	181	26	26	27	27
	28	04-10	10-30	204	26	26	26	26
	24	03-26	11-20	240	26	26	23	23
	20	03-15	11-30	260	26	26	23	23
	16	03-10	12-07	272	26	26	23	22
SETAUKET	32	04-10	11-09	214	30	30	28	28
	28	03-26	11-23	242	30	30	28	28
	24	03-15	12-05	265	30	30	28	27
	20	03-09	12-09	275	30	30	28	27
	16	02-28	12-19	294	30	29	28	21
SODUS CENTER	32	05-06	10-09	155	16	16	11	11
	28	04-22	10-27	187	16	16	9	9
	24	04-12	11-05	207	16	16	9	9
	20	03-28	11-22	239	15	15	9	9
	16	03-21	12-03	257	15	15	9	8
SOUTH WALES EMERY PK	32	05-17	10-01	137	19	19	20	20
	28	05-05	10-14	163	19	19	20	20
	24	04-16	11-02	200	19	19	20	20
	20	04-03	11-15	226	19	19	20	20
	16	03-25	11-24	244	19	19	20	20
SPIER FALLS	32	05-10	10-07	150	29	29	27	27
	28	04-21	10-20	182	29	29	27	27
	24	04-12	11-04	206	29	29	27	27
	20	03-29	11-16	232	29	29	27	27
	16	03-20	11-26	252	29	29	27	27
STILLWATER RES	32	05-28	09-19	114	24	24	23	23
	28	05-15	09-28	136	23	23	23	23
	24	05-05	10-11	160	23	23	23	23
	20	04-22	10-27	187	23	23	23	23
	16	04-12	11-07	210	23	23	23	23
SYRACUSE WSO AP	32	04-30	10-15	168	29	29	29	29
	28	04-15	10-29	197	29	29	29	29
	24	04-04	11-14	224	29	29	29	29
	20	03-21	11-23	247	29	29	29	29
	16	03-14	12-03	265	29	29	29	28
TUPPER LAKE SUNMOUNT	32	05-28	09-16	111	26	26	24	24
	28	05-14	09-28	137	26	26	24	24
	24	04-26	10-10	166	25	25	24	24
	20	04-17	10-28	194	25	25	24	23
	16	04-06	11-10	217	25	25	22	22
UTICA FAA AP	32	05-14	10-01	140	20	20	18	18
	28	04-28	10-17	172	20	20	18	18
	24	04-12	11-01	203	20	20	18	18
	20	03-28	11-10	227	20	20	18	18
	16	03-20	11-25	251	20	20	18	18
UTICA 3 W	32	05-14	10-02	141	23	23	21	21
	28	04-30	10-19	173	23	23	21	21
	24	04-12	11-01	203	23	23	21	21
	20	03-28	11-13	230	23	23	21	21
	16	03-20	11-27	252	23	23	21	21
WANAKENA R SCHOOL	32	06-02	09-13	103	29	29	26	26
	28	05-14	09-28	137	29	29	26	26
	24	05-02	10-07	158	28	28	26	26
	20	04-22	10-22	183	28	28	26	26
	16	04-11	11-05	208	28	28	26	26
WATERTOWN	32	05-07	10-04	151	30	30	30	30
	28	04-26	10-18	176	30	30	30	30
	24	04-13	10-28	198	30	30	30	30
	20	03-31	11-13	228	30	30	30	30
	16	03-23	11-24	246	30	30	30	30
WEST POINT	32	04-18	10-26	191	25	25	26	26
	28	04-03	11-08	219	25	25	26	26
	24	03-21	11-25	249	24	24	26	26
	20	03-14	12-04	266	24	24	26	25
	16	03-09	12-08	274	24	24	26	25

Data in the above table are based on the period 1921-1950, or that portion of this period for which data are available.

Means have been adjusted to take into account years of non-occurrence.

A freeze is a numerical substitute for the former term "killing frost" and is the occurrence of a minimum temperature at or below the threshold temperature of 32°, 28°, etc.

Freeze data tabulations in greater detail are available and can be reproduced at cost.

*NORMALS BY CLIMATOLOGICAL DIVISIONS

Taken from "Climatography of the United States No. 81-4, Decennial Census of U. S. Climate"

TEMPERATURE (°F) / PRECIPITATION (In.)

STATIONS (By Divisions)	JAN	FEB	MAR	APR	MAY	JUNE	JULY	AUG	SEPT	OCT	NOV	DEC	ANN	JAN	FEB	MAR	APR	MAY	JUNE	JULY	AUG	SEPT	OCT	NOV	DEC	ANN
WESTERN PLATEAU																										
ALFRED	23.5	23.4	30.9	43.5	54.7	63.6	67.4	65.7	59.2	49.0	37.1	25.9	45.3	2.28	2.08	3.28	3.05	3.78	3.69	3.46	3.29	2.99	2.85	2.83	2.33	35.91
ALLEGANY STATE PARK	25.6	25.3	32.6	44.8	55.4	63.8	67.5	66.0	59.8	49.9	38.3	27.6	46.4	2.84	2.75	3.46	3.49	4.27	4.27	4.35	3.62	4.05	3.44	3.77	3.07	43.38
ANGELICA	24.5	24.2	32.0	44.5	55.3	64.5	68.6	66.7	59.9	49.5	37.9	26.7	46.2	2.11	1.87	2.84	2.72	3.37	3.46	3.38	2.89	2.88	2.52	2.59	2.09	32.72
CORNING	.	.	.	.	.	.	.	.	.	.	.	.	.	1.84	1.69	2.74	2.98	3.92	3.18	3.71	3.85	2.95	2.73	2.41	2.19	34.19
ELMIRA	27.1	27.0	34.7	47.2	58.4	67.6	72.0	69.7	62.4	51.7	40.5	29.7	49.0	1.88	1.92	2.98	3.00	3.94	3.37	3.58	3.97	2.84	2.70	2.49	2.22	34.89
HASKINVILLE	.	.	.	.	.	.	.	.	.	.	.	.	.	1.87	1.92	2.81	2.85	3.64	3.46	3.82	3.28	2.81	2.71	2.48	2.06	33.71
OLEAN	.	.	.	.	.	.	.	.	.	.	.	.	.	2.40	2.22	3.05	3.39	3.85	3.77	3.92	3.00	3.30	2.87	2.88	2.46	37.11
DIVISION	24.7	24.5	32.1	44.7	55.7	64.7	68.7	66.9	60.2	49.8	38.2	27.1	46.4	2.33	2.26	3.13	3.14	3.82	3.68	3.69	3.44	3.20	2.94	2.91	2.51	37.05
EASTERN PLATEAU																										
BAINBRIDGE	.	.	.	.	.	.	.	.	.	.	.	.	.	2.77	2.57	3.20	3.34	3.77	3.87	4.56	3.71	3.44	3.26	3.08	3.08	40.65
BINGHAMTON WSO	23.8	23.8	31.3	43.5	55.1	63.5	68.4	66.5	59.5	49.8	38.0	26.8	45.8	2.50	2.18	2.89	2.94	3.49	3.85	3.71	3.57	2.95	3.09	2.51	2.56	36.24
COOPERSTOWN	22.7	23.0	31.0	44.1	55.3	64.4	68.7	66.6	59.7	49.4	38.0	25.7	45.7	2.96	2.74	3.22	3.33	3.93	3.77	4.27	3.98	3.70	3.49	3.32	3.07	41.78
CORTLAND	23.7	23.5	31.3	44.3	55.8	65.2	70.1	68.1	60.6	50.3	38.6	26.7	46.5	2.81	2.83	3.54	3.21	3.78	3.70	4.16	3.74	3.33	3.36	2.99	3.21	40.66
DELHI 2 SW	23.7	23.9	31.6	44.2	55.4	64.2	68.4	66.6	59.8	49.5	38.1	26.2	46.0	2.75	2.66	3.09	3.32	3.89	3.92	4.62	4.21	3.90	3.45	3.56	2.86	42.23
FREEHOLD 2 NW	26.0	27.2	35.8	48.5	59.7	68.9	73.4	71.1	63.3	52.5	41.2	29.3	49.7	2.46	2.33	3.09	3.10	3.58	3.29	3.63	3.13	3.92	3.65	3.38	3.03	38.59
MORRISVILLE	20.8	20.7	29.0	42.3	53.5	62.9	67.4	65.4	58.1	47.7	36.6	24.0	44.0	2.53	2.55	2.98	3.07	3.65	3.57	3.74	3.81	3.44	3.42	2.96	2.80	38.72
NORWICH 1 NE	22.3	22.3	30.8	44.0	54.7	64.0	68.4	66.8	59.2	48.6	37.5	25.1	45.3	2.81	2.57	3.39	3.41	3.71	3.89	4.25	3.56	3.52	3.45	3.23	3.00	40.79
ROXBURY	23.4	23.9	31.8	44.1	55.2	63.8	68.3	66.5	59.2	49.0	38.0	26.2	45.8	3.01	2.62	3.44	3.58	3.98	3.87	4.04	3.90	3.75	3.63	3.51	2.94	42.27
SHERBURNE 2 S	.	.	.	.	.	.	.	.	.	.	.	.	.	2.39	2.16	2.95	3.20	3.45	3.59	4.13	3.50	3.67	3.26	2.94	2.83	38.07
DIVISION	23.7	24.0	32.1	44.8	56.0	65.0	69.5	67.6	60.3	50.0	38.5	26.5	46.5	2.81	2.61	3.24	3.38	3.78	3.70	4.17	3.89	3.60	3.40	3.24	2.97	40.79
NORTHERN PLATEAU																										
BIG MOOSE 3 E	.	.	.	.	.	.	.	.	.	.	.	.	.	4.02	3.65	3.74	3.79	4.00	3.88	4.59	3.76	4.73	4.39	4.48	4.48	49.51
HIGHMARKET	.	.	.	.	.	.	.	.	.	.	.	.	.	4.12	3.77	4.09	4.04	4.44	3.62	4.53	4.07	4.98	5.11	4.70	4.55	52.02
HOFFMEISTER 3 W	.	.	.	.	.	.	.	.	.	.	.	.	.	4.12	3.52	4.42	4.28	4.33	4.56	5.04	4.07	4.93	4.62	4.66	4.40	52.95
HOPE	.	.	.	.	.	.	.	.	.	.	.	.	.	3.81	3.32	3.78	3.80	3.67	3.74	4.46	3.53	4.07	3.68	4.08	3.95	45.89
INDIAN LAKE 2 SW	16.5	16.9	25.4	30.3	50.8	59.9	64.1	62.2	55.0	44.5	32.8	19.6	40.5	3.16	2.78	3.29	3.11	3.35	3.72	4.32	3.18	4.28	3.62	3.69	3.44	41.94
LAKE PLACID CLUB	14.9	15.6	24.8	38.6	51.3	60.7	64.9	62.7	55.2	44.8	32.1	18.4	40.3	3.13	2.97	3.24	2.70	3.25	3.67	3.96	3.41	3.74	2.92	2.98	3.19	39.16
LOWVILLE	18.3	19.2	28.3	42.7	54.9	64.3	68.7	66.8	59.2	48.5	35.9	22.3	44.1	2.86	2.51	3.01	3.05	3.25	2.80	3.27	3.08	3.25	3.41	3.54	3.36	37.39
LYONS FALLS	.	.	.	.	.	.	.	.	.	.	.	.	.	3.53	2.92	3.30	3.30	3.72	3.12	4.11	3.27	4.18	4.09	4.12	3.88	43.54
SOUTH EDWARDS 1 E	.	.	.	.	.	.	.	.	.	.	.	.	.	3.03	2.88	3.05	3.35	3.89	3.47	4.10	3.79	4.26	3.94	3.67	3.43	43.12
STILLWATER RESERVOIR	14.4	14.5	23.9	38.5	51.7	61.1	65.5	63.9	56.3	45.4	32.8	18.2	40.5	3.81	3.63	4.08	3.97	4.06	3.91	4.82	3.89	4.56	4.45	4.32	4.41	49.91
TUPPER LAKE SUNMOUNT	16.3	17.0	25.7	39.3	52.1	61.3	65.4	63.4	55.8	45.4	33.1	19.6	41.2	2.44	2.29	2.60	2.64	3.36	3.38	3.97	3.83	3.65	3.12	2.79	2.67	36.74
WANAKENA RANGER SCHOOL	16.8	17.4	26.2	40.3	52.9	61.6	65.7	63.9	56.6	46.1	33.9	20.2	41.8	3.06	2.83	3.17	3.24	3.61	3.53	4.18	3.48	4.02	3.81	3.50	3.50	41.93
DIVISION	16.4	17.0	25.8	39.6	52.3	61.4	65.7	63.7	56.3	45.8	33.5	19.9	41.5	3.26	2.97	3.35	3.24	3.58	3.58	4.21	3.59	4.06	3.71	3.64	3.63	42.82
COASTAL																										
BRIDGEHAMPTON	32.0	31.9	37.6	46.6	56.1	65.3	71.3	70.7	64.4	55.1	45.3	34.8	50.9	4.20	3.59	4.61	3.62	3.44	2.88	2.92	4.42	3.67	3.55	4.66	4.10	45.66
NEW YORK CNTRL PK WSO	33.2	33.4	40.5	51.4	62.4	71.4	76.8	75.1	68.5	58.3	47.0	35.9	54.5	3.31	2.84	4.01	3.43	3.67	3.31	3.70	4.44	3.87	3.14	3.39	3.26	42.37
NY JOHN F KENNEDY INAP	31.8	31.6	38.7	49.0	60.2	70.1	75.9	74.5	67.8	57.6	46.2	34.9	53.2	3.23	2.93	4.15	3.48	3.67	3.35	4.04	4.97	4.16	3.21	3.51	3.23	43.93
N Y LA GUARDIA WSO	33.6	33.6	40.8	51.2	62.1	71.5	76.8	75.4	68.8	58.6	47.4	36.4	54.7	3.31	3.09	4.23	3.57	3.58	3.38	3.71	5.08	3.92	3.37	3.59	3.39	44.22
SCARSDALE	30.5	31.2	38.5	49.7	60.5	69.3	74.3	72.7	65.6	55.3	44.2	33.1	52.1	3.36	2.78	4.39	4.10	4.21	3.79	4.51	4.90	4.40	3.81	4.10	3.73	48.08
SETAUKET	33.0	32.8	39.2	49.5	59.8	68.4	73.8	72.5	66.3	57.1	46.8	35.8	52.9	3.87	3.19	4.26	3.70	3.55	3.40	3.55	4.10	3.91	3.36	4.12	3.64	44.65
DIVISION	32.2	32.4	39.2	49.4	59.9	69.1	74.5	73.1	66.5	56.5	45.9	35.0	52.8	3.64	3.14	4.37	3.75	3.71	3.39	3.78	4.68	3.91	3.46	3.97	3.67	45.47
HUDSON VALLEY																										
ALBANY WSO	22.7	23.7	33.0	46.2	57.9	67.3	72.1	70.0	61.6	50.8	39.1	26.5	47.6	2.47	2.20	2.72	2.77	3.47	3.25	3.49	3.07	3.58	2.77	2.70	2.59	35.08
ALBANY	25.7	26.7	35.7	48.4	59.9	69.0	74.0	71.7	63.7	53.1	41.7	29.4	49.9	2.51	2.26	2.86	2.90	3.62	3.74	4.29	3.30	4.00	2.84	2.90	2.73	37.95
BEDFORD HILLS	29.6	30.5	38.2	49.8	60.8	69.3	74.3	72.4	65.1	54.8	43.1	32.1	51.7	3.32	2.85	4.00	4.02	4.07	3.99	5.00	4.57	4.11	3.70	3.92	3.99	47.54
CARMEL 1 SW	26.4	27.0	35.1	47.2	57.8	66.5	71.7	70.0	62.8	52.8	41.4	29.4	49.0	3.34	2.75	3.76	3.70	4.35	3.90	4.71	4.61	4.24	3.63	4.12	3.59	46.70
CONKLINGVILLE DAM	.	.	.	.	.	.	.	.	.	.	.	.	.	3.45	3.13	3.60	3.47	3.28	3.55	3.97	3.46	3.56	3.33	3.73	3.63	42.16
MECHANICVILLE 2 S	.	.	.	.	.	.	.	.	.	.	.	.	.	2.51	1.97	2.55	3.01	3.47	3.82	3.97	3.17	3.94	3.06	2.93	2.68	37.08
MOHONK LAKE	26.0	26.8	34.5	46.7	58.1	66.1	70.8	68.9	61.9	52.0	40.6	28.8	48.4	3.48	3.08	3.91	4.44	4.33	3.98	4.71	4.20	4.26	3.88	4.03	3.66	47.96
POUGHKEEPSIE	27.3	28.5	37.3	49.5	60.7	69.6	74.6	72.4	64.6	54.0	42.6	30.5	51.0	2.89	2.45	3.09	3.65	3.57	3.48	4.13	3.89	3.67	3.04	3.35	3.00	40.21
SCHENECTADY	23.6	24.6	33.7	46.9	59.0	68.3	73.2	70.8	62.4	51.4	39.8	27.6	48.4	2.57	2.29	2.87	2.84	3.28	3.46	3.26	3.55	2.91	2.68	2.62	2.60	35.83
SMITHS BASIN	.	.	.	.	.	.	.	.	.	.	.	.	.	2.64	2.14	2.71	3.25	3.19	3.95	4.52	3.03	3.64	2.87	3.29	2.60	37.83
SPIER FALLS	22.1	23.2	32.7	46.0	58.1	67.4	71.8	69.9	62.1	51.2	39.2	26.2	47.5	3.09	2.64	3.14	3.33	3.26	3.42	3.88	3.17	3.23	2.95	3.38	3.19	38.68
WARWICK	.	.	.	.	.	.	.	.	.	.	.	.	.	3.00	2.53	3.68	3.73	3.79	4.04	4.36	4.50	3.94	3.30	4.03	3.39	44.29
WEST POINT	28.2	29.3	37.5	49.6	60.6	69.6	74.9	73.0	65.5	54.9	43.2	31.3	51.5	3.34	2.96	4.16	4.08	4.20	3.85	4.40	4.08	4.18	3.56	4.10	3.80	46.71
DIVISION	25.2	26.2	34.7	47.3	58.7	67.5	72.4	70.3	62.6	52.0	40.6	28.5	48.8	3.06	2.62	3.42	3.62	3.82	3.79	4.30	3.83	3.92	3.29	3.61	3.23	42.51

*NORMALS BY CLIMATOLOGICAL DIVISIONS

Taken from "Climatography of the United States No. 81-4, Decennial Census of U. S. Climate"

	TEMPERATURE (°F)													PRECIPITATION (In.)												
STATIONS (By Divisions)	JAN	FEB	MAR	APR	MAY	JUNE	JULY	AUG	SEPT	OCT	NOV	DEC	ANN	JAN	FEB	MAR	APR	MAY	JUNE	JULY	AUG	SEPT	OCT	NOV	DEC	ANN
MOHAWK VALLEY																										
CANAJOHARIE	•	•	•	•	•	•	•	•	•	•	•	•	•	2.31	1.98	2.45	2.82	3.50	3.49	3.84	3.51	3.53	2.86	2.61	2.44	35.34
DELTA	•	•	•	•	•	•	•	•	•	•	•	•	•	3.72	3.28	3.42	3.68	3.76	3.63	3.90	3.61	3.74	3.94	3.83	3.82	44.33
DOLGEVILLE	•	•	•	•	•	•	•	•	•	•	•	•	•	3.25	2.78	3.13	3.46	3.92	3.94	4.74	3.69	4.39	3.61	3.36	3.12	43.39
FRANKFORT LOCK 19	•	•	•	•	•	•	•	•	•	•	•	•	•	2.77	2.37	2.61	3.27	3.50	3.63	4.17	3.50	4.20	3.59	3.42	2.70	39.73
GLOVERSVILLE	21.4	22.6	31.2	44.7	56.8	66.1	70.5	68.3	60.3	49.5	37.7	25.0	46.2	3.41	2.89	3.35	3.61	3.83	3.90	4.27	3.66	4.07	3.64	3.44	3.17	43.24
HINCKLEY	•	•	•	•	•	•	•	•	•	•	•	•	•	4.26	3.46	3.99	4.06	4.14	3.84	4.36	3.99	4.46	4.34	4.32	4.16	49.38
LITTLE FALLS CITY RES	21.0	21.7	30.3	44.1	56.2	65.3	70.1	68.2	60.7	49.8	37.5	24.4	45.8	2.85	2.30	2.82	3.37	3.64	3.92	4.51	3.93	4.36	3.56	3.22	2.83	41.31
LITTLE FALLS MILL ST	•	•	•	•	•	•	•	•	•	•	•	•	•	3.32	2.89	3.19	3.58	3.70	3.79	4.42	3.77	4.27	3.59	3.48	3.24	43.24
NEW LONDON LOCK 22	•	•	•	•	•	•	•	•	•	•	•	•	•	3.13	2.98	3.03	3.44	3.65	3.42	3.96	3.54	3.57	3.90	3.64	3.29	41.55
SALISBURY	18.4	18.9	27.8	41.4	53.2	62.4	66.9	64.9	57.3	46.8	35.1	21.8	42.9	3.71	3.18	3.33	3.94	4.14	4.10	4.94	4.08	4.61	4.23	4.01	3.53	47.80
TRENTON FALLS	•	•	•	•	•	•	•	•	•	•	•	•	•	4.00	3.35	3.70	4.06	4.11	4.20	4.73	3.92	4.47	4.38	4.32	4.17	49.41
TRIBES HILL	•	•	•	•	•	•	•	•	•	•	•	•	•	3.21	2.53	3.03	3.16	3.65	3.23	3.82	3.20	3.77	3.20	3.00	2.90	38.70
DIVISION	20.7	21.5	30.2	43.8	55.8	65.2	69.8	67.8	60.2	49.3	37.5	24.4	45.5	3.32	2.76	3.14	3.55	3.74	3.92	4.46	3.88	4.13	3.73	3.54	3.23	43.40
CHAMPLAIN VALLEY																										
DANNEMORA	17.9	19.1	27.9	42.0	55.1	64.5	69.1	67.3	59.7	48.5	35.4	21.6	44.0	2.23	2.18	2.28	2.75	3.26	3.45	3.35	3.09	3.28	2.88	2.49	2.48	33.72
WHITEHALL	•	•	•	•	•	•	•	•	•	•	•	•	•	2.89	2.55	2.86	3.07	3.09	3.42	3.90	3.20	3.43	2.84	3.27	2.92	37.44
DIVISION	18.4	19.6	29.1	43.3	55.9	65.3	69.9	67.9	59.9	48.7	36.5	22.7	44.8	2.26	2.11	2.34	2.72	3.05	3.27	3.39	3.00	3.25	2.72	2.60	2.44	33.15
ST. LAWRENCE VALLEY																										
CANTON 4 SE	17.3	18.6	28.7	43.3	55.6	65.1	69.7	67.7	59.6	48.7	36.4	21.9	44.4	2.41	2.32	2.50	3.01	3.35	3.11	3.59	3.18	3.46	3.03	2.90	2.79	35.65
LAWRENCEVILLE	17.2	18.5	28.4	43.0	55.8	65.2	69.9	67.9	60.0	48.7	36.3	21.5	44.4	1.70	1.82	1.96	2.71	3.39	3.13	3.49	3.23	3.50	3.12	2.36	2.10	32.51
OGDENSBURG 3 NE	17.9	19.7	29.2	44.0	55.9	65.6	70.7	69.0	60.9	50.2	37.7	23.1	45.3	2.09	1.97	2.26	2.54	2.92	2.60	2.93	2.86	2.96	2.55	2.58	2.44	30.70
DIVISION	17.5	18.9	28.7	43.1	55.3	64.9	69.6	67.6	59.7	48.8	36.6	22.2	44.4	2.32	2.26	2.49	2.93	3.32	2.96	3.49	3.06	3.48	3.11	2.86	2.74	35.02
GREAT LAKES													—													
BATAVIA	•	•	•	•	•	•	•	•	•	•	•	•	•	2.25	2.24	2.68	2.93	3.08	2.62	2.94	2.96	2.73	2.59	2.59	2.30	31.91
BUFFALO WSO	24.5	24.1	31.5	43.5	54.8	64.8	69.8	68.4	61.4	50.8	39.1	27.7	46.7	2.84	2.72	3.24	3.01	2.95	2.54	2.57	3.05	3.13	3.00	3.60	3.00	35.65
CLYDE LOCK 26	•	•	•	•	•	•	•	•	•	•	•	•	•	2.33	2.59	3.02	3.18	3.71	3.06	3.40	3.00	3.20	3.61	3.07	2.69	36.86
FREDONIA	28.0	27.5	34.3	46.5	57.4	67.9	72.2	71.0	64.7	53.9	42.1	31.4	49.7	2.66	2.13	3.02	3.15	3.12	3.16	3.00	3.27	3.80	3.47	3.33	2.64	36.75
LOCKPORT 2 NE	25.5	25.3	32.6	44.9	55.9	66.2	71.2	69.6	62.5	51.8	40.0	28.9	47.9	2.44	2.38	2.54	2.69	3.19	2.37	2.61	3.02	2.93	2.73	2.67	2.40	31.97
MACEDON	•	•	•	•	•	•	•	•	•	•	•	•	•	2.03	2.40	2.73	2.88	3.25	2.75	2.98	2.84	2.73	3.25	2.49	2.33	32.66
NEWARK	•	•	•	•	•	•	•	•	•	•	•	•	•	1.99	2.18	2.62	2.71	3.35	2.79	3.13	2.72	2.70	3.39	2.60	2.23	32.41
OSWEGO EAST	25.1	25.4	32.6	44.0	54.6	64.5	70.5	69.4	62.3	52.2	40.8	29.0	47.5	2.70	2.62	2.80	2.72	2.97	2.28	2.74	2.51	2.78	3.26	3.01	3.17	33.56
ROCHESTER WSO	25.2	24.9	32.3	45.1	56.7	66.9	71.6	69.7	62.4	51.8	40.1	28.7	48.0	2.40	2.53	3.01	2.67	2.77	2.56	2.84	2.72	2.53	2.58	2.51	2.38	31.50
SOUTH WALES EMERY PARK	24.2	23.9	31.5	44.1	54.9	64.7	69.3	67.5	60.5	50.2	38.4	27.3	46.4	3.33	2.93	3.61	3.50	3.47	3.27	3.29	2.86	3.55	3.50	3.91	3.53	40.75
WATERTOWN	20.4	21.4	30.8	44.5	56.2	66.1	70.5	69.4	61.6	50.6	38.7	24.8	46.3	3.03	2.57	2.67	3.11	3.39	2.72	3.33	3.21	3.71	3.68	3.78	3.65	38.85
DIVISION	25.5	25.4	32.8	45.2	56.3	66.4	71.1	69.6	62.6	52.0	40.3	28.9	48.0	2.67	2.49	2.93	2.97	3.16	2.76	2.94	2.88	3.15	2.99	3.06	2.83	34.83
CENTRAL LAKES																										
AUBURN WATER WORKS	26.0	25.3	32.2	44.4	55.7	66.4	72.1	70.8	63.7	52.9	42.4	30.8	48.6	2.27	2.38	2.98	2.69	2.93	2.86	3.18	2.45	2.50	2.91	2.69	2.58	32.42
BALDWINSVILLE	•	•	•	•	•	•	•	•	•	•	•	•	•	2.94	3.16	3.46	3.04	3.36	3.22	3.19	3.16	3.21	3.52	3.25	3.09	38.60
CAYUGA LOCK 1	•	•	•	•	•	•	•	•	•	•	•	•	•	2.11	2.29	2.80	2.98	3.31	2.93	3.49	2.74	2.94	3.26	2.78	2.43	34.06
HEMLOCK	25.4	24.9	32.2	44.9	56.7	66.8	71.2	69.3	62.0	51.6	40.2	29.0	47.9	1.84	1.88	2.93	2.69	3.25	3.03	2.87	2.86	2.58	2.73	2.39	1.99	31.04
MAYS POINT LOCK 25	•	•	•	•	•	•	•	•	•	•	•	•	•	2.22	2.39	2.83	2.74	3.28	2.98	3.54	2.66	2.91	3.34	2.74	2.54	34.17
PENN YAN 2 SW	26.3	26.2	33.6	46.4	57.7	67.4	72.2	70.6	63.4	52.5	40.8	29.4	48.9	2.06	1.91	2.80	2.81	3.32	2.99	3.72	2.85	2.31	2.79	2.27	1.99	31.82
SKANEATELES	•	•	•	•	•	•	•	•	•	•	•	•	•	2.43	2.61	3.22	3.10	3.53	3.33	3.64	2.90	3.19	3.44	3.02	2.77	37.18
SYRACUSE WSO	24.0	24.3	32.6	46.0	57.7	67.3	72.2	70.2	62.4	51.8	40.2	27.8	48.0	3.15	3.13	3.60	3.08	3.27	2.96	3.09	3.25	2.84	3.18	2.90	3.15	37.60
WATERLOO	•	•	•	•	•	•	•	•	•	•	•	•	•	2.20	2.27	2.82	2.89	3.27	2.87	3.49	2.65	2.70	3.11	2.51	2.21	32.99
DIVISION	25.7	25.5	33.0	45.7	56.9	66.8	71.4	69.6	62.5	51.9	40.5	29.0	48.2	2.08	2.12	2.83	2.82	3.16	3.08	3.30	3.01	2.69	2.84	2.51	2.22	32.66

* Normals for the period 1931-1960. Divisional normals may not be the arithmetical average of individual stations published, since additional data for shorter period stations are used to obtain better areal representation.

CONFIDENCE - LIMITS

In absence of trend or record changes, the chances are 9 out of 10 that the true mean will lie in the interval formed by adding and subtracting the values in the following table from the means for any station in the State. Because of the wider variation in mean precipitation, the corresponding monthly means and annual mean must be substituted for "p" in the precipitation table below to obtain mean precipitation confidence limits.

| 1.6 | 1.4 | 1.5 | 1.1 | .9 | .7 | .6 | .8 | .9 | 1.0 | 1.0 | 1.2 | .5 | .21√p | .17√p | .21√p | .21√p | .25√p | .27√p | .26√p | .27√p | .31√p | .31√p | .26√p | .24√p | .25√p |

COMPARATIVE DATA

Data in the following table are the mean temperature and average precipitation for Geneva Experiment Station, New York, for the period 1906-1930 and are included in this publication for comparative purposes.

| 25.2 | 24.6 | 34.4 | 45.8 | 56.7 | 66.3 | 71.6 | 69.5 | 63.7 | 51.7 | 40.1 | 28.9 | 48.2 | 2.01 | 1.65 | 2.24 | 3.01 | 3.27 | 3.56 | 3.27 | 3.12 | 2.80 | 2.62 | 2.40 | 2.02 | 31.97 |

NORMALS, MEANS AND EXTREMES
(Table Revised 1972. Base Period for Climatological Normals: 1931-1960)

Station: ALBANY COUNTY AIRPORT — ALBANY, NEW YORK
Standard time used: EASTERN Latitude: 42° 45' N Longitude: 73° 48' W Elevation (ground): 275 feet

Month	Normal Daily max	Normal Daily min	Normal Monthly	Rec. highest	Yr	Rec. lowest	Yr	Normal heating degree days (Base 65°)	Precip. Normal total	Max. monthly	Yr	Min. monthly	Yr	Max. 24 hrs	Yr	Snow Mean total	Max. monthly	Yr	Max. 24 hrs	Yr	RH 01	RH 07	RH 13	RH 19	Mean wind spd	Prev. dir	Fastest spd	Dir	Yr	Mean sky cover	Pct. sunshine	Clear	Partly cloudy	Cloudy	Precip ≥.01	Snow ≥1.0	Tstms	Heavy fog	Max 90°+	Max 32°-	Min 32°-	Min 0°-
J	31.0	14.4	22.7	57	1966	-28	1971	1311	2.47	4.12	1958	0.73	1968	1.68	1952	15.3	28.8	1966+	15.4	1964	80	79	64	74	9.7	W	57	W	1952	6.9	46	8	8	17	13	4	*	1	0	6	30	11
F	32.5	14.8	23.7	54	1966	-18	1973	1156	2.20	4.14	1950	0.36	1968	1.42	1953	15.2	34.5	1962	17.9	1958	74	76	59	68	10.3	NW	71	NW	1953	6.9	51	8	8	15	12	3	*	1	0	3	27	5
M	41.9	24.0	33.0	77	1968	-10	1967	992	2.72	5.74	1953	1.07	1949	1.53	1964	12.8	34.7	1956	15.9	1956	71	74	55	63	10.4	WNW	55	W	1941	6.9	53	8	10	16	13	3	1	1	0	1	27	*
A	56.7	35.7	46.2	88	1970	10	1965	564	2.77	4.71	1953	1.14	1963	2.20	1968	2.3	8.4	1957	7.2	1955	59	69	46	55	10.4	WNW	49	W	1954+	6.9	53	6	10	16	13	1	1	1	0	0	16	0
M	69.5	46.3	57.9	92	1970	26	1968	239	3.47	8.96	1953	1.22	1968	2.17	1968	0.1	1.4	1966	1.4	1966	69	59	55	59	8.1	WNW	50	W	1950	6.5	60	5	12	13	11	*	4	2	*	0	3	0
J	78.7	55.8	67.3	98	1968	36	1967	45	3.25	5.30	1969	0.65	1952	3.48	1952	0.0	0.0		0.0		82	77	55	63	7.3	S	57	NW	1971	6.7	64	6	13	11	11	0	6	1	2	0	0	0
J	83.7	60.5	72.1	98	1966	43	1969	19	3.49	5.89	1968	0.49	1968	2.70	1970	0.0	0.0		0.0		82	77	53	62	7.3	S	43	NW	1938	6.0	64	8	13	10	11	0	7	1	4	0	0	0
A	81.4	58.5	70.0	98	1970+	37	1965	30	3.07	7.33	1950	0.73	1947	4.52	1971	0.0	0.0		0.0		85	81	57	67	7.0	S	38	S	1950+	6.0	62	8	12	11	10	0	5	3	4	0	*	0
S	72.9	50.3	61.6	91	1970+	30	1965	138	3.58	7.89	1960	0.40	1964	3.66	1955	0.0	0.0		0.0		86	85	54	73	7.9	S	48	NW	1961	6.0	57	8	10	12	9	0	3	4	1	0	1	0
O	61.8	39.8	50.8	84	1968	16	1969	440	2.77	8.83	1955	0.40	1963	2.65	1959	0.1	2.0	1952	2.0	1952	82	84	55	74	8.6	S	45	NE	1939	5.4	55	8	10	13	10	*	1	5	*	0	8	0
N	47.5	30.6	39.1	75	1971	5	1969	777	2.70	5.56	1969	1.17	1960	2.01	1971	3.8	24.0	1971	21.9	1971	80	82	63	63	8.8	S	70	E	1950	6.1	37	6	8	18	12	1	*	2	0	1	18	*
D	34.5	18.5	26.5	65	1969	-22	1969	1194	2.59	6.51	1969	0.64	1969	4.02	1958	16.2	57.5	1969	18.3	1966	79	81	67	75	9.1	S	54	W	1944	7.2	40	5	8	18	12	4	*	3	0	5	28	3
YR	57.7	37.4	47.6	98	JUN 1968+	-28	JAN 1971	6875	35.08	8.96	MAY 1953	0.20	OCT 1963	4.52	AUG 1971	65.8	57.5	DEC 1969	21.9	NOV 1971	79	73	57	67	8.5	S	71	NW	FEB 1953	6.7	54	72	114	179	133	16	28	23	8	51	158	18

Ø For period February 1965 through current year.
Means and extremes above are from existing and comparable exposures. Annual extremes have been exceeded at other sites in the locality as follows: Highest temperature 104 in July 1911; maximum monthly precipitation 13.48 in October 1869; minimum monthly precipitation 0.08 in January 1860; maximum precipitation in 24 hours 4.75 in October 1903; maximum snowfall in 24 hours 30.4 in March 1888.

NORMALS, MEANS AND EXTREMES
(Table Revised 1975. Base Period for Climatological Normals: 1941-1970)

Station: ALBANY COUNTY AIRPORT — ALBANY, NEW YORK (Elev. 292 feet m.s.l.; Average station pressure)

Month	Normal Daily max	Normal Daily min	Normal Monthly	Rec. highest	Yr	Rec. lowest	Yr	Heating DD	Cooling DD	Precip. Normal	Max. monthly	Yr	Min. monthly	Yr	Max. 24 hrs	Yr	Snow Mean total	Max. monthly	Yr	Max. 24 hrs	Yr	RH 01	RH 07	RH 13	RH 19	Mean wind spd	Prev. dir	Fastest spd	Dir	Yr	Mean sky cover	Pct. sunshine	Clear	Partly cloudy	Cloudy	Precip ≥.01	Snow ≥1.0	Tstms	Heavy fog	Max 90°+	Max 32°-	Min 32°-	Min 0°-	Sta. press mb
J	30.4	12.5	21.5	62	1974	-28	1971	1349	0	2.20	4.12	1958	0.73	1955	1.68	1952	28.8			15.4	1964	78	79	63	73	9.8	WNW	57	W	1952	7.0	46	6	8	17	12	4	*	1	0	9	30	9	1008.5
F	32.7	14.3	23.5	57	1974	-21	1973	1162	0	2.11	4.14	1950	0.36	1968	1.42	1953	34.5			17.9	1958	76	76	59	68	10.3	WNW	71	W	1953	6.9	51	6	8	11	11	3	*	1	0	5	27	5	1007.6
M	42.6	24.2	33.4	77	1967	-10	1967	980	0	2.58	5.74	1953	1.07	1949	1.53	1964	34.7			15.9	1956	72	74	55	62	10.5	WNW	55	W	1941	6.9	52	6	9	16	12	3	1	1	0	*	26	*	1003.8
A	58.0	35.7	46.9	92	1970	10	1965	543	0	2.70	4.71	1953	1.22	1965	2.20	1968	8.4			10.7	1955	70	70	44	56	10.4	WNW	49	W	1954+	6.9	55	6	9	15	11	1	1	2	*	0	15	0	1003.2
M	69.7	45.7	57.7	92	1970	26	1968	253	27	3.26	8.96	1953	0.65	1952	2.17	1968	1.4			1.4	1966	74	60	52	60	9.1	S	50	W	1950	6.6	55	5	11	15	12	*	4	2	2	0	2	0	1002.7
J	78.?	55.5		98	1968	36	1967	39	114	3.00	7.36	1973	0.49	1960	3.48	1952	0.0			0.0		83	79	54	63	7.3	S	43	NW	1938	6.0	64	7	13	10	11	0	6	4	2	0	0	0	1004.7
J	83.9	60.1	72.0	98	1973	43	1969	22	226	3.12	5.89	1947	0.73	1947	2.70	1970	0.0			0.0		83	81	55	75	7.3	S	38	S	1950+	6.0	61	8	12	11	11	0	7	2	4	0	0	0	1007.5
A	81.5	58.5	69.6	98	1973	37	1965	135	165	3.12	7.33	1950	0.40	1960	4.52	1971	0.0			0.0		84	84	57	75	6.9	S	48	S	1961	6.0	57	8	12	11	10	0	5	3	2	0	1	0	1007.1
S	73.7	51.4	62.6	93	1973	30	1965	422	42	2.63	7.89	1960	0.20	1963	3.66	1972	0.0			0.0		84	85	55	73	7.9	S	64	NE	1939	6.1	55	8	9	13	9	*	3	4	1	0	0	0	1005.9
O	62.8	40.0	51.4	84	1958	16	1969	762	0	2.84	8.07	1973	1.17	1960	2.65	1959	2.0			2.01		84	82	55	77	7.5	S	45	E	1950	5.4	56	7	9	15	8	*	1	5	0	0	9	0	1005.4
N	48.1	31.1	39.6	75	1974	5	1974	762	0	2.84	6.73	1969	0.64	1958	2.01	1966	24.6			21.9	1971	81	82	64	77	8.9	S	70	E	1950	6.5	38	4	8	18	13	2	*	2	0	2	18	0	1005.4
D	34.1	17.7	25.9	65	1970	-22	1970	1212	0	2.93	6.73	1973	0.64	1958	4.02	1948	57.5			18.3	1966	82	82	68	—	9.1	S	54	W	1944	7.3	—	5	7	19	13	3	*	2	0	9	28	2	1007.0
YR	58.1	37.1	47.6	98	JUN 1968+	-28	JAN 1971	6888	574	33.36	8.96	MAY 1953	0.20	OCT 1963	4.52	AUG 1971	57.5		DEC 1969	21.9	NOV 1971	79	79	57	68	8.5	S	71	NW	FEB 1953	6.7	53	72	109	184	135	16	28	24	8	48	156	17	1006.2

Ø For period February 1965 through current year.
Means and extremes above are from existing and comparable exposures. Annual extremes have been exceeded at other sites in the locality as follows: Highest temperature 104 in July 1911; maximum monthly precipitation 13.48 in October 1869; minimum monthly precipitation 0.28 in January 1860; maximum precipitation in 24 hours 4.75 in October 1903; maximum snowfall in 24 hours 30.4 in March 1888.

REFERENCE NOTES APPLYING TO TABLES APPEAR ON THE PAGE FOLLOWING LAST TABLE.
(Caution: Letters and symbols may have different meanings in 1941-1970 tables than in earlier tables. See notes.)

NORMALS, MEANS AND EXTREMES
(Table Revised 1972. Base Period for Climatological Normals: 1931-1960)

Station: BINGHAMTON, NEW YORK BROOME COUNTY AIRPORT   Standard time used: EASTERN   Latitude: 42° 13' N   Longitude: 75° 59' W   Elevation (ground): 1590 feet

Month	Temperature Normal Daily max (b)	Daily min (b)	Monthly (b)	Extremes Record highest	Year	Record lowest	Year	Normal heating degree days (Base 65°)	Precip Normal total (b)	Max monthly	Year	Min monthly	Year	Max in 24 hrs	Year	Snow Mean total	Max monthly	Year	Max in 24 hrs	Year	RH 01	RH 07	RH 13	RH 19	Wind mean speed	Prevailing dir	Fastest mile speed	Dir	Year	Mean sky cover	Pct sunshine	Clear	Partly cloudy	Cloudy	Precip .01+	Snow 1.0+	Thunderstorms	Heavy fog	Max 90°+	Max 32°-	Min 32°-	Min 0°-	Solar langleys
(yrs)	20	(b)	(b)	20		20		20	(b)	20		20		20		20	20		20		20	20	20	20	20	12	20	20	20	20	37	20	20	20	20	20	20	20	20	20	20	20	20
J	30.1	17.4	23.8	63	1957	-20	1957	1277	2.50	4.31	1958	0.76	1970	1.80	1958	19.9	36.6	1966	18.4	1964	81	79	75	81	11.7	WSW	59	NW	1959	8.0	37	3	6	22	17	6	*	4	0	*	30	30	4
F	31.0	16.0	23.5	66	1961	-15	1961	1154	2.18	4.36	1971	0.51	1968	2.16	1966	19.6	39.4	1966	23.0	1961	81	78	73	83	11.7	SSE	66	NW	1956	7.7	43	4	7	17	16	5	*	4	0	*	27	27	3
M	39.0	23.0	31.3	72	1968	-6	1967	1045	2.89	5.11	1956	0.69	1969	1.95	1964	16.1	33.5	1971	15.8	1971	81	78	65	76	11.8	NW	61	NW	1955	7.4	45	5	6	20	16	4	1	5	0	*	21	10	*
A	52.9	34.1	43.5	84	1962	11	1954	645	2.94	5.09	1961	1.61	1960	1.77	1961	5.0	16.4	1957	11.5	1960	78	70	57	66	11.8	NNW	54	W	1954	7.2	50	5	11	14	13	1	3	4	0	*	7	1	0
M	65.3	44.8	55.1	87	1966+	25	1966+	313	3.49	6.46	1968	1.78	1962	1.75	1952	0.3	3.4	1966	3.4	1966	79	56	50	56	10.3	NNW	54	W	1954	6.7	57	5	11	15	13	0	4	5	*	2	0	0
J	73.4	53.6	63.5	94	1952	36	1958	99	3.85	4.85	1960	1.15	1960	3.11	1960	0.0	0.0		0.0		81	58	52	56	9.5	WSW	60	SW	1958	6.6	58	5	11	14	11	0	7	3	1	0	0	0
J	78.3	58.5	68.4	95	1966+	39	1952	22	3.71	7.40	1956	0.83	1952	3.01	1952	0.0	0.0		0.0		84	60	52	58	8.5	WSW	58	N	1956	6.5	62	5	13	13	10	0	7	4	1	0	0	0
A	76.2	56.8	66.5	96	1955	37	1965	65	3.57	7.48	1955	0.61	1953	3.19	1959	0.0	0.0		0.0		85	62	53	60	8.3	WSW	58	N	1958	6.1	59	6	15	10	10	0	5	4	1	0	0	0
S	68.6	49.5	59.5	96	1953	26	1963	201	2.95	5.57	1963	0.61	1961	3.40	1961	0.0	T	1970+	T	1970+	90	62	57	62	9.0	SSW	42	NW	1960	5.9	59	7	10	13	9	*	2	5	*	0	1	0
O	58.7	40.8	49.8	82	1971	21	1969	471	2.51	5.62	1955	0.26	1963	2.10	1955	0.4	1.9	1952	1.3	1953	85	61	61	61	9.8	WSW	72	SE	1954	5.6	56	6	8	16	11	*	1	4	0	*	6	18	1
N	44.7	31.3	38.0	74	1971	-1	1958	810	3.09	5.42	1955	1.01	1960	1.90	1967	8.0	10.1	1954	10.1	1953	80	75	67	71	11.1	WSW	57	SE	1951	7.7	30	2	6	22	14	2	*	5	0	*	18	28	1
D	32.7	20.8	26.8	64	1966	-8	1969	1184	2.56	4.85	1969	0.94	1960	1.51	1952	18.9	59.6	1969	15.6	1969	82	84	78	78	11.4	WSW	59	NW	1956	8.4	27	2	6	23	18	6	*	5	0	1	28	28	8
YR	54.2	37.4	45.8	96	SEP. 1953	-20	JAN. 1957	7286	36.24	9.46	JUN. 1960	0.26	OCT. 1963	3.88	OCT. 1955	88.2	59.6	DEC. 1969	23.0	FEB. 1961	80	83	64	70	10.4	WSW	72	S	OCT. 1954	7.3	51	50	100	215	161	25	31	52	3	67	148	8

Means and extremes above are from existing and comparable exposures. Annual extremes have been exceeded at other sites in the locality as follows:
Highest temperature 103 in July 1936; lowest temperature -28 in January 1893; maximum monthly precipitation 9.59 in October 1955; minimum monthly
precipitation 0.16 in December 1928; maximum precipitation in 24 hours 4.55 in September 1924.

NORMALS, MEANS AND EXTREMES
(Table Revised 1975. Base Period for Climatological Normals: 1941-1970)

Station: BINGHAMTON, NEW YORK   Elev. 1638 feet m.s.l.   Average station pressure mb.

Month	Temp Normal Daily max	Daily min	Monthly	Extremes Record highest	Year	Record lowest	Year	Normal Degree days Heating	Cooling	Precip Water equiv Normal	Max monthly	Year	Min monthly	Year	Max in 24 hrs	Year	Snow Max monthly	Year	Max in 24 hrs	Year	RH 01	RH 07	RH 13	RH 19	Wind mean speed	Prevailing dir	Fastest speed	Dir	Year	Pct sunshine	Mean sky cover	Clear	Partly cloudy	Cloudy	Precip .01+	Snow 1.0+	Thunderstorms	Heavy fog	Max 90°+	Max 32°-	Min 32°-	Min 0°-	Avg station pressure mb
(yrs)	23		23	23		23		23	23	23	23		23		23		23		23		23	23	23	23	23	12	23	23	23	23	23	23	23	23	23	23	23	23	23	23	23	23	2
J	28.7	15.2	22.0	63	1967	-20	1957	1333	0	2.32	4.31	1958	0.76	1970	1.80	1958	36.6	1966	18.4	1964	80	80	72	74	11.6	WSW	59	NW	1959	38	8.0	2	8	21	16	5	*	4	0	*	30	30	4	957.9
F	30.1	15.4	22.8	66	1954	-15	1961	1182	0	2.25	4.36	1971	0.51	1968	2.16	1966	44.3	1972	23.0	1971	80	79	69	73	11.8	SSE	66	NW	1956	42	7.8	2	7	19	16	4	*	3	0	*	27	27	3	956.7
M	38.9	23.1	31.3	72	1968	-6	1967	1045	0	2.87	5.11	1956	0.69	1969	1.95	1964	33.5	1971	15.8	1971	77	77	61	69	11.8	NW	61	NW	1955	47	7.4	4	7	19	16	5	1	5	0	*	21	10	*	955.8
A	54.0	35.3	44.7	84	1962	11	1954	609	0	3.18	5.09	1961	1.61	1968	1.77	1961	16.4	1957	11.5	1960	74	65	53	63	11.4	NNW	57	W	1954	51	7.2	4	10	16	14	1	2	4	0	*	7	1	0	955.0
M	65.0	45.2	55.1	87	1969+	25	1962	320	13	3.83	6.46	1968	1.78	1962	1.75	1952	0.0		3.4	1966	79	57	49	58	10.2	NNW	54	W	1954	56	6.7	5	12	14	13	0	4	5	*	2	0	0	955.0
J	74.5	55.1	64.8	94	1952	35	1972	75	69	3.59	4.46	1960	1.15	1960	3.19	1966	0.0		0.0		81	58	51	58	9.1	WSW	60	SW	1957	63	6.8	5	12	13	12	0	7	5	1	0	0	0	957.5
J	78.5	59.6	69.1	95	1966	39	1966	21	148	3.83	7.40	1956	0.83	1955	3.01	1956	0.0		0.0		82	58	50	56	8.3	WSW	58	N	1956	66	6.5	5	13	13	11	0	7	4	1	0	0	0	958.7
A	76.8	57.9	67.3	96	1953	37	1965	40	111	3.61	7.48	1953	0.61	1953	3.40	1960	0.0		0.0		86	60	53	60	8.0	WSW	58	N	1960	60	6.6	5	14	12	11	0	5	4	1	0	0	0	960.7
S	69.4	50.4	60.2	96	1953	25	1963	172	28	3.02	5.57	1960	0.61	1961	3.88	1968	T	1970	T	1970	86	61	55	61	9.4	SSW	42	NW	1954	58	6.5	7	10	13	11	*	2	5	*	0	1	0	960.4
O	59.2	41.8	50.4	82	1974	7	1974	456	0	3.00	5.81	1972	0.26	1963	2.66	1955	2.4	1972	2.4	1972	80	61	57	61	9.6	NNW	57	SE	1951	50	6.2	6	9	16	11	1	1	5	0	*	6	18	1	961.5
N	44.5	31.8	38.2	74	1958	-7	1958	804	0	3.10	5.81	1973	1.01	1960	2.66	1972	24.4	1954	10.1	1953	80	75	75	79	10.9	WSW	57	SE	1951	30	8.2	2	6	22	15	2	*	5	0	*	18	28	6	956.5
D	31.3	19.4	25.4	64	1952	-8	1973	1228	0	2.75	4.85	1973	0.94	1960	1.59	1973	59.6	1969	15.6	1969	81	83	76	77	11.3	WSW	59	NW	1956	26	8.5	2	6	23	18	6	*	5	0	1	28	28	8	956.7
YR	54.3	37.6	46.0	96	SEP 1953	-20	JAN 1957	7285	369	37.35	9.46	JUN 1960	0.26	OCT 1963	3.88	FEB 1955	59.6	DEC 1969	23.0	FEB 1961	80	83	64	70	10.3	WSW	72	S	OCT 1954	51	7.3	46	104	215	162	25	31	54	2	67	148	8	957.7

Means and extremes above are from existing and comparable exposures. Annual extremes have been exceeded at other sites in the locality as follows:
Highest temperature 103 in July 1936; lowest temperature -28 in January 1893; maximum monthly precipitation 9.59 in October 1955; minimum monthly
precipitation 0.16 in December 1928; maximum precipitation in 24 hours 4.55 in September 1924.

REFERENCE NOTES APPLYING TO TABLES APPEAR ON THE PAGE FOLLOWING LAST TABLE.
(Caution: Letters and symbols may have different meanings in 1941-1970 tables than in earlier tables. See notes.)

NORMALS, MEANS AND EXTREMES
(Table Revised 1972. Base Period for Climatological Normals: 1931-1960)

Station: BUFFALO, NEW YORK — GREATER BUFFALO INTL AIRPORT
Standard time used: EASTERN Latitude: 42° 56' N Longitude: 78° 44' W Elevation (ground): 705 feet

Month	Temp Normal Daily max	Temp Normal Daily min	Temp Normal Monthly	Extremes Record highest	Year	Record lowest	Year	Normal heating degree days (Base 65°)	Precip Normal total	Precip Max monthly	Year	Precip Min monthly	Year	Precip Max 24 hrs	Year	Snow Mean total	Snow Max monthly	Year	Snow Max 24 hrs	Year
(a)	(b)	(b)	(b)	11		11		(b)	(b)	28		28		28		28	28		28	
J	30.8	18.2	24.5	61	1967	-11	1968+	1256	2.84	6.47	1959	1.03	1946	2.40	1957	22.3	50.6	1945	18.2	1966
F	31.0	17.2	24.1	60	1965	-20	1961	1145	2.72	5.80	1960	0.20	1968	2.31	1954	17.8	54.2	1958	14.5	1958
M	38.6	24.4	31.5	78	1966	-4	1967	1020	3.24	5.25	1956	1.04	1967	2.14	1956	12.6	34.2	1959	15.8	1959
A	52.9	34.0	43.5	83	1970+	17	1969	645	3.01	5.90	1961	1.27	1946	1.49	1944	2.6	13.1	1961	5.1	1961
M	65.5	44.5	55.0	88	1963	29	1970+	329	2.95	6.48	1953	0.11	1965	2.03	1957	0.2	2.0	1945	2.0	1945
J	75.1	54.8	64.8	94	1963	38	1964	78	2.54	4.48	1955	0.11	1955	3.04	1958	0.0	0.0		0.0	
J	80.1	59.4	69.8	94	1968+	46	1965	19	2.57	6.43	1963	1.04	1944	3.38	1963	0.0	0.0		0.0	
A	78.6	58.1	68.4	93	1962	38	1965	37	3.05	6.36	1963	1.10	1948	3.88	1963	0.0	0.0		0.0	
S	71.5	51.2	61.4	90	1960	32	1963	141	3.13	6.36	1967	0.30	1964	3.63	1967	0.0	T	1956+	T	1956+
O	60.1	41.4	50.8	82	1965	20	1965	440	3.00	9.13	1954	0.30	1963	3.49	1954	0.2	2.0	1962	19.9	1962
N	46.5	31.7	39.1	80	1961	9	1971+	777	3.60	5.83	1954	1.44	1944	2.51	1949	12.6	28.6	1949	19.9	1955
D	34.3	21.1	27.7	66	1966	-4	1962	1156	3.00	5.83	1954	0.69	1943	2.16	1945	20.2	51.1	1945	24.3	1945
YR	55.4	37.9	46.7	94	JUL 1968+	-20	FEB 1961	7062	35.65	9.13	OCT 1954	0.11	JUN 1955	3.88	AUG 1963	88.0	54.2	FEB 1958	24.3	DEC 1945

Month	RH 01	RH 07	RH 13	RH 19	Wind Mean speed	Prevailing direction	Fastest mile Speed	Direction	Year	Mean sky cover sunrise to sunset	Pct. of possible sunshine	Days Clear	Partly cloudy	Cloudy	Precip .01 in or more	Snow 1.0 in or more	Thunderstorms	Heavy fog	Max 90° and above	Max 32° and below	Min 32° and below	Min 0° and below	Avg daily solar radiation langleys
	11	11	11	11	32	14	28	28	28	28	28	28	28	28	28	28					11	11	11
J	75	77	72	75	14.4	WSW	91	SW	1950+	8.3	33	2	6	23	20	7	*	2	0	18	18	3
F	78	79	71	74	14.2	SW	70	SW	1959	8.1	40	6	6	20	17	5	*	2	0	16	16	3
M	78	77	64	65	14.0	SW	67	W	1957	7.5	47	6	8	19	16	4	1	3	0	8	8	2
A	75	74	58	61	13.1	SW	67	W	1957	7.1	52	8	8	17	15	1	2	3	0	*	*	*
M	75	72	56	61	12.0	SW	63	SW	1950	6.7	57	8	8	15	14	*	3	1	0	0	0	0
J	78	75	56	61	10.7	SW	59	NW	1953	6.1	67	7	10	13	11	0	5	1	1	0	0	0
J	80	79	56	61	10.7	SW	56	SW	1944	5.8	69	14	14	11	10	0	6	1	1*	0	0	0
A	79	83	59	61	10.2	SW	56	SW	1954	5.8	67	12	12	11	10	0	6	1	0	0	0	0
S	81	83	60	74	11.5	SW	63	NW	1954	6.2	60	10	8	13	10	*	4	1	*	0	3	3
O	80	81	62	74	11.5	SW	66	S	1948	6.3	54	8	8	15	11	1	2	1	0	0	3	1
N	80	82	71	77	13.6	SW	60	S	1945	8.3	30	2	6	22	16	6	1	1	0	3	14	4
D	79	81	73	77	13.6	WSW				8.4	29	2	7	23	19	9	1	1	0	14	26	6
YR	79	80	63	70	12.4	SW	91	SW	JAN 1950+	7.1	53	58	108	199	165	26	30	20	3	58	137	5

Means and extremes above are from existing and comparable exposures. Annual extremes have been exceeded at other sites in the locality as follows:
Highest temperature 99 in August 1948; lowest temperature -21 in February 1934; maximum monthly precipitation 1C.63 in August 1885; minimum monthly precipitation .05 in August 1876; maximum precipitation in 24 hours 4.28 in August 1893.

NORMALS, MEANS AND EXTREMES
(Table Revised 1975. Base Period for Climatological Normals: 1941-1970)

Month	Temp Normal Daily max	Temp Normal Daily min	Temp Normal Monthly	Extremes Record highest	Year	Record lowest	Year	Degree days Heating	Degree days Cooling	Precip Normal	Max monthly	Year	Min monthly	Year	Max 24 hrs	Year	Snow Max monthly	Year	Snow Max 24 hrs	Year
(a)	14	14		14		14					31		31		31		31		31	
J	29.8	17.6	23.7	61	1967	-11	1968	1280	0	2.90	6.47	1946	1.03	1959	2.40	1957	50.6	1945	18.2	1966
F	31.0	17.7	24.4	60	1965	-20	1961	1137	0	2.55	5.80	1968	0.81	1960	2.31	1954	54.2	1958	14.5	1958
M	39.4	25.2	32.1	78	1966	-4	1967	1020	0	2.85	5.25	1956	1.20	1956	2.14	1954	34.2	1959	15.8	1959
A	53.3	36.4	44.9	83	1970	17	1964	603	0	3.15	5.90	1961	1.27	1961	1.49	1944	13.1	1961	5.1	1961
M	64.3	45.9	55.1	88	1944	29	1953	321	14	2.97	6.06	1955	1.21	1963	2.03	1957	2.0	1945	2.0	1945
J	75.1	56.3	65.7	94	1963	36	1972	58	79	2.23	6.06	1955	0.11	1955	3.04	1968	0.0		0.0	
J	79.5	60.7	70.1	94	1968	46	1965	12	170	2.93	8.04	1972	0.99	1972	3.38	1963	0.0		0.0	
A	77.9	59.1	68.4	93	1962	38	1965	33	138	3.53	8.04	1963	1.08	1963	3.88	1963	T	1956	T	1956
S	70.8	51.6	61.6	90	1971	32	1963	138	36	3.25	9.13	1967	0.77	1967	3.63	1967	2.5	1972	2.5	1972
O	60.2	42.7	51.5	82	1963	20	1964	419	0	3.01	9.13	1954	0.30	1964	3.49	1954	19.9	1949	19.9	1949
N	46.1	33.5	39.8	80	1961	9	1971	756	0	3.74	6.37	1949	1.44	1944	2.51	1949	28.6	1949	24.3	1955
D	33.6	22.2	27.9	66	1966	-4	1962	1150	0	3.00	5.83	1954	0.59	1943	2.16	1945	51.1	1945	24.3	1945
YR	55.0	39.1	47.1	94	JUL 1968	-20	FEB 1961	6927	437	36.11	9.13	OCT 1954	0.11	JUN 1955	3.88	AUG 1963	54.2	FEB 1958	24.3	DEC 1945

Month	RH 01	RH 07	RH 13	RH 19	Wind Mean speed m.p.h.	Prevailing direction	Fastest mile speed m.p.h.	Direction	Year	Mean sky cover tenths	Pct. of possible sunshine	Days Clear	Partly cloudy	Cloudy	Precip .01 or more	Snow 1.0 or more	Thunderstorms	Heavy fog ¼ mi or less	Max 90° above	Max 32° below	Min 32° below	Min 0° below	Avg station pressure mb
	14	14	14	14	35	14	31	31	31	31	31	31	31	31	31	31	31	31	14	14	14	14	2
J	76	77	72	75	14.5	WSW	91	SW	1950	8.4	34	2	6	23	20	7	*	2	0	17	29	2	992.2
F	78	79	70	73	14.1	SW	70	SW	1946	8.2	40	4	7	20	17	6	*	3	0	16	27	2	992.3
M	77	77	65	68	13.8	SW	80	W	1959	7.6	47	5	8	17	15	4	1	3	0	7	25	2*	989.7
A	75	74	58	61	13.1	SW	67	W	1957	7.6	52	6	9	16	14	1	3	2	0	1	12	0	988.5
M	75	72	56	62	11.9	SW	63	SW	1950	6.8	57	6	12	12	11	*	4	1	0	0	1	0	987.2
J	77	75	57	62	11.3	SW	56	NW	1954	6.2	66	6	12	12	10	0	5	1	1	0	0	0	988.9
J	79	77	57	62	11.3	SW	56	SW	1954	5.9	67	7	13	11	10	0	6	1	1*	0	0	0	990.4
A	79	83	55	60	10.6	SW				5.9	67	7	15	11	10	0	6	1	*	0	0	0	992.4
S	79	83	60	67	10.1	SW				6.3	63	7	9	14	11	0	3	1	*	0	3	2*	992.4
O	80	80	61	73	11.4	SW	63	S	1948	6.4	53	7	9	15	11	1	2	1	0	0	3	3	993.4
N	79	81	73	79	11.4	SW	60	S	1945	8.3	29	2	5	24	16	6	1	1	0	3	14	3	989.9
D	79	80	75	80	13.5	WSW				8.5	29	2	5	24	20	6	1	1	0	13	26	*	990.7
YR	79	80	63	70	12.3	SW	91	SW	JAN 1950	7.1	53	56	105	204	168	26	30	20	2	56	136	5	990.7

Means and extremes above are from existing and comparable exposures. Annual extremes have been exceeded at other sites in the locality as follows:
Highest temperature 99 in August 1948; lowest temperature -21 in February 1934; maximum monthly precipitation 10.63 in August 1885; minimum monthly precipitation .05 in August 1876; maximum precipitation in 24 hours 4.28 in August 1893.

REFERENCE NOTES APPLYING TO TABLES APPEAR ON THE PAGE FOLLOWING LAST TABLE.
(Caution: Letters and symbols may have different meanings in 1941-1970 tables than in earlier tables. See notes.)

NORMALS, MEANS AND EXTREMES
(Table Revised 1972. Base Period for Climatological Normals: 1931-1960)

Station: NEW YORK, NEW YORK   CENTRAL PARK OBSERVATORY

Standard time used: EASTERN   Latitude: 40° 47' N   Longitude: 73° 58' W   Elevation (ground): 132 feet

Temperature / Extremes / Degree days / Precipitation / Snow

Month	Normal Daily max	Normal Daily min	Normal Monthly	Record highest	Year	Record lowest	Year	Normal heating degree days (Base 65°)	Precip Normal total	Precip Max monthly	Year	Precip Min monthly	Year	Precip Max in 24 hrs	Year	Snow Mean total	Snow Max monthly	Year	Snow Max in 24 hrs	Year
(yrs)	(b)	(b)	(b)	103		103		(b)	(b)	103		103		60		103	103		103	
J	39.5	26.9	33.2	72	1950	-6	1882	986	3.21	7.94	1915	0.66	1970	3.33	1915	7.7	27.4	1925	13.0	1925
F	40.2	26.4	33.4	75	1930	-15	1934	885	2.84	6.87	1869	0.45	1895	2.96	1941	8.6	27.9	1934	17.4	1961
M	48.4	33.2	40.8	86	1945	3	1872	760	4.01	8.77	1876	0.90	1885	4.25	1876	5.4	30.5	1896	18.1	1896
A	60.8	43.1	52.0	92	1915	12	1923	408	3.43	8.51	1874	0.95	1881	2.67	1886	1.0	13.5	1875	10.2	1875
M	71.4	53.4	62.4	99	1962	32	1891	118	3.67	9.78	1908	0.60	1903	4.88	1946	T	1.8	1946	T	1946
J	80.2	62.5	71.4	101	1966+	44	1945		3.31	10.02	1903	0.02	1949	4.74	1884	0.0	0.0		0.0	
J	85.3	68.2	76.8	106	1936	52	1943	0	3.70	11.89	1910	0.49	1910	3.60	1971	0.0	0.0		0.0	
A	83.3	66.8	75.1	104	1918	50	1965+	0	4.44	10.86	1955	0.24	1964	5.78	1971	0.0	0.0		0.0	
S	76.8	60.1	68.5	102	1953	39	1912	30	3.87	16.85	1882	0.21	1884	8.30	1882	0.0	0.0		0.0	
O	65.3	50.3	57.8	94	1941	28	1936	233	3.14	13.31	1903	0.14	1963	11.17	1903	0.8	0.8	1925	0.8	1925
N	53.7	40.3	47.0	84	1950	5	1875	540	3.39	9.97	1931	0.60	1931	3.81	1963	1.0	19.0	1898	10.0	1898
D	42.1	29.7	35.9	70	1946+	-13	1917	902	3.26	7.53	1936	0.25	1955	3.21	1909	6.1	29.6	1947	26.4	1947
YR	62.2	46.7	54.5	106	JUL. 1936	-15	FEB. 1934	4871	42.37	16.85	SEP. 1882	0.02	JUN. 1949	11.17	OCT. 1903	29.8	30.5	MAR. 1896	26.4	DEC. 1947

Relative humidity / Wind / Sky / Sunshine / Days / Temperatures / Solar

Month	RH 01	RH 07	RH 13	RH 19	Wind Mean speed	Prevailing direction	Fastest mile Speed	Direction	Year	Mean sky cover (sunrise to sunset)	Pct. of possible sunshine	Days Clear	Days Partly cloudy	Days Cloudy	Precip .01"+	Snow 1.0"+	Thunderstorms	Heavy fog	Max 90° & above	Max 32° & below	Min 32° & below	Min 0° & below	Avg. daily solar radiation (langleys)
(yrs)	40	52	52	52	52	45	25	25		42	95	42	42	42	42	101	11		58	58	58	58	44
J	73	73	60	65	10.9	NW	47	NW	1959	5.8	51	8	9	14	11	2	*		0	1	23	**	136
F	73	73	58	63	11.0	NW	47	NW	1961+	5.8	55	9	9	11	10	2	*		0	1	20	*	205
M	72	71	55	58	11.1	NW	60	NW	1950	5.7	57	8	10	12	12	2	1		0	*	13	*	295
A	71	71	53	58	11.0	NW	45	NE	1961	5.7	57	8	10	12	11	*	1		*	0	1	0	376
M	75	75	55	63	8.9	SW	38	NE	1967+	5.6	62	8	12	11	11	0	2		1	0	*	0	440
J	75	77	55	66	8.2	SW	65	NE	1952	5.6	65	8	12	10	10	0	4		3	0	0	0	478
J	75	77	58	66	7.7	SW	43	NW	1949	5.5	65	8	13	10	10	0	4		6	0	0	0	465
A	77	79	58	67	7.5	SW	36	NE	1954	5.5	63	9	12	10	9	0	4		4	0	0	0	402
S	76	79	59	69	8.2	NE	40	NE	1960	5.2	63	11	10	9	8	0	1		2	0	*	0	335
O	73	76	55	63	9.0	NW	70	NE	1955+	5.2	61	12	9	9	8	*	1		*	0	5	*	246
N	70	72	59	60	10.4	NW	70	NW	1950	5.8	54	9	9	12	9	2	*		0	0	19	*	149
D	67	70	60	63	10.4	NW	43	NE	1962+	5.9	50	9	9	13	10	2	*		0	5	19	*	117
YR	70	72	56	62	9.5	NW	70	NE	NOV. 1950	5.6	59	107	125	133	121	18	8	18	17	20	82	*	304

Means and extremes above are from existing and comparable exposures. Annual extremes have been exceeded at other sites in the locality as follows:
Maximum monthly snowfall 37.9 inches in February 1894; fastest mile of wind 113 from the SE in October 1954.

NORMALS, MEANS AND EXTREMES
(Table Revised 1975. Base Period for Climatological Normals: 1941-1970)

Station: NEW YORK, NEW YORK   CENTRAL PARK OBSERVATORY

Average station pressure 1013.5 mb   Elev. 87 feet m.s.l.

Temperature / Extremes / Degree days / Precipitation / Snow

Month	Normal Daily max	Normal Daily min	Normal Monthly	Record highest	Year	Record lowest	Year	Heating deg. days	Cooling deg. days	Precip Normal	Precip Max monthly	Year	Precip Min monthly	Year	Precip Max 24 hrs	Year	Snow Max monthly	Year	Snow Max 24 hrs	Year
(yrs)	(b)	(b)	(b)	106		106					106		106		63		106		106	
J	38.5	25.9	32.2	72	1950	-6	1882	1017	0	2.71	7.94	1915	0.66	1970	3.33	1915	27.4	1925	13.0	1925
F	40.2	26.5	33.4	75	1930	-15	1934	885	0	2.92	6.37	1889	0.46	1895	3.04	1941	27.9	1934	17.4	1961
M	48.7	33.7	41.1	86	1945	3	1872	741	0	3.73	8.79	1876	0.90	1885	4.25	1876	30.5	1896	18.1	1896
A	61.4	43.5	52.5	92	1915	12	1923	387	0	3.30	8.71	1874	0.95	1881	2.67	1886	13.5	1875	10.2	1875
M	71.6	53.2	62.4	99	1962	32	1891	137	54	3.47	8.51	1968	0.30	1903	4.88	1946	1.8	1946	T	1946
J	80.5	62.6	71.6	101	1966	44	1945	0	202	2.96	9.78	1903	0.02	1949	4.74	1884	0.0		0.0	
J	85.2	68.0	76.6	106	1966	52	1943	0	360	3.68	11.89	1910	0.49	1910	3.60	1971	0.0		0.0	
A	83.4	66.4	74.9	104	1964	50	1965	0	307	4.01	10.86	1955	0.24	1964	5.78	1971	0.0		0.0	
S	76.8	59.9	68.4	102	1953	39	1912	29	131	3.27	16.85	1882	0.21	1884	8.30	1882	0.0		0.0	
O	66.8	50.6	58.7	94	1941	28	1936	209	14	2.85	13.31	1903	0.14	1972	11.17	1903	0.8	1925	0.8	1925
N	54.0	40.8	47.4	84	1950	5	1875	528	0	3.12	9.73	1972	0.60	1931	3.81	1963	19.0	1898	10.0	1898
D	41.4	29.5	35.5	70	1962	-13	1917	915	0	3.53	9.98	1973	0.25	1955	3.21	1909	29.6	1947	26.4	1947
YR	62.3	46.7	54.5	106	JUL. 1966	-15	FEB. 1934	4848	1068	40.19	16.85	SEP. 1882	0.02	JUN. 1949	11.17	OCT. 1903	30.5	MAR. 1896	26.4	DEC. 1947

Relative humidity / Wind / Sky / Sunshine / Days / Temperatures / Pressure

Month	RH 01	RH 07	RH 13	RH 19	Wind Mean speed m.p.h.	Prevailing direction	Fastest mile m.p.h.	Direction	Year	Mean sky cover (tenths) sunrise to sunset	Pct. of possible sunshine	Days Clear	Days Partly cloudy	Days Cloudy	Precip .01"+	Snow 1.0"+	Thunderstorms	Heavy fog ¼ mile or less	Max 90° & above	Max 32° & below	Min 32° & below	Min 0° & below	Avg. station pressure mb
(yrs)	43	55	55	55	55	45	28	28		42	98	42	42	42	106	104	14		61	61	61	61	2
J	73	74	60	65	10.8	NW	47	NW	1959	5.8	51	8	9	14	11	2	*		0	1	22	**	1016.1
F	74	74	58	63	10.9	NW	47	NW	1961	5.8	56	9	9	11	10	2	*		0	1	21	*	1013.9
M	72	70	56	57	11.1	NW	60	NW	1950	5.7	56	8	10	12	12	2	1		0	*	13	*	1012.7
A	70	67	51	57	10.9	NW	45	NE	1961	5.7	59	8	11	11	11	*	1		*	0	1	0	1010.7
M	75	74	51	62	8.9	SW	38	NE	1967	5.6	64	8	12	11	11	0	2		1	0	*	0	1010.3
J	74	75	55	62	8.1	SW	49	SW	1952	5.6	64	8	12	10	10	0	4		3	0	0	0	1010.2
J	75	77	55	63	7.7	SW	43	NW	1949	5.5	65	8	13	10	10	0	4		6	0	0	0	1012.4
A	78	79	57	66	7.7	SW	36	NE	1954	5.5	64	9	12	10	9	0	4		4	0	0	0	1014.9
S	77	78	57	67	8.2	SW	40	NE	1960	5.2	63	11	10	9	8	0	1		2	0	*	0	1014.9
O	76	76	59	70	9.0	NW	70	NE	1955	5.2	61	12	9	9	8	*	1		*	0	5	*	1017.3
N	73	73	59	61	10.0	NW	70	NW	1950	5.8	52	9	9	12	9	2	*		0	0	18	*	1013.1
D	67	70	61	63	10.4	NW	43	NE	1962	5.9	49	9	9	13	10	2	*		0	5	18	*	1013.5
YR	70	72	56	62	9.4	NW	70	NE	NOV. 1950	5.6	59	107	125	133	121	19	8		17	20	81	*	1013.5

Means and extremes above are from existing and comparable exposures. Annual extremes have been exceeded at other sites in the locality as follows:
Maximum monthly snowfall 37.9 inches in February 1894; fastest mile of wind 113 from the SE in October 1954.

REFERENCE NOTES APPLYING TO TABLES APPEAR ON THE PAGE FOLLOWING LAST TABLE.
(Caution: Letters and symbols may have different meanings in 1941-1970 tables than in earlier tables. See notes.)

NORMALS, MEANS AND EXTREMES

(Table Revised 1972. Base Period for Climatological Normals: 1931-1960)

Station: YORK, NEW YORK   JOHN F. KENNEDY INTL AIRPORT   Standard time used: EASTERN   Latitude: 40° 39' N   Longitude: 73° 47' W   Elevation (ground): 13 feet

Ø For the period June 1961 through the current year.

$ Greatest calendar day through March 1969.
# Greatest calendar day August 1966 through March 1969.

NORMALS, MEANS AND EXTREMES

(Table Revised 1975. Base Period for Climatological Normals: 1941-1970)

REFERENCE NOTES APPLYING TO TABLES APPEAR ON THE PAGE FOLLOWING LAST TABLE.
(Caution: Letters and symbols may have different meanings in 1941-1970 tables than in earlier tables. See notes.)

NORMALS, MEANS AND EXTREMES

(Table Revised 1972. Base Period for Climatological Normals: 1931-1960)

Station: NEW YORK, NEW YORK — LA GUARDIA AIRPORT

Standard time used: EASTERN Latitude: 40° 46' N Longitude: 73° 54' W Elevation (ground): 11 feet

Month	Normal Daily maximum	Normal Daily minimum	Normal Monthly (b)	Extremes Record highest	Year	Extremes Record lowest (∅)	Year	Normal heating degree days (Base 65°) (b)	Precip Normal total	Precip Max monthly	Year	Precip Min monthly	Year	Precip Max in 24 hrs	Year	Snow Mean total	Snow Max monthly	Year	Snow Max in 24 hrs	Year
J	39.6	27.5	33.6	68	1967	-1	1968	973	3.31	5.77	1949	0.76	1970	2.12	1944	6.8	18.3	1948	9.0	1948
F	40.4	27.0	33.6	63	1971	-2	1963	879	3.09	5.76	1960	1.37	1960	2.90	1961	8.4	21.5	1947	17.4	1947
M	47.8	33.8	40.8	77	1963	10	1967	750	4.23	8.73	1953	0.87	1966	3.25	1953	5.4	18.9	1958	15.3	1956
A	59.3	43.0	51.2	84	1964	25	1972	414	3.57	7.36	1961	1.21	1942	2.52	1958	0.8	6.4	1956	T	1961
M	70.7	53.4	62.1	95	1962	38	1966	124	3.58	7.42	1968	0.43	1964	3.02	1968	T	0.0	1961	0.0	
J	80.0	62.1	71.5	97	1964	47	1967	0	3.38	6.16	1968	0.03	1949	3.26	1946	0.0	0.0		0.0	
J	84.8	68.8	76.8	107	1966	57	1969+	0	3.71	9.27	1960	0.69	1954	3.82	1971	0.0	0.0		0.0	
A	82.9	67.8	75.4	94	1968+	53	1965	0	5.08	16.05	1955	0.24	1964	7.11	1955	0.0	0.0		0.0	
S	76.2	61.3	68.8	94	1965	44	1963	27	3.37	8.04	1944	0.62	1941	4.52	1969	0.0	0.0		0.0	
O	65.9	51.0	58.6	85	1967	30	1969	223	3.37	9.09	1943	0.06	1963	3.36	1943	0.5	1.2	1962	1.2	1962
N	53.2	41.3	47.4	76	1971	20	1967	528	3.39	7.92	1951	1.03	1949	3.68	1963	6.7	4.0	1953	4.0	1953
D	42.4	30.3	36.4	66	1970+	3	1962	887	3.39	5.82	1955	0.31	1955	3.44	1947	6.7	26.8	1947	22.8	1947
YR	62.0	47.3	54.7	107 JUL. 1966		-2 FEB. 1963		4811	44.22	16.05 AUG. 1955		0.03 JUN. 1949		7.11 AUG. 1955		28.6	26.8 DEC. 1947		22.8 DEC. 1947	

Month	Rel. humidity Hour 01	Hour 07	Hour 13	Hour 19	Wind Mean speed	Prevailing direction	Fastest mile Speed	Direction	Year	Pct. of possible sunshine	Mean sky cover	Clear	Partly cloudy	Cloudy	Precip .01"+	Snow 1.0"+	Thunderstorms	Heavy fog	Max 90°+	Max 32° & below	Min 32° & below	Min 0° & below	Avg daily solar radiation
(years)	9	9	9	9	23	14	22	22		22	23	23	23	23	31	27	23	23				9	
J	62	64	56	57	14.2	WNW	68	NE	1958		6.4	8	7	15	11	2	*	2	0	1	24	0*	0
F	61	64	54	56	14.2	WNW	64	NW	1958		6.3	7	6	14	11	2	*	1	0	1	21	0	0
M	64	67	49	54	14.5	NW	59	NW	1959		6.3	8	7	11	11	2	1	1	0	1	16	*	0
A	63	66	50	56	13.7	NE	52	W	1963+		6.1	7	10	13	11	1	2	1	0	0	4	0	0
M	66	71	50	58	10.9	S	54	NW	1952		5.9	8	11	12	11	*	3	2	1	0	0	0	0
J	70	70	51	59	10.4	S	59	NW	1954		6.1	7	12	11	10	0	5	1	3	0	0	0	0
J	71	73	52	60	10.3	S	63	NE	1958		5.8	7	12	12	9	0	5	1	5	0	0	0	0
A	73	75	53	62	9.7	S	66	SE	1960		5.5	9	12	10	9	0	5	3	3	0	0	0	0
S	71	73	53	61	11.7	SW	66	SE	1954		5.3	10	11	9	8	*	2	4	1	0	*	0	0
O	69	70	51	57	11.7	NW	56	NW	1951		5.5	11	9	11	8	*	1	5	0	0	3	0*	0
N	67	67	57	62	12.8	WNW	70	NW			5.4	9	9	12	10	1	*	3	0	0	8	3	0
D	65	60	58	61	13.7	WNW	56	NW			6.3	8	7	16	10	2	*	1	0	0	18	1	*
YR	67	69	53	59	12.4	WNW	70	NE SEP. 1960		6.1	95	117	153	119	8	24	14	22	78			

∅ For period May 1962 through the current year.

Means and extremes above are from existing and comparable exposures. Annual extremes have been exceeded at other sites in the locality as follows: Lowest temperature -7 in February 1943.

NORMALS, MEANS AND EXTREMES

(Table Revised 1975. Base Period for Climatological Normals: 1941-1970)

Month	Normal Daily maximum	Normal Daily minimum	Normal Monthly	Extremes Record highest	Year	Record lowest	Year	Degree days Base 65°F Heating	Cooling	Water equiv. Max monthly	Year	Min monthly	Year	Precip Normal	Max in 24 hrs	Year	Snow Mean	Max monthly	Year	Max in 24 hrs	Year
J	37.7	26.4	32.1	68	1967	-1	1968	1020	0	5.77	1949	0.76	1970	2.88	2.12	1944	18.3	18.3	1948	9.0	1948
F	39.2	27.0	33.1	63	1971	-2	1963	893	0	5.76	1960	1.28	1960	3.10	2.90	1961	21.5	21.5	1947	17.4	1947
M	47.1	34.1	40.6	77	1963	10	1967	756	0	8.73	1953	0.87	1966	3.70	3.25	1953	19.8	18.9	1958	15.3	1956
A	59.3	43.0	51.2	84	1961	25	1972	399	0	6.66	1942	1.21	1942	3.56	2.52	1958	6.4	6.4	1956	6.4	1956
M	70.4	53.7	62.1	95	1962	38	1966	145	46	7.42	1968	0.43	1964	3.47	3.02	1968	T	0.0	1961	T	1961
J	79.4	63.6	71.5	97	1964	47	1967	0	199	6.16	1949	0.03	1949	3.63	3.67	1973	0.0	0.0		0.0	
J	84.1	69.3	76.6	107	1966	56	1972	0	363	9.27	1960	0.69	1954	3.85	3.82	1971	0.0	0.0		0.0	
A	82.1	67.6	74.9	97	1973	53	1965	0	307	16.05	1955	0.24	1964	4.48	4.52	1955	0.0	0.0		0.0	
S	75.2	60.9	68.1	94	1965	44	1944	30	123	8.04	1944	0.62	1941	3.15	3.58	1969	0.0	0.0		0.0	
O	65.2	51.0	58.1	85	1967	30	1969	227	10	9.09	1943	0.06	1963	2.96	3.58	1972	0.06	1.2	1962	1.2	1962
N	53.2	41.3	47.3	76	1974	20	1967	531	0	7.92	1972	1.01	1974	3.77	4.43	1972	4.0	4.0	1953	4.0	1953
D	41.0	30.1	35.6	66	1970	3	1962	911	0	5.82	1955	0.31	1955	3.63	3.44	1947	26.8	26.8	1947	22.8	1947
YR	61.1	47.4	54.3	107 JUL. 1966		-2 FEB. 1963		4909	1048	16.05 AUG. 1955		0.03 JUN. 1949		41.61	7.11 AUG. 1955			26.8 DEC. 1947		22.8 DEC. 1947	

Month	Rel. humidity Hour 01	07	13	19	Wind Mean speed	Prevailing direction	Fastest mile Speed	Direction	Year	Pct. of possible sunshine	Mean sky cover	Clear	Partly cloudy	Cloudy	Precip .01"+	Snow 1.0"+	Thunderstorms	Heavy fog, vis. ½ mi or less	Max 90°+	Max 32° & below	Min 32° & below	Min 0° & below	Avg station pressure mb	Elev. feet m.s.l.
(years)	12	12	12	12	26	14	25	25		26	26	26	26	26	34	30	26	26	12	12	12	12		31 / 2
J	62	65	56	57	13.9	WNW	68	NE	1958		6.4	8	7	15	11	2	*	2	0	1	22	0*	1018.6	
F	61	63	54	54	14.2	WNW	64	NW	1958		6.3	8	6	14	10	2	*	1	0	1	21	0*	1018.8	
M	63	65	49	54	14.3	NW	60	W	1959		6.4	7	7	13	11	2	1	1	0	0	16	*	1019.5	
A	63	65	50	58	13.2	NE	59	W	1963		6.4	7	10	13	11	1	2	1	0	0	3	0	1013.6	
M	65	69	54	61	11.0	S	52	NW	1955		6.0	8	11	12	11	*	3	2	1	0	0	0	1012.9	
J	72	73	55	61	10.8	S	54	NW	1952		6.1	7	11	11	10	0	5	1	3	0	0	0	1014.7	
J	70	71	52	60	10.3	S	59	NW	1954		6.0	7	11	13	9	0	5	1	5	0	0	0	1014.4	
A	74	74	53	62	10.3	S	63	NE	1958		5.8	9	11	11	9	0	5	3	3	0	0	0	1016.8	
S	72	72	53	61	11.6	SW	70	SE	1960		5.7	10	9	11	8	*	2	4	1	0	*	0	1017.1	
O	72	72	57	63	12.6	WNW	66	SE	1954		5.3	11	9	11	7	1	1	4	0	0	4	0	1019.7	
N	67	68	61	63	13.5	WNW	68	N	1953		6.4	9	9	14	10	1	1	4	0	0	16	0	1016.5	
D	67	68	61	61	12.3	WNW	56	NW	1951		6.5	8	9	14	11	2	*	5	0	0	16	0	1016.4	
YR	67	69	54	59	12.3	WNW	70	NE SEP. 1960			6.1	93	118	154	120	7	24	14	22	75				1016.0	

Means and extremes above are from existing and comparable exposures. Annual extremes have been exceeded at other sites in the locality as follows: Lowest temperature -7 in February 1943.

REFERENCE NOTES APPLYING TO TABLES APPEAR ON THE PAGE FOLLOWING LAST TABLE.

(Caution: Letters and symbols may have different meanings in 1941-1970 tables than in earlier tables. See notes.)

NORMALS, MEANS AND EXTREMES

(Table Revised 1972. Base Period for Climatological Normals: 1931-1960)

Station: ROCHESTER, NEW YORK — ROCHESTER-MONROE COUNTY AP
Standard time used: EASTERN Latitude: 43° 07' N Longitude: 77° 40' W Elevation (ground): 547 feet

Month	Normal Daily maximum (b)	Normal Daily minimum (b)	Normal Monthly (b)	Extremes Record highest	Year	Extremes Record lowest	Year Ø	Normal heating degree days (Base 65°) (b)	Precipitation Normal total	Precipitation Maximum monthly	Year	Precipitation Minimum monthly	Year	Precipitation Maximum in 24 hrs.	Year	Snow, ice pellets Mean total	Snow Maximum monthly	Year	Snow Maximum in 24 hrs.	Year	Rel. humidity Hour 01	07	13	19	Mean speed	Prevailing direction	Fastest mile Speed	Direction	Year	Pct. possible sunshine	Mean sky cover
J	32.6	17.8	25.2	68	1967	-11	1968	1234	2.40	4.10	1966	0.81	1946	1.64	1950	22.0	60.2	1966	18.2	1966	78	73	71	78	11.6	WSW	73	W	1950	37	8.1
F	25.2	16.9	24.0	64	1965	-8	1970+	1123	2.53	5.07	1950	0.74	1968	2.43	1958	22.0	64.8	1958	18.7	1958	79	73	67	78	11.6	WSW	66	W	1956	43	7.7
M	32.3	24.3	32.3	79	1966	-1	1967	1014	3.01	5.42	1942	0.47	1942	2.21	1959	14.7	40.3	1959	17.6	1959	75	74	65	73	11.6	W	65	SW	1956	51	7.3
A	54.5	35.7	45.1	93	1970	14	1964	597	2.67	4.90	1944	1.28	1944	1.99	1943	4.0	7.6	1957	6.0	1971	75	75	53	60	10.6	W	63	W	1946	57	6.9
M	67.3	46.1	56.7	96	1970	26	1966	279	2.77	5.51	1943	0.50	1969	1.76	1962	0.1	2.0	1962	2.0	1945	75	53	53	58	9.3	WSW	59	W	1950	60	6.5
J	78.8	56.0	66.9	96	1964	36	1965	48	2.56	4.69	1969	0.22	1963	2.86	1963	0.0	0.0		0.0		77	77	58	58	8.3	SW	61	SW	1949	68	5.8
J	82.5	60.7	71.6	98	1966+	42	1963	9	2.84	9.70	1947	0.98	1955	2.94	1947	0.0	0.0	1965	0.0	1965	82	81	53	59	7.8	SW	56	NE	1956	70	5.5
A	80.4	59.0	69.7	98	1965	36	1965	31	2.72	5.95	1968	0.76	1951	2.39	1968	0.0	T	1956	T	1956	85	86	55	65	7.6	SW	59	SW	1955	68	5.7
S	73.1	51.6	62.4	96	1973	30	1963	126	2.53	5.21	1945	0.28	1959	2.61	1959	0.0	T	1956	T	1956	87	87	58	73	7.9	SW	50	SW	1941	61	6.3
O	61.7	41.6	51.8	86	1972	22	1971	415	2.58	7.85	1955	0.23	1963	2.52	1945	0.2	1.4	1960+	1.4	1960+	80	83	59	73	8.7	SW	68	E	1950	51	6.3
N	47.5	32.5	40.1	81	1971	5	1971	747	2.51	5.51	1950	0.73	1965	3.11	1953	6.7	16.3	1953	11.2	1953	81	81	63	72	10.1	W	61	SW	1950	32	8.1
D	35.5	21.9	28.7	68	1966	-5	1969+	1125	2.38	5.05	1944	0.62	1958	1.59	1944	17.3	44.2	1970	17.8	1970	79	79	77	77	10.8	WSW	56	W	1949	32	8.3
YR	57.2	38.6	48.0	98 1966+	JUL	-11 1968	JAN	6748	31.50	9.70	JUL 1947	0.22	JUN 1963	3.13	NOV 1945	85.2	64.8	FEB 1958	18.7	FEB 1960	79	78	60	67	9.6	WSW	73	W	JAN 1950	55	6.8

Ø For period July 1963 through the current year.
Means and extremes above are from existing and comparable exposures. Annual extremes have been exceeded at other sites in the locality as follows: Highest temperature 102 in July 1936; lowest temperature -22 in February 1934; minimum monthly precipitation 0.08 in October 1924; maximum precipitation in 24 hours 4.19 in August 1893; maximum snowfall in 24 hours 29.8 in March 1900.

NORMALS, MEANS AND EXTREMES

(Table Revised 1975. Base Period for Climatological Normals: 1941-1970)

Elev. 555 feet m.s.l. Average station pressure mb.: 2

Month	Temperature °F Normal Daily maximum (a)(b)	Normal Daily minimum	Normal Monthly	Extremes Record highest	Year	Extremes Record lowest	Year	Normal Degree days Base 65°F Heating	Cooling	Precipitation Water equivalent Normal	Maximum monthly	Year	Minimum monthly	Year	Maximum in 24 hrs.	Year	Snow, ice pellets Maximum monthly	Year	Maximum in 24 hrs.	Year	Rel. humidity Hour 01	07	13	19	Mean speed m.p.h.	Prevailing direction	Fastest mile Speed m.p.h.	Direction	Year	Pct. possible sunshine	Mean sky cover
J	31.3	16.7	24.0	68	1967	-11	1968	1271	0	2.25	4.10	1966	0.81	1946	1.64	1966	60.2	1966	18.2	1966	82	74	67	74	11.7	WSW	73	W	1950	37	8.1
F	32.6	16.9	24.0	64	1965	-8	1970	1126	0	2.42	5.07	1950	0.74	1968	2.43	1950	64.8	1958	18.7	1958	81	75	65	75	11.5	WSW	66	W	1956	43	7.3
M	41.0	24.3	33.0	79	1966	-1	1967	992	0	2.57	5.42	1942	0.47	1942	2.21	1959	40.3	1959	17.6	1959	77	74	62	73	11.7	WSW	65	SW	1958	55	7.3
A	56.0	36.1	46.1	93	1970	14	1970	567	0	2.74	4.90	1944	0.50	1974	1.99	1943	8.2	1972	2.4	1974	76	73	53	63	10.6	W	59	SW	1950	55	6.6
M	67.3	46.0	56.5	96	1970	26	1966	248	22	2.80	6.62	1974	0.22	1963	3.85	1974	2.0	1945	2.0	1945	77	76	54	60	9.3	SW	63	W	1950	60	6.6
J	78.0	55.8	66.9	96	1964	36	1972	46	103	2.54	6.56	1972	0.22	1963	2.86	1945	0.0		0.0		81	78	54	60	8.4	SW	61	SW	1949	67	5.9
J	82.2	60.2	71.2	98	1966	42	1963	9	202	2.89	9.70	1947	0.98	1955	2.94	1947	0.0	1965	0.0	1965	82	81	52	57	7.9	SW	56	NE	1956	70	5.6
A	80.1	59.3	69.3	98	1965	36	1968	26	159	2.35	5.95	1968	0.76	1951	2.39	1968	T	1956	T	1956	86	87	55	65	7.6	SW	59	SW	1955	68	5.7
S	73.1	51.5	62.3	96	1973	30	1959	126	45	2.62	7.85	1945	0.28	1963	2.61	1959	1.4	1960	1.4	1960	86	87	59	73	8.0	SW	50	SW	1941	61	5.4
O	62.4	41.5	52.3	86	1972	22	1972	398	0	2.83	7.85	1955	0.28	1963	2.52	1945	1.6	1972	1.4	1972	80	81	59	73	8.7	SW	68	E	1950	51	6.4
N	47.9	33.1	40.5	81	1971	5	1971	735	0	2.35	5.51	1950	0.73	1965	3.13	1953	16.3	1953	11.2	1953	81	80	69	75	10.2	W	61	SW	1950	32	8.2
D	34.9	21.7	28.3	68	1966	-5	1969	1138	0	2.35	5.05	1944	0.62	1958	1.59	1944	44.2	1970	17.8	1970	81	81	77	77	10.7	WSW	56	W	1949	31	8.3
YR	57.2	38.6	47.9	98 JUL 1966		-11 JAN 1968		6719	536	31.33	9.70 JUL 1947		0.22 JUN 1963		3.85 MAY 1974		64.8 FEB 1958		18.7 FEB 1960		80	79	60	63	9.6	WSW	73	W	JAN 1950	54	6.9

Means and extremes above are from existing and comparable exposures. Annual extremes have been exceeded at other sites in the locality as follows: Highest temperature 102 in July 1936; lowest temperature -22 in February 1934; minimum monthly precipitation 0.08 in October 1924; maximum precipitation in 24 hours 4.19 in August 1893; maximum snowfall in 24 hours 29.8 in March 1900.

REFERENCE NOTES APPLYING TO TABLES APPEAR ON THE PAGE FOLLOWING LAST TABLE.
(Caution: Letters and symbols may have different meanings in 1941-1970 tables than in earlier tables. See notes.)

NORMALS, MEANS AND EXTREMES
(Table Revised 1972. Base Period for Climatological Normals: 1931-1960)

Station: SYRACUSE, NEW YORK   HANCOCK AIRPORT

Latitude: 43° 07' N   Longitude: 76° 07' W   Standard time used: EASTERN   Elevation (ground): 410 feet

Month	Temperature Normal Daily maximum	Normal Daily minimum	Normal Monthly	Extremes Record highest	Year	Extremes Record lowest	Year	Precipitation Normal total	Normal heating degree days (Base 65°)	Precipitation Max monthly	Year	Min monthly	Year	Max in 24 hrs.	Year	Snow, Ice pellets Mean total	Max monthly	Year	Max in 24 hrs.	Year
J	31.5	16.5	24.0	70	1967	-26	1966	3.15	1271	4.59	1959	1.02	1970	1.47	1958	26.9	71.0	1966	24.5	1966
F	31.9	16.7	24.3	70	1965	-22	1967	3.19	1140	5.38	1951	1.10	1968	1.99	1961	27.4	72.6	1958	21.4	1961
M	40.1	24.8	32.6	85	1968	-5	1967	3.60	1004	6.84	1955	1.08	1964	1.33	1964	17.4	37.2	1971	14.7	1971
A	56.5	36.4	46.5	92	1971	16	1956	3.68	570	4.75	1954	1.57	1950	2.42	1970	2.4	4.0	1961	5.9	1961
M	67.9	46.0	57.0	94	1971+	25	1966	3.27	248	4.35	1953	1.05	1969	3.13	1969	0.1	0.0		0.0	
J	77.7	56.8	67.3	96	1966+	35	1968	2.96	45	6.14	1968	1.10	1962	2.96	1968	0.0	0.0		0.0	
J	82.4	62.0	72.2	96	1965	45	1965	3.09	6	6.49	1971	0.90	1969	3.09	1969	0.0	0.0		0.0	
A	80.5	59.9	70.2	97	1965	40	1965	3.25	28	8.41	1956	1.77	1969	4.27	1956	0.0	0.0		0.0	
S	72.4	52.3	62.4	93	1964	28	1965	2.64	132	4.89	1958	0.75	1964	3.09	1960	0.7	4.4	1952	2.1	1952
O	61.1	42.4	51.8	87	1963	20	1965	3.18	414	8.29	1955	0.21	1963	3.60	1955	0.9	9.0	1954	8.9	1954
N	48.3	33.2	40.2	76	1971	10	1967	2.90	744	5.65	1963	1.68	1960	2.09	1963	9.0	22.1	1958	8.9	1958
D	34.6	20.9	27.8	70	1966	-7	1966	3.15	1153	5.01	1959	1.73	1958	2.18	1952	25.2	52.5	1969	15.5	1969
YR	56.9	39.2	48.0	97 AUG. 1965		-26 JAN. 1966		37.60	6756	8.41 AUG. 1956		0.21 OCT. 1963		4.27 AUG. 1954		109.1	72.6 FEB. 1958		24.5 JAN. 1966	

Ø For period July 1963 through the current year.

Means and extremes above are from existing and comparable exposures. Annual extremes have been exceeded at other sites in the locality as follows:
Highest temperature 102 in July 1936; maximum monthly precipitation 15.92 in June 1922; minimum monthly precipitation 0.19 in May 1920; maximum precipitation in 24 hours 4.79 in June 1922; maximum snowfall in 24 hours 27.2 in January 1925; fastest mile wind 69 SW in December 1921.

NORMALS, MEANS AND EXTREMES
(Table Revised 1975. Base Period for Climatological Normals: 1941-1970)

Month	Temperature °F Normal Daily maximum	Normal Daily minimum	Normal Monthly	Extremes Record highest	Year	Extremes Record lowest	Year	Normal Degree days Base 65° F Heating	Cooling	Precipitation Normal	Max monthly	Year	Min monthly	Year	Max in 24 hrs.	Year	Snow, Ice pellets Max monthly	Year	Max in 24 hrs.	Year
J	31.4	15.8	23.6	70	1967	-26	1966	1283	0	2.68	4.59	1959	1.02	1970	1.47	1958	71.0	1966	24.5	1966
F	32.7	16.5	24.6	59	1974	-22	1967	1131	0	2.79	5.38	1951	1.10	1968	1.99	1961	72.6	1958	21.4	1961
M	41.5	24.8	33.2	85	1968	-5	1967	986	0	3.03	6.86	1955	1.08	1964	1.34	1964	37.2	1971	14.7	1971
A	56.5	36.4	46.5	85	1974	16	1950	555	18	3.03	6.91	1973	1.57	1950	2.42	1970	8.4	1961	5.9	1961
M	68.0	46.0	56.8	92	1971	25	1966	272	103	3.08	6.19	1973	1.05	1969	3.13	1969	1.2	1973	1.2	1973
J	77.7	56.1	66.9	94	1974	35		46	212	3.09	12.30	1972	1.10	1962	3.88	1972	0.0		0.0	
J	82.0	61.0	71.5	96	1966	45	1965	11	212	3.08	9.52	1974	0.90	1969	4.07	1974	0.0		0.0	
A	80.2	59.2	69.7	97	1965	40	1965	18	164	3.50	8.41	1956	1.64	1954	4.27	1956	0.0		0.0	
S	73.4	52.5	62.5	93	1973	28	1965	120	54	2.71	4.89	1958	0.75	1964	3.09	1960	4.4	1952	2.1	1952
O	62.4	42.5	52.5	87	1963	21	1972	392	0	2.89	8.29	1955	0.21	1967	3.60	1955	12.1	1958	8.9	1954
N	48.3	33.6	41.0	76	1971	10	1967	720	0	3.25	6.79	1972	1.68	1960	2.09	1963	22.1	1958	8.9	1958
D	35.0	21.2	28.1	70	1966	-7	1973	1144	0	3.09	5.01	1952	1.73	1958	2.18	1952	52.5	1969	15.5	1969
YR	57.4	38.8	48.1	97 AUG. 1965		-26 JAN. 1966		6678	551	36.41	12.30 OCT. 1972		0.21 OCT. 1963		4.27 AUG. 1954		72.6 FEB. 1958		24.5 JAN. 1966	

Average station pressure mb. 1001.3   Elev. 408 feet m.s.l.

Means and extremes above are from existing and comparable exposures. Annual extremes have been exceeded at other sites in the locality as follows:
Highest temperature 102 in July 1936; maximum monthly precipitation 15.92 in June 1922; minimum monthly precipitation 0.19 in May 1920; maximum precipitation in 24 hours 4.79 in June 1922; maximum snowfall in 24 hours 27.2 in January 1925; fastest mile wind 69 SW in December 1921.

REFERENCE NOTES APPLYING TO TABLES APPEAR ON THE PAGE FOLLOWING LAST TABLE.
(Caution: Letters and symbols may have different meanings in 1941-1970 tables than in earlier tables. See notes.)

Reference notes applying to Normals, Means, and Extremes tables for 1931–1960 base period.

(a) Length of record, years, based on January data. Other months may be for more or fewer years if there have been breaks in the record. Climatological standard normals (1931-1960).

(b) Less than one half.

+ Also on earlier dates, months, or years.

T Trace, an amount too small to measure.

Below zero temperatures are preceded by a minus sign. The prevailing direction for wind in the Normals, Means, and Extremes table is from records through 1963.

‡ ≥ 70° at Alaskan stations.

Unless otherwise indicated, dimensional units used in this bulletin are: temperature in degrees F.; precipitation, including snowfall, in inches; wind movement in miles per hour; and relative humidity in percent. Heating degree day totals are the sums of negative departures of average daily temperatures from 65° F. Cooling degree day totals are the sums of positive departures of average daily temperatures from 65° F. Sleet was included in snowfall totals beginning with July 1948. The term "ice pellets" includes solid grains of ice (sleet) and particles consisting of snow pellets encased in a thin layer of ice. Heavy fog reduces visibility to 1/4 mile or less.

Sky cover is expressed in a range of 0 for no clouds or obscuring phenomena to 10 for complete sky cover. The number of clear days is based on average cloudiness 0-3, partly cloudy days 4-7, and cloudy days 8-10 tenths.

Solar radiation data are the averages of direct and diffuse radiation on a horizontal surface. The langley denotes one gram calorie per square centimeter.

& Figures instead of letters in a direction column indicate direction in tens of degrees from true North; i.e., 09 - East, 18 - South, 27 - West, 36 - North, and 00 - Calm. Resultant wind is the vector sum of wind directions and speeds divided by the number of observations. If figures appear in the direction column under "Fastest mile" the corresponding speeds are fastest observed 1-minute values.

To 8 compass points only.

Reference notes applying to Normals, Means, and Extremes tables for 1941–1970 base period.

(a) Length of record, years, through the current year unless otherwise noted, based on January data.

(b) 70° and above at Alaskan stations.

* Less than one half.

T Trace.

NORMALS - Based on record for the 1941-1970 period.

DATE OF AN EXTREME - The most recent in cases of multiple occurrence.

PREVAILING WIND DIRECTION - Record through 1963.

WIND DIRECTION - Numerals indicate tens of degrees clockwise from true north. 00 indicates calm.

FASTEST MILE WIND - Speed is fastest observed 1-minute value when the direction is in tens of degrees.

$ Greatest calendar day through March 1969.

Greatest calendar day August 1966 through March 1969.

MEAN SEASONAL SNOWFALL, INCHES

Data are based on the period 1931-68. Isolines are drawn through points of approximately equal value. Caution should be used in interpolating on these maps, particularly in mountainous areas.

MEAN ANNUAL PRECIPITATION, INCHES

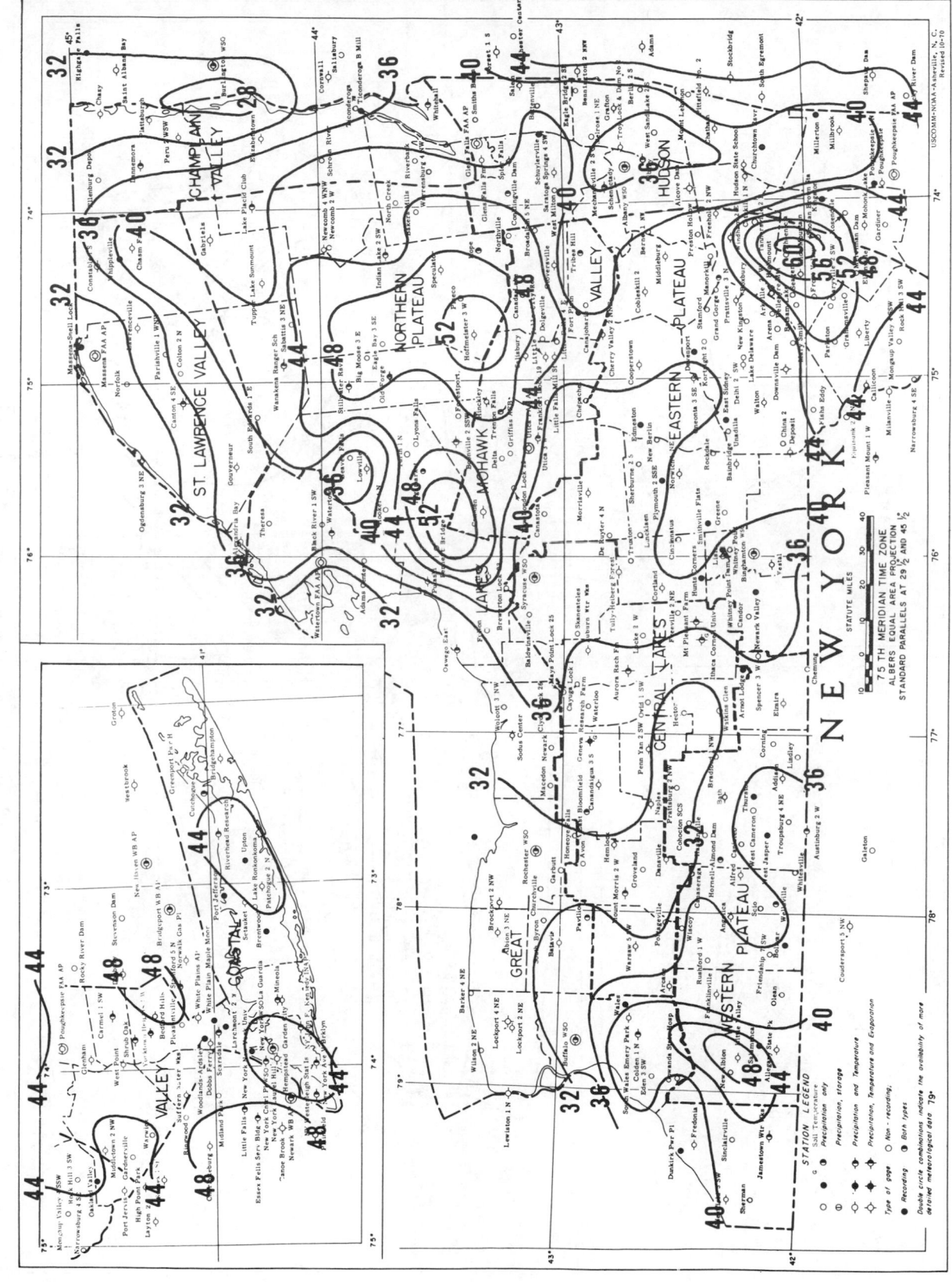

Data are based on the period 1931-55. Isolines are drawn through points of approximately equal value. Caution should be used in interpolating on these maps, particularly in mountainous areas.

MEAN MAXIMUM TEMPERATURE (°F.), JANUARY

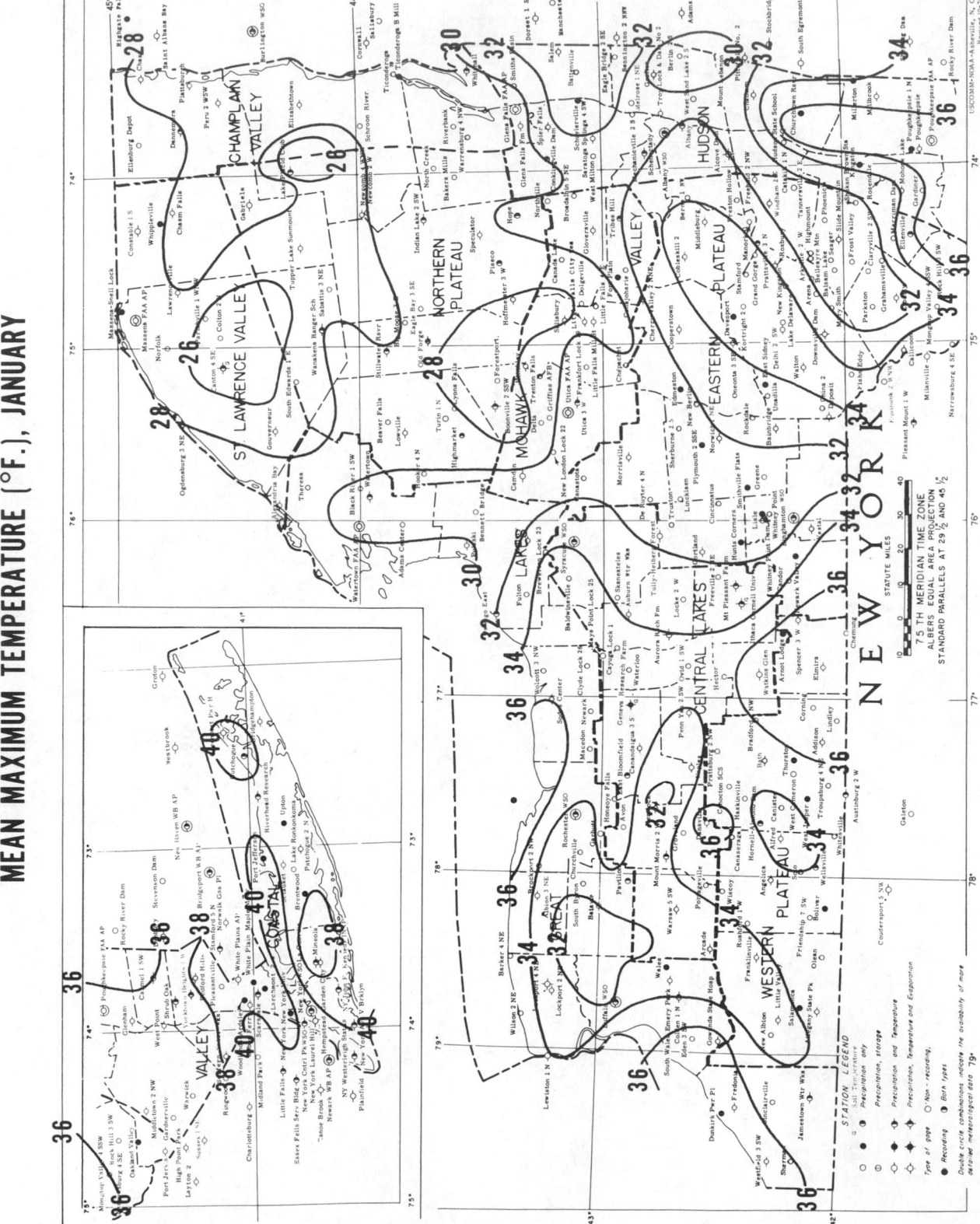

Data are based on the period 1931-52. Isolines are drawn through points of approximately equal value. Caution should be used in interpolating on these maps, particularly in mountainous areas.

MEAN MINIMUM TEMPERATURE (°F.), JANUARY

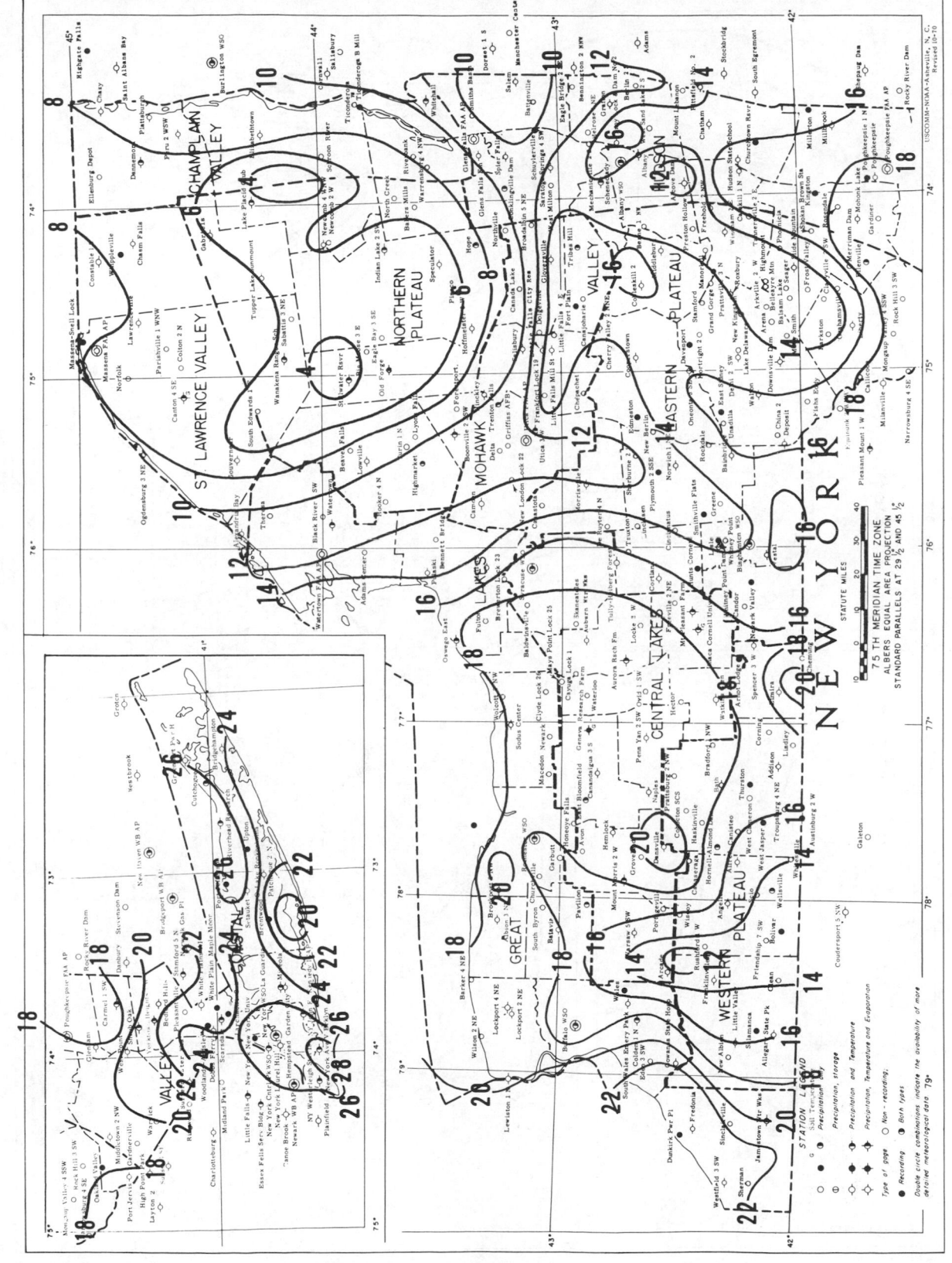

Data are based on the period 1931-52. Isolines are drawn through points of approximately equal value. Caution should be used in interpolating on these maps, particularly in mountainous areas.

MEAN MAXIMUM TEMPERATURE (°F.), JULY

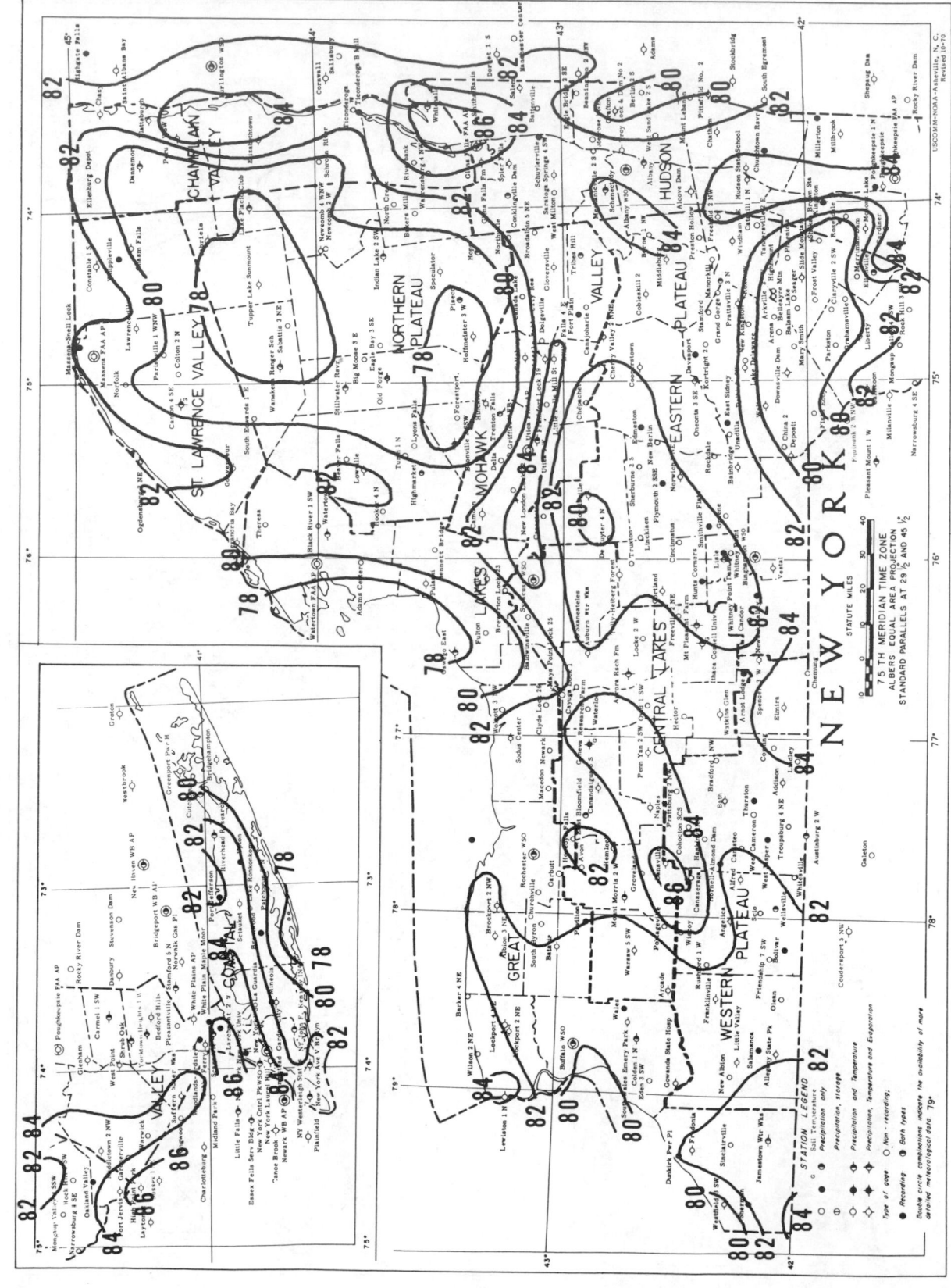

Data are based on the period 1931-52. Isolines are drawn through points of approximately equal value. Caution should be used in interpolating on these maps, particularly in mountainous areas.

MEAN MINIMUM TEMPERATURE (°F.), JULY

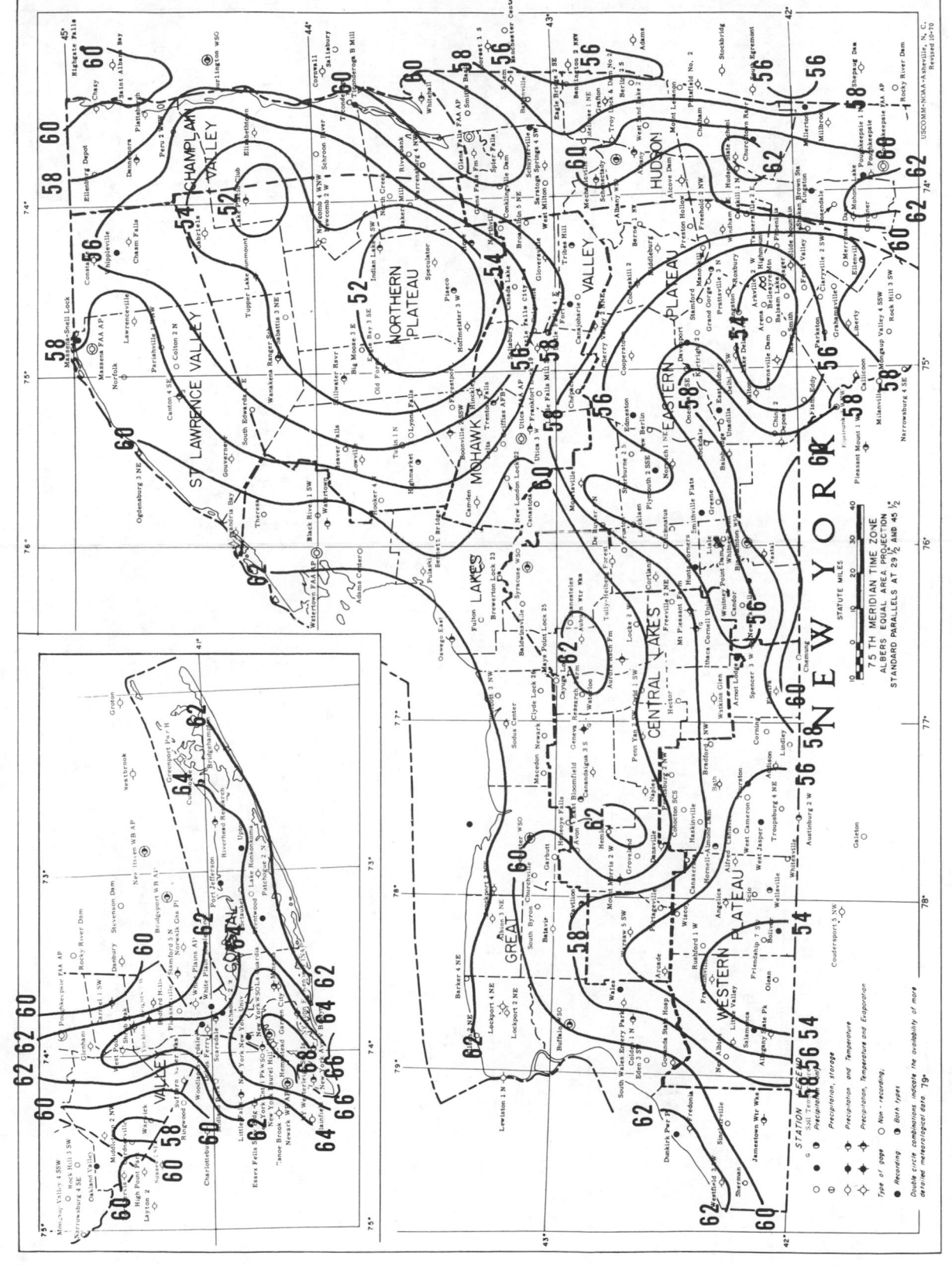

Data are based on the period 1931-52. Isolines are drawn through points of approximately equal value. Caution should be used in interpolating on these maps, particularly in mountainous areas.

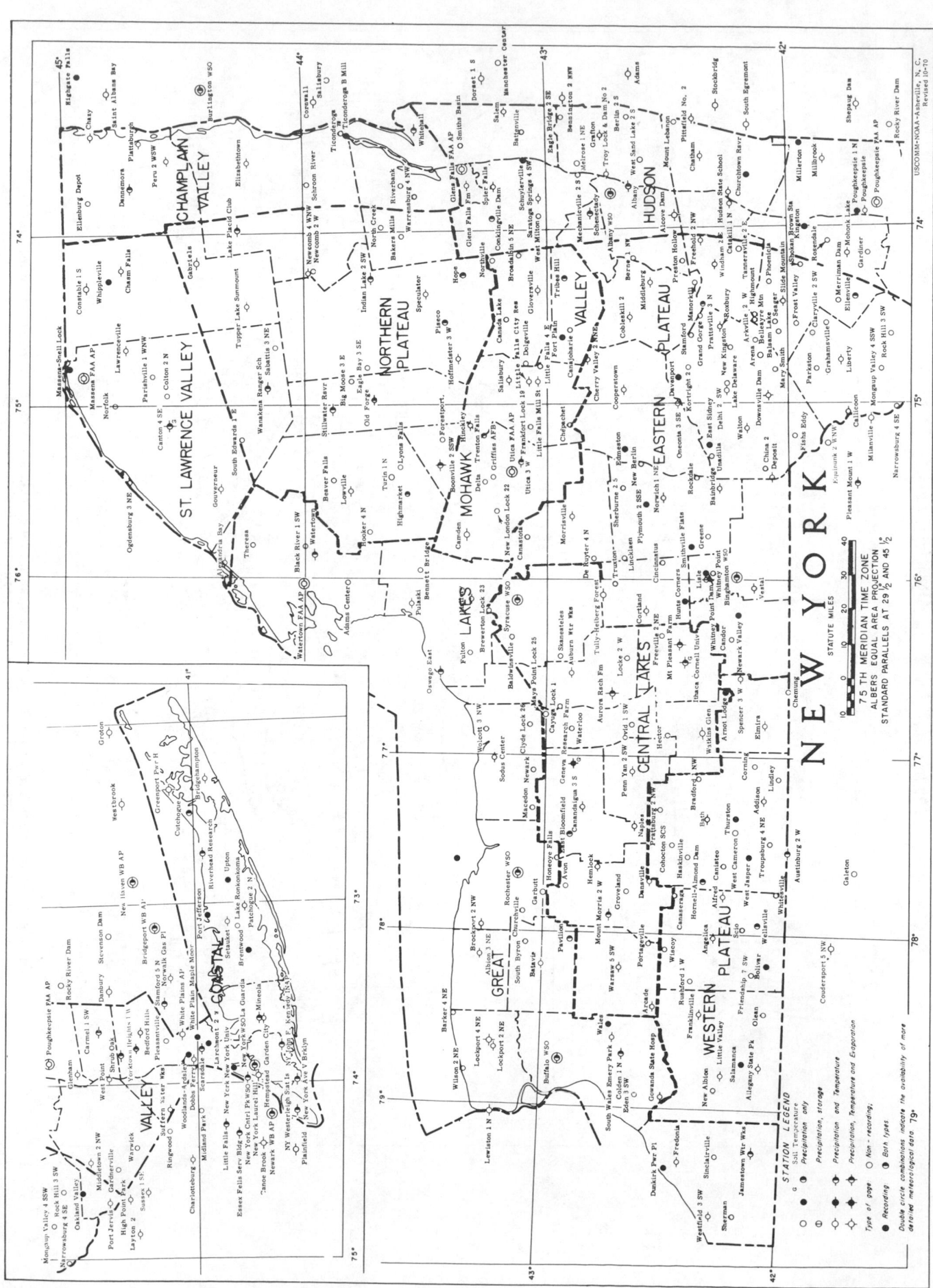

NEW YORK

75 TH MERIDIAN TIME ZONE
ALBERS EQUAL AREA PROJECTION
STANDARD PARALLELS AT 29½ AND 45½

STATUTE MILES

STATION LEGEND

○ Precipitation only
⊙ Precipitation, storage
● Precipitation and Temperature
⊕ Precipitation, Temperature and Evaporation
⬦ Type of gage   ● Recording   ○ Non - recording;
⊕ Both types
Ⓖ Soil Temperature
Double circle combinations indicate the availability of more
detailed meteorological data

USCOMM-NOAA-Asheville, N. C.
Revised 10-70

CLIMATES OF THE STATES

North Carolina

(Normals, Means and Extremes tables revised 1970 and 1975. Basic report revised June 1970.)

Albert V. Hardy, ESSA State Climatologist

North Carolina lies between 33 1/2° and 37° north latitude and between 75° and 84 1/2° west longitude. The span of longitude is greater than that of any other state east of the Mississippi River. The greatest length from east to west is 503 miles. The greatest breadth from north to south is 187 miles. The total area is 52,712 square miles; 49,142 square miles of land and 3,570 square miles of water.

The range of altitude is also the greatest of any eastern state. North Carolina rises from sea level along the Atlantic coast to 6,684 feet at the summit of Mount Mitchell, the highest peak in the eastern United States. Mount Mitchell is in the heart of the Blue Ridge Range. This Range, along with the Great Smokies, lies partly in North Carolina and partly in Tennessee and forms the highest part of the Appalachian Mountains.

The three principal physiographic Divisions of the eastern United States are particularly well developed in North Carolina. Beginning in the east, they are: The Coastal Plain, the Piedmont, and the Mountains. For assembly and study of climate and crop statistics, the Mountain Division is subdivided into a northern and a southern sector, while the Coastal Plain and Piedmont Divisions are each subdivided into northern, central, and southern sectors. Thus a total of eight climatological subdivisions are recognized.

The land and water areas of the Coastal Plain Division comprise nearly half the area of the State. The tidewater portion is generally flat and swampy, while the interior is gently sloping and, for the most part, naturally well drained. Throughout the Coastal Plain the soils consist of soft sediment, with little or no underlying hard rock near the surface. The average slope is from about 200 feet above mean sea level at the "fall line", or western boundary, to generally 30 feet or less over the tidewater area.

The Piedmont Division rises gently from about 200 feet at the fall line to near 1,500 feet at the base of the Mountains; its area is about one-third of the State. The land is mostly gently rolling, with a great deal of hard rock near the surface. There are several ranges of steeper hills. The principal of these are the Uwharrie Range in and around Randolph County, and the Kings Mountain Range in Cleveland and Gaston Counties.

The Mountain Division is the smallest of the three, little more than one-fifth of the State's area. In elevation it ranges downward from Mount Mitchell's peak to about 1,000 feet above mean sea level in the lowest valleys. There are more than 40 peaks higher than 6,000 feet and about 80 others over 5,000 feet high. The surface of the Mountain Division is rocky, the soils being mainly of weathered and eroded rocky materials.

North Carolina has the most varied climate of any eastern state. This is due mainly to its wide range in elevation and distance from the ocean. Lesser influences are latitude, inland bodies of water, soil surface, and plant cover. In all seasons of the year the average temperature varies more than 20° from the lower coast to the highest mountain elevations. The average annual temperature at Southport on the lower coast is nearly as high as that of interior northern Florida, while the average on the summit of Mount Mitchell is lower than that of Buffalo, New York.

Altitude also has an important effect on rainfall. The rainiest part of the eastern United States, with an annual average of more than 80 inches, is in southwestern North Carolina where moist southerly winds are forced upward in passing over the mountain barrier. Less than 50 miles north, in the valley of the French Broad River, is the driest place south of Virginia and east of the Mississippi. Sheltered by mountain ranges on all sides, this point has an average annual precipitation of only 37 inches. East of the Mountains, average annual rainfall ranges mostly 40 to 55 inches.

In winter the greater part of North Carolina is partially protected by the mountain ranges from the frequent outbreaks of cold which move southeastward across the Central States. Such outbreaks often spread southward all the way to the Gulf of Mexico without attaining strength and depth to cross the Appalachian Range. When cold waves do break across they are usually modified by the crossing and the descent on the eastern slopes. The temperature drops to around 10° over central North Carolina once or twice during an average winter. Near the coast a comparable figure is some 10 degrees higher, and in the upper mountains 10 degrees lower. Temperatures as low as 0° are rare outside the mountains, but have occurred at one time or another throughout the western part of the State. The lowest temperature of record is -29°, recorded January 30, 1966, at Mount Mitchell 2 SSW.

Winter temperatures in the eastern Coastal Plain are modified by the proximity of the Atlantic Ocean. This effect raises the average winter temperature and reduces the average day-to-night range. The Gulf Stream, contrary to popular opinion, has little direct effect on North Carolina temperatures, even on the immediate coast. The Stream lies some 50 miles offshore at its nearest point. The southern reaches of the cold Labrador Current pass between the Gulf Stream and the North Carolina coast. This offsets any warming effect the Stream might otherwise have on coastal temperatures. The meeting of the two opposing currents does provide a breeding ground for rough weather. Not infrequently low pressure storms having their origin there develop major proportions, causing rain on the North Carolina coast and over states to the north.

In spring the storm systems that bring cold weather southward reach North Carolina less forcefully than in winter, and temperatures begin to modify. Day-to-day variations in temperature are less pronounced, and warm weather is more likely to occur in conjunction with fair weather.

The rise in average temperatures is greater in May than in any other month. Artificial heating is generally discontinued at some time during May.

Occasional mild invasions of air from the north continue to occur during the summer, but their effect on temperatures is slight and of short duration. Ordinarily, such outbreaks serve principally to clear the air of excess humidity. The increase in sunshine which follows usually brings temperatures back up quickly. When the drying of the air is sufficient to keep cloudiness at a minimum for several days, temperatures may occasionally reach 100° or a little higher in interior sections at elevations below 1,500 feet. The highest of record is 109°. Ordinarily, however, summer cloudiness develops to limit the sun's heating while temperatures are still in the 90° range. An entire summer, or even several consecutive summers, may pass without a high of 100° being recorded in the State. The average daily maximum reading in midsummer is below 90° for most localities.

Differences in temperatures over the various parts of the State are no less pronounced in summer than in winter. The warmest days, however, are found in the interior rather than on the coast in summer. The average daily maximum at midsummer is near 92° at Fayetteville and Goldsboro. On the extreme south coast it is only 86° at that season. The average mid-July maximum is only 67° atop Mount Mitchell, while over widely populated areas in the Mountain Division the figure is about 75°. Lowest morning temperatures average about 20° lower than afternoon highs except along the immediate coast, where the daily range is most often between 10° and 15°.

Autumn is the season of most rapidly changing temperature, the daily downward trend being greater than the corresponding rise in spring. The dropoff is most rapid in October and continues almost as fast in November. Average daily temperatures by the end of that month are generally within 5° of the annual low.

There are no distinct wet and dry seasons in North Carolina. There is some seasonal variation in average precipitation. Summer rainfall is normally the greatest, and July the wettest month. Since the rain at this time of year comes mostly with thunderstorms and convective showers, it is also more variable than at other seasons. Daily showers are not uncommon, nor are periods of one or two weeks without rain. Autumn is the driest season, and October the driest month. Precipitation in winter and spring occurs mostly with migratory low pressure storms. It appears with greater regularity and more even distribution than summer showers.

Winter precipitation usually occurs with southerly through easterly winds, and is seldom associated with very cold weather. Snow and sleet occur on an average of once or twice a year near the coast, and not much more often over the southeastern half of the State. Such occurrences are nearly always connected with northeasterly winds, generated when high pressure over the interior or the northeastern United States causes a flow of cold air down parallel to the coastline, while offshore low pressure brings in cool, moist air from the North Atlantic. Over the Mountains and western Piedmont frozen precipitation sometimes occurs with interior low pressure storms. In the extreme west it can happen with a cold front passage from northwest. Average winter snowfall ranges from about 1 inch per year on the Outer Banks and the lower coast, to about 9 inches in the northern Piedmont and southern Mountains. Some of the higher mountain peaks and upper slopes receive an average of nearly 50 inches a year.

The greatest North Carolina 24-hour snowfall of record was 31 inches at Nashville in March 1927. The record fall for an entire season was 148 inches at Mount Mitchell in 1930-31. The greatest 24-hour rainfall of record, 22.22 inches, fell at Altapass in July 1916. The greatest annual total of 129.60 inches was recorded at Rosman in 1964.

Relative humidity may vary greatly from day to day and even from hour to hour, especially in winter. Observed percentages range from 100 down to 10 or lower. The average relative humidity, however, does not vary greatly from season to season, there being a slight tendency for highest averages in winter and lowest in spring. The lowest relative humidities are found over the southern Piedmont, where the year around average is about 65 percent. The highest are along the immediate coast, where the average may be as high as 75 to 80 percent. The lowest amount of actual moisture is found in the higher mountain areas, but the lower temperatures there bring the relative humidity up as high or higher than other interior areas.

Sunshine is abundant, the average annual percentage of possible ranging from 60 to 65 percent at most recording points. Observations of sky condition taken at airport locations in recent years indicate an average of about 112 days per year clear, 105 partly cloudy, and 148 cloudy. Measurable rain falls on about 120 days. Prevailing winds blow from southwest 10 months of the year, and from northeast during September and October. The average wind speed for interior locations is about 8 m.p.h., for coastal points about 12. The highest wind of record (fastest mile) is 110 m.p.h. at Hatteras in September 1944.

North Carolina rivers fall into two groups: Those that flow into the Atlantic Ocean and those that drain westward into the Mississippi River system. The two are separated by a ridge averaging 2,200 feet above mean sea level. A second chain of mountains ranging up to 6,000 feet marks the western boundary of the State.

Most of the State, including the Coastal Plain, the Piedmont, and the eastern and southern slopes, drain into the Atlantic Ocean. The principal rivers involved are the Roanoke, Tar, Neuse, Cape Fear, Yadkin, and Catawba. The Roanoke rises in the Allegheny Mountains west of Roanoke, Virginia, and flows southeasterly through Virginia and North Carolina a distance of 400 miles, about 150 miles of which is in North Carolina. Flood peaks which used to beset the lower Roanoke are now much subdued by John H. Kerr Dam just upstream from the North Carolina-Virginia State border, and other dams both upstream and downstream. The Tar, Neuse, and Cape Fear Rivers rise in the North Carolina Piedmont and flow southeastwardly to the Atlantic. The Yadkin and Catawba Rivers rise in the Blue Ridge Mountains of western North Carolina, but reach the Atlantic via the Pee Dee and Santee Rivers of South Carolina.

The main stream draining the extreme western part of North Carolina is the French Broad River, which rises in the mountains southwest of Asheville. It flows first northward, then westward through the Great Smoky Mountains into Tennessee, there to join the Holston River near Knoxville to form the Tennessee River, which in turn feeds into the Ohio. Other parts of the southwestern mountain area drain into the Little Tennessee. The northern mountains are drained by streams flowing into the Ohio River system. All eventually reach the Mississippi.

Intense rainstorms occur in the precipitous mountain terrain, especially in the southern portion. Streams here rise quickly to flood, and almost as quickly subside when rain ends. Because of the unusually abundant rainfall and infrequent freezing of the rivers, North Carolina streams when properly dammed furnish a reliable flow of water for hydroelectric development. Extensive use has been made of this resource in the area drained by tributaries of the Tennessee River as developed by the Tennessee Valley Authority. Waters draining eastward have been harnessed less extensively. A survey taken several years ago showed North Carolina as the fourth state in the nation in installed capacity of hydroelectric generators. In addition to public power projects, there were estimated to be nearly 100 private hydroelectric generating plants.

Floods occur frequently, affecting some part of North Carolina each year. Loss of life is rare, and the economic loss not generally large, but the cost of floods is increasing as river lowgrounds are developed. Floods may occur at any season, but are most frequent in early spring, summer, and early fall. Rains associated with West Indian hurricanes are the main cause of summer and fall floods. In mid-August 1940, severe floods occurred as a result of hurricane rains. Later in the same month intense rains of local origin caused severe flooding in western North Carolina. Major floods also occurred in September 1945, October 1929, August 1928, and July 1916.

The greatest economic loss entailed in North Carolina because of stormy weather is that due to summer thunderstorms. These usually affect only limited areas, but hail and wind occurring with some of them account for an average yearly loss of about $5 million. Three to five people are killed in the State by lightning during the average year. Farm livestock, especially cattle, are killed in larger numbers, and there is a considerable loss of property due to fires set by lightning. In any given locality, 40 to 50 days with thunderstorms may be expected in a year.

North Carolina is outside the principal tornado area of the United States. A total of 192 tornadoes were reported during the 53-year period 1916-1968, an average of less than 4 per year. In recent years the number reported has increased to almost twice that figure, probably due to rising population and more effective reporting. A large percentage of North Carolina tornadoes affect only a small area.

Tropical hurricanes come close enough to influence North Carolina weather about twice in an average year. Only about once in 10 years, on the average, does this type storm strike the State with sufficient force to do much damage to inland property. Coastal properties occasionally suffer damage from associated high tides and seas.

The wide variety of climate in North Carolina leads to a wide variety of vegetation. The average annual freeze-free period or "growing season" ranges from near 130 days in the highest mountain areas to around 290 days on the central section of the Outer Banks. At Hatteras entire seasons may pass without either frost or freezing temperature occurring. Tropical fruits can be grown there with a little care. Both orange and grapefruit trees have reached bearing age on the island.

Corn and hay are the most widely planted farm crops, growing in every North Carolina county. Most of these crops are used as feed for livestock, but an increasing percentage of corn is being grown for market. Small grains are almost as widely planted as corn, but the acreage is smaller. The most widely planted of all "money crops" is tobacco. Three-fourths of the North Carolina counties grow some tobacco, and in about one-half the counties it is the chief cash crop. The gross value of tobacco produced in the State is nearly twice that of all other field crops combined, and is the greatest of any single crop in any state except cotton in Texas. The widespread growing of tobacco in North Carolina is partly the result of suitable soils and partly of climate. While some types of tobacco are grown from Florida to Canada, the particular quality of tobacco which abounds in North Carolina can be produced over a rather limited area.

Other important field crops, listed in order of the total value of production in North Carolina in 1968, are: Corn, peanuts, soybeans, cotton, and hay. Sweet potatoes, Irish potatoes, and various fruits and vegetables are also grown commercially in some areas.

Livestock production plays an increasingly important part in the farm economy of North Carolina. The mild climate lends itself to the economical production of poultry and eggs, which account for more than half the total value of livestock products. Cattle and dairy products run second in value. Hogs account for nearly one-fifth of the total. Altogether, livestock and livestock products total nearly one-third of the entire value of farm output.

Favorable climatic factors are causing many manufacturers to establish plants in North Carolina. Manufacturing interests in the State employ more people than agriculture, and the gross value of output is greater. Textiles are the most important of the manufactured products. Tobacco is second, followed by food, furniture, and lumber in that order.

The importance of North Carolina as a vacation area is increasing. The mountains, because of their mild summer temperatures, provide a welcome escape from the heat of lower lands of both this and other states. Midsummer afternoon temperatures average below 80° at elevations of 3,000 feet or higher. Nights are crisp and cool, but seldom too cold for light camping. Mountain streams and forests furnish plenty of fishing and hunting, and there are lakes with boating facilities. Hiking is coming into increased favor, with numerous mountain trails to follow. There are peaks which provide challenge to amateur mountain climbers. Ski slopes are operating with increasing success at several higher mountain locations in midwinter.

Summer temperatures on the North Carolina beaches are considerably modified from the heat of the interior. The entire coastline has developed into a summer vacation area. Fishing, ranging from deep sea to inland lake and stream, is one of the most popular of sports. Midsummer ocean water temperatures range near 80° along the North Carolina coast, and practically the entire shoreline is suitable for ocean bathing. For those who prefer to swim in protected inland waters, hundreds of miles of beaches are found bordering the "sounds", or waters lying within the offshore islands.

The North Carolina beach country, especially the southern portion, is becoming increasingly popular for year-round vacationing. Both air and water temperatures are generally mild. The average ocean water temperature at Southport in January is higher than that along the northern coast of Maine in July.

The best known winter vacation area, however, is the "sandhills" section of the southern Piedmont. Here, due to the sandy character of the soils, the average winter daytime temperature is higher and the corresponding relative humidity lower than in other similarly exposed sections of the State. Winter rains are readily absorbed by the sandy soils without "mucking", and snow is a rarity, so that horse racing, golf and other outdoor sports are widely enjoyed. Pinehurst and Southern Pines, in the center of this area, have become popular places for retirement of those seeking a moderate climate.

Another favored area for winter living is the

"thermal belt" section of the southern mountains. Thermal belts are small areas along the mountain slopes that tend to have higher minimum temperatures than other adjacent or nearby areas less favorably situated. This is because cold air on quiet nights tends to drain into lower valleys, while higher elevations above the thermals belts are cold because of their elevation. In this particular area the situation is augmented by the existence of many long, southward-facing slopes which absorb much winter sunshine, at the same time being protected to a degree from cold air outbreaks by higher mountains to the north and west. As a result "thermal belts" part way up the mountain sides may support vulnerable vegetation long after frost has killed all green both above and below. Often in the dead of winter the contrast of a belt of green flanked by brown both upslope and in the valley is most striking. The term "thermal belt" is sometimes mistakenly applied to the entire area where this phenomenon is most common.

REFERENCES

(1) Carney, C. B., Climate of North Carolina, Climates of the States, Climatography of the United States No. 60-31. ESSA, Environmental Data Service, 1960.

(2) Carney, C. B., and A. V. Hardy. Weather and Climate in North Carolina. N. C. Agricultural Experiment Station Bull. No. 396. Reprint, Raleigh, 1964.

(3) Goerch and Ehringhaus. North Carolina Almanac and State Industrial Guide. Almanac Publishing Company, Raleigh, 1953.

(4) Hardy, A. V. Low Temperature Probabilities in North Carolina. N. C. Agricultural Experiment Station Bull. No. 423. N. C. State Univ., Raleigh, 1964.

(5) Hardy, A. V., C. B. Carney, and H. V. Marshall, Jr. Climate of North Carolina Research Stations. N. C. Agr. Exp. Sta. Bull. No. 433. N.C.S.U., Raleigh, 1967.

(6) North Carolina Department of Agriculture. North Carolina Agricultural Statistics; Annuals, 1953 and 1968, and Special Livestock Issue, 1968.

(7) North Carolina Department of Conservation and Development. Topography, Geology, and Mineral Resources of North Carolina; Educational Series No. 2, 1952.

(8) _____ Geology and Mineral Resources of North Carolina; Educational Series No. 3, 1953.

(9) North Carolina State Planning Board. Report on Water Resources. Raleigh, October 1937.

(10) Weather Bureau Technical Paper No. 15 - Maximum Station Precipitation for 1, 2, 3, 6, 12, and 24 Hours. ESSA, Weather Bureau.

(11) Weather Bureau Technical Paper No. 16 - Maximum 24-Hour Precipitation in the U. S. ESSA, Weather Bureau.

(12) Weather Bureau Technical Paper No. 25 - Rainfall Intensity - Duration - Frequency Curves. ESSA, Weather Bureau.

(13) Weather Bureau Technical Paper No. 29 - Rainfall Intensity - Frequency Regime. ESSA, Weather Bureau.

BIBLIOGRAPHY

(A) Climatic Summary of the United States (Bulletin W) 1930 Edition, Sections 95, 96 and 97. ESSA, Weather Bureau.

(B) Climatic Summary of the United States, North Carolina - Supplement for 1931 through 1952. (Bulletin W Supplement). ESSA, Weather Bureau.

(C) Climatic Summary of the United States, North Carolina - Supplement for 1951 through 1960. (Bulletin W Supplement). ESSA, Weather Bureau.

(D) Climatological Data, North Carolina. ESSA, Environmental Data Service.

(E) Climatological Data, National Summary. ESSA, Environmental Data Service.

(F) Hourly Precipitation Data, North Carolina. ESSA, Environmental Data Service.

(G) Local Climatological Data for Asheville, Charlotte, Greensboro, Hatteras, Raleigh, Wilmington, and Winston-Salem, North Carolina. ESSA, Environmental Data Service.

FREEZE DATA

STATION	Freeze threshold temperature	Mean date of last Spring occurrence	Mean date of first Fall occurrence	Mean No. of days between dates	Years of record Spring	No. of occurrences in Spring	Years of record Fall	No. of occurrences in Fall
ALBEMARLE	32	04-14	10-23	192	30	30	30	30
	28	03-31	11-05	219	30	30	30	30
	24	03-13	11-16	249	30	30	30	30
	20	02-27	12-01	277	30	30	30	27
	16	02-06	12-16	313	30	24	30	21
ANDREWS 2 E	32	04-29	10-15	169	21	21	25	25
	28	04-14	10-24	193	21	21	25	25
	24	03-30	11-05	219	21	21	25	25
	20	03-10	11-15	250	21	21	25	25
	16	02-27	11-30	276	20	20	25	23
ASHEBORO 2 W	32	04-07	11-02	209	25	25	25	25
	28	03-23	11-16	238	25	25	25	24
	24	03-05	11-29	270	25	25	24	23
	20	02-18	12-08	293	25	23	24	20
	16	02-03	12-20	320	25	20	24	15
ASHEVILLE WB CITY	32	04-12	10-24	195	30	30	30	30
	28	03-29	11-04	220	30	30	30	30
	24	03-14	11-21	252	30	30	30	30
	20	02-27	12-04	280	30	30	30	26
	16	02-10	12-16	309	30	26	30	17
BANNER ELK	32	05-14	10-03	142	29	29	30	30
	28	04-30	10-10	163	29	29	29	29
	24	04-11	10-19	191	29	29	27	27
	20	03-28	11-04	221	29	29	27	27
	16	03-17	11-17	245	29	29	27	26
BREVARD	32	04-25	10-15	173	12	12	13	13
	28	04-14	10-21	190	12	12	13	13
	24	04-01	10-31	214	12	12	12	12
	20	03-04	11-20	262	11	11	12	12
	16	02-11	12-05	297	10	8	10	9
CAROLEEN	32	04-10	10-30	202	30	30	29	29
	28	03-24	11-12	233	30	30	29	29
	24	03-09	11-24	260	30	30	29	28
	20	02-24	12-07	286	30	30	29	24
	16	01-29	12-15	320	28	20	27	19
CHAPEL HILL 2 W	32	03-31	11-04	218	30	30	30	30
	28	03-17	11-18	246	30	30	30	30
	24	03-04	11-29	269	30	30	30	30
	20	02-16	12-12	299	30	28	30	23
	16	01-30	12-23	328	30	22	30	13
CHARLOTTE WB CITY	32	03-21	11-15	239	30	30	30	30
	28	03-10	11-29	263	30	30	30	30
	24	02-21	12-11	293	30	28	30	24
	20	02-07	12-21	317	30	26	30	17
	16	01-17	12-26	342	30	15	30	10
CLINTON	32	03-25	11-02	223	14	14	14	14
	28	03-14	11-16	248	14	14	14	14
	24	02-27	12-02	278	14	14	14	13
	20	02-09	12-17	311	14	12	14	9
	16	01-14	12-26	346	14	6	14	4
CONCORD	32	04-04	10-31	211	18	18	18	18
	28	03-20	11-15	240	18	18	18	18
	24	03-05	11-25	266	18	18	18	18
	20	02-15	12-07	295	17	15	18	16
	16	01-31	12-19	322	17	13	18	10
CULLOWHEE	32	04-27	10-16	172	30	30	30	30
	28	04-15	10-23	191	30	30	30	30
	24	03-30	10-31	215	30	30	29	29
	20	03-13	11-14	246	30	30	29	29
	16	03-01	11-24	269	30	30	29	28
DURHAM	32	04-13	10-24	195	29	29	28	28
	28	03-25	11-08	228	29	29	28	28
	24	03-06	11-22	261	28	28	28	28
	20	02-22	12-04	285	28	28	28	25
	16	02-13	12-15	306	28	26	27	19
EDENTON	32	03-29	11-11	227	30	30	30	30
	28	03-12	11-19	252	30	30	30	30
	24	02-28	12-03	278	30	29	29	28
	20	02-15	12-17	305	30	28	29	20
	16	01-22	12-27	339	29	15	29	8
ELIZABETH CITY	32	03-29	11-10	226	29	29	30	30
	28	03-12	11-20	253	29	29	30	30
	24	02-26	12-08	285	29	28	30	26
	20	02-17	12-18	304	28	26	29	19
	16	01-25	12-27	336	28	18	29	9
FAYETTEVILLE	32	03-28	11-06	222	30	30	29	29
	28	03-13	11-16	248	30	30	29	29
	24	03-01	11-30	274	30	30	29	28
	20	02-07	12-17	313	30	27	29	17
	16	01-18	12-25	341	30	16	29	11
GOLDSBORO	32	03-29	11-07	224	30	30	30	30
	28	03-17	11-15	243	30	30	29	29
	24	02-27	11-30	276	30	30	28	27
	20	02-13	12-13	303	30	28	27	19
	16	01-27	12-27	334	30	20	26	8
GREENVILLE NO. 2	32	03-28	11-05	222	30	30	30	30
	28	03-14	11-17	248	30	30	30	30
	24	03-01	12-03	277	30	30	30	30
	20	02-13	12-12	302	30	29	30	24
	16	01-26	12-26	334	30	21	30	11
HATTERAS WB CITY	32	02-25	12-18	296	30	29	30	20
	28	02-04	12-26	324	30	25	30	9
	24	01-14	●	●	29	11	30	4
	20	01-08	●	●	29	8	30	2
	16	●	●	●	29	1	30	0
HENDERSON 3 SW	32	04-14	10-24	193	29	29	29	29
	28	04-03	11-05	216	29	29	28	28
	24	03-14	11-22	253	29	29	28	28
	20	02-28	12-05	280	29	28	28	26
	16	02-12	12-13	304	29	25	28	21
HENDERSONVILLE	32	04-24	10-19	177	30	30	30	30
	28	04-07	10-29	205	30	30	30	30
	24	03-23	11-09	231	30	30	30	30
	20	03-06	11-25	264	30	30	30	29
	16	02-21	12-05	287	30	29	29	27
HICKORY	32	04-04	10-31	210	30	30	30	30
	28	03-22	11-13	236	30	30	30	30
	24	03-09	11-30	266	30	30	30	29
	20	02-19	12-08	292	30	29	30	25
	16	02-05	12-19	317	30	26	30	17
HIGHLANDS	32	04-22	10-21	182	29	29	29	29
	28	04-12	10-30	202	29	29	29	29
	24	04-04	11-14	225	28	28	28	28
	20	03-16	11-26	255	28	28	27	24
	16	03-04	12-08	278	27	27	27	20
HOT SPRINGS 2	32	04-16	10-23	190	30	30	28	28
	28	04-05	10-31	209	30	30	28	28
	24	03-17	11-14	241	30	30	28	27
	20	03-05	11-27	267	30	29	27	25
	16	02-15	12-12	300	28	25	26	19
KINSTON	32	03-25	11-08	228	26	26	25	25
	28	03-12	11-19	252	25	25	24	23
	24	02-24	11-30	279	25	24	24	23
	20	02-11	12-13	305	25	23	24	18
	16	01-17	12-25	342	25	12	24	10
LENOIR	32	04-20	10-18	181	29	29	30	30
	28	04-07	10-31	208	29	29	30	30
	24	03-19	11-15	240	29	29	30	30
	20	03-04	11-29	271	29	29	30	29
	16	02-15	12-07	296	29	26	30	24
LEXINGTON	32	04-07	11-03	211	12	12	13	13
	28	03-28	11-20	237	12	12	13	13
	24	02-28	12-03	278	12	12	13	13
	20	02-14	12-14	303	12	10	13	10
	16	02-01	12-23	325	12	10	13	6
LOUISBURG	32	04-12	10-23	195	29	29	30	30
	28	03-31	11-06	221	29	29	30	30
	24	03-11	11-18	251	29	29	30	30
	20	02-25	12-02	280	29	28	30	29
	16	02-10	12-13	306	29	26	30	23

FREEZE DATA

STATION	Freeze threshold temperature	Mean date of last Spring occurrence	Mean date of first Fall occurrence	Mean No. of days between dates	Years of record Spring	No. of occurrences in Spring	Years of record Fall	No. of occurrences in Fall	STATION	Freeze threshold temperature	Mean date of last Spring occurrence	Mean date of first Fall occurrence	Mean No. of days between dates	Years of record Spring	No. of occurrences in Spring	Years of record Fall	No. of occurrences in Fall
LUMBERTON	32	03-26	11-06	225	30	30	30	30	RALEIGH WB CITY	32	03-24	11-16	237	30	30	30	30
	28	03-13	11-18	249	30	30	30	30		28	03-09	11-27	262	30	30	30	29
	24	02-22	12-03	284	30	28	30	27		24	02-27	12-08	284	30	30	30	25
	20	02-07	12-16	312	30	24	30	19		20	02-10	12-20	313	30	28	30	18
	16	01-13	12-27	347	30	12	30	9		16	01-23	12-25	336	30	18	30	11
MANTEO	32	03-13	11-26	258	22	22	24	23	REIDSVILLE	32	04-06	11-03	211	30	30	30	30
	28	02-28	12-07	282	22	22	24	21		28	03-21	11-16	240	30	30	30	30
	24	02-10	12-22	315	22	18	24	12		24	03-08	11-27	264	30	30	30	30
	20	01-20	12-25	340	21	11	24	9		20	02-24	12-07	287	30	29	30	24
	16	01-13	@	@	21	8	24	2		16	02-07	12-19	315	30	26	30	19
MARION	32	04-16	10-18	186	24	24	29	29	ROCK HOUSE	32	04-12	10-27	198	30	30	29	29
	28	03-28	11-05	221	23	23	29	29		28	04-05	11-04	213	30	30	29	29
	24	03-14	11-16	247	20	20	28	28		24	03-19	11-23	249	29	29	29	27
	20	03-05	11-28	268	19	19	27	25		20	03-11	12-04	268	29	29	29	24
	16	02-14	12-08	296	17	16	27	21		16	02-21	12-14	297	29	28	29	17
MARSHALL 2 NE	32	04-24	10-15	174	30	30	30	30	SALISBURY	32	04-07	10-28	205	30	30	30	30
	28	04-11	10-25	198	30	30	29	29		28	03-27	11-08	226	30	30	30	30
	24	03-26	11-05	224	30	30	29	29		24	03-12	11-19	252	30	30	30	29
	20	03-06	11-21	260	30	29	29	27		20	02-27	12-02	278	30	29	30	27
	16	02-21	12-04	286	30	28	29	25		16	02-10	12-17	310	30	27	30	20
MONCURE	32	04-21	10-21	183	30	30	30	30	SHELBY	32	04-05	11-04	213	14	14	15	15
	28	04-06	11-02	210	30	30	30	30		28	03-22	11-15	238	14	14	15	15
	24	03-17	11-14	242	30	30	30	30		24	03-03	11-27	269	14	14	15	15
	20	03-04	11-24	265	30	30	30	30		20	02-10	12-10	303	14	12	15	12
	16	02-17	12-09	295	30	28	30	25		16	01-23	12-21	332	14	9	15	7
MONROE 4 SE	32	04-10	10-27	201	30	30	30	30	SLOAN 3 S	32	04-08	11-03	209	28	28	26	26
	28	03-27	11-05	223	30	30	30	30		28	03-22	11-13	235	28	28	25	25
	24	03-12	11-19	252	30	30	30	30		24	03-06	11-24	264	27	27	25	24
	20	02-22	11-30	282	30	29	30	28		20	02-11	12-10	302	26	22	23	19
	16	01-28	12-17	324	30	22	30	17		16	01-20	12-21	336	25	11	23	9
MORGANTON	32	04-14	10-23	192	29	29	28	28	SMITHFIELD	32	04-08	10-25	200	28	28	30	30
	28	03-31	11-01	215	29	29	27	27		28	03-23	11-10	232	28	28	30	30
	24	03-16	11-17	246	29	29	27	27		24	03-10	11-19	254	28	28	30	30
	20	02-28	12-01	276	28	28	26	25		20	02-23	12-08	287	28	27	30	26
	16	02-11	12-13	304	26	23	26	20		16	02-05	12-19	317	28	23	30	17
MT AIRY	32	04-23	10-18	178	29	29	28	28	SOUTHPORT	32	03-15	11-23	254	30	30	30	30
	28	04-07	10-28	204	29	28	28	28		28	02-25	12-03	281	30	30	30	27
	24	03-20	11-10	234	29	29	28	28		24	02-09	12-18	312	30	28	30	17
	20	03-08	11-26	263	29	29	28	26		20	01-16	12-25	343	30	15	30	11
	16	02-21	12-06	289	29	28	28	25		16	01-08	@	@	30	9	30	4
NASHVILLE	32	04-06	10-28	205	29	29	29	29	STATESVILLE 2 W	32	04-07	11-05	212	25	25	26	26
	28	03-21	11-10	234	29	29	28	28		28	03-21	11-19	243	25	25	26	26
	24	03-07	11-21	259	29	29	28	28		24	03-09	11-30	266	25	25	25	24
	20	02-22	12-08	290	29	28	28	24		20	02-18	12-09	294	25	23	25	21
	16	02-06	12-19	317	29	25	28	17		16	02-11	12-20	312	24	20	25	15
NEW BERN	32	03-20	11-16	241	30	30	29	29	TARBORO	32	03-30	11-01	216	30	30	30	30
	28	03-06	11-26	265	30	30	29	29		28	03-17	11-16	244	30	30	30	30
	24	02-18	12-11	297	30	29	29	23		24	03-01	11-30	274	30	30	30	29
	20	01-30	12-22	326	30	22	29	15		20	02-17	12-13	299	30	29	30	24
	16	01-13	12-28	350	30	12	29	7		16	01-26	12-23	331	30	18	30	12
NEW HOLLAND	32	03-23	11-13	235	29	29	30	30	TRYON	32	04-10	10-30	203	30	30	29	29
	28	03-08	11-23	260	29	29	30	30		28	03-26	11-08	228	28	28	28	28
	24	02-21	12-11	292	28	28	29	25		24	03-09	11-25	261	26	26	25	25
	20	01-27	12-24	330	28	21	29	12		20	02-20	12-05	288	26	26	24	22
	16	01-15	12-27	346	27	11	29	7		16	01-30	12-15	319	26	19	22	14
OXFORD 2 SW	32	04-09	10-28	202	30	30	30	30	WADESBORO	32	03-31	11-11	225	11	11	11	11
	28	03-23	11-15	237	30	30	30	30		28	03-15	11-26	256	11	11	8	8
	24	03-08	11-25	262	30	30	30	30		24	02-24	12-03	282	8	8	8	7
	20	02-23	12-08	288	30	29	30	25		20	02-09	12-10	304	7	6	8	6
	16	02-07	12-19	315	30	27	30	18		16	01-29	12-22	327	7	5	7	4
PARKER 1 E	32	05-02	10-09	160	30	30	29	29	WATERVILLE	32	04-02	10-31	212	20	20	20	20
	28	04-23	10-18	179	30	30	29	29		28	03-20	11-14	239	20	20	20	20
	24	04-08	10-28	203	30	30	29	29		24	03-10	11-29	265	20	20	20	18
	20	03-27	11-13	231	30	30	28	28		20	02-24	12-10	289	20	19	20	15
	16	03-10	11-26	261	30	30	28	25		16	02-10	12-17	311	20	17	20	11
PINEHURST	32	04-03	11-01	212	30	30	30	30	WAYNESVILLE 1 E	32	05-01	10-11	163	25	25	25	25
	28	03-22	11-15	239	30	30	30	30		28	04-15	10-21	188	24	24	24	24
	24	03-08	11-30	267	30	30	30	29		24	04-01	11-02	215	23	23	22	22
	20	02-17	12-09	294	30	30	30	24		20	03-17	11-11	238	22	22	21	21
	16	02-01	12-20	323	30	22	30	16		16	03-05	11-29	270	22	22	21	20

FREEZE DATA

STATION	Freeze threshold temperature	Mean date of last Spring occurrence	Mean date of first Fall occurrence	Mean No. of days between dates	Years of record Spring	No. of occurrences in Spring	Years of record Fall	No. of occurrences in Fall	STATION	Freeze threshold temperature	Mean date of last Spring occurrence	Mean date of first Fall occurrence	Mean No. of days between dates	Years of record Spring	No. of occurrences in Spring	Years of record Fall	No. of occurrences in Fall
WELDON	32	04-05	11-02	210	30	30	30	30	WILMINGTON WB CITY	32	03-08	11-24	262	30	30	30	30
	28	03-22	11-13	237	30	30	30	30		28	02-22	12-08	289	30	29	30	26
	24	03-06	11-25	264	30	30	30	30		24	02-03	12-22	322	30	22	30	15
	20	02-21	12-08	291	30	29	30	26		20	01-16	12-27	345	30	14	30	8
	16	02-04	12-24	323	30	25	30	13		16	01-07	⊕	⊕	30	8	30	4
WILLARD 1 N	32	04-10	11-05	209	28	28	24	24	WINSTON SALEM WB AP	32	04-09	10-29	203	30	30	30	30
	28	03-21	11-13	237	26	26	21	21		28	03-26	11-09	229	30	30	30	30
	24	03-06	11-22	262	26	26	21	21		24	03-09	11-21	257	30	30	30	30
	20	02-24	12-14	294	24	24	19	15		20	02-26	12-06	283	30	29	30	25
	16	01-31	12-21	324	20	16	19	10		16	02-09	12-15	310	30	27	30	21

Data in the above table are based on the period 1921-1950, or that portion of this period for which data are available.

⊕When the frequency of occurrence in either spring or fall is one year in ten, or less, mean dates are not given.

Means have been adjusted to take into account years of non-occurrence.

A freeze is a numerical substitute for the former term "killing frost" and is the occurrence of a minimum temperature at or below the threshold temperature of 32°, 28°, etc.

Freeze data tabulations in greater detail are available and can be reproduced at cost.

TROPICAL CYCLONE DATA HAVING IMPORTANCE FOR THIS STATE IS INCLUDED IN STATISTICS AND CHARTS ON PAGES 1161 THROUGH 1164.

Taken from "Climatography of the United States No. 81-4, Decennial Census of U. S. Climate"

STATIONS (By Divisions)	TEMPERATURE (°F)													PRECIPITATION (In.)												
	JAN	FEB	MAR	APR	MAY	JUNE	JULY	AUG	SEPT	OCT	NOV	DEC	ANN	JAN	FEB	MAR	APR	MAY	JUNE	JULY	AUG	SEPT	OCT	NOV	DEC	ANN
SOUTHERN MOUNTAINS																										
ANDREWS 2 E	39.3	40.6	46.3	55.5	63.7	71.0	73.8	72.8	67.6	57.2	46.0	39.5	56.1	6.45	6.67	6.47	5.27	4.24	5.00	6.32	4.91	3.34	3.14	4.11	6.10	62.02
ASHEVILLE WB AP	37.6	38.9	44.6	54.6	63.0	70.1	72.4	71.5	65.9	55.8	44.6	37.9	54.7	4.17	4.03	4.82	4.02	3.66	3.52	5.85	4.94	3.62	3.13	2.77	3.62	48.15
ASHEVILLE	39.7	40.6	46.2	56.0	64.4	71.8	74.4	73.5	67.8	57.6	46.5	40.0	56.5	3.17	3.04	3.74	3.19	2.87	3.52	4.31	3.63	2.78	2.49	2.22	2.92	37.88
BREVARD	•	•	•	•	•	•	•	•	•	•	•	•	•	5.75	5.55	6.18	5.03	4.83	4.94	6.51	6.05	4.46	4.31	4.25	5.63	63.49
CANTON 1 SW	•	•	•	•	•	•	•	•	•	•	•	•	•	3.46	3.53	4.00	3.30	3.00	3.13	4.70	4.40	2.83	2.61	2.35	3.00	40.31
CAROLEEN	42.8	44.6	50.7	60.6	69.2	77.0	79.3	78.2	72.5	61.6	50.0	42.3	60.7	4.34	4.25	5.12	4.04	3.55	3.68	4.98	5.02	3.95	3.25	3.18	4.39	49.75
CULLOWHEE	40.7	42.0	47.1	55.9	63.8	70.9	73.9	73.2	67.9	57.7	46.4	40.3	56.7	4.59	4.66	4.94	3.79	3.60	3.55	5.07	4.31	2.91	2.76	2.88	4.40	47.46
ENKA	•	•	•	•	•	•	•	•	•	•	•	•	•	3.19	3.17	3.91	3.20	3.21	3.42	4.89	4.09	2.81	2.70	2.49	2.91	39.99
FLETCHER 3 W	39.8	40.8	46.3	55.6	63.6	70.8	73.4	72.5	66.7	56.6	45.8	39.7	56.0	4.17	4.03	4.82	4.02	3.66	3.52	5.85	4.94	3.62	3.13	2.77	3.62	48.15
HENDERSONVILLE 1 NE	39.9	41.2	46.9	56.0	63.9	71.2	73.7	72.8	67.1	56.9	46.1	39.6	56.3	4.76	4.58	5.48	4.54	4.15	4.82	6.16	5.52	4.04	3.75	3.60	4.72	56.12
HIGHLANDS 2 S	39.3	39.9	45.0	54.2	61.9	68.3	70.1	69.4	65.1	56.2	46.1	39.9	54.6	7.17	6.76	7.66	6.28	5.40	6.55	9.32	7.51	4.93	5.06	5.16	7.33	79.13
HOT SPRINGS 2	40.3	41.7	47.2	57.2	65.7	73.2	76.0	75.3	70.2	59.6	47.5	40.6	57.9	3.43	3.69	4.23	3.69	3.37	4.39	5.41	4.43	2.96	2.48	2.64	2.95	43.67
MARSHALL	37.7	38.7	44.3	54.1	62.9	70.4	73.4	72.4	66.8	56.1	44.6	37.7	54.9	3.22	3.29	3.84	3.18	2.98	3.64	4.72	3.56	2.52	2.19	2.29	2.79	38.22
MORGANTON	42.3	43.5	49.1	58.7	66.8	74.7	77.1	76.0	70.2	59.9	49.2	42.0	59.1	4.27	3.95	4.89	3.90	3.64	3.69	5.12	5.42	4.46	3.55	3.03	4.11	50.03
MURPHY	•	•	•	•	•	•	•	•	•	•	•	•	•	5.65	5.85	5.93	4.95	3.72	4.45	5.61	4.52	3.01	2.83	3.65	5.07	55.24
SWANNANOA 2 E	•	•	•	•	•	•	•	•	•	•	•	•	•	3.54	3.43	4.08	3.34	2.96	3.54	4.34	4.56	3.51	3.19	2.65	3.24	42.38
TAPOCO	•	•	•	•	•	•	•	•	•	•	•	•	•	5.38	5.71	5.76	4.74	4.15	4.67	5.70	5.03	3.04	2.94	3.64	4.94	55.70
TRYON	44.3	45.7	51.4	60.1	67.9	74.9	77.1	76.0	70.7	61.2	51.2	44.0	60.4	5.29	5.34	6.35	4.83	4.33	4.41	6.21	6.48	4.52	4.35	3.90	5.48	61.49
WAYNESVILLE 1 E	38.9	39.9	45.2	54.2	61.8	68.6	71.3	70.5	64.7	55.1	44.7	38.8	54.5	4.65	4.32	4.95	3.58	3.51	3.41	5.01	4.02	2.74	2.86	2.88	3.99	45.92
DIVISION	39.5	40.5	46.0	55.2	63.3	70.5	73.0	72.2	66.9	57.0	46.1	39.5	55.8	4.83	4.71	5.49	4.45	4.00	4.47	5.94	5.45	3.84	3.60	3.54	4.71	55.03
NORTHERN MOUNTAINS																										
BANNER ELK	34.3	34.9	40.0	49.1	57.0	64.0	66.7	65.8	60.8	51.5	40.9	34.8	50.0	3.99	4.12	4.63	4.33	4.13	4.52	6.11	5.34	3.89	3.27	3.44	3.82	51.59
BOONE	35.1	35.8	40.8	50.2	58.8	66.2	68.9	68.2	62.4	52.9	42.3	36.1	51.5	4.06	4.04	5.10	4.65	4.51	4.39	6.34	5.41	4.38	3.83	4.02	4.07	54.80
ELKIN	•	•	•	•	•	•	•	•	•	•	•	•	•	3.58	3.37	4.36	3.78	3.95	4.29	5.07	5.61	3.86	3.28	3.02	3.57	47.74
JEFFERSON	•	•	•	•	•	•	•	•	•	•	•	•	•	3.48	3.44	4.11	3.83	4.09	3.82	5.50	5.82	4.00	3.23	3.22	3.52	48.06
LENOIR	41.7	42.9	48.6	58.1	66.2	73.4	76.1	75.2	69.9	59.6	48.7	41.5	58.5	3.90	4.03	4.63	4.14	3.93	4.33	5.04	5.65	3.85	3.63	2.85	3.77	49.75
MOUNT AIRY	39.6	40.9	47.2	57.1	65.7	73.3	76.2	75.0	69.3	58.9	47.4	39.6	57.5	3.66	3.44	4.33	3.83	4.18	4.12	5.39	4.80	3.56	3.28	3.00	3.47	47.06
NORTH WILKESBORO	•	•	•	•	•	•	•	•	•	•	•	•	•	3.92	3.75	4.66	3.78	3.96	4.08	5.80	5.55	3.78	3.72	3.06	3.72	49.78
DIVISION	36.8	37.8	43.2	52.7	61.0	68.2	70.9	69.9	64.5	54.9	44.1	37.2	53.4	3.89	3.99	4.79	4.24	4.26	4.40	5.85	5.60	4.11	3.52	3.37	3.82	51.84
NORTHERN PIEDMONT																										
CHAPEL HILL 2 W	43.3	44.7	51.0	60.7	69.0	76.5	79.0	77.8	72.7	62.0	51.6	43.4	61.0	3.67	3.66	3.82	3.78	3.23	3.48	5.71	5.01	3.91	2.79	3.18	3.55	45.79
DURHAM	42.2	43.4	49.9	59.6	68.2	75.8	78.4	77.2	71.4	60.7	50.0	41.9	59.9	3.30	3.36	3.57	3.37	3.17	3.74	5.35	4.73	3.44	2.68	2.90	3.04	42.65
GRAHAM 2 ENE	•	•	•	•	•	•	•	•	•	•	•	•	•	3.50	3.51	3.74	3.67	3.50	3.86	5.08	5.01	3.94	3.02	2.88	3.24	44.95
GREENSBORO WB AIRPORT	39.7	41.0	47.4	57.4	66.9	74.8	77.3	76.2	70.1	59.3	47.9	39.9	58.2	3.40	3.30	3.69	3.43	3.29	3.47	4.79	4.61	3.66	2.71	2.68	3.13	42.16
HENDERSON 2 SW	41.7	43.0	49.6	59.6	67.9	75.7	78.4	77.1	71.3	60.6	50.4	41.9	59.8	3.51	3.22	3.99	3.76	3.80	4.03	5.95	4.94	3.45	2.68	3.10	3.15	45.58
HIGH POINT	•	•	•	•	•	•	•	•	•	•	•	•	•	3.60	3.57	3.92	3.80	3.57	3.68	5.52	5.06	4.03	2.59	2.88	3.47	45.69
LAKE MICHIE	•	•	•	•	•	•	•	•	•	•	•	•	•	3.44	3.48	3.82	3.61	3.69	3.54	5.52	4.82	3.79	2.86	2.88	2.93	44.38
LOUISBURG	41.7	43.1	49.6	59.6	68.4	76.2	79.1	77.7	72.0	60.9	50.1	41.7	60.0	4.59	4.69	3.83	3.58	3.61	3.93	5.85	5.35	4.05	2.63	3.25	3.28	46.59
MANGUMS STORE 4 WSW	•	•	•	•	•	•	•	•	•	•	•	•	•	3.38	3.14	3.80	3.68	3.18	3.86	5.09	4.71	3.30	2.59	3.09	3.16	42.98
OXFORD 2 SW	41.9	43.1	49.3	59.4	67.8	75.4	78.1	76.8	71.3	60.6	50.5	42.1	59.7	3.45	3.31	3.70	3.77	3.93	4.45	5.44	4.77	3.56	2.74	3.07	3.17	45.36
REIDSVILLE 2 NW	41.9	43.1	45.9	59.6	68.1	75.8	78.4	77.3	71.5	61.2	50.6	42.3	59.9	3.37	3.26	3.80	3.74	3.79	3.51	4.34	4.24	4.09	3.21	2.94	3.08	43.37
ROUGEMONT	•	•	•	•	•	•	•	•	•	•	•	•	•	3.41	3.21	3.86	3.92	3.72	4.21	5.72	4.91	3.68	3.01	3.16	3.09	45.90
ROXBORO	•	•	•	•	•	•	•	•	•	•	•	•	•	3.48	3.26	3.57	3.71	3.78	4.00	4.97	4.56	4.12	2.78	2.95	3.08	44.26
DIVISION	41.6	42.9	49.3	59.3	67.9	75.6	78.2	76.9	71.3	60.6	50.0	41.7	59.6	3.51	3.44	3.83	3.67	3.58	3.81	5.27	4.73	3.82	2.85	2.96	3.25	44.72
CENTRAL PIEDMONT																										
ASHEBORO 2 W	43.3	44.9	51.2	60.9	68.7	76.0	78.4	77.2	71.7	61.6	51.4	43.5	60.7	3.67	3.49	4.07	3.64	3.69	3.81	5.54	4.85	3.65	2.97	2.69	3.31	45.38
HICKORY	41.3	42.4	48.6	58.6	67.3	75.1	77.4	76.3	70.9	60.6	49.0	41.1	59.1	4.11	4.05	4.73	3.78	3.62	3.88	4.99	5.65	3.81	3.40	3.08	4.05	49.15
MONCURE 3 SE	41.5	43.1	49.5	59.5	67.8	75.9	78.5	77.4	71.7	60.5	49.6	41.3	59.7	3.56	3.68	3.96	3.79	3.90	3.62	6.84	5.51	4.29	2.89	3.16	3.45	48.65
NEUSE 2 NE	•	•	•	•	•	•	•	•	•	•	•	•	•	3.53	3.68	3.68	3.77	3.53	4.07	5.49	5.58	4.14	2.89	3.20	3.45	47.01
RALEIGH DURHAM WB AP	41.6	43.0	49.5	59.3	67.6	75.1	77.9	76.9	71.2	60.5	50.0	41.9	59.5	3.22	3.23	3.35	3.52	3.52	3.70	5.49	5.20	3.85	2.71	2.77	3.02	43.58
RALEIGH 3 W	•	•	•	•	•	•	•	•	•	•	•	•	•	3.42	3.39	3.85	3.63	3.86	4.07	5.56	4.98	4.36	2.85	3.07	3.24	46.28
RALEIGH STATE COLLEGE	43.3	44.3	50.7	60.6	69.2	76.9	79.4	78.0	72.5	62.3	51.8	43.6	61.1	3.33	3.49	3.72	3.78	3.80	3.94	5.90	5.35	4.57	2.77	3.01	3.20	46.86
RANDLEMAN	•	•	•	•	•	•	•	•	•	•	•	•	•	3.62	3.72	3.83	3.65	3.68	3.66	5.78	5.12	4.09	2.90	2.86	3.36	46.27
SALISBURY	42.2	43.8	50.1	60.3	69.2	76.9	79.2	78.0	72.0	61.0	49.8	41.8	60.4	3.91	3.65	4.39	3.76	3.79	3.77	5.59	4.89	3.91	3.12	2.96	3.61	47.35
STATESVILLE 2 NNE	42.3	44.0	49.7	59.7	68.2	75.9	78.0	76.8	71.7	61.6	50.5	42.2	60.1	4.06	3.86	4.69	3.73	3.52	3.64	5.07	4.63	3.73	3.30	3.04	3.81	47.08
DIVISION	42.4	43.9	50.1	60.0	68.3	76.0	78.5	77.3	71.7	61.3	50.4	42.4	60.2	3.73	3.66	4.17	3.71	3.71	3.79	5.60	5.07	4.00	3.01	2.99	3.52	46.96

*NORMALS BY CLIMATOLOGICAL DIVISIONS

Taken from "Climatography of the United States No. 81-4, Decennial Census of U. S. Climate"

TEMPERATURE (°F) PRECIPITATION (In.)

STATIONS (By Divisions)	JAN	FEB	MAR	APR	MAY	JUNE	JULY	AUG	SEPT	OCT	NOV	DEC	ANN	JAN	FEB	MAR	APR	MAY	JUNE	JULY	AUG	SEPT	OCT	NOV	DEC	ANN
SOUTHERN PIEDMONT																										
ALBEMARLE 4 N	43.5	45.0	50.9	60.4	69.1	76.8	78.9	77.7	72.1	61.5	50.9	42.9	60.8	3.65	3.59	4.13	3.87	3.20	3.82	6.22	4.61	4.43	2.77	2.80	3.46	46.55
CHARLOTTE WB AP	42.7	44.2	50.0	60.3	69.0	77.1	79.2	78.7	72.9	62.5	50.4	42.7	60.8	3.53	3.55	4.39	3.49	3.11	3.61	4.88	4.22	3.49	2.96	2.53	3.62	43.38
GASTONIA	44.2	45.9	52.0	62.0	70.5	78.0	79.8	78.8	73.2	62.4	51.7	44.0	61.9	3.91	4.09	4.66	3.95	5.64	3.21	4.93	4.57	4.10	3.51	2.87	3.94	47.38
MONROE 4 SE	44.8	46.3	52.1	61.1	69.7	77.4	79.5	78.3	72.9	62.4	52.1	44.4	61.8	3.53	3.57	3.85	3.74	3.08	3.67	5.31	5.04	3.81	2.64	2.68	3.41	44.33
MOUNT GILEAD 4 W	•	•	•	•	•	•	•	•	•	•	•	•	•	3.20	3.16	3.61	3.59	2.87	3.78	5.72	4.58	4.00	2.66	2.76	3.19	43.12
PINEHURST SRN-PINES	44.1	45.6	51.8	61.7	70.0	77.2	79.2	78.2	72.7	62.4	51.9	43.9	61.6	3.55	3.79	4.18	4.01	3.66	4.30	6.90	5.74	4.04	3.09	3.07	3.55	49.88
SHELBY 2 NNE	•	•	•	•	•	•	•	•	•	•	•	•	•	4.04	4.11	4.60	3.69	3.52	3.64	5.09	4.68	3.46	3.31	3.02	4.15	47.31
SOUTHERN PINES	44.9	46.3	52.2	61.5	69.8	77.4	78.9	77.7	72.9	62.7	52.6	45.1	61.8	3.47	3.71	4.27	4.08	3.80	4.78	7.11	5.50	4.35	3.30	2.92	3.38	50.67
DIVISION	43.8	45.3	51.4	61.0	69.6	77.2	79.3	78.2	72.8	62.3	51.6	43.7	61.4	3.60	3.66	4.28	3.79	3.42	3.86	5.71	5.00	4.03	2.97	2.79	3.53	46.64
SOUTHERN COASTAL PLAIN																										
ELIZABETHTOWN LOCK 2	•	•	•	•	•	•	•	•	•	•	•	•	•	2.78	3.25	3.61	3.23	2.98	4.63	6.14	6.02	4.58	3.03	2.75	3.03	46.03
FAYETTEVILLE	45.3	46.0	52.4	62.2	70.4	78.0	80.3	79.1	74.2	63.4	53.0	44.6	62.4	3.02	3.39	3.89	3.80	3.64	4.16	6.06	6.00	4.39	2.25	2.83	3.01	46.44
LUMBERTON 6 NW	45.9	47.3	53.7	62.8	70.6	77.4	79.5	78.5	73.6	63.2	52.8	45.4	62.6	2.86	3.50	3.87	3.78	3.25	4.56	6.21	4.91	4.23	2.52	2.69	3.10	45.48
RED SPRINGS	•	•	•	•	•	•	•	•	•	•	•	•	•	2.90	3.39	3.69	3.65	3.37	4.54	5.47	5.37	4.07	2.35	2.83	3.29	44.92
SLOAN 3 S	46.0	46.8	52.9	61.6	68.9	75.9	78.9	77.7	73.1	62.9	53.3	45.7	62.0	3.12	3.52	3.82	2.95	3.61	5.17	7.22	5.71	5.04	2.57	2.83	3.21	48.77
SOUTHPORT 5 N	48.7	49.1	54.2	62.7	70.8	77.6	80.2	79.7	75.9	66.5	57.0	49.5	64.3	3.00	3.65	3.94	2.57	3.17	3.87	6.56	5.71	7.08	3.27	3.17	3.50	49.49
WILLARD 1 N	47.3	48.5	54.1	62.5	70.3	77.2	79.6	78.5	74.1	64.0	54.3	46.9	63.1	2.86	3.54	3.67	2.76	3.66	4.73	7.56	5.98	5.28	2.78	2.88	3.27	48.97
WILMINGTON WB AP	47.9	48.7	54.2	62.5	70.5	77.7	80.0	79.4	75.2	65.4	55.4	48.2	63.8	2.85	3.42	4.03	2.86	3.52	4.76	7.68	6.86	6.29	3.01	3.09	3.42	51.29
DIVISION	46.8	47.8	53.5	62.4	70.3	77.3	79.7	78.7	74.1	64.0	54.2	46.6	63.0	3.06	3.51	3.90	3.23	3.56	4.50	6.94	5.94	5.23	2.77	2.96	3.32	48.92
CENTRAL COASTAL PLAIN																										
GOLDSBORO 1 SSW	44.7	46.0	52.5	62.3	70.6	78.0	80.5	79.1	74.0	63.3	53.0	44.6	62.4	3.32	3.37	3.85	3.79	3.85	4.78	7.50	5.62	4.35	2.87	3.10	3.21	49.61
GREENVILLE	•	•	•	•	•	•	•	•	•	•	•	•	•	3.34	3.40	3.60	3.47	3.38	4.08	6.79	5.58	4.84	2.77	3.03	3.26	47.54
KINSTON 5 SE	45.3	46.4	52.9	62.1	70.2	77.4	80.0	79.4	74.2	63.8	53.5	45.2	62.5	3.03	3.46	3.51	3.18	3.64	4.62	7.09	5.83	4.79	2.58	2.95	3.17	47.84
NEW BERN 3 NW	46.6	47.7	53.5	62.8	70.8	77.8	80.2	79.3	74.7	64.7	54.7	46.8	63.3	3.27	3.85	3.96	3.10	3.91	4.54	8.17	6.98	6.68	3.17	3.65	4.13	55.41
NEW HOLLAND	47.0	47.2	52.8	61.0	68.9	75.9	78.8	78.2	73.9	64.6	55.1	47.2	62.6	3.52	3.80	3.65	3.07	3.77	4.74	7.18	6.71	5.86	3.35	3.71	4.01	53.37
SMITHFIELD	44.1	45.4	51.8	61.3	69.4	77.0	79.4	78.3	72.9	62.3	52.0	43.9	61.5	3.42	3.57	3.86	3.97	3.56	4.57	6.04	6.08	4.19	2.81	3.03	3.24	48.34
DIVISION	45.5	46.5	52.5	61.7	70.0	77.3	79.8	79.0	74.2	64.0	54.0	45.8	62.5	3.39	3.64	3.79	3.35	3.69	4.46	7.04	6.16	5.33	3.12	3.38	3.55	50.90
NORTHERN COASTAL PLAIN																										
CAPE HATTERAS WB	46.6	46.5	51.0	59.3	68.0	75.2	78.0	77.6	74.1	65.4	56.2	48.2	62.2	3.90	3.93	4.16	2.99	3.98	4.14	6.15	6.42	5.89	4.24	4.09	4.58	54.47
EDENTON	43.5	44.7	50.9	60.5	69.2	76.9	79.5	78.0	72.8	62.1	52.0	43.8	61.2	3.63	3.40	3.87	3.46	3.39	4.13	6.84	6.28	4.75	2.81	3.31	3.01	48.88
ELIZABETH CITY	44.6	45.2	51.2	60.4	68.9	76.4	79.5	78.2	73.3	63.1	53.3	45.0	61.6	3.61	3.76	3.84	3.42	3.29	3.82	7.11	6.41	5.08	3.18	3.50	3.25	50.27
ENFIELD	•	•	•	•	•	•	•	•	•	•	•	•	•	3.36	3.38	3.69	3.30	3.54	3.74	5.61	5.20	3.50	2.62	2.64	2.89	43.47
HATTERAS	47.6	47.4	52.0	60.1	68.5	75.8	78.9	78.8	75.4	66.6	57.5	49.2	63.2	3.92	4.01	4.21	2.99	4.07	4.06	6.28	6.58	6.02	4.47	4.31	4.87	55.79
MANTEO	45.3	45.6	51.0	59.5	68.4	75.9	79.2	78.3	74.0	64.2	54.8	46.7	61.9	3.25	3.49	3.31	2.59	2.95	3.27	5.84	5.52	4.66	3.08	3.22	3.13	44.31
NASHVILLE	42.8	44.2	50.8	60.4	69.0	76.6	79.2	77.8	72.4	61.8	51.2	42.9	60.8	3.46	3.44	3.69	3.61	3.87	3.86	5.88	6.02	4.31	2.71	2.99	3.17	47.01
ROCKY MOUNT POWER PL	•	•	•	•	•	•	•	•	•	•	•	•	•	3.25	3.29	3.69	3.53	3.59	4.10	6.55	5.80	4.45	2.82	3.10	3.00	47.17
ROCKY MOUNT 8 ESE	•	•	•	•	•	•	•	•	•	•	•	•	•	3.44	3.78	3.84	3.36	3.49	4.39	6.12	5.68	4.18	2.75	3.22	3.04	47.64
SCOTLAND NECK	•	•	•	•	•	•	•	•	•	•	•	•	•	3.43	3.67	3.80	3.54	3.60	4.76	5.98	5.44	4.22	3.04	3.02	3.14	47.64
TARBORO 1 S	43.2	44.3	51.3	61.2	69.9	77.2	80.0	78.9	73.5	62.4	51.7	43.3	61.4	3.47	3.49	3.93	3.40	3.57	4.49	6.08	6.17	4.28	2.87	3.16	3.20	48.11
WELDON	42.4	43.4	50.0	60.0	69.0	77.0	79.9	78.3	72.6	61.5	51.1	42.5	60.6	3.29	3.37	3.43	3.36	3.71	3.38	5.66	5.26	3.97	2.51	2.82	2.89	43.65
WILLIAMSTON 1 ENE	•	•	•	•	•	•	•	•	•	•	•	•	•	3.25	3.45	3.80	3.54	3.55	4.28	6.74	6.04	4.41	2.69	3.10	2.98	47.83
DIVISION	44.0	44.9	50.9	60.2	68.8	76.3	79.2	78.1	73.2	62.8	52.8	44.5	61.3	3.50	3.57	3.78	3.30	3.55	4.00	6.28	6.02	4.69	3.00	3.29	3.32	48.30

* Normals for the period 1931-1960. Divisional normals may not be the arithmetical averages of individual stations published, since additional data for shorter period stations are used to obtain better areal representation.

TEMPERATURE PRECIPITATION

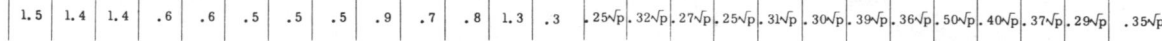

JAN	FEB	MAR	APR	MAY	JUNE	JULY	AUG	SEPT	OCT	NOV	DEC	ANN	JAN	FEB	MAR	APR	MAY	JUNE	JULY	AUG	SEPT	OCT	NOV	DEC	ANN

CONFIDENCE - LIMITS

In the absence of trend or record changes, the chances are 9 out of 10 that the true mean will lie in the interval formed by adding and subtracting the values in the following table from the means for any station in the State. Because of the wider variation in mean precipitation, the corresponding monthly means and annual mean must be substituted for "p" in the precipitation table below to obtain mean precipitation confidence limits.

1.5	1.4	1.4	.6	.6	.5	.5	.5	.9	.7	.8	1.3	.3	.25√p	.32√p	.27√p	.25√p	.31√p	.30√p	.39√p	.36√p	.50√p	.40√p	.37√p	.29√p	.35√p

COMPARATIVE DATA

Data in the following table are the mean temperature and average precipitation for Salisbury, North Carolina, for the period 1906-1930 and are included in this publication for comparative purposes.

41.9	43.7	51.1	59.6	67.6	75.2	78.3	77.2	72.3	61.1	50.2	42.6	60.1	4.20	4.07	4.65	3.86	4.16	4.76	5.28	5.18	3.83	3.13	2.45	4.04	49.61

(Table Revised 1970. Base Period for Climatological Normals: 1931-1960)

Station: ASHEVILLE, NORTH CAROLINA ASHEVILLE AIRPORT
Standard time used: EASTERN Latitude: 35° 25' N Longitude: 82° 32' W Elevation (ground): 2140 feet

Month	Temperature Normal Ø Daily maximum	Daily minimum	Monthly	Extremes Record highest	Year	Record lowest	Year	Normal heating degree days (Base 65°)	Precipitation Normal total	Maximum monthly	Year	Minimum monthly	Year	Maximum in 24 hrs.	Year	Snow, ice pellets Mean total	Maximum monthly	Year	Maximum in 24 hrs.	Year
J	47.5	27.7	37.6	71	1966	-7	1966	849	4.17	11.28	1966	2.02	1967	1.40	1969	6.4	17.6	1965	7.6	1965
F	49.3	27.8	38.9	68	1965	-2	1965	731	4.03	6.56	1966	0.62	1968	3.17	1966	9.2	25.5	1969	11.7	1969
M	56.1	33.0	44.6	81	1965	14	1965	633	4.82	6.65	1968	2.59	1968	5.13	1969	3.7	13.0	1968	10.9	1965
A	67.1	42.1	54.6	88	1967	24	1968+	312	4.02	5.47	1967	1.11	1967	2.17	1969	T	0.0		0.0	
M	75.6	50.3	63.0	91	1969	30	1966	111	3.66	6.79	1967	2.92	1968	2.23	1967	0.0	0.0		0.0	
J	82.1	58.1	70.1	96	1969	35	1966	0	3.52	5.06	1968	2.46	1966	2.43	1967	0.0	0.0		0.0	
J	83.5	61.3	72.4	94	1967	46	1967	0	5.85	7.53	1969	3.24	1966	4.02	1969	0.0	0.0		0.0	
A	82.5	60.5	71.5	94	1969	43	1968	0	4.94	11.28	1967	3.31	1968	4.12	1967	0.0	0.0		0.0	
S	77.5	55.2	66.3	85	1969	30	1967	75	3.62	4.69	1965	2.53	1967	2.65	1969	0.0	0.0		0.0	
O	68.3	43.2	55.8	83	1969	24	1965	294	3.13	5.37	1966	1.30	1969	2.95	1966	0.2	2.2	1966	5.7	1968
N	56.3	34.3	44.6	77	1968	13	1968	612	2.77	2.94	1965	1.30	1966	1.38	1966	1.0	10.9	1968	11.7	1968
D	48.4	27.4	37.9	68	1965	14	1968+	840	3.62	6.13	1965	0.16	1965	1.36	1969	2.7	25.5	1969	5.6	1969
YR	66.2	43.2	54.7	96 JUN 1969		-7 JAN 1966		4466	48.15	11.28 AUG 1967		0.16 DEC 1965		5.13 MAR 1968		24.2	25.5 FEB 1969		11.7 FEB 1969	

Ø Corrected after 1968 issue.

Means and extremes above are from existing and comparable exposures. Annual extremes have been exceeded at other sites in the locality as follows: Highest temperature 99 in July 1936; maximum monthly precipitation 13.75 in August 1940; minimum monthly precipitation T in October 1963; maximum precipitation in 24 hours 7.92 in October 1918; maximum monthly snowfall 28.9 in March 1960; maximum snowfall 28.9 in 24 hours 15.8 in March 1942.

NORMALS, MEANS AND EXTREMES
(Table Revised 1975. Base Period for Climatological Normals: 1941-1970)

ASHEVILLE, NORTH CAROLINA Elev. 2170 feet m.s.l. Average station pressure 942.3 mb.

Month	Temperatures °F Normal Daily maximum	Daily minimum	Monthly	Extremes Record highest	Year	Record lowest	Year	Normal Degree days Base 65°F Heating	Cooling	Precipitation Water equivalent Normal	Maximum monthly	Year	Minimum monthly	Year	Maximum in 24 hrs.	Year	Snow, ice pellets Maximum monthly	Year	Maximum in 24 hrs.	Year
J	48.4	27.3	37.9	73	1974	-7	1966	840	0	3.39	4.26	1973	1.75	1970	1.54	1970	17.6	1966	7.6	1966
F	50.6	28.2	39.4	77	1972	-2	1967	717	0	3.60	6.56	1966	0.62	1968	3.17	1966	25.5	1969	10.9	1969
M	56.9	33.5	45.2	84	1974	14	1968	592	6	4.66	8.91	1973	2.59	1966	3.06	1971	13.0	1960	10.2	1960
A	67.4	42.4	55.0	89	1972	24	1973	279	6	3.53	5.71	1973	1.11	1967	3.06	1972	0.2	1971	0.2	1971
M	76.8	50.6	63.7	91	1969	30	1971	100	60	3.11	8.83	1972	1.72	1970	3.06	1970	0.0		0.0	
J	82.5	58.7	70.6	96	1969	35	1966	14	182	3.97	6.54	1968	2.46	1968	3.54	1968	0.0		0.0	
J	84.3	62.6	73.5	95	1970	46	1967	0	264	4.87	7.53	1969	3.24	1966	4.02	1969	0.0		0.0	
A	83.8	61.8	72.8	94	1968	43	1968	0	244	4.50	11.28	1967	1.88	1972	4.12	1967	0.0		0.0	
S	78.0	55.4	66.7	89	1973	30	1967	50	101	3.57	5.11	1972	1.17	1970	2.65	1968	0.0		0.0	
O	68.1	44.5	56.3	84	1974	24	1974	289	15	3.25	7.05	1971	1.30	1965	1.74	1968	0.0		0.0	
N	58.2	34.3	46.3	78	1974	8	1970	561	0	2.94	4.42	1973	1.30	1965	2.66	1973	9.6	1968	5.7	1968
D	49.3	28.1	38.7	73	1971	11	1972	815	0	3.59	8.48	1973	0.16	1965			16.3	1971	16.3	1971
YR	67.4	44.0	55.7	96 JUN 1969		-7 JAN 1966		4237	872	45.18	11.28 AUG 1967		0.16 DEC 1965		5.13 MAR 1968		25.5 FEB 1969		16.3 DEC 1971	

Means and extremes above are from existing and comparable exposures. Annual extremes have been exceeded at other sites in the locality as follows: Highest temperature 99 in July 1936; maximum monthly precipitation 13.75 in August 1940; minimum monthly precipitation T in October 1963; maximum precipitation in 24 hours 7.92 in October 1918; maximum monthly snowfall 28.9 in March 1960.

REFERENCE NOTES APPLYING TO TABLES APPEAR ON THE PAGE FOLLOWING LAST TABLE.
(Caution: Letters and symbols may have different meanings in 1941-1970 tables than in earlier tables. See notes.)

NORMALS, MEANS AND EXTREMES

(Table Revised 1970. Base Period for Climatological Normals: 1931-1960)

Station: CAPE HATTERAS, NORTH CAROLINA   WEATHER BUREAU BUILDING   Standard time used: EASTERN   Latitude: 35° 16' N   Longitude: 75° 33' W   Elevation (ground): 7 feet

Means and extremes above are from existing and comparable exposures. Annual extremes have been exceeded at other sites in the locality as follows:
Highest temperature 97 in June 1952; lowest temperature 8 in December 1880; maximum monthly precipitation 20.95 in June 1949; minimum monthly precipitation T in November 1890; maximum precipitation in 24 hours 14.73 in June 1949; maximum snowfall 12.0 in December 1917; maximum snowfall in 24 hours 12.0 in December 1917; fastest mile wind 110 W in September 1944.

NORMALS, MEANS AND EXTREMES

(Table Revised 1975. Base Period for Climatological Normals: 1941-1970)

Means and extremes above are from existing and comparable exposures. Annual extremes have been exceeded at other sites in the locality as follows:
Highest temperature 97 in June 1952; lowest temperature 8 in December 1880; maximum monthly precipitation 20.95 in June 1949; minimum monthly precipitation T in November 1890; maximum precipitation in 24 hours 14.73 in June 1949; maximum snowfall 12.0 in December 1917; maximum snowfall in 24 hours 12.0 in December 1917; fastest mile wind 110 W in September 1944.

REFERENCE NOTES APPLYING TO TABLES APPEAR ON THE PAGE FOLLOWING LAST TABLE.

(Caution: Letters and symbols may have different meanings in 1941-1970 tables than in earlier tables. See notes.)

NORMALS, MEANS AND EXTREMES
(Table Revised 1970. Base Period for Climatological Normals: 1931-1960)

Station: CHARLOTTE, NORTH CAROLINA — DOUGLAS MUNICIPAL AIRPORT
Standard time used: EASTERN   Latitude 35° 13' N   Longitude 80° 56' W   Elevation (ground) 736 feet

Month	Normal Daily maximum	Normal Daily minimum	Normal Monthly	Record highest	Year	Record lowest	Year	Normal heating degree days (Base 65°)	Precipitation Normal total	Max monthly	Year	Min monthly	Year	Max in 24 hrs.	Year	Snow, Ice pellets Mean total	Max monthly	Year	Max 24 hrs.	Year
J	51.4	34.6	42.7	74	1966	4	1966	691	3.53	7.44	1949	1.24	1956	3.57	1962	2.3	11.7	1962	10.2	1965
F	53.7	34.7	44.2	78	1967+	4	1967	582	3.55	6.86	1961	0.87	1968	2.92	1955	1.5	5.2	1969	8.0	1969
M	60.0	40.6	50.0	86	1965	18	1960	461	4.39	8.69	1944	2.11	1944	3.64	1962	1.1	19.3	1960	12.0	1960
A	70.0	49.6	60.3	91	1963+	28	1961	156	3.49	7.64	1958	0.97	1942	3.20	1962	T	T	1954	T	1954
M	79.4	58.6	69.0	99	1963	32	1963	22	3.11	5.29	1957+	0.11	1941	2.67	1965	0.0	0.0		0.0	
J	87.6	66.6	77.1	99	1964	46	1964	0	3.61	8.26	1961	0.67	1954	3.77	1949	0.0	0.0		0.0	
J	88.6	69.5	79.2	99	1966	53	1961	0	4.88	9.12	1941	1.26	1948	4.40	1948	0.0	0.0		0.0	
A	87.8	69.5	78.7	100	1963	53	1965	0	4.40	9.98	1967	0.88	1968	4.58	1967	0.0	0.0		0.0	
S	82.6	63.8	72.9	94	1966+	39	1967	64	3.49	10.89	1945	0.02	1959	8.07	1945	0.0	0.0		0.0	
O	72.8	52.2	62.5	87	1962	24	1962	124	2.69	10.89	1954	T	1953	4.84	1954	T	2.5	1962	2.9	1958
N	60.2	40.1	50.4	85	1961	13	1969	438	2.53	8.17	1948	0.60	1960	2.79	1962	0.1	2.5	1968	2.5	1968
D	51.3	34.1	42.7	77	1962	2	1962	691	3.62	7.41	1965	0.43	1955	2.59	1958	0.3	4.4	1945	2.9	1945
YR	70.5	51.1	60.5	100	AUG. 1963	2	DEC. 1962	3191	43.38	10.89	SEP. 1945	T	OCT. 1953	4.84	OCT. 1955	5.4	19.3	MAR. 1960	12.0	1960

Means and extremes above are from existing and comparable exposures. Annual extremes have been exceeded at other sites in the locality as follows: Highest temperature 104 in September 1954; lowest temperature −5 in February 1899; maximum monthly precipitation 16.55 in July 1916; maximum precipitation in 24 hours 6.59 in July 1944; maximum snowfall in 24 hours 14.0 in February 1902.

NORMALS, MEANS AND EXTREMES
(Table Revised 1975. Base Period for Climatological Normals: 1941-1970)

Month	Normal Daily maximum	Normal Daily minimum	Normal Monthly	Record highest	Year	Record lowest	Year	Degree days Heating	Degree days Cooling	Precipitation Normal	Max monthly	Year	Min monthly	Year	Max in 24 hrs.	Year
J	52.1	32.1	42.1	74	1974	14	1970	710	0	3.51	7.44	1956	1.24	1948	3.57	1962
F	54.3	33.1	43.7	78	1962	14	1967	588	0	3.83	8.69	1958	0.87	1972	2.92	1955
M	62.2	39.0	50.6	86	1967	18	1960	461	15	4.52	8.69	1944	2.11	1949	3.64	1962
A	72.7	48.9	60.8	91	1964	25	1972	145	19	3.40	7.64	1961	0.97	1942	3.20	1962
M	80.4	57.4	68.9	95	1964	32	1963	34	152	2.90	5.51	1972	0.11	1941	2.67	1965
J	86.4	65.51	75.9	99	1964	45	1972	0	327	3.70	8.26	1961	0.67	1954	3.77	1949
J	88.3	68.7	78.5	99	1966	53	1961	0	419	4.57	9.12	1941	1.26	1948	3.00	1949
A	87.0	67.9	77.7	100	1963	53	1965	0	394	3.96	9.98	1967	0.88	1972	4.40	1967
S	82.4	61.9	72.1	94	1970+	39	1967	50	220	2.46	10.89	1945	0.02	1959	4.58	1954
O	73.1	50.3	61.7	87	1962	24	1962	152	20	2.69	7.66	1947	T	1953	4.84	1953
N	62.7	40.6	51.0	85	1961	13	1970	420	0	2.74	8.17	1948	0.46	1973	2.79	1973
D	52.5	32.4	42.5	77	1971	2	1962	698	0	3.44	7.41	1965	0.43	1955	2.87	1972
YR	71.2	49.7	60.5	100	AUG. 1963	2	DEC. 1962	3218	1596	42.72	10.89	SEP. 1945	T	OCT. 1953	4.84	OCT. 1955

Means and extremes above are from existing and comparable exposures. Annual extremes have been exceeded at other sites in the locality as follows: Highest temperature 104 in September 1954; lowest temperature −5 in February 1899; maximum monthly precipitation 16.55 in July 1916; maximum precipitation in 24 hours 6.59 in July 1944; maximum snowfall in 24 hours 14.0 in February 1902.

REFERENCE NOTES APPLYING TO TABLES APPEAR ON THE PAGE FOLLOWING LAST TABLE.
(Caution: Letters and symbols may have different meanings in 1941-1970 tables than in earlier tables. See notes.)

NORMALS, MEANS AND EXTREMES

(Table Revised 1970. Base Period for Climatological Normals: 1931-1960)

Station: GREENSBORO, HIGH POINT, WINSTON SALEM AP, N. C. Standard time used: EASTERN Latitude: 36° 05' N Longitude: 79° 57' W Elevation (ground): 897 feet

Month	Temperature Normal Daily maximum	Daily minimum	Monthly	Extremes Record highest	Year	Record lowest	Year	Normal heating degree days (Base 65°)	Precipitation Normal total	Maximum monthly	Year	Minimum monthly	Year	Maximum in 24 hrs	Year	Snow, Ice pellets Mean total	Maximum monthly	Year	Maximum in 24 hrs	Year	Relative humidity Hour 01	07	13	19	Mean sky cover	Pct. of possible sunshine	Mean number of days

(a) and (b) columns with monthly data J through D; YR 69.3 | 47.0 | 58.2 | 100 | JUL. 1966 | 3 | JAN. 1966 | 3805 | 42.16 | 13.26 | SEP. 1947 | 0.13 | SEP. 1939 | 7.49 | SEP. 1947 | 8.8 | 22.9 | JAN. 1966 | 14.3 | DEC. 1930 | 80 | 84 | 54 | 64 | 5.7 | 62 | ...

Ø For period July 1963 through the current year.
Means and extremes above are from existing and comparable exposures. Annual extremes have been exceeded at other sites in the locality as follows:
Highest temperature 102 in July 1954+; lowest temperature -7 in January 1940.

NORMALS, MEANS AND EXTREMES

(Table Revised 1975. Base Period for Climatological Normals: 1941-1970)

Month	Temperatures °F		Average station pressure mb. Elev. 886 feet m.s.l.

YR 69.2 | 47.0 | 58.1 | 100 | JUL. 1972 | -1 | JAN. 1972 | 3825 | 1341 | 41.36 | 13.26 | SEP. 1947 | 0.13 | SEP. 1939 | 7.49 | SEP. 1947 | 8.8 | 22.9 | JAN. 1966 | 14.3 | DEC. 1930 | 81 | 84 | 56 | 65 | 5.8 | 61 | ... 988.5

Means and extremes above are from existing and comparable exposures. Annual extremes have been exceeded at other sites in the locality as follows:
Highest temperature 102 in July 1954+; lowest temperature -7 in January 1940.

REFERENCE NOTES APPLYING TO TABLES APPEAR ON THE PAGE FOLLOWING LAST TABLE.
(Caution: Letters and symbols may have different meanings in 1941-1970 tables than in earlier tables. See notes.)

NORMALS, MEANS AND EXTREMES
(Table Revised 1970. Base Period for Climatological Normals: 1931-1960)

Station: RALEIGH, NORTH CAROLINA — RALEIGH-DURHAM AIRPORT
Standard time used: EASTERN Latitude: 35° 52′ N Longitude: 78° 47′ W Elevation (ground): 434 feet

Temperature °F and Heating Degree Days

Month	Normal Daily maximum	Normal Daily minimum	Normal Monthly	Extremes Record highest	Year	Extremes Record lowest	Year	Normal heating degree days (Base 65°)
J	51.9	31.3	41.6	77	1967	3	1966	725
F	54.0	31.9	43.0	79	1968+	5	1968+	616
M	61.4	37.8	49.5	89	1968	17	1965	487
A	71.8	46.8	59.3	93	1967	23	1965	180
M	79.4	55.9	67.6	95	1969+	28	1966	34
J	86.3	63.9	75.1	98	1964	43	1966	0
J	88.1	67.6	77.9	97	1966	55	1965	0
A	87.1	66.7	76.9	98	1968	46	1965	0
S	81.9	60.4	71.2	92	1966+	42	1967	21
O	72.8	48.2	60.5	89	1968	24	1965	164
N	62.2	37.7	50.0	84	1974	11	1967	450
D	52.3	31.4	41.9	77	1971	3	1964	716
YR	70.8	48.3	59.5	98 AUG 1968		3 JAN 1966		3393

Precipitation (inches)

Month	Normal total	Max monthly	Year	Min monthly	Year	Max in 24 hrs	Year
J	3.22	7.52	1954	1.05	1956	2.79	1954
F	3.23	5.75	1961	1.00	1968	2.40	1946
M	3.35	4.94	1962	1.48	1949	2.51	1952
A	3.52	5.83	1959	1.43	1969	2.02	1958
M	3.52	6.69	1959	1.92	1954	4.44	1957
J	3.70	8.32	1965	1.12	1965	3.44	1967
J	5.49	10.05	1945	0.80	1953	3.89	1952
A	5.20	10.49	1955	0.81	1950	5.20	1955
S	3.85	12.94	1945	0.57	1954	5.20	1944
O	2.71	6.53	1954	0.44	1963+	4.70	1954
N	2.77	8.22	1948	0.88	1960	4.48	1960
D	3.02	6.20	1945	0.25	1955	3.18	1958
YR	43.58	12.94 SEP 1945		0.25 DEC 1965		5.20 AUG 1955	

Snow, Ice pellets (inches)

Month	Mean total	Max monthly	Year	Max in 24 hrs	Year
J	3.0	14.4	1955	9.0	1955
F	2.3	13.5	1948	9.1	1948
M	1.7	14.5	1960	9.3	1960
A	T	T	1950	T	1950
M	0.0	0.0		0.0	
J	0.0	0.0		0.0	
J	0.0	0.0		0.0	
A	0.0	0.0		0.0	
S	0.0	0.0		0.0	
O	0.0	0.0		0.0	
N	0.1	1.3	1962	1.3	1962
D	0.9	10.6	1958	9.1	1958
YR	7.4	14.4 JAN 1955		9.3 MAR 1969	

Relative humidity, Wind, Sky cover, Sunshine, Mean number of days

(Relative humidity Hour 01 07 13 19; Wind Mean speed / Prevailing direction / Fastest mile Speed, Direction, Year; Mean sky cover sunrise to sunset; Pct. of possible sunshine)

Footnote:

∅ For period November 1964 through the current year.
Means and extremes above are from existing and comparable exposures. Annual extremes have been exceeded at other sites in the locality as follows: Highest temperature 105° in July 1952; lowest temperature -2 in February 1899; maximum monthly precipitation 13.63 in August 1908; minimum monthly precipitation 0.06 in November 1931+; maximum precipitation in 24 hours 6.66 in September 1929; maximum monthly snowfall 20.0 in January 1893; maximum snowfall in 24 hours 17.8 in March 1927.

NORMALS, MEANS AND EXTREMES
(Table Revised 1975. Base Period for Climatological Normals: 1941-1970)

Average station pressure: Elev. 441 feet m.s.l.

Temperature °F and Degree Days

Month	Normal Daily maximum	Normal Daily minimum	Normal Monthly	Extremes Record highest	Year	Extremes Record lowest	Year	Heating (Base 65°)	Cooling (Base 65°)
J	51.0	30.0	40.5	77	1967	0	1970	760	0
F	53.1	31.1	42.1	79	1965	0	1966	638	0
M	61.0	37.4	49.2	89	1968	17	1960	502	12
A	72.2	46.7	59.5	93	1967	23	1972	180	15
M	79.4	55.4	67.4	95	1969+	28	1966	48	123
J	85.6	63.1	74.4	98	1973	43	1972	0	282
J	87.7	67.2	77.5	97	1966	55	1974	0	388
A	86.6	66.2	76.6	98	1968	46	1968	0	357
S	81.5	59.7	70.6	92	1973	39	1965	18	180
O	72.4	48.0	60.2	89	1968	24	1965	186	0
N	62.1	37.8	50.0	84	1974	11	1970	450	0
D	51.9	30.5	41.2	77	1971	0	1970	738	0
YR	70.4	47.8	59.1	98 AUG 1968		0 JAN 1970		3514	1394

Precipitation in inches

Month	Normal	Max monthly	Year	Min monthly	Year	Water equiv. Max in 24 hrs	Year
J	3.22	7.52	1956	1.05	1956	2.79	1954
F	3.32	5.75	1961	1.00	1968	3.22	1973
M	3.44	4.94	1960	1.48	1949	2.51	1952
A	3.07	5.83	1959	1.32	1967	2.09	1958
M	3.32	7.67	1974	0.92	1974	4.40	1957
J	3.67	9.38	1970	0.87	1970	3.44	1967
J	5.08	10.05	1953	0.81	1952	3.89	1952
A	4.93	10.49	1955	0.80	1955	5.20	1955
S	3.78	7.53	1944	0.57	1944	5.16	1944
O	2.81	6.53	1954	0.44	1963	4.10	1954
N	2.82	8.22	1948	0.61	1973	4.70	1948
D	3.08	6.38	1973	0.25	1965	3.18	1958
YR	42.54	12.94 SEP 1945		0.25 DEC 1965		5.20 AUG 1955	

Snow, Ice pellets (inches)

Month	Max monthly	Year	Max in 24 hrs	Year
J	14.4	1955	9.0	1955
F	13.5	1948	9.1	1948
M	14.5	1960	9.3	1960
A	T	1950	T	1950
M	0.0		0.0	
J	0.0		0.0	
J	0.0		0.0	
A	0.0		0.0	
S	0.0		0.0	
O	0.0		0.0	
N	1.3	1962	1.3	1962
D	10.6	1958	9.1	1958
YR	14.4 JAN 1955		9.3 MAR 1969	

Average station pressure (mb)

Month	Station pressure mb
J	1004.6
F	1004.1
M	1000.7
A	1000.7
M	998.7
J	1000.4
J	1001.4
A	1002.6
S	1002.2
O	1005.1
N	1003.8
D	1002.4
YR	1002.1

Footnote:

Means and extremes above are from existing and comparable exposures. Annual extremes have been exceeded at other sites in the locality as follows: Highest temperature 105° in July 1952; lowest temperature -2 in February 1899; maximum monthly precipitation 13.63 in August 1908; minimum monthly precipitation 0.06 in November 1931+; maximum precipitation in 24 hours 6.66 in September 1929; maximum monthly snowfall 20.0 in January 1893; maximum snowfall in 24 hours 17.8 in March 1927.

REFERENCE NOTES APPLYING TO TABLES APPEAR ON THE PAGE FOLLOWING LAST TABLE.
(Caution: Letters and symbols may have different meanings in 1941-1970 tables than in earlier tables. See notes.)

NORMALS, MEANS AND EXTREMES

(Table Revised 1970. Base Period for Climatological Normals: 1931-1960)

Station: WILMINGTON, NORTH CAROLINA   NEW HANOVER COUNTY AIRPORT   Standard time used: EASTERN   Latitude: 34° 16' N   Longitude: 77° 55' W   Elevation (ground): 28 feet

Ø For period July 1963 through the current year.
Means and extremes above are from existing and comparable exposures. Annual extremes have been exceeded at other sites in the locality as follows:
Highest temperature 104 in June 1952; lowest temperature 5 in February 1899; maximum monthly precipitation 21.12 in July 1886; minimum monthly precipitation .02 in October 1943; maximum precipitation in 24 hours 9.52 in September 1938; maximum snowfall 12.1 in February 1896; maximum snowfall in 24 hours 11.1 in February 1896.

NORMALS, MEANS AND EXTREMES

(Table Revised 1975. Base Period for Climatological Normals: 1941-1970)

Means and extremes above are from existing and comparable exposures. Annual extremes have been exceeded at other sites in the locality as follows:
Highest temperature 104 in June 1952; lowest temperature 5 in February 1899; maximum monthly precipitation 21.12 in July 1886; minimum monthly precipitation .02 in October 1943; maximum precipitation in 24 hours 9.52 in September 1938.

REFERENCE NOTES APPLYING TO TABLES APPEAR ON THE PAGE FOLLOWING LAST TABLE.
(Caution: Letters and symbols may have different meanings in 1941-1970 tables than in earlier tables. See notes.)

Reference notes applying to Normals, Means, and Extremes tables for 1931–1960 base period.

(a) Length of record, years, based on January data. Other months may be for more or fewer years if there have been breaks in the record.

(b) Climatological standard normals (1931-1960).

* Less than one half.

+ Also on earlier dates, months, or years.

T Trace, an amount too small to measure.

Below zero temperatures are preceded by a minus sign.

The prevailing direction for wind in the Normals, Means, and Extremes table is from records through 1963.

‡ ≥ 70° at Alaskan stations.

Unless otherwise indicated, dimensional units used in this bulletin are: temperature in degrees F.; precipitation, including snowfall, in inches; wind movement in miles per hour; and relative humidity in percent. Heating degree day totals are the sums of negative departures of average daily temperatures from 65° F. Cooling degree day totals are the sums of positive departures of average daily temperatures from 65° F. Sleet was included in snowfall totals beginning with July 1948. The term "Ice pellets" includes solid grains of ice (sleet) and particles consisting of snow pellets encased in a thin layer of ice. Heavy fog reduces visibility to 1/4 mile or less.

Sky cover is expressed in a range of 0 for no clouds or obscuring phenomena to 10 for complete sky cover. The number of clear days is based on average cloudiness 0-3, partly cloudy days 4-7, and cloudy days 8-10 tenths.

Solar radiation data are the averages of direct and diffuse radiation on a horizontal surface. The langley denotes one gram calorie per square centimeter.

& Figures instead of letters in a direction column indicate direction in tens of degrees from true North; i.e., 09 - East, 18 - South, 27 - West, 36 - North, and 00 - Calm. Resultant wind is the vector sum of wind directions and speeds divided by the number of observations. If figures appear in the direction column under "Fastest mile" the corresponding speeds are fastest observed 1-minute values.

\# To 8 compass points only.

Reference notes applying to Normals, Means, and Extremes tables for 1941–1970 base period.

(a) Length of record, years, through the current year unless otherwise noted, based on January data.

(b) 70° and above at Alaskan stations.

* Less than one half.

T Trace.

NORMALS - Based on record for the 1941-1970 period.
DATE OF AN EXTREME - The most recent in cases of multiple occurrence.
PREVAILING WIND DIRECTION - Record through 1963.
WIND DIRECTION - Numerals indicate tens of degrees clockwise from true north. 00 indicates calm.
FASTEST MILE WIND - Speed is fastest observed 1-minute value when the direction is in tens of degrees.

∅ Through January 1971.

Mean Maximum Temperature (°F.), January

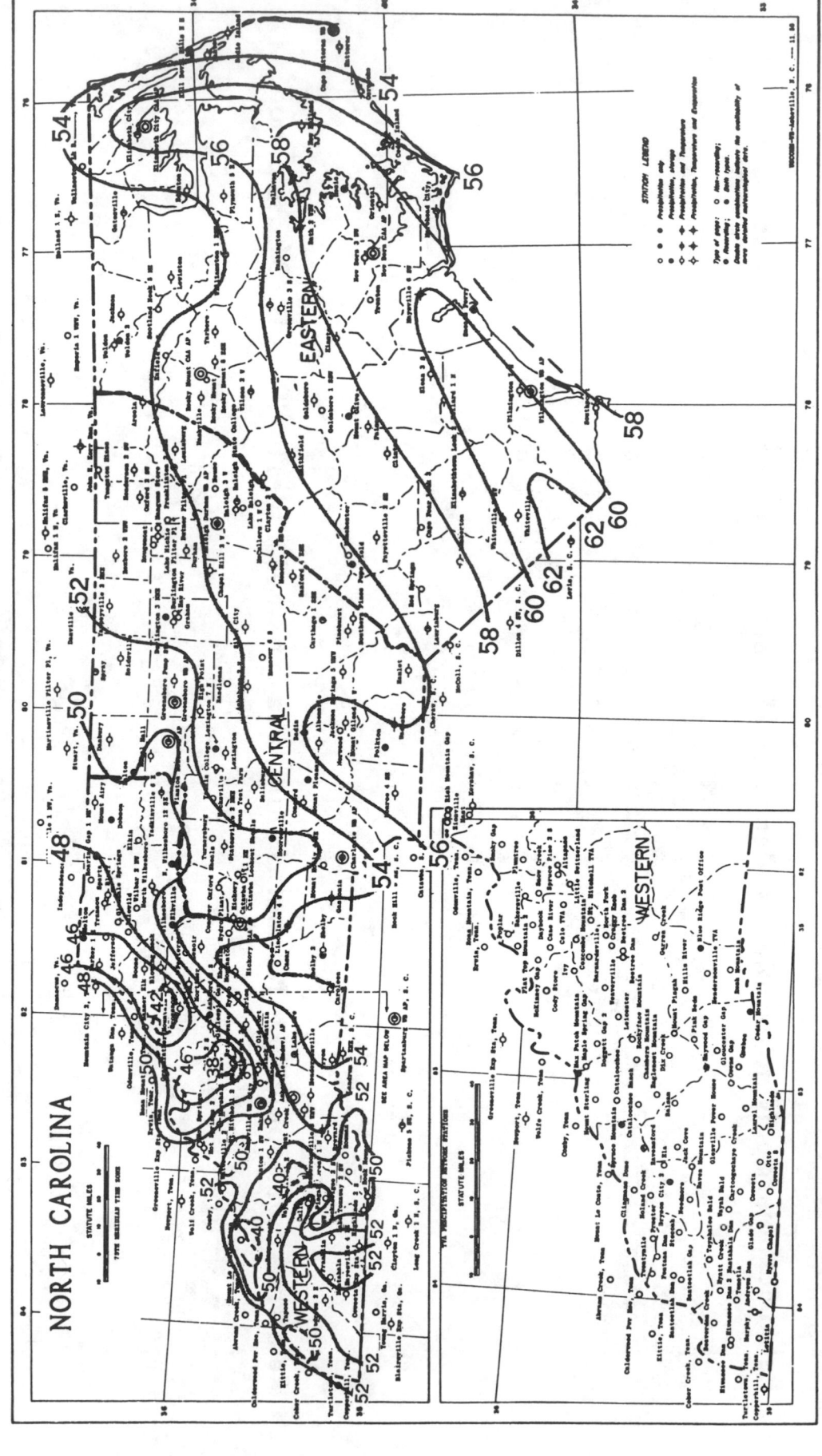

Based on period 1931-52

Isolines are drawn through points of approximately equal value. Caution should be used in interpolating on these maps, particularly in mountainous areas.

742

Mean Minimum Temperature (°F.), January

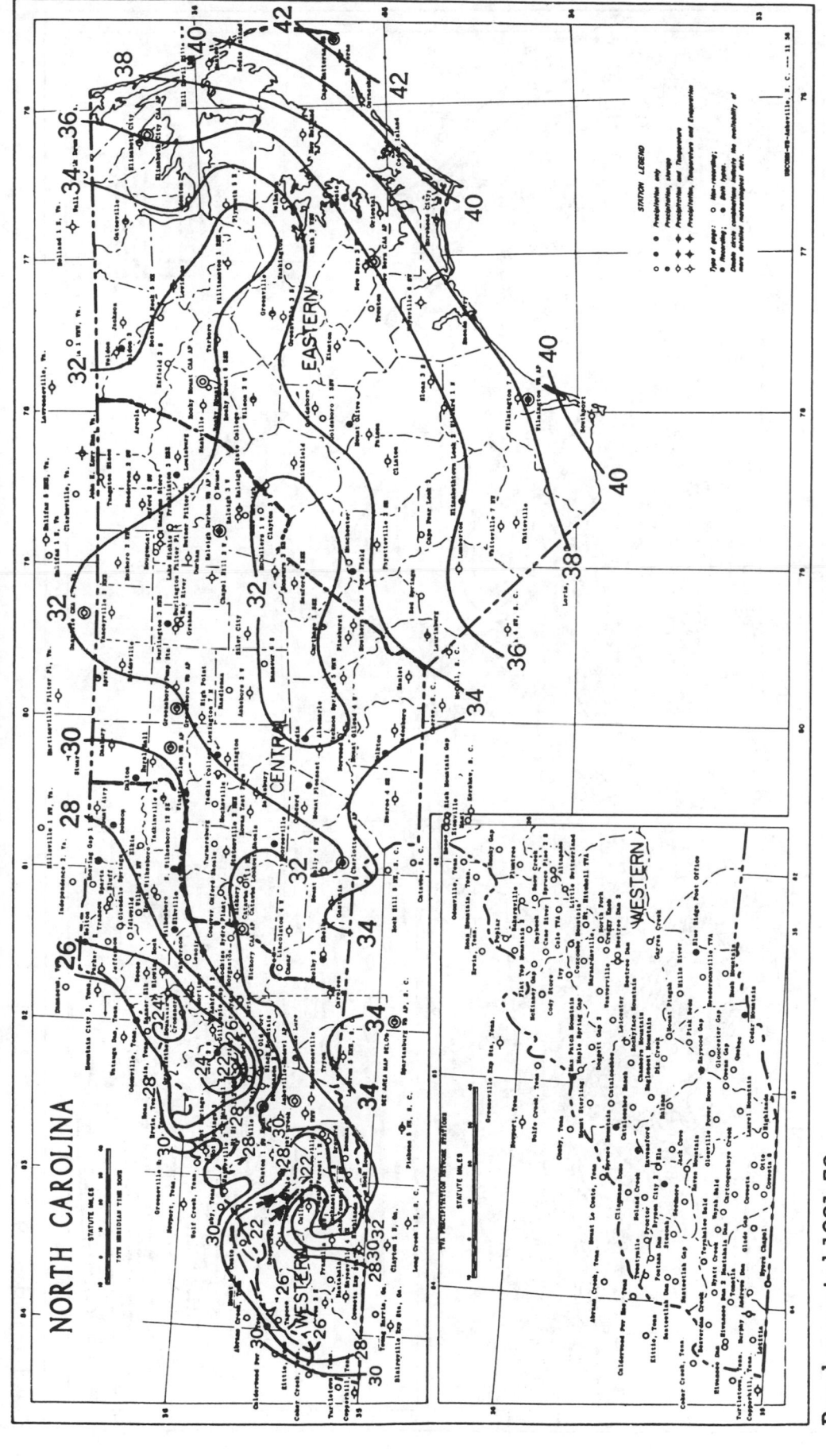

Based on period 1931-52

Isolines are drawn through points of approximately equal value. Caution should be used in interpolating on these maps, particularly in mountainous areas.

Mean Maximum Temperature (°F.), July

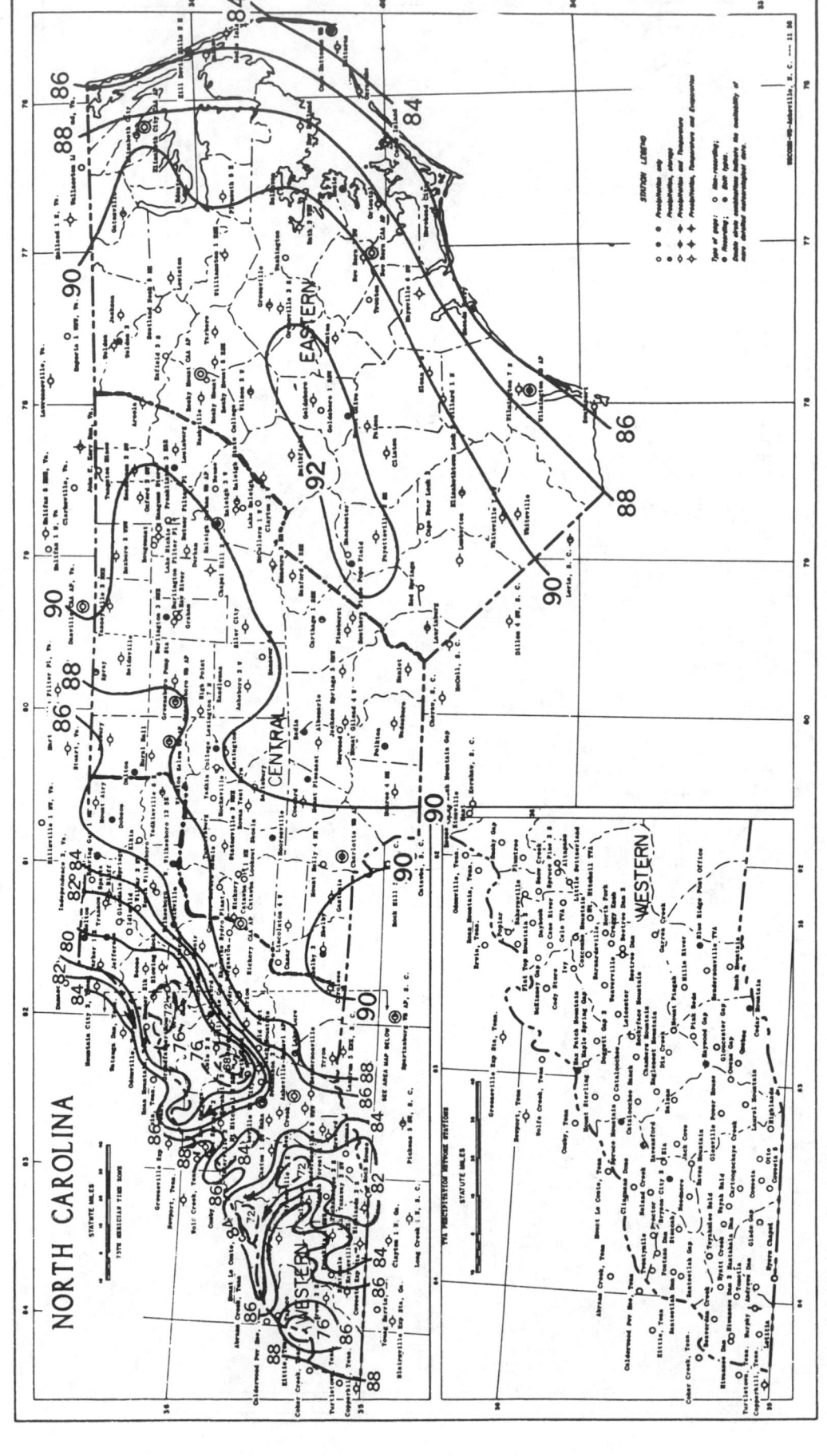

Based on period 1931-52

Isolines are drawn through points of approximately equal value. Caution should be used in interpolating on these maps, particularly in mountainous areas.

Mean Minimum Temperature (°F.), July

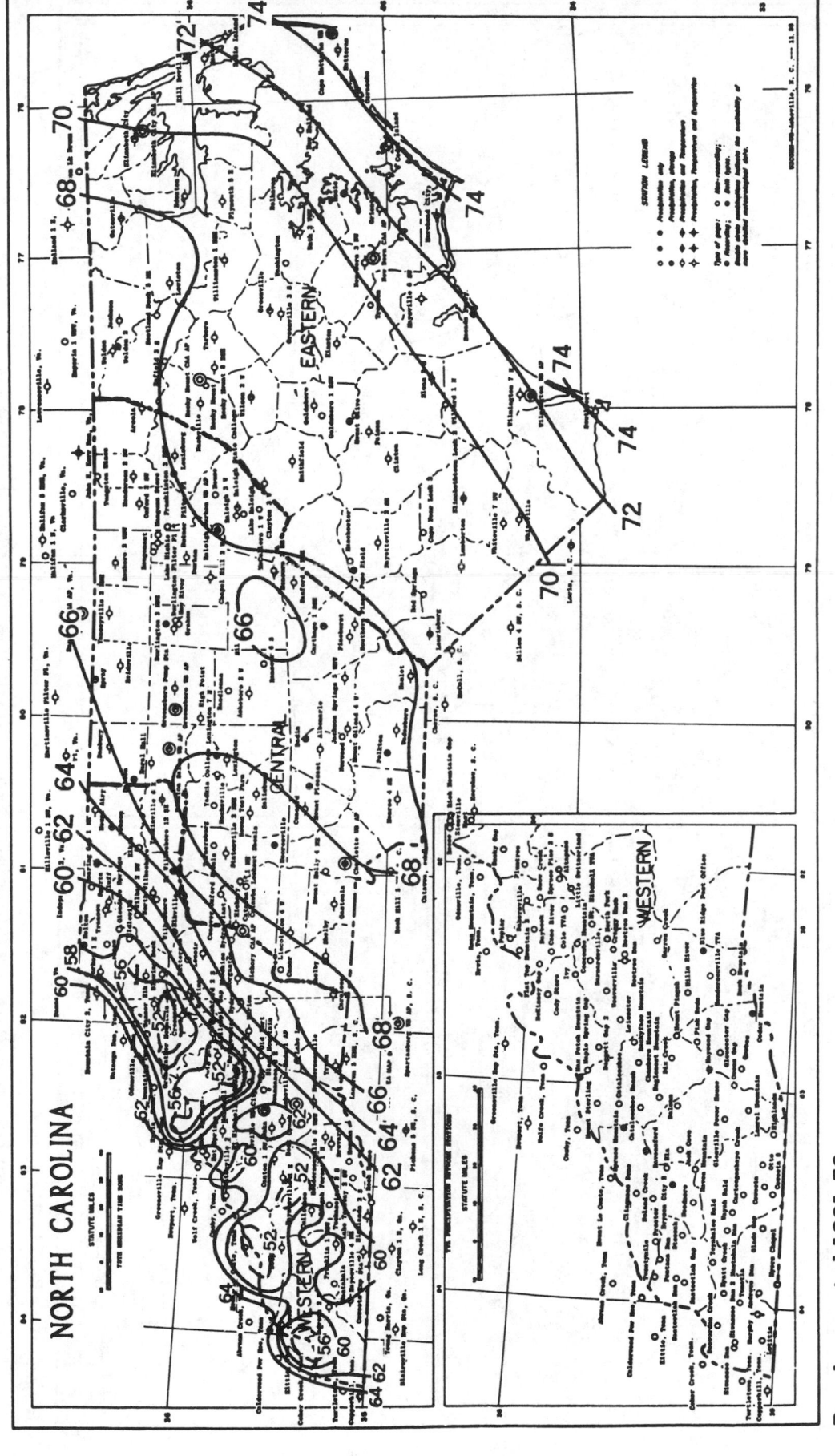

NORTH CAROLINA

Based on period 1931-52

Isolines are drawn through points of approximately equal value. Caution should be used in interpolating on these maps, particularly in mountainous areas.

745

Mean Annual Precipitation, Inches

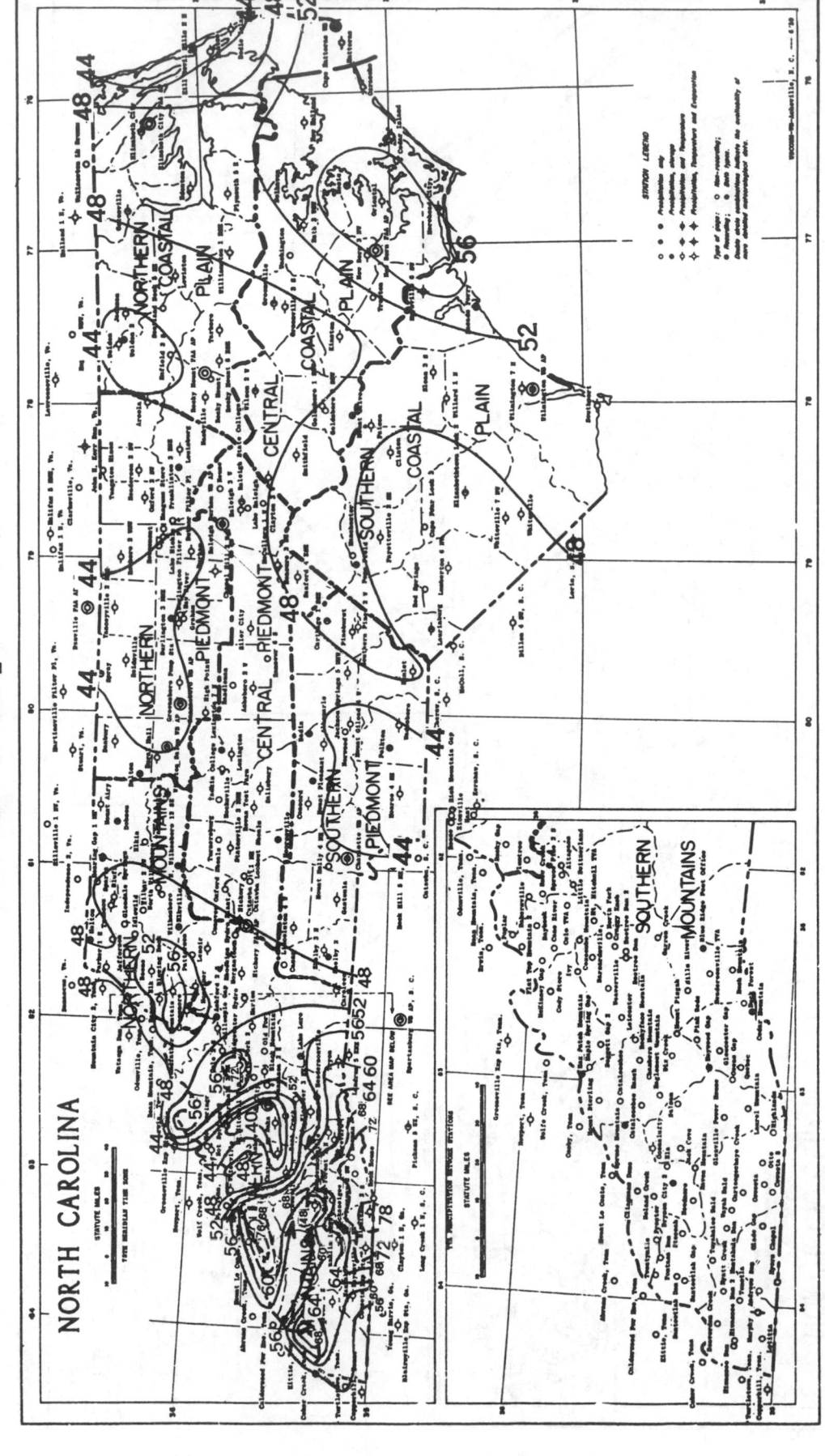

Based on period 1931-55

Isolines are drawn through points of approximately equal value. Caution should be used in interpolating on these maps, particularly in mountainous areas.

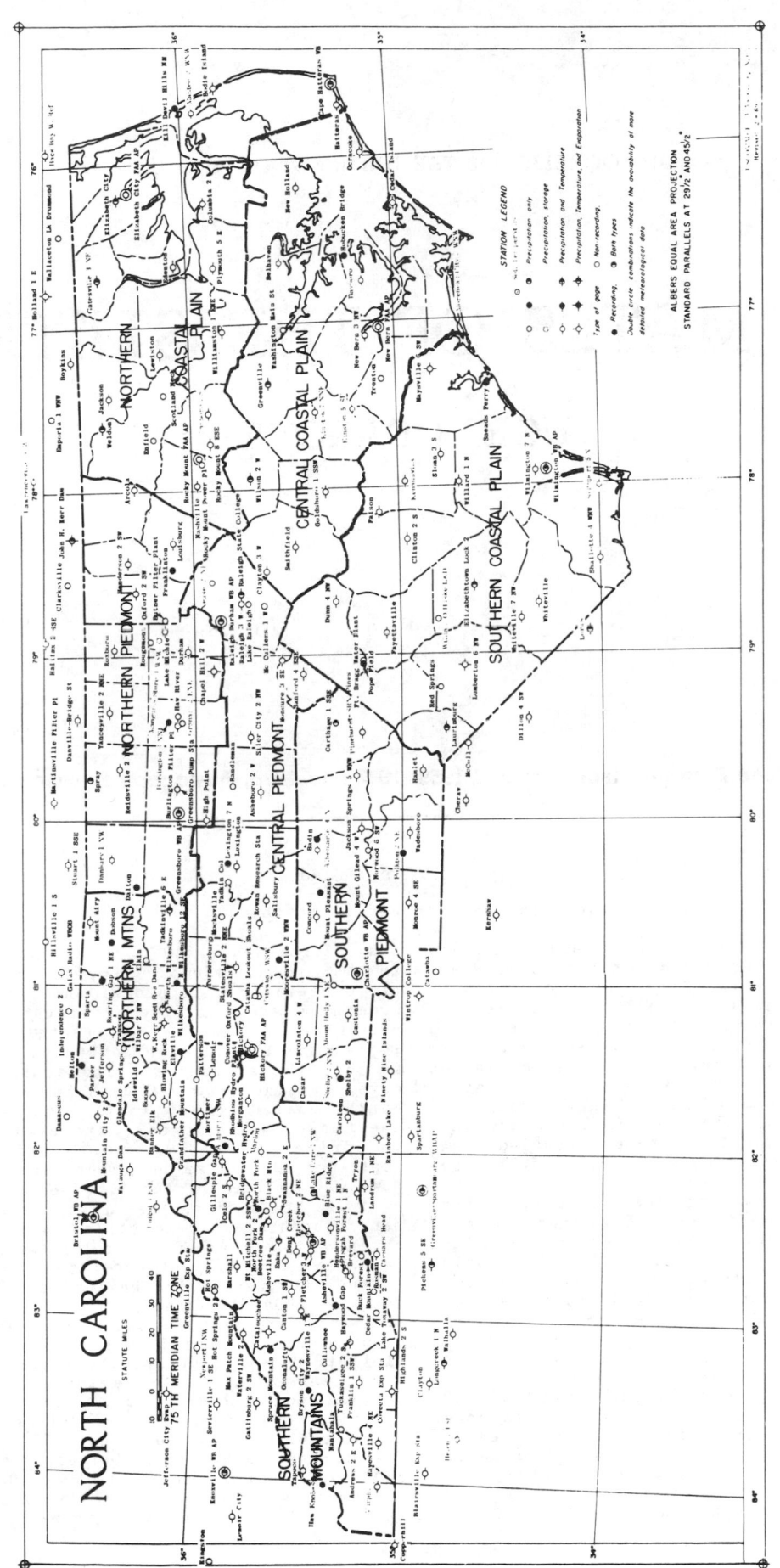

CLIMATES OF THE STATES

NORTH DAKOTA

(Normals, Means and Extremes tables revised 1959, 1973 and 1975. Basic report revised October 1959.)

Climate of North Dakota

Frank J. Bavendick, Weather Bureau State Climatologist

North Dakota is typically plains country located near the center of the North American Continent. The eastern part of the State is flat, with an elevation in the Red River Valley of 780 feet at Pembina in the north to 962 feet above sea level at Wahpeton in the south. To the westward there is a gradual rise of terrain until an elevation of 3,468 is reached at Black Butte in the southwestern part of the State. The Turtle Mountains in the north-central part of the State are only about 500 feet higher than the surrounding area, with the highest elevation about 2,300 feet above sea level.

Most of the eastern part has fine agricultural lands, the soil consisting of a black loam of varying depth, underlaid with a subsoil of clay. The west is gently rolling, with soil ranging from poor in the Badlands to excellent in some other sections. Cereals and feed crops are grown in large quantities. Potatoes and sugar beets are also grown in the Red River Valley. There is much grazing land in the western part of the State. Fruit is grown only in small amounts.

Summers are usually very pleasant, but hot winds and periods of prolonged high temperatures occur occasionally. However, minimum temperatures are seldom above 70°F., so it is unusual to have uncomfortable nights. Winters are usually cold with occasional ones that are open and mild.

The annual mean temperature for North Dakota ranges from about 36°F. in the northeast to 43°F. in the extreme south. Temperatures above 100°F. are occasionally recorded, and zero readings are common in winter. The average number of days a year when the temperature reaches 90°F. or higher is 14, and the average number with zero or lower is 53. The average growing season is about 121 days, ranging from 110 days in the northeast and north-central to 135 in the extreme south. For the State, the average date of the last freeze in spring is May 19, and the first in fall is September 18. Freezing temperatures have occurred, however, as late as the first part of June and as early in the fall as the first few days of September. The highest temperature of record is 121°F. observed July 6, 1936 at Steele; the lowest, -60°F., February 15, 1936, at Parshall.

Precipitation in the eastern third of the State averages about 19 inches, in the middle third about 16 inches, and in the western third about 15 inches. On an average, about 77 percent of the annual precipitation occurs during the crop-growing freeze-free season, April to September, and almost 50 percent falls during May, June, and July. The normal precipitation for the driest months, November to February, is about one-half an inch a month. The greatest amount falls between 5 p.m. and 8 p.m. and again about midnight. In North Dakota, precipitation is considered the most important climatic factor. Because the soil is well supplied with all the necessary plant nutrients, it is of prime importance that the rains come principally during the crop season. Most of the rain in the summer months occurs in storms accompanied by thunder and lightning, often with heavy falls for a short time. The average number of thunderstorm days is 30, mostly in June, July, and August. In most years at least some part of the State is visited by a storm that brings a rainfall of 2 or 3 inches in 24 hours, and occasionally 5 or 6 inches falls in 1 day. On an average, rain falls about 1 day in 4 during the summer months. The annual number of days with measurable precipitation averages 66, ranging from about 50 in the west to 90 in the east. During the 4 years, 1933-1936, North Dakota's precipitation averaged slightly more than 12 inches per year. During the 4 years, 1941-1944, the State's precipitation averaged slightly more than 20 inches. The first light snow in autumn occasionally falls in September, but usually very little occurs until after October. The average number of days with 0.1 inch or more of snow is 23. The average annual snowfall is 32 inches with the greatest amount in the northeast and least in the southwest. Occasionally there is heavy snowfall in winter, and the amount of snow on the ground accumulates to a considerable depth. Much of the North Dakota soil is capable of yielding surprisingly well on little moisture, provided moisture is present when it is most needed. With enough moisture to start the crop, 6 inches of rain from the middle of May to the middle of July will usually produce a good crop.

The streams of North Dakota fall into two main groups -- those in the west and south-central portions draining into the Missouri Basin and those in the east and north-central portions draining into the Red River of the North.

Some of the important tributaries which drain into the Missouri in North Dakota are: The Cannonball, Grand, Heart, Knife, Little Missouri, and the James. Local floods occur occasionally on all the tributaries, mainly associated with ice breakup, notably on the Heart River where serious floods have occurred from ice jams in 1923, 1935, 1939, 1943, 1947, 1948, 1949, 1950, and 1952. A serious flood due to heavy rainfall occurred in the Knife River in July 1938.

Floods along the main stem of the Missouri in the past have been caused primarily by snowmelt in the high plains. The resulting flooding has been almost invariably aggravated by ice jams. The most recent flood of this nature occurred in April 1952. Others occurred in 1947, 1943, 1917, 1910, 1887, and 1881. The completion of Garrison Dam since 1952 has greatly reduced the danger of Missouri River flooding.

The streams draining the east and north-central portions of North Dakota flow into the Red River of the North, which flows in a northerly direction between Minnesota and North Dakota into Canada. The most important tributaries in the eastern portion of North Dakota are the Sheyenne and the Pembina, the latter rises in the province of Manitoba, Canada. In the north-central portion the Souris River originates in the province of Saskatchewan, Canada, flows southeastward into North Dakota, and then curves back into Canada, and flows in a northerly direction into the Assiniboine River which empties into the Red River of the North above the International Boundary. Floods in the Red River of the North Basin occur primarily during the spring season (April and May) and are caused chiefly by melting snow. Ice conditions, particularly on the northward flowing streams, increase flood crests and occasionally cause extremely high flood stages due to jams. Early freeze-up in the fall before snow occurs is also a contributing factor in producing flood conditions in the spring. Considerably higher crests result along the tributaries and the main stem of the river if the snowmelt is accompanied by a period of prolonged heavy rains. Major rainstorms of sufficient magnitude to cause more than local flooding (without snowmelt) are extremely rare.

Flooding in the Red River of the North during the spring of 1950 was one of the worst floods of record in the Valley. The crests at Grand Forks, N. Dak., were the highest since 1897 and at Pembina, N. Dak., the highest on record. From below Grand Forks, N. Dak., to the Canadian Border the flood was generally the highest in the past 100 years. Major floods occurred also in 1952, 1947, 1943, 1916, and 1882.

The prevailing direction of the wind in all months of the year is from the northwest, unless it is influenced by local conditions. More southerly winds are observed during the summer than during the winter. The average annual wind speed is about 11 miles an hour. The highest speeds are in spring and the lowest in late summer. High winds frequently accompany severe thunderstorms. A total of 71 tornadoes were reported during the 7-year period 1952-58. Loss of life from tornadoes has been small.

The average relative humidity is about 68 percent, slightly higher in the east than in the west. Humidity is frequently low during the afternoon in summer, sometimes below 20 percent. Dense fogs are experienced, on an average, on only 8 days of the year.

The average number of clear days is 160, partly cloudy 100, and cloudy 105. On a clear day the sun shines for more than 15 hours from the middle of May to the end of July. These long hours of sunshine make it possible to grow many crops in what appears to be a comparatively short growing season. The yearly average amount of sunshine is 59 percent of the possible amount, with 74 percent in July and 72 percent in August.

BIBLIOGRAPHY

A. Climatic Summary of the United States (Bulletin W) 1930 edition, Sections 34 and 35. U. S. Weather Bureau

B. Climatic Summary of the United States, North Dakota - Supplement for 1931 through 1952 Bulletin W Supplement). U. S. Weather Bureau.

C. Climatological Data - North Dakota. U. S. Weather Bureau

D. Climatological Data National Summary. U. S. Weather Bureau

E. Hourly Precipitation Data - North Dakota. U. S. Weather Bureau

F. Local Climatological Data, U. S. Weather Bureau, for Bismarck, Devils Lake, Fargo, and Williston, North Dakota

REFERENCES

(a) Weather Bureau Technical Paper No. 25 - Rainfall Intensity-Duration-Frequency Curves. For selected stations in the United States, Alaska, Hawaiian Islands, and Puerto Rico.

(b) Weather Bureau Technical Paper No. 16 - Maximum 24-Hour Precipitation in the United States. Washington, D. C. 1952.

FREEZE DATA

STATION	Freeze threshold temperature	Mean date of last Spring occurrence	Mean date of first Fall occurrence	Mean No. of days between dates	Years of record Spring	No. of occurrences in Spring	Years of record Fall	No. of occurrences in Fall
AMENIA	32	05-22	09-20	121	19	19	20	20
	28	05-14	10-01	141	18	18	20	20
	24	05-02	10-09	161	18	18	20	20
	20	04-26	10-18	176	18	18	20	20
	16	04-12	10-31	202	18	18	20	20
AMIDON	32	05-18	09-18	123	20	20	19	19
	28	05-10	10-02	145	20	20	19	19
	24	05-01	10-10	162	20	20	19	19
	20	04-26	10-21	178	20	20	19	19
	16	04-16	10-29	196	20	20	19	19
ASHLEY	32	05-21	09-16	118	30	30	30	30
	28	05-14	09-28	137	28	28	30	30
	24	05-02	10-03	154	28	28	30	30
	20	04-24	10-15	174	28	28	29	29
	16	04-07	10-25	201	29	29	29	29
BEACH	32	05-17	09-19	125	24	24	25	25
	28	05-09	10-01	145	24	24	25	25
	24	04-28	10-13	168	24	24	25	25
	20	04-15	10-21	189	24	24	25	25
	16	04-09	10-30	204	24	24	25	25
BISMARCK WB AP	32	05-11	09-24	136	30	30	30	30
	28	05-04	10-04	152	30	30	30	30
	24	04-23	10-17	177	30	30	30	30
	20	04-11	10-22	194	30	30	30	30
	16	03-31	11-03	217	30	30	30	30
BOTTINEAU	32	05-23	09-14	114	29	29	30	30
	28	05-16	09-25	132	29	29	30	30
	24	05-05	10-02	150	28	28	30	30
	20	04-28	10-13	168	28	28	30	30
	16	04-15	10-22	189	30	30	30	30
BOWBELLS	32	05-30	09-10	103	15	15	16	16
	28	05-16	09-24	131	15	15	16	16
	24	05-07	10-02	149	15	15	16	16
	20	04-26	10-14	170	15	15	16	16
	16	04-12	10-26	198	15	15	16	16
BOWMAN COURT HOUSE	32	05-16	09-23	130	27	27	25	25
	28	05-07	09-30	146	27	27	25	25
	24	04-26	10-12	170	27	27	24	24
	20	04-17	10-20	187	27	27	23	23
	16	04-11	10-27	199	26	26	23	23
CARSON	32	05-20	09-14	117	30	30	30	30
	28	05-11	09-27	140	30	30	30	30
	24	04-29	10-05	159	30	30	30	30
	20	04-26	10-20	178	30	30	30	30
	16	04-16	10-26	193	30	30	29	29
COOPERSTOWN	32	05-25	09-19	116	16	16	16	16
	28	05-10	09-25	138	16	16	16	16
	24	05-02	10-07	158	16	16	16	16
	20	04-23	10-14	174	16	16	16	16
	16	04-12	10-24	195	16	16	16	16
CROSBY	32	05-20	09-14	117	30	30	30	30
	28	05-11	09-25	137	30	30	30	30
	24	05-02	10-04	155	30	30	30	30
	20	04-19	10-14	178	30	30	30	30
	16	04-10	10-23	196	30	30	30	30
DEVILS LAKE WB CITY	32	05-18	09-22	127	30	30	30	30
	28	05-09	10-02	146	30	30	30	30
	24	04-30	10-11	163	30	30	30	30
	20	04-19	10-19	183	30	30	30	30
	16	04-08	10-30	206	30	30	30	30
DICKINSON EXP STA	32	05-23	09-11	111	30	30	30	30
	28	05-11	09-25	137	30	30	30	30
	24	05-03	10-06	156	30	30	30	30
	20	04-25	10-16	174	30	30	30	30
	16	04-13	10-26	196	30	30	30	30
DRAKE	32	05-20	09-22	124	21	21	21	21
	28	05-09	10-01	145	21	21	21	21
	24	04-28	10-11	166	21	21	20	20
	20	04-21	10-18	180	21	21	20	20
	16	04-09	10-26	201	19	19	20	20

STATION	Freeze threshold temperature	Mean date of last Spring occurrence	Mean date of first Fall occurrence	Mean No. of days between dates	Years of record Spring	No. of occurrences in Spring	Years of record Fall	No. of occurrences in Fall
DUNN CENTER	32	05-20	09-12	115	30	30	30	30
	28	05-14	09-23	132	30	30	30	30
	24	05-01	10-01	154	30	30	30	30
	20	04-24	10-12	171	30	30	30	30
	16	04-16	10-23	190	30	30	30	30
DUNSEITH ST SANITARIUM	32	05-26	09-15	112	30	30	29	29
	28	05-16	09-25	132	30	30	30	30
	24	05-08	10-03	148	30	30	30	30
	20	04-30	10-16	169	30	30	30	30
	16	04-16	10-21	188	30	30	30	30
ECKMAN	32	05-30	09-08	100	29	29	27	27
	28	05-18	09-17	122	28	28	27	27
	24	05-07	09-29	145	28	28	27	27
	20	04-30	10-09	162	28	28	27	27
	16	04-20	10-15	178	28	28	27	27
EDGELEY EXP FARM	32	05-24	09-18	117	30	30	30	30
	28	05-13	09-26	136	30	30	30	30
	24	05-03	10-09	158	30	30	30	30
	20	04-22	10-16	177	30	30	30	30
	16	04-07	10-31	207	30	30	30	30
EDMORE 1 W	32	06-02	09-10	100	17	17	18	18
	28	05-18	09-18	123	17	17	18	18
	24	05-03	09-28	148	17	17	18	18
	20	04-25	10-07	165	17	17	18	18
	16	04-15	10-16	184	17	17	18	18
ELBOWOODS	32	05-24	09-13	112	30	30	30	30
	28	05-11	09-21	133	30	30	30	30
	24	05-04	10-02	151	30	30	30	30
	20	04-25	10-09	168	30	30	30	30
	16	04-12	10-24	195	30	30	30	30
ELLENDALE	32	05-16	09-24	131	30	30	30	30
	28	05-06	10-03	151	30	30	30	30
	24	04-26	10-13	170	30	30	30	30
	20	04-13	10-21	192	30	30	30	30
	16	03-31	11-02	215	30	30	30	30
FARGO WB AP	32	05-13	09-27	137	30	30	30	30
	28	05-04	10-06	155	30	30	30	30
	24	04-24	10-15	174	30	30	30	30
	20	04-11	10-29	201	30	30	30	30
	16	03-31	11-05	219	30	30	30	30
FESSENDEN	32	05-21	09-16	118	30	30	30	30
	28	05-10	09-25	138	30	30	30	30
	24	04-28	10-07	163	30	30	30	30
	20	04-17	10-16	182	30	30	30	30
	16	04-04	10-26	205	30	30	30	30
FORMAN	32	05-16	09-21	127	14	14	14	14
	28	05-09	10-01	145	14	14	14	14
	24	04-25	10-11	169	14	14	14	14
	20	04-11	10-23	196	14	14	14	14
	16	03-29	11-03	218	14	14	14	14
FORT YATES	32	05-19	09-16	120	22	22	22	22
	28	05-11	09-28	140	22	22	22	22
	24	04-27	10-09	164	22	22	22	22
	20	04-17	10-22	188	22	22	22	22
	16	04-06	10-29	206	22	22	22	22
FOXHOLM 7 N	32	05-22	09-17	118	30	30	30	30
	28	05-13	09-29	139	30	30	30	30
	24	05-02	10-05	156	30	30	30	30
	20	04-21	10-17	179	30	30	30	30
	16	04-12	10-27	198	30	30	30	30
FULLERTON	32	05-19	09-22	127	30	30	30	30
	28	05-08	10-02	147	30	30	30	30
	24	04-27	10-09	166	30	30	30	30
	20	04-16	10-22	189	30	30	30	30
	16	04-04	11-03	213	30	30	30	30
GACKLE	32	05-16	09-20	126	13	13	13	13
	28	05-09	09-29	143	13	13	13	13
	24	04-23	10-14	174	13	13	13	13
	20	04-14	10-17	186	13	13	13	13
	16	04-07	10-22	199	13	13	12	12

FREEZE DATA

STATION	Freeze threshold temperature	Mean date of last Spring occurrence	Mean date of first Fall occurrence	Mean No. of days between dates	Years of record Spring	No. of occurrences in Spring	Years of record Fall	No. of occurrences in Fall
GARRISON	32	05-21	09-16	119	30	30	30	30
	28	05-06	09-26	142	30	30	30	30
	24	04-30	10-08	161	30	30	30	30
	20	04-21	10-17	179	30	30	30	30
	16	04-10	10-26	199	30	30	30	30
GRAFTON STATE SCHOOL	32	05-22	09-18	119	30	30	30	30
	28	05-10	09-28	141	30	30	30	30
	24	04-30	10-08	161	30	30	30	30
	20	04-19	10-21	185	30	30	30	30
	16	04-08	10-31	206	30	30	30	30
GRAND FORKS UNIVER	32	05-20	09-23	126	30	30	30	30
	28	05-08	09-30	146	30	30	30	30
	24	04-27	10-09	165	30	30	30	30
	20	04-16	10-24	190	30	30	30	30
	16	04-05	11-02	211	30	30	30	30
GRANVILLE	32	05-24	09-13	112	29	29	30	30
	28	05-14	09-25	134	29	29	30	30
	24	05-02	10-06	157	29	29	30	30
	20	04-25	10-14	172	29	29	30	30
	16	04-14	10-25	195	29	29	30	30
GRENORA	32	05-23	09-11	112	28	28	28	28
	28	05-13	09-24	135	28	28	28	28
	24	05-05	10-03	151	28	28	28	28
	20	04-24	10-09	168	28	28	28	28
	16	04-11	10-23	195	28	28	28	28
HANNAH	32	05-27	09-15	111	18	18	18	18
	28	05-19	09-24	128	18	18	18	18
	24	05-08	10-03	148	17	17	18	18
	20	04-27	10-15	171	17	17	18	18
	16	04-08	10-23	198	17	17	18	18
HANSBORO	32	06-04	08-29	86	30	30	30	30
	28	05-20	09-16	118	30	30	29	29
	24	05-12	09-30	142	30	30	28	28
	20	05-02	10-07	159	30	30	28	28
	16	04-20	10-17	180	30	30	28	28
HETTINGER	32	05-19	09-17	121	28	28	29	29
	28	05-11	09-29	141	28	28	29	29
	24	04-28	10-05	160	28	28	29	29
	20	04-21	10-17	179	28	28	29	29
	16	04-10	10-26	199	28	28	29	29
HILLSBORO	32	05-15	09-25	133	30	30	28	28
	28	05-05	10-04	152	30	30	27	27
	24	04-26	10-10	167	30	30	27	27
	20	04-14	10-25	194	30	30	27	27
	16	04-02	11-03	215	30	30	27	27
JAMESTOWN STATE HOS	32	05-18	09-18	123	30	30	30	30
	28	05-06	09-27	144	30	30	30	30
	24	04-28	10-08	163	30	30	30	30
	20	04-19	10-16	181	30	30	30	30
	16	04-06	10-31	207	30	30	30	30
LANGDON EXP FARM	32	06-03	09-13	102	30	30	30	30
	28	05-17	09-24	130	30	30	30	30
	24	05-07	10-05	151	30	30	30	30
	20	04-29	10-14	168	30	30	30	30
	16	04-18	10-25	190	30	30	30	30
LARIMORE	32	05-20	09-20	123	29	29	29	29
	28	05-12	09-29	140	29	29	29	29
	24	05-01	10-12	164	29	29	29	29
	20	04-24	10-15	174	29	29	29	29
	16	04-11	10-29	201	29	29	29	29
LEEDS	32	05-23	09-17	117	10	10	10	10
	28	05-15	09-20	129	10	10	10	10
	24	05-06	10-03	150	10	10	9	9
	20	04-25	10-19	177	10	10	9	9
	16	03-31	10-26	210	10	10	9	9
LINTON	32	05-26	09-15	113	29	29	30	30
	28	05-14	09-23	133	29	29	30	30
	24	05-03	10-04	154	30	30	30	30
	20	04-22	10-12	173	30	30	30	30
	16	04-10	10-24	197	30	30	29	29
LISBON	32	05-22	09-18	119	30	30	30	30
	28	05-09	09-28	142	30	30	30	30
	24	04-28	10-06	161	30	30	30	30
	20	04-20	10-17	180	30	30	30	30
	16	04-05	10-29	208	30	30	30	30
MADDOCK AG SCHOOL	32	05-27	09-10	106	30	30	30	30
	28	05-15	09-21	129	30	30	30	30
	24	05-07	10-01	148	30	30	30	30
	20	04-27	10-13	169	30	30	30	30
	16	04-13	10-23	193	30	30	30	30
MARMARTH	32	05-17	09-15	121	27	27	25	25
	28	05-07	09-27	143	27	27	25	25
	24	04-29	10-09	163	27	27	24	24
	20	04-19	10-15	179	27	27	24	24
	16	04-08	10-27	202	27	27	24	24
MC LEOD 3 E	32	05-19	09-20	124	30	30	29	29
	28	05-10	09-29	142	30	30	29	29
	24	04-30	10-09	161	30	30	29	29
	20	04-22	10-17	177	30	30	29	29
	16	04-06	10-27	204	30	30	29	29
MINOT CAA AP	32	05-24	09-12	111	30	30	29	29
	28	05-13	09-27	138	30	30	29	29
	24	05-04	10-04	153	30	30	29	29
	20	04-23	10-15	175	30	30	29	29
	16	04-08	10-25	200	30	30	29	29
MOTT	32	05-21	09-15	117	30	30	30	30
	28	05-08	09-26	141	30	30	30	30
	24	04-29	10-07	161	30	30	30	30
	20	04-21	10-18	180	30	30	30	30
	16	04-08	10-29	205	30	30	30	30
NAPOLEON 1 SE	32	05-26	09-13	110	30	30	30	30
	28	05-13	09-25	135	30	30	30	30
	24	05-05	10-02	150	30	30	30	30
	20	04-27	10-12	167	30	30	30	30
	16	04-14	10-22	192	30	30	30	30
NEW ENGLAND	32	05-21	09-15	117	30	30	29	29
	28	05-12	09-25	137	30	30	29	29
	24	04-29	10-06	160	29	29	29	29
	20	04-22	10-16	177	29	29	29	29
	16	04-11	10-26	198	28	28	29	29
NEW SALEM 1 S	32	05-19	09-21	125	30	30	29	29
	28	05-10	09-26	140	30	30	29	29
	24	05-01	10-08	160	30	30	29	29
	20	04-26	10-20	177	30	30	29	29
	16	04-10	10-28	201	30	30	29	29
PARK RIVER	32	05-23	09-20	119	30	30	30	30
	28	05-12	09-30	141	30	30	30	30
	24	05-01	10-13	165	30	30	30	30
	20	04-23	10-20	181	30	30	30	30
	16	04-10	10-29	203	30	30	30	30
PEMBINA CAA AP	32	05-25	09-17	115	27	27	26	26
	28	05-15	09-26	134	27	27	26	26
	24	05-03	10-08	158	27	27	26	26
	20	04-26	10-19	176	27	27	26	26
	16	04-12	10-29	201	27	27	26	26
PETTIBONE	32	05-27	09-08	104	29	29	27	27
	28	05-15	09-20	128	29	29	27	27
	24	05-04	10-02	150	29	29	27	27
	20	04-28	10-13	168	29	29	27	27
	16	04-19	10-23	187	29	29	26	26
PORTAL	32	05-23	09-08	108	22	22	22	22
	28	05-15	09-22	131	22	22	22	22
	24	05-01	10-02	154	22	22	22	22
	20	04-24	10-12	171	22	22	22	22
	16	04-16	10-20	188	22	22	22	22
POWERS LAKE	32	05-22	09-11	112	23	23	21	21
	28	05-07	09-29	145	23	23	21	21
	24	04-25	10-13	171	23	23	21	21
	20	04-16	10-14	181	23	23	21	21
	16	04-05	10-18	196	23	23	21	21

FREEZE DATA

STATION	Freeze threshold temperature	Mean date of last Spring occurrence	Mean date of first Fall occurrence	Mean No. of days between dates	Years of record Spring	No. of occurrences in Spring	Years of record Fall	No. of occurrences in Fall
SANISH	32	05-25	09-12	109	20	20	21	21
	28	05-14	09-20	129	20	20	21	21
	24	05-02	10-03	154	20	20	21	21
	20	04-22	10-10	171	20	20	20	20
	16	04-09	10-28	202	19	19	20	20
STANLEY	32	05-27	09-11	107	13	13	13	13
	28	05-16	09-17	124	13	13	13	13
	24	05-09	09-30	145	13	13	13	13
	20	04-28	10-10	165	13	13	13	13
	16	04-10	10-22	196	13	13	13	13
STEELE	32	05-24	09-13	112	30	30	30	30
	28	05-12	09-25	136	29	29	30	30
	24	05-04	10-04	152	29	29	30	30
	20	04-28	10-14	168	29	29	30	30
	16	04-18	10-21	187	29	29	30	30
TOWNER	32	05-26	09-11	109	30	30	29	29
	28	05-17	09-21	127	30	30	29	29
	24	05-05	10-01	148	29	29	29	29
	20	04-26	10-10	167	29	29	28	28
	16	04-13	10-22	192	29	29	27	27
TURTLE LAKE	32	05-22	09-19	119	24	24	24	24
	28	05-09	09-27	140	24	24	24	24
	24	05-03	10-08	158	24	24	23	23
	20	04-21	10-18	180	24	24	23	23
	16	04-14	10-29	198	24	24	23	23
VALLEY CITY	32	05-20	09-19	122	29	29	28	28
	28	05-08	09-27	142	29	29	28	28
	24	04-29	10-08	161	29	29	28	28
	20	04-19	10-20	184	29	29	28	28
	16	04-08	10-28	203	29	29	28	28
VELVA	32	05-22	09-16	116	23	23	23	23
	28	05-10	09-27	140	23	23	23	23
	24	05-01	10-07	159	23	23	23	23
	20	04-24	10-16	175	23	23	23	23
	16	04-12	10-29	199	23	23	23	23
WAHPETON ST SCHOOL	32	05-17	09-22	128	30	30	30	30
	28	05-08	10-02	148	30	30	30	30
	24	04-27	10-12	168	30	30	30	30
	20	04-16	10-22	189	30	30	30	30
	16	04-04	11-02	213	30	30	30	30
WATFORD CITY	32	05-20	09-17	120	15	15	15	15
	28	05-13	09-30	140	14	14	15	15
	24	04-25	10-07	165	14	14	15	15
	20	04-14	10-22	191	14	14	15	15
	16	04-07	10-29	205	14	14	15	15
WESTHOPE	32	05-27	09-14	110	29	29	28	28
	28	05-13	09-24	134	29	29	28	28
	24	05-03	10-05	154	29	29	28	28
	20	04-25	10-14	172	28	28	28	28
	16	04-13	10-23	193	28	28	28	28
WILLISTON WB CITY	32	05-14	09-23	132	30	30	30	30
	28	05-04	10-02	151	30	30	30	30
	24	04-21	10-15	177	30	30	30	30
	20	04-10	10-25	197	30	30	30	30
	16	04-04	11-04	214	30	30	30	30
WILLOW CITY	32	06-01	09-09	101	30	30	30	30
	28	05-17	09-17	123	30	30	30	30
	24	05-08	09-24	139	30	30	28	28
	20	04-30	10-04	157	30	30	28	28
	16	04-18	10-19	183	30	30	28	28
WILTON R R STATION	32	05-19	09-18	122	16	16	16	16
	28	05-09	09-28	143	16	16	16	16
	24	04-30	10-11	164	16	16	16	16
	20	04-23	10-21	181	16	16	16	16
	16	04-07	10-28	204	16	16	16	16
WISHEK	32	05-26	09-09	107	22	22	22	22
	28	05-14	09-22	131	22	22	22	22
	24	05-06	09-28	146	22	22	22	22
	20	04-23	10-08	168	22	22	22	22
	16	04-15	10-17	185	21	21	22	22

Data in the above table are based on the period 1921-1950, or that portion of this period for which data are available.

Means have been adjusted to take into account years of non-occurrence.

A freeze is a numerical substitute for the former term "killing frost" and is the occurrence of a minimum temperature at or below the threshold temperature of 32°, 28°, etc.

Freeze data tabulations in greater detail are available and can be reproduced at cost.

*MEAN TEMPERATURE AND PRECIPITATION

STATION	JAN Temp	JAN Precip	FEB Temp	FEB Precip	MAR Temp	MAR Precip	APR Temp	APR Precip	MAY Temp	MAY Precip	JUN Temp	JUN Precip	JUL Temp	JUL Precip	AUG Temp	AUG Precip	SEP Temp	SEP Precip	OCT Temp	OCT Precip	NOV Temp	NOV Precip	DEC Temp	DEC Precip	ANN Temp	ANN Precip
NORTHWEST DIVISION																										
CROSBY	6.0	.36	10.3	.34	21.5	.62	39.7	.79	52.9	1.64	60.7	3.29	68.5	2.13	65.7	1.85	55.0	1.09	44.0	.83	25.8	.40	13.8	.41	38.7	13.75
FOXHOLM 7 N	7.3	.50	11.4	.36	22.6	.77	41.0	1.19	54.2	1.84	62.1	3.52	69.6	1.98	67.3	2.10	56.9	1.24	45.5	.78	26.6	.57	14.4	.41	39.9	15.26
KENMARE		.50		.49		.83		1.07		1.86		3.52		2.25		1.90		1.36		.80		.53		.46		15.57
MOHALL	5.3	.50	9.4	.47	21.2	.89	40.4	1.31	54.0	2.01	61.9	4.01	69.1	2.55	66.8	2.18	56.4	1.42	44.7	.83	25.5	.58	12.6	.44	38.9	17.19
NEWTOWN	6.0	.50	9.3	.44	20.6	.57	41.3	1.21	53.4	1.66	61.7	3.61	69.4	1.97	67.0	1.68	56.1	1.22	44.6	.74	26.6	.60	14.6	.39	39.2	14.59
PARSHALL		.40		.31		.47		1.15		2.02		3.54		2.41		1.67		1.44		.72		.49		.28		14.90
PORTAL	4.1	.43	8.2	.44	19.4	.67	39.5	.83	52.7	1.88	60.7	3.78	68.2	2.10	65.3	2.27	55.1	1.24	43.3	.75	24.5	.47	11.8	.44	37.7	15.30
WILDROSE		.54		.48		.69		.90		1.63		3.26		1.80		1.85		1.13		.76		.45		.48		13.97
WILLISTON WB CITY	10.0	.49	13.5	.46	26.5	.75	42.9	1.07	54.6	1.66	63.0	3.59	70.9	2.13	68.1	1.41	57.2	1.21	45.5	.77	28.3	.58	15.7	.54	41.3	14.66
DIVISION	6.3	.46	10.5	.42	21.4	.67	40.5	1.09	53.2	1.90	61.2	3.59	69.0	2.15	66.5	1.94	56.0	1.28	44.6	.77	26.0	.50	13.6	.40	39.1	15.17
NORTH CENTRAL DIV																										
BOTTINEAU	3.3	.46	7.6	.35	20.1	.74	39.4	.96	53.4	1.97	61.5	3.80	68.9	2.55	66.5	2.57	55.8	1.38	43.6	.78	23.8	.39	10.2	.40	37.8	16.35
DRAKE	6.0	.45	9.9	.38	21.7	.70	40.6	1.21	54.0	1.84	62.6	3.73	70.0	2.63	67.5	2.09	56.9	1.41	44.8	.86	26.0	.63	13.3	.37	39.4	16.30
ECKMAN 2 SE	3.0		7.6		20.1		40.0		53.3		61.4		68.7		65.9		55.5		43.6		24.2		11.2		37.9	
GRANVILLE	6.7	.55	10.9	.37	23.2	.67	41.6	1.22	55.3	2.04	62.4	3.93	70.0	2.42	67.5	2.14	57.4	1.23	45.5	.75	26.5	.51	14.0	.41	40.1	16.24
MADDOCK AGR SCHOOL	5.2	.54	9.6	.46	22.3	.74	41.0	1.08	54.2	2.01	62.6	3.57	69.6	2.79	67.7	2.09	57.5	1.47	45.0	.90	25.7	.50	12.4	.40	39.4	16.55
RUGBY		.58		.39		.78		1.03		1.76		3.37		2.65		2.21		1.31		.81		.53		.53		15.95
SAN HAVEN	4.2	.49	8.8	.40	19.7	.58	39.4	.87	53.1	1.92	60.9	3.51	68.5	2.85	66.0	2.38	55.6	1.39	43.6	.72	24.0	.43	11.4	.40	37.9	15.94
TOWNER	5.0	.57	9.3	.45	21.7	.96	41.5	1.35	54.5	1.98	62.5	3.69	69.8	2.26	67.0	2.24	54.3	1.42	44.6	.79	25.6	.63	12.5	.53	39.0	16.87
VELVA	8.2	.57	12.6	.43	24.3	.71	42.6	1.29	55.3	2.16	63.2	3.65	70.0	2.56	67.6	2.11	57.6	1.33	46.0	.80	27.3	.69	15.3	.46	40.8	16.76
WESTHOPE	3.7		7.7		20.4		40.2		54.3		62.1		69.2		66.9		56.3		43.8		24.3		10.9		38.3	
WILLOW CITY	2.5	.55	6.8	.39	19.8	.70	39.7	1.07	53.1	1.81	61.5	3.70	68.6	2.48	66.0	2.04	55.7	1.26	43.1	.71	23.9	.54	10.0	.51	37.6	15.76
DIVISION	4.5	.54	9.2	.41	21.0	.79	40.5	1.13	53.7	1.97	61.9	3.77	69.1	2.52	66.7	2.26	56.1	1.39	44.3	.79	25.1	.56	11.8	.46	38.7	16.59
NORTHEAST DIVISION																										
DEVILS LAKE WB CITY	4.8	.40	8.7	.41	22.6	.73	40.0	1.44	53.0	2.13	62.4	3.18	69.3	2.67	66.8	2.17	56.3	2.07	43.9	1.26	25.1	.67	11.2	.48	38.7	17.61
GRAFTON	4.7	.72	9.7	.58	22.9	.90	41.6	1.46	55.4	1.86	64.3	3.19	70.9	3.20	68.7	2.65	58.1	2.05	46.2	1.18	26.4	.90	11.3	.60	40.0	19.29
GRAND FORKS UNIVERSITY	4.4	.60	8.9	.55	22.5	.91	41.3	1.51	54.9	2.33	64.0	3.57	70.7	2.96	68.2	3.13	57.4	1.73	45.3	1.22	26.0	.92	11.5	.62	39.6	20.05
HANNAH		.54		.45		.77		1.09		2.06		3.66		2.70		2.89		1.91		1.06		.70		.42		18.25
HANSBORO	2.9	.42	7.4	.26	19.6	.66	38.6	.92	51.9	1.85	60.3	3.73	67.2	2.59	65.0	2.56	55.0	1.37	43.3	.82	23.7	.52	10.5	.33	37.1	16.03
LANGDON EXP FARM	1.9	.79	6.5	.58	19.5	.92	38.7	1.29	52.5	1.95	61.3	3.06	68.0	2.62	65.9	2.63	55.5	2.05	43.0	1.11	22.8	.75	9.3	.56	37.1	18.31
LARIMORE	5.3	.45	9.9	.46	22.0	.86	40.8	1.37	54.6	2.21	63.4	3.09	70.4	2.53	68.3	2.61	57.8	1.73	45.7	1.20	26.5	.78	12.5	.47	39.9	17.76
PARK RIVER	4.4	.36	9.0	.36	21.8	.74	40.1	1.32	54.2	1.79	63.1	3.06	69.6	2.89	67.7	2.19	57.2	2.13	45.3	1.11	25.6	.63	11.6	.40	39.1	16.98
PEMBINA 2 N		.58		.44		.72		1.21		1.89		3.07		2.92		2.88		2.45		1.18		.72		.40		18.46
PETERSBURG		.42		.43		.69		1.13		2.03		3.15		2.68		2.50		1.78		1.17		.57		.43		16.98
DIVISION	3.3	.56	7.9	.46	20.8	.80	39.9	1.27	53.4	1.99	62.2	3.32	69.0	2.81	66.9	2.71	56.3	1.92	44.4	1.09	24.9	.72	10.6	.49	38.3	18.14
WEST CENTRAL DIVISION																										
DUNN CENTER 2 SW	10.4	.43	14.2	.39	24.1	.65	40.8	1.26	53.1	1.99	61.4	3.85	69.5	2.34	66.5	1.97	55.4	1.24	44.9	.81	27.8	.46	17.7	.27	40.5	15.66
GARRISON	8.3	.59	12.4	.54	23.7	.75	41.6	1.27	54.2	2.05	62.8	3.48	70.4	2.35	67.7	1.82	57.4	1.27	45.2	.78	27.0	.62	15.0	.44	40.5	15.96
MAX	6.0	.45	10.0	.40	21.3	.73	40.2	1.25	53.2	2.19	62.1	3.98	69.7	2.66	67.1	2.16	56.1	1.30	43.9	.72	25.4	.59	13.2	.36	39.0	16.79
WASHBURN	10.2	.40	14.4	.40	25.6	.69	43.6	1.32	55.9	1.75	64.3	3.69	72.0	2.35	69.2	1.78	59.2	1.32	47.1	.82	28.9	.49	16.9	.32	42.3	15.33
WATFORD CITY	11.2	.61	15.5	.55	25.6	.85	43.0	1.33	55.1	1.85	62.9	3.58	71.4	2.36	68.9	1.55	58.0	1.25	46.5	.96	29.0	.56	17.7	.49	42.1	15.94
DIVISION	8.8	.47	12.7	.44	23.6	.75	41.8	1.30	54.1	2.01	62.3	3.79	70.1	2.38	67.7	1.93	57.1	1.36	45.6	.83	27.2	.55	15.7	.36	40.6	16.17
CENTRAL DIVISION																										
CARRINGTON	6.2	.52	10.1	.48	22.2	.74	40.1	1.40	53.2	2.29	62.5	3.64	69.7	2.64	67.2	1.90	56.4	1.46	45.0	1.15	25.8	.58	12.8	.41	39.3	17.21
COURTENAY		.62		.48		.88		1.25		2.15		3.29		2.51		2.28		1.57		1.11		.60		.59		17.33
FESSENDEN	6.3	.57	10.8	.51	22.9	.78	41.2	1.34	54.2	2.14	62.8	3.36	70.2	2.82	67.9	1.97	57.5	1.47	45.3	1.04	26.2	.68	13.2	.53	39.9	17.21
HARVEY		.36		.39		.71		1.23		1.92		3.48		2.69		1.81		1.43		1.00		.50		.30		15.82
JAMESTOWN STATE HOSP	8.2	.46	12.3	.57	24.9	.88	42.3	1.64	55.2	2.19	64.6	3.32	71.8	2.73	69.2	2.13	58.5	1.32	46.5	1.17	28.0	.67	15.0	.48	41.4	17.56
MC CLUSKY	7.5	.63	11.6	.51	23.4	.84	41.8	1.49	54.3	2.32	62.8	4.07	70.7	2.14	68.3	1.91	58.0	1.48	45.8	1.02	27.0	.65	14.6	.45	40.5	17.57
PETTIBONE	6.5	.47	10.6	.50	22.7	.79	41.2	1.39	54.1	2.11	62.8	3.01	70.5	2.72	68.2	2.15	57.9	1.64	45.2	1.15	26.4	.50	13.4	.38	40.0	16.81
STEELE	7.6	.44	11.6	.37	23.8	.60	41.8	1.32	54.3	2.26	63.0	3.88	70.5	2.51	68.3	2.04	58.1	1.71	45.9	1.20	27.4	.45	14.7	.32	40.6	17.10
DIVISION	6.8	.51	11.0	.48	23.0	.78	41.3	1.40	53.8	2.27	62.6	3.57	70.0	2.66	67.9	2.14	57.4	1.57	45.6	1.14	26.7	.59	13.6	.44	40.0	17.55

*MEAN TEMPERATURE AND PRECIPITATION

STATION	JAN Temp	JAN Precip	FEB Temp	FEB Precip	MAR Temp	MAR Precip	APR Temp	APR Precip	MAY Temp	MAY Precip	JUN Temp	JUN Precip	JUL Temp	JUL Precip	AUG Temp	AUG Precip	SEP Temp	SEP Precip	OCT Temp	OCT Precip	NOV Temp	NOV Precip	DEC Temp	DEC Precip	ANN Temp	ANN Precip
EAST CENTRAL DIVISION																										
COOPERSTOWN		.47		.63		.78		1.17		2.36		3.38		2.51		2.57		1.75		1.14		.65		.57		17.98
FARGO WB AP	7.1	.60	10.8	.66	25.3	.89	42.1	1.88	55.0	2.17	64.6	3.04	71.3	2.31	69.1	2.73	58.9	1.72	46.1	1.26	27.6	.87	12.9	.60	40.9	18.73
HILLSBORO	6.3	.53	10.8	.54	24.1	.83	42.2	1.70	55.4	2.42	64.7	3.77	71.6	3.00	69.0	2.82	58.8	1.88	46.3	1.13	27.3	.76	13.0	.57	40.8	19.95
MAYVILLE	6.8	.53	11.3	.53	24.5	.68	42.4	1.27	55.7	2.27	64.7	3.52	71.8	2.56	69.8	2.69	59.6	1.74	47.1	1.06	27.7	.76	13.5	.49	41.2	17.98
SHARON	4.8	.52	9.5	.47	22.4	.88	40.5	1.44	53.8	2.64	62.7	3.36	70.1	2.66	67.7	2.74	57.1	1.83	44.9	1.25	25.8	.74	11.7	.52	39.3	19.05
VALLEY CITY	8.5	.46	13.1	.49	26.0	.75	43.4	1.42	56.1	2.35	65.1	3.30	71.7	2.81	69.4	2.33	59.1	1.50	47.2	1.18	28.6	.74	15.4	.51	42.0	17.84
DIVISION	5.9	.50	11.0	.52	24.1	.78	41.9	1.42	55.2	2.44	64.2	3.45	71.2	2.67	68.9	2.54	58.5	1.67	46.2	1.13	27.2	.70	13.0	.53	40.6	18.35
SOUTHWEST DIVISION																										
BOWMAN COURT HOUSE	14.8		17.7		26.5		42.7		54.2		62.2		71.1		68.8		60.4		46.5		30.0		21.2		43.0	
DICKINSON EXP STATION	10.7	.48	13.9	.45	23.9	.76	41.3	1.36	53.1	2.01	61.4	3.98	69.8	2.00	67.1	1.76	56.4	1.13	44.8	.88	27.9	.55	15.5	.32	40.7	15.68
HETTINGER	15.2	.37	18.3	.29	27.0	.72	43.5	1.25	55.0	2.11	63.5	3.47	72.0	1.98	72.5	1.82	59.1	1.13	47.1	.83	30.8	.36	21.3	.23	43.8	14.56
MARMARTH	14.9	.56	18.7	.45	28.0	.80	44.4	1.12	55.5	1.92	63.6	3.59	72.1	1.92	69.1	1.65	58.3	1.00	47.1	.92	30.1	.45	20.6	.38	43.5	14.76
MOTT	13.6	.43	17.1	.41	26.6	.79	43.5	1.07	55.0	1.90	63.4	3.65	71.8	1.97	69.1	1.55	58.9	1.27	46.8	.68	29.9	.44	19.8	.30	43.0	14.46
NEW ENGLAND	14.3	.63	17.4	.59	27.0	1.11	43.0	1.51	54.8	1.88	63.4	4.00	72.1	2.39	69.4	1.58	58.7	1.15	46.2	.85	30.2	.52	19.9	.38	41.0	16.59
RICHARDTON ABBEY	12.9	.56	16.1	.55	25.5	1.05	42.7	1.46	54.4	2.02	63.0	3.99	70.8	2.59	68.8	1.79	58.7	1.14	47.2	.83	29.6	.82	19.2	.36	42.4	17.16
TROTTERS 6 SE		.54		.44		.75		1.05		1.84		3.01				2.06		1.68		1.16		.74		.54		14.19
DIVISION	13.7	.50	17.2	.46	26.0	.85	42.8	1.33	54.3	1.95	62.5	3.77	71.0	2.11	68.7	1.76	58.1	1.18	46.5	.86	29.7	.49	19.8	.32	42.5	15.58
SOUTH CENTRAL DIV																										
BISMARCK WB AP	9.2	.36	12.7	.43	26.7	.76	43.1	1.39	54.8	1.94	64.3	3.33	72.1	2.33	69.3	1.50	58.5	1.43	45.7	1.00	28.4	.53	15.5	.40	41.7	15.40
CARSON NO 2	11.8	.53	15.1	.55	25.4	1.11	42.8	1.41	54.8	2.36	63.5	3.87	71.7	2.09	69.4	1.51	59.0	1.34	46.8	1.02	29.3	.55	18.6	.35	42.4	16.69
FORT YATES	12.1	.39	16.5	.45	27.9	.80	44.9	1.37	56.7	2.10	65.4	3.58	73.1	2.01	70.5	1.74	60.4	1.18	48.1	1.12	30.9	.36	19.1	.25	43.8	15.35
LINTON	10.1	.40	14.7	.39	26.7	.72	44.3	1.36	56.6	2.05	65.0	4.09	72.8	2.47	70.9	1.85	60.3	1.43	47.4	1.10	29.3	.40	16.8	.31	42.9	16.57
MANDAN EXP STATION	9.4	.50	13.0	.48	24.6	.90	42.9	1.42	55.2	1.95	63.9	3.78	71.8	2.33	69.2	1.76	58.9	1.34	46.6	.92	28.3	.37	16.2	.37	41.7	16.34
NEW SALEM 1 S	11.0	.45	14.5	.48	25.0	.81	42.6	1.29	54.6	2.02	63.1	3.54	71.1	2.21	68.9	1.80	58.9	1.30	46.7	.76	28.8	.47	17.6	.30	41.9	15.43
DIVISION	10.9	.44	15.2	.47	25.6	.83	43.5	1.32	55.5	2.06	64.0	3.89	71.9	2.26	69.7	1.85	59.4	1.36	47.3	1.01	29.4	.50	17.7	.32	42.5	16.31
SOUTHEAST DIVISION																										
ASHLEY	9.0	.51	12.5	.44	24.7	.77	42.0	1.26	54.4	1.98	63.3	3.83	70.6	2.65	68.4	2.19	58.2	1.23	45.9	.94	28.1	.42	15.6	.28	41.1	16.50
EDGELEY EXP FARM	9.0	.44	13.1	.45	25.4	.76	42.7	1.53	54.9	2.38	63.9	3.76	71.0	2.57	69.0	1.94	58.8	1.43	46.7	.95	28.0	.46	15.1	.34	41.5	17.01
ELLENDALE	9.8	.54	13.9	.62	26.6	1.00	43.5	1.94	54.7	2.22	63.6	3.85	71.9	2.84	69.1	2.49	58.8	1.36	47.5	1.08	29.3	.63	16.3	.51	42.4	19.08
FORMAN		.49		.57		.92		1.91		2.55		3.92		3.13		2.19		1.42		1.17		.64		.52		19.43
FULLERTON	10.1	.87	14.1	.95	26.7	1.26	43.7	2.15	56.1	2.51	65.3	3.97	72.2	2.47	70.0	2.21	59.6	1.36	47.4	1.16	29.4	.86	16.5	.78	42.6	20.55
HANKINSON R R STATION	9.2	.56	13.4	.58	25.8	1.10	43.0	1.89	56.5	2.41	65.9	3.91	72.5	2.46	69.7	2.43	59.3	1.69	46.9	1.04	28.7	.69	15.6	.59	42.2	19.35
LISBON	8.7	.51	13.1	.58	26.2	.90	42.4	1.92	55.6	2.35	65.0	3.86	71.7	2.53	69.0	2.87	58.3	1.47	47.3	1.04	29.0	.67	15.2	.51	41.8	19.36
MC LEOD 3 E	8.5	.53	13.0	.47	26.3	.99	43.4	1.67	56.6	2.51	65.2	3.44	72.0	2.85	69.8	2.65	59.9	1.54	47.7	1.12	29.2	.70	14.9	.47	41.8	18.94
NAPOLEON	8.4	.54	12.2	.45	24.0	.95	42.2	1.37	54.7	2.31	63.4	3.72	70.8	2.89	69.1	1.95	58.8	1.40	47.1	1.20	27.8	.63	15.3	.37	41.1	17.78
OAKES	8.4	.49	12.6	.59	25.9	.93	42.7	1.73	55.6	2.32	65.1	3.68	72.0	2.49	69.7	2.04	59.0	1.25	46.8	1.09	28.6	.63	15.0	.50	41.8	17.74
WAHPETON STATE SCHOOL	9.6	.60	13.8	.57	26.9	.86	42.9	1.99	57.5	2.64	66.8	3.57	73.1	3.04	71.2	3.03	61.3	1.61	49.0	1.15	30.1	.71	16.0	.61	43.3	20.38
WISHEK	7.7	.45	11.6	.38	23.5	.71	41.4	1.38	53.8	2.16	63.0	3.76	70.1	2.76	67.6	1.78	59.5	1.30	44.7	.96	28.7	.41	15.2	.28	40.4	16.33
DIVISION	8.8	.53	13.1	.55	25.5	.92	42.9	1.70	55.6	2.37	64.6	3.79	71.6	2.73	69.4	2.29	59.1	1.42	47.0	1.08	28.7	.65	15.6	.47	41.8	18.46

* Averages for period 1931-1955, except for stations marked WB which are "normals" based on period 1921-1950. Divisional means may not be the arithmetical average of individual stations published, since additional data from shorter period stations are used to obtain better areal representation.

CONFIDENCE LIMITS

In the absence of trend or record changes, the chances are 9 out of 10 that the true mean will lie in the interval formed by adding and subtracting the values in the following table from the means for any station in the State.

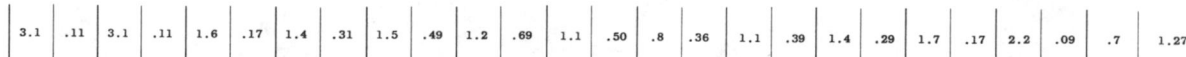

3.1	.11	3.1	.11	1.6	.17	1.4	.31	1.5	.49	1.2	.69	1.1	.50	.8	.36	1.1	.39	1.4	.29	1.7	.17	2.2	.09	.7	1.27

COMPARATIVE DATA

Data in the following table are the mean temperature and average precipitation for Dickinson Experiment Station, North Dakota for the period 1906-1930 and are included in this publication for comparative purposes:

10.2	.46	14.1	.40	26.0	.66	41.8	1.21	51.4	2.50	61.9	3.49	68.1	2.21	65.7	1.82	55.9	1.32	43.5	1.02	29.2	.48	14.8	.50	40.2	16.07

NORMALS, MEANS AND EXTREMES
(Table Revised 1973. Base Period for Climatological Normals: 1931-1960)

Station: BISMARCK, NORTH DAKOTA — MUNICIPAL AIRPORT
Standard time used: CENTRAL
Latitude: 45° 46' N Longitude: 100° 45' W Elevation (ground): 1647 feet

Month	Normal Daily max	Normal Daily min	Normal Monthly	Record highest	Year	Record lowest	Year	Normal heating degree days (Base 65°)	Precip. Normal total	Precip. Max monthly	Year	Min monthly	Year	Max in 24 hrs	Year	Snow Mean total	Max monthly	Year	Max in 24 hrs	Year
J	19.6	.1	9.9	54	1961	-41	1970+	1708	0.44	1.29	1969	0.02	1940	0.67	1952	7.4	16.9	1950	7.9	1950
F	23.3	3.7	13.5	61	1963	-37	1971	1442	0.43	1.17	1969	0.12	1968+	0.73	1958	6.2	17.4	1969	8.7	1969
M	35.1	17.3	26.2	80	1964+	-28	1962	1203	0.78	2.84	1950	0.11	1961	1.30	1950	7.8	29.7	1950	15.5	1950
A	54.9	32.1	43.6	91	1962	15	1962	645	1.22	4.05	1970	0.31	1952	1.97	1964	4.0	18.2	1970	8.7	1970
M	68.2	42.6	55.9	95	1969	19	1967	329	1.97	5.18	1965	0.85	1952	1.95	1972	1.2	10.3	1969	11.0	1969
J	76.5	52.4	64.5	100	1961	30	1969	117	3.40	8.22	1947	0.85	1967	3.25	1969+	T	T	1969	T	1969+
J	85.7	57.7	71.7	108	1960	35	1971	34	2.19	5.24	1969	0.18	1968	2.33	1969	0.0	0.0		0.0	
A	83.7	54.8	69.3	106	1960	33	1964	28	1.73	5.05	1944	0.03	1971	2.68	1965	0.0	0.0		0.0	
S	72.6	44.1	58.7	100	1971	11	1965	222	1.19	3.91	1941	0.05	1948	1.99	1941	0.3	4.1	1942	4.1	1942
O	59.4	34.0	46.7	95	1963	13	1960	577	0.85	3.74	1971	0.05	1968	1.62	1971	1.8	7.6	1958	7.5	1958
N	38.6	18.9	28.9	74	1965	-29	1964	1083	0.59	2.55	1944	T	1963	0.99	1944	4.8	16.1	1959	7.5	1959
D	26.9	8.6	17.8	62	1969+	-43	1967	1463	0.36	0.95	1967	T	1944	0.59	1960	6.1	13.7	1969	7.6	1969
YR	53.7	30.7	42.2	108	JUL. 1960	-43	DEC. 1967	8851	15.15	8.22	JUN. 1947	T	NOV. 1963+	3.25	JUN. 1969+	39.1	29.7	MAR. 1950	15.5	MAR. 1950

Means and extremes above are from existing and comparable exposures. Annual extremes have been exceeded at other sites in the locality as follows: Highest temperature 114 in July 1936; lowest temperature -45 in February 1936 and earlier; maximum monthly precipitation 9.90 in June 1914; maximum precipitation in 24 hours 3.76 in June 1914; maximum snowfall 31.0 in November 1896.

NORMALS, MEANS AND EXTREMES
(Table Revised 1975. Base Period for Climatological Normals: 1941-1970)

Elev. 1660 feet m.s.l. Average station pressure 2 mb

Month	Normal Daily max	Normal Daily min	Normal Monthly	Record highest	Year	Record lowest	Year	Heating degree days Base 65°F	Cooling degree days Base 65°F	Precip. Normal	Max monthly	Year	Min monthly	Year	Max in 24 hrs	Year	Snow Max monthly	Year	Max in 24 hrs	Year
J	19.7	-2.8	8.2	54	1961	-42	1971	1761	0	0.51	1.29	1969	0.02	1969	0.67	1952	16.9	1950	7.9	1950
F	24.5	3.5	13.5	61	1963	-37	1962	1442	0	0.44	1.17	1969	0.08	1969	0.73	1958	17.4	1969	8.7	1969
M	34.7	14.7	25.1	80	1962	-28	1962	1237	0	0.73	2.84	1950	0.1	1970	1.30	1950	29.7	1950	15.5	1950
A	54.8	31.1	43.0	91	1962	11	1973	660	0	1.44	4.05	1970	0.31	1952	1.97	1964	18.2	1970	8.7	1970
M	68.8	43.0	54.4	95	1969	15	1967	339	11	2.17	5.18	1965	0.50	1947	1.95	1972	10.3	1969	11.0	1969
J	75.8	51.8	63.8	100	1961	30	1969	122	86	3.58	8.29	1947		1947	3.25	1947	T	1969	T	1969
J	84.3	57.3	70.8	108	1973	35	1971	18	198	2.20	5.24	1969	0.18	1968	2.33	1969	0.0		0.0	
A	84.3	54.9	69.2	107	1973	33	1964	35	165	1.32	5.05	1944	0.03	1941	2.68	1965	0.0		0.0	
S	71.3	43.7	57.5	100	1971	11	1974	252	27	1.30	3.91	1941	0.02	1948	1.99	1941	4.1	1942	4.1	1942
O	60.3	33.2	46.8	95	1963	11	1971	577	0	0.86	3.74	1971	0.05	1971	1.62	1971	7.6	1958	7.5	1958
N	39.4	18.3	28.9	74	1965	-29	1964	1083	0	0.56	2.55	1944	0.00	1944	0.99	1944	16.1	1959	7.5	1959
D	26.0	5.2	15.6	62	1969+	-43	1967	1531	0	0.45	0.95	1967	0.95	1967	0.59	1960	13.7	1969	7.6	1969
YR	53.5	29.3	41.4	108	JUL. 1973	-43	DEC. 1967	9044	487	16.16	8.29	JUN. 1947	T	NOV. 1963	3.25	JUN. 1947	29.7	MAR. 1950	15.5	MAR. 1950

Means and extremes above are from existing and comparable exposures. Annual extremes have been exceeded at other sites in the locality as follows: Highest temperature 114 in July 1936; lowest temperature -45 in February 1936 and earlier; maximum monthly precipitation 9.90 in June 1914; maximum precipitation in 24 hours 3.76 in June 1914; maximum snowfall 31.0 in November 1896.

REFERENCE NOTES APPLYING TO TABLES APPEAR ON THE PAGE FOLLOWING LAST TABLE.
(Caution: Letters and symbols may have different meanings in 1941-1970 tables than in earlier tables. See notes.)

DEVILS LAKE, NORTH DAKOTA
715 FIFTH AVENUE
LATITUDE 48° 07' N
LONGITUDE 98° 52' W
ELEVATION (ground) 1471 Feet

NORMALS, MEANS AND EXTREMES

(Table Revised 1959. Base Period for Climatological Normals: 1921-1950)

Temperature and Precipitation

Month	Normal Daily max	Normal Daily min	Normal Monthly	Record highest	Year	Record lowest	Year	Normal degree days	Precip Normal total	Precip Max monthly	Year	Precip Min monthly	Year	Max in 24 hrs	Year
(a)	(b)	(b)	(b)	53		53		(b)	(b)	53		53		53	
J	14.2	-4.7	4.8	51	1908	-44	1916	1866	.40	1.67	1907	.04	1942	.70	1956
F	18.0	-0.7	8.7	56	1932	-46	1936	1576	.41	2.20	1908	.01	1928	.80	1908
M	31.8	13.3	22.6	85	1910	-32	1948	1314	.73	2.29	1956	.09	1912#	1.36	1956
A	50.3	29.6	40.0	93	1952	-4	1908	750	1.44	4.96	1924	.04	1952	2.43	1938
M	64.7	41.2	53.0	106	1934	6	1907	394	2.13	5.18	1917	.04	1917	2.86	1909
J	73.7	51.1	62.4	103	1933	29	1910	137	3.18	8.55	1954	.96	1933	4.08	1929
J	81.7	56.9	69.3	112	1936	37	1917	47	2.67	7.44	1912	.67	1930	4.82	1905
A	79.7	53.9	66.8	103	1925#	31	1915	61	2.17	6.55	1944	.27	1933	4.53	1921
S	68.3	44.2	56.3	103	1906#	15	1906#	276	2.07	8.34	1941	.02	1938	4.55	1941
O	54.6	33.2	43.9	89	1914	-1	1919	654	1.26	3.94	1944	.01	1944	2.14	1949
N	33.6	16.7	25.1	69	1939	-21	1905	1197	.67	2.59	1906	.08	1942	1.73	1928
D	19.8	2.5	11.2	64	1939	-37	1916	1668	.48	1.55	1906	.04	1954	.72	1906
Year	49.2	28.1	38.7	112 JULY 1936		-46 FEB 1936		9940	17.61	8.55 JUNE 1954		T MAY 1917		4.82 JULY 1905	

Snow/Sleet, Relative Humidity, and Wind

Month	Snow Mean total	Snow Max monthly	Year	Snow Max 24 hrs	Year	RH 6:00 A.M. CST	RH Noon CST	RH 6:00 P.M. CST	Wind Mean hourly speed	Prevailing direction	Fastest mile Speed	Fastest mile Direction	Fastest mile Year
	53	53		53		20	20	20	53		53		
J	6.1	19.4	1954	7.7	1956	76	72	75	9.9	NW	41	NW	1923
F	5.0	18.2	1908	7.8	1911	78	72	77	10.0	NW	54	N	1917
M	6.2	25.6	1956	13.4	1956	82	70	74	10.7	NW	54	N	1920
A	3.2	18.9	1950	8.6	1950	82	56	55	11.3	NW	47	NW	1914
M	.9	12.7	1935	7.4	1935	80	52	49	10.9	N	50	W	1920
J	T					85	57	56	9.6	SE			
J	.0	.0		.0		86	53	53	8.5	NW	56	NW	1936
A	.0	.0		.0		87	52	50	8.6	SE	47	NW	1940
S	T					86	53	55	9.3	NW	46	NW	1933
O	1.6	1.5	1934	1.5	1934	83	56	59	9.3	NW	47	N	1913#
N	5.6	19.9	1906	9.2	1945	83	72	76	10.0	NW	57	NW	1905
D	6.4	20.1	1949	9.5	1945	78	73	77	10.2	NW	42	NW	1936
Year	35.0	25.6 MAR 1956		13.4 MAR 1956		82	62	63	9.9	NW	57	N	NOV 1905

Sky, Sunshine, and Mean Number of Days

Month	Pct of possible sunshine	Mean sky cover sunrise to sunset	Clear	Partly cloudy	Cloudy	Precip .01 inch or more	Snow/Sleet 1.0 inch or more	Thunderstorms	Heavy fog	Max 90° and above	Max 32° and below	Min 32° and below	Min 0° and below
	53	53	53	53	53	53	53	53	53	53	53	53	53
J	53	6.1	9	8	14	8	2	0	1	0	28	31	21
F	59	6.0	8	8	12	6	2	0	1	0	23	28	15
M	59	6.1	9	7	14	8	2	1	1	0	17	29	7
A	60	5.8	9	9	12	8	1	1	*	0	3	19	*
M	59	5.6	9	10	12	10	*	4	*	*	0	3	0
J	61	5.6	9	11	10	10	0	7	*	1	0	*	0
J	71	4.5	13	11	7	10	0	8	*	4	0	0	0
A	67	4.6	13	10	8	9	0	7	1	4	0	*	0
S	59	5.3	13	10	9	9	*	3	1	1	0	3	0
O	56	5.6	10	9	12	7	*	1	1	0	*	15	0
N	44	6.6	8	8	15	7	2	*	1	0	13	28	4
D	45	6.3	8	7	16	8	2	0	1	0	24	30	13
Year	59	5.7	113	110	142	101	11	30	9	11	111	189	60

Station moved or discontinued. See other stations for this state.

REFERENCE NOTES APPLYING TO TABLES APPEAR ON THE PAGE FOLLOWING LAST TABLE.

NORMALS, MEANS AND EXTREMES
(Table Revised 1973. Base Period for Climatological Normals: 1931-1960)

Station: FARGO, NORTH DAKOTA   HECTOR AIRPORT   Standard time used: CENTRAL   Latitude: 46° 54' N   Longitude: 96° 48' W   Elevation (ground): 896 feet

Month	Temperature Normal Daily maximum	Normal Daily minimum	Normal Monthly	Extremes Record highest	Year	Record lowest	Year	Normal heating degree days (Base 65°)	Precipitation Normal total	Max monthly	Year	Min monthly	Year	Max in 24 hrs	Year	Snow, Ice pellets Mean total	Max monthly	Year	Max in 24 hrs	Year
J	17.3	-2.8	7.3	46	1964	-35	1965	1789	0.53	1.36	1950	0.09	1961	0.68	1949	6.7	16.5	1972	7.0	1972
F	20.5	.7	10.7	51	1963	-34	1963	1520	0.51	1.60	1948	0.03	1954	1.22	1946	5.4	15.8	1948	11.2	1951
M	33.4	15.1	24.3	78	1967	-23	1962	1262	0.75	2.21	1950	0.03	1958	1.16	1950	5.2	15.4	1966	8.6	1966
A	52.7	31.3	42.0	88	1962	-8	1963	690	1.72	4.24	1962	0.02	1962	1.91	1963	3.0	12.8	1970	8.0	1970
M	67.8	43.0	55.4	98	1964	20	1964	332	2.03	5.95	1962	0.47	1952	2.35	1944	0.1	1.0	1950	1.0	1950
J	76.1	53.3	64.8	98	1964	30	1969	99	3.04	5.64	1943	0.58	1972	2.77	1957	0.0	0.0		0.0	
J	83.8	59.0	71.4	100	1960	36	1967	28	2.91	8.42	1952	0.42	1950	3.93	1952	0.0	0.0		0.0	
A	81.9	57.0	69.5	100	1960	33	1964	37	2.95	8.52	1944	0.38	1969	4.72	1943	0.0	0.0		0.0	
S	71.0	46.5	58.8	93	1960	10	1965	219	1.48	6.13	1971	0.15	1952	3.97	1957	T	0.6	1942	0.6	1942
O	58.8	35.0	46.5	88	1963	-10	1965	574	0.84	4.42	1957	0.08	1952	2.06	1971	0.8	8.1	1951	8.1	1951
N	37.5	18.8	28.1	70	1964	-22	1964	1107	0.58	2.01	1960	0.04	1960	1.52	1960	4.3	19.0	1947	10.9	1947
D	23.9	4.8	14.4	57	1962	-32	1967	1569	0.58?	2.19	1951	0.04	1958	0.87	1958	6.8	20.3	1951	8.0	1951
YR	52.0	30.1	41.1	100 AUG 1965*		-35 JAN 1965		9226	18.45	8.52 APR 1949		0.02 APR 1949		4.72 AUG 1943		34.2	20.3 DEC 1951		11.2 FEB 1951	

Means and extremes above are from existing and comparable exposures. Annual extremes have been exceeded at other sites in the locality as follows: Highest temperature 114 in July 1936; lowest temperature -48 in January 1887; maximum monthly precipitation 9.58 in August 1900; minimum monthly precipitation T in November 1901; maximum precipitation in 24 hours 5.17 in July 1886; maximum monthly snowfall 30.4 in November 1896; maximum snowfall in 24 hours 19.2 in December 1927.

NORMALS, MEANS AND EXTREMES
(Table Revised 1975. Base Period for Climatological Normals: 1941-1970)

Month	Temperature Normal Daily maximum	Normal Daily minimum	Normal Monthly	Extremes Record highest	Year	Record lowest	Year	Normal Degree days Base 65°F Heating	Cooling	Precipitation (inches) Water equiv. Normal	Max monthly	Year	Min monthly	Year	Max in 24 hrs	Year	Snow, Ice pellets Max monthly	Year	Max in 24 hrs	Year
J	15.4	-3.6	5.9	46	1964	-35	1965	1832	0	0.50	1.36	1950	0.09	1961	0.68	1949	16.5	1972	7.0	1972
F	20.6	.8	10.7	51	1963	-34	1963	1520	0	0.44	1.60	1954	0.03	1954	1.22	1946	15.8	1948	11.2	1951
M	33.0	14.9	24.3	78	1963	-23	1962	1265	0	0.83	2.21	1958	0.03	1958	1.91	1949	15.4	1966	8.6	1966
A	52.6	31.9	42.2	88	1962	-8	1963	681	11	2.08	4.24	1962	0.02	1962	1.91	1963	12.8	1970	8.0	1970
M	66.8	42.6	56.6	98	1964	20	1954	334	88	2.58?	5.95	1962	0.47	1952	2.77	1943	1.0	1950	1.0	1950
J	75.9	53.4	64.7	98	1964	30	1969	97		3.20	5.64	1943	0.58	1972	2.77	1957	0.0		0.0	
J	82.8	58.6	70.7	100	1960	36	1967	13	190	3.16	8.42	1952	0.42	1950	3.93	1952	0.0		0.0	
A	81.0	54.8	67.9	100	1960	33	1964	13	161	2.84	8.52	1944	0.38	1969	4.72	1943	0.0		0.0	
S	58.2	46.5	57.0	93	1960	10	1965	234	20	1.84	6.13	1974	0.15	1952	3.97	1957	0.6	1942	0.6	1942
O	58.2	35.2	46.2	88	1963	-10	1965	588	0	1.00	4.42	1957	0.13	1967	2.06	1947	8.1	1951	8.1	1951
N	37.2	20.0	28.6	70	1964	-22	1964	1092	0	0.72	2.01	1960	0.04	1960	1.95	1947	19.0	1947	10.9	1947
D	21.9	4.1	13.0	57	1962	-32	1967	1612	0	0.62	2.19	1951	0.04	1958	0.87	1958	20.3	1951	8.0	1951
YR	51.4	30.1	40.8	100 AUG 1965		-35 JAN 1965		9271	473	19.62	8.52 APR 1944		0.02 APR 1944		4.72 AUG 1943		20.3 DEC 1951		11.2 FEB 1951	

Means and extremes above are from existing and comparable exposures. Annual extremes have been exceeded at other sites in the locality as follows: Highest temperature 114 in July 1936; lowest temperature -48 in January 1887; maximum monthly precipitation 9.58 in August 1900; minimum monthly precipitation T in November 1901; maximum precipitation in 24 hours 5.17 in July 1886; maximum monthly snowfall 30.4 in November 1896; maximum snowfall in 24 hours 19.2 in December 1927.

Average station pressure (mb.): 982.2   Elev. 899 feet m.s.l.

REFERENCE NOTES APPLYING TO TABLES APPEAR ON THE PAGE FOLLOWING LAST TABLE.
(Caution: Letters and symbols may have different meanings in 1941-1970 tables than in earlier tables. See notes.)

NORMALS, MEANS AND EXTREMES

(Table Revised 1973. Base Period for Climatological Normals: 1931-1960)

Station: WILLISTON, NORTH DAKOTA — SLOULIN FIELD INT'L AIRPORT
Standard time used: CENTRAL Latitude: 48°11'N Longitude: 103°38'W Elevation (ground): 1899 feet

Month	Norm Daily Max	Norm Daily Min	Norm Monthly	Rec High	Yr	Rec Low	Yr	Heat Deg Days (Base 65°)	Precip Norm Total	Precip Max Mon	Yr	Precip Min Mon	Yr	Precip Max 24h	Yr	Snow Mean Total	Snow Max Mon	Yr	Snow Max 24h	Yr
J	19.4	-2.9	8.3	50	1968	-40	1966	1758	0.55	1.42	1967	0.35	1967	0.49	1967	7.6	16.6	1966	5.1	1967
F	24.1	.7	12.4	55	1973	-41	1962	1473	0.48	1.48	1967	0.01	1970	0.43	1967	5.7	16.5	1972	4.7	1972
M	35.0	13.0	24.3	75	1966	-28	1962	1262	0.71	0.94	1964	0.01	1966	0.47	1964	4.6	22.2	1964	7.6	1964
A	55.9	29.7	42.3	90	1962	3	1967	681	0.94	1.43	1967	0.38	1967	2.04	1967	2.0	7.38	1970	4.6	1970
M	68.8	40.5	54.7	98	1964	20	1967	357	1.43	3.25	1965	0.18	1965	2.05	1965	0.3	1.4	1965	1.4	1965
J	76.1	50.1	63.1	99	1966	30	1969	141	3.31	5.92	1964	0.91	1964	2.20	1964	0.0	0.0			
J	85.9	56.9	71.3	107	1966	34	1967	31	1.87	6.20	1963	0.63	1968	5.03	1963	0.0	0.0			
A	84.2	53.8	69.0	103	1969+	36	1964	43	1.48	3.38	1968	0.07	1971	2.45	1972	0.0	0.0	1972+	3.0	1972+
S	71.2	43.1	57.2	97	1963	21	1965	261	1.12	3.06	1971	0.11	1963	2.24	1971	0.5	3.0	1962	5.1	1962
O	58.7	32.4	45.6	93	1963	3	1972	601	0.73	3.56	1969	T	1965	2.21	1971	1.4	6.4	1970	4.1	1970
N	37.6	17.6	27.6	69	1965	-23	1964	1122	0.58	0.81	1970	0.04	1969	0.31	1970	3.0	6.4	1970	6.4	1970
D	26.8	5.6	16.2	56	1969	-35	1968	1513	0.50	1.28	1962	0.27	1964	0.52	1972	6.7	12.8	1964	6.4	1964
YR	53.6	28.3	40.9	107 JUL 1966		-41 FEB 1962		9243	13.70	7.38 MAY 1965		T OCT 1965+		5.03 JUL 1963		36.0	22.2 APR 1970		7.6 APR 1970	

Month	RH 00	RH 06	RH 12	RH 18	Wind Mean Speed	Prev Dir %	Fastest Speed	Dir	Yr	Mean Sky Cover	Pct Poss Sun	Clear	Partly Cloudy	Cloudy	Precip .01+	Snow 1.0+	Tstorms	Heavy fog	Max 90+	Max 32−	Min 32−	Min 0−	Avg Solar
J	78	77	73	73	10.4	W	70	NW	1964	7.1	49	5	8	18	10	3	0	0	0	24	31	20
F	82	82	72	73	9.4	NE	66	NW	1965	6.6	58	8	7	13	8	2	0	1	0	31	28	13
M	82	83	65	73	10.2	NW	52	NW	1970	6.7	62	8	7	16	8	1	0	3	*	24	28	0
A	74	80	53	60	10.2	SE	56	SW	1962	6.7	57	7	11	12	8	1	1	3	0	3	12	0
M	69	78	48	54	11.3	SE	56	SE	1963	6.5	64	9	11	11	11	0	3	7	1	0	1	0
J	74	84	52	61	9.7	SE	61	NW	1965	6.1	64	6	12	12	11	0	7	2	*	0	0	0
J	73	83	40	40	9.2	SE	64	NW	1966	4.9	75	13	11	7	8	0	7	1	7	0	0	0
A	63	78	35	35	9.8	SW	47	NW	1967	4.6	77	14	9	8	8	0	6	0	1	0	0	0
S	69	81	43	50	10.4	SW	50	NW	1962	5.0	63	9	9	11	7	*	3	*	1	0	3	3
O	75	79	56	70	10.4	SW	57	SW	1971	5.9	57	8	8	15	6	1	1	1	0	8	15	9
N	82	82	67	74	9.6	SW	47	W	1970	6.2	40	4	7	17	6	2	0	2	0	18	29	3
D	81	79	74	74	9.5	SW	56	SW	1964	6.7	51	7	7	17	8	2	0	0	0	22	31	15
YR	75	81	55	58	10.1	SW	70 NW JAN 1964			6.3	61	89	108	168	94	12	25	10	21	87	191	56

Means and extremes above are from existing and comparable exposures. Annual extremes have been exceeded at other sites in the locality as follows: Highest temperature 110 in July 1936; lowest temperature -50 in February 1936; maximum monthly precipitation 8.84 in June 1901; maximum snowfall 26.6 in April 1896; maximum snowfall in 24 hours 10.6 in April 1947.

NORMALS, MEANS AND EXTREMES

(Table Revised 1975. Base Period for Climatological Normals: 1941-1970)

Station: WILLISTON, NORTH DAKOTA Elevation 1905 feet m.s.l.

Month	Norm Daily Max	Norm Daily Min	Norm Monthly	Rec High	Yr	Rec Low	Yr	Heat DD	Cool DD	Precip Norm	Precip Max Mon	Yr	Precip Min Mon	Yr	Precip Max 24h	Yr	Snow Max Mon	Yr	Snow Max 24h	Yr
J	19.3	-2.8	8.3	50	1968	-40	1966	1758	0	0.59	1.42	1967	0.03	1973	0.49	1967	16.6	1967	5.1	1967
F	25.7	-.6	12.4	55	1973	-41	1962	1422	0	0.53	1.48	1967	0.10	1970	0.43	1967	16.5	1972	4.7	1972
M	35.5	13.8	24.7	75	1966	-28	1962	1244	0	0.53	1.94	1966	0.01	1966	0.47	1964	22.2	1964	7.6	1964
A	54.9	29.9	42.4	90	1962	3	1967	678	7	1.24	3.31	1967	0.38	1967	2.04	1967	7.38	1970	4.6	1970
M	67.0	41.4	54.1	98	1964	20	1967	345	66	1.62	3.56	1969	0.81	1969	2.20	1965	1.4	1965	1.4	1965
J	75.0	50.4	62.7	99	1966	30	1969	135	180	3.25	5.92	1964	0.91	1964	2.20	1964	0.0			
J	84.0	56.1	70.1	107	1966	34	1967	22	144	2.04	6.20	1963	0.63	1968	5.03	1963	0.0			
A	82.4	53.8	68.2	103	1969	36	1964	35	25	1.56	3.38	1968	0.07	1972	2.45	1971	3.0	1972	3.0	1972
S	70.3	43.1	56.7	97	1966	17	1974	298	0	1.21	3.06	1971	0.11	1963	2.24	1971	3.0	1962	5.1	1962
O	58.3	31.7	45.7	93	1963	3	1964	601	0	0.84	3.56	1969	0.04	1969	2.21	1971	6.4	1973	4.1	1973
N	38.3	17.8	28.1	69	1965	-23	1964	1107	0	0.53	0.81	1970	0.04	1969	0.80	1970	6.4	1974	6.4	1974
D	25.8	5.0	15.4	56	1969	-35	1968	1538	0	0.49	1.28	1962	0.27	1965	0.52	1964	12.8	1964	6.4	1964
YR	53.1	28.6	40.6	107 JUL 1966		-41 FEB 1962		9161	422	14.33	7.38 MAY 1965		T OCT 1965		5.03 JUL 1963		22.2 APR 1970		7.6 APR 1970	

Month	RH 00	RH 06	RH 12	RH 18	Wind Mean Speed	Prev Dir	Fastest Speed m.p.h.	Dir	Yr	Mean Sky Cover	Pct Poss Sun	Clear	Partly Cloudy	Cloudy	Precip .01+	Snow 1.0+	Tstorms	Heavy fog	Max 90+	Max 32−	Min 32−	Min 0−	Avg Station Pressure mb
J	79	77	71	73	10.1	W	70	NW	1964	7.1	48	5	8	18	9	7	0	*	0	23	31	19	945.8
F	81	81	73	73	9.6	NE	66	NW	1965	6.9	57	5	7	16	8	6	0	1	0	31	28	12	947.5
M	83	83	66	65	11.0	NW	56	NW	1966	7.0	61	5	7	16	8	4	1	1	*	24	29	5	945.5
A	74	80	53	55	11.3	SE	56	SW	1962	6.9	55	6	8	16	8	2	1	1	0	4	20	0	945.5
M	71	82	49	57	10.9	SE	61	SE	1963	6.4	60	6	13	12	10	1	3	*	2	0	4	0	943.7
J	74	83	48	61	10.0	SE	61	W	1965	6.6	66	7	12	11	11	0	7	1	1	0	0	0	942.3
J	71	83	45	46	9.3	SE	64	NW	1966	4.7	76	12	12	7	8	0	7	1	7	0	0	0	946.5
A	64	79	44	44	9.8	SW	47	NW	1967	4.7	75	13	11	8	8	0	6	1	1	0	0	0	945.3
S	70	81	43	51	10.3	SW	50	NW	1962	4.9	64	9	8	13	6	*	3	*	1	0	3	0	946.8
O	74	81	51	62	10.2	SW	57	SW	1971	5.9	55	7	8	14	6	2	1	1	0	9	29	0	947.2
N	82	82	68	72	9.4	SW	47	W	1974	6.1	41	4	6	17	8	2	0	2	0	22	31	3	947.1
D	80	82	75	75	10.0	SW	56	W	1964	6.8	49	7	7	17	8	2	0	1	0	22	31	13	945.9
YR	75	81	55	58	10.1	SW	70 NW JAN 1964			6.3	61	91	105	169	93	12	25	10	21	85	191	53	945.7

Means and extremes above are from existing and comparable exposures. Annual extremes have been exceeded at other sites in the locality as follows: Highest temperature 110 in July 1936; lowest temperature -50 in February 1936; maximum monthly precipitation 8.84 in June 1901; maximum snowfall 26.6 in April 1896; maximum snowfall in 24 hours 10.6 in April 1947.

REFERENCE NOTES APPLYING TO TABLES APPEAR ON THE PAGE FOLLOWING LAST TABLE.
(Caution: Letters and symbols may have different meanings in 1941-1970 tables than in earlier tables. See notes.)

Reference notes applying to Normals, Means, and Extremes tables for 1921–1950 base period.

(a) Length of record, years.
(b) Normal values are based on the period 1921-1950, and are means adjusted to represent observations taken at the present standard location.
* Less than one-half.

- No record.
† Airport data.
‡ City Office data.
# Also on earlier dates, months, or years.
T Trace, an amount too small to measure.

Sky cover is expressed in a range of 0 for no clouds or obscuring phenomena to 10 for complete sky cover. The number of clear days is based on average cloudiness 0-3 tenths; partly cloudy days on 4-7 tenths; and cloudy days on 8-10 tenths. Monthly degree day totals are the sum of the negative departures of average daily temperatures from 65°F. Sleet was included in snowfall totals beginning with July 1948. Heavy fog also includes data referred to at various times in the past as "Dense" or "Thick". The upper limit for heavy fog is 1/4 mile. Data in these tables are based on records through 1957.

Reference notes applying to Normals, Means, and Extremes tables for 1931–1960 base period.

(a) Length of record, years, based on January data. Other months may be for more or fewer years if there have been breaks in the record.
(b) Climatological standard normals (1931-1960).
* Less than one half.
+ Also on earlier dates, months, or years.
T Trace, an amount too small to measure.
  Below zero temperatures are preceded by a minus sign. The prevailing direction for wind in the Normals, Means, and Extremes table is from records through 1964.
‡ ≥70° at Alaskan stations.

Unless otherwise indicated, dimensional units used in this bulletin are: temperature in degrees F.; precipitation, including snowfall, in inches; wind movement in miles per hour; and relative humidity in percent. Heating degree day totals are the sums of negative departures of average daily temperatures from 65° F. Cooling degree day totals are the sums of positive departures of average daily temperatures from 65° F. Sleet was included in snowfall totals beginning with July 1948. The term "Ice pellets" includes solid grains of ice (sleet) and particles consisting of snow pellets encased in a thin layer of ice. Heavy fog reduces visibility to 1/4 mile or less.

Sky cover is expressed in a range of 0 for no clouds or obscuring phenomena to 10 for complete sky cover. The number of clear days is based on average cloudiness 0-3, partly cloudy days 4-7, and cloudy days 8-10 tenths.

Solar radiation data are the averages of direct and diffuse radiation on a horizontal surface. The langley denotes one gram calorie per square centimeter.

& Figures instead of letters in a direction column indicate direction in tens of degrees from true North; i.e., 09 - East, 18 - South, 27 - West, 36 - North, and 00 - Calm. Resultant wind is the vector sum of wind directions and speeds divided by the number of observations. If figures appear in the direction column under "Fastest mile" the corresponding speeds are fastest observed 1-minute values.

% Based on available record. The station did not operate 24 hours daily prior to March 1967.
# To 8 compass points only.
** The National Weather Service considers the accuracy of solar radiation data questionable; therefore, publication is suspended pending determination of corrected values.

Reference notes applying to Normals, Means, and Extremes tables for 1941–1970 base period.

(a) Length of record, years, through the current year unless otherwise noted, based on January data.
(b) 70° and above at Alaskan stations.
* Less than one half.
T Trace.

NORMALS - Based on record for the 1941-1970 period.
DATE OF AN EXTREME - The most recent in cases of multiple occurrence.
PREVAILING WIND DIRECTION - Record through 1963.
WIND DIRECTION - Numerals indicate tens of degrees clockwise from true north. 00 indicates calm.
FASTEST MILE WIND - Speed is fastest observed 1-minute value when the direction is in tens of degrees.

% Based on available record. The station did not operate 24 hours daily prior to March 1967.

Mean Maximum Temperature (°F.), January

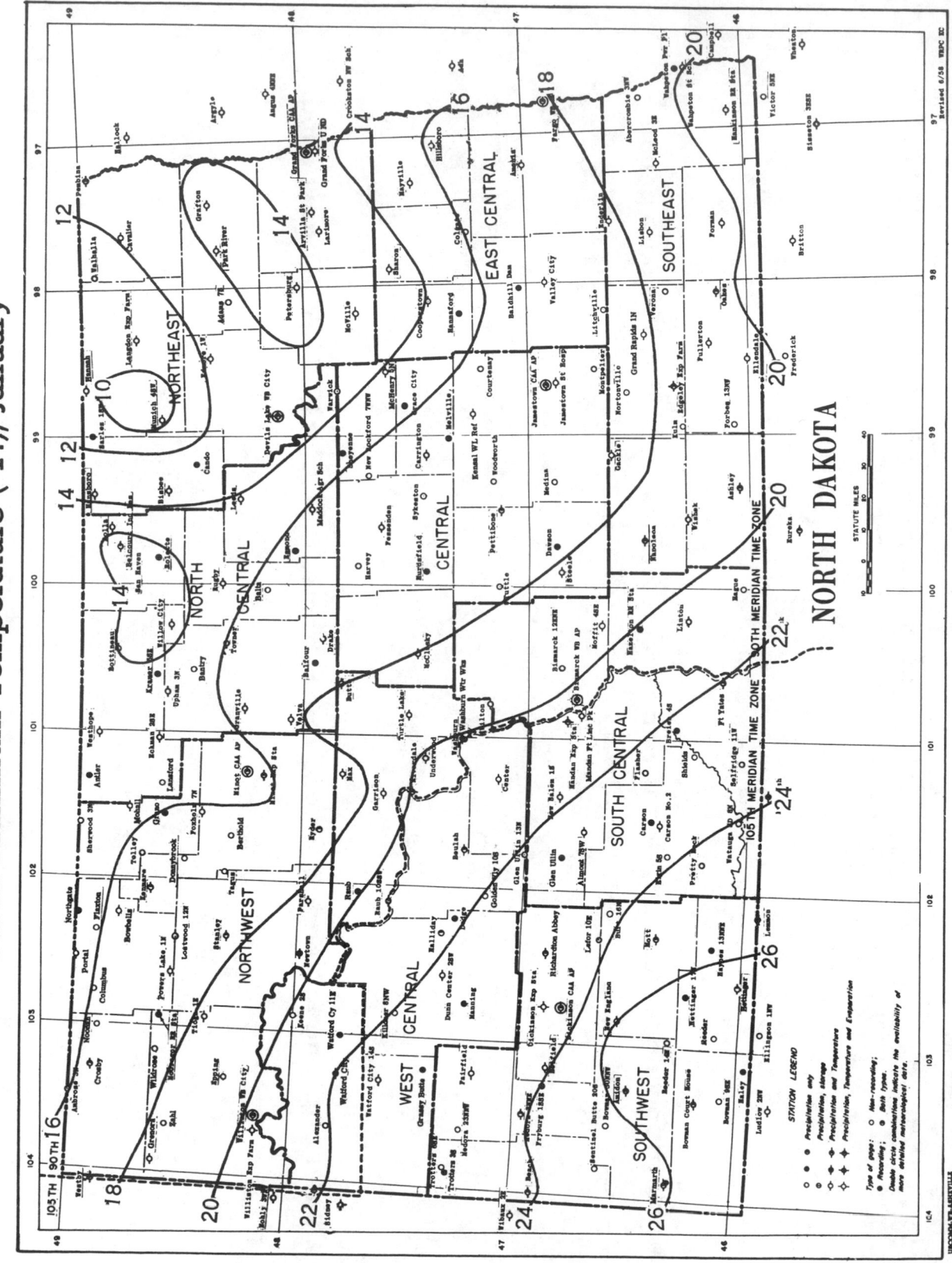

Based on period 1931-52

Isolines are drawn through points of approximately equal value. Caution should be used in interpolating on these maps,

760

Mean Minimum Temperature (°F.), January

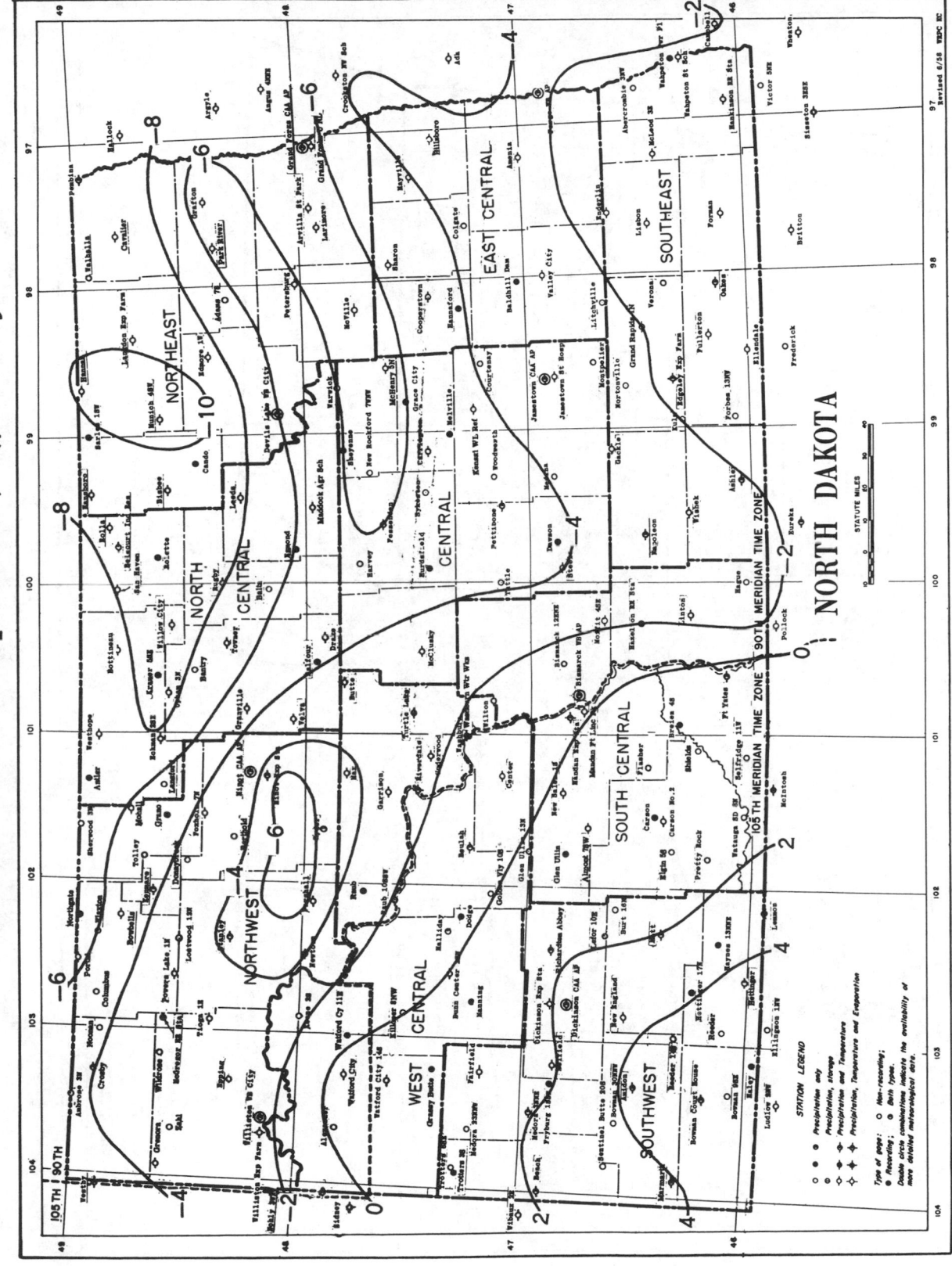

NORTH DAKOTA

Based on period 1931-52

Isolines are drawn through points of approximately equal value. Caution should be used in interpolating on these maps.

STATION LEGEND

Precipitation only
Precipitation, storage
Precipitation and Temperature
Precipitation, Temperature and Evaporation

Type of gage:
○ ● ◑ Non-recording;
◇ ◆ Recording;
◇ ◆ Both types.

Double circle combinations indicate the availability of more detailed meteorological data.

Mean Maximum Temperature (°F.), July

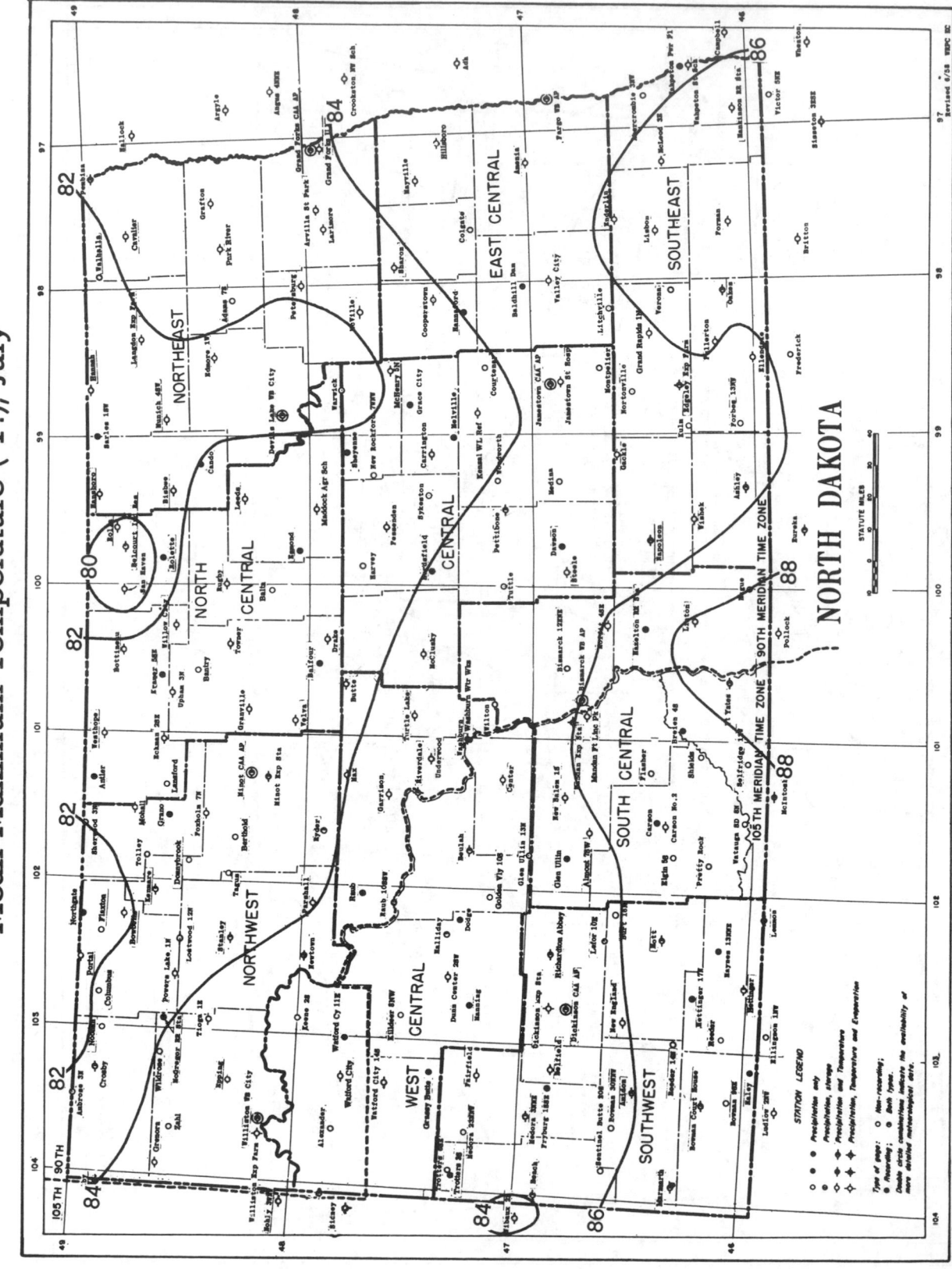

NORTH DAKOTA

Based on period 1931-52

Isolines are drawn through points of approximately equal value. Caution should be used in interpolating on these maps.

STATUTE MILES

Mean Minimum Temperature (°F.), July

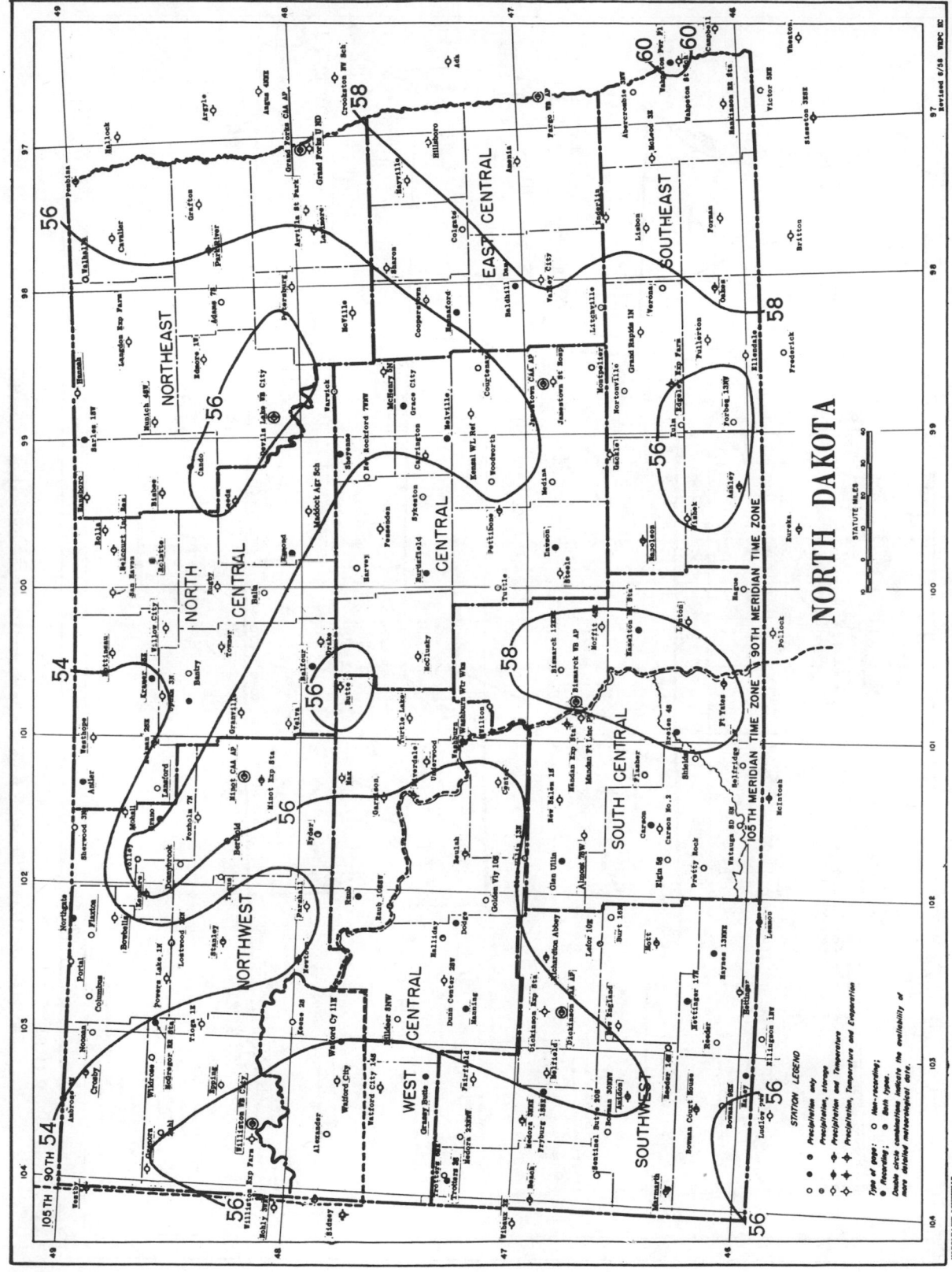

Isolines are drawn through points of approximately equal value. Caution should be used in interpolating on these maps.

Based on period 1931-52

Mean Annual Precipitation, Inches

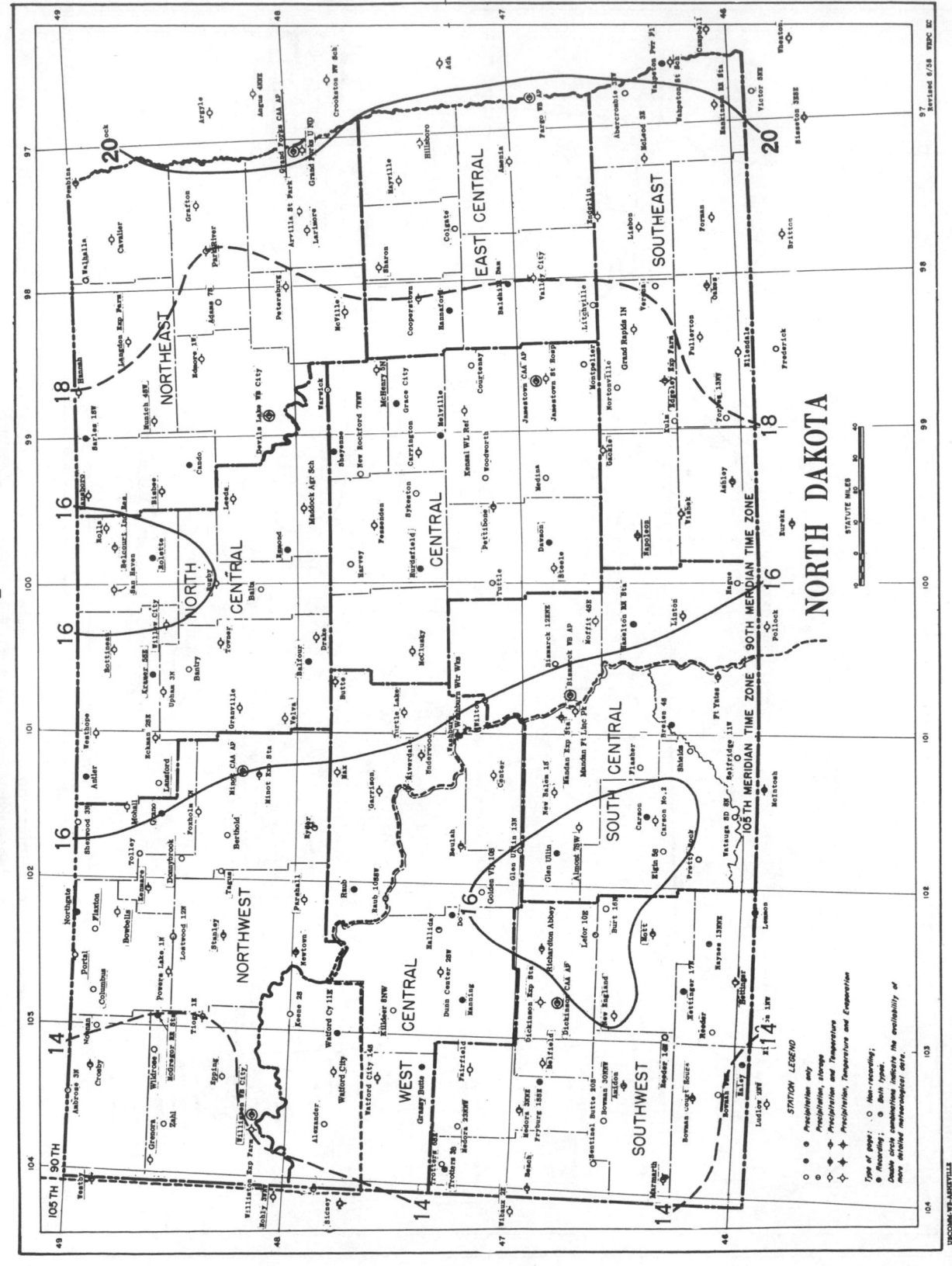

NORTH DAKOTA

STATION LEGEND

Type of gage: ○ Non-recording;

                 ● Recording;   ◐ Both types.

Double circle combinations indicate the availability of
more detailed meteorological data.

- ● Precipitation only
- ◐ Precipitation, storage
- ◆ Precipitation and Temperature
- ✦ Precipitation, Temperature and Evaporation

STATUTE MILES

UCOMM.WB-ASHEVILLE

Based on period 1931-55

Isolines are drawn through points of approximately equal value. Caution should be used
in interpolating on these maps.

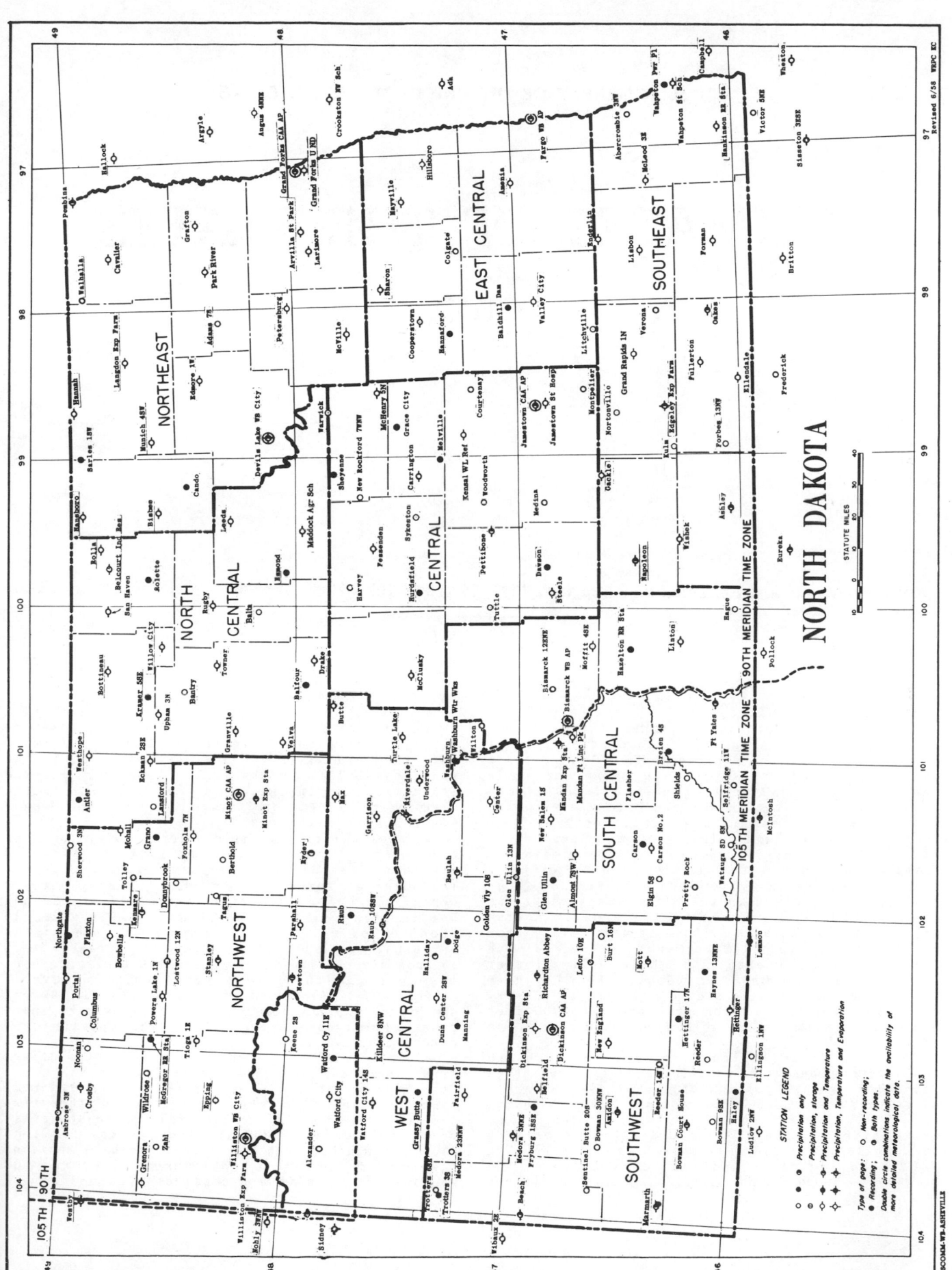

NORTH DAKOTA

STATUTE MILES

Revised 6/58   WBPC KC

CLIMATES OF THE STATES

OHIO

(Normals, Means and Extremes tables revised 1959, 1973 and 1975. Basic report revised December 1959.)

Climate of Ohio

L. T. Pierce, Weather Bureau State Climatologist

The climate of Ohio is remarkably varied. Less than one-half of its area is occupied by typical plains, while most of eastern and much of southern Ohio is decidedly hilly. Topography ranges in elevation from 430 feet above sea level at the junction of the Great Miami and Ohio Rivers up to 1,550 feet on a summit near Bellefontaine. In addition to this high point there are innumerable other hills which rise above 1,400 feet (mean sea level). These are located mainly along the dividing line between the Ohio River and Lake Erie drainage basins. Large areas in the State have elevations above 1,000 feet. An extensive area in northwestern Ohio is occupied by a flat lake plain--once the bottom of glacial Lake Maumee which was much larger than the present Lake Erie. The greater part of eastern Ohio is within the Allegheny Plateau, an unglaciated area consisting of picturesque hills many of which rise above 1,300 feet and between which there are many winding rivers and streams.

The Ohio River, which forms the southern and southeastern boundaries of Ohio, and its tributaries drain the greater portion of the State. A number of streams drain northward into Lake Erie. Although this area comprises nearly a third of the State, the divide between the two drainages is only 20 to 40 miles from the lake shore for a distance of more than 100 miles until it dips south of the arrowhead-shaped Maumee Basin. The largest streams in this region are the Maumee, Sandusky, and Cuyahoga Rivers. Principal tributaries flowing southward into the Ohio River include the Muskingum in the east, the Scioto in the central section, and the Great Miami in the west. A small portion in the west-central region drains westward into the Wabash River basin of Indiana.

The availability of ample water supply, together with Ohio's favorable geographical location and its large mineral resources, have favored the development of large industries. In fact, Ohio is one of the major industrial states of the Union. Due in large part to the accessibility of ample supplies of coal and water, this is one of the principal steel producing states. It ranks first in such diversified lines as machine tools, rubber, business machines, clay products, and several others. While Cleveland, Cincinnati, Toledo, Akron, and Dayton are the largest industrial centers, many other cities have become sites for important and diversified manufacturing plants.

Ohio's agriculture is characterized by its diversity. Much of the western half lies in the great Corn Belt which includes the bed of the Old Lake Maumee in the northwest. Central and northeastern sections have rolling terrain and are consequently devoted largely to dairing and general farming. The hilly southeastern and extreme southern section, however, is principally a general farming and woodland area, although considerable tobacco and truck crops are grown along the Ohio River. The largest sources of farm income are dairy, hogs, and poultry, followed by wheat, soybeans, and beef cattle. Tobacco is grown extensively in southern counties, sugar beets in the northwest, and grapes principally along the shore of Lake Erie. Truck crops are grown mostly in the west and the north, their distribution being controlled primarily by climatic and topographic considerations.

Located west of the Appalachian Mountains, Ohio has a climate essentially continental in nature, characterized by moderate extremes of heat and cold, and wetness and dryness. Summers are moderately warm and humid, with occasional days when temperatures exceed 100°F.; winters are reasonably cold, with an average of about 2 days of subzero weather; and autumns are predominately cool, dry, and invigorating. Spring is the wettest season and vegetation grows luxuriantly. Annual precipitation is slightly in excess of the national average and is well distributed, though with peaks in early spring and summer. In spite of the relatively small range in latitude and the compact shape of Ohio, rainfall varies considerably in amount and seasonal distribution. This is accounted for not only by the presence of Lake Erie on the north, but also by its topography and proximity to rain producing storm paths. Annual precipitation averages about 38 inches, being most generous in spring (about 4 inches in April) and least in the fall (about 2.5 inches in October). Greatest amounts are measured in the southwest where Wilmington has an average of 44.36 inches; and the lake shore is driest, Gilbralter Island having a normal of only 29.06 inches. The southern half of the State is visited more frequently by productive rainstorms which, together with the general roughness of terrain, accounts for the larger total precipitation. The lifting of moist air masses over the hills tends to increase the yield of rainfall, especially in winter and spring. There is a marked tendency during the cold season for northeastern counties to receive snowfall amounts substantially in excess of those measured elsewhere. Northerly winds have a long fetch across Lake Huron and the widest part of Lake Erie, thus picking up moisture and heat from the lakes. This moisture is then forced to condense as the air is lifted abruptly over the divide a short distance from the lake. Average snowfall ranges from 60 inches in parts of Lake and adjoining counties down to 16 inches or less along the Ohio River.

The normal annual temperature for the State ranges from 49.6°F. at Hiram in Portage County up to 56.9°F. at Portsmouth on the Ohio River. Variations over the State are due mainly to differences in Latitude and topography, but the immediate Lake shore area experiences a moderating effect due to its proximity to a large body of water. This latter effect accounts in large part for the concentration of grape culture along the lake shore. Widest temperature ranges are found generally among the eastern hills. The lowest ever recorded in the State was -39°F. at Milligan in Perry County on February 10, 1899, while the record high of 113°F. has been observed at scattered stations, among them Wilmington in the southwest. In an average year, 90-degree heat may be expected about 20 times in summer with 100°F. or more once or twice. Readings of zero or lower are generally to be expected on 2 to 4 days each winter, and these are just as likely to occur in the south as the north. However, 1 winter out of 6 or 8 will pass without experiencing zero readings anywhere in the State.

The growing season, as defined by the period 32°F. or higher, ranges widely because of latitude and proximity to Lake Erie. The longest is about 200 days on the lake shore and the shortest is in the northeastern valleys within the Ohio River drainage. Dates of the average last freezing temperature in spring range from April 15 to May 18 and the mean first freeze date in fall varies from September 30 to November 6, the latter being on the western lake shore.

Damaging windstorms are mostly associated with heavy thunderstorms or line squalls. Three or four tornadoes may be expected to strike in Ohio each year. Most tornadoes, however, are of limited effect having paths that are short and narrow.

Most floods in Ohio are caused by unusual precipitation. The storms causing floods may bring rainfall of unusual intensity or of unusual duration and extent. Some floods may be caused by a series of ordinary storms which follow one another in rapid succession. Others may result from rain falling at relatively high temperatures on snow-covered areas. At times, though infrequent, flood conditions are caused or aggravated by ice gorges, especially in the tributary streams. Severe thunderstorms frequently cause local flash flooding. General flooding occurs most frequently during January to March and rarely occurs during August to October.

The frequency of floods in Ohio is about 1 in 3 years. Notable flood years in Ohio have been 1832, 1882, 1883, 1884, 1898, 1907, 1913, 1927, 1929, 1937, 1943, and 1959.

A system of flood control reservoirs in the Muskingum and Miami River basins controls the flow of those rivers. The lower 93 miles of the Muskingum is canalized by a system of 11 locks and dams. There is a series of 30 navigational locks in the Ohio River from East Liverpool to North Bend, Ohio.

In general Ohio's climate is a good one in which to live, and one favorable for agricultural and industrial development. While this is not a "vacation land" in the usual sense, people are coming more and more to enjoy recreational facilities afforded by the numerous parks, flood control reservoirs, and the shore of Lake Erie.

REFERENCES

(1) "Ohio, An Empire Within An Empire", published by the Ohio Development and Publicity Commission.

(2) "Climatological History of Ohio" - 1924 W. H. Alexander: Bulletin No. 26 Engineering Experiment Station Ohio State University.

(3) Weather Bureau Technical Paper No. 16 - Maximum 24-Hour Precipitation in the United States. Washington, D. C. 1952.

(4) Weather Bureau Technical Paper No. 29 - Rainfall Intensity-Duration-Frequency Regime. Washington, D. C.

(5) "Climatic Features of Ohio", G. W. Mindling (manuscript).

(6) Weather Bureau Technical Paper No. 15 - Maximum Station Precipitation for 1, 2, 3, 6, 12, and 24 Hours.

(7) Weather Bureau Technical Paper No. 25 - Rainfall Intensity-Duration-Frequency Curves. For selected stations in the United States, Alaska, Hawaiian Islands, and Puerto Rico.

BIBLIOGRAPHY

(A) Climatic Summary of the United States (Bulletin W) 1930 edition, Section 68, 69, 70 and 71. U. S. Weather Bureau

(B) Climatic Summary of tne United States Ohio - Supplement for 1931 through 1952 (Bulletin W Supplement). U. S. Weather Bureau

(C) Climatological Data - Ohio. U. S. Weather Bureau

(D) Climatological Data National Summary. U. S. Weather Bureau

(E) Hourly Precipitation Data - Ohio. U. S. Weather Bureau

(F) Local Climatological Data, U. S. Weather Bureau, for Akron-Canton, Cincinnati, Cleveland, Columbus, Dayton, Sandusky, Toledo and Youngstown, Ohio.

FREEZE DATA

STATION	Freeze threshold temperature	Mean date of last Spring occurrence	Mean date of first Fall occurrence	Mean No. of days between dates	Years of record Spring	No. of occurrences in Spring	Years of record Fall	No. of occurrences in Fall
AKRON CANTON WB AP	32	04-29	10-20	173	30	30	30	30
	28	04-14	11-05	205	30	30	30	30
	24	04-01	11-17	231	30	30	30	30
	20	03-20	11-28	253	30	30	30	30
	16	03-11	12-05	269	30	30	30	28
ASHLAND	32	05-07	10-11	157	30	30	30	30
	28	04-19	10-24	187	30	30	30	30
	24	04-05	11-04	213	30	30	30	30
	20	03-26	11-21	240	30	30	30	30
	16	03-15	12-02	262	30	30	30	29
ATHENS 1 E	32	04-28	10-10	165	20	20	19	19
	28	04-15	10-24	193	19	19	19	19
	24	03-25	11-07	227	18	18	19	19
	20	03-16	11-20	249	18	18	19	19
	16	03-09	12-02	267	18	18	19	18
BATAVIA 4 N	32	04-28	10-13	168	30	30	30	30
	28	04-14	10-26	195	30	30	30	30
	24	03-27	11-07	225	30	30	30	30
	20	03-15	11-18	249	30	30	30	30
	16	03-04	12-03	274	30	30	30	28
BELLEFONTAINE SEWAGE	32	05-04	10-15	164	28	20	29	29
	28	04-22	10-23	184	28	28	28	28
	24	04-06	11-06	214	28	28	28	28
	20	03-24	11-17	238	28	28	28	28
	16	03-12	11-26	259	28	28	28	27
BOWLING GREEN SWG PL	32	05-06	10-08	155	30	30	30	30
	28	04-24	10-26	185	30	30	30	30
	24	04-08	11-05	211	30	30	30	30
	20	03-25	11-19	239	30	30	30	30
	16	03-12	11-30	263	30	30	30	29
BUCYRUS	32	05-03	10-09	159	30	30	30	30
	28	04-22	10-22	184	30	30	30	30
	24	04-06	11-05	213	30	30	30	30
	20	03-25	11-17	237	30	30	30	30
	16	03-11	11-29	263	30	30	30	29
CADIZ	32	05-01	10-14	166	28	28	28	28
	28	04-18	10-26	191	28	28	28	28
	24	04-03	11-10	220	28	28	27	27
	20	03-24	11-19	240	28	28	27	27
	16	03-14	11-29	260	27	27	27	27
CALDWELL 4 W	32	04-29	10-15	169	16	16	17	17
	28	04-16	10-28	195	16	16	17	17
	24	04-04	11-12	222	16	16	17	17
	20	03-26	11-23	242	16	16	17	17
	16	03-10	11-30	265	16	16	17	17
CAMBRIDGE ST HOSP	32	05-04	10-10	159	30	30	30	30
	28	04-18	10-23	188	30	30	30	30
	24	04-03	11-07	219	30	30	30	30
	20	03-16	11-19	249	30	30	30	30
	16	03-08	12-03	271	30	30	30	28
CANFIELD 1 S	32	05-15	10-04	142	29	29	30	30
	28	04-28	10-16	170	30	30	30	30
	24	04-15	10-30	198	30	30	30	30
	20	04-02	11-20	232	30	30	30	30
	16	03-18	11-30	256	29	29	30	30
CANTON	32	05-05	10-14	162	28	28	27	27
	28	04-18	10-28	193	28	28	27	27
	24	04-05	11-16	225	28	28	27	27
	20	03-19	11-22	247	28	28	27	27
	16	03-10	12-06	271	28	28	27	25
CATAWBA ISLAND 1 SW	32	04-18	10-30	195	30	30	30	30
	28	04-06	11-12	220	30	30	30	30
	24	03-28	11-23	240	29	29	30	30
	20	03-17	12-03	261	29	29	30	29
	16	03-07	12-09	277	29	29	30	28
CHILLICOTHE	32	04-25	10-15	173	28	28	27	27
	28	04-10	10-27	200	28	28	27	27
	24	03-25	11-09	230	27	27	27	27
	20	03-12	11-20	253	27	27	27	27
	16	03-05	12-06	276	27	26	27	25
CHIPPEWA LAKE	32	05-14	10-07	147	30	30	30	30
	28	04-29	10-22	175	30	30	30	30
	24	04-14	11-07	206	30	30	30	30
	20	03-31	11-21	235	30	30	30	30
	16	03-16	11-29	258	30	30	30	29
CINCINNATI ABBE OBS	32	04-15	10-25	192	30	30	30	30
	28	03-29	11-07	223	30	30	30	30
	24	03-19	11-21	247	30	30	30	30
	20	03-08	12-02	269	30	30	30	29
	16	02-28	12-11	286	30	30	30	26
CIRCLEVILLE	32	04-28	10-12	166	27	27	27	27
	28	04-18	10-24	188	26	26	27	27
	24	03-29	11-08	224	25	25	27	27
	20	03-12	11-19	252	25	25	27	27
	16	03-03	11-30	273	24	24	26	24
CLEVELAND WB AP	32	04-21	11-02	195	30	30	30	30
	28	04-07	11-13	220	30	30	30	30
	24	03-25	11-24	245	30	30	30	30
	20	03-15	12-04	263	30	30	30	28
	16	03-05	12-11	281	30	30	30	27
COLUMBUS WB CITY	32	04-17	10-30	196	30	30	30	30
	28	04-01	11-12	225	30	30	30	30
	24	03-22	11-22	245	30	30	30	30
	20	03-10	12-06	271	30	30	30	28
	16	03-05	12-10	280	30	30	30	27
COSHOCTON	32	05-04	10-11	160	30	30	30	30
	28	04-18	10-25	190	30	30	30	30
	24	04-01	11-09	222	30	30	30	30
	20	03-19	11-20	246	30	30	30	30
	16	03-10	12-03	268	30	30	30	29
DELAWARE	32	05-01	10-09	161	30	30	29	29
	28	04-19	10-23	187	30	30	29	29
	24	04-04	11-04	214	30	30	29	29
	20	03-21	11-16	239	30	30	29	29
	16	03-10	11-30	265	30	30	29	28
DENNISON	32	05-12	10-03	144	30	30	30	30
	28	04-29	10-19	172	30	30	30	30
	24	04-14	11-02	203	29	29	30	30
	20	03-27	11-11	229	29	29	30	30
	16	03-13	11-26	258	29	29	30	30
FINDLAY SWG PLANT	32	05-03	10-11	160	30	30	30	30
	28	04-22	10-27	188	30	30	30	30
	24	04-06	11-07	215	30	30	30	30
	20	03-22	11-19	242	30	30	30	30
	16	03-12	12-01	264	30	30	30	29
GERMANTOWN 3 NE	32	04-27	10-13	168	30	30	30	30
	28	04-14	10-27	196	30	30	30	30
	24	03-25	11-09	229	30	30	30	30
	20	03-14	11-22	252	30	30	30	30
	16	03-06	12-05	274	29	29	30	28
GREENVILLE SWG PLT	32	04-30	10-14	167	29	29	30	30
	28	04-17	10-28	194	29	29	30	30
	24	03-30	11-09	224	30	30	30	30
	20	03-19	11-18	244	30	30	30	30
	16	03-07	12-04	272	30	30	30	27
HAMILTON WATER WORKS	32	04-30	10-14	167	29	29	29	29
	28	04-14	10-26	195	29	29	29	29
	24	03-25	11-11	231	29	29	29	29
	20	03-12	11-21	255	29	29	29	29
	16	03-04	12-08	280	29	29	29	26
HILLSBORO	32	04-29	10-14	168	30	30	30	30
	28	04-16	10-28	195	30	30	30	30
	24	04-03	11-08	219	30	30	30	30
	20	03-17	11-19	247	30	30	30	30
	16	03-07	12-01	269	30	30	30	29
HIRAM	32	05-06	10-14	161	29	29	30	30
	28	04-22	10-31	192	29	29	29	29
	24	04-11	11-13	216	29	29	29	29
	20	03-29	11-20	236	29	29	29	29
	16	03-17	12-01	259	29	29	29	28

STATION	Freeze threshold temperature	Mean date of last Spring occurrence	Mean date of first Fall occurrence	Mean No. of days between dates	Years of record Spring	No. of occurrences in Spring	Years of record Fall	No. of occurrences in Fall
IRONTON	32	04-23	10-16	176	30	30	30	30
	28	04-11	10-30	202	30	30	30	30
	24	03-26	11-11	231	30	30	30	30
	20	03-10	11-24	258	30	30	30	30
	16	03-02	12-07	280	30	30	30	27
JACKSON	32	04-30	10-10	162	29	29	29	29
	28	04-17	10-21	187	29	29	29	29
	24	04-01	11-03	216	29	29	29	29
	20	03-19	11-12	238	29	29	28	28
	16	03-09	11-25	261	29	29	28	27
JEFFERSON	32	05-03	10-16	166	11	11	11	11
	28	04-21	10-30	192	11	11	11	11
	24	04-07	11-14	222	11	11	11	11
	20	03-28	11-28	244	12	12	11	11
	16	03-14	12-04	265	12	12	11	10
KENTON 2 W	32	05-07	10-08	155	30	30	30	30
	28	04-23	10-22	182	30	30	30	30
	24	04-05	11-03	212	30	30	30	30
	20	03-24	11-16	237	30	30	30	30
	16	03-11	11-30	264	30	30	30	29
LANCASTER	32	05-07	10-05	152	30	30	30	30
	28	04-24	10-20	179	30	30	30	30
	24	04-09	10-30	205	30	30	30	30
	20	03-21	11-15	239	30	30	30	30
	16	03-11	11-22	256	30	30	30	30
LIMA SWG PLT	32	05-03	10-11	161	29	29	29	29
	28	04-19	10-26	190	29	29	29	29
	24	04-01	11-06	219	29	29	29	29
	20	03-20	11-20	245	29	29	29	29
	16	03-09	12-04	269	29	29	29	27
LONDON 4 W	32	05-02	10-14	165	30	30	30	30
	28	04-19	10-25	189	30	30	30	30
	24	04-02	11-10	222	29	29	30	30
	20	03-22	11-17	239	29	29	30	30
	16	03-10	11-29	263	29	29	30	30
MANSFIELD 6 W	32	05-09	10-06	149	30	30	29	29
	28	04-27	10-18	175	30	30	29	29
	24	04-13	11-03	205	30	30	29	29
	20	03-20	11-14	229	30	30	29	29
	16	03-16	11-27	256	30	30	29	28
MARIETTA WATER WORKS	32	04-27	10-14	171	28	28	29	29
	28	04-11	10-30	202	28	28	29	29
	24	03-27	11-12	230	28	28	29	29
	20	03-10	11-27	261	28	28	29	29
	16	03-06	12-07	276	28	28	29	27
MARION WATER WKS	32	04-28	10-19	174	30	30	30	30
	28	04-13	10-31	201	30	30	30	30
	24	03-30	11-13	228	30	30	30	30
	20	03-20	11-24	250	30	30	30	29
	16	03-09	12-08	274	30	30	30	27
MARYSVILLE	32	05-05	10-10	157	30	30	30	30
	28	04-24	10-22	181	30	30	30	30
	24	04-04	11-03	213	30	30	30	30
	20	03-22	11-15	238	30	30	30	30
	16	03-10	11-27	262	30	30	30	29
MC CONNELSVILLE LK 7	32	05-03	10-13	163	29	29	28	28
	28	04-18	10-26	191	29	29	28	28
	24	04-03	11-08	219	29	29	28	28
	20	03-18	11-18	245	29	29	28	28
	16	03-08	12-01	268	29	29	28	28
MILLERSBURG	32	05-12	10-04	145	28	28	30	30
	28	04-29	10-16	170	29	29	30	30
	24	04-13	10-30	200	29	29	30	30
	20	03-24	11-15	235	29	29	30	30
	16	03-12	11-28	262	29	29	30	29
MILLPORT 2 NW	32	05-18	09-30	135	29	29	29	29
	28	05-02	10-13	164	29	29	30	30
	24	04-22	10-27	187	29	29	30	30
	20	04-06	11-10	218	30	30	30	30
	16	03-18	11-23	250	30	30	30	30
MONTPELIER	32	05-04	10-11	159	19	19	19	19
	28	04-17	10-23	189	20	20	19	19
	24	04-04	11-10	220	20	20	19	19
	20	03-27	11-24	242	20	20	19	19
	16	03-15	11-28	258	18	18	19	19
MT HEALTHY EXP FARM	32	04-27	10-18	175	30	30	30	30
	28	04-10	11-02	206	30	30	30	30
	24	03-25	11-15	235	30	30	30	30
	20	03-13	11-23	254	30	30	29	29
	16	03-04	12-05	276	30	30	29	26
NAPOLEON	32	05-03	10-12	162	29	29	29	29
	28	04-18	10-25	190	29	29	29	29
	24	04-03	11-07	218	29	29	29	29
	20	03-20	11-19	245	29	29	29	28
	16	03-10	12-02	267	29	29	29	28
NEWARK WATR WKS	32	05-06	10-01	149	16	16	16	16
	28	04-27	10-12	168	16	16	16	16
	24	04-14	10-29	197	16	16	16	16
	20	03-26	11-12	230	16	16	16	16
	16	03-14	11-25	256	16	16	16	16
NEW BREMEN	32	04-30	10-15	169	19	19	20	20
	28	04-12	10-30	200	19	19	20	20
	24	03-29	11-09	225	19	19	20	20
	20	03-20	11-19	244	19	19	20	20
	16	03-06	12-02	271	19	19	20	17
NORWALK	32	05-10	10-10	153	29	29	29	29
	28	04-26	10-21	178	29	29	29	29
	24	04-10	11-03	208	30	30	29	29
	20	03-25	11-18	238	30	30	29	28
	16	03-14	12-02	263	30	30	30	29
OBERLIN	32	05-07	10-12	158	30	30	30	30
	28	04-24	10-28	187	30	30	30	30
	24	04-06	11-09	217	30	30	30	30
	20	03-22	11-23	246	30	30	30	30
	16	03-14	12-04	265	29	29	30	28
OTTAWA	32	05-02	10-14	165	28	28	29	29
	28	04-18	10-28	193	28	28	29	29
	24	04-03	11-09	220	28	28	29	29
	20	03-20	11-21	247	28	28	29	29
	16	03-12	12-01	264	28	28	29	28
PAULDING	32	05-07	10-07	153	28	28	30	30
	28	04-26	10-22	179	29	29	30	30
	24	04-10	11-04	209	28	28	30	30
	20	03-26	11-14	233	28	28	30	30
	16	03-14	11-26	258	28	28	30	29
PEEBLES 1 S	32	05-04	10-07	156	30	30	30	30
	28	04-23	10-18	178	30	30	30	30
	24	04-06	10-30	207	30	30	30	30
	20	03-21	11-12	236	30	30	30	30
	16	03-08	11-23	261	30	30	29	29
PLYMOUTH	32	05-07	10-05	152	18	18	18	18
	28	04-25	10-22	180	18	17	18	17
	24	04-04	11-04	214	18	18	18	18
	20	03-28	11-19	236	18	18	18	18
	16	03-16	12-01	260	18	18	18	18
PORTSMOUTH	32	04-18	10-21	186	30	30	30	30
	28	03-29	11-04	221	30	30	30	30
	24	03-13	11-18	249	30	30	30	30
	20	03-06	12-02	271	30	30	30	29
	16	02-25	12-12	290	30	30	30	25
PUT IN BAY STONE LAB	32	04-16	11-06	205	30	30	30	30
	28	04-03	11-20	231	30	30	30	30
	24	03-26	11-28	247	30	30	30	30
	20	03-13	12-07	269	30	30	30	28
	16	03-08	12-15	282	30	30	30	26
SANDUSKY WB CITY	32	04-17	10-30	197	30	30	30	30
	28	04-05	11-10	219	30	30	30	30
	24	03-24	11-22	244	30	30	30	30
	20	03-13	12-02	265	30	30	30	29
	16	03-04	12-10	281	30	30	30	28

FREEZE DATA

STATION	Freeze threshold temperature	Mean date of last Spring occurrence	Mean date of first Fall occurrence	Mean No. of days between dates	Years of record Spring	No. of occurrences in Spring	Years of record Fall	No. of occurrences in Fall	STATION	Freeze threshold temperature	Mean date of last Spring occurrence	Mean date of first Fall occurrence	Mean No. of days between dates	Years of record Spring	No. of occurrences in Spring	Years of record Fall	No. of occurrences in Fall
SIDNEY (2)	32	05-01	10-12	164	27	27	28	28	VAN WERT	32	05-06	10-09	156	26	26	27	27
	28	04-16	10-27	194	27	27	28	28		28	04-26	10-24	181	26	26	27	27
	24	03-29	11-07	223	27	27	28	28		24	04-11	11-05	208	26	26	27	27
	20	03-17	11-19	247	27	27	28	28		20	03-23	11-15	236	26	26	27	27
	16	03-08	11-30	267	26	26	27	26		16	03-14	11-28	259	26	26	27	27
TIFFIN	32	05-02	10-12	163	30	30	29	29	VICKERY 2 NW	32	05-04	10-14	163	30	30	30	30
	28	04-19	10-26	190	30	30	29	29		28	04-21	10-28	190	30	30	30	30
	24	04-01	11-08	222	30	30	29	29		24	04-03	11-10	220	30	30	30	30
	20	03-23	11-23	245	30	30	29	29		20	03-18	11-26	253	29	29	30	30
	16	03-12	12-04	267	30	30	29	28		16	03-10	12-05	270	29	29	30	29
TOLEDO WB AP	32	04-24	10-25	184	30	30	30	30	WARREN	32	05-12	10-07	148	29	29	30	30
	28	04-09	11-07	213	30	30	30	30		28	04-23	10-25	184	29	29	30	30
	24	03-29	11-19	235	30	30	30	30		24	04-09	11-10	215	29	29	30	30
	20	03-20	11-29	255	30	30	30	29		20	03-26	11-26	245	30	30	30	30
	16	03-08	12-07	274	30	30	30	28		16	03-13	12-06	268	30	30	30	27
UPPER SANDUSKY	32	05-04	10-07	156	30	30	30	30	WASHINGTON C HOUSE	32	04-28	10-12	167	25	25	24	24
	28	04-22	10-24	185	30	30	30	30		28	04-20	10-25	188	25	25	24	24
	24	04-05	11-08	217	30	30	30	30		24	03-31	11-08	223	25	25	25	25
	20	03-23	11-16	238	30	30	30	30		20	03-17	11-15	242	25	25	25	25
	16	03-11	12-01	264	30	30	30	29		16	03-07	11-28	266	25	25	25	24
URBANA JR COLLEGE	32	05-02	10-12	163	30	30	30	30	WAUSEON SEWAGE	32	05-06	10-13	160	30	30	30	30
	28	04-19	10-26	190	30	30	30	30		28	04-20	10-26	189	30	30	30	30
	24	04-01	11-07	220	30	30	29	29		24	04-07	11-08	214	30	30	30	30
	20	03-18	11-20	247	30	30	29	29		20	03-23	11-20	242	30	30	30	30
	16	03-08	12-02	269	30	30	29	27		16	03-12	11-30	263	30	30	30	29
									WAVERLY	32	04-29	10-09	162	27	27	28	28
										28	04-18	10-18	189	27	27	28	28
										24	03-31	11-03	217	27	27	28	28
										20	03-17	11-16	244	27	27	28	28
										16	03-08	11-28	265	27	27	28	27
									WILMINGTON	32	05-02	10-09	160	30	30	30	30
										28	04-24	10-21	181	30	30	30	30
										24	04-02	11-06	218	30	30	30	30
										20	03-20	11-15	240	30	30	30	30
										16	03-09	11-27	264	30	30	30	29
									WOOSTER EXP STA	32	05-11	10-05	147	29	29	29	29
										28	04-25	10-19	177	29	29	29	29
										24	04-10	11-01	205	29	29	29	29
										20	03-24	11-16	236	29	29	29	29
										16	03-12	12-01	263	29	29	29	28

Data in the above table are based on the period 1921-1950, or that portion of this period for which data are available.

Means have been adjusted to take into account years of non-occurrence.

A freeze is a numerical substitute for the former term "killing frost" and is the occurrence of a minimum temperature at or below the threshold temperature of 32°, 28°, etc.

Freeze data tabulations in greater detail are available and can be reproduced at cost.

*MEAN TEMPERATURE AND PRECIPITATION

STATION	JANUARY		FEBRUARY		MARCH		APRIL		MAY		JUNE		JULY		AUGUST		SEPTEMBER		OCTOBER		NOVEMBER		DECEMBER		ANNUAL	
	Temperature	Precipitation	Temperature	Precipitation	Temperature	Precipitation	Temperature	Precipitation	Temperature	Precipitation	Temperature	Precipitation	Temperature	Precipitation	Temperature	Precipitation	Temperature	Precipitation	Temperature	Precipitation	Temperature	Precipitaion	Temperature	Precipitation	Temperature	Precipitation
NORTHWEST																										
BOWLING GREEN SEWAGE P	28.9	2.07	29.5	1.69	37.7	2.76	48.9	3.11	60.4	3.62	70.8	3.70	74.8	3.11	72.7	2.50	65.9	2.72	54.9	2.48	41.1	2.04	30.5	1.86	51.3	31.66
FINDLAY SEWAGE PLANT	28.5	2.38	29.1	2.03	37.2	3.35	48.2	3.22	59.7	3.71	70.2	4.18	74.2	3.57	72.3	2.86	65.1	2.57	53.7	2.74	40.4	2.31	30.1	2.15	50.7	35.07
LIMA SEWAGE PLANT	29.5	2.57	30.2	1.96	38.5	3.40	49.4	3.32	60.5	3.45	70.8	4.15	74.5	3.29	72.5	3.41	65.6	3.00	54.5	2.88	41.3	2.65	31.1	2.20	51.5	36.28
MONTPELIER	27.5	2.28	28.2	2.03	36.7	3.00	48.6	3.34	59.9	3.62	70.4	3.89	74.4	3.04	72.4	3.10	65.0	2.91	53.9	2.59	40.0	2.39	29.3	2.01	50.5	34.20
NAPOLEON	28.3	2.52	29.0	2.24	37.4	3.10	48.8	3.37	60.4	3.79	70.8	4.20	74.8	3.29	72.6	2.98	65.3	3.15	54.2	2.62	40.6	2.29	29.9	2.28	51.0	35.83
PAULDING	28.2	2.42	29.1	2.04	37.7	3.24	49.0	3.38	59.9	3.64	70.5	4.40	74.5	3.41	72.6	2.46	65.6	2.92	54.1	2.75	40.3	2.37	29.5	2.13	50.9	35.16
TOLEDO WB EXPRESS AP	26.4	2.25	27.3	1.86	35.8	2.86	46.5	3.25	58.2	2.95	68.6	3.55	73.1	2.65	71.0	2.69	64.2	3.02	52.7	2.32	39.8	2.15	28.9	2.29	49.4	31.84
VAN WERT	29.1	2.57	30.1	1.91	38.6	3.49	49.8	3.45	60.9	4.05	71.2	4.19	75.0	3.47	72.8	2.60	66.0	3.07	54.9	2.94	41.0	2.48	30.6	2.12	51.7	36.34
WAUSEON SEWAGE	27.3	2.28	27.8	1.96	36.2	2.86	47.8	3.27	59.3	3.65	70.0	3.81	73.8	2.93	71.7	2.62	64.4	2.98	53.4	2.58	39.6	2.28	29.1	2.07	50.0	33.29
DIVISION	28.4	2.35	29.0	1.95	37.3	3.06	48.6	3.20	60.0	3.60	70.6	3.99	74.5	3.14	72.5	3.72	65.4	2.83	54.2	2.65	40.5	2.27	30.0	2.08	50.9	33.85
NORTH CENTRAL																										
BUCYRUS SEWAGE PLANT	29.1	2.85	29.6	2.25	38.2	3.42	48.9	3.00	60.1	3.21	70.3	4.37	73.7	3.03	71.4	3.29	64.4	2.68	53.4	2.43	40.4	2.37	30.5	2.22	50.8	35.12
CATAWBA ISLAND 1 SW		2.16		1.77		2.76		3.11		3.26		3.47		2.83		2.76		2.40		2.25		2.22		2.07		31.06
FREMONT		2.50		1.99		2.91		3.32		3.54		4.10		3.48		2.88		2.82		2.43		2.26		2.05		34.28
NORWALK	29.2	2.41	29.5	1.87	37.6	2.94	48.1	3.24	59.6	3.67	69.8	4.05	73.3	3.61	71.2	3.43	64.8	3.22	54.1	2.30	41.2	2.32	30.9	2.00	50.8	35.06
OBERLIN	29.7	2.48	29.7	2.03	37.6	2.98	48.4	3.24	59.6	3.41	69.8	3.73	73.5	3.05	71.6	3.23	65.2	3.05	54.4	2.40	41.4	2.26	31.3	2.17	51.0	34.03
SANDUSKY WB CITY	28.8	2.29	29.4	1.92	37.5	2.89	47.9	2.96	59.3	3.32	69.9	3.73	74.6	3.45	72.8	2.81	66.5	3.26	55.0	2.10	42.2	2.27	31.5	2.16	51.3	33.16
TIFFIN	29.5	2.54	30.0	2.07	38.2	3.29	49.2	3.18	60.4	3.67	70.7	4.04	74.4	3.24	72.2	3.34	65.6	2.91	54.6	2.44	41.3	2.42	31.0	2.10	51.4	35.24
UPPER SANDUSKY	29.6	2.74	30.3	2.20	38.8	3.42	49.6	3.07	60.9	3.40	71.0	4.25	74.5	3.22	72.6	3.22	65.8	2.80	54.6	2.38	41.2	2.43	31.1	2.21	51.7	35.34
DIVISION	29.3	2.46	29.6	1.98	37.7	3.02	48.4	3.14	59.8	3.45	70.3	3.96	74.3	3.14	72.3	3.12	65.7	2.80	54.7	2.35	41.4	2.25	31.1	2.06	51.2	33.73
NORTHEAST																										
AKRON CANTON WB AP	27.4	2.74	28.1	2.13	36.5	3.16	47.1	3.20	58.5	3.75	68.4	3.84	72.4	4.20	70.6	3.26	64.6	3.63	53.0	2.39	40.4	2.38	30.1	2.58	49.7	37.26
CHIPPEWA LAKE		2.88		2.36		3.49		3.49		3.71		4.14		3.80		3.03		2.89		2.61		2.60		2.41		37.41
CLEVELAND WB AIRPORT	28.5	2.38	28.6	2.12	36.8	2.89	47.3	2.73	59.1	2.73	69.4	3.05	73.7	3.04	71.9	2.64	65.5	3.13	54.4	2.42	41.7	2.66	30.9	2.29	50.6	32.0[illegible]
HIRAM	28.1	2.88	27.7	2.19	35.8	3.40	47.1	3.52	58.2	3.81	67.8	3.83	71.7	3.31	70.0	3.24	63.8	2.89	53.2	2.86	40.1	2.78	29.9	2.49	49.5	37.2[illegible]
MINERAL RIDGE WATER WK		2.58		1.87		2.93		3.21		3.44		3.27		3.55		3.15		2.54		2.61		2.27		2.05		33.4[illegible]
WARREN	29.9	2.87	29.6	2.19	37.6	3.28	48.5	3.50	59.6	3.64	69.2	3.55	73.1	3.38	71.2	3.27	64.6	2.83	53.8	2.72	41.6	2.69	31.6	2.34	50.9	36.2[illegible]
YOUNGSTOWN WB AIRPORT	27.5	3.32	28.2	2.72	36.7	3.45	47.2	3.64	58.3	4.09	68.0	3.71	72.3	4.16	70.4	3.46	64.7	3.58	53.9	2.77	40.6	3.49	30.0	2.94	49.8	41.3[illegible]
DIVISION	29.1	2.75	28.9	2.22	36.8	3.27	47.7	3.36	58.9	3.61	68.8	3.57	72.8	3.45	71.1	3.19	64.6	2.95	53.9	2.80	41.1	2.68	31.0	2.40	50.4	36.2[illegible]
WEST CENTRAL																										
BELLEFONTAINE SEWAGE		2.98		1.94		3.29		3.24		3.31		4.38		3.37		3.48		2.62		2.60		2.31		2.30		35.8[illegible]
GREENVILLE SEWAGE PL	29.7	3.02	30.3	2.28	38.8	3.69	49.5	3.43	60.4	4.00	70.2	4.23	73.9	3.46	71.9	2.86	64.9	3.32	53.9	2.58	40.7	2.72	30.8	2.38	51.3	37.0[illegible]
KENTON 2 W	28.7	2.49	29.5	1.80	37.6	3.13	48.4	3.00	59.5	3.22	69.6	4.05	73.5	3.23	71.4	2.85	64.9	2.98	53.7	2.42	40.6	2.08	30.2	2.01	50.6	33.2[illegible]
LAKEVIEW 3 NE		2.59		2.00		3.10		3.13		3.31		4.64		3.34		3.30		2.95		2.44		2.23		2.11		35.1[illegible]
NEW CARLISLE		3.25		2.56		3.52		3.32		3.80		4.27		3.26		3.54		2.76		2.48		2.83		2.48		38.0[illegible]
PIQUA		3.06		2.36		3.61		3.45		3.67		4.30		3.52		2.82		3.05		2.53		2.54		2.41		37.3[illegible]
PLEASANT HILL 1 NW		3.00		2.18		3.42		3.36		3.52		3.77		3.28		3.11		2.78		2.39		2.48		2.34		35.63
TIPP CITY		3.19		2.34		3.34		3.30		3.77		4.13		3.33		3.13		3.05		2.41		2.62		2.37		36.98
URBANA GRIMES FIELD	30.3	3.25	31.3	2.36	39.6	3.53	50.1	3.46	60.8	3.64	70.6	4.14	74.3	3.52	72.0	3.58	65.2	2.97	54.2	2.53	41.3	2.55	31.4	2.38	51.8	37.91
VERSAILLES		2.94		2.37		3.57		3.68		3.85		4.49		3.15		3.22		2.99		2.48		2.68		2.40		37.82
DIVISION	29.9	2.96	30.8	2.21	39.1	3.40	49.8	3.32	60.8	3.59	70.8	4.19	74.5	3.42	72.5	3.14	65.7	3.05	54.5	2.54	41.2	2.46	31.1	2.28	51.7	36.56
CENTRAL																										
CIRCLEVILLE		3.46		2.55		4.02		3.83		3.61		4.11		4.52		3.19		2.78		1.93		2.77		2.51		39.28
COLUMBUS VALLEY CROSS	31.3	3.16	32.1	2.24	40.4	3.36	50.6	3.63	61.3	3.54	70.6	3.92	74.0	3.89	72.1	3.10	65.7	2.55	54.6	1.99	41.6	2.50	32.1	2.39	52.2	36.27
COLUMBUS WB AIRPORT	29.7	2.94	31.2	2.27	39.8	3.43	50.2	3.44	60.8	3.97	70.7	4.33	74.4	3.85	72.4	3.21	66.5	2.91	54.5	2.18	41.9	2.86	31.7	2.49	52.0	37.88
COLUMBUS WB CITY	31.1	2.81	32.6	2.15	41.1	3.22	51.4	3.19	62.2	3.23	72.0	3.66	75.8	3.53	73.8	3.01	67.8	2.58	56.1	2.00	43.2	2.54	33.3	2.44	53.4	34.36
DELAWARE	30.7	3.07	31.3	2.31	39.9	3.48	50.2	3.41	61.0	3.44	71.0	4.36	74.4	3.36	72.5	3.09	65.7	2.81	54.6	2.16	41.4	2.43	31.5	2.46	52.0	36.38
LANCASTER 2 NW	32.6	3.35	33.3	2.68	41.3	4.07	51.4	3.72	61.7	3.81	71.3	4.19	74.4	4.22	72.7	3.73	66.1	3.13	55.2	2.17	42.7	2.66	33.4	2.81	53.0	40.54
LA RUE		2.78		2.08		3.45		3.12		3.35		4.19		3.05		3.16		2.29		2.39		2.12		2.19		34.17
LONDON WATER WORKS	30.1	3.36	31.0	2.53	39.5	3.68	49.9	3.86	60.7	4.07	70.8	4.00	74.3	4.08	72.6	3.05	66.1	2.81	55.0	2.31	41.0	2.76	31.1	2.69	51.8	39.20
MARION WATER WORKS	30.5	2.89	31.2	2.18	39.7	3.40	50.3	3.21	61.3	3.31	71.1	4.42	74.7	3.10	72.7	3.46	66.2	2.65	55.4	2.54	42.1	2.41	31.8	2.25	52.3	35.82
MARYSVILLE	29.7	3.12	30.5	2.27	39.0	3.47	49.6	3.43	60.6	3.51	70.3	4.27	73.9	3.28	72.0	3.59	65.2	2.82	54.3	2.37	40.7	2.37	30.7	2.53	51.4	37.03
PROSPECT 3 N		2.82		2.14		3.22		3.22		3.29		4.18		2.89		2.93		2.57		2.29		2.21		2.11		33.87
WASHINGTON COURT HOUSE	31.5	3.38	32.3	2.44	40.5	3.99	50.9	3.84	61.8	3.81	71.5	3.97	74.9	3.72	73.1	3.00	66.6	2.71	55.3	2.00	41.8	2.70	32.1	2.48	52.7	38.04
DIVISION	31.4	3.20	32.1	2.36	40.4	3.59	50.8	3.60	61.7	3.65	71.4	4.16	74.8	3.77	73.0	3.25	66.3	2.74	55.3	2.18	42.0	2.55	32.2	2.48	52.6	37.53
CENTRAL HILLS																										
ASHLAND 2 ENE		2.60		1.96		3.30		3.06		3.47		3.53		3.29		3.54		2.75		2.15		2.23		2.05		33.93
ASHLAND 3 NW	29.3	2.79	29.6	2.08	37.9	3.50	48.7	3.21	60.0	3.50	69.8	3.74	73.5	3.57	71.8	3.67	65.1	2.76	54.4	2.27	40.7	2.40	30.6	2.23	51.0	35.72
COSHOCTON SEWAGE PL	31.6	3.49	32.0	2.44	40.4	3.79	51.2	3.65	62.2	3.97	71.9	4.37	75.0	4.13	73.1	4.00	66.5	2.86	55.5	2.33	42.4	2.88	32.6	2.54	52.9	40.45
WOOSTER EXP STATION	28.9	2.81	29.1	2.04	37.3	3.29	47.8	3.11	58.7	3.75	68.7	3.92	72.4	3.76	70.7	3.51	63.8	2.70	53.0	2.32	40.1	2.19	30.3	2.21	50.1	35.61
WOOSTER 2 SE	29.4	2.76	29.6	1.99	37.6	3.25	48.4	3.23	59.6	3.72	69.6	3.99	73.3	3.76	71.5	3.51	64.[illegible]	2.67	53.9	2.33	40.7	2.23	30.7	2.25	50.8	35.69
DIVISION	29.7	3.01	29.9	2.15	38.0	3.42	48.6	3.26	59.7	3.61	69.5	4.17	73.1	3.77	71.3	3.54	64.7	2.83	53.8	2.30	40.6	2.45	30.9	2.36	50.8	36.87
NORTHEAST HILLS																										
CANFIELD 1 S		2.50		1.81		3.13		3.05		3.65		3.62		3.41		3.00		2.50		2.61		2.30		2.12		33.70
DENNISON	31.9	3.22	32.2	2.36	40.2	3.72	50.7	3.61	61.0	3.92	70.3	4.27	73.3	4.04	71.5	3.75	65.0	2.89	54.2	2.55	42.0	2.76	32.6	2.60	52.1	39.69
ELLSWORTH		2.53		1.84		3.28		3.58		3.92		3.80		3.41		3.05		2.88		2.82		2.49		2.14		35.74
MILLPORT 2 NW	29.0	3.10	29.1	2.15	37.2	3.57	47.7	3.29	58.2	3.58	67.6	3.71	71.1	3.98	69.7	3.33	62.9	3.13	52.0	2.58	39.9	2.40	30.4	2.30	49.6	37.12
NEWCOMERSTOWN		3.34		2.29		3.64		3.50		3.82		4.26		3.91		4.09		2.69		2.36		2.64		2.30		38.84
DIVISION	30.3	3.04	30.4	2.20	38.5	3.58	49.1	3.40	60.0	3.68	69.3	4.01	72.8	4.01	71.1	3.43	64.5	2.93	53.6	2.56	41.1	2.52	31.4	2.43	51.0	37.79
SOUTHWEST																										
CHILO DAM 34		3.98		2.97		4.58		3.93		3.43		3.96		3.61		3.41		2.56		2.07		2.95		2.86		40.31
CINCINNATI ABBE OBS	33.1	3.44	34.8	2.55	43.3	4.07	53.6	3.64	63.6	3.54	72.8	4.05	76.6	3.70	74.8	3.38	68.9	2.88	57.4	2.19	44.6	3.06	35.0	2.84	54.9	39.34
CINCINNATI WB CITY	34.6	3.44	36.0	2.55	44.5	4.07	54.9	3.64	64.9	3.54	74.4	4.05	78.1	3.70	76.2	3.38	70.1	2.88	59.1	2.19	46.1	3.06	36.6	2.84	56.3	39.34
DAYTON WB AIRPORT	29.7	2.96	31.4	2.11	39.8	3.24	50.5	3.07	61.0	3.54	70.9	3.90	75.0	3.29	72.9	2.86	66.9	2.80	55.1	2.30	41.9	2.61	31.7	2.47	52.2	35.15
EATON		3.61		2.64		3.81		3.51		3.61		4.29		3.37		3.15		3.18		2.53		2.87		2.52		39.09
FERNBANK DAM 37		3.73		2.78		4.27		3.60		3.45		3.97		3.53		2.92		2.86		2.26		2.93		2.59		38.92
FRANKLIN		3.65		2.66		3.78		3.58		3.88		4.18		3.44		3.17		3.26		2.38		2.92		2.68		39.58
HAMILTON		3.70		2.68		4.04		3.64		3.65		4.06		3.77		2.85		3.54		2.33		2.90		2.69		39.85
HAMILTON WTR WKS SOUTH		3.70		2.68		4.05		3.62		3.60		3.99		3.68		2.80		3.63		2.34		2.81		2.69		39.59
HILLSBORO	33.0	4.11	33.7	3.16	42.0	4.77	52.1	4.08	62.1	3.92	70.9	4.38	74.3	4.04	72.7	4.03	66.6	3.23	56.0	2.35	43.1	2.96	33.7	3.10	53.4	44.13
KINGS MILLS		3.73		2.79		4.09		3.95		3.98		4.29		3.84		3.34		3.33		2.18		3.12		2.77		41.41
MIAMISBURG		3.78		2.68		3.95		3.77		4.01		4.47		3.25		3.10		3.44		2.64		3.12		2.79		41.00
MIDDLETOWN		3.78		2.85		4.00		3.59		3.98		4.14		3.76		3.02		3.25		2.39		2.92		2.78		40.46
WEST MANCHESTER 3 SW		3.46		2.55		3.63		3.42		3.53		4.26		3.21		3.24		2.98		2.61		2.84		2.66		38.39
WILMINGTON		4.26		3.33		4.71		4.29		3.88		4.17		4.25		3.47		3.30		2.29		3.16		3.25		44.36
XENIA 4 SSW	31.7	3.15	32.7	2.42	41.1	3.44	51.2	3.55	61.0	3.71	71.5	4.43	74.6	3.66	72.5	3.77	66.4	2.90	55.3	2.24	42.5	2.83	32.4	2.40	52.7	39.50
DIVISION	33.0	3.66	34.0	2.76	42.2	4.10	52.6	3.69	62.9	3.71	72.5	4.13	76.0	3.63	74.2	3.21	67.7	3.03	56.6	2.35	43.4	2.92	33.9	2.73	54.1	39.92

*MEAN TEMPERATURE AND PRECIPITATION

STATION	JAN Temp	JAN Precip	FEB Temp	FEB Precip	MAR Temp	MAR Precip	APR Temp	APR Precip	MAY Temp	MAY Precip	JUN Temp	JUN Precip	JUL Temp	JUL Precip	AUG Temp	AUG Precip	SEP Temp	SEP Precip	OCT Temp	OCT Precip	NOV Temp	NOV Precip	DEC Temp	DEC Precip	ANN Temp	ANN Precip
SOUTH CENTRAL																										
CHILLICOTHE	33.9	3.54	34.5	2.71	42.7	3.94	53.4	3.71	63.7	3.73	72.9	4.01	76.2	4.08	74.3	3.57	67.8	2.44	56.6	1.82	43.8	2.56	34.6	2.67	54.5	39.28
GALLIPOLIS 5 W		3.85		2.98		4.27		3.59		3.90		3.98		4.15		3.66		2.45		2.12		2.62		2.90		40.97
IRONTON		4.08		2.85		4.57		3.76		3.97		4.57		4.51		3.74		2.86		2.09		2.76		2.92		42.68
JACKSON 2 NW	34.5	3.43	35.1	2.88	43.3	4.39	53.6	3.77	63.1	3.91	71.5	4.07	74.7	3.44	73.1	3.54	66.7	2.42	55.4	2.23	43.5	2.73	34.6	3.05	54.1	41.36
NORTH KENOVA DAM 28		4.02		2.91		4.28		3.48		3.82		4.23		4.47		3.29		2.66		2.03		2.67		2.82		40.68
PORTSMOUTH	36.8	4.22	37.7	3.08	45.6	4.63	55.9	3.67	65.9	4.10	74.6	3.94	77.8	4.73	76.2	3.62	69.9	2.54	58.6	2.11	46.1	2.64	37.3	2.88	56.9	42.16
PORTSMOUTH US GRANT BR		4.14		3.00		4.57		3.65		4.07		3.83		4.88		3.68		2.61		2.16		2.92		2.92		42.33
RACINE DAM 23		3.89		2.98		4.31		3.58		3.76		4.27		4.52		3.88		2.87		2.09		2.71		2.86		41.74
WAVERLY		3.78		2.79		4.13		3.54		3.87		3.89		4.06		3.54		2.52		1.99		2.54		2.79		39.54
DIVISION	35.3	3.91	36.1	2.87	44.2	4.36	54.4	3.69	64.1	3.93	73.0	4.10	76.1	4.23	74.6	3.67	68.2	2.85	57.0	2.09	44.5	2.57	35.6	2.90	55.3	41.27
SOUTHEAST																										
ATHENS		3.42		2.65		3.93		3.49		4.01		4.23		4.44		4.04		2.87		1.92		2.63		2.71		40.34
CAMBRIDGE STATE HOSP		3.16		2.36		3.75		3.46		3.62		4.10		4.15		3.84		2.97		2.36		2.75		2.51		39.03
CLARINGTON LOCK 14		3.41		2.66		3.69		3.59		4.23		4.76		4.36		4.22		2.67		2.49		2.61		2.85		41.54
MARIETTA LOCK 1		3.42		2.74		3.94		3.38		3.79		4.19		4.56		4.42		2.74		2.08		2.76		2.76		40.69
MC CONNELSVILLE LOCK 7	34.1	3.35	34.3	2.62	42.1	3.96	52.2	3.67	62.5	4.16	71.4	4.62	74.7	4.12	73.4	3.43	67.1	2.98	56.4	2.27	43.3	2.44	34.1	2.62	53.8	40.14
PHILO	33.4	3.16	33.9	2.48	42.0	3.89	52.3	3.64	62.7	4.07	71.9	4.41	75.0	4.38	73.3	3.78	66.9	2.91	56.1	2.27	43.2	2.62	33.8	2.71	53.7	40.32
PHILO 3 SW	33.1	2.72	33.6	2.07	41.8	3.32	52.5	3.01	62.8	3.63	71.8	4.31	75.1	3.71	73.2	3.45	67.3	2.74	56.8	2.10	43.7	2.31	34.0	2.39	53.8	35.57
ZANESVILLE LOCK 10		2.96		2.21		3.46		3.34		3.66		4.19		4.06		3.50		2.92		2.12		2.55		2.30		37.28
DIVISION	33.2	3.22	33.5	2.46	41.6	3.75	52.0	3.47	62.3	3.91	71.3	4.28	74.5	4.17	72.9	3.77	66.5	2.85	55.7	2.20	42.9	2.56	33.6	2.62	53.3	39.26

* Averages for period 1931-1955, except for stations marked WB which are "normals" based on period 1921-1950. Divisional means may not be the arithmetical average of individual stations published, since additional data from shorter period stations are used to obtain better areal representation.

CONFIDENCE LIMITS

In the absence of trend or record changes, the chances are 9 out of 10 that the true mean will lie in the interval formed by adding and subtracting the values in the following table from the means for any station in the State. Because of the wider variation in mean precipitation, the corresponding monthly means and annual mean must be substituted for "p" in the precipitation table below to obtain mean precipitation confidence limits.

1.8	$.38\sqrt{p}$	1.6	$.25\sqrt{p}$	1.8	$.32\sqrt{p}$	1.1	$.26\sqrt{p}$	1.1	$.34\sqrt{p}$	1.0	$.34\sqrt{p}$	.8	$.35\sqrt{p}$	.9	$.31\sqrt{p}$	1.0	$.30\sqrt{p}$	1.1	$.32\sqrt{p}$	1.1	$.30\sqrt{p}$	1.4	$.24\sqrt{p}$	.4	$.32\sqrt{p}$

COMPARATIVE DATA

Data in the following table are the mean temperature and average precipitation for Wooster Experiment Station, Ohio for the period 1906-1930 and are included in this publication for comparative purposes:

27.5	3.33	28.4	2.32	37.9	3.61	48.8	3.29	58.4	3.65	67.2	3.86	71.6	3.86	70.1	3.71	64.3	3.24	52.3	2.75	40.7	2.57	30.1	2.81	49.8	39.00

NORMALS, MEANS AND EXTREMES
(Table Revised 1973. Base Period for Climatological Normals: 1931-1960)

Station: AKRON, OHIO   AKRON-CANTON AIRPORT   Standard time used: EASTERN   Latitude: 40° 55' N   Longitude: 81° 26' W   Elevation (ground): 1208 feet

Month	Normal Daily maximum	Normal Daily minimum	Normal Monthly	Record highest	Year	Record lowest	Year	Normal heating degree days (Base 65°)	Normal total precip.	Max monthly precip.	Year	Min monthly precip.	Year	Max in 24 hrs	Year	Snow mean total	Snow max monthly	Year	Snow max 24 hrs	Year
J	35.6	20.9	28.3	66	1967	-15	1972	1138	2.86	8.70	1950	0.71	1961	2.99	1959	9.8	17.6	1964	10.9	1966
F	36.6	20.7	28.7	62	1971+	-7	1972	1016	2.30	5.24	1956	0.48	1968	2.57	1959	9.5	19.7	1967	10.7	1951
M	46.3	27.5	36.9	78	1967	3	1964	871	3.32	8.83	1964	1.04	1958	3.29	1964	9.7	15.8	1960	8.5	1952+
A	59.6	37.8	48.7	85	1970	10	1964	489	3.34	6.06	1964	0.91	1957	1.98	1971	2.1	3.2	1961	3.2	1961
M	71.2	48.1	59.7	88	1967	24	1966	202	3.85	9.60	1964	1.52	1971	2.70	1956	0.2	0.0	1966	2.0	1966
J	80.6	58.2	69.4	97	1971	32	1972	39	4.26	7.08	1970	1.01	1967	2.87	1967	0.0	0.0		0.0	
J	83.2	61.9	72.6	94	1965	43	1963	9	3.77	11.43	1958	1.56	1958	4.18	1958	0.0	0.0		0.0	
A	82.0	61.0	71.5	93	1968	36	1965	9	3.11	6.11	1956	1.11	1964	2.77	1964	0.0	0.0		0.0	
S	75.1	52.9	64.0	93	1964	22	1966	96	2.62	6.47	1972	0.20	1960	2.33	1967	0.0	T	1965	T	1965
O	63.3	42.5	52.9	85	1963	10	1964	381	2.48	8.42	1954	0.45	1953	2.77	1954	0.5	6.8	1952	3.9	1952
N	48.6	32.7	40.8	77	1968	-6	1971+	726	2.25	5.05	1966	1.07	1954	2.00	1955	5.4	22.3	1950	7.4	1950
D	37.4	23.5	30.5	67	1971+	-6	1963	1070	2.27	4.89	1951	0.31	1955	1.53	1968	9.6	25.4	1960	8.5	1960
YR	60.0	40.6	50.3	97	JUN 1971	-15	JAN 1972	6037	36.43	11.43	JUL 1958	0.20	SEP 1958	4.18	JUL 1958	46.8	25.4	DEC 1960	10.9	JAN 1966

Ø For the period July 1963 through the current year.

Means and extremes above are from existing and comparable exposures. Annual extremes have been exceeded at other sites in the locality as follows:
Highest temperature 104 in August 1918; lowest temperature -21 in January 1963; maximum monthly precipitation 11.98 in September 1926; maximum precipitation in 24 hours 5.96 in July 1943; maximum snowfall 26.6 in April 1901; maximum snowfall in 24 hours is 15.6 in April 1901.

NORMALS, MEANS AND EXTREMES
(Table Revised 1975. Base Period for Climatological Normals: 1941-1970)

Month	Normal Daily maximum	Normal Daily minimum	Normal Monthly	Record highest	Year	Record lowest	Year	Heating degree days Base 65°F	Cooling degree days	Normal precip.	Max monthly water equiv.	Year	Min monthly water equiv.	Year	Max in 24 hrs	Year	Snow max monthly	Year	Snow max 24 hrs	Year
J	33.9	18.0	26.0	66	1967	-15	1972	1200	0	2.69	8.70	1950	0.71	1961	2.99	1959	17.6	1964	10.9	1966
F	36.0	19.4	27.7	62	1971	-8	1974	1044	0	2.16	5.24	1956	0.48	1968	2.57	1959	19.7	1967	10.7	1951
M	46.2	26.9	36.6	78	1967	3	1964	893	0	3.15	8.03	1964	1.04	1971	3.48	1964	20.9	1960	5.9	1961
A	59.3	37.8	48.6	85	1970	10	1964	495	9	3.27	6.06	1964	0.91	1957	1.98	1971	3.2	1961	3.2	1961
M	70.4	48.3	59.4	88	1967	24	1966	231	36	3.82	9.03	1964	1.52	1971	2.70	1956	0.0		0.0	
J	79.4	57.1	68.3	97	1971	32	1972	33	132	3.50	7.06	1970	1.01	1967	2.87	1967	0.0		0.0	
J	82.6	60.8	71.7	94	1965	43	1963	9	217	3.80	11.43	1958	1.56	1965	4.18	1958	0.0		0.0	
A	81.7	59.3	70.3	93	1973	35	1974	16	181	2.60	8.19	1974	0.49	1970	2.77	1964	T	1965	T	1965
S	74.7	52.7	63.7	93	1964	21	1967	101	72	2.60	6.47	1972	0.20	1954	2.33	1967	6.8	1952	3.9	1952
O	63.7	42.8	53.3	85	1963	10	1964	369	6	2.50	8.42	1954	0.45	1953	2.77	1954	7.4	1954	7.4	1950
N	48.6	32.7	40.7	77	1968	-6	1971	729	0	2.35	5.05	1966	1.07	1955	2.00	1955	17.9	1950	17.9	1974
D	36.5	22.2	29.4	67	1971	-6	1963	1104	0	2.40	4.89	1951	0.31	1955	1.66	1974	29.4	1974	17.9	1974
YR	59.3	39.8	49.6	97	JUN 1971	-15	JAN 1972	6224	634	35.13	11.43	JUL 1958	0.20	SEP 1954	4.18	JUL 1958	29.4	DEC 1974	17.9	DEC 1974

Means and extremes above are from existing and comparable exposures. Annual extremes have been exceeded at other sites in the locality as follows:
Highest temperature 104 in August 1918; lowest temperature -21 in January 1963; maximum monthly precipitation 11.98 in September 1926; maximum precipitation in 24 hours 5.96 in July 1943.

REFERENCE NOTES APPLYING TO TABLES APPEAR ON THE PAGE FOLLOWING LAST TABLE.
(Caution: Letters and symbols may have different meanings in 1941-1970 tables than in earlier tables. See notes.)

(Table Revised 1973. Base Period for Climatological Normals: 1931-1960)

Station: CINCINNATI, OHIO   ABBE OBSERVATORY   Standard time used: EASTERN   Latitude: 39° 09' N   Longitude: 84° 31' W   Elevation (ground): 761 feet

Ø Wind observation record at Abbe Observatory began April 1, 1915; transferred to Lunken Airport January 2, 1946; transferred to Federal Building March 6, 1947; transferred back to Abbe Observatory April 18, 1951. Wind record for 6 years through 1951 not used due to poor instrumental exposure.

NORMALS, MEANS AND EXTREMES

(Table Revised 1975. Base Period for Climatological Normals: 1941-1970)

Ø Wind observation record at Abbe Observatory began April 1, 1915; transferred to Lunken Airport January 2, 1946; transferred to Federal Building March 6, 1947; transferred back to Abbe Observatory April 18, 1951. Wind record for 6 years through 1951 not used due to poor instrumental exposure.

REFERENCE NOTES APPLYING TO TABLES APPEAR ON THE PAGE FOLLOWING LAST TABLE.

(Caution: Letters and symbols may have different meanings in 1941-1970 tables than in earlier tables. See notes.)

NORMALS, MEANS AND EXTREMES

(Table Revised 1973. Base Period for Climatological Normals: 1931-1960)

Station: CLEVELAND, OHIO — CLEVELAND HOPKINS INTL AIRPORT
Standard time used: EASTERN Latitude: 41° 24' N Longitude: 81° 51' W Elevation (ground): 777 feet

Temperature / Precipitation / Degree days

Month	Daily max	Daily min	Monthly	Rec. highest	Year	Rec. lowest	Year	Heating deg days (Base 65°)	Precip. normal total	Max monthly	Year	Min monthly	Year	Max 24 hrs	Year
J	35.4	21.3	28.4	68	1961	-19	1963	1135	2.67	7.01	1950	0.36	1961	2.33	1959
F	36.7	20.8	28.5	69	1961	-15	1963	1022	2.33	6.40	1950	0.73	1963	2.33	1959
M	43.9	26.3	35.1	80	1967+	10	1964	927	3.13	6.07	1954	0.78	1958	2.76	1948
A	57.3	36.7	47.0	85	1962	25	1964	540	3.41	6.18	1961	1.18	1946	2.24	1961
M	68.9	47.0	58.0	91	1962	31	1966	251	3.44	9.04	1947	1.10	1963	3.73	1955
J	78.3	57.2	67.8	95	1971	35	1972	59	3.52	9.06	1972	1.17	1967	4.00	1972
J	82.4	61.3	71.9	98	1965	41	1968	0	3.43	6.47	1969	1.23	1952	2.87	1969
A	80.8	60.0	70.4	93	1968	41	1965	21	3.31	6.95	1960	0.53	1969	3.07	1947
S	74.5	53.8	64.2	86	1961	34	1964	95	3.28	6.37	1945	0.74	1964	2.26	1969
O	63.4	43.7	53.4	79	1963	23	1964	365	2.90	9.50	1954	0.61	1952	3.44	1954
N	48.8	34.0	41.4	69	1970+	13	1970+	711	2.42	5.60	1950	0.92	1964	2.23	1958
D	37.0	24.0	30.5	69	1963+	-4	1963+	1070	2.34	5.60	1951	0.71	1958	1.62	1968
YR	58.9	40.5	49.7	98	JUL 1965	-19	JAN 1963	6196	35.35	9.50	OCT 1954	0.36	JAN 1961	4.00	JUN 1972

Snow, Ice pellets / Relative humidity / Wind / Sky

Month	Snow mean total	Max monthly	Year	Max 24 hrs	Year	RH 01	RH 07	RH 13	RH 19	Wind mean speed	Prev. dir	Fastest speed	Dir	Year	Mean sky cover	Pct. poss. sunshine
J	10.7	18.7	1945	9.3	1945	73	75	69	68	12.5	SW	68	SW	1959	8.1	31
F	10.7	20.7	1942	9.3	1942	75	75	67	68	12.5	SW	68	W	1956	7.8	37
M	8.6	26.3	1954	14.9	1954	77	77	65	60	12.3	W	74	W	1948	7.6	45
A	2.4	14.5	1943	7.6	1943	74	72	57	60	11.8	W	64	W	1951	7.0	53
M	T	2.1	1963	0.1	1963	75	71	57	58	10.5	S	65	S	1957	6.6	60
J	0.0	0.0		0.0		78	72	61	60	9.3	S	65	W	1958	6.0	65
J	0.0	0.0		0.0		81	82	57	57	8.5	S	61	S	1956	5.6	67
A	0.0	T	1970+	T	1970+	84	84	60	61	8.4	S	61	S	1956	5.4	65
S	T	0.8	1962	6.7	1962	80	80	59	60	9.1	S	45	W	1953	5.7	61
O	0.7	8.0	1962	15.0	1962	76	80	67	65	10.5	S	43	W	1946	6.6	55
N	5.7	22.3	1950	9.3	1950	74	77	71	71	12.4	S	59	W	1948	7.9	31
D	10.6	30.3	1962			76	76	71	73	12.4	S	49	SW	1971	8.2	27
YR	50.5	30.3	DEC 1962	15.0	DEC 1962	76	78	62	66	10.9	S	74	W	MAR 1948	6.8	52

Mean number of days

Month	Clear	Partly cloudy	Cloudy	Precip .01"+	Snow 1.0"+	Thunderstorms	Heavy fog	Max 90°+	Max 32°-	Min 32°-	Min 0°-	Avg daily solar radiation
J	3	5	23	16	4	*	2	0	10	28	3	12
F	3	5	20	13	4	*	2	0	6	23	2	3
M	4	8	20	16	3	2	1	0	2	16	0	0
A	5	11	14	15	1	4	1	*	0	5	0	0
M	6	14	11	13	0	5	1	*	0	1	0	3
J	7	11	12	11	0	7	1	2	0	0	0	0
J	9	12	10	10	0	6	1	3	0	0	0	0
A	10	11	10	10	0	5	1	2	0	0	0	0
S	9	9	12	10	0	3	1	1	0	0	0	0
O	9	9	13	11	*	2	1	0	0	1	0	0
N	6	6	18	15	2	1	1	0	1	11	0	0
D	3	5	23	16	4	*	1	0	1	24	2	*
YR	71	100	194	154	18	36	13	8	48	127	6	

Means and extremes above are from existing and comparable exposures. Annual extremes have been exceeded at other sites in the locality as follows: Highest temperature 103 in July 1941; maximum monthly precipitation 9.77 in June 1902; minimum monthly precipitation 0.17 in August 1881; maximum precipitation in 24 hours 4.97 in September 1901; maximum monthly snowfall 30.5 in February 1908; maximum snowfall 30.5 in 24 hours 17.4 in November 1913; fastest mile of wind 78 from Southwest in May 1940.

NORMALS, MEANS AND EXTREMES

(Table Revised 1975. Base Period for Climatological Normals: 1941-1970)

Elev.: 805 feet m.s.l. Average station pressure: 987.6 mb

Temperature / Degree days / Precipitation

Month	Daily max	Daily min	Monthly	Rec. highest	Year	Rec. lowest	Year	Heating DD	Cooling DD	Precip. normal	Max monthly	Year	Min monthly	Year	Max 24 hrs	Year
J	33.4	20.3	26.9	68	1967	-19	1963	1181	0	2.56	7.01	1950	0.36	1950	2.33	1959
F	35.0	20.8	27.9	69	1961	-15	1963	1039	0	2.18	6.40	1950	0.73	1969	2.33	1959
M	44.1	28.1	36.1	80	1967	10	1974	896	0	3.05	6.07	1954	0.78	1945	2.76	1948
A	57.3	38.5	48.3	85	1962	10	1964	501	0	3.49	6.37	1961	1.18	1954	2.24	1943
M	68.4	48.1	58.3	91	1955	31	1961	244	37	3.49	9.50	1943	1.00	1950	3.73	1955
J	78.3	57.5	67.8	95	1971	40	1972	40	127	3.28	9.06	1972	0.71	1951	4.00	1972
J	81.6	61.2	71.4	98	1965	41	1968	9	208	3.45	6.47	1969	1.23	1952	2.87	1969
A	80.4	59.6	70.0	93	1973	41	1965	17	172	3.00	6.95	1960	0.53	1969	3.07	1947
S	74.2	53.5	63.9	86	1973	34	1964	95	62	2.80	6.37	1945	0.74	1954	2.26	1969
O	63.6	43.4	53.5	79	1963	23	1952	354	7	2.57	9.50	1954	0.61	1958	3.44	1954
N	48.8	34.6	41.6	79	1961	13	1958	702	0	2.76	5.60	1950	0.92	1950	2.23	1958
D	36.4	24.1	30.3	69	1971	-4	1963	1076	0	2.36	5.60	1951	0.71	1951	2.06	1974
YR	58.5	40.8	49.7	98	JUL 1965	-19	JAN 1963	6154	613	34.99	9.50	OCT 1954	0.36	JAN 1961	4.00	JUN 1972

Snow, Ice pellets / Relative humidity / Wind / Sky

Month	Snow mean total	Max monthly	Year	Max 24 hrs	Year	RH 01	RH 07	RH 13	RH 19	Wind mean speed	Prev. dir	Fastest speed	Dir	Year	Mean sky cover	Pct. poss. sunshine
J	10.7	18.7	1945	9.3	1945	73	75	68	70	12.4	SW	68	SW	1959	8.1	32
F	10.7	20.7	1942	10.6	1962	75	75	65	67	12.5	SW	68	W	1956	7.8	37
M	8.6	26.3	1954	14.9	1954	74	77	65	61	12.5	W	74	W	1948	7.5	44
A	2.4	14.5	1943	7.6	1943	75	72	57	59	11.9	W	64	W	1951	7.0	53
M	T	2.1	1974	2.1	1957	76	71	57	58	10.5	S	65	S	1957	6.6	65
J	0.0	0.0		0.0		79	72	61	59	9.1	S	65	W	1958	6.0	65
J	0.0	0.0		0.0		81	82	58	57	8.7	S	61	S	1956	5.6	67
A	0.0	T	1970	T	1970	85	84	60	61	8.4	S	61	S	1956	5.4	65
S	T	T	1962	6.7	1962	80	84	59	60	9.1	S	45	W	1953	5.9	55
O	0.8	8.0	1962	15.0	1962	76	80	67	65	10.0	S	43	W	1946	5.7	50
N	5.7	22.3	1950	12.1	1950	76	77	72	71	12.1	S	59	W	1948	5.9	30
D	10.6	30.3	1962	12.2	1974	75	79	71	73	12.3	S	49	SW	1971	8.3	26
YR	50.4	30.3	DEC 1962	15.0	DEC 1962	77	79	62	67	10.8	S	74	W	MAR 1948	6.8	52

Mean number of days

Month	Clear	Partly cloudy	Cloudy	Precip .01"+	Snow 1.0"+	Thunderstorms	Heavy fog ¼ mi or less	Max 90°+	Max 32°-	Min 32°-	Min 0°-
J	3	5	23	16	4	*	2	0	15	28	3
F	3	5	20	15	4	*	2	0	11	26	2
M	3	8	20	16	3	2	1	0	6	21	0
A	6	11	14	15	1	4	1	*	0	6	0
M	7	12	11	13	*	5	1	0	0	0	0
J	9	11	12	11	0	7	1	2	0	0	0
J	10	12	10	10	0	6	1	3	0	0	0
A	10	11	12	10	0	5	1	2	0	0	0
S	9	9	13	10	0	3	1	1	0	0	0
O	9	9	21	11	*	2	1	0	0	1	0
N	3	6	15	15	2	1	1	0	1	12	0
D	3	5	23	16	4	*	1	0	1	24	2
YR	71	98	196	156	18	36	13	8	46	124	5

Means and extremes above are from existing and comparable exposures. Annual extremes have been exceeded at other sites in the locality as follows: Highest temperature 103 in July 1941; maximum monthly precipitation 9.77 in June 1902; minimum monthly precipitation 0.17 in August 1881; maximum precipitation in 24 hours 4.97 in September 1901; maximum monthly snowfall 30.5 in February 1908; maximum snowfall 30.5 in 24 hours 17.4 in November 1913; fastest mile of wind 78 from Southwest in May 1940.

REFERENCE NOTES APPLYING TO TABLES APPEAR ON THE PAGE FOLLOWING LAST TABLE.

(Table Revised 1973. Base Period for Climatological Normals: 1931-1960)

Station: COLUMBUS, OHIO — PORT COLUMBUS INTL AIRPORT
Standard time used: EASTERN Latitude: 40° 00' N Longitude: 82° 53' W Elevation (ground): 812 feet

Month	Daily maximum	Daily minimum	Monthly	Record highest	Year	Record lowest	Year	Normal heating degree days (Base 65°)	Normal total precip	Snow/Ice pellets mean total
J	37.8	22.0	29.9	68	1967	-15	1963	1088	3.16	7.2
F	39.5	22.7	31.1	72	1961	-11	1963	949	2.31	6.4
M	48.4	29.3	38.9	79	1967	-2	1967+	809	3.49	5.0
A	62.2	39.4	50.8	88	1960	18	1972+	426	4.00	0.6
M	72.9	49.4	61.5	93	1962	31	1966	171	4.16	0.0
J	82.9	58.6	70.8	96	1966	43	1972	27	3.93	0.0
J	86.6	62.9	74.8	97	1966+	39	1965	0	2.86	0.0
A	85.0	61.4	73.2	98	1964	31	1963+	6	2.65	0.0
S	77.6	54.2	65.9	96	1963	20	1962	84	2.11	T
O	65.8	42.5	54.2	86	1963	11	1964	347	2.50	0.1
N	50.6	31.8	41.2	79	1961	-11	1963	714	2.34	3.1
D	39.6	23.4	31.5	72	1971	-10	1963	1039		5.9
YR	62.5	41.5	52.0	98 AUG.1964		-15 JAN.1963		5660	36.67	28.5

Means and extremes above are from existing and comparable exposures. Annual extremes have been exceeded at other sites in the locality as follows:
Highest temperature 106 in July 1936; lowest temperature -20 in February 1899+; maximum monthly precipitation 13.71 in January 1937; minimum monthly precipitation .10 in October 1924; maximum monthly snowfall 29.2 in February 1910; fastest mile wind 84 NW in July 1916.

NORMALS, MEANS AND EXTREMES
(Table Revised 1975. Base Period for Climatological Normals: 1941-1970)

Elev. 833 feet m.s.l. 812 feet Average station pressure 987.3 mb

Month	Daily maximum	Daily minimum	Monthly	Record highest	Year	Record lowest	Year	Normal degree days Heating	Normal degree days Cooling	Precip Normal
J	36.4	20.4	28.4	68	1967	-15	1963	1135	0	2.87
F	39.2	21.4	30.3	72	1961	-11	1963	972	0	2.32
M	49.3	29.1	39.2	80	1974	-2	1967	800	0	3.44
A	62.8	39.5	51.2	88	1960	18	1972	418	0	3.71
M	72.9	49.4	61.1	93	1962	31	1966	176	55	4.10
J	81.9	58.9	70.4	96	1966	35	1972	13	175	4.13
J	83.7	62.9	73.6	97	1966+	43	1972	0	267	4.21
A	83.7	60.1	71.9	98	1964	39	1965	8	222	2.86
S	77.6	54.2	65.2	96	1963	31	1963	76	82	2.41
O	66.4	42.0	54.2	86	1963	20	1964	342	0	1.99
N	50.9	32.4	41.7	79	1961	11	1964	699	0	2.68
D	38.7	22.7	30.7	72	1971	-10	1963	1063	0	2.39
YR	62.1	40.9	51.5	98 AUG.1964		-15 JAN.1963		5702	809	37.01

Means and extremes above are from existing and comparable exposures. Annual extremes have been exceeded at other sites in the locality as follows:
Highest temperature 106 in July 1936; lowest temperature -20 in February 1899+; maximum monthly precipitation 10.71 in January 1937; minimum monthly precipitation .10 in October 1924; maximum monthly snowfall 29.2 in February 1910; fastest mile wind 84 NW in July 1916.

REFERENCE NOTES APPLYING TO TABLES APPEAR ON THE PAGE FOLLOWING LAST TABLE.
(Caution: Letters and symbols may have different meanings in 1941-1970 tables than in earlier tables. See notes.)

NORMALS, MEANS AND EXTREMES

(Table Revised 1973. Base Period for Climatological Normals: 1931-1960)

Station: DAYTON, OHIO   MUNICIPAL AIRPORT   Standard time used: EASTERN   Latitude: 39° 54' N   Longitude: 84° 13' W   Elevation (ground): 1002 feet

Month	Temperature Normal Daily maximum	Normal Daily minimum	Normal Monthly	Extremes Record highest	Year	Record lowest	Year	Normal heating degree days (Base 65°)	Precipitation Normal total	Maximum monthly	Year	Minimum monthly	Year	Maximum in 24 hrs.	Year	Snow, Ice pellets Mean total	Maximum monthly	Year	Maximum in 24 hrs.	Year	Relative humidity Hour 01	Hour 07	Hour 13	Hour 19	Wind Mean speed	Prevailing direction	Fastest mile Speed	Direction	Year	Pct. of possible sunshine	Mean sky cover sunrise to sunset	Clear	Partly cloudy	Cloudy	Precipitation .01 inch or more	Snow, Ice pellets 1.0 inch or more	Thunderstorms	Heavy fog	Temperatures Max. 90° and above	32° and below	Min. 32° and below	0° and below	Average daily solar radiation - langleys
(a)	(b)	(b)	(b)	9		9		(b)	(b)	29		29		29		29	29		29		9	9	9	9	15	15	29	29	29	29	29	29	29	29	29	29	29	29	9	9	9	9	9
J	36.9	22.2	29.6	69	1967	-16	1972	1097	3.18	9.86	1950	0.62	1944	4.30	1959	6.7	19.4	1964	9.8	1964	72	74	65	60	11.8	S	59	SW	1959+	42	7.5	5	6	20	12	2	1	4	0	13	26	3	9
F	38.8	22.9	30.9	67	1972	-7	1965	955	3.02	4.57	1950	0.14	1947	2.79	1951	5.9	12.5	1964	7.7	1947	75	76	61	58	12.1	WNW	60	SW	1946	45	7.2	5	6	17	10	2	1	2	0	8	25	2	3
M	47.5	29.9	38.7	77	1967	3	1960	809	3.32	7.65	1964	1.07	1966	2.87	1964	5.0	13.8	1968	11.3	1968	73	77	61	55	12.4	WNW	70	S	1948	49	7.4	5	7	19	13	2	2	1	1	3	20	0	2
A	60.5	40.8	50.7	86	1960	15	1957	429	3.52	6.69	1961	0.56	1962	2.61	1957	0.5	3.2	1963	4.7	1953	74	77	54	51	11.9	SSW	61	SW	1957	53	7.1	6	8	17	12	1	3	*	1	*	7	0	0
M	70.7	51.4	61.6	90	1962	29	1966	167	3.73	7.76	1968	1.55	1964	2.93	1964	T	T	1966	T	1966	74	76	54	53	10.1	SSW	57	SW	1970	60	6.7	6	11	14	12	1	0	4	1	0	*	0	0
J	81.4	61.5	71.5	98	1971	40	1972	30	4.10	10.89	1958	0.32	1962	3.25	1962	0.0	0.0		0.0		77	76	53	53	9.1	SSW	67	NW	1950	67	6.1	7	11	12	10	1	0	7	1	4	0	0	0
J	85.3	65.1	75.2	100	1972	44	1972	0	3.53	7.21	1955	0.79	1955	2.98	1955	0.0	0.0		0.0		78	80	55	58	8.0	SSW	52	W	1972	68	5.8	8	12	11	9	1	0	7	2	7	0	0	0
A	83.7	63.7	73.7	100	1965	40	1965	6	2.88	4.35	1947	0.34	1967	2.48	1947	0.0	0.0		0.0		80	84	55	60	7.5	SSW	49	SW	1950	71	5.5	8	13	10	9	1	0	5	2	4	0	0	0
S	77.3	56.2	66.8	97	1954	36	1964	78	2.68	5.69	1950	0.27	1963	2.45	1950	0.0	0.0		0.0		80	85	55	60	8.3	SSW	55	SW	1955	67	5.5	10	11	9	8	1	0	3	2	3	0	0	0
O	66.0	45.4	55.6	87	1963	25	1972+	310	2.23	4.28	1954	0.10	1954	1.99	1954	0.1	2.0	1962	2.0	1962	75	80	56	64	9.0	SSW	56	SW	1954	63	5.3	11	8	12	8	1	*	1	2	0	4	0	0
N	50.1	33.5	41.8	77	1968	8	1964	696	2.67	5.67	1949	0.48	1949	2.93	1949	2.6	12.7	1950	10.0	1950	77	80	60	67	11.4	SSW	56	W	1952	42	7.2	7	7	16	11	2	1	1	1	7	14	*	0
D	39.0	23.5	31.3	67	1971+	-12	1963	1045	2.37	5.07	1951	0.36	1955	1.71	1945	5.9	15.6	1960	7.0	1960	77	79	67	75	11.5	SSW	47	W	1951	37	7.5	4	7	20	12	2	1	1	4	12	23	1	*
YR	61.5	43.0	52.3	100	JUL. 1965+	-16	JAN. 1972	5622	36.04	10.89	JUN. 1958	0.10	OCT. 1944	4.30	JAN. 1959	27.3	19.4	JAN. 1964	11.3	MAR. 1968	75	79	64	64	10.3	SSW	78	NW	JUN. 1950	57	6.6	79	104	182	128	8	40	23	17	32	117	5	5

Ø For period July 1963 through the current year.
Means and extremes above are from existing and comparable exposures. Annual extremes have been exceeded at other sites in the locality as follows: Highest temperature 108 in July 1901; lowest temperature -28 in February 1899; maximum monthly precipitation 12.41 in January 1937; maximum precipitation in 24 hours 4.56 in September 1925; maximum monthly snowfall 24.4 in January 1918.

NORMALS, MEANS AND EXTREMES

(Table Revised 1975. Base Period for Climatological Normals: 1941-1970)

Station: DAYTON, OHIO   MUNICIPAL AIRPORT   Standard time used: EASTERN   Latitude: 39° 54' N   Longitude: 84° 12' W   Elevation (ground): 995 feet

Month	Temperatures °F Normal Daily maximum	Normal Daily minimum	Normal Monthly	Extremes Record highest	Year	Record lowest	Year	Normal Degree days Base 65°F Heating	Cooling	Precipitation in inches Water equivalent Normal	Maximum monthly	Year	Minimum monthly	Year	Maximum in 24 hrs.	Year	Snow, Ice pellets Maximum monthly	Year	Maximum in 24 hrs.	Year	Relative humidity pct. Hour 01	Hour 07	Hour 13	Hour 19	Wind Mean speed m.p.h.	Prevailing direction	Fastest mile Speed m.p.h.	Direction	Year	Pct. of possible sunshine	Mean sky cover, tenths, sunrise to sunset	Clear	Partly cloudy	Cloudy	Precipitation .01 inch or more	Snow, Ice pellets 1.0 inch or more	Thunderstorms	Heavy fog, visibility ¼ mile or less	Temperatures °F Max. 90° and above	32° and below	Min. 32° and below	0° and below	Average station pressure mb. Elev. 1003 feet m.s.l.
(a)				11		11				31	31		31		31		31		31		11	11	11	11	31	15	31	31	31	31	31	31	31	31	31	31	31	31	11	11	11	11	2
J	35.8	20.4	28.1	69	1967	-16	1972	1144	0	2.76	9.86	1950	0.62	1944	4.30	1959	19.4	1964	9.8	1964	73	75	66	69	11.8	S	59	SW	1959	42	7.6	5	5	20	12	2	1	1	0	12	26	3	982.9
F	38.6	21.1	30.4	67	1972	-7	1965	969	0	2.24	4.57	1950	0.14	1947	2.79	1951	12.5	1964	7.7	1947	75	76	63	68	12.1	WNW	66	SW	1946	45	7.4	5	6	17	10	2	1	1	0	8	25	2	982.5
M	48.1	29.0	39.0	84	1974	-1	1960	806	0	3.21	7.65	1964	1.07	1966	2.87	1964	13.8	1968	11.3	1968	73	78	62	66	12.0	WNW	70	S	1964	53	7.1	5	7	19	13	2	2	1	*	4	21	1	978.5
A	61.7	41.0	51.4	90	1970	15	1957	413	5	3.34	6.69	1961	0.56	1962	2.61	1957	4.9	1974	4.7	1953	72	76	54	62	11.4	SSW	61	SW	1957	53	7.0	6	8	16	13	1	3	1	1	*	7	0	978.3
M	72.0	51.2	61.6	90	1962	28	1966	166	61	3.76	7.76	1968	1.55	1964	2.93	1964	T	1966	T	1966	74	76	54	60	10.1	SSW	57	SW	1970	60	6.7	6	11	14	12	1	0	5	1	0	*	0	977.7
J	81.6	61.0	71.3	98	1971	40	1972	7	202	3.88	10.89	1958	0.32	1962	3.25	1958	0.0		0.0		76	75	53	57	9.1	SSW	67	NW	1950	67	6.2	7	11	12	10	1	0	7	1	4	0	0	978.7
J	84.7	64.6	74.6	100	1972	44	1972	0	298	3.54	7.21	1955	0.47	1974	3.29	1974	0.0		0.0		78	80	54	57	8.0	SSW	52	W	1972	68	5.8	8	12	11	11	1	0	7	1	7	0	0	980.9
A	83.4	62.5	73.0	100	1965	40	1965	9	255	2.55	8.03	1974	0.34	1967	3.62	1974	0.0		0.0		80	85	55	61	7.5	SSW	49	SW	1950	71	5.5	10	13	8	11	1	0	6	1	4	0	0	980.3
S	77.0	55.5	66.3	97	1954	25	1963	63	102	2.28	5.69	1950	0.27	1963	2.45	1950	0.0		0.0		80	85	55	66	8.3	SSW	55	SW	1955	63	5.3	11	8	11	8	1	0	3	1	1	3	0	982.8
O	66.0	45.5	55.8	87	1963	23	1972	307	13	1.94	4.28	1954	0.10	1954	1.99	1954	2.0	1962	2.0	1962	76	80	56	72	9.0	SSW	56	SW	1954	63	5.3	11	7	13	8	1	*	1	2	0	4	0	984.6
N	50.1	33.4	41.8	77	1968	8	1964	696	0	2.29	5.67	1949	0.48	1949	2.93	1949	12.7	1950	7.6	1950	76	80	60	72	11.4	SSW	56	W	1952	42	7.2	7	7	16	11	2	1	2	1	7	13	*	981.6
D	38.3	23.4	30.9	67	1971	-12	1963	1057	0	2.29	5.07	1951	0.36	1955	1.71	1945	15.6	1960	7.6	1960	77	79	67	75	11.5	SSW	47	W	1973	37	7.6	5	7	20	12	2	1	3	*	12	23	1	980.5
YR	61.5	42.5	52.0	100	JUL 1965	-16	JAN 1972	5641	936	34.36	10.89	JUN 1958	0.10	OCT 1944	4.30	JAN 1959	19.4	JAN 1964	11.3	MAR 1968	76	79	60	64	10.3	SSW	78	NW	JUN 1950	57	6.6	79	103	183	130	8	41	23	17	31	115	4	980.8

Means and extremes above are from existing and comparable exposures. Annual extremes have been exceeded at other sites in the locality as follows: Highest temperature 108 in July 1901; lowest temperature -28 in February 1899; maximum monthly precipitation 12.41 in January 1937; maximum precipitation in 24 hours 4.56 in September 1925; maximum monthly snowfall 24.4 in January 1918.

REFERENCE NOTES APPLYING TO TABLES APPEAR ON THE PAGE FOLLOWING LAST TABLE.
(Caution: Letters and symbols may have different meanings in 1941-1970 tables than in earlier tables. See notes.)

SANDUSKY, OHIO
POST OFFICE BUILDING
LATITUDE 41° 27' N
LONGITUDE 82° 43' W
ELEVATION (ground) 603 feet

NORMALS, MEANS AND EXTREMES

(Table Revised 1959. Base Period for Climatological Normals: 1921-1950)

Month	Temperature Normal Daily maximum	Temperature Normal Daily minimum	Temperature Normal Monthly	Temp Extremes Record highest	Year	Temp Extremes Record lowest	Year	Normal degree days	Precip Normal total	Precip Max monthly	Year	Precip Min monthly	Year	Precip Max in 24 hrs	Year	Snow Mean total	Snow Max monthly	Year	Snow Max in 24 hrs	Year	RH 7:00 a.m. EST	RH 1:00 p.m. EST	RH 7:00 p.m. EST	Wind Mean hourly speed	Wind Prevailing direction	Wind Fastest speed	Wind Fastest direction	Wind Fastest year	Pct possible sunshine	Mean sky cover sunrise to sunset	Days Clear	Days Partly cloudy	Days Cloudy	Precip .01+	Snow/Sleet 1.0+	Thunderstorms	Heavy fog	90° & above (max)	32° & below (max)	32° & below (min)	0° & below (min)
(a)	(b)	(b)	(b)	80	80	80	80	(b)	(b)	81	81	81	81	81	81	73	73	73	73	73	23	25	20	80	80	80	80	80	47	69	80	80	80	81	66	74	74	80	80	80	80
J	35.7	21.8	28.8	73	1950	-16	1879	1122	2.29	6.58	1937	0.60	1902	1.71	1937	8.2	29.8	1893	11.8	1910	81	68	78	10.8	SW	56	SW	1952	35	7.4	5	8	18	14	4	0	1	0	13	26	2
F	38.4	22.4	29.4	72	1944	-15	1899	997	1.92	8.53	1887	0.27	1920	2.98	1887	6.5	22.1	1893	10.1	1900	81	70	76	10.9	SW	64	NE	1881	41	7.0	5	8	15	12	4	0	1	0	11	24	1
M	45.2	29.8	37.5	85	1910	-3	1885	853	2.89	8.69	1913	0.28	1910	2.96	1913	4.5	16.1	1916	8.7	1955	79	64	74	11.3	SW	56	NW	1918	49	6.7	6	9	16	13	3	2	1	0	5	21	0
A	56.3	39.4	47.9	90	1942	14	1923	513	2.96	6.24	1910	0.35	1915	2.21	1929	1.2	12.0	1957	9.0	1957	73	61	68	10.6	SW	58	SW	1919	52	6.3	7	9	14	12	0	3	0	0	0	5	0
M	68.3	50.2	59.3	93	1941+	32	1923+	217	3.32	9.04	1943	0.64	1934	3.83	1938	0.0	0.0		0.0		72	53	67	9.0	SW	48	SW	1917	62	5.8	8	11	12	13	0	5	0	1	0	0	0
J	79.0	60.7	69.9	104	1934	40	1894	41	3.73	12.51	1937	0.91	1919	5.95	1937	0.0	0.0		0.0		73	53	68	8.1	SW	77	NW	1924	69	5.3	9	13	8	11	0	7	0	3	0	0	0
J	83.7	65.5	74.6	105	1936	50	1918	0	3.45	9.71	1943	0.26	1916	3.87	1935	0.0	0.0		0.0		72	53	65	7.6	SW	69	N	1879	71	4.7	11	14	6	10	0	7	0	5	0	0	0
A	81.7	63.9	72.8	105	1918	45	1946	0	2.81	8.02	1882	0.23	1894	4.20	1906	0.0	0.0		0.0		74	57	67	7.6	SW	63	NE	1885	68	4.8	11	13	7	9	0	5	0	3	0	0	0
S	75.4	57.5	66.5	99	1953+	34	1956+	66	3.26	7.72	1950	0.73	1928	4.28	1950	0.0	0.0		0.0		74	53	68	8.4	SW	52	NW	1897	63	5.0	11	11	8	9	0	3	1	2	0	0	0
O	63.5	46.5	55.0	93	1953	22	1925	327	2.10	6.22	1917	0.43	1897	2.76	1920	T	1.6	1917	1.5	1917	77	59	70	9.3	SW	54	N	1885	55	5.5	10	9	12	10	0	1	1	0	0	1	0
N	49.2	35.2	42.2	82	1950	0	1880	684	2.27	6.43	1927	0.09	1904	2.26	1927	1.9	12.3	1950	12.3	1950	79	65	73	10.9	SW	68	SW	1919	37	7.1	5	8	17	12	1	1	1	0	2	11	0
D	37.6	25.3	31.5	70	1899	-13	1880	1039	2.16	6.27	1881	0.63	1934	1.74	1927	6.1	20.2	1951	9.0	1886	81	71	75	10.6	SW	56	NW	1895	29	7.7	3	8	20	13	3	0	1	0	10	23	0
Year	59.3	43.2	51.3	105	July 1936+	-16	Jan. 1879	5859	33.16	12.51	June 1937	0.09	Nov. 1904	5.95	June 1937	28.2	29.8	Jan. 1893	12.3	Nov. 1950	76	62	71	9.6	SW	77	NW	June 1924	53	6.1	91	121	153	138	13	34	6	13	41	111	3

Station moved or discontinued. See other stations for this state.

REFERENCE NOTES APPLYING TO TABLES APPEAR ON THE PAGE FOLLOWING LAST TABLE.

(Caution: Letters and symbols may have different meanings in 1941-1970 tables than in earlier tables. See notes.)

NORMALS, MEANS AND EXTREMES
(Table Revised 1973. Base Period for Climatological Normals: 1931-1960)

Station: TOLEDO, OHIO   TOLEDO EXPRESS AIRPORT   Standard time used: EASTERN   Latitude: 41°36'N   Longitude: 83°48'W   Elevation (ground): 669 feet

Month	Temperature Normal Daily maximum	Normal Daily minimum	Normal Monthly	Extremes Record highest	Year	Record lowest	Year	Normal heating degree days (Base 65°)	Precipitation Normal total	Maximum monthly	Year	Minimum monthly	Year	Maximum in 24 hrs.	Year
J	34.1	18.4	26.3	62	1967+	-17	1972+	1200	2.33	4.61	1965	0.27	1961	1.78	1959
F	35.7	18.9	27.3	68	1957	-14	1960	1056	1.88	3.13	1960	0.27	1960	1.35	1959
M	44.7	26.6	35.2	80	1963	-7	1960	924	2.26	4.88	1964	0.58	1958	1.56	1964
A	59.6	37.2	48.4	87	1960	11	1964	543	2.77	4.94	1961	0.88	1962	2.39	1956
M	70.4	46.1	58.3	95	1962	26	1968	242	3.04	5.13	1968	0.96	1964	1.96	1970
J	80.3	56.3	68.3	99	1971	32	1972	60	3.79	4.86	1960	1.89	1964	2.50	1956
J	85.1	60.2	72.7	96	1966+	43	1972+	8	2.59	6.75	1969	1.58	1969	4.39	1969
A	83.0	58.8	70.9	98	1964	37	1965	16	3.33	8.47	1965	0.81	1967	2.42	1967
S	75.5	51.3	63.4	95	1960	29	1961	117	2.13	8.10	1972	0.58	1963	3.97	1972
O	63.8	40.3	52.1	91	1963	16	1965	406	2.39	3.72	1959	0.28	1959	1.71	1957
N	47.3	29.8	38.6	78	1968	-11	1958	792	2.04	4.63	1966	0.77	1958	2.06	1966
D	35.8	20.8	28.3	67	1971	-11	1960	1138	1.95	6.81	1967	0.54	1958	3.53	1967
YR	59.5	38.5	49.0	99	JUN. 1971	-17	JAN. 1972+	6494	30.50	8.47	AUG. 1965	0.27	FEB. 1969+	4.39	JUL. 1969

Means and extremes above are from existing and comparable exposures. Annual extremes have been exceeded at other sites in the locality as follows:
Highest temperature 105 in July 1936; maximum monthly precipitation 8.49 in October 1881; minimum monthly precipitation 0.04 in November 1904;
maximum precipitation in 24 hours 5.98 in September 1918; maximum snowfall 26.2 in January 1918; maximum monthly snowfall 26.2 in
February 1900; fastest mile wind 87 in March 1948.

NORMALS, MEANS AND EXTREMES
(Table Revised 1975. Base Period for Climatological Normals: 1941-1970)

Means and extremes above are from existing and comparable exposures. Annual extremes have been exceeded at other sites in the locality as follows:
Highest temperature 105 in July 1936; maximum monthly precipitation 8.49 in October 1881; minimum monthly precipitation 0.04 in November 1904;
maximum precipitation in 24 hours 5.98 in September 1918; maximum snowfall 26.2 in January 1918; maximum monthly snowfall 26.2 in
February 1900; fastest mile wind 87 in March 1948.

REFERENCE NOTES APPLYING TO TABLES APPEAR ON THE PAGE FOLLOWING LAST TABLE.
(Caution: Letters and symbols may have different meanings in 1941-1970 tables than in earlier tables. See notes.)

NORMALS, MEANS AND EXTREMES
(Table Revised 1973. Base Period for Climatological Normals: 1931-1960)

Station: YOUNGSTOWN, OHIO
MUNICIPAL AIRPORT
Standard time used: EASTERN Latitude: 41° 16' N Longitude: 80° 40' W Elevation (ground): 1178 feet

Month	Daily maximum	Daily minimum	Monthly	Record highest	Year	Record lowest	Year	Normal heating degree days (Base 65°)	Precip. Normal total
J	34.4	20.2	27.3	71	1950	-18	1963	1169	3.16
F	35.4	19.7	27.6	67	1961	-11	1963+	1047	2.50
M	44.0	26.5	35.3	79	1945	-4	1948	921	3.35
A	57.6	36.3	47.0	86	1948	11	1950	540	3.70
M	68.9	46.8	57.9	92	1962	24	1970+	248	4.07
J	78.0	56.1	67.1	99	1952	30	1972	60	3.58
J	81.7	59.3	70.5	100	1954	42	1968	6	4.32
A	80.3	59.7	70.0	97	1953	41	1965	19	3.50
S	73.9	51.7	62.5	99	1954	29	1957	120	2.89
O	61.4	42.3	51.9	87	1953+	20	1969	412	2.71
N	46.6	31.9	39.3	80	1961	-9	1958	771	2.81
D	36.0	22.8	29.4	69	1971	-9	1951	1104	2.68
YR	58.1	39.4	48.8	100	JUL. 1954	-18	JAN. 1963	6417	39.27

Month	Precip max monthly	Year	Precip min monthly	Year	Max in 24 hrs	Year	Snow mean total	Snow max monthly	Year	Snow max 24 hrs	Year
J	7.64	1950	0.82	1961	2.79	1959	12.7	30.1	1948	17.5	1948
F	5.26	1950	0.71	1968	2.76	1959	11.5	22.7	1965	9.5	1951
M	6.20	1964	1.34	1960+	2.47	1964	11.4	23.2	1961	9.4	1964
A	6.43	1957	1.30	1946	2.85	1957	2.5	12.2	1957	6.4	1957
M	9.87	1946	1.38	1957	2.96	1946	0.2	5.0	1966	5.4	1966
J	6.97	1957	1.35	1952	3.82	1957	0.0	0.0		0.0	
J	7.41	1958	1.57	1958	2.79	1967	0.0	0.0		0.0	
A	7.86	1956	0.51	1969+	4.31	1956	0.0	0.0		0.0	
S	6.17	1950	0.27	1960	4.31	1959	T	T	1970+	T	1970+
O	8.59	1954	0.43	1953	2.27	1954	0.5	7.4	1962	4.9	1962
N	5.52	1966	1.38	1950	2.47	1955	7.0	30.6	1950	20.7	1950
D	5.52	1971	0.88	1958	1.51	1971+	12.8	23.0	1963	14.8	1963
YR	9.87	MAY 1946	0.27	SEP. 1960	4.31	OCT. 1954	58.6	30.6	NOV. 1950	20.7	NOV. 1950

NORMALS, MEANS AND EXTREMES
(Table Revised 1975. Base Period for Climatological Normals: 1941-1970)

Month	Daily maximum	Daily minimum	Monthly	Record highest	Year	Record lowest	Year	Heating	Cooling	Precip Normal
J	33.0	18.3	25.7	71	1950	-18	1963	1218	0	2.94
F	34.7	18.7	26.7	67	1961	-11	1963	1072	0	2.42
M	44.3	26.3	35.3	79	1945	-14	1948	1021	0	3.64
A	57.6	36.3	47.0	86	1948	11	1950	519	0	3.67
M	68.9	46.3	57.6	92	1962	24	1966	258	29	3.67
J	78.3	56.1	67.0	99	1952	30	1972	42	102	3.59
J	81.8	59.6	70.7	100	1954	42	1968	9	185	3.90
A	80.4	58.0	69.2	97	1953	41	1965	22	153	3.23
S	73.9	51.4	62.7	99	1954	29	1957	118	40	2.64
O	62.6	42.6	52.6	87	1969	20	1969	384	0	2.94
N	47.9	32.2	40.3	80	1961	-9	1958	741	0	2.97
D	35.6	22.0	28.8	69	1971	-9	1951	1122	0	2.55
YR	58.4	39.0	48.7	100	JUL 1954	-18	JAN 1963	6426	518	37.99

Average station pressure: Elev. 1186 feet m.s.l.

REFERENCE NOTES APPLYING TO TABLES APPEAR ON THE PAGE FOLLOWING LAST TABLE.
(Caution: Letters and symbols may have different meanings in 1941-1970 tables than in earlier tables. See notes.)

Reference notes applying to Normals, Means, and Extremes tables for 1931–1960 base period.

(a) Length of record, years, based on January data, Other months may be for more or fewer years if there have been breaks in the record.

(b) Climatological standard normals (1931-1960).
* Less than one half.
+ Also on earlier dates, months, or years.
T Trace, an amount too small to measure.
Below zero temperatures are preceded by a minus sign. The prevailing direction for wind in the Normals, Means, and Extremes table is from records through 1963.
‡ > 70° at Alaskan stations.

Unless otherwise indicated, dimensional units used in this bulletin are: temperature in degrees F.; precipitation, including snowfall, in inches; wind movement in miles per hour; and relative humidity in percent. Heating degree day totals are the sums of negative departures of average daily temperatures from 65° F. Cooling degree day totals are the sums of positive departures of average daily temperatures from 65° F. Sleet was included in snowfall totals beginning with July 1948. The term "Ice pellets" includes solid grains of ice (sleet) and particles consisting of snow pellets encased in a thin layer of ice. Heavy fog reduces visibility to 1/4 mile or less.

Sky cover is expressed in a range of 0 for no clouds or obscuring phenomena to 10 for complete sky cover. The number of clear days is based on average cloudiness 0-3, partly cloudy days 4-7, and cloudy days 8-10 tenths.

Solar radiation data are the averages of direct and diffuse radiation on a horizontal surface. The langley denotes one gram calorie per square centimeter.

& Figures instead of letters in a direction column indicate direction in tens of degrees from true North; i.e., 09 - East, 18 - South, 27 - West, 36 - North, and 00 - Calm. Resultant wind is the vector sum of wind directions and speeds divided by the number of observations. If figures appear in the direction column under "Fastest mile" the corresponding speeds are fastest observed 1-minute values.
¢ Through 1964.
% Through 1964. The station did not operate 24 hours daily. Thunderstorm data therefore may be incomplete.
# To 8 compass points.
** The National Weather Service considers the accuracy of solar radiation data questionable; therefore, publication is suspended pending determination of corrected values.

Reference notes applying to Normals, Means, and Extremes tables for 1941–1970 base period.

(a) Length of record, years, through the current year unless otherwise noted, based on January data.
(b) 70° and above at Alaskan stations.
* Less than one half.
T Trace.

NORMALS - Based on record for the 1941-1970 period.
DATE OF AN EXTREME - The most recent in cases of multiple occurrence.
PREVAILING WIND DIRECTION - Record through 1963.
WIND DIRECTION - Numerals indicate tens of degrees clockwise from true north. 00 indicates calm.
FASTEST MILE WIND - Speed is fastest observed 1-minute value when the direction is in tens of degrees.

¢ Through 1964.

% Through 1964. The station did not operate 24 hours daily. Thunderstorm data therefore may be incomplete.

$ Record through 1972.

Mean Annual Precipitation, Inches

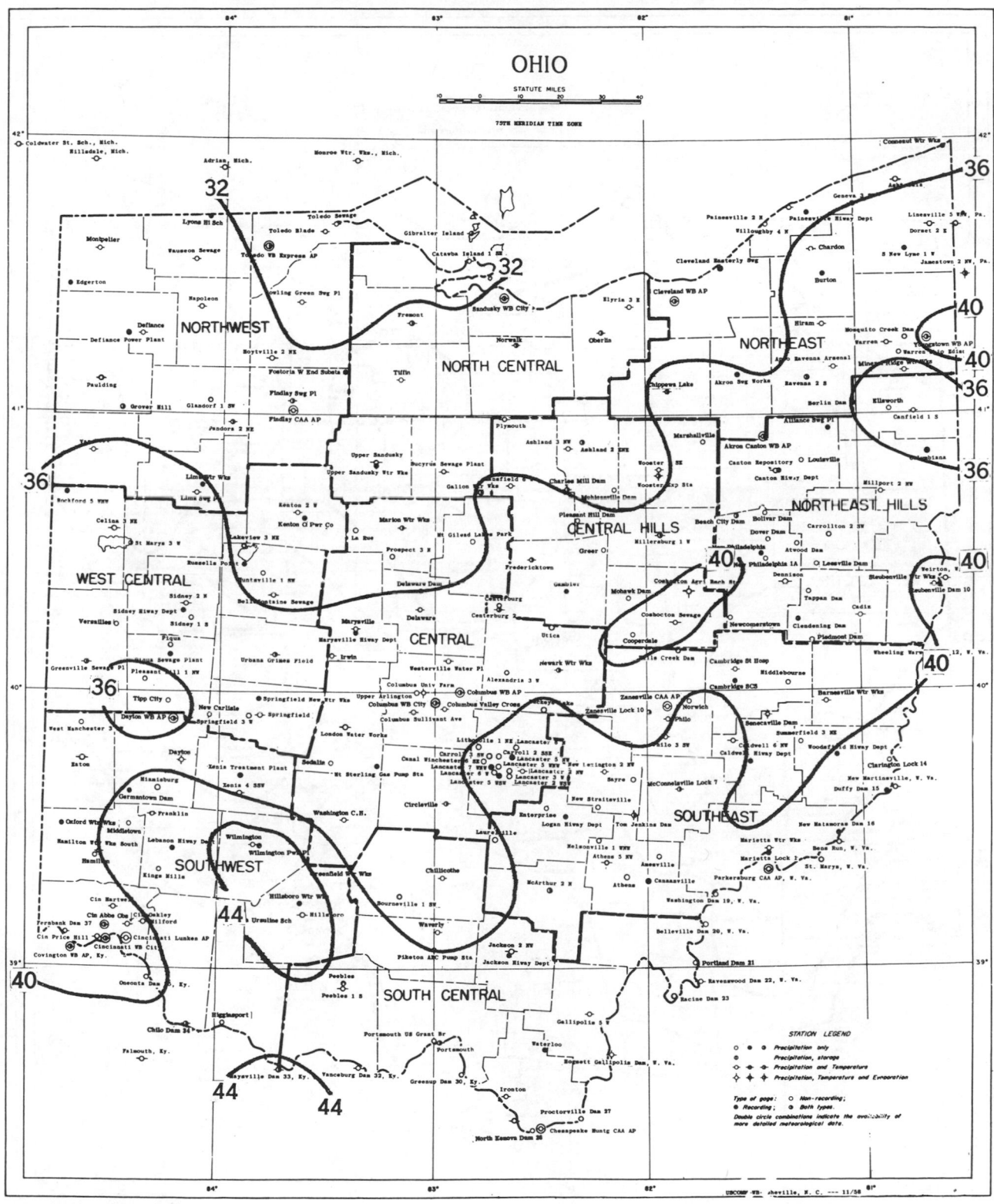

Based on period 1931-55

Isolines are drawn through points of approximately equal value. Caution should be used in interpolating on these maps, particularly in mountainous areas.

Mean Maximum Temperature (°F.), January

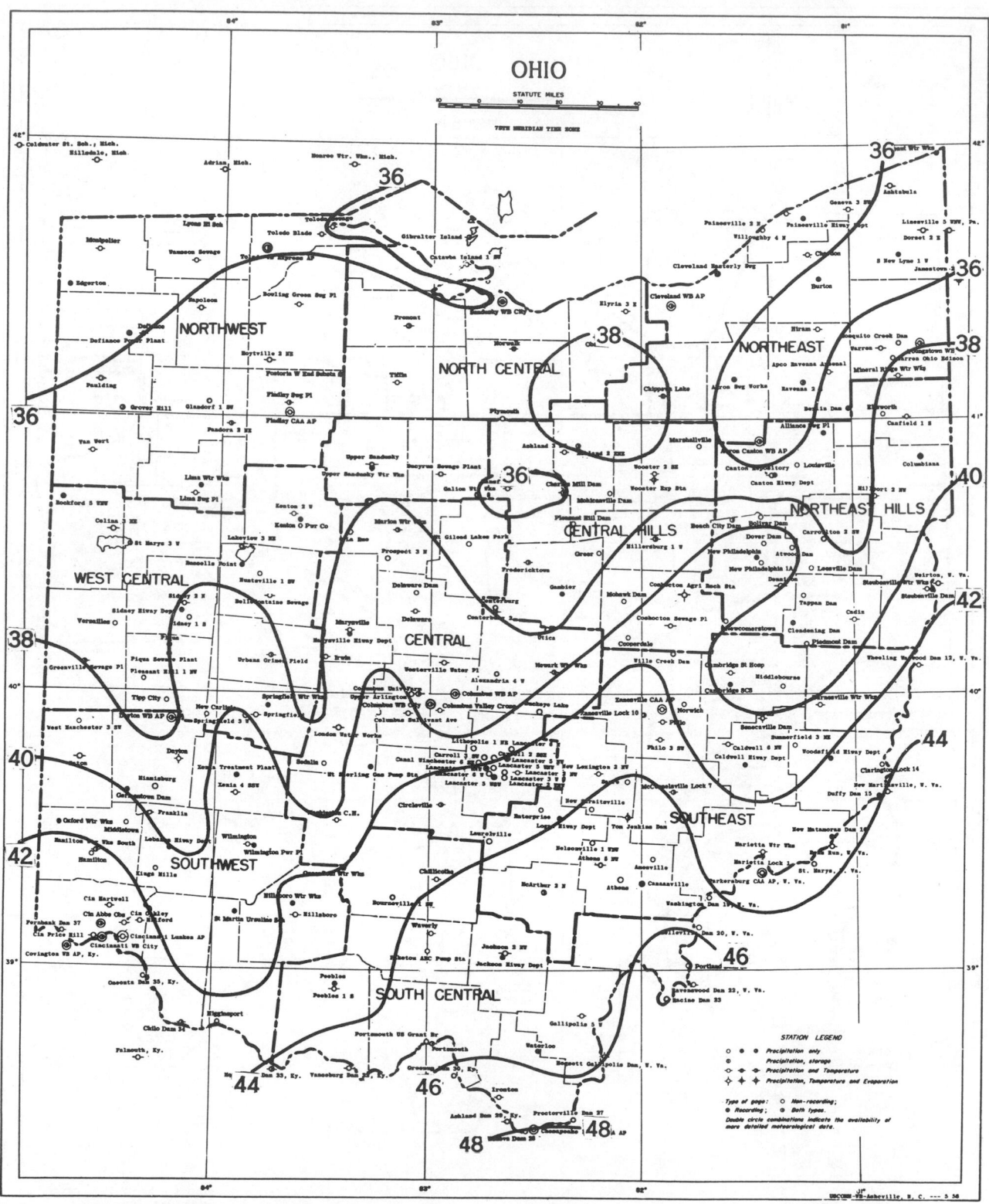

Based on period 1931-52

Isolines are drawn through points of approximately equal value. Caution should be used in interpolating on these maps, particularly in mountainous areas.

784

Mean Minimum Temperature (°F.), January

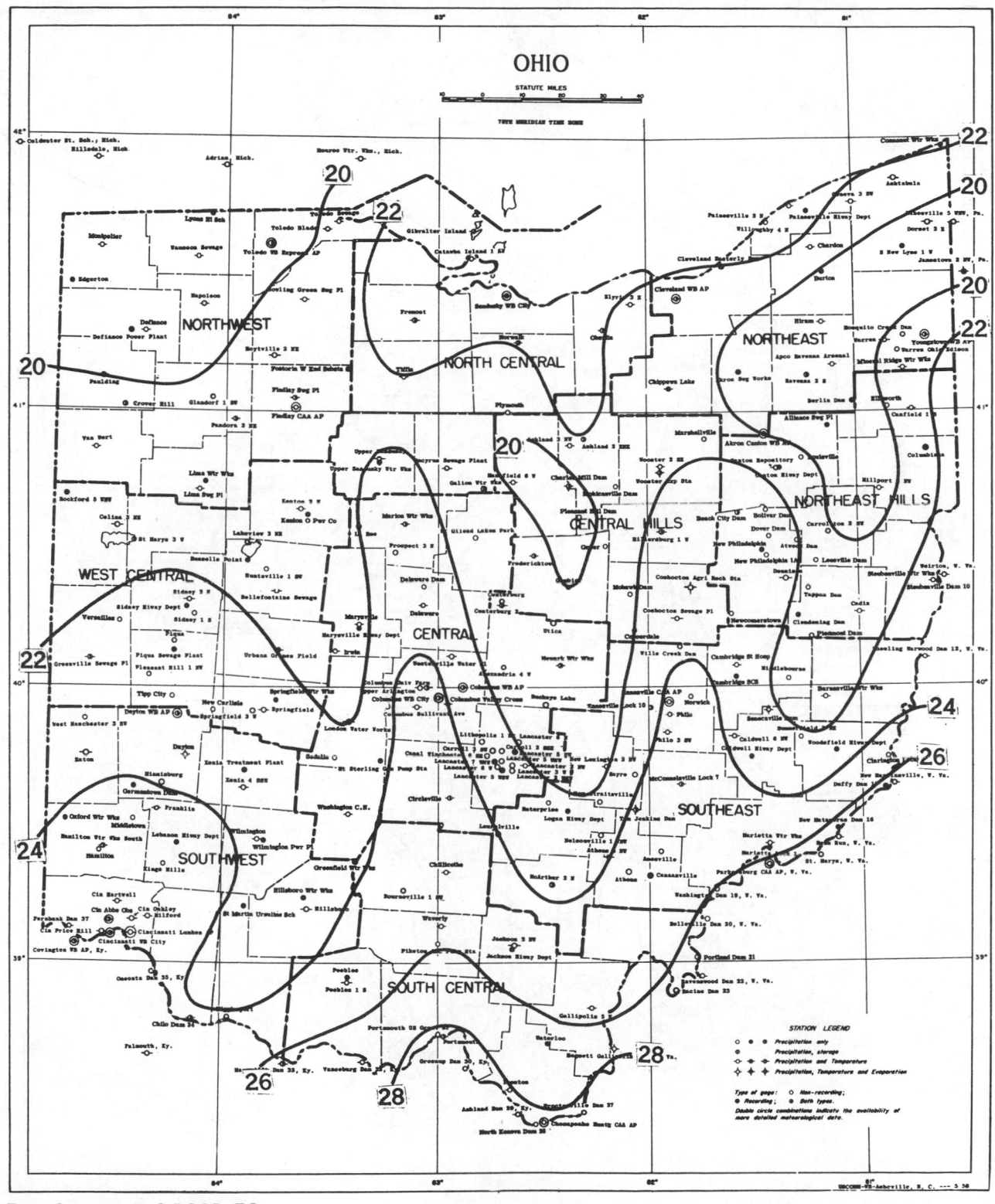

Based on period 1931-52

Isolines are drawn through points of approximately equal value. Caution should be used in interpolating on these maps, particularly in mountainous areas.

Mean Maximum Temperature (°F.), July

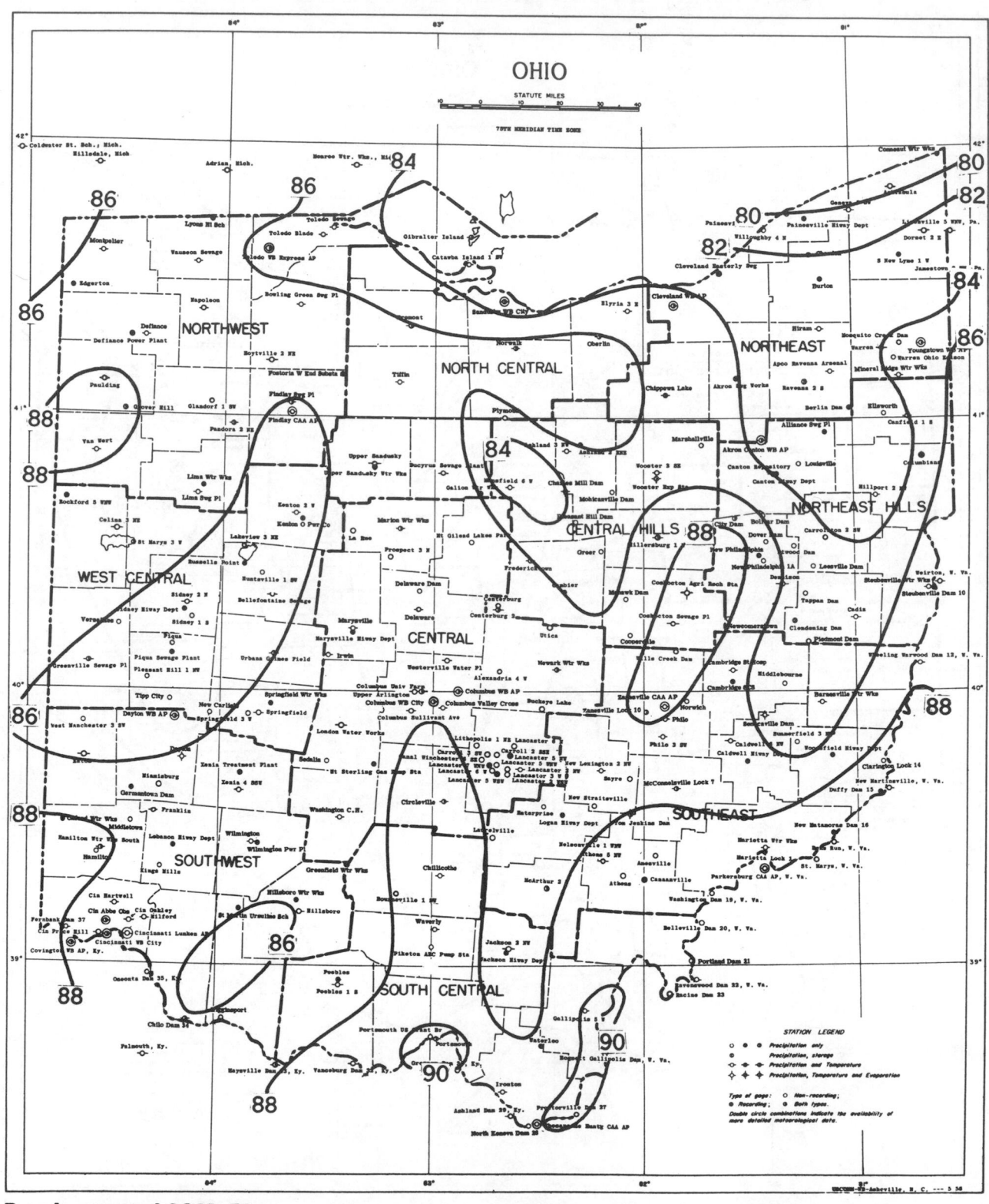

Based on period 1931-52

Isolines are drawn through points of approximately equal value. Caution should be used in interpolating on these maps, particularly in mountainous areas.

Mean Minimum Temperature (°F.), July

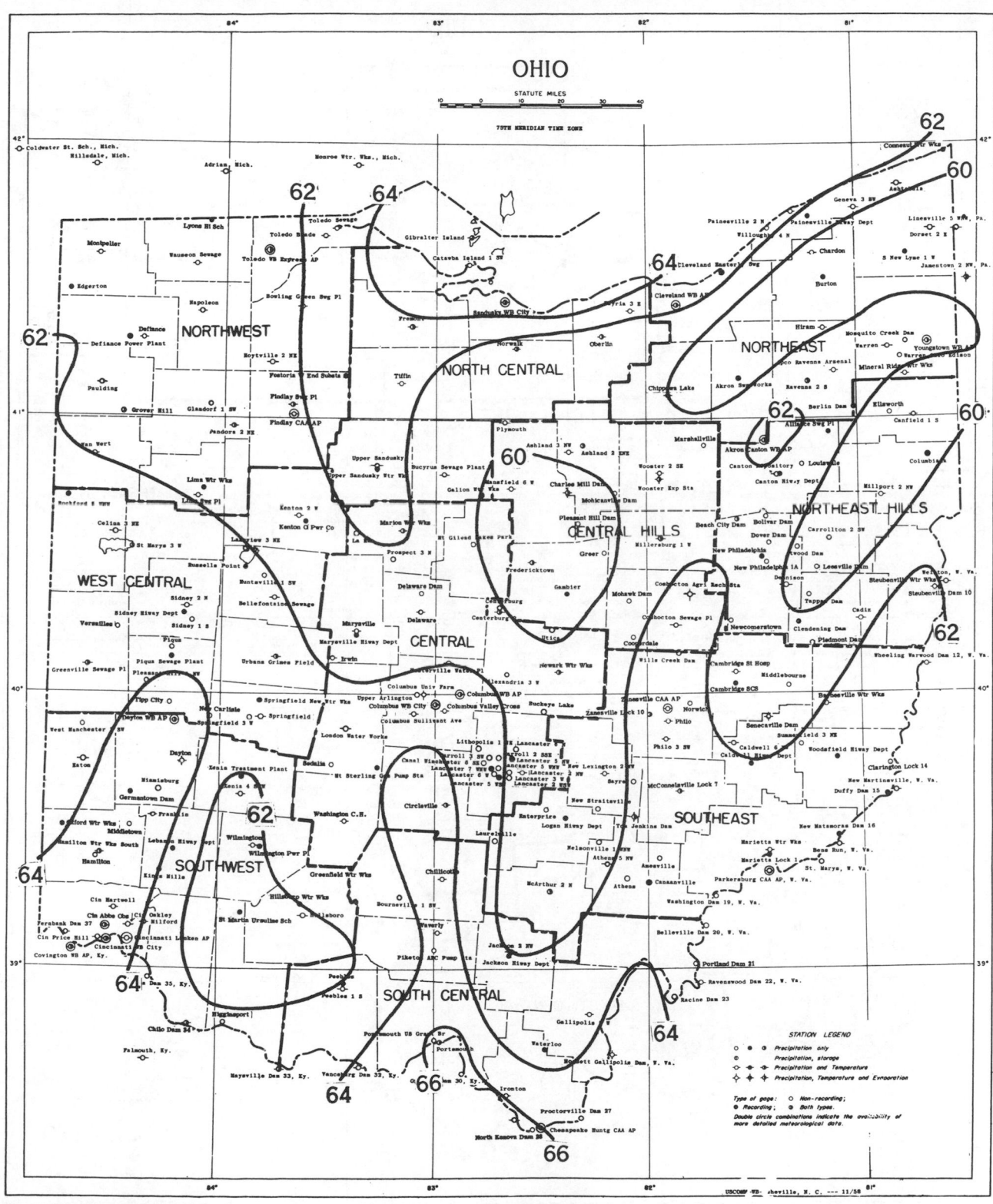

Based on period 1931-52

Isolines are drawn through points of approximately equal value. Caution should be used in interpolating on these maps, particularly in mountainous areas.

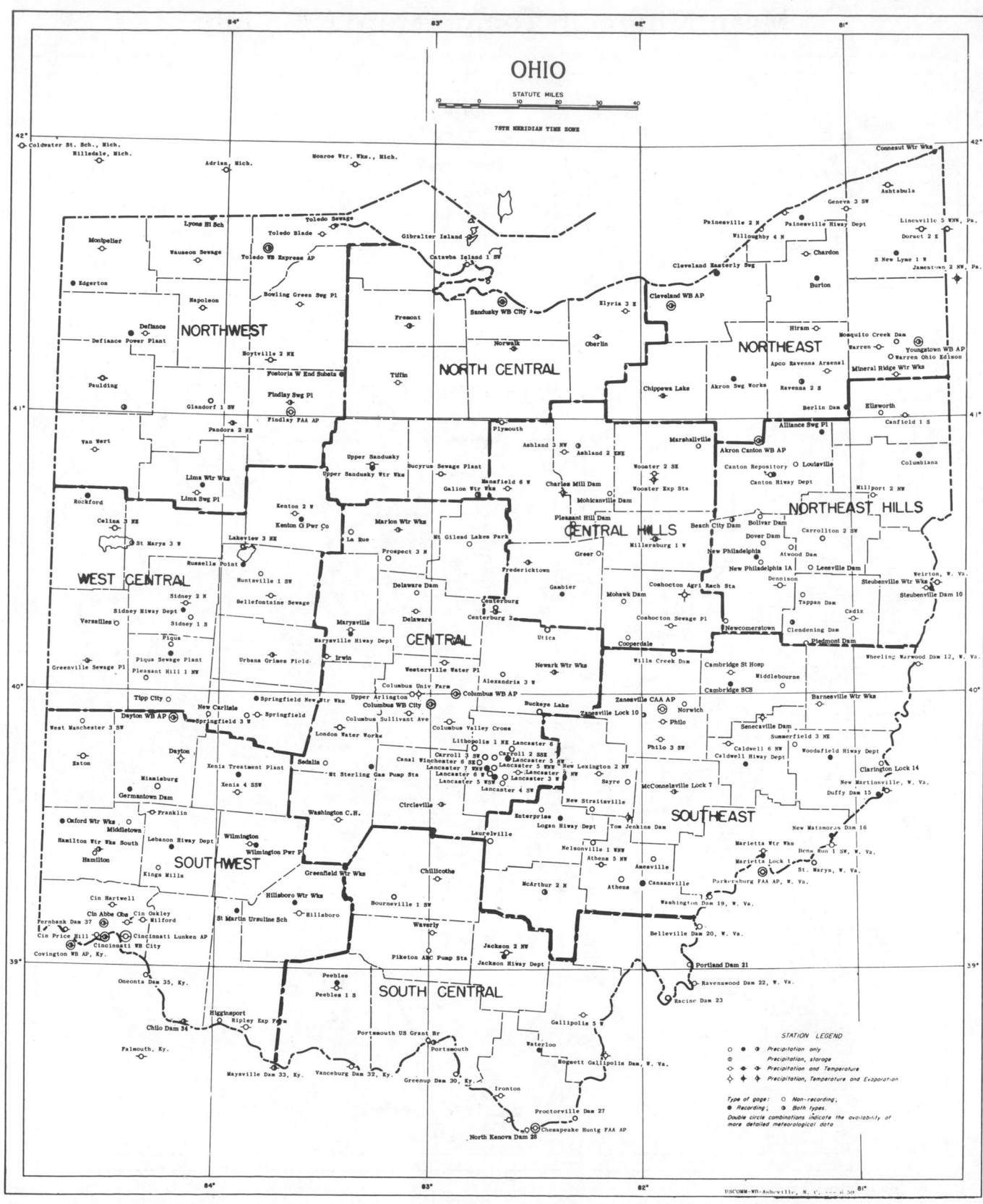

OHIO

STATUTE MILES

79TH MERIDIAN TIME ZONE

STATION LEGEND

○ ◑ ● *Precipitation only*

⊙ ◉ *Precipitation, storage*

◇ ◈ ◆ *Precipitation and Temperature*

◇ ◈ ◆ *Precipitation, Temperature and Evaporation*

Type of gage: ○ *Non-recording;*

● *Recording;* ◉ *Both types.*

Double circle combinations indicate the availability of
more detailed meteorological data

USCOMM-WB-Asheville, N. C. - - - 9 59

CLIMATES OF THE STATES

Oklahoma

(Normals, Means and Extremes tables revised 1970 and 1975. Basic report revised June 1970.)

Billy R. Curry, ESSA State Climatologist

Oklahoma is located in the southern Great Plains. Of the 50 states, it ranks 18th in size with an area of approximately 70,000 square miles, only 935 of which are covered by lakes and ponds. Its northern boundary is about 465 miles in length and its southern boundary 315 miles in length. Greatest depth is 222 miles.

The terrain is mostly rolling plains, sloping downward from west to east. The plains are broken by scattered hilly areas where most points are 600 feet or less above the adjacent countryside, and by a mountainous area in the southeast where some peaks rise more than 2000 feet above their base. The hilly areas consist of the Wichita Mountains, with some isolated peaks, in the southwest; the Arbuckle Mountains in the south-central; and, an extension of the Ozarks in the northeast. The Ouachita Mountains occupy much of the southeast. Elevations in the State range from 4,976 feet above sea level on Black Mesa in the northwestern corner of the Panhandle, to about 305 feet above sea level in the bed of the Red River where it leaves Oklahoma at the southeastern corner of the State.

Oklahoma lies entirely within the drainage basin of the Mississippi River. The two main rivers in the State are the Arkansas which drains the northern two-thirds of Oklahoma and the Red River which drains the southern third and forms the State's southern boundary. Principal tributaries of the Arkansas are the Verdigris, Grand (Neosho), Illinois, Cimarron, North Canadian, and Canadian

Rivers. The Red draws largely from the North Fork of the Red, Washita, Boggy, and Little Rivers.

In western Oklahoma, rivers tend to be broad, shallow, sand choked, and dry or nearly dry much of the time. Basins are mostly long and narrow. In the east, rivers are fairly swift and clear and basins more oval in form. Most lakes are man-made and were built for flood control, irrigation, municipal water storage, recreational, and in a few cases hydro-electric power purposes. The largest lakes are Texoma on the Red River and Eufaula Reservoir on the Canadian.

Agriculture, mining, manufacturing, trade, and government are all important sectors of Oklahoma's economy. Leading agricultural crops and their main areas of production are: Wheat, western half of the State; cotton, southern two-thirds; corn, eastern half; peanuts, south; broomcorn, central and west; and milo, western half, especially in the Panhandle where a tremendous agricultural economy, based on irrigation from ground water, is being developed. The livestock industry is of great importance to all sections of the State. Minerals produced in the State include petroleum, natural gas, coal, lead, and zinc. Leading manufactured products include food products, transportation equipment, primary and fabricated metal products, machinery, and petroleum and coal products. Lumbering is important in the southeast.

The climate of Oklahoma is mostly continental in type, as in all of the central Great Plains. Warm, moist air moving northward from the Gulf of Mexico exerts much influence at times, particularly over the southern and more eastern sections of the State where, as a result, humidities and cloudiness are generally greater and precipitation considerably heavier than in the western and northern sections. Summers are long and occasionally very hot. Winters are shorter and less rigorous than those of the more northern Plain States. Periods of extreme cold are infrequent.

The mean annual temperature over the State ranges from 64° along the southern border to about 60° along the northern border. It then decreases westward across the Panhandle to about 57° in Cimarron County. Temperatures of 90° or higher occur, on an average, about 85 days per year in the western Panhandle and in the northeast corner of the State. In the southwest, the average is about 120 days, and in the southeast from 95 to 100 days. Temperatures of 100° or higher are common over the State from May well into September. In the southwest part of the State the average number of 100° days is 20 to 25 per year. Other sections of the State will average somewhat less, but very seldom will any location in the State not reach a 100° temperature sometime during the summer months.

Low humidities and good southerly breezes usually accompany the high summer temperatures and somewhat lessen their discomforting effect. Occasionally strong, hot winds accompany the high daytime temperatures; this combination produces rapid evaporation and often injures crops. When these conditions persist for long periods of time, droughts develop and occasionally become severe. Nights are generally comfortable because the clear skies and dry air allows for rapid cooling after sunset.

The highest temperature ever recorded in the State was 120°. This reading was observed at Alva on July 18, at Altus on July 19 and August 12, and at Poteau on August 10--all during the extremely hot summer of 1936. Tishomingo also observed a 120° temperature on July 26, 1943.

Temperatures of 32° or less occur on an average of 55 to 65 days per year along the southern tier of counties and from 90 to 100 days per year along the Kansas border in the north-central and northeastern sections of the State. In the Panhandle, days with 32° or less occur, on an average, 125 to 140 days per year. The lowest temperature of record is -27° and was observed at Watts on January 18, 1930, and at Vinita on February 13, 1905.

The average length of the growing season, or freeze-free period, ranges from 168 days at Kenton, in the northwestern corner of the Panhandle, to about 225 days along the Red River in the south-central and southeastern sections of Oklahoma. Along the northern border of the State, the average date of the last spring freeze varies from April 5 in the northeast to April 27 in the western end of the Panhandle. Along the southern border, the average date varies from March 27 to April 5. The average date of the first fall freeze varies from October 12 to October 27 in the north and from November 5 to November 10 along the southern border, the latest dates occurring in the south-central area. Freezing temperatures have occurred as late as April 20 along the southern border and as late as May 15 in the extreme northwest and in the Panhandle. Fall freezes have occurred as early as September 20 in the Panhandle and as early as October 9 along the southern border.

Frozen soil is not a major problem, nor much of a deterrent to seasonal activities. Its occurrence is rather infrequent, of very limited extent, and of brief duration. The average maximum depth that frost penetrates the soil ranges from less than 3 inches in the southeastern corner of the State to more than 10 inches in the extreme northwestern portion. Extreme frost penetration ranges from 10 inches in the southeast corner to about 30 inches in the northwest corner of the Panhandle. Factors having an important bearing on frost penetration are severity and duration of temperature below freezing; condition, character, and moisture content of the soil; and amount and character of protective cover of the soil including snow cover.

The geographical distribution of rainfall decreases sharply from east to west. Average annual precipitation ranges from about 56 inches in southern LeFlore County, in the southeastern corner of the State, to 15 inches in the extreme western Panhandle. The greatest annual precipitation recorded at an official reporting station was 84.47 inches at Kiamichi Tower in southern LeFlore County in 1957. The least annual amount was 6.53 inches at Regnier in northwestern Cimarron County in 1956.

Frequency of rainfall, as determined from the average number of days with 0.01 inch or more, varies from 95 to 100 days a year in the extreme east to from 70 to 80 days a year over the western third of the State.

Precipitation is usually adequate for successful production of the State's principal crops. Spring and early summer rains are of more general and abundant character. Late summer and early fall rainfall is more localized and less abundant as a rule. Fall precipitation, except in some western districts where occurrence at this season is uncertain, is usually adequate for putting soils in good workable condition and for giving fall-sown grains a good start. Average summer rainfall of less than 2 inches per month is unfavorable to crops normally maturing during that season of the year, and poor yields of such crops usually result even on good soils.

Excessively heavy rains occur at times. Amounts of 10 inches or more within a 24-hour period, have been recorded. The greatest official rainfall, within a 24-hour period, was 15.50 inches at Sapulpa on September 3-4, 1940. Larger unofficial amounts have been recorded and in a shorter period of time; for instance, 24 inches in a 10-hour period in the Hallett-Maramec area in southern Pawnee County, also on September 3-4, 1940, and 23 inches within a 12-hour period on

April 3-4, 1934, near Cheyenne in Roger Mills County.

Floods may occur during any season. They occur with greater frequency, however, from May to July and in September and October, representing periods when storms are of greater magnitude and rains of greatest intensities. In general, floods in other seasons are the result of more abnormal and persistent buildup of soil moisture plus a concurrent increase in streamflow due to prolonged rains. The number of lakes built in Oklahoma within the past 10 years has done much in reducing flood damage.

Some notable flood years in Oklahoma are as follows: Verdigris--1904, 1922, 1927, 1941, 1951, 1957, 1959; Cimarron--1926, 1932, 1943, 1945, 1949, 1955, 1957; Grand (Neosho)--1943, 1948, 1951; Illinois--1950, 1951; Canadian--1904, 1914, 1941, 1943; North Canadian--1923, 1927, 1945; Washita--1908, 1923, 1927, 1936, 1938, 1949, 1951, 1955, 1957; Arkansas--1943, 1945, 1957.

Relative humidity averages about 10 percent higher in the eastern portion of the State because of lower elevations and more frequent inflow of Gulf moisture. Summer afternoon and early evening relative humidities are considerably lower than those of winter.

The geographical distribution of annual snowfall is usually almost the reverse of the annual precipitation pattern and ranges, on an average, from approximately 2 inches in the southeastern corner of the State to approximately 20 inches in the western sections of the.Panhandle. Snow rarely remains on the ground more than a few days. At times, strong winds with heavy snowfalls cause bad drifting and occasionally produce blizzard conditions which restrict highway traffic and endangers livestock. The greatest seasonal snowfall ever recorded in Oklahoma was 87.3 inches at Beaver, in Beaver County, during the season of 1911-12. The greatest daily snowfall was 22 inches at Beaver on December 19, 1911.

Oklahoma, along with other states in the southern Great Plains, has at times been subject to droughts of varying degree and duration, although drought years have been far less frequent than dry summers and falls. Most notable of the drought periods in Oklahoma were the dry years which occurred in the late 1890's, the drought of 1910 to 1919, the very severe drought of the 1930's, and the most recent drought which persisted from July 1951 to March 1957. While little can be done at this time to correct deficient rainfall, which is the major contributing cause of droughts, much has been done since the late 1930's in adapting land use and cultivation practices to climatic deficiencies. The tre-

mendous increase in the past 20 years in irrigation farming has also played a major role in reducing drought conditions in Oklahoma. Since 1947, farmland under irrigation has increased from 50,000 acres to approximately 600,000 acres--a 12-fold increase. Most of the irrigation land is in the western third of the State.

Average annual lake evaporation varies from about 48 inches in the extreme eastern sections of the State to as high as 65 inches in the southwestern corner. In the western Panhandle approximately 58 inches of water is evaporated each year. The importance of these evaporative losses from reservoirs and other surface water supplies has prompted a good deal of research to find ways of retarding the evaporation process.

Prevailing winds are southerly although northerly winds predominate during the winter months. Average yearly wind speeds vary from 9 m.p.h. in the east to approximately 14 m.p.h. in the west. March and April are the windiest months, and July and August the calmest.

Thunderstorms occur, on an average, of 50 to 60 days per year in the eastern half of the State and from 40 to 50 days per year in the western half. Some of the more severe thunderstorms are accompanied by tornadoes and damaging hail, and approximately 75 percent of these occur during the spring season. Since 1875, over 1600 tornadoes have struck the State. One of the best known and most destructive moved out of the Texas Panhandle on the evening of April 9, 1947, striking Ellis, Woodward, and Woods Counties before moving into Kansas. One hundred and one people lost their lives in Oklahoma, 95 of these at Woodward, Woodward County. Property damage in the State was a little over $8 million. Since 1924, estimated damage from hailstorms has averaged a little over $3 million per year. The most destructive hailstorm known to have hit the State struck Oklahoma City during the night of May 23-24, 1968, causing over $20 million property damage.

Skies are preponderantly clear in western and central sections and about equally clear and cloudy in eastern sections. Sunshine records show an annual average of 68 percent of the possible amount of Oklahoma City and 63 percent at Tulsa. Summer is the period of greatest possible sunshine and winter the least.

The climate of Oklahoma is favorable for a long vacation season and a wide variety of recreational activities. The wooded and more watered and mountainous eastern sections of the State are particularly attractive to the vacationer, especially those who enjoy boating, water-skiing, camping, hiking, and fishing.

REFERENCES

(1) Holbrook, Stanley G., ESSA, Weather Bureau, Oklahoma's Severe Hail, 1924-1961, 1962 (unpublished).

(2) Klein, John J., et al., Department of Economics, Oklahoma State University, Stillwater, Oklahoma, The Oklahoma Economy, 1963.

(3) Kohler, M. A., et al., ESSA, Weather Bureau, Evaporation Maps for the United States, Weather Bureau Technical Paper No. 37, 1959.

(4) Lehrer, Hugo V., ESSA, Weather Bureau, "Climate of Oklahoma", published in Climates of the States, Oklahoma, Climatography of the United States No. 60-34, ESSA, Environmental Data Service, 1960.

(5) Bureau of Business Research, University of Oklahoma, Norman, Oklahoma, Oklahoma Data Book, 1968.

(6) ESSA, Weather Bureau Climatological Services Division, Mean Number of Thunderstorm Days in the United States, Weather Bureau Technical Paper No. 19, 1952.

(7) Oklahoma Almanac, Incorporated, Norman, Oklahoma, The Oklahoma Almanac, 1961 edition.

(8) Oklahoma Water Resources Board, Oklahoma City, Oklahoma, Reported Water Use in Oklahoma, 1968.

BIBLIOGRAPHY

(A) Climatic Summary of the United States (Bulletin W), 1930 edition, Sections 42 and 43. ESSA, Weather Bureau.

(B) Climatic Summary of the United States, Oklahoma - Supplement for 1931 through 1952 (Bulletin W Supplement), ESSA, Weather Bureau.

(C) Climatic Summary of the United States, Oklahoma - Supplement for 1951 through 1960 (Bulletin W Supplement), ESSA, Weather Bureau.

(D) Climatological Data - Oklahoma, ESSA, Environmental Data Service.

(E) Climatological Data - National Summary, ESSA, Environmental Data Service.

(F) Hourly Precipitation Data - Oklahoma, ESSA, Environmental Data Service.

(G) Local Climatological Data - Oklahoma City and Tulsa, Oklahoma, ESSA, Environmental Data Service.

(H) Selected Climatic Maps of the United States, ESSA, Environmental Data Service

(I) Storm Data, ESSA, Environmental Data Service.

FREEZE DATA

STATION	Freeze threshold temperature	Mean date of last Spring occurrence	Mean date of first Fall occurrence	Mean No. of days between dates	Years of record Spring	No. of occurrences in Spring	Years of record Fall	No. of occurrences in Fall
ADA	32	03-31	11-06	220	30	30	30	30
	28	03-21	11-17	241	30	30	30	30
	24	03-09	11-30	266	30	30	30	30
	20	02-19	12-13	297	30	30	30	25
	16	02-07	12-20	315	30	27	30	17
ALTUS	32	04-01	11-07	219	30	30	29	29
	28	03-25	11-16	236	30	30	29	29
	24	03-13	11-29	260	29	29	28	28
	20	02-27	12-08	284	29	28	28	27
	16	02-19	12-17	301	29	27	28	17
ALVA	32	04-09	10-30	204	30	30	30	30
	28	03-29	11-07	223	29	29	30	30
	24	03-21	11-19	243	29	29	30	30
	20	03-13	11-27	259	29	29	29	29
	16	03-01	12-08	282	29	28	29	24
ANTLERS	32	03-27	11-05	223	28	28	28	28
	28	03-17	11-14	242	28	28	28	28
	24	03-01	11-28	273	28	28	26	25
	20	02-16	12-15	302	28	28	26	21
	16	02-03	12-21	321	28	25	26	13
APACHE	32	04-05	10-30	208	29	29	29	29
	28	03-27	11-10	228	29	29	28	28
	24	03-15	11-23	253	29	29	28	28
	20	03-04	12-05	276	29	28	28	26
	16	02-19	12-17	302	28	27	28	18
ARDMORE CAA AP	32	03-29	11-11	227	30	30	30	30
	28	03-15	11-23	253	30	30	30	30
	24	03-02	12-07	280	30	30	30	28
	20	02-14	12-17	306	30	30	30	22
	16	02-02	12-21	321	30	25	30	17
BARTLESVILLE 2 W	32	04-02	10-28	210	29	29	30	30
	28	03-27	11-04	221	29	29	30	30
	24	03-17	11-16	244	29	29	30	30
	20	03-11	11-28	262	29	28	30	30
	16	02-23	12-09	289	29	28	30	24
BEAVER	32	04-20	10-21	184	30	30	29	29
	28	04-05	10-29	208	30	30	29	29
	24	03-29	11-08	225	30	30	29	29
	20	03-20	11-15	240	29	29	29	29
	16	03-12	11-25	259	29	29	29	28
BOISE CITY	32	04-23	10-13	174	24	24	23	23
	28	04-13	10-25	194	23	23	22	22
	24	04-03	11-05	216	23	23	22	22
	20	03-29	11-11	227	23	23	21	21
	16	03-16	11-18	247	23	23	20	20
BRISTOW	32	04-03	11-02	212	28	28	24	24
	28	03-26	11-07	226	27	27	23	23
	24	03-16	11-19	248	26	26	23	23
	20	03-07	12-01	269	24	24	21	21
	16	02-19	12-15	299	24	24	21	15
BUFFALO	32	04-17	10-24	190	27	27	30	30
	28	04-02	11-03	215	27	27	30	30
	24	03-27	11-12	230	27	27	30	30
	20	03-16	11-21	250	27	27	29	29
	16	03-02	12-03	276	27	27	29	27
CARNEGIE	32	04-08	10-26	201	30	30	29	29
	28	03-31	11-03	217	30	30	29	29
	24	03-17	11-20	247	30	30	28	28
	20	03-09	11-29	265	30	30	28	28
	16	02-23	12-11	291	30	30	28	23
CHANDLER 1	32	04-01	10-29	210	30	30	29	29
	28	03-26	11-11	230	30	30	29	29
	24	03-15	11-24	254	30	30	29	29
	20	03-05	12-05	273	30	30	29	29
	16	02-15	12-18	306	30	29	29	20
CHATTANOOGA	32	04-04	11-03	212	29	29	27	27
	28	03-25	11-18	238	29	29	27	27
	24	03-12	11-29	262	28	28	26	26
	20	02-27	12-10	286	28	27	24	21
	16	02-13	12-22	311	28	26	24	13
CHEROKEE	32	04-10	10-29	202	27	27	28	28
	28	04-01	11-06	220	26	26	26	26
	24	03-23	11-18	240	25	25	25	25
	20	03-12	11-27	260	24	24	25	25
	16	02-27	12-12	288	23	21	24	18
CHICKASHA	32	04-04	10-31	209	30	30	30	30
	28	03-25	11-10	230	30	30	30	30
	24	03-11	11-23	257	30	30	30	30
	20	03-03	12-05	277	30	29	30	27
	16	02-16	12-19	306	30	29	30	18
CLAREMORE	32	04-03	10-29	209	30	30	30	30
	28	03-28	11-03	220	29	29	30	30
	24	03-12	11-18	251	29	29	30	30
	20	03-01	12-05	279	29	28	29	28
	16	02-15	12-16	304	29	28	29	23
CLEVELAND 1	32	04-05	10-29	206	30	30	29	29
	28	03-28	11-04	221	30	30	29	29
	24	03-16	11-16	245	30	30	29	29
	20	03-06	12-01	270	30	30	29	29
	16	02-22	12-15	296	30	29	29	24
CLINTON	32	04-04	10-29	207	14	14	15	15
	28	03-27	11-06	223	14	14	15	15
	24	03-21	11-16	240	14	14	15	15
	20	03-06	11-24	263	14	14	15	15
	16	02-23	12-13	293	14	13	15	12
CLOUD CHIEF	32	04-06	11-01	209	28	28	27	27
	28	03-30	11-11	226	28	28	26	26
	24	03-18	11-23	250	28	28	26	26
	20	03-09	12-06	272	28	28	26	25
	16	02-25	12-18	296	28	27	26	17
COALGATE	32	03-31	11-01	215	13	13	12	12
	28	03-13	11-10	242	13	13	12	12
	24	02-27	11-18	264	13	13	12	12
	20	02-16	12-10	297	13	13	11	8
	16	02-02	12-20	321	12	10	11	5
DURANT	32	03-30	11-07	223	30	30	30	30
	28	03-13	11-20	251	30	30	30	30
	24	02-24	12-05	284	30	30	30	29
	20	02-09	12-17	310	30	28	30	22
	16	01-29	12-20	325	30	26	30	16
ELK CITY	32	04-06	10-31	208	25	25	25	25
	28	03-27	11-10	228	25	25	25	25
	24	03-17	11-22	250	24	24	24	24
	20	03-09	12-02	268	24	24	24	23
	16	02-23	12-14	294	24	23	24	19
ENID	32	04-02	11-03	214	30	30	28	28
	28	03-25	11-15	235	30	30	27	27
	24	03-16	11-25	253	29	29	27	27
	20	03-05	12-07	277	29	29	27	22
	16	02-21	12-17	299	29	27	27	20
ERICK	32	04-07	10-31	207	29	29	27	27
	28	03-27	11-11	228	29	29	25	25
	24	03-20	11-15	240	28	28	25	25
	20	03-09	11-27	263	28	28	24	24
	16	02-21	12-13	295	28	26	24	19
EUFAULA	32	04-02	10-29	211	21	21	27	27
	28	03-22	11-10	233	21	21	26	26
	24	03-04	11-23	264	20	20	25	25
	20	02-19	12-12	296	19	19	24	22
	16	02-09	12-19	313	19	19	24	15
FORT RENO	32	04-05	11-02	211	29	29	29	29
	28	03-28	11-13	231	29	29	29	29
	24	03-18	11-25	252	29	29	29	29
	20	03-10	12-04	269	29	29	29	27
	16	03-01	12-15	290	29	29	29	18
FREDERICK	32	03-31	11-09	224	30	30	29	29
	28	03-22	11-21	244	30	30	28	28
	24	03-09	12-06	273	30	30	27	26
	20	02-23	12-15	295	30	29	27	21
	16	02-13	12-23	314	29	27	26	13

STATION	Freeze threshold temperature	Mean date of last Spring occurrence	Mean date of first Fall occurrence	Mean No. of days between dates	Years of record Spring	No. of occurrences in Spring	Years of record Fall	No. of occurrences in Fall
GEARY	32	04-05	11-04	219	28	28	30	30
	28	03-28	11-13	231	28	28	30	30
	24	03-17	11-24	252	28	28	30	30
	20	03-08	12-07	274	28	28	30	25
	16	02-22	12-15	297	28	28	30	21
GOODWELL	32	04-23	10-20	180	29	29	29	29
	28	04-15	10-29	197	29	29	29	29
	24	04-05	11-07	216	29	29	29	29
	20	03-28	11-15	231	29	29	28	28
	16	03-21	11-22	246	29	29	28	28
GUTHRIE	32	04-07	10-29	205	30	30	30	30
	28	03-28	11-05	222	30	30	30	30
	24	03-17	11-21	249	30	30	30	30
	20	03-10	11-30	266	29	29	30	29
	16	02-25	12-14	292	29	28	30	23
HAMMON	32	04-12	10-27	198	30	30	28	28
	28	03-30	11-06	221	30	30	26	26
	24	03-24	11-18	240	30	30	26	26
	20	03-14	11-25	256	29	29	26	26
	16	02-27	12-10	286	29	29	24	20
HENNESSEY	32	04-06	10-31	209	29	29	29	29
	28	03-28	11-10	227	29	29	29	29
	24	03-17	11-22	249	28	28	29	29
	20	03-06	12-04	272	28	28	29	27
	16	02-27	12-15	292	28	27	29	23
HOBART CAA AP	32	04-05	11-01	210	29	29	28	28
	28	03-26	11-13	232	29	29	28	28
	24	03-17	11-24	252	29	29	28	28
	20	03-06	12-07	276	29	28	28	28
	16	02-24	12-17	296	29	28	28	20
HOLDENVILLE	32	03-31	11-02	217	30	30	30	30
	28	03-21	11-15	239	30	30	30	30
	24	03-07	11-29	267	30	30	30	30
	20	02-22	12-13	294	30	30	30	26
	16	02-09	12-19	313	30	28	30	18
HOLLIS	32	04-07	11-01	209	26	26	28	28
	28	03-28	11-11	228	25	25	26	26
	24	03-21	11-19	243	25	25	24	24
	20	03-03	12-05	276	25	25	24	24
	16	02-25	12-18	296	25	24	23	15
HOOKER	32	04-24	10-18	177	30	30	30	30
	28	04-14	10-28	197	30	30	30	30
	24	03-31	11-08	221	30	30	30	30
	20	03-27	11-14	232	29	29	30	30
	16	03-14	11-21	251	29	29	30	30
HUGO	32	03-29	11-06	222	30	30	30	30
	28	03-14	11-22	253	30	30	30	30
	24	02-21	12-06	288	30	29	30	29
	20	02-07	12-16	312	30	29	30	22
	16	01-24	12-20	331	30	20	30	16
IDABEL	32	03-27	11-03	222	24	24	25	25
	28	03-13	11-16	248	24	24	24	24
	24	02-24	12-02	281	24	24	24	22
	20	02-11	12-12	305	24	23	24	18
	16	01-30	12-21	326	24	18	24	12
JEFFERSON	32	04-14	10-26	195	28	28	30	30
	28	04-01	11-03	216	28	28	29	29
	24	03-26	11-15	234	27	27	28	28
	20	03-16	11-27	256	27	27	28	28
	16	03-02	12-10	283	27	26	27	20
KENTON	32	04-27	10-12	168	30	30	30	30
	28	04-14	10-26	195	30	30	30	30
	24	04-05	11-03	212	30	30	30	30
	20	03-28	11-10	227	30	30	30	30
	16	03-19	11-20	246	30	30	30	30
KINGFISHER	32	04-05	10-30	208	30	30	28	28
	28	03-28	11-09	226	30	30	28	28
	24	03-17	11-24	253	30	30	29	29
	20	03-10	12-06	271	30	30	29	28
	16	02-27	12-16	292	30	29	29	20

STATION	Freeze threshold temperature	Mean date of last Spring occurrence	Mean date of first Fall occurrence	Mean No. of days between dates	Years of record Spring	No. of occurrences in Spring	Years of record Fall	No. of occurrences in Fall
LAWTON 2 N	32	04-02	11-03	215	30	30	30	30
	28	03-23	11-13	236	30	30	30	30
	24	03-14	11-28	259	30	30	30	30
	20	02-26	12-12	289	30	30	29	25
	16	02-13	12-19	309	30	28	30	20
MANGUM	32	04-03	11-04	216	29	29	26	26
	28	03-24	11-14	235	29	29	26	26
	24	03-18	11-26	254	28	28	26	26
	20	03-06	12-12	281	26	26	25	24
	16	02-22	12-19	299	26	26	24	19
MARLOW	32	04-01	11-03	217	30	30	29	29
	28	03-22	11-16	239	30	30	29	29
	24	03-11	11-30	263	30	30	28	28
	20	02-28	12-14	289	30	29	28	22
	16	02-12	12-19	310	30	28	28	15
MC ALESTER	32	03-31	11-02	216	28	28	29	29
	28	03-22	11-14	238	27	27	28	28
	24	03-10	11-29	264	27	27	27	27
	20	02-17	12-10	297	27	27	26	22
	16	02-07	12-19	315	27	25	26	14
MEEKER	32	04-03	10-30	210	27	27	29	29
	28	03-26	11-02	221	26	26	28	28
	24	03-18	11-21	247	26	26	27	27
	20	03-09	12-01	268	26	26	27	27
	16	02-18	12-17	302	26	25	27	19
MIAMI	32	04-04	10-27	205	29	29	28	28
	28	03-27	11-03	221	29	29	28	28
	24	03-15	11-16	246	29	29	28	28
	20	03-03	12-04	275	29	29	28	27
	16	02-17	12-14	300	29	29	28	24
MUSKOGEE	32	03-31	11-01	216	30	30	30	30
	28	03-22	11-12	234	30	30	30	30
	24	03-04	11-29	270	30	30	30	30
	20	02-19	12-13	298	30	30	30	25
	16	02-09	12-20	314	30	29	30	17
MUTUAL	32	04-14	10-30	199	28	28	27	27
	28	03-28	11-08	225	28	28	27	27
	24	03-22	11-19	242	28	28	27	27
	20	03-16	11-29	258	28	28	27	27
	16	03-04	12-06	277	27	27	27	25
NEWKIRK	32	04-10	10-29	203	30	30	30	30
	28	03-27	11-08	225	30	30	30	30
	24	03-20	11-19	244	30	30	30	30
	20	03-11	11-30	264	29	29	30	30
	16	02-27	12-14	290	29	28	29	22
NORMAN	32	04-02	11-04	216	28	28	25	25
	28	03-24	11-17	238	28	28	23	23
	24	03-16	11-26	256	27	27	22	22
	20	03-05	12-09	279	27	27	21	18
	16	02-16	12-18	305	26	25	21	13
NOWATA	32	04-08	10-27	202	12	12	12	12
	28	03-27	11-06	224	12	12	12	12
	24	03-22	11-09	232	12	12	12	12
	20	03-02	12-02	275	12	12	12	11
	16	02-24	12-12	292	12	12	11	8
OAKWOOD 1 W	32	04-14	10-24	193	28	28	29	29
	28	04-02	11-04	216	27	27	29	29
	24	03-24	11-13	234	27	27	29	29
	20	03-17	11-23	250	27	27	29	29
	16	03-04	12-05	277	27	26	29	28
OKEENE	32	04-11	10-30	202	27	27	28	28
	28	03-29	11-09	225	25	25	28	28
	24	03-22	11-21	244	25	25	27	27
	20	03-11	12-02	266	25	25	27	25
	16	02-28	12-15	290	25	24	27	20
OKEMAH	32	03-30	11-05	220	29	29	30	30
	28	03-19	11-17	242	29	29	30	30
	24	03-10	11-29	264	29	29	30	30
	20	02-25	12-13	292	29	29	30	26
	16	02-09	12-20	314	29	28	30	17

FREEZE DATA

STATION	Freeze threshold temperature	Mean date of last Spring occurrence	Mean date of first Fall occurrence	Mean No. of days between dates	Years of record Spring	No. of occurrences in Spring	Years of record Fall	No. of occurrences in Fall
OKLAHOMA CITY WB CTY	32	03-28	11-07	223	30	30	30	30
	28	03-19	11-20	246	30	30	30	30
	24	03-11	12-03	267	30	30	30	29
	20	03-01	12-14	288	30	30	29	23
	16	02-17	12-20	306	30	28	29	16
OKMULGEE	32	04-01	10-30	212	30	30	27	27
	28	03-24	11-08	229	30	30	27	27
	24	03-11	11-21	255	30	30	27	27
	20	02-23	12-07	287	30	29	27	25
	16	02-11	12-15	307	30	28	27	21
PAULS VALLEY	32	04-04	10-29	207	30	30	30	30
	28	03-24	11-09	230	30	30	30	30
	24	03-14	11-23	254	30	30	29	29
	20	02-25	12-04	282	30	30	29	27
	16	02-13	12-18	308	30	29	29	20
PAWHUSKA	32	04-08	10-25	200	30	30	27	27
	28	03-29	11-02	218	30	30	27	27
	24	03-21	11-17	241	30	30	27	27
	20	03-12	11-29	262	29	29	27	27
	16	02-27	12-12	288	29	28	26	23
PERRY	32	04-07	10-30	206	29	29	25	25
	28	03-25	11-09	229	29	29	25	25
	24	03-18	11-26	252	28	28	23	23
	20	03-06	12-03	273	27	26	23	22
	16	02-24	12-15	294	26	23	23	15
POTEAU	32	04-03	11-01	212	29	29	29	29
	28	03-20	11-08	234	27	27	28	28
	24	03-07	11-23	261	25	25	28	28
	20	02-18	12-10	296	25	25	26	23
	16	02-05	12-19	317	24	20	26	14
PRYOR	32	04-03	10-31	210	23	23	25	25
	28	03-29	11-05	221	22	22	25	25
	24	03-13	11-17	249	22	22	25	25
	20	03-01	12-05	279	22	22	25	23
	16	02-12	12-16	307	22	21	25	16
SHAWNEE	32	04-01	11-01	214	29	29	29	29
	28	03-27	11-13	231	29	29	29	29
	24	03-13	11-25	257	29	29	28	28
	20	03-05	12-03	273	28	28	28	27
	16	02-14	12-17	306	28	27	28	19
SMITHVILLE	32	04-15	10-22	190	27	27	25	25
	28	03-30	10-30	214	27	27	25	25
	24	03-18	11-12	239	27	27	25	25
	20	03-01	11-24	268	26	26	25	24
	16	02-13	12-12	302	26	25	24	18
STILLWATER	32	04-04	10-28	207	30	30	30	30
	28	03-27	11-09	227	30	30	30	30
	24	03-19	11-23	250	30	30	30	30
	20	03-08	12-04	271	30	30	30	27
	16	02-25	12-14	292	30	30	30	22
SULPHUR	32	04-04	10-30	209	30	30	30	30
	28	03-26	11-11	230	30	30	30	30
	24	03-12	11-22	255	30	30	30	30
	20	02-26	12-06	283	30	30	28	27
	16	02-10	12-16	309	30	28	28	21
TAHLEQUAH	32	04-09	10-25	199	28	28	29	29
	28	03-28	11-03	220	28	28	29	29
	24	03-17	11-14	242	28	28	27	27
	20	03-04	12-02	273	27	27	26	24
	16	02-11	12-13	305	27	27	26	20
TISHOMINGO	32	04-01	11-02	215	25	25	26	26
	28	03-21	11-10	234	25	25	25	25
	24	03-02	11-23	267	25	25	25	25
	20	02-18	12-04	289	25	24	24	23
	16	02-04	12-20	319	25	23	24	13
TULSA WB AP	32	03-31	11-02	216	30	30	30	30
	28	03-22	11-14	237	30	30	30	30
	24	03-12	11-30	263	30	30	30	30
	20	02-24	12-11	290	30	30	30	26
	16	02-12	12-19	311	30	29	30	18
VINITA	32	04-09	10-25	199	30	30	30	30
	28	03-29	11-01	217	30	30	30	30
	24	03-19	11-13	239	30	30	30	30
	20	03-05	11-28	269	30	29	30	30
	16	02-15	12-11	300	30	29	30	25
WATTS	32	04-17	10-22	188	25	25	21	21
	28	04-06	10-31	208	24	24	19	19
	24	03-23	11-11	233	24	24	19	19
	20	03-07	11-19	257	23	23	18	18
	16	02-17	12-10	296	22	22	17	16
WAUKOMIS	32	04-08	10-31	206	29	29	28	28
	28	03-30	11-08	223	29	29	28	28
	24	03-21	11-21	244	29	29	28	28
	20	03-10	12-02	267	29	29	28	26
	16	02-28	12-13	287	29	29	28	23
WAURIKA	32	04-01	11-05	218	30	30	30	30
	28	03-19	11-18	243	30	30	30	30
	24	03-03	12-05	276	30	30	30	30
	20	02-17	12-14	300	30	29	30	22
	16	02-04	12-20	319	30	26	30	17
WAYNOKA	32	04-13	10-29	199	12	12	12	12
	28	03-31	10-31	214	12	12	12	12
	24	03-27	11-11	229	12	12	11	11
	20	03-17	11-22	251	11	11	11	11
	16	02-25	11-28	276	10	9	11	11
WEATHERFORD	32	04-03	11-01	212	30	30	30	30
	28	03-25	11-10	230	30	30	30	30
	24	03-18	11-23	250	30	30	30	30
	20	03-10	12-04	270	30	30	30	29
	16	02-22	12-15	296	30	29	30	24
WEBBERS FALLS	32	03-30	11-01	216	28	28	28	28
	28	03-23	11-10	232	28	28	27	27
	24	03-09	11-22	259	28	28	27	27
	20	02-14	12-12	301	28	28	27	22
	16	02-09	12-18	312	28	27	27	17
WICHITA MT WLR	32	04-12	10-29	200	27	27	27	27
	28	04-04	11-08	218	27	27	27	27
	24	03-23	11-15	237	25	25	26	26
	20	03-11	12-01	265	25	25	25	24
	16	02-25	12-10	289	24	24	25	22
WILBURTON	32	03-30	10-24	208	15	15	13	13
	28	03-27	11-09	227	14	14	13	13
	24	03-04	11-20	261	13	13	12	12
	20	02-25	12-03	282	13	13	12	12
	16	02-07	12-12	308	12	11	12	12
WOODWARD	32	04-14	10-21	190	30	30	30	30
	28	04-01	11-01	214	30	30	30	30
	24	03-25	11-12	232	30	30	30	30
	20	03-16	11-23	253	30	30	30	29
	16	03-06	12-02	271	29	29	30	29

Data in the above table are based on the period 1921-1950, or that portion of this period for which data are available.

Means have been adjusted to take into account years of non-occurrence.

A freeze is a numerical substitute for the former term "killing frost" and is the occurrence of a minimum temperature at or below the threshold temperature of 32°, 28°, etc.

Freeze data tabulations in greater detail are available and can be reproduced at cost.

* NORMALS BY CLIMATOLOGICAL DIVISIONS

Taken from "Climatography of the United States No. 81-4, Decennial Census of U. S. Climate"

TEMPERATURE (°F)   PRECIPITATION (In.)

STATIONS (By Divisions)	JAN	FEB	MAR	APR	MAY	JUNE	JULY	AUG	SEPT	OCT	NOV	DEC	ANN	JAN	FEB	MAR	APR	MAY	JUNE	JULY	AUG	SEPT	OCT	NOV	DEC	ANN
PANHANDLE																										
ARNETT	35.8	39.8	46.7	57.5	66.2	76.2	80.7	80.3	72.3	61.2	46.5	38.9	58.5	.66	.96	1.17	2.12	3.68	3.42	2.28	2.38	1.66	2.07	.74	.69	21.83
BEAVER	•	•	•	•	•	•	•	•	•	•	•	•	•	.56	.71	1.01	1.52	3.17	2.67	2.34	2.24	1.46	1.41	.65	.61	18.35
BUFFALO	35.7	40.3	47.6	58.6	67.4	78.2	83.3	82.4	73.9	62.5	47.2	38.8	59.7													
GOODWELL	35.2	39.2	45.0	55.5	64.8	75.3	79.8	78.8	70.9	59.5	44.6	37.6	57.2	.41	.54	.77	1.25	2.75	2.20	2.78	2.38	1.57	1.42	.60	.48	17.15
HOOKER 1 N	33.8	38.1	44.5	55.5	64.7	75.5	80.6	79.2	71.4	59.3	44.2	36.8	57.0	.59	.66	1.08	1.36	3.18	2.89	2.69	2.55	1.44	1.43	.81	.51	19.19
KENTON	34.7	38.4	43.8	53.9	63.1	73.6	78.3	77.6	70.0	58.1	43.9	37.2	56.1	.40	.34	.89	1.46	2.50	1.74	2.41	2.20	1.66	.93	.48	.37	15.38
DIVISION	34.8	38.9	45.1	55.9	65.1	75.5	80.4	79.4	71.4	59.8	45.0	37.6	57.4	.55	.69	1.03	1.63	3.15	2.63	2.49	2.34	1.62	1.46	.67	.57	18.83
NORTH CENTRAL																										
ALVA	36.3	40.2	47.8	58.9	68.0	78.6	83.9	83.4	74.6	63.1	47.9	39.1	60.2	.96	1.16	1.57	2.66	4.23	3.61	2.02	2.67	2.58	1.95	1.22	1.01	25.64
BLACKWELL 1 W	•	•	•	•	•	•	•	•	•	•	•	•	•	1.04	1.22	1.60	2.80	3.93	3.78	3.25	3.00	2.90	2.28	1.60	1.26	28.66
CHEROKEE	36.5	40.6	48.3	59.3	68.1	78.9	83.7	83.2	74.5	63.0	47.7	39.1	60.2													
ENID	37.4	41.4	48.8	59.8	68.5	78.7	83.4	83.2	74.8	63.7	48.9	40.4	60.8	1.02	1.20	1.58	2.93	4.37	3.86	2.76	3.46	2.96	2.27	1.43	1.31	29.15
JEFFERSON	36.0	40.2	47.8	58.9	67.8	78.7	83.2	82.9	74.3	62.9	47.5	39.1	59.9	.98	1.14	1.60	2.76	4.32	3.77	3.15	2.74	2.79	2.25	1.48	1.29	28.27
MUTUAL	36.0	40.1	47.2	58.1	66.9	77.9	82.4	82.0	73.3	62.0	47.2	38.9	59.3	.81	1.03	1.42	2.65	4.43	3.33	1.95	2.41	2.35	2.09	.98	.82	24.27
NEWKIRK	35.3	39.7	47.4	58.8	67.6	77.6	82.4	81.9	73.7	62.5	47.4	38.7	59.4	.95	1.20	1.77	3.11	4.58	4.20	3.16	3.27	3.25	2.37	1.53	1.21	30.60
PERRY	38.0	42.2	49.9	61.0	68.8	78.2	82.9	82.7	74.8	64.1	49.4	41.2	61.1	1.15	1.39	2.06	3.33	4.95	4.22	2.97	2.98	3.29	2.33	1.80	1.38	31.85
PONCA CITY	37.1	41.5	49.1	60.5	69.2	78.8	83.5	83.2	74.9	63.9	48.9	40.2	60.9	1.04	1.23	1.92	3.13	4.71	4.43	3.60	3.09	3.52	2.41	1.70	1.33	32.11
SUPPLY 1 E	•	•	•	•	•	•	•	•	•	•	•	•	•	.80	1.23	1.47	2.12	3.75	3.18	2.40	1.98	1.67	1.81	.88	.85	22.14
WOODWARD	36.9	40.7	48.0	59.3	67.9	78.1	82.7	82.1	73.5	62.2	47.4	39.5	59.9	.84	1.20	1.53	2.49	4.10	3.62	2.35	2.36	2.12	2.18	1.10	.90	24.79
DIVISION	36.5	40.6	48.1	59.3	68.1	78.4	83.1	82.8	74.3	63.0	47.9	39.4	60.1	.95	1.19	1.66	2.76	4.37	3.76	2.77	2.84	2.73	2.20	1.35	1.13	27.71
NORTHEAST																										
BARTLESVILLE 2 W	36.4	40.6	48.3	59.7	68.1	77.3	81.8	81.2	73.1	62.1	47.9	39.4	59.7	1.56	1.56	2.25	3.55	5.20	4.62	3.29	2.68	4.21	3.14	2.02	1.47	35.55
CLAREMORE 2 ENE	37.9	41.9	49.5	60.3	68.4	77.3	81.9	81.6	73.9	63.1	48.8	40.7	60.4	1.85	1.87	2.54	3.95	5.52	4.91	3.16	3.03	3.89	3.36	2.21	1.79	38.08
CLEVELAND	38.3	42.5	50.1	61.2	69.2	78.5	82.9	82.7	74.7	63.5	49.4	41.3	61.2	1.51	1.61	2.34	3.31	5.21	4.29	3.60	3.36	4.00	3.44	2.32	1.35	36.34
MIAMI	37.2	41.5	48.9	59.9	68.0	77.6	82.1	81.8	74.0	63.0	48.6	40.1	60.2	1.75	2.01	2.83	3.99	6.01	5.19	3.43	3.13	4.65	3.86	2.36	2.01	41.22
PAWHUSKA	•	•	•	•	•	•	•	•	•	•	•	•	•	1.26	1.45	2.26	3.39	5.35	4.42	3.06	3.12	3.92	3.05	1.98	1.38	34.64
PRYOR	37.5	41.5	48.7	59.8	68.3	77.4	81.9	81.6	73.6	62.8	48.5	40.5	60.2	1.99	2.16	2.81	4.29	5.73	4.98	3.06	3.13	4.06	3.86	2.47	1.89	40.43
RALSTON	•	•	•	•	•	•	•	•	•	•	•	•	•	1.08	1.16	1.91	3.08	5.06	4.56	3.05	3.21	3.58	2.50	1.94	1.38	32.51
SPAVINAW	•	•	•	•	•	•	•	•	•	•	•	•	•	2.12	2.30	2.92	4.80	6.02	5.05	3.49	3.25	4.63	4.58	2.57	2.06	43.79
TULSA WB AIRPORT	36.2	40.6	48.1	58.9	67.8	77.3	82.2	81.6	73.8	62.8	47.6	39.6	59.7	1.71	1.77	2.43	4.02	5.26	4.69	2.94	3.04	4.01	3.31	2.28	1.62	37.08
VINITA 3 NNE	37.1	41.1	48.4	59.4	67.9	77.3	82.2	81.7	73.5	62.5	48.1	40.0	59.9	1.94	2.00	2.85	4.28	6.16	5.81	3.42	3.46	4.94	3.92	2.51	2.12	43.41
DIVISION	37.4	41.5	49.0	60.1	68.2	77.4	82.0	81.7	73.7	62.8	48.6	40.4	60.2	1.79	1.94	2.60	4.08	5.64	4.97	3.28	3.12	4.14	3.63	2.32	1.81	39.32
WEST CENTRAL																										
CHEYENNE	•	•	•	•	•	•	•	•	•	•	•	•	•	.82	1.09	1.40	2.95	4.71	3.34	1.86	2.53	1.86	2.46	.93	.86	24.81
CLOUD CHIEF 2 SE	38.6	42.8	49.6	60.1	68.9	79.0	83.4	83.1	75.0	63.8	49.1	41.0	61.2	1.02	1.02	1.47	2.75	4.78	3.37	2.43	2.14	2.13	2.37	.99	1.16	25.63
ELK CITY	38.0	41.8	49.1	59.7	68.1	77.9	81.9	81.4	73.3	62.4	48.4	40.6	60.2	.84	.89	1.35	2.30	4.82	2.75	2.29	1.70	1.88	2.19	.87	.97	22.85
GEARY	38.4	42.2	49.4	60.2	68.7	78.6	83.3	83.1	75.0	64.1	49.4	41.0	61.1	1.08	1.17	1.72	2.78	4.40	3.86	2.35	2.55	2.81	2.53	1.26	1.27	27.78
HAMMON	37.8	41.6	49.1	59.6	68.2	78.4	83.0	82.3	73.8	62.5	48.1	40.1	60.4	•	•	•	•	•	•	•	•	•	•	•	•	•
OKEENE	•	•	•	•	•	•	•	•	•	•	•	•	•	.87	1.03	1.54	2.61	4.06	3.35	2.46	2.69	2.59	2.02	1.16	1.11	25.49
WATONGA	•	•	•	•	•	•	•	•	•	•	•	•	•	1.02	1.13	1.57	2.61	4.47	3.79	2.37	2.30	2.46	2.41	1.27	1.43	26.83
WEATHERFORD	38.4	42.6	49.9	60.4	68.8	78.7	82.9	82.7	74.5	63.4	49.1	41.1	61.0	1.00	1.03	1.53	2.75	4.54	4.28	2.52	2.52	2.57	2.59	1.13	1.24	27.70
DIVISION	37.9	41.9	49.2	59.9	68.4	78.4	82.8	82.5	74.1	63.0	48.6	40.4	60.6	.90	1.00	1.46	2.59	4.53	3.39	2.23	2.22	2.28	2.29	1.02	1.09	25.00
CENTRAL																										
BRISTOW	38.1	42.3	49.6	60.5	68.4	77.5	81.8	81.9	74.2	63.3	49.3	41.2	60.7	1.71	1.86	2.45	4.11	5.46	4.95	3.59	2.66	3.75	3.19	2.25	1.56	37.54
CHANDLER NO 1	39.0	43.2	50.6	61.4	68.9	78.0	82.5	82.6	74.9	64.1	50.1	41.9	61.4	1.58	1.70	2.21	3.62	5.14	4.51	3.16	2.68	3.46	2.77	2.11	1.50	34.44
CHICKASHA	39.3	43.7	50.9	61.2	69.2	78.5	83.0	82.9	75.0	63.8	49.7	41.8	61.6	1.35	1.47	1.90	3.07	5.36	4.16	2.42	2.57	3.13	3.03	1.63	1.51	31.60
EL RENO 1 N	•	•	•	•	•	•	•	•	•	•	•	•	•	1.15	1.31	1.62	2.85	4.79	3.83	2.48	2.51	2.71	2.85	1.65	1.33	29.08
GUTHRIE 2 WNW	38.3	42.2	49.7	60.8	69.0	78.2	82.9	82.8	74.6	63.8	49.3	41.1	61.1	1.22	1.37	1.77	3.09	4.51	4.06	2.69	2.99	3.45	2.83	1.71	1.28	30.97
HENNESSEY 1 N	37.4	41.6	49.1	59.9	68.8	79.0	83.5	83.3	75.1	63.9	48.7	40.0	60.9	1.17	1.24	1.71	2.84	4.60	4.01	2.76	2.89	2.40	2.30	1.46	1.33	28.71
KINGFISHER	38.0	42.1	49.6	60.5	69.0	79.2	83.6	83.4	75.1	63.7	48.8	40.4	61.1	1.06	1.14	1.71	2.60	4.25	3.66	2.63	2.49	2.69	2.37	1.31	1.28	27.19
LAKE OVERHOLSER	•	•	•	•	•	•	•	•	•	•	•	•	•	1.17	1.34	1.91	3.14	4.80	4.14	2.48	2.41	3.17	2.82	1.58	1.33	30.29
NORMAN 3 S	39.7	44.0	51.3	61.6	69.4	78.4	82.8	82.9	75.2	64.5	50.6	42.6	61.9													
OKEMAH	39.6	43.8	50.9	61.2	68.9	77.5	82.0	82.0	74.5	64.1	50.4	42.6	61.5	1.88	2.04	2.65	4.55	5.82	5.19	3.10	2.64	3.85	3.36	2.32	2.07	39.47
OKLAHOMA CITY PENN AVE	38.5	42.5	49.8	60.8	70.0	78.4	82.8	82.6	74.8	63.9	49.4	41.3	61.2	1.39	1.45	1.98	3.42	5.35	4.51	2.62	2.68	3.10	2.97	1.70	1.41	32.58
OKLAHOMA CITY WB AP	37.0	41.3	48.5	59.9	68.4	78.0	82.5	82.8	73.8	62.9	48.4	40.3	60.3	1.31	1.37	1.97	3.12	5.19	4.47	2.37	2.52	3.02	2.51	1.56	1.41	30.82
PERKINS 1 SSE	•	•	•	•	•	•	•	•	•	•	•	•	•	1.53	1.46	2.20	3.16	5.09	4.58	3.45	3.19	3.81	3.21	1.90	1.42	35.00
SHAWNEE	39.7	43.8	51.1	61.6	69.2	78.1	82.4	82.3	74.7	64.0	50.3	42.4	61.6	1.63	1.86	2.37	4.11	5.81	4.93	3.22	2.92	3.48	3.16	2.07	1.66	37.22
STILLWATER 2 W	37.9	42.2	49.6	60.6	68.5	77.9	82.5	82.3	74.2	63.5	49.1	40.8	60.8	1.16	1.35	1.86	2.86	4.62	4.24	3.53	3.38	2.78	3.21	1.85	1.34	32.18
UNION CITY 1 SSE	•	•	•	•	•	•	•	•	•	•	•	•	•	1.33	1.69	2.05	3.59	5.53	4.57	2.71	2.57	3.02	3.02	1.79	1.55	33.42
DIVISION	38.7	42.8	50.1	60.9	68.9	78.2	82.6	82.5	74.6	63.8	49.5	41.5	61.2	1.44	1.59	2.12	3.43	5.34	4.44	3.01	2.75	3.37	2.94	1.85	1.54	33.82

*NORMALS BY CLIMATOLOGICAL DIVISIONS

Taken from "Climatography of the United States No. 81-4, Decennial Census of U. S. Climate"

TEMPERATURE (°F) PRECIPITATION (In.)

STATIONS (By Divisions)	JAN	FEB	MAR	APR	MAY	JUNE	JULY	AUG	SEPT	OCT	NOV	DEC	ANN	JAN	FEB	MAR	APR	MAY	JUNE	JULY	AUG	SEPT	OCT	NOV	DEC	ANN
EAST CENTRAL																										
CALVIN	•	•	•	•	•	•	•	•	•	•	•	•	•	1.98	2.85	3.06	3.98	6.42	4.60	3.52	3.16	3.76	3.30	2.45	2.25	41.33
HOLDENVILLE	40.8	44.7	51.8	62.1	69.6	78.4	82.9	82.9	75.6	65.0	51.3	43.6	62.4	2.02	2.50	3.06	4.27	6.34	5.23	3.97	3.10	3.75	3.26	2.41	2.30	42.21
MCALESTER FAA AIRPORT	40.7	44.6	51.8	61.4	69.5	78.4	82.9	82.9	75.3	64.6	50.8	43.3	62.2	2.13	3.03	3.28	4.33	6.10	4.57	3.72	3.17	3.66	3.29	2.60	2.81	42.69
MUSKOGEE	39.9	43.9	51.2	61.8	69.6	78.3	82.8	82.7	75.2	64.3	50.4	42.5	61.9	2.30	2.63	3.23	4.54	5.75	5.48	3.15	2.89	3.49	3.42	2.79	2.35	42.02
OKMULGEE WATER WORKS	39.4	43.4	50.6	61.4	68.9	77.8	81.9	81.5	73.9	63.2	49.9	42.3	61.2	1.97	2.21	2.84	4.58	5.63	5.63	3.26	2.49	3.94	3.62	2.46	2.05	40.68
SALLISAW	40.0	43.8	50.8	61.4	69.3	77.8	82.0	81.7	74.8	63.8	50.1	42.5	61.5	2.57	3.22	3.35	4.43	5.68	4.87	2.92	3.31	3.80	3.07	2.86	2.68	42.76
TAHLEQUAH	38.6	42.7	49.8	60.5	67.7	76.5	81.0	80.7	73.2	62.1	48.7	41.4	60.2	2.33	2.79	3.17	4.63	6.14	5.10	3.00	3.05	4.00	3.67	2.89	2.33	43.10
WEBBERS FALLS	39.5	43.2	50.7	61.7	69.6	78.3	82.8	82.3	74.9	63.8	49.9	42.1	61.6	2.49	3.06	3.19	4.67	6.09	5.28	3.28	3.22	3.57	3.55	2.83	2.69	43.92
DIVISION	39.8	43.7	50.9	61.4	69.0	77.8	82.2	82.0	74.7	63.9	50.2	42.5	61.5	2.26	2.81	3.23	4.54	5.97	5.11	3.35	3.09	3.76	3.45	2.73	2.48	42.78
SOUTHWEST																										
ALTUS IRR. RESCH STN	41.0	44.7	52.2	62.7	71.1	80.6	84.4	84.1	76.1	65.4	51.4	43.3	63.1	.98	1.02	1.27	2.25	4.47	3.49	1.90	1.95	2.36	2.78	.85	1.17	24.49
APACHE	39.9	44.2	51.2	61.4	69.4	78.6	83.1	83.2	75.4	64.3	50.3	42.3	61.9	1.39	1.63	2.08	2.97	5.87	3.79	2.75	2.30	3.34	2.98	1.55	1.65	32.30
CARNEGIE 4 ENE	38.8	43.3	50.7	61.4	69.8	79.7	83.6	83.4	75.1	63.7	49.3	41.2	61.7	1.19	1.23	1.53	2.68	4.80	3.63	2.49	2.38	2.57	2.35	1.06	1.49	27.40
CHATTANOOGA 3 NE	40.8	45.1	52.1	61.9	70.4	79.9	83.9	84.1	76.4	65.2	51.0	43.0	62.8	1.23	1.44	1.53	2.24	5.52	3.17	2.36	1.73	2.57	2.88	1.31	1.39	27.37
FREDERICK	42.2	46.5	53.4	63.3	71.4	80.8	85.0	85.0	77.3	66.4	52.3	44.4	64.0	1.21	1.39	1.78	2.26	4.98	3.29	2.18	1.97	2.39	2.67	1.36	1.35	26.83
HOBART FAA AIRPORT	38.5	42.5	49.5	60.3	69.0	79.4	83.6	83.4	74.9	63.4	48.9	40.7	61.2	1.02	1.00	1.30	2.56	4.90	3.47	2.13	1.96	2.01	2.60	.96	1.22	25.13
HOLLIS	40.2	44.4	51.7	62.4	71.3	81.3	85.3	84.9	76.3	64.9	50.4	42.4	63.0	.78	.87	1.01	2.50	5.07	3.08	2.04	1.72	2.15	2.21	.70	1.07	23.20
LAWTON	40.7	45.1	52.3	62.4	70.4	79.5	83.6	83.7	76.1	64.8	50.9	43.1	62.7	1.39	1.58	1.78	2.37	5.95	3.67	2.61	1.96	2.71	3.01	1.62	1.53	30.18
MANGUM RESEARCH STA	•	•	•	•	•	•	•	•	•	•	•	•	•	.97	.94	1.22	2.42	4.89	2.70	1.98	2.05	2.16	2.28	.82	1.07	23.50
WALTERS	41.7	45.9	53.2	63.2	71.0	80.4	84.6	84.7	77.0	65.9	52.0	44.0	63.6	1.26	1.53	1.83	2.56	5.40	3.42	2.33	2.35	2.84	3.11	1.44	1.62	29.69
WICHITA MT WL REF	39.3	43.2	50.5	61.0	68.5	77.6	82.0	82.1	74.3	62.9	49.0	41.5	61.0	1.37	1.47	1.71	2.46	5.26	3.77	2.68	2.33	2.81	2.64	1.31	1.38	29.19
DIVISION	40.2	44.4	51.5	62.0	70.2	79.8	83.9	83.8	75.8	64.6	50.4	42.5	62.4	1.16	1.28	1.55	2.46	5.14	3.44	2.31	2.10	2.55	2.71	1.19	1.36	27.25
SOUTH CENTRAL																										
ADA	41.6	45.6	52.4	62.3	69.8	78.4	82.9	82.8	75.5	65.3	51.6	43.9	62.7	1.92	2.55	2.77	3.79	6.38	4.47	3.13	3.12	3.43	3.45	2.40	2.35	39.76
ARDMORE	43.2	46.9	54.0	63.9	71.1	79.9	84.0	84.2	77.3	67.0	53.6	45.7	64.2	2.00	2.46	2.95	4.18	5.74	3.97	2.95	2.54	2.87	3.16	2.06	2.26	37.14
DURANT SE STATE COL	43.0	46.7	53.7	63.4	71.0	79.4	83.6	83.4	76.3	65.7	52.8	45.4	63.7	2.26	3.08	2.85	4.61	5.44	3.76	3.04	2.37	3.16	3.23	2.81	2.59	39.00
MARLOW 1 WSW	40.5	44.8	51.9	62.1	69.5	78.2	82.6	82.8	75.4	65.0	51.2	43.2	62.3	1.40	1.70	2.07	2.99	6.24	4.56	2.37	2.23	2.88	3.36	1.74	1.72	33.26
PAULS VALLEY	40.5	44.8	52.1	62.7	70.6	79.5	83.5	83.3	75.7	64.5	50.7	42.7	62.6	1.68	2.18	2.39	3.26	5.87	4.27	2.72	2.79	3.47	3.31	2.10	1.87	35.91
SULPHUR PLATT NATL PK	41.7	45.8	53.0	62.8	70.0	78.9	83.1	83.3	76.2	65.5	51.6	43.8	63.0	2.08	2.54	2.89	3.97	6.08	4.07	2.77	2.73	3.31	3.22	2.43	2.31	38.40
TISHOMINGO NATIONAL WL	42.5	46.3	53.3	63.0	70.4	78.8	82.7	82.7	75.8	65.3	52.2	44.7	63.1	2.20	2.79	3.07	4.77	5.79	3.96	2.87	2.21	3.07	3.25	2.29	2.58	38.85
WAURIKA	42.6	46.8	53.7	63.8	71.3	80.0	83.9	84.1	76.8	66.3	52.8	44.9	63.9	1.53	1.80	2.04	2.84	5.67	3.47	2.52	2.03	2.74	3.19	1.72	1.93	31.48
DIVISION	42.1	46.1	53.1	63.1	70.6	79.2	83.3	83.4	76.2	65.6	52.1	44.4	63.3	1.89	2.44	2.69	3.88	5.83	4.13	2.82	2.53	3.14	3.28	2.19	2.22	37.04
SOUTHEAST																										
ANTLERS 2 ENE	•	•	•	•	•	•	•	•	•	•	•	•	•	3.40	3.83	3.29	5.20	6.66	4.05	3.15	3.19	3.82	3.17	3.53	3.15	46.44
HUGO	43.4	47.1	54.2	63.6	71.0	79.3	83.1	83.0	76.3	65.9	52.9	45.8	63.8	3.54	3.31	3.83	5.15	5.94	4.34	3.65	3.54	3.59	3.44	3.50	3.25	47.08
IDABEL	44.1	47.4	54.0	63.2	70.3	78.5	82.3	82.3	75.7	65.3	52.6	45.5	63.4	4.05	3.80	4.19	5.23	6.35	3.50	3.79	2.42	3.25	3.17	3.96	3.78	47.49
POTEAU	42.6	46.1	53.0	62.8	70.3	79.0	83.5	83.2	76.1	65.1	52.0	44.7	63.2	3.04	3.53	3.84	4.46	6.00	3.88	3.42	3.40	3.53	3.22	3.47	2.80	44.59
DIVISION	42.8	46.4	53.2	62.8	70.2	78.5	82.6	82.4	75.5	64.9	52.1	44.8	63.0	3.38	3.69	3.85	5.06	6.14	4.16	3.84	3.34	3.66	3.26	3.57	3.34	47.29

* Normals for the period 1931-1960. Divisional normals may not be the arithmetical averages of individual stations published, since additional data for shorter period stations are used to obtain better areal representation.

TEMPERATURE PRECIPITATION

JAN	FEB	MAR	APR	MAY	JUNE	JULY	AUG	SEPT	OCT	NOV	DEC	ANN	JAN	FEB	MAR	APR	MAY	JUNE	JULY	AUG	SEPT	OCT	NOV	DEC	ANN

CONFIDENCE - LIMITS

In the absence of trend or record changes, the chances are 9 out of 10 that the true mean will lie in the interval formed by adding and subtracting the values in the following table from the means for any station in the State. Because of the wider variation in mean precipitation, the corresponding monthly means and annual mean must be substituted for "p" in the precipitation table below to obtain mean precipitation confidence limits.

| 1.8 | 1.5 | 1.5 | 1.0 | .7 | .7 | .8 | .9 | 1.0 | 1.2 | 1.0 | 1.1 | .4 | $.29\sqrt{p}$ | $.26\sqrt{p}$ | $.36\sqrt{p}$ | $.38\sqrt{p}$ | $.38\sqrt{p}$ | $.44\sqrt{p}$ | $.41\sqrt{p}$ | $.35\sqrt{p}$ | $.41\sqrt{p}$ | $.43\sqrt{p}$ | $.36\sqrt{p}$ | $.26\sqrt{p}$ | $.36\sqrt{p}$ |

COMPARATIVE DATA

Data in the following table are the mean temperature and average precipitation for Chandler, Oklahoma, for the period 1906-1930 and are included in this publication for comparative purposes.

| 38.0 | 42.8 | 51.8 | 61.1 | 68.6 | 78.0 | 82.5 | 82.3 | 75.0 | 62.3 | 51.5 | 40.4 | 61.2 | 1.40 | 1.16 | 2.43 | 4.14 | 4.65 | 3.70 | 2.55 | 3.02 | 3.02 | 3.55 | 2.09 | 1.62 | 33.33 |

NORMALS, MEANS AND EXTREMES

(Table Revised 1970. Base Period for Climatological Normals: 1931-1960)

Station: OKLAHOMA CITY, OKLAHOMA   WILL ROGERS WORLD AIRPORT   Standard time used: CENTRAL   Latitude: 35° 24' N   Longitude: 97° 36' W   Elevation (ground): 1285 feet

Ø For period November 1965 through current year.
Means and extremes above are from existing and comparable exposures. Annual extremes have been exceeded at other sites in the locality as follows:
Highest temperature 113 in August 1936; lowest temperature -17 in February 1899; maximum monthly precipitation 14.12 in June 1932; maximum precipitation in 24 hours 7.87 in October 1927; maximum monthly snowfall 20.7 in March 1924; maximum snowfall in 24 hours 11.3 in March 1924.

NORMALS, MEANS AND EXTREMES

(Table Revised 1975. Base Period for Climatological Normals: 1941-1970)

Elev. 1304 feet m.s.l.   Average station pressure mb.

Ø For period November 1965 through current year.
Means and extremes above are from existing and comparable exposures. Annual extremes have been exceeded at other sites in the locality as follows:
Highest temperature 113 in August 1936; lowest temperature -17 in February 1899; maximum monthly precipitation 14.12 in June 1932; maximum precipitation in 24 hours 7.87 in October 1927; maximum monthly snowfall 20.7 in March 1924; maximum snowfall in 24 hours 11.3 in March 1924.

REFERENCE NOTES APPLYING TO TABLES APPEAR ON THE PAGE FOLLOWING LAST TABLE.

(Caution: Letters and symbols may have different meanings in 1941-1970 tables than in earlier tables. See notes.)

798

(Table Revised 1970. Base Period for Climatological Normals: 1931-1960)

Station: **TULSA, OKLAHOMA** — INTERNATIONAL AIRPORT
Standard time used: CENTRAL Latitude: 36° 12' N Longitude: 95° 54' W Elevation (ground): 650 feet

Month	Temperature Normal Daily max	Daily min	Monthly	Extremes Record highest	Year	Record lowest	Year	Normal heating degree days (Base 65°)	Precip Normal total	Max monthly	Year	Min monthly	Year	Max in 24 hrs	Year	Snow, ice pellets Mean total	Max monthly	Year	Max in 24 hrs	Year	Rel. humidity Hr 00	Hr 06	Hr 12	Hr 18	Wind Mean speed	Prevailing dir.	Fastest mile Speed	Dir.	Year	Pct possible sunshine	Mean sky cover	Clear	Partly cloudy	Cloudy	Precip .01" or more	Snow 1.0" or more	Thunderstorms	Heavy fog	Max 90°+	Max 32°-	Min 32°-	Min 0°-	Avg daily solar radiation
(a)	(b)	(b)	(b)	10		10		(b)	(b)	31		31		31		31	31		31		9	9	9	9	21	15	27	27	27	27	27	31	31	31	31	31	31	31	31	31	31	31	9
J	45.9	26.5	36.2	77	1965	-2	1963	893	1.71	6.65	1949	T	1944	2.25	1946	3.3	10.8	1949	9.0	1944	71	77	58	58	11.4	N	55	SW	1947	52	6.2	9	9	13	6?	2	1	3	0	5	25	9	1
F	51.6	29.6	40.6	86	1962	6	1962	683	1.77	3.95	1951	0.40	1947	1.77	1964	2.4	10.1	1960	6.3	1964	69	75	53	53	11.4	N	50	SW	1948	56	6.0	8	8	12	7	1	1	2	0	1	20		
M	60.0	36.2	48.1	94	1967	11	1960	539	2.43	6.14	1949	0.25	1940	2.67	1969	1.8	11.8	1968	1.7	1968	66	75	47	47	12.5	S	70	SW	1950	56	6.0	8	9	14	7	1	3	*	0	*	12		
A	71.9	47.2	59.6	94	1960	26	1960	213	4.02	9.23	1947	0.51	1950	4.58	1964	0.1	1.7	1957	1.7	1957	70	79	52	49	12.5	S	60	N	1963	58	6.1	8	10	12	9	*	5	*	1	*	1		
M	79.5	57.7	68.6	93	1964+	35	1961+	47	5.26	18.00	1943	1.33	1964	5.26	1950	0.0	0.0		0.0		72	82	54	56	11.4	S	65	NW	1961	61	6.4	8	10	13	11	0	6	*	1	0	0		
J	87.2	67.8	77.5	100	1966+	51	1964	(b)	4.69	11.17	1948	0.53	1948	5.01	1941	0.0	0.0		0.0		79	86	60	57	10.4	S	65	N	1948	65	5.5	10	11	10	10	0	9	1	3	0	0		
J	92.9	71.4	82.2	106	1966	54	1967	0	2.94	10.88	1961	0.03	1954	7.54	1963	0.0	0.0		0.0		73	85	55	51	9.3	S	56	N	1963	70	5.0	12	12	8	7	0	6	*	23	0	0		
A	93.1	70.0	81.6	110	1964	52	1967	0	3.04	7.47	1942	0.21	1945	4.16	1942	0.0	0.0		0.0		75	85	50	51	9.3	S	48	E	1954	73	4.5	13	11	7	7	0	7	1	20	0	0		
S	85.0	61.9	73.5	103	1965	39	1967	18	4.01	10.50	1962	T	1940	6.39	1940	0.0	0.0		0.0		81	88	59	58	9.3	S	45	SW	1944	67	4.4	14	9	7	6	0	5	1	8	0	0		
O	75.0	50.6	62.8	97	1963	31	1961	158	3.31	16.51	1941	T	1952	5.46	1959	T	0.3	1951	2.5	1951	74	86	52	52	10.5	S	62	SW	1949	67	4.4	14	8	9	6	0	3	1	3	0	*		
N	58.6	36.5	47.6	84	1965	13	1964	522	2.28	7.57	1946	T	1949	2.77	1946	0.3	4.0	1958	2.5	1958	74	85	45	62	10.5	S	62	SW	1964	62	5.4	11	7	11	6	*	1	1	0	3	9		
D	49.0	30.2	39.6	80	1966+	-3	1963	787	1.62	4.29	1965	0.16	1950	3.19	1965	1.6	9.9	1958	8.8	1958	79	80	60	61	10.7	S	56	NW	1948	54	5.9	10	8	13	7	1	1	1	0	3	22		
YR	70.6	48.8	59.7	110	AUG. 1964	-3	DEC. 1963	3860	37.08	18.00	MAY 1943	T	OCT. 1952+	7.54	JUL. 1963	9.5	11.8	MAR. 1968	9.8	MAR. 1968	73	81	55	54	10.7	S	75	SW	MAY 1952+	62	5.4	125	104	136	90	4.53	53?	9	71	10	89	1

Means and extremes above are from existing and comparable exposures. Annual extremes have been exceeded at other sites in the locality as follows: Highest temperature 115 in August 1936; lowest temperature -16 in January 1930; minimum monthly precipitation 0.00 in December 1896 and earlier; maximum monthly snowfall 19.2 in March 1924; maximum snowfall in 24 hours 11.5 in March 1924.

NORMALS, MEANS AND EXTREMES
(Table Revised 1975. Base Period for Climatological Normals: 1941-1970)

Month	Temp Normal Daily max	Daily min	Monthly	Extremes Record highest	Year	Record lowest	Year	Normal Degree days Heating	Cooling	Precip Normal	Max monthly	Year	Min monthly	Year	Max 24 hrs	Year	Snow Max monthly	Year	Max 24 hrs	Year	Rel. hum. Hr 00	Hr 06	Hr 12	Hr 18	Wind Mean mph	Prevailing dir.	Fastest mile mph	Dir.	Year	Pct poss. sunshine	Mean sky cover	Clear	Partly cloudy	Cloudy	Precip .01"+	Snow 1.0"+	Thunder	Heavy fog	Max 90°+	Max 32°-	Min 32°-	Min 0°-	Avg station pressure mb	Elev ft
(a)				15		15		15		36	36		36		36		36		36		14	14	14	14	26	15	32	32	32	32	32	36	36	36	36	36	36	36	14	14	14	14		676
J	47.0	26.1	36.6	77	1965	-2	1963	880	0	1.43	6.65	1949	T	1944	2.25	1946	10.8	1949	9.0	1944	79	80	60	60	10.8	N	55	SW	1947	52	6.1	9	7	15	7	1	1	2	0	6	25	14	995.3
F	52.2	30.2	41.2	86	1962	6	1962	666	0	1.72	4.18	1951	T	1947	2.27	1964	10.1	1960	6.3	1964	76	76	55	54	11.3	N	50	SW	1948	56	6.0	8	6	14	7	1	1	2	*	2	19		994.4
M	59.7	36.9	48.3	94	1974	4	1964	528	10	2.52	6.14	1969?	0.08	1940	2.67	1969	11.8	1968	1.7	1968	75	76	49	51	12.5	S	51	SW	1950	57	6.0	8	8	15	8	1	3	*	0	*	11		987.5
A	71.8	49.7	60.8	102	1972	26	1962	176	50	4.17	9.23	1947	0.51	1950	4.58	1964	1.7	1957	1.7	1957	78	78	52	55	12.4	S	60	NW	1943	60	6.0	8	9	13	9	*	5	*	1	0	0		989.0
M	79.2	58.4	68.8	93	1973	35	1964	28	176	5.11	11.17	1943	1.33	1964	7.30	1950	0.0		0.0		82	86	55	56	10.9	S	65	NW	1949	62	6.1	8	11	12	11	0	6	*	1	0	0		988.2
J	87.3	67.3	77.3	100	1966	51	1971	0	369	4.69	11.17	1948	0.53	1948	5.01	1941	0.0		0.0		79	86	59	57	10.3	S	65	N	1949	65	5.4	11	11	8	7	0	9	1	13	0	0		988.9
J	92.8	71.4	82.1	109	1974	51	1971	0	530	3.51	10.88	1961	0.03	1954	7.54	1963	0.0		0.0		73	82	54	50	9.0	S	56	N	1963	71	5.5	12	12	8	7	0	6	*	23	0	0		991.0
A	92.7	70.0	81.4	110	1970	51	1967	0	508	2.95	7.47	1942	0.21	1945	4.16	1942	0.0		0.0		74	85	50	50	9.0	S	48	E	1954	73	4.4	13	11	8	7	0	5	1	21	0	0		991.6
S	84.8	61.7	73.3	103	1965	31	1961	143	259	4.07	18.81	1971	T	1940	6.39	1940	0.0		0.0		83	89	53	57	9.4	S	45	NW	1948	65	4.6	14	8	10	6	0	6	1	8	0	0		994.8
O	75.0	50.8	62.9	97	1963	31	1961	468	78	3.22	16.51	1941	T	1952	5.50	1971	5.6	1971	4.0	1972	77	84	51	57	10.0	S	62	SW	1949	65	4.2	14	8	9	6	*	5	1	2	0	*		993.4
N	60.8	38.0	49.4	84	1966	13	1963	781	0	1.87	7.57	1946	T	1949	5.14	1974	5.6	1958	2.5	1958	80	85	46	62	10.6	S	62	SW	1948	60	5.2	12	7	12	6	*	1	1	0	3	8		992.9
D	50.1	29.5	39.8	80	1966	-3	1963	781	0	1.64	6.34	1971	0.16	1971	3.19	1965	9.9	1958	8.8	1958	79	80	62	63	10.5	S	56	SW	1948	53	5.9	11	6	14	7	1	1	2	0	3	21	1	992.9
YR	71.1	49.2	60.2	110	AUG 1970	-3	DEC 1963	3680	1949	36.90	18.81	SEP 1971	T	OCT 1952	7.54	JUL 1963	11.8	MAR 1968	9.8	MAR 1968	82	75	53	56	10.6	S	75	SW	MAY 1949	62	5.4	124	103	138	90	4	52	10	71	11	84	1	991.5

Means and extremes above are from existing and comparable exposures. Annual extremes have been exceeded at other sites in the locality as follows: Highest temperature 115 in August 1936; lowest temperature -16 in January 1930; minimum monthly precipitation 0.00 in December 1896 and earlier; maximum monthly snowfall 19.2 in March 1924; maximum snowfall in 24 hours 11.5 in March 1924.

REFERENCE NOTES APPLYING TO TABLES APPEAR ON THE PAGE FOLLOWING LAST TABLE.
(Caution: Letters and symbols may have different meanings in 1941-1970 tables than in earlier tables. See notes.)

Reference notes applying to Normals, Means, and Extremes tables for 1931–1960 base period.

(a) Length of record, years, based on January data.
Other months may be for more or fewer years if
there have been breaks in the record.

(b) Climatological standard normals (1931-1960).
* Less than one half.
T Also on earlier dates, months, or years.
Trace, an amount too small to measure.
Below zero temperatures are preceded by a minus sign.
The prevailing direction for wind in the Normals,
Means, and Extremes table is from records through
1963.

‡ ≥ 70° at Alaskan stations.

Unless otherwise indicated, dimensional units used in this bulletin are: temperature in degrees F.;
precipitation, including snowfall, in inches; wind movement in miles per hour; and relative humidity
in percent. Heating degree day totals are the sums of negative departures of average daily temperatures from 65° F. Cooling degree day totals are the sums of positive departures of average daily
temperatures from 65° F. Sleet was included in snowfall totals beginning with July 1948. The term
"Ice pellets" includes solid grains of ice (sleet) and particles consisting of snow pellets encased
in a thin layer of ice. Heavy fog reduces visibility to 1/4 mile or less.

Sky cover is expressed in a range of 0 for no clouds or obscuring phenomena to 10 for complete sky
cover. The number of clear days is based on average cloudiness 0-3, partly cloudy days 4-7, and
cloudy days 8-10 tenths.

Solar radiation data are the averages of direct and diffuse radiation on a horizontal surface. The langley
denotes one gram calorie per square centimeter.

Δ Figures instead of letters in a direction column indicate direction in tens of degrees from true North;
i.e., 09 -East, 18 -South, 27 -West, 36 -North, and 00 -Calm. Resultant wind is the vector sum of
wind directions and speeds divided by the number of observations. If figures appear in the direction
column under "Fastest mile" the corresponding speeds are fastest observed 1-minute values.

To 8 compass points only.

Reference notes applying to Normals, Means, and Extremes tables for 1941–1970 base period.

(a) Length of record, years, through the
current year unless otherwise noted,
based on January data.

(b) 70° and above at Alaskan stations.
* Less than one half.
T Trace.

NORMALS - Based on record for the 1941-1970 period.
DATE OF AN EXTREME - The most recent in cases of multiple
  occurrence.
PREVAILING WIND DIRECTION - Record through 1963.
WIND DIRECTION - Numerals indicate tens of degrees clockwise
  from true north. 00 indicates calm.
FASTEST MILE WIND - Speed is fastest observed 1-minute value
  when the direction is in tens of degrees.

Mean Annual Precipitation, Inches

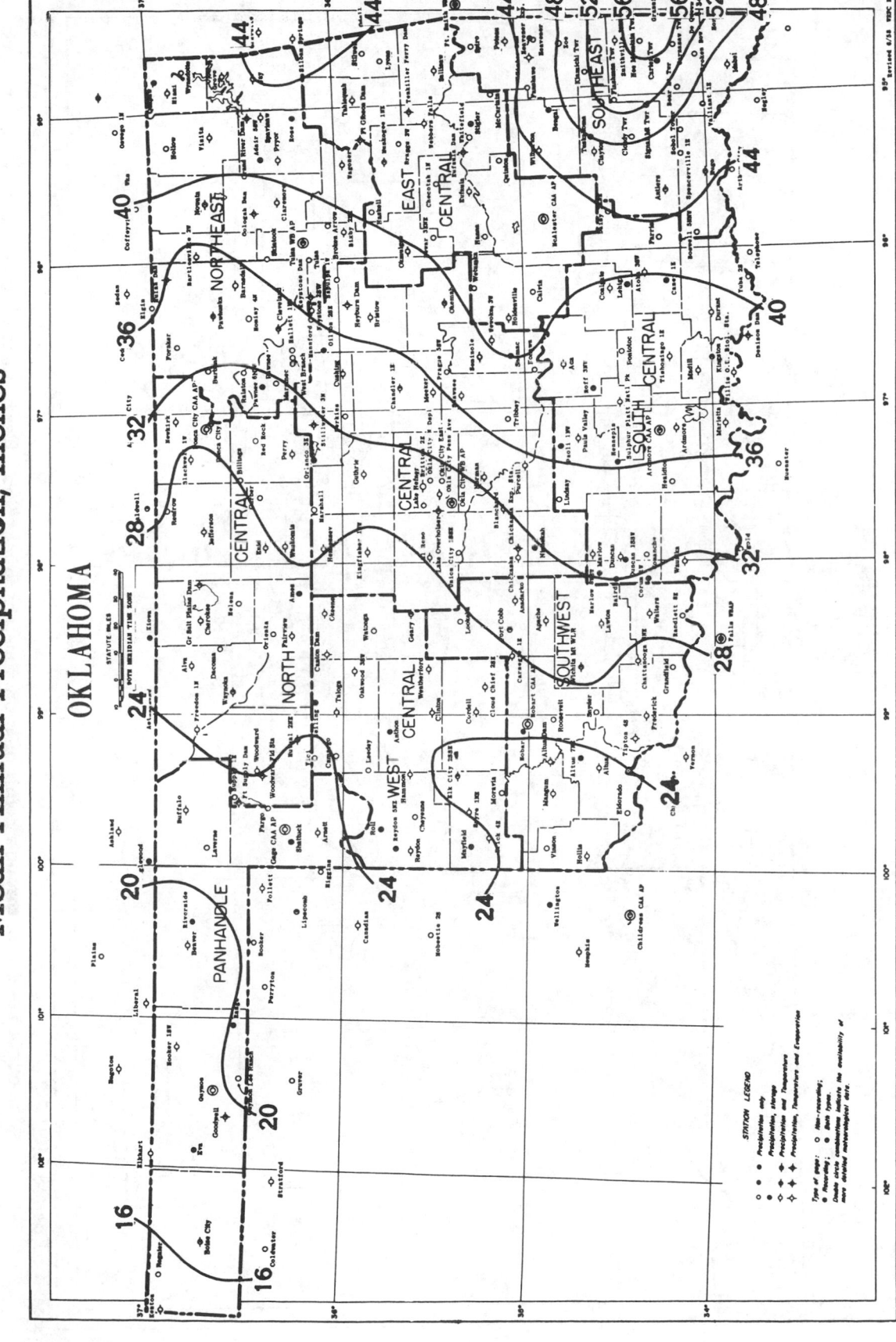

Based on period 1931-55

Isolines are drawn through points of approximately equal value. Caution should be used in interpolating on these maps.

Mean Maximum Temperature (°F.), January

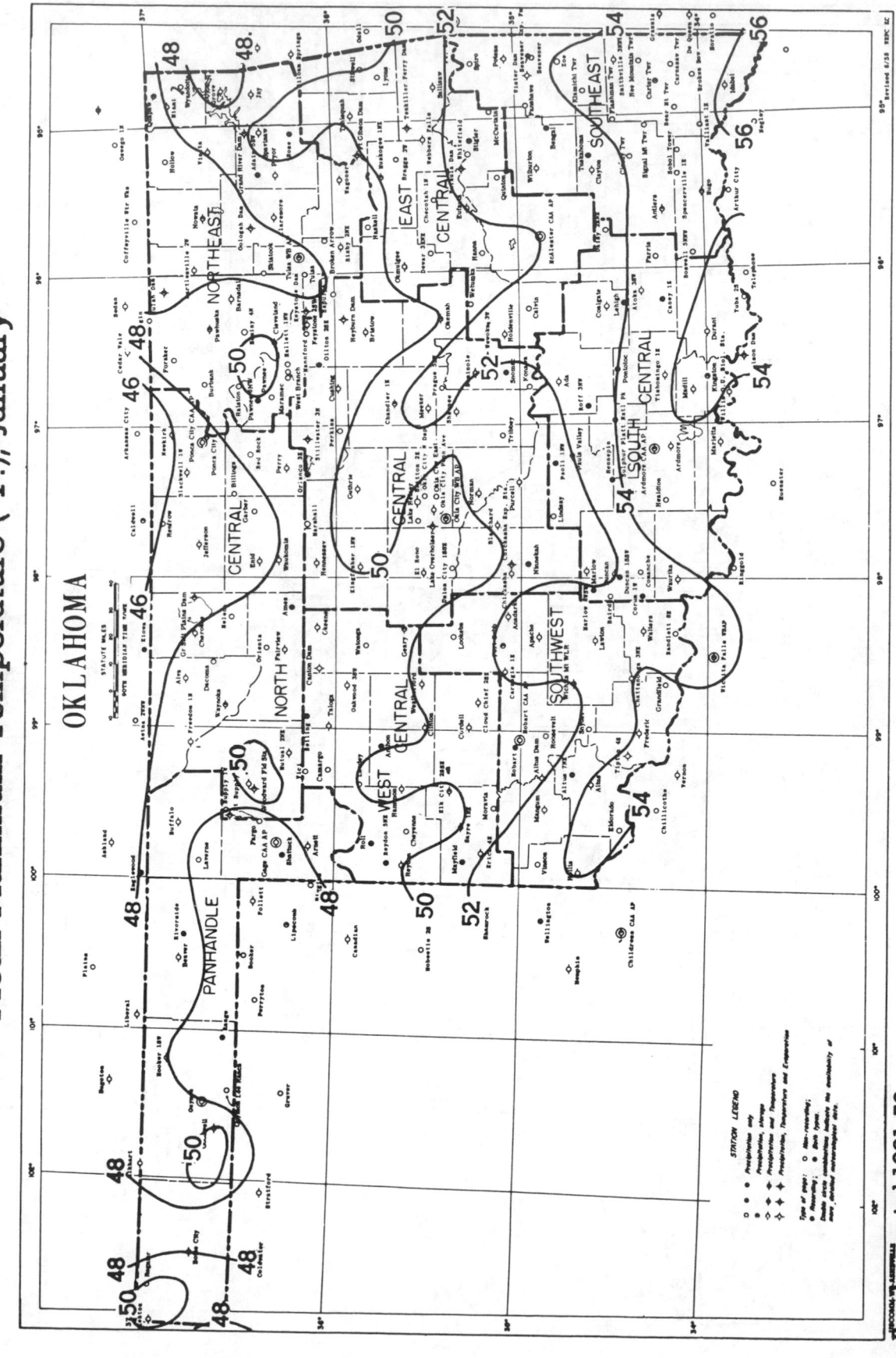

Isolines are drawn through points of approximately equal value. Caution should be used in interpolating on these maps.

Mean Minimum Temperature (°F.), January

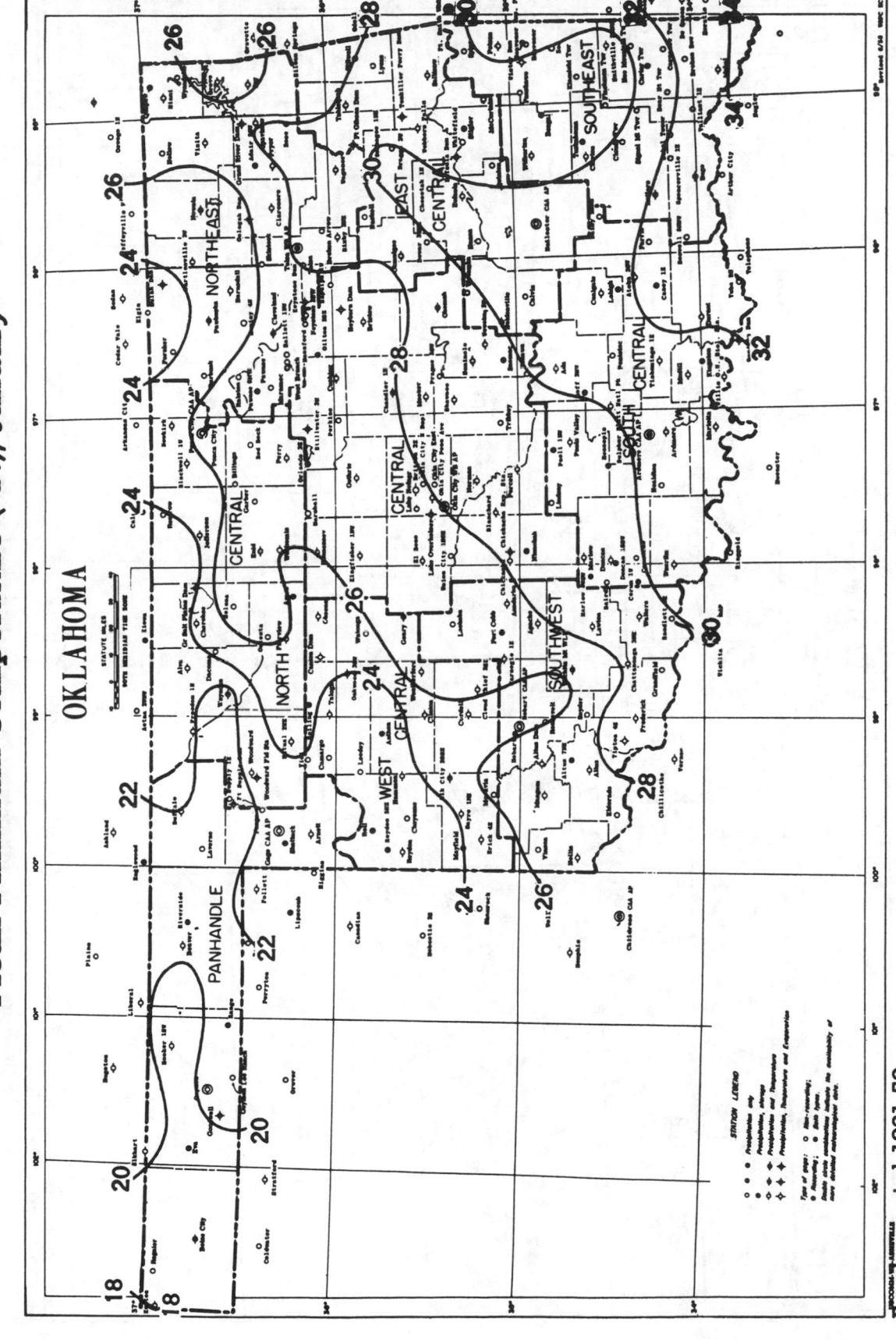

Based on period 1931-52

Isolines are drawn through points of approximately equal value. Caution should be used in interpolating on these maps.

Mean Maximum Temperature (°F.), July

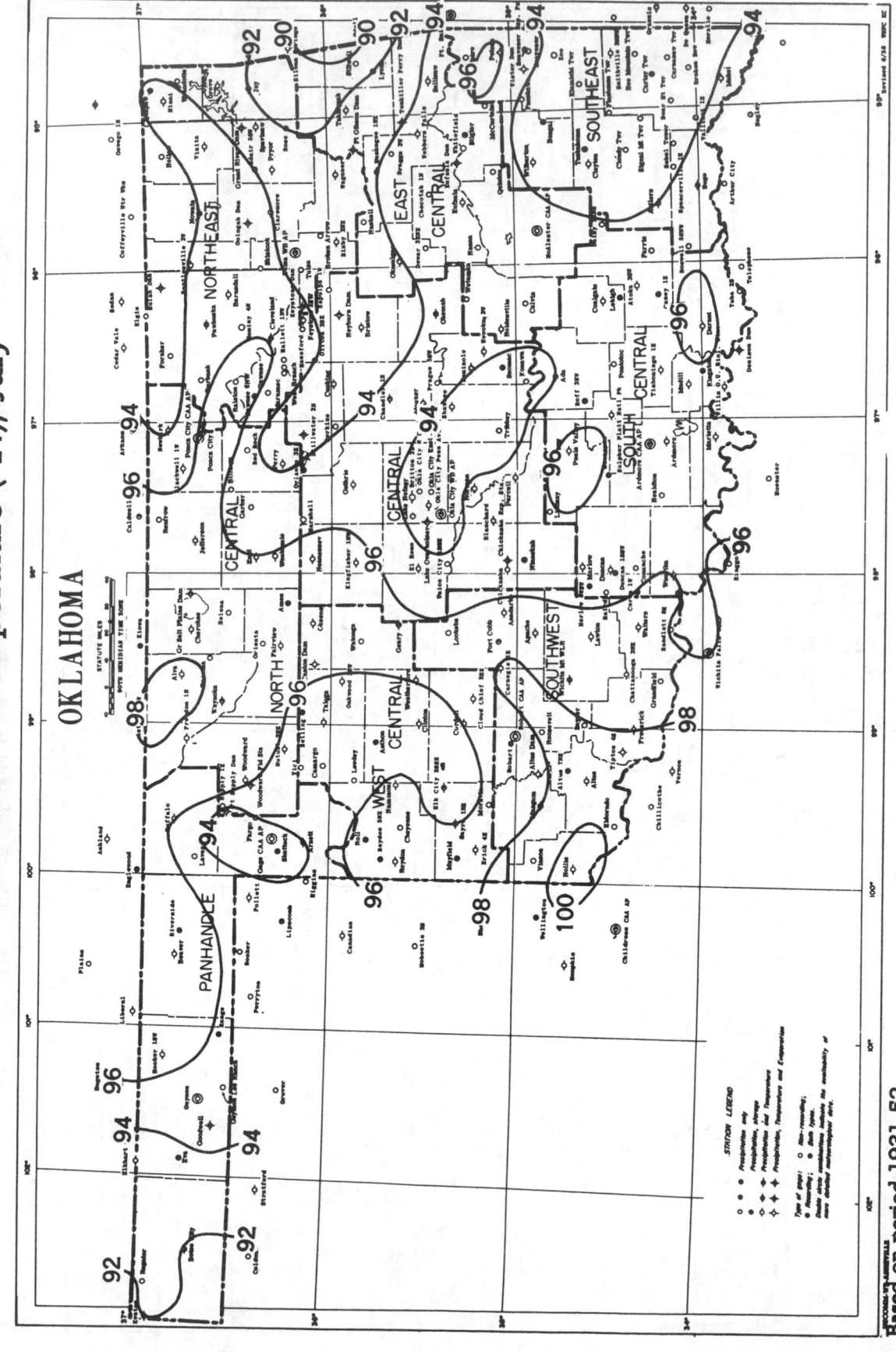

Based on period 1931-52.

Isolines are drawn through points of approximately equal value. Caution should be used in interpolating on these maps.

804

Mean Minimum Temperature (°F.), July

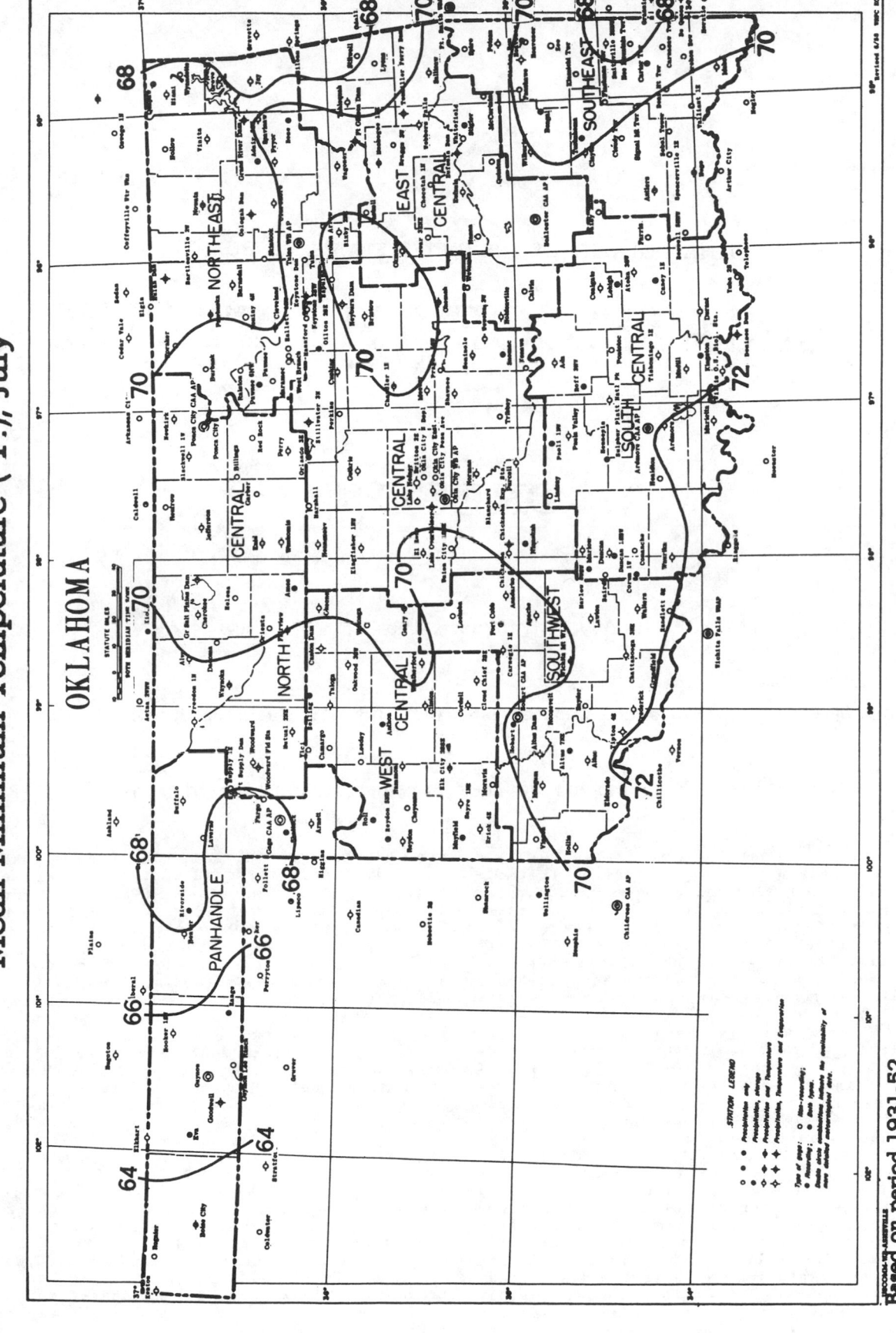

Based on period 1931-52

Isolines are drawn through points of approximately equal value. Caution should be used in interpolating on these maps.

CLIMATES OF THE STATES

OREGON

February 1960

(Normals, Means and Extremes tables revised 1960, 1972 and 1975. Basic report revised February 1960.)

Gilbert L. Sternes, Weather Bureau State Climatologist

Oregon generally enjoys a mild though varied climate with only a rare occurrence of the more devastating weather elements such as cloudbursts, tornadoes, and hailstorms, severe enough to cause serious widespread damage. This discussion of Oregon's climate includes the weather elements that comprise it, the topographical features most important in producing it, and the agricultural, industrial, and recreational interests that result from it.

The most important single geographic feature in the climate of western Oregon and possibly the whole State is the Pacific Ocean, 429 miles of whose coastline make up the western border. Because of the normal movement of air masses from west to east, most of those moving across Oregon have been conditioned by from one to several days over the Pacific. As a result, winter minimum and summer maximum temperatures in the west and to a lesser extent in the eastern portion are greatly moderated. The occurrence of the more extreme low or high temperatures is generally associated with the occasional invasion of the continental air masses. The unlimited supply of moisture available for absorption by those air masses that move across the Pacific is largely responsible for the abundance of rainfall over western Oregon and extending into the higher elevations of the eastern portion.

Beginning at or near the coast and extending the full north-south length of Oregon, the Coast Range is the farthest west of the three mountain ranges exerting important influence on Oregon's climate. This range rises to a crest of between 2,000 and 3,000 feet above sea level in the northern two-thirds of the State and between 3,000 and 4,000 feet in the remaining southern portion. There are occasional peaks extending from 1,000 to 1,500 feet higher. This range, athwart the path of the moisture laden marine air masses moving in from the Pacific, forces their ascent as they pass inland. The resultant cooling produces some of the heaviest annual rainfalls in the United States along these higher western slopes, and materially lessens the moisture available for distribution further inland.

The crest of the Cascade Mountains parallels that of the Coast Range along a path about 75 miles to the east and to within 50 to 75 miles of the California border where the two ranges merge forming a fairly broad, rugged mountain chain known as the Siskiyou Mountains. The Cascades slope upward from the broad valley of the Willamette eastward to an average height of about 5,000 feet above sea level, with a few mountain peaks about twice that elevation. One of these, Mount Hood, with an elevation of 11,245 feet is the highest point in the State. Again the forced ascent of the air masses from the west causes them to give up additional moisture. The rain bearing potential of the marine winds, however, was greatly reduced by passage over the Coast Range and the rainfall on the west slopes of the Cascades is only about one-half to two-thirds as great and drops off rapidly once the crest is crossed and descent down the eastward side begins

The portion of the Blue Mountains in Oregon extend from its northeast corner southwestward to the valleys of the John Day and Deschutes Rivers in the central portion with several branches projecting southeastward to the Snake River Valley. In the extreme northeast corner a separate branch of the Blue Mountains is known as the Wallowa Mountains. The elevations of these mountains, roughly between 5,000 and 6,000 feet above sea level with peaks between 7,000 and 9,000 feet, exert a major influence on the climate in their immediate area including several sizable valleys, particularly those of the Umatilla and Grand Ronde Rivers. However, their effects on Oregon's climate are much less significant and for far smaller areas than that of either the Coast or the Cascade Range. These mountains are, in addition to the rivers mentioned, the source of a number of smaller streams useful to irrigation and power production. They form an area of fairly heavy snowfall which provides excellent skiing and maintains the water flow necessary for some of Oregon's better fishing.

The Steens Mountains are a short range of mountains in the southeast corner of the State less than 25 miles in length and only of very local climatic significance. The main crest is slightly more than 8,000 feet above sea level, with one peak of 9,354 feet. They are important as a snow shed which feeds several small streams useful to local irrigation. Most important of these is the Donner und Blitzen River.

The Columbia River is of vital economic importance to the State since the large dams along its course generate most of the cheap hydro-electric power of the Northwest. Its waters are used in the irrigation of several thousand acres of rich agricultural land in the Boardman and Portland areas. As a major waterway it carries millions of tons of shipping hundreds of miles inland each year. The ports of Vancouver on the Columbia, and Portland on the Willamette at its confluence with the Columbia, are among the finest and largest fresh water ports in the world. However, about its only real importance climatically is its gorge. Cutting through both the Cascade and the Coast Ranges, it offers ready passage of marine air from the Pacific to materially moderate the temperatures to the east in both summer and winter and through which continental air occasionally passes in reverse to produce the more extreme effects in the western valleys.

Scattered through the rugged terrain that makes up much of Oregon are the Columbia and Snake River Basins, the valleys of the many streams that head in the mountains and several very wide plateau regions. The valleys, particularly those of the Columbia, Snake, Willamette, Rouge, and Hood Rivers, produce most of Oregon's agricultural wealth; however, the mountain and plateau regions are used extensively for livestock grazing and considerable dryland farming. This is discussed in more detail under the paragraphs dealing with agriculture. The plateau region, covering approximately one-third of the State's total area, extends from the eastern border west to the eastern slopes of the Cascade Mountains and from the southern border north to the Columbia River. Its elevation is mostly between 4,000 and 6,000 feet above sea level. Because of its arid nature and the very scant vegetation over much of it, summer heating and winter cooling often become quite extreme.

Few states have experienced greater temperature extremes than Oregon. They have ranged from a low of 54°F. below zero to a high of 119°F. Seldom, however, do daily extremes occur even closely approaching these absolute records. In 80 percent of the past 60 years the highest temperature recorded in the State has not exceeded 114°F., nor has the absolute minimum been lower than -37°F., and 50 percent of those years has seen no temperature recorded in Oregon higher than 110°F. or lower than -27°F. The more extreme temperatures generally occur east of the Cascades. In the coastal sections they never drop as low as zero and on very few occasions pass the 100° mark. Here the mean of the coldest month, January, is 45°F. only 15° less than that of July, the warmest month. In the Willamette Valley most stations have not had a maximum temperature greater than 98°F. or a minimum temperature lower than 16°F. for over half their years of record. Temperatures of 90°F. or more occur here only about 6 to 8 days a year and those below zero occur on an average only about once every 25 years. Here the mean temperatures average 38°F. in January and 66°F. in July. In the inland valleys of the southwest the average summer temperatures are about 5° higher than in the northwest and maximums of 90°F. or more occur 40 to 50 days a year. In south-central Oregon the median annual maximum temperatures over a period of years have been between 95° and 100°F., varying, of course, with the different stations; in most other areas east of the Cascades this variance is between 100° and 105°F. Median annual minimum temperatures for eastern Oregon vary from near zero in the more protected areas of the Columbia Basin to -26.F. in the high mountain and plateau regions. This figure for a large majority of these stations, however, lies in the range of -1° to -10°F. The normal mean January temperature in southeast Oregon is 25° to 28°F. and in the northeast 29° to 33°F.; July normal means range between 65° and 70°F. in the central valleys and plateau regions and 70° to 78°F. along the eastern border.

The average annual rainfall in Oregon varies from less than 8 inches in drier plateau regions to as much as 130 inches at points along the upper west slopes of the Coast Range. As can be expected by this wide precipitation variation, vegetation ranges from the heavily wooded Coast Range and west slopes of the Cascades with their dense undergrowth to only a very sparse growth of sagebrush and desert type grasses over the wide plateau areas of central Oregon. This has also resulted in a similar variance in the amount and type of agriculture, though rapidly expanding irrigation projects of both the Federal Government and local districts in recent years have converted many thousands of acres of semidesert areas into highly productive farmland.

The State as a whole has a very definite winter rainfall climate. West of the Cascades between 40 and 50 percent of the annual total precipitation falls during the 3 winter months, December through February; about 20 to 25 percent each in the spring and fall and 6 percent or less during the summer months. East of the Cascades the differences are not as pronounced. Here the winter fall makes up about 30 to 40 percent of the total, spring 25 to 30 percent, fall 20 to 25 percent, and summer between 10 and 15 percent. Along the coast the normal annual total is from 75 to 90 inches, and gradually increases up the west slopes of the Coast Range to between 115 and 130 inches near the crest. This decreases on the east slopes to between 40 and 45 inches along the valley floor of the Willamette and 15 to 30 inches in valleys of the southwest. Again on the west slopes of the Cascades there is a marked increase with elevation as annual averages here range from 50 to 75 inches. The decrease on the east side is very rapid; the annual average for the great plateau area extending from the foothills of the Cascades over the southeastern third of the State is only 8 to 10 inches. In the north-central and northeastern divisions, composed largely of the Columbia Basin and the Blue Mountains, totals are slightly heavier. However, with the exception of the mountain regions where averages as high as 35 inches are attained, this entire area has an average of only 8 to 13 inches annually.

In the higher levels of the Cascades where the State's heaviest snowfalls occur there are few official Weather Bureau Observing stations. In preparing this discussion considerable attention was given to measurements obtained on various snow courses (1). From the records available it appears that here the annual average totals range from 300 to 550 inches. A maximum annual fall of 879 inches and a depth on the ground of 242 inches has been officially recorded at Crater Lake National Park headquarters. Winter precipitation in the Coast Range, due to its lower elevation, occurs largely in the form of rain, though it too i occasionally subject to very heavy snows. In the Blue Mountains seasonal totals range between 150 and 300 inches and depths on the ground may occasionally exceed 120 inches, but during most years greatest recorded depths are less than 100 inches. The periods of continuous snow, of course, vary with elevation. On the peaks of the Cascades higher than 7,000 feet above sea level it persists. in glacial form the year around. In most mountain areas above 4,500 feet there is usually a snow cover from early December until the latter part of April. Snow course averages show that above 4,500 feet snow depth (again varying with elevation) are approximately 60 to 100 inches in the Cascades, 25 to 65 inches in the Blue Mountains at the end of January; 60 to 125 inches and 25 to 70 inches respectively at end of February; 75 to 135 inches and 25 to 80 inches at the end of March; 40 to 120 inches and 5 to 45 inches at the end of April.

Along the Oregon coast the average annual snowfall is only 1 to 3 inches, with many years in which there is no measurable amount. In the inland western valleys most average yearly totals range between 10 and 15 inches, with ground coverage seldom lasting more than 2 or 3 days at a time. In north-central Oregon the annual average fall is 15 to 30 inches, while over the higher plateau region that makes up the south-central portion it ranges up to as much as 60 inches. In the valleys of the northeast 40 to 75 inches is normal, but in the Snake Basin which makes up most of the southeast it is only 15 to 40 inches. Once in every several years the State, with the possible exception of the coastal areas, will be visited by heavy snowstorms which even in the Willamette Valley have on occasion produced 20 to 25 inches in a 24-hour period.

Over the State there are a number of hailstorms each year, but these are generally light and cover only very small areas. They cause several thousand dollars damage annually to crops and a much lesser amount to buildings, but the overall total is insignificant. Practically all of these storms severe enough to damage either buildings or crops occur east of the Cascades. Several times each year winds of hurricane speeds (74 m.p.h. and over) strike the Oregon coast and occasionally move inland to the western valleys and up the Columbia Gorge in considerable strength. At Portland gusts with instantaneous speeds of 75 to 80 miles per hour are occasionally observed. Damage is usually confined to power and communication lines, in a minor way to growing hay and grain crops, outdoor signs, and timber. Very rarely does loss of life or major structural damage to buildings result. The few tornadoes that have been reported have been short lived in the incipient stage, never more than a few hundred feet in width and a fraction of a mile in length.

In the western portion of the State thunderstorms occur in the valleys on an average of only 4 or 5 days a year. Here they are not usually severe and cause little damage. In the eastern portion, the average is 12 to 15 days and the accompanying precipitation and resulting damage is somewhat greater; however, even here total losses are not extensive. It is in the mountain areas that these storms occur most frequently, and each year a considerable number of forest fires are started by the accompanying lightning.

Most of Oregon is drained into the Pacific Ocean through the Columbia River which forms much of the northern boundary of the State. Major tributaries of the Columbia are the Willamette River which flows through the broad valley between the Coast Range and the Cascades; the Deschutes and John Day Rivers in north central Oregon, and the Snake River which forms more than half of the eastern boundary of the State, but which has its confluence with the Columbia near Burbank, Wash. The Powder, Malheur, and Owyhee Rivers, principal western tributaries of the Snake, drain practically all of eastern Oregon except for a sizeable area in the southeast where the drainage is into brackish lakes, or depressions, some of which hold water only during wet periods. A number of streams in the rugged southwestern region drain directly into the Ocean. The largest of these are the Umpqua and the Rogue Rivers which rise in the Cascades.

Major floods in the Willamette Basin and in the coastal streams usually result from a combiantion of (1) a period of several days of moderate to heavy rain, (2) moderate to heavy snowpack on the west slope and summit of the Cascades, and (3) warm southerly winds. Major flooding during the period October to April in the Willamette has a frequency of near 1 year in 4. In the coastal drainage where the climate is mild, most of the precipitation occurs during November through March and flooding due to heavy rains occurs about once in two years and usually may be expected in the late fall and winter.

West of the Cascades, annual peak discharge occurs between April and June, the period of the most rapid melting of snow. Moderate to heavy rains falling on a heavy snowpack during this period constitutes a major flood threat.

Major floods have occurred in Oregon in 1894, 1910, 1933, 1948 and 1956, in the Columbia Basin. In addition, the Willamette Valley experienced serious flooding in 1890, 1923, 1943, 1945, 1953 and 1955. In the coastal drainage notable flood years have been 1861, 1890, 1909, 1927, 1953 and 1955.

During the early morning hours (the period when relative humidity is greatest) there is little variation between winter and summer and between eastern and western Oregon. The 4:30 a.m. average for practically every station in the State for which relative humidity figures are available is between 82 and 92 percent in January and only about 5 percent less in each case in July. In contrast, the 4:30 p.m. averages (the time of near minimum relative humidities) show a very marked difference between summer and winter and a considerable difference, particularly in the summer months, between the areas east and the areas west of the Cascades. In January over the State generally the afternoon averages range between 75 and 85 percent. In July this drops to between 25 and 30 percent east of the Cascades and 35 to 45 percent west.

Under extreme conditions that occur occasionally in the west and rather often in the east during the summer and early fall this will drop down to 10 to 20 percent during the warmest part of the day.

In recent years the Weather Bureau has placed a much larger emphasis on dates of occurrence of specific temperature values in various tabulations and publications. The publication of dates of killing frosts has been discontinued. The reason for this is the indefinite nature of "killing frost". A temperature that is very critical to one type of vegetation is of little consequence to another. It is never certain by the general term "killing frost" which type of vegetation is being referred to. Elsewhere in this publication will be found a tabulation of last occurrence in spring and first in fall of certain selected temperatures for fifty-two Oregon stations with 30 years of complete records for the period 1921-1950.

Droughts are not the problem in Oregon that they are to the Middle West and other parts of the country as there is a normal dry period during the summer months of each year. There are many years when the average rainfall over the entire State for July and August combined is only a few hundredths of an inch. As a result the agricultural program is planned accordingly. The drought condition that is most damaging is the result of marked deficiency in moisture during the spring and early summer months when it is particularly needed for completing maturity of hay and grain and sizing of potatoes and other row crops. Seldom if ever are these severe enough to induce a complete crop failure over sizeable areas, but frequently yields and quality of important crops are materially reduced, and untimely deterioration of pastures and ranges force livestock onto the markets before reaching prime condition.

The heavy winter rains in the Coast Range and rains and snow in the Cascade Mountains provide an abundant water supply for irrigation in all areas at present able to take advantage of it. In recent years there has been a tremendous growth in irrigation in land adjacent to the Cascades. This has been done in several ways. A number of very large dams have been built by the Government for storing water during the heavy rainfall months for use during the summer. Large irrigation districts have been created and canals built leading from the larger rivers to previously semidesert areas. Along the rivers individuals have installed pumping stations and sprinkler systems. In central Oregon successful experimentation has been carried on with drilling wells and pumping water for irrigation. There is still a tremendous natural flow in the streams of Oregon that can be utilized if irrigation systems for bringing it to the land where it is needed can be devised at a reasonable cost.

Several evaporation stations are at present being operated in Oregon, largely at agricultural experiment stations and U. S. Engineer reservoir installations. Most of these records begin May 1 and continue through September 30. In general this period encompasses approximately 70 to 80 percent of their annual total evaporation. A summary of the average total evaporation to the closest inch for these 5 months at representative stations follows: Corvallis, 26; Cottage Grove Dam, 30; Dorena Dam, 31; Fern Ridge Dam, 31; Malheur Experiment Station, 36; Medford Experiment Station, 33; Odell Lake, 18; Wickiup Dam, 34; Madras Airport, 37; Malheur Experiment Station (near Ontario) 36; Warm Springs reservoir, 46.

Agriculturally the State can be divided into six distinct areas; Coastal and Lower Columbia, Willamette Valley, Southwestern Oregon, Columbia Basin, South Central Area, and the Snake Basin. The type of agriculture carried on in each area is briefly summarized as follows:

Coastal and Lower Columbia (Clatsop, Columbia, Coos, Curry, Lincoln, and Tillamook Counties): This is a heavily timbered area with copious rainfall much of the year resulting in lush pastures, largely responsible for the predominance of dairying in these counties. The raising of meat animals, both sheep and cattle, is rapidly becoming a close second in importance to timber. Along the lower Columbia and in the northern Coastal valleys production of Bent grass seed, narcissus bulbs, cranberries, and mink add significantly to the farm income. In the southern coastal counties, in addition to dairy and meat production, substantial supplements to the farm income include wool, cranberries, legume crops and seeds, nuts, fruit, strawberries, and a large lily bulb and cut flower industry.

Willamette Valley (Benton, Clackamas, Lane, Linn, Marion, Multnomah, Polk, Washington, and Yamhill Counties): This is by far the most diversified agricultural area in the State. Its climate is relatively free of extremes in temperatures; growing seasons are long; and moisture is abundant most of the year with the use of summer irrigation expanding rapidly. There are extensive commercial productions of hay, grain, a wide variety of truck crops, apples, pears, sweet cherries, prunes, walnuts, filberts, strawberries, all types of cane berries, hops, onions, potatoes, mint, fiber flax, numerous vegetable and specialty grass seed crops, and many other miscellaneous crops. Other multimillion dollar farm operations include dairying; poultry raising, particularly turkeys; and extensive nursery businesses.

Southwestern Oregon (Douglas, Jackson, and Josephine Counties): This region is best known for the Rogue River pear industry with 10 to 12 thousand acres devoted to that crop. Here also is produced a considerable quantity of peaches and apples. Following fruit raising in importance is dairying, then the raising of hay, grain, seed crops, beef cattle, poultry, and sheep. In local areas the growing of bulbs and hops adds substantially to the agricultural economy. The development of sprinkler irrigation is greatly extending pasture use and materially aiding certain horticulture interests.

Columbia Basin (Gilliam, Hood River, Morrow, Sherman, Umatilla, Wasco, and Wheeler Counties): This is Oregon's major wheat producing area. Grain, raised on dryland farms, is very predominately the source of farm income in all of these counties except Hood River where fruit, particularly pears, apples, and sweet cherries, leads. These fruits are also raised commercially in other counties immediately adjacent to the Columbia River. Umatilla County realizes a sizable income from fruit as well as from a multimillion dollar green pea raising and processing industry, and a sizable annual return from truck crops raised under irrigation. In addition, beef cattle, sheep, alfalfa, and poultry all make substantial contributions, to the agricultural wealth of the region.

South Central Oregon (Crook, Deschutes, Grant, Harney, Jefferson, Klamath, and Lake Counties): This is primarily a livestock country with vast spring, summer, and fall ranges. In addition to beef cattle, which are the dominant livestock interest, there is also extensive raising of sheep, dairy herds, horses, and swine. In the three counties of Deschutes, Crook, and Jefferson, which make up the literal center of the State, are 160,000 acres of land irrigated by Federal and locally owned irrigation systems. Another quite extensive operation supplied by the Klamath Lake is in operation in Klamath County. Field crops grown on a large commercial basis include potatoes, particularly those for certified seed stock; alfalfa as both a hay and seed crop, as well as large acreages of other hay crops; a major portion of the State's Ladino clover seed; and wheat, oats, barley, and onions.

Snake River Basin (Baker, Malheur, Union, and Wallowa Counties): With over 5 million acres of Federal range being utilized by ranchers in these counties, it follows that livestock raising is the major industry. The one exception is Union County where field crops hold a slight lead. While beef cattle predominate, sheep, dairy herds, hogs, and poultry are all important contributors to the farm income. In most of these counties the principal crops are wheat, potatoes, barley, oats, and grass seed. In recent years irrigation has greatly increased in Malheur County, and the most diversified agriculture in this portion of the State is carried on here. In addition to crops already mentioned, this county produces large acreages of sugar beets, onions, peas, tomatoes, berries, sweet corn, and other miscellaneous crops.

Industry and climate in Oregon are very closely related. Much of the industrial development is tied to the vast timber resources or to the abundant supply of power. Both of these are direct products of the heavy rainfall and winter snowpack. A large percent of the remaining industry is concerned in one way or another with the extremely large and widely diversified agricultural production. This too is the result of a favorable climate. In the timber industries not only are billions of board feet of construction lumber sawed each year, but there are numerous large furniture plants scattered over the State. A considerable number of persons are employed in the manufacture of various wooden novelty and art ware. Some of the largest paper manufacturing and plywood plants in the world are located here. There is a wide variety of items manufactured as byproducts of these major industries, including various compressed wood fiber boards, insulating material, and fuels. Very recently tree farming itself has become of major importance. Millions of small trees have been planted by various means on logged-off land of large corporations for harvest, which will begin with thinning about 30 years after planting and will continue until the trees are 80 to 100 years old. As a result of abundance of now existing and future potential power, large metallurgical, chemical, and other lesser industries have developed. The many dams now under construction or in the planning stage will undoubtedly provide for their much greater future expansion. The large industries that exist in Oregon for the purposes of processing its fruits, vegetables, meats, and dairy products have been mentioned under the previous section dealing with agriculture. Fishing is also a major industry in Oregon, with an annual cash income of between $30 and $40 Million. Since many of these fish spawn in fresh water, the large number of coastal streams and their tributaries fed by the snowpack in the Oregon mountains are an important factor in this industry.

Oregon affords almost every conceivable type of outdoor recreation. It has over 400 miles of coast line, three major mountain ranges, thousands of miles of fishing streams, wide range of geological formations, abundance of forest and plateau refuges for many types of game, and hundreds of square miles of lakes and reservoirs providing a natural habitat for all kinds of waterfowl. There are three major ski centers, two in the Cascades and one in the Blue Mountains of the north, besides many smaller ones. One of these, Mt. Hood, is most popular for national ski meets. Tens of thousands of persons ply the streams, lakes, and coast each year in fishing for trout, steelhead, salmon, and many less known varieties. Forests in all areas of the State abound with deer and elk. Large herds of antelope are once more to be found on the semidesert plateaus of central Oregon. Large bodies of water from the Columbia River on the north to Klamath and Goose Lakes to the south and from the coastal bays of the Pacific to the Snake River on the eastern border provide very fine duck and goose hunting. The mild climate and beautiful wooded mountains and streams have placed the finest possible camping and hiking to be found anywhere within the financial reach of everyone. Scores of State parks now have very complete camping facilities available at little if any cost. In central Oregon the varied lava formations and opal, agate, and jasper deposits provide a large number of professional and amateur "rock hounds" with excellent hunting and some profit. Water sports of every type are widely followed here. The mild climate and abundance of rainfall produce beautiful golf courses, and these too are very plentiful and very popular even in the smaller metropolitan areas.

REFERENCES

(1) Summary of Snow Survey Measurements in Oregon 1926-1951. - Division of Irrigation, Soil Conservation Service, Oregon Agricultural Experiment Station and Oregon State Engineer.

(2) Agriculture in Oregon (Revised July, 1952) - issued under direction of E. L. Peterson, Director, Oregon State Department of Agriculture, Salem, Oregon

(3) Fisheries Statistics of Oregon - Issued under direction of Oregon Fish Commission, Portland, Oregon

(4) Weather Bureau Technical Paper No. 15 - Maximum Station Precipitation for 1, 2, 3, 6, 12 and 24 Hours.

(5) Weather Bureau Technical Paper No. 16 - Maximum 24-Hour Precipitation in the United States. Washington, D. C. 1952.

(6) Weather Bureau Technical Paper No. 25 - Rainfall Intensity-Duration-Frequency Curves. For selected stations in the United States, Alaska, Hawaiian Islands, and Puerto Rico.

BIBLIOGRAPHY

(A) Climatic Summary of the United States (Bulletin W) 1930 edition, Sections 3 and 4. U. S. Weather Bureau

(B) Climatic Summary of the United States Oregon - Supplement for 1931 through 1952 (Bulletin W Supplement). U. S. Weather Bureau

(C) Climatological Data - Oregon. U. S. Weather Bureau

(D) Climatological Data National Summary, U. S. Weather Bureau

(E) Hourly Precipitation Data Oregon. U. S. Weather Bureau

(F) Local Climatological data, U. S. Weather Bureau, for Astoria, Burns, Eugene, Meacham, Medford, Pendleton, Portland, Roseburg, Salem and Sexton Summit, Oregon.

FREEZE DATA

STATION	Freeze threshold temperature	Mean date of last Spring occurrence	Mean date of first Fall occurrence	Mean No. of days between dates	Years of record Spring	No. of occurrences in Spring	Years of record Fall	No. of occurrences in Fall
ADRIAN	32	05-03	09-29	148	30	30	30	30
	28	04-22	10-13	174	29	29	30	30
	24	04-01	10-25	207	29	29	30	30
	20	03-15	11-06	236	29	29	29	29
	16	02-18	11-22	277	29	28	28	27
ALBANY	32	03-24	11-06	227	30	30	30	30
	28	02-18	11-24	279	30	29	30	28
	24	01-23	12-16	327	30	23	30	20
	20	01-15	12-24	343	30	17	30	10
	16	01-09	@	@	30	10	30	3
ASTORIA	32	03-18	11-24	251	30	30	30	29
	28	01-31	12-20	323	30	22	29	15
	24	01-17	12-27	344	30	18	29	6
	20	01-08	@	@	30	11	29	3
	16	01-05	@	@	30	6	29	1
BAKER WB CITY	32	05-22	09-23	124	30	30	30	30
	28	04-27	10-05	161	30	30	30	30
	24	04-10	10-25	198	30	30	30	30
	20	03-19	11-04	230	30	30	30	30
	16	02-24	11-20	269	30	29	30	30
BEND	32	06-17	08-17	62	30	30	30	30
	28	05-28	09-19	114	30	30	30	30
	24	05-06	10-11	158	30	30	30	30
	20	04-17	10-28	194	30	30	30	30
	16	03-18	11-17	244	30	29	30	28
BIG EDDY	32	03-30	10-30	214	30	30	30	30
	28	03-07	11-13	251	30	30	30	29
	24	02-09	12-02	296	30	27	30	25
	20	01-28	12-17	323	30	21	30	17
	16	01-20	12-22	336	30	19	30	13
BROOKINGS	32	02-20	12-17	300	30	25	29	18
	28	01-07	@	@	30	11	29	3
	24	@	@	@	30	4	29	2
	20	@	@	@	30	0	29	2
	16	@	@	@	30	0	29	0
CASCADE LOCKS	32	04-02	11-10	222	30	30	30	30
	28	02-22	12-09	291	30	28	30	24
	24	01-29	12-20	325	30	23	30	17
	20	01-19	12-23	338	30	14	30	10
	16	01-16	12-25	343	30	17	30	10
CASCADIA R S	32	05-03	10-13	163	26	26	26	26
	28	03-22	11-10	233	26	26	26	24
	24	02-11	12-07	299	25	23	25	19
	20	01-22	12-19	331	25	18	25	11
	16	01-17	12-27	344	25	16	24	4
CHILOQUIN	32	06-21	07-22	31	30	30	30	30
	28	06-12	08-24	73	30	30	30	30
	24	05-20	09-13	116	29	29	29	29
	20	04-29	10-10	164	28	28	28	28
	16	04-07	10-23	199	28	28	28	28
CONDON	32	05-28	09-30	125	30	30	30	30
	28	05-05	10-15	163	30	30	30	30
	24	04-02	10-26	207	30	30	30	30
	20	03-08	11-18	255	30	29	30	30
	16	02-10	12-03	296	30	28	30	24
CORVALLIS O S C	32	04-12	10-31	202	30	30	30	30
	28	02-25	11-23	271	30	29	30	27
	24	01-25	12-15	324	30	21	30	20
	20	01-17	12-23	340	30	18	30	13
	16	01-08	@	@	30	9	30	4
COTTAGE GROVE 1 S	32	05-06	10-14	161	30	30	30	30
	28	03-23	11-16	240	30	30	30	27
	24	02-07	12-09	305	30	24	30	21
	20	01-15	12-21	340	30	17	30	13
	16	01-11	12-27	350	30	15	30	5
COVE 1 ENE	32	05-19	09-19	123	30	30	30	30
	28	04-25	10-08	166	30	30	30	30
	24	03-31	11-02	215	30	30	30	30
	20	03-15	11-12	243	30	30	30	30
	16	02-21	11-27	279	30	29	30	27
DUFUR	32	05-20	10-04	137	30	30	29	29
	28	04-27	10-22	178	30	30	29	29
	24	03-20	11-07	232	30	29	29	29
	20	02-21	11-27	279	29	28	29	25
	16	02-08	12-12	307	29	25	29	20
ECHO	32	04-21	10-09	171	30	30	30	30
	28	04-07	10-25	201	30	30	30	30
	24	03-09	11-08	244	30	30	30	30
	20	02-18	11-30	284	30	28	30	25
	16	02-02	12-12	312	30	26	30	21
ESTACADA 2 SE	32	04-21	10-21	184	30	30	30	30
	28	03-10	11-22	257	30	29	30	28
	24	02-02	12-11	313	30	27	30	23
	20	01-18	12-22	338	30	18	30	17
	16	01-11	12-27	350	30	13	30	7
EUGENE WB AP	32	04-09	10-31	204	30	30	30	30
	28	03-01	11-22	266	30	29	30	28
	24	01-22	12-12	324	30	20	30	20
	20	01-15	12-21	340	30	17	30	14
	16	01-10	@	@	30	13	30	4
FALLS CITY 1 SW	32	04-23	10-25	186	27	27	27	27
	28	03-14	11-14	245	27	27	27	27
	24	02-05	12-10	308	27	24	27	20
	20	01-18	12-22	339	27	17	27	11
	16	01-12	12-26	348	27	12	27	6
FOREST GROVE	32	04-26	10-18	175	30	30	30	30
	28	03-24	11-12	233	30	30	30	30
	24	01-31	12-07	310	30	26	30	25
	20	01-18	12-19	335	30	18	30	18
	16	01-15	12-24	343	30	17	30	8
FREMONT	32	06-28	07-08	11	30	30	30	30
	28	06-24	07-15	21	30	30	30	30
	24	06-15	08-06	52	30	30	30	30
	20	06-01	09-05	96	30	30	30	30
	16	05-08	09-24	139	29	29	30	30
GRANTS PASS	32	04-29	10-08	162	30	30	30	30
	28	04-03	10-28	208	30	30	30	29
	24	02-23	12-03	283	30	25	30	23
	20	01-24	12-18	328	30	20	30	17
	16	01-08	12-23	350	30	10	30	8
HDWKS PTLAND WTR BUR	32	04-09	11-11	216	30	30	30	30
	28	02-28	12-01	275	30	29	30	27
	24	01-29	12-19	324	30	22	30	18
	20	01-16	12-26	343	30	17	30	10
	16	01-12	12-27	350	30	15	30	5
HEPPNER	32	05-08	10-07	152	30	30	30	30
	28	04-11	10-24	195	30	30	30	30
	24	03-14	11-13	245	30	30	30	30
	20	02-22	11-30	281	29	28	30	25
	16	02-08	12-12	307	29	26	30	20
HERMISTON 2 S	32	05-03	10-03	153	30	30	30	30
	28	04-11	10-15	187	30	30	30	30
	24	03-30	10-30	214	30	30	30	30
	20	03-14	11-17	248	30	30	30	27
	16	02-19	12-04	288	30	28	30	24
HOOD RIVER EXP STA	32	04-30	10-11	165	30	30	30	30
	28	04-04	10-29	207	30	30	30	30
	24	02-27	11-24	270	30	29	30	25
	20	02-06	12-11	308	30	25	30	23
	16	01-22	12-19	332	30	19	30	15
JOSEPH	32	06-11	08-31	81	30	30	30	30
	28	05-24	09-26	126	30	30	30	30
	24	05-02	10-10	161	30	30	29	29
	20	04-10	10-25	198	30	30	29	29
	16	03-27	11-07	224	30	30	29	29
KLAMATH FALLS	32	05-22	09-25	126	30	30	30	30
	28	05-05	10-17	164	30	30	30	30
	24	04-12	10-31	202	30	30	30	30
	20	03-18	11-17	244	30	30	29	29
	16	02-14	12-04	293	30	26	29	26

FREEZE DATA

STATION	Freeze threshold temperature	Mean date of last Spring occurrence	Mean date of first Fall occurrence	Mean No. of days between dates	Years of record Spring	No. of occurrences in Spring	Years of record Fall	No. of occurrences in Fall	STATION	Freeze threshold temperature	Mean date of last Spring occurrence	Mean date of first Fall occurrence	Mean No. of days between dates	Years of record Spring	No. of occurrences in Spring	Years of record Fall	No. of occurrences in Fall
LA GRANDE	32	05-09	09-29	143	30	30	30	30	PRINEVILLE 2 NW	32	06-14	08-13	59	29	29	30	30
	28	04-15	10-22	190	30	30	30	30		28	05-28	09-12	107	29	29	30	30
	24	03-27	11-06	224	30	30	29	29		24	05-03	09-28	148	29	29	30	30
	20	03-08	11-22	259	30	29	29	29		20	04-17	10-16	182	29	29	30	30
	16	02-16	12-03	290	29	28	29	25		16	03-26	11-03	222	29	29	30	30
LAKE CREEK 8 SE	32	05-14	10-05	144	30	30	30	30	PROSPECT 2 SW	32	06-06	09-17	102	30	30	30	30
	28	04-14	10-28	197	30	30	30	30		28	05-05	10-05	153	30	30	30	30
	24	02-28	11-29	275	30	28	27	25		24	03-30	11-05	220	30	30	30	30
	20	01-30	12-15	319	29	24	27	16		20	02-20	11-30	283	30	28	30	26
	16	01-13	12-26	347	28	13	27	6		16	01-30	12-17	321	30	25	29	16
LAKEVIEW	32	06-10	09-03	85	30	30	30	30	RIDDLE	32	04-21	10-17	179	30	30	30	30
	28	05-24	09-23	122	30	30	30	30		28	03-17	11-16	244	30	29	30	28
	24	04-29	10-12	166	30	30	30	30		24	01-30	12-13	317	30	22	30	16
	20	04-12	10-29	200	30	30	30	30		20	01-09	12-26	351	30	12	30	8
	16	03-19	11-12	238	30	30	30	30		16	01-05	⊕	⊕	30	7	30	4
MADRAS	32	06-17	08-07	51	29	29	30	30	ROSEBURG WB CITY	32	03-27	11-13	232	30	30	30	29
	28	06-04	09-13	101	29	29	30	30		28	02-15	12-09	297	30	25	30	22
	24	05-18	10-02	137	29	29	30	30		24	01-15	12-22	341	30	16	30	12
	20	04-30	10-15	168	29	29	30	29		20	01-09	12-27	352	30	12	30	5
	16	04-08	10-29	203	29	29	30	29		16	01-05	⊕	⊕	30	6	30	2
MEDFORD EXP STATION	32	04-25	10-20	178	30	30	30	30	SALEM WB AP	32	04-14	10-27	197	30	30	30	30
	28	03-26	11-01	220	30	30	30	30		28	03-06	11-17	256	30	30	30	29
	24	02-19	11-23	277	30	29	30	28		24	01-31	12-11	314	30	25	30	25
	20	01-21	12-13	326	30	20	30	19		20	01-17	12-20	337	30	19	30	17
	16	01-09	12-26	351	30	12	30	9		16	01-11	12-26	349	30	13	30	5
MILTON	32	04-10	10-21	194	30	30	30	30	TALENT	32	05-02	10-12	164	30	30	30	30
	28	03-20	11-03	227	30	30	30	30		28	03-31	10-29	212	30	30	30	30
	24	02-26	11-21	268	30	28	30	29		24	03-02	11-22	266	30	28	30	29
	20	02-08	12-10	305	30	27	30	21		20	01-26	12-12	320	30	24	30	19
	16	01-29	12-14	319	30	24	30	19		16	01-12	12-25	347	30	14	30	10
MODOC ORCHARD	32	05-10	10-03	146	30	30	30	30	THE DALLES	32	04-03	10-23	204	30	30	30	30
	28	04-04	10-21	200	30	30	30	30		28	03-12	11-07	241	30	30	30	29
	24	03-05	11-10	249	30	29	30	30		24	02-16	12-07	294	30	28	29	23
	20	02-03	12-03	302	30	25	30	24		20	01-26	12-16	324	29	21	29	17
	16	01-15	12-19	337	30	16	30	13		16	01-21	12-21	334	29	20	29	13
MORO	32	05-08	10-06	152	30	30	30	30	UMATILLA	32	04-14	10-18	188	30	30	29	29
	28	04-13	10-28	198	30	30	30	30		28	03-30	10-30	213	30	30	29	29
	24	03-18	11-14	241	30	30	30	29		24	03-10	11-17	252	30	29	29	28
	20	02-13	12-02	292	30	27	30	26		20	02-18	12-09	294	29	26	29	19
	16	02-01	12-14	316	30	26	30	19		16	01-31	12-17	320	28	21	29	16
NEWPORT	32	03-01	12-10	284	29	26	29	23	VALLEY FALLS	32	06-16	08-14	58	30	30	29	29
	28	01-22	12-26	337	29	16	29	9		28	05-29	09-04	98	30	30	29	29
	24	01-10	12-27	351	29	12	29	8		24	05-12	09-22	132	30	30	29	29
	20	01-05	⊕	⊕	29	8	29	3		20	04-21	10-08	170	30	30	29	29
	16	⊕	⊕	⊕	29	3	29	1		16	04-03	10-23	203	30	30	29	29
PARKDALE	32	05-30	09-17	110	29	29	29	29	WALLOWA	32	06-08	08-23	75	30	30	30	30
	28	05-02	10-16	167	29	29	29	29		28	05-21	09-15	117	30	30	30	30
	24	03-25	11-04	224	29	29	29	29		24	04-27	10-01	156	30	30	30	30
	20	02-24	12-03	283	29	28	29	24		20	04-05	10-25	203	30	30	30	30
	16	02-09	12-13	307	29	26	29	20		16	03-11	11-06	240	30	30	30	30
PENDLETON ROUNDUP	32	04-27	10-08	163	30	30	30	30	WASCO	32	05-02	10-13	164	30	30	30	30
	28	04-06	10-20	198	30	30	30	30		28	04-07	10-27	203	30	30	30	30
	24	03-13	11-08	240	30	30	30	29		24	03-06	11-11	250	30	28	30	29
	20	02-20	11-28	281	30	28	30	26		20	02-16	12-02	289	29	27	29	23
	16	01-31	12-12	315	30	24	30	20		16	01-30	12-13	317	29	25	29	20
PORTLAND WB CITY	32	02-25	12-01	279	30	29	30	26	WESTON 2 SE	32	04-22	10-23	184	28	28	29	29
	28	01-23	12-21	332	30	24	30	18		28	03-28	11-07	224	28	28	29	29
	24	01-17	12-25	342	30	18	30	9		24	03-09	11-20	256	28	27	29	28
	20	01-13	12-29	350	30	14	30	6		20	02-18	12-01	287	28	27	28	25
	16	01-11	⊕	⊕	30	13	30	3		16	02-11	12-11	303	28	25	28	20

Data in the above table are based on the period 1921-1950, or that portion of this period for which data are available.

⊕ When the frequency of occurrence in either spring or fall is one year in ten, or less, mean dates are not given.

Means have been adjusted to take into account years of non-occurrence.

A freeze is a numerical substitute for the former term "killing frost" and is the occurrence of a minimum temperature at or below the threshold temperature of 32°, 28°, etc.

Freeze data tabulations in greater detail are available and can be reproduced.

*MEAN TEMPERATURE AND PRECIPITATION

STATION	JANUARY Temperature	JANUARY Precipitation	FEBRUARY Temperature	FEBRUARY Precipitation	MARCH Temperature	MARCH Precipitation	APRIL Temperature	APRIL Precipitation	MAY Temperature	MAY Precipitation	JUNE Temperature	JUNE Precipitation	JULY Temperature	JULY Precipitation	AUGUST Temperature	AUGUST Precipitation	SEPTEMBER Temperature	SEPTEMBER Precipitation	OCTOBER Temperature	OCTOBER Precipitation	NOVEMBER Temperature	NOVEMBER Precipitation	DECEMBER Temperature	DECEMBER Precipitation	ANNUAL Temperature	ANNUAL Precipitation
COASTAL AREA																										
ASTORIA	41.0	11.60	43.3	9.96	46.0	8.83	50.1	5.20	54.0	3.19	57.6	2.86	60.7	1.38	61.4	1.47	59.4	3.08	54.1	7.75	46.8	11.24	43.3	14.18	51.5	80.74
ASTORIA WB AP	40.1	10.66	43.1	9.44	45.3	8.27	49.7	5.15	53.5	3.53	57.6	2.66	60.8	1.07	61.8	1.45	60.3	3.04	54.1	7.15	47.1	10.36	42.7	13.21	51.4	75.99
BANDON		9.18		6.98		6.67		3.76		2.50		1.52		.45		.39		1.56		5.03		7.95		9.55		55.54
BROOKINGS	47.0	12.98	47.9	10.49	48.9	9.89	50.9	5.76	54.6	4.03	57.5	2.77	59.1	.62	58.9	.61	59.4	1.76	56.0	7.30	51.8	11.25	48.7	14.40	53.4	81.86
CANARY	43.0	12.02	45.0	9.76	46.3	9.68	49.2	5.30	53.0	3.51	56.8	2.42	59.9	.95	60.6	.90	59.0	2.57	54.4	7.03	48.4	11.05	44.9	13.49	51.7	78.68
FALLS CITY	37.8	12.35	41.2	9.90	44.5	8.39	49.6	3.97	54.8	2.72	59.2	1.51	64.1	.38	64.6	.68	61.4	1.68	53.4	5.78	44.4	10.80	40.5	14.50	51.3	72.66
GOLD BEACH RS		13.60		9.92		10.17		5.77		3.81		2.28		.53		.67		1.68		7.07		11.31		14.75		81.56
LANGLOIS 2		13.62		10.30		10.29		5.22		4.01		2.68		.51		.62		2.24		7.35		11.82		15.05		83.71
NEWPORT	43.7	9.48	44.9	7.96	46.2	8.07	49.4	4.08	52.3	2.78	55.8	2.38	57.6	.89	58.0	.87	57.0	2.25	53.8	5.94	49.1	8.84	45.4	11.33	51.1	64.87
NORTH BEND CAA AP	45.2	9.89	46.5	8.00	47.6	7.60	50.0	4.12	53.5	2 67	56.8	1.77	59.1	.44	59.6	.57	58.2	1.69	55.1	5.42	50.3	8.99	47.0	11.14	52.4	62.30
POWERS		10.55		7.75		7.96		4.10		2.89		1.74		.29		.38		1.21		4.99		8.51		11.30		61.67
SEASIDE	43.3	11.79	45.0	9.90	46.1	9.08	49.2	5.35	52.9	3.34	56.9	3.10	59.4	1.33	60.1	1.52	58.6	2.95	54.5	7.48	48.4	10.38	45.2	13.64	51.6	79.86
SUMMIT		10.32		8.52		7.77		3.98		2.93		1.90		.56		.74		1.80		6.04		9.37		12.21		66.14
TILLAMOOK		13.05		11.38		10.71		5.67		4.14		3.28		1.19		1.30		3.12		7.81		12.14		15.72		89.51
TIMBER		9.95		8.10		6.76		3.34		2.43		1.69		.51		.58		1.59		5.08		9.16		12.13		61.32
DIVISION	42.5	11.71	44.9	9.73	46.6	8.90	50.1	4.91	54.1	3.30	57.9	2.31	61.2	.72	61.4	.86	59.3	2.21	54.4	6.59	48.2	10.56	44.8	13.62	52.1	75.42
WILLAMETTE VALLEY																										
ALBANY	39.3	6.01	42.8	4.94	46.7	4.34	52.2	2.31	57.5	1.98	62.0	1.48	67.0	.42	66.6	.51	62.3	1.47	53.9	4.14	45.4	5.98	41.7	7.08	53.1	40.66
CASCADIA RS		8.30		6.87		7.20		4.90		3.82		3.07		.63		.83		2.26		5.76		8.17		9.44		61.25
CORVALLIS STATE COL	39.6	6.05	43.1	4.58	46.8	3.88	51.9	2.01	57.2	1.67	61.8	1.22	66.5	.35	66.7	.41	62.9	1.26	54.6	3.60	45.8	5.39	41.9	6.85	53.2	37.27
COTTAGE GROVE 1 S	39.5	6.91	42.6	5.27	45.8	5.27	50.1	3.22	55.1	2.40	59.7	1.91	64.9	.32	64.8	.51	60.5	1.61	54.0	4.36	45.8	6.70	41.6	7.82	52.0	46.30
DISSTON 1 NE LAYNG CR		7.54		5.86		5.91		4.26		3.33		2.66		.44		.58		1.95		4.83		6.97		8.41		52.74
ESTACADA 2 SE	39.0	7.42	42.0	6.18	45.5	6.76	50.8	4.07	55.9	3.38	59.9	2.85	64.9	.71	65.1	1.08	60.9	2.38	53.3	5.78	45.0	7.78	41.4	8.90	52.0	57.29
EUGENE WB AP	38.2	5.41	42.7	4.84	46.7	3.86	50.9	2.54	56.4	1.99	61.3	1.37	66.6	.26	66.0	.38	60.9	1.60	52.7	3.57	45.3	5.69	40.6	6.00	52.4	37.51
FOREST GROVE	37.4	7.36	41.1	5.88	45.4	5.11	51.0	2.33	56.9	1.83	61.4	1.33	66.3	.46	66.3	.52	62.0	1.55	53.0	3.95	44.1	7.04	40.2	9.03	52.1	46.39
HASKINS DAM		12.18		10.45		8.55		4.24		2.88		1.56		.41		.63		1.86		6.14		11.48		14.81		75.19
HEADWORKS PTLND WATER	36.9	11.68	39.9	9.15	43.8	9.65	49.7	5.84	55.3	5.31	59.3	4.69	64.6	1.24	64.2	1.45	60.4	3.81	52.8	8.02	43.9	12.19	39.5	13.38	50.9	86.41
HILLSBORO	37.8	5.79	41.4	4.49	45.4	3.95	51.3	1.89	56.8	1.70	61.2	1.47	66.2	.41	65.6	.47	61.7	1.50	53.1	3.45	44.0	5.76	40.4	7.45	52.1	38.33
MEHAMA		9.34		7.86		8.01		5.01		4.26		3.22		.81		.90		2.68		6.07		9.25		10.90		68.31
PORTLAND WB AP	37.4	4.59	41.5	4.15	46.6	3.47	51.8	2.11	57.6	1.76	62.6	1.72	67.3	.42	67.3	.61	62.3	1.84	54.7	3.17	45.5	5.22	40.8	6.17	53.0	35.23
PORTLAND WB CITY	39.5	5.43	43.8	4.87	48.4	4.15	53.6	2.43	59.1	1.87	63.8	1.62	68.5	.42	68.4	.61	63.9	1.83	56.1	3.53	47.2	6.05	42.4	7.10	54.6	39.91
SALEM WB AP	38.4	5.72	42.8	5.32	46.8	4.19	51.4	2.39	57.2	1.93	62.8	1.22	67.7	.32	67.0	.49	62.4	1.49	54.5	3.70	45.4	5.96	41.0	7.12	53.1	39.85
WATERLOO		6.29		5.44		4.95		3.32		2.81		2.09		.39		.62		1.76		4.54		6.70		7.32		46.23
DIVISION	38.0	7.71	41.3	6.25	44.4	5.59	49.8	3.30	55.3	2.66	59.9	2.04	64.9	.53	64.9	.69	61.3	1.80	53.1	4.93	44.7	7.62	40.5	8.64	51.5	51.76
SOUTHWESTERN VALLEYS																										
ASHLAND 1 N	37.4	2.82	41.5	2.01	45.4	1.97	51.0	1.43	56.7	1.66	62.0	1.25	69.0	.22	68.3	.25	63.0	.83	53.4	1.89	44.2	2.66	39.2	3.16	52.6	20.15
FISH LAKE		6.32		5.10		5.53		3.57		3.37		2.41		.27		.36		1.56		3.77		5.33		7.00		44.59
GRANTS PASS	38.9	5.45	43.3	3.83	48.0	2.82	53.5	1.78	59.1	1.56	64.1	.95	70.1	.24	69.2	.19	64.3	.62	54.8	2.71	44.6	4.17	40.8	5.81	54.2	30.13
MEDFORD WB AP	37.2	2.51	42.6	2.02	47.2	1.52	52.3	1.19	58.7	1.22	64.9	.97	71.8	.17	71.1	.19	64.4	.65	54.5	1.91	44.2	2.67	38.5	3.13	54.0	18.15
MODOC ORCHARD	37.3	3.78	42.1	2.53	46.7	1.95	52.5	1.32	58.3	1.36	63.8	1.16	70.7	.22	69.9	.20	64.0	.60	54.1	2.07	43.9	3.19	39.3	4.24	53.6	22.62
PROSPECT 2 SW	35.2	6.59	38.3	4.71	42.0	4.35	47.9	3.05	53.9	2.64	58.9	1.80	65.9	.27	65.1	.32	60.3	1.02	51.3	3.97	42.2	5.80	37.1	7.12	49.8	41.64
RIDDLE 2 NNE		5.19		3.70		3.24		2.01		1.56		1.29		.19		.25		.87		2.80		4.38		6.06		31.54
ROSEBURG WB AP	39.5	4.60	43.1	3.76	46.9	2.93	51.5	2.13	56.8	1.73	62.1	1.35	67.2	.20	66.8	.32	61.8	1.11	54.3	2.93	46.2	4.51	41.2	4.93	53.1	30.50
SEXTON SUMMIT WB	35.8	4.02	36.4	3.59	39.3	2.96	44.3	2.08	50.5	1.95	56.0	1.50	63.2	.32	63.7	.29	59.8	1.17	50.3	3.84	41.2	4.54	36.7	4.70	48.1	30.96
TALENT	37.3	2.69	41.3	1.79	45.8	1.68	51.7	1.36	57.8	1.55	63.8	1.26	70.7	.27	69.4	.17	63.3	.66	53.2	1.95	43.7	2.49	39.2	3.08	53.1	18.95
DIVISION	36.9	4.81	40.4	3.41	44.2	3.09	49.9	2.06	55.7	1.90	61.0	1.44	67.9	.25	67.2	.26	62.3	.87	53.0	2.78	43.5	4.17	38.8	5.25	51.7	30.29
NORTHERN CASCADES																										
DETROIT		11.20		8.92		8.36		5.11		3.72		2.84		.63		.88		2.45		7.16		11.03		13.04		75.34
MC KENZIE BRIDGE RS		10.62		8.15		8.14		4.76		3.64		2.95		.53		.69		2.23		6.54		9.74		12.00		69.99
OAKRIDGE SALMON HATCH		6.78		4.86		4.87		3.23		2.50		2.14		.47		.52		1.34		3.87		6.08		6.98		43.64
THREE LYNX	35.9	9.61	38.8	7.86	42.9	7.87	49.3	4.64	54.8	4.00	59.3	3.05	64.7	.65	64.4	.81	60.7	2.38	52.5	6.43	42.9	9.28	38.3	11.17	50.4	67.75
DIVISION	33.0	8.50	36.0	7.28	40.0	7.40	45.9	4.43	51.7	4.04	56.0	3.23	62.9	.78	62.5	.83	58.1	2.46	49.3	6.22	40.0	8.79	35.5	10.21	47.6	64.17
HIGH PLATEAU																										
FREMONT	25.6	1.38	30.4	.92	35.5	.75	41.3	.57	47.8	1.06	53.0	.99	60.5	.35	58.7	.24	52.4	.37	44.0	.82	34.6	1.07	28.9	1.68	42.7	10.20
GERBER DAM		2.52		2.05		1.99		1.38		1.65		1.40		.36		.39		.64		1.43		2.07		2.76		18.64
ROUND GROVE		1.72		1.55		1.67		1.33		1.78		1.47		.32		.37		.66		1.44		1.63		2.20		16.14
DIVISION	25.3	5.40	29.2	3.82	32.5	3.55	39.5	2.11	46.2	1.90	51.8	1.64	59.9	.41	58.2	.48	52.8	.91	43.4	2.85	34.8	4.54	28.6	5.97	41.9	33.58
NORTH CENTRAL																										
ANTELOPE 1 N	30.3	1.48	34.7	1.05	39.9	1.04	46.5	.94	53.1	1.23	58.8	1.17	66.7	.31	65.5	.39	59.3	.64	49.9	1.15	39.3	1.60	33.7	1.69	48.1	12.69
ARLINGTON	32.9	1.35	38.2	.94	46.2	.75	53.9	.49	61.7	.54	67.7	.74	75.3	.14	73.8	.18	66.7	.31	55.2	.90	41.9	1.30	36.4	1.43	54.2	9.07
BIG EDDY	33.7	2.35	38.4	1.80	45.9	1.29	53.8	.59	60.7	.66	66.4	.68	72.7	.09	72.0	.15	66.3	.47	55.0	1.23	42.7	2.11	37.6	2.44	53.8	13.86
CONDON	29.2	1.32	33.2	1.08	39.6	1.07	45.9	.97	52.7	1.15	58.5	1.36	66.4	.38	65.2	.33	59.0	.64	49.2	1.17	38.3	1.47	32.7	1.45	47.5	12.39
DUFUR	31.2	1.96	36.0	1.34	42.8	1.05	49.3	.66	56.0	.76	61.5	.90	67.7	.18	67.0	.16	61.6	.46	51.7	1.02	39.8	1.70	34.6	1.81	49.9	12.00
ECHO	33.1	1.21	38.3	1.05	46.1	.92	53.8	.86	61.6	.67	67.6	.92	74.3	.22	72.2	.17	64.5	.56	54.1	.97	41.8	1.24	36.7	1.37	53.7	10.16
FRIEND 1 W		2.82		2.06		1.37		.78		1.00		1.06		.19		.18		.62		1.08		2.49		2.96		16.61
HEPPNER	32.5	1.16	36.5	1.11	42.7	1.17	48.7	1.32	55.0	1.21	60.6	1.53	67.7	.34	66.6	.37	60.6	.74	51.6	1.26	41.0	1.50	35.9	1.43	50.0	13.14
HERMISTON 2 S	31.6	1.07	36.9	.90	45.5	.73	53.7	.63	61.4	.58	67.5	.80	74.3	.21	72.1	.11	64.2	.44	53.3	.79	40.6	1.10	35.4	1.17	53.0	8.53
HOOD RIVER EXP STA	33.2	4.98	37.4	3.54	43.7	3.17	50.4	1.41	56.6	1.06	61.4	.96	66.6	.16	65.8	.23	60.7	.86	52.0	2.62	41.1	4.75	36.5	5.71	50.5	29.45
KENT	29.6	1.22	33.5	.99	39.6	.93	46.3	.84	53.6	1.07	59.8	1.11	67.7	.23	66.6	.26	60.3	.57	50.4	.94	38.5	1.33	32.8	1.34	48.2	10.83
MIKKALO 6 W		1.30		.95		.74		.71		.89		.92		.21		.18		.47		.96		1.24		1.37		9.94
MILTON FREEWATER	33.2	1.41	38.0	1.26	46.2	1.41	54.1	1.34	61.1	1.19	66.8	1.45	74.4	.31	72.2	.21	65.0	.62	54.4	1.30	41.8	1.57	37.0	1.76	53.7	13.83
MORGAN		1.25		1.00		.79		.67		.71		.87		.14		.20		.42		.97		1.13		1.24		9.39
MORO	30.1	1.75	34.6	1.23	41.6	.98	48.4	.78	55.6	.76	61.3	.90	68.7	.18	67.7	.16	61.3	.49	50.8	1.15	38.9	1.71	33.6	1.74	49.4	11.83
PARKDALE	30.4	8.08	34.3	5.83	40.3	5.00	47.0	2.27	53.1	1.85	58.0	1.33	63.2	.28	62.4	.31	57.6	1.26	48.8	3.87	38.6	6.86	33.7	8.97	47.3	45.91
PENDLETON WB AP	30.6	1.48	36.6	1.40	45.2	1.18	52.3	1.04	59.8	.97	66.7	1.21	74.5	.26	72.7	.33	64.6	.71	53.6	1.18	41.1	1.51	35.3	1.69	52.7	12.96
PILOT ROCK 1 SE		1.32		1.18		1.32		1.43		1.36		1.55		.37		.40		.72		1.17		1.38		1.53		13.73
THE DALLES	34.1	2.45	38.8	1.76	46.4	1.23	54.1	.51	61.1	.64	67.0	.70	73.2	.09	72.1	.16	65.8	.47	55.1	1.21	42.8	2.10	37.8	2.48	54.0	13.80
UMATILLA	32.4	.97	37.3	.74	45.6	.70	54.0	.56	61.7	.53	67.7	.72	74.7	.20	72.8	.10	64.7	.40	53.2	.71	41.1	1.06	36.0	1.14	53.4	7.83
WASCO		1.78		1.22		1.03		.66		.73		.83		.19		.19		.46		1.15		1.62		1.86		11.72
DIVISION	31.6	1.91	36.3	1.52	43.1	1.35	50.2	1.04	57.1	1.01	62.7	1.12	70.1	.23	68.7	.25	62.2	.61	52.1	1.33	40.1	1.89	35.1	2.13	50.8	14.39
SOUTH CENTRAL																										
BEND	30.5	1.85	34.0	1.03	38.5	.76	44.8	.68	50.7	1.02	56.2	1.25	63.7	.50	62.4	.51	56.5	.44	48.1	.93	38.7	1.45	32.9	1.83	46.4	12.25
BURNS WB CITY	23.9	1.47	31.2	1.23	38.9	.77	46.6	.82	54.1	.75	60.5	.85	69.8	.25	67.2	.27	58.3	.43	47.2	.80	36.5	1.17	27.7	1.44	46.8	10.25
CHILOQUIN		2.66		1.79		1.51		1.15		1.47		1.11		.26		.27		.60		1.42		2.34		2.91		17.49
KENO		3.02		2.10		1.78		1.13		1.42		.93		.28		.30		.60		1.70		2.42		3.16		18.84
KLAMATH FALLS	29.2	2.09	33.6	1.49	39.6	1.16	47.1	.92	53.9	1.13	60.0	.94	68.7	.29	67.3	.31	60.8	.50	50.3	1.11	38.6	1.67	32.0	2.33	48.4	13.94
LAKEVIEW	26.8	1.73	30.4	1.61	36.5	1.49	44.7	1.17	51.7	1.45	57.7	1.38	66.5	.18	64.7	.16	57.8	.52	48.2	1.14	37.2	1.43	30.7	1.99	46.1	14.25
MADRAS	30.4	1.06	35.0	.74	39.9	.68	45.9	.64	52.6	.95	58.6	1.00	65.4	.24	63.6	.28	57.4	.52	47.9	.82	38.2	1.22	33.5	1.14	47.4	9.29
PAISLEY		1.19		.90		.91		.63		1.14		1.23		.30		.27		.42		.76		.90		1.27		9.92
PRINEVILLE 2 NW	30.3	.94	34.7	.73	39.2	.62	45.5	.71	51.6	1.08	56.8	1.24	63.2	.30	61.2	.38	55.7	.50	47.7	.85	38.7	1.13	33.3	1.11	46.5	9.59
REDMOND	31.8	1.04	36.2	.70	40.8	.56	47.0	.58	53.0	.82	58.4	1.08	65.6	.32	63.8	.28	58.3	.46	49.8	.69	39.9	.90	34.6	1.11	48.3	8.54
VALLEY FALLS	29.5	1.39	33.1	1.14	38.0	1.13	45.0	.94	51.5	1.34	57.4	1.44	65.8	.26	64.0	.29	56.7	.55	48.0	1.04	37.8	1.05	32.3	1.60	46.6	12.17
DIVISION	28.6	1.48	33.1	1.14	38.4	1.03	45.5	.92	52.0	1.21	57.8	1.25	66.0	.29	63.9	.31	57.3	.54	48.2	1.00	37.7	1.36	32.0	1.61	46.7	12.14

* Averages for period 1931 - 1955, except for stations marked WB which are "normals" based on period 1921 - 1950. Divisional means may not be the arithmetical average of individual stations published, since additional data from shorter period stations are used to obtain better areal representation.

*MEAN TEMPERATURE AND PRECIPITATION

STATION	JANUARY		FEBRUARY		MARCH		APRIL		MAY		JUNE		JULY		AUGUST		SEPTEMBER		OCTOBER		NOVEMBER		DECEMBER		ANNUAL	
	Temperature	Precipitation	Temperature	Precipitation	Temperature	Precipitation	Temperature	Precipitation	Temperature	Precipitation	Temperature	Precipitation	Temperature	Precipitation	Temperature	Precipitation	Temperature	Precipitation	Temperature	Precipitation	Temperature	Precipitation	Temperature	Precipitation	Temperature	Precipitation
NORTHEAST																										
AUSTIN 3 S		2.65		2.13		1.83		1.09		1.35		1.65		.50		.50		.65		1.52		2.21		2.83		18.91
COVE 1 ENE	29.1	1.76	33.1	1.96	38.9	2.19	46.6	2.38	53.1	2.56	58.4	2.82	66.4	.57	64.8	.54	58.0	1.13	49.1	1.93	38.5	2.38	32.7	2.22	47.4	22.44
ENTERPRISE	23.5	.89	28.0	.84	35.0	1.23	43.7	1.26	50.3	1.52	55.7	2.25	62.9	.58	60.7	.51	54.3	.97	45.8	1.17	34.0	1.06	27.3	.98	43.4	13.26
HUNTINGTON	29.2	1.57	34.3	1.56	43.0	.93	52.4	.76	61.8	.68	68.9	.84	79.3	.30	77.7	.16	67.2	.38	54.7	.64	39.7	1.28	32.2	1.68	53.4	10.78
LA GRANDE	30.2	2.00	33.7	2.05	39.6	2.16	48.1	1.94	55.7	1.76	61.3	2.13	70.3	.53	68.8	.46	60.8	.86	50.6	1.59	39.3	2.26	33.9	2.35	49.4	20.09
MEACHAM WB AP	24.9	4.24	29.0	4.06	34.0	3.66	41.1	2.73	48.0	2.37	54.1	2.61	63.2	.46	62.6	.51	55.7	1.46	45.5	2.94	34.9	3.87	29.1	4.38	43.5	33.29
MINAM 7 NE		2.33		2.26		2.07		1.68		1.89		2.16		.52		.54		1.03		1.88		2.43		2.59		21.38
ROCK CREEK		2.60		2.38		1.88		1.49		1.75		2.09		.60		.62		.67		1.62		2.19		2.97		20.86
UKIAH		1.84		1.76		1.72		1.58		1.74		2.18		.41		.48		.89		1.47		1.82		2.27		18.16
UNION	29.8	.83	33.8	1.03	39.9	1.29	47.6	1.36	53.8	1.64	58.7	1.85	66.3	.42	64.6	.44	57.9	.75	49.4	1.14	39.2	1.20	33.5	1.17	47.2	13.12
WALLOWA		1.62		1.53		1.65		1.54		1.61		2.07		.65		.47		.98		1.60		1.91		1.90		17.53
DIVISION	25.6	2.29	29.8	1.96	35.8	1.87	44.3	1.56	51.5	1.73	57.1	1.94	65.6	.49	63.6	.46	56.7	.83	47.3	1.52	35.9	2.15	29.2	2.56	45.2	19.36
SOUTHEAST																										
ADRAIN	28.3	.97	34.3	.96	42.4	.86	51.7	.78	59.7	.82	66.2	.84	75.5	.21	72.7	.19	62.8	.35	51.9	.67	38.3	.99	32.4	1.06	51.3	8.70
DANNER	25.4	1.12	30.4	1.05	37.6	1.04	46.1	1.03	53.5	1.22	59.8	1.13	69.1	.30	66.8	.12	58.0	.55	48.0	.84	36.2	1.13	29.6	1.37	46.7	10.90
VALE	27.2	1.01	33.2	.95	42.2	.76	50.8	.67	58.6	.90	64.8	.88	73.7	.18	70.7	.22	61.0	.40	50.7	.71	37.6	.89	31.5	1.05	50.2	8.62
WARM SPRINGS RES	26.4	.93	32.0	.82	40.0	.60	48.8	.57	56.2	.79	62.7	1.14	72.6	.34	70.6	.20	62.0	.40	51.0	.54	37.7	.72	31.1	1.02	49.3	8.07
DIVISION	27.3	1.10	32.2	.93	40.3	.85	49.2	.82	57.0	.97	63.6	.99	72.8	.23	70.3	.18	61.2	.37	50.4	.68	37.4	.96	30.7	1.18	49.4	9.26

* Averages for period 1931 - 1955, except for stations marked WB which are "normals" based on period 1921 - 1950. Divisional means may not be the arithmetical average of individual stations published, since additional data from shorter period stations are used to obtain better areal representation.

*MEAN TEMPERATURE AND PRECIPITATION

STATION	JANUARY		FEBRUARY		MARCH		APRIL		MAY		JUNE		JULY		AUGUST		SEPTEMBER		OCTOBER		NOVEMBER		DECEMBER		ANNUAL	
	Temperature	Precipitation	Temperature	Precipitation	Temperature	Precipitation	Temperature	Precipitation	Temperature	Precipitation	Temperature	Precipitation	Temperature	Precipitation	Temperature	Precipitation	Temperature	Precipitation	Temperature	Precipitation	Temperature	Precipitation	Temperature	Precipitation	Temperature	Precipitation
NORTHEAST																										
--BAKER WB	24.1	1.01	30.3	.89	38.0	.79	45.3	.91	52.8	1.09	58.9	1.31	67.6	.33	66.0	.40	57.2	.60	43.3	.70	36.6	.95	28.3	1.04	45.1	10.02

CONFIDENCE LIMITS

In the absence of trend or record changes, the chances are 9 out of 10 that the true mean will lie in the interval formed by adding and subtracting the values in the following table from the means for any station in the State. Because of the wider variation in mean precipitation, the corresponding monthly means and annual mean must be substituted for "p" in the precipitation table below to obtain mean precipitation confidence limits.

1.8	.30 $\sqrt{p}$	1.5	.34 $\sqrt{p}$	1.0	.33 $\sqrt{p}$	1.0	.26 $\sqrt{p}$	.9	.29 $\sqrt{p}$	.7	.32 $\sqrt{p}$	.6	.23 $\sqrt{p}$	.6	.24 $\sqrt{p}$	.9	.31 $\sqrt{p}$	.8	.43 $\sqrt{p}$	1.1	.46 $\sqrt{p}$	1.2	.34 $\sqrt{p}$	.5	.33 $\sqrt{p}$

COMPARATIVE DATA

Data in the following table are the mean temperature and average precipitation for Albany, Oregon for the period 1906-1930 and are included in this publication for comparative purposes:

38.7	6.43	42.6	4.91	46.7	3.56	51.0	2.52	55.8	1.91	61.5	1.25	66.9	0.42	66.5	0.52	60.5	1.85	52.9	2.83	45.2	6.63	39.6	5.95	52.3	38.78

NORMALS, MEANS AND EXTREMES

(Table Revised 1972. Base Period for Climatological Normals: 1931-1960)

Station: ASTORIA, OREGON CLATSOP COUNTY AIRPORT Standard time used: PACIFIC Latitude: 46° 09' N Longitude: 123° 53' W Elevation (ground): 8 feet

Month	Temperature: Normal: Daily maximum	Daily minimum	Monthly	Extremes: Record highest	Year	Record lowest	Year	Normal heating degree days (Base 65°)	Precipitation: Normal total	Maximum monthly	Year	Minimum monthly	Year	Maximum in 24 hrs.	Year
(a)	(b)	(b)	(b)	18		18		(b)	(b)	18		18		18	
J	46.5	34.8	40.7	65	1961	11	1969	753	11.71	18.94	1954	4.76	1963+	4.32	1964
F	49.9	35.7	42.8	72	1968	19	1962	622	9.89	21.89	1961	4.06	1964	2.86	1961
M	51.9	37.1	44.5	73	1964	22	1971	636	8.92	13.47	1956	0.93	1965	2.66	1956
A	56.9	41.0	49.0	83	1956	29	1968+	480	5.18	8.04	1955	1.33	1956	2.26	1965
M	61.6	44.9	53.3	86	1959+	30	1954	363	3.20	6.60	1960	1.03	1958	1.74	1968
J	65.1	49.5	57.3	93	1955	38	1966+	231	3.02	5.48	1954	0.75	1965	2.42	1968
J	68.6	52.5	60.6	100	1961	39	1971	146	1.27	3.42	1955	0.01	1960	1.43	1955
A	69.2	52.7	61.0	89	1966	42	1969	130	1.49	5.22	1968	0.08	1970	1.65	1968
S	67.2	48.8	58.0	92	1955	33	1970	210	3.13	6.55	1969	0.51	1965	2.63	1953
O	61.1	44.7	52.9	81	1971+	26	1971	375	7.78	12.24	1955	2.61	1964	3.47	1958
N	53.5	39.1	46.3	71	1970+	15	1955	561	11.20	14.64	1955	2.57	1956	3.48	1959
D	49.4	36.8	43.1	63	1963	15	1964	679	13.65	16.57	1955	6.12	1960	3.03	1955
YR	58.4	43.1	50.8	100	JUL. 1961	11	JAN. 1969	5186	80.44	21.89	FEB. 1961	0.01	JUL. 1960	4.32	JAN. 1964

Month	Snow, Ice pellets: Mean total	Maximum monthly	Year	Maximum in 24 hrs.	Year	Relative humidity: Hour 04	Hour 10	Hour 16	Hour 22 (Local time)	Wind: Mean speed	Prevailing direction	Fastest mile: Speed	Direction	Year	Pct. of possible sunshine	Mean sky cover sunrise to sunset	Mean number of days: Sunrise to sunset: Clear	Partly cloudy	Cloudy	Precipitation .01 inch or more	Snow, Ice pellets 1.0 inch or more	Thunderstorms	Heavy fog	Temperatures Max.: 90° and above ‡	32° and below	Min.: 32° and below	0° and below	Average daily solar radiation - langleys
(a)	18	18		18		18	18	18	18	18	10	18	18			18	18	18	18	18	18	18	18	18	18	18	18	18
J	4.0	26.3	1969	10.8	1971	88	85	80	87	9.1	E	55	17	1971		8.6	2	3	26	24	1	1	4	0	1	9	0	87
F	0.4	4.0	1962	4.0	1962	87	82	74	86	8.7	ESE	46	20	1963		8.1	3	3	22	19	*	*	3	0	0	7	0	163
M	1.2	6.7	1966	5.9	1960	88	77	70	86	8.8	SE	44	19	1964		8.0	3	5	23	20	1	1	2	0	0	8	0	274
A	T	0.2	1971	0.2	1971	89	74	70	85	8.5	WNW	52	20	1962		8.1	3	5	22	18	0	1	2	0	0	3	0	383
M	T	T	1971+	T	1971+	90	73	69	85	8.4	NW	35	24	1971		7.7	3	8	20	15	0	*	2	0	0	*	0	490
J	0.0	0.0		0.0		91	76	72	86	8.1	NW	29	20	1962		7.7	3	7	20	14	0	*	2	*	0	0	0	495
J	0.0	0.0		0.0		91	75	70	87	8.4	NW	26	32	1964		6.6	6	10	15	7	0	*	2	*	0	0	0	521
A	0.0	0.0		0.0		93	77	70	89	7.8	NW	28	23	1961+		6.5	7	10	14	9	0	*	5	0	0	0	0	457
S	0.0	0.0		0.0		92	76	70	89	7.3	SE	35	20	1959		6.3	9	6	15	11	0	1	6	*	0	0	0	356
O	T	T	1971+	T	1971+	91	82	74	90	7.5	SE	44	20	1962		7.4	5	6	20	18	0	1	7	0	0	*	0	210
N	0.1	2.2	1955	2.0	1955	89	83	78	88	8.2	SE	46	20	1962		8.1	3	5	22	21	*	1	4	0	*	4	0	112
D	2.0	19.0	1964	7.2	1964	88	86	82	88	8.9	ESE	52	25	1961+		8.7	2	4	25	24	1	1	5	0	*	7	0	72
YR	7.7	26.3	JAN. 1969	10.8	JAN. 1971	90	79	73	87	8.3	SE	55	17	JAN. 1971		7.7	49	72	244	200	2	9	43	1	1	39	0	302

Means and extremes above are from existing and comparable exposures. Annual extremes have been exceeded at other sites in the locality as follows: Highest temperature 101 in July 1942; lowest temperature 10 in January 1946 and earlier; maximum monthly precipitation 36.07 in December 1933; maximum precipitation in 24 hours 6.98 in January 1919; maximum monthly snowfall 27.3 in January 1916.

NORMALS, MEANS AND EXTREMES

(Table Revised 1975. Base Period for Climatological Normals: 1941-1970)

Month	Temperatures °F: Normal: Daily maximum	Daily minimum	Monthly	Extremes: Record highest	Year	Record lowest	Year	Normal Degree days Base 65 °F: Heating	Cooling	Precipitation in inches: Water equivalent: Normal	Maximum monthly	Year	Minimum monthly	Year	Maximum in 24 hrs.	Year
(a)				21		21					21		21		21	
J	46.5	34.6	40.6	65	1961	11	1969	756	0	9.73	18.94	1954	4.76	1963	4.32	1964
F	50.6	36.6	43.6	72	1968	19	1962	599	0	7.82	21.89	1961	2.60	1973	2.86	1961
M	52.1	36.7	44.4	73	1964	22	1971	639	0	6.62	13.47	1956	0.93	1965	2.66	1956
A	55.6	40.0	47.8	83	1956	29	1968	516	0	4.61	8.04	1955	1.33	1956	2.26	1965
M	60.3	44.3	52.3	86	1959	30	1954	394	0	2.72	6.60	1960	1.03	1958	1.74	1968
J	63.8	49.1	56.5	93	1955	38	1966	255	0	2.45	5.48	1954	0.75	1965	2.42	1968
J	67.7	52.2	60.0	100	1961	39	1971	163	8	0.96	4.20	1974	0.01	1960	1.98	1974
A	68.3	52.2	60.3	93	1972	39	1973	151	5	1.46	5.22	1968	0.08	1970	1.65	1968
S	67.6	49.1	58.4	95	1972	33	1972	201	0	2.83	6.55	1969	0.51	1965	2.63	1953
O	61.0	44.5	52.8	81	1971	26	1971	378	0	6.80	12.24	1955	1.85	1974	3.47	1958
N	53.4	39.6	46.5	71	1970	15	1955	555	0	9.78	14.93	1973	2.57	1956	3.48	1959
D	48.6	36.9	42.8	63	1963	6	1972	688	0	10.57	16.57	1955	6.12	1960	3.61	1974
YR	58.0	43.0	50.5	100	JUL 1961	6	DEC 1972	5295	13	66.34	21.89	FEB 1961	0.01	JUL 1960	4.32	JAN 1964

Month	Snow, Ice pellets: Maximum monthly	Year	Maximum in 24 hrs.	Year	Relative humidity pct.: Hour 04	Hour 10	Hour 16	Hour 22 (Local time)	Wind: Mean speed m.p.h.	Prevailing direction	Fastest mile: Speed m.p.h.	Direction	Year	Pct. of possible sunshine	Mean sky cover, tenths, sunrise to sunset	Mean number of days: Sunrise to sunset: Clear	Partly cloudy	Cloudy	Precipitation .01 inch or more	Snow, Ice pellets 1.0 inch or more	Thunderstorms	Heavy fog, visibility ¼ mile or less	Temperatures °F Max.: 90° and above (b)	32° and below	Min.: 32° and below	0° and below	Average station pressure mb. Elev. 22 feet m.s.l.
(a)	21		21		21	21	21	21	21	10	21	21			21	21	21	21	21	21	21	21	21	21	21	21	2
J	26.3	1969	10.8	1971	87	84	79	86	9.4	E	55	17	1971		8.5	3	3	25	24	1	1	4	0	1	10	0	1015.6
F	4.0	1962	4.0	1962	87	82	74	86	8.8	ESE	46	20	1963		8.1	3	4	21	19	*	*	3	0	0	7	0	1016.6
M	6.7	1966	5.9	1960	88	78	71	86	8.8	SE	44	19	1964		8.1	3	5	23	21	1	1	2	0	0	7	0	1013.4
A	0.2	1972	0.2	1972	88	74	69	85	8.6	WNW	52	20	1962		8.1	3	5	22	18	0	1	2	0	0	2	0	1020.5
M	T	1974	T	1974	90	73	70	85	8.4	NW	35	24	1971		7.7	3	8	20	15	0	*	2	0	0	*	0	1018.5
J	0.0		0.0		90	76	72	86	8.2	NW	29	20	1962		7.7	3	7	20	13	0	*	2	*	0	0	0	1017.8
J	0.0		0.0		90	75	69	87	8.4	NW	26	32	1964		6.6	6	10	15	7	0	*	2	*	0	0	0	1018.3
A	0.0		0.0		92	77	70	88	7.8	NW	28	23	1961		6.4	7	10	14	8	0	*	5	*	0	0	0	1018.0
S	T	1972	T	1972	92	76	70	89	7.3	SE	35	20	1959		6.2	9	6	15	11	0	1	6	*	0	0	0	1016.6
O	T	1971	T	1971	91	81	74	89	7.4	SE	44	20	1962		7.2	5	7	19	17	0	1	7	0	0	*	0	1018.5
N	2.2	1955	2.0	1955	89	83	78	87	8.3	SE	46	20	1962		8.1	3	5	22	21	*	1	4	0	*	4	0	1012.7
D	19.0	1964	7.2	1964	87	85	81	87	9.1	ESE	52	25	1961		8.7	2	4	25	24	1	1	4	0	1	7	0	1016.4
YR	26.3	JAN 1969	10.8	JAN 1971	89	79	73	87	8.4	SE	55	17	JAN 1971		7.6	50	74	241	199	2	8	43	1	2	38	0	1016.9

Means and extremes above are from existing and comparable exposures. Annual extremes have been exceeded at other sites in the locality as follows:
Highest temperature 101 in July 1942; maximum monthly precipitation 36.07 in December 1933; maximum precipitation in 24 hours 6.98 in January 1919; maximum monthly snowfall 27.3 in January 1916.

REFERENCE NOTES APPLYING TO TABLES APPEAR ON THE PAGE FOLLOWING LAST TABLE.
(Caution: Letters and symbols may have different meanings in 1941-1970 tables than in earlier tables. See notes.)

NORMALS, MEANS AND EXTREMES
(Table Revised 1972. Base Period for Climatological Normals: 1931-1960)

Station: BURNS, OREGON   FEDERAL BUILDING   Standard time used: PACIFIC   Latitude: 43° 35' N   Longitude: 119° 03' W   Elevation (ground): 4151 feet

Month	Temperature Normal — Daily maximum	Normal — Daily minimum	Normal — Monthly	Extremes — Record highest	Year	Record lowest	Year	Normal heating degree days (Base 65°)
J	35.2	14.3	24.8	58	1971+	-26	1962	1246
F	40.4	18.9	29.7	64	1954+	-18	1956	988
M	48.8	25.9	37.4	75	1966	-3	1951	856
A	59.8	32.2	46.0	80	1968	14	1968+	570
M	68.0	38.8	53.4	93	1954	19	1964	366
J	75.0	44.6	59.8	98	1961	25	1962	177
J	86.8	52.1	69.5	101	1968+	34	1966+	12
A	84.7	49.6	67.2	99	1961	31	1951	37
S	76.6	40.9	58.8	99	1955	18	1970	210
O	64.2	32.6	48.4	87	1962	-17	1971	515
N	48.3	23.9	36.1	70	1962	-10	1955	867
D	39.1	19.0	29.1	61	1958	...	1964	1113
YR	60.6	32.7	46.7	103 AUG. 1961		-26 JAN. 1962		6957

Month	Precipitation Normal total	Max monthly	Year	Min monthly	Year	Max in 24 hrs	Year	Snow/Ice pellets Mean total	Max monthly	Year	Max in 24 hrs	Year
J	1.62	5.73	1970	0.59	1955	1.02	1969	15.4	36.6	1956	7.1	1956
F	1.27	2.31	1957	0.13	1955	0.93	1960	6.0	17.5	1952	4.8	1952
M	0.97	2.44	1957	0.21	1968	0.91	1967	7.0	23.1	1967	11.0	1967
A	0.75	2.10	1963	0.09	1968	0.97	1951	1.8	12.1	1955	4.9	1955
M	0.89	2.69	1953	0.18	1966	1.00	1971	0.3	2.3	1970	1.8	1970
J	0.88	1.93	1965+	T	1960	0.74	1955	T	1.0	1954	1.0	1954
J	0.34	1.61	1955	0.00	1953	1.20	1955	0.0	0.0	1960	0.0	1960
A	0.29	1.73	1968	0.00	1970+	0.88	1968	0.0	0.2	1971	T	1971
S	0.50	1.93	1959	T	1965+	1.16	1959	T	0.2	1971	0.2	1971
O	0.86	3.68	1962	T	1962	1.41	1962	0.6	4.9	1962	4.9	1962
N	1.16	3.20	1970	0.13	1959	1.57	1953	4.7	17.9	1955	10.9	1955
D	1.43	5.47	1964	0.67	1959	2.15	1964	13.3	30.2	1951	8.1	1951
YR	10.96	5.73 JAN. 1970		0.00 AUG. 1970+		2.15 DEC. 1964		49.1	36.6 JAN. 1956		11.0 MAR. 1967	

Month	Relative humidity Hour 04	10	16	22	Wind Mean speed	Mean sky cover sunrise to sunset
J	83	76	69	81	6.1	7.6
F	82	69	57	78	6.4	7.1
M	78	60	46	70	7.4	6.7
A	72	47	35	60	8.6	6.4
M	72	45	31	53	8.0	5.9
J	68	44	31	53	7.9	5.1
J	53	35	21	37	7.7	2.6
A	54	36	22	39	7.7	2.9
S	61	40	25	47	7.0	3.4
O	73	52	37	67	7.0	5.6
N	81	76	56	77	5.5	6.6
D	84	77	69	82	5.9	7.4
YR	72	54	42	62	7.1	5.6

Mean number of days:

Month	Clear	Partly cloudy	Cloudy	Precip .01 in or more	Snow/ice pellets 1.0 in or more	Thunderstorms	Heavy fog	Max 90°+	Max 32°−	Min 32°−	Min 0°−
J	5	5	21	14	5	0	4	0	8	29	3
F	5	6	17	12	3	0	2	0	2	26	1
M	7	7	17	11	2	0	1	0	1	28	*
A	7	8	15	7	1	1	*	0	0	20	0
M	10	8	13	8	*	3	*	*	0	6	0
J	11	10	10	7	*	3	*	1	0	1	0
J	21	7	3	3	0	3	*	11	0	0	0
A	19	8	4	3	0	3	*	8	0	*	0
S	17	8	5	3	0	2	*	1	0	3	0
O	12	7	11	6	*	1	1	0	*	17	*
N	7	6	17	10	2	*	4	0	1	26	1
D	5	7	19	12	4	0	4	0	7	30	
YR	124	91	150	93	17	14	12	22	20	188	4

NORMALS, MEANS AND EXTREMES
(Table Revised 1975. Base Period for Climatological Normals: 1941-1970)

Average station pressure: mb.   Elev. 4170 feet m.s.l.

Month	Temperature Normal — Daily maximum	Daily minimum	Monthly	Extremes — Record highest	Year	Record lowest	Year	Normal degree days Base 65°F Heating	Cooling
J	35.4	15.0	25.2	58	1971+	-26	1962	1234	0
F	41.5	21.0	31.0	64	1954+	-18	1956	952	0
M	49.4	24.2	36.8	75	1966	-3	1951	896	0
A	58.1	31.2	44.7	80	1968	14	1968+	624	0
M	66.5	37.8	52.2	93	1954	19	1964	402	5
J	73.9	44.1	59.0	98	1961	25	1962	205	25
J	85.6	51.1	68.4	101	1968+	34	1966+	30	135
A	83.4	48.8	66.1	101	1961	31	1951	68	102
S	75.4	40.9	58.2	99	1955	19	1971	220	22
O	64.0	32.0	48.0	87	1962	-17	1971	549	0
N	47.6	24.3	36.0	70	1962	-26	1955	876	0
D	37.5	18.3	27.9	61	1958	-26	1972	1150	0
YR	59.6	32.3	46.0	103 AUG. 1961		-26 DEC. 1972		7212	289

Month	Precipitation Normal	Max monthly	Year	Min monthly	Year	Max in 24 hrs	Year	Snow/Ice pellets Max monthly	Year	Max in 24 hrs	Year
J	1.76	5.73	1970	0.32	1955	1.06	1956	36.6	1956	7.1	1956
F	1.18	2.31	1957	0.44	1968	0.93	1972	17.5	1952	4.8	1952
M	0.92	2.10	1967	0.46	1962	0.91	1967	23.1	1967	11.0	1967
A	0.70	2.10	1963	0.13	1968	0.91	1951	12.1	1955	4.9	1955
M	1.03	2.69	1953	0.09	1966	1.00	1971	2.3	1970	1.8	1970
J	0.97	1.93	1965	T	1960	0.74	1955	1.0	1954	1.0	1954
J	0.32	1.61	1955	0.00	1953	1.20	1955	0.0	1960	0.0	1960
A	0.44	1.73	1968	0.00	1970	0.88	1965	0.2	1971	T	1971
S	0.46	1.93	1959	T	1962	1.41	1962	0.2	1971	0.2	1971
O	0.83	3.68	1962	T	1964	1.41	1962	4.9	1962	4.9	1962
N	1.23	5.47	1959	0.11	1959	1.57	1955	17.9	1955	10.9	1955
D	1.73	5.47	1964	0.67	1964	2.15	1970	30.2	1951	8.1	1970
YR	11.83	5.73 JAN. 1970		0.00 AUG. 1970		2.15 DEC. 1964		36.6 JAN. 1956		11.0 MAR. 1967	

Month	Relative humidity Hour 04	10	16	22	Wind Mean speed m.p.h.	Mean sky cover tenths sunrise to sunset
J	82	76	68	81	6.2	7.6
F	82	72	60	78	6.3	6.8
M	78	60	47	70	7.4	6.6
A	70	46	33	57	8.7	5.8
M	70	45	30	53	8.0	5.8
J	67	45	32	53	7.9	5.1
J	53	35	21	37	7.7	2.6
A	54	36	24	39	7.1	3.0
S	61	41	26	47	7.0	3.4
O	67	52	37	67	6.4	5.8
N	81	76	56	77	5.4	6.6
D	84	77	70	82	5.8	7.4
YR	71	54	42	62	7.0	5.6

Mean number of days:

Month	Clear	Partly cloudy	Cloudy	Precip .01 in or more	Snow/ice pellets 1.0 in or more	Thunderstorms	Heavy fog, visibility ¼ mile or less	Max 90°+	Max 32°−	Min 32°−	Min 0°−
J	5	6	20	13	5	0	4	0	9	29	3
F	5	8	15	11	2	0	2	0	2	26	1
M	6	8	17	11	3	0	1	0	1	28	*
A	7	9	14	7	1	1	*	0	0	20	0
M	11	8	12	6	*	3	*	*	0	7	0
J	11	10	9	6	*	3	*	1	0	1	0
J	21	7	3	3	0	3	*	11	0	0	0
A	18	8	5	3	0	3	*	8	0	*	*
S	16	8	6	3	0	2	*	1	0	3	0
O	12	7	12	6	*	1	1	0	*	17	*
N	6	6	18	10	2	*	4	0	1	26	1
D	5	6	20	12	4	0	4	0	7	30	
YR	124	91	150	91	16	14	12	22	20	187	5

REFERENCE NOTES APPLYING TO TABLES APPEAR ON THE PAGE FOLLOWING LAST TABLE.
(Caution: Letters and symbols may have different meanings in 1941-1970 tables than in earlier tables. See notes.)

NORMALS, MEANS AND EXTREMES
(Table Revised 1972. Base Period for Climatological Normals: 1931-1960)

Station: EUGENE, OREGON MAHLON SWEET FIELD
Standard time used: PACIFIC Latitude: 44° 07' N Longitude: 123° 13' W Elevation (ground): 359 feet

Month	Normal Daily max	Normal Daily min	Normal Monthly	Extremes Record highest	Year	Extremes Record lowest	Year	Normal heating degree days (Base 65°)	Precip Normal total	Max monthly	Year	Max in 24 hrs	Year	Min monthly	Year	Snow Mean total	Max monthly	Year	Max in 24 hrs	Year	RH 04	RH 10	RH 16	RH 22	Wind mean speed	Prevailing dir	Fastest speed	dir	Year	Mean sky cover	Clear	Partly cloudy	Cloudy	Precip .01+	Snow 1.0+	Tstm	Heavy fog	Max 90+	Max 32-	Min 32-	Min 0-
J	45.2	32.8	39.0	65	1971	-4	1957	803	6.33	14.83	1964	3.98	1966	1.39	1962	5.8	47.1	1969	22.9	1969	92	87	81	91	8.3		58	20	1961	8.6	2	4	25	19	1	*	8	0	*	15	*
F	50.2	34.9	42.6	71	1968	-3	1950	627	4.97	11.58	1961	4.17	1961	0.86	1964	0.5	4.8	1949	2.5	1949	87	72	64	80	7.7		54	24	1963	8.2	3	4	21	17	*	*	6	0	*	10	*
M	55.2	36.8	46.0	75	1969	20	1956	589	4.32	9.81	1960	2.44	1963	0.79	1965	0.8	10.8	1951	4.9	1951	90	64	56	81	7.8		48	18	1963	8.2	3	7	21	17	*	1	3	0	0	8	0
A	61.2	39.6	50.4	86	1957+	27	1953	426	2.38	5.23	1963	2.25	1971	0.52	1966	T	T	1971+	T	1971+	85	57	53	78	7.4		40	27	1971	7.3	4	8	18	12	0	1	2	0	0	2	0
M	68.3	44.0	56.2	91	1963+	28	1954	279	2.14	4.44	1952	1.79	1963	T	1951	0.0	0.0		0.0		81	49	49	74	7.4		29	23	1952	7.3	4	9	18	10	0	1	1	*	0	*	0
J	73.6	48.1	60.9	100	1961+	35	1956	135	1.42	4.76	1951	2.36	1952	0.00	1967	0.0	0.0		0.0		78	46	45	70	7.7		29	23	1961	6.2	8	8	15	7	0	1	1	1	0	0	0
J	82.6	51.0	66.8	105	1961+	39	1955+	34	0.27	2.63	1947	1.39	1947	0.00	1967	0.0	0.0		0.0		71	37	39	71	8.0		26	36	1960+	3.6	17	9	6	2	0	*	5	0	0	0	0
A	81.3	50.2	66.0	101	1961	38	1968	39	0.57	3.04	1968	1.46	1968	0.00	1967	0.0	0.0		0.0		77	39	42	73	7.4		32	20	1959	4.3	14	8	8	4	0	1	3	5	0	0	0
S	76.5	46.3	61.5	101	1944	26	1945	129	1.27	3.04	1950	1.88	1963	0.62	1952	0.0	0.0		0.0		89	63	42	77	7.6		63	18	1962	5.0	12	8	10	6	0	1	2	5	0	0	0
O	64.2	42.1	53.2	88	1970	14	1971	366	3.83	12.66	1950	3.85	1950	1.20	1952	T	T	1971	5.0	1971	94	86	65	90	7.0		49	23	1957	7.1	5	7	17	12	0	*	*	12	0	*	7	0
N	52.5	38.4	45.5	67	1949	10	1955	585	5.42	12.02	1973	4.23	1960	2.69	1960	0.3	6.0	1955	4.0	1955	88	78	84	92	7.0		52	24	1958	8.4	2	6	22	16	*	1	*	10	0	*	7	*
D	47.5	36.0	41.8	67	1958	-4	1964	719	6.61	20.99	1964	4.82	1955	2.69	1944	1.0	10.2	1964	4.0	1964	92	88	84	92	7.7		40	25	1961	8.9	2	4	26	17	1	*	1	9	0	*	11	*
YR	63.2	41.7	52.5	105	JUL. 1961+	-4	JAN. 1957	4726	39.56	20.99	DEC. 1964	4.82	DEC. 1955	0.00	AUG. 1967+	8.4	47.1	JAN. 1969	22.9	JAN. 1969	91	73	60	84	7.6	18	63	18	OCT. 1962	6.8	77	83	205	137	3	5	59	15	3	55	*

NORMALS, MEANS AND EXTREMES
(Table Revised 1975. Base Period for Climatological Normals: 1941-1970)

Elev. 373 feet m.s.l. Average station pressure 1004.0 mb

Month	Normal Daily max	Normal Daily min	Normal Monthly	Extremes Record highest	Year	Extremes Record lowest	Year	Heating DD Base 65	Cooling DD Base 65	Precip Normal	Max monthly	Year	Max in 24 hrs	Year	Min monthly	Year	Water equiv Max monthly	Year	Max in 24 hrs	Year	Snow Mean total	Max monthly	Year	Max 24 hrs	Year	RH 04	RH 10	RH 16	RH 22	Wind mean speed	Prev dir	Fastest speed	dir	Year	Mean sky cover	Clear	Partly cloudy	Cloudy	Precip .01+	Snow 1.0+	Tstm	Heavy fog	Max 90+	Max 32-	Min 32-	Min 0-
J	45.6	33.1	39.4	65	1971	-4	1957	794	0	7.54	14.83	1964	4.88	1974	1.39	1962	14.83	1964	4.88	1974	5.8	47.1	1969	22.9	1969	91	86	80	90	8.4		58	20	1961	8.6	2	4	25	19	1	*	8	0	*	15	*
F	51.7	35.2	43.5	71	1968	-3	1950	602	0	4.67	11.58	1961	4.17	1961	0.86	1964	11.58	1961	4.17	1961	0.5	4.8	1949	2.5	1949	92	73	65	85	7.8		54	20	1963	8.3	3	5	21	17	*	*	6	0	*	10	*
M	55.2	36.8	45.9	75	1969	20	1956	592	0	4.43	9.81	1960	2.44	1963	0.79	1965	9.81	1960	2.44	1963	0.8	10.8	1951	4.9	1951	91	77	64	86	8.5		48	18	1963	8.2	4	7	21	17	*	1	3	0	0	9	0
A	61.2	39.7	50.3	86	1957	27	1972	441	0	2.31	5.80	1972	2.25	1971	0.52	1966	5.80	1972	2.25	1971	T	T	1971+	T	1971+	90	69	56	83	7.8		40	18	1972	7.3	4	8	18	12	0	1	2	0	0	3	0
M	67.1	43.7	55.4	91	1963+	28	1954	289	25	2.06	4.44	1952	2.37	1973	0.29	1951	4.44	1952	2.37	1973	0.0	0.0		0.0		90	65	53	81	7.5		29	23	1961	7.3	6	9	18	10	0	1	2	1	0	*	0
J	74.1	48.7	61.4	100	1961	35	1952	133	100	1.28	4.76	1952	2.36	1952	T	1952	4.76	1952	2.36	1952	0.0	0.0		0.0		78	62	49	78	7.7		29	23	1961	6.2	8	8	15	7	0	1	1	7	0	0	0
J	82.6	51.1	66.9	105	1961+	39	1968	41	100	0.26	2.63	1947	1.29	1947	0.00	1967	2.63	1947	1.29	1947	0.0	0.0		0.0		71	37	38	71	8.1		26	36	1960	3.6	17	8	6	3	0	*	7	0	0	0	0
A	81.3	50.0	66.1	106	1972	38	1969	51	85	0.58	3.04	1968	1.46	1968	0.00	1968	3.04	1968	1.46	1968	0.0	0.0		0.0		77	39	42	77	7.6		32	20	1959	4.3	14	8	8	3	0	1	3	6	0	0	0
S	76.5	47.4	62.0	105	1944	31	1972	119	29	1.00	3.04	1950	1.88	1963	0.12	1963	3.04	1950	1.88	1963	0.0	0.0		0.0		89	63	42	79	7.4		63	18	1962	5.0	12	8	10	6	0	1	2	12	0	0	0
O	64.2	42.0	53.2	88	1970	14	1970	366	0	4.53	12.66	1950	3.85	1950	1.00	1960	12.66	1950	3.85	1950	T	T	1971	5.0	1971	94	87	64	92	6.1		49	23	1957	7.0	5	6	17	12	*	*	1	10	0	*	7	0
N	53.0	38.1	45.6	74	1949	14	1955	582	0	6.53	20.48	1973	4.23	1960	2.69	1960	20.48	1973	4.23	1960	0.3	6.0	1955	4.0	1955	92	79	79	92	6.3		52	24	1958	8.4	2	6	22	16	*	1	1	10	0	*	7	*
D	47.4	35.6	41.5	67	1961	-12	1972	729	0	7.04	20.99	1964	4.82	1955	2.69	1944	20.99	1964	4.82	1955	1.0	10.2	1964	6.3	1972	92	88	84	92	7.7		40	25	1961	8.9	2	4	26	19	1	*	1	9	0	*	10	*
YR	63.4	41.8	52.6	106	AUG. 1972	-12	DEC. 1972	4739	239	42.56	20.99	DEC. 1964	4.88	JAN. 1974	0.00	AUG. 1967	20.99	DEC. 1964	4.88	JAN. 1974	8.4	47.1	JAN. 1969	22.9	JAN. 1969	91	73	60	84	7.7	18	63	18	OCT. 1962	6.9	76	84	205	138	3	5	58	16	3	54	*

REFERENCE NOTES APPLYING TO TABLES APPEAR ON THE PAGE FOLLOWING LAST TABLE.
(Caution: Letters and symbols may have different meanings in 1941-1970 tables than in earlier tables. See notes.)

NORMALS, MEANS AND EXTREMES

Station: MEACHAM, OREGON   MEACHAM EMERGENCY AIRPORT   Standard time used: PACIFIC   Latitude: 45° 30' N   Longitude: 118° 24' W   Elevation (ground): 4050 feet

(Table Revised 1972. Base Period for Climatological Normals: 1931-1960)

Month	Normal Daily max	Normal Daily min	Normal Monthly	Extremes Record highest	Year	Extremes Record lowest	Year	Normal heating degree days (Base 65°)	Precip. Normal total	Max monthly	Year	Max 24 hrs.	Year	Min monthly	Year	Snow Mean total	Max monthly	Year	Max 24 hrs.	Year
(a)	(b)		(b)	27	27	27	27	(b)	(b)	27		26		27		27	27		26	
J	32.7	19.3	26.0	57	1971	-23	1963	1209	4.20	10.30	1970	2.37	1946	1.48	1949	32.6	67.7	1946	22.9	1947
F	36.3	21.8	29.1	61	1968	-14	1956	1005	4.00	6.71	1949	2.34	1945	0.83	1947	24.2	62.0	1949	14.2	1948
M	41.6	24.9	33.3	68	1960	-14	1955	983	3.96	5.54	1965	1.40	1950	1.10	1965	13.3	55.0	1945	17.2	1951
A	50.7	30.8	40.8	81	1946	11	1968	726	2.80	5.87	1958	1.46	1957	0.40	1956	0.5	34.9	1945	15.1	1970
M	59.1	36.8	48.0	87	1947	20	1954	527	2.45	5.68	1956	1.51	1946	0.40	1964	T	17.6	1954	7.2	1960
J	65.7	41.7	53.7	93	1961	29	1966+	339	2.45	4.02	1969	1.54	1950	0.29	1962	0.0	6.1	1954	4.3	1954
J	77.6	49.2	63.4	103	1960	32	1955	84	0.50	1.78	1955	1.12	1955	T	1962+	0.0	0.2	1955	0.2	1955
A	75.1	48.2	61.7	105	1961	35	1965+	124	0.53	1.89	1965	1.11	1965	0.00	1969+	0.0	0.0		0.0	
S	67.5	43.8	55.7	97	1950	24	1970	288	1.46	5.80	1959+	1.45	1968	0.09	1953	T	2.7	1971	2.3	1971
O	55.6	35.0	46.3	68	1952	-12	1971	580	2.84	6.49	1951	1.77	1967	0.09	1952	0.3	19.1	1971	10.8	1971
N	41.6	27.1	34.4	68	1955	-22	1955	918	4.15	7.77	1945	2.81	1963	0.79	1952	17.4	66.8	1945	19.3	1945
D	35.5	24.1	29.8	53	1958	-22	1964	1091	4.53	8.13	1964	3.00	1946	1.17	1959	28.8	68.8	1948	18.4	1954
YR	53.3	33.7	43.5	105	AUG. 1961	-23	JAN. 1963	7874	33.87	10.30	DEC. 1970	3.00	DEC. 1946	0.00	AUG. 1969+	149.8	68.8	DEC. 1948	22.9	JAN. 1947

Month	RH Hr 04	RH Hr 10	RH Hr 16	RH Hr 22	Wind Fastest mile speed	Direction	Year	Pct. sunshine	Mean sky cover	Clear	Partly cloudy	Cloudy	Precip .01"+	Snow 1.0"+	Thunderstorms	Heavy fog	Max 90&above	Max 32&below	Min 32&below	Min 0&below	Avg daily solar
	26	26	26	22	5	5			26	26	26	26	27	27	20	20	27	27	27	27	27
J	81	78	80	80	35	26	1967		8.5	3	3	25	20	9	0	4	0	15	28	2
F	74	69	74	74	37	26	1967		8.5	3	3	22	17	7	0	4	0	5	25	1
M	70	59	72	72	32	23	1967		8.0	4	5	22	16	8	*	5	0	1	27	*
A	62	61	62		38	03	1967		7.8	4	6	20	13	4	1	3	0	0	21	0
M	60	60	48		32	27	1967		7.6	6	8	17	14	1	4	2	0	0	8	0
J	55	43	57	57	26	10	1970		6.1	8	8	14	10	*	3	1	*	0	1	0
J	42	24	24	45	26	27	1967		3.7	19	7	5	5	0	3	*	2	0	0	0
A	43	35	45	51	31	24	1967		3.8	17	7	7	5	0	4	*	1	0	0	0
S	48	34	31	52	30	24	1967		4.7	13	6	10	7	*	1	1	*	0	2	0
O	62	54	54	77	39	24	1967		7.9	8	4	17	11	2	*	2	0	*	10	0
N	78	80	77	80	37	25	1967		8.4	4	4	22	16	6	2	4	0	5	22	*
D	81	85	80		32	19	1967		8.4	3		24	18	8	0	4	0	12	27	*
YR	63	57	64		39	24	OCT. 1967		6.6	92	68	205	151	45	16	31	4	46	170	3

NORMALS, MEANS AND EXTREMES

(Table Revised 1975. Base Period for Climatological Normals: 1941-1970)

Elev. 4056 feet m.s.l.

Month	Normal Daily max	Normal Daily min	Normal Monthly	Extremes Record highest	Year	Extremes Record lowest	Year	Normal Degree days Heating	Cooling	Precip. Normal	Max monthly	Year	Min monthly	Year	Max 24 hrs.	Year	Snow Max monthly	Year	Max 24 hrs.	Year
(a)				30		30				30	30		30		29		30		29	
J	32.2	20.1	26.2	57	1971	-23	1963	1203	0	4.34	10.30	1970	1.48	1949	2.37	1946	67.7	1946	22.9	1947
F	37.3	24.3	30.8	61	1968	-14	1956	958	0	3.52	6.71	1949	0.83	1947	2.34	1945	62.0	1949	14.2	1948
M	41.1	25.6	40.1	68	1960	-14	1955	747	0	2.90	5.54	1965	1.10	1965	1.78	1972	55.0	1945	17.2	1951
A	49.3	30.8	40.1	81	1946	11	1968	536	0	2.83	5.87	1956	0.40	1956	1.46	1957	34.9	1945	15.1	1957
M	57.7	37.0	47.4	87	1947	20	1954	323	5	2.58	5.68	1956	0.40	1964	1.54	1946	17.6	1960	7.2	1960
J	65.3	43.4	54.4	93	1961	29		46	45	2.15	4.02	1969	0.29	1962	1.54	1950	6.1	1954	4.3	1954
J	77.2	49.9	63.6	103	1960	32	1955	88	45	0.56	1.82	1974	T	1962	1.12	1955	0.2	1955	0.2	1955
A	75.1	49.2	62.2	105	1961	35	1965	132	46	0.90	4.89	1965	0.00	1969	1.11	1965	0.0	1968	0.0	1968
S	67.2	43.9	55.6	97	1950	24	1970	294	12	1.52	5.80	1959	0.26	1953	1.45	1967	2.7	1971	2.3	1971
O	55.1	36.6	45.8	68	1952	-12	1971	595	0	2.70	6.49	1951	0.79	1952	1.77	1967	19.1	1971	19.3	1971
N	41.4	28.6	35.0	68	1955	-22	1955	900	0	4.01	10.09	1973	0.79	1973	2.81	1963	66.8	1945	19.3	1945
D	34.5	24.0	29.3	53	1972	-22	1964	1107	0	4.61	8.96	1959	1.17	1959	3.00	1946	68.8	1948	18.4	1954
YR	52.8	34.5	43.7	105	AUG. 1961	-23	JAN. 1963	7863	108	32.68	10.30	JAN. 1970	0.00	AUG. 1969	3.00	DEC. 1946	68.8	DEC. 1948	22.9	JAN. 1947

Month	RH Hr 04	RH Hr 10	RH Hr 16	RH Hr 22	Wind Fastest mile m.p.h.	Direction	Year	Pct. sunshine	Mean sky cover tenths	Clear	Partly cloudy	Cloudy	Precip .01"+	Snow 1.0"+	Thunderstorms	Heavy fog, vis. ¼ mi or less	Max 90&above	Max 32&below	Min 32&below	Min 0&below	Avg station pressure mb
	27	3	3	3	7	7			28	28	28	28	30	30	20	20	30	30	30	30	2
J	87	81	79	84	44	25	1972		8.5	3	3	25	19	9	0	4	0	15	28	2	874.9
F	88	75	73	80	37	26	1967		8.5	3	3	22	16	7	0	4	0	5	25	1	876.6
M	84	69	61	79	32	23	1967		8.0	4	6	22	17	8	*	5	0	1	27	*	872.4
A	62	60	60	75	38	03	1967		7.6	4	6	20	14	4	1	3	0	0	21	0	877.3
M	60	55	50	66	32	27	1967		6.8	6	9	17	13	1	4	2	0	0	8	0	877.1
J					26	27	1967		6.0	8	8	14	10	*	3	1	*	2	0	1	876.9
J	62	39	26	45	28	10	1970		3.6	20	7	5	5	0	3	*	2	0	0	0	878.8
A	72	42	34	44	31	27	1967		4.7	18	7	7	5	0	4	*	1	0	0	0	877.9
S	82	48	40	62	30	24	1967		4.7	14	6	10	7	*	1	1	*	0	2	0	879.3
O	93	78	55	74	39	24	1967		6.5	8	4	17	11	2	*	2	0	*	10	*	879.0
N	80	82	85	87	37	25	1967		8.0	4	4	22	16	6	2	4	0	5	22	1	874.1
D	86	82	85	87	32	19	1967		8.5	3	4	24	18	8	0	4	0	12	27		877.0
YR	81	63	58	71	44	25	JAN. 1972		6.6	94	67	204	150	45	16	31	4	46	170	4	876.8

REFERENCE NOTES APPLYING TO TABLES APPEAR ON THE PAGE FOLLOWING LAST TABLE.
(Caution: Letters and symbols may have different meanings in 1941-1970 tables than in earlier tables. See notes.)

NORMALS, MEANS AND EXTREMES
(Table Revised 1972. Base Period for Climatological Normals: 1931-1960)

Station: MEDFORD, OREGON   MEDFORD-JACKSON COUNTY AIRPORT   Standard time used: PACIFIC   Latitude: 42° 22' N   Longitude: 122° 52' W   Elevation (ground): 1298 feet

Month	Temperature Normal Daily maximum (b)	Daily minimum	Monthly	Extremes Record highest	Year	Record lowest	Year	Normal heating degree days (Base 65°) (b)	Precip. Normal total (b)
J	42.4	28.4	35.4	70	1962	11	1962	918	3.14
F	49.5	30.6	40.1	77	1958	10	1962	697	2.40
M	55.7	32.7	44.3	81	1956	21	1966+	642	1.78
A	63.5	37.7	50.6	88	1958	28	1968+	432	1.06
M	71.4	43.5	57.5	94	1970	28	1968	242	1.47
J	78.1	50.3	64.2	98	1961	34	1966+	78	1.02
J	88.3	55.7	72.0	108	1961	38	1962	0	0.21
A	87.3	54.1	70.7	109	1971	39	1962	0	0.18
S	80.0	47.4	64.2	104	1963	31	1965+	78	0.60
O	66.1	39.9	53.0	94	1970	18	1971	372	1.94
N	51.0	33.7	42.4	75	1962	14	1961	678	2.60
D	43.0	30.8	36.9	72	1962	12	1965	871	3.38
YR	64.7	40.5	52.6	109	AUG. 1971+	0	JAN. 1962	5008	19.78

Means and extremes above are from existing and comparable exposures. Annual extremes have been exceeded at other sites in the locality as follows:
Highest temperature 115 in July 1946; lowest temperature -10 in December 1919; maximum snowfall in 24 hours 11.0 in December 1919.

NORMALS, MEANS AND EXTREMES
(Table Revised 1975. Base Period for Climatological Normals: 1941-1970)

Month	Temperature °F Normal Daily maximum (a)	Daily minimum	Monthly	Extremes Record highest	Year	Record lowest	Year	Degree days Base 65°F Heating	Cooling	Precip. Normal
J	44.2	29.0	36.6	70	1962	0	1962	880	0	3.54
F	51.8	30.7	41.3	77	1968	10	1962	664	0	2.15
M	56.4	32.8	44.6	81	1966	21	1966	626	0	1.64
A	63.8	36.6	50.2	88	1968	24	1972	444	11	1.01
M	71.4	37.3	54.3	98	1973	28	1968	250	73	1.44
J	79.4	49.1	64.3	109	1961	34	1966	94	218	0.89
J	89.5	53.8	71.7	108	1961	38	1962	21	189	0.25
A	87.8	52.9	70.4	106	1972	39	1962	11	171	0.33
S	82.1	46.7	64.4	105	1973	31	1965	71	90	0.55
O	66.4	39.4	53.4	94	1970	18	1971	360	0	2.05
N	52.7	34.2	43.5	75	1961	14	1961	645	0	3.10
D	44.2	31.1	37.7	72	1962	14	1972	846	0	3.67
YR	66.0	39.9	53.0	109	AUG. 1972	-6	DEC. 1972	4930	562	20.64

Means and extremes above are from existing and comparable exposures. Annual extremes have been exceeded at other sites in the locality as follows:
Highest temperature 115 in July 1946; lowest temperature -10 in December 1919; maximum snowfall in 24 hours 11.0 in December 1919.

REFERENCE NOTES APPLYING TO TABLES APPEAR ON THE PAGE FOLLOWING LAST TABLE.
(Caution: Letters and symbols may have different meanings in 1941-1970 tables than in earlier tables. See notes.)

NORMALS, MEANS AND EXTREMES
(Table Revised 1972. Base Period for Climatological Normals: 1931-1960)

Station: PENDLETON, OREGON PENDLETON FIELD Standard time used: PACIFIC Latitude: 45°41'N Longitude: 118°51'W Elevation (ground): 1482 feet

Month	Temperature — Normal Daily maximum	Daily minimum	Monthly	Extremes Record highest	Year	Record lowest	Year	Normal heating degree days (Base 65°)	Precipitation Normal total	Max monthly	Year	Min monthly	Year	Max in 24 hrs.	Year	Snow, Ice pellets Mean total	Max monthly	Year	Max in 24 hrs.	Year
J	39.3	25.1	32.2	67	1968+	-22	1957	1017	1.42	3.92	1970	0.21	1949	1.29	1956	8.2	41.6	1950	13.3	1950
F	45.4	29.4	37.4	66	1968+	-18	1950	773	1.18	3.03	1940	0.07	1964	1.09	1959	3.8	15.8	1936	9.7	1949
M	54.2	36.0	45.1	79	1964	10	1964	617	1.20	2.31	1957	0.24	1941	0.65	1970	1.0	4.9	1971	4.0	1971
A	63.5	40.5	52.0	89	1946	18	1936	396	1.09	2.45	1958	0.01	1956	0.98	1969	0.1	1.9	1950	T	1950+
M	72.1	47.1	59.6	99	1954	25	1954	205	1.12	2.70	1962	0.03	1964	1.14	1956	0.0	0.0	1971+	0.0	1971+
J	78.9	52.6	65.8	108	1961	36	1966+	63	1.17	2.70	1947	0.12	1940	1.49	1947	0.0	0.0		0.0	
J	89.3	57.6	73.6	110	1939	42	1971+	0	0.22	1.26	1948	T	1967+	1.19	1948	0.0	0.0		0.0	
A	87.3	56.5	71.9	113	1961	41	1964	0	0.28	0.34	1941	T	1963+	0.63	1941	0.0	0.0		0.0	
S	78.1	50.2	64.2	102	1955	30	1970+	111	0.63	2.34	1947	0.00	1956+	0.95	1955	0.1	1.9	1971	1.7	1971
O	64.7	41.3	53.7	86	1960	11	1935	350	1.08	2.75	1945	0.04	1939	1.09	1961	1.2	9.2	1961	6.6	1961
N	49.4	33.3	41.3	74	1965	-6	1955	711	1.40	3.23	1964	0.27	1965	1.35	1971	3.2	11.9	1968	9.9	1948
D	43.0	29.9	36.5	67	1959+	-12	1957	884	1.49		1970			1.23	1958					
YR	63.8	41.7	52.8	113	AUG 1961	-22	JAN 1957	5127	12.38	3.92	JAN 1970	0.00		1.49	JUN 1947	17.6	41.6	JAN 1950	13.3	JAN 1950

May - December 1935 data considered in extracting extremes in above table.
Means and extremes above are from existing and comparable exposures. Annual extremes have been exceeded at other sites in the locality as follows:
Highest temperature 119 in August 1898; lowest temperature -28 in December 1919; maximum monthly precipitation 4.24 in November 1897.

NORMALS, MEANS AND EXTREMES
(Table Revised 1975. Base Period for Climatological Normals: 1941-1970)

Month	Temperatures °F — Normal Daily maximum	Daily minimum	Monthly	Extremes Record highest	Year	Record lowest	Year	Normal Degree days Base 65°F Heating	Cooling	Precipitation in inches — Water equivalent Normal	Max monthly	Year	Min monthly	Year	Max in 24 hrs.	Year	Snow, Ice pellets Mean total	Max monthly	Year	Max in 24 hrs.	Year
J	38.6	25.3	32.0	68	1957	-22	1957	1023	0	1.60	3.92	1970	0.21	1949	1.29	1949	8.2	41.6	1950	13.3	1950
F	46.5	31.3	38.9	68	1950	-18	1950	731	0	1.07	3.03	1940	0.07	1964	1.09	1964	3.8	15.8	1936	9.7	1936
M	53.2	34.3	43.8	79	1964	10	1964	657	0	1.01	2.31	1957	0.24	1941	1.01	1941	4.9	4.9	1971	4.0	1971
A	63.0	39.8	50.9	89	1946	18	1936	423	18	1.01	2.45	1958	0.01	1956	1.24	1956	0.1	1.9	1969	1.9	1969
M	70.4	46.5	58.5	99	1954	25	1954	220	88	1.24	3.02	1962	0.03	1964	1.01	1964	T	T	1972	T	1972
J	78.9	52.8	65.8	108	1961	36	1966	70	269	1.01	2.70	1947	0.12	1940	1.52	1947	0.0	0.0		0.0	
J	88.2	58.8	73.5	110	1939	42	1971	6	214	0.26	1.26	1948	T	1967	1.19	1948	0.0	0.0		0.0	
A	85.5	57.5	71.5	113	1961	41	1964	13	67	0.34	2.34	1941	T	1963	0.95	1941	0.0	0.0		0.0	
S	76.9	51.1	64.0	102	1955	30	1970	97	0	0.64	2.34	1947	0.04	1936	1.11	1947	0.2	3.2	1957	1.7	1957
O	64.8	41.8	52.6	86	1974	11	1935	384	0	1.11	2.75	1945	0.04	1939	1.50	1936	3.2	9.2	1973	3.2	1973
N	48.9	33.8	41.3	74	1974	-6	1955	708	0	1.50	3.76	1973	0.04	1973	1.35	1939	3.2	12.6	1972	9.9	1972
D	41.8	29.6	35.7	67	1959	-12	1957	908	0	1.53	4.68	1973	0.27	1965	1.23	1958	9.9	11.9	1972	9.9	
YR	62.8	41.9	52.4	113	AUG 1961	-22	JAN 1957	5240	656	12.31	4.68	DEC 1973	0.00	AUG 1969	1.52	MAY 1972	41.6	41.6	JAN 1950	13.3	JAN 1950

May - December 1935 data considered in extracting extremes in above table.
Means and extremes above are from existing and comparable exposures. Annual extremes have been exceeded at other sites in the locality as follows:
Highest temperature 119 in August 1898; lowest temperature -28 in December 1919.

REFERENCE NOTES APPLYING TO TABLES APPEAR ON THE PAGE FOLLOWING LAST TABLE.
(Caution: Letters and symbols may have different meanings in 1941-1970 tables than in earlier tables. See notes.)

NORMALS, MEANS AND EXTREMES
(Table Revised 1972. Base Period for Climatological Normals: 1931-1960)

Station: PORTLAND, OREGON   INTERNATIONAL AIRPORT   Standard time used: PACIFIC   Latitude: 45° 36' N   Longitude: 122° 36' W   Elevation (ground): 21 feet

Month	Temperature Normal — Daily max	Daily min	Monthly	Extremes — Record highest	Year	Record lowest	Year	Normal heating degree days (Base 65°)	Precip. Normal total	Max monthly	Year	Min monthly	Year	Max in 24 hrs	Year	Snow, Ice pellets Mean total	Max monthly	Year	Max in 24 hrs	Year
J	43.7	33.0	38.4	62	1958+	-2	1950	825	5.37	12.83	1953	1.02	1949	2.49	1956	4.9	41.4	1950	10.6	1950
F	46.8	35.1	42.0	70	1968	-3	1950	644	4.22	9.46	1949	0.78	1964	2.00	1949	0.7	13.2	1949	3.2	1962
M	51.9	37.0	46.1	80	1947	19	1955	586	3.83	7.52	1957	1.10	1965	1.83	1943	T	12.9	1951	7.7	1951+
A	61.7	41.8	51.8	87	1957	29	1954	396	2.29	4.57	1945	0.57	1950	1.47	1962	0.0	0.6	1953	0.5	1953
M	67.5	46.3	57.4	92	1957	29	1954	245	1.99	4.53	1945	0.53	1951	1.47	1968	0.0	0.0		0.0	
J	71.6	52.2	62.2	100	1942	39	1966	105	1.67	3.58	1954	0.03	1958	1.82	1958	0.0	0.0		0.0	
J	78.6	55.7	67.2	107	1965	43	1955	25	0.41	1.40	1942	0.00	1967	0.91	1966	0.0	0.0		0.0	
A	77.9	55.2	66.6	102	1967	44	1951	28	0.65	4.53	1968	T	1970+	1.38	1968	0.0	0.0		T	1949
S	73.7	50.7	62.2	101	1944	34	1965	114	1.63	3.96	1945	0.01	1969	2.18	1969	0.0	T	1949	0.2	1950
O	63.2	45.1	54.2	90	1970	26	1971+	335	3.61	8.04	1947	0.72	1952	2.60	1941	T	0.2	1950	4.5	1955
N	51.7	38.4	45.1	70	1969	13	1955	597	5.33	11.57	1942	1.44	1946	2.17	1942	0.3	8.2	1955	8.0	1968
D	46.5	36.1	41.3	64	1950+	6	1964	735	6.38	11.12	1968	1.90	1944	2.17	1968	1.6	15.7	1968		
YR	61.6	44.0	52.9	107	JUL 1965	-3	FEB 1950	4635	37.18	12.83	JAN 1953	0.00	JUL 1967	2.60	NOV 1946	8.4	41.4	JAN 1950	10.6	JAN 1950

Month	Rel. hum. Hour 04	Hour 10	Hour 16	Hour 22	Wind Mean speed	Prevailing direction	Fastest mile Speed	Direction	Year	Mean sky cover	Pct possible sunshine	Days Clear	Partly cloudy	Cloudy	Precip .01+	Snow 1.0+	Thunderstorms	Heavy fog	Max 90°+	Max 32°-	Min 32°-	Min 0°-	Avg daily solar radiation
J	86	80	77	83	10.0	ESE	54	S	1951	8.7	23	2	3	26	20	2	*	4	0	*	13	31
F	86	77	68	78	8.7	ESE	61	SW	1958	8.1	36	3	3	22	16	*	*	4	0	3	6	
M	86	73	60	78	8.1	ESE	57	SW	1963	8.1	41	3	4	24	17	*	1	2	0	*	1	
A	86	66	57	74	7.1	W	42	NW	1957	7.2	48	4	5	21	14	0	1	1	0	*	*	
M	85	66	53	74	6.7	NW	40	NW	1958	6.8	53	6	7	19	12	0	1	1	0	0	0	
J	85	65	53	72	6.7	NW					50			17	10	0	1	*	0	0	0	
J	83	62	49	65	7.3	NW	31	S	1956	4.5	69	14	9	8	8	0	1	3	0	0	0	*
A	85	64	47	69	6.9	NW	29	SW	1961+	5.1	63	11	10	10	10	0	1	3	0	0	0	*
S	87	67	52	76	6.2	NW	61	SW	1963	6.2	55	10	8	12	12	0	1	1	0	0	0	0
O	90	82	65	85	6.4	ESE	88	SW	1962	7.3	40	5	4	19	17	*	*	*	0	*	1	0
N	88	84	79	85	9.5	ESE	56	SW	1951	9.0	29	1	3	23	18	*	3	0	*	5	9	0
D	87	84	79	85			57	S	1951		20	1	3	27	19	1	3	6	0	1	9	*
YR	86	73	60	78	7.7	NW	88	S	OCT 1962	7.2	47	67	70	228	153	3	7	33	9	4	45	

Means and extremes above are from existing and comparable exposures. Annual extremes have been exceeded at other sites in the locality as follows: Maximum monthly precipitation 20.14 in December 1882; maximum snowfall in 24 hours 16.0 in January 1937.

NORMALS, MEANS AND EXTREMES
(Table Revised 1975. Base Period for Climatological Normals: 1941-1970)

Average station pressure: 1016.3 mb   Elev. 39 feet m.s.l.

Month	Temperature Normal — Daily max	Daily min	Monthly	Extremes — Record highest	Year	Record lowest	Year	Degree days heating	cooling	Precip. Normal	Max monthly	Year	Min monthly	Year	Max in 24 hrs	Year	Water equiv. Max monthly	Year	Min monthly	Year	Snow, Ice pellets Mean total	Max monthly	Year	Max in 24 hrs	Year
J	43.6	32.5	38.1	62	1950	-2	1950	834	0	5.88	12.83	1953	1.02	1949	2.61	1974	12.83	1953	0.47	1974	4.9	41.4	1950	10.6	1950
F	50.1	35.5	42.8	70	1950	-3	1950	622	0	4.06	9.46	1949	0.78	1964	2.00	1949	9.46	1949	0.90	1964	0.7	13.2	1949	3.2	1962
M	54.3	37.0	45.6	80	1947	19	1955	598	0	4.04	7.52	1957	1.10	1965	2.00	1956	8.04	1947	1.80	1965	T	12.9	1951	7.7	1951
A	60.3	40.8	50.6	87	1957	29	1955	432	0	2.22	4.72	1955	0.53	1950	1.47	1962	4.72	1955	2.09		0.0	0.6	1953	0.5	1953
M	67.0	46.3	56.7	92	1957	29	1954	264	7	2.09	4.53	1954	0.09	1951	1.47	1968	4.53	1954			0.0	0.0		0.0	
J	72.1	51.8	62.0	100	1942	39	1966	128	38	1.59	3.58	1966	0.03		1.82		3.58	1966			0.0	0.0		0.0	
J	79.0	55.2	67.1	107	1965	43	1955	48	114	0.47	2.01	1974	0.00	1967	0.91	1967	2.01	1974	0.00	1967	0.0	0.0		0.0	
A	78.1	55.0	66.6	104	1967	44	1951	56	106	0.90	2.80	1970	T	1970	1.38	1968	2.80	1970	0.01	1970	0.0	0.0		T	1949
S	73.9	50.5	62.2	101	1944	34	1965	119	35	1.80	3.96	1945	0.01	1969	2.23	1965	3.96	1947	0.72	1969	0.0	T	1949	0.2	1950
O	62.1	44.7	53.8	90	1970	26	1971	347	0	3.59	8.04	1947	0.72	1952	2.18	1973	8.04	1947	1.44	1952	T	0.2	1950	4.5	1955
N	52.1	38.5	45.3	70	1969	13	1955	591	0	5.61	11.57	1942	1.44	1942	2.62	1942	11.12	1942	1.90	1944	0.3	8.2	1955	8.0	1968
D	46.0	35.3	40.7	64	1950	6	1964	753	0	6.04	11.12	1968	1.90	1944	2.17	1968	11.12	1968			1.6	15.7	1968		
YR	61.6	43.6	52.6	107	JUL 1965	-3	FEB 1950	4792	300	37.61	12.83	JAN 1953	0.00	JUL 1967	2.62	NOV 1962	12.83	JAN 1953	0.00	JUL 1967	8.4	41.4	JAN 1950	10.6	JAN 1950

Month	Rel. hum. Hour 04	Hour 10	Hour 16	Hour 22	Wind Mean speed	Prevailing direction	Fastest mile m.p.h.	Direction	Year	Pct possible sunshine	Mean sky cover	Days Clear	Partly cloudy	Cloudy	Precip .01+	Snow 1.0+	Thunderstorms	Heavy fog ¼ mile or less	Max 90°+	Max 32°-	Min 32°-	Min 0°-
J	86	82	76	83	10.2	ESE	54	S	1951	24	8.6	2	4	25	16	2	*	4	0	*	13
F	87	79	68	81	8.7	ESE	61	SW	1958	35	8.3	3	4	24	14	1	*	4	0	3	6
M	86	68	60	78	8.1	ESE	57	SW	1963	41	8.1	3	5	21	17	*	1	2	0	*	1
A	85	66	55	74	6.8	NW	42	NW	1957	48	7.7	4	7	19	14	*	1	1	0	*	*
M	84	65	49	72	6.8	NW	40	NW	1958	51	6.8	6	7	17	12	*	1	1	0	0	0
J	83	61	45	69	7.4	NW	31	S	1956	64	4.5	14	9	8	9	*	1	4	0	0	0
J	84	61	45	69	7.4	NW	31	S	1956	64	4.5	14	9	8	9	*	1	4	0	0	0
A	87	74	57	75	6.3	NW	29	SW	1961	59	5.2	12	10	8	8	*	1	5	0	0	0
S	90	82	64	84	6.3	ESE	61	SW	1963	57	5.7	10	7	12	10	*	1	3	0	0	1
O	88	82	74	85	8.3	ESE	88	SW	1962	28	8.2	5	5	23	18	*	*	*	0	1	5
N	87	84	79	85	9.6	ESE	56	SW	1951	20	9.0	3	3	27	19	1	3	*	0	1	9
D	87	84	79	85			57	S	1951			1	3	27	19	1	3	4	0	1	9
YR	86	72	60	78	7.7	NW	88	S	OCT 1962	47	7.2	68	69	228	153	2	7	33	10	4	43

Means and extremes above are from existing and comparable exposures. Annual extremes have been exceeded at other sites in the locality as follows: Maximum monthly precipitation 20.14 in December 1882; maximum snowfall in 24 hours 16.0 in January 1937.

REFERENCE NOTES APPLYING TO TABLES APPEAR ON THE PAGE FOLLOWING LAST TABLE.
(Caution: Letters and symbols may have different meanings in 1941-1970 tables than in earlier tables. See notes.)

NORMALS, MEANS AND EXTREMES

(Table Revised 1960. Base Period for Climatological Normals: 1921-1950)

LATITUDE 43° 14' N
LONGITUDE 123° 22' W
ELEVATION (ground) 505 Feet

Temperature, Precipitation, Snow/Sleet

Month	Norm Daily max	Norm Daily min	Norm Monthly	Record highest	Year	Record lowest	Year	Normal degree days	Precip Normal total	Precip Max monthly	Year	Precip Min monthly	Year	Precip Max in 24 hrs	Year	Snow Mean total	Snow Max monthly	Year	Snow Max 24 hrs	Year
(a)	(b)	(b)	(b)	5		5		(b)	(b)	5		5		5		5	5		5	
J	47.1	31.8	39.5	65	1954	9	1957	791	4.60	10.16	1953	2.42	1957	2.80	1956	5.4	13.3	1954	9.1	1954
F	51.9	34.3	43.1	66	1957	13	1956	613	3.76	7.07	1956	1.79	1955	3.24	1956	0.6	1.2	1956	0.9	1956
M	57.7	36.1	46.9	79	1953	19	1956	561	2.93	5.80	1957	2.23	1954	1.23	1954	1.6	7.0	1956	6.7	1956
A	64.0	38.9	51.5	90	1957+	27	1955	405	2.13	3.41	1955	0.71	1956	0.84	1957	0.0	2.4	1953	2.4	1953
M	70.3	43.2	56.8	95	1956	26	1954	262	1.73	3.22	1953	0.30	1953	0.56	1956	0.0	0.0		0.0	
J	76.1	48.1	62.1	96	1955	34	1954	112	1.35	1.60	1953	0.23	1957			0.0	0.0		0.0	
J	83.2	51.2	67.2	101	1956	40	1955+	29	0.20	0.32	1957	T	1956+	0.30	1957	0.0	0.0		0.0	
A	82.9	50.7	66.8	99	1953	41	1953	24	0.32	1.29	1953	T	1955	0.61	1953	0.0	0.0		0.0	
S	77.2	46.3	61.8	102	1955	32	1954	125	1.11	1.98	1957	0.46	1956	0.98	1957	0.0	0.0		0.0	
O	66.1	41.8	54.0	86	1953	26	1953	332	2.93	7.00	1956	1.34	1954	2.16	1956	0.0	0.0		0.0	
N	55.1	37.2	46.2	73	1955+	15	1955	564	4.51	8.44	1953	1.02	1956	3.71	1953	1.1	5.4	1955	3.3	1955
D	48.6	33.8	41.2	66	1955	23	1955	738	4.93	15.74	1955	3.30	1956	4.03	1955	0.3	0.8	1956	0.8	1956
Yr.	65.1	41.1	53.1	102	SEP. 1955	9	JAN. 1957	4556	30.50	15.74	DEC. 1955	T	JUL. 1956+	4.03	DEC. 1955	9.2	13.3	JAN. 1954	9.1	JAN. 1954

Relative Humidity, Wind, Sky Cover, Mean Number of Days

Month	RH 10:00 A. PST	RH 4:00 P. PST	Wind Mean hourly speed	Wind Prevailing direction	Wind Fastest speed	Dir.	Year	Mean sky cover sunrise to sunset	Pct. of possible sunshine	Clear	Partly cloudy	Cloudy	Precip .01 in or more	Snow/Sleet 1.0 in or more	Thunderstorms	Heavy fog	90° & above	Max 32° & below	Min 32° & below	Min 0° & below
	5	5			5	5	5		27	80	80	80					5	5	5	5
J	84	75	4.4		34	SW	1956	8.9	24	3	9	19	19	2	*	9	0	1	13	0
F	82	65	4.1		26	NW	1953	8.2	33	3	10	15	15	*	0	6	0	0	14	0
M	69	56	4.8		27	SW	1953	7.8	40	5	11	15	15	1	1	7	0	0	11	0
A	62	50	4.9		27	S	1957+	7.4	52	7	10	13	13	*	1	3	*	0	5	0
M	59	49	5.0		22	S	1957+	6.6	57	9	11	11	12	0	1	1	1	0	1	0
J	59	48	5.3		22	N	1956	6.2	60	11	10	9	9	0	1	1	1	0	0	0
J	54	36	6.1		25	NW	1954	3.2	78	19	9	3	4	0	*	0	4	0	0	0
A	57	39	5.5		25	N	1953	4.1	77	19	8	4	5	0	1	*	3	0	0	0
S	61	40	4.7		21	N	1956+	4.4	69	15	9	6	6	0	1	3	4	0	0	0
O	78	59	3.5		33	S	1954	7.3	41	8	11	12	12	0	*	11	0	0	2	0
N	85	74	3.5		26	SE	1953	8.5	28	4	10	16	14	*	0	12	0	*	8	0
D	87	80	4.0		31	SW	1954	8.9	17	2	9	20	18	*	0	10	0	*	12	0
Yr.	70	56	4.7		34	SW	JAN. 1956	6.8	51	105	118	142	137	3	6	55	13	1	66	0

Means and extremes in the above table are from the existing or comparable locations. Annual extremes have been exceeded at prior locations as follows: Highest Temperature 109 in July 1946; lowest temperature -6 in January 1888; maximum monthly precipitation 0.00 in June 1951 and earlier dates; maximum precipitation in 24 hours 4.15 in February 1899; maximum monthly snowfall 28.0 in January 1950; fastest mile of wind 40 from Southwest in December 1951.

Station moved or discontinued. See other stations for this state.

REFERENCE NOTES APPLYING TO TABLES APPEAR ON THE PAGE FOLLOWING LAST TABLE.
(Caution: Letters and symbols may have different meanings in 1941-1970 tables than in earlier tables. See notes.)

NORMALS, MEANS AND EXTREMES
(Table Revised 1972. Base Period for Climatological Normals: 1931-1960)

Station: SALEM, OREGON MC NARY FIELD Standard time used: PACIFIC Latitude: 44° 55' N Longitude: 123° 01' W Elevation (ground): 196 feet

Month	Temperature Normal Daily maximum	Normal Daily minimum	Normal Monthly	Extremes Record highest	Year	Extremes Record lowest	Year	Normal heating degree days (Base 65°)	Precipitation Normal total	Maximum monthly	Year	Minimum monthly	Year	Maximum in 24 hrs.	Year	Snow, Ice pellets Mean total	Maximum monthly	Year	Maximum in 24 hrs.	Year	Relative humidity 04	10	16	22	Wind Mean speed	Prevailing direction	Fastest mile Speed	Direction	Year	Pct. of possible sunshine	Mean sky cover sunrise to sunset	Days Clear	Partly cloudy	Cloudy	Precip. .01 in. or more	Snow 1.0 in. or more	Thunderstorms	Heavy fog	Max 90° & above	Max 32° & below	Min 32° & below	Min 0° & below	Avg daily solar radiation
	(b)	(b)	(b)	10		10		(b)	(b)	34		34		34		34	34		34		9	9	9	9	23	15	22	22	22	27	34	34	34	34	34	34	34	34	9	9	9	9	9
J	45.5	31.5	38.5	64	1971	8	1969	822	6.70	15.40	1953	0.57	1949	2.91	1956	4.2	32.4	1950	10.8	1943	88	86	79	87	8.9	S	40	18	1966		8.4	3	4	24	19	1	*	7	0	1	13	0
F	50.3	33.5	41.9	72	1968	15	1971	647	5.31	12.31	1949	0.78	1964	3.16	1949	0.9	8.4	1962	5.4	1962	88	83	68	84	7.9	S	40	19	1958		8.1	4	5	20	17	*	*	4	0	0	13	0
M	54.0	35.3	45.3	75	1969	12	1971	611	4.68	8.42	1938	0.87	1965	3.03	1943	T	10.9	1951	8.5	1951	84	74	60	71	7.9	S	40	19	1971		7.8	3	6	21	18	0	*	2	0	0	7	0
A	62.5	39.6	51.1	84	1968	23	1968	417	2.33	5.18	1955	0.39	1939	2.22	1971	T	T	1971+	T	1971+	80	65	57	70	7.4	S	44	21	1962		7.3	4	8	18	14	0	*	1	*	0	4	0
M	66.9	44.2	55.6	92	1963	30	1966	273	2.11	4.58	1942	0.18	1963	1.84	1963	0.0	0.0		0.0		83	63	52	65	7.1	S	28	21	1951		6.9	6	8	17	13	0	1	*	1	0	1	0
J	73.4	47.8	60.6	100	1969+	35	1965	144	1.45	3.60	1947	0.01	1951	1.60	1950	0.0	0.0		0.0		82	58	49	60	6.5	S	25	25	1971		6.5	7	9	15	8	0	1	*	1	0	0	0
J	81.7	50.4	66.1	104	1965	37	1962	37	0.35	1.41	1947	0.00	1967+	0.87	1961	0.0	0.0		0.0		87	58	40	72	6.0	N	26	29	1957		4.0	15	9	7	3	0	1	*	7	0	0	0
A	80.8	51.3	66.1	102	1967	38	1969+	51	0.45	1.25	1968	T	1970+	1.25	1968	0.0	0.0		0.0		87	59	40	70	6.0	N	24	24	1943		4.6	11	9	8	4	0	1	*	6	0	0	0
S	76.8	48.1	62.5	98	1967	28	1970+	111	1.38	3.98	1971	0.01	1942	1.86	1971	0.0	0.0		T	1970	87	65	47	79	6.3	S	31	18	1969		5.1	9	9	12	7	0	1	2	4	0	0	0
O	62.5	42.5	54.1	93	1970+	23	1971	338	3.91	11.17	1947	0.83	1952	2.84	1955	T	T	1970	T	1955	91	76	64	86	6.3	S	58	18	1951		5.8	5	6	20	13	0	1	7	0	0	4	0
N	53.8	36.6	45.2	72	1970+	17	1969	594	5.71	13.38	1942	0.84	1939	2.82	1942	0.2	5.8	1955	4.0	1955	90	86	78	87	7.5	S	38	21	1969		8.1	3	5	22	18	*	*	7	0	0	7	0
D	48.6	34.3	41.5	61	1969+	8	1969+	729	7.37	12.40	1964	2.67	1964	2.72	1964	1.2	12.6	1968	6.7	1968	88	88	82	87	8.4	S	45	23	1953		8.8	2	3	26	20	1	*	8	0	1	11	0
YR	63.5	41.3	52.4	104	JUL 1965	8	JAN 1969+	4754	41.75	15.40	JAN 1953	0.00	JUL 1967+	3.16	FEB 1949	7.4	32.8	JAN 1950	10.8	1943	88	73	60	81	7.3	S	58	18	OCT 1962		6.9	76	84	205	150	2	6	39	17	2	70	0

NORMALS, MEANS AND EXTREMES
(Table Revised 1975. Base Period for Climatological Normals: 1941-1970)

(Second table for the same station, base period 1941-1970, with Average station pressure Elev. 201 feet m.s.l. 1010.1 mb.)

Month	Temperature Normal Daily maximum	Normal Daily minimum	Normal Monthly	Record highest	Year	Record lowest	Year	Degree days Heating	Cooling	Precipitation Normal	Maximum monthly	Year	Minimum monthly	Year	Maximum in 24 hrs.	Year	Snow, Ice pellets Maximum monthly	Year	Maximum in 24 hrs.	Relative humidity 04	10	16	22	Wind Mean speed m.p.h.	Prevailing direction	Fastest mile m.p.h.	Direction	Year	Pct. possible sunshine	Mean sky cover	Clear	Partly cloudy	Cloudy	Precip. .01 in. or more	Snow 1.0 in. or more	Thunderstorms	Heavy fog vis. ¼ mi or less	Max 90° & above	Max 32° & below	Min 32° & below	Min 0° & below
				13		13					37		37		37		37		37	12	12	12	12	26	15	25	25	25	30	37	37	37	37	37	37	37	37	12	12	12	12
J	45.3	32.2	38.8	64	1971	6	1974	812	0	6.90	15.40	1953	0.57	1949	3.07	1949	32.8	1950	10.8	86	83	76	85	8.9	S	40	18	1966		8.4	5	5	21	19	1	*	5	0	0	14	0
F	51.4	34.4	42.9	72	1968	15	1971	619	0	4.79	12.31	1949	0.78	1964	3.16	1949	8.4	1962	5.4	83	76	68	80	7.8	S	40	19	1958		7.9	5	6	17	17	1	*	4	0	0	12	0
M	54.0	34.5	44.3	75	1969	12	1971	614	0	4.33	8.42	1938	0.87	1965	3.03	1943	10.9	1951	8.5	80	70	60	75	7.8	S	40	19	1971		7.4	4	8	18	16	*	*	2	0	0	12	0
A	60.5	38.5	49.8	84	1968	23	1968	456	7	2.29	5.18	1955	0.39	1939	2.22	1971	T	1972	0.1	77	65	56	73	7.1	S	44	21	1962		7.4	6	8	14	14	0	*	1	*	0	7	0
M	68.1	43.6	55.7	92	1963	30	1966	295	7	2.09	4.58	1942	0.18	1963	1.84	1963	0.0		0.0	77	62	49	67	6.7	S	28	21	1951		6.9	8	8	15	11	0	1	*	1	0	1	0
J	74.0	48.4	61.2	100	1969	35	1973	133	19	1.39	3.60	1947	0.01	1951	1.60	1950	0.0		0.0	85	62	49	73	6.6	S	25	25	1974		6.4	9	9	12	8	0	1	*	3	0	0	0
J	82.4	50.7	66.6	104	1965	37	1962	43	92	0.35	1.80	1974	0.00	1967	0.87	1961	0.0		0.0	84	56	39	67	6.6	N	26	29	1957		3.9	15	9	7	3	0	1	*	7	0	0	0
A	81.4	50.9	66.1	104	1972	38	1970	53	87	0.57	1.25	1968	T	1970	1.25	1968	0.0		0.0	80	57	40	70	6.3	N	24	24	1943		4.5	14	9	8	4	0	1	*	7	0	0	0
S	76.1	47.3	62.5	99	1973	26	1971	120	27	1.46	3.98	1971	0.01	1942	1.86	1971	T	1974	T	87	63	47	80	6.3	S	31	18	1969		5.1	10	7	13	7	0	1	3	4	0	0	0
O	64.1	42.3	53.2	93	1971	23	1971	366	0	3.98	11.17	1947	0.83	1952	2.84	1955	T	1955	4.0	90	76	60	86	6.3	S	58	18	1951		5.6	7	5	17	13	0	1	7	*	0	4	0
N	53.0	37.4	45.2	72	1970	17	1972	594	0	6.08	15.23	1973	0.84	1939	2.82	1942	5.8	1955	4.0	90	86	78	89	7.5	S	38	21	1969		8.8	5	5	22	18	*	*	7	0	0	7	*
D	47.1	34.7	40.9	61	1970	-12	1972	747	0	6.85	12.40	1964	2.57	1964	2.72	1964	14.5	1972	9.4	87	86	81	86	8.4	S	45	23	1953		8.8	3	3	26	20	1	*	11	0	*	11	2
YR	63.3	41.3	52.3	104	AUG 1972	-12	DEC 1972	4852	232	41.08	15.40	JAN 1953	0.00	JUL 1967	3.16	FEB 1949	32.8	JAN 1950	10.8	86	72	58	79	7.3	S	58	18	OCT 1962		6.9	83	77	205	150	2	6	38	19	3	68	2

Ø For period July 1962 through the current year.
Means and extremes above are from existing and comparable exposures. Annual extremes have been exceeded at other sites in the locality as follows: Highest temperature 108 in July 1941; lowest temperature -10 in January 1950; maximum monthly precipitation 17.54 in December 1933; maximum precipitation in 24 hours 4.30 in December 1933; maximum snowfall in 24 hours 25.0 in February 1937.

Means and extremes above are from existing and comparable exposures. Annual extremes have been exceeded at other sites in the locality as follows: Highest temperature 108 in July 1941; maximum monthly precipitation 17.54 in December 1933; maximum precipitation in 24 hours 4.30 in December 1933; maximum snowfall in 24 hours 25.0 in February 1937.

REFERENCE NOTES APPLYING TO TABLES APPEAR ON THE PAGE FOLLOWING LAST TABLE. (See notes.)

(Note: 1941-1970 tables than in earlier tables)

Station: SEXTON SUMMIT, OREGON TOP OF SEXTON MOUNTAIN

Standard time used: PACIFIC **Latitude:** 42° 37' N **Longitude:** 123° 22' W **Elevation (ground):** 3836 feet

Month	Normal Daily maximum	Normal Daily minimum	Normal Monthly	Record highest	Year	Record lowest	Year	Normal heating degree days (Base 65°)	Normal total Precipitation	Max monthly	Year	Min monthly	Year	Max in 24 hrs	Year	Snow, Ice pellets Mean total	Max monthly	Year	Max in 24 hrs	Year
J	39.3	28.9	34.1	65	1971	-2	1962	958	5.59	19.03	1970	1.11	1949	3.76	1948	26.7	106.7	1969	34.1	1969
F	41.8	30.4	36.1	66	1962	-9	1962+	809	4.12	10.32	1961	0.81	1964	3.16	1961	19.7	57.8	1956	22.6	1956
M	45.8	31.4	38.6	74	1966	15	1971	818	3.47	8.57	1960	0.26	1965	2.19	1960	19.7	45.5	1948	18.3	1948
A	53.8	35.6	44.7	82	1947	21	1964+	609	2.85	7.54	1963	0.37	1966	2.03	1963	8.7	43.5	1948	13.0	1948
M	60.8	39.2	50.0	87	1963	22	1965	465	2.19	7.26	1963	0.19	1966	3.07	1963	2.2	12.8	1953	9.5	1953
J	66.5	44.8	55.7	95	1961	27	1966	279	1.53	4.65	1947	0.00	1960+	1.79	1950	0.2	1.5	1954+	1.5	1954+
J	75.5	51.6	63.6	100	1946	36	1955	81	0.32	3.78	1947	0.00	1947	2.23	1947	0.0	0.0		0.0	
A	75.9	50.5	63.2	97	1971	36	1947	81	0.25	2.42	1968	0.00	1968	1.32	1962	0.0	0.0		0.0	
S	70.7	44.6	57.7	97	1955	26	1965	171	1.25	2.81	1959	0.01	1965	2.07	1957	0.1	3.1	1971	3.1	1971
O	58.8	42.6	50.7	85	1952	20	1971	443	3.41	13.87	1950	0.35	1952	2.81	1950	1.6	17.3	1956	8.8	1956
N	48.9	42.8	42.8	75	1955	11	1955	666	4.72	11.80	1973	0.57	1959	4.72	1961	9.0	36.4	1945	33.5	1945
D	41.6	32.0	36.8	64	1956	13	1968	874	4.76	20.03	1964	1.37	1964	4.43	1964	22.4	80.3	1971	33.5	1971
YR	56.6	39.4	48.0	100	JUL. 1946	-2	JAN. 1962	6254	33.12	20.03	DEC. 1964	0.00	AUG. 1970+	5.20	OCT. 1950	104.9	106.7	JAN. 1969	34.1	JAN. 1969

NORMALS, MEANS AND EXTREMES

Month	Normal Daily maximum	Normal Daily minimum	Normal Monthly	Record highest	Year	Record lowest	Year	Heating Degree days Base 65°F	Cooling	Precipitation Normal	Max monthly	Year	Min monthly	Year	Max in 24 hrs	Year	Snow Max monthly	Year	Max in 24 hrs	Year
J	39.8	29.2	34.5	65	1971	-5	1962	946	0	6.54	19.03	1970	1.11	1949	5.98	1949	106.7	1969	34.1	1969
F	43.9	30.7	37.3	66	1962	-9	1962	776	0	4.05	10.35	1961	1.05	1964	3.96	1961	57.8	1956	22.6	1956
M	43.0	30.7	37.8	74	1966	19	1971	843	0	4.43	8.57	1960	0.26	1965	2.64	1960	45.5	1948	18.3	1948
A	51.5	34.1	42.8	82	1947	19	1972	666	0	2.06	7.54	1963	0.37	1966	2.03	1963	42.6	1948	13.1	1948
M	60.3	38.7	49.5	87	1963	21	1965	481	0	2.22	7.28	1963	0.19	1966	3.07	1963	12.8	1953	10.0	1953
J	66.3	44.7	55.5	95	1961	27	1966	292	7	1.26	4.65	1947	0.00	1960	1.79	1950	1.5	1954	1.5	1954
J	75.7	51.4	63.6	100	1946	36	1955	97	53	0.33	3.78	1947	0.00	1947	2.23	1947	0.0		0.0	
A	74.7	51.4	63.1	97	1971	36	1947	115	57	0.44	2.42	1968	0.00	1968	1.32	1962	0.0		0.0	
S	70.2	49.2	59.7	97	1955	26	1965	179	20	1.03	2.81	1959	0.00	1973	2.07	1957	3.1	1971	3.1	1971
O	58.7	42.0	50.5	85	1952	20	1971	450	0	3.63	13.87	1950	0.35	1952	5.20	1973	42.6	1973	8.8	1956
N	47.1	35.2	41.2	75	1955	11	1955	714	0	5.61	24.09	1973	0.57	1959	5.20	1959	36.4	1945	33.5	1945
D	41.7	32.0	36.9	64	1956	2	1972	871	0	6.19	20.03	1964	1.37	1943	4.43	1943	80.3	1971	33.5	1971
YR	56.1	39.2	47.7	100	JUL. 1946	-9	JAN. 1962	6430	137	36.79	24.09	NOV. 1973	0.00	SEP. 1974	5.98	JAN. 1974	106.7	JAN. 1969	34.1	JAN. 1969

REFERENCE NOTES APPLYING TO TABLES APPEAR ON THE PAGE FOLLOWING LAST TABLE.

(Caution: Letters and symbols may have different meanings in 1941-1970 tables than in earlier tables. See notes.)

Reference notes applying to Normals, Means, and Extremes tables for 1931–1960 base period.

(a) Length of record, years, based on January data. Other months may be for more or fewer years if there have been breaks in the record.
(b) Climatological standard normals (1931-1960).
* Less than one half.
+ Also on earlier dates, months, or years.
T Trace, an amount too small to measure.
Below zero temperatures are preceded by a minus sign. The prevailing direction for wind in the Normals, Means, and Extremes table is from records through 1963.
‡ ≥ 70° at Alaskan stations.

Unless otherwise indicated, dimensional units used in this bulletin are: temperature in degrees F.; precipitation, including snowfall, in inches; wind movement in miles per hour; and relative humidity in percent. Heating degree day totals are the sums of negative departures of average daily temperatures from 65° F. Cooling degree day totals are the sums of positive departures of average daily temperatures from 65° F. Sleet was included in snowfall totals beginning with July 1948. The term "ice pellets" includes solid grains of ice (sleet) and particles consisting of snow pellets encased in a thin layer of ice. Heavy fog reduces visibility to 1/4 mile or less.

Sky cover is expressed in a range of 0 for no clouds or obscuring phenomena to 10 for complete sky cover. The number of clear days is based on average cloudiness 0-3, partly cloudy days 4-7, and cloudy days 8-10 tenths.

Solar radiation data are the averages of direct and diffuse radiation on a horizontal surface. The langley denotes one gram calorie per square centimeter.

& Figures instead of letters in a direction column indicate direction in tens of degrees from true North; i.e., 09 - East, 18 - South, 27 - West, 36 - North, and 00 - Calm. Resultant wind is the vector sum of wind directions and speeds divided by the number of observations. If figures appear in the direction column under "Fastest mile" the corresponding speeds are fastest observed 1-minute values.

% Through 1964. The station did not operate 24 hours daily. Fog and thunderstorm data may be incomplete.

∅ For period beginning with 1965, records incomplete. Summary based on available data.
$ Through 1964.
c Data through 1968 and 1972 to date.

Reference notes applying to Normals, Means, and Extremes tables for 1941–1970 base period.

(a) Length of record, years, through the current year unless otherwise noted, based on January data.
(b) 70° and above at Alaskan stations.
* Less than one half.
T Trace.

NORMALS - Based on record for the 1941-1970 period.
DATE OF AN EXTREME - The most recent in cases of multiple occurrence.
PREVAILING WIND DIRECTION - Record through 1963.
WIND DIRECTION - Numerals indicate tens of degrees clockwise from true north. 00 indicates calm.
FASTEST MILE WIND - Speed is fastest observed 1-minute value when the direction is in tens of degrees.

% Through 1964. The station did not operate 24 hours daily. Fog and thunderstorm data may be incomplete.
∅ For period beginning with 1965 records incomplete. Summary based on available data.
$ Through 1964.
c Data through 1968 and 1972 to date.

Mean Maximum Temperature (°F.), January

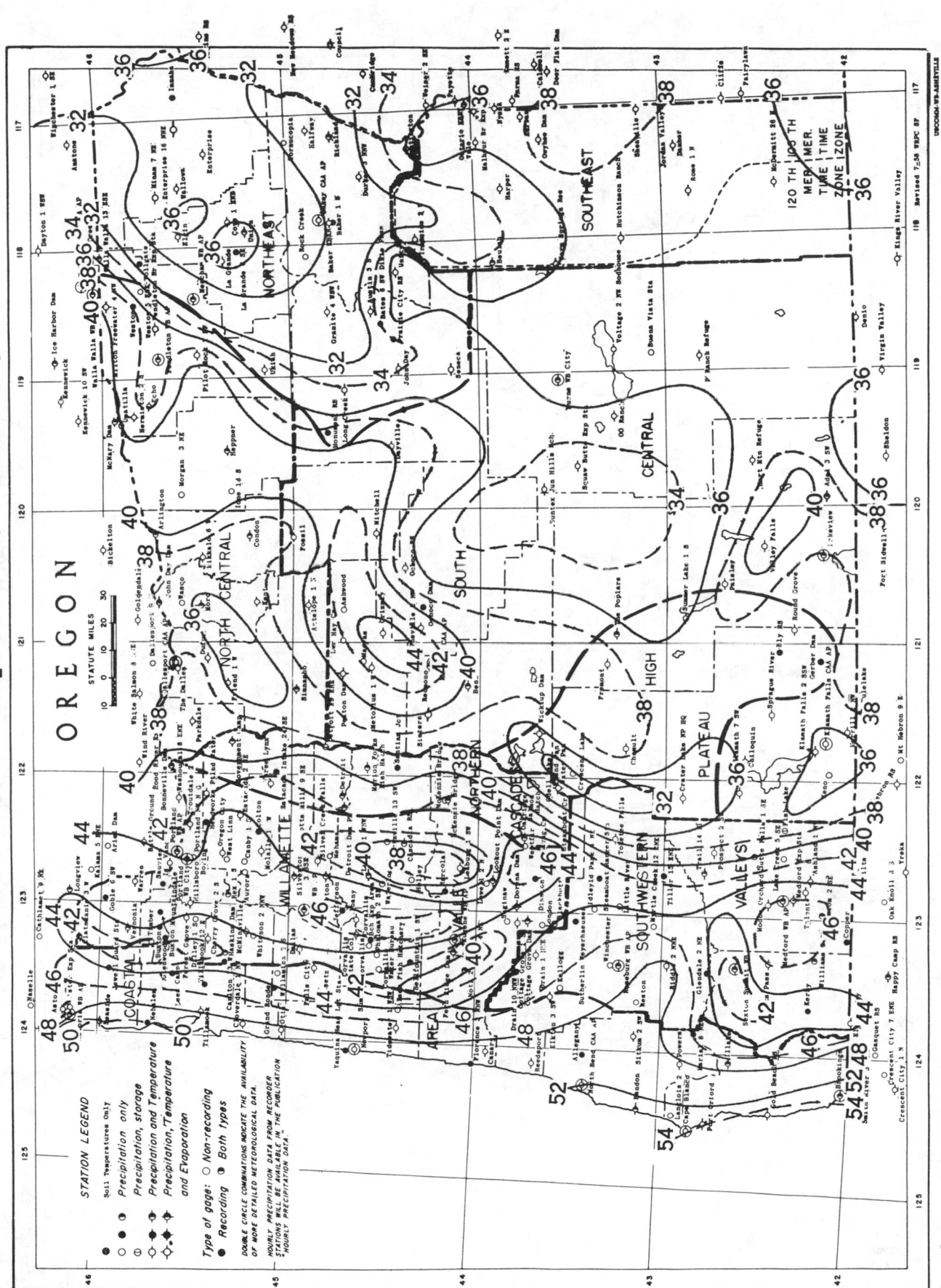

Based on period 1931-52

Isolines are drawn through points of approximately equal value. Caution should be used in interpolating on these maps, particularly in mountainous areas.

Mean Minimum Temperature (°F.), January

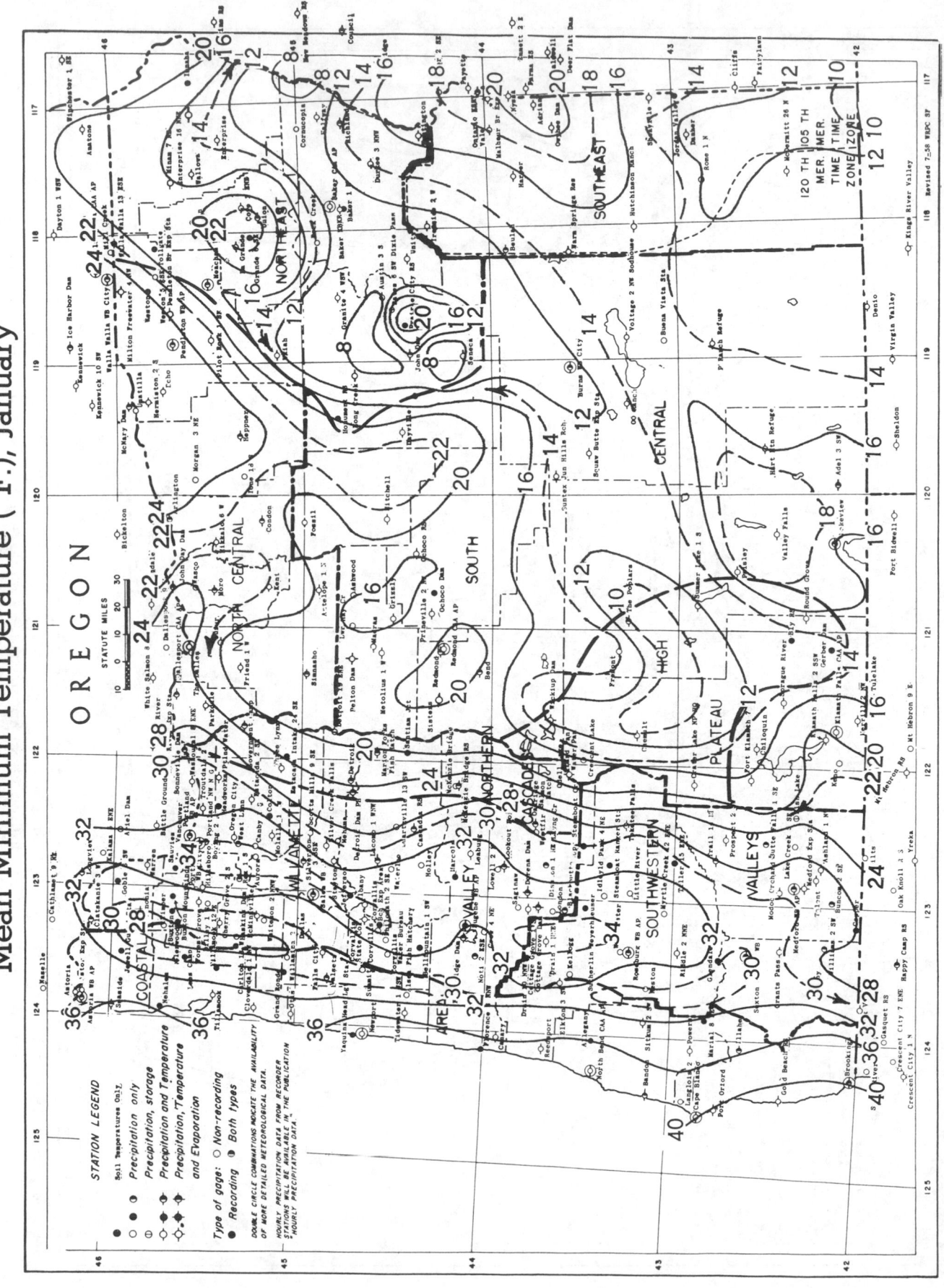

Based on period 1931-52

Isolines are drawn through points of approximately equal value. Caution should be used in interpolating on these maps, particularly in mountainous areas.

Mean Maximum Temperature (°F), July

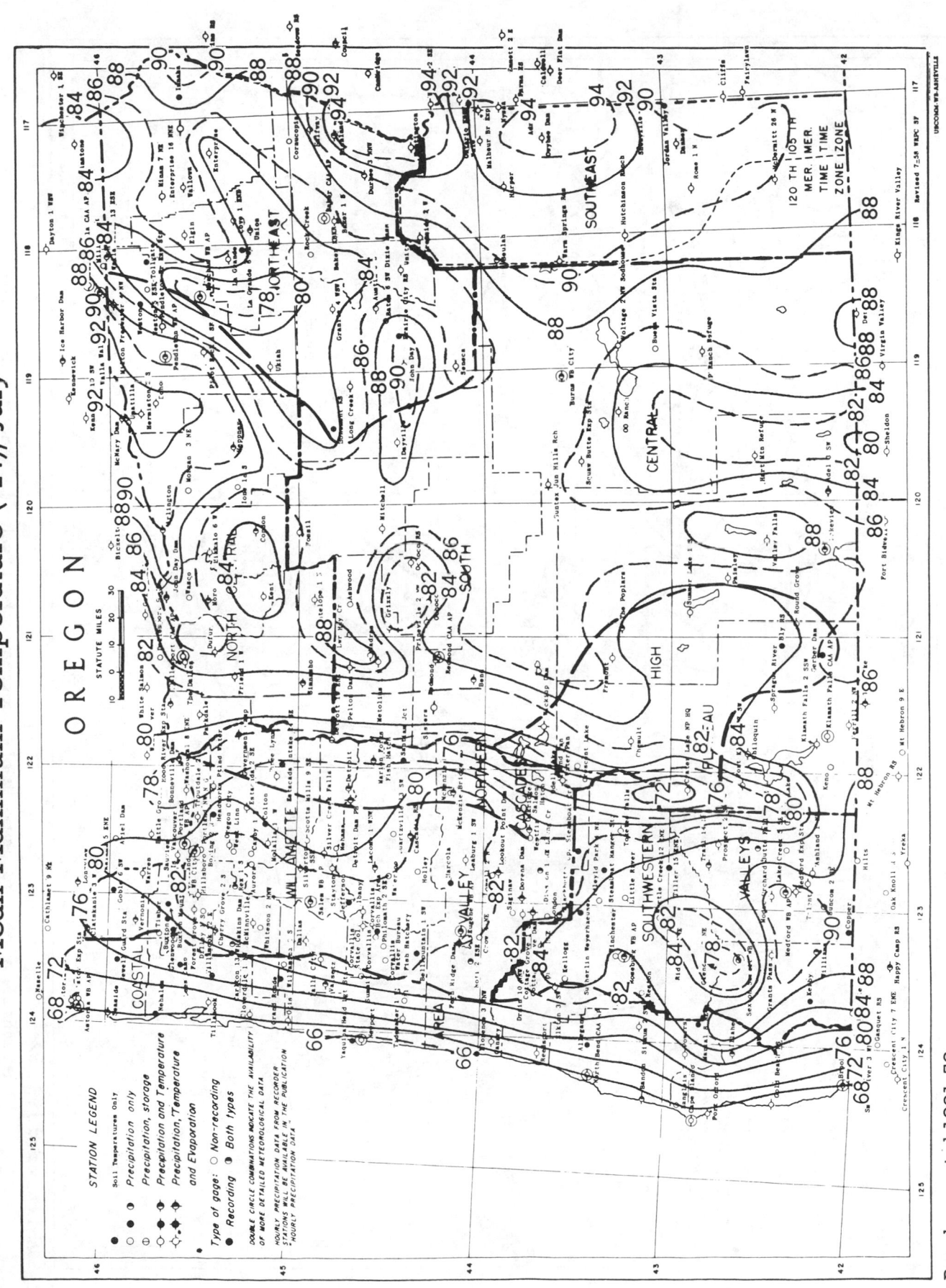

Based on period 1931-52

Isolines are drawn through points of approximately equal value. Caution should be used in interpolating on these maps, particul ly in mountainous areas.

Mean Minimum Temperature (°F), July

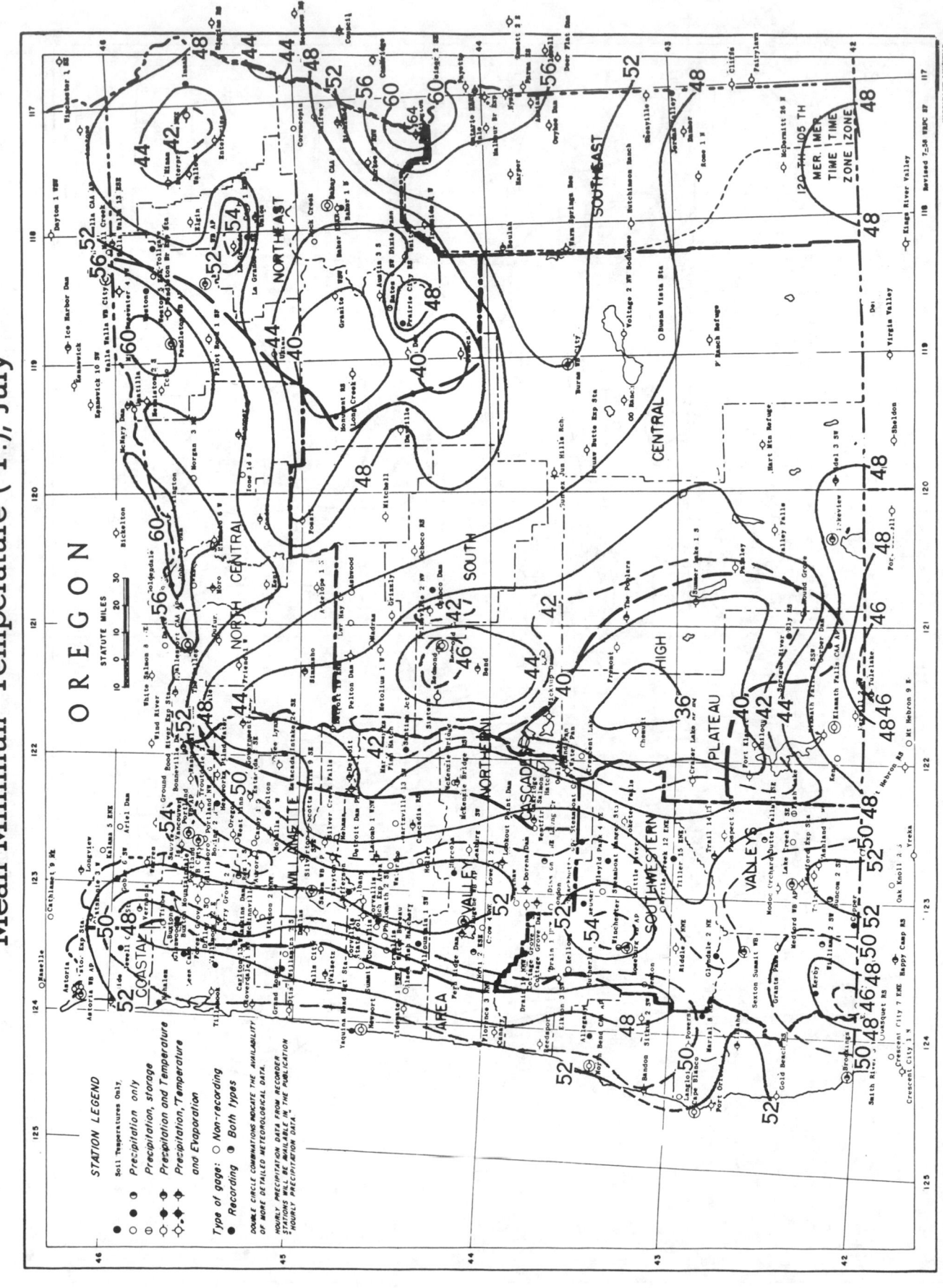

Based on period 1931-52

Isolines are drawn through points of approximately equal value. Caution should be used in interpolating on these maps, particularly in mountainous areas.

Mean Annual Precipitation, Inches

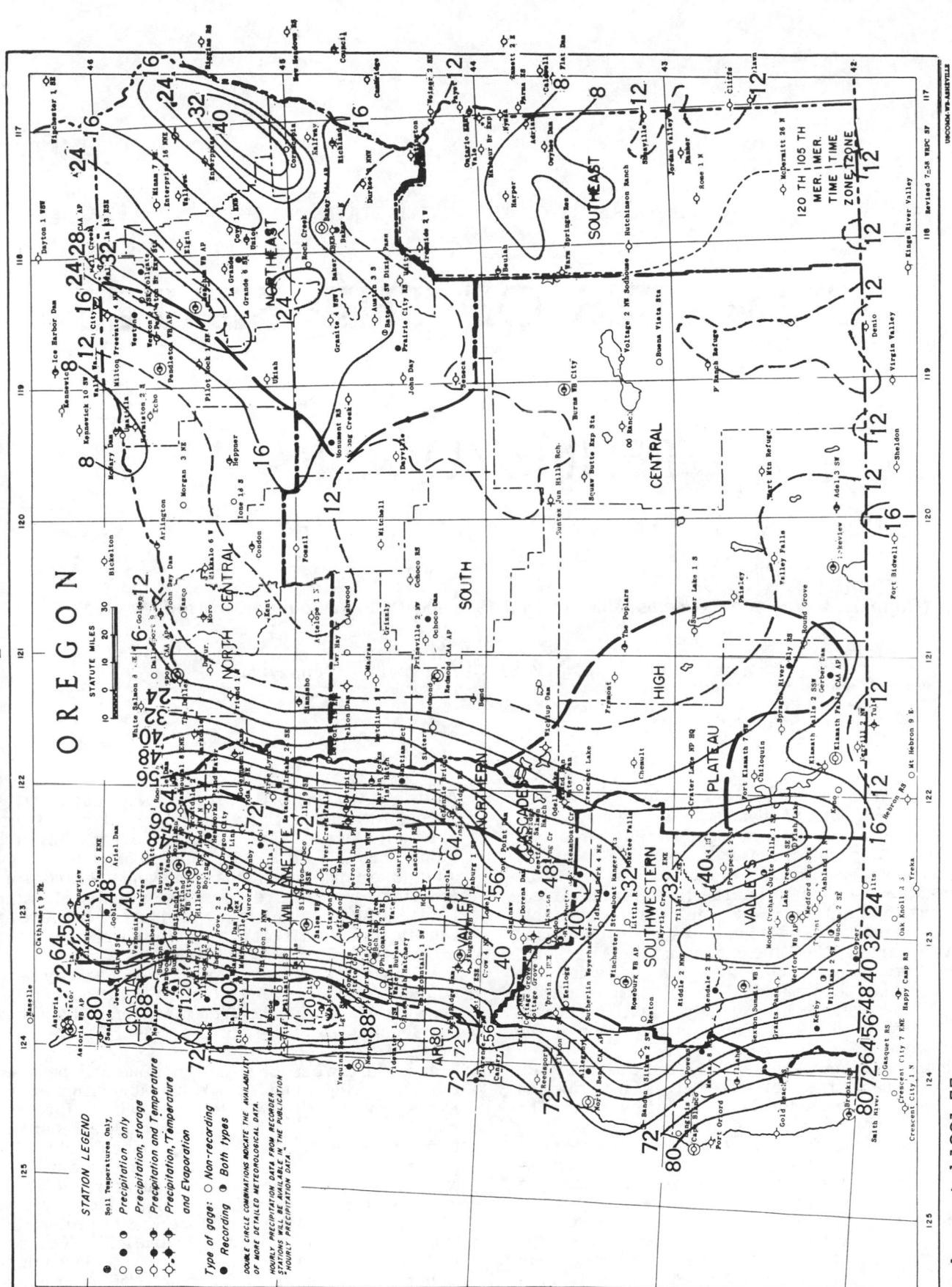

Based on period 1931-55

Isolines are drawn through points of approximately equal value. Caution should be used in interpolating on these maps, particul   in mountainous areas.

CLIMATES OF THE STATES

PENNSYLVANIA

(Normals, Means and Extremes tables revised 1971 and 1975. Basic report revised March 1971.)

Paul W. Dailey, Jr., NOAA Climatologist - Pennsylvania

The erratic course of the Delaware River is the only natural boundary of Pennsylvania. All others are arbitrary boundaries that do not conform to physical features. Notable contrasts in topography, climate, and soils exist. Within this 45,126-square-mile area lies a great variety of physical land forms of which the most notable is the Appalachian Mountain system composed of two ranges, the Blue Ridge and the Allegheny. These mountains divide the Commonwealth into three major topographical sections. In addition, two plain areas of relatively small size also exist, one in the southeast and the other in the northwest.

In the extreme southeast is the Coastal Plain situated along the Delaware River and covering an area 50 miles long and 10 miles wide. The land is low, flat, and poorly drained, but has been improved for industrial and commercial use because of its proximity to ocean transportation via the Delaware River. Philadelphia lies almost in the center of this area.

Bordering the Coastal Plain and extending 60 to 80 miles northwest to the Blue Ridge is the Piedmont Plateau, with elevations ranging from 100 to 500 feet and including rolling or undulating uplands, low hills, fertile valleys, and well-drained soils. These features, combined with the prevailing climate, have aided this area in becoming the leading agricultural section of the State. Good pastures, productive land, and short distances to markets have resulted in dairy farm-

ing becoming one of the leading agricultural activities. Another activity is the growing of fruit, primarily apples and peaches. Gentle hillside slopes provide an excellent place for fruit trees, as cold air drainage helps to prevent unseasonable freezing temperatures on these slightly elevated lands. The area has many orchards, with Adams County leading all others within the region in the production of apples. The climate and soils in the Lancaster County area are especially well suited for the growing of cigar leaf tobacco, as is pointed up by the fact that Pennsylvania is the leading producer of cigar leaf of any type in the Nation.

Just northwest of the Piedmont and between the Blue Ridge and Allegheny Mountains is the Ridge and Valley Region, in which forested ridges alternate with fertile and extensively farmed valleys. Vegetables, grown primarily for canning, are the leading crop. This has led to a well-developed canning industry, which is concentrated in the middle Susquehanna Valley. The Ridge and Valley Province is 80 to 100 miles wide and characterized by parallel ridges and valleys oriented northeast-southwest. The mountain ridges vary from 1,300 to 1,600 feet above sea level, with local relief 600 to 700 feet.

North and west of the Ridge and Valley Region and extending to the New York and Ohio borders is the area known as the Allegheny Plateau. This is the largest natural division of the State and occupies more than half the area. It is crossed by many deep narrow valleys and drained by the Delaware, Susquehanna, Allegheny, and Monongahela River systems. Elevations are generally 1,000 to 2,000 feet above sea level; however, some mountain peaks extend to 3,000 feet. The area is heavily wooded and among the most rugged in the State. Numerous lakes and swamps characterize this once glaciated area, creating a very picturesque landscape; this is particularly outstanding in the more northerly counties. The combination of lakes and forests at elevations high enough to keep summer temperatures comfortable and its location close to heavily populated cities have made the Pocono Mountain area the leading tourist and recreational center in Pennsylvania.

Bordering Lake Erie is a narrow 40-mile strip of flat, rich land 3 to 4 miles wide called the Lake Erie Plain. Fine alluvial soils and favorable climate permit intensive vegetable and fruit cultivation, which is typical of the much larger area surrounding Lake Erie.

Eastern and central Pennsylvania drains into the Atlantic Ocean, while the western portion of the State lies in the Ohio River Basin, except the Lake Erie Plain in the northwest, which is drained by a number of small streams into Lake Erie. The Delaware River, which forms the eastern boundary, drains the eastern portion and flows into Delaware Bay. The Susquehanna River drains the central portion and flows into Chesapeake Bay. In the western portion, the Allegheny and the Monongahela Rivers have their confluence at Pittsburgh and form the Ohio River.

Floods may occur during any month of the year in Pennsylvania, although they occur with greater frequency in the spring months of March and April. They may result from heavy rains during any season. Generally, the most widespread flooding occurs during the winter and spring when associated with heavy rains, or heavy rains combined with snowmelt. Serious local flooding sometimes results from ice jams during the spring thaw. Heavy local thunderstorm rains cause severe flash flooding in many areas. Storms of tropical origin sometimes deposit flood-producing rains, especially in the eastern portion of the State.

Floods may be expected at least once in most years. For instance, flood stage at Pittsburgh is exceeded on the average of 1.3 times per year, based on the long-term record. However, floods of notable severity and magnitude for the State occur about once in 8 years.

Some years in which major flooding occurred along principal rivers are as follows: Schuylkill, 1902, 1935, 1942, 1955, 1969; Delaware, 1903, 1936, 1955, 1967; Susquehanna, 1865, 1889, 1894, 1902, 1904, 1936, 1964; Allegheny, 1865, 1889, 1892, 1905, 1907, 1910, 1913, 1936, 1942, 1947, 1964; Monongahela, 1888, 1907, 1918, 1936; Ohio, 1907, 1936, 1942, 1954.

Pennsylvania is generally considered to have a humid continental type of climate, but the varied physiographic features have a marked effect on the weather and climate of the various sections within the State. The prevailing westerly winds carry most of the weather disturbances that affect Pennsylvania from the interior of the continent, so that the Atlantic Ocean has only limited influence upon the climate of the State. Coastal storms do, at times, affect the day-to-day weather, especially in eastern sections. It is here that storms of tropical origin have the greatest effect within the State, causing floods in some instances.

Throughout the State temperatures generally remain between 0° and 100° and average from near 47° annually in the north-central mountains to 57° annually in the extreme southeast. The highest temperature of record in Pennsylvania of 111° was observed at Phoenixville July 9 and 10, 1936, while the record low of -42° occurred at Smethport January 5, 1904.

Summers are generally warm, averaging about 68° along Lake Erie to 74° in southeastern counties. High temperatures, 90° or above, occur on the average of 10 to 20 days per year in most sections; but occasionally southeastern localities may experience a season with as many as 30 days, while the extreme northwest averages as few as 4 days annually. Only rarely does a summer pass without excessive temperatures being reported somewhere in the State. However, there are places such as immediately adjacent to Lake Erie and at some higher elevations where readings of 100° have never been recorded. Daily temperatures during the warm season usually have a range of about 20° over much of the State, while the daily range in winter is several degrees less. During the coldest months temperatures average near the freezing point with daily minimum readings sometimes near 0° or below. Freezing temperatures occur on the average of 100 or more days annually with the greatest number of occurrences in mountainous regions. Records show that freezing temperatures have occurred somewhere in the State during all months of the year and below 0° readings from November to April, inclusive.

Precipitation is fairly evenly distributed throughout the year. Annual amounts generally range between 34 to 52 inches, while the majority of places receive 38 to 46 inches. Greatest amounts usually occur in spring and summer months, while February is the driest month, having about 2 inches less than the wettest months. Precipitation tends to be somewhat greater in eastern sections due primarily to coastal storms which occasionally frequent the area. During the warm season these storms bring heavy rain, while in winter heavy snow or a mixture of rain and snow may be produced. Thunderstorms, which average between 30 to 35 per year, are concentrated in the warm months and are responsible for most of the summertime rainfall, which averages from 11 inches in the northwest to 13 inches in the east. Occasionally dry spells may develop and persist for several months during which time monthly precipitation may total less than one-quarter inch. These periods almost never affect all sections of the State at the same time, nor are they confined to any particular season of the year. Winter precipitation is usually 3 to 4 inches less than summer rainfall and is produced most frequently from northeastward-moving storms. When temperatures are low enough these storms sometimes cause heavy snow which may accumulate to 20 inches or more. Annual snowfall ranges between wide limits from year to year and place to place. Some years are quite lean as snowfall may total less than 10 inches while other years may produce upwards to 100 inches mostly in northern and mountainous areas. Annual snowfall averages from about 20 inches in the extreme southeast to 90 inches in parts of McKean County. Measurable snow generally occurs between November 20 and March 15 although snow has been observed as early as the beginning of October and as late as May, especially in northern counties. Greatest monthly amounts usually fall in December and January, however, greatest amounts from individual storms generally occur in March as the moisture supply increases with the annual march of temperature.

As mentioned earlier, hurricanes or low pressure systems with a tropical origin seldom affect the State. Damages, as a result of hurricane winds, are rare and usually confined to extreme eastern portions. However, nature's most violent storm, the tornado, does occur in Pennsylvania. At least one tornado has been noted in almost all counties since the advent of severe storms records in 1854. On the average, 5 or 6 tornadoes are observed annually in Pennsylvania, and the State ranks 27th nationally. June is the month of highest frequency, followed closely by July and August. Principal areas of tornado concentration are in the extreme northwest, the Southwest Plateau, and the Southeastern Piedmont. The frequency in the latter area is the highest in the State per square mile, similar to what is observed in portions of Midwestern United States. Many of the tornadoes in Pennsylvania have caused relatively minor damages. However, several have claimed lives and dealt severe local economic setbacks. The most destructive activity occurred June 23, 1944, when 3 tornadoes raked the southwestern portion of the Commonwealth, killing 45 persons, injuring another 362, and causing over $2 million in property damage.

The topographic features of Pennsylvania divide the State into four rather distinct climatic areas:

(1) The Southeastern Coastal Plain and Piedmont Plateau,

(2) The Ridge and Valley Province,

(3) The Allegheny Plateau, and

(4) The Lake Erie Plain.

In the Southeastern Coastal Plain and Piedmont Plateau summers are long and at times uncomfortably hot. Daily temperatures reach 90°

or above on the average of 25 days during the summer season; however, readings of 100° or above are comparatively rare. From about July 1 to the middle of September this area occasionally experiences uncomfortably warm periods, 4 to 5 days to a week in length, during which light wind movement and high relative humidity make conditions oppressive. In general, the winters are comparatively mild, with an average of less than 100 days with minimum temperatures below the freezing point. Temperatures 0° or lower occur at Philadelphia, on an average, 1 winter in 4, and at Harrisburg 1 in 3. The freeze-free season averages 170 to 200 days.

Average annual precipitation in the area ranges from about 30 inches in the lower Susquehanna Valley to about 46 in Chester County. Under the influence of an occasional severe coastal storm, a normal month's rainfall, or more, may occur within a period of 48 hours. The average seasonal snowfall is about 30 inches, and fields are ordinarily snow covered about one-third of the time during the winter season.

The Ridge and Valley Province is not rugged enough for a true mountain type of climate, but it does have many of the characteristics of such a climate. The mountain-and-valley influence on the air movements causes somewhat greater temperature extremes than are experienced in the southeastern part of the State where the modifying coastal and Chesapeake Bay influence hold them relatively constant, and the daily range of temperature increases somewhat under the valley influences.

The effects of nocturnal radiation in the valleys and the tendency for cool airmasses to flow down them at night result in a shortening of the growing season by causing freezes later in spring and earlier in fall than would otherwise occur. The growing (freeze-free) season in this section is longest in the middle Susquehanna Valley, where it averages about 165 days, and shortest in Schuylkill and Carbon Counties, averaging less than 130 days.

The annual precipitation in this area has a mean value of 3 or 4 inches more than in the southeastern part of the State, but its geographic distribution is less uniform. The mountain ridges are high enough to have some deflecting influence on general storm winds, while summer showers and thunderstorms are often shunted up the valleys.

Seasonal snowfall of the Ridge and Valley Province varies considerably within short distances. It is greatest in Somerset County, averaging 88 inches in the vicinity of Somerset, and least in Huntingdon, Mifflin, and Juniata Counties, averaging about 37 inches.

The Allegheny Plateau is fairly typical of a continental type of climate, with changeable temperatures and more frequent precipitation than other parts of the State. In the more northerly sections the influence of latitude, together with higher elevation and radiation conditions, serve to make this the coldest area in the State. Occasionally, winter minimum temperatures are severe. The daily temperature range is fairly large, averaging about 20° in midwinter and 26° in midsummer. In the southern counties the daily temperature range is a few degrees higher and the same may be said of the normal annual range. Because of the rugged topography the freeze-free season is variable, ranging between 130 days in the north to 175 days in the south.

Annual precipitation has a mean of about 41 inches, ranging from less than 35 inches in the northern parts of Tioga and Bradford Counties to more than 45 inches in parts of Crawford, Warren, and Wayne Counties. The seasonal snowfall averages 54 inches in northern areas, while southern sections receive several inches less. Fields are normally snow covered three-fourths of the time during the winter season. With rapidly flowing streams in the Ohio Drainage system (except the Monongahela), it is fortunate that this part of the State is not subject to torrential rains such as sometimes occur along the Atlantic slope. Although average annual precipitation is about equal to that for the State as a whole, it usually occurs in smaller amounts at more frequent intervals; 24-hour rains exceeding 2.5 inches are comparatively rare.

Although the Lake Erie Plain is of relatively small size, it has a unique and agriculturally advantageous climate typical of the coastal areas surrounding much of the Great Lakes. Both in spring and autumn the lake water exerts a retarding influence on the temperature regime and the freeze-free season is extended about 45 days. In the autumn this prevents early freezing temperatures, which is a critical factor in the growing of fruit and vegetables.

Annual precipitation totals about 34.5 inches, which is fairly evenly distributed throughout the year. Snowfall exceeds 54 inches per year, with heavy snows sometimes experienced late in April.

BIBLIOGRAPHY

Benfer, Neil A., "An Economic Geography of the Canning Industry in the Middle Susquehanna Valley Lowland," Master's thesis, Pennsylvania State University, University Park, Pa., 1951.

Bingham, Christopher, "Probabilities of Weekly Averages of the Daily Temperature Maximum, Minimum, and Range," The Climate of the Northeast, Bulletin 659, The Connecticut Agricultural Experiment Station, New Haven, Conn., Sept. 1963, 28 pp.

Chester County Planning Commission, "Chester County Natural Environment and Planning," West Chester, Pa., July 1963.

Chestnutwood, Charles M., "The Geographical Bases of Pennsylvania's Tourist Industry," Master's thesis, Pennsylvania State University, University Park, Pa., 1954.

Cornell University Agricultural Station, "Precipitation Amounts in the Northeast Region of the U.S.," The Climate of the Northeast, Agronomy Mimeo 61-4, Ithaca, N.Y., Nov. 1961, 302 pp.

Dailey, Paul W., Jr., "Tornadoes in Pennsylvania," Information Report No. 63, Institute for Research on Land and Water Resources, Pennsylvania State University, University Park, Pa., June 1970.

Dethier, B. E. and Vittum, M. T., "Growing Degree Days," The Climate of the Northeast, Bulletin 801, New York State Agricultural Experiment Station, Geneva, N. Y., Aug. 1963.

Dickerson, W. H., "Heating Degree Days," The Climate of the Northeast, Bulletin 483T, West Virginia University Agricultural Experiment Station, Morgantown, W. Va., June 1963.

Fuller, Theodore E., "A Geographical Analysis of the Agricultural Regions of Adams County, Pennsylvania," Master's thesis, Pennsylvania State University, University Park, Pa., 1956.

Havens, A. V. and McGuire, J. K., "Spring and Fall Low-Temperature Probabilities," The Climate of the Northeast, Bulletin 801, New Jersey Agricultural Experiment Station, Rutgers University, New Brunswick, N. J., June 1961, 32 pp.

Heppell, Roger C., "Agricultural Geography of the Cigar Tobacco Industry of the Lancaster, Pennsylvania Region," Master's thesis, Pennsylvania State University, University Park, Pa., 1953.

Kauffman, Nelson M., "Climates of the States--Pennsylvania," Climatography of the United States No. 60-36, U.S. Weather Bureau, U.S. Government Printing Office, Washington, D. C., 1960, 19 pp.

_____, and Butler, R. G., "In Pennsylvania--Late Spring and Early Fall Freezes," Pennsylvania Crop Reporting Service, Harrisburg, Pa., Aug. 1961, 11 pp.

Murphy, Raymond E. and Marion, Pennsylvania--A Regional Geography, The Pennsylvania Book Service, Harrisburg, Pa., 1937.

NOAA, Environmental Data Service, Climatological Data--National Summary, Silver Spring, Md., monthly plus annual summary.

_____, Environmental Data Service, "Climatological Data--Pennsylvania," Silver Spring, Md., monthly plus annual summary.

_____, Environmental Data Service, Hourly Precipitation Data, Silver Spring, Md., monthly plus annual summary.

_____, "Local Climatological Data" for Allentown, Erie, Harrisburg, Philadelphia, Pittsburgh, Reading, Scranton, Shippingport, and Williamsport, Pennsylvania, Silver Spring, Md., monthly plus annual summary.

_____, "Substation Climatological Summary," Climatography of the United States, No. 20-36, in cooperation with the Pennsylvania Department of Commerce for Altoona, Bethlehem, Brookville, Chambersburg, Emporium, Franklin, Freeland, George School, Johnstown, Lancaster, Lawrenceville, Lebanon, Madera, Montrose, Mount Pocono, New Castle, Selinsgrove-Sunbury, Slippery Rock, Somerset, State College, Towanda, Uniontown, Warren, and York, Penna.;Silver Spring, Md., irregular.

BIBLIOGRAPHY

Pennsylvania Crop Reporting Service, "Pennsylvania Crop and Livestock Annual Summary," Harrisburg, Pa., annual, 1959-

Pennsylvania Writers Project, Pennsylvania--A Guide to the Keystone State, Oxford University Press, New York, N. Y., 1940.

Sharpe, William E.; Lee, Richard; and Jones, E. Bruce, "Summary of Temperature Means of State College, Pennsylvania, for 1931-1967," Information Report No. 59, Institute for Research on Land and Water Resources, Pennsylvania State University, University Park, Pa., October 1968.

Tukey, L. D.; Kauffman, N. M.; and Weiser, E. V., Jr., "Regional Weather Summary of Pennsylvania," Progress Report 254, Climatic Series 1, Southeastern Area, Pennsylvania State University Agricultural Experiment Station, University Park, Pa., Jan. 1965, 39 pp.

_____, "Regional Weather Summary of Pennsylvania," Progress Report 260, Climatic Series 2, Lower Susquehanna Area, Pennsylvania State University Agricultural Experiment Station, University Park, Pa., July 1965, 40 pp.

_____, "Regional Weather Summary of Pennsylvania," Progress Report 266, Climatic Series 3, Northeastern Area, Pennsylvania State University Agricultural Experiment Station, University Park, Pa., June 1966, 32 pp.

U.S. Department of Agriculture, Soil Survey, Soil Conservation Service, county summaries for Adams (May 1967), Carbon (Nov. 1962), Chester and Delaware (May 1963), Clinton (Aug. 1966), Columbia (Mar. 1967), Erie (Dec. 1960), Fulton (Nov. 1969), Indiana (Jan. 1968), Jefferson (Aug. 1964), Lancaster (Oct. 1959), Lehigh (Nov. 1963), Montgomery (Apr. 1967), Pike (June 1969), Potter (July 1958), Westmoreland (Nov. 1968), and York (May 1963), Pennsylvania. Washington, D. C.

_____, Yearbook of Agriculture, 1941: Climate and Man, U.S. Government Printing Office, Washington, D. C., 1248 pp.

U.S. Weather Bureau, "Maximum Station Precipitation for 1, 2, 3, 6, 12, and 24 hours, Part XVI, Pennsylvania," Technical Paper No. 15, Washington, D. C., 1956, 146 pp.

U.S. Weather Bureau, "Maximum 24-Hour Precipitation in the United States," its Technical Paper No. 16, Washington, D. C., 1952, 284 pp.

_____, Climatic Summary of the United States (Bulletin W), Sections 87, 88, and 89, Washington, D. C., 1930, 69 pp.

_____, Climatic Summary of the United States, Pennsylvania, (Bulletin W Supplement), 1931-1952, Washington, D. C., 1953, 77 pp.

_____, Climatic Summary of the United States, Pennsylvania, (Bulletin W Supplement), 1951-1960, Washington, D. C., 1961, 95 pp.

_____, "Rainfall Intensity-Duration-Frequency Curves for Selected Stations in the United States, Alaska, Hawaiian Islands, and Puerto Rico," its Technical Paper No. 25, Washington, D. C., 1955, 53 pp.

_____, "Rainfall Intensity-Frequency Regime, Part 3--the Middle Atlantic Region," its Technical Paper No. 29, Washington, D. C., 1958, 38 pp.

FREEZE DATA

STATION	Freeze threshold temperature	Mean date of last Spring occurrence	Mean date of first Fall occurrence	Mean No. of days between dates	Years of record Spring	No. of occurrences in Spring	Years of record Fall	No. of occurrences in Fall
ALTOONA HORSESHOE CV	32	05-06	10-08	155	30	30	30	30
	28	04-21	10-21	183	30	30	30	30
	24	04-12	11-05	207	30	30	30	30
	20	03-31	11-18	232	30	30	30	30
	16	03-16	12-01	259	30	30	29	29
BETHLEHEM LEHI UNIV	32	04-14	10-23	192	28	28	29	29
	28	04-01	11-07	221	28	28	28	28
	24	03-21	11-23	247	28	28	28	28
	20	03-10	12-03	268	28	28	28	28
	16	03-06	12-11	280	28	28	28	25
BROOKVILLE AP	32	05-27	09-20	116	27	27	30	30
	28	05-14	10-03	143	27	27	30	30
	24	04-29	10-18	172	27	27	29	29
	20	04-15	11-05	203	27	27	29	29
	16	03-29	11-17	234	28	28	29	29
BUTLER	32	05-21	10-03	135	23	23	24	24
	28	04-29	10-15	170	23	23	24	24
	24	04-15	11-01	200	23	23	24	24
	20	04-08	11-12	218	23	23	24	24
	16	03-17	11-28	256	23	23	24	24
CARLISLE	32	04-27	10-11	167	30	30	30	30
	28	04-17	10-24	190	30	30	30	30
	24	03-30	11-09	224	30	30	30	30
	20	03-16	11-26	255	30	30	30	30
	16	03-08	12-06	273	30	30	30	29
CHAMBERSBURG	32	05-01	10-12	164	30	30	30	30
	28	04-15	10-26	195	30	30	30	30
	24	03-28	11-14	231	30	30	30	30
	20	03-16	11-28	257	30	30	30	30
	16	03-08	12-09	276	30	30	30	26
CLAYSVILLE 3 W	32	05-15	09-30	138	30	30	29	29
	28	04-30	10-13	166	30	30	29	29
	24	04-17	10-28	194	30	30	29	29
	20	04-03	11-14	225	30	30	29	29
	16	03-20	11-25	250	30	30	29	29
COATESVILLE 1 SW	32	04-29	10-16	170	30	30	30	30
	28	04-12	10-28	200	30	30	30	30
	24	03-28	11-07	224	30	30	30	30
	20	03-18	11-24	250	30	30	30	30
	16	03-09	12-07	273	30	30	30	29
CORRY	32	05-24	09-23	123	28	28	28	28
	28	05-09	10-11	155	28	28	28	28
	24	04-23	10-27	188	28	28	28	28
	20	04-10	11-13	217	28	28	28	28
	16	03-29	11-27	244	28	28	28	27
COUDERSPORT	32	05-29	09-12	107	18	18	18	18
	28	05-17	09-30	136	17	17	17	17
	24	05-02	10-10	161	17	17	17	17
	20	04-15	10-22	190	17	17	17	17
	16	04-01	11-11	224	16	16	18	18
DERRY	32	05-08	10-08	153	30	30	30	30
	28	04-21	10-20	183	30	30	29	29
	24	04-12	11-05	207	30	30	30	30
	20	03-27	11-20	238	29	29	30	30
	16	03-16	12-02	262	29	29	30	30
EBENSBURG	32	05-15	10-01	139	29	29	29	29
	28	05-01	10-16	168	29	29	28	28
	24	04-19	10-25	189	29	30	29	29
	20	04-04	11-10	220	30	30	30	30
	16	03-26	11-24	243	30	30	30	30
EMPORIUM 1 E	32	05-19	09-29	133	29	29	30	30
	28	05-08	10-16	160	29	29	29	29
	24	04-24	10-24	183	29	29	29	29
	20	04-12	11-08	209	29	29	29	29
	16	03-26	11-19	238	29	29	29	29
ERIE WSO	32	04-20	11-07	200	30	30	30	30
	28	04-06	11-14	223	30	30	30	30
	24	03-28	11-30	247	30	30	30	30
	20	03-16	12-07	267	30	30	28	28
	16	03-10	12-10	276	30	30	27	27
FRANKLIN	32	05-14	10-05	144	30	30	30	30
	28	04-27	10-22	178	30	30	30	30
	24	04-16	11-06	204	30	30	30	30
	20	04-06	11-18	226	30	30	30	30
	16	03-18	11-30	257	30	30	30	30
FREELAND	32	05-07	10-09	155	30	30	30	30
	28	04-25	10-23	181	30	30	30	30
	24	04-15	11-03	201	30	30	30	30
	20	04-06	11-16	224	30	30	30	30
	16	03-26	11-25	244	30	30	30	30
GEORGE SCHOOL	32	04-25	10-16	174	30	30	30	30
	28	04-12	10-31	202	30	30	30	30
	24	03-25	11-11	231	30	30	30	30
	20	03-15	11-27	258	30	30	30	30
	16	03-05	12-07	278	30	30	30	30
GETTYSBURG	32	04-23	10-19	179	30	30	30	30
	28	04-08	11-03	209	30	30	30	30
	24	03-21	11-18	241	30	30	30	30
	20	03-13	12-01	263	30	30	30	30
	16	03-04	12-11	282	30	30	30	28
GORDON	32	05-17	09-27	133	30	30	30	30
	28	05-04	10-09	158	30	30	30	30
	24	04-20	10-25	187	30	30	30	30
	20	04-04	11-06	217	30	30	30	30
	16	03-20	11-20	245	30	30	30	30
GREENSBURG 2 S	32	05-07	10-06	152	30	30	30	30
	28	04-22	10-19	180	30	30	30	30
	24	04-10	11-05	208	30	30	30	30
	20	03-29	11-21	237	30	30	30	30
	16	03-14	11-29	260	30	30	30	30
GREENVILLE	32	05-22	09-30	131	29	29	30	30
	28	05-03	10-12	162	30	30	30	30
	24	04-21	10-31	193	29	29	30	30
	20	04-07	11-13	220	29	29	30	30
	16	03-23	12-01	253	29	29	28	26
HANOVER	32	04-20	10-21	184	30	30	30	30
	28	04-07	11-04	212	30	30	30	30
	24	03-22	11-21	244	30	30	30	30
	20	03-13	12-03	265	30	30	29	29
	16	03-06	12-11	281	29	29	29	26
HAWLEY	32	05-19	09-26	130	30	30	31	31
	28	05-05	10-08	155	30	30	31	31
	24	04-21	10-21	183	30	30	30	30
	20	04-09	11-01	206	30	30	30	30
	16	03-25	11-18	238	30	30	30	30
HAWLEY 1 S DAM	32	05-18	09-18	123	14	14	15	15
	28	05-11	10-01	143	11	11	15	15
	24	04-21	10-15	177	6	6	12	12
	20	04-07	10-21	196	5	5	10	10
	16	03-30	11-06	221	5	5	7	7
HOLTWOOD	32	04-05	11-08	217	30	30	30	30
	28	03-23	11-22	243	30	30	30	30
	24	03-11	12-03	267	30	30	29	29
	20	03-06	12-10	280	30	30	30	29
	16	02-24	12-19	299	30	29	30	22
HUNTINGDON	32	05-14	10-05	145	30	30	30	30
	28	04-28	10-18	172	30	30	30	30
	24	04-15	10-28	197	30	30	30	30
	20	03-29	11-13	229	30	30	30	30
	16	03-14	11-25	255	30	30	30	30
INDIANA	32	05-14	10-07	145	25	25	24	24
	28	04-30	10-11	164	25	25	24	24
	24	04-17	10-30	196	25	25	25	25
	20	04-06	11-13	222	24	24	24	24
	16	03-20	11-28	253	24	24	22	22
IRWIN	32	05-11	10-07	149	28	28	29	29
	28	05-01	10-16	168	28	28	29	29
	24	04-17	10-30	195	28	28	27	27
	20	03-29	11-15	231	27	27	27	27
	16	03-14	11-29	261	27	27	26	26

FREEZE DATA

STATION	Freeze threshold temperature	Mean date of last Spring occurrence	Mean date of first Fall occurrence	Mean No. of days between dates	Years of record Spring	No. of occurrences in Spring	Years of record Fall	No. of occurrences in Fall
JOHNSTOWN 1	32	04-28	10-14	169	29	29	30	30
	28	04-20	10-27	190	29	29	30	30
	24	03-30	11-13	228	28	28	30	30
	20	03-21	11-26	250	28	28	30	30
	16	03-10	12-06	271	28	28	30	30
KANE 1 NNE	32	06-02	09-13	104	18	18	19	19
	28	05-16	10-01	138	18	18	19	19
	24	05-01	10-11	163	18	18	19	19
	20	04-22	10-19	180	18	18	19	19
	16	04-11	11-09	212	18	18	19	19
LANCASTER 2 NE PUMP	32	05-04	10-09	158	30	30	30	30
	28	04-24	10-22	180	29	29	30	30
	24	04-07	11-05	212	29	29	30	30
	20	03-21	11-19	242	29	29	30	30
	16	03-08	12-07	274	30	30	30	29
LAWRENCEVILLE 1 S	32	05-22	09-28	129	30	30	30	30
	28	05-05	10-11	159	29	29	30	30
	24	04-20	10-22	186	29	29	30	30
	20	04-06	11-05	214	30	30	30	30
	16	03-20	11-23	247	30	30	30	30
LEBANON	32	04-22	10-17	179	30	30	30	30
	28	04-08	10-30	205	30	30	30	30
	24	03-22	11-16	239	30	30	30	30
	20	03-12	12-02	265	30	30	30	29
	16	03-06	12-10	279	30	30	30	28
LEWISTOWN	32	04-25	10-16	174	11	11	12	12
	28	04-12	10-21	192	11	11	12	12
	24	04-04	11-09	220	9	9	12	12
	20	03-22	11-29	252	9	9	12	12
	16	03-12	12-11	274	9	9	12	11
LOCK HAVEN	32	05-04	10-12	162	25	25	27	27
	28	04-18	10-24	189	25	25	27	27
	24	04-03	11-08	219	25	25	27	27
	20	03-20	11-21	246	25	25	27	27
	16	03-12	12-02	265	25	25	27	27
MAUCH CHUNK 1 SW	32	05-09	10-01	145	28	28	28	28
	28	04-25	10-16	174	29	29	29	29
	24	04-07	10-28	204	29	29	29	29
	20	03-27	11-12	230	28	28	29	29
	16	03-15	11-28	258	28	28	29	29
MEADVULLE 1 S	32	05-18	10-04	139	22	22	23	23
	28	04-30	10-21	174	22	22	23	23
	24	04-17	11-07	204	22	22	23	23
	20	04-04	11-20	230	22	22	23	23
	16	03-23	12-02	254	22	22	23	22
MERCER AP	32	05-19	10-08	142	11	11	11	11
	28	05-05	10-17	165	11	11	11	11
	24	04-22	11-08	200	11	11	11	11
	20	04-03	11-18	230	11	11	11	11
	16	03-23	11-28	250	11	11	11	11
MIDDLETOWN OLMSTED F	32	04-24	10-24	184	10	10	10	10
	28	04-01	11-08	221	10	10	10	10
	24	03-22	11-22	245	10	10	10	10
	20	03-12	12-01	264	10	10	10	10
	16	03-05	12-14	284	10	10	10	10
MIDLAND DAM LOWER 7	32	05-02	10-19	170	13	13	13	13
	28	04-13	11-05	205	13	13	13	13
	24	03-27	11-21	239	13	13	13	13
	20	03-16	11-28	256	13	13	13	13
	16	03-10	12-10	275	13	13	13	12
MONTROSE 3 E	32	05-15	10-02	139	28	28	27	27
	28	04-30	10-17	170	28	28	26	26
	24	04-18	10-30	195	28	28	27	27
	20	04-09	11-11	216	28	28	28	28
	16	03-29	11-22	238	27	27	28	28
MT POCONO	32	05-17	09-30	136	24	24	27	27
	28	05-05	10-14	162	23	23	27	27
	24	04-15	10-28	196	23	23	24	24
	20	04-04	11-06	216	24	24	24	24
	16	03-24	11-15	237	23	23	23	23
NEW CASTLE 1 N	32	05-15	10-05	143	30	30	30	30
	28	04-29	10-21	175	30	30	30	30
	24	04-15	11-05	204	30	30	30	30
	20	03-28	11-21	238	30	30	30	30
	16	03-14	12-03	264	30	30	30	29
PALMERTON	32	05-04	10-09	159	30	30	30	30
	28	04-17	10-19	186	30	30	30	30
	24	04-01	11-07	221	30	30	30	30
	20	03-18	11-21	248	30	30	30	30
	16	03-09	12-02	268	30	30	30	29
PHILADELPHIA SHAWMNT	32	04-18	10-26	192	30	30	30	30
	28	03-31	11-09	223	30	30	30	30
	24	03-18	11-23	250	30	30	30	30
	20	03-09	12-06	272	30	30	30	29
	16	03-01	12-12	287	30	30	30	28
PHILADELPHIA CITY	32	03-30	11-17	232	30	30	30	30
	28	03-19	11-29	255	30	30	30	30
	24	03-11	12-06	270	30	30	29	29
	20	03-04	12-13	284	30	30	30	26
	16	02-17	12-20	306	30	27	30	20
PHOENIXVILLE 1 E	32	05-03	10-11	161	30	30	30	30
	28	04-16	10-23	190	30	30	28	28
	24	03-30	11-06	221	30	30	28	28
	20	03-17	11-24	252	30	30	28	28
	16	03-08	12-04	271	30	30	27	26
PITTSBURGH ALLEG CO AP	32	04-20	10-23	187	15	15	16	16
	28	04-07	11-10	217	15	15	16	16
	24	03-28	11-21	238	15	15	16	16
	20	03-21	11-29	254	15	15	16	16
	16	03-14	12-06	267	15	15	16	16
PITTSBURGH WSO	32	04-16	11-03	200	30	30	30	30
	28	03-31	11-16	230	30	30	30	30
	24	03-21	11-26	250	30	30	30	30
	20	03-13	12-07	270	30	30	30	27
	16	03-03	12-14	286	30	30	30	26
PORT CLINTON 1 S	32	05-06	10-10	158	28	28	29	29
	28	04-19	10-22	186	28	28	28	28
	24	04-02	11-05	217	29	29	28	28
	20	03-17	11-19	247	29	29	27	27
	16	03-08	12-02	269	29	29	27	27
QUAKERTOWN	32	05-08	10-04	149	30	30	30	30
	28	04-21	10-18	180	30	30	30	30
	24	04-04	10-27	207	30	30	30	30
	20	03-18	11-16	243	30	30	29	29
	16	03-13	12-02	264	30	30	29	28
READING 3 N	32	04-13	10-29	198	30	30	30	30
	28	03-27	11-17	235	30	30	30	30
	24	03-16	11-30	260	30	30	30	30
	20	03-09	12-08	274	30	30	29	28
	16	03-04	12-15	286	30	30	27	25
RETREAT 1 SW	32	05-07	10-09	155	16	16	18	18
	28	04-22	10-20	180	16	16	18	18
	24	04-04	11-03	214	16	16	18	18
	20	03-28	11-15	231	17	17	18	18
	16	03-17	11-30	258	17	17	18	18
RIDGWAY	32	05-28	09-24	119	29	29	28	28
	28	05-14	10-06	145	28	28	28	28
	24	04-30	10-19	172	28	28	28	28
	20	04-17	10-30	196	28	28	28	28
	16	04-03	11-18	229	26	26	28	28
SCRANTON	32	04-24	10-14	174	30	30	30	30
	28	04-11	10-31	203	30	30	30	30
	24	03-26	11-19	237	30	30	30	30
	20	03-18	11-29	257	30	30	30	30
	16	03-10	12-06	271	30	30	30	29
SELINSGROVE FAA AP	32	05-04	10-09	158	29	29	30	30
	28	04-20	10-22	185	29	29	30	30
	24	04-03	11-07	218	28	28	30	30
	20	03-21	11-19	244	28	28	29	29
	16	03-12	12-02	264	28	28	29	29

FREEZE DATA

STATION	Freeze threshold temperature	Mean date of last Spring occurrence	Mean date of first Fall occurrence	Mean No. of days between dates	Years of record Spring	No. of occurrences in Spring	Years of record Fall	No. of occurrences in Fall	STATION	Freeze threshold temperature	Mean date of last Spring occurrence	Mean date of first Fall occurrence	Mean No. of days between dates	Years of record Spring	No. of occurrences in Spring	Years of record Fall	No. of occurrences in Fall
SOMERSET WATER WORKS	32	05-24	09-21	120	29	29	28	28	WARREN	32	05-14	10-06	146	30	30	30	30
	28	05-14	10-04	143	28	28	28	28		28	04-30	10-22	175	30	30	30	30
	24	04-30	10-15	168	28	28	28	28		24	04-15	11-09	208	30	30	30	30
	20	04-15	10-27	195	28	28	28	28		20	04-04	11-19	229	30	30	30	30
	16	03-27	11-12	230	28	28	28	28		16	03-21	11-28	252	30	30	30	30
SPRINGS 1 SW	32	05-25	09-29	126	25	25	27	27	WELLSBORO 3 S	32	05-27	09-22	118	30	30	29	29
	28	05-06	10-09	156	25	25	27	27		28	05-12	10-05	146	30	30	30	30
	24	04-23	10-22	183	26	26	27	27		24	04-24	10-16	176	30	30	30	30
	20	04-13	11-04	206	26	26	27	27		20	04-14	10-27	196	30	30	30	30
	16	03-29	11-18	234	26	26	27	27		16	03-27	11-14	232	30	30	29	29
STATE COLLEGE	32	04-29	10-12	166	30	30	30	30	WEST CHESTER	32	04-18	10-25	189	30	30	29	29
	28	04-17	10-21	187	30	30	30	30		28	04-04	11-05	215	30	30	29	29
	24	04-05	11-10	219	30	30	30	30		24	03-21	11-23	247	30	30	29	29
	20	03-23	11-19	241	30	30	30	30		20	03-12	12-02	265	30	30	29	29
	16	03-12	12-05	268	30	30	30	30		16	03-05	12-12	282	30	30	29	26
TOWANDA	32	05-12	10-02	143	30	30	30	30	WILLIAMSPORT WSO	32	05-03	10-13	164	30	30	30	30
	28	04-29	10-15	169	30	30	30	30		28	04-17	10-25	191	30	30	30	30
	24	04-14	10-25	194	30	30	30	30		24	03-30	11-09	224	30	30	30	30
	20	03-30	11-13	228	30	30	30	30		20	03-19	11-24	250	30	30	30	30
	16	03-17	11-27	254	30	30	30	30		16	03-10	12-05	271	30	30	30	29
UNIONTOWN	32	04-30	10-15	167	30	30	30	30	YORK 3 SW PUMP STA	32	05-02	10-09	160	30	30	30	30
	28	04-18	10-27	193	30	30	30	30		28	04-20	10-23	186	30	30	30	30
	24	04-04	11-12	222	30	30	30	30		24	04-02	11-08	220	30	30	30	30
	20	03-19	11-26	252	30	30	30	30		20	03-17	11-21	249	30	30	30	30
	16	03-11	12-04	268	30	30	30	29		16	03-08	12-07	274	30	30	30	30

Data in the above table are based on the period 1921-1950, or that portion of this period for which data are available.

Means have been adjusted to take into account years of non-occurrence.

A freeze is a numerical substitute for the former term "killing frost" and is the occurrence of a minimum temperature at or below the threshold temperature of 32°, 28°, etc.

Freeze data tabulations in greater detail are available and can be reproduced at cost.

*NORMALS BY CLIMATOLOGICAL DIVISIONS

Taken from "Climatography of the United States No. 81-4, Decennial Census of U. S. Climate"

TEMPERATURE (°F)   **PRECIPITATION (In.)**

STATIONS (By Divisions)	T JAN	T FEB	T MAR	T APR	T MAY	T JUNE	T JULY	T AUG	T SEPT	T OCT	T NOV	T DEC	T ANN	P JAN	P FEB	P MAR	P APR	P MAY	P JUNE	P JULY	P AUG	P SEPT	P OCT	P NOV	P DEC	P ANN
POCONO MOUNTAINS																										
FREELAND	25.4	25.8	33.4	45.8	57.7	63.9	70.3	68.4	61.6	51.1	38.8	27.7	47.5	3.09	2.72	3.82	4.10	4.84	4.20	5.38	4.27	3.95	3.68	4.15	3.77	47.97
GOULDSBORO														3.10	2.89	3.84	4.09	4.60	3.89	4.86	4.41	3.81	3.83	4.10	3.48	46.90
HAWLEY														2.79	2.62	3.42	3.58	4.03	3.47	4.35	4.28	3.70	3.38	3.33	3.02	41.97
HOLLISTERVILLE														2.93	2.85	3.66	3.97	4.08	3.93	4.81	4.00	3.50	3.49	3.56	3.20	43.98
LAKEVILLE 2 NNE														2.72	2.76	3.59	3.81	4.12	3.84	4.91	4.35	3.73	3.69	3.56	3.12	44.20
MATAMORAS														3.02	2.65	3.45	3.92	3.96	3.91	4.29	4.24	3.99	3.42	3.65	3.29	43.79
PAUPACK 2 WNW														2.89	2.77	3.61	3.83	4.07	3.67	4.62	4.38	3.64	3.60	3.60	3.20	43.88
PLEASANT MOUNT 1 W														3.34	2.95	3.83	4.09	4.32	4.06	4.63	4.12	4.18	3.82	3.96	3.50	46.80
SCRANTON	28.4	28.6	36.5	48.5	59.6	68.3	70.5	70.7	63.4	53.0	41.8	30.8	50.0	2.24	2.12	2.88	3.43	3.62	3.73	4.81	3.86	3.03	2.98	3.01	2.44	38.15
STROUDSBURG	28.0	28.8	36.6	48.7	59.3	68.0	72.7	70.5	63.0	52.5	41.4	30.6	50.0	3.36	2.86	4.12	4.15	4.01	4.21	4.70	4.57	4.29	3.65	4.15	3.89	47.96
TOBYHANNA														3.60	3.09	4.46	4.34	4.52	4.33	5.23	5.25	4.40	4.32	4.30	3.97	51.81
WILKES-BARRE 4 NE														2.31	2.27	2.89	3.38	3.73	3.66	4.64	4.03	3.36	3.28	3.07	2.75	39.37
W-BARRE-SCRANTON WSO	27.7	28.3	36.2	48.4	59.6	68.2	72.4	70.0	62.5	51.0	39.6	29.4	49.4	2.29	1.99	2.82	3.27	3.95	3.91	4.79	3.58	2.97	3.50	2.94	2.47	38.48
DIVISION	25.9	26.2	34.0	46.3	57.3	65.6	70.0	68.1	60.9	50.8	39.4	28.3	47.7	2.93	2.61	3.67	3.90	4.16	3.97	4.81	4.31	3.81	3.61	3.70	3.26	44.74
EAST CENTRAL MOUNTAIN																										
ALLENTOWN WSO	29.0	29.2	37.6	49.3	60.4	69.4	74.1	72.0	64.7	53.8	41.9	31.3	51.1	3.17	2.64	3.79	3.76	4.08	4.07	4.82	4.47	3.75	2.97	3.33	3.27	44.17
BETHLEHEM LEHIGH UNIV	31.1	32.1	39.8	51.5	62.2	71.1	75.5	73.5	66.8	56.2	44.5	33.5	53.2	3.11	2.59	3.60	3.63	3.85	3.72	4.66	4.44	3.56	2.99	3.36	3.25	42.76
PALMERTON	29.2	29.6	37.4	49.0	59.7	68.4	73.1	70.9	63.5	52.9	41.6	31.3	50.6	2.96	2.49	3.56	3.49	3.88	3.88	5.10	4.75	3.99	3.24	3.49	3.24	44.07
PORT CLINTON	29.6	30.0	37.5	49.0	59.7	68.2	72.7	70.6	63.4	52.8	41.8	31.4	50.6	3.29	2.81	4.20	4.11	4.51	3.66	4.74	4.45	4.32	3.57	3.75	3.58	46.99
TAMAQUA 4 N DAM														3.32	2.89	4.17	4.26	4.74	4.12	5.03	4.94	4.19	3.96	4.18	3.71	49.51
DIVISION	29.2	29.7	37.3	49.1	59.9	68.5	73.0	70.9	63.9	53.3	41.9	31.2	50.7	3.16	2.67	3.87	3.85	4.32	3.88	5.01	4.54	3.85	3.34	3.67	3.44	45.60
SOUTHEASTERN PIEDMONT																										
COATESVILLE 1 SW	31.4	31.6	39.2	50.7	61.3	70.4	74.7	72.8	65.7	54.7	43.3	32.9	52.4	3.30	2.78	4.11	3.49	4.17	4.42	4.26	5.06	3.49	3.19	3.55	3.39	45.21
CONSHOHOCKEN														3.25	2.86	3.99	3.52	3.97	3.77	4.45	4.62	3.68	3.12	3.61	3.20	44.04
DOYLESTOWN														3.09	2.54	3.60	3.36	4.10	3.81	4.70		3.57	3.12	3.48	3.15	43.21
EPHRATA	32.1	32.9	40.7	52.0	62.8	71.4	75.8	73.7	66.6	55.6	44.3	33.8	53.5	2.98	2.52	3.65	3.41	3.83	4.20	4.74	4.71	3.76	3.20	3.30	3.07	43.37
GEORGE SCHOOL	31.8	32.4	39.8	51.0	61.5	70.1	74.8	72.9	66.2	55.6	44.4	33.5	52.8	3.34	2.70	4.07	3.64	3.95	3.75	4.94	4.74	3.66	3.18	3.76	3.35	45.08
GRATERFORD														3.25	2.71	3.99	3.49	4.26	3.84	4.85	4.43	3.47	3.18	3.55	3.28	44.30
HOLTWOOD	32.8	33.2	40.7	52.1	63.4	72.5	77.5	75.7	68.8	57.4	45.3	34.7	54.5	2.63	2.31	3.34	3.09	3.25	3.50	3.79	4.21	3.21	2.66	2.94	2.73	37.66
LANCASTER 2 NE PMP STA	31.6	32.5	40.6	51.6	62.2	70.3	74.4	72.4	65.4	54.0	43.1	32.8	52.6	2.89	2.48	3.81	3.60	3.90	3.90	4.89	4.94	3.51	3.22	3.16	2.99	43.29
LEBANON 3 W	30.3	30.9	38.7	50.4	61.5	70.4	75.1	72.9	65.5	54.5	42.6	31.9	52.1	3.10	2.59	3.75	3.65	4.23	3.93	4.39	4.27	3.89	3.55	3.36	3.28	43.99
MARCUS HOOK	35.8	36.5	43.9	54.6	65.6	74.6	79.0	77.1	70.2	59.8	47.9	38.0	56.9	3.36	2.82	3.98	3.64	3.73	3.54	3.88	5.06	3.22	2.87	3.51	3.12	42.73
NESHAMINY FALLS														3.26	2.83	4.27	3.64	3.97	3.96	4.71	4.98	3.56	3.33	3.66	3.17	45.34
PHILADELPHIA WSO	32.3	33.2	41.0	52.0	62.6	71.0	75.6	73.6	66.7	55.7	44.3	33.9	53.5	3.32	2.80	3.80	3.40	3.74	4.05	4.16	4.63	3.46	2.78	3.40	2.94	42.48
PHOENIXVILLE 1 E	32.9	33.5	41.5	52.8	63.2	71.7	76.3	74.1	67.3	56.3	45.1	34.4	54.1	3.44	2.82	4.22	3.50	4.17	3.82	4.47	4.98	3.55	3.21	3.79	3.40	45.37
QUAKERTOWN	30.0	30.6	38.6	50.0	60.4	68.7	73.4	70.6	63.8	53.2	42.5	31.9	51.1	3.38	2.71	4.00	3.89	4.16	4.19	4.63	4.85	3.74	3.47	3.71	3.45	46.18
READING 3 N	32.7	33.4	41.3	52.6	63.3	72.2	76.9	74.8	67.5	57.2	45.1	34.7	54.3	3.07	2.64	3.78	3.42	3.79	3.72	4.26	4.05	3.32	2.84	3.40	3.14	41.43
WEST CHESTER 1 W	33.0	33.5	41.0	52.1	62.5	71.0	75.5	73.5	66.9	56.4	45.2	34.7	53.8	3.43	2.90	4.18	3.43	4.17	3.85	4.73	5.13	3.80	3.16	3.88	3.37	46.03
WEST GROVE 1 SE														3.39	2.75	4.24	3.56	4.21	3.69	4.59	5.27	3.46	3.00	3.52	3.35	45.03
YORK HAVEN														2.72	2.36	3.47	3.47	4.06	3.68	3.84	3.94	3.04	3.24	3.16	2.91	39.89
DIVISION	32.4	33.0	40.6	51.8	62.6	71.1	75.7	73.7	66.8	56.1	44.7	34.2	53.6	3.21	2.67	3.88	3.51	3.99	3.90	4.40	4.66	3.52	3.16	3.48	3.15	43.53
LOWER SUSQUEHANNA																										
ARENDTSVILLE	30.7	31.3	39.0	50.8	61.5	70.2	74.6	72.8	65.5	54.4	42.5	32.1	52.1	2.96	2.49	3.91	3.74	4.29	3.94	3.85	4.34	3.51	3.57	3.30	3.15	43.05
BLOSERVILLE 1 N														2.87	2.26	3.68	3.87	4.48	4.08	4.27	4.85	3.46	3.52	3.52	2.98	43.84
CARLISLE	31.6	32.7	40.9	52.6	63.3	71.7	75.8	73.7	66.5	55.4	43.5	33.0	53.4	3.07	2.52	3.78	3.73	4.13	3.87	3.96	4.05	3.11	3.31	3.26	5.13	41.92
CHAMBERSBURG 1 ESE	31.6	32.6	40.1	51.4	62.1	70.8	74.9	73.0	65.9	54.7	43.0	33.0	52.8	3.02	2.32	3.77	3.46	4.10	4.08	3.95	4.08	3.30	3.15	3.12	2.98	41.33
GETTYSBURG	32.7	33.6	41.2	52.6	63.0	71.5	75.7	73.8	66.9	56.1	44.6	34.2	53.8	2.92	2.48	3.84	3.51	4.07	3.52	4.15	4.22	3.32	3.31	3.26	2.95	41.55
HANOVER	32.8	33.4	40.8	52.3	63.0	71.5	75.8	73.8	66.9	56.1	44.8	34.3	53.8	3.01	2.61	3.82	3.52	4.03	3.76	4.09	4.14	3.08	3.27	3.07	3.10	41.50
HARRISBURG WSO	31.3	32.6	40.3	51.8	62.7	71.3	76.2	74.1	66.9	55.7	43.4	33.0	53.3	2.76	2.31	3.43	3.02	3.90	3.42	3.51	3.65	2.82	2.97	2.95	2.91	37.65
MERCERSBURG														2.73	2.35	3.63	3.35	4.27	4.00	3.57	3.56	3.11	3.33	2.98	2.87	39.75
NEW PARK														3.15	2.60	3.94	3.77	4.39	4.10	4.35	5.22	3.64	3.46	3.46	3.27	45.35
SPRING GROVE														2.97	2.55	3.76	3.53	4.09	3.71	3.88	4.56	3.24	3.30	3.08	3.08	41.75
YORK 3 SSW PUMP STA	32.8	33.7	41.5	52.6	63.3	71.5	75.7	73.9	67.1	55.8	44.4	34.0	53.9	2.97	2.38	3.62	3.52	4.34	3.81	3.84	4.75	3.35	3.38	3.17	2.87	42.00
DIVISION	31.8	32.7	40.3	51.8	62.5	71.0	75.3	73.4	66.3	55.3	43.6	33.3	53.1	2.92	2.39	3.72	3.50	4.17	3.78	3.86	4.16	3.18	3.26	3.18	3.03	41.15
MIDDLE SUSQUEHANNA																										
BEAR GAP														2.80	2.47	3.73	3.55	4.41	4.01	4.83	4.01	3.50	3.51	3.71	3.22	43.75
NEWPORT														2.90	2.37	3.55	3.56	4.07	3.71	4.19	4.11	3.03	3.44	3.37	3.05	41.35
SHAMOKIN														2.82	2.63	3.73	3.49	4.49	4.58	3.91		3.43	3.32	3.54	3.27	43.06
SUNBURY														3.04	2.58	3.62	3.65	4.60	3.45	3.85	3.86	3.27	3.23	3.36	3.22	41.73
WILLIAMSPORT WSO	28.8	29.2	37.4	49.4	60.4	69.2	73.6	71.6	64.1	53.1	41.1	30.4	50.7	2.67	2.51	3.73	3.55	4.08	3.23	4.18	3.62	3.27	3.32	3.45	3.04	40.65
DIVISION	29.4	30.2	38.0	49.8	60.7	69.2	73.4	71.6	64.5	53.5	41.8	31.1	51.1	2.70	2.28	3.60	3.54	4.28	3.48	3.87	3.88	3.19	3.23	3.31	2.90	40.26

*NORMALS BY CLIMATOLOGICAL DIVISIONS

Taken from "Climatography of the United States No. 81-4, Decennial Census of U. S. Climate"

TEMPERATURE (°F)   PRECIPITATION (In.)

STATION	JAN	FEB	MAR	APR	MAY	JUNE	JULY	AUG	SEPT	OCT	NOV	DEC	ANN	JAN	FEB	MAR	APR	MAY	JUNE	JULY	AUG	SEPT	OCT	NOV	DEC	ANN
UPPER SUSQUEHANNA																										
LAWRENCEVILLE	26.1	26.2	33.9	46.4	57.6	66.7	70.6	68.6	61.7	51.0	39.0	28.2	48.0	1.77	1.67	2.79	2.68	3.91	3.30	3.97	3.95	3.03	2.79	2.36	1.97	34.19
TOWANDA 1 ESE	27.4	27.6	35.4	47.5	58.3	67.0	71.3	69.5	62.5	51.8	40.6	29.5	49.0	1.86	1.93	2.86	3.09	4.02	3.18	3.84	3.33	3.49	2.95	2.62	2.17	35.34
DIVISION	25.1	25.2	32.9	45.3	56.5	65.3	69.7	67.8	60.7	50.2	38.6	27.4	47.1	2.22	2.15	3.07	3.28	4.10	3.38	3.98	3.93	3.43	3.11	2.91	2.51	38.07
CENTRAL MOUNTAINS																										
CLEARFIELD	•	•	•	•	•	•	•	•	•	•	•	•	•	3.19	2.62	4.00	3.97	4.39	4.13	4.30	3.78	3.04	3.04	3.11	3.09	42.66
LOCK HAVEN	29.4	30.1	38.0	50.8	61.4	69.7	73.5	71.7	64.6	53.5	41.3	31.0	51.3	2.57	2.21	3.67	3.39	4.31	3.48	4.33	3.89	3.14	3.14	3.12	2.77	40.02
RENOVO 1 W	•	•	•	•	•	•	•	•	•	•	•	•	•	2.46	2.19	3.46	3.38	4.31	3.72	4.20	3.18	2.67	3.03	2.66		38.28
RIDGWAY	25.9	25.0	32.5	45.1	56.1	64.8	68.2	66.8	59.9	49.2	37.8	27.3	46.6	2.94	2.39	3.44	3.75	4.17	3.99	4.44	3.50	3.25	3.00	3.06	2.87	40.80
STATE COLLEGE	28.7	29.3	36.6	48.7	59.9	68.2	72.1	70.2	62.9	52.6	41.0	30.5	50.1	2.67	2.19	3.70	3.51	4.39	3.57	3.78	3.53	2.62	2.93	3.05	2.76	38.70
DIVISION	26.8	27.0	34.5	46.8	57.8	66.1	70.0	68.3	61.3	50.8	39.1	28.5	48.1	2.82	2.38	3.71	3.60	4.41	3.80	4.26	3.77	3.03	3.06	3.09	2.86	40.79
SOUTH CENTRAL MOUNTAI																										
ALTOONA HORSESHOE CRVE	28.5	29.1	36.4	48.6	59.1	67.3	70.9	69.4	62.8	52.7	40.6	30.2	49.6	3.23	2.47	4.27	4.20	4.47	4.57	4.54	3.72	3.06	3.21	3.06	3.03	43.83
BUFFALO MILLS	•	•	•	•	•	•	•	•	•	•	•	•	•	2.69	2.03	3.53	3.28	3.94	3.65	3.68	3.62	2.77	2.97	2.53	2.40	37.09
EBENSBERG	26.9	27.2	34.2	46.1	56.8	65.1	68.8	67.3	61.1	50.8	38.6	28.6	47.6	3.49	2.90	4.06	4.01	4.61	4.71	4.50	4.09	3.09	2.97	3.11	3.46	45.00
HUNTINGDON	30.0	30.6	37.8	49.6	60.4	68.8	72.8	71.1	64.1	53.2	41.7	31.4	51.0	2.69	2.11	3.72	3.54	4.14	3.76	4.14	3.96	3.00	2.83	3.02	2.73	39.64
HYNDMAN	•	•	•	•	•	•	•	•	•	•	•	•	•	2.55	1.97	3.42	3.10	3.59	3.55	3.41	3.83	2.57	2.65	2.41	2.42	35.47
JOHNSTOWN	30.8	31.4	38.6	50.2	60.9	69.6	73.1	71.5	62.7	54.3	42.5	32.4	51.5	3.64	3.07	4.24	4.25	4.52	4.45	4.70	4.10	2.99	2.74	2.77	3.30	44.77
DIVISION	29.5	30.1	37.4	49.1	59.7	68.0	71.8	70.2	63.3	52.8	41.0	31.0	50.3	2.98	2.41	3.85	3.68	4.22	4.11	4.20	3.84	2.88	2.92	2.84	2.92	40.85
SOUTHWEST PLATEAU																										
ACMETONIA LOCK 3	•	•	•	•	•	•	•	•	•	•	•	•	•	3.20	2.66	3.70	3.76	3.98	4.20	4.12	3.85	2.76	2.79	2.69	2.79	40.50
BEAVER FALLS	•	•	•	•	•	•	•	•	•	•	•	•	•	2.67	2.18	3.10	3.21	3.72	3.74	3.65	3.45	2.67	2.42	2.31	2.24	35.36
BUTLER	30.3	30.5	37.9	49.7	59.9	68.1	72.6	70.6	64.1	53.3	41.6	31.6	50.9	2.96	2.57	3.45	3.76	3.98	4.06	3.95	3.46	3.08	2.95	2.77	2.82	39.81
CHARLEROI LOCK 4	•	•	•	•	•	•	•	•	•	•	•	•	•	3.00	2.44	3.68	3.69	3.90	3.84	3.90	3.78	2.89	2.47	2.58	2.62	38.79
CLAYSVILLE 3 W	31.1	31.6	38.8	49.9	60.1	68.7	71.9	70.4	64.3	53.1	41.4	32.0	51.1	3.07	2.51	3.66	3.49	4.08	4.28	4.29	3.73	3.20	2.53	2.61	2.79	40.24
CONFLUENCE 1 NW	•	•	•	•	•	•	•	•	•	•	•	•	•	3.62	2.87	4.06	3.68	4.41	4.68	4.65	4.40	3.05	2.89	2.75	3.17	44.23
CONNELLSVILLE	•	•	•	•	•	•	•	•	•	•	•	•	•	2.96	2.35	3.37	3.59	4.46	4.46	4.84	4.30	3.21	2.95	2.45	2.45	41.37
CORAOPOLIS NEVILLE IS	•	•	•	•	•	•	•	•	•	•	•	•	•	2.66	2.13	3.23	3.33	3.81	3.77	3.88	3.38	2.69	2.50	2.37	2.31	36.06
CREEKSIDE	•	•	•	•	•	•	•	•	•	•	•	•	•	3.24	2.71	3.67	3.98	4.12	4.59	4.81	3.92	3.20	3.03	2.98	3.05	43.30
DONORA	34.3	34.9	41.8	53.3	63.4	72.3	75.7	74.2	67.9	57.0	45.2	35.8	54.7	2.75	2.30	3.57	3.51	3.82	3.92	4.04	3.91	2.76	2.52	2.35	2.42	37.87
GREENSBORO LOCK 7	•	•	•	•	•	•	•	•	•	•	•	•	•	3.10	2.52	3.64	3.55	4.01	4.26	4.64	4.20	2.96	2.58	2.52	2.57	40.55
MC KEESPORT	•	•	•	•	•	•	•	•	•	•	•	•	•	2.85	2.34	3.42	3.30	3.59	4.08	3.67	3.64	2.76	2.38	2.39	2.48	36.90
NATRONA LOCK 4	•	•	•	•	•	•	•	•	•	•	•	•	•	3.11	2.60	3.71	3.72	3.98	4.29	4.26	3.96	2.83	2.86	2.64	2.67	40.63
NEW CASTLE 1 N	30.0	30.0	37.6	49.4	59.8	69.1	72.6	71.1	64.7	53.7	42.0	31.8	51.0	2.86	2.38	3.38	3.61	4.00	4.05	4.27	3.57	2.88	2.73	2.66	2.56	38.95
PITTSBURGH WSO 2	28.9	29.2	36.8	49.0	59.8	68.4	72.1	70.8	64.2	53.1	40.8	30.7	50.3	2.97	2.19	3.32	3.08	3.91	3.78	3.88	3.31	2.54	2.52	2.24	2.40	36.14
PITTSBURGH CITY WSO	32.1	32.9	40.2	52.1	62.7	71.4	74.9	73.0	66.2	54.7	42.8	33.4	53.0	2.82	2.31	3.52	3.37	3.75	3.95	3.60	3.50	2.67	2.50	2.34	2.54	36.87
SCHENLEY LOCK 5	•	•	•	•	•	•	•	•	•	•	•	•	•	3.19	2.78	4.10	3.80	4.10	4.18	3.89	3.73	2.98	2.77	2.71	2.85	41.08
UNIONTOWN	34.0	34.6	41.4	52.7	62.5	70.7	73.9	72.4	66.5	55.8	44.3	35.1	53.7	3.27	2.58	3.76	3.70	4.52	4.60	4.30	4.00	3.00	2.90	2.71	2.72	42.06
VANDERGRIFT	•	•	•	•	•	•	•	•	•	•	•	•	•	2.68	2.38	3.46	3.55	3.87	4.34	4.09	4.04	3.02	3.04	2.53	2.54	39.54
WAYNESBURG 1 E	32.6	33.1	40.2	51.2	61.0	69.4	72.7	71.3	64.9	54.0	42.6	33.3	52.2	3.08	2.47	3.65	3.59	4.24	4.21	4.32	3.91	3.08	2.66	2.46	2.61	40.28
DIVISION	30.7	31.1	38.2	49.8	60.0	68.6	72.1	70.6	64.2	53.5	41.8	32.1	51.1	3.12	2.58	3.74	3.75	4.22	4.38	4.33	3.95	3.02	2.84	2.68	2.80	41.41
NORTHWEST PLATEAU																										
CORRY	26.0	25.8	33.3	45.9	56.8	66.1	69.9	66.4	61.7	50.9	38.9	28.5	47.5	3.39	2.93	3.80	4.06	4.03	4.36	3.89	3.43	3.58	3.47	4.00	3.52	44.46
ERIE WSO	27.3	26.4	33.6	45.5	56.4	66.7	71.1	69.8	63.3	52.6	41.2	30.7	48.7	2.67	2.32	2.88	3.56	3.54	3.05	3.67	2.98	3.56	3.30	3.36	2.61	37.50
FARRELL SHAPON	30.0	30.3	38.5	50.7	61.7	71.3	74.8	72.9	66.1	54.8	42.4	32.2	52.1	2.83	2.29	2.95	3.39	3.84	3.71	3.58	3.89	2.71	2.45	2.51	2.41	36.56
FRANKLIN	27.9	27.2	34.8	46.4	58.0	66.9	70.8	69.2	62.6	51.8	40.2	30.0	48.8	3.06	2.43	3.24	3.82	4.35	3.87	4.33	3.11	3.13	3.03	3.07	2.57	40.01
GALETON	•	•	•	•	•	•	•	•	•	•	•	•	•	2.59	2.15	3.47	3.29	4.06	3.51	3.90	3.74	3.16	3.12	3.01	2.61	38.61
GREENVILLE	28.4	28.7	36.4	48.2	59.0	68.4	72.1	70.2	63.7	52.7	40.8	30.7	49.9	3.08	2.52	3.29	3.85	4.06	4.13	4.11	3.79	2.96	3.17	3.00	2.58	40.54
JAMESTOWN 2 NW	•	•	•	•	•	•	•	•	•	•	•	•	•	2.76	2.26	2.97	3.63	4.01	3.82	3.92	3.54	3.21	2.97	2.77	2.36	38.22
MEADVILLE 1 S	26.8	26.2	34.2	46.4	57.2	66.7	70.4	68.7	62.1	51.4	39.5	29.1	48.2	2.91	2.52	3.27	3.59	4.01	4.31	4.25	3.46	3.01	3.31	3.09	2.71	40.44
WARREN	27.6	27.3	34.4	46.4	57.6	66.7	70.5	69.0	62.6	52.0	40.1	29.8	48.7	2.86	2.56	3.51	3.61	4.27	4.67	4.29	3.50	3.43	3.41	3.59	3.03	42.73
DIVISION	27.0	26.7	34.3	46.4	57.3	66.5	70.3	68.6	62.1	51.5	39.5	29.1	48.3	3.04	2.58	3.38	3.75	4.15	4.09	4.14	3.46	3.22	3.15	3.24	2.84	41.04

* Normals for the period 1931-1960. Divisional normals may not be the arithmetical average of individual stations published, since additional data for shorter period stations are used to obtain better areal representation.

*NORMALS BY CLIMATOLOGICAL DIVISIONS

Taken from "Climatography of the United States No. 81-4,   Decennial Census of U. S. Climate"

TEMPERATURE (°F)													PRECIPITATION (In.)												
JAN	FEB	MAR	APR	MAY	JUNE	JULY	AUG	SEPT	OCT	NOV	DEC	ANN	JAN	FEB	MAR	APR	MAY	JUNE	JULY	AUG	SEPT	OCT	NOV	DEC	ANN

CONFIDENCE - LIMITS

In the absence of trend or record changes, the chances are 9 out of 10 that the true mean will lie in the interval formed by adding and subtracting the values in the following table from the means for any station in the State.   Because of the wider variation in mean precipitation, the corresponding monthly means and annual mean must be substituted for "p" in the precipitation table below to obtain mean precipitation confidence limits.

| 1.6 | 1.4 | 1.5 | 1.0 | .9 | .7 | .5 | .7 | .9 | .9 | .8 | 1.1 | .4 | $.22\sqrt{p}$ | $.19\sqrt{p}$ | $.22\sqrt{p}$ | $.24\sqrt{p}$ | $.28\sqrt{p}$ | $.27\sqrt{p}$ | $.26\sqrt{p}$ | $.28\sqrt{p}$ | $.35\sqrt{p}$ | $.38\sqrt{p}$ | $.29\sqrt{p}$ | $.25\sqrt{p}$ | $.29\sqrt{p}$ |

COMPARATIVE DATA

Data in the following table are the mean temperature and average precipitation for Huntingdon, Pennsylvania, for the period 1906-1930 and are included in this publication for comparative purposes.

| 28.7 | 30.0 | 39.4 | 49.8 | 59.9 | 68.1 | 72.3 | 70.3 | 64.9 | 53.2 | 41.8 | 31.2 | 50.8 | 3.30 | 2.77 | 3.14 | 3.73 | 3.62 | 4.16 | 3.88 | 3.67 | 3.39 | 2.87 | 2.19 | 2.83 | 39.55 |

NORMALS, MEANS AND EXTREMES

(Table Revised 1971. Base Period for Climatological Normals: 1931-1960)

Station: ALLENTOWN, PENNSYLVANIA   ALLENTOWN-BETHLEHEM-EASTON AP   Standard time used: EASTERN   Latitude: 40°39'N   Longitude: 75°26'W   Elevation (ground): 387 feet

Month	Normal Daily maximum	Daily minimum	Monthly	Extremes Record highest	Year	Record lowest	Year	Normal total heating degree days (Base 65°)	Precipitation Normal total	Maximum monthly	Year	Minimum monthly	Year	Maximum in 24 hrs.	Year	Snow, Ice pellets Mean total	Maximum monthly	Year	Maximum in 24 hrs.	Year	Relative humidity Hour 01	Hour 07	Hour 13	Hour 19	Wind Mean speed	Prevailing direction	Fastest mile Speed	Direction	Year	Pct. of possible sunshine	Mean sky cover sunrise to sunset	Clear	Partly Cloudy	Cloudy	Precipitation .01 inch or more	Snow, Ice pellets 1.0 inch or more	Thunderstorms	Heavy fog	Temperatures Max 90° and above	32° and below	Min 32° and below	0° and below	Average daily solar radiation - langleys

(yrs) 27

Month				27		27		(b)	(b)			27		27		27			27		20	20	20	20	21	14	22	22	22		26	27	27	27	27	27	27	27	27	27	27	27	27
J	36.7	21.3	29.0	72	1950	-12	1961	1116	3.17	6.16	1953	0.69	1970	2.47	1961	8.0	24.1	1966	16.0	1961	74	74	62	69	11.0	W	55	29	1959		6.4	7	8	16	11	2	*	3	0	10	28	1	
F	38.0	20.4	29.2	72	1954	-17	1967	1002	2.94	4.90	1956	0.69	1968	2.05	1966	8.9	22.4	1964	16.4	1961	76	77	55	62	11.4	WNW	55	29	1958		6.4	7	8	13	10	2	*	2	0	7	26	1	
M	47.2	26.5	37.0	86	1945	-18	1967	834	3.76	7.21	1953	1.39	1949	3.02	1952	6.6	30.5	1958	17.5	1958	77	77	55	53	12.1	NW	58	29	1952		6.5	8	11	12	11	2	1	2	*	1	22	1	*
A	61.3	38.1	49.3	89	1962	18	1964	471	3.78	6.83	1952	1.64	1964	3.08	1968	0.9	9.9	1957	3.1	1957	78	78	50	55	11.6	NW	60	30	1956		6.6	10	11	9	11	*	2	1	1	0	7	0	0
M	72.0	48.4	60.1	97	1962	28	1947	167	4.08	7.88	1948	0.09	1964	3.02	1953	T	T	1966+	T	1966+	79	78	52	60	10.2	WSW	58	30	1954		6.6	11	12	8	12	0	5	2	4	0	*	0	0
J	80.7	58.1	69.4	100	1966	39	1945	24	4.07	6.59	1946	0.34	1949	3.55	1967	0.0	0.0		0.0		80	80	53	61	9.2	SW	81	27	1964		5.9	7	12	11	10	0	6	1	7	0	0	0	0
J	85.4	63.0	74.1	105	1966	48	1963	0	4.82	10.42	1969	0.42	1955	4.54	1969	0.0	0.0		0.0		84	82	55	67	8.2	W	55	23	1951		5.7	8	12	11	10	0	7	3	7	0	0	0	0
A	82.6	61.7	72.2	100	1955	43	1944	90	4.79	12.10	1955	1.26	1964	4.79	1955	0.0	0.0		0.0		87	87	57	72	7.3	WSW	55	23	1949		5.5	10	11	10	9	0	6	3	4	0	0	0	0
S	75.7	53.7	64.7	99	1953	30	1947	353	3.75	6.84	1967	0.94	1952	4.06	1952	0.0	0.0		0.0		87	87	56	70	7.0	SW	46	25	1956		5.3	11	10	9	8	0	3	3	3	0	*	0	0
O	65.1	42.6	53.8	90	1951	22	1969	359	2.97	6.84	1955	0.15	1963	2.70	1955	0.2	1.4	1972	1.4	1972	84	83	55	70	7.5	WSW	49	14	1954		5.5	11	9	11	7	0	1	3	0	8	0	0
N	51.2	32.5	41.9	81	1950	11	1955	693	3.33	6.84	1972	1.00	1946	2.70	1951	1.4	7.8	1967	6.4	1967	83	83	60	72	8.4	NW	49	14	1950		6.1	8	9	13	10	1	1	3	0	15	1	0
D	39.3	23.2	31.3	70	1946	-8	1950	1045	3.27	6.43	1969	0.39	1955	2.85	1953	8.3	28.4	1966	13.3	1966	80	80	63	72	10.5	NW	52	29	1956		6.5	7	8	16	11	2	*	3	0	10	27	*
YR	61.2	40.9	51.1	105 JUL. 1966		-12 JAN. 1961		5810	44.12	12.10 AUG. 1955		0.09 MAY 1964		4.79 AUG. 1955		34.2	30.5 MAR. 1958		17.5 MAR. 1958		80	81	56	66	9.5	W	81	27	JUN. 1964		6.1	94	115	156	122	9	32	29	17	130	3

Means and extremes above are from existing and comparable exposures. Annual extremes have been exceeded at other sites in the locality as follows:
Maximum monthly snowfall 43.2 in January 1925.

NORMALS, MEANS AND EXTREMES

(Table Revised 1975. Base Period for Climatological Normals: 1941-1970)

Elev. 385 feet m.s.l.   Average station pressure mb.

Month	Normal Daily maximum	Daily minimum	Monthly	Extremes Record highest	Year	Record lowest	Year	Normal Degree days Base 65°F Heating	Cooling	Precipitation in inches Water equivalent Normal	Maximum monthly	Year	Minimum monthly	Year	Maximum in 24 hrs.	Year	Snow, Ice pellets Maximum monthly	Year	Maximum in 24 hrs.	Year	Relative humidity pct. Hour 01	Hour 07	Hour 13	Hour 19	Wind Mean speed m.p.h.	Prevailing direction	Fastest mile Speed m.p.h.	Direction	Year	Pct. of possible sunshine	Mean sky cover, tenths, sunrise to sunset	Clear	Partly cloudy	Cloudy	Precipitation .01 inch or more	Snow, Ice pellets 1.0 inch or more	Thunderstorms	Heavy fog, visibility ¼ mile or less	Temperatures °F Max 90° and above	32° and below	Min 32° and below	0° and below	Average station pressure mb.

(yrs) 31

Month				31		31				31	31		31		31		31		31		24	24	24	24	25	14	26	26	26	30	31	31	31	31	31	31	31	31	31	31	31	31	2
J	35.7	19.8	27.8	72	1950	-12	1961	1153	0	3.02	6.16	1953	0.69	1970	2.47	1961	24.1	1966	16.0	1961	75	77	62	69	10.8	W	55	29	1959		6.6	7	8	16	11	2	*	3	1	0	10	28	1	1005.6
F	37.5	20.4	29.0	72	1954	-7	1967	997	0	2.78	5.44	1958	1.31	1958	2.05	1966	22.4	1964	16.4	1958	77	77	55	62	12.0	WNW	58	29	1956		6.4	7	8	13	10	2	*	2	1	0	7	26	1	1003.7
M	47.7	28.5	38.1	86	1945	-18	1967	834	0	3.61	7.21	1953	1.39	1949	3.08	1952	30.5	1958	17.5	1958	73	77	52	59	12.2	NW	58	29	1952		6.5	8	11	12	11	2	1	2	*	*	1	22	1003.2
A	61.3	38.1	49.7	89	1962	18	1964	453	0	3.75	10.00	1948	1.64	1964	2.52	1963	9.9	1957	3.1	1957	74	74	50	54	11.3	NW	60	30	1956		6.5	8	12	10	11	*	2	1	1	0	7	1000.6
M	71.7	48.4	60.1	97	1962	28	1947	190	38	3.78	7.88	1948	0.09	1964	3.24	1964	T	1966	T	1966	78	78	52	61	10.3	WSW	58	30	1964		6.6	8	12	11	12	0	5	2	1	4	0	999.9
J	81.0	58.1	69.5	100	1966	39	1945	21	156	3.59	8.58	1972	0.34	1949	3.55	1967	0.0		0.0		80	80	54	61	9.4	SW	81	27	1964		6.0	8	12	10	10	0	6	1	*	7	0	1001.9
J	85.4	62.7	74.1	105	1966	48	1963	0	282	4.36	10.42	1969	0.42	1955	4.54	1969	0.0		0.0		83	82	55	66	8.3	W	55	23	1951		5.7	8	12	11	11	0	7	3	1	7	0	1002.2
A	82.8	60.6	71.7	100	1955	43	1944	6	214	4.18	12.10	1955	1.26	1964	4.79	1955	0.0		0.0		86	87	57	70	7.9	WSW	55	23	1949		5.7	10	11	10	11	0	7	3	1	4	0	1004.3
S	75.7	53.7	64.7	99	1953	30	1947	85	76	2.73	6.84	1952	0.94	1952	4.06	1952	0.0		0.0		87	89	55	72	8.3	SW	46	25	1956		5.3	11	11	9	7	0	3	3	*	0	1004.6
O	65.1	42.3	53.7	90	1951	22	1969	344	0	2.53	6.84	1967	0.15	1963	2.96	1963	1.4	1972	1.4	1972	84	84	60	72	7.3	WSW	49	14	1950		5.3	11	10	10	8	0	1	3	0	15	1002.9
N	51.7	32.9	42.3	81	1950	11	1955	681	0	3.59	6.84	1972	1.00	1946	3.40	1966	7.8	1967	6.4	1967	81	83	60	72	8.3	W	49	11	1950		6.6	7	8	15	10	1	1	3	0	15	1002.9
D	38.7	22.6	30.7	70	1946	-8	1950	1003	0	3.59	6.43	1973	0.39	1955	2.85	1973	28.4	1966	13.3	1966	82	80	64	71	10.3	W	52				6.7	7	8	16	11	2	*	3	0	26	1003.6
YR	61.3	40.7	51.0	105 JUL. 1966		-12 JAN. 1961		5827	772	42.49	12.10 AUG. 1955		0.09 MAY 1964		4.79 AUG. 1955		30.5 MAR. 1958		17.5 MAR. 1958		80	81	57	66	9.4	W	81	27	JUN. 1964		6.2	93	115	157	123	9	32	29	17	129	2	1003.2

Means and extremes above are from existing and comparable exposures. Annual extremes have been exceeded at other sites in the locality as follows:
Maximum monthly snowfall 43.2 in January 1925.

REFERENCE NOTES APPLYING TO TABLES APPEAR ON THE PAGE FOLLOWING LAST TABLE.
(Caution: Letters and symbols may have different meanings in 1941-1970 tables than in earlier tables. See notes.)

NORMALS, MEANS AND EXTREMES
(Table Revised 1971. Base Period for Climatological Normals: 1931-1960)

Station: ERIE, PENNSYLVANIA   ERIE INTERNATIONAL AIRPORT   Standard time used: EASTERN   Latitude: 42° 05' N   Longitude: 80° 11' W   Elevation (ground): 731 feet

Month	Normal Daily maximum	Normal Daily minimum	Normal Monthly	Normal heating degree days (Base 65°)
J	34.0	20.5	27.3	1169
F	34.0	18.8	26.4	1081
M	41.1	25.4	33.6	973
A	54.7	36.3	45.5	585
M	66.1	46.5	56.4	288
J	76.1	57.2	66.7	60
J	80.1	62.0	71.1	0
A	78.5	61.1	69.8	25
S	72.1	54.5	63.3	102
O	61.0	44.2	52.6	391
N	48.0	34.3	41.2	714
D	36.8	23.3	30.7	1063
YR	56.9	40.5	48.7	6451

Mean and extremes above are from existing and comparable exposures. Annual extremes have been exceeded at other sites in the locality as follows: Highest temperature 99 in September 1953; lowest temperature -16 in February 1875; maximum monthly precipitation 13.27 in July 1947; minimum monthly precipitation 0.02 in October 1924; maximum precipitation in 24 hours 10.42 in July 1947; maximum snowfall in 24 hours 26.5 in December 1944.

NORMALS, MEANS AND EXTREMES
(Table Revised 1975. Base Period for Climatological Normals: 1941-1970)

Month	Normal Daily maximum	Normal Daily minimum	Normal Monthly	Normal Degree days Base 65°F Heating	Normal Degree days Base 65°F Cooling
J	31.7	18.5	25.1	1237	0
F	32.5	17.9	25.2	1114	0
M	40.5	25.4	32.9	995	0
A	53.7	36.1	44.8	606	0
M	64.5	45.5	54.6	336	13
J	73.5	55.6	64.6	80	68
J	77.4	60.0	68.7	24	139
A	78.0	58.9	67.5	43	120
S	70.2	52.6	61.4	141	33
O	60.0	43.0	51.6	415	0
N	46.0	33.7	40.1	747	0
D	34.8	23.3	29.1	1113	0
YR	55.0	39.2	47.1	6851	373

Average station pressure mb. Elev. 737 feet m.s.l.
991.9
991.6
988.7
987.8
987.9
989.9
991.4
993.4
989.7
990.2
990.0

Means and extremes above are from existing and comparable exposures. Annual extremes have been exceeded at other sites in the locality as follows: Highest temperature 99 in September 1953; lowest temperature -16 in February 1875; maximum monthly precipitation 13.27 in July 1947; maximum precipitation in 24 hours 10.42 in July 1947; maximum snowfall in 24 hours 26.5 in December 1944.

REFERENCE NOTES APPLYING TO TABLES APPEAR ON THE PAGE FOLLOWING LAST TABLE.
(Caution: Letters and symbols may have different meanings in 1941-1970 tables than in earlier tables. See notes.)

NORMALS, MEANS AND EXTREMES
(Table Revised 1971. Base Period for Climatological Normals: 1931-1960)

Station: HARRISBURG, PENNSYLVANIA   HARRISBURG STATE AIRPORT   Standard time used: EASTERN   Latitude: 40° 13' N   Longitude: 76° 51' W   Elevation (ground): 338 feet

Month	Normal Daily max	Normal Daily min	Normal Monthly	Extremes Record highest	Year	Extremes Record lowest	Year	Normal heating degree days (Base 65°)	Precip Normal total	Precip Max monthly	Year	Precip Min monthly	Year	Precip Max 24 hrs	Year	Snow Mean total	Snow Max monthly	Year	Snow Max 24 hrs	Year	RH 01	RH 07	RH 13	RH 19	Wind Mean speed	Prevailing dir	Fastest speed	Fastest dir#	Fastest Year	Mean sky cover	Pct sunshine	Clear	Partly cloudy	Cloudy	Precip .01"+	Snow 1.0"+	T-storms	Heavy fog	Max 90°+	Max 32°-	Min 32°-	Min 0°-	Solar
(yrs)	(b)	(b)	(b)	32		32		(b)	(b)	32		32		32		32	32		32		32	32	32	32	14	32	32		32	20	32	32	32	32	32	32	32	32	32	32	32	32	32
J	38.9	23.6	31.3	73	1950	-5	1968	1045	2.76	4.78	1964	0.70	1945	1.94	1964	9.5	34.0	1961	21.0	1945	69	71	58	64	8.4	WNW	47	W	1956	6.6	48	7	7	17	11	2	*	3	0	9	26	1	32
F	41.4	23.7	32.6	75	1954	-2	1967	907	2.31	5.62	1960	0.53	1964	1.93	1966	9.5	30.2	1964	19.6	1964	71	74	54	67	9.3	WNW	68	INV	1939	6.6	54	7	7	14	10	2	*	2	0	6	23	*	23
M	50.0	30.5	40.6	86	1945	8	1943	766	3.43	5.47	1949	1.20	1944	2.11	1960	7.1	22.6	1960	10.7	1960	67	71	49	55	9.7	WNW	68	W	1955	6.7	57	7	8	16	12	2	1	1	0	1	18	0	18
A	63.4	40.5	51.8	92	1957+	21	1965+	396	3.02	6.23	1952	0.57	1942	3.1	1966+	0.3	3.1	1966+	3.1	1966+	65	70	46	50	9.2	WNW	60	SW	1952	6.7	55	8	9	13	12	*	2	1	*	0	3	0	*
M	74.5	51.6	62.7	97	1942+	31	1966	124	3.90	8.06	1964	0.51	1953	3.11	1964	0.0	0.0		0.0		73	76	49	57	7.8	WNW	46	S	1940	6.5	60	7	10	14	13	0	6	1	1	0	*	0	0
J	82.7	61.0	72.0	100	1966+	41	1961	12	3.42	8.00	1946	0.07	1966	2.81	1966+	0.0	0.0		0.0		76	78	50	61	6.9	W	61	NW	1959	6.1	65	6	11	13	11	0	7	*	3	0	0	0	*
J	87.0	65.3	76.2	107	1966	49	1945	0	3.51	9.72	1969	0.78	1969	5.36	1969	0.0	0.0		0.0		78	81	51	65	6.3	WNW	47	NW	1954	6.1	68	8	11	12	10	0	7	3	7	0	0	0	*
A	84.6	63.5	74.1	101	1944	46	1944	0	3.65	9.07	1955	0.93	1955	3.35	1955	0.0	0.0		0.0		81	84	53	67	6.0	WNW	45	NW	1939	5.8	67	8	11	12	10	0	6	3	2	0	0	0	*
S	77.0	56.2	66.9	102	1953	30	1963	63	2.82	6.12	1966	0.72	1941	4.36	1966	0.0	0.0		0.0		82	84	54	67	6.1	WNW	50	SE	1954	5.5	65	11	9	13	9	0	3	3	1	0	*	*	0
O	66.7	44.6	55.7	97	1941	23	1969	298	2.97	6.82	1943	0.04	1963	2.61	1943	T	T	1954	T	1954	79	82	52	65	6.7	WNW	58	W	1954	5.6	63	11	7	13	8	*	1	3	0	0	2	0	0
N	52.9	34.5	43.4	84	1950	13	1955	648	2.95	5.93	1963	0.53	1939	2.93	1963	1.9	15.4	1953	12.9	1953	74	78	56	65	7.9	WNW	50	NW	1950	6.5	47	7	7	16	10	1	*	2	0	0	12	0	0
D	40.7	25.3	33.0	71	1946	-8	1960	992	2.91	4.46	1969	0.23	1955	2.01	1969	8.3	28.3	1969	12.9	1969	71	73	59	63	8.2	WNW	61	NW	1953	6.9	45	6	7	18	12	1	*	2	0	1	24	1	1
YR	63.3	43.2	53.3	107	JUL. 1966	-8	DEC. 1960	5251	37.65	9.72	JUL. 1969	0.04	OCT. 1963	5.36	JUL. 1969	36.6	34.0	JAN. 1961	21.0	JAN. 1945	74	75	54	62	7.7	WNW	68	MAR. W 1955		6.4	59	86	106	173	124	10	33	21	26	22	108	1

Means and extremes above are from existing and comparable exposures. Annual extremes have been exceeded at other sites in the locality as follows: Lowest temperature -14 in January 1912; maximum monthly precipitation 10.67 in August 1933; minimum monthly precipitation 0.02 in October 1924; maximum precipitation in 24 hours 7.46 on May 31-June 1, 1889.

\# To 8 compass points only.

NORMALS, MEANS AND EXTREMES
(Table Revised 1975. Base Period for Climatological Normals: 1941-1970)

Month	Normal Daily max	Normal Daily min	Normal Monthly	Extremes Record highest	Year	Extremes Record lowest	Year	Normal Degree days Base 65°F Heating	Cooling	Precip Normal	Precip Max monthly	Year	Precip Min monthly	Year	Precip Max 24 hrs	Year	Snow Max monthly	Year	Snow Max 24 hrs	Year	RH 01	RH 07	RH 13	RH 19	Wind Mean speed	Prevailing dir	Fastest speed	Fastest dir	Fastest Year	Mean sky cover	Pct sunshine	Clear	Partly cloudy	Cloudy	Precip .01"+	Snow 1.0"+	T-storms	Heavy fog	Max 90°+	Max 32°-	Min 32°-	Min 0°-	Elev. station pressure mb.
(yrs)	36			36		36				36	36		36		36		36		36		35	36	35	36	36	14	36		36	24	36	36	36	36	36	36	36	36	36	36	36	36
J	37.7	22.5	30.1	73	1950	-5	1968	1082	0	2.57	4.78	1964	0.70	1955	1.94	1955	34.0	1961	21.0	1945	69	71	58	64	8.4	WNW	47	W	1956	6.6	48	7	7	17	11	2	*	3	0	9	26	1	1007.4
F	40.5	24.0	32.3	75	1954	-2	1967	916	0	2.42	5.62	1971	0.53	1968	1.93	1966	30.2	1964	19.6	1964	71	74	54	67	9.3	WNW	68	INV	1939	6.7	54	7	8	14	10	2	*	2	0	6	23	*	1005.4
M	50.7	31.2	41.0	86	1945	8	1943	744	0	3.22	5.47	1949	1.20	1944	2.11	1949	22.6	1960	10.7	1960	67	71	45	57	9.7	WNW	68	W	1955	6.7	56	8	8	15	12	2	1	1	0	1	18	0	1003.8
A	64.1	41.5	52.8	92	1957	21	1965	370	0	2.98	6.23	1952	0.57	1957	2.18	1973	3.1	1966	3.1	1966	65	70	42	55	9.4	WNW	56	SW	1952	6.5	59	7	11	12	12	*	2	1	*	0	3	0	1002.1
M	74.5	51.6	63.1	97	1942	31	1966	128	69	3.76	8.06	1964	0.51	1953	3.11	1964	T	1966	T	1966	73	76	51	57	7.8	WNW	48	W	1940	6.5	59	7	11	13	11	0	6	1	1	0	*	0	1002.1
J	83.0	61.0	72.0	100	1966	41	1961	0	214	3.11	12.55	1972	0.07	1966	12.55	1972	0.0		0.0		76	78	52	60	6.8	W	61	NW	1963	6.2	64	6	11	13	11	0	7	*	5	0	0	0	1002.9
J	86.8	65.4	76.1	107	1966	49	1945	0	344	3.70	9.72	1969	0.78	1969	5.36	1969	0.0		0.0		78	81	52	65	6.3	WNW	47	NW	1954	6.1	68	8	11	12	10	0	7	3	9	0	0	0	1003.6
A	84.6	63.2	73.9	101	1944	46	1944	0	279	3.66	9.07	1955	0.93	1955	3.35	1955	0.0		0.0		81	82	54	65	6.0	WNW	45	NW	1939	5.7	67	8	12	11	9	0	6	3	6	0	0	0	1005.4
S	78.0	56.0	67.0	102	1953	30	1963	51	111	3.22	6.79	1974	0.72	1941	4.36	1943	0.0		0.0		82	84	54	67	6.1	WNW	50	SE	1954	5.5	65	11	9	10	9	0	3	3	1	0	*	*	1008.3
O	66.9	44.7	55.8	97	1941	23	1969	293	0	2.57	6.82	1943	0.04	1963	2.91	1943	T		T		79	82	55	66	6.6	WNW	58	W	1954	5.6	62	11	10	10	8	*	1	3	0	0	2	0	1008.8
N	52.9	34.7	43.8	84	1950	13	1955	636	0	3.19	5.93	1963	0.53	1939	2.93	1963	15.4	1953	12.9	1953	74	78	56	65	7.9	WNW	50	NW	1950	6.4	46	7	9	13	10	1	*	2	0	0	12	0	1008.1
D	40.1	25.0	32.6	71	1946	-8	1960	1004	0	3.07	6.52	1973	0.23	1955	2.01	1973	28.3	1969	12.9	1969	73	76	60	64	8.1	WNW	61	NW	1953	6.0	44	6	8	17	10	1	*	3	0	1	23	1	1005.1
YR	63.3	43.4	53.4	107	JUL. 1966	-8	DEC. 1960	5224	1025	36.47	18.55	JUN 1972	0.04	OCT. 1963	12.55	JUN 1972	34.0	JAN 1961	21.0	JAN 1945	74	75	54	62	7.7	WNW	68	MAR. W 1955		6.4	58	86	105	174	124	9	33	20	25	21	107	1	1004.6

Means and extremes above are from existing and comparable exposures. Annual extremes have been exceeded at other sites in the locality as follows: Lowest temperature -14 in January 1912; minimum monthly precipitation 0.02 in October 1924.

REFERENCE NOTES APPLYING TO TABLES APPEAR ON THE PAGE FOLLOWING LAST TABLE.
(Caution: Letters and symbols may have different meanings in 1941-1970 tables than in earlier tables. See notes.)

NORMALS, MEANS AND EXTREMES
(Table Revised 1971. Base Period for Climatological Normals: 1931-1960)

Station: PHILADELPHIA, PENNSYLVANIA INTERNATIONAL AIRPORT   Standard time used: EASTERN   Latitude: 39° 53' N   Longitude: 75° 15' W   Elevation (ground): 5 feet

Means and extremes above are from existing and comparable exposures. Annual extremes have been exceeded at other sites in the locality as follows:
Highest temperature 106 in August 1918; lowest temperature -11 in February 1934; maximum monthly precipitation 12.10 in August 1911; maximum precipitation in 24 hours 5.89 in August 1898; maximum snowfall 31.5 in February 1899; maximum snowfall 31.5 in December 1909; fastest mile of wind 88 from North in July 1931.

\# To 8 compass points only.

NORMALS, MEANS AND EXTREMES
(Table Revised 1975. Base Period for Climatological Normals: 1941-1970)

Month	Normal Daily maximum	Normal Daily minimum	Normal Monthly	Extremes Record highest	Year	Extremes Record lowest	Year	Degree days Heating	Degree days Cooling	Precip. Normal	Snow/Ice pellets Mean total
J	40.1	24.4	32.3	69	1961	-5	1963	1014	0	2.81	5.6
F	42.2	25.5	33.9	69	1972	-5	1961	871	0	2.62	6.3
M	51.2	32.5	41.9	80	1960	9	1960	716	0	3.69	3.4
A	63.1	41.9	52.5	90	1960	24	1960	367	7	3.35	0.1
M	74.0	52.3	63.2	96	1964	28	1966	122	67	3.70	0.0
J	83.0	61.6	72.3	100	1964	44	1972	0	223		0.0
J	86.8	66.7	76.8	104	1966	51	1966	0	366		0.0
A	84.4	65.1	74.8	104	1973	45	1965	0	304		0.0
S	78.4	57.8	68.1	97	1970	35	1963	38	131		T
O	67.5	46.9	57.4	88	1974	25	1964	249	13		T
N	54.4	36.9	45.7	81	1961	17	1964	564	0		0.8
D	43.2	27.2	35.2	71	1966	3	1962	924	0	3.32	4.7
YR	64.2	44.9	54.6	104	JUL 1966	-5	JAN 1963	4865	1104	39.93	21.6

Means and extremes above are from existing and comparable exposures. Annual extremes have been exceeded at other sites in the locality as follows:
Highest temperature 106 in August 1918; lowest temperature -11 in August 1918; lowest temperature -11 in February 1934; maximum monthly precipitation 12.10 in August 1911; maximum precipitation in 24 hours 5.89 in August 1898; maximum snowfall 31.5 in February 1899; maximum snowfall 31.5 in December 1909; fastest mile of wind 88 from North in July 1931.

REFERENCE NOTES APPLYING TO TABLES APPEAR ON THE PAGE FOLLOWING LAST TABLE.
(Caution: Letters and symbols may have different meanings in 1941-1970 tables than in earlier tables. See notes.)

NORMALS, MEANS AND EXTREMES
(Table Revised 1971. Base Period for Climatological Normals: 1931-1960)

Station: PITTSBURGH, PENNSYLVANIA   GREATER PITTSBURGH AIRPORT   Standard time used: EASTERN   Latitude: 40° 30' N   Longitude: 80° 13' W   Elevation (ground): 1137 feet

Month	Normal Daily maximum	Normal Daily minimum	Normal Monthly	Record highest	Year	Record lowest	Year	Normal heating degree days (Base 65°)	Precip Normal total	Max monthly	Year	Min monthly	Year	Max 24 hrs	Year	Snow Mean total	Max monthly	Year	Max 24 hrs	Year	RH 01	RH 07	RH 13	RH 19	Wind Mean speed	Prevailing dir	Fastest speed	Dir	Year	Mean sky cover	Pct sunshine
J	36.5	21.2	28.9	66	1967	-18	1963	1119	2.97	4.52	1966	1.06	1967	1.42	1966	11.0	24.6	1966	14.0	1966	73	76	67	66	10.7	WSW	44	26	1964	67	37
F	37.6	20.7	29.2	66	1961	-9	1963	1002	2.19	5.93	1950	0.51	1969	1.72	1961	10.0	22.5	1961	12.3	1960	73	77	64	65	11.1	WSW	58	26	1967	66	40
M	46.1	27.4	36.8	80	1960	-1	1960	874	3.32	6.10	1964	1.14	1969	2.00	1969	10.0	21.3	1960	14.7	1954	73	77	60	64	11.0	WSW	46	25	1954	69	48
A	60.0	37.9	49.0	87	1970	15	1964	480	3.08	7.61	1964	1.37	1960	2.15	1964	1.7	5.9	1961	3.1	1953	68	74	52	55	10.7	WSW	46	25	1970+	64	49
M	71.4	48.1	59.8	91	1962	26	1966	195	3.91	6.36	1968	1.21	1965	2.18	1968	0.3	5.9	1966	3.1	1966	74	79	51	59	9.6	WSW	40	27	1957	62	55
J	79.9	56.9	68.4	96	1966	36	1966	39	3.78	4.61	1970	0.90	1967	1.93	1955	0.0	0.0				78	79	51	57	8.2	WSW	40	27	1957	62	62
J	83.3	60.9	72.1	98	1966	42	1963	0	3.88	7.43	1958	1.82	1965	2.73	1957	0.0	0.0				81	83	52	59	7.5	WSW	51	25	1956	59	63
A	81.9	59.6	70.8	98	1966+	40	1965	9	3.06	9.95	1956	0.78	1957	3.06	1956	0.0	0.0				83	86	55	62	7.1	WSW	46	02	1960	55	63
S	75.5	52.8	64.2	95	1959	31	1959	105	2.54	5.42	1962	0.74	1964	2.10	1962	0.0	0.0				82	86	55	65	7.7	WSW	32	02	1960	60	65
O	63.7	42.0	52.9	87	1963	16	1965	375	2.52	8.20	1954	0.74	1963	3.56	1954	0.2	1.8	1962	1.8	1962	80	84	53	61	8.5	WSW	35	29	1959	58	60
N	49.5	32.0	41.3	82	1961	-5	1959	726	2.24	4.70	1961	1.00	1954	1.38	1961+	3.9	11.0	1958	6.9	1958	78	80	62	69	10.2	WSW	45	30	1968	58	41
D	38.1	23.0	30.7	68	1966	-11	1962	1063	2.40	4.26	1964	0.40	1955	1.25	1964	9.4	20.8	1960	6.9	1958	71	76	66	71	10.6	WSW	48	27	1968	60	33
YR	60.3	40.3	50.3	98	JUL 1966	-18	JAN 1963	5987	36.14	8.20	OCT 1954	0.16	OCT 1963	3.56	MAR 1962	46.5	24.6	JAN 1966	14.7	MAR 1962	76	79	57	62	9.4	WSW	58	26	FEB 1967	53	7.1

Means and extremes above are from existing and comparable exposures. Annual extremes have been exceeded at other sites in the locality as follows:
AIRPORT - Highest temperature 102 in July 1936; maximum monthly precipitation 10.25 in June 1951; maximum monthly snowfall 32.3 in November 1950; and maximum snowfall in 24 hours 17.5 in November 1950.
CITY OFFICE - Highest temperature 103 in July 1936; lowest temperature -20 in February 1899; minimum monthly precipitation 0.06 in October 1874; maximum precipitation in 24 hours 4.08 in September 1876; and maximum monthly snowfall 36.3 in December 1890.

NORMALS, MEANS AND EXTREMES
(Table Revised 1975. Base Period for Climatological Normals: 1941-1970)

Average station pressure: Elev. 1225 feet m.s.l.

Month	Normal Daily maximum	Normal Daily minimum	Normal Monthly	Record highest	Year	Record lowest	Year	Heating deg days	Cooling deg days	Precip Normal	Max monthly	Year	Min monthly	Year	Max 24 hrs	Year	Snow Mean total	Max monthly	Year	Max 24 hrs	Year	RH 01	RH 07	RH 13	RH 19	Wind Mean speed	Prevailing dir	Fastest mph	Dir	Year	Mean sky cover	Pct sunshine
J	35.3	20.8	28.1	68	1972	-18	1963	1144	0	2.79	4.52	1966	1.06	1967	1.42	1966	11.4	24.6	1966	14.0	1966	73	76	66	67	10.8	WSW	48	27	1971	67	37
F	37.2	21.3	29.3	66	1961	-9	1963	1000	0	2.35	5.10	1966	0.51	1969	1.72	1961	10.6	22.5	1972	12.3	1960	73	75	64	65	11.1	WSW	58	26	1967	66	39
M	47.2	29.0	38.1	80	1966	-1	1960	834	0	3.60	6.10	1964	0.44	1967	2.00	1969	11.0	21.0	1960	14.7	1954	72	75	58	61	11.0	WSW	46	27	1954	69	49
A	60.9	39.4	50.2	87	1970	15	1964	444	0	3.40	7.61	1964	0.88	1971	2.15	1964	2.4	5.9	1961	3.1	1953	67	73	51	54	10.6	WSW	46	25	1974	64	49
M	70.8	49.0	59.9	91	1962	26	1966	208	46	3.63	6.36	1968	0.90	1965	2.44	1968	0.2	5.9	1966	3.1	1966	73	73	51	55	9.8	WSW	45	27	1957	62	60
J	79.5	57.7	68.6	96	1971	36	1966	46	134	3.48	5.08	1974	0.90	1967	1.93	1955	0.0	0.0				78	80	53	58	8.2	WSW	40	27	1957	62	64
J	82.5	61.3	71.9	98	1966	42	1963	16	221	3.84	7.43	1958	1.82	1965	2.97	1966	0.0	0.0				80	83	53	59	7.6	WSW	51	25	1956	60	62
A	80.9	59.4	70.2	98	1966	40	1965	17	177	3.15	9.95	1956	0.78	1957	3.06	1956	0.0	0.0				82	86	55	62	7.3	WSW	46	09	1963	61	61
S	74.9	52.7	63.8	95	1959	31	1959	98	62	2.52	5.42	1962	0.74	1964	2.18	1962	0.0	0.0				82	86	57	66	7.3	WSW	32	02	1960	56	60
O	63.8	41.9	52.9	87	1963	16	1972	372	7	2.47	8.20	1954	1.00	1963	3.56	1954	0.4	1.8	1972	1.8	1972	80	85	53	61	8.5	WSW	35	29	1959	54	50
N	49.3	33.3	41.3	81	1961	-5	1959	711	0	2.47	4.70	1972	0.65	1972	1.47	1974	4.0	11.0	1958	10.5	1958	76	80	63	69	10.4	WSW	45	25	1968	55	40
D	37.3	23.6	30.5	72	1971	-5	1962	1070	0	2.48	4.26	1964	0.40	1955	1.25	1964	9.8	21.2	1974	12.5	1974	76	78	69	71	10.6	WSW	48	25	1968	60	30
YR	60.0	40.8	50.4	98	JUL 1966	-18	JAN 1963	5930	647	36.23	8.20	OCT 1954	0.16	OCT 1963	3.56	MAR 1962	51	24.6	JAN 1966	14.7	MAR 1962	75	79	58	62	9.4	WSW	58	26	FEB 1967	51	7.1

Means and extremes above are from existing and comparable exposures. Annual extremes have been exceeded at other sites in the locality as follows:
AIRPORT - Highest temperature 102 in July 1936; maximum monthly precipitation 10.25 in June 1951; maximum monthly snowfall 32.3 in November 1950; and maximum snowfall in 24 hours 17.5 in November 1950.
CITY OFFICE - Highest temperature 103 in July 1936; lowest temperature -20 in February 1899; minimum monthly precipitation 0.06 in October 1874; maximum precipitation in 24 hours 4.08 in September 1876; and maximum monthly snowfall 36.3 in December 1890.

REFERENCE NOTES APPLYING TO TABLES ON THE PAGE FOLLOWING LAST TABLE.
(Caution: Letters and symbols may have different meanings in 1941-1970 tables than in earlier tables. See notes.)

NORMALS, MEANS AND EXTREMES
(Table Revised 1971. Base Period for Climatological Normals: 1931-1960)

Station: PITTSBURGH, PENNSYLVANIA   FEDERAL BUILDING   Standard time used: EASTERN   Latitude: 40° 27' N   Longitude: 80° 00' W   Elevation (ground): 747 feet

Month	Temperature Normal Daily maximum	Daily minimum	Monthly	Extremes Record highest	Year	Record lowest	Year	Precipitation Normal total	Normal heating degree days (Base 65°)
J	39.5 (b)	24.7	32.1	77	1950	-13	1963	2.82 (b)	1020
F	41.0	24.9	32.9	74	1937	-5	1958	2.31	899
M	49.0	31.4	40.2	83	1950	5	1943	3.52	777
A	62.5	41.6	52.1	90	1970	18	1950	3.17	394
M	73.4	51.9	62.7	92	1962	29	1966	3.75	131
J	81.6	61.1	71.4	99	1952	39	1966	3.45	14
J	85.0	64.8	74.6	103	1936	50	1965	3.60	0
A	83.0	62.9	73.0	100	1948	45	1965	3.50	0
S	77.6	55.7	66.7	100	1953	37	1967	2.50	74
O	64.7	44.5	54.6	91	1951	24	1958	2.60	330
N	50.5	36.5	42.8	83	1961	-7	1958	2.54	666
D	40.5	26.2	33.4	71	1952	-7	1951	2.54	980
YR	62.3	43.7	53.0	103	JUL 1936	-13	JAN 1963	36.87	5291

Means and extremes above are from existing and comparable exposures. Annual extremes have been exceeded at other sites in the locality as follows: Lowest temperature -20 in February 1899; maximum monthly precipitation 9.51 in July 1887; minimum monthly precipitation 0.06 in October 1874; maximum precipitation in 24 hours 4.08 in September 1876; maximum monthly snowfall 36.3 in December 1890.

NORMALS, MEANS AND EXTREMES
(Table Revised 1975. Base Period for Climatological Normals: 1941-1970)

Month	Temperature Normal Daily maximum	Daily minimum	Monthly	Extremes Record highest	Year	Record lowest	Year	Precipitation Normal	Heating Degree days (Base 65°F)	Cooling Degree days
J	37.4	23.4	30.6	77	1950	-13	1963	2.61	1066	0
F	41.0	23.9	32.0	74	1937	-5	1958	2.29	924	0
M	48.5	30.6	40.4	83	1950	5	1943	3.58	763	7
A	62.2	40.4	52.2	90	1970	18	1950	3.44	382	74
M	72.6	50.2	62.3	92	1962	29	1966	3.59	161	199
J	81.0	59.2	72.3	99	1952	39	1972	3.74	10	298
J	84.0	65.2	74.6	103	1936	50	1972	3.78	0	298
A	82.0	63.3	72.7	100	1948	45	1956	3.18	6	254
S	76.7	56.2	66.5	100	1953	35	1962	2.53	58	103
O	65.2	44.1	54.7	91	1951	24	1963	2.47	298	13
N	51.8	35.8	44.1	83	1971	-7	1958	2.49	627	0
D	39.8	26.8	33.3	71	1952	-7	1951	2.52	983	0
YR	61.9	44.1	53.0	103	JUL 1936	-13	JAN 1963	36.22	5278	948

Means and extremes above are from existing and comparable exposures. Annual extremes have been exceeded at other sites in the locality as follows: Lowest temperature -20 in February 1899; maximum monthly precipitation 9.51 in July 1887; minimum monthly precipitation 0.06 in October 1874; maximum precipitation in 24 hours 4.08 in September 1876; maximum monthly snowfall 36.3 in December 1890.

REFERENCE NOTES APPLYING TO TABLES APPEAR ON THE PAGE FOLLOWING LAST TABLE.
(Caution: Letters and symbols may have different meanings in 1941-1970 tables than in earlier tables. See notes.)

NORMALS, MEANS AND EXTREMES
(Table Revised 1971. Base Period for Climatological Normals: 1931-1960)

Station: WILLIAMSPORT-LYCOMING CO. AP

WILLIAMSPORT, PENNSYLVANIA

Standard time used: EASTERN Latitude: 41° 15' N Longitude: 76° 55' W Elevation (ground): 524 feet

NORMALS, MEANS AND EXTREMES
(Table Revised 1975. Base Period for Climatological Normals: 1941-1970)

	Elev. 525 feet m.s.l.	Average station pressure mb. 2

REFERENCE NOTES APPLYING TO TABLES APPEAR ON THE PAGE FOLLOWING LAST TABLE.
(Caution: Letters and symbols may have different meanings in 1941-1970 tables than in earlier tables. See notes.)

Reference notes applying to Normals, Means, and Extremes tables for 1931–1960 base period.

(a) Length of record, years, based on January data. Other months may be for more or fewer years if there have been breaks in the record.

(b) Climatological standard normals (1931-1960).

° Less than one half.
+ Also on earlier dates, months, or years.
T Trace, an amount too small to measure.
Below zero temperatures are preceded by a minus sign. The prevailing direction for wind in the Normals, Means, and Extremes table is from records through 1963.
‡ >70° at Alaskan stations.

Unless otherwise indicated, dimensional units used in this bulletin are: temperature in degrees F.; precipitation, including snowfall, in inches; wind movement in miles per hour; and relative humidity in percent. Heating degree day totals are the sums of negative departures of average daily temperatures from 65° F. Cooling degree day totals are the sums of positive departures of average daily temperatures from 65° F. Sleet was included in snowfall totals beginning with July 1948. The term "Ice pellets" includes solid grains of ice (sleet) and particles consisting of snow pellets encased in a thin layer of ice. Heavy fog reduces visibility to 1/4 mile or less.

Sky cover is expressed in a range of 0 for no clouds or obscuring phenomena to 10 for complete sky cover. The number of clear days is based on average cloudiness 0-3, partly cloudy days 4-7, and cloudy days 8-10 tenths.

Solar radiation data are the averages of direct and diffuse radiation on a horizontal surface. The langley denotes one gram calorie per square centimeter.

& Figures instead of letters in a direction column indicate direction in tens of degrees from true North; i.e., 09 - East, 18 - South, 27 - West, 36 - North, and 00 - Calm. Resultant wind is the vector sum of wind directions and speeds divided by the number of observations. If figures appear in the direction column under "Fastest mile" the corresponding speeds are fastest observed 1-minute values.

Reference notes applying to Normals, Means, and Extremes tables for 1941–1970 base period.

(a) Length of record, years, through the current year unless otherwise noted, based on January data.
(b) 70° and above at Alaskan stations.
* Less than one half.
T Trace.

NORMALS - Based on record for the 1941-1970 period.
DATE OF AN EXTREME - The most recent in cases of multiple occurrence.
PREVAILING WIND DIRECTION - Record through 1963.
WIND DIRECTION - Numerals indicate tens of degrees clockwise from true north. 00 indicates calm.
FASTEST MILE WIND - Speed is fastest observed 1-minute value when the direction is in tens of degrees.

Mean Annual Precipitation, Inches

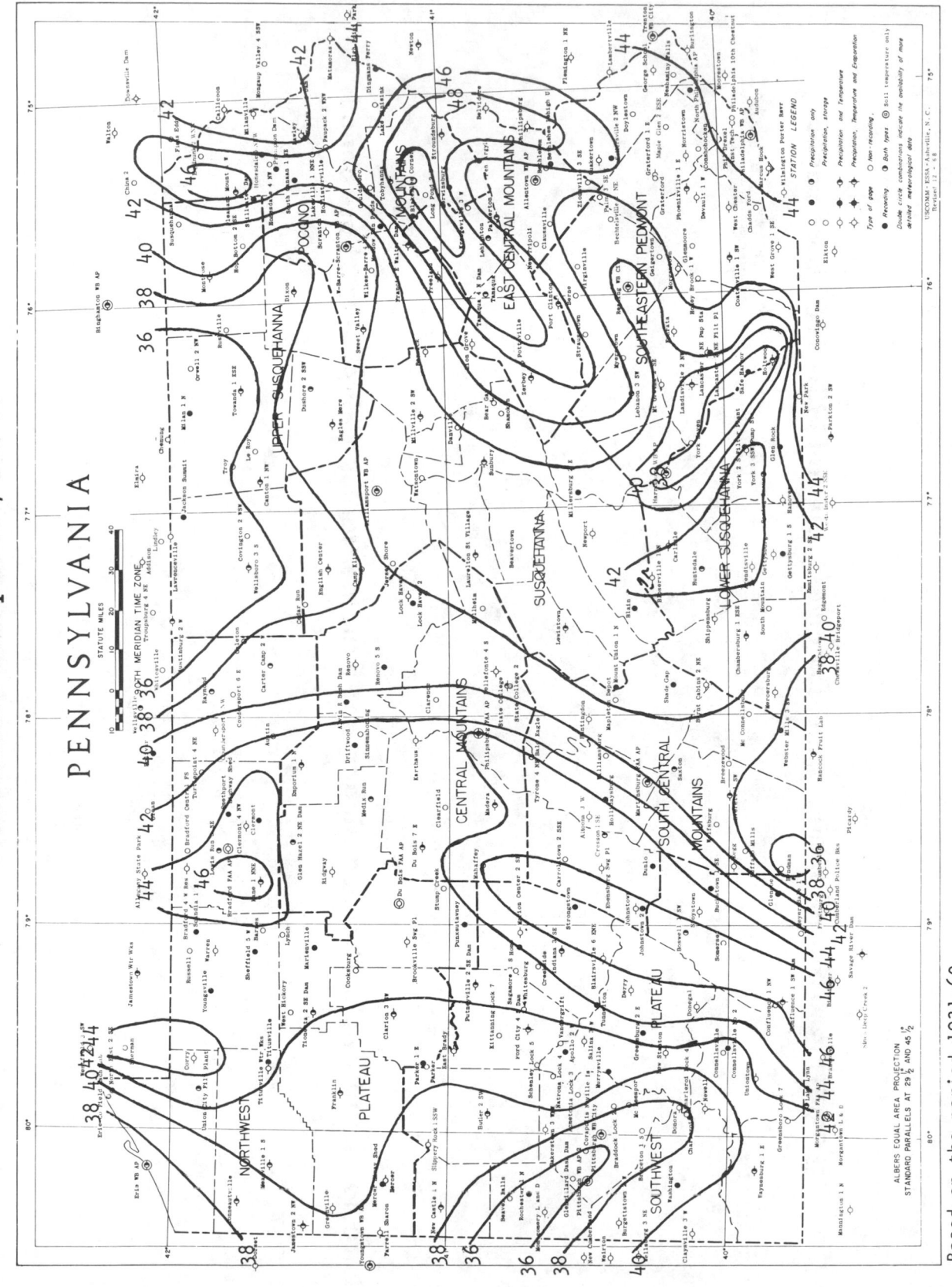

Based on the period 1931-60

Isolines are drawn through points of approximately equal value. Caution should be used in interpolating on these maps, particularly in mountainous areas.

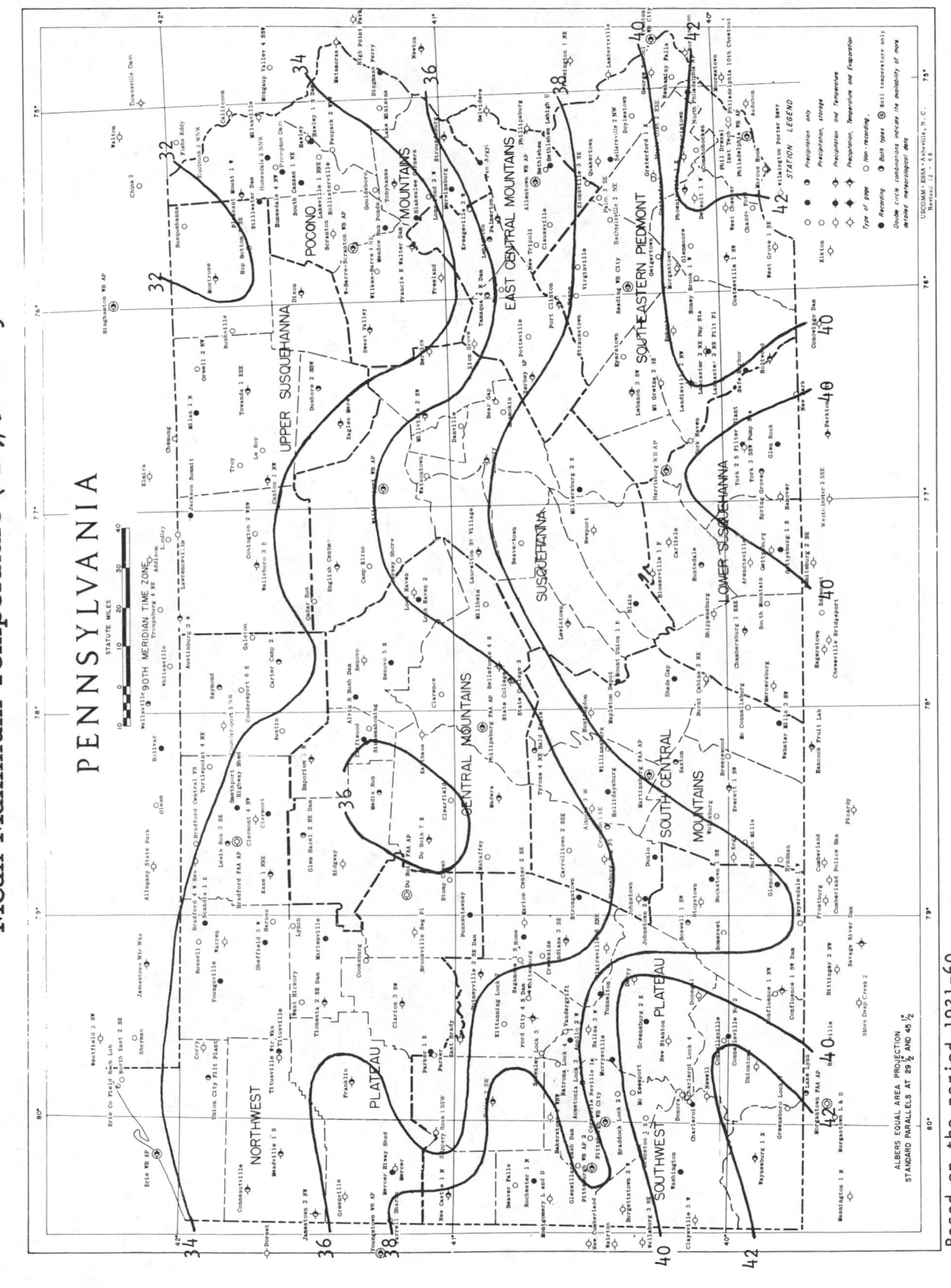

Mean Maximum Temperature (°F), January

PENNSYLVANIA

Based on the period 1931-60

Isolines are drawn through points of approximately equal value. Caution should be used in interpolating on these maps, particularly in mountainous areas.

ALBERS EQUAL AREA PROJECTION
STANDARD PARALLELS AT 29½ AND 45½

USCOMM - ESSA - Asheville, N. C.
Revised 12 - 68

Mean Minimum Temperature (°F), January

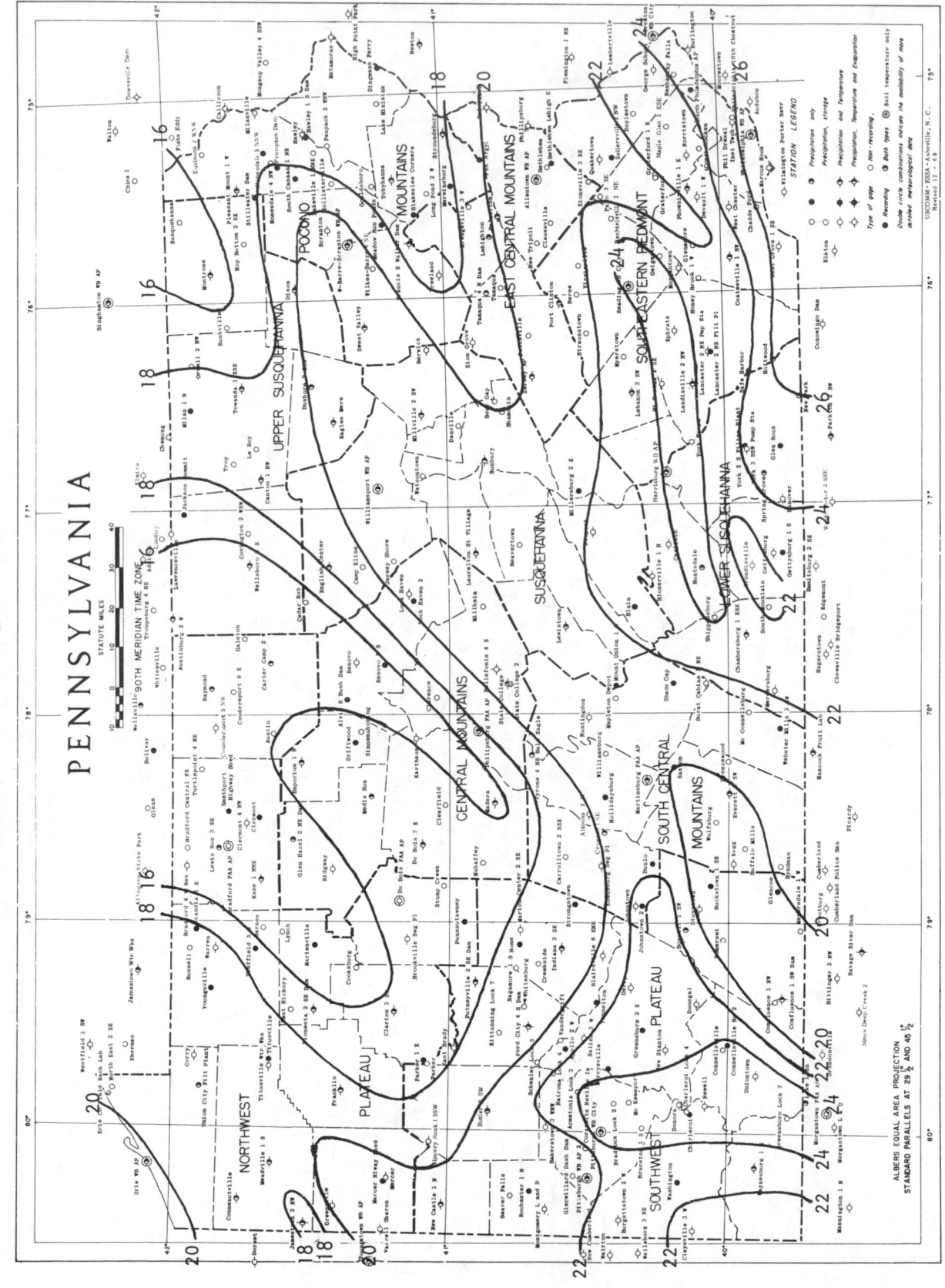

PENNSYLVANIA

Based on the period 1931-60

Isolines are drawn through points of approximately equal value. Caution should be used in interpolating on these maps, particularly in mountainous areas.

854

Mean Maximum Temperature (°F), July

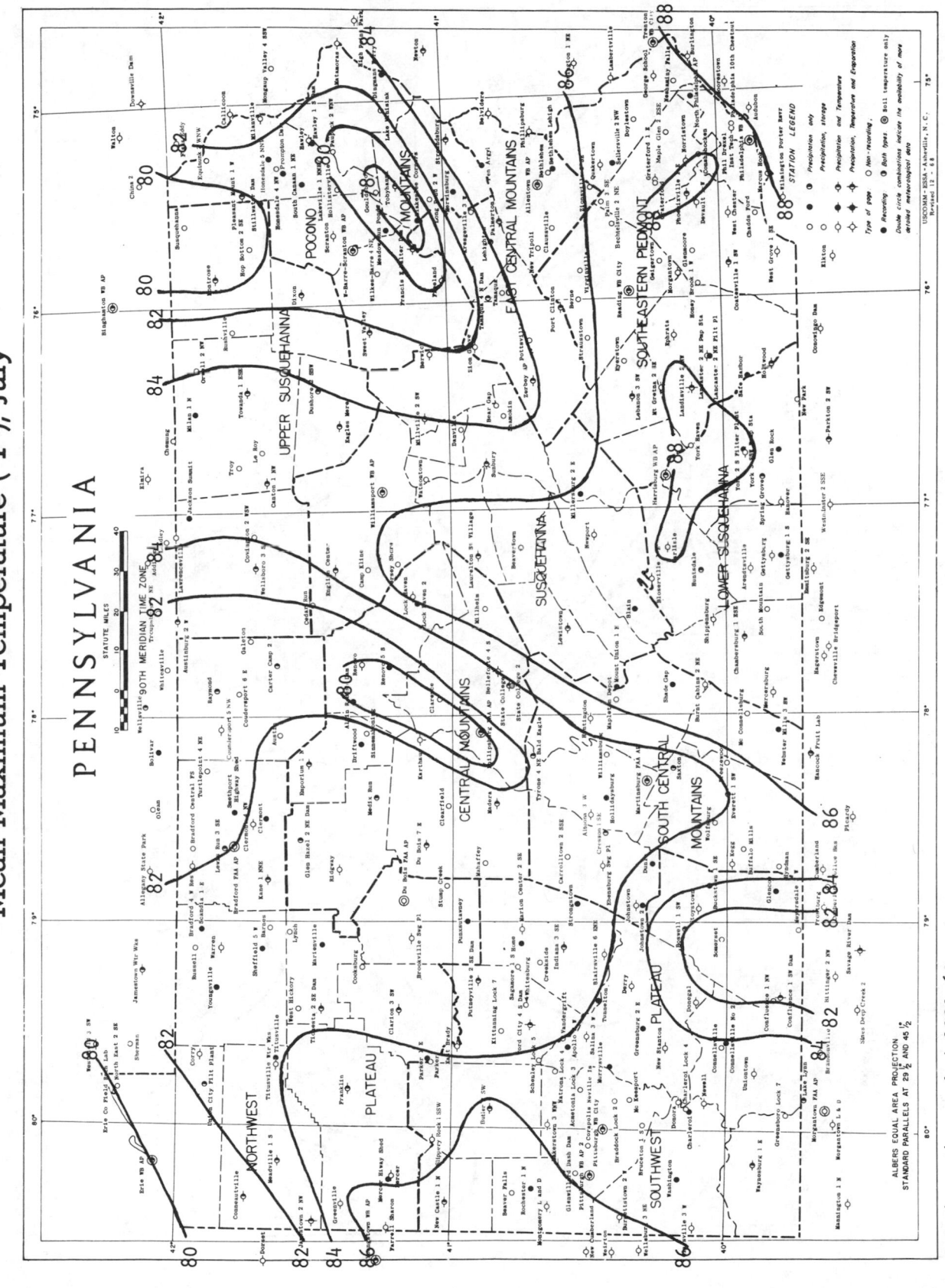

PENNSYLVANIA

STATUTE MILES

ALBERS EQUAL AREA PROJECTION
STANDARD PARALLELS AT 29½ AND 45½

Based on the period 1931-60

Isolines are drawn through points of approximately equal value.   Caution should be used
in interpolating on these maps, particularly in mountainous areas.

Mean Minimum Temperature (°F), July

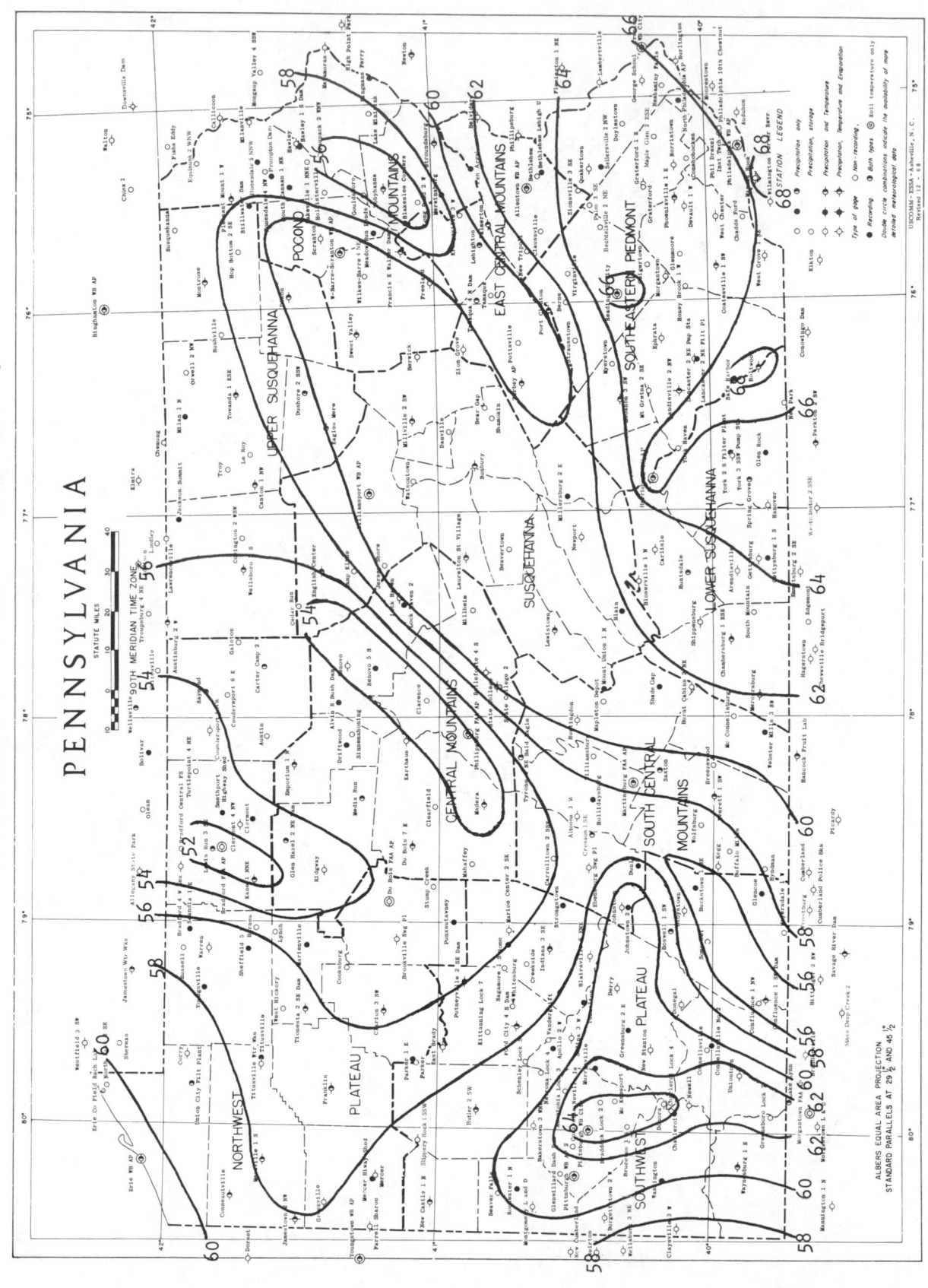

Based on the period 1931-60

Isolines are drawn through points of approximately equal value. Caution should be used in interpolating on these maps, particularly in mountainous areas.

856

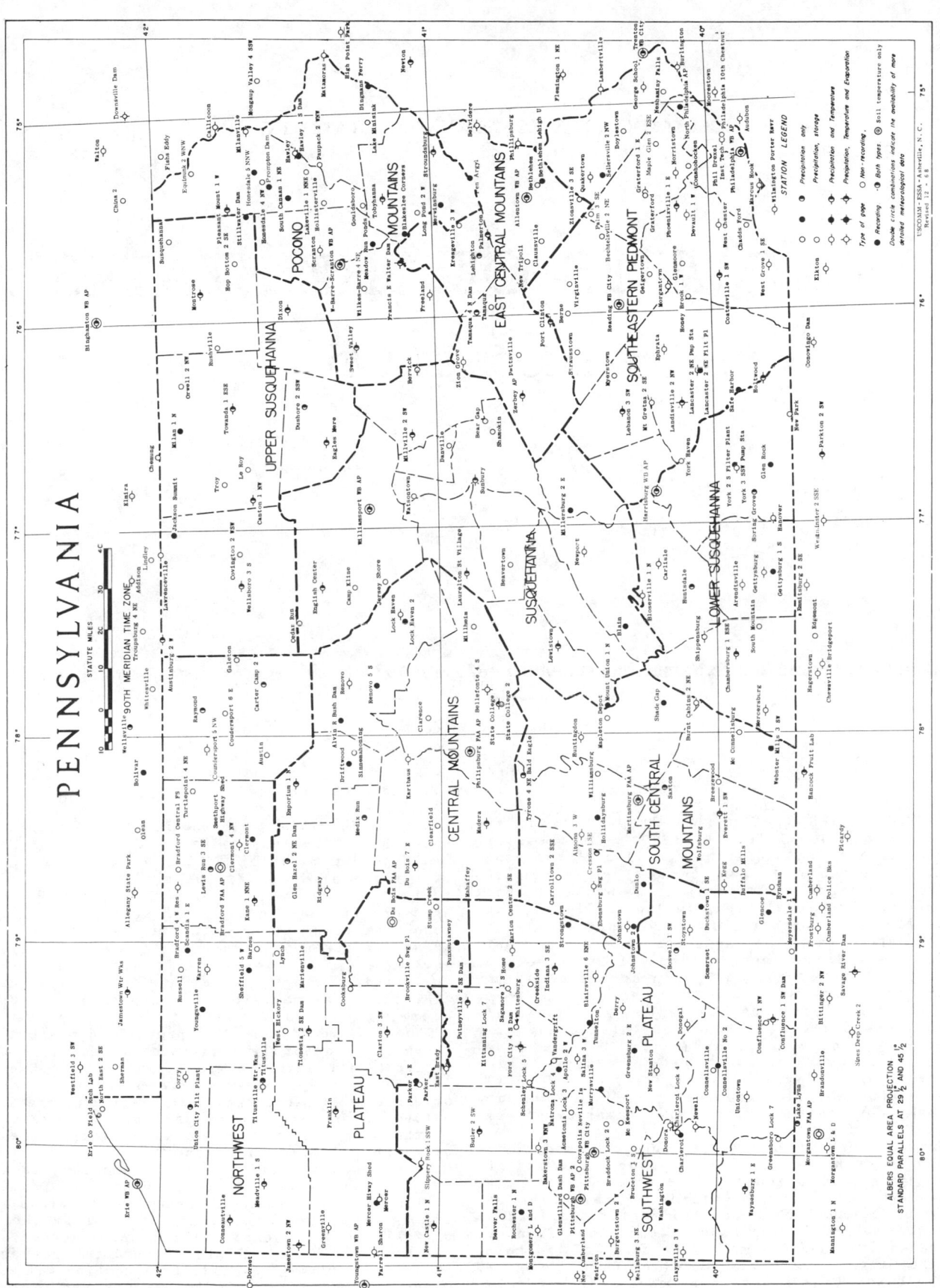

PENNSYLVANIA

STATUTE MILES

90TH MERIDIAN TIME ZONE

ALBERS EQUAL AREA PROJECTION
STANDARD PARALLELS AT 29½ AND 45½

USCOMM - ESSA - Asheville, N. C.
Revised 12 - 68

STATION LEGEND

Precipitation only
Precipitation, storage
Precipitation and Temperature
Precipitation, Temperature and Evaporation

Type of gage
○ Non - recording
◑ Recording
● Both types

Double circle combinations indicate the availability of more
detailed meteorological data
⊙ soil temperature only

NORTHWEST
PLATEAU
SOUTHWEST
CENTRAL MOUNTAINS
SOUTH CENTRAL MOUNTAINS
POCONO MOUNTAINS
UPPER SUSQUEHANNA
SUSQUEHANNA
LOWER SUSQUEHANNA
EAST CENTRAL MOUNTAINS
SOUTHEASTERN PIEDMONT

CLIMATES OF THE STATES

RHODE ISLAND

(Normals, Means and Extremes tables revised 1973 and 1975. Basic report revised November 1959.)

Climate of Rhode Island

A. Boyd Pack, Weather Bureau State Climatologist

PHYSICAL DESCRIPTION: -- Rhode Island, the smallest of the states, shares the southeastern corner of New England with a portion of Massachusetts. The State extends for 50 miles in a north-south direction and has an average width of about 30 miles. The total area, including Block Island some 10 miles offshore, is 1,497 square miles of which Narragansett Bay occupies about 25 percent.

There are three topographical divisions of the State. A narrow coastal plain occurs along the south shore and around Narragansett Bay with an elevation of less than 100 feet. A second division lies to the north and east of the Bay with gently rolling uplands of up to 200 feet elevation. The western two-thirds of Rhode Island consists of predominantly hilly uplands of mostly 200 to 600 feet elevation but rising to a maximum of 800 feet above sea level in the northwest corner of the State.

Narragansett Bay has a very irregular shoreline, indented by numerous small bays or coves and the mouths of the Taunton and Blackstone Rivers. The Bay contains several islands of which the one known as Aquidneck, or Rhode Island, is the largest. The shore line facing Long Island Sound is about 20 miles long and has many fine beaches. No point in the State is more than 25 miles from the ocean.

The Blackstone River in northeastern Rhode Island is the principal river. A number of smaller rivers or brooks originating in the western up-lands of the State or in southeastern Massachusetts empty into Narragansett Bay or Long Island Sound.

GENERAL CLIMATIC FEATURES: -- The chief characteristics of Rhode Island's climate may be summarized as follows: (1) Equable distribution of precipitation among the four seasons; (2) large ranges of temperature both daily and annual; (3) great differences in the same season of different years; and (4) considerable diversity of the weather over short periods of time. These characteristics are modified by nearness to the Bay or ocean, elevation, and nature of the terrain.

Rhode Island lies in the "prevailing westerlies", the belt of generally eastward air movement which encircles the globe in middle latitudes. Embedded in this circulation are extensive masses of air originating in higher and lower latitudes and interacting to produce storm systems. A large number of these systems and air-mass fronts pass near or over Rhode Island in a year.

Air masses affecting the State belong to three types: (1) Cold, dry air pouring down from sub-arctic North America; (2) warm, moist air streaming up on a long overland journey from the Gulf of Mexico and adjacent waters; and (3) cool, damp air moving from the North Atlantic. Because the atmospheric flow is usually from continental areas, Rhode Island is more influenced by the first two types than it is by the third. The ocean constitutes an important modifying factor, particularly in southeast sections of the State, but does not dominate the climate as it would if the prevailing circulation was onshore.

The procession of contrasting air masses and the relatively frequent passage of "Lows" bring about a roughly twice-weekly alternation from fair to cloudy or stormy weather, usually attended by abrupt changes in temperature, moisture, sunshine, wind direction, and speed. There is no regular or persistent rhythm to this sequence, and it is sometimes interrupted by periods of several days, or infrequently of a few weeks with the same weather pattern.

Day-to-day variety rather than monotony is the main feature of Rhode Island's weather. Changeability is also one of its features on a longer time scale. That is, the same month or season will exhibit varying characteristics over the years, sometimes in close alternation and sometimes arranged in similar groups for successive years. A "Normal" month, season, or year is the exception rather than the rule.

As just outlined, the basic climate does not result from the predominance of any single controlling weather regime. It is composed of a large variety of weather patterns. Hence, weather averages in Rhode Island are not useful for important planning purposes and should be supplemented by more detailed climatological analysis.

TEMPERATURE: -- The mean annual temperature ranges from 48° to 49°F., except near the south shore, Narragansett Bay, and in the large built-up area around Providence, where it is 50° to 51°F. Southwestern Rhode Island, from 4 to 10 miles inland, exhibits a coolness not suggested by the nearness to the ocean or the general elevation of 50 to 150 feet. Here the annual mean temperature is not more than 48°F., making the section as cool as the cooler areas of the northwest interior.

The average daily minimum temperature in January and February is 19° to 20°F. over about two-thirds of the State, increasing to near 25°F. in immediate coastal sections. The number of days with minimum temperature of zero or below average one or less per year in the Bay and coastal areas. The number increases to about 5 per year in most of the interior. In a particularly cold winter month as many as 6 to 8 days with such temperatures are observed in southwestern Rhode Island, a few miles inland from the coast.

A maximum temperature of 32°F. or lower occurs on an average of 20 to 25 days per year along the shoreline and 30 to 40 days in the remainder of the state.

Summer temperatures are considerably influenced by proximity to the coastal waters and the frequent onshore flow of air during the warmer months. The average July maximum temperature is about 80°F., except in the northwestern interior where it is a few degrees higher. The greatest number of hot days occur in the metropolitan areas and in parts of the northern interior. Here, about 8 to 10 days of temperatures 90°F. or higher may be expected per year with a variation of from 2 to 5 days in cool summers to 20 or more in exceptionally warm summers. Near the immediate coast the occurrence of 90°F. temperatures is limited to 1 day in the average summer if it occurs at all. Temperatures of 100°F. or higher have been recorded in the northern interior in an occasional year. Among eight weather stations scattered over the State and operated for 14 years by the Rhode Island Agricultural Experiment Station, a temperature of 100°F. was noted only once.

The length of the freeze-free season, as limited by the occurrence of temperatures of 32° or lower, averages from 155 to 180 days in most of the State.

Exceptions are in the southwestern interior with an average length of 130 to 145 days and in the immediate Bay area with 200 days or more. Near the southeastern shore of the Bay the first autumn freezing temperature is considerably delayed compared to the rest of the State. From year to year there is a good deal of variation in the length of period free of temperatures 32°F. or lower.

Grasses and hardy crops begin growing about mid-April in the interior and by early April in areas modified by oceanic influence. The growing season for freeze-sensitive crops starts about a month later in both areas. It comes to a conclusion by early October in the interior and by the end of October or early November in the southeastern corner. Climatic differences of temperature in this small State are very striking in the fall season. Autumnal coloration of foliage will be past its peak of brilliance in the northwestern interior before leaves have begun to noticeably turn color in the Newport area of the southeast.

PRECIPITATION: -- The climate of Rhode Island is characterized by the rather even distribution of precipitation throughout the year. Storm centers and their accompanying fronts are the principal year-round producers of precipitation. Storms moving up the Atlantic coast generally yield the heaviest amounts of rain and snow. Bands and patches of thunderstorms or convective showers contribute considerable precipitation in the summer and make up the difference resulting from decreased activity of the storm centers. In comparison with the general storms, these are of brief duration, but they yield the heaviest local rainfall.

Variations in precipitation from month to month are sometimes extreme in Rhode Island. A month having 5 inches or more may be preceded or followed by one with less than 2 inches of precipitation, in any season. Months with less than 1 inch generally over the State are known to occur as well as those with precipitation in excess of 8 inches. Such large fluctuations, however, are not characteristic of the precipitation supply in the State. Consequently, prolonged droughts are infrequent. So are widespread floods.

Annual precipitation averages 42 to 46 inches over most of the State, with a tendency for decreasing amounts from west to east. It varies from about 40 inches in the immediate southeastern Bay area and on Block Island to 48 inches in the western uplands. Total precipitation in the freeze-free season of April through October shows similar differences over the State with an average of 22 to 24 inches near the Bay and 26 to 29 inches in the western interior.

While there are no pronounced wet and dry months as in other climates, the months of May through July are relatively dry in proximity of the Bay. The average total precipitation for each of these months is 2.5 to 3.0 inches. October and February have an average total of slightly more than 3 inches over most of the State. The remaining months each yield from 3.5 to 4.0 inches.

Measurable precipitation falls on an average of 1 day in 3 or on approximately 120 days per year. Periods of 5 days or more of successive daily precipitation occur a few times during most years. On the other hand, extended periods of little or no precipitation are observed nearly every summer or early fall. Such a period may last from 10 to 20 days.

Twenty-four hour periods with rainfall of more than 4 inches have been observed in all parts of Rhode Island on rare occasions. Such heavy rainfalls have occurred most frequently during the summer and early fall months.

SNOWFALL: -- The average annual snowfall in Rhode Island increases from about 20 inches on

859

Block Island and along the southeast shores of Narragansett Bay to from 40 to 55 inches in the western third of the State. Areas near the western and northern shores of the Bay, including greater Providence, have an average of from 25 to a little more than 30 inches of snow per year. In mild winters the snowfall in the southeast may total only 10 to 15 inches, while a light annual total in the western interior would amount to 25 to 30 inches. The more snowy winters will yield a total of 35 to 40 inches to areas like Block Island and 60 inches or more to the western uplands.

Most of the snow falls in January and February, with each month averaging about 10 inches in the Providence area, for example, and 5 inches at Block Island. However, there are occasional winters when in coastal sections, particularly, heavier monthly amounts will occur in December or March.

In the western and northern portions of the State the first snowfall of 1 inch or more usually occurs in mid or late November. The southeastern Bay area does not observe measurable snow before December in the great majority of years. In about half of the years of record the total winter snowfall has been less than 1.5 inches by January 1 on Block Island. The last measurable snowfall usually occurs by late March in the populous areas of the State, although an April snowstorm is by no means rare.

The average number of days with 1 inch or more of snow on the ground also increases from the shore areas to the western interior. In the latter, a snow cover prevails most of the time from mid or late December to about mid-March. Near the Bay a snow cover does not last more than a few days unless a heavy snowstorm is followed by prolonged cold temperatures.

WINDS AND STORMS: -- The prevailing wind in Rhode Island is northwesterly from December through March and southwesterly in the remaining months. An important feature of the climate is the sea breeze which affects a considerable portion of the State's area. From approximately late spring to midautumn this cool onshore wind blows during the afternoon hours and penetrates from 5 to 10 miles inland. Since much of Rhode Island is within 10 miles of the Sound or Bay, the relatively cool summer maximum temperatures can be accounted for.

Thunderstorms occur on an average of about 20 days per year, although the number in some years may be as many as 30 in the western third of the State. Often these storms are accompanied by high winds and occasionally by destructive hail. They may cause considerable damage to property and crops. Storms of glaze or freezing rain are usually a part of the winter's weather, especially in the Bay area. Extensive disruption of telephone and power lines result, as well as the serious crippling of surface transportation.

Coastal storms or "northeasters", aside from hurricanes, are the most serious weather hazard in Rhode Island. They generate very strong winds and heavy rains, and produce the greatest snowfalls in the winter. Heavy water damage results along the shores of Long Island Sound and Narragansett Bay when these storms occur at the time of high or rising tide.

Hurricanes or storms of tropical origin occasionally affect the State during the summer or fall months as they move on a path well out to sea. However, within the past 20 years 4 hurricanes (1938, 1944, and two in 1954) have passed over Rhode Island. They have caused enormous damage to property and heavy loss of life. In contrast, tornadoes have been recorded only twice in the State within the last 150 years.

While the frequency of floods is not high, a few major ones have occurred in the last 35 years, particularly along the Blackstone River. Localized thunderstorms with heavy and intense rainfall on occasions cause damaging flash floods in the small as well as the larger streams of the State. Major floods have occurred in November 1927, March 1936, July and September 1938, and August 1955.

OTHER CLIMATIC ELEMENTS: -- The percentage of possible sunshine averages 55 to 60 percent, ranging from about 50 percent in the winter months to a little over 60 percent during the summer. The average number of clear and cloudy days per year are about equal with 125 to 130 each. The highest number of clear days per month usually occur in September or October, while the maximum number of cloudy days are noted in December and January.

Heavy fog is observed on an average of about 50 days per year on Block Island and in the southeastern areas of the Bay. This number decreases to 30 or 35 along the western and northern shores of the Bay and to about 25 days in the western interior. Fog occurs on an average of 1 day in 4 during the late spring and early summer months on Block Island. In an especially foggy month as many as 20 days of heavy fog may be observed at this island outpost.

The humidity tends to be lowest in the early spring season and highest in late summer or early fall. While an occasional summer day may be uncomfortable from a combination of high humidity and temperature in the interior and urban areas of Rhode Island, the frequency of such days is much less than in the Southern or Midwestern States.

CLIMATE AND THE STATE ECONOMY: -- The long coastline with numerous beaches and harbors make Rhode Island a strong attraction for bathers, fishermen, and sailing enthusiasts. These recreational activities, favored by the climate, contribute greatly to the economy of this State. Rhode Island is primarily an industrial state with agriculture secondary in importance. The agriculture is of an intensive type and about 10 percent of the land area is devoted to crop and pasture production. An index of crop production shows that crop yields in Rhode Island are above the national average.

The climate plays a significant role in the State's agriculture. A summer mean temperature of close to 70°F. is favorable for the production of alfalfa, mixed grass hays, and pasture. Thus dairying is the leading agricultural enterprise. Poultry raising ranks second and represents about 20 percent of the total farm income.

Potatoes are the major single crop as southern Rhode Island ranks among the important potato producing areas of New England. The cool summer temperatures and adequate precipitation are ideal for this crop. The mean maximum temperature in July, for example, is only slightly warmer than it is in the great potato area of Aroostook County, Maine.

A considerable variety of truck and fruit crops are produced, and together they account for about 10 percent of the total farm income. The truck crops are grown mostly in close proximity to Narragansett Bay. The mild nights, delayed fall freezes, and a reduced freeze hazard in the mid and late spring make this area very satisfactory for these crops. Central Rhode Island contains most of the commercial apple and peach orchards, while various small fruits are produced in scattered areas of the eastern half of the State. Field corn and a small acreage of small grains are grown for livestock feeding.

Forests cover about 75 percent of the total land area. While the income from wood and wood products is relatively small, the forests are a valuable resource in erosion and flood control. They are also a tourist attraction during the autumnal coloration. A wide range of deciduous and coniferous trees are supported by the temperature and precipitation features of the climate. The greenhouse and nursery industry is becoming increasingly important in the agricultural economy of the State. A climate favorable to a wide variety of ornamental trees, shrubs, and flowers has helped to bring this enterprise to third rank in the agricultural economy of Rhode Island.

The growth and development of industry has been aided by the mild, temperate climate. Comfortable summer temperatures, relatively mild winters, and ample rainfall make the climate tolerable to the many aspects of the industrial economy.

SELECTED REFERENCES

General:

1. National Planning Association: The Economic State of New England. 1954.

2. U. S. Dept. of Agriculture: Atlas of American Agriculture. 1936.

3. -------------------------: Climate and Man. Yearbook of Agriculture for 1941.

4. -------------------------: Local Climatological Data with Comparative Data. Issued monthly and annually for Bridgeport, New Haven, and Hartford, Conn.

5. Upton, W.: Characteristics of the New England Climate. Annals Harvard Astron. Observatory. 1890.

Specialized:

1. Brooks, C. F.: New England Snowfall. Monthly Weather Review, Vol. 45, 1917.

2. ------------: The Rainfall of New England., General Statement. Journal New Eng. Water Works Assoc., Vol. 44, 1930.

3. Church, P. E.: A Geographical Study of New England Temperatures. Geographical Review, Vol. 26, 1936.

4. Eustis, R. S.: Winds over New England in Relation to Topography. Bulletin Amer. Met. Soc., Vol. 23, 1942.

5. Goodnough, X. H.: Rainfall in New England, Jour. New Eng. Water Works Assoc., Vols. 29, 1915; 35, 1921; 40, 1926.

6. Harris, B. K., and Odland, T. E.: Rhode Island Weather. Rhode Island Agricultural Experiment Station, Bull. 299 (1948)

7. Perley, S.: Historic Storms of New England. 1891.

8. Stone, R. G.: Distribution of snow depths over New York and New England. Trans. Amer. Geophy. Union. 1940.

9. --------------: The average length of the season with snow cover of various depths in New England. Trans. Amer. Geophy. Union. 1944.

10. Westveld, Marinus, et al: Natural Forest Vegetation Zones in New England. Jour. of Forestry, Vol. 54, 332-338. 1956.

11. Weber, J. H.: The Rainfall of New England. Historical Statement. Annual Rainfall. Seasonal Rainfall. Mean Monthly Rainfall of Southern New England. Maximum and Minimum Rainfall of Southern New England. Jour. New Eng. Water Works Assoc., Vol. 44, 1930.

12. White, G. V.: Rainfall in New England. Jour. New Eng. Water Works Assoc. Vols. 56, 1942; 57, 1943.

13. Brown, R. A.: Twisters in New England. Unpublished compilation of historical records of tornadoes. 1957.

14. Weather Bureau Technical Paper No. 15 - Maximum Station Precipitation for 1, 2, 3, 6, 12, and 24-Hours.

15. Weather Bureau Technical Paper No. 16 - Maximum 24-Hour Precipitation in the United States. Washington, D. C. 1952.

16. Weather Bureau Technical Paper No. 25 - Rainfall Intensity-Duration-Frequency Curves. For selected stations in the United States. Alaska, Hawaiian Islands, and Puerto Rico.

BIBLIOGRAPHY

(A) Climatic Summary of the United States (Bulletin W) 1930 edition, Massachusetts, Rhode Island and Connecticut. Section 86. U. S. Weather Bureau

(B) Climatic Summary of the United States, New England. Supplement for 1931 through 1952 (Bulletin W Supplement). U. S. Weather Bureau

(C) Climatological Data - Rhode Island. U. S. Weather Bureau

(D) Climatological Data National Summary. U. S. Weather Bureau

(E) Hourly Precipitation Data - New England. U. S. Weather Bureau

(F) Local Climatological Data, U. S. Weather Bureau, for Block Island and Providence, Rhode Island.

FREEZE DATA

STATION	Freeze threshold temperature	Mean date of last Spring occurrence	Mean date of first Fall occurrence	Mean No. of days between dates	Years of record Spring	No. of occurrences in Spring	Years of record Fall	No. of occurrences in Fall	STATION	Freeze threshold temperature	Mean date of last Spring occurrence	Mean date of first Fall occurrence	Mean No. of days between dates	Years of record Spring	No. of occurrences in Spring	Years of record Fall	No. of occurrences in Fall
	RHODE ISLAND									CONNECTICUT							
BLOCK ISLAND WB CITY	32	04-09	11-16	221	30	30	30	30	HARTFORD	32	04-22	10-19	180	30	30	30	30
	28	03-23	11-27	249	30	30	30	30		28	04-06	11-02	210	30	30	30	30
	24	03-14	12-03	264	30	30	30	30		24	03-24	11-17	239	30	30	30	30
	20	03-08	12-09	276	30	30	30	27		20	03-13	11-29	261	30	30	30	30
	16	02-28	12-15	290	30	30	30	24		16	03-08	12-07	273	30	30	30	29
KINGSTON	32	05-08	10-05	150	30	30	30	30	NEW HAVEN WB AP	32	04-15	10-27	195	30	30	30	30
	28	04-24	10-16	176	30	30	30	30		28	03-28	11-10	226	30	30	30	30
	24	04-08	10-28	203	30	30	29	29		24	03-18	11-23	249	30	30	30	30
	20	03-24	11-16	237	30	30	29	29		20	03-10	12-03	268	30	30	30	29
	16	03-12	11-26	259	30	30	29	29		16	03-05	12-10	280	30	30	30	27
PROVIDENCE WB CITY	32	04-13	10-27	197	30	30	30	30	PUTNAM	32	05-15	09-24	133	15	15	17	17
	28	04-01	11-11	224	29	29	30	30		28	04-29	10-08	162	15	15	17	17
	24	03-18	11-24	251	29	29	30	30		24	04-13	10-19	189	15	15	17	17
	20	03-12	12-02	265	29	29	30	30		20	03-30	11-06	221	15	15	16	16
	16	03-05	12-07	276	29	29	30	29		16	03-19	11-18	244	15	15	16	16

Data in the above table are based on the period 1921-1950, or that portion of this period for which data are available.

Means have been adjusted to take into account years of non-occurrence.

A freeze is a numerical substitute for the former term "killing frost" and is the occurrence of a minimum temperature at or below the threshold temperature of 32°, 28°, etc.

Freeze data tabulations in greater detail are available and can be reproduced at cost.

TROPICAL CYCLONE DATA HAVING IMPORTANCE FOR THIS STATE IS INCLUDED IN STATISTICS AND CHARTS ON PAGES 1161 THROUGH 1164.

*MEAN TEMPERATURE AND PRECIPITATION

STATION	JANUARY Temperature	JANUARY Precipitation	FEBRUARY Temperature	FEBRUARY Precipitation	MARCH Temperature	MARCH Precipitation	APRIL Temperature	APRIL Precipitation	MAY Temperature	MAY Precipitation	JUNE Temperature	JUNE Precipitation	JULY Temperature	JULY Precipitation	AUGUST Temperature	AUGUST Precipitation	SEPTEMBER Temperature	SEPTEMBER Precipitation	OCTOBER Temperature	OCTOBER Precipitation	NOVEMBER Temperature	NOVEMBER Precipitation	DECEMBER Temperature	DECEMBER Precipitation	ANNUAL Temperature	ANNUAL Precipitation
RHODE ISLAND																										
BLOCK ISLAND WB AP	31.9	3.67	30.9	3.25	37.1	3.54	44.9	3.37	54.2	2.96	63.0	2.86	69.1	2.55	69.0	3.46	63.5	2.98	54.5	3.10	45.3	3.53	35.1	3.36	49.9	38.63
KINGSTON	29.5	4.12	29.7	3.27	36.7	4.26	45.6	3.83	55.7	3.49	64.2	3.06	70.1	2.57	68.9	4.66	62.2	3.74	52.5	3.21	42.2	4.74	31.7	3.75	49.1	44.70
PROVIDENCE WB AIRPORT	28.7	3.75	28.6	2.84	36.8	3.58	46.0	3.37	56.8	3.02	65.6	3.17	71.0	3.06	69.4	3.63	62.7	3.19	52.7	2.83	42.6	3.74	31.6	3.45	49.4	39.63
DIVISION	30.7	3.99	30.5	3.15	37.4	4.16	46.6	3.78	56.7	3.31	65.4	3.02	71.5	2.55	70.3	4.33	63.6	3.50	54.3	3.15	43.9	4.44	33.3	3.61	50.4	42.99
CONNECTICUT																										
NORTHWEST																										
CREAM HILL	25.0	3.55	25.4	2.99	33.7	3.88	45.4	3.76	56.9	4.23	65.4	4.70	70.4	4.59	68.5	4.24	61.2	4.24	51.6	3.34	39.3	4.14	27.5	3.60	47.5	47.26
FALLS VILLAGE	25.5	3.09	26.0	2.38	34.6	3.39	45.7	3.55	57.2	3.97	65.7	4.58	70.3	4.01	68.4	3.92	60.9	4.14	50.5	3.03	39.8	3.74	27.7	2.93	47.7	42.73
DIVISION	24.7	3.56	25.1	2.91	33.6	3.91	45.1	3.78	56.6	4.31	64.9	4.78	69.8	4.20	67.8	4.33	60.4	4.21	50.5	3.37	38.7	4.26	27.3	3.57	47.0	47.19
CENTRAL																										
COLLINSVILLE 1 S		4.05		3.25		4.66		4.14		4.30		4.30		3.87		4.91		3.96		3.37		4.55		3.97		49.33
HARTFORD BRAINARD FLD	27.0	3.74	27.5	3.03	36.9	3.53	47.4	3.55	58.9	3.77	67.7	3.76	72.7	3.93	70.4	3.67	63.1	3.41	52.6	2.70	41.7	3.85	30.1	3.49	49.7	42.43
HARTFORD WB AIRPORT	27.0	3.15	28.1	2.57	37.2	3.81	48.0	3.56	59.7	3.66	68.9	3.62	73.8	3.56	71.4	3.54	63.8	3.44	52.9	2.80	41.3	3.48	29.6	3.29	50.1	40.48
MIDDLETOWN WB	26.8	3.91	27.1	3.01	36.0	3.92	46.3	3.96	57.0	4.14	66.7	3.80	71.6	3.23	69.7	3.54	63.0	3.49	52.3	2.81	41.3	4.40	29.8	3.93	49.0	44.14
MIDDLETOWN 4 W	28.3	4.19	28.7	3.21	36.5	4.71	47.4	4.38	58.7	4.48	67.3	4.18	72.7	3.72	70.6	4.51	63.2	4.40	53.8	3.48	42.4	5.06	31.0	4.20	50.1	50.52
STORRS	26.5	3.62	26.7	2.82	34.5	4.34	45.2	3.84	56.4	4.04	65.1	3.65	70.2	3.56	68.4	5.13	61.2	4.07	51.7	3.35	40.6	4.23	29.1	3.56	48.0	46.21
DIVISION	28.0	3.80	28.4	2.92	36.5	4.33	47.1	3.92	58.1	4.03	66.8	3.82	72.1	3.69	70.0	4.62	62.7	4.04	53.0	3.29	41.7	4.47	30.4	3.79	49.6	46.72
COASTAL																										
BRIDGEPORT WB AP	29.2	3.43	29.0	2.97	36.9	3.60	46.3	3.49	57.2	3.60	66.9	3.47	72.8	3.97	71.7	4.43	65.2	3.55	54.4	2.83	43.5	3.59	32.3	3.08	50.5	42.01
LAKE KONOMOC		4.36		3.44		4.99		4.18		4.19		3.55		3.73		4.83		4.49		3.86		4.97		4.29		50.88
NEW HAVEN WB AIRPORT	29.1	3.89	29.1	3.30	37.1	4.12	46.1	3.89	56.7	3.87	65.8	3.81	71.2	3.66	69.8	4.11	63.6	3.46	53.3	3.00	42.9	3.94	31.9	3.94	49.7	44.99
DIVISION	30.1	3.98	30.4	3.09	37.8	4.71	47.9	3.91	58.4	4.00	67.3	3.61	72.8	3.59	71.1	4.95	64.3	4.08	54.4	3.42	43.4	4.30	32.4	3.85	50.9	47.49

* Averages for period 1931-1955, except for stations marked WB which are "normals" based on period 1921-1950. Divisional means may not be the arithmetical average of individual stations published, since additional data from shorter period stations are used to obtain better areal representation.

CONFIDENCE LIMITS

In the absence of trend or record changes, the chances are 9 out of 10 that the true mean will lie in the interval formed by adding and subtracting the values in the following table from the means for any station in the State:

1.5	.45	1.3	.29	1.4	.50	.9	.52	.8	.54	.7	.68	.5	.63	.7	.78	.7	.64	.9	.61	.9	.80	1.1	.55	.4	2.41

COMPARATIVE DATA

Data in the following table are the mean temperature and average precipitation for Kingston, Rhode Island, for the period 1906 - 1930 and are included in this publication for comparative purposes :

27.7	4.79	26.9	4.24	35.1	4.22	44.1	4.47	54.3	3.99	63.2	3.41	68.8	3.44	67.3	4.50	61.6	3.46	52.1	3.80	40.8	4.13	30.6	5.04	47.7	49.49

NORMALS, MEANS AND EXTREMES
(Table Revised 1973. Base Period for Climatological Normals: 1931-1960)

Station: BLOCK ISLAND, RHODE ISLAND   STATE AIRPORT   Standard time used: EASTERN   Latitude: 41° 10' N   Longitude: 71° 35' W   Elevation (ground): 110 feet

Month	Temperature Normal Daily maximum	Normal Daily minimum	Normal Monthly	Extremes Record highest	Year	Record lowest	Year	Normal heating degree days (Base 65°)	Precipitation Normal total	Max monthly	Year	Min monthly	Year	Max in 24 hrs	Year	Snow, Ice pellets Mean total	Max monthly	Year	Max in 24 hrs.	Year	Relative humidity Hour 01	Hour 07	Hour 13	Hour 19	Mean sky cover	Clear	Partly cloudy	Cloudy	Precip. .01 in+	Snow 1.0 in+	Thunder-storms	Heavy fog	Max 90°+	Max 32°-	Min 32°-	Min 0°-	Avg daily solar radiation
(a)	(b)	(b)	(b)	22		22		(b)	(b)	22		22		22		22	22		22		14	14	14	14	18	18	18	18	22	22	14	14	22	22	22	22	22
J	37.9	26.3	32.1	57	1968	-2	1968	1020	3.84	6.74	1958	0.27	1970	4.06	1962	5.1	21.5	1952	9.0	1965		73	65	65	6.5	8	9	14	11	2	*	4	0	8	24	**
F	36.7	25.1	30.9	55	1951	-2	1961	955	3.29	6.88	1971	1.20	1968	2.86	1972	5.2	16.9	1961	16.5	1961		73	65	65	6.0	8	8	14	11	2	*	4	0	8	22	*
M	42.2	31.2	36.7	60	1960	6	1967	877	4.07	8.52	1959	1.16	1966	3.63	1968	6.3	24.1	1956	11.5	1960		75	65	67	5.8	9	9	13	13	2	*	5	0	4	17	0
A	51.7	38.4	45.4	73	1954	21	1954	612	3.61	6.24	1961	1.15	1960	2.50	1960	0.3	1.0	1965	1.0	1965		79	80	66	5.3	8	11	13	11	*	0	1	0	1	3	0
M	60.1	47.6	53.9	82	1969	34	1972+	344	3.01	5.98	1967	0.72	1955	2.35	1961	0.0	0.0		0.0			80	84	67	6.2	7	11	13	10	1	0	0	0	0	0	0
J	69.7	56.2	62.9	90	1952	41	1967	99	2.56	6.81	1962	T	1957	2.39	1972	0.0	0.0		0.0			84	69		6.2	9	10	11	8	2	0	0	0	0	0	*
J	75.6	63.2	69.6	91	1972+	51	1962	0	2.69	6.15	1959	0.39	1959	2.66	1959	0.0	0.0		0.0			87	72		6.8	7	11	13	8	2	0	0	0	0	0	0
A	75.2	63.6	69.0	88	1962+	48	1971	16	3.86	9.73	1954	0.26	1964	4.86	1964	0.0	0.0		0.0			86	71		6.6	7	11	13	7	2	0	0	0	0	0	0
S	69.8	57.8	64.0	87	1953	43	1969+	38	3.20	11.51	1961	0.33	1971	8.52	1960	0.0	0.0		0.0			84	70		5.5	10	9	11	6	1	*	0	0	0	0	*
O	61.1	49.2	55.2	77	1967	30	1967	307	3.02	8.74	1952	0.81	1952	6.63	1969	T	T		T			80	66		5.4	11	8	12	8	1	1	0	0	5	0
N	52.1	40.1	45.9	70	1950	20	1969+	594	3.71	8.06	1955	1.27	1955	6.39	1955	0.2	2.5	1955	2.5	1955		76	65		6.1	7	11	12	11	1	1	0	1	11	5	*
D	42.1	29.6	35.9	64	1953	-4	1962	902	3.59	8.12	1967	0.83	1969	4.39	1967	3.2	10.4	1963	6.3	1960		72	65		6.1	9	10	14	12	1	3	0	4	18	5	*
YR	56.2	43.9	50.1	91 JUL. 1972+		-4 DEC. 1962		5804	40.45	11.51 SEP. 1961		T JUN. 1957		8.52 SEP. 1960		21.3	24.1 MAR. 1956		16.9 FEB. 1961		79	67		6.2	98	114	153	112	7	17	82	20	90	22	

NORMALS, MEANS AND EXTREMES
(Table Revised 1975. Base Period for Climatological Normals: 1941-1970)

Elev.: 118 feet m.s.l.   Average station pressure mb.

Month	Temperature Normal Daily maximum	Normal Daily minimum	Normal Monthly	Extremes Record highest	Year	Record lowest	Year	Normal Degree days Base 65°F Heating	Cooling	Precipitation Water equivalent Normal	Max monthly	Year	Min monthly	Year	Max in 24 hrs.	Year	Snow, Ice pellets Max monthly	Year	Max in 24 hrs.	Year	Relative humidity Hour 01	Hour 07	Hour 13	Hour 19	Mean sky cover	Clear	Partly cloudy	Cloudy	Precip. .01 in+	Snow 1.0 in+	Thunder-storms	Heavy fog	Max 90°+	Max 32°-	Min 32°-	Min 0°-
(a)				24		24				24	24		24		24		24		24		14	14	14	14	19	19	19	19	24	24	14	14	24	24	24	24
J	37.3	25.4	31.4	57	1967	-2	1968	1042	0	3.41	6.74	1958	0.27	1970	4.06	1962	21.5	1965	9.0	1965		73	65	65	6.5	8	9	14	11	2	*	4	0	8	24	**
F	37.2	25.1	31.3	55	1974	-2	1961	944	0	3.32	6.88	1971	1.20	1968	2.86	1972	16.9	1961	16.5	1961		73	65	65	6.0	8	9	13	11	2	*	4	0	8	22	*
M	43.0	31.0	36.8	60	1973	6	1967	871	0	3.88	8.52	1959	1.16	1966	3.63	1968	24.1	1956	11.5	1956		75	65	67	5.8	9	9	13	12	2	*	5	0	4	17	0
A	51.7	38.8	45.3	73	1973	21	1954	591	0	3.81	7.78	1973	1.15	1963	2.67	1973	0.0		0.0			79	80	66	5.3	8	11	11	10	*	0	1	0	1	3	0
M	60.4	47.4	53.8	82	1969	34	1972	347	10	3.25	5.98	1967	1.15	1955	2.35	1961	0.0		0.0			80	84	67	6.2	7	11	13	10	1	0	0	0	0	0	*
J	69.7	56.2	63.0	90	1952	41	1967	82	25	2.20	6.81	1962	0.72	1957	2.39	1957	0.0		0.0			84	69		6.3	9	10	11	8	2	0	0	*	0	0	*
J	75.6	63.3	69.5	91	1972	51	1962	9	149	2.74	6.15	1959	0.39	1959	3.16	1952	0.0		0.0			87	72		6.6	7	11	13	8	2	4	0	*	0	0	0
A	75.3	63.7	69.8	88	1973	48	1971	11	142	3.86	9.73	1954	0.26	1964	4.86	1964	0.0		0.0			86	70		6.0	7	11	12	7	2	4	0	*	0	0	0
S	69.8	57.7	63.8	87	1953	43	1969	79	43	3.00	11.51	1961	0.83	1955	6.53	1955	0.0		0.0			84	70		5.5	10	8	12	7	1	1	0	0	0	0	0
O	61.8	49.2	55.3	77	1956	30	1967	301	0	2.88	8.74	1952	0.23	1972	8.06	1969	2.5	1970	2.5	1970		80	66		5.4	11	9	12	7	1	1	0	0	5	0	*
N	51.8	40.1	46.0	70	1955	20	1957	570	0	3.86	8.06	1955	1.23	1955	6.39	1955	10.4	1955	6.3	1960		76	65		6.1	7	10	14	10	1	1	0	1	11	5	0
D	41.2	29.1	35.2	64	1953	-4	1962	924	0	4.11	8.12	1967	0.63	1969	4.39	1967	10.4	1963	6.3	1960		72	65		6.2	9	10	14	11	1	*	0	4	18	5	*
YR	56.2	43.9	50.1	91 AUG 1973		-4 DEC 1962		5771	359	40.51	11.51 SEP 1961		T		8.52 SEP 1960		24.1 MAR 1956		16.9 FEB 1961		79	67		6.2	97	114	154	111	6	17	82	20	89	24	

REFERENCE NOTES APPLYING TO TABLES APPEAR ON THE PAGE FOLLOWING LAST TABLE.
(Caution: Letters and symbols may have different meanings in 1941-1970 tables than in earlier tables. See notes.)

NORMALS, MEANS AND EXTREMES
(Table Revised 1973. Base Period for Climatological Normals: 1931-1960)

Station: PROVIDENCE, RHODE ISLAND THEO FRANCIS GREEN STATE AP Standard time used: EASTERN Latitude: 41° 44' N Longitude: 71° 26' W Elevation (ground): 51 feet

Month	Temperature Normal Daily maximum	Normal Daily minimum	Normal Monthly	Extremes Record highest	Year	Record lowest	Year	Normal heating degree days (Base 65°)	Precipitation Normal total	Maximum monthly	Year	Minimum monthly	Year	Maximum in 24 hrs	Year	Snow, ice pellets Mean total	Maximum monthly	Year	Maximum in 24 hrs	Year	Relative humidity Hour 01	Hour 07	Hour 13	Hour 19	Wind Mean speed	Prevailing direction	Fastest mile Speed	Direction	Year
(a)	(b)	(b)	(b)	9	9	9	9	(b)	(b)	19		19		19		19	19		19		9	9	13	19	19	10	19	19	19
J	37.3	21.0	29.2	65	1967	-5	1971+	1110	3.81	7.12	1958	0.50	1970	3.34	1962	9.6	28.7	1965	10.5	1965	68	68	55	62	11.7	NW	46	18	1957
F	38.3	21.1	29.7	56	1967	-5	1967	988	3.10	5.63	1960	1.31	1968	2.72	1970	9.0	30.9	1962	18.3	1962	68	68	55	61	12.0	NNW	46	16	1972+
M	45.3	28.6	37.0	73	1964	1	1967	868	4.14	7.83	1968	1.72	1965	4.53	1968	10.5	31.6	1956	16.9	1956	68	70	54	59	12.6	NNW	60	20	1959
A	56.6	37.7	47.2	84	1964	19	1964	534	3.75	7.32	1961	1.48	1965	2.82	1970	0.7	3.7	1958	3.7	1958	70	68	47	59	12.6	SW	51	20	1956
M	67.7	47.2	57.5	94	1972+	32	1967	236	3.35	7.27	1967	0.71	1964	3.76	1967	0.0	0.0		0.0		75	70	50	61	11.3	SW	42	20	1957
J	76.1	56.3	66.2	94	1964	41	1967	51	2.76	6.83	1972	0.39	1957	2.09	1972	0.0	0.0		0.0		82	75	56	66	10.2	SW	40	20	1954
J	81.4	62.7	72.1	97	1964	49	1965	0	2.91	6.29	1958	1.00	1970	2.80	1958	0.0	0.0		0.0		81	77	57	67	9.7	SW	35	34	1964
A	80.0	60.9	70.5	97	1968	40	1965	16	3.96	11.01	1955	0.91	1956	5.47	1955	0.0	0.0		0.0		81	79	58	73	9.7	SSW	90	11	1954
S	73.1	53.3	63.2	93	1968	35	1965	96	3.52	7.92	1961	0.77	1961	4.89	1961	0.0	0.0		0.0		83	80	56	73	9.9	SW	58	18	1950
O	63.4	43.2	53.3	85	1968+	21	1969	372	3.10	11.89	1962	0.77	1963	6.63	1962	0.1	1.6	1969	1.6	1969	80	80	53	70	9.9	SW	52	18	1954
N	52.2	33.8	43.0	73	1971	14	1972	660	4.11	8.65	1972	2.08	1963	3.04	1963	3.6	19.8	1955	11.2	1955	76	73	59	70	10.7	SW	48	14	1957
D	40.4	23.5	32.0	69	1965	-2	1968	1023	3.62	10.75	1969	0.58	1955	3.85	1969	8.4			11.2	1963	74	76	61	69	11.2	NNW			
YR	59.3	40.8	50.1	97	JUL. 1964	-5	JAN. 1971+	5954	42.13	11.89	OCT. 1962	0.39	JUN. 1957	6.63	OCT. 1962	40.1	31.6	MAR. 1956	18.3	FEB. 1961	76	74	55	66	10.9	SW	90	11	AUG. 1954

Mean number of days:

Month	Mean sky cover sunrise to sunset	Sunrise to sunset Clear	Partly cloudy	Cloudy	Precipitation .01 inch or more	Snow, ice pellets 1.0 inch or more	Thunderstorms	Heavy fog	Temperatures Max. 90° and above	Max. 32° and below	Min. 32° and below	Min. 0° and below	Average daily solar radiation - langleys	Pct. of possible sunshine
	19	19	19	19	19	19	19	19	9	9	9	9	9	19
J	6.2	11	8	12	14	3	*	2	0	1	28	2	2	57
F	6.5	7	8	13	11	3	*	2	0	0	26	2	1	56
M	6.6	8	8	15	13	3	1	2	0	0	22	2	0	56
A	6.0	8	10	12	11	*	1	2	0	0	8	0	0	56
M	6.6	7	10	14	11	0	3	2	*	0	*	0	0	58
J	6.2	7	12	11	10	0	4	2	2	0	0	0	0	59
J	6.4	9	12	11	9	0	5	2	2	0	0	0	0	59
A	6.0	8	12	11	9	0	4	2	1	0	0	0	*	58
S	5.7	12	8	10	8	0	2	3	1	0	0	0	*	55
O	5.5	12	7	12	8	0	1	3	0	0	4	0	4	48
N	6.4	8	8	13	12	1	*	2	0	0	14	0	4	52
D	6.1	8	6	15	13	2	1	2	0	1	24	7	3	52
YR	6.2	106	103	156	124	11	21	26	7	28	127			56

Ø For period November 1963 through the current year.
Means and extremes above are from existing and comparable exposures. Annual extremes have been exceeded at other sites in the locality as follows:
Highest temperature 102 in August 1948; lowest temperature -17 in February 1934; maximum monthly precipitation 12.24 in August 1946; minimum monthly precipitation .04 in June 1949; maximum monthly snowfall 31.9 in January 1948; fastest mile wind 95 SW in September 1938.

NORMALS, MEANS AND EXTREMES
(Table Revised 1975. Base Period for Climatological Normals: 1941-1970)

Month	Temperatures °F Normal Daily maximum	Normal Daily minimum	Normal Monthly	Extremes Record highest	Year	Record lowest	Year	Normal Degree days Base 65°F Heating	Cooling	Precipitation in inches Normal	Maximum monthly	Year	Minimum monthly	Year	Maximum in 24 hrs	Year	Water equivalent Maximum monthly	Year	Minimum monthly	Year	Maximum in 24 hrs	Year	Snow, ice pellets Maximum monthly	Year	Maximum in 24 hrs	Year
(a)	(b)	(b)	(b)	11	11	11	11			21	21		21		21		21		21		21		21		21	
J	36.2	20.6	28.4	66	1974	-5	1971	1135	0	3.52	7.12	1958	0.50	1970	3.34	1962	2.85	1958	0.91	1970	2.80	1958	28.7	1965	10.6	1965
F	37.6	21.2	29.4	59	1974	-5	1967	997	0	3.45	5.63	1960	1.31	1968	2.72	1970	2.90	1960	1.12	1968	5.47	1955	30.9	1962	18.3	1962
M	44.7	29.0	36.9	73	1964	1	1967	871	0	3.99	7.83	1968	1.72	1965	4.53	1968	3.26	1968	1.56	1961	4.63	1961	31.6	1956	16.9	1956
A	56.6	37.8	47.3	94	1974	19	1964	531	0	3.72	7.32	1961	1.48	1965	2.82	1970	3.27	1961	1.48	1965	6.63	1959	3.7	1958	3.7	1958
M	66.8	46.9	56.9	94	1964	32	1967	259	8	3.49	7.27	1967	0.71	1964	3.76	1967	4.52	1972	1.56	1972	3.04	1963	0.0		0.0	
J	76.5	56.5	66.4	94	1973	41	1967	36	78	2.49	6.83	1972	0.39	1957	2.09	1972	4.13	1955	0.58	1955	3.85	1969	0.0		0.0	
J	81.1	63.0	72.1	97	1964	49	1965	0	224	2.80													0.0		0.0	
A	79.1	61.0	70.4	97	1973	40	1965	0	177	2.90													0.0		0.0	
S	71.6	53.6	62.6	93	1968	34	1974	93	45	3.26													0.0		0.0	
O	63.0	43.7	53.4	85	1969	21	1969	350	0	3.27													0.1	1962	1.6	1962
N	52.6	34.6	43.6	73	1974	14	1972	651	0	4.52													3.6	1955	3.6	1955
D	39.6	23.4	31.5	69	1965	-2	1968	1039	0	4.13													19.8	1969	11.2	1969
YR	59.0	40.9	50.0	97	JUL 1964	-5	JAN 1971	5972	532	42.75	11.89	OCT 1962	0.39	JUN 1957	6.63	OCT 1962	11.89	OCT 1962	0.39	JUN 1957	6.63	OCT 1962	31.6	MAR 1956	18.3	FEB 1961

Mean number of days:

Month	Mean sky cover, tenths, sunrise to sunset	Sunrise to sunset Clear	Partly cloudy	Cloudy	Precipitation .01 inch or more	Snow, ice pellets 1.0 inch or more	Thunderstorms	Heavy fog, visibility ¼ mile or less	Temperatures °F Max. 90° and above	Max. 32° and below	Min. 32° and below	Min. 0° and below	Average station pressure mb. Elev. 62 feet m.s.l.	Relative humidity pct. Hour 01	Hour 07	Hour 13	Hour 19	Wind Mean speed m.p.h.	Prevailing direction	Fastest mile Speed m.p.h.	Direction	Year	Pct. of possible sunshine
	21	21	21	21	21	21	21	21	11	11	11	11	2	11	11	11	11	21	10	21	21	21	21
J	6.2	10	7	14	11	3	*	2	0	1	28	2	1015.8	69	71	56	63	11.6	NW	46	18	1957	57
F	6.5	9	7	13	11	3	*	2	0	0	26	1	1014.3	70	65	55	61	12.0	NNW	46	16	1959	56
M	6.6	8	8	14	11	2	1	2	0	0	21	*	1013.4	70	70	54	60	12.4	NW	60	20	1959	55
A	6.7	8	10	13	11	*	1	2	0	0	7	0	1011.4	68	69	52	58	12.5	SW	51	20	1956	56
M	6.4	7	10	14	11	0	3	2	*	0	*	0	1011.1	75	71	56	63	11.2	SW	42	20	1957	58
J	6.3	7	13	11	10	0	4	2	2	0	0	0	1013.6	82	76	57	67	10.1	SW	40	20	1954	58
J		9	11	11	9	0	5	2	3	0	0	0	1012.5	83	78	57	69	9.6	SW	35	34	1964	59
A	6.4	8	11	12	10	0	5	3	2	0	0	0	1015.3	82	78	54	69	9.5	SSW	90	11	1954	59
S	5.8	11	8	11	8	0	3	3	1	0	0	0	1015.3	84	79	56	70	9.7	SW	58	18	1950	59
O	5.3	12	7	12	8	0	1	3	0	0	4	0	1017.8	82	79	52	70	9.8	NW	52	18	1954	48
N	6.4	8	7	15	12	1	1	2	0	0	13	0	1012.9	79	78	57	69	10.7	NW	52	18	1957	48
D	6.2	8	8	15	12	2	*	2	0	1	24	*	1014.6	76	76	61	69	11.1	WNW	48	14	1957	52
YR	6.2	103	101	161	124	10	21	26	8	27	123	2	1014.0	76	75	55	66	10.8	SW	90	11	AUG 1954	56

Means and extremes above are from existing and comparable exposures. Annual extremes have been exceeded at other sites in the locality as follows:
Highest temperature 102 in August 1948; lowest temperature -17 in February 1934; maximum monthly precipitation 12.24 in August 1946; minimum monthly precipitation .04 in June 1949; maximum monthly snowfall 31.9 in January 1948; fastest mile wind 95 SW in September 1938.

REFERENCE NOTES APPLYING TO TABLES APPEAR ON THE PAGE FOLLOWING LAST TABLE.
(Caution: Letters and symbols may have different meanings in 1941-1970 tables than in earlier tables. See notes.)

Reference notes applying to Normals, Means, and Extremes tables for 1931–1960 base period.

(a) Length of record, years, based on January data. Other months may be for more or fewer years if there have been breaks in the record.

(b) Climatological standard normals (1931-1960).

* Less than one half.

+ Also on earlier dates, months, or years.

T Trace, an amount too small to measure.

Below zero temperatures are preceded by a minus sign. The prevailing direction for wind in the Normals, Means, and Extremes table is from records through 1963.

‡ ≥70° at Alaskan stations.

Unless otherwise indicated, dimensional units used in this bulletin are: temperature in degrees F.; precipitation, including snowfall, in inches; wind movement in miles per hour; and relative humidity in percent. Heating degree day totals are the sums of negative departures of average daily temperatures from 65° F. Cooling degree day totals are the sums of positive departures of average daily temperatures from 65° F. Sleet was included in snowfall totals beginning with July 1948. The term "Ice pellets" includes solid grains of ice (sleet) and particles consisting of snow pellets encased in a thin layer of ice. Heavy fog reduces visibility to 1/4 mile or less.

Sky cover is expressed in a range of 0 for no clouds or obscuring phenomena to 10 for complete sky cover. The number of clear days is based on average cloudiness 0-3, partly cloudy days 4-7, and cloudy days 8-10 tenths.

Solar radiation data are the averages of direct and diffuse radiation on a horizontal surface. The langley denotes one gram calorie per square centimeter.

& Figures instead of letters in a direction column indicate direction in tens of degrees from true North; i.e., 09 - East, 18 - South, 27 - West, 36 - North, and 00 - Calm. Resultant wind is the vector sum of wind directions and speeds divided by the number of observations. If figures appear in the direction column under "Fastest mile" the corresponding speeds are fastest observed 1-minute values.

% Through 1964. The station did not operate 24 hours daily. Fog and thunderstorm data may be incomplete.

$ Through 1964.

c Through 1968.

Reference notes applying to Normals, Means, and Extremes tables for 1941–1970 base period.

(a) Length of record, years, through the current year unless otherwise noted, based on January data.

(b) 70° and above at Alaskan stations.

T Trace.

NORMALS - Based on record for the 1941-1970 period.

DATE OF AN EXTREME - The most recent in cases of multiple occurrence.

PREVAILING WIND DIRECTION - Record through 1963.

WIND DIRECTION - Numerals indicate tens of degrees clockwise from true north. 00 indicates calm.

FASTEST MILE WIND - Speed is fastest observed 1-minute value when the direction is in tens of degrees.

% Through 1964. The station did not operate 24 hours daily. Fog and thunderstorm data may be incomplete.

$ Through 1964.

c Through 1968.

Mean Maximum Temperature (°F.), January

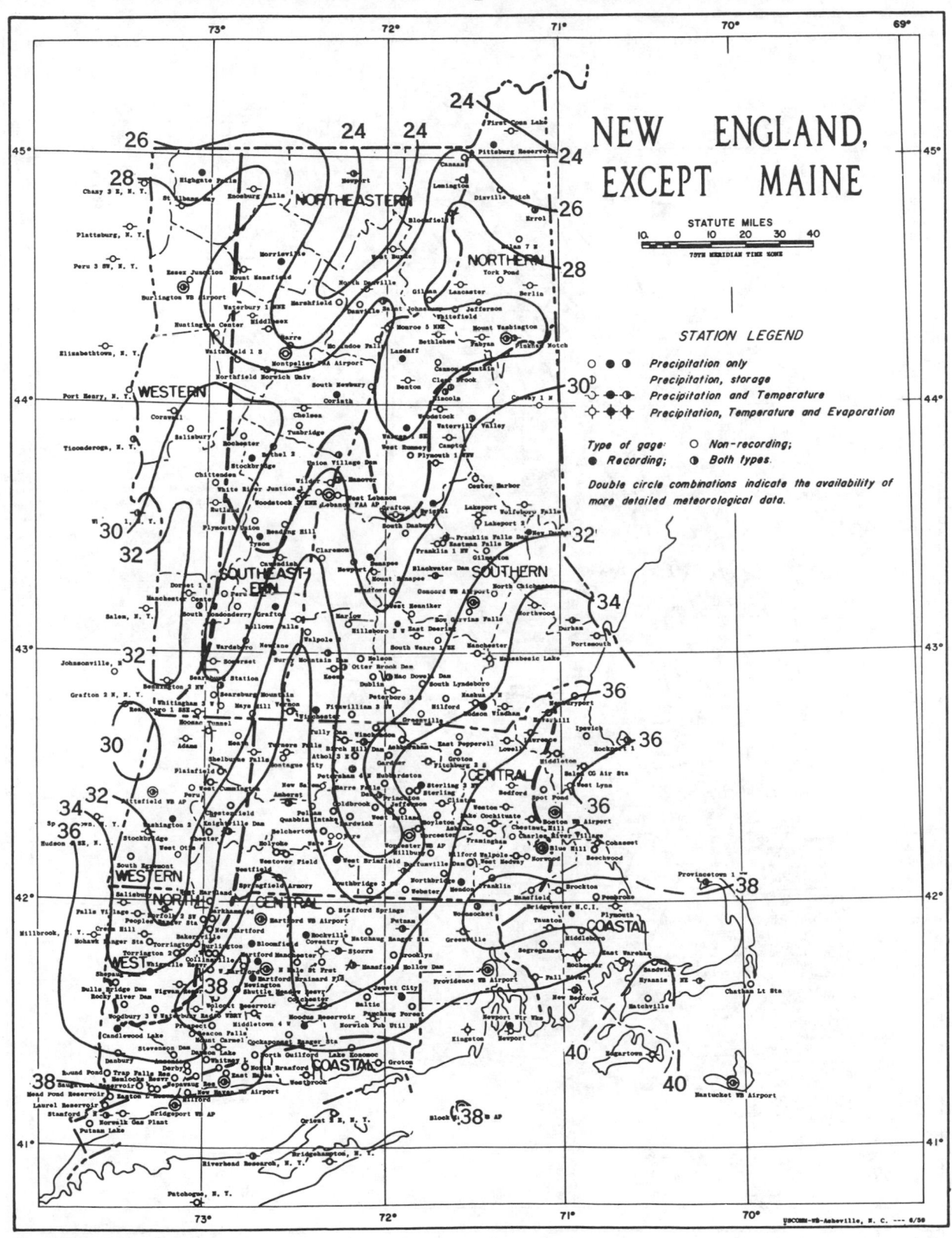

NEW ENGLAND, EXCEPT MAINE

Based on period 1931-52

Isolines are drawn through points of approximately equal value. Caution should be used in interpolating on these maps, particularly in mountainous areas.

Mean Minimum Temperature (°F.), January

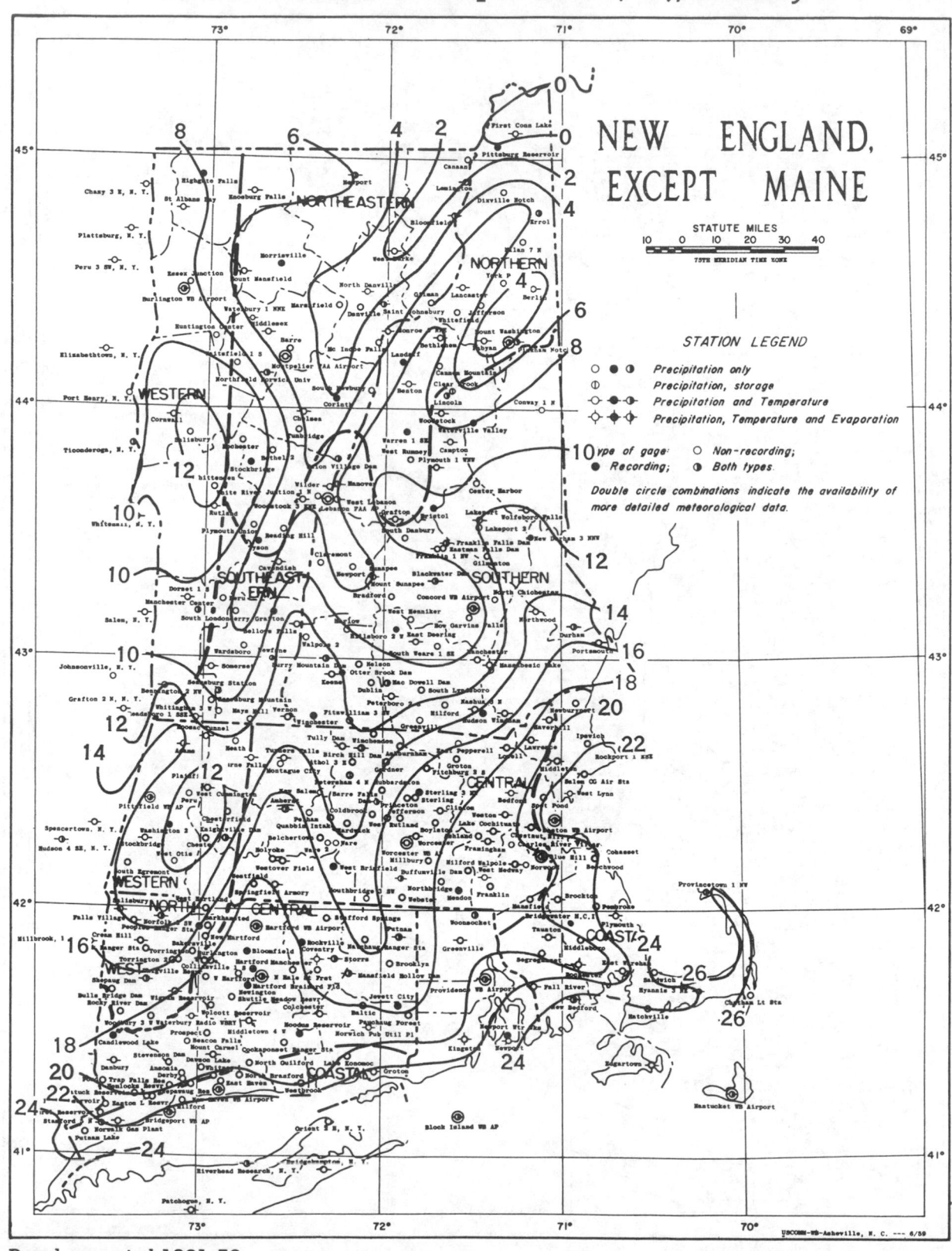

NEW ENGLAND,
EXCEPT MAINE

STATUTE MILES

STATION LEGEND

Precipitation only
Precipitation, storage
Precipitation and Temperature
Precipitation, Temperature and Evaporation

Type of gage: Non-recording;
Recording;   Both types.

Double circle combinations indicate the availability of
more detailed meteorological data.

Based on period 1931-52

Isolines are drawn through points of approximately equal value.   Caution should be used
in interpolating on these maps, particularly in mountainous areas.

Mean Maximum Temperature (°F.), July

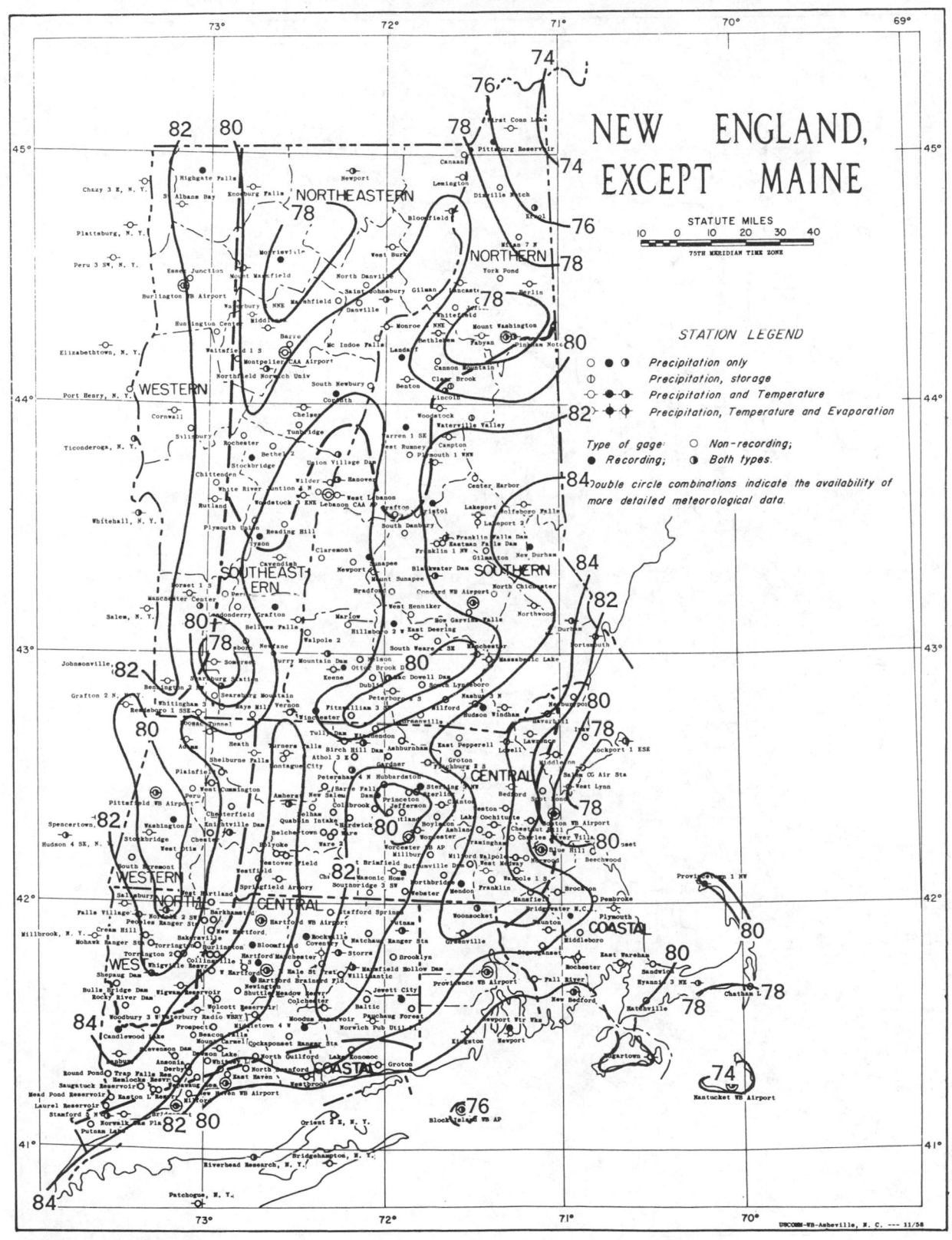

Based on period 1931-52

Isolines are drawn through points of approximately equal value. Caution should be used in interpolating on these maps, particularly in mountainous areas.

Mean Minimum Temperature (°F.), July

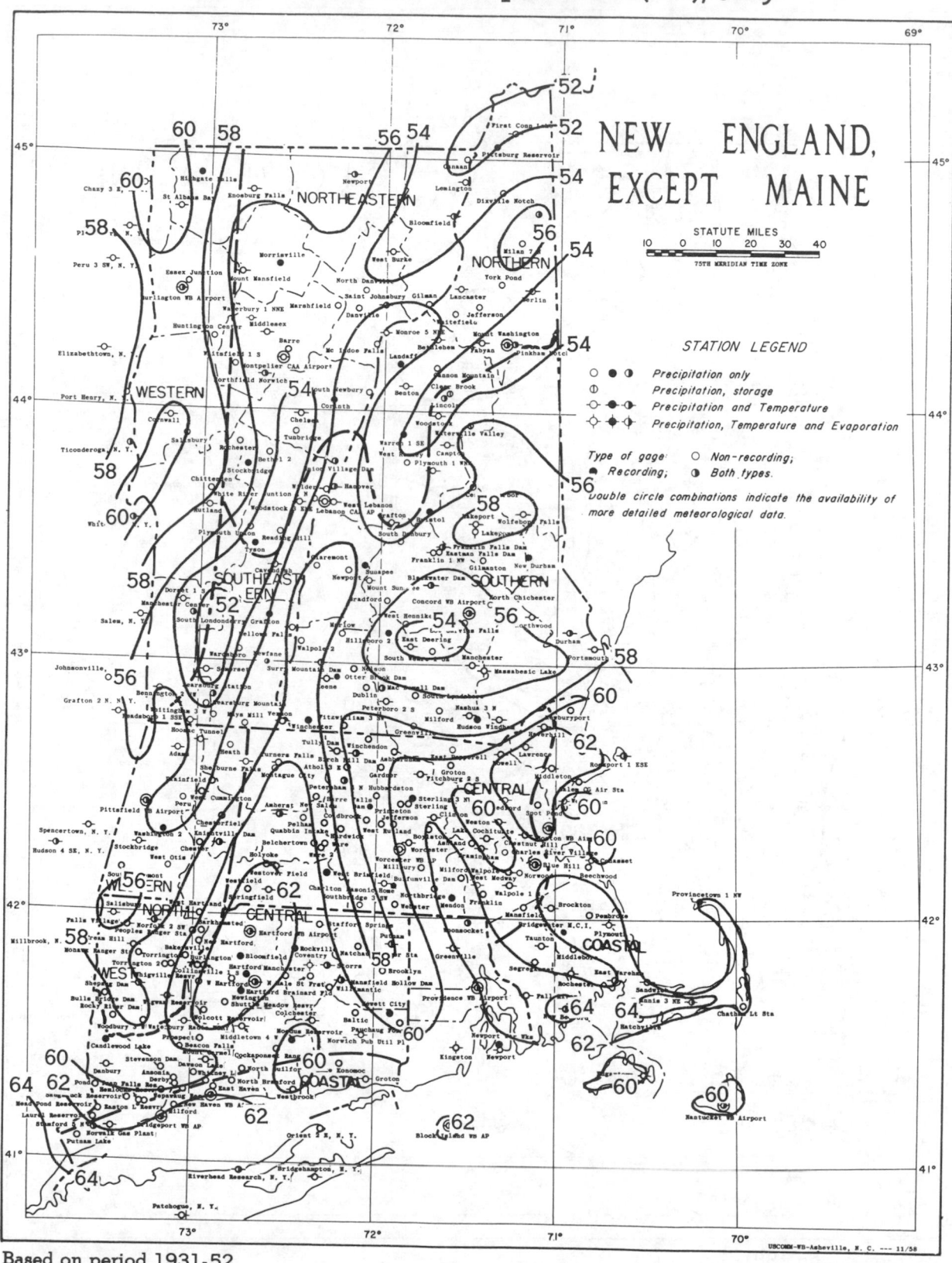

NEW ENGLAND, EXCEPT MAINE

STATUTE MILES
10 0 10 20 30 40
75TH MERIDIAN TIME ZONE

STATION LEGEND

○ ◐ ● — *Precipitation only*
◍ — *Precipitation, storage*
◌─◑─● — *Precipitation and Temperature*
◌─◈─◆ — *Precipitation, Temperature and Evaporation*

Type of gage: ○ *Non-recording;*
● *Recording;* ◐ *Both types.*

Double circle combinations indicate the availability of more detailed meteorological data.

Based on period 1931-52

UBCOMM-WB-Asheville, N. C. --- 11/58

Isolines are drawn through points of approximately equal value. Caution should be used in interpolating on these maps, particularly in mountainous areas.

Mean Annual Precipitation, Inches

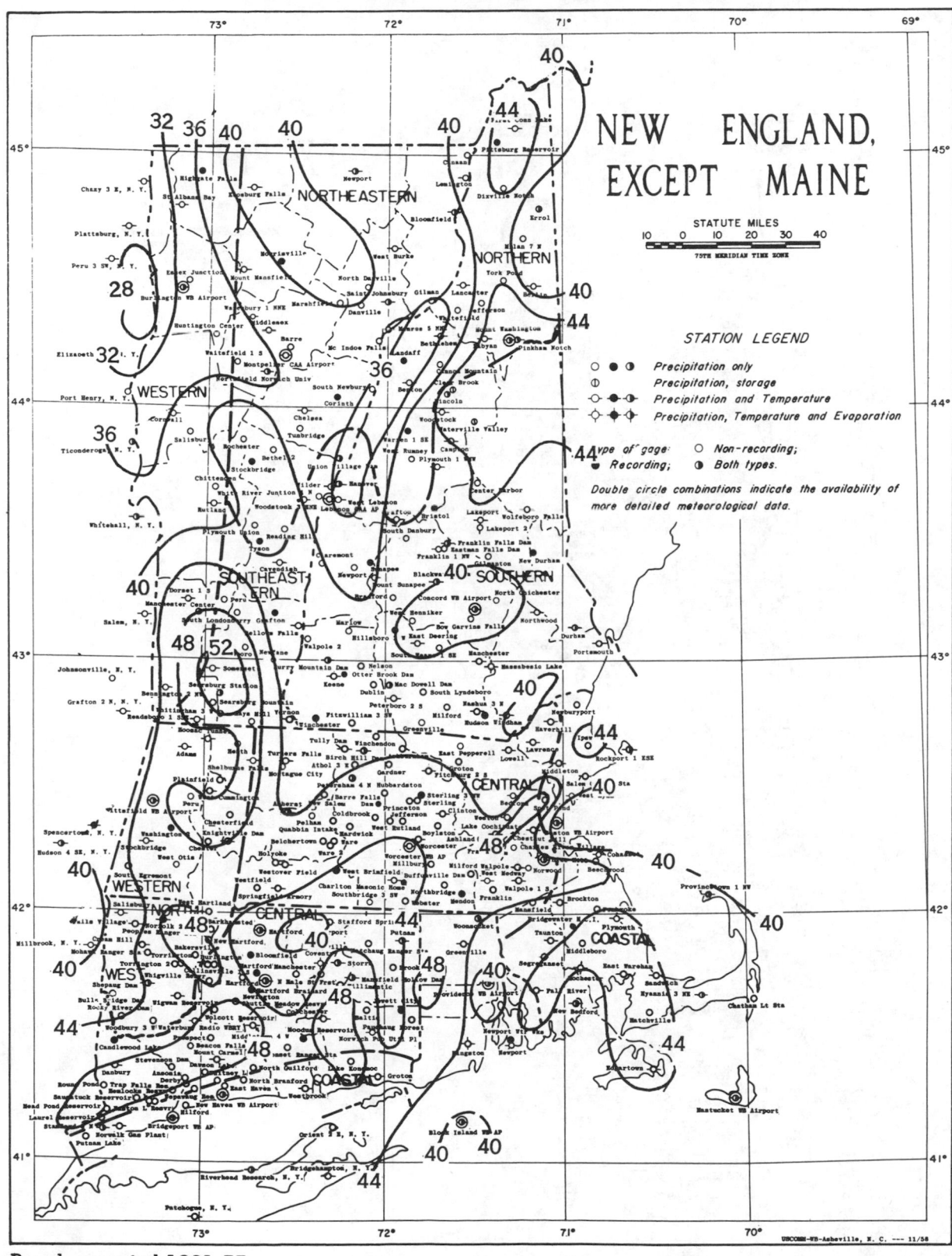

Based on period 1931-55

Isolines are drawn through points of approximately equal value. Caution should be used in interpolating on these maps, particularly in mountainous areas.

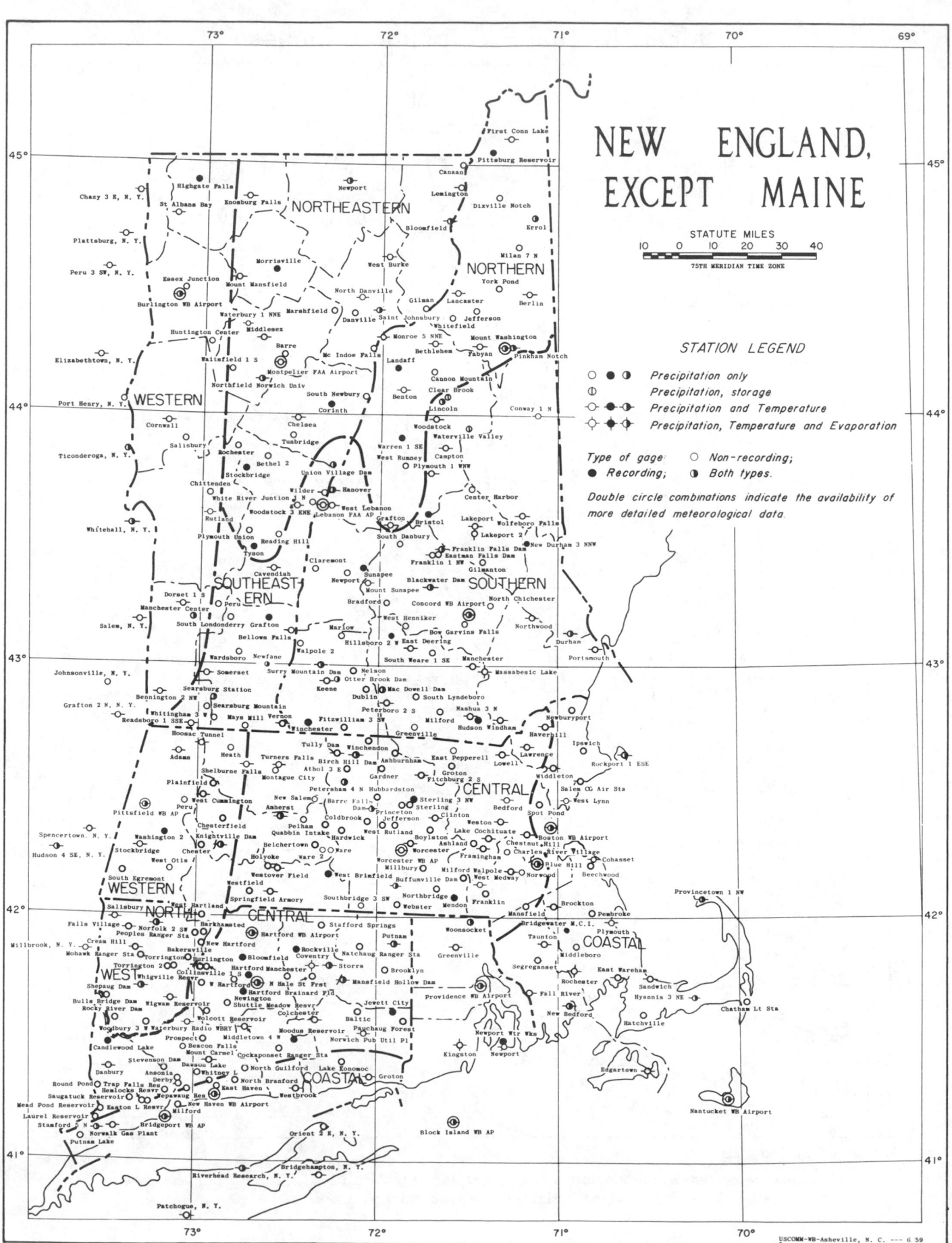

NEW ENGLAND, EXCEPT MAINE

STATUTE MILES
10   0   10   20   30   40
75TH MERIDIAN TIME ZONE

STATION LEGEND

○ ● ◑	*Precipitation only*
◐	*Precipitation, storage*
○- ●- ◑-	*Precipitation and Temperature*
○- ●- ◑-	*Precipitation, Temperature and Evaporation*

Type of gage: ○ *Non-recording;*
● *Recording;*   ◐ *Both types.*

Double circle combinations indicate the availability of more detailed meteorological data.

CLIMATES OF THE STATES

South Carolina

(Normals, Means and Extremes tables revised 1970 and 1975. Basic report revised June 1970.)

H. Landers, ESSA State Climatologist

South Carolina is located on the southeastern coast of the United States between the southern part of the Appalachian Mountains and the Atlantic Ocean. Its north-south extent is 220 miles, from 32° to 35.2° north latitude. The mountains in the extreme northwestern part of the State are 240 miles from the coastline. The coastline is 185 miles long and oriented southwest to northeast.

South Carolina shares some common topographic features with several eastern seaboard states. All of these features have a southwest to northeast orientation and extend across the whole State. The Blue Ridge Range of the Appalachian mountains lies in the extreme northwestern part of the State. Elevations range from 1,000 to 2,000 feet with several peaks going over 3,000 feet. Sassafras Mountain, at 3,554 feet elevation, is the highest point in the State. The Mountain Region covers less than 10 percent of the State's area and to its southeast lies the Piedmont Plateau. The Plateau extends nearly to the center of the State with elevations decreasing northwest to southeast from 1,000 to 500 feet. There is a narrow hilly region where the Plateau descends to the Coastal Plain. In South Carolina this "fall line" region is known as the "Sand Hills", elevations range from 500 to 200 feet. The width of the Sand Hills area is about 30 to 40 miles. Between the Sand Hills and the Atlantic Ocean lies the Coastal Plain. The Plain is broad and nearly level with elevations mostly between 50 and 200 feet. About 40 percent of the area of the State lies in the Coastal Plain.

All of the State's rivers drain southeast from the Mountain Region or Piedmont Plateau toward the ocean. There are three major and one minor river-basin systems. The Santee is the largest and drains the entire center portion of the State. The Savannah drains the western part of the State. Both of these systems extend all the way from the mountains to the ocean. The third major system is the Pee Dee, located in the northeastern section. Its tributaries drain parts of the Piedmont area of South Carolina and North Carolina. The Edisto is a lesser river system lying between the Santee and Savannah. It drains a part of the Piedmont Plateau in western South Carolina known as the "ridge" which extends southeastward to northern Aiken county. Several large lakes and reservoirs have been created by damming the major rivers, mostly in the Plateau area. In addition to the large hydroelectric power plants located at these dam sites, nuclear power plants are now being constructed on the Keowee River near the edge of the Mountain Region.

The major coast indentations are Winyah Bay, Charleston Harbor, St. Helena Sound, Port Royal Sound, and Tybee Roads at the mouth of the Savannah River. There are many low sea islands separated from the mainland by shallow straits, sounds, and coastal streams. The Intracoastal Waterway can be found along much of the coastline.

Several major factors combine to give South Carolina a pleasant, mild, and humid climate. It is located at a relatively low latitude (32° to 35° N.) and most of the State is under 1,000 feet in elevation. It has a long coastline along which moves the warm Gulf Stream current. The mountains to the north and west block or delay many cold air masses approaching from those directions. Even the deep cold air masses which cross the mountains rapidly are warmed somewhat as the air is heated by compression when it descends on the southeastern side. This effect can be seen on the maps of minimum temperature in January and to a lesser degree in July, where a fairly large area of relatively higher temperature appears just southeast of the mountains. It is convenient for climatic discussion to divide the State into areas coinciding closely with the topographic features already discussed. Six areas can be defined, each of which is closely associated with the existing temperature and rainfall patterns:

1. The Outer Coastal Plain, elevations 0-50 feet, width 25-30 miles.

2. The Inner Coastal Plain, elevations 51-200 feet, width 40-45 miles.

3. The Sand Hills, elevations 201-500 feet, width 30-40 miles.

4. The Lower Piedmont Plateau, elevations 501-700 feet, width 45-50 miles.

5. The Upper Piedmont Plateau, elevations 701-1,000 feet, width 30 miles.

6. The Mountain Region, elevations greater than 1,000 feet, width 15 miles.

Some major factors affecting temperatures are elevation, latitude, and distance inland from the coast. All three of these factors work together in South Carolina. Lower temperatures can be expected in the Upper Piedmont and Mountain Region, where latitude, elevation and distance inland all have large values. Higher temperatures will result from smaller values of the three factors, as are found along the southern coast. Annual average temperatures are 10° lower in the extreme Upper Piedmont than along the coast between Charleston and Savannah. Except for small-scale and local irregularities, there is a gradual decrease in annual average temperature northwestward from 68° at the coast to 58° at the edge of the mountains. Within the Mountain Region variations in elevation are great over short horizontal distances. Thus, variations in temperature are due almost entirely to elevation differences. All of the record low wintertime temperatures were set in the Mountain Region or extreme Upper Piedmont. The lowest temperature on record was -13° on January 6, 1940, at

Long Creek, which has an elevation of nearly 2,000 feet. But the highest summertime temperatures ever observed are not found along the south coast. The ocean waters have very small daily and annual changes in temperature when compared with the land surface. The air over the coastal water is cooler than the air over the land in summer and warmer than the air over land in winter and this has a controlling effect on the temperatures of locations on and very near the coast. The July average maximum temperature map reveals that the highest temperatures of 92° to 93° are found in the central part of the State with the coast being 4° to 5° cooler. The July maximum and minimum temperature maps viewed together show a daily range along the coast of about 13°, while the range is about 21° in the center of the State. Even in January at the time of maximum temperatures, the air over the land a short distance from the coast is 2° or 3° warmer than at the coastal stations. The daily range along the coast in January is about 16° and about 23° in the center of the State. The highest temperature on record was 111°. This reading occurred three times; at Calhoun Falls on September 8, 1925, Blackville on September 4, 1925, and Camden on June 28, 1954. The record highest July and August temperatures are 110° and 109°, respectively. Clouds and rainfall have a minor effect on temperature. Maximum temperatures in summer are reduced slightly in areas where afternoon cloudiness and rain are persistent. Such an area is found along the Outer Coastal Plain where sea breezes produce clouds and rain nearly every summer day and dissipate at night. Another minor effect is drainage of cold air, mostly in winter, into some of the river and stream valleys causing minimum temperatures to be somewhat lower than they would be otherwise. One example of this takes place in a rather deep section of the Broad River valley from Lockhart to a little north of Columbia. Generally, not enough temperature stations are available for this small-scale effect to be clearly defined on minimum temperature maps.

The growing season for most cultivated crops is limited by the fall and spring freezes. The freeze-free period, the time elapsing between the last temperature of 32° or less in the spring and the first in the fall, is quite important to agriculture. The average length of the freeze-free period varies from about 200 days in the coldest area to about 280 days along the south coast. In the area where most of the major crops are grown, it is from 210 to 235 days, or 7 to 8 months. The average date of the last freezing temperature in spring ranges from March 10 in the south to April 1 in the north. The fall dates range from late October in the north to November 20 in the south. Freezes have occurred as much as four weeks later than the average date in spring and three weeks earlier than the average date in the fall. The minimum temperature is 32° or less on 50 to 70 winter days in the Upper Piedmont and 10 days near the coast. Counties in the Inner Coastal Plain and Sand Hills area have maximum temperatures of 90° or more on 80 summer days. There are 30 such days along the coast and 10 to 30 in the mountains.

Summers are rather hot and air conditioning is desirable at elevations below 500 feet. Fall and spring are mild and winters are rather cool at elevations above 500 feet. Heating of homes and places of business in varying amounts is necessary in all parts of the State.

Rainfall is adequate in all parts of the State. Annual rainfall averages up to 80 inches in the highest part of the Mountain Region and less than 42 inches in parts of the Inner Coastal Plain and the Sand Hills. The general features on the annual rainfall map are related to the topographical features defined earlier. The Mountain Region is wet with amounts of 56 inches or more, the Upper Piedmont is relatively wet with amounts of 48 to 55 inches, the Lower Piedmont is relatively dry with amounts of 43 to 47 inches, the Outer Coastal Plain is relatively wet with amounts of 48 to 53 inches, and the Inner Coastal Plain is relatively dry with amounts of 38 to 47 inches. The Sand Hills area is less clear cut but is in general a relatively wet strip with a small dry area imbedded in it a few miles south of Columbia. The immediate south coast is also on the dry side. In winter (Dec.-Jan.-Feb.), rainfall decreases from 6 or 7 inches per month in the mountains to 3 or 3 1/2 inches along the coast. In summer (June-July-August) there is a maximum of 6 to 8 inches per month along the Outer Coastal Plain, less in the Inner Coastal Plain, a maximum again of 5 to 7 inches in the Sand Hills, only 3 1/2 to 4 1/2 inches per month in the Lower Piedmont, and the Upper Piedmont and Mountain Region receiving 4 to 7 inches per month. In September and early October the northeast coast gets some additional rain from occasional tropical storms, but fall is actually a dry season with amounts dropping to less than 2 inches along the south coast by November. March is a month of heavy rain in all parts of the State ranging from 4 inches in the Coastal Plain to 7 1/2 inches in the Mountain Region. The pattern of a relatively dry Lower Piedmont and a relatively wet Sand Hills begins to appear in April and continues into May at which time the sea breeze maximum in the Outer Coastal Plain begins to appear. This arrangement persists through the summer and begins to break down in early October. May, however, is the driest month in the spring dry period with less than 3 1/2 inches everywhere except the mountains. In summary, the driest period is in October and November when there is little cyclonic storm activity and "Indian Summer" prevails. Rainfall increases gradually and reaches a peak in March when cyclone and cold front activity are at a maximum. There is a general decrease again to a dry period from late April through early June. From the latter part of June through early September is a wet period primarily due to thunderstorm and shower activity which reaches its peak in July, the wettest summer month. The summer maximum stretches a little into the fall along the coast due to occasional tropical storm activity. The greatest monthly rainfall on record was 31.13 inches at Kingstree in July 1916. The greatest in 24 hours was 13.25 inches in the same month and year at nearby Effingham.

Solid forms of precipitation include snow, sleet, and hail. Hail is not too frequent but does occur with spring thunderstorms from March through early May. These thunderstorms usually accompany squall lines or cold fronts. Snow and sleet may occur separately or together or mixed with rain during the winter months of December through February. Snow may occur from one to three times in winter. Seldom do accumulations remain very long on the ground except in the mountains. Statewide snows of notable amounts can occur when a cyclonic storm moves northeastward along or just off the coast. Two intense storms of this type brought record snowfalls with a belt of greatest amounts running along the Sand Hills and the southern edge of the Lower Piedmont. These two storms, in February 1899 and in February 1914, blanketed the State with depths of 2 inches along the coast and from 8 to 18 inches along the heavy snow belt, with amounts of 4 to 6 inches in the Upper Piedmont. Nearly all of the State's snowfall records were established on Caesars Head Mountain at an elevation of 3,200 feet. The greatest snowfall in 24 hours at Caesars Head was 22 inches, the greatest in one month was 34 inches, and the greatest annual was 48 inches. All of these records were set in 1969 and the greatest month was February. Freezing rain also occurs from one to three times per winter in the northern half of the State. This rain, which freezes on contact with the ground and other objects, can cause hazardous driving conditions, breakage of limbs and tree tops, and breakage of various types of wires and the poles on which they are strung. One of the most severe cases of ice accumulation from freezing rain was in February 1969 in several north central and northeastern counties. Timber losses were tremendous and power and telephone services were seriously disrupted over a large area.

Severe drouths occur about once in 15 years with less severe and less widespread drouths about once in 7 or 8 years. Irrigation is used for most of the truck vegetable crops and tobacco, and by most of the large peach growers. The field crops of corn, cotton, soy beans, and others are largely unirrigated and dry weather takes a heavy toll when it extends over several weeks during the growing season.

The percent of possible sunshine received varies over the State in a way similar to the variation in cloudiness and precipitation. Values in winter range from 50 to 60 percent, in summer from 60 to 70, with the dry periods in spring and fall receiving 70 to 75 percent. The variation in relative humidity with time of day is considerably greater than day to day and month to month variations. Highest values of 80 to 90 percent or more are reached at about sunrise and the lowest values of 45 to 50 percent occur an hour or two after local noon. There is about a 10 percent difference between winter and summer, with summer being the higher of the two seasons. The prevailing surface winds tend to be either from northeast or southwest due to the presence and orientation of the Appalachian Mountains. Winds of all directions occur throughout the State during the year, but the prevailing directions by seasons are; spring--southwest, summer--south and southwest, autumn--northeast, and in winter--northeast and southwest have almost the same frequency. Average surface wind speeds for all months range between 6 and 10 m.p.h. Prevailing winds at levels above the mountain effect are between southwest and northwest in winter and spring, from south to southwest in summer, and from southwest to west in autumn.

Severe weather comes to South Carolina occasionally in the form of violent thunderstorms, tornadoes and hurricanes. Although thunderstorms are common in the summer months, the really violent ones generally accompany the squall lines and active cold fronts of spring. Generally, they bring high winds, hail and considerable lightning and sometimes spawn a tornado. In the 57-year period prior to 1970 there were 252 tornadoes, which is an average of 4 or 5 per year. Many tornadoes were not detected in the early part of that period due to a much smaller population and lack of the organized effort launched in recent years to cut down on loss of life due to these storms. In the 20-year period from 1950 through 1969 the tornado count was 148, an average of 7 or 8 tornadoes per year. Since a tornado is very small and affects a very small area, the probability of any given locality having a tornado in any given year is close to zero. Sixty percent of the tornadoes occur from March through June with April being the peak month with 25 percent. A smaller maximum is found in August and September which accounts for 21 percent of the total. Many are waterspouts or tornadoes that accompany tropical storms and are detected near the coast with some never reaching land. Tornadoes are rather rare from October through February. Only 13 percent of the total are experienced during this five-month period. The worst tornado day in South Carolina's history was April 30, 1924 when two tornadoes struck. The paths of both were unusually long, each over 100 miles, and together they killed 77 persons, injured 778 more, destroyed 465 homes and many other buildings, resulting in many millions of dollars of damage. Tropical storms or hurricanes affect the State about one year out of two. Most of the occurrences are tropical storms which do little damage, frequently bringing rains at a time when they are needed. Most of the hurricanes affect only the Outer Coastal Plain. If they do come far inland they decrease in intensity quite rapidly. The most devastating hurricane, as far as loss of life is concerned, was the one that struck south of Savannah on August 27, 1893. The fast moving storm had piled up vast amounts of water to the east of its center and the south coast and sea islands were badly inundated. Winds of 120 m.p.h. were measured at Charleston and were probably higher between Charleston and Beaufort. More than 1,000 persons were drowned and damage estimates exceeded $10 million. The greatest amount of property damage, $27 million, was done at Myrtle Beach by a hurricane on October 15, 1954. No lives were lost and the highest windspeed was measured at 100 m.p.h. A hurricane that crossed the State moved inland between Charleston and Savannah on September 29, 1959, and continued on a northerly track. Coastal winds were estimated at 140 m.p.h., damage estimated at $20 million and seven lives were lost. Considerable flooding accompanies hurricanes which come very far inland and high tides occur along the coast to the north and east of the storm centers.

There is minor flooding somewhere in the State every year. It can occur on any of the many streams and rivers. A certain amount of control can be effected on the larger rivers which have dams. There is a major flood about once every 7 or 8 years. They are not generally a threat to human life but are very damaging to crops and livestock, disrupting logging operations and damaging homes, stores and other structures. The most extensive flooding took place in August 1908 when all the major rivers in the State rose from 9 to 22 feet above flood stage. Flooding in July 1916 did not affect the western rivers, the Saluda and Savannah, but broke several of the 1908 records in the eastern two thirds of the State. At the six points where the records were broken, heights above flood stage ranged from 13 to 30 feet.

There have been many earth tremors in South Carolina over the years. The southern part of the Coastal Plain is indicated as earthquake prone on a recently published seismic map. There has been only one major quake recorded in the area in the last 100 years. After having light tremors on the 27th, 28th and 30th, a major shock hit the Charleston area on August 31, 1886, just before 10 o'clock in the evening. The area of greatest intensity of the shock was about 10 miles west-northwest of the center of the city of Charleston. More than 100 buildings, including frame houses, were demolished. Ninety percent of all brick structures suffered some damage and 14,000 chimneys tumbled down. The pulverized mortar from the demolished masonry structures filled the city air with a fine white dust. There were many cracks in the earth and numerous small craters up to 20 feet in diameter which spewed water and sand in the air to heights of 20 feet. There was considerable loud moaning and praying as many thought the day of judgement had surely arrived. Eighty-three persons were killed and many injured. The major shock was felt over an area of 2,800,000 square miles, from Canada to the Gulf of Mexico and from Bermuda to Iowa, Missouri and Arkansas, and it broke windows in Milwaukee.

South Carolina is an agricultural State although there is a steady decrease of farm acreage with time. At the same time the size of the remaining farms is increasing. Industry is increasing steadily and many textile mills and other manufacturing plants dot the countryside mostly in the Plateau section. Important crops are tobacco, cotton, corn, soybeans, peaches, truck crops, and small grains. There are many wooded areas

and lumbering is an important activity. The port of Charleston is very important in shipping the State's products abroad and bringing in needed items from other countries. The State is naturally suited to many kinds of recreational activities with its beautiful beaches on the coast, abundant streams, rivers, lakes and forests in the interior, and mountains in the northwest. Parks and recreation areas have always been ample and are now being increased in number and improved. About the only kinds of recreation not available are those which require ice or snow.

REFERENCES

(1) Kronberg, N. and Purvis, J. C., "Climate of South Carolina", published in Climates of the States, South Carolina, Climatography of the United States, No. 60-38, 1959, ESSA, Environmental Data Service.

(2) Agricultural Weather Research Series. Data for Agricultural Weather Stations and Reports on Agriculture-Weather Relationships. ESSA, Weather Bureau and South Carolina Agricultural Experiment Station at Clemson University, Clemson, South Carolina.

(3) Climatic Research Series. Reports on various aspects of South Carolina Climate. ESSA, Weather Bureau and South Carolina Agricultural Experiment Station at Clemson University, Clemson, South Carolina.

BIBLIOGRAPHY

A. Climatic Summary of the United States (Bulletin W), 1930 Edition, Sections 98 and 99. ESSA, Weather Bureau.

B. Climatic Summary of the United States, South Carolina-Supplement for 1931 through 1952 (Bulletin W Supplement), ESSA, Weather Bureau.

C. Climatic Summary of the United States, South Carolina-Supplement for 1951 through 1960 (Bulletin W Supplement), ESSA, Weather Bureau.

D. Climatic Summaries of Resort Areas, Myrtle Beach and Lake Harwell, South Carolina, Climatography of the United States, No. 21-38-1 and 2. ESSA, Environmental Data Service.

E. Climatological Data-South Carolina, Vols. 38 through 67, ESSA, Environmental Data Service.

F. Climatological Data - National Summary, ESSA, Environmental Data Service.

G. Climatological Summaries for 21 South Carolina Substations. Climatography of the United States, No. 20-33, 1965, ESSA Weather Bureau.

H. Hourly Precipitation Data-South Carolina, ESSA, Environmental Data Service.

I. Local Climatological Data for Charleston, Columbia, Florence, Greenville and Spartanburg, South Carolina. ESSA, Environmental Data Service.

J. Storm Data. ESSA, Environmental Data Service

K. Summary of the Climatological Data by Sections (Bulletin W), Volume II, Sections 87 and 88. ESSA, Weather Bureau.

TROPICAL CYCLONE DATA HAVING IMPORTANCE FOR THIS STATE IS INCLUDED IN STATISTICS AND CHARTS ON PAGES 1161 THROUGH 1164 .

FREEZE DATA

STATION	Freeze threshold temperature	Mean date of last Spring occurrence	Mean date of first Fall occurrence	Mean No. of days between dates	Years of record Spring	No. of occurrences in Spring	Years of record Fall	No. of occurrences in Fall
AIKEN	32	03-21	11-18	242	30	30	30	30
	28	03-04	12-01	271	30	30	30	27
	24	02-12	12-16	307	30	26	30	17
	20	01-30	12-23	326	30	22	30	13
	16	01-10	12-27	352	30	11	30	6
ANDERSON	32	03-26	11-13	232	30	30	30	30
	28	03-12	11-27	260	30	30	30	29
	24	02-20	12-09	292	30	30	30	24
	20	02-01	12-17	319	30	22	30	18
	16	01-19	12-24	340	30	13	30	10
BEAUFORT 7 SW	32	03-05	11-25	265	25	25	24	23
	28	02-15	12-09	297	25	24	23	18
	24	01-21	12-19	333	25	17	23	10
	20	01-08	@	@	25	8	23	3
	16	01-03	@	@	25	4	23	2
BISHOPVILLE	32	03-27	11-12	230	16	16	15	15
	28	03-09	11-20	256	15	15	15	15
	24	02-20	12-04	287	13	13	15	13
	20	02-02	12-20	321	13	11	15	7
	16	01-18	12-25	342	13	8	15	4
BLACKVILLE	32	03-17	11-18	247	29	29	30	30
	28	03-01	11-28	271	29	30	30	29
	24	02-04	12-12	311	29	22	29	19
	20	01-19	12-21	337	29	16	29	12
	16	01-08	12-28	354	29	9	29	6
CAESARS HEAD	32	04-11	10-30	201	26	26	26	26
	28	03-31	11-06	220	26	26	25	25
	24	03-18	11-20	247	26	26	25	24
	20	03-11	11-28	262	25	25	25	24
	16	02-21	12-09	291	25	25	25	19
CALHOUN FALLS	32	03-29	11-09	225	29	29	28	28
	28	03-14	11-21	252	29	29	27	27
	24	02-25	12-01	278	29	29	27	24
	20	02-08	12-13	308	29	25	27	19
	16	01-25	12-24	333	29	19	27	10
CAMDEN	32	03-26	11-11	231	29	29	30	30
	28	03-14	11-21	252	29	29	30	29
	24	02-28	12-02	276	29	29	30	27
	20	02-03	12-14	314	29	22	30	20
	16	01-17	12-26	343	29	15	29	9
CHARLESTON WB CITY	32	02-19	12-10	294	30	29	30	21
	28	02-01	12-19	321	30	21	30	15
	24	01-10	12-26	350	30	11	30	7
	20	01-04	@	@	30	6	30	4
	16	@	@	@	30	0	30	0
CHERAW	32	04-01	11-03	217	30	30	30	30
	28	03-15	11-16	247	30	30	30	30
	24	03-04	11-29	271	30	30	30	28
	20	02-14	12-12	301	30	28	30	23
	16	01-15	12-23	342	30	14	30	13
CHESTER	32	04-06	11-03	211	19	19	17	17
	28	03-25	11-15	235	19	19	17	17
	24	03-06	11-26	265	19	19	17	17
	20	02-17	12-10	296	19	18	17	13
	16	01-29	12-19	324	17	13	17	8
CLEMSON COLLEGE	32	04-04	10-30	209	30	30	30	30
	28	03-20	11-12	237	30	29	30	30
	24	03-07	11-25	263	29	29	30	30
	20	02-18	12-07	292	29	27	30	25
	16	02-05	12-19	317	29	24	30	17
COLUMBIA WB CITY	32	03-14	11-21	252	30	30	30	30
	28	02-25	12-02	281	30	29	30	27
	24	02-05	12-17	315	30	23	30	18
	20	01-21	12-26	339	30	16	30	9
	16	01-06	@	@	30	10	30	4
CONWAY	32	03-18	11-12	240	30	30	30	30
	28	03-05	11-24	264	30	30	30	30
	24	02-14	12-13	302	30	27	30	24
	20	01-19	12-24	339	30	17	30	12
	16	01-08	12-29	355	30	9	30	5
DARLINGTON	32	03-25	11-10	229	28	28	27	27
	28	03-11	11-19	253	28	28	26	26
	24	02-25	12-02	279	29	28	26	26
	20	02-01	12-16	318	27	21	26	16
	16	01-14	12-25	344	27	12	26	10
EUTAWVILLE	32	03-19	11-08	234	27	27	28	28
	28	03-03	11-21	263	26	25	27	27
	24	02-10	12-06	299	26	22	27	24
	20	01-18	12-21	338	26	15	26	12
	16	01-06	12-27	355	26	7	26	5
FLORENCE 2 N	32	03-17	11-13	241	28	28	28	28
	28	03-02	11-25	267	28	28	28	27
	24	02-16	12-08	295	28	26	28	22
	20	01-21	12-21	334	28	16	28	14
	16	01-11	12-27	350	28	11	28	6
GEORGETOWN	32	03-19	11-15	242	24	24	21	21
	28	02-24	11-25	274	23	22	21	21
	24	02-06	12-10	308	22	17	21	17
	20	01-20	12-25	339	21	12	21	7
	16	01-07	@	@	21	6	21	3
GREENVILLE WB AP	32	03-23	11-17	239	30	30	30	30
	28	03-05	11-29	269	30	30	30	30
	24	02-22	12-11	291	30	30	30	21
	20	02-05	12-20	318	30	23	30	17
	16	01-16	12-27	344	30	14	30	7
GREENWOOD	32	03-22	11-13	236	30	30	30	30
	28	03-09	11-29	265	30	30	30	30
	24	02-19	12-09	294	30	29	30	22
	20	02-01	12-21	323	30	23	30	14
	16	01-15	12-27	346	30	14	30	7
HEATH SPRINGS	32	03-30	11-08	222	30	30	28	28
	28	03-15	11-26	256	30	30	27	27
	24	02-27	12-03	279	30	30	26	22
	20	02-11	12-16	308	29	27	26	18
	16	01-25	12-25	334	28	17	26	9
KERSHAW	32	03-28	11-11	228	30	30	28	28
	28	03-13	11-20	253	30	30	28	27
	24	02-28	12-05	280	30	30	27	23
	20	02-11	12-14	306	30	26	27	20
	16	01-21	12-23	337	28	15	26	11
KINGSTREE	32	03-17	11-10	239	30	30	30	30
	28	03-03	11-21	263	30	30	30	30
	24	02-10	12-04	297	30	24	30	28
	20	01-17	12-20	337	30	16	30	18
	16	01-09	12-27	352	30	11	30	7
LAKE CITY	32	03-21	11-16	240	12	12	13	13
	28	03-04	11-24	265	12	12	13	13
	24	02-08	12-02	298	12	10	13	12
	20	01-20	12-21	335	12	7	13	7
	16	01-11	12-27	351	12	4	13	2
LANDRUM	32	04-01	11-02	216	30	30	29	29
	28	03-18	11-20	247	30	30	29	29
	24	03-03	11-30	272	30	30	29	29
	20	02-17	12-10	296	30	28	29	23
	16	01-30	12-20	324	29	23	29	15
LAURENS	32	04-01	11-06	219	30	30	28	28
	28	03-19	11-19	245	30	30	28	28
	24	03-02	12-05	278	29	29	28	25
	20	02-08	12-12	307	28	24	28	21
	16	01-24	12-22	332	28	16	28	13
LITTLE MOUNTAIN	32	03-29	11-07	223	30	30	30	30
	28	03-15	11-24	254	30	30	30	30
	24	02-26	12-08	284	30	30	30	24
	20	02-05	12-19	317	30	24	30	18
	16	01-22	12-26	338	30	17	30	8
MARION	32	03-26	11-09	228	17	17	17	17
	28	03-18	11-18	245	16	16	16	16
	24	02-25	12-02	280	15	14	16	14
	20	02-01	12-21	324	15	11	16	8
	16	01-11	12-26	350	14	5	16	5

STATION	Freeze threshold temperature	Mean date of last Spring occurrence	Mean date of first Fall occurrence	Mean No. of days between dates	Years of record Spring	No. of occurrences in Spring	Years of record Fall	No. of occurrences in Fall	STATION	Freeze threshold temperature	Mean date of last Spring occurrence	Mean date of first Fall occurrence	Mean No. of days between dates	Years of record Spring	No. of occurrences in Spring	Years of record Fall	No. of occurrences in Fall
NEWBERRY	32	04-01	11-05	219	30	30	29	29	SOCIETY HILL	32	03-31	11-07	221	30	30	28	28
	28	03-17	11-19	247	30	30	29	28		28	03-17	11-16	245	29	29	28	28
	24	02-28	12-03	278	30	30	29	26		24	03-02	11-28	271	29	28	28	26
	20	02-04	12-14	313	30	24	29	19		20	02-15	12-14	302	28	26	28	18
	16	01-16	12-25	343	30	13	29	9		16	01-20	12-25	339	28	17	28	10
ORANGEBURG 2 SE	32	03-18	11-14	241	30	30	30	30	SPARTANBURG WB AP	32	03-29	11-11	227	30	30	29	29
	28	02-28	11-26	271	30	30	30	28		28	03-11	11-24	258	30	30	29	28
	24	02-04	12-11	310	30	23	30	25		24	02-28	12-04	279	30	30	29	24
	20	01-16	12-25	343	30	15	30	12		20	02-13	12-16	305	30	27	29	18
	16	01-06	12-29	357	30	8	30	6		16	01-22	12-24	336	30	16	29	9
PINOPOLIS DAM	32	03-13	11-20	252	22	22	17	17	TRENTON 1 NNE	32	03-22	11-15	238	27	27	27	27
	28	02-22	12-01	282	22	21	17	15		28	03-05	11-28	268	26	26	27	26
	24	01-31	12-16	318	21	15	17	11		24	02-15	12-13	301	25	23	27	16
	20	01-13	12-24	345	20	8	17	5		20	01-31	12-23	325	25	19	27	11
	16	01-03	⊕	⊕	20	3	17	1		16	01-11	12-27	350	25	9	27	5
SALUDA	32	03-29	11-03	219	25	25	29	29	WALHALLA	32	04-11	10-28	200	30	30	30	30
	28	03-18	11-14	242	25	25	29	29		28	03-27	11-06	225	30	30	30	30
	24	02-27	11-28	274	25	25	29	28		24	03-13	11-22	254	30	30	30	30
	20	02-04	12-14	312	23	18	29	20		20	03-01	12-07	281	30	29	30	25
	16	01-18	12-25	341	22	13	29	9		16	02-07	12-17	312	30	26	30	16
SANTUCK 4 SE	32	04-09	10-30	204	30	30	30	30	WALTERBORO	32	03-21	11-18	242	13	13	13	13
	28	03-27	11-07	224	29	29	30	30		28	03-06	11-26	265	13	13	13	13
	24	03-07	11-20	258	29	29	30	30		24	02-05	12-10	308	13	10	13	10
	20	02-22	12-04	286	30	28	30	28		20	01-28	12-23	329	13	8	13	6
	16	01-31	12-15	318	30	21	30	19		16	01-10	12-28	352	13	5	13	2
									WEDGEFIELD	32	03-19	11-17	243	27	27	27	27
										28	03-06	11-24	262	26	26	27	27
										24	02-18	12-12	297	24	24	27	17
										20	01-29	12-25	331	24	17	27	9
										16	01-14	12-28	349	24	12	27	5
									WINTHROP COLLEGE	32	04-01	11-07	220	30	30	30	30
										28	03-18	11-22	249	30	30	30	30
										24	03-03	12-01	274	30	30	30	26
										20	02-19	12-11	296	30	28	30	22
										16	01-26	12-23	331	30	18	30	12
									YEMASSEE 4 W	32	03-18	11-17	244	30	30	30	30
										28	02-25	11-24	271	30	30	30	30
										24	02-07	12-10	306	30	26	30	21
										20	01-19	12-23	339	30	17	30	10
										16	01-06	12-28	356	30	8	30	5

Data in the above table are based on the period 1921-1950, or that portion of this period for which data are available.

⊕ When the frequency of occurrence in either spring or fall is one year in ten, or less, mean dates are not given.

Means have been adjusted to take into account years of non-occurrence.

A freeze is a numerical substitute for the former term "killing frost" and is the occurrence of a minimum temperature at or below the threshold temperature of 32°, 28°, etc.

Freeze data tabulations in greater detail are available and can be reproduced at cost.

*NORMALS BY CLIMATOLOGICAL DIVISIONS

Taken from "Climatography of the United States No. 81-4, Decennial Census of U. S. Climate"

TEMPERATURE (°F)   PRECIPITATION (In.)

STATIONS (By Divisions)	JAN	FEB	MAR	APR	MAY	JUNE	JULY	AUG	SEPT	OCT	NOV	DEC	ANN	JAN	FEB	MAR	APR	MAY	JUNE	JULY	AUG	SEPT	OCT	NOV	DEC	ANN
MOUNTAIN																										
CAESARS HEAD	39.0	39.9	45.3	55.0	62.8	69.6	71.5	70.5	65.8	56.6	46.4	39.6	55.2	6.74	6.15	7.22	5.83	5.40	5.42	7.55	7.74	5.53	5.09	5.21	7.19	75.07
CAESARS HEAD 1 NE	39.0	39.9	45.3	55.0	62.8	69.6	71.5	70.5	65.8	56.6	46.4	39.6	55.2	6.74	6.15	7.22	5.83	5.40	5.42	7.55	7.74	5.53	5.09	5.21	7.19	75.07
DIVISION	40.4	41.3	46.6	56.3	63.9	70.8	72.7	72.0	67.0	57.9	47.8	40.7	56.5	6.34	5.91	6.70	5.43	5.03	5.12	7.14	7.18	5.03	4.97	4.68	6.66	70.19
NORTHWEST																										
ANDERSON	45.1	46.6	52.3	61.7	70.0	77.9	79.5	78.8	73.6	63.7	52.8	45.0	62.3	4.60	4.35	5.15	3.87	2.86	3.52	4.69	4.48	3.19	2.96	3.08	4.62	47.37
CLEMSON UNIVERSITY	44.6	46.2	52.0	61.3	69.4	77.1	79.0	78.3	73.3	62.7	51.6	44.5	61.7	4.83	4.98	5.54	4.36	3.67	3.33	5.13	4.95	3.23	3.29	3.17	5.10	51.58
CRESCENT 1 S														5.00	5.07	5.30	4.08	3.33	2.73	4.95	4.78	3.97	3.58	3.27	4.91	50.97
GRNVLE-SPTNBRG WBAP	43.7	45.1	51.4	60.9	69.4	77.0	79.0	78.2	72.7	62.4	51.3	43.6	61.2	4.28	4.07	4.75	3.94	3.22	2.89	4.65	4.51	3.88	3.36	2.84	4.03	46.42
LANDRUM 1 NE	43.6	45.0	50.8	60.1	68.1	75.9	78.0	76.5	71.4	61.3	50.6	43.4	60.4	5.04	4.47	5.75	4.63	4.10	4.24	5.94	5.94	4.29	3.71	3.90	4.77	56.78
LAURENS	45.1	47.0	53.0	62.3	70.4	78.3	80.3	79.3	73.8	63.2	52.0	44.7	62.5	4.04	4.05	4.83	3.72	3.52	3.77	4.89	3.92	3.52	2.64	2.88	4.20	45.98
RAINBOW LAKE	43.1	44.6	50.4	59.7	68.1	75.9	78.2	77.4	71.8	61.2	50.1	42.6	60.3	4.43	4.48	5.20	3.93	3.53	3.52	5.00	4.57	4.01	3.43	3.05	4.54	49.69
SANTUCK	44.9	46.3	52.3	61.4	69.8	77.4	79.5	78.6	73.1	62.9	52.0	44.5	61.9	3.95	3.84	4.45	3.57	3.38	3.15	5.15	4.84	3.43	2.84	2.71	3.77	45.08
WALHALLA	43.4	44.6	50.1	59.4	67.8	75.6	77.5	76.6	71.7	61.2	50.2	42.9	60.1	5.38	5.62	6.44	4.87	3.96	4.61	5.57	5.44	3.92	4.15	3.78	5.47	59.21
WEST PELZER	•	•	•	•	•	•	•	•	•	•	•	•	•	4.70	4.54	5.26	3.88	3.04	3.43	4.93	4.72	3.50	3.53	3.02	4.46	49.01
DIVISION	44.1	45.6	51.4	60.8	69.2	77.0	79.0	78.0	72.7	62.4	51.3	43.9	61.3	4.56	4.42	5.22	4.06	3.48	3.53	5.02	4.80	3.62	3.24	3.14	4.52	49.61
NORTH CENTRAL																										
BLAIR	•	•	•	•	•	•	•	•	•	•	•	•	•	3.67	3.64	4.29	3.73	3.30	2.91	5.69	4.40	3.52	2.37	2.61	3.85	43.98
CAMDEN 2 WSW	45.4	46.6	52.7	61.6	70.6	78.4	80.6	79.6	74.4	63.6	52.4	44.8	62.6	3.60	3.83	4.46	4.09	3.51	4.40	5.86	5.04	4.07	2.74	2.99	3.63	48.22
CATAWBA	•	•	•	•	•	•	•	•	•	•	•	•	•	3.77	3.70	3.93	3.74	3.27	3.29	6.04	4.50	3.13	2.76	2.68	3.77	45.74
KERSHAW	45.1	46.4	52.7	61.9	70.4	77.6	79.7	78.6	73.6	63.5	52.7	44.8	62.3	3.43	3.45	3.82	4.09	2.95	3.94	5.88	5.32	4.02	2.61	2.72	3.51	45.74
WINNSBORO	45.7	46.9	53.1	62.6	71.0	78.4	80.4	79.0	73.9	63.7	53.1	45.2	62.8	3.72	3.56	4.55	3.91	3.17	3.47	5.71	5.29	3.65	2.37	2.47	3.38	45.25
WINTHROP COLLEGE	44.7	46.4	52.4	61.9	70.3	77.6	79.5	78.2	73.2	63.2	52.5	44.3	62.0	4.16	3.88	4.40	3.87	3.27	3.11	6.04	4.64	3.51	3.05	2.78	4.01	46.72
DIVISION	45.2	46.6	52.6	61.9	70.4	77.8	79.9	78.7	73.6	63.5	52.6	44.7	62.3	3.66	3.64	4.21	3.89	3.18	3.60	5.62	4.95	3.77	2.63	2.69	3.57	45.41
NORTHEAST																										
CHERAW	44.6	45.8	51.9	61.5	70.6	78.2	80.7	79.5	74.3	63.6	52.4	44.4	62.3	3.36	3.47	4.14	3.84	3.36	4.09	5.62	5.35	4.26	2.68	2.80	3.27	46.24
CONWAY	48.9	50.1	55.9	64.3	72.2	79.0	81.1	80.3	75.6	65.7	55.7	48.4	64.8	2.75	3.70	4.28	3.06	3.31	5.12	7.52	6.59	5.77	2.70	2.41	3.15	50.36
DARLINGTON	46.7	47.7	54.0	63.0	71.3	78.3	80.4	79.3	74.2	63.7	53.4	46.1	63.2	2.80	3.27	3.76	3.81	3.02	4.03	5.78	4.59	4.54	2.52	2.60	3.19	43.91
EFFINGHAM	•	•	•	•	•	•	•	•	•	•	•	•	•	2.57	3.33	3.92	3.39	3.22	4.05	5.36	5.36	4.27	2.35	2.18	3.11	43.11
FLORENCE FAA AP	47.5	48.8	54.7	63.5	71.9	78.8	80.7	79.9	74.9	65.0	54.6	47.2	64.0	2.64	3.17	3.64	3.61	2.89	4.28	6.24	4.57	4.10	2.18	2.31	3.06	42.69
FLORENCE 2 N	•	•	•	•	•	•	•	•	•	•	•	•	•	2.90	3.39	3.89	3.58	3.27	4.33	6.21	4.66	4.45	2.23	2.57	2.97	44.45
KINGSTREE 1 SE	49.1	50.3	56.4	64.8	72.7	79.6	81.5	80.5	75.8	65.7	55.7	48.5	65.1	2.64	3.55	4.11	3.16	3.21	4.34	5.85	6.42	5.14	2.35	2.35	3.07	46.19
PEE DEE	•	•	•	•	•	•	•	•	•	•	•	•	•	2.68	3.33	3.68	3.79	3.13	3.98	5.81	5.99	4.95	2.80	2.37	3.10	45.61
DIVISION	47.4	48.5	54.5	63.3	71.5	78.4	80.5	79.5	74.7	64.5	54.4	46.9	63.7	2.84	3.43	3.92	3.43	3.17	4.47	6.18	5.42	4.87	2.53	2.52	3.14	45.92
WEST CENTRAL																										
AIKEN	48.9	50.5	56.1	64.5	72.3	79.2	80.8	80.1	75.3	65.7	55.6	48.6	64.8	3.24	3.89	4.33	3.82	3.35	3.45	4.34	4.71	3.79	2.33	2.60	3.71	43.56
CALHOUN FALLS	45.2	47.0	52.7	62.1	70.6	78.2	80.3	79.5	74.1	63.6	52.5	45.1	62.6	4.22	4.09	4.84	3.81	3.47	3.70	4.52	3.73	3.50	2.74	2.56	4.32	45.50
CHAPPELLS	•	•	•	•	•	•	•	•	•	•	•	•	•	3.93	3.94	4.73	3.96	3.42	3.20	4.37	4.52	3.51	2.71	2.54	3.79	44.62
EDGEFIELD 1 ENE	•	•	•	•	•	•	•	•	•	•	•	•	•	3.52	3.87	4.49	4.44	3.16	3.25	4.92	4.14	3.89	2.15	2.32	3.58	43.73
GREENWOOD	44.7	46.0	51.9	61.5	70.5	78.2	80.1	79.2	73.4	63.3	52.2	44.5	62.1	4.51	4.24	4.99	3.87	3.49	3.27	4.65	4.43	3.65	2.71	2.63	3.90	46.34
LITTLE MOUNTAIN	47.3	48.8	54.5	63.5	71.8	79.2	81.0	80.0	75.2	65.3	55.0	47.0	64.1	3.70	3.93	4.36	3.96	3.33	3.24	5.19	4.60	3.80	2.63	2.45	3.61	44.80
NEWBERRY	46.0	47.6	53.3	62.5	71.0	78.5	80.6	79.6	74.2	63.6	52.6	45.2	62.9	3.89	3.81	4.47	3.80	3.41	3.57	5.06	3.93	3.60	2.59	2.47	3.88	44.48
SALUDA 2 W	46.9	48.5	54.0	63.0	71.6	78.8	81.6	80.1	74.7	64.0	53.1	46.1	63.5	3.89	4.05	4.87	4.20	3.09	3.67	4.61	4.72	4.14	2.42	2.44	3.84	45.94
DIVISION	46.6	48.2	53.8	62.9	71.4	78.6	80.5	79.6	74.5	64.4	53.7	46.2	63.4	3.83	3.99	4.59	3.92	3.36	3.48	4.80	4.33	3.70	2.55	2.47	3.86	44.88
CENTRAL																										
COLUMBIA WB AIRPORT	46.9	48.4	54.4	63.6	72.2	79.7	81.6	80.5	75.3	64.7	53.7	46.4	64.0	3.02	3.74	4.26	4.01	3.54	3.85	6.09	5.74	4.31	2.38	2.36	3.52	46.82
COLUMBIA UNI OF S C	48.2	49.7	55.4	64.3	72.6	79.6	81.3	80.3	75.3	65.5	55.0	47.7	64.6	•	•	•	•	•	•	•	•	•	•	•	•	•
ORANGEBURG 2	48.4	49.9	55.8	64.4	72.6	79.2	81.2	80.5	75.4	65.2	54.8	47.6	64.6	2.70	3.72	3.84	3.55	3.34	4.29	4.75	5.58	4.33	2.55	2.35	3.21	44.21
RIMINI	•	•	•	•	•	•	•	•	•	•	•	•	•	2.85	3.84	3.80	3.56	3.28	4.81	4.63	4.72	4.37	2.52	2.32	3.51	44.21
SUMTER	48.3	50.1	56.0	64.6	72.3	79.0	81.0	79.9	75.1	65.2	55.1	47.8	64.5	2.71	3.43	3.88	3.68	3.32	4.17	5.56	5.27	4.39	2.40	2.29	3.28	44.38
DIVISION	47.8	49.4	55.2	64.0	72.2	79.1	80.9	80.0	75.1	64.9	54.6	47.3	64.2	2.69	3.48	3.83	3.53	3.25	3.94	5.47	5.30	4.16	2.40	2.27	3.20	43.52

*NORMALS BY CLIMATOLOGICAL DIVISIONS

Taken from "Climatography of the United States No. 81-4, Decennial Census of U. S. Climate"

TEMPERATURE (°F)   PRECIPITATION (In.)

STATIONS (By Divisions)	JAN	FEB	MAR	APR	MAY	JUNE	JULY	AUG	SEPT	OCT	NOV	DEC	ANN	JAN	FEB	MAR	APR	MAY	JUNE	JULY	AUG	SEPT	OCT	NOV	DEC	ANN
SOUTHERN																										
BEAUFORT 7 SW	51.4	52.8	57.6	65.6	73.2	79.3	81.0	80.6	76.5	67.4	58.1	51.3	66.2	2.62	2.97	3.79	2.54	3.48	4.61	6.08	5.91	5.59	2.28	1.88	2.66	44.41
BLACKVILLE 3 W	48.9	50.2	56.0	64.5	72.8	79.1	80.9	80.0	75.6	66.0	55.7	48.3	64.8	2.73	3.65	4.23	3.74	3.53	4.29	5.01	4.91	4.06	2.69	2.45	3.33	44.62
CHARLESTON WB AP	49.8	51.5	56.7	64.8	72.9	79.2	80.6	79.7	75.6	66.2	55.9	50.0	65.2	2.54	3.29	3.93	2.88	3.61	4.98	7.71	6.61	5.83	2.84	2.09	2.85	49.16
CHARLESTON WB CITY	51.5	52.3	57.5	65.6	73.3	79.6	81.5	81.0	77.1	68.1	58.6	51.7	66.5	2.40	3.07	3.62	2.54	3.09	4.25	7.78	6.07	6.32	2.74	1.92	2.74	46.54
SUMMERVILLE 2 WNW	49.3	50.5	56.3	64.4	72.3	78.7	80.7	80.1	75.8	66.2	55.8	48.8	64.9	2.44	3.25	3.74	2.82	3.51	5.03	7.15	6.45	5.22	2.30	2.05	2.73	46.69
YEMASSEE 4 W	50.6	52.0	57.4	65.2	72.6	78.9	80.7	80.3	76.0	66.5	56.6	50.1	65.6	2.57	3.49	3.87	3.41	3.78	5.15	6.78	6.49	4.99	2.92	2.25	2.74	48.44
DIVISION	50.5	51.7	57.1	65.1	72.9	79.2	81.0	80.4	76.3	66.8	56.9	50.1	65.7	2.57	3.35	3.89	3.00	3.56	4.67	6.54	6.05	5.23	2.71	2.13	2.81	46.51

* Normals for the period 1931-1960.   Divisional normals may not be the arithmetical averages of individual stations published, since additional data for shorter period stations are used to obtain better areal representation.

TEMPERATURE   PRECIPITATION

JAN	FEB	MAR	APR	MAY	JUNE	JULY	AUG	SEPT	OCT	NOV	DEC	ANN	JAN	FEB	MAR	APR	MAY	JUNE	JULY	AUG	SEPT	OCT	NOV	DEC	ANN

CONFIDENCE - LIMITS

In the absence of trend or record changes, the chances are 9 out of 10 that the true mean will lie in the interval formed by adding and subtracting the values in the following table from the means for any station in the State. Because of the wider variation in mean precipitation, the corresponding monthly means and annual mean must be substituted for "p" in the precipitation table below to obtain mean precipitation confidence limits.

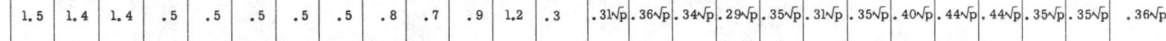

| 1.5 | 1.4 | 1.4 | .5 | .5 | .5 | .5 | .5 | .8 | .7 | .9 | 1.2 | .3 | $.31\sqrt{p}$ | $.36\sqrt{p}$ | $.34\sqrt{p}$ | $.29\sqrt{p}$ | $.35\sqrt{p}$ | $.31\sqrt{p}$ | $.35\sqrt{p}$ | $.40\sqrt{p}$ | $.44\sqrt{p}$ | $.44\sqrt{p}$ | $.35\sqrt{p}$ | $.35\sqrt{p}$ | $.36\sqrt{p}$ |

COMPARATIVE DATA

Data in the following table are the mean temperature and average precipitation for Winthrop College, South Carolina, for the period 1906-1930 and are included in this publication for comparative purposes.

| 44.1 | 45.6 | 53.0 | 61.5 | 69.4 | 76.7 | 79.3 | 78.1 | 73.4 | 62.6 | 51.7 | 44.3 | 61.6 | 3.93 | 4.07 | 4.05 | 3.11 | 3.93 | 4.66 | 5.50 | 4.75 | 3.60 | 2.95 | 2.39 | 4.05 | 46.99 |

NORMALS, MEANS AND EXTREMES
(Table Revised 1970. Base Period for Climatological Normals: 1931-1960)

Station: CHARLESTON, SOUTH CAROLINA MUNICIPAL AIRPORT Standard time used: EASTERN Latitude: 32° 54' N Longitude: 80° 02' W Elevation (ground): 40 feet

Means and extremes above are from existing and comparable exposures. Annual extremes have been exceeded at other sites in the locality as follows:
Highest temperature 104 in June 1944; lowest temperature 7° in February 1899; maximum monthly precipitation 23.75 in July 1964 at City Office; minimum monthly precipitation 0.01 in October 1942; maximum precipitation in 24 hours 10.57 in September 1933; maximum snowfall 3.9 in February 1899; maximum snowfall in 24 hours 3.9 in February 1899.

NORMALS, MEANS AND EXTREMES
(Table Revised 1975. Base Period for Climatological Normals: 1941-1970)

Means and extremes above are from existing and comparable exposures. Annual extremes have been exceeded at other sites in the locality (East Bay Street) as follows:
highest temperature 104 in June 1944; lowest temperature 7° in February 1899; maximum precipitation in 24 hours 10.57 in September 1933.

REFERENCE NOTES APPLYING TO TABLES APPEAR ON THE PAGE FOLLOWING LAST TABLE.
(Caution: Letters and symbols may have different meanings in 1941-1970 tables than in earlier tables. See notes.)

(Table Revised 1970. Base Period for Climatological Normals: 1931-1960)

Station: COLUMBIA, SOUTH CAROLINA — COLUMBIA METROPOLITAN AIRPORT
Standard time used: EASTERN Latitude: 33° 57′ N Longitude: 81° 07′ W Elevation (ground): 213 feet

Month	Temperature Normal — Daily maximum	Daily minimum	Monthly	Normal heating degree days (Base 65°)
J	58.2	35.6	46.9	570
F	60.5	36.3	48.4	470
M	67.4	41.4	54.4	357
A	76.3	50.6	63.6	81
M	84.7	59.6	72.2	0
J	91.5	67.5	79.7	0
J	92.5	70.7	81.6	0
A	91.2	69.7	80.5	0
S	86.1	64.5	75.3	84
O	77.1	52.0	64.6	345
N	66.7	40.7	53.7	577
D	58.2	34.5	46.4	—
YR	75.8	52.1	64.0	2484

Means and extremes above are from existing and comparable exposures. Annual extremes have been exceeded at other sites in the locality as follows: Highest temperature 107 in June 1954+; lowest temperature -2 in February 1899; maximum monthly snowfall 11.8 in February 1899; maximum snowfall in 24 hours 11.7 in February 1914.

NORMALS, MEANS AND EXTREMES

(Table Revised 1975. Base Period for Climatological Normals: 1941-1970)

Month	Temperatures °F Normal — Daily maximum	Daily minimum	Monthly	Precipitation Normal (in.)
J	56.9	33.9	45.4	3.44
F	59.7	35.5	47.6	3.67
M	66.5	41.9	54.2	4.07
A	76.9	51.3	64.1	3.51
M	84.5	59.6	72.1	3.35
J	90.3	67.2	78.8	3.82
J	92.0	70.3	81.2	5.65
A	91.2	70.4	80.2	5.63
S	85.4	64.5	74.9	4.32
O	77.1	51.3	64.2	2.58
N	66.9	40.6	53.8	2.34
D	57.9	34.1	46.0	3.38
YR	75.4	51.5	63.5	46.36

Means and extremes above are from existing and comparable exposures. Annual extremes have been exceeded at other sites in the locality as follows: Highest temperature 107 in June 1954+; lowest temperature -2 in February 1899.

REFERENCE NOTES APPLYING TO TABLES APPEAR ON THE PAGE FOLLOWING LAST TABLE.
(Caution: Letters and symbols may have different meanings in 1941-1970 tables than in earlier tables. See notes.)

NORMALS, MEANS AND EXTREMES
(Table Revised 1970. Base Period for Climatological Normals: 1931-1960)

Station: GREENVILLE-SPARTANBURG AP **GREER, S. C.** **Standard time used:** EASTERN **Latitude:** 34° 54' N **Longitude:** 82° 13' W **Elevation (ground):** 957 feet

Temperature / Precipitation / Degree Days

Month	Normal Daily max	Normal Daily min	Normal Monthly	Record highest	Year	Record lowest	Year	Normal heating degree days (Base 65°)	Precip. Normal total
J	52.4	35.0	43.7	76	1966	-6	1966	665	4.28
F	54.4	35.8	45.1	75	1967	8	1967	557	4.07
M	61.4	41.0	51.4	88	1969	17	1969	441	4.75
A	71.4	50.3	60.9	91	1966	28	1966	145	3.94
M	79.4	59.3	69.4	97	1963	38	1963	23	3.22
J	86.1	67.8	77.0	98	1964	46	1964	0	2.89
J	87.0	71.0	79.0	98	1966	58	1967+	0	4.65
A	86.4	70.2	78.2	98	1968	52	1968	0	4.51
S	81.0	64.4	72.7	97	1965+	36	1967	130	3.88
O	71.5	52.3	62.4	91	1967	27	1965	411	3.16
N	61.5	41.0	51.3	81	1964	17	1965	663	2.84
D	52.4	34.7	43.6	78	1967+	14	1963	663	4.03
YR	70.5	51.9	61.2	99 AUG 1968		-6 JAN 1966		3044	46.42

Precipitation detail / Snow

Month	Max monthly	Year	Min monthly	Year	Max 24 hrs	Year	Snow mean total	Snow max monthly	Year	Snow max 24 hrs	Year
J	5.44	1965	2.39	1965	2.61	1969	2.7	9.9	1966	5.3	1965
F	6.78	1966	1.00	1968	2.33	1968	2.6	6.9	1969	5.3	1969
M	9.66	1963	1.98	1963	3.45	1969	0.5	1.0	1969	1.0	1969
A	11.30	1964	1.36	1967	3.75	1963	0.0	0.0		0.0	
M	4.97	1964	1.09	1968	4.21	1969	0.0	0.0		0.0	
J	9.59	1969	3.84	1966	4.21	1965+	0.0	0.0		0.0	
J	7.44	1964	2.46	1963	3.49	1964	0.0	0.0		0.0	
A	7.51	1963	1.16	1963	4.49	1967	0.0	0.0		0.0	
S	7.98	1966	0.24	1963	4.28	1964	0.0	0.0		0.0	
O	10.24	1963	0.24	1963	4.28	1964	0.3	1.9	1968	1.9	1968
N	5.07	1966	1.93	1966	2.83	1964	0.3	2.1	1963	1.4	1963
D	7.40	1967	0.37	1965	1.91	1967	—	—		—	
YR	11.30 APR 1964		0.24 OCT 1963		4.49 AUG 1967		6.4	9.1 JAN 1966		5.7 JAN 1965	

Relative humidity / Wind / Sunshine / Sky cover

Month	RH 01	RH 07	RH 13	RH 19	Mean speed	Prevailing dir.	Fastest speed	Fastest dir.	Year	Pct possible sunshine	Mean sky cover
J	70	74	52	61	7.4	NE	44	SW	1967	54	6.0
F	67	72	49	56	8.5	NE	44	SW	1966	56	6.0
M	66	73	45	50	8.5	SW	38	NW	1969	58	5.9
A	71	76	41	50	8.3	SW	40	NW	1964	60	5.8
M	79	81	50	54	6.5	NE	36	N	1969	59	5.9
J	81	84	56	60	6.2	WSW	35	N	1969	59	5.9
J	85	87	59	69	6.1	NE	52	NE	1966	64	6.5
A	85	88	58	71	6.0	NE	31	NE	1963	65	5.7
S	85	86	57	67	6.1	NE	36	NE	1964	64	5.2
O	73	78	48	61	7.1	NE	32	N	1964+	60	4.2
N	73	78	54	64	7.3	SW	31	S	1968	54	5.7
D	—	—	—	—	7....	NE	47	NE	1963	58	5.7
YR	77	80	52	62	7.2	NE	52 NE JUL 1966			60	5.6

Mean number of days

Month	Clear	Partly cloudy	Cloudy	Precip .01"+	Snow 1.0"+	Thunderstorms	Heavy fog	Max 90°+	Max 32°-	Min 32°-	Min 0°-	Avg daily solar radiation (langleys)
J	10	5	16	11	1	*	3	0	1	20	*	*
F	11	5	12	10	1	*	3	0	*	18	0	0
M	11	7	13	11	1	3	3	0	0	8	0	0
A	13	9	12	10	*	3	3	*	0	*	0	0
M	13	9	12	10	0	6	2	3	0	0	0	0
J	12	10	13	10	0	7	1	7	0	0	0	0
J	8	13	13	12	0	12	2	10	0	0	0	0
A	10	11	11	10	0	9	3	10	0	0	0	0
S	11	10	10	7	0	4	2	5	0	0	0	0
O	16	5	11	7	0	1	2	1	0	1	0	0
N	12	5	11	7	*	1	2	0	*	6	0	0
D	12	5	14	9	*	2	5	0	1	17	*	*
YR	124	95	146	112	3	41	31	31	1	71	*	

NORMALS, MEANS AND EXTREMES
(Table Revised 1975. Base Period for Climatological Normals: 1941-1970)

Station: GREENVILLE-SPARTANBURG **Elev.:** 971 feet m.s.l. **Average station pressure:** 983.4 mb

Temperature / Degree Days

Month	Normal Daily max	Normal Daily min	Normal Monthly	Record highest	Year	Record lowest	Year	Heating DD	Cooling DD
J	51.6	33.0	42.3	77	1974	-6	1966	704	0
F	54.1	34.7	44.4	79	1972	8	1967	577	0
M	61.0	40.2	51.0	86	1972	17	1969	450	13
A	72.0	49.9	61.0	91	1963	25	1973	144	24
M	79.9	58.3	69.1	97	1964	33	1972	29	156
J	85.9	65.9	75.9	98	1972	40	1972	0	327
J	87.6	69.0	78.3	99	1970	58	1972	0	412
A	86.8	68.1	77.5	99	1968	52	1968	0	388
S	81.4	62.3	71.7	93	1970	36	1965	145	210
O	71.8	50.2	61.0	91	1970	27	1965	420	43
N	61.8	40.1	51.0	85	1971	12	1970	685	0
D	52.4	33.3	42.9	75	1973	12	1973	—	0
YR	70.6	50.5	60.6	99 JUL 1970		-6 JAN 1966		3163	1573

Precipitation / Snow

Month	Normal	Max monthly	Year	Min monthly	Year	Max 24 hrs	Year	Snow max monthly	Year	Snow max 24 hrs	Year
J	4.07	6.14	1971	1.74	1970	2.61	1969	9.1	1966	5.7	1966
F	4.43	7.43	1971	1.00	1968	2.98	1973	6.9	1969	5.3	1969
M	5.33	11.30	1964	1.98	1963	3.76	1963	6.0	1971	6.0	1971
A	4.31	9.59	1964	2.28	1964	3.58	1972	0.0		0.0	
M	2.95	9.59	1969	2.19	1965	4.66	1971	0.0		0.0	
J	4.05				1971		1972	0.0		0.0	
J	4.18	7.44	1964	2.31	1970	3.89	1964	0.0		0.0	
A	4.51	7.51	1968	1.16	1967	4.49	1967	0.0		0.0	
S	3.79	7.98	1966	0.93	1964	6.21	1973	0.0		0.0	
O	3.10	10.24	1964	0.24	1973	6.21	1964	0.0		0.0	
N	3.10	5.31	1972	1.34	1973	2.83	1964	1.9	1968	1.9	1968
D	4.09	7.55	1971	0.37	1965	3.00	1971	11.4	1971	11.4	1971
YR	47.54	11.30 APR 1964		0.24 OCT 1974		6.21 SEP 1973		11.4 DEC 1971		11.4 DEC 1971	

Relative humidity / Wind / Sunshine / Sky cover

Month	RH 01	RH 07	RH 13	RH 19	Mean speed	Prevailing dir.	Fastest speed	Fastest dir.	Year	Pct possible sunshine	Mean sky cover (tenths)
J	74	78	55	64	7.2	NE	44	SW	1967	51	6.2
F	69	74	50	57	8.0	NE	44	SW	1966	58	5.5
M	69	75	46	50	8.1	SW	38	NW	1969	64	5.6
A	69	76	45	52	7.1	SW	40	NW	1970	63	5.3
M	83	84	55	64	6.3	NE	36	NW	1967	58	6.0
J	85	87	58	68		WSW	52	NE	1966	58.5	6.5
J	86	88	59	72	5.7	NE	31	NE	1966	59	6.5
A	86	88	58	72		NE	31	NE	1963	62	5.5
S	85	85	53	68	6.6	NE	36	NE	1964	64	5.1
O	80	83	50	64	6.8	NE	34	SW	1974	62	4.8
N	75	80	48	68	7.3	SW	34	SW	1971	52	4.6
D	76	80	57	68		NE	47	NE	1963	6.0	6.0
YR	78	81	54	63	6.9	NE	52 NE JUL 1966			59	5.7

Mean number of days

Month	Clear	Partly cloudy	Cloudy	Precip .01"+	Snow 1.0"+	Thunderstorms	Heavy fog, visibility ¼ mile or less	Max 90°+	Max 32°-	Min 32°-	Min 0°-
J	10	5	16	12	1	1	2	0	1	18	*
F	11	6	12	11	1	1	3	0	*	18	0
M	11	8	11	11	1	3	2	0	0	10	0
A	11	9	10	10	*	6	1	*	0	1	0
M	13	9	10	10	0	6	1	2	0	0	0
J	11	10	13	10	0	7	1	6	0	0	0
J	7	13	11	12	0	12	2	11	0	0	0
A	10	11	11	10	0	7	3	8	0	0	0
S	15	7	8	8	0	3	2	2	0	0	0
O	13	7	10	8	*	1	3	0	0	*	0
N	10	6	15	11	*	1	6	0	*	8	*
D										15	
YR	118	102	145	118	2	44	35	30	2	68	*

REFERENCE NOTES APPLYING TO TABLES APPEAR ON THE PAGE FOLLOWING LAST TABLE.
(Caution: Letters and symbols may have different meanings in 1941-1970 tables than in earlier tables. See notes.)

Reference notes applying to Normals, Means, and Extremes tables for 1931–1960 base period.

(a) **Length of record, years, based on January data.** Other months may be for more or fewer years if there have been breaks in the record.

(b) Climatological standard normals (1931-1960).

* Less than one half.

+ Also on earlier dates, months, or years.

T Trace, an amount too small to measure.

Below zero temperatures are preceded by a minus sign.

The prevailing direction for wind in the Normals, Means, and Extremes table is from records through 1963.

‡ ≥ 70° at Alaskan stations.

Unless otherwise indicated, dimensional units used in this bulletin are: temperature in degrees F.; precipitation, including snowfall, in inches; wind movement in miles per hour; and relative humidity in percent. Heating degree day totals are the sums of negative departures of average daily temperatures from 65° F. Cooling degree day totals are the sums of positive departures of average daily temperatures from 65° F. Sleet was included in snowfall totals beginning with July 1948. The term "Ice pellets" includes solid grains of ice (sleet) and particles consisting of snow pellets encased in a thin layer of ice. Heavy fog reduces visibility to 1/4 mile or less.

Sky cover is expressed in a range of 0 for no clouds or obscuring phenomena to 10 for complete sky cover. The number of clear days is based on average cloudiness 0-3, partly cloudy days 4-7, and cloudy days 8-10 tenths.

Solar radiation data are the averages of direct and diffuse radiation on a horizontal surface. The langley denotes one gram calorie per square centimeter.

‡ Figures instead of letters in a direction column indicate direction in tens of degrees from true North; i.e., 09 - East, 18 - South, 27 - West, 36 - North, and 00 - Calm. Resultant wind is the vector sum of wind directions and speeds divided by the number of observations. If figures appear in the direction column under "Fastest mile" the corresponding speeds are fastest observed 1-minute values.

To 8 compass points only.

Reference notes applying to Normals, Means, and Extremes tables for 1941–1970 base period.

(a) Length of record, years, through the current year unless otherwise noted, based on January data.

(b) 70° and above at Alaskan stations.

T Trace.

NORMALS - Based on record for the 1941-1970 period.

DATE OF AN EXTREME - The most recent in cases of multiple occurrence.

PREVAILING WIND DIRECTION - Record through 1963.

WIND DIRECTION - Numerals indicate tens of degrees clockwise from true north. 00 indicates calm.

FASTEST MILE WIND - Speed is fastest observed 1-minute value when the direction is in tens of degrees.

January Average Maximum Temperature (°F)

SOUTH CAROLINA

75 TH MERIDIAN TIME ZONE

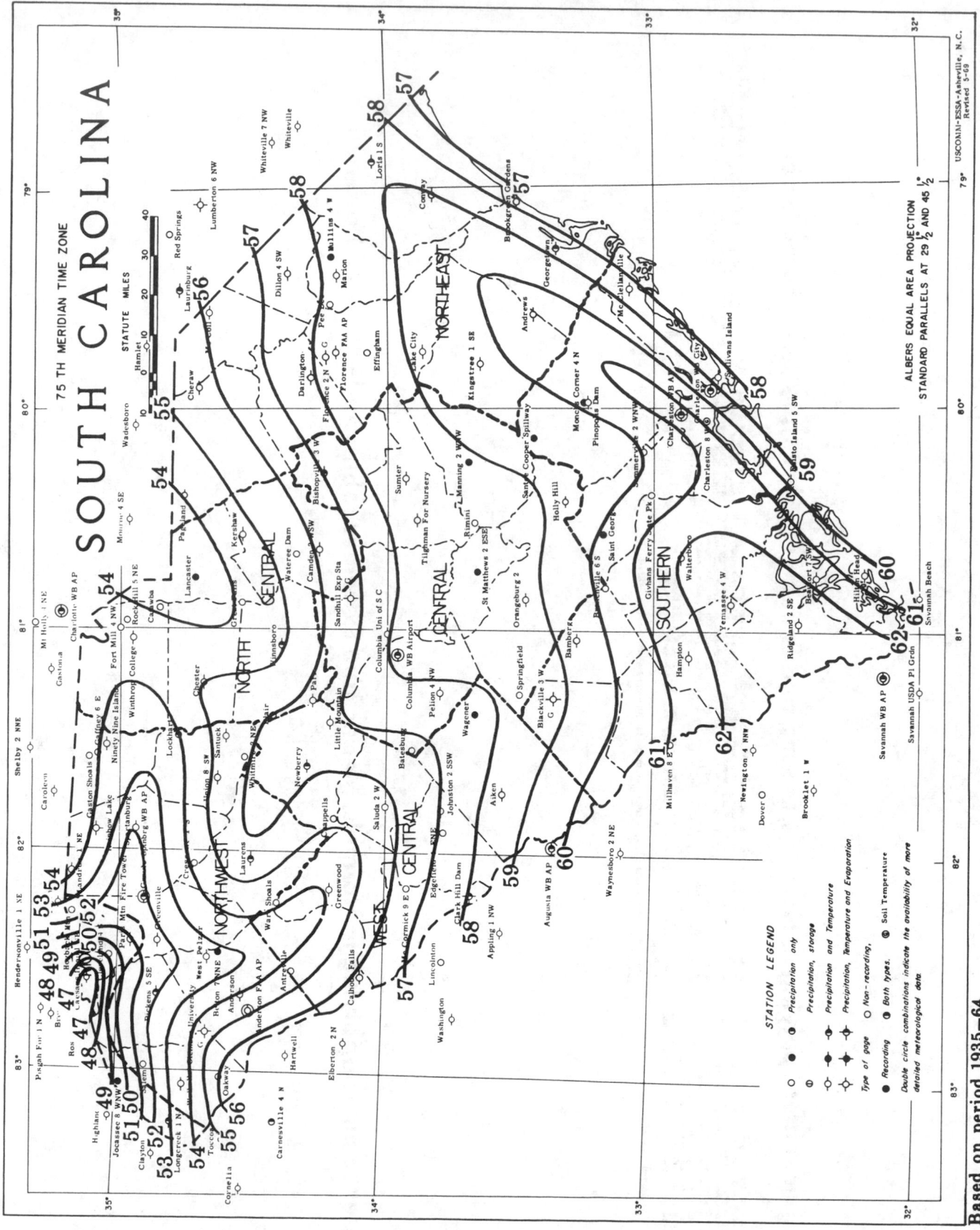

STATUTE MILES

STATION LEGEND

○ Precipitation only
⊖ Precipitation, storage
◐ Precipitation and Temperature
⊕ Precipitation, Temperature and Evaporation

Type of page ○ Non-recording;
● Recording ● Both types.
⊙ Soil Temperature

Double circle combinations indicate the availability of more detailed meteorological data.

ALBERS EQUAL AREA PROJECTION
STANDARD PARALLELS AT 29½ AND 45½

USCOMM–ESSA–Asheville, N.C.
Revised 3–69

Based on period 1935–64

Isolines are drawn through points of approximately equal value. Caution should be used in interpolating on these maps, particularly in mountainous areas.

886

January Average Minimum Temperature (°F)

SOUTH CAROLINA

75 TH MERIDIAN TIME ZONE

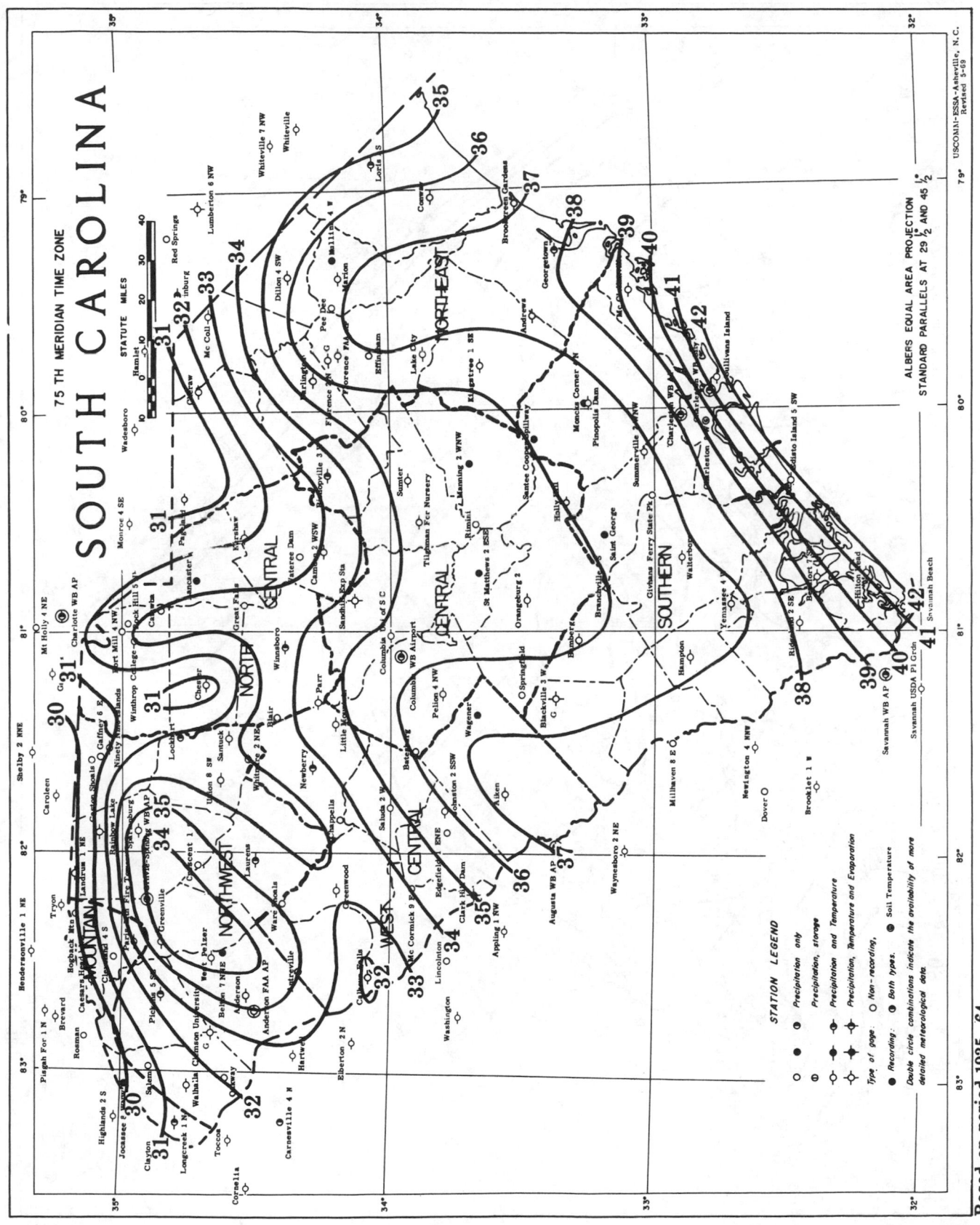

STATUTE MILES

STATION LEGEND

○ Precipitation only

◑ Precipitation, storage

◐ Precipitation and Temperature

◑ Precipitation, Temperature and Evaporation

Type of gage: ○ Non-recording,

● Recording;   ◑ Both types.

⊛ Soil Temperature

Double circle combinations indicate the availability of more
detailed meteorological data

Based on period 1935–64

Isolines are drawn through points of approximately equal value. Caution should be used
in interpolating on these maps, particularly in mountainous areas.

USCOMM-ESSA-Asheville, N.C.
Revised 5-69

ALBERS EQUAL AREA PROJECTION
STANDARD PARALLELS AT 29½° AND 45½°

July Average Maximum Temperature (°F)

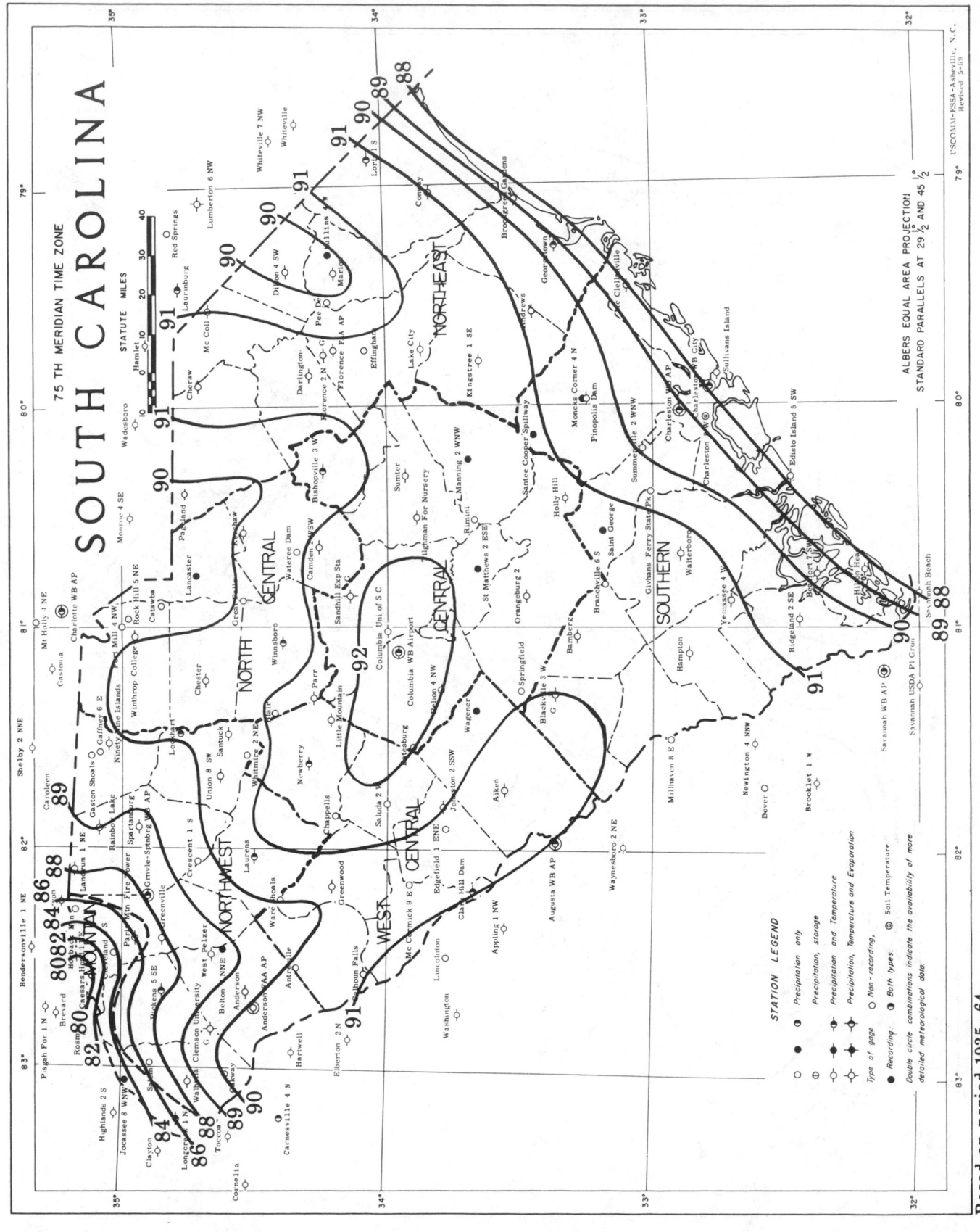

SOUTH CAROLINA

75 TH MERIDIAN TIME ZONE

STATUTE MILES

Based on period 1935–64
Isolines are drawn through points of approximately equal value. Caution should be used
in interpolating on these maps, particularly in mountainous areas.

ALBERS EQUAL AREA PROJECTION
STANDARD PARALLELS AT 29½ AND 45½

STATION LEGEND

○ Precipitation only
● Precipitation, storage
◐ Precipitation and Temperature
◒ Precipitation, Temperature and Evaporation
○ Type of gage   ○ Non - recording
● Recording :   ◑ Both types.
⊙ Soil Temperature
● Double circle combinations indicate the availability of more
   detailed meteorological data

USCOMM-ESSA-Asheville, N. C.
Revised 5-69

888

July Average Minimum Temperature (°F)

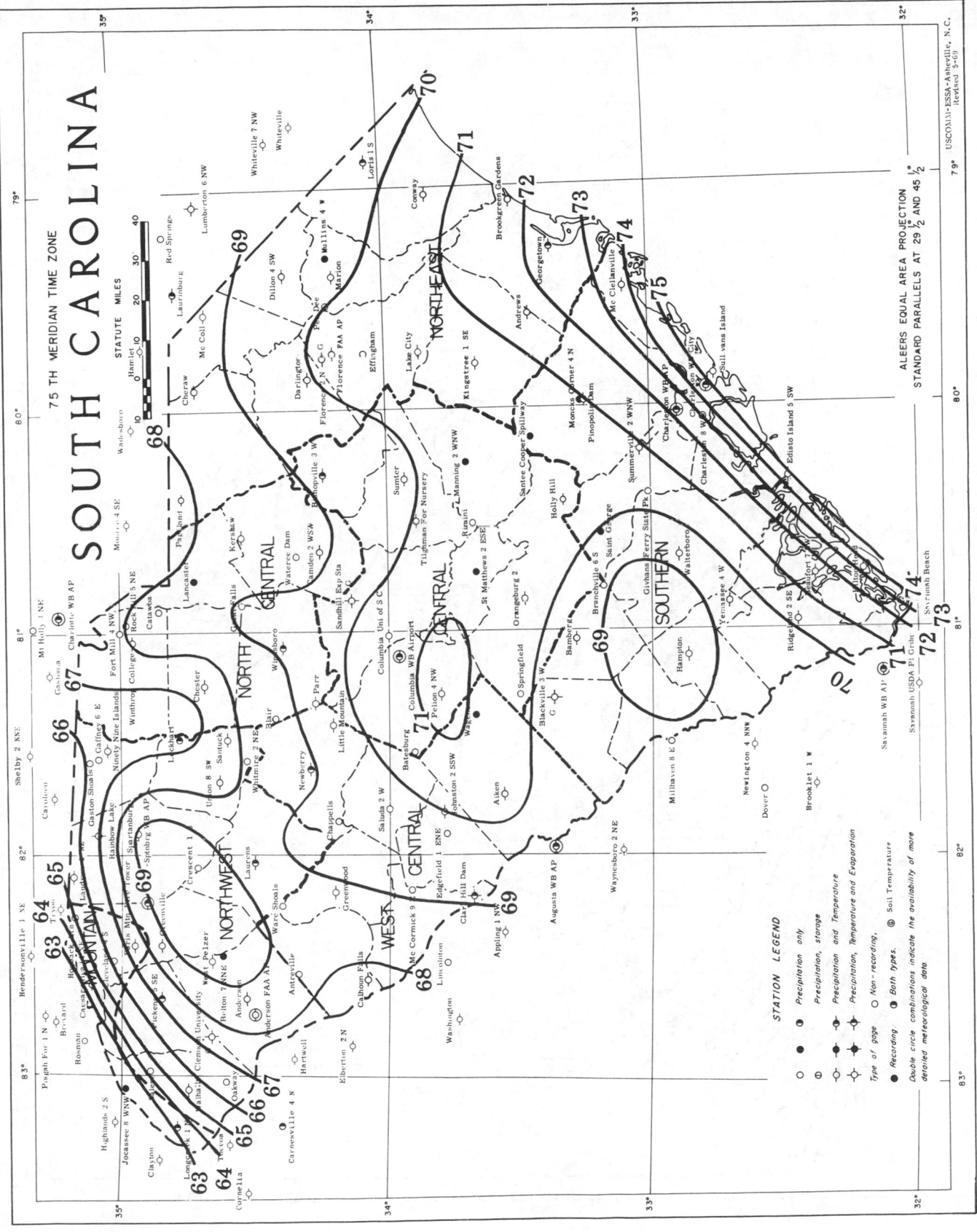

SOUTH CAROLINA

75 TH MERIDIAN TIME ZONE

STATUTE MILES

Based on period 1935—64
Isolines are drawn through points of approximately equal value. Caution should be used
in interpolating on these maps, particularly in mountainous areas.

STATION LEGEND

○ Precipitation only
● Precipitation, storage
◐ Precipitation and Temperature
◓ Precipitation, Temperature and Evaporation

Type of gage ○ Non - recording
● Recording ● Both types. ◉ Soil Temperature

Double circle combinations indicate the availability of more
detailed meteorological data

ALBERS EQUAL AREA PROJECTION
STANDARD PARALLELS AT 29 ½ AND 45 ½

USCOMM-ESSA-Asheville, N.C.
Revised 5-69

Annual Average Rainfall (Inches)

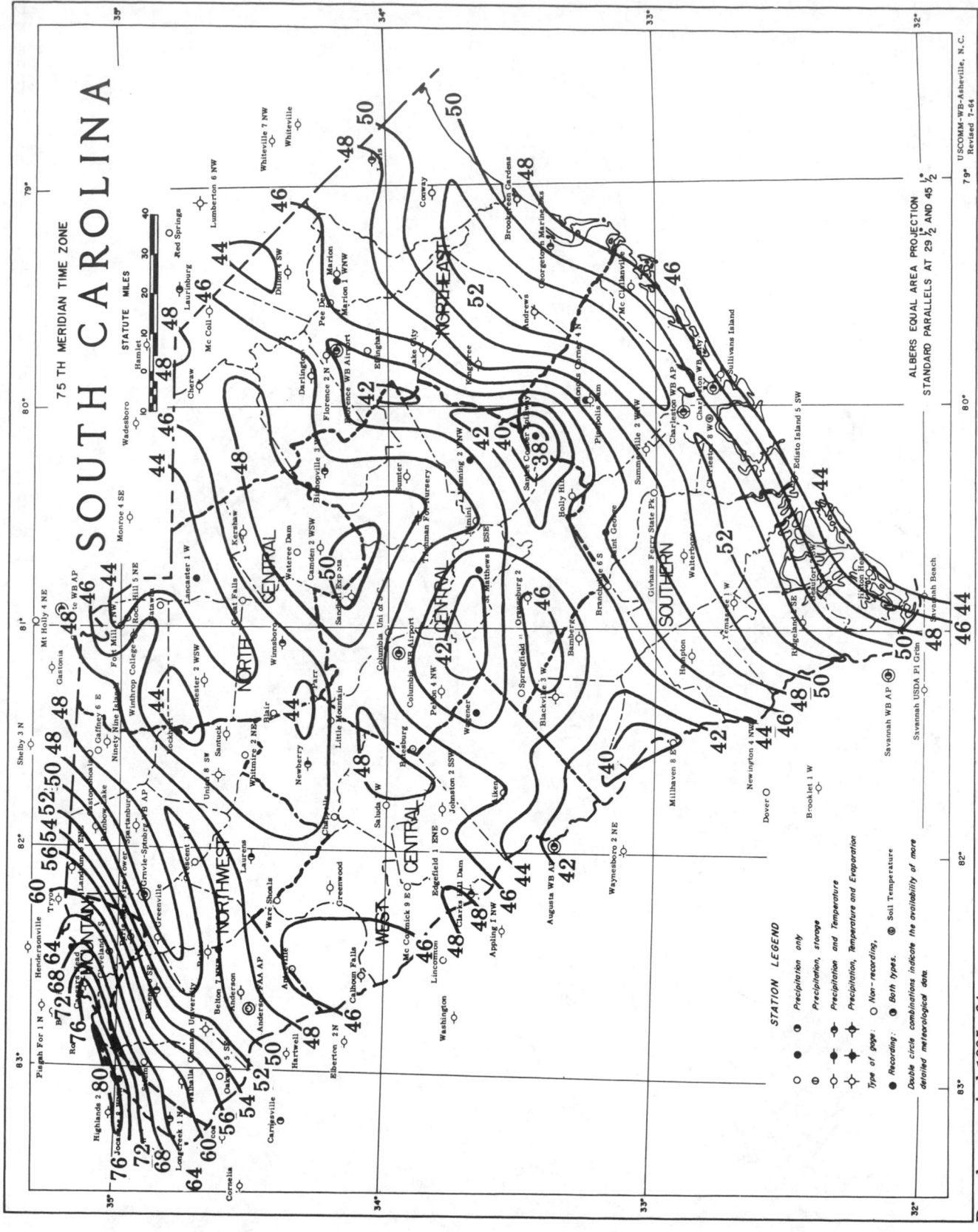

SOUTH CAROLINA

75 TH MERIDIAN TIME ZONE

STATUTE MILES

STATION LEGEND

Type of gage: ○ Non-recording;
● Recording; ⊕ Both types.

Double circle combinations indicate the availability of more
detailed meteorological data.

○ Precipitation only
⊖ Precipitation, storage
⊕ Precipitation and Temperature
⊕ Precipitation, Temperature and Evaporation
⊕ Soil Temperature

ALBERS EQUAL AREA PROJECTION
STANDARD PARALLELS AT 29 ½ AND 45 ½

U.S.COMM-WB-Asheville, N.C.
Revised 7-64

Based on period 1935–64
Isolines are drawn through points of approximately equal value. Caution should be used
in interpolating on these maps, particularly in mountainous areas.

890

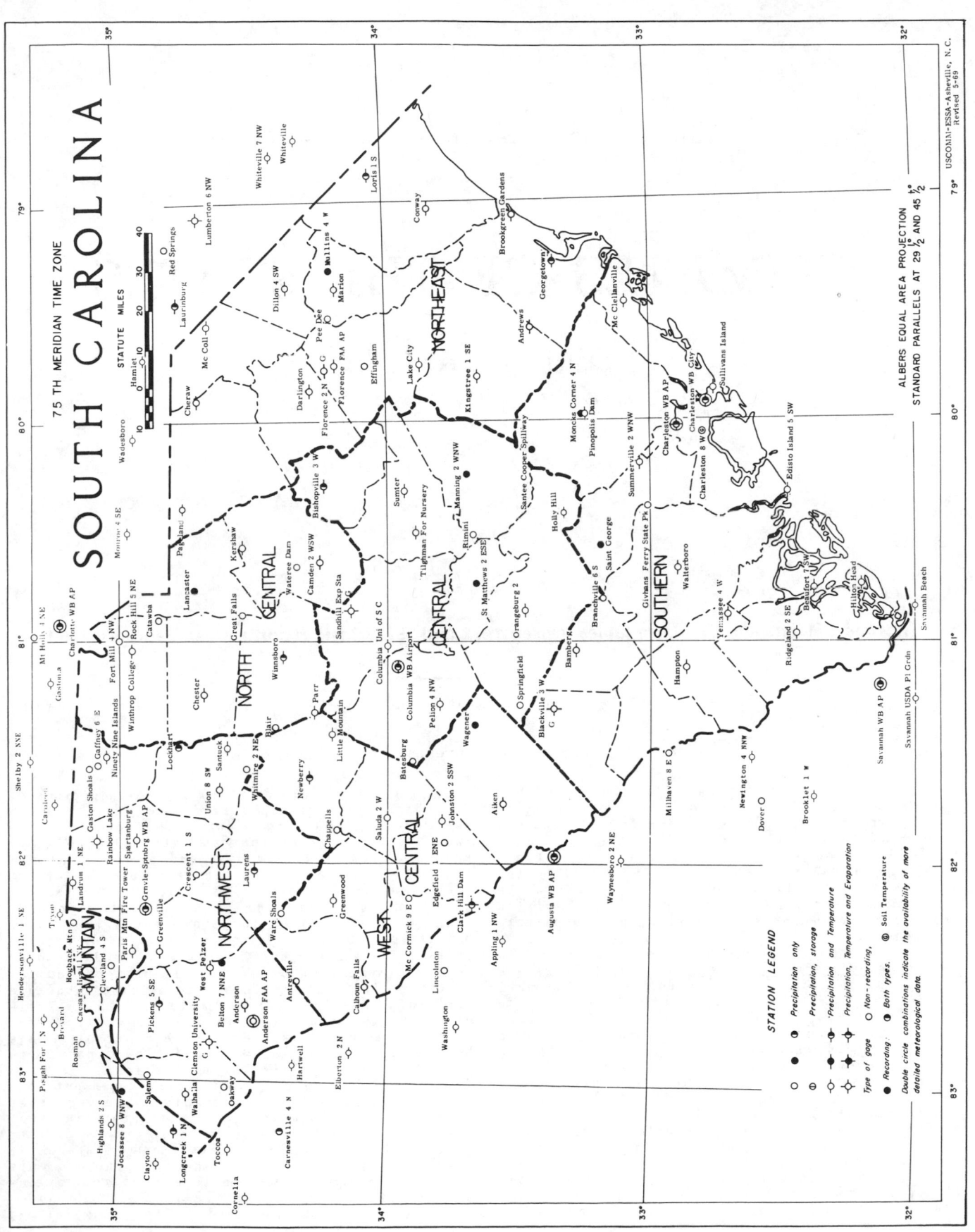

SOUTH CAROLINA

75 TH MERIDIAN TIME ZONE

STATUTE MILES

ALBERS EQUAL AREA PROJECTION
STANDARD PARALLELS AT 29½° AND 45½°

USCOMM–ESSA–Asheville, N.C.
Revised 5-69

STATION LEGEND

○ Precipitation only
⊖ Precipitation, storage
✦ Precipitation and Temperature
✦ Precipitation, Temperature and Evaporation

Type of page   ○ Non - recording
● Recording   ◐ Both types.   ⊙ Soil Temperature

Double circle combinations indicate the availability of more
detailed meteorological data.

891

CLIMATES OF THE STATES

SOUTH DAKOTA

(Normals, Means and Extremes tables revised 1973 and 1975. Basic report revised February 1960.)

Climate of South Dakota

William T. Hodge, Weather Bureau State Climatologist

Rolling plains are the main feature of South Dakota, varying from nearly level land to hilly ridges, and increasing in elevation from the eastern border to the western edge of the State. The general elevation above sea level in the extreme eastern portion is about 1,500 feet, and in the extreme west is about 3,000 feet, except in the Black Hills area. The Black Hills, an isolated group of forest-covered mountains, have a climate of their own.

The soil covering the State was laid down in past ages by glaciers, water, and wind. There are occasional outcroppings of bedrock. The Missouri River and its tributaries drain all of South Dakota except for a small portion of the northeastern part of the State. Some of this small drainage area is in the headwaters of the Red river of the North in the Hudson Bay Drainage, and the remainder is in the headwater area of the Minnesota River which forms a part of the upper Mississippi River Basin.

South Dakota is bisected by the Missouri River which flows in a southerly direction to Pierre and then turns to the south-southeast to where it forms the South Dakota-Nebraska State line. West of the Missouri lies a country of canyons, broad upland flats, and buttes. The principal tributaries which drain this region are the Grand, and the Moreau and Cheyenne, which drain the Black Hills, and the White. To the east of the Missouri which is mostly prairie land, there are numerous small

ponds and lakes, some of which dry up in periods of droughts. The principal rivers in this area are the James and the Big Sioux. The larger of the two, the James River, has an extremely low slope and consequently is sluggish and meanders. Water falling on much of the eastern area does not reach the stream valleys at all, but lies in depressions until it evaporates or soaks into the ground.

The eastern half of the State has a limy soil, but it is arable, fertile, and suitable for growing crops. The western half, because of limited moisture and more rugged terrain, is best suited for stock grazing, although other forms of agriculture are profitable in years of ample moisture. Irrigation is practiced to some extent, particularly near the Black Hills. Extensive reservoirs have been created along the Missouri River, partly to provide water supplies for irrigation.

There are few, if any, states more dependent upon agriculture than is South Dakota. Since grasses flourish in the climate, it is not surprising that livestock and their products account for the greater part of the farm income. Cattle and hogs are the principal livestock, while wheat, corn, and oats are the main crops. Hunting is a major recreational activity. Pheasant thrive in the east; antelope, deer, and elk, are found in the west. Buffalo are maintained in preserves.

Since South Dakota is situated in the heart of the North American Continent, it is near the paths

892

of many cyclones and anticyclones, and has the extremes of summer heat and winter cold that are characteristic of continental climates. The highest temperature of record in the State is 120°F., observed July 5, 1936, at Gannvalley; the lowest, -58°F., February 17, 1936, at McIntosh. Rapid fluctuations in temperature are common. Partly because of the great distance from any large body of water, the ranges of daily, monthly, and annual temperatures are very large. Temperatures of 100°F., or higher, are experienced in some part of the State each summer, and on rare occasions such readings have been noted as early as April and as late as October. These high temperatures are usually attended by low humidity, which greatly reduces the oppressiveness of the heat. Below-zero temperatures occur frequently on midwinter mornings, but it is not often that the temperature stays below zero during the entire day. In the north, subzero temperatures can occur in October and April.

Warm, "chinook" winds and frequent sunny skies make the Black Hills area the warmest part of the State in winter. Also, because of the tendency for very cold air masses to stay at low elevations, some of the arctic air outbreaks that blanket the eastern counties do not reach the higher counties in the west. During summer, the higher elevation of the Black Hills results in that section having cooler temperatures than the rest of the State. At this season, the central and southeastern counties are warmest. The freeze-free season is shortest high in the Black Hills where brief freezing has been known to occur at any time of summer. Elsewhere, the first autumn freeze generally occurs in mid-September in the northwest, in late September in the central and east, and in the first week of October in the extreme southeastern corner. The average time of the last freeze in spring ranges from early May in the southeast to late May in the northwest.

The annual precipitation decreases northwestward from about 25 inches in the extreme southeast to less than 13 inches in part of the northwest. The Black Hills are again an exception, varying from 16 inches in their southern portion to almost 25 inches in the northern, where rain and snow are often formed when the prevailing winds are abruptly forced up the mountainsides. Most of the State's precipitation occurs during the crop season, April through September. On the average, it reaches a maximum during June, and decreases sharply in early July. In the east, there is a small secondary increase in August, followed by an overall diminishing during autumn. The least precipitation is received during winter.

Occasionally there is heavy snowfall in winter and the amount of snow on the ground accumulates to a considerable depth, but as a rule, the snow cover is not great. Wind usually accompanies the snow, causing a large proportion of it to collect in gullies and behind windbreaks. In the worst storms, isolated drifts many feet deep may block roads, while windswept fields nearby are nearly bare of snow. Accurate measurements of the snow are difficult, since irregularities are introduced by the presence of buildings, fences, trees, and weeds; and by variations in the terrain, wind, and the snow itself. Snow that falls early in the season seldom stays on the ground very long. After the ground has frozen deeply and the days become very short, it remains longer. Once a snow cover is present, there is a tendency for it to continue, since the temperature falls to much lower levels over snow than over bare ground. Snowfall reaches a maximum in February and early March, and decreases markedly near the end of March. Violent, cold winds carrying snow picked up from the ground, commonly called "blizzards", are not very frequent and are a hazard only to those who are unprepared for a winter storm.

Rainstorms occur most frequently in early summer, hailstorms are most frequent in midsummer, and lightning does its worst damage in late summer. In dry seasons, and particularly in the west in late summer, thunderstorm bases may be as high as 2 miles above the ground; consequently, the rain showers may evaporate before reaching ground. When the thundershowers do reach the ground during summer, there is a high incidence of hail. A total of 75 tornadoes were reported during the 7-year period 1952-58. They are not as frequent as in states farther south and east. Partly because the centers of population are widely dispersed, the State's tornadoes have seldom caused great damage or many deaths. Much more damage is caused by straight-line thunderstorm winds. Such winds are not impeded by trees or other obstacles on the open prairie, and speeds near the ground become very high. Farm buildings not having the protection of a shelterbelt are especially vulnerable to this type of damage.

The most serious flooding (e.g., the record April 1952 flood on the Missouri River) has been caused by rapid melting of a heavy snow pack and aggravated by ice jams. Heavy rainfall alone causes severe floods on tributary streams, especially in the eastern part of the State. Intense local storms result in flash flooding along minor tributaries. Since 1956 the flow of the Missouri River through the State has been controlled by Fort Randall and Garrison Dams which were constructed for purposes of flood control, power, navigation, and irrigation. Major floods in recent years occurred in 1951, 1952, and 1957.

Drought effects usually become evident soon after the rainfall drops below normal. Much of the State lies in a semiarid region, so crop returns are very sensitive in relation to the rainfall. Slightly below-normal rainfall can reduce yields sharply, especially if the rains do not fall at opportune times. Regarding corn and similar crops, a somewhat common experience is for enough moisture to fall early in the season so that the plants get a start. Then hot, dry spells occur in July and August, causing the crops to develop poorly. If a crop is in a critical stage with depleted soil moisture, a hot dry wind can do great damage within a day or two.

South Dakota has considerable fair weather. The air is generally clear with excellent visibility, since much of it arrives by way of the Rocky Mountains and Canada. The wind is most frequently from the south and southeast during the summer, and from the north and northwest during the winter. Wind speeds are often moderate to fresh at midday and almost calm at night, averaging 11 or 12 miles per hour on a year-round basis.

REFERENCES

(1) Weather Bureau Technical Paper No. 16-Maximum 24-Hour precipitation in the United States. Washington, D. C. 1952.

(2) Weather Bureau Technical Paper No. 25-Rainfall Intensity-Duration-Frequency Curves. For selected stations in the United States, Alaska, Hawaiian Islands, and Puerto Rico.

(3) South Dakota State College Agricultural Economics Phamphlet 68, Average Weekly Temperatures, Precipitation and New Snow Received at 60 Weather Bureau Stations through 1954, by Ray F. Pengra, Brookings, S. D., 1956.

(4) South Dakota State College Agricultural Economics Department Pamphlet 77, Temperature Summary of Data for 60 South Dakota Weather Stations, by Ray F. Pengra, Brookings, S. D., 1957.

BIBLIOGRAPY

(A) Climatic Summary of the United States (Bulletin W) 1930 edition, Sections 36 and 37. U. S. Weather Bureau

(B) Climatic Summary of the United States South Dakota - Supplement for 1931 through 1952 (Bulletin W Supplement). U. S. Weather Bureau

(C) Climatological Data - South Dakota. U. S. Weather Bureau

(D) Climatological Data National Summary. U. S. Weather Bureau

(E) Hourly Precipitation Data South Dakota. U. S. Weather Bureau

(F) Local Climatological Data, U. S. Weather Bureau, for Huron, Rapid City, and Sioux Falls, South Dakota.

STATION	Freeze threshold temperature	Mean date of last Spring occurrence	Mean date of first Fall occurrence	Mean No. of days between dates	Years of record Spring	No. of occurrences in Spring	Years of record Fall	No. of occurrences in Fall
ABERDEEN CAA AP	32	05-13	09-27	137	30	30	30	30
	28	05-04	10-04	153	30	30	30	30
	24	04-20	10-16	180	30	30	29	29
	20	04-11	10-27	199	30	30	29	29
	16	03-29	11-04	221	30	30	29	29
ACADEMY	32	05-11	09-26	138	28	28	30	30
	28	04-29	10-09	163	28	28	30	30
	24	04-21	10-20	183	28	28	30	30
	20	04-10	10-26	200	27	27	28	28
	16	03-29	11-01	217	27	27	28	28
ARDMORE	32	05-15	09-22	130	28	28	30	30
	28	05-03	10-06	156	28	28	29	29
	24	04-25	10-12	170	28	28	29	29
	20	04-13	10-19	189	28	28	29	29
	16	04-06	10-29	206	28	28	29	29
BELLE FOURCHE 2 NE	32	05-13	09-23	133	30	30	30	30
	28	05-03	10-05	155	30	30	30	30
	24	04-26	10-16	173	30	30	30	30
	20	04-14	10-23	192	30	30	30	30
	16	04-05	10-30	208	30	30	30	30
BISON	32	05-15	09-25	133	18	18	13	13
	28	05-04	10-07	157	16	16	13	13
	24	04-22	10-14	176	16	16	13	13
	20	04-18	10-25	190	16	16	13	13
	16	04-14	11-02	203	15	15	12	12
BRITTON	32	05-20	09-20	123	29	29	30	30
	28	05-07	10-03	149	29	29	30	30
	24	04-28	10-10	165	29	29	30	30
	20	04-17	10-24	190	29	29	29	29
	16	04-04	11-02	212	29	29	29	29
BROOKINGS	32	05-13	09-27	137	30	30	30	30
	28	05-04	10-05	154	30	30	30	30
	24	04-25	10-15	173	30	30	30	30
	20	04-12	10-25	196	30	30	30	30
	16	03-27	11-03	221	30	30	30	30
CAMP CROOK	32	05-20	09-17	120	30	30	28	28
	28	05-10	09-27	140	30	30	26	26
	24	04-29	10-09	163	30	30	26	26
	20	04-23	10-17	177	30	30	26	26
	16	04-13	10-29	198	30	30	26	26
CASTLEWOOD	32	05-21	09-20	122	29	29	29	29
	28	05-11	09-30	142	28	28	29	29
	24	04-30	10-08	161	28	28	29	29
	20	04-18	10-18	183	28	28	29	29
	16	04-03	10-31	211	28	28	29	29
CHEYENNE AGENCY	32	05-15	09-28	136	10	10	11	11
	28	05-02	10-05	156	10	10	11	11
	24	04-29	10-19	173	10	10	11	11
	20	04-09	11-02	206	10	10	11	11
	16	03-22	11-10	233	9	9	11	11
CLARK	32	05-16	09-24	131	29	29	30	30
	28	05-07	10-03	149	29	29	30	30
	24	04-29	10-13	167	28	28	30	30
	20	04-17	10-22	188	28	28	30	30
	16	04-04	10-29	209	28	28	30	30
COTTONWOOD 3 E	32	05-19	09-21	125	30	30	30	30
	28	05-06	10-02	149	30	30	30	30
	24	04-26	10-11	168	30	30	30	30
	20	04-18	10-22	187	30	30	30	30
	16	04-08	10-26	201	30	30	30	30
DE SMET	32	05-11	09-25	136	27	27	26	26
	28	05-03	10-03	153	27	27	26	26
	24	04-24	10-10	169	26	26	25	25
	20	04-13	10-20	190	26	26	25	25
	16	04-05	10-29	207	26	26	24	24
DUPREE	32	05-17	09-25	131	29	29	29	29
	28	05-04	10-05	154	29	29	29	29
	24	04-26	10-13	171	29	29	29	29
	20	04-14	10-22	191	29	29	29	29
	16	04-03	10-30	210	29	29	29	29
EUREKA	32	05-17	09-20	126	30	30	30	30
	28	05-09	09-29	143	30	30	30	30
	24	05-02	10-07	159	30	30	30	30
	20	04-20	10-16	179	30	30	30	30
	16	04-06	10-28	206	30	30	30	30
FAIRFAX	32	05-07	10-03	149	30	30	30	30
	28	04-28	10-14	169	29	29	30	30
	24	04-15	10-24	193	29	29	29	29
	20	04-04	10-30	209	29	29	29	29
	16	03-28	11-07	224	29	29	29	29
FAITH	32	05-12	09-29	140	20	20	21	21
	28	04-29	10-07	161	20	20	21	21
	24	04-17	10-18	184	20	20	20	20
	20	04-11	10-25	197	20	20	20	20
	16	04-02	11-09	220	20	20	19	19
FAULKTON	32	05-12	09-26	137	30	30	30	30
	28	05-01	10-03	155	30	30	30	30
	24	04-22	10-13	174	30	30	30	30
	20	04-11	10-27	198	30	30	30	30
	16	03-30	11-02	217	30	30	30	30
GANNVALLEY	32	05-08	09-28	144	29	29	29	29
	28	05-03	10-09	160	29	29	29	29
	24	04-24	10-20	179	29	29	29	29
	20	04-12	10-28	199	29	29	29	29
	16	03-30	11-05	220	29	29	29	29
GELHAUS FARM	32	05-19	09-22	126	18	18	18	18
	28	05-02	09-29	150	18	18	18	18
	24	04-26	10-13	170	18	18	18	18
	20	04-17	10-19	185	18	18	18	18
	16	04-02	11-01	213	18	18	18	18
GETTYSBURG	32	05-16	09-27	134	20	20	20	20
	28	05-04	10-03	152	20	20	20	20
	24	04-24	10-15	175	20	20	20	20
	20	04-14	10-24	193	19	19	20	20
	16	04-04	10-31	211	19	19	20	20
GREGORY	32	05-05	10-02	150	20	20	21	21
	28	04-26	10-12	169	20	20	21	21
	24	04-21	10-21	183	20	20	21	21
	20	04-07	10-29	206	20	20	21	21
	16	03-29	11-08	223	20	20	21	21
HIGHMORE 1 W	32	05-15	09-22	129	30	30	30	30
	28	05-06	10-02	149	30	30	30	30
	24	04-27	10-12	168	30	30	30	30
	20	04-20	10-24	187	30	30	30	30
	16	04-06	10-31	208	30	30	30	30
HOT SPRINGS	32	05-12	09-27	139	30	30	30	30
	28	05-02	10-09	160	30	30	30	30
	24	04-19	10-17	181	30	30	30	30
	20	04-09	10-29	204	30	30	30	30
	16	04-01	11-03	216	30	30	30	30
HOWARD	32	05-12	09-27	138	30	30	30	30
	28	05-02	10-06	157	30	30	30	30
	24	04-22	10-17	178	30	30	30	30
	20	04-12	10-26	197	29	29	30	30
	16	04-02	11-01	213	29	29	29	29
* HURON	32	05-04	09-30	149	30	30	30	30
	28	04-24	10-11	169	30	30	30	30
	24	04-16	10-25	192	30	30	30	30
	20	04-08	11-01	207	30	30	30	30
	16	03-28	11-07	224	30	30	30	30

FREEZE DATA

STATION	Freeze threshold temperature	Mean date of last Spring occurrence	Mean date of first Fall occurrence	Mean No. of days between dates	Years of record Spring	No. of occurrences in Spring	Years of record Fall	No. of occurrences in Fall
KENNEBEC	32	05-15	09-23	131	28	28	29	29
	28	05-05	10-03	152	28	28	29	29
	24	04-25	10-14	172	28	28	29	29
	20	04-18	10-23	188	28	28	29	29
	16	04-03	11-01	211	28	28	29	29
LEMMON	32	05-16	09-23	130	30	30	29	29
	28	05-07	10-01	148	30	30	29	29
	24	04-28	10-12	167	30	30	29	29
	20	04-16	10-23	190	30	30	29	29
	16	04-08	10-27	202	30	30	29	29
LONGVALLEY	32	05-15	09-24	132	13	13	14	14
	28	05-04	10-07	157	13	13	13	13
	24	04-23	10-19	180	12	12	13	13
	20	04-13	10-27	197	12	12	12	12
	16	04-01	11-07	220	12	12	12	12
LUDLOW 2 NW	32	05-21	09-19	121	27	27	26	26
	28	05-12	09-27	138	27	27	26	26
	24	05-02	10-10	161	27	27	26	26
	20	04-22	10-18	179	27	27	26	26
	16	04-10	10-27	200	27	27	26	26
MARTIN	32	05-16	09-28	135	14	14	14	14
	28	05-04	10-12	161	14	14	14	14
	24	04-21	10-23	186	13	13	14	14
	20	04-09	10-31	205	12	12	14	14
	16	04-01	11-06	219	12	12	14	14
MC INTOSH	32	05-19	09-22	126	24	24	24	24
	28	05-10	09-28	141	24	24	24	24
	24	05-03	10-16	166	23	23	23	23
	20	04-24	10-21	180	23	23	23	23
	16	04-07	10-30	205	22	22	21	21
MELLFTTE	32	05-16	09-23	130	29	29	30	30
	28	05-04	10-01	150	29	29	29	29
	24	04-24	10-11	170	29	29	29	29
	20	04-12	10-21	192	29	29	29	29
	16	04-02	10-29	209	29	29	29	29
MENNO	32	05-07	09-27	143	30	30	26	26
	28	04-27	10-08	165	30	30	25	25
	24	04-16	10-20	187	30	30	24	24
	20	04-06	10-29	206	30	30	24	24
	16	03-31	11-06	219	30	30	24	24
MILBANK	32	05-07	09-30	147	29	29	30	30
	28	05-01	10-09	160	29	29	30	30
	24	04-20	10-19	182	30	30	30	30
	20	04-07	10-31	208	30	30	30	30
	16	03-28	11-07	224	30	30	30	30
MILLER	32	05-14	09-26	136	30	30	30	30
	28	05-02	10-06	157	30	30	30	30
	24	04-26	10-18	175	30	30	30	30
	20	04-11	10-28	200	30	30	30	30
	16	04-01	11-04	217	30	30	30	30
MITCHELL 2 SE	32	05-05	10-01	149	30	30	30	30
	28	04-29	10-11	165	30	30	30	30
	24	04-13	10-25	195	30	30	30	30
	20	04-06	10-29	206	30	30	30	30
	16	03-30	11-08	224	30	30	30	30
MOBRIDGE	32	05-13	09-22	132	24	24	24	24
	28	05-02	10-01	151	24	24	24	24
	24	04-26	10-13	171	24	24	24	24
	20	04-17	10-24	190	24	24	24	24
	16	03-29	11-03	219	24	24	24	24
MURDO	32	05-24	09-22	121	30	30	30	30
	28	05-10	10-01	144	30	30	30	30
	24	05-01	10-14	165	30	30	30	30
	20	04-20	10-22	186	30	30	29	29
	16	04-09	10-31	205	30	30	29	29
* PIERRE CAA AP	32	05-04	10-06	155	30	30	30	30
	28	04-22	10-16	177	30	30	30	30
	24	04-13	10-26	196	30	30	30	30
	20	04-02	11-02	215	30	30	30	30
	16	03-26	11-12	231	30	30	30	30
PINE RIDGE	32	05-23	09-20	121	19	19	19	19
	28	05-07	09-27	143	19	19	19	19
	24	04-26	10-11	168	19	19	18	18
	20	04-16	10-24	190	18	18	18	18
	16	04-05	11-03	212	17	17	18	18
POLLOCK	32	05-23	09-14	114	30	30	29	29
	28	05-12	09-22	132	30	30	29	29
	24	05-02	10-04	155	29	29	28	28
	20	04-26	10-11	169	29	29	28	28
	16	04-11	10-27	198	29	29	27	27
PUKWANA	32	05-12	09-23	134	26	26	27	27
	28	05-02	10-05	156	26	26	27	27
	24	04-22	10-16	177	26	26	27	27
	20	04-08	10-26	201	26	26	27	27
	16	03-26	11-02	221	26	26	27	27
* RAPID CITY	32	05-07	10-04	150	30	30	30	30
	28	04-26	10-13	169	30	30	30	30
	24	04-16	10-21	187	30	30	30	30
	20	04-07	10-27	203	30	30	30	30
	16	03-31	11-06	219	30	30	30	30
REDFIELD	32	05-10	09-25	138	30	30	30	30
	28	05-03	10-06	156	30	30	30	30
	24	04-26	10-15	172	30	30	30	30
	20	04-11	10-24	197	30	30	30	30
	16	03-30	10-31	215	30	30	30	30
REDIG 9 NE	32	05-20	09-16	119	30	30	29	29
	28	05-13	09-25	136	30	30	29	29
	24	05-02	10-04	155	29	29	29	29
	20	04-24	10-18	178	29	29	29	29
	16	04-10	10-25	198	29	29	29	29
SIOUX FALLS WB CITY	32	05-05	10-03	152	30	30	30	30
	28	04-26	10-13	171	30	30	30	30
	24	04-14	10-26	195	30	30	30	30
	20	04-05	11-04	212	30	30	30	30
	16	03-26	11-12	232	30	30	30	30
SISSETON	32	05-10	09-30	143	18	18	19	19
	28	04-28	10-10	165	18	18	19	19
	24	04-18	10-14	179	18	18	19	19
	20	04-10	10-31	204	18	18	19	19
	16	03-29	11-05	220	18	18	19	19
TIMBER LAKE	32	05-19	09-23	127	24	24	24	24
	28	05-04	10-03	152	24	24	24	24
	24	04-26	10-11	168	24	24	23	23
	20	04-18	10-22	187	24	24	23	23
	16	04-09	11-01	206	24	24	23	23
VALE	32	05-16	09-24	131	30	30	30	30
	28	05-05	10-05	153	30	30	30	30
	24	04-24	10-13	172	30	30	30	30
	20	04-13	10-23	193	30	30	30	30
	16	04-04	10-30	209	30	30	30	30
VERMILLION	32	05-04	10-07	156	30	30	28	28
	28	04-21	10-19	181	30	30	28	28
	24	04-08	10-26	201	30	30	27	27
	20	03-31	11-05	219	30	30	27	27
	16	03-26	11-14	234	30	30	27	27
VIVIAN	32	05-18	09-25	130	30	30	30	30
	28	05-02	10-04	155	29	29	30	30
	24	04-28	10-14	170	29	29	30	30
	20	04-19	10-24	189	28	28	29	29
	16	04-05	10-30	208	28	28	29	29

FREEZE DATA

STATION	Freeze threshold temperature	Mean date of last Spring occurrence	Mean date of first Fall occurrence	Mean No. of days between dates	Years of record Spring	No. of occurrences in Spring	Years of record Fall	No. of occurrences in Fall
WAGNER	32	05-06	09-30	147	29	29	28	28
	28	04-27	10-14	170	29	29	28	28
	24	04-16	10-26	193	28	28	26	26
	20	04-04	11-03	213	27	27	25	25
	16	03-27	11-07	225	27	27	24	24
WATERTOWN CAA AP	32	05-17	09-27	133	30	30	30	30
	28	05-05	10-04	152	30	30	30	30
	24	04-24	10-13	173	30	30	30	30
	20	04-13	10-27	197	30	30	30	30
	16	03-31	11-03	218	30	30	30	30
WEBSTER	32	05-18	09-23	128	28	28	23	23
	28	05-07	10-01	147	28	28	23	23
	24	04-29	10-14	168	29	29	23	23
	20	04-15	10-22	190	29	29	22	22
	16	04-03	11-01	211	29	29	22	22
WENTWORTH	32	05-14	09-29	137	29	29	26	26
	28	05-02	10-06	157	29	29	26	26
	24	04-21	10-18	180	29	29	24	24
	20	04-11	10-26	198	29	29	24	24
	16	03-30	11-04	219	29	29	24	24
WHITE LAKE	32	05-10	09-29	142	28	28	27	27
	28	04-29	10-11	165	28	28	27	27
	24	04-17	10-19	186	28	28	27	27
	20	04-04	10-29	208	27	27	27	27
	16	03-26	11-08	227	27	27	27	27
WINNER	32	05-07	09-29	145	26	26	22	22
	28	04-29	10-09	163	25	25	22	22
	24	04-16	10-17	184	25	25	21	21
	20	04-09	10-28	201	25	25	20	20
	16	04-02	11-09	221	25	25	20	20
WOOD	32	05-14	09-25	135	30	30	29	29
	28	05-03	10-04	154	30	30	29	29
	24	04-24	10-17	175	30	30	29	29
	20	04-14	10-23	192	30	30	29	29
	16	04-05	11-02	211	30	30	29	29
* YANKTON	32	04-29	10-08	162	30	30	30	30
	28	04-18	10-20	185	30	30	30	30
	24	04-10	10-29	202	30	30	30	30
	20	03-29	11-06	222	30	30	30	30
	16	03-25	11-11	231	30	30	30	30

Data in the above table are based on the period 1921-1950, or that portion of this period for which data are available.

Means have been adjusted to take into account years of non-occurrence.

A freeze is a numerical substitute for the former term "killing frost" and is the occurrence of a minimum temperature at or below the threshold temperature of 32°, 28°, etc.

Freeze data tabulations in greater detail are available and can be reproduced at cost.

* Freeze Data for Huron, Pierre, Rapid City and Yankton should be used with caution, since temperatures for part of the period were taken from instruments mounted above the roofs of buildings.

*MEAN TEMPERATURE AND PRECIPITATION

STATION	JAN Temp	JAN Precip	FEB Temp	FEB Precip	MAR Temp	MAR Precip	APR Temp	APR Precip	MAY Temp	MAY Precip	JUN Temp	JUN Precip	JUL Temp	JUL Precip	AUG Temp	AUG Precip	SEP Temp	SEP Precip	OCT Temp	OCT Precip	NOV Temp	NOV Precip	DEC Temp	DEC Precip	ANN Temp	ANN Precip
CENTRAL DIVISION																										
HIGHMORE 1 W	14.8	.42	18.4	.53	29.2	1.20	45.8	1.71	57.2	2.18	66.7	3.71	74.9	1.80	72.8	1.99	62.8	1.21	50.1	1.13	32.4	.49	20.1	.40	45.4	16.77
KENNEBEC	19.0	.52	22.7	.56	32.2	1.08	48.0	1.69	59.0	2.36	68.7	3.28	77.3	1.54	75.1	2.03	64.8	1.38	51.5	1.04	34.9	.55	23.5	.43	48.1	16.46
MURDO	21.1	.60	24.4	.52	33.1	1.30	48.7	1.79	59.4	2.75	68.8	3.34	77.7	1.19	75.7	1.95	65.5	1.22	53.0	1.14	36.4	.47	25.8	.36	49.1	16.63
PIERRE CAA AP	17.5	.43	21.0	.63	31.3	1.15	47.7	1.49	59.1	2.45	68.8	3.15	77.6	1.69	75.2	2.03	64.3	1.13	51.6	.90	36.4	.46	22.7	.47	47.6	15.98
PUKWANA 3 W	18.2		22.3		32.7		48.5		59.6		69.3		77.4		74.8		64.7		51.3		34.3		22.6		48.0	
VIVIAN	19.0	.46	22.5	.56	33.2	1.30	47.8	1.76	59.1	2.79	68.2	3.60	77.3	1.60	74.9	1.93	64.6	1.17	51.7	.91	34.5	.55	23.5	.43	48.0	17.06
DIVISION	17.9	.45	21.6	.54	31.2	1.15	47.5	1.63	58.5	2.46	67.9	3.53	76.7	1.62	74.5	1.94	64.1	1.22	51.2	1.01	34.3	.47	22.8	.39	47.4	16.41
EAST CENTRAL DIVISION																										
ARLINGTON		.59		.66		1.49		2.33		2.76		3.98		2.31		2.67		2.15		1.33		.80		.59		21.66
BROOKINGS 1 NE	13.9	.40	18.3	.50	29.7	1.07	45.4	1.81	57.6	2.65	67.4	3.99	73.7	2.06	71.3	2.90	61.6	2.10	49.6	1.24	31.7	.67	19.2	.47	45.0	19.86
CASTLEWOOD	12.4	.50	16.9	.48	28.6	1.04	44.3	2.09	56.3	2.60	65.9	4.14		2.69		2.57		1.98		1.31		.69		.48		20.57
CLARK	12.2		16.5		28.1		44.4		56.7		65.9		73.0		70.8		60.9		48.6		30.4		17.9		43.8	
DE SMET	13.8		18.5		29.5		45.4		57.4		67.3		74.1		71.8		61.2		49.2		31.3				44.9	
FLANDREAU	14.3	.58	18.6	.78	29.5	1.32	45.5	2.03	57.6	2.93	67.2	4.31	73.3	2.60	71.0	2.99	61.6	2.63	49.7	1.32	32.1	.86	19.4	.57	45.0	22.92
FORESTBURG 3 NE	15.9	.48	20.3	.55	31.2	1.44	47.2	2.38	58.4	2.61	67.8	3.90	75.0	2.26	72.9	2.37	63.1	1.66	50.7	1.29	33.4	.70	21.0	.69	46.4	20.31
HOWARD	15.5	.50	19.9	.58	30.7	1.27	46.8	2.18	58.6	2.80	68.4	4.05	75.6	2.43	73.1	2.72	63.2	2.20	50.9	1.19	33.0	.73	20.6	.54	46.4	21.19
HURON WB AP	13.5	.57	17.6	.49	31.7	1.09	46.4	1.92	58.0	2.25	68.2	3.06	75.4	2.08	72.9	1.98	62.8	1.73	50.0	1.26	32.5	.66	19.6	.45	45.7	17.54
LA DELLE 7 NE	13.0	.47	17.8	.44	29.8	1.03	45.8	1.88	57.6	2.28	67.2	3.85	74.4	2.54	71.9	2.21	62.1	1.64	49.4	1.14	31.4	.57	18.5	.50	44.9	18.55
MILLER	16.0	.56	19.9	.50	30.5	1.13	46.9	2.13	58.4	2.40	67.9	4.07	75.5	2.02	73.1	1.83	63.2	1.28	50.5	1.25	33.4	.57	21.3	.47	46.4	18.21
REDFIELD	13.8	.50	18.3	.47	30.4	.99	46.3	1.82	58.8	2.19	68.4	3.71	75.7	2.04	73.3	2.03	63.1	1.34	50.3	1.22	32.4	.56	19.8	.48	45.9	17.35
WATERTOWN CAA AP	11.3	.51	15.7	.60	27.3	1.08	43.4	2.06	56.1	2.80	65.8	3.81	72.6	2.84	70.1	2.65	60.1	1.93	47.7	1.16	29.7	.72	17.4	.51	43.1	20.67
WENTWORTH		.55		.68		1.50		2.21		3.00		4.04		2.79		3.16		2.48		1.36		.89		.61		23.27
DIVISION	14.2	.52	18.5	.57	29.7	1.21	45.7	2.09	57.6	2.65	67.2	4.05	74.2	2.51	71.9	2.57	62.0	1.92	49.7	1.27	32.0	.72	19.5	.53	45.2	20.61
SOUTH CENTRAL DIV																										
ACADEMY	20.3	.52	23.6	.67	33.1	1.23	47.9	2.35	58.8	2.95	69.0	4.11	77.2	2.23	74.7	2.56	64.6	1.52	52.0	1.24	36.0	.56	24.6	.46	48.5	20.40
GREGORY	21.5	.50	24.8	.74	33.3	1.40	47.9	2.22	59.0	3.01	68.7	4.24	76.7	2.08	74.4	2.48	65.0	1.66	52.8	1.29	36.5	.78	26.0	.60	49.1	21.00
WINNER	21.8	.49	25.6	.58	33.4	1.24	48.5	2.00	59.4	2.71	68.9	3.67	77.5	1.77	75.1	2.55	65.1	1.41	52.5	1.11	36.3	.52	26.0	.46	49.2	18.51
WOOD	23.0	.59	26.1	.67	34.0	1.39	49.0	1.73	59.2	2.96	68.5	3.60	77.5	1.80	76.0	1.75	66.0	1.44	53.6	1.13	37.4	.61	27.4	.45	49.8	18.12
DIVISION	21.6	.52	25.0	.66	33.4	1.32	48.3	2.08	59.1	2.91	68.8	3.90	77.2	1.97	75.0	2.33	65.2	1.51	52.7	1.19	36.6	.62	26.0	.49	49.1	19.50
SOUTHEAST DIVISION																										
ALEXANDRIA	17.2	.42	21.3	.60	32.3	1.28	48.4	2.18	59.9	2.72	69.6	3.90	76.7	2.24	74.3	2.64	64.6	2.11	52.4	1.24	34.9	.66	22.3	.46	47.8	20.45
ARMOUR	19.6	.53	23.3	.70	33.7	1.45	49.2	2.05	60.7	2.70	70.7	4.18	76.7	1.98	75.7	3.33	65.0	1.66	52.8	1.29	36.5	.74	24.0	.58	49.1	21.41
BONESTEEL	20.5	.58	23.9	.77	33.1	1.75	48.1	2.29	59.1	3.30	68.7	3.90	76.3	2.34	74.2	3.00	64.3	2.09	52.3	1.37	35.9	.72	25.0	.54	48.5	22.65
CANISTOTA 2 N		.60		.84		1.55		2.06		2.87		4.36		2.64		3.13		2.48		1.26		.78		.59		23.16
CANTON	17.6		21.8		32.8		48.5		60.1		69.9		75.9		73.7		64.3		52.2		34.9		22.3		47.8	
MARION	17.3	.53	21.4	.91	32.4	1.63	48.2	2.02		2.15	69.6	4.34	76.3	2.80	74.0	2.87	64.4	2.34	52.0	1.42	34.2	.82	21.9	.63	47.7	23.46
MENNO	18.4	.66	22.5	.88	33.3	1.62	49.2	2.08	60.0	3.24	70.1	4.28	76.7	2.38	74.0	3.06	64.8	1.98	52.6	1.26	34.9	.79	23.0	.63	48.3	22.86
MITCHELL 2 SE	18.2	.47	22.0	.60	32.9	1.41	48.8	2.24	60.0	2.73	69.6	3.93	76.8	2.26	74.3	2.71	64.5	2.16	52.3	1.25	35.1	.69	22.9	.50	48.1	20.95
PARKSTON 5 E		.52		.68		1.34		1.99		2.65		4.17		2.35		2.91		2.31		1.13		.69		.55		21.29
SIOUX FALLS WB AP	14.2	.72	19.5	.74	32.0	1.35	46.4	2.05	58.1	3.38	68.0	4.25	74.8	3.00	72.4	3.28	62.4	2.93	50.0	1.46	32.2	1.11	19.4	.67	45.8	25.24
TYNDALL	19.6	.54	23.2	.81	33.6	1.52	48.8	2.09	60.0	3.16	70.0	4.02	77.1	2.42	74.8	3.10	64.9	2.20	52.9	1.17	35.8	.78	24.1	.53	48.7	22.34
VERMILLION 2 N	20.8	.51	24.5	.84	35.0	1.23	50.4	2.26	61.7	3.09	71.5	4.34	77.5	3.14	75.3	3.13	66.2	2.68	54.6	1.30	37.2	.96	25.6	.60	50.0	24.08
WHITE LAKE	18.1	.46	22.0		32.4		48.0		59.3		69.1		76.8		74.5		65.2		52.2		35.0		22.8		47.9	
YANKTON 3 NNW	18.6	.46	22.0	.69	32.2	1.22	47.5	2.08	59.5	3.23	69.6	4.58	76.5	2.50	73.8	3.05	64.0	2.29	52.1	1.36	34.9	.77	23.6	.49	47.9	22.72
DIVISION	18.8	.54	22.8	.80	33.1	1.45	48.7	2.17	60.2	3.04	69.9	4.22	76.5	2.47	74.8	3.02	65.2	2.18	52.7	1.26	35.4	.78	23.5	.57	48.5	22.50

* Averages for period 1931-1955, except for stations marked WB which are "normals" based on period 1921-1950. Divisional means may not be the arithmetical average of individual stations published, since additional data from shorter period stations are used to obtain better areal representation.

*MEAN TEMPERATURE AND PRECIPITATION

STATION	JANUARY Temperature	JANUARY Precipitation	FEBRUARY Temperature	FEBRUARY Precipitation	MARCH Temperature	MARCH Precipitation	APRIL Temperature	APRIL Precipitation	MAY Temperature	MAY Precipitation	JUNE Temperature	JUNE Precipitation	JULY Temperature	JULY Precipitation	AUGUST Temperature	AUGUST Precipitation	SEPTEMBER Temperature	SEPTEMBER Precipitation	OCTOBER Temperature	OCTOBER Precipitation	NOVEMBER Temperature	NOVEMBER Precipitation	DECEMBER Temperature	DECEMBER Precipitation	ANNUAL Temperature	ANNUAL Precipitation
NORTHWEST DIVISION																										
BELLE FOURCHE 2 NE	22.1	.26	24.5	.23	31.4	.64	45.9	1.86	56.1	2.10	64.5	3.15	73.1	1.33	70.5	1.21	60.3	1.18	49.1	.77	35.0	.42	26.1	.22	46.6	13.37
CAMP CROOK	17.4	.50	20.3	.31	28.6	.79	43.5	1.27	54.6	2.03	63.5	2.85	73.0	1.80	70.1	1.39	59.5	.95	46.9	.69	31.4	.30	22.7	.24	44.3	13.12
DUPREE	16.7	.59	20.1	.59	29.1	1.15	45.8	1.48	57.0	1.92	66.1	3.46	75.5	1.52	73.0	1.58	62.4	.94	46.9	.99	32.9	.53	22.4	.35	45.9	15.10
LEMMON	13.7	.47	16.7	.52	26.3	.98	42.7	1.24	54.3	2.05	63.1	3.91	72.0	1.78	69.6	1.76	59.0	1.18	46.8	.81	30.0	.48	19.9	.26	42.8	15.44
LUDLOW 2 NW	17.2	.33	19.5	.31	27.2	.51	42.7	1.15	53.8	1.83	62.4	3.59	71.1	1.98	68.8	1.40	58.4	.94	46.9	.77	31.4	.41	22.5	.20	43.5	13.42
NEWELL 2 NW	18.5	.44	21.6	.39	29.2	.92	44.6	1.75	55.4	2.58	64.4	3.23	73.5	1.75	71.2	1.27	60.6	1.13	48.5	.87	33.1	.53	23.3	.32	45.3	15.18
ORMAN DAM	21.4	.32	24.3	.29	31.3	.82	46.2	1.87	56.8	2.42	65.3	3.44	74.8	1.60	72.5	1.17	62.1	1.15	50.4	.80	35.5	.52	26.0	.26	47.2	14.66
REDIG 9 NE	16.9	.44	19.9	.33	27.7	.75	43.4	1.29	54.1	1.80	62.9	2.93	72.1	1.58	69.6	1.17	58.9	.99	47.0	.74	31.7	.39	22.1	.26	43.9	12.67
VALE	20.4	.35	23.7	.37	31.5	.93	46.3	1.81	56.6	2.69	65.2	3.48	74.0	1.88	71.4	1.27	60.9	1.17	49.2	.99	34.3	.50	24.7	.32	46.5	15.76
DIVISION	17.9	.41	21.0	.38	29.0	.85	44.5	1.48	55.4	2.18	64.0	3.39	73.2	1.67	70.8	1.38	60.2	1.09	48.3	.85	32.5	.44	23.1	.26	45.0	14.38
NORTH CENTRAL DIV																										
BOWDLE		.58		.50	26.7	.95	43.9	1.72		2.45		4.03		2.09		1.93		1.25		1.01		.53		.30	43.0	17.34
EUREKA	10.6	.38	14.8	.39	26.7	.69	43.9	1.38	56.3	2.40	65.1	4.02	72.8	2.37	70.6	2.20	60.2	1.30	47.6	.99	29.9	.38	17.1	.25	43.0	16.75
FAULKTON 1 NW	13.7	.48	17.8	.45	29.3	.97	45.5	1.86	57.2	2.41	66.4	3.55	73.9	2.10	71.6	2.09	61.5	1.38	49.3	1.10	32.3	.52	19.7	.36	44.9	17.27
GETTYSBURG		.38		.48		.95		1.58		2.22				1.92		1.72		1.19		.90		.42		.34		15.75
MOBRIDGE	13.3	.42	17.4	.47	29.0	.94	46.0	1.29	58.0	2.22	67.0	3.51	75.1	2.17	73.0	1.87	62.1	1.12	49.1	1.13	31.7	.38	19.6	.31	45.1	15.83
ONAKA		.38		.45		.88		1.57		2.13		3.65		2.03		1.97		1.68		.93		.40		.49		16.16
POLLOCK	10.2	.37	14.5	.37	26.9	.61	44.0	1.21	56.1	1.96	64.8	3.61	72.5	2.05	70.5	1.73	59.8	1.12	47.1	1.04	29.4	.34	17.3	.20	42.8	14.61
TIMBER LAKE	13.9		17.3		27.6		44.6		56.5		65.8		74.2		71.5		60.8		48.1		31.0		20.0		44.3	
DIVISION	12.7	.48	16.6	.48	28.0	.92	44.9	1.49	56.8	2.20	65.7	3.81	73.8	2.08	71.1	1.90	61.0	1.23	48.5	1.03	31.1	.45	19.0	.32	44.1	16.40
NORTHEAST DIVISION																										
ABERDEEN CAA AP	11.0	.68	15.8	.63	28.5	1.20	45.2	2.12	57.5	2.24	66.8	4.04	73.9	2.61	71.6	2.16	61.0	1.43	48.4	1.18	30.5	.71	17.6	.63	44.0	19.63
BRITTON	10.4	.44	14.8	.56	27.6	.79	44.8	1.77	57.3	2.22	66.4	3.95	73.4	2.47	71.4	2.82	61.4	1.50	49.0	1.01	30.0	.55	17.0	.41	43.6	18.49
LEOLA		.44		.43		.92		1.72		2.05		3.89		2.73		2.14		1.41		.92		.49		.35		17.49
MELLETTE	12.3	.60	17.0	.63	29.8	1.18	46.0	1.97	57.9	2.33	67.2	4.04	74.5	2.07	72.4	2.04	62.0	1.42	49.4	1.17	31.5	.67	18.4	.48	44.9	18.60
MILBANK	13.5	.57	17.9	.55	29.4	1.19	45.7	2.22	58.7	2.78	68.0	4.25	74.7	2.43	72.3	2.63	62.5	1.87	50.7	1.33	32.3	.78	19.6	.58	45.4	21.18
VICTOR 5 NE		.62		.74		1.34		2.01		2.69		4.31		3.02		2.40		1.69		1.21		.90		.61		21.54
DIVISION	11.7	.58	16.3	.63	28.6	1.08	45.3	2.11	57.5	2.47	66.7	4.19	73.6	2.53	71.6	2.45	61.3	1.59	49.3	1.18	31.1	.70	18.1	.53	44.3	20.04
BLACK HILLS DIVISION																										
BUSKALA RANCH		1.27		.87		1.76		3.11		4.01		4.25		2.59		1.80		1.48		1.26		1.11		.98		24.49
CUSTER		.41		.38		1.00		1.88		3.09		3.09		1.97		2.52		1.05		1.05		.39		.36		17.19
DUMONT 2 ENE		1.39		.87		1.74		2.53		3.31				2.12		1.76		1.08		1.13		1.23		.98		22.00
HOT SPRINGS	25.1	.41	27.8	.52	34.3	1.12	46.8	1.91	56.8	3.04	66.3	3.01	75.3	2.12	73.2	1.71	62.9	1.33	51.0	.77	36.9	.37	29.0	.36	48.8	16.67
LEAD	24.3	1.11	25.7	.87	29.7	1.70	41.4	2.96	51.1	3.90	60.2	4.04	69.7	2.08	67.9	1.68	58.4	1.57	48.0	1.32	34.4	1.14	28.1	.95	44.9	23.37
SPEARFISH 9 WNW		.60		.51		.98		2.25		2.74		3.61		1.66		1.41		1.57		.95		.74		.50		17.52
DIVISION	24.6	.76	26.7	.67	31.2	1.43	43.9	2.28	53.1	3.49	62.1	3.70	71.3	1.98	69.4	1.90	59.6	1.45	48.9	1.11	35.2	.86	27.5	.64	46.1	20.27
SOUTHWEST DIVISION																										
COTTONWOOD 2 E	20.0	.51	23.3	.40	31.8	.95	46.8	1.62	57.3	2.57	66.8	2.90	75.9	1.39	73.7	1.25	63.0	.99	50.3	.89	34.8	.36	24.3	.36	47.3	14.19
LONGVALLEY		.45		.47		1.14		1.85		2.91		3.24		1.40		1.75		1.05		1.05		.45		.28		16.16
OELRICHS	23.4	.33	26.7	.29	33.8	1.02	46.8	1.79	56.6	3.09	66.1	2.69	75.6	1.98	73.5	1.30	63.0	1.16	50.8	.78	35.9	.39	26.8	.36	48.3	15.18
RAPID CITY WB AP	21.1	.48	23.9	.32	31.3	1.06	44.5	2.01	54.7	3.05	63.7	3.36	72.3	2.06	70.9	1.56	60.5	1.18	49.1	1.13	35.3	.55	25.7	.34	46.1	17.10
RAPID CITY	24.4	.42	26.7	.43	32.1	1.08	45.7	2.00	56.2	2.77	65.1	3.45	74.3	1.85	72.0	1.67	62.2	.98	51.2	.95	37.3	.39	29.0	.31	48.0	16.31
DIVISION	21.8	.40	25.5	.40	32.5	.95	46.4	1.73	56.7	2.76	66.0	3.11	75.5	1.66	73.2	1.60	62.9	1.17	50.8	.94	35.4	.40	25.9	.32	47.7	15.44

* Averages for period 1931-1955, except for stations marked WB which are "normals" based on period 1921-1950. Divisional means may not be the arithmetical average of individual stations published, since additional data from shorter period stations are used to obtain better areal representation.

CONFIDENCE LIMITS

In the absence of trend or record changes, the chances are 9 out of 10 that the true mean will lie in the interval formed by adding and subtracting the values in the following table from the means for any station in the State. Because of the wider variation in mean precipitation, the corresponding monthly means and annual mean must be substituted for "p" in the precipitation table below to obtain mean precipitation confidence limits.

2.8	$.22\sqrt{p}$	2.8	$.16\sqrt{p}$	1.6	$.27\sqrt{p}$	1.3	$.39\sqrt{p}$	1.5	$.41\sqrt{p}$	1.2	$.32\sqrt{p}$	1.2	$.29\sqrt{p}$	1.0	$.32\sqrt{p}$	1.2	$.31\sqrt{p}$	1.4	$.31\sqrt{p}$	1.4	$.24\sqrt{p}$	1.8	$.18\sqrt{p}$	.9	$.30\sqrt{p}$

COMPARATIVE DATA

Data in the following table are the mean temperature and average precipitation for Redfield, South Dakota for the period 1906-1930 and are included in this publication for comparative purposes.

11.3	0.57	16.7	0.64	30.5	0.89	45.8	2.00	56.4	2.78	66.6	3.21	72.3	2.51	70.4	2.33	61.1	1.66	47.6	1.39	33.1	0.74	17.9	0.53	44.1	19.25

NORMALS, MEANS AND EXTREMES
(Table Revised 1973. Base Period for Climatological Normals: 1931-1960)

Station: HURON, SOUTH DAKOTA   W. W. HOWES MUNICIPAL AIRPORT   Standard time used: CENTRAL   Latitude: 44° 23' N   Longitude: 98° 13' W   Elevation (ground): 1281 feet

Month	Normal Daily maximum	Normal Daily minimum	Normal Monthly	Record highest	Year	Record lowest	Year	Normal heating degree days (Base 65°)	Precip. Normal total	Max monthly	Year	Min monthly	Year	Max in 24 hrs	Year	Snow Mean total	Max monthly	Year	Max in 24 hrs	Year	RH 00	06	12	18	Wind Mean speed	Prevailing direction	Fastest Speed	Direction	Year	Pct. possible sunshine	Mean sky cover	Clear	Partly cloudy	Cloudy	Precip .01"+	Snow 1.0"+	Thunderstorms	Heavy fog	Max 90°+	Max 32°-	Min 32°-	Min 0°-	Solar radiation
(yrs)	(b)	(b)	(b)	13		13		(b)	(b)	33		33		33		33	33		33		13	13	13	13	33	14	33	33	33	33	29	33	33	33	33	33	29	33	33	33	13	13	13
J	23.1	1.9	12.5	61	1963	-35	1970	1628	0.48	1.66	1944	0.04	1942+	1.58	1944	6.2	17.1	1949	7.4	1949	74	73	70	70	11.8	SSE	57	NW	1967	55	6.7	10	8	13	7	6	*	2	0	26	31	16	160
F	27.7	5.5	16.6	59	1963	-39	1962	1355	0.60	3.87	1962	0.03	1939	2.16	1962	9.7	39.9	1962	17.5	1962	78	78	71	71	11.9	NW	56	NW	1955	57	6.7	8	8	12	7	6	*	2	0	22	28	10	240
M	39.6	17.7	28.7	86	1963	-24	1960	1125	1.11	3.41	1942	0.12	1942	1.87	1939	8.7	24.2	1962	11.1	1960+	80	79	64	65	12.8	NW	68	W	1955	61	7.2	6	9	16	8	6	1	2	0	14	18	2	400
A	58.2	31.7	45.0	93	1962	6	1961	600	1.84	5.39	1968	0.30	1952	2.40	1952	2.5	12.8	1970	5.0	1970	77	78	55	55	14.2	SSE	73	SE	1955	64	6.1	6	10	14	9	2	3	1	0	1	2	0	440
M	71.5	43.1	57.3	94	1964	17	1967+	288	2.36	6.44	1962	0.50	1945	3.49	1967+	0.2	1.6	1954	1.6	1954	77	84	52	56	12.2	SSE	72	NW	1949	63	6.0	7	11	13	10	*	6	1	*	0	*	0	560
J	81.5	53.6	67.6	102	1970	32	1964	87	3.14	8.30	1967	0.67	1950	5.48	1950	0.0	0.0		0.0		82	87	54	58	11.6	SSE	65	SE	1960	68	5.8	9	11	10	11	0	8	1	3	0	0	0	580
J	90.3	59.6	75.0	112	1966	37	1971	9	1.81	4.93	1972	0.42	1941	2.19	1950	0.0	0.0		0.0		79	86	48	52	10.8	SSE	72	NW	1957	77	4.6	12	13	6	6	0	8	1	10	0	0	0	600
A	88.0	57.6	72.8	110	1965	36	1964	12	2.07	5.47	1976	0.45	1958	4.11	1956	0.0	0.0		0.0		81	87	48	53	10.5	SSE	64	NW	1968	75	4.7	12	11	8	6	0	8	1	7	0	0	0	510
S	77.5	46.3	61.9	106	1970	19	1965	165	1.53	3.57	1965	0.10	1950	4.20	1961	T	5.3	1970	5.0	1970	78	85	54	56	11.6	SSE	64	W	1951	67	5.3	11	11	8	6	*	4	1	3	0	*	0	400
O	63.3	34.0	48.8	102	1963	11	1969	508	1.15	6.44	1946	T	1961	4.20	1961	0.4	28.2	1947	10.0	1953	75	81	56	60	11.7	SSE	72	W	1949	63	5.3	11	7	11	6	1	2	1	1	5	5	0	290
N	43.3	19.1	31.2	77	1962	-21	1964	1014	0.68	3.01	1947	T	1939	2.00	1972	4.7	28.2	1972	10.7	1968	78	78	63	65	12.2	SSE	73	NW	1954	51	6.0	7	6	17	6	5	*	1	0	12	27	1	170
D	29.2	8.4	18.8	62	1969	-26	1964+	1432	0.56	1.53	1968	T	1943	1.24	1949	6.4	26.0	1968	10.7	1968	82	77	77	73	11.4	SSE	77	NW	1963	49	6.8	7	6	17	7	6	*	2	0	20	31	9	40
YR	57.8	31.5	44.7	112	JUL 1966	-39	FEB 1962	8223	17.33	8.30	JUN 1967	T	OCT 1952+	5.48	JUN 1967	39.0	39.9	FEB 1962	17.5	FEB 1962	78	82	59	59	12.0	SSE	77	NW	JUL 1957	64	6.0	102	109	154	94	42	15	12	28	73	175	40	

Means and extremes above are from existing and comparable exposures. Annual extremes have been exceeded at other sites in the locality as follows:
Lowest temperature -43 in January 1912; maximum monthly precipitation 11.56 in June 1914.

NORMALS, MEANS AND EXTREMES
(Table Revised 1975. Base Period for Climatological Normals: 1941-1970)

Average station pressure mb.   Elev. 1289 feet m.s.l.

Month	Normal Daily maximum	Normal Daily minimum	Normal Monthly	Record highest	Year	Record lowest	Year	Degree days Heating	Cooling	Precip. Normal	Max monthly	Year	Min monthly	Year	Max in 24 hrs	Year	Snow Max monthly	Year	Max in 24 hrs	Year	RH 00	06	12	18	Wind Mean speed	Prevailing direction	Fastest Speed	Direction	Year	Pct. possible sunshine	Mean sky cover	Clear	Partly cloudy	Cloudy	Precip .01"+	Snow 1.0"+	Thunderstorms	Heavy fog	Max 90°+	Max 32°-	Min 32°-	Min 0°-	Station pressure mb.
(yrs)	(a)			15		15					35		35		35		35		35		15	15	15	15	35	14	35	35	35	35	31	35	35	35	35	31	35	35	15	15	15	15	
J	23.4	1.6	12.5	61	1963	-35	1970	1627	0	0.43	1.66	1944	0.04	1942	1.58	1944	17.1	1949	7.4	1949	74	73	67	70	11.6	SSE	57	NW	1967	56	6.7	8	9	16	7	6	*	2	0	22	31	15	970.1
F	28.7	7.0	17.9	59	1963	-39	1962	1319	0	0.57	3.87	1962	0.03	1968	2.16	1962	39.9	1962	17.5	1962	78	78	69	71	11.7	NW	56	NW	1955	60	6.7	6	9	14	6	3	*	3	0	16	28	9	971.2
M	39.2	18.8	29.0	86	1963	-24	1960	1116	0	1.07	3.41	1942	0.12	1939	1.87	1939	24.6	1962	11.1	1968	79	79	64	64	12.8	NW	68	W	1955	61	7.2	6	9	16	8	4	1	2	0	16	18	3	965.3
A	57.9	33.6	45.8	93	1962	6	1961	576	0	1.96	5.39	1968	0.30	1952	2.40	1952	12.8	1970	5.0	1970	77	77	52	54	14.1	SSE	73	SE	1955	64	6.3	6	10	14	10	1	3	1	*	1	2	0	965.5
M	71.0	44.2	57.6	94	1964	17	1954	273	25	2.76	6.44	1962	0.50	1945	3.49	1967	1.6	1954	1.6	1954	77	84	55	58	12.8	SSE	72	NW	1942	68	6.2	7	11	13	11	*	6	1	*	0	*	0	964.1
J	79.0	55.2	67.1	102	1970	32	1964	72	135	3.76	8.30	1967	0.67	1950	5.48	1950	0.0		0.0		82	87	54	58	11.6	SSE	65	SE	1960	68	5.8	9	11	10	11	0	8	1	3	0	0	0	964.4
J	86.8	60.6	73.7	112	1966	37	1971	9	278	2.23	4.93	1972	0.42	1968	2.19	1950	0.0		0.0		78	86	48	52	10.7	SSE	72	NW	1957	76	4.6	12	13	6	7	0	8	1	11	0	0	0	967.2
A	85.1	57.7	71.4	110	1965	36	1964	29	239	2.19	5.47	1976	0.45	1956	4.11	1956	0.0		0.0		79	86	48	54	10.9	SSE	64	NW	1968	74	4.7	11	11	9	6	0	8	1	10	0	0	0	966.9
S	74.0	46.1	60.1	106	1970	19	1965	169	40	1.78	3.57	1965	0.10	1950	2.60	1961	5.3	1970	5.0	1970	78	85	54	56	11.6	SSE	64	W	1951	67	5.1	11	11	8	6	*	4	1	5	*	1	0	970.1
O	63.4	34.0	48.7	102	1963	11	1969	482	5	0.87	6.44	1946	T	1952	4.20	1961	28.2	1947	10.7	1953	78	82	56	63	12.1	SSE	73	W	1949	63	5.3	11	7	11	6	1	2	1	1	5	5	1	969.9
N	44.9	21.3	33.1	77	1962	-21	1964	978	0	0.67	3.01	1947	T	1939	2.00	1943	28.2	1972	10.7	1968	78	78	63	69	11.4	SSE	73	NW	1954	51	6.0	7	6	17	6	5	*	2	0	17	27	7	969.6
D	29.4	9.0	19.2	62	1969	-26	1964	1420	0	0.53	1.53	1968	T	1943	1.24	1949	26.0	1968	10.7	1968	77	78	69	73	11.4	SSE	59	NW	1963	49	6.8	7	6	17	6	2	*	2	0	20	31	9	967.9
YR	56.7	32.8	44.8	112	JUL 1966	-39	FEB 1962	8054	716	19.44	8.30	JUN 1967	T	OCT 1952	5.48	JUN 1967	39.9	FEB 1962	17.5	FEB 1962	77	82	59	59	11.9	SSE	77	NW	JUL 1957	63	6.0	102	109	154	93	41	15	12	29	71	173	37	967.9

Means and extremes above are from existing and comparable exposures. Annual extremes have been exceeded at other sites in the locality as follows:
Lowest temperature -43 in January 1912; maximum monthly precipitation 11.56 in June 1914.

REFERENCE NOTES APPLYING TO TABLES APPEAR ON THE PAGE FOLLOWING LAST TABLE.
(Caution: Letters and symbols may have different meanings in 1941-1970 tables than in earlier tables. See notes.)

NORMALS, MEANS AND EXTREMES
(Table Revised 1973. Base Period for Climatological Normals: 1931-1960)

Station: RAPID CITY, SOUTH DAKOTA   RAPID CITY REGIONAL AIRPORT   Standard time used: MOUNTAIN   Latitude: 44° 03' N   Longitude: 103° 04' W   Elevation (ground): 3162 feet

Temperature (°F)

Month	Normal Daily max (b)	Normal Daily min (b)	Normal Monthly (b)	Record highest	Year	Record lowest	Year
J	34.1	9.8	22.0	74	1953	-27	1950+
F	35.5	12.3	24.1	74	1954+	-22	1971
M	42.7	18.7	31.1	82	1946+	-17	1962
A	56.7	32.3	44.5	89	1962	2	1960
M	67.8	43.6	55.7	98	1969	18	1950+
J	77.0	52.8	64.9	106	1961	31	1951
J	87.7	59.9	73.8	109	1954	39	1959
A	86.0	58.0	72.0	106	1947	38	1966
S	75.6	47.6	61.6	104	1960	10	1972
O	63.1	36.8	50.0	94	1963+	-10	1972
N	47.0	23.2	35.1	77	1965+	-19	1959
D	38.7	15.7	27.2	75	1965	-23	1972+
YR	59.3	34.3	46.8	109	JUL 1954	-27	JAN 1950+

Normal heating degree days (Base 65°), Precipitation (inches), Snow/Ice pellets

Month	Heating deg. days (b)	Precip Normal total (b)	Precip Max monthly	Year	Precip Min monthly	Year	Precip Max 24 hrs	Year	Snow Mean total	Snow Max monthly	Year	Snow Max 24 hrs	Year
J	1333	0.36	1.77	1944	0.01	1952	1.26	1944	5.3	24.0	1949	16.3	1944
F	1145	0.48	2.46	1953	0.06	1950	1.00	1953	5.6	20.7	1953	10.7	1953
M	1051	1.07	3.02	1945	0.14	1948	2.19	1963	8.7	30.7	1950	14.7	1970
A	615	1.67	5.16	1946	0.27	1954	3.01	1946	6.7	30.6	1970	13.4	1967
M	326	2.66	7.35	1965	0.33	1966	3.40	1965	1.0	11.6	1950	8.0	1951
J	126	3.08	7.00	1968	0.76	1961	4.01	1963	0.1	3.6	1951	3.6	1951
J	22	1.78	6.13	1969	0.60	1965	2.51	1954	0.0	0.0		0.0	
A	12	1.22	4.40	1945	.01	1943	1.65	1943	0.0	0.0		0.0	
S	165	0.95	3.94	1946	0.03	1958	2.13	1966	0.1	2.0	1970	2.0	1970
O	481	0.79	2.25	1968	T	1945	1.45	1962	1.8	10.2	1971	7.8	1971
N	897	0.41	2.09	1944	0.04	1945	1.09	1944	3.8	12.6	1959	8.2	1959
D	1172	0.30	0.89	1967	0.30	1957+	0.50	1960	4.3	9.1	1967	5.0	1967
YR	7345	14.71	7.35	MAY 1965	T	OCT 1945	4.01	JUN 1963	38.4	30.7	MAR 1950	16.3	JAN 1944

Relative humidity, Wind, Sky cover, Sunshine

Month	RH 05	RH 11	RH 17	RH 23	Wind Mean speed	Prevailing direction	Fastest mile Speed	Direction	Year	Mean sky cover	Pct possible sunshine
J	59	60	64	69	10.5	NNW	65	SW	1967	6.6	53
F	73	61	64	72	10.8	NNW	59	NW	1970	6.7	58
M	74	58	61	67	11.2	NNW	59	W	1953	6.5	61
A	72	48	46	67	13.2	NNW	61	NW	1967	6.5	57
M	77	52	50	73	10.7	NNW	56	NW	1967	5.7	59
J	73	46	41	65	9.9	NNW	66	SW	1959	4.3	71
A	68	41	36	59	10.1	NNW	66	NW	1951	4.2	73
S	64	39	40	60	11.0	NNW	61	NW	1964	4.0	66
O	64	44	44	66	11.1	NNW	59	NW	1971	4.8	65
N	68	51	58	68	11.0	NNW	59	NW	1966+	6.2	55
D	68	59	59	68	10.4	NNW	60	W	1955	6.3	53
YR	70	50	51	67	11.1	NNW	66	SW	JUL 1959	5.8	61

Mean number of days

Month	Clear	Partly cloudy	Cloudy	Precip .01"+	Snow 1.0"+	Thunder-storms	Heavy fog	Max 90°+	Max 32°-	Min 32°-	Min 0°-	Solar radiation (langleys)
J	7	8	16	8	3	*	0	0	13	30	8	
F	6	8	14	7	3	0	0	0	10	27	4	
M	6	9	15	9	3	*	1	0	8	28	3	
A	6	9	15	9	2	1	*	0	1	15	0	
M	6	10	11	12	*	6	*	*	0	3	0	
J	9	10	11	13	0	11	1	3	0	*	0	
J	14	12	5	9	0	12	*	11	0	0	0	
A	14	12	5	7	0	9	*	13	0	0	0	
S	13	9	8	6	*	3	1	3	0	1	0	
O	13	8	10	6	*	*	2	*	0	10	1	
N	7	8	14	6	5	*	2	0	5	24	4	
D	8	8	15	6	6	*	2	0	11	30	0	
YR	110	112	143	96	12	43	15	31	48	169	20	

Means and extremes above are from existing and comparable exposures. Annual extremes have been exceeded at other sites in the locality as follows: Lowest temperature -33 in February 1936; maximum monthly precipitation 9.66 in July 1905; maximum precipitation in 24 hours 5.57 in May 1926; maximum monthly snowfall 38.5 in April 1927; maximum snowfall in 24 hours 18.3 in March 1927; fastest mile of wind 75 from Northwest in February 1947.

NORMALS, MEANS AND EXTREMES
(Table Revised 1975. Base Period for Climatological Normals: 1941-1970)

Station: RAPID CITY, SOUTH DAKOTA — Average station pressure 903.8 mb — Elev. 3168 feet m.s.l.

Temperature (°F)

Month	Normal Daily max	Normal Daily min	Normal Monthly	Record highest	Year	Record lowest	Year
J	34.2	9.6	21.9	74	1953	-27	1950
F	37.6	13.9	25.8	74	1954+	-22	1971
M	42.7	19.7	31.2	82	1946+	-17	1962
A	57.2	32.0	44.6	89	1962	2	1966
M	67.4	42.9	55.2	98	1969	18	1950
J	76.3	52.0	64.2	106	1961	31	1951
J	86.3	58.8	72.6	110	1973	39	1959
A	85.7	57.2	71.6	106	1947	38	1966
S	74.7	46.3	60.5	104	1960	10	1947
O	63.6	36.4	50.0	94	1963	-10	1972
N	47.5	23.2	35.4	77	1965+	-19	1959
D	38.0	14.9	26.5	75	1965	-23	1972
YR	59.3	33.9	46.6	110	JUL 1973	-27	JAN 1950

Normal Degree days (Base 65°F), Precipitation (inches), Snow/Ice pellets

Month	Heating	Cooling	Precip Normal	Water equiv. Max monthly	Year	Water equiv. Min monthly	Year	Precip Max 24 hrs	Year	Snow Max monthly	Year	Snow Max 24 hrs	Year
J	1336	0	0.47	1.77	1944	0.01	1952	1.26	1944	24.0	1949	16.3	1944
F	1098	0	0.59	2.46	1953	0.06	1950	1.00	1953	20.7	1953	10.7	1953
M	1048	0	0.99	3.02	1945	0.14	1948	2.19	1945	30.7	1950	14.9	1970
A	612	0	2.09	5.16	1946	0.27	1954	3.01	1946	30.6	1970	13.4	1967
M	319	15	2.81	7.00	1963	0.33	1966	3.40	1963	11.6	1950	16.0	1951
J	134	110	3.67	7.00	1968	0.64	1961	4.01	1963	3.6	1951	3.6	1951
J	13	249	2.10	6.13	1969	0.60	1965	2.51	1944	0.0		0.0	
A	17	222	1.47	4.40	1945	.01	1943	1.65	1943	0.0		0.0	
S	191	56	1.22	3.94	1946	0.03	1958	2.13	1966	2.0	1970	2.0	1970
O	474	0	0.86	2.25	1946	T	1945	1.45	1962	10.2	1971	7.8	1971
N	888	0	0.48	2.09	1944	0.03	1945	1.09	1944	12.6	1959	8.2	1959
D	1194	0	0.39	0.89	1957	0.03	1957	0.50	1960	9.1	1967	5.0	1967
YR	7324	661	17.12	7.35	MAY 1946	T	OCT 1960	4.01	JUN 1963	30.7	MAR 1950	16.3	JAN 1944

Relative humidity pct., Wind, Sky cover, Sunshine

Month	RH 05	RH 11	RH 17	RH 23	Wind Mean speed	Prevailing direction	Fastest mile Speed	Direction	Year	Mean sky cover	Pct possible sunshine
J	68	59	63	68	10.5	NNW	65	SW	1967	6.6	54
F	74	56	55	72	12.5	NNW	59	W	1970	6.5	61
M	74	56	57	67	12.5	NNW	59	W	1953	6.7	61
A	74	49	46	70	12.4	NNW	61	NW	1967	6.4	57
M	76	49	49	74	10.7	NNW	56	NW	1967	5.7	59
J	72	44	41	65	9.9	NNW	66	SW	1959	4.3	71
A	68	41	35	59	9.2	NNW	66	NW	1951	4.2	73
S	64	40	39	60	11.0	NNW	61	NW	1964	4.6	66
O	64	41	40	64	11.1	NNW	59	NW	1971	5.2	65
N	68	59	59	68	11.0	NNW	59	NW	1966	6.2	55
D	68	59	64	68	10.4	NNW	60	W	1955	6.3	54
YR	70	50	50	66	11.1	NNW	66	SW	JUL 1959	5.8	62

Mean number of days, Average station pressure

Month	Clear	Partly cloudy	Cloudy	Precip .01"+	Snow 1.0"+	Thunder-storms	Heavy fog, vis. ¼ mi or less	Max 90°+	Max 32°-	Min 32°-	Min 0°-	Avg station pressure (mb)
J	7	8	16	8	3	0	2	0	13	30	8	903.0
F	7	8	14	6	3	0	2	0	10	27	4	900.5
M	6	9	16	9	3	*	2	0	8	28	3	900.1
A	6	9	14	10	2	1	1	0	1	15	0	902.3
M	7	10	14	12	*	6	1	*	0	3	0	902.5
J	7	11	10	12	0	11	1	3	0	*	0	905.5
J	13	13	5	9	0	12	*	11	0	0	0	907.2
A	14	12	5	7	0	9	*	12	0	0	0	907.2
S	13	9	8	6	*	3	*	3	0	1	0	906.2
O	13	8	10	6	*	*	1	*	0	10	1	903.0
N	8	8	14	6	5	*	2	0	5	25	4	
D	8	8	15	6	6	*	2	0	11	30	0	
YR	110	113	142	95	12	42	16	31	47	169	20	903.8

Means and extremes above are from existing and comparable exposures. Annual extremes have been exceeded at other sites in the locality as follows: Lowest temperature -33 in February 1936; maximum monthly precipitation 9.66 in July 1905; maximum precipitation in 24 hours 5.57 in May 1926; maximum monthly snowfall 38.5 in April 1927; maximum snowfall in 24 hours 18.3 in March 1927; fastest mile of wind 75 from Northwest in February 1947.

REFERENCE NOTES APPLYING TO TABLES APPEAR ON THE PAGE FOLLOWING LAST TABLE.
(Caution: Letters and symbols may have different meanings in 1941-1970 tables than in earlier tables. See notes.)

NORMALS, MEANS AND EXTREMES
(Table Revised 1973. Base Period for Climatological Normals: 1931-1960)

Station: SIOUX FALLS, SOUTH DAKOTA   FOSS FIELD
Standard time used: CENTRAL   Latitude: 43° 34' N   Longitude: 96° 44' W   Elevation (ground): 1418 feet

Temperature (°F)

Month	Normal Daily max	Normal Daily min	Normal Monthly	Record highest	Year	Record lowest	Year
J	25.1	5.2	15.2	57	1964	-36	1970
F	29.1	9.0	19.1	59	1964	-30	1965
M	39.4	20.8	30.1	87	1968	-10	1971+
A	57.0	34.7	45.9	92	1964	18	1971+
M	70.1	46.5	58.3	100	1967	27	1967
J	79.4	56.8	68.1	101	1968	33	1969
J	86.0	62.5	74.3	106	1966	38	1971+
A	83.3	60.3	71.8	103	1947	37	1964
S	73.5	50.0	61.8	101	1970	26	1970
O	61.9	38.6	50.3	94	1963	3	1965
N	42.5	22.7	32.6	72	1965	-17	1964
D	30.3	11.9	21.1	60	1970	-22	1972+
YR	56.5	34.9	45.7	106 JUL.1966		-36 JAN.1970	

Normal heating degree days (Base 65°): 1544, 1285, 1082, 573, 270, 78, 19, 25, 168, 472, 972, 1361; YR 7839

Precipitation

Normal total: 0.62, 0.93, 1.54, 2.31, 3.13, 4.35, 2.84, 3.59, 2.61, 1.25, 1.00, 0.74; YR 25.16

Maximum monthly precipitation: 1.71, 4.05, 3.33, 5.23, 7.29, 7.94, 7.79, 7.47, 6.34, 4.57, 2.62; JUN. 7.94 (1951)

Maximum in 24 hrs: 1.61, 2.00, 1.96, 2.94, 4.32, ... ; JUN. 4.32 (1957)

Snow, Ice pellets

Mean total: 5.9, 10.1, 10.5, 1.8, T, 0.0, 0.0, 0.0, T, 0.4, 3.8, 7.8; YR 40.3
Maximum monthly: 19.6, 48.4, 31.5, 10.7, 0.2, 0.0, 0.0, 0.0, T, 5.1, 17.0, 41.1; FEB. 48.4
Maximum in 24 hrs: 11.8, 26.0, 18.9, 0.0, 0.0, 0.0, 0.0, 0.0, T, 5.0, 9.1, 16.6; FEB. 26.0 (1962)

Relative humidity pct. (Local time) / Wind

Mean sky cover (tenths) sunrise to sunset: 6.5, 6.5, 6.9, 6.5, 6.3, 5.8, 4.8, 4.8, 5.1, 5.1, 5.3, 6.7; YR 6.0

Fastest mile (Speed / Direction / Year):
45 32 1967, 44 30 1972+, 60 02 1950, 48 25 1950, 46 11 1950, 70 23 1952, 60 27 1955, 52 16 1956, 50 16 1953, 60 27 1949, 52 32 1954, 46 29 1956; JUN. 70 23 1952

Means and extremes above are from existing and comparable exposures. Annual extremes have been exceeded at other sites in the locality as follows:
Highest temperature 110 in July 1936; lowest temperature -42 in February 1899; maximum monthly precipitation 9.42 in May 1898; minimum monthly precipitation 0.00 in November 1914.

NORMALS, MEANS AND EXTREMES
(Table Revised 1975. Base Period for Climatological Normals: 1941-1970)

Temperature (°F)

Month	Normal Daily max	Normal Daily min	Normal Monthly	Record highest	Year	Record lowest	Year
J	24.6	3.7	14.2	57	1964	-36	1970
F	29.7	9.0	19.1	59	1964	-30	1965
M	39.7	20.2	30.0	87	1968	-10	1971+
A	57.8	34.4	46.1	92	1964	18	1971+
M	69.7	45.7	57.7	100	1957	27	1967
J	78.9	56.3	67.6	101	1954	33	1969
J	85.1	61.5	73.3	106	1966	38	1971+
A	83.8	59.8	71.8	103	1973	37	1964
S	73.0	47.6	60.9	104	1963	22	1974
O	62.7	37.0	49.9	94	1963	3	1965
N	43.5	22.7	33.1	72	1965	-17	1964
D	29.6	10.4	20.0	60	1970	-26	1973
YR	55.5	34.2	45.4	108 AUG.1973		-36 JAN.1970	

Normal Degree days Base 65°F

Heating: 1575, 1277, 1085, 567, 259, 65, 10, 18, 165, 465, 957, 1395; YR 7838
Cooling: 0, 0, 0, 0, 32, 143, 267, 229, 42, 0, 0, 0; YR 719

Precipitation in inches

Water equivalent Normal: 0.57, 1.04, 1.40, 2.30, 3.17, 4.32, 2.94, 2.84, 2.85, 1.50, 0.85, 0.74; YR 24.72

Maximum monthly: 1.71, 4.05, 3.52, 5.23, 7.29, 7.94, 7.79, 7.47, 6.34, 5.73, 2.45, 2.62; JUN. 7.94 (1951)
Year: 1969, 1962, 1973, 1953, 1951, ..., 1948, 1959, 1966, 1973, 1971, 1968
Minimum monthly: 0.05, 0.05, 0.14, 0.17, 0.72, 1.48, 0.25, 0.53, 0.29, T, 0.03, 0.09; OCT. T (1952)
Maximum in 24 hrs: 1.61, 2.00, 2.64, 2.92, 4.32, ..., 2.83, 3.94, 4.02, 1.62, 1.44; OCT. 4.54 (1973)

Snow, Ice pellets

Maximum monthly: 19.6, 48.4, 31.5, 10.7, 0.2, 0.0, 0.0, 0.0, T, 5.1, 17.0, 41.1; FEB. 48.4 (1962)
Maximum in 24 hrs: 11.8, 26.0, 9.0, 0.0, 0.0, 0.0, 0.0, 0.0, T, 5.1, 9.1, 16.6; FEB. 26.0 (1962)

Mean sky cover sunrise to sunset / Wind

Mean sky cover (tenths): 6.5, 6.5, 7.0, 6.5, 6.3, 5.7, 4.8, 4.8, 5.1, 5.1, 5.6, 6.7; YR 6.0

Fastest mile (Speed / Direction / Year):
45 32 1967, 44 30 1972, 60 02 1950, 48 25 1955, 46 23 1952, 70 23 1952, 60 27 1955, 52 16 1956, 50 16 1953, 60 27 1949, 52 32 1954, 46 29 1956; JUN. 70 23 1952

Average station pressure: Elev. 1427 feet m.s.l.

Means and extremes above are from existing and comparable exposures. Annual extremes have been exceeded at other sites in the locality as follows:
Highest temperature 110 in July 1936; lowest temperature -42 in February 1899; maximum monthly precipitation 9.42 in May 1898; minimum monthly precipitation 0.00 in November 1914.

REFERENCE NOTES APPLYING TO TABLES APPEAR ON THE PAGE FOLLOWING LAST TABLE.
(Caution: Letters and symbols may have different meanings in 1941-1970 tables than in earlier tables. See notes.)

Reference notes applying to Normals, Means, and Extremes tables for 1931–1960 base period.

(a) Length of record, years, based on January data. Other months may be for more or fewer years if there have been breaks in the record.

(b) Climatological standard normals (1931-1960).

* Less than one half.

+ Also on earlier dates, months, or years.

T Trace, an amount too small to measure.

Below zero temperatures are preceded by a minus sign. The prevailing direction for wind in the Normals, Means, and Extremes table is from records through 1963.

‡ >70° at Alaskan stations.

Unless otherwise indicated, dimensional units used in this bulletin are: temperature in degrees F.; precipitation, including snowfall, in inches; wind movement in miles per hour; and relative humidity in percent. Heating degree day totals are the sums of negative departures of average daily temperatures from 65° F. Cooling degree day totals are the sums of positive departures of average daily temperatures from 65° F. Sleet was included in snowfall totals beginning with July 1948. The term "ice pellets" includes solid grains of ice (sleet) and particles consisting of snow pellets encased in a thin layer of ice. Heavy fog reduces visibility to 1/4 mile or less.

Sky cover is expressed in a range of 0 for no clouds or obscuring phenomena to 10 for complete sky cover. The number of clear days is based on average cloudiness 0-3, partly cloudy days 4-7, and cloudy days 8-10 tenths.

Solar radiation data are the averages of direct and diffuse radiation on a horizontal surface. The langley denotes one gram calorie per square centimeter.

& Figures instead of letters in a direction column indicate direction in tens of degrees from true North; i.e., 09 - East, 18 - South, 27 - West, 36 - North, and 00 - Calm. Resultant wind is the vector sum of wind directions and speeds divided by the number of observations. If figures appear in the direction column under "Fastest mile" the corresponding speeds are fastest observed 1-minute values.

$ Per observational day prior to 1948.

To 8 compass points only.

** The National Weather Service considers the accuracy of solar radiation data questionable; therefore, publication is suspended pending determination of corrected values.

Reference notes applying to Normals, Means, and Extremes tables for 1941–1970 base period.

(a) Length of record, years, through the current year unless otherwise noted, based on January data.

(b) 70° and above at Alaskan stations.

* Less than one half.

T Trace.

NORMALS - Based on record for the 1941-1970 period.

DATE OF AN EXTREME - The most recent in cases of multiple occurrence.

PREVAILING WIND DIRECTION - Record through 1963.

WIND DIRECTION - Numerals indicate tens of degrees clockwise from true north. 00 indicates calm.

FASTEST MILE WIND - Speed is fastest observed 1-minute value when the direction is in tens of degrees.

Mean Annual Precipitation, Inches

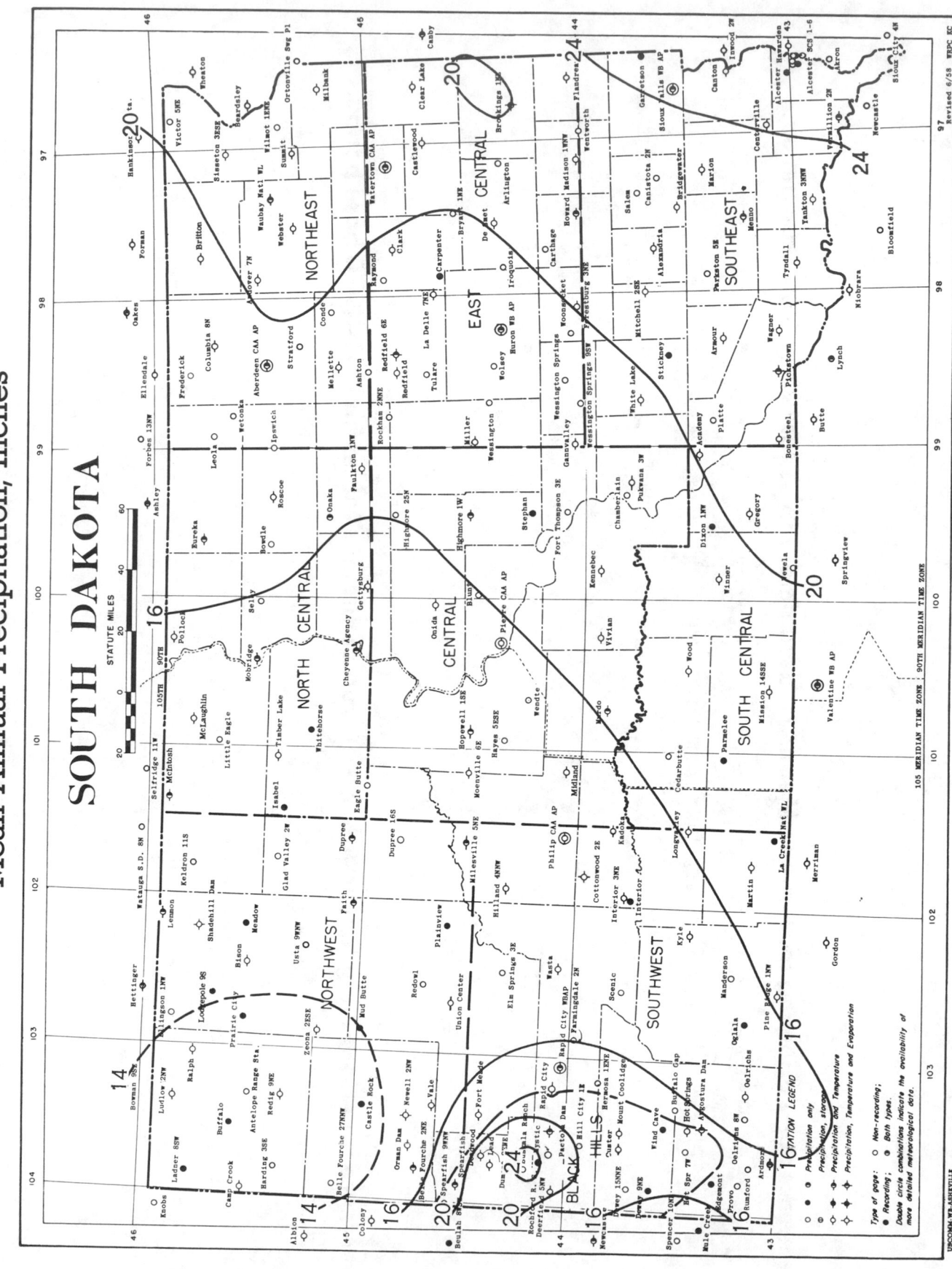

SOUTH DAKOTA

STATUTE MILES

Based on period 1931-55

Isolines are drawn through points of approximately equal value. Caution should be used in interpolating on these maps, particularly in mountainous areas.

UDCOMM-WB-ASHEVILLE

LEGEND

Type of gage: ○ Non-recording;
● Recording;   ⊙ Both types.

Double circle combinations indicate the availability of more detailed meteorological data.

Precipitation only
Precipitation, storage
Precipitation and Temperature
Precipitation, Temperature and Evaporation

105 MERIDIAN TIME ZONE   90TH MERIDIAN TIME ZONE

Revised 6/58 WBPC KC

Mean Maximum Temperature (°F.), January

SOUTH DAKOTA

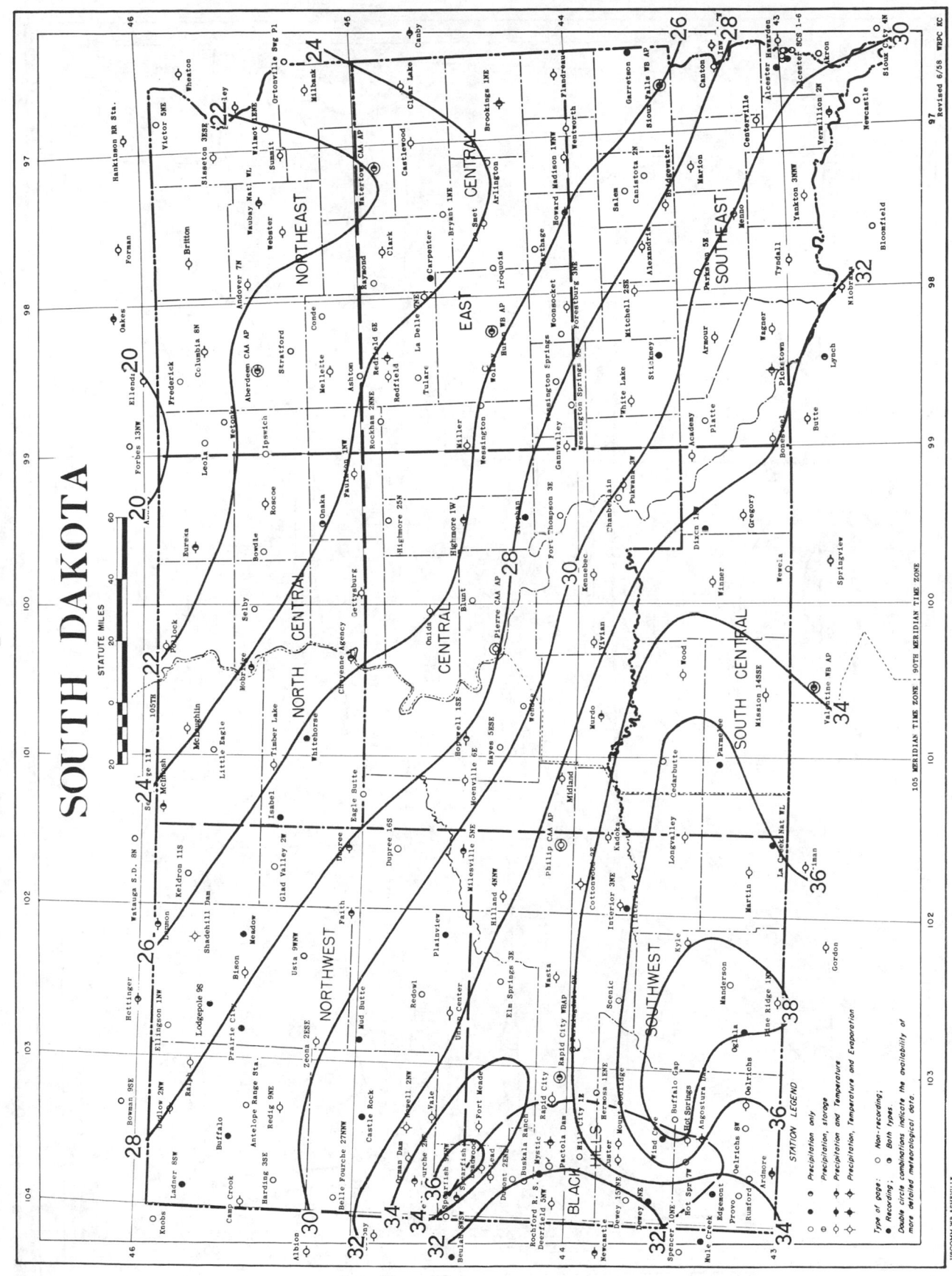

USCOMM WB-ASHEVILLE

Based on period 1931-52

Isolines are drawn through points of approximately equal value. Caution should be used in interpolating on these maps, particularly in mountainous areas.

Mean Minimum Temperature (°F.), January

SOUTH DAKOTA

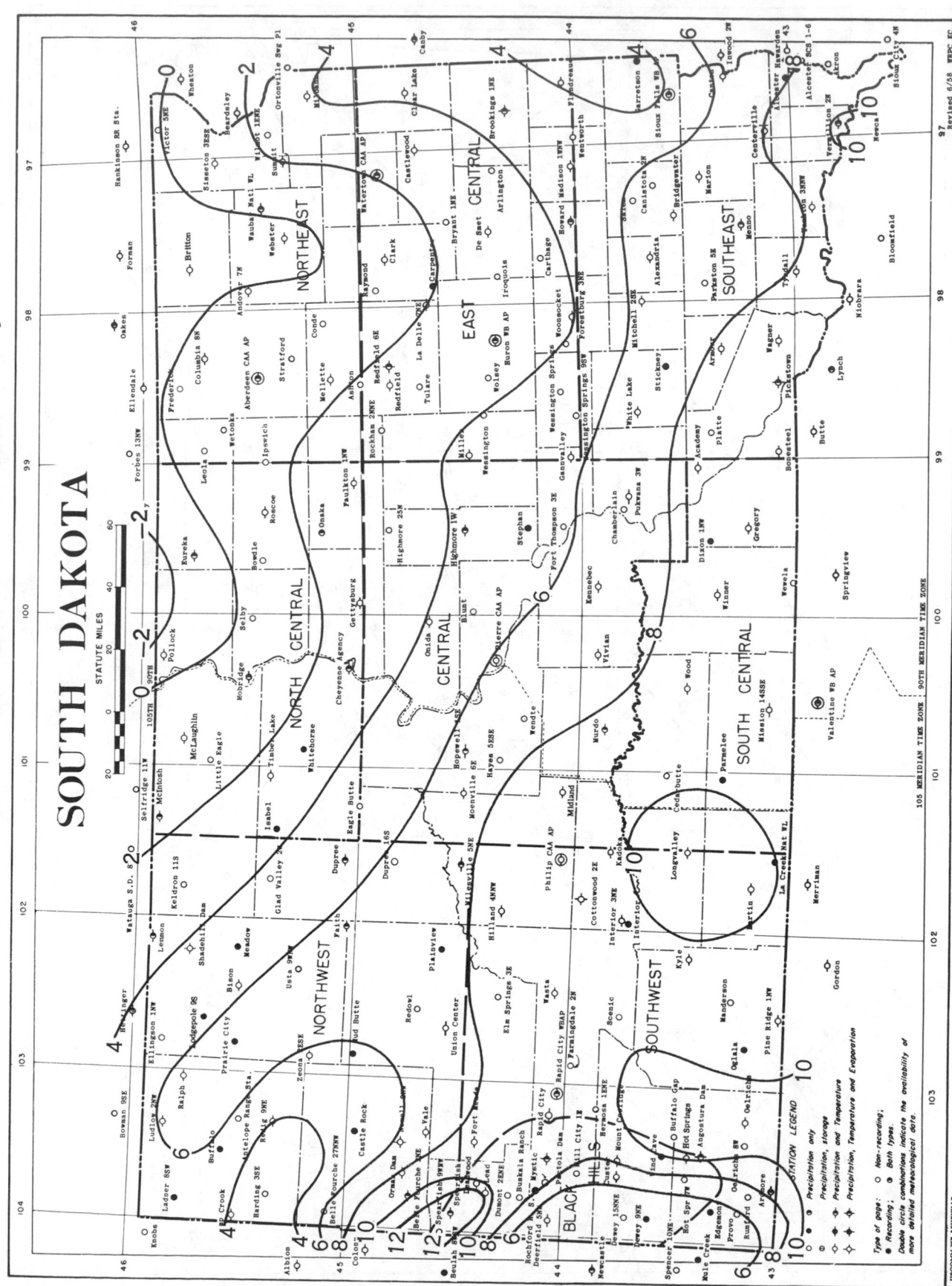

Based on period 1931-52

Isolines are drawn through points of approximately equal value. Caution should be used in interpolating on these maps, particularly in mountainous areas.

Mean Maximum Temperature (°F.), July

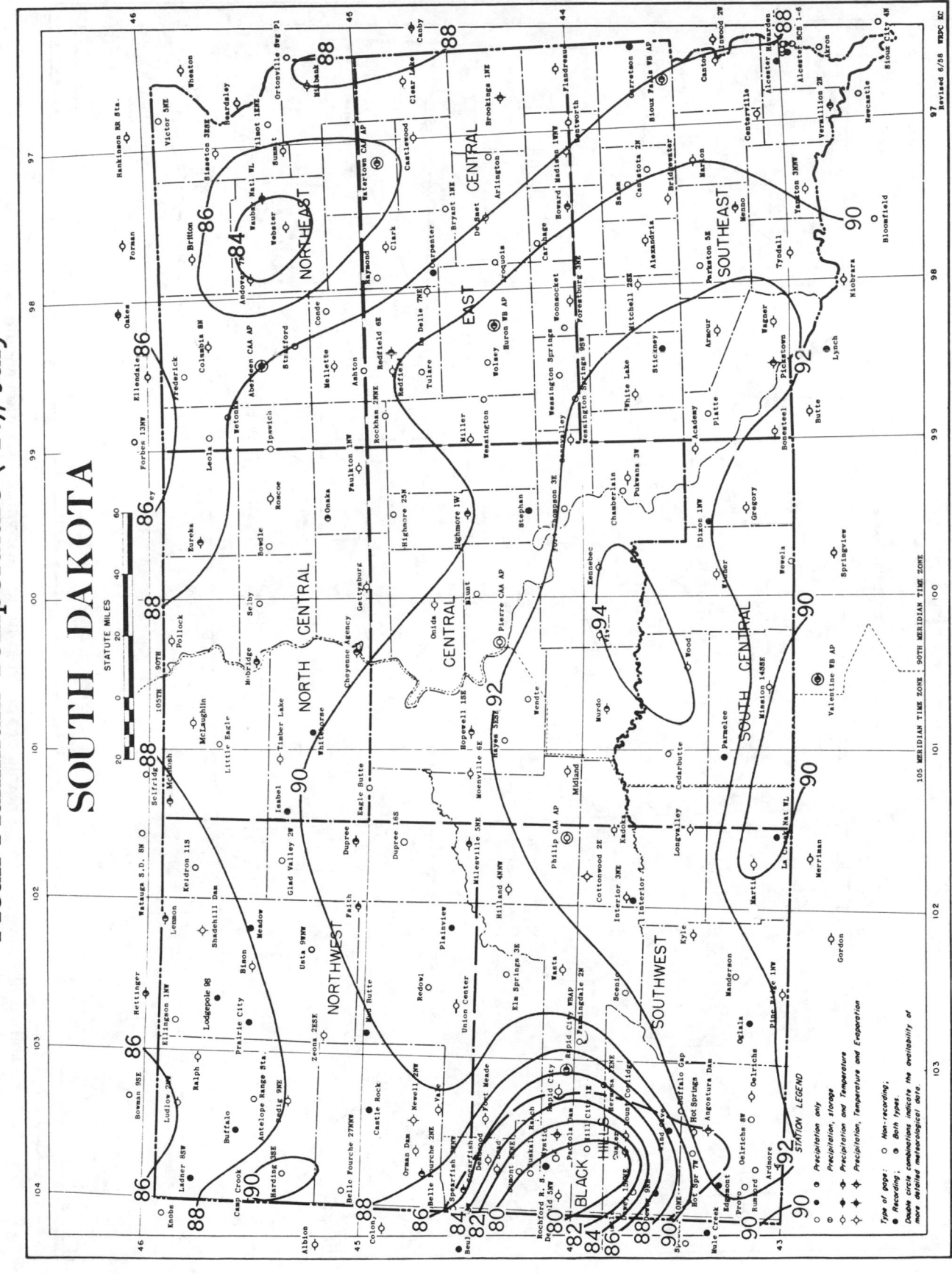

SOUTH DAKOTA

STATUTE MILES

Based on period 1931-52

Isolines are drawn through points of approximately equal value. Caution should be used in interpolating on these maps, particularly in mountainous areas.

STATION LEGEND

Type of page:　● Non-recording;
　　　　　　　○ Precipitation only
　◔ Precipitation, storage
　◑ Precipitation and Temperature
　◕ Precipitation, Temperature and Evaporation

Type of page:　○ Non-recording;
　● Recording;　　● Both types.
Double circle combinations indicate the availability of more detailed meteorological data.

USCOMM-WB-ASHEVILLE

Revised 6/58 WBPC KC

Mean Minimum Temperature (°F.), July

SOUTH DAKOTA

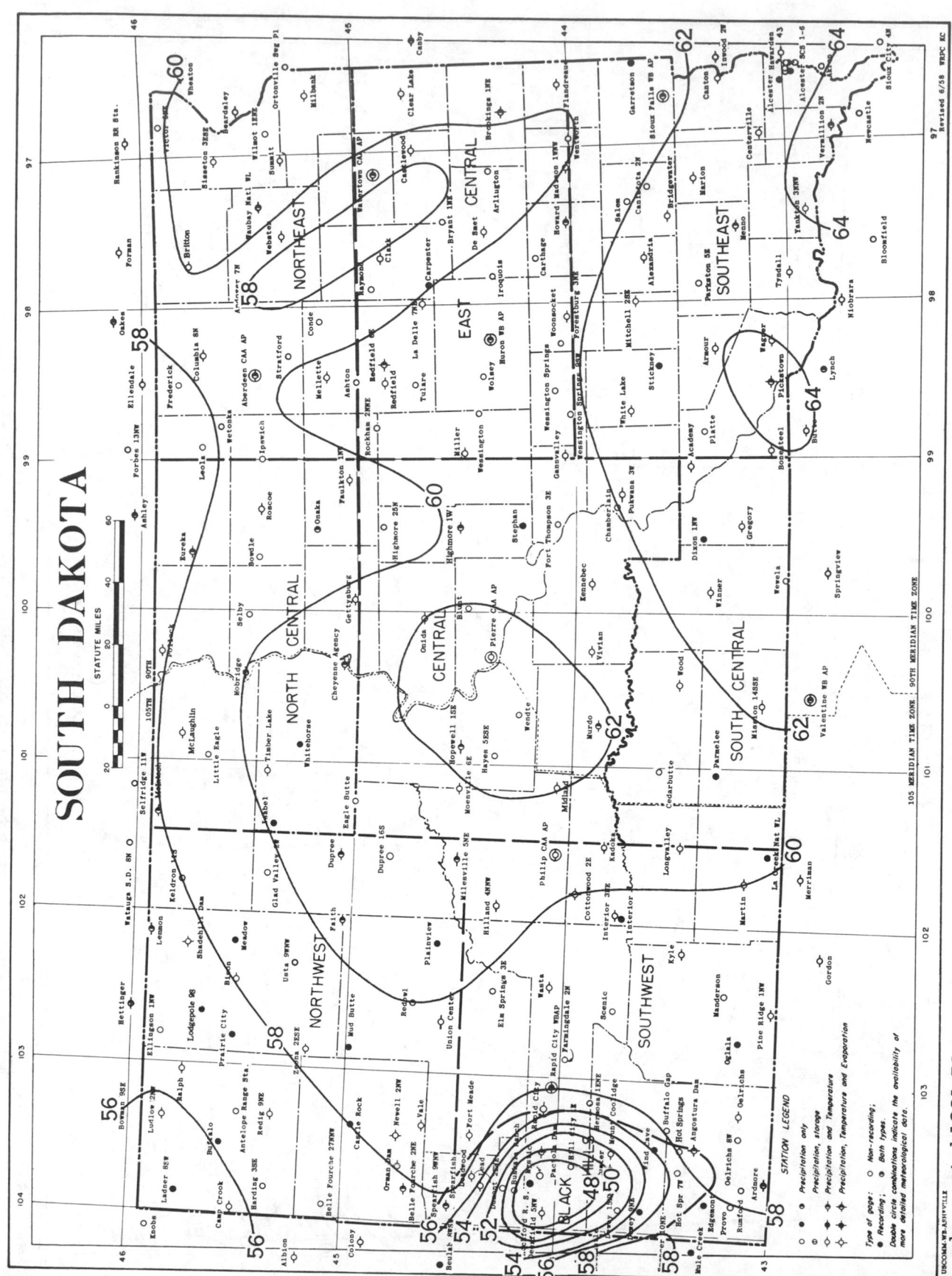

Based on period 1931-52

Isolines are drawn through points of approximately equal value. Caution should be used in interpolating on these maps, particularly in mountainous areas.

STATION LEGEND

Type of gage: O Non-recording; ⊕ Recording; ◉ Both types.

Double circle combinations indicate the availability of more detailed meteorological data.

O Precipitation only
⊕ Precipitation, storage
◉ Precipitation and Temperature
✦ Precipitation, Temperature and Evaporation

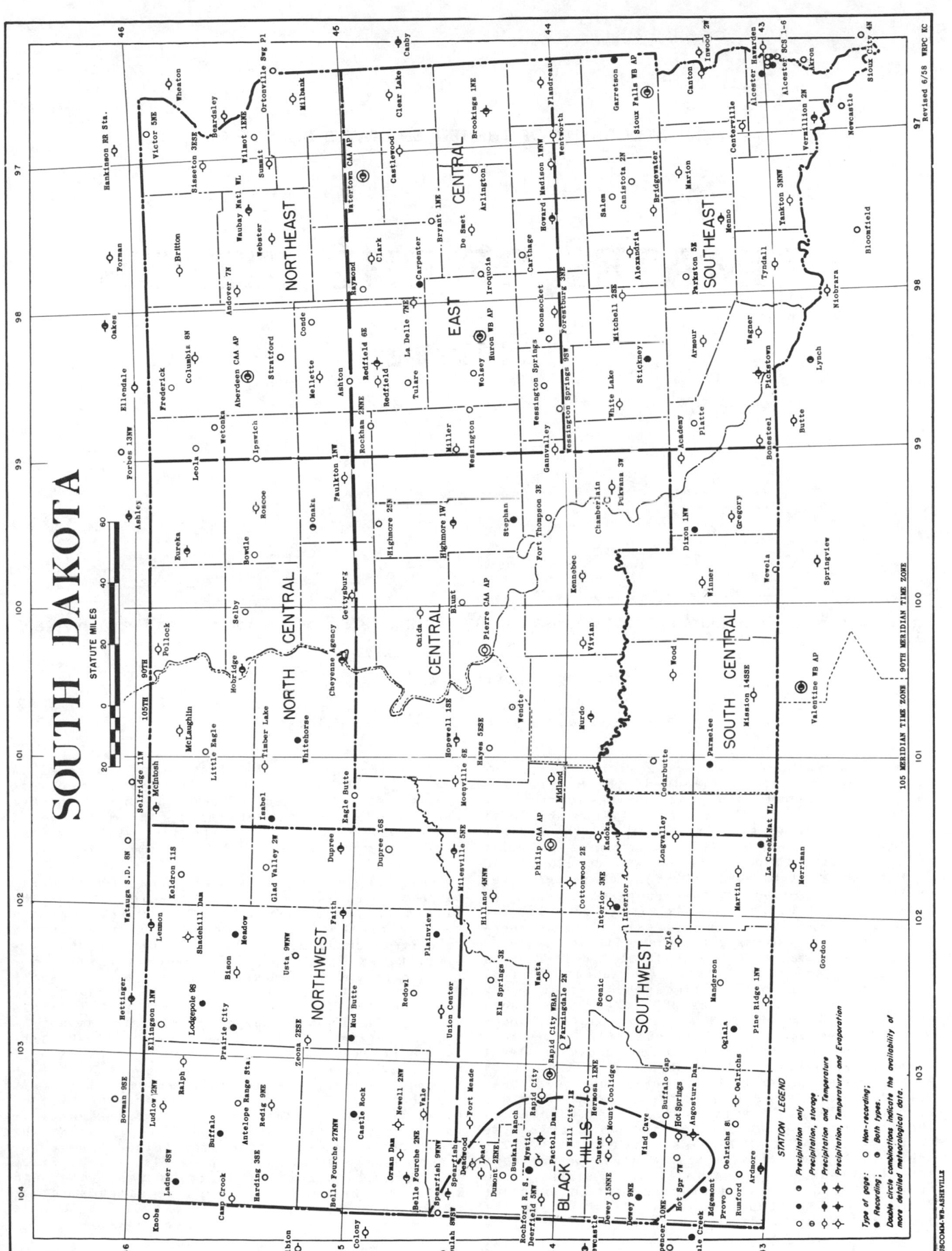

SOUTH DAKOTA

STATUTE MILES

STATION LEGEND

Precipitation only
Precipitation, storage
Precipitation and Temperature
Precipitation, Temperature and Evaporation

Type of gage: ○ Non-recording;
● Recording; ● Both types.
Double circle combinations indicate the availability of
more detailed meteorological data.

USCOMM-WB-ASHEVILLE

Revised 6/58   WBPC KC

105 MERIDIAN TIME ZONE   90TH MERIDIAN TIME ZONE

CLIMATES OF THE STATES

TENNESSEE

(Normals, Means and Extremes tables revised 1960, 1973 and 1975. Basic report revised February 1960.)

Climate of Tennessee

Robert R. Dickson, Weather Bureau State Climatologist

The topography of Tennessee is quite varied, stretching from the lowlands of the Mississippi Valley to the mountain peaks in the east. The westernmost part of the State, between the bluffs overlooking the Mississippi River and the western valley of the Tennessee River, is a region of gently rolling plains sloping gradually from 200 to 250 feet in the west to about 600 feet above sea level in the hills overlooking the Tennessee River. The hilly Highland Rim, in a wide circle touching the Tennessee River Valley on the west and the Cumberland Plateau on the east, together with the enclosed Central Basin makes up the whole of Middle Tennessee. The Highland Rim ranges from about 600 feet in elevation along the Tennessee River to 1,000 feet in the east and rises 300 to 400 feet above the Central Basin which is a rolling plain of about 600 feet average elevation, but with a crescent of hills reaching to over 1,000 feet above sea level south of Nashville. The Cumberland Plateau, with an average elevation of 2,000 feet above sea level, extends roughly northeast-southwest across the State in a belt 30 to 50 miles wide, being bounded on the west by the Highland Rim and overlooking the Great Valley of East Tennessee on the east. The Great Valley, parelleling the Plateau to the west and the Great Smoky Mountains to the east, is a funnel shaped valley varying in width from about 30 miles in the south to about 90 miles in the north. Within the valley which slopes from 1,500 feet in the north to 700 feet above sea level in the south are a series of northeast-southwest ridges. Along the Tennessee-North Carolina border lie the Great Smoky Mountains, the most rugged and elevated portion of Tennessee, with numerous peaks from 4,000 to 6,000 feet above sea level.

Tennessee, except for a small area east of Chattanooga, lies entirely within the drainage of the Mississippi River system. The extreme western section of the State is drained through several relatively small rivers directly into the Mississippi River. Otherwise drainage is into either the Cumberland or Tennessee Rivers, both of which flow northward near the end of their courses to join the Ohio River along the Kentucky - Illinois border. The Cumberland River which drains north-central portions of Tennessee rises in the Cumberland Mountains in Kentucky, flows southwestward, then south into Tennessee reaching the Nashville area before turning northward to re-enter Kentucky. The Tennessee River is formed by the juncture of the Holston and French Broad Rivers at Knoxville in the east. It flows southwesterly along the Great Valley of East Tennessee into Alabama, re-enters Tennessee at the Alabama-Mississippi line, and then flows northward across the State into Kentucky. Besides the headwater streams, other important tributaries include the Clinch, Little Tennessee, Hiwassee, Elk and Duck Rivers.

Most aspects of the State's climate are related to the widely varying topography within its borders. The decrease of temperature with elevation is quite apparent, amounting to, on the average, 3°F. per 1000 feet increase in elevation. Thus higher portions of the State, such as the Cumberland Plateau and the mountains of the east, have lower average temperatures than the Great Valley of East Tennessee, which they flank, and other lower parts of the State. In the Great Valley temperature increases from north to south, reaching a value at the low end comparable to that of Middle and West Tennessee where elevation variations are a generally minor consideration. Across the State, the average annual temperature varies from over 62°F. in the extreme southwest to near 45°F. atop the highest peaks of the east. Temperatures at elevated locations (e.g. Bristol and Crossville) can be compared with those at lower elevations (e.g. Nashville and Union City) in the accompanying tables. It is of interest to note that average January temperature atop a 6,000-foot peak in the Great Smokies is equivalent to that in central Ohio, while average July temperature is duplicated along the southern edge of the Hudson Bay in Canada. While most of the State can be described as having a warm, humid summer and a mild winter, this must be qualified to include variations with elevation. Thus with increasing elevation, summers become cooler and more pleasant while winters become colder and more blustery.

This dependence of temperature upon elevation is of considerable importance to a variety of interests. Temperature, together with precipitation, plays an important role in determining what plant and animal life are adaptable to the area. In the Great Smoky Mountains, for example, the variations in elevation from 1,000 to 6,000 feet above sea level with attendant variations in temperature contribute to the remarkable variety of plant life found. The relative coolness of the mountain area has also contributed to the popularity of that area during the warmer part of the year.

Length of growing season (freeze-free period) is linked to topography in a way similar to temperature, varying from 237 days at low-lying Memphis to near 130 days on the highest mountains in the east. Most of the State is included in the range 180 to 220 days. Shorter growing seasons than this are confined to the mountains forming the State's eastern border and to the northern part of the Cumberland Plateau. Longer growing seasons are found in counties bordering the Mississippi River, parts of the Central Basin of the Middle Tennessee, and the southern end of the Great Valley of East Tennessee.

Since the principal source of moist air for this area is the Gulf region, there exists a gradual decrease of average precipitation from south to north. This effect is largely obscured, however, by the overruling influence of topography. Air forced to ascend cools and condenses out a portion of its moisture charge; thus average precipitation is generally greater at higher elevations. This is apparent in all parts of the State. In West Tennessee average annual precipitation ranges from 46 to 54 inches, increasing from Mississippi bottomlands to the slight hills farther east. In Middle Tennessee the variation is from a minimum of 45 inches in the Central Basin to 50 to 55 inches in the surrounding hilly Highland Rim. Over the elevated Cumberland Plateau average annual precipitation is generally from 50 to 55 inches. In contrast, average annual precipitation in the Great Valley of East Tennessee increases from near 40 inches in northern portions to over 50 inches in

the south. The northern minimum, lowest for the entire state, results from the shielding influence of the Great Smoky Mountains to the southeast and the Cumberland Plateau to the northwest. The mountainous eastern border of the State is its wettest part, having average annual precipitation ranging up to about 80 inches on the higher, well-exposed peaks of the Smokies.

Over most of the State greatest precipitation occurs during the winter and early spring due to the more frequent passage of large scale storms over and near the State during those months. A secondary maximum of precipitation occurs in midsummer in response to shower and thundershower activity. This is especially pronounced in the mountains of the east where July rainfall exceeds the precipitation of any other month. Lightest precipitation, observed in the fall, is brought about by the maximum occurrence of slow moving, rain suppressing high pressure areas during that season. Although all parts of Tennessee are generally well supplied with precipitation, there occurs on the average one or more prolonged dry spells each year during summer and fall. Current studies illustrate the beneficial effects of supplemental irrigation of crops, despite the usually bountiful annual precipitation.

The most important flood season is during the winter and early spring (December through March) when the frequent migratory storms bring general rains of high intensity. During this period both widespread flooding and local flash floods can occur. During summer, heavy thunderstorm rains frequently result in local flash flooding. In the fall, while flood producing rains are rare, decadent hurricanes on occasion cause serious floods in the east. The numerous dams constructed along the Tennessee and Cumberland Rivers are major features in the control of flood waters in the State.

Some of the more notable flood years in Tennessee are 1793, 1867, 1886, 1901, 1902, 1917, 1926, 1927, 1929, 1937 and 1955.

The dams of the Tennessee and Cumberland River Systems and the lakes so formed, in addition to vastly reducing flood damage have facilitated water transportation, provided abundant low cost hydroelectric power and created extensive recreation areas. Fishing, boating, swimming, and camping along the "Great Lakes of the South", together with utilization of the several state and national parks, has made the tourist industry one of major proportions in the State.

Average annual snowfall varies from 4 to 6 inches in southern and western parts of the State and in most of the Great Valley of East Tennessee to more than 10 inches over the northern Cumberland Plateau and the mountains of the east. Over most of the State, due to relatively mild winter temperatures, a snow cover rarely persists for more than a few days.

Water resources of Tennessee have been a major factor in the recent rapid industrial growth. The bountiful and good quality water supply has influenced the location of industry, especially chemical processing plants. Three major waterways, the Mississippi, Cumberland, and Tennessee Rivers, are suitable for commercial traffic. Finally, the availability of low cost hydroelectric power from the multipurpose dams of the Cumberland and Tennessee Rivers and tributaries has been a stimulus to industry of all types. The principal types of manufacturing industries in Tennessee, in order of value added, are chemical and allied products, food and kindred products, textile mill products, primary metals, fabricated metal products, and lumber and lumber products.

Although surpassed in monetary value by industrial activity, agriculture remains a vital feature of Tennessee's economic life. The wide range of climates in Tennessee, from river bottom to mountaintop, coupled with a wide range of soils, has resulted in a large number of crops which prosper in the State. Cotton, in addition to being the foremost money crop, also serves as raw material for the State's extensive textile industry. Primary crops (and major areas in which grown) in order of value of production, are cotton (west), tobacco (north-central and northeast), corn (west and south-central), and hay (middle and east). Livestock and livestock products constitute nearly one-half of the entire farm cash income. Major activity concerns hogs, cattle, calves, dairy products, poultry, and eggs.

Forests represent an additional important segment of Tennessee's natural resources related to the climate of the State. Timberland, containing principally hardwood types, covers approximately one-half the total area of Tennessee. This has led to a highly diversified woodworking industry in which Memphis has become a hardwood flooring

center of the United States. The temperate climate of the State is very favorable for logging operations, allowing full scale activity during 9 months of the year and to a lesser extent during the winter months. The abundant wood supply is utilized by a number of chemical processing industries in the State.

Severe storms are relatively infrequent in Tennessee, being east of the center of tornado activity, south of most blizzard conditions, and too far inland to be often affected by hurricanes. On the average four or five tornadoes are observed in the State each year, with greatest frequency in March when one or two usually occur. Tornado occurrence is not evenly distributed throughout the State, being largely confined to areas west of the Cumberland Plateau. Annual expectancy is less than in bordering locations to the south and west. Damage from tropical storms is rare, occurring only about once every 18 years and blizzard conditions during a recent 40-year period were non-existent. Hailstorms at a given locality are observed two or three times a year and damaging glaze storms occur in the State every 5 or 6 years. Thunderstorms are frequent and severe thunderstorms with damaging winds are experienced at scattered locations throughout the State each year during the warm season.

REFERENCES

(1) Federal Disaster Insurance -- Staff Study for the Committee on Banking and Currency, United States Senate, "Storms", pages 123 - 137, November, 1955.

(2) Klein, W. H., "Principal Tracks and Mean Frequencies of Cyclones & Anticyclones in the Northern Hemisphere", U. S. Weather Bureau Research Paper No. 40, Washington, D. C., 1957.

(3) Tennessee Deaprtment of Agriculture, Agricultural Trends in Tennessee, November 1958.

(4) Tennessee State Planning Commission, Vol. I., Industrial Trends in Tennessee, December 1949; Vol. II Forests, Agriculture and Minerals, December 1948; Vol. III Water Supplies, Fuels, Electric Power and Transportation Facilities, December 1948.

(5) Tharp, Max M. & C. W. Crickman, "Supplemental Irrigation in Humid Regions", Water, 1955 Yearbook of Agriculture, pp 252 - 258, Washington, D. C.

(6) U. S. Weather Bureau, "Tornado Occurrences in the United States", Technical Paper No. 20, Washington, D. C. September 1952.

(7) U. S. Weather Bureau, "Mean Temperature and Precipitation, Tennessee", L. S. 5733, Washington, D. C., August, 1957.

(8) Van Horn, A. G., W. M. Whitaker, R. H. Lush and John R. Carreker, "Irrigation of Pastures for Dairy Cows", University of Tennessee Agricultural Experiment Station Bulletin No. 248, Knoxville, Tennessee, June 1956.

(9) Wessenauer, A. M. "Success of the TVA multiple-purpose river development", Civil Engineering, July 1956.

(10) Weather Bureau Technical Paper No. 15 - Maximum Station Precipitation for 1,2,3,6,12, and 24 hours. Washington, D.C. September 1956.

(11) Weather Bureau Technical Paper No. 16 - Maximum 24-Hour Precipitation in the United States - Washington, D. C. January, 1952.

(12) Weather Bureau Technical Paper No. 25 - Rainfall Intensity-Duration-Frequency Curves. For selected stations in the United States, Alaska, Hawaiian Islands, and Puerto Rico. Washington, D. C. December 1955.

(13) Weather Bureau Technical Paper No. 29 - Rainfall Intensity-Frequency Regime, Part I - The Ohio Valley. Washington, D. C. June 1957.

(14) Hoyt, William G. and Langbern, Walter B., Floods, Princeton University Press, Princeton, N. J. 1947.

BIBLIOGRAPHY

(A) Climatic Summary of the United States (Bulletin W) 1930 edition, Sections 76 and 77. U. S. Weather Bureau

(B) Climatic Summary of the United States, Tennessee - Supplement for 1931 through 1952 (Bulletin W Supplement). U. S. Weather Bureau

(C) Climatological Data - Tennessee. U. S. Weather Bureau

(D) Climatological Data National Summary. U. S. Weather Bureau

(E) Hourly Precipitation Data - Tennessee. U. S. Weather Bureau

(F) Local Climatological Data, U. S. Weather Bureau for Bristol, Chattanooga, Knoxville, Memphis, Nashville, and Oak Ridge, Tennessee.

FREEZE DATA

STATION	Freeze threshold temperature	Mean date of last Spring occurrence	Mean date of first Fall occurrence	Mean No. of days between dates	Years of record Spring	No. of occurrences in Spring	Years of record Fall	No. of occurrences in Fall	STATION	Freeze threshold temperature	Mean date of last Spring occurrence	Mean date of first Fall occurrence	Mean No. of days between dates	Years of record Spring	No. of occurrences in Spring	Years of record Fall	No. of occurrences in Fall
ASHWOOD	32	04-08	10-24	199	30	30	30	30	DOVER 1 NW	32	04-15	10-19	187	30	30	30	30
	28	03-29	11-07	223	30	30	30	30		28	03-31	11-03	217	30	30	30	30
	24	03-13	11-20	252	30	30	30	30		24	03-15	11-12	242	30	30	30	30
	20	02-28	12-01	276	29	29	30	27		20	03-03	11-26	268	30	30	30	28
	16	02-13	12-10	300	29	27	27	23		16	02-18	12-09	294	30	28	28	25
BOLIVAR	32	04-03	10-23	203	26	26	25	25	FRANKLIN	32	04-12	10-21	192	30	30	30	30
	28	03-21	11-04	229	26	26	25	25		28	03-28	11-02	219	30	30	30	30
	24	03-04	11-17	258	25	25	25	24		24	03-13	11-14	246	30	30	30	30
	20	02-15	12-02	290	25	23	24	24		20	02-27	11-29	275	30	30	30	28
	16	02-08	12-14	309	23	21	24	16		16	02-14	12-11	301	30	27	28	22
BRISTOL WB AP	32	04-16	10-23	190	13	13	13	13	GATLINBURG 2 SW	32	04-29	10-16	171	24	24	25	25
	28	04-02	11-02	213	13	13	13	13		28	04-13	10-27	197	24	24	25	25
	24	03-24	11-15	236	13	13	13	13		24	03-27	11-09	227	24	24	25	25
	20	03-05	11-24	264	13	13	13	13		20	03-11	11-23	257	24	24	25	25
	16	02-19	12-12	296	13	12	13	11		16	03-01	12-02	276	24	24	24	23
BROWNSVILLE	32	03-28	10-31	217	30	30	30	30	GREENEVILLE EXP STA	32	04-20	10-17	180	18	18	18	18
	28	03-16	11-14	243	30	30	30	30		28	04-12	10-28	198	18	18	18	18
	24	03-01	11-30	274	30	30	30	29		24	03-27	11-08	226	18	18	18	18
	20	02-16	12-11	298	30	28	29	25		20	03-10	11-22	257	18	18	18	18
	16	01-31	12-18	321	28	24	25	16		16	02-27	12-05	281	18	18	18	18
CARTHAGE	32	04-07	10-28	205	30	30	30	30	JACKSON CAA AP	32	04-06	10-23	200	30	30	30	30
	28	03-22	11-11	233	30	30	30	30		28	03-25	11-05	226	30	30	30	30
	24	03-10	11-23	257	30	30	30	30		24	03-05	11-22	262	30	30	30	29
	20	02-24	12-04	283	30	29	29	26		20	02-19	12-07	291	30	28	29	25
	16	02-15	12-14	302	29	27	26	20		16	02-04	12-17	316	28	24	25	16
CHATTANOOGA WB CITY	32	03-26	11-10	229	30	30	30	29	KENTON	32	04-01	10-24	206	25	25	24	24
	28	03-14	11-20	252	30	30	29	28		28	03-19	11-05	231	25	25	25	25
	24	02-23	12-03	283	30	28	28	25		24	03-07	11-21	259	25	25	25	24
	20	02-17	12-11	297	28	27	24	20		20	02-23	12-03	283	25	25	24	21
	16	01-26	12-18	326	26	18	20	13		16	02-06	12-12	309	25	22	21	18
CLARKSVILLE	32	04-04	10-29	208	30	30	30	30	KNOXVILLE WB CO	32	03-31	11-06	220	30	30	30	30
	28	03-22	11-09	232	30	30	30	30		28	03-18	11-20	247	30	30	30	29
	24	03-06	11-22	261	30	30	30	29		24	03-03	12-03	275	30	30	29	28
	20	02-24	12-03	282	30	29	29	26		20	02-20	12-14	297	30	29	28	21
	16	02-12	12-14	304	29	26	26	23		16	01-30	12-19	323	29	21	21	15
COLDWATER 1 E	32	04-11	10-21	192	29	29	30	30	LEWISBURG EXP STA	32	04-14	10-20	189	30	30	30	30
	28	03-30	10-31	215	29	29	29	29		28	03-31	10-31	214	30	30	30	30
	24	03-11	11-15	249	29	29	29	28		24	03-15	11-15	245	30	30	30	30
	20	02-26	11-26	273	29	28	27	25		20	03-04	11-25	266	30	30	30	28
	16	02-13	12-07	296	28	27	25	21		16	02-16	12-10	296	30	28	28	25
COPPERHILL	32	04-11	10-23	196	30	30	30	30	LOUDON	32	04-10	10-24	197	29	29	29	29
	28	03-28	11-04	221	30	30	30	30		28	03-31	11-07	222	29	29	29	29
	24	03-14	11-19	250	30	30	30	29		24	03-15	11-19	249	29	29	29	28
	20	02-27	12-02	278	30	29	29	28		20	03-02	11-27	270	29	29	27	27
	16	02-11	12-12	304	29	27	28	20		16	02-13	12-12	303	29	26	27	20
COVINGTON	32	03-27	11-03	222	30	30	30	30	LYNNVILLE 4 SW	32	04-13	10-20	190	30	30	30	30
	28	03-14	11-12	244	30	30	30	30		28	03-29	11-01	217	29	29	30	30
	24	03-01	11-27	271	30	30	30	29		24	03-18	11-14	240	29	29	29	28
	20	02-14	12-10	300	30	27	29	24		20	03-04	11-25	267	29	28	28	28
	16	01-31	12-18	321	27	23	24	16		16	02-15	12-10	298	29	28	28	22
CROSSVILLE	32	04-21	10-14	176	29	29	29	29	MC MINNVILLE	32	04-08	10-28	203	30	30	30	30
	28	04-13	10-25	195	29	29	29	29		28	03-28	11-09	226	29	29	30	30
	24	04-01	11-04	217	29	29	30	30		24	03-10	11-22	257	29	29	30	29
	20	03-18	11-18	245	29	29	29	29		20	02-28	12-03	279	29	29	29	27
	16	03-06	11-29	269	29	29	29	28		16	02-13	12-14	304	29	27	27	19
DALE HOLLOW DAM (Celina)	32	04-13	10-21	191	16	16	16	16	MEMPHIS WB CO	32	03-20	11-12	237	30	30	30	30
	28	04-05	11-02	211	16	16	16	16		28	03-05	11-29	269	30	30	30	29
	24	03-16	11-13	242	16	16	16	16		24	02-16	12-09	296	30	28	29	27
	20	03-04	11-27	268	14	14	16	16		20	02-09	12-17	311	28	26	27	19
	16	02-27	12-08	284	14	14	15	13		16	01-28	12-19	325	26	22	19	12
DECATUR	32	04-14	10-24	192	30	30	30	30	MILAN	32	04-04	10-26	206	30	30	30	30
	28	04-03	11-04	216	30	30	30	30		28	03-25	11-05	226	30	30	30	30
	24	03-19	11-16	242	30	30	30	30		24	03-06	11-21	260	30	30	30	29
	20	03-03	11-30	272	30	30	30	27		20	02-19	12-03	287	30	29	29	26
	16	02-19	12-08	291	30	28	27	23		16	02-07	12-14	310	29	26	26	19
DICKSON	32	04-10	10-23	197	30	30	30	30	MOSCOW	32	04-02	10-24	205	30	30	30	30
	28	03-28	11-07	224	30	30	30	30		28	03-23	11-07	230	30	30	30	30
	24	03-13	11-18	251	30	30	30	29		24	03-04	11-19	260	30	30	30	30
	20	02-28	11-29	275	30	29	29	29		20	02-15	12-09	297	30	27	30	24
	16	02-19	12-13	296	29	28	29	24		16	02-03	12-16	317	27	24	24	17

STATION	Freeze threshold temperature	Mean date of last Spring occurrence	Mean date of first Fall occurrence	Mean No. of days between dates	Years of record Spring	No. of occurrences in Spring	Years of record Fall	No. of occurrences in Fall
MURFREESBORO	32	04-05	10-25	203	30	30	30	30
	28	03-25	11-05	225	30	30	30	30
	24	03-09	11-18	255	30	30	30	29
	20	02-23	12-02	282	30	28	30	27
	16	02-11	12-16	309	29	26	30	22
NASHVILLE WB CO	32	03-28	11-07	224	30	30	30	30
	28	03-18	11-17	245	30	30	30	29
	24	03-04	11-28	269	30	30	29	28
	20	02-18	12-11	296	30	29	28	23
	16	02-05	12-17	315	29	26	23	18
NEWBERN	32	03-27	10-31	217	30	30	30	30
	28	03-20	11-13	238	30	30	30	30
	24	03-07	11-26	265	29	29	29	28
	20	02-20	12-07	290	29	27	29	27
	16	02-07	12-19	315	29	26	29	17
NEWPORT	32	04-11	10-24	197	30	30	30	30
	28	03-30	11-05	220	30	30	30	30
	24	03-14	11-18	249	30	30	30	30
	20	02-28	12-02	277	30	29	30	28
	16	02-19	12-12	297	29	27	28	20
NORRIS	32	04-16	10-27	194	16	16	16	16
	28	04-04	11-02	211	16	16	16	16
	24	03-28	11-21	238	16	16	16	16
	20	03-10	11-29	264	16	16	16	16
	16	02-18	12-10	295	16	14	16	13
PALMETTO	32	04-10	10-24	197	30	30	29	29
	28	03-27	11-06	224	30	30	29	29
	24	03-11	11-21	255	30	30	29	28
	20	02-20	11-30	283	30	28	28	26
	16	02-14	12-15	304	28	27	26	19
PARIS	32	04-12	10-31	202	17	17	16	16
	28	03-29	11-06	222	16	16	16	16
	24	03-11	11-18	252	15	15	16	16
	20	02-27	11-30	276	15	15	16	16
	16	02-09	12-14	309	15	14	16	11
ROGERSVILLE	32	04-17	10-24	190	28	28	30	30
	28	03-30	11-03	218	28	28	30	30
	24	03-14	11-18	250	28	28	30	30
	20	03-02	12-01	274	28	28	30	27
	16	02-20	12-08	290	28	27	27	23
RUGBY	32	05-03	10-08	158	30	30	29	29
	28	04-16	10-20	187	30	30	29	29
	24	04-02	10-31	212	30	30	29	29
	20	03-19	11-11	237	30	30	29	29
	16	03-08	11-25	262	29	29	29	27
SAMBURG WL REF (Tiptonville)	32	03-29	11-02	218	22	22	23	23
	28	03-19	11-11	237	22	22	23	23
	24	03-03	11-23	266	22	22	23	22
	20	02-18	12-04	288	22	22	22	21
	16	02-10	12-19	312	22	20	21	14
SAVANNAH	32	04-04	10-28	207	29	29	30	30
	28	03-22	11-12	235	29	29	30	30
	24	03-08	11-27	264	29	28	29	28
	20	02-21	12-02	284	28	27	28	25
	16	02-09	12-15	309	27	25	25	18
SPRINGFIELD EXP STA (Cedar Hill)	32	04-07	10-30	206	30	30	30	30
	28	03-27	11-09	227	30	30	30	30
	24	03-13	11-22	254	30	30	30	29
	20	03-02	12-03	276	30	30	30	27
	16	02-13	12-12	302	30	27	29	24
TULLAHOMA	32	04-10	10-23	196	30	30	30	30
	28	03-25	11-06	226	29	29	30	30
	24	03-13	11-16	248	30	30	30	30
	20	02-26	12-03	280	30	30	30	26
	16	02-12	12-14	305	29	27	30	21
UNION CITY	32	03-31	10-28	211	30	30	30	30
	28	03-22	11-09	232	30	30	30	30
	24	03-05	11-22	262	30	30	30	30
	20	02-21	12-07	289	30	30	30	26
	16	02-09	12-15	309	30	27	26	19
WAYNESBORO	32	04-18	10-15	180	28	28	28	28
	28	04-05	10-29	208	28	28	28	28
	24	03-24	11-07	227	28	28	28	28
	20	03-08	11-22	260	28	28	28	27
	16	02-23	12-07	287	28	27	28	22

Data in the above table are based on the period 1921-1950, or that portion of this period for which data are available.

Means have been adjusted to take into account years of non-occurrence.

A freeze is a numerical substitute for the former term "killing frost" and is the occurrence of a minimum temperature at or below the threshold temperature of 32°, 28°, etc.

Freeze data tabulations in greater detail are available and can be reproduced at cost.

*MEAN TEMPERATURE AND PRECIPITATION

STATION	January Temperature	January Precipitation	February Temperature	February Precipitation	March Temperature	March Precipitation	April Temperature	April Precipitation	May Temperature	May Precipitation	June Temperature	June Precipitation	July Temperature	July Precipitation	August Temperature	August Precipitation	September Temperature	September Precipitation	October Temperature	October Precipitation	November Temperature	November Precipitation	December Temperature	December Precipitation	Annual Temperature	Annual Precipitation
EASTERN																										
BLUFF CITY		4.13		3.65		4.42		3.30		3.72		3.66		5.84		4.02		2.88		2.15		2.69		3.53		43.99
BRISTOL WB AIRPORT	38.6	3.50	40.1	3.41	46.7	3.77	56.0	3.25	64.2	3.44	72.3	3.67	74.8	5.14	73.6	3.89	68.5	2.79	57.9	2.39	45.8	2.56	38.7	3.43	56.4	41.24
CHARLESTON		5.70		5.13		5.84		4.25		3.77		3.95		4.63		4.13		2.55		2.72		3.78		5.26		51.71
CHATTANOOGA WB AP	41.6	5.23	44.0	5.11	50.7	6.05	59.7	4.53	67.7	4.16	75.8	4.21	78.3	5.34	77.3	3.70	72.5	2.69	60.8	3.24	49.1	4.03	42.1	5.31	60.0	53.60
CLINTON		5.68		5.44		5.43		3.87		4.17		3.59		5.28		3.75		3.34		2.63		3.99		5.22		52.39
COPPERHILL	40.4	5.88	41.2	5.38	47.7	6.35	56.8	4.26	65.6	3.66	73.4	4.22	76.1	5.77	75.1	4.99	69.3	2.89	58.0	2.97	46.3	3.66	40.1	5.34	57.5	55.37
DANDRIDGE		4.40		4.33		4.87		3.36		3.45		3.18		4.75		3.82		2.59		2.19		2.84		3.72		43.50
EMBREEVILLE		3.65		3.72		4.62		3.46		4.04		3.55		6.36		4.00		3.04		2.48		2.70		3.25		44.87
ERWIN		3.67		3.76		4.86		3.48		4.11		3.90		6.55		4.23		3.07		2.51		2.68		3.27		46.09
KINGSPORT 3 SE		4.03		3.80		4.56		3.31		3.40		3.36		5.05		3.98		2.64		2.19		2.69		3.33		42.34
KINGSTON		5.35		5.28		5.39		3.91		3.64		3.53		4.46		4.10		2.55		2.40		3.64		4.85		49.10
KNOXVILLE WB AP	40.5	4.54	42.5	4.73	49.4	4.83	59.0	3.64	67.4	3.58	75.8	3.47	78.4	4.72	77.0	3.43	72.1	2.53	60.3	2.63	48.4	3.15	41.0	4.26	59.3	45.51
LOUDON 1 E	41.3	5.57	42.5	5.25	47.7	5.54	58.9	3.84	67.7	3.78	76.2	3.42	79.0	5.35	78.0	3.57	72.4	2.74	61.0	2.61	48.0	3.56	40.9	4.80	59.5	50.03
NEWPORT	40.9	4.29	41.7	4.14	48.5	4.82	58.1	3.40	67.1	3.56	75.4	3.73	78.1	5.25	77.0	4.02	71.3	2.45	60.0	2.42	47.2	2.65	40.2	3.46	58.8	44.19
OAK RIDGE	39.6	5.11	41.4	5.21	48.3	5.64	57.7	4.29	66.0	3.78	74.2	4.18	76.7	4.83	75.4	4.45	71.1	2.97	59.3	2.77	47.5	3.94	40.1	5.21	58.1	52.38
OAK RIDGE WB	39.0	6.06	41.0	5.79	48.1	5.69	57.6	4.08	66.0	4.18	74.2	3.91	76.7	5.34	75.3	4.06	70.8	2.55	58.8	2.42	47.0	3.98	39.6	5.66	57.8	53.72
PARKSVILLE		5.46		5.16		5.72		4.27		3.78		3.90		5.73		4.38		2.87		2.74		3.53		4.89		52.43
ROGERSVILLE 1 NE	40.9	4.47	42.2	4.24	49.1	4.65	58.2	3.28	66.5	3.50	74.3	3.10	77.0	4.78	76.3	3.48	71.3	2.60	60.5	2.17	47.8	3.01	40.3	3.63	58.7	42.91
TELLICO PLAINS		5.88		5.16		6.11		4.03		4.34		4.24		5.24		4.73		2.81		2.77		3.84		4.72		53.87
DIVISION	40.9	4.94	42.0	4.67	48.7	5.25	57.9	3.75	66.5	3.74	74.5	3.75	77.2	5.52	76.2	4.13	70.7	2.74	59.6	2.58	47.3	3.19	40.4	4.31	58.5	48.57
CUMBERLAND PLATEAU																										
CROSSVILLE EXP STA	37.8	5.97	38.6	5.61	45.7	5.63	54.9	4.23	63.6	3.90	70.8	3.95	73.8	4.53	73.0	4.55	67.5	3.18	57.2	2.70	45.3	3.82	37.9	5.46	55.5	53.53
MC MINNVILLE	42.5	5.74	43.9	5.65	50.7	5.62	59.4	3.77	67.5	3.83	75.5	4.24	78.1	4.36	77.3	4.15	71.9	3.22	61.2	2.49	49.1	3.50	42.5	5.04	60.0	51.61
ROCK ISLAND 2 NW		5.79		5.59		5.78		3.83		3.91		3.69		4.62		4.03		3.31		2.58		3.70		5.16		51.99
SEWANEE		6.12		5.80		6.29		4.87		4.10		4.20		5.69		4.13		3.13		2.92		4.35		5.39		56.99
TULLAHOMA	41.5	6.19	43.0	5.85	49.9	6.06	58.8	4.39	66.9	3.62	74.9	3.79	77.6	5.12	76.8	3.54	71.1	3.12	60.3	2.56	48.4	3.85	41.8	5.39	59.3	53.48
DIVISION	39.9	6.03	41.1	5.65	48.1	5.87	57.1	4.22	65.5	3.87	73.3	4.20	76.1	5.01	75.2	4.14	69.8	3.15	59.3	2.62	47.1	3.89	40.1	5.38	57.7	54.03
MIDDLE																										
ASHWOOD	41.5	5.92	43.0	5.34	50.1	5.92	58.9	4.35	67.2	4.04	75.8	3.40	78.4	4.39	77.4	3.54	71.4	3.23	60.8	2.51	48.9	3.78	42.0	4.74	59.6	51.16
CARTHAGE	41.9	6.09	43.4	5.06	50.7	5.58	59.8	3.93	68.4	3.96	76.7	4.54	79.6	4.36	78.8	3.89	72.9	3.29	62.0	2.38	49.4	4.04	42.2	4.68	60.5	51.90
CLARKSVILLE	40.5	6.13	42.7	4.14	50.4	5.41	59.9	4.11	68.3	3.61	77.2	3.33	80.3	3.67	79.4	3.34	72.7	3.13	61.7	2.55	49.1	3.72	41.3	4.32	60.3	47.46
COLDWATER 1 E	42.9	6.27	44.5	5.87	51.3	6.43	59.9	4.23	67.9	4.26	76.6	3.15	79.3	4.53	78.5	3.66	72.5	3.25	61.6	2.82	49.4	3.94	43.2	5.33	60.6	53.74
DICKSON	40.8	6.14	42.7	4.60	50.5	6.03	59.9	4.39	67.8	4.26	76.4	3.71	79.1	4.56	78.2	3.47	71.8	2.07	61.2	2.66	48.9	4.02	41.5	4.73	59.9	51.54
DOVER 1 NW	39.9	5.80	42.0	3.92	49.7	5.47	59.2	3.85	66.9	3.71	75.5	3.36	78.6	4.01	77.7	3.38	71.0	3.37	60.3	2.84	48.2	3.79	40.5	4.25	59.1	47.75
FAYETTEVILLE 1 NE		5.70		5.51		5.74		4.04		3.99		3.13		5.12		3.79		3.10		2.61		3.98		5.18		51.89
FRANKLIN SEWAGE PLANT	40.6	5.82	42.4	4.96	49.6	5.53	58.6	4.04	67.2	3.76	76.0	3.27	78.6	3.85	77.9	3.49	71.4	2.67	60.5	2.20	48.2	3.43	40.9	4.27	59.3	47.29
LEWISBURG EXP STA	40.7	5.86	42.2	5.98	49.5	5.92	58.6	4.37	67.2	3.73	75.9	3.60	78.8	4.65	78.1	3.05	71.8	2.94	60.6	2.54	48.0	3.98	41.2	4.64	59.4	51.26
MURFREESBORO	41.7	5.67	43.5	5.29	50.7	5.46	59.7	3.96	68.3	3.86	76.8	2.94	79.5	4.25	78.6	3.74	72.6	3.15	61.5	2.31	49.1	3.97	42.2	4.41	60.4	49.01
NASHVILLE WB AP	39.9	4.93	42.3	4.16	49.8	5.28	59.7	3.69	68.2	3.78	76.9	3.19	80.0	3.96	78.7	3.31	73.2	2.74	61.8	2.52	49.3	3.41	41.6	4.06	60.1	45.03
NASHVILLE		5.63		4.42		5.38		3.83		3.67		3.09		4.17		2.87		2.78		2.10		3.26		4.12		45.40
PALMETTO	41.6	5.63	43.1	5.68	50.5	5.47	59.0	4.00	67.7	3.99	76.2	3.56	78.7	4.98	78.0	3.22	72.3	2.80	61.5	2.64	48.9	3.84	42.2	4.58	60.0	50.39
PERRYVILLE		7.08		5.05		5.81		4.41		4.04		3.67		3.97		3.47		2.90		2.45		4.27		4.81		51.93
SAVANNAH	42.8	6.83	44.7	5.39	52.2	5.91	61.2	4.45	69.1	4.03	77.6	3.86	80.6	3.88	80.0	3.91	73.5	3.12	62.5	2.64	50.4	4.28	43.1	5.04	61.5	53.34
WAYNESBORO	40.2	6.27	41.6	5.74	49.5	6.06	58.4	4.93	66.0	4.07	74.7	4.04	77.5	4.60	76.9	3.91	70.5	3.08	59.5	2.70	47.4	4.54	40.6	5.12	58.6	55.06
DIVISION	41.0	6.16	42.7	5.07	50.1	5.70	59.3	4.19	67.6	3.92	76.3	3.65	79.1	4.22	78.3	3.44	72.0	3.04	61.1	2.56	48.7	3.89	41.5	4.64	59.8	50.48
WESTERN																										
BOLIVAR 2	41.8	6.92	44.0	5.10	51.6	5.67	60.8	4.64	68.5	3.70	77.0	3.88	80.0	4.40	79.2	3.59	72.3	3.70	61.6	2.76	49.4	4.54	42.5	4.72	60.7	53.62
BROWNSVILLE	42.0	6.64	44.3	4.78	52.1	5.74	61.4	4.47	69.6	4.19	78.1	3.78	80.7	4.12	79.7	2.93	72.9	3.62	63.0	2.59	50.7	4.44	43.1	4.76	61.5	52.06
COVINGTON	41.9	6.39	44.2	4.55	51.6	5.94	60.9	4.51	69.5	4.57	78.3	3.70	81.0	3.78	80.0	2.79	73.2	3.37	62.9	2.98	50.5	4.31	42.9	4.79	61.4	51.68
DRESDEN		5.86		4.36		5.59		4.22		4.31		3.77		3.57		2.97		3.22		2.83		3.87		4.59		49.16
JACKSON EXP STA	41.1	6.55	43.5	4.74	50.9	5.56	60.0	4.60	68.2	4.21	76.9	4.09	79.6	4.63	78.8	3.65	69.1	3.58	61.6	2.60	49.2	4.22	42.1	4.53	60.1	52.06
MC KENZIE		6.14		4.29		5.50		3.74		3.89		3.35		3.53		2.49		3.38		2.85		3.91		4.39		47.46
MEMPHIS WB AIRPORT	41.6	5.55	44.5	4.59	52.0	5.59	61.8	4.80	70.1	3.92	78.3	3.33	81.2	3.23	80.3	2.94	74.3	2.55	63.6	3.27	50.6	4.56	43.3	5.09	61.8	49.42
MEMPHIS WB CITY	41.9	5.38	44.5	4.22	52.4	5.20	62.2	4.70	70.5	3.65	78.7	3.25	81.3	3.10	80.3	2.53	74.8	2.50	65.0	3.18	52.1	4.31	44.4	4.79	62.4	46.91
MILAN	41.0	6.60	43.5	4.50	51.0	6.15	60.3	4.63	68.5	4.23	77.3	4.30	80.0	4.58	79.2	3.45	72.5	3.61	61.9	3.03	49.5	4.18	40.3	4.69	60.4	53.95
MOSCOW	42.6	6.18	44.8	5.44	52.4	5.76	61.2	4.53	69.2	4.11	77.3	4.21	80.3	4.01	79.6	3.51	72.9	3.08	62.4	2.78	50.4	4.38	43.3	5.13	61.4	53.12
NEWBERN	40.2	5.27	42.6	4.18	50.5	5.62	60.4	4.00	69.3	3.84	77.9	3.93	80.6	3.89	79.8	3.10	73.2	3.20	62.7	3.05	49.5	3.78	41.5	4.44	60.7	48.30
SAMBURG W L REFUGE	39.1	5.25	41.4	4.27	49.8	5.38	59.9	4.18	69.0	4.38	77.9	3.89	80.8	3.83	79.8	2.80	72.4	3.34	61.6	3.19	48.4	3.85	40.3	4.36	60.0	48.72
SELMER		6.37		5.67		5.45		4.50		3.80		4.17		3.78		3.44		3.16		2.70		4.25		4.87		52.16
UNION CITY	38.2	5.15	40.2	4.23	48.3	5.49	58.6	4.43	68.0	4.14	77.3	3.78	80.3	3.93	79.2	2.87	71.7	3.47	60.8	3.17	47.6	3.97	39.5	4.38	59.1	49.01
DIVISION	40.8	6.08	43.1	4.55	50.9	5.61	60.4	4.32	68.9	4.07	77.7	3.84	80.4	4.01	79.6	3.08	72.7	3.31	62.1	2.89	49.4	4.09	41.7	4.58	60.6	50.43

* Averages for period 1931-1955, except for stations marked WB which are "normals" based on period 1921-1950. Divisional means may not be the arithmetical average of individual stations published, since additional data from shorter period stations are used to obtain better areal representation.

CONFIDENCE LIMITS

In the absence of trend or record changes, the chances are 9 out of 10 that the true mean will lie in the interval formed by adding and subtracting the values in the following table from the means for any station in the State. Because of the wider variation in mean precipitation, the corresponding monthly means and annual mean must be substituted for "p" in the precipitation table below to obtain mean precipitation confidence limits.

1.8	.48 $\sqrt{p}$	1.6	.38 $\sqrt{p}$	1.7	.36 $\sqrt{p}$	.9	.35 $\sqrt{p}$	.9	.37 $\sqrt{p}$	.9	.40 $\sqrt{p}$	.7	.42 $\sqrt{p}$	.8	.38 $\sqrt{p}$	1.2	.37 $\sqrt{p}$	1.1	.44 $\sqrt{p}$	1.1	.37 $\sqrt{p}$	1.5	.39 $\sqrt{p}$	.3	.39 $\sqrt{p}$

COMPARATIVE DATA

Data in the following table are the mean temperature and average precipitation for Lewisburg Experiment Station, Tennessee for the period 1906 - 1930 and are included in this publication for comparative purposes :

40.7	4.94	42.3	4.37	50.5	5.81	59.2	4.73	66.8	4.77	75.2	4.37	78.4	4.11	77.6	4.61	72.9	3.09	60.3	3.18	49.3	3.63	41.2	5.20	59.5	52.81

NORMALS, MEANS AND EXTREMES
(Table Revised 1973. Base Period for Climatological Normals: 1931-1960)

Station: BRISTOL-JOHNSON CITY-KINGSPORT, TENN. TRI-CITY AIRPORT Standard time used: EASTERN Latitude: 36° 29' N Longitude: 82° 24' W Elevation (ground): 1507 feet

Month	Temperature Normal Daily maximum (b)	Daily minimum (b)	Monthly (b)	Extremes Record highest	year	Record lowest	year Ø	Normal heating degree days (Base 65°) (b)	Precipitation Normal total (b)	Max. monthly	year	Min. monthly	year	Max. in 24 hrs	year
J	46.8	29.7	38.3	78	1972	-15	1966	828	3.69	9.18	1957	1.85	1955	2.34	1950
F	50.0	30.0	40.0	76	1965	-14	1965	700	3.56	7.29	1956	0.75	1968	1.87	1954
M	56.5	35.5	46.0	81	1963	12	1963	598	3.98	9.56	1955	1.33	1957	3.10	1963
A	68.2	44.2	56.6	89	1970	21	1964	261	3.16	5.85	1970	1.38	1963	3.16	1956
M	77.1	53.7	65.7	92	1969+	30	1963	68	3.45	9.71	1950	1.31	1963	2.44	1958
J	84.5	61.9	73.2	95	1966+	38	1966	8	3.38	6.68	1957	1.14	1954	3.10	1954
J	86.4	65.4	75.9	95	1970	48	1967	0	5.55	9.73	1949	0.79	1957	2.90	1946
A	85.0	63.1	74.9	95	1965	48	1968	0	3.80	7.07	1966	0.82	1954	2.50	1963
S	80.6	56.6	69.1	92	1971+	34	1967	51	2.62	6.60	1972	0.07	1963	3.61	1972
O	70.7	45.3	58.0	85	1970	20	1962	236	2.15	5.65	1959	0.07	1963	2.55	1964
N	56.7	34.5	45.9	80	1971+	10	1970	573	2.51	5.90	1948	1.07	1963	3.55	1957
D	47.1	29.5	38.3	76	1966	-9	1962	828	3.21	6.75	1961	0.21	1965	2.95	1969
YR	67.5	46.1	56.8	95	JUL 1970+	-15	JAN 1966	4143	41.06	9.73	JUL 1949	0.07	OCT 1963	3.65	OCT 1964

Month	Snow, Ice pellets Mean total	Max. monthly	year	Max. in 24 hrs	year	Relative humidity Hour 01	07	13	19	Wind Mean speed	Prevailing direction	Fastest mile Speed	Direction	Year	Mean sky cover sunrise to sunset
J	4.9	22.1	1955	9.7	1955	76	80	61	65	6.9	WSW	40	25	1965	7.1
F	4.5	17.2	1947	10.7	1969	78	77	56	60	7.1	WSW	46	25	1961	6.9
M	2.8	27.9	1960	13.0	1960	70	77	51	54	7.1	NNW	46	25	1952	6.7
A	T	T	1962	0.6	1962	70	77	51	57	6.9	NW	40	25	1970	6.7
M	T	0.0		T	1963	83	81	55	64	5.3	SW	50	32	1970	6.0
J	0.0	0.0		0.0		87	87	61	68	4.1	WSW	30	31	1973	6.0
J	0.0	0.0		0.0		88	91	68	70	3.8	SW	40	23	1961	6.4
A	0.0	0.0		0.0		88	92	65	70	3.6	E	29	34	1962	5.4
S	0.0	0.0		0.0		87	91	63	65	4.3	NE	31	31	1967	5.4
O	T	T	1968	T	1968+	82	84	57	63	4.7	NE	28	28	1965	5.0
N	1.4	18.1	1952	16.2	1952	78	81	61	68	5.9	NE	35	29	1965	6.9
D	2.9	12.9	1963	9.6	1969	78	81	59	68	6.0	WSW	40	24	1968+	6.9
YR	16.5	27.9	MAR 1960	16.2	NOV 1952	80	85	57	63	5.5	WSW	50	32	MAY 1951	6.2

Month	Mean number of days Clear	Partly cloudy	Cloudy	Precip. .01 in or more	Snow 1.0 in or more	Thunderstorms	Heavy fog	Temp. Max 90° and above	Max 32° and below	Min 32° and below	Min 0° and below
J	6	7	18	14	2	*	3	0	5	24	1
F	6	6	16	14	1	1	3	0	3	20	*
M	7	7	16	14	1	2	1	0	*	15	0
A	7	7	14	11	0	3	2	0	0	4	0
M	7	11	13	11	0	7	3	1	0	*	0
J	6	11	13	11	0	9	4	4	0	0	0
J	5	13	13	12	0	10	5	5	0	0	0
A	7	14	10	11	0	8	7	3	0	0	0
S	11	9	10	8	0	4	5	3	0	0	0
O	13	11	8	8	0	1	5	0	0	*	0
N	7	7	14	11	1	1	3	0	*	13	0
D	7	7	17	11	1	*	3	0	3	20	*
YR	91	111	163	134	5	46	43	15	12	100	2

Ø For period May 1961 through the current year.
Means and extremes above are from existing and comparable exposures. Annual extremes have been exceeded at other sites in the locality as follows: Highest temperature 102 in July 1952.

NORMALS, MEANS AND EXTREMES
(Table Revised 1975. Base Period for Climatological Normals: 1941-1970)

Month	Temperatures °F Normal Daily maximum (b)	Daily minimum (b)	Monthly (b)	Extremes Record highest	year	Record lowest	year	Normal Degree days Base 65°F Heating	Cooling	Precipitation in inches Normal	Max. monthly	year	Min. monthly	year	Max. in 24 hrs	year
J	46.0	26.7	36.4	78	1972	-15	1966	887	0	3.62	9.18	1957	1.80	1973	2.34	1950
F	48.7	28.4	38.7	76	1965	-4	1965	736	0	3.71	7.29	1956	0.75	1968	1.87	1954
M	57.1	34.5	45.8	86	1973	12	1964	604	9	3.90	9.56	1955	1.33	1957	3.35	1973
A	69.4	44.2	56.8	89	1970	21	1964	270	9	3.30	5.85	1970	1.38	1963	3.16	1956
M	77.1	52.6	64.9	92	1970	30	1963	90	87	3.29	9.71	1950	1.31	1963	2.32	1996
J	84.0	60.7	72.4	95	1969+	38	1966	8	230	3.52	6.68	1957	1.34	1964	3.10	1954
J	85.9	64.4	75.2	95	1970	48	1967	0	316	4.98	9.73	1949	0.79	1957	2.90	1946
A	85.0	63.1	74.0	95	1965	48	1968	0	285	3.72	7.07	1966	0.82	1954	2.50	1963
S	80.4	56.6	68.5	93	1973+	34	1967	37	142	2.88	6.60	1972	0.93	1972	3.61	1972
O	70.1	45.3	57.7	85	1970	20	1962	249	29	2.25	5.65	1959	1.07	1963	2.55	1964
N	56.7	34.5	45.7	81	1974	10	1970	579	0	2.78	5.90	1948	1.07	1953	3.55	1957
D	47.3	28.0	37.7	76	1971	-9	1962	846	0	3.51	6.75	1961	0.21	1965	2.95	1969
YR	67.3	44.9	56.1	95	JUL 1970	-15	JAN 1966	4306	1107	41.47	9.73	JUL 1949	0.07	OCT 1963	3.65	OCT 1964

Month	Snow, Ice pellets Water equivalent Max. monthly	year	Max. in 24 hrs	year	Snow Ice pellets Max. monthly	year	Max. in 24 hrs	year	Relative humidity pct. Hour 01	07	13	19	Wind Mean speed m.p.h.	Prevailing direction	Fastest mile Speed m.p.h.	Direction	Year	Mean sky cover, tenths, sunrise to sunset
J	22.1	1955	9.7	1955	22.1	1966	9.7	1955	77	80	62	60	6.4	WSW	40	25	1965	7.1
F	17.2	1947	10.7	1969	17.2	1947	10.7	1969	72	77	58	55	7.0	WSW	46	25	1961	6.8
M	27.9	1960	13.0	1960	27.9	1960	13.0	1960	72	78	52	51	7.2	WNW	40	25	1952	6.8
A	T	1962	0.6	1962	T	1962	0.6	1962	72	81	51	52	6.8	WSW	40	25	1970	6.6
M	0.0		T	1963	0.0		T	1963	81	90	52	65	5.4	NE	50	32	1970	6.2
J	0.0		0.0		0.0		0.0		87	93	55	68	4.6	WSW	35	26	1973	6.4
J	0.0		0.0		0.0		0.0		89	93	61	68	4.1	WSW	40	23	1961	6.4
A	0.0		0.0		0.0		0.0		88	91	58	71	3.8	NE	29	34	1962	5.5
S	0.0		0.0		0.0		0.0		88	93	56	69	4.6	NE	31	31	1967	5.4
O	T	1968	T	1968	T	1968	T	1968	87	84	56	65	4.6	NE	28	28	1965	4.9
N	18.1	1952	16.2	1952	18.1	1952	16.2	1952	81	78	63	69	5.9	NE	35	29	1965	6.2
D	12.9	1963	9.6	1969	12.9	1963	9.6	1969	78	84	64	69	6.0	WSW	40	24	1968	7.0
YR	27.9	MAR 1960	16.2	NOV 1952	27.9	MAR 1960	16.2	NOV 1952	81	85	57	64	5.6	WSW	50	32	MAY 1951	6.3

Month	Pct. of possible sunshine	Mean number of days Sunrise to sunset Clear	Partly cloudy	Cloudy	Precip. .01 in or more	Snow 1.0 in or more	Thunderstorms	Heavy fog, visibility ¼ mile or less	Temp. °F Max 90° and above	Max 32° and below	Min 32° and below	Min 0° and below	Average station pressure mb. Elev. 1525 feet m.s.l.
J		6	7	18	14	2	*	3	0	5	22	1	965.3
F		7	7	15	12	1	1	4	0	3	20	*	963.1
M		7	7	17	14	1	2	1	0	*	14	0	961.4
A		7	8	14	11	0	3	1	0	0	3	0	961.6
M		6	11	13	11	0	7	3	1	0	*	0	962.9
J		5	13	13	12	0	9	5	4	0	0	0	962.8
J		5	13	13	11	0	10	5	4	0	0	0	964.1
A		7	14	10	10	0	8	5	3	0	0	0	965.0
S		10	12	8	8	0	4	5	3	0	0	0	965.2
O		13	11	8	8	0	1	5	0	0	*	0	965.4
N		6	7	14	10	*	*	3	0	*	13	0	965.3
D		7	7	17	12	1	*	3	0	3	20	*	963.1
YR		90	112	163	134	5	46	44	13	12	96	2	963.8

Means and extremes above are from existing and comparable exposures. Annual extremes have been exceeded at other sites in the locality as follows: Highest temperature 102 in July 1952.

REFERENCE NOTES APPLYING TO TABLES APPEAR ON THE PAGE FOLLOWING LAST TABLE.
(Caution: Letters and symbols may have different meanings in 1941-1970 tables than in earlier tables. See notes.)

NORMALS, MEANS AND EXTREMES

(Table Revised 1973. Base Period for Climatological Normals: 1931-1960)

Station: CHATTANOOGA, TENNESSEE   LOVELL FIELD   Standard time used: EASTERN   Latitude: 35° 02' N   Longitude: 85° 12' W   Elevation (ground): 665 feet

Means and extremes above are from existing and comparable exposures. Annual extremes have been exceeded at other sites in the locality as follows:
Maximum monthly precipitation 15.29 in April 1911; minimum monthly precipitation .04 in September 1919; maximum precipitation in 24 hours 7.61 in
March 1886; maximum monthly snowfall 15.8 in January 1893; maximum snowfall in 24 hours 12.0 in December 1886.

NORMALS, MEANS AND EXTREMES

(Table Revised 1975. Base Period for Climatological Normals: 1941-1970)

Means and extremes above are from existing and comparable exposures. Annual extremes have been exceeded at other sites in the locality as follows:
Maximum monthly precipitation 15.29 in April 1911; minimum monthly precipitation .04 in September 1919; maximum precipitation in 24 hours 7.61 in
March 1886; maximum monthly snowfall 15.8 in January 1893; maximum snowfall in 24 hours 12.0 in December 1886.

REFERENCE NOTES APPLYING TO TABLES APPEAR ON THE PAGE FOLLOWING LAST TABLE.
(Caution: Letters and symbols may have different meanings in 1941-1970 tables than in earlier tables. See notes.)

917

NORMALS, MEANS AND EXTREMES
(Table Revised 1973. Base Period for Climatological Normals: 1931-1960)

Station: KNOXVILLE, TENNESSEE   MC GHEE TYSON AIRPORT   Standard time used: EASTERN   Latitude: 35°49'N   Longitude: 83°59'W   Elevation (ground): 980 feet

Month	Normal Daily maximum	Normal Daily minimum	Normal Monthly	Extremes Record highest	Year	Extremes Record lowest	Year	Normal heating degree days (Base 65°)	Precipitation Normal total	Max monthly	Year	Min monthly	Year	Max in 24 hrs	Year	Snow Mean total	Snow Max monthly	Year	Max in 24 hrs	Year
J	50.1	32.6	41.4	72	1972+	-9	1966	732	4.88	11.74	1954	1.63	1955	3.89	1946	3.9	15.1	1962	12.0	1962
F	52.9	33.4	43.1	75	1972	0	1970	613	4.81	9.38	1944	0.74	1968	2.89	1944	3.3	23.3	1960	17.5	1960
M	59.5	39.2	49.6	86	1963	18	1969+	493	4.73	9.92	1963	2.26	1957	4.82	1963	2.0	7.0	1960	7.0	1960
A	70.9	48.9	58.9	89	1965	27	1964	198	3.70	9.92	1964	0.84	1957	2.32	1957	T	2.0	1971	T	1971
M	78.6	56.4	67.7	94	1964	34	1963	43	3.50	6.40	1960	0.74	1946	2.35	1953	0.0	0.0		0.0	
J	86.1	64.9	75.5	96	1964	44	1972	0	3.33	7.58	1969	0.20	1969	3.57	1972	0.0	0.0		0.0	
J	88.5	68.2	78.4	98	1966	51	1961	0	4.82	10.09	1967	0.70	1957	4.69	1942	0.0	0.0		0.0	
A	87.1	67.2	77.4	99	1968	53	1968+	0	3.46	8.88	1942	0.77	1954	3.25	1959	0.0	0.0		0.0	
S	83.8	60.6	72.2	96	1962	36	1967	30	2.54	8.61	1944	0.50	1961	5.08	1944	0.0	0.0		0.0	
O	73.0	48.8	60.9	89	1962	25	1962	171	2.61	6.67	1949	T	1963	2.44	1961	T	T	1962	T	1962
N	59.2	38.1	48.7	83	1961	21	1970	489	3.24	6.35	1948	0.97	1942	4.06	1948	1.0	18.2	1952	18.2	1952
D	50.3	32.8	41.6	75	1970	-2	1962	725	4.23	11.63	1961	0.45	1965	4.89	1961	2.1	12.2	1963	8.9	1963
YR	70.0	49.2	59.6	99 AUG. 1968		-9 JAN. 1966		3494	45.85	11.74 JAN. 1954		T	OCT. 1963	5.08 SEP. 1944		13.0	23.3 FEB. 1960		18.2 NOV. 1952	

Means and extremes above are from existing and comparable exposures. Annual extremes have been exceeded at other sites in the locality as follows: Highest temperature 104 in July 1930; lowest temperature -16 in January 1884; maximum monthly precipitation 17.32 in April 1874; maximum precipitation in 24 hours 6.30 in July 1917; maximum monthly snowfall 25.7 in February 1895.

NORMALS, MEANS AND EXTREMES
(Table Revised 1975. Base Period for Climatological Normals: 1941-1970)

Average station pressure: Elev. 980 feet m.s.l. — 982.9 mb

Month	Normal Daily maximum	Normal Daily minimum	Normal Monthly	Extremes Record highest	Year	Extremes Record lowest	Year	Normal Degree days Heating	Cooling	Precipitation Normal	Max monthly	Year	Min monthly	Year	Max in 24 hrs	Year	Snow Max monthly	Year	Max in 24 hrs	Year
J	48.9	32.2	40.6	72	1972+	-9	1966	756	0	4.67	11.74	1954	1.63	1955	3.89	1946	15.1	1962	12.0	1962
F	52.0	33.5	42.8	75	1972	0	1970	630	8	4.71	9.38	1944	0.74	1968	2.89	1956	23.3	1960	17.5	1960
M	59.4	39.4	49.4	86	1963	18	1969+	484	16	4.86	10.24	1973	2.26	1957	4.85	1957	20.2	1960	7.0	1960
A	72.0	48.6	60.3	89	1965	27	1964	173	64	3.61	7.20	1970	0.84	1970	2.92	1957	T	1971	T	1971
M	79.8	56.6	68.4	94	1964	34	1963	47	152	3.56	10.98	1974	0.74	1946	3.57	1974	0.0		0.0	
J	86.1	64.9	75.5	96	1964	44	1972	0	315	3.63	7.58	1969	0.20	1969	3.57	1972	0.0		0.0	
J	88.0	68.3	78.2	98	1966	51	1961	0	409	4.70	10.09	1967	0.70	1957	4.69	1942	0.0		0.0	
A	87.3	67.3	77.3	99	1968	53	1968+	0	381	3.24	8.88	1942	0.77	1954	3.25	1959	0.0		0.0	
S	82.0	61.2	71.6	96	1962	36	1967	10	208	2.78	8.61	1944	0.50	1961	5.08	1944	0.0		0.0	
O	71.8	50.0	60.9	89	1962	25	1962	175	48	2.67	6.67	1949	T	1963	2.44	1961	T	1962	T	1962
N	58.9	39.4	49.2	83	1961	21	1970	474	0	3.56	6.35	1948	0.97	1942	4.06	1948	18.2	1952	18.2	1952
D	49.8	33.1	41.5	75	1970	-2	1962	729	0	4.47	11.63	1961	0.45	1965	4.89	1961	12.2	1963	8.9	1963
YR	69.8	49.5	59.7	99 AUG. 1968		-9 JAN. 1966		3478	1569	46.18	11.74 JAN. 1954		T OCT. 1963		5.08 SEP. 1944		23.3 FEB. 1960		18.2 NOV. 1952	

Means and extremes above are from existing and comparable exposures. Annual extremes have been exceeded at other sites in the locality as follows: Highest temperature 104 in July 1930; lowest temperature -16 in January 1884; maximum monthly precipitation 17.32 in April 1874; maximum precipitation in 24 hours 6.30 in July 1917; maximum monthly snowfall 25.7 in February 1895.

REFERENCE NOTES APPLYING TO TABLES APPEAR ON THE PAGE FOLLOWING LAST TABLE.
(Caution: Letters and symbols may have different meanings in 1941-1970 tables than in earlier tables. See notes.)

NORMALS, MEANS AND EXTREMES
(Table Revised 1973. Base Period for Climatological Normals: 1931-1960)

Station: MEMPHIS, TENNESSEE   INTERNATIONAL AIRPORT   Standard time used: CENTRAL   Latitude: 35° 03' N   Longitude: 89° 59' W   Elevation (ground): 258 feet

Month	Temperature Normal Daily maximum	Normal Daily minimum	Normal Monthly	Extremes Record highest	Year	Record lowest	Year	Precipitation Normal total	Normal heating degree days (Base 65°)	Max monthly	Year	Min monthly	Year	Max in 24 hrs	Year	Snow, Ice pellets Mean total	Max monthly	Year	Max 24 hrs	Year	Relative humidity Hour 00	06	12	18	Mean speed	Prevailing direction	Fastest mile Speed	Dir	Year	Pct. possible sunshine	Mean sky cover	Mean number of days
J	50.6	32.4	41.5	78	1972+	-4	1962	6.07	729	12.21	1951	0.84	1951	3.87	1954	2.1	12.2	1966	6.5	1960	76	76	71	68	10.6	S	43	NW	1956	48	6.9
F	53.9	34.3	44.1	81	1962	-11	1951	4.69	585	9.99	1956	1.98	1968	3.63	1966	1.4	7.3	1960	5.8	1960	74	79	60	60	10.7	S	49	SW	1951	54	6.3
M	61.4	40.7	51.1	85	1963+	-12	1943	5.07	456	8.91	1953	1.50	1953	3.59	1955	1.4	17.3	1968	16.1	1968	71	78	56	55	11.3	S	57	W	1952	63	6.1
A	72.1	50.7	61.4	91	1951	29	1944	4.63	147	12.29	1955	2.05	1965	3.50	1955	T	0.0		T	1971	71	78	54	54	10.9	S	50	N	1953	69	5.8
M	80.1	60.4	70.3	97	1953	38	1944	4.23	22	11.58	1953	0.84	1962	4.94	1958	0.0	0.0		0.0		76	82	55	56	8.9	S	50	NW	1951	70	5.4
J	88.3	68.6	78.5	104	1954	48	1966	3.68	0	6.88	1951	0.04	1953	2.91	1968	0.0	0.0		0.0		79	83	55	57	8.1	S	56	NW	1964	73	5.0
J	91.1	71.5	81.3	106	1952	52	1947	3.54	0	8.84	1959	0.43	1954	3.31	1959	0.0	0.0		0.0		80	84	57	59	7.6	S	56	NW	1967+	75	5.6
A	90.7	70.3	80.5	105	1943	48	1946	2.97	0	7.84	1960	0.43	1953	3.35	1969	0.0	0.0		0.0		80	86	55	60	7.1	S	42	NE	1967+	76	5.0
S	85.6	62.1	73.9	103	1954	36	1949+	2.82	18	9.37	1958	T	1953	4.63	1957	0.0	0.0		0.0		78	84	56	62	7.5	S	47	SE	1968	73	4.5
O	75.7	50.5	63.1	95	1954+	25	1952	2.72	80	6.20	1970	0.19	1963	2.47	1968	T	0.8	1971	0.8	1971	78	81	51	55	7.8	S	47	W	1955	71	4.5
N	61.5	38.7	50.1	85	1955	-9	1950	4.38	447	8.56	1972	0.75	1965	3.64	1972	0.3	0.8	1963	0.8	1963	76	80	61	64	9.4	S	43	S	1952+	58	4.5
D	52.5	32.5	42.5	79	1951	-13	1963	4.93	698	9.37	1951	1.05	1955	4.15	1964	1.1	14.3	1963	14.3	1963	76	80	62	69	10.0	S	43	N	1954	49	6.4
YR	72.0	51.1	61.5	106	JUL 1952	-13	DEC 1963	49.73	3232	12.29	DEC 1963	T	OCT 1963	4.94	MAY 1958	6.0	17.3	MAR 1968	16.1	MAR 1968	76	81	56	61	9.2	S	57	W	APR 1955	65	5.7

Means and extremes above are from existing and comparable exposures. Annual extremes have been exceeded at other sites in the locality as follows:
Maximum monthly precipitation 18.16 in June 1877; minimum monthly precipitation 0.00 in September 1897; maximum precipitation in 24 hours 10.48 in
November 1934; maximum snowfall 18.5 in March 1892; maximum snowfall in 24 hours 18.0 in March 1892.

NORMALS, MEANS AND EXTREMES
(Table Revised 1975. Base Period for Climatological Normals: 1941-1970)

Average station pressure mb.   Elev. 284 feet m.s.l.

Month	Temperatures °F Normal Daily maximum	Normal Daily minimum	Normal Monthly	Extremes Record highest	Year	Record lowest	Year	Degree days Base 65°F Heating	Cooling	Precipitation in inches Water equivalent Normal	Max monthly	Year	Min monthly	Year	Max 24 hrs	Year	Snow, Ice pellets Max monthly	Year	Max 24 hrs	Year	Relative humidity pct. Hour 00	06	12	18	Wind Mean speed m.p.h.	Prevailing direction	Fastest mile Speed m.p.h.	Direction	Year	Pct. possible sunshine	Mean sky cover	Mean number of days
J	49.4	31.6	40.5	78	1972	-4	1962	760	0	4.93	12.21	1951	0.84	1961	3.89	1974	12.2	1966	6.5	1960	76	80	64	69	10.5	S	43	NW	1956	48	6.9
F	51.1	34.4	43.8	81	1962	-11	1951	594	0	4.73	9.39	1956	1.98	1968	3.63	1966	7.6	1960	5.8	1960	74	79	60	63	10.6	S	49	SW	1951	54	6.3
M	60.8	41.5	51.1	85	1963	-12	1943	457	23	5.10	8.91	1953	1.50	1953	3.59	1955	17.3	1968	16.1	1968	71	78	55	58	11.3	S	57	W	1952	57	6.4
A	72.7	52.3	62.5	91	1951	29	1944	131	56	5.42	12.29	1955	2.05	1965	3.50	1955	T	1971	T	1971	72	79	54	55	10.9	S	50	N	1953	69	5.8
M	81.2	60.6	70.9	97	1953	38	1944	0	205	4.39	11.58	1953	0.84	1962	4.94	1958	0.0		0.0		76	82	55	56	8.9	S	50	NW	1951	73	5.4
J	88.7	68.5	78.6	104	1954	48	1966	0	408	3.46	6.88	1951	0.04	1953	2.91	1968	0.0		0.0		79	83	56	57	8.1	S	50	NW	1951	73	5.4
J	91.6	71.5	81.6	106	1952	52	1947	0	515	3.53	8.84	1959	0.43	1954	3.31	1959	0.0		0.0		80	85	57	59	7.6	S	56	NW	1964	75	5.6
A	90.6	70.1	80.4	105	1943	48	1946	0	477	3.33	7.84	1960	0.43	1953	3.35	1969	0.0		0.0		80	86	56	63	7.1	S	42	NE	1967	75	5.1
S	84.3	62.1	73.6	103	1954	36	1957	80	265	3.01	9.37	1958	0.19	1963	4.63	1957	0.0		0.0		78	84	51	56	7.5	S	47	SE	1968	69	5.1
O	74.9	50.9	62.9	95	1954	25	1952	142	19	2.58	6.20	1970	0.19	1972	2.47	1968	0.8	1971	0.8	1971	78	81	55	56	7.8	S	47	W	1955	71	4.5
N	61.5	40.3	50.9	85	1955	-9	1950	423	0	3.92	8.56	1972	0.75	1965	3.64	1972	0.8	1963	0.8	1963	75	80	55	62	9.4	S	43	S	1952	58	4.5
D	51.7	33.7	42.7	79	1951	-13	1963	691	0	4.70	9.37	1951	1.05	1964	4.15	1964	14.3	1963	14.3	1963	76	80	62	69	10.1	S	43	N	1954	49	6.4
YR	71.7	51.5	61.6	106	JUL 1952	-13	DEC 1963	3227	2029	49.10	12.29	APR 1955	T	OCT 1963	4.94	APR 1955	17.3	MAR 1968	16.1	MAR 1968	77	82	57	61	9.2	S	57	W	APR 1970	64	5.7

Means and extremes above are from existing and comparable exposures. Annual extremes have been exceeded at other sites in the locality as follows:
Maximum monthly precipitation 18.16 in June 1877; minimum monthly precipitation 0.00 in September 1897; maximum precipitation in 24 hours 10.48 in
November 1934; maximum snowfall 18.5 in March 1892; maximum snowfall in 24 hours 18.0 in March 1892.

Average station pressure: 1010.4, 1010.5, 1009.9, 1005.6, 1005.1, 1004.4, 1005.8, 1007.0, 1007.5, 1010.5, 1009.3, 1008.3, 1007.1 mb.

REFERENCE NOTES APPLYING TO TABLES APPEAR ON THE PAGE FOLLOWING LAST TABLE.
(Caution: Letters and symbols may have different meanings in 1941-1970 tables than in earlier tables. See notes.)

NORMALS, MEANS AND EXTREMES

(Table Revised 1973. Base Period for Climatological Normals: 1931-1960)

Station: NASHVILLE, TENNESSEE   METROPOLITAN AIRPORT   Standard time used: CENTRAL   Latitude: 36° 07′ N   Longitude: 86° 41′ W   Elevation (ground): 590 feet

Month	Normal Daily maximum	Normal Daily minimum	Normal Monthly	Record highest	Year	Record lowest	Year	Normal heating degree days (Base 65°)	Precip. Normal total	Max. monthly	Year	Min. monthly	Year	Max. in 24 hrs	Year	Snow Mean total	Max. monthly	Year	Max. in 24 hrs	Year	Mean sky cover	Pct. possible sunshine	Fastest mile Speed	Dir.	Year
J	48.8	30.9	39.9	78	1972	-6	1966	778	5.49	13.92	1950	1.16	1970	4.40	1946	3.7	18.8	1948	7.5	1966	7.1	40	56	SW	1949+
F	51.4	32.5	42.0	79	1972	-5	1971	644	4.51	10.31	1960	0.64	1968	4.04	1948	3.4	16.0	1960	7.4	1960	6.6	45	52	NW	1960
M	59.2	38.7	49.1	86	1967	14	1968	512	5.19	10.03	1963	1.39	1966	4.65	1955	2.0	11.1	1960	8.8	1951	6.5	52	70	NW	1953
A	70.8	48.4	59.6	88	1965	28	1965	189	3.74	6.91	1962	1.15	1948	3.57	1944	0.1	1.1	1971	1.1	1971	6.5	59	61	SW	1958
M	79.8	57.4	68.6	91	1970	35	1971	40	3.72	8.23	1957	0.78	1962	3.57	1957	0.0	0.0		0.0		6.3	63	43	SW	1961
J	88.4	66.3	77.4	98	1952	42	1960	0	3.25	9.37	1960	0.78	1952	4.91	1960	0.0	0.0		0.0		5.9	66	73	NW	1953
J	90.7	69.6	80.2	103	1966	54	1954	0	3.72	7.75	1950	0.71	1954	3.56	1950	0.0	0.0		0.0		5.7	64	52	NW	1956
A	89.5	68.4	79.2	99	1968	51	1967	0	2.86	8.31	1942	0.69	1968	5.34	1963	0.0	0.0		0.0		5.3	64	45	NW	1955
S	84.5	61.5	72.8	99	1962	37	1962	30	2.87	8.03	1962	0.28	1956	5.09	1962	0.0	0.0		0.0		4.8	63	42	NE	1952
O	73.7	49.3	61.5	90	1968	27	1968	158	2.32	6.13	1959	T	1963	2.39	1968	T	0.7	1950	2.1	1950	4.0	61	38	SE	1947
N	59.1	37.9	48.5	84	1951	12	1959	495	3.28	9.04	1945	0.54	1949	3.61	1949	0.7	9.2	1950	9.2	1950	5.6	51	43	S	1947
D	50.3	32.4	41.4	74	1966	-6	1966	732	4.19	10.66	1951	1.01	1965	3.88	1951	2.1	13.2	1963	10.2	1963	6.8	46	47	W	1946
YR	70.6	49.4	60.0	103	JUL 1966	-6	JAN 1966	3578	45.14	13.92	JAN 1950	T	OCT 1963	5.34	AUG 1963	11.4	18.8	JAN 1948	10.2	DEC 1963	6.0	58	73	NW	JUN 1953

ø For period March 1965 through current year.

Means and extremes above are from existing and comparable exposures. Annual extremes have been exceeded at other sites in the locality as follows: Highest temperature 107 in July 1952; lowest temperature -15 in January 1937; maximum monthly precipitation 14.75 in January 1937; maximum precipitation in 24 hours 6.05 in November 1900; maximum snowfall 21.5 in March 1892; maximum snowfall in 24 hours 17.0 in March 1892.

NORMALS, MEANS AND EXTREMES

(Table Revised 1975. Base Period for Climatological Normals: 1941-1970)

Average station pressure: 998.6 mb.   Elev. 605 feet m.s.l.

Month	Normal Daily maximum	Normal Daily minimum	Normal Monthly	Record highest	Year	Record lowest	Year	Normal Heating Degree days	Normal Cooling Degree days	Water equiv. Normal	Max. monthly	Year	Min. monthly	Year	Max. in 24 hrs	Year	Snow Max. monthly	Year	Max. in 24 hrs	Year	Mean sky cover	Pct. possible sunshine	Fastest mile Speed	Dir.	Year
J	47.6	29.0	38.3	78	1972	-6	1966	828	0	4.75	13.92	1950	1.16	1970	4.40	1946	18.8	1948	7.5	1966	7.2	40	56	SW	1949
F	50.9	31.0	41.0	78	1972	-5	1971	672	0	4.43	10.31	1956	0.64	1968	4.04	1948	16.0	1960	7.4	1951	6.6	45	52	NW	1960
M	58.1	38.1	48.7	86	1967	14	1968	524	19	5.00	10.03	1955	1.39	1966	4.65	1955	11.1	1971	8.8	1971	6.2	52	70	NW	1953
A	71.3	48.8	60.1	88	1965	24	1973	176	29	4.11	7.00	1973	1.15	1948	3.25	1962	0.0		0.0		6.2	59	61	SW	1958
M	80.1	57.3	68.5	98	1974	35	1971	45	153	4.10	8.23	1957	0.78	1962	3.57	1944	0.0		0.0		5.9	62	43	SW	1961
J	87.5	65.7	76.6	98	1971	42	1966	0	348	3.38	9.37	1960	0.78	1952	4.91	1960	0.0		0.0		5.6	67	73	NW	1953
J	90.2	69.0	79.6	103	1966	54	1972	0	453	3.83	7.75	1950	0.71	1954	3.56	1950	0.0		0.0		5.7	66	52	NW	1956
A	89.2	67.7	78.5	99	1968	51	1967	0	419	3.24	8.31	1942	0.69	1968	5.34	1963	0.0		0.0		5.3	66	45	NW	1952
S	83.2	60.5	72.0	95	1973	37	1967	100	220	3.09	10.44	1974	0.28	1956	5.09	1962	0.0		0.0		4.7	63	42	NE	1952
O	73.2	48.4	60.8	90	1971	27	1968	180	16	2.16	6.13	1955	T	1963	2.39	1968	T	1954	T	1954	4.7	62	38	SE	1947
N	59.0	37.7	48.4	84	1971	12	1970	498	0	3.46	9.04	1945	0.54	1949	3.74	1973	9.2	1950	9.2	1950	5.4	50	43	W	1947
D	49.6	31.1	40.4	74	1971	-6	1966	763	0	4.45	10.66	1951	1.01	1965	3.88	1951	13.2	1963	10.2	1963	6.8	40	47	NW	1946
YR	70.1	47.8	59.4	103	JUL 1966	-6	JAN 1966	3696	1694	46.00	13.92	JAN 1950	T	OCT 1963	5.34	AUG 1963	18.8	JAN 1948	10.2	DEC 1963	6.0	57	73	NW	JUN 1953

ø For period March 1965 through current year.

Means and extremes above are from existing and comparable exposures. Annual extremes have been exceeded at other sites in the locality as follows: Highest temperature 107 in July 1952; lowest temperature -15 in January 1937; maximum monthly precipitation 14.75 in January 1937; maximum precipitation in 24 hours 6.05 in November 1900; maximum snowfall 21.5 in March 1892; maximum snowfall in 24 hours 17.0 in March 1892.

REFERENCE NOTES APPLYING TO TABLES APPEAR ON THE PAGE FOLLOWING LAST TABLE.

(Caution: Letters and symbols may have different meanings in 1941-1970 tables than in earlier tables. See notes.)

NORMALS, MEANS AND EXTREMES
(Table Revised 1973. Base Period for Climatological Normals: 1931-1960)

Station: OAK RIDGE, TENNESSEE WEATHER SERVICE OFFICE Standard time used: EASTERN Latitude: 36° 01' N Longitude: 84° 14' W Elevation (ground): 905 feet

Month	Normal Daily max	Normal Daily min	Normal Monthly	Extreme Record highest	Year	Extreme Record lowest	Year	Normal heating degree days (Base 65°)	Precip Normal total	Precip Max monthly	Year	Precip Min monthly	Year	Precip Max in 24 hrs	Year	Snow Mean total	Snow Max monthly	Year	Snow Max 24 hrs	Year
J	49.2	30.6	39.9	75	1952	-9	1966	778	5.94	13.27	1954	1.86	1961	4.25	1954	3.0	9.6	1966	8.3	1966
F	51.0	31.2	41.1	76	1972	1	1965+	669	5.80	10.47	1956	0.84	1968	2.94	1954	3.5	11.3	1960	9.1	1960
M	58.4	36.6	47.5	85	1963	8	1960	552	5.59	9.92	1965	1.39	1957	4.40	1965	1.5	21.3	1960	12.0	1960
A	69.6	45.9	57.8	91	1970	24	1950	228	4.20	9.71	1956	1.80	1950	3.74	1956	0.0	0.3	1971	0.3	1971
M	78.6	54.4	66.5	93	1962	31	1963	56	3.73	7.17	1971	0.67	1971	3.99	1971	0.0	0.0		0.0	
J	85.8	63.2	74.5	101	1954	40	1972	0	3.24	8.09	1962	0.86	1962	3.70	1959	0.0	0.0		0.0	
J	87.7	66.7	77.1	105	1952	50	1967	0	4.95	9.27	1967	1.55	1970	4.91	1967	0.0	0.0		0.0	
A	86.8	65.3	76.1	102	1954	51	1965+	0	4.62	10.46	1960	0.54	1953	7.48	1990	0.0	0.0		0.0	
S	83.2	58.8	71.0	100	1954	33	1967	39	3.30	9.10	1957	0.41	1961	3.43	1972+	0.0	0.0		0.0	
O	72.5	47.3	59.9	90	1961	21	1952	192	2.67	6.95	1972+	T	1963	2.13	1957	T	T		T	
N	58.7	35.9	47.3	83	1951	0	1950	531	3.77	12.22	1948	1.37	1949	2.97	1959	0.5	6.5	1962	6.5	1962
D	49.5	30.7	40.1	74	1951	-3	1962	772	5.90	10.31	1956	0.67	1956	5.12	1959	2.3	14.8	1963	10.8	1963
YR	69.3	47.2	58.2	105	JUL 1952	-9	JAN 1966	3817	54.71	19.27	JUL 1967	T	OCT 1963	7.48	AUG 1950	10.3	21.0	MAR 1960	12.0	MAR 1960

Month	Mean sky cover	Sunrise to sunset Clear	Partly cloudy	Cloudy	Precip .01 in+	Snow 1.0 in+	Thunderstorms	Heavy fog	Max 90°+	Max 32° below	Min 32° below	Min 0° below	Wind Mean speed	Prevailing direction	Peak gust Speed	Direction	Year
J	6.7	7	6	18	13	1	1	1	0	2	19	*	4.8	SW	59		1959
F	6.6	7	5	16	13	1	2	1	0	1	16	0	5.0	SW	47		1957
M	6.5	8	6	17	13	1	3	1	0	*	12	0	5.3	ENE	48		1952
A	6.1	9	6	15	11	1	5	1	*	0	2	0	5.7	SW	50		1959
M	5.7	9	8	14	10	*	8	2	2	0	*	0	4.5	SW	45		1972
J	5.7	9	10	10	10	0	9	2	7	0	0	0	4.2	SW	49		1953
J	6.1	8	11	12	12	0	11	3	11	0	0	0	3.9	SW	50		1961
A	5.5	11	8	11	10	0	9	4	9	0	0	0	3.7	SW	53		1954
S	5.0	14	6	11	8	0	3	8	3	0	0	0	3.6	E	39		1959
O	5.0	14	7	11	8	0	1	6	*	0	1	0	3.6	E	39		1967
N	6.9	9	6	14	10	0	1	2	0	*	12	*	4.1	E	45		1968
D	6.9	7	6	18	11	1	1	2	0	1	18	*	4.5	SW	50		1964
YR	6.0	109	90	166	128	3	53	34	32	5	81	1	4.4	SW	59	SW	JAN 1959

NORMALS, MEANS AND EXTREMES
(Table Revised 1975. Base Period for Climatological Normals: 1941-1970)

Month	Normal Daily max	Normal Daily min	Normal Monthly	Extreme Record highest	Year	Extreme Record lowest	Year	Heating degree days Base 65°	Cooling degree days Base 65°	Precip Normal	Precip Max monthly	Year	Precip Min monthly	Year	Precip Max 24 hrs	Year	Snow Max monthly	Year	Snow Max 24 hrs	Year
J	47.1	29.1	38.1	75	1952	-9	1966	834	0	5.25	13.27	1954	1.86	1961	4.25	1954	9.6	1966	8.3	1966
F	50.3	30.4	40.4	75	1972	1	1965+	689	0	5.24	10.47	1956	0.84	1968	2.94	1954	11.3	1960	9.1	1960
M	58.5	36.4	47.6	85	1963	8	1960	551	12	5.45	9.71	1965	2.13	1957	4.74	1965	21.0	1960	12.0	1960
A	70.8	46.2	58.5	90	1970	24	1950	220	22	4.21	9.71	1956	1.39	1950	3.74	1956	0.3	1971	0.3	1971
M	78.8	54.4	66.7	93	1962	31	1963	77	129	3.54	7.17	1971	0.80	1971	4.41	1973	0.0		0.0	
J	85.5	62.9	74.2	101	1954	40	1972	5	281	3.94	8.09	1962	0.86	1962	3.70	1969	0.0		0.0	
J	87.5	66.4	77.0	105	1952	50	1967	0	372	5.67	9.27	1967	1.55	1970	4.91	1967	0.0		0.0	
A	86.8	66.3	76.1	100	1954	51	1965+	0	344	3.85	10.46	1960	0.54	1953	7.48	1960	0.0		0.0	
S	81.1	58.8	70.0	100	1954	34	1962	20	173	3.34	9.10	1961	0.41	1961	3.43	1972	0.0		0.0	
O	71.8	47.1	59.1	90	1961	21	1952	214	34	2.72	6.95	1972	T	1963	2.13	1948	T		T	
N	57.9	37.2	47.1	83	1951	0	1950	537	0	4.13	12.22	1973	1.37	1956	5.29	1956	6.5	1962	6.5	1962
D	48.3	30.1	39.2	74	1954	-3	1962	800	0	5.36	10.31	1969	0.67	1956	5.12	1969	14.8	1963	10.8	1963
YR	68.6	47.0	57.8	105	JUL 1952	-9	JAN 1966	3949	1367	52.60	19.27	JUL 1967	T	OCT 1963	7.48	AUG 1960	21.0	MAR 1960	12.0	MAR 1960

Month	Mean sky cover tenths	Clear	Partly cloudy	Cloudy	Precip .01 in+	Snow 1.0 in+	Thunderstorms	Heavy fog, vis. 1/4 mile or less	Max 90°+	Max 32° below	Min 32° below	Min 0° below	Wind Mean speed m.p.h.	Prevailing direction	Peak gust Speed m.p.h.	Direction	Year
J	6.8	7	6	18	13	1	1	1	0	2	19	*	4.8	SW	59		1959
F	6.6	7	5	16	13	1	2	1	0	1	16	0	5.0	SW	47		1957
M	6.7	8	6	17	13	1	3	1	0	*	11	0	5.3	ENE	48		1962
A	6.1	9	6	15	11	1	5	1	*	0	2	0	5.7	SW	50		1959
M	5.8	9	8	14	10	*	8	2	2	0	*	0	4.5	SW	46		1973
J	5.7	9	10	10	10	0	9	2	7	0	0	0	4.2	SW	49		1963
J	6.1	8	11	12	12	0	11	3	10	0	0	0	3.9	SW	50		1961
A	5.5	11	8	11	10	0	9	4	8	0	0	0	3.7	SW	53		1964
S	5.0	14	6	11	8	0	3	8	3	0	0	0	3.6	E	39		1959
O	5.0	14	7	11	8	0	1	6	*	0	1	0	3.6	E	39		1968
N	6.1	9	6	14	10	0	1	2	0	*	12	*	4.1	E	45		1968
D	6.9	7	6	18	11	1	1	2	0	1	19	*	4.5	SW	50		1964
YR	6.1	109	88	168	128	3	53	34	30	5	81	*	4.4	SW	59	SW	JAN 1959

Average station pressure mb.   Elev. feet m.s.l.

REFERENCE NOTES APPLYING TO TABLES APPEAR ON THE PAGE FOLLOWING LAST TABLE.
(Caution: Letters and symbols may have different meanings in 1941-1970 tables than in earlier tables. See notes.)

NORMALS, MEANS AND EXTREMES
(Table Revised 1960. Base Period for Climatological Normals: 1921-1950)

OAK RIDGE, TENNESSEE
AREA STATION
ELEVATION (ground) 886 Feet

Month	Temperature Normal Daily maximum	Normal Daily minimum	Normal Monthly	Extremes Record highest	Year	Record lowest	Year	Normal degree days	Precip. Normal total	Max. monthly	Year	Min. monthly	Year	Max. in 24 hrs	Year	Wind Mean hourly speed	Prevailing direction	Mean no. days Precip .01 in or more	Temp Max 90° and above	Temp Max 32° and below	Temp Min 32° and below	Temp Min 0° and below
(a)	(b)	(b)	(b)	13		13		(b)	(b)	14		14		14		10	10	11	13	13	13	13
J	48.7	30.4	39.6	77	1952	1	1948	787	5.11	12.37	1954	1.11	1944	3.96	1954	5.7	NE	14	0	1	17	0
F	51.2	31.5	41.4	77	1948	3	1951	661	5.21	10.01	1956	1.89	1947	3.23	1948	5.9	NE	11	0	1	13	0
M	59.3	37.3	48.3	87	1945	14	1955	534	5.64	8.91	1951	2.06	1957	2.85	1952	6.8	SW	12	0	*	10	0
A	69.5	45.9	57.7	89	1948+	24	1950	227	4.29	8.54	1956	1.25	1952	2.96	1956	6.8	SW	10	0	0	2	0
M	77.7	54.2	66.0	94	1956	34	1954	64	3.78	7.01	1950	0.90	1951	2.09	1955	4.9	SW	10	2	0	0	0
J	85.7	62.7	74.2	99	1952	41	1946	0	4.18	5.87	1945	1.18	1956	3.08	1945	4.1	SW	9	10	0	0	0
J	87.8	65.6	76.7	103	1952	49	1947	0	4.83	8.13	1950	2.14	1954	2.40	1948	3.8	NE	10	13	0	0	0
A	86.6	64.2	75.4	99	1957+	44	1946	0	4.45	10.31	1950	0.50	1953	2.34	1950	3.2	NE	9	13	0	0	0
S	83.0	59.1	71.1	103	1954+	33	1949	41	2.97	12.84	1944	1.07	1945	7.75	1944	4.0	NE	7	6	0	1	0
O	71.8	46.7	59.3	91	1954	21	1952	206	2.77	6.43	1949	0.63	1953	2.32	1949	4.2	NE	7	*	*	1	0
N	58.4	36.6	47.5	82	1950	4	1950	531	3.94	12.00	1948	1.01	1949	3.20	1948	4.9	SW	10	0	*	12	0
D	49.3	30.9	40.1	76	1951	4	1957	772	5.21	10.28	1954	3.00	1947	4.38	1954	4.9	NE	10	0	1	19	0
Year	69.1	47.1	58.1	103	Sept. 1954+	1	Jan. 1948	3823	52.38	12.84	Sept. 1944	0.50	Aug. 1953	7.75	Sept. 1944	4.9	NE	119	44	3	74	0

Reference notes applying to Normals, Means, and Extremes tables for 1921–1950 base period.

(a) Length of record, years.

(b) Normal values are based on the period 1921-1950, and are means adjusted to represent observations taken at the present standard location.

* Less than one-half.

. No record.

† Airport data.

‡ City Office data.

+ Also on earlier dates, months, or years.

T Trace, an amount too small to measure.

Sky cover is expressed in a range of 0 for no clouds or obscuring phenomena to 10 for complete sky cover. The number of clear days is based on average cloudiness 0-3 tenths; partly cloudy days on 4-7 tenths; and cloudy days on 8-10 tenths. Monthly degree day totals are the sum of the negative departures of average daily temperatures from 65°F. Sleet was included in snowfall totals beginning with July 1948. Heavy fog also includes data referred to at various times in the past as "Dense" or "Thick". The upper visibility limit for heavy fog is 1/4 mile. Data in these tables are based on records through 1957.

Reference notes applying to Normals, Means, and Extremes tables for 1931–1960 base period.

(a) Length of record, years, based on January data. Other months may be for more or fewer years if there have been breaks in the record.

(b) Climatological standard normals (1931-1960).

* Less than one half.

+ Also on earlier dates, months, or years.

T Trace, an amount too small to measure.

Below zero temperatures are preceded by a minus sign. The prevailing direction for wind in the Normals, Means, and Extremes table is from records through 1963.

‡ ≥ 70° at Alaskan stations.

Unless otherwise indicated, dimensional units used in this bulletin are: temperature in degrees F.; precipitation, including snowfall, in inches; wind movement in miles per hour; and relative humidity in percent. Heating degree day totals are the sums of negative departures of average daily temperatures from 65° F. Cooling degree day totals are the sums of positive departures of average daily temperatures from 65° F. Sleet was included in snowfall totals beginning with July 1948. The term "Ice pellets" includes solid grains of ice (sleet) and particles consisting of snow pellets encased in a thin layer of ice. Heavy fog reduces visibility to 1/4 mile or less.

Sky cover is expressed in a range of 0 for no clouds or obscuring phenomena to 10 for complete sky cover. The number of clear days is based on average cloudiness 0-3, partly cloudy days 4-7, and cloudy days 8-10 tenths.

Solar radiation data are the averages of direct and diffuse radiation on a horizontal surface. The langley denotes one gram calorie per square centimeter.

& Figures instead of letters in a direction column indicate direction in tens of degrees from true North; i.e., 09 - East, 18 - South, 27 - West, 36 - North, and 00 - Calm. Resultant wind is the vector sum of wind directions and speeds divided by the number of observations. If figures appear in the direction column under "Fastest mile" the corresponding speeds are fastest observed 1-minute values.

* Through 1964. The station did not operate 24 hours daily. Fog and thunderstorm data therefore may be incomplete.

Through 1964.

** The National Weather Service considers the accuracy of solar radiation data questionable; therefore, publication is suspended pending determination of corrected values.

Reference notes applying to Normals, Means, and Extremes tables for 1941–1970 base period.

(a) Length of record, years, through the current year unless otherwise noted, based on January data.

(b) 70° and above at Alaskan stations.

* Less than one half.

T Trace.

NORMALS - Based on record for the 1941-1970 period.

DATE OF AN EXTREME - The most recent in cases of multiple occurrence.

PREVAILING WIND DIRECTION - Record through 1963.

WIND DIRECTION - Numerals indicate tens of degrees clockwise from true north. 00 indicates calm.

FASTEST MILE WIND - Speed is fastest observed 1-minute value when the direction is in tens of degrees.

% Through 1964. The station did not operate 24 hours daily. Fog and thunderstorm data therefore may be incomplete.

Through 1964.

Mean Annual Precipitation, Inches

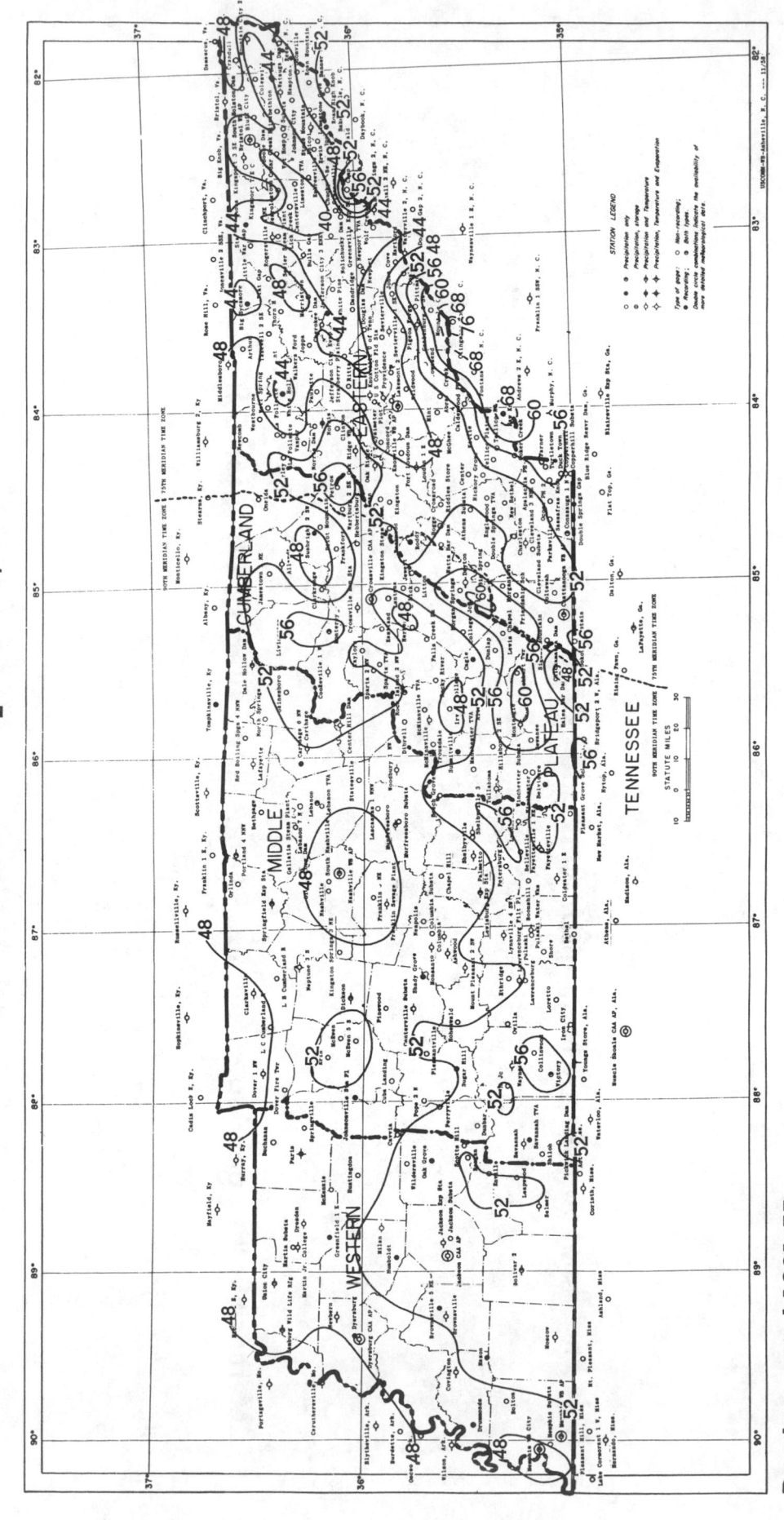

Based on period 1931-55

Isolines are drawn through points of approximately equal value. Caution should be used in interpolating on these maps, particularly in mountainous areas.

Mean Maximum Temperature (°F.), January

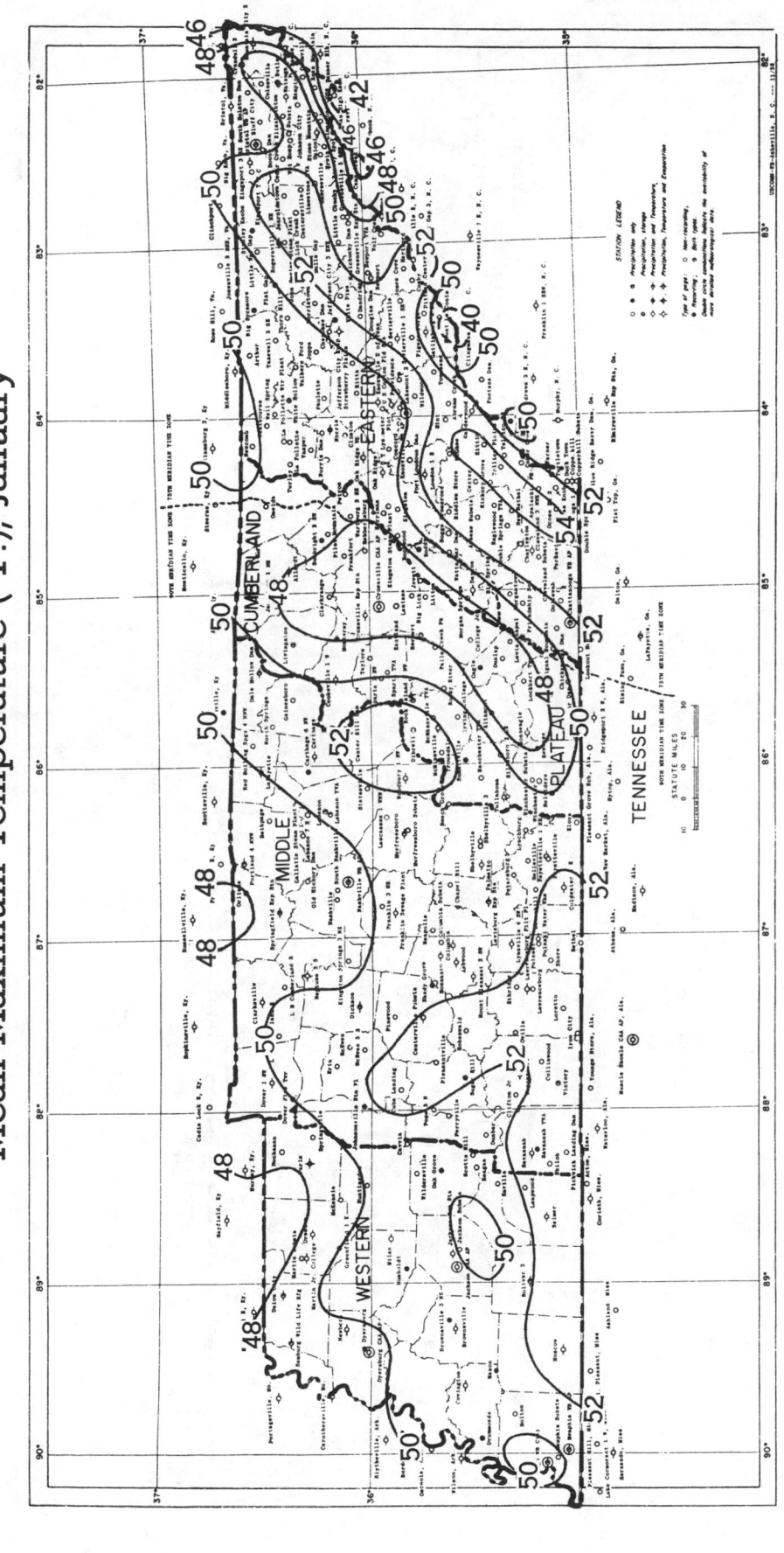

Based on period 1931-52

Isolines are drawn through points of approximately equal value. Caution should be used in interpolating on these maps, particularly in mountainous areas.

Mean Minimum Temperature (°F.), January

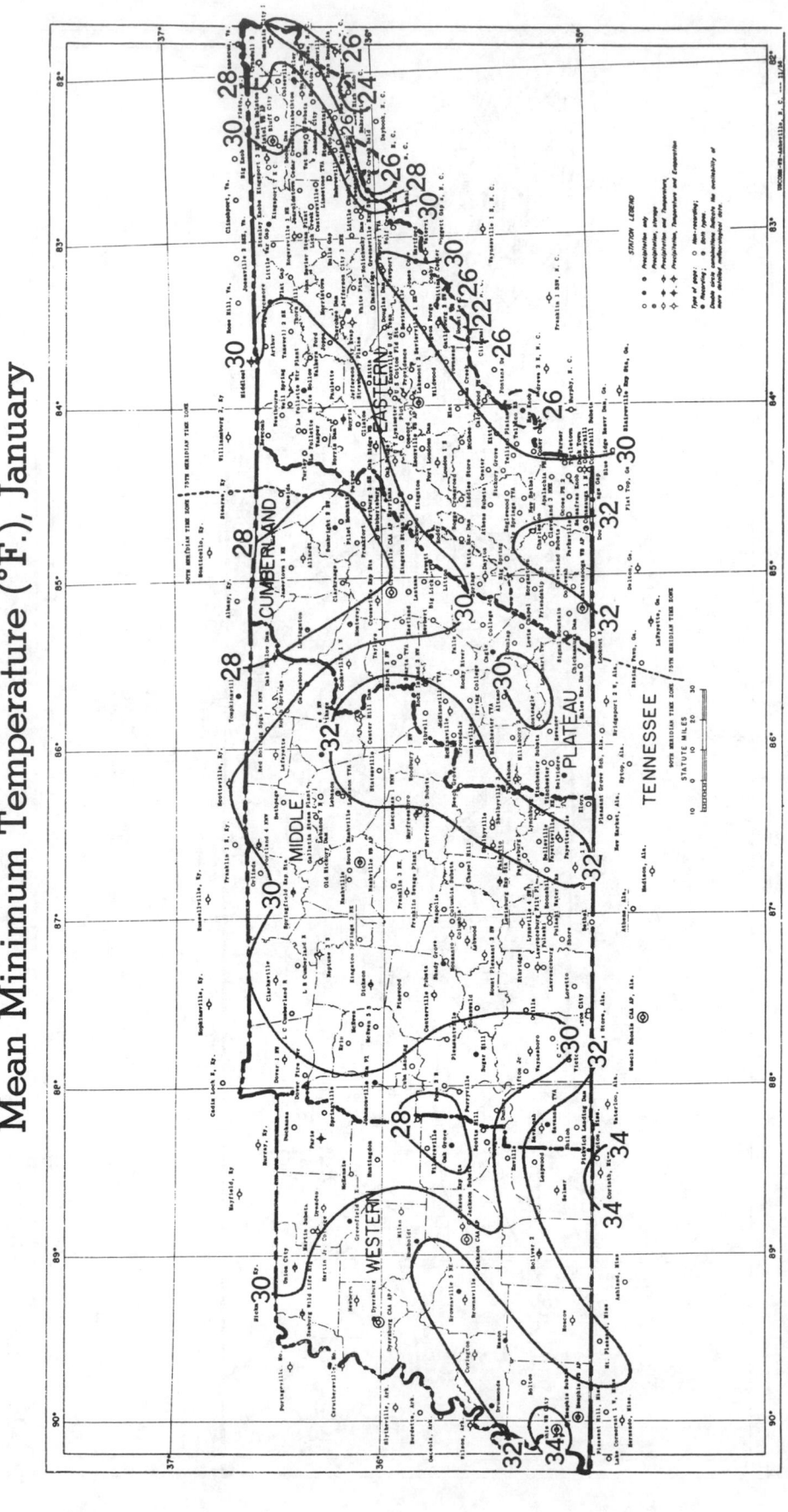

Based on period 1931-52

Isolines are drawn through points of approximately equal value. Caution should be used in interpolating on these maps, particulary in mountainous areas.

Mean Maximum Temperature (°F.), July

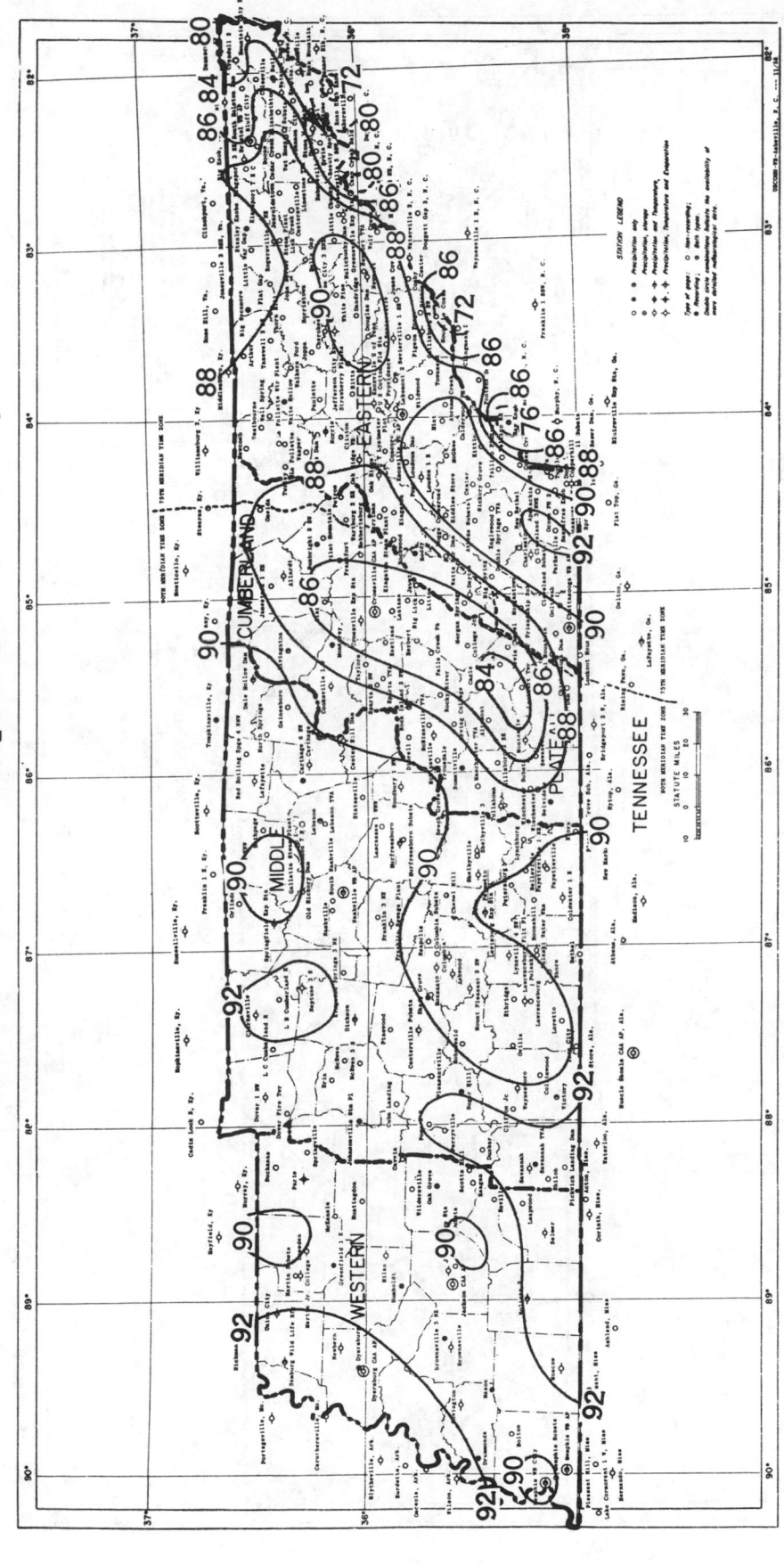

Based on period 1931-52

Isolines are drawn through points of approximately equal value. Caution should be used in interpolating on these maps, particulary in mountainous areas.

Mean Minimum Temperature (°F.), July

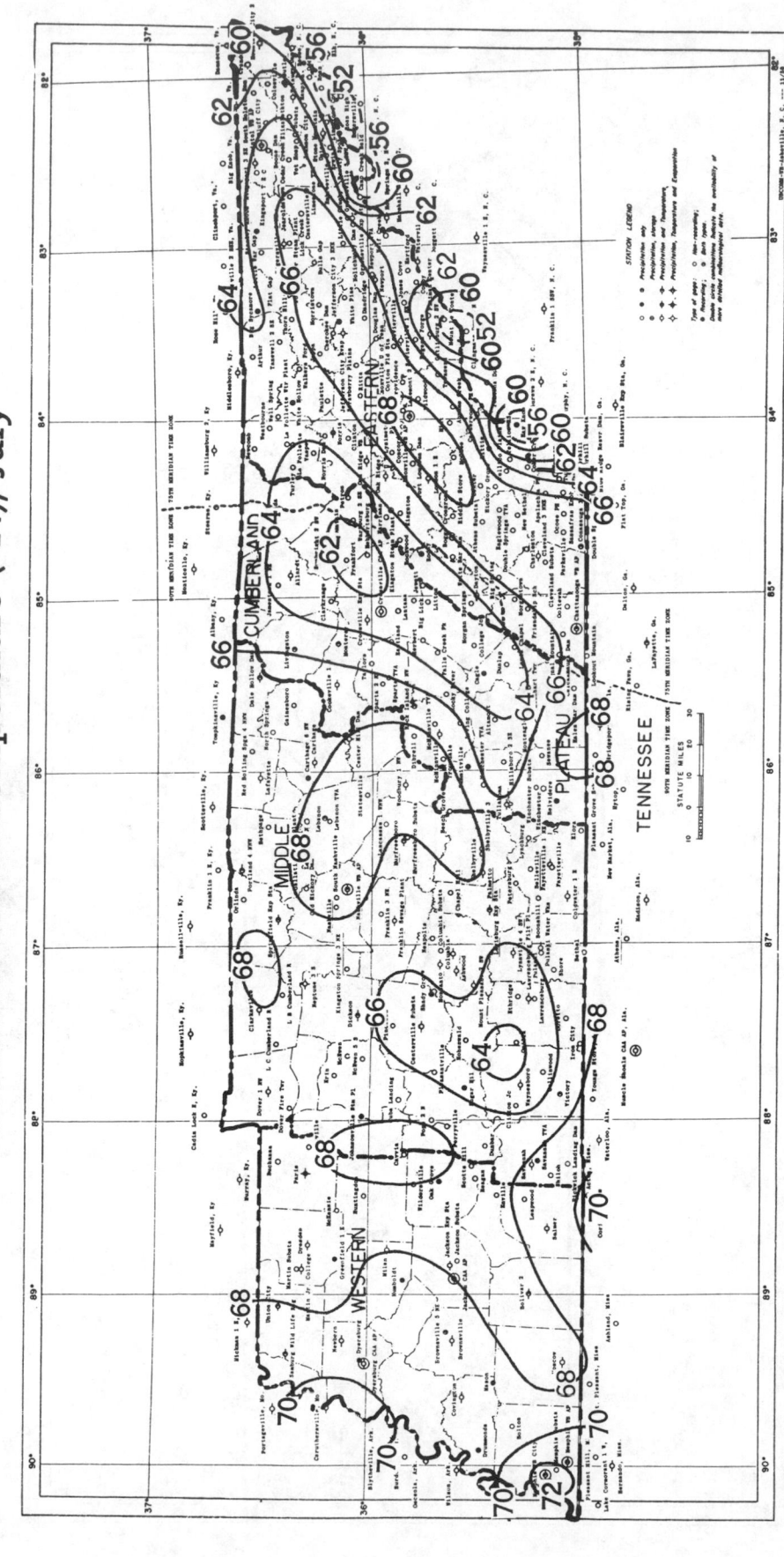

Based on period 1931-52

Isolines are drawn through points of approximately equal value. Caution should be used in interpolating on these maps, particulary in mountainous areas.

CLIMATES OF THE STATES

TEXAS

(Normals, Means and Extremes tables revised 1969 and 1975. Basic report revised June 1969.)

Climate of Texas

Robert B. Orton, ESSA State Climatologist

Texas, from "tejas", an Indian word meaning friendly, has been called "the crossroads of North American geology." Within the State's boundaries four great physiographic subdivisions of the North American Continent come together. These are: the Gulf Coastal Forested Plain, Great Western Lower Plains, Great Western High Plains, and the Rocky Mountain Region. Texas may be described as a vast amphitheater, sloping upward from sea level along the coast of the Gulf of Mexico to more than 4,000 feet general elevation along the Texas-New Mexico line. While much of the State is relatively flat, there are 90 mountains a mile or more high, all of them in the Trans-Pecos region. Guadalupe Peak, at 8,751 feet, is the State's highest. The most spectacular physical features of Texas are found in the western portion of the State, from the Palo Duro Canyon in the north to the Big Bend in the south, and westward from the Balcones Escarpment and the Hill Country to the Guadalupe Mountains. The State is subdivided into 12 geographic regions; from east to west, these are: Pine Woods Region, Gulf Coast Plain, Post Oak Belt, Blackland Prairie, Grand Prairie, South Texas Plain, Cross Timbers, Llano Basin, Edwards Plateau, Lower Plains, High Plains, and Mountain and Basin Region.

Texas contains 267,339 square miles or 7.4 percent of the Nation's total area. It is large enough to accomodate 15 of the 50 states within its borders. In straight-line distance, Texas extends 801 miles from north to south and 773 miles from east to west. The boundary of Texas extends 4,137 miles. The Rio Grande forms the longest segment of the boundary, 1,569 miles. The second longest segment, 726 miles, is formed by the Red River. The tidewater coastline extends 624 miles. Brewster, in Southwest Texas, is the largest of the State's 254 counties with 6,208 square miles. The smallest county is Rockwall in Northeast Texas with 147 square miles.

Texas ranks second only to Alaska among the 50 states in the volume of its inland water with nearly 6,000 square miles of lakes and streams. Toledo Bend Reservoir, situated on the Sabine River between Texas and Louisiana, is the largest reservoir in Texas, or on its border, with a capacity of 4,477,000 acre-feet, and is one of the largest artificially constructed reservoirs in the United States. Most Texas rivers parallel each other and flow directly into the Gulf, but the Canadian, Red, and Sulphur Rivers are part of the Mississippi River system. The Brazos is the largest river between the Rio

Grande and the Red and third in size of all rivers flowing either partly or wholly in Texas. Other principal rivers are the Colorado, Trinity, Sabine, Nueces, Neches, and Guadalupe.

Included in Texas' 26 million acres of woodland are four national forests with 775,265 acres and four state forests with 6,306 acres. The most important forest area of the State, producing nearly all the commercial timber, is the East Texas pine-hardwood belt, known locally as the "Piney Woods." It extends over all or parts of 43 counties.

Wedged between the warm waters of the Gulf of Mexico and the high plateaus and mountain ranges of the North American Continent, Texas has diverse meteorological and climatological conditions. Continental, marine, and mountain types of climates are all found in Texas; the continental and mountain types in true form, but the marine climate modified by surges of continental air. The High Plains, separated from the Lower Plains by the Cap Rock Escarpment, lies in a Cool-Temperature climatic zone. Except for some small areas in the Trans-Pecos, the remainder of the State lies in a Warm-Temperature Subtropical Zone. Within these broad zones, six subclasses appropriately identify the climates of Texas. The proximity to the Gulf of Mexico, the persistent southerly and southeasterly flow of warm Tropical Maritime air into Texas from around the westward extension of the Azores High, and adequate rainfall, combine to produce a humid subtropical climate with hot summers across the eastern third of the State. The Gulf moisture supply gradually decreases westward and is cut off more frequently during the colder months by intrusions of drier polar air from the north and west; as a result, most of Central Texas, as far north as the High Plains has a subtropical climate with dry winters and humid summers. This region is semi-arid. As the distance from the Gulf increases westward, the summer moisture supply continues to decrease gradually, producing a subtropical steppe climate across a broad section along the middle Rio Grande Valley that extends as far west as the Pecos Valley, where rainfall is most often inadequate for agriculture without supplemental irrigation. Except for "islands" of cool-temperate, mountain type climates at the higher elevations in the Guadalupe, Davis, and Chisos Mountains, the area west of the Pecos is mostly arid subtropical, and rainfall is inadequate for other than desert or semi-desert types of vegetation. The mountain climates in the Trans-Pecos are cooler throughout the year than those of the adjacent lowlands. Temperatures decrease with altitude and average about 1° F. lower for each 300 feet of increased elevation. The rate of change varies with the season, being more rapid in summer and greatest during the warmer hours of the day.

Stretching over the largest level plain of its kind in the United States, the High Plains rise gradually from about 2,700 feet on the east to more than 4,000 feet in spots along the New Mexico border. The combination of high elevation, remoteness from moisture source regions, and frequent intrusions of dry polar airmasses, result in a dry steppe climate with relatively mild winters. This region is semi-arid and rainfall is often inadequate for profitable agricultural production without supplemental irrigation.

While the changes in climate across Texas are considerable, they are nevertheless gradual; no natural boundary separates the moist East from the dry West or the cool North from the warm South.

Rainfall in Texas is not evenly distributed over the State and varies greatly from year to year. Average annual rainfall along the Louisiana border exceeds 56 inches, and in the western extremity of the State, is less than 8 inches. In the way of extremes, Clarksville, in Northeast Texas, recorded 109.38 inches in 1873, while Wink, in extreme West Texas, recorded only 1.76 inches in 1956. The number of days with measurable precipitation follows the general trend of rainfall totals so that seasonal frequencies are lowest where amounts are lowest. At a single location amounts for any one month will nearly always vary widely from the mean or normal precipitation. Except along the upper Texas coast, it is possible for one or two thunderstorms to account for the entire month's rainfall at a station. Torrential rains of 10 to 20 inches or more may accompany a tropical storm as it moves inland across the Texas coast. These infrequent but excessive amounts are reflected in mean rainfall data and seriously limit the usefulness of this type of statistic in describing rainfall.

Patterns of seasonal precipitation in Texas vary considerably for different areas of the State. Rains occur most frequently in late spring as a result of squall-line thunderstorms; consequently, most areas of the State show a peak in May. This includes most of the High and Low Rolling Plains, the Edwards Plateau, North Central, and South Central Texas. Rainfall in the Pecos Valley, most of southern Texas, the lower Rio Grande Valley, and in the coastal section, shows a peak in September, with a secondary peak in May. On the High Plains, particularly the northern portion, a significant percentage of the total annual precipitation occurs during the summer months (following the May peak). Throughout the central part of the State, July and August are relatively dry months. In the mountainous Trans-Pecos area of West Texas, afternoon thundershowers during July, August, and September account for most of the annual rainfall. Throughout most of East Texas (east of about 95° west longitude) rainfall is fairly evenly distributed throughout the year. East of about 96° west longitude annual rainfall exceeds average potential evapotranspiration. West of this meridian, average potential evapotranspiration exceeds annual average rainfall.

In most of Texas a large portion of the annual rainfall occurs within short periods of time, resulting in excessive run-off and frequently producing damaging floods. Some 320 Texas cities have flood problems resulting from stream overflow, local drainage, or coastal floods. One hundred of these cities have stream overflow flood problems, 112 have local drainage problems, 20 have coastal flood problems, and another 88 have some combination of these.

Due to the variety in the physiographic features of the State, the types of floods experienced vary in character. In the east, where the average annual rate of rainfall is highest, there are broad flat valleys and large areas of timber and brushland. The natural drainage channels have gentle slopes, limited capacity, and follow meandering courses from their headwater areas to the Gulf. Runoff is comparatively slow. During periods of intense general rainfall, large volumes of water are temporarily stored in the valleys of the basins and then slowly released to the streams. This produces a broad flat crested slow-moving flood which in the lower basin regions results in protracted periods of inundation.

In the west, ground and tree cover is sparse and stream slopes vary in the portions of the basins from steep to moderately steep, and tend to flatten in the coastal strip. During periods of intense general rain the time of concentration of runoff is more rapid in the western portion of the State than in the eastern section. This results in the production of higher peaks and more rapidly moving floods with shorter periods of inundation of flood areas.

The amount of water flowing in Texas streams ranges widely from east to west. The average annual runoff is about 39 million acre-feet, with about three-fourths of this total coming from the eastern one-fourth of the State. The annual amount of runoff varies widely also. For the period 1940-46, the average annual amount was approximately 59 million acre-feet, dropping to about 24 million acre-feet annually for the dry period 1950-56.

Flood stage is reached on some Texas streams nearly every year. The worst general floods of recent years were in 1957 when every major river and tributary in the State flooded during the spring months between April and June. In late April 1966, intense flooding occurred in northeast Texas where 20 to 26 inches of rain fell in some areas in a relatively short period of time. Flash flooding in the Sanderson area in 1965 cost 23 lives in a period of hours. Severe flooding occurred in South Texas from heavy rainfall accompanying Hurricane Beulah in September 1967.

The greatest rainfall recorded in Texas during 24 consecutive hours was at Thrall in Williamson County where 38.2 inches fell during parts of September 9 and 10, 1921.

From the early days of Texas history recorded by Spaniards exploring the Southwest, drought has been a re-occurring problem. A drought in Central Texas dried up the San Gabriel River in 1756, forcing the abandonment of a settlement of missionaries and Indians. Stephen F. Austin's first colonists also were hurt by drought. Their initial corn crop was snuffed out in 1822, forcing the once ambitious farmers into desperate hunters. Drought means various things to various people. The agriculturist, hydrologist, economist, and the meteorologist each have a different concept because of his specific interest and experiences. In Texas agricultural drought is probably the most important. Agricultural drought may be defined as a condition in which sufficient soil moisture is not available in the root zone for plant growth and development. Thus, an evaluation of drought on a purely agricultural basis must take into account the kind of crop, its stages of growth and root depths, the characteristics of the soil, and the degree of plant wilting, as well as the various meteorological factors that daily affect moisture supply and demand, the duration of drought, and the size of the affected area. Obviously, such an evaluation of agricultural drought becomes rather involved and very unwieldy on a regional scale.

Meteorological drought may be defined as a prolonged and abnormal moisture deficiency. More specifically, a drought period may be defined as an interval of time, generally of the order of months or years in duration, during which the actual moisture supply at a given place rather consistently falls short of the climatically expected or climatically appropriate moisture supply. Thus, drought is a relative rather than an absolute condition. A prescribed set of weather events could provide unusually wet weather in semi-desert regions such as is found in the Trans-Pecos, while the identical set would be regarded as drought in East Texas. Therefore, drought severity depends not only on current weather, but on antecedent weather as well.

The worst drought in recent years, by any classification, agricultural, economic, hydrologic, or meteorological, began in 1950 in the western part of the State and spread until about 94 percent of the counties (244 of 254) were classified as disaster areas by the end of 1956. In terms of financial loss, necessary human adjustments, and deterioration of physical resources, this was the worst drought in recorded Texas history. Also, it was the longest. Other severe droughts occurred in 1909-1910, 1916-1917, and 1933-1934. Departures from normal precipitation, based on the period 1931-1960, show the year 1917 to be the driest on record in Texas.

In 9 out of the 10 climatic divisions within the State, precipitation was less than 60 percent of normal; in 5 climatic divisions precipitation was less than 50 percent of normal. The year 1956 is the second driest year on record. The two wettest years on record are 1919 and 1941.

In most years, some sections of the State receive less than normal rainfall, while other sections receive a greater than normal supply. Severe drought or excessively wet conditions rarely exist over the entire State at the same time. While the Great Plains drought of the early 1930's received considerable publicity as the "dust bowl days", its presence in Texas was confined largely to the western one-third of the State, and to the years, 1933-1934.

Ground water is a significant resource throughout much of the State, supplying about 75 percent of the total water used for municipal, industrial, and irrigation purposes. Many areas now supplied by ground water are depleting the available supply because the rate of pumping grossly exceeds the rate of replenishment.

The present level of Texas irrigation - 8.36 million acres in 1968, has developed rapidly, mostly since World War II. Nearly 83 percent of all present irrigation is supplied with ground water. However, many presently irrigated areas - the High Plains, Lower Rio Grande Valley, Winter Garden, Trans-Pecos, and elsewhere - face the prospect of returning to dryland farming as available water supplies are exhausted. By 1985, according to Texas Water Development Board estimates, if a supplemental surface supply of water has not reached the High Plains, this vast area will have begun an area-wide retrogression to dryland farming.

Major water uses in Texas are for domestic and municipal supply, industry, and irrigated agriculture. Other beneficial uses of water include mining and secondary oil recovery, hydro-electric power generation, navigation, and recreation. Demands for recreation, particularly for water oriented recreation, are increasing rapidly in Texas. There is not enough water available in Texas to supply future water needs. By the year 2020, the State's population is expected to exceed 30.5 million. At that time, according to Texas Water Development Board estimates, an import of as much as 12 to 13 million acre-feet of water per year will be necessary to prevent economic loss to the State of major geographic areas where ground water supplies are now being depleted and other sources of supply do not occur.

The vast land area of Texas experiences a wide range of temperatures. The High Plains experiences rather low temperatures in winter, while there are several separate areas within the State that experience very high temperatures in summer. The average January temperature in Amarillo, in the Panhandle is 36.7° F. and at Brownsville, in the lower Rio Grande Valley 61.4° F. From November through March, surges of cold air from the north are frequent. These cold fronts, or "northers" as they are called locally, modify rapidly as they reach warmer latitudes. Fast moving cold fronts, followed by rapid warming, result in frequent and pronounced temperature changes from day to day, and sometimes from hour to hour, during the colder months of the year.

Extended periods of subfreezing temperatures are rare, even on the High Plains. Cold spells that seriously interfere with outdoor activities usually do not last more than 48 to 72 hours at the most. In South Texas, subfreezing temperatures associated with arctic airmasses ordinarily are confined to several hours prior to sunrise, and seasons may pass with no subfreezing temperatures at all. Temperatures of 32° F. or lower occur only about three years out of four, on an average, in the Lower Rio Grande Valley. Extremely cold spells were experienced in South Texas in 1951 and again, 11 years later, in 1962. Brownsville experienced 65 consecutive hours of subfreezing temperatures January 29-February 1, 1951, and 64 consecutive hours January 9-12, 1962.

In summer, the temperature contrast is much less pronounced from north to south with daily highs generally in the 90's. August is the hottest month. Extremely hot daytime temperatures occur with greatest frequency in the triangle in South Texas bounded by Rio Grande City, Cotulla and Eagle Pass, and in an area along the Upper Rio Grande from about Presidio northward to Candelaria. Almost as hot, is a small area along the Red River bounded by Childress and Chillicothe, Texas, and Hollis and Altus, Oklahoma.

Because of the amazing adaptability of the human body, it is very difficult to evaluate the effect of either extremely low or extremely high temperatures on the productivity or efficiency of outdoor activity. People in South Texas perform work out-of-doors at temperatures that would seem intolerably high to people accustomed to more moderate climates. As a matter of fact, low temperatures in the Lower Rio Grande Valley in winter are as likely to cause a loss of efficiency among workers as do extremely high temperatures in summer, although these low temperatures would seem mild to people living in the northern United States. It is conceded generally that in most of Texas the number of days of 100° F. temperatures or above provides a more useful criterion for evaluating the effects of temperature on outdoor worker activity than does the number of days of a lower value such as 90° F. or 95° F.

The highest temperature recorded in Texas was 120° F. at Seymour, southwest of Wichita Falls, on August 12, 1936. The lowest temperature ever recorded was 23° F. below zero at Seminole, southwest of Lubbock, on February 8, 1933, and at Tulia, north of Lubbock, on February 12, 1899. The coldest winter in Texas history was that of 1898-99, according to best available records.

Relative humidity is highest in the coastal region, and decreased gradually inland, as the distance from the Gulf of Mexico increases. Mean annual relative humidity at noon, Central Standard Time, varies from slightly more than 60 percent near the coast to around 35 percent in the El Paso area. On the whole, there is a range of approximately 10 percent between the high summer and low winter averages. As temperatures increase relative humidities generally decrease, and when temperatures fall the relative humidity tends to rise. The lowest relative humidities are found generally in the daytime, especially in the afternoon, while the highest values usually occur in the early morning.

Heating engineers and fuel distributors use an index based on the assumption that buildings require heating when the average temperature for a day falls below 65° F. The greater the accumulation of heating degree days (difference between temperature for a day below 65° and 65°) the more heat is required to produce a comfortable indoor temperature, and more fuel will be consumed. In the northern Panhandle an average season will exceed 4,000 heating degree days, yet in the southernmost section of the State, an average season has less than 1,000 heating degree days.

The climate of Texas is such that summer air conditioning is desirable in all parts of the State. While it is true that residents of the High Plains area enjoy cool summers occasionally when little air conditioning is needed, more often than not, July and August temperatures rise to the point where some cooling adds measurably to human comfort. During the summer the temperature and humidity conditions are such as to make the use of evaporative cooling practical for home use west and north of a line from Wichita Falls southward to San Antonio and westward to Del Rio. The most satisfactory results from evaporative cooling are obtained when the outside dry bulb temperature during the daytime is 90° F. or above and the outside wet bulb temperature is below 75° F. East and south of the above line, refrigerated type air conditioning is recommended for maximum comfort.

Sunshine is abundant in the extreme southwestern section of the State, decreasing gradually eastward. On an average, the western Trans-Pecos receives 80 percent of the total possible sunshine annually, while the Upper Coast receives only 60 percent.

Except for "freak" occurrences, significant amounts of snowfall are confined almost entirely to the mountainous Trans-Pecos region and the High Plains. Measurable snow falls south of the High Plains but usually melts almost as fast as it falls. Falling snow rarely interferes with outdoor operations more than an hour or two at a time, except when it is associated with an intense extratropical cyclone. Heavy snows of 4 inches or more rarely occur. Vega, elevation 4,000 feet, in the western Panhandle receives an average of 23.3 inches annually, while no measurable snow has fallen at Brownsville during the 73-year period, 1896-1968. Blizzards may occur in extreme West Texas or Northwest Texas during the winter or early spring months, but are rare. Blizzards are characterized by subfreezing temperatures, very strong winds, and considerable blowing or drifting snow. From 1950 through the winter of 1967-1968 there have been only two blizzards worthy of note in Texas: the blizzard of February 1-5, 1956, on the High Plains, which resulted in 20 deaths and dumped a record 33.0 inches of snow at Hale Center; and the late March blizzard of 1957 (March 22-25) which resulted in 10 deaths and 4,000 persons being marooned.

Tropical cyclones are a threat to all sections of the Texas coast during the summer and fall. From 1885 to 1968, inclusive, 69 tropical storms of all intensities have affected the Texas coast. Those tropical cyclones with sustained wind speeds of 64 knots (74 m.p.h.) or greater are known as hurricanes. Virtually all tropical cyclones which have affected the Texas coast originated in the Gulf of Mexico, the Caribbean Sea or the southern part of the North Atlantic Ocean. The season extends from June to October; storms are most frequent in August and September and rarely affect the Texas coast after the first days of October. The average frequency for the entire Texas coast is approximately one per year. Tropical cyclones were most frequent in 1886 and 1933, with four in each of these years.

Over inland areas with higher elevations, the greatest concern regarding hurricanes is possible damage from winds (including tornadoes) or damage from flooding due to excessive rainfall. Near, and along the immediate coast, the additional hazard of hurricane tides (or storm tides) must be considered. Persons and property along the immediate shore line, without some type of natural or man-made protective barrier, are exposed to the direct forces of hurricane waves and swells. These forces reach magnitude almost beyond comprehension. Although hurricane winds and tornadoes spawned by hurricanes cause a great amount of damage and sometimes loss of life, surveys of past hurricanes indicate that storm tides cause by far the greatest destruction and largest number of deaths.

The Great Galveston Storm of September 8-9, 1900, was the worst natural disaster in United States history. Loss of life at Galveston has been estimated at 6,000 to 8,000, but the exact number has never been definitely ascertained. The island was completely inundated, and not a single structure escaped damage. Most of the loss of life was due to drowning by storm tides that reached 15 feet or more. The anemometer blew away when the wind reached 100 m.p.h. at 6:15 p.m. on the 8th. The wind reached an estimated maximum velocity of 120 m.p.h. between 7:30 and 8:30 p.m. Property damage has been estimated at $30 to $40 million.

A severe hurricane struck Galveston again, August 16-17, 1915, resulting in at least 275 lives lost and property damage estimated at $50 million. A new sea wall prevented a repetition of the 1900 disaster. The center of a severe hurricane moved inland south of Corpus Christi on September 14, 1919. Tides were 16.0 feet above normal in that area and 8.8 feet above normal at Galveston. Two hundred eighty-four persons lost their lives in Texas, and property damage was estimated at $20.3 million.

The most severe hurricanes in recent years were Carla, September 8-14, 1961, and Beulah, September 18-23, 1967. Great Hurricane Carla, the largest hurricane of record, moved inland over Port O'Connor on September 11. The highest wind gusts, estimated at 175 m.p.h., occurred at Port Lavaca. The highest tide, 18.5 feet, was at Port Lavaca also. The most damage was inflicted to coastal counties between Corpus Christi and Port Arthur and to inland counties of Jackson, Harris, and Wharton. In Texas, 34 persons died; seven of these deaths were attributed to a vicious tornado that swept across Galveston Island during the storm. Four hundred sixty-five persons were injured. Hurricane Carla caused property and crop damage conservatively estimated at $300 million.

Hurricane Beulah, the third largest hurricane of record, after Hurricane Carla and the great New England storm of 1938, moved inland just east of Brownsville, near the mouth of the Rio Grande on September 20, 1967. The SHIRLEY LYKES, in Port Brownsville, reported a 136 m.p.h. wind gust during Beulah's passage. More damaging than the wind or tides was the torrential rainfall which resulted in record-breaking floods on streams and and rivers south of San Antonio. Storm rainfall amounts ranged from 10 to 20 inches over much of the area south of San Antonio, and totals exceeded 30 inches in some areas. An unofficial gaging station in Falfurrias, in Brooks County, registered the highest accumulated rainfall, 36 inches. The resultant stream overflow and surface runoff inundated over 1.4 million acres. Beulah spawned 115 tornadoes; all occurred in Texas. This number surpassed the previous high of 26 hurricane-induced tornadoes during Carla in 1961. Tornadoes resulted in five deaths and 34 injuries. Eight deaths, by drowning, and three injuries were attributed to Hurricane Beulah directly, for a storm total of 13 deaths and 37 injuries. Property losses over the State were estimated at $100 million and crop losses at $50 million.

Tornadoes have occurred in Texas during all seasons; however, they have occurred with greatest frequency during April, May, and June. On an average, three-fifths of the total annual number of occurrences fall within this 3-month period. Approximately one-fourth of the total annual number of tornadoes occur in the month of May alone. Because of the record number of tornadoes spawned by Hurricane Beulah, the greatest number of these violent storms ever reported in Texas in a single month was 124, in September 1967; the greatest number in a single year was 232, in 1967.

For the 17-year period 1951-1967, the average annual number of tornado occurrences in Texas is 89. An annual total of 77 tornadoes occurred in each of three years: 1954, 1960, and 1966. The worst outbreak of spring tornadoes occurred in April 1957, when 69 were reported. More tornadoes have occurred in Texas than in any other state -- a total 2,000 known occurrences during the 52-year period, 1916-1967. This is due partly to the State's size, whose 262,840 square-mile land area is considerably larger than that of the average state. Actually, in total number of tornadoes per 10,000 square-mile area, Texas ranked eleventh among the States, with an average of 3.4 occurrences per year, during the period 1951-1967. Tornadoes occur with greatest frequency on the Low Rolling Plains and in the northern half of North Central Texas. They are quite rare in the Trans-Pecos and in southwestern Texas.

The greatest outbreak of tornadoes of record was associated with Hurricane Beulah in September 1967. One hundred fifteen tornadoes, all in Texas, are known to have occurred with this great hurricane within a 5-day period, September 19-23. Most likely, the actual total was considerably greater than 115. Sixty-seven of these occurred on September 20, a record number for a single day.

The most disastrous tornadoes to strike in Texas occurred more than 50 years apart. On May 18, 1902, a tornado struck the picturesque little town of Goliad. One hundred fourteen persons were killed, and more than twice this number were injured. On May 11, 1953, a tornado moved through the city of Waco; 114 persons were killed, 597 injured, and property damage exceed $41 million.

Hailstorms occur in all parts of the State. The most frequent and most damaging of these occur in spring and early summer. From November through March hail occurrences are closely associated with active cold fronts. Areal patterns of maximum hail frequency, determined from long periods of weather records, coincide with the mean position of the polar front. The lowest hail frequency is in January. Hail occurs on one day in 20 years, on an average, throughout the central section of the State in January. The frequency is 2 hail days in 20 years in small areas around El Paso, north of Dallas, and in the Longview-Marshall area of East Texas. The northward shift of the polar front in March in association with higher temperatures, and more frequent intrusions of moist air from the Gulf, result in a general increase in hail frequency. Hail occurs in about 7 out of 10 years at points in the Dallas area, and in about 4 out of 10 years in the Waco-Austin-San Antonio area, and in the vicinity of Childress, during March.

Hail occurrences in April, May, and June are closely associated with macro-scale weather conditions that are relatively short lived; these are the lines of thunderstorms, popularly called "squall lines", that form, move generally eastward, then dissipate, usually within a period of

12 hours. Peak hail frequency is reached in May. Most points on the High Plains, the Low Rolling Plains, and in North Central Texas, average one hail day per year during May. Hail occurrences at points in the Davis Mountains during May average one day per year also. Hail activity decreases in June as the Azores Anticyclone begins to establish anticyclonic flow across Texas and temperatures continue to rise. The occurrence of hail decreases sharply in July as the westward extension of the Azores Anticyclone becomes firmly established and temperatures rise. Maximum occurrence is one hail day every four years, on an average, at points on the southern High Plains and in the Davis Mountains. After January, hail occurs with lowest frequencies in August and November.

Thunderstorms, from which most damaging local weather develops (tornadoes, hail, windstorms, and high intensity showers) occur on about 60 days each year in the extreme eastern section of the State. The mean annual number of thunderstorm days decreases to about 40 in extreme West Texas, and to 30 in the lower Rio Grande Valley.

Blowing dust or sand may occur occasionally in West Texas where strong winds are more frequent and vegetation is sparse. Duststorms are rare; those that reduce the visibility to less than 1 mile are associated only with the strongest pressure gradients, such as those that accompany intense extratropical cyclones. These low centers form generally east of the Rockies, along the western edge of the High Plains in Kansas, Colorado, New Mexico, or Texas during winter and early spring. Winds of 50 to 60 m.p.h. and higher may persist for several days if these lows become stationary. These winds produce severe duststorms in West Texas when no moist air from the Gulf is carried northward into the system, and blizzards or heavy snow conditions when moisture is available. While blowing dust or sand may reduce visibility to less than 5 miles over an area of thousands of square miles, restrictions to less than 1 mile are quite localized and depend on soil type, soil condition, cultivation practices and vegetation in the immediate area.

Texas growing seasons (freeze-free period) range from 329 days in the Lower Rio Grande Valley to 178 days on the northern High Plains. Normal precipitation ranges from more than 50 inches in southeast Texas to less than 10 inches on the extreme west side of the State. The many combinations of these two climatic variables produce a wide range of season and crop adaptations. An example of this wide range is reflected in the fact that cotton and sorghums are maturing and being harvested in the Lower Rio Grande Valley when planting is still underway on the High Plains. During all 12 months of the year there is cotton in the fields somewhere in Texas. In the Lower Rio Grande Valley, a 32° F. freeze will occur before January 1, every other year, on an average; while after January 1, a 32° F.

freeze will occur in three out of four years, on an average.

Texas farming and ranching changed greatly in the two decades after World War II. Prewar Texas farms and ranches were largely self-contained, producing much of the supplies, labor and power they required. The State was predominantly rural and agricultural. In 1940, one out of three Texans lived on farms. In 1960, Texas farm population had declined to 694,482 out of a total population that numbered 9,581,508. By 1966, it was estimated that only 560,000 out of 10,500,000 lived on farms and ranches. The number and nature of farms changed appreciably also. There were 418,000 farms in the State in 1940. Twenty years later, there were only 227,000, and in 1969 there were 191,000. Farms are much larger, however, and average much higher investment in land, buildings, and equipment. Total land in farms, 145,000,000 acres in 1969, has not changed appreciably in recent years while the average size of farm shows a steady increase. The average size of a Texas farm in 1969 is 759 acres compared to the national average of 377 acres.

In 1968 total value of production of Texas' principal crops was $1,443,544. The value of all livestock and poultry on Texas' farms, January 1, 1969, was $1,696,781.

Wild cotton grew in Texas before Europeans arrived. Indians and Spaniards used it some, but cotton planted by a member of Stephen F. Austin's colony marked the real beginning of commercial production. Cotton has long been Texas' leading crop, although in recent years, reductions in cotton plantings under federal control programs, have allowed sorghum grain to challenge it in total value. In 1968 the total value of cotton lint produced in Texas was $367,410,000 and cottonseed, $74,052,000. Cotton was harvested from 4,125,000 acres, yielding 404 pounds of lint per acre. The largest acreage ever harvested in the State was in 1926 when 17,749,000 acres produced 4,626,000 500-pound gross weight bales. The largest crop was in 1949 when 6,040,000 500-pound gross weight bales were harvested from 10,900,000 acres. Almost all Texas cotton is mechanically harvested. The southern High Plains, with Lubbock as its hub, is the principal cotton-producing area of the State. Cameron and Hidalgo Counties in the Lower Rio Grande Valley are heavy cotton producers also.

Sorghum grain, Texas' second most important crop, had a total value of $320,333,000 in 1968. More acreage, 6,196,000 acres in 1968, was planted to grain sorghums in Texas than to any other crop, as a result of recent reductions in cotton plantings under federal control programs. Average yield in 1968 was 55.0 bushels per acre. Principal sorghum grain-producing areas are the High Plains, Blacklands, Coastal Bend, and the Lower Rio Grande Valley.

Rice and winter wheat rank third and fourth,

respectively, in importance among Texas' leading crops. Rice acreage is small and concentrated in a few Gulf Coast counties, stretching from about Calhoun and Victoria Counties, eastward to the Louisiana border; however, the value of 1968 production was $142,802,200, from 597,000 acres that produced an average of 4,600 pounds per acre. Two-thirds of Texas' wheat is produced in the Panhandle, but the crop is grown for grain or grazing over most of the State. In 1968, an average of 22.0 bushels of winter wheat was produced from 3,825,000 acres, with a production value of $106,029,000, at $1.26 per bushel.

Hay, including all kinds, occupying 2,376,000 acres in 1968, is the State's sixth most important crop. The value of 1968 production was $105,501,000.

Other important crops are: corn, oats, peanuts, soybeans, potatoes, pecans, citrus, and vegetables. Produced also, but in smaller quantities are: barley, rye, flaxseed, sweet potatoes, broomcorn, sugarbeets, castorbeans, sesame, and peaches. Peanuts is the second most important oilseed, after cottonseed. The 1968 peanut crop, grown on 296,000 acres, was valued at $48,757,000.

Most of Texas' 254 counties grow vegetables for sale, but commercial production is largely concentrated in 112 counties of 11 major areas. The value of commercial vegetables (for fresh market and processing), potatoes, sweet potatoes, and strawberries produced during 1968 totaled $118,558,000. Total acreage of these crops harvested in 1968 was 271,000 acres, 9 percent below the 5-year average (1962-66). The heavy rains of Hurricane Beulah prevented a substantial amount of the 1968 winter crops acreage from being planted. Leading vegetable crops in value of production are onions, carrots, potatoes, cantaloups, and cabbage. In acreage harvested, watermelons is the leader. Nationally, Texas ranks third in harvested acreage, production and value of fresh market vegetables. Texas ranks first in the nation in harvested acreage for fresh market for onions, watermelons, spinach, and beets. Other important fresh vegetables produced are: snap beans, broccoli, cauliflower, sweet corn, cucumbers, honeydew melons, lettuce, green peppers and tomatoes.

Citrus is the principal commercial fruit of Texas, which ranks among the three leading citrus states. Production is concentrated in the Lower Rio Grande Valley (Cameron, Hidalgo, Willacy, and Starr Counties). The value of Texas citrus has varied widely, partly as a result of severe freezes in the Lower Valley on January 29-31, 1949, January 29-February 3, 1951, and January 9-12, 1962. The freeze of January 29-February 3, 1951, was the most disastrous of this century, as far as the Lower Valley is concerned. At Brownsville temperatures were 32° F. or below for 93 hours during the cold period; 65 of these were consecutive between January 29-February 1. The severe

freeze was preceded by 52 days of mild temperatures. Buds on trees were actively growing; as a result, an estimated 75 percent of Valley citrus trees were killed. Citrus production, which had not recovered from the 1949 freeze, dropped from 10.2 million boxes in the 1950-51 crop year to only 0.5 million boxes for the 1951-52 year. For the next 10 consecutive years, 1952 through 1961, the Lower Rio Grande Valley did not experience a severe freeze. Most of the Texas citrus is marketed fresh, between September and March, but there is increasing processing for juice and fruit. The 1968 crop totalled 4,100,000 boxes of oranges and 6,500,000 boxes of grapefruit.

An estimated 56 percent of the total value of crops grown in Texas is produced through irrigation, although only about 25 percent of the harvested cropland is irrigated.

Historically, Texas is the nation's leading livestock state. The State continues to rank first in total cattle numbers, beef cattle, sheep and goats. According to the Texas Crop and Livestock Reporting Service, total value of livestock and poultry was $1,696,781,000 on January 1, 1969. The 11,521,000 cattle in Texas accounted for 92.3 percent of the total value. All sheep accounted for 3.9 percent, hogs 1.3 percent, goats 1.2 percent and poultry 1.3 percent of total value of livestock. The number of cattle on Texas farms and ranches set a new record for the fourth consecutive year. Milk cows, sheep, goats, and chickens are in a downward trend while hog and turkey production is increasing.

Texas is rapidly becoming one of the nation's most highly industrialized states. Almost 709,000 persons were estimated by the Texas Employment Commission to be at work in the State's factories in December 1968. Value added to the economy is estimated by the Business and Defense Services Administration to reach $11.8 billion. The value of Texas mineral production in 1968 is estimated at $5.67 billion by the Bureau of Mines. In 1967 transportation equipment surpassed food processing as Texas' largest employment category in manufacture, and it still retains this lead. In December 1968 an estimated 101,400 persons were working in the Texas transportation-equipment industry compared to 86,800 food-processing employees. Texas employment in the production of chemicals and allied products totalled 62,300 in 1968. Texas leads the other forty-nine states in petroleum refining capacity, in value added by manufacture of refinery products, and in refining employment. The Texas lumber and wood products industry includes slightly over 8 percent of our national pulping capacity. Other major industries include: electrical and non-electrical supplies, electronic systems and equipment, chemicals and allied products, apparel, primary metals (steel, aluminum and magnesium), stone, clay, glass, and concrete. Texas accounts for about 90 percent of the magnesium manufacturing capacity in the nation, while steel companies produce thirteen of the twenty-five major product categories of the steel industry within the State.

In 1968 Texas ranked second among all fifty states in the amount of money spent on prime military contracts by the federal government. The transportation-equipment industry, chiefly the aircraft segment, accounted for the largest share.

Tonnages and values of commerce through Texas ports have been at peak levels in recent years. Texas ranks among the leading states in total export values, value of agricultural exports, petroleum chemicals, sulphur and food. Thirteen major ports handle almost all of the total. The value of Texas manufactured products exported in 1968 was estimated at $1.5 billion. Houston, Texas' largest city, also is its leading port and usually ranks second or third nationally in tonnage among deepwater ports. Intracoastal commerce is handled by the Gulf Intracoastal Waterway, a 12-feet-deep by 125-feet-wide channel that parallels the Texas Coast for 423 miles between Brownsville and Port Arthur. In addition to waterborne commerce, increasing tonnages move to and from Texas by truck, air, and rail.

The increases in urbanization, industrialization, and agricultural activities in Texas have increased air-pollution problems also. Many of the atmospheric pollution problems in Texas arise from the fact that industrial plants which once were remote from homes and business institutions now have been surrounded by the sprawling towns and cities in which most Texans live. Most recently, the recognition of these problems by a concerned citizenry has led to a more concentrated effort toward abatement and control of the contamination of the atmosphere.

The weather is a large-scale creator of air pollution problems. When serious pollution episodes occur they happen not so much because of a great or sudden increase in the output of pollutants as because of adverse weather elements which trap the pollutants in a mass of stagnant air. One of the most important factors in the dispersion of air pollution is wind. Wind direction determines the part of the city that may be affected by the transport of pollutants from given sources or areas. Wind speed is important since the concentration of pollutants is inversely proportional to wind speed. Horizontal wind speeds of 7 m.p.h. or less are generally considered conducive to high air pollution potential; the weakest transport and dilution effects occur at low wind speeds.

A second important meteorological factor in the potential for air pollution is stability, which may be described simply as resistance to change. In the atmosphere it may be measured by the vertical variations (lapse rate) of temperature.

Air unsaturated by water vapor is said to have neutral stability if its temperature decreases at the rate of 5.4° F. for each 1,000-foot increase in elevation. A stable condition exists when the temperature decreases with height less rapidly than the above rate, which is called the dry-adiabatic lapse rate. A more rapid decrease of temperature with height through a layer of air is an unstable condition. Unstable lapse rates favor vertical motions and atmospheric mixing, and accelerate the diffusion of air pollutants, while stable lapse rates oppose vertical motions and inhibit the diffusion of air pollutants.

The Gulf of Mexico plays an important role in the Texas potential for air pollution. Its 624 miles of coastline are subjected to land and sea breezes which affect the transport of pollutants in coastal areas. The sea breeze is a localized coastal circulation which has the surface winds blowing from sea to land. It is most frequently a daytime phenomenon. This alternates with a usually weaker nighttime circulation of the opposite direction which is called a land breeze. The sea breeze circulation extends about 25 miles inland. Thus, the strength and diurnal variation of the sea breeze is quite important in determining the transport of air pollutants in coastal areas. The Gulf of Mexico, being a warm body of water, contributes to the instability of the air passing over it. This is in direct contrast to the cold California Current, which stabilizes the off-shore circulation along the southern California coast. In the fall and winter the surface-water temperature of the Gulf of Mexico off the Texas coast averages about 3° F. warmer than the temperature of the layer of air near the water surface. This results in a decrease in the lapse rate of the layer of air near the warm-water surface and makes it less stable. This tends to inhibit the formation of temperature inversions over the land at night. Thus, the influence of the Gulf of Mexico is to decrease the air-pollution potential of the adjacent land areas during those seasons when other meteorological factors contribute to a higher potential.

From about the first of May to the end of September, Maritime Tropical air from the Gulf of Mexico and southern North Atlantic source regions largely controls the Texas weather. Because of its long trajectory across a warm ocean, Maritime Tropical air is warm, humid, and conditionally stable. This airmass is responsible for almost all of the thundershower activity in Texas, and for most of the State's precipitation. As it moves across the hot land surface during the day the airmass becomes warmer in the lower layer; it becomes unstable and rises to form the cumulus clouds so characteristic of Texas' summer skies. The vertical motions generated by this dynamic process dilute any concentration of air pollutants present, mixing them into a larger volume of air. The climatic controls imposed on the State by the Gulf of Mexico favor a low air-pollution potential, and from this point of view, Texas is fortunate indeed to be so closely associated with this warm body of water.

Three factors which tend to increase air pollution--light winds, atmospheric stability, and photochemical reaction--are associated with high-pressure systems (anticyclones). Major air-pollution episodes are related to the incidence of stagnating cells of high atmospheric pressure over urban areas. In such cases the anticyclone lingers over an area for a protracted period of four days or longer. Under these meteorological conditions the air-pollution potential of the area reaches its maximum. Most anticyclones that enter Texas are transitory and continue their movement across the State without pausing long enough to meet the criteria prescribed for stagnating anticyclones. These migratory systems provide a change in the airmass over a particular area. The "dirty" air, contaminated by local pollutant sources, moves eastward or southward out of the area, and a fresh, clean airmass replaces it. Thus, the air-pollution potential of the area is determined partially by the frequency and speed at which these migratory anticyclones move across the area.

Since few high-pressure systems stagnate over Texas for any length of time, the meteorological situation most favorable for serious air-pollution episodes is rarely present. In Texas stagnation is more likely to occur over the East Texas Pine Belt than elsewhere. Thus, the "restless" Texas climate, characterized by frequent changes in airmass and by numerous local and regional-scale weather disturbances, does not favor objectionable concentrations of air pollutants much of the time. However, the frequency of occurrence of light winds and temperature inversions at night may create pollution problems at some locations during certain periods of the year.

The variety of recreation available in Texas is unsurpassed among the 50 states. Those seeking fun and relaxation have almost endless choices. For vacationers who have not seen Texas and expect a flat, barren land limited to cowboys and oil fields, a visit to Texas is as surprising as it is delightful. Inland water, more square miles of it than in any other state except Alaska, plus hundreds of miles of sandy beach along the Gulf of Mexico, offer every kind of water sport. Towering pines and hardwoods cover millions of acres, and there are 65 State Parks and two National Parks to visit.

Travel is made easy by the largest all-paved toll-free highway network in the world. The State's vast highway system encompasses approximately 68,000 miles of designated highways, of which 39,000 miles are farm-to-market and ranch-to-market roads. On this system are more than 700,000 signs and markers and 1,100 safety rest areas maintained by the Texas Highway Department for the convenience and safety of the traveling public. Texas is America's most popular gateway to Mexico, with six principal entryways into that republic.

TROPICAL CYCLONE DATA HAVING IMPORTANCE FOR THIS STATE IS INCLUDED IN STATISTICS AND CHARTS ON PAGES 1161 **THROUGH** 1164.

Normal Annual Temperature

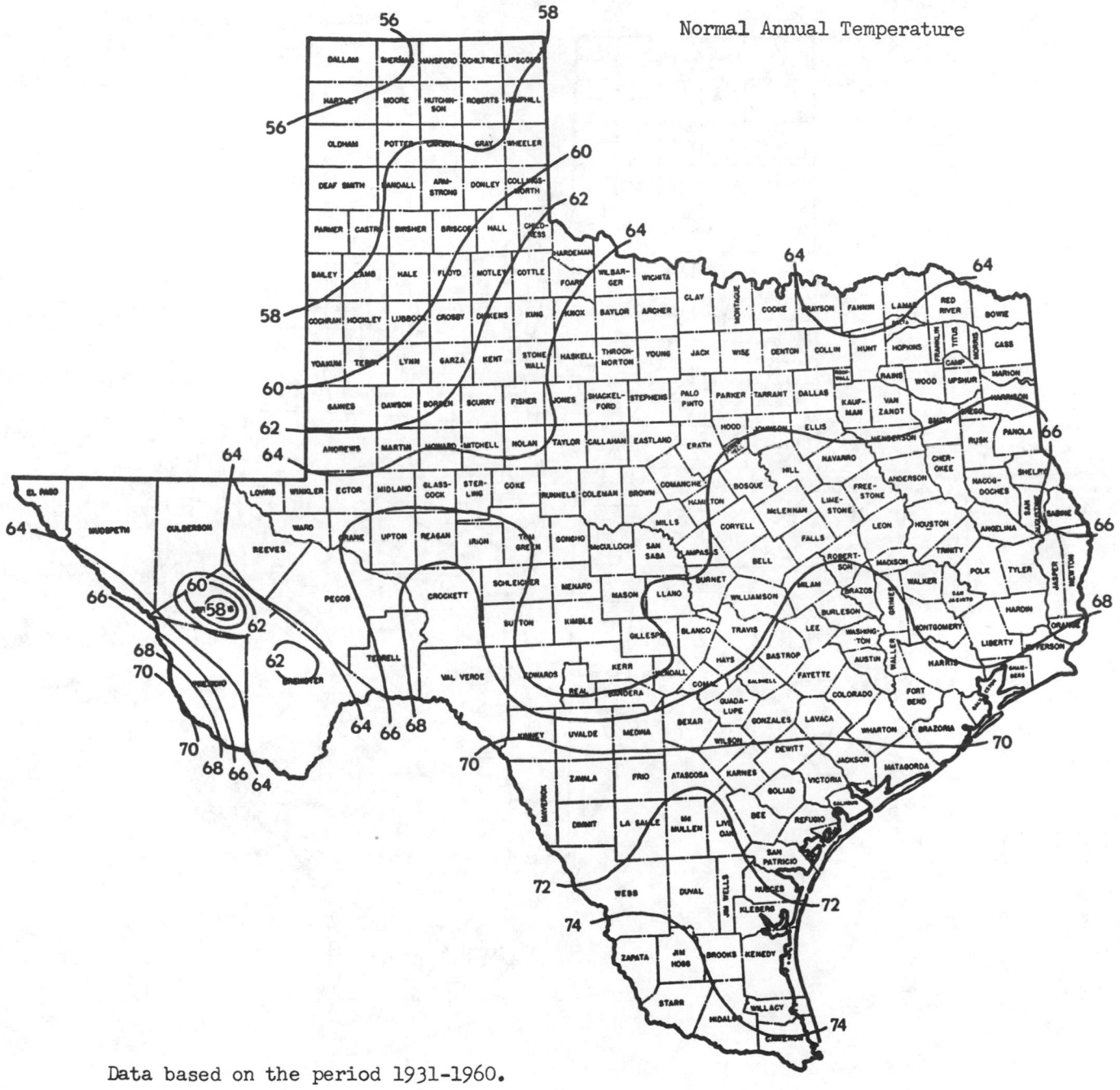

Data based on the period 1931-1960.

October 1968   577.21.1

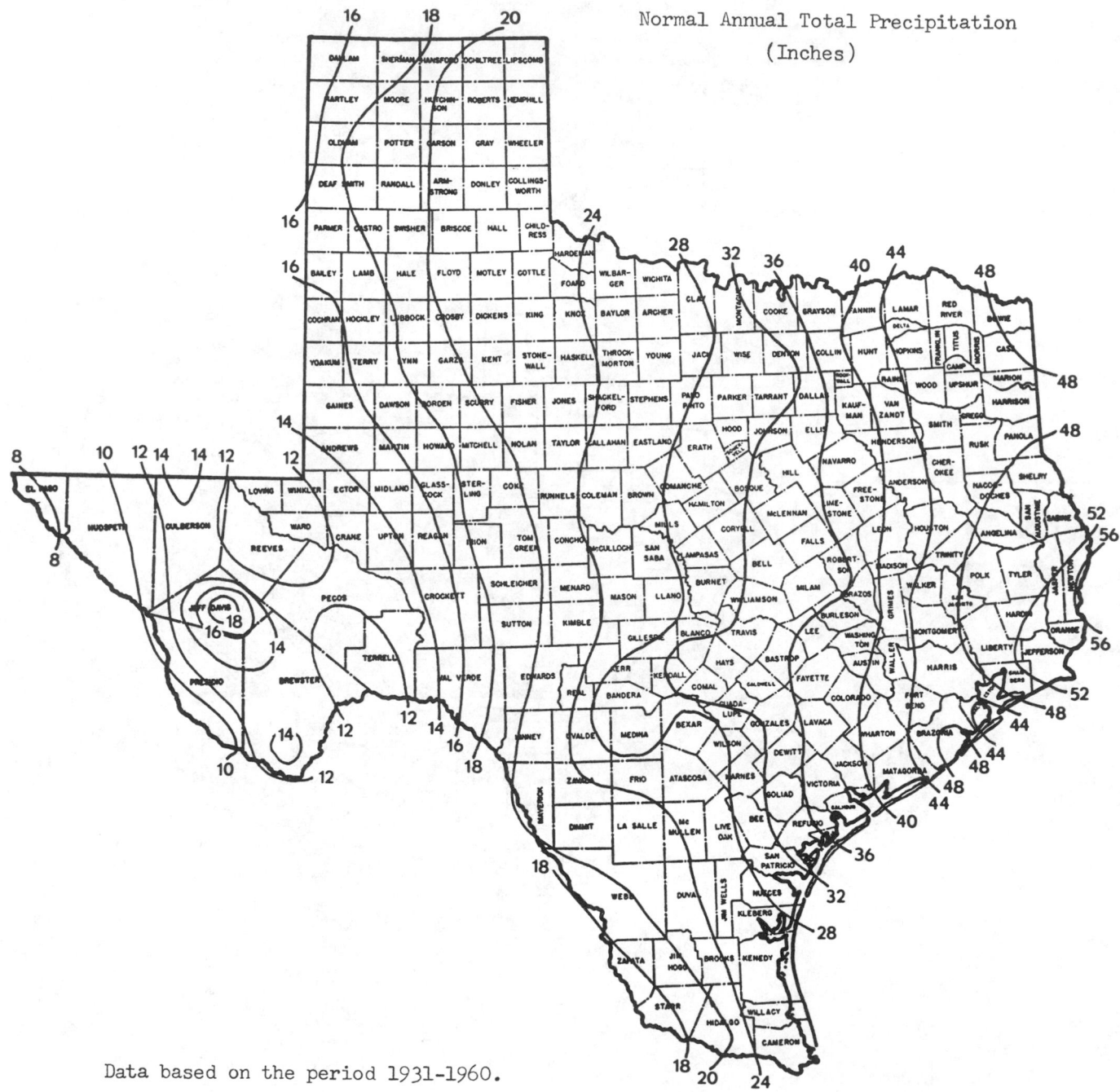

Normal Annual Total Precipitation
(Inches)

Data based on the period 1931-1960.

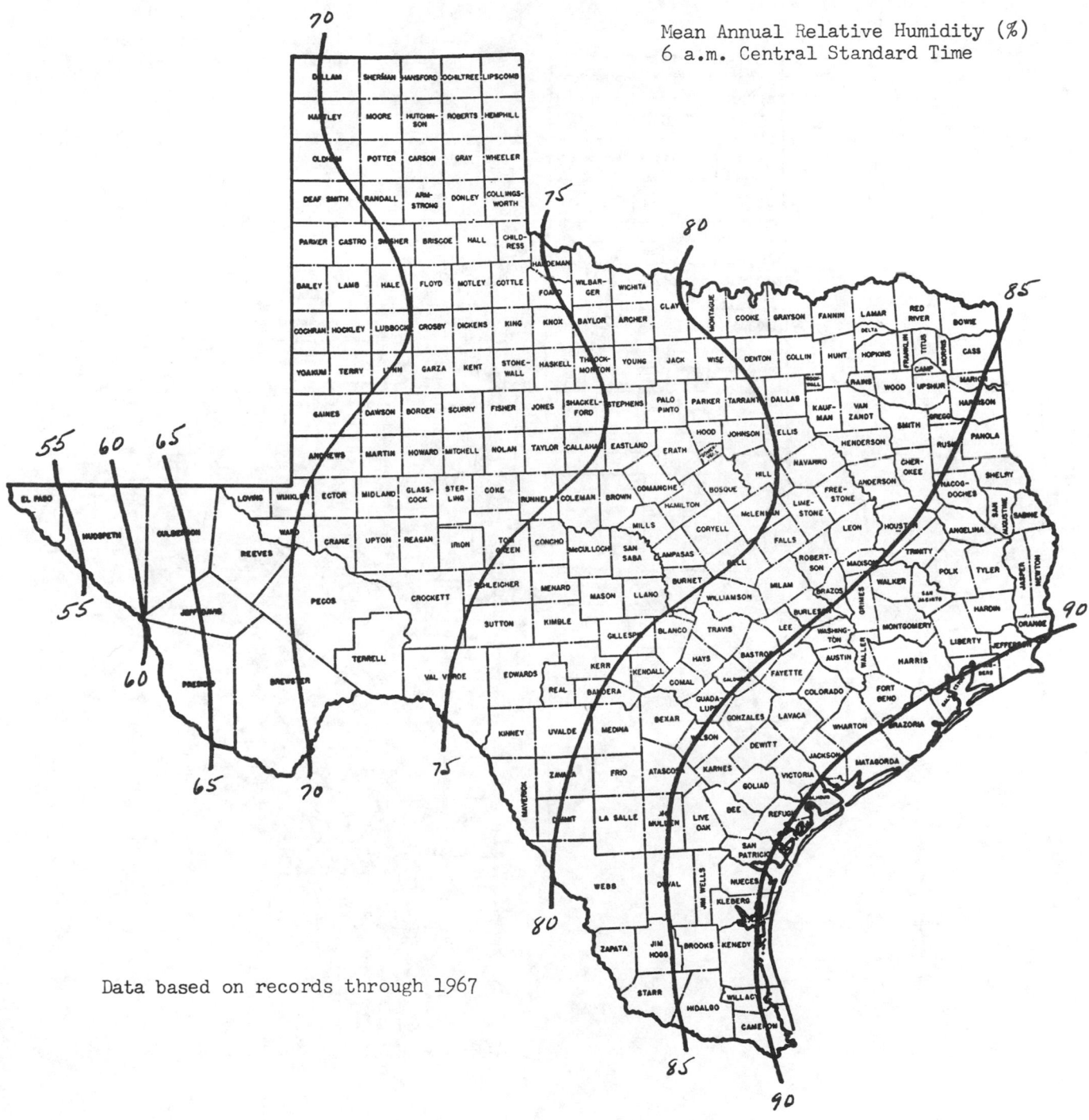

Mean Annual Relative Humidity (%)
6 a.m. Central Standard Time

Data based on records through 1967

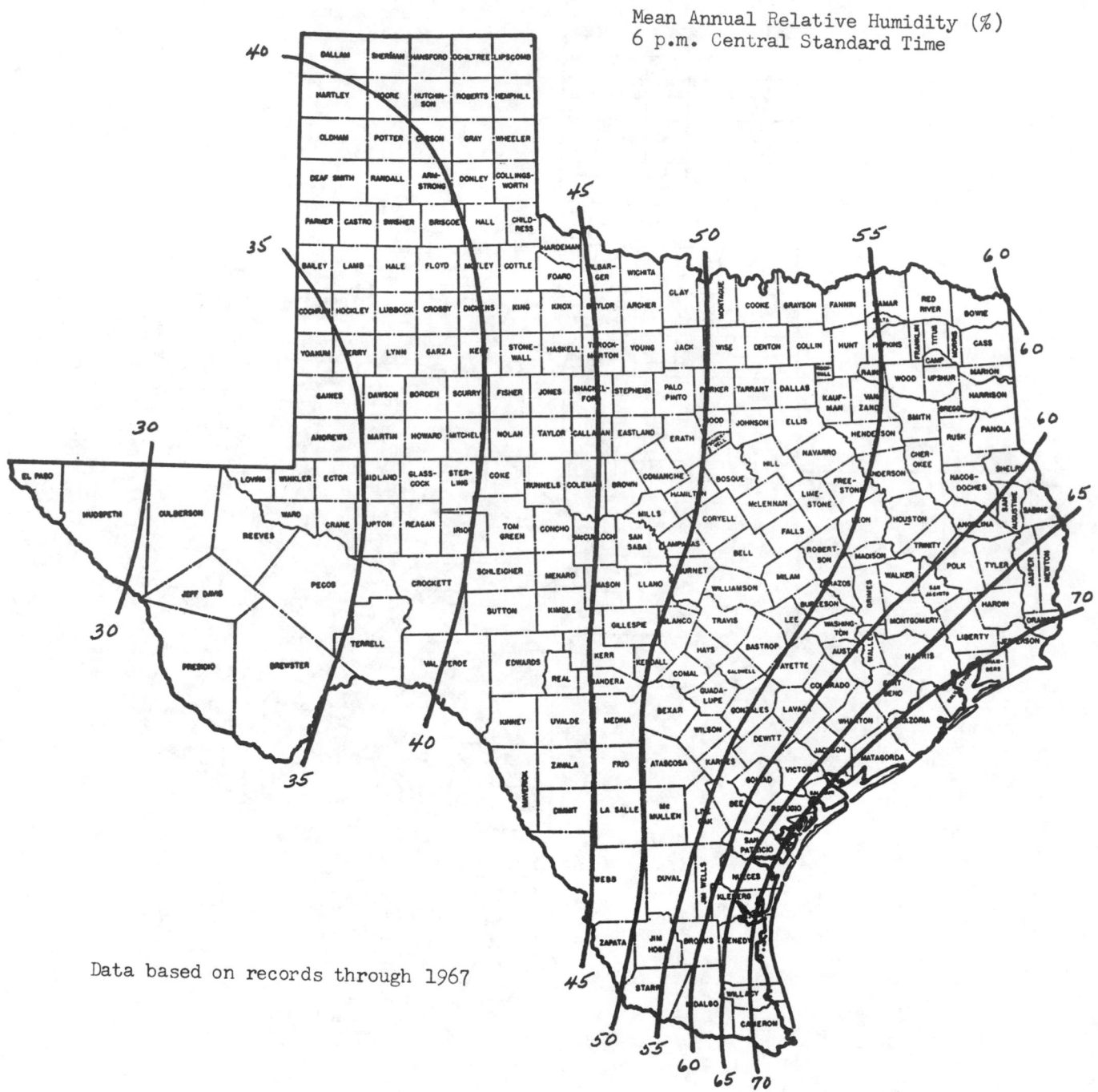

Mean Annual Relative Humidity (%)
6 p.m. Central Standard Time

Data based on records through 1967

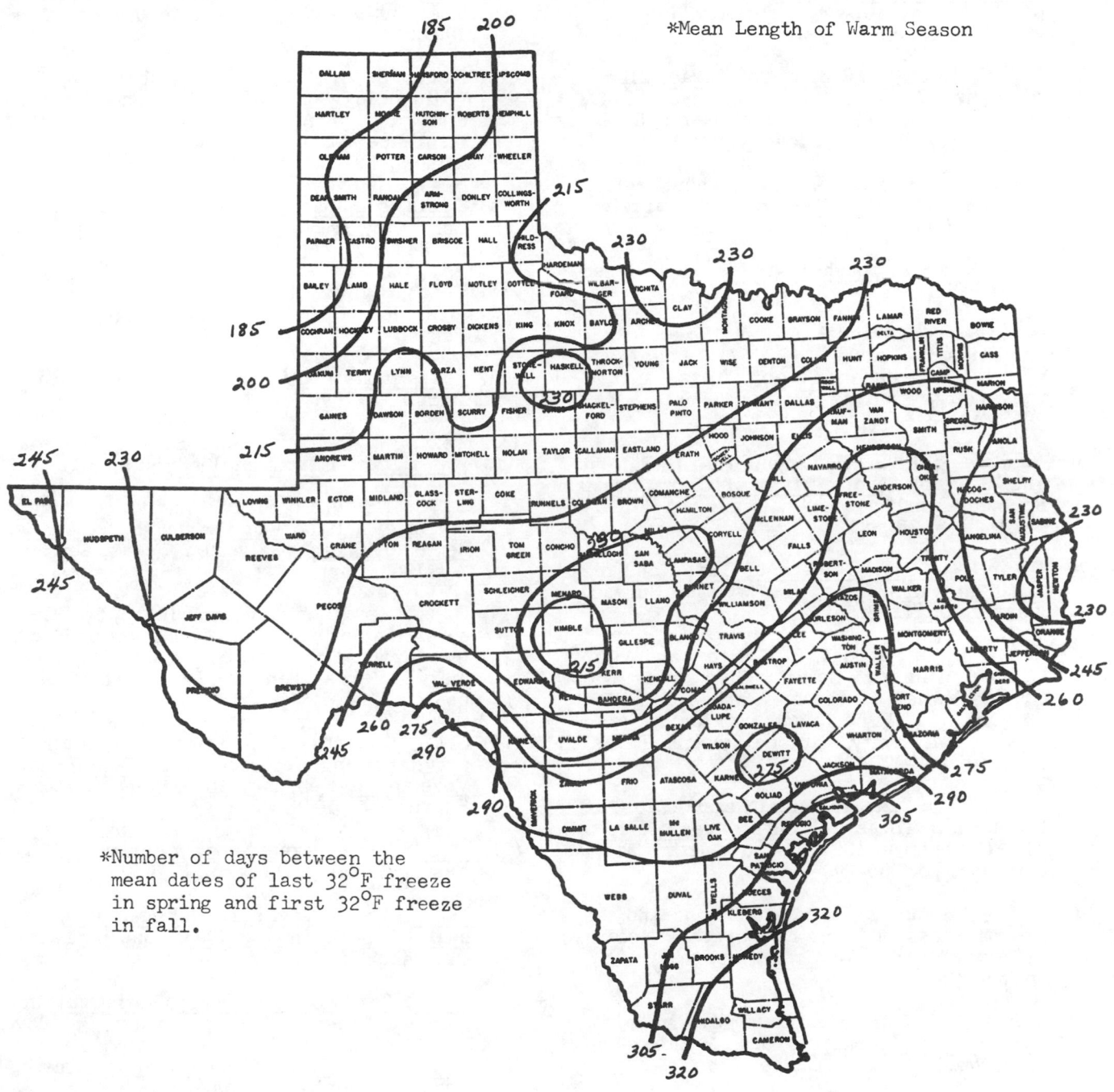

*Mean Length of Warm Season

*Number of days between the
mean dates of last 32°F freeze
in spring and first 32°F freeze
in fall.

REFERENCES

(1) Arbingast, S. A., "Texas Industry, 1968", _Texas Business Review_, The University of Texas at Austin, February 1969, pp. 29-38.

(2) Blood, R. D. W., _Climate of Texas_, Climatography of the United States No. 60-41, U. S. Weather Bureau, Washington, D. C., February 1960, 27 pp. (out of print).

(3) Chambers, William T., "Geographic Regions of Texas", _Texas Geographic Magazine_, Spring, 1948.

(4) Dunn, Roy S., _Drought in Texas_, unpublished manuscript.

(5) Hildreth, R. J., and R. Orton, "Freeze Probabilities in Texas", _MP-657_, Texas Agricultural Experiment Station, Texas A & M University, College Station, Texas, May 1963, 19 pp.

(6) Landsberg, H. E., H. Lippman, K. H. Paffen, and C. Troll, _World Maps of Climatology_, Springer-Verlag, Berlin-Gottinger-Heidelberg, 1963, 28 pp.

(7) Orton, R., _The Climate of Texas and the Adjacent Gulf Waters_, U. S. Weather Bureau, Washington, D. C., 1964, 195 pp.

(8) ____, D. J. Haddock, E. G. Bice, and A. C. Webb, "Climatic Guide, The Lower Rio Grande Valley of Texas, " _MP-841_, Texas Agricultural Experiment Station, Texas A & M University, College Station, Texas, September 1967, 108 pp.

(9) ____, "The Meteorological Potential for Air Pollution in Texas", _Texas Business Review_, The University of Texas at Austin, October 1968, pp. 285-290.

(10) ____, "Summer Air Conditioning and the Effectiveness of Home Evaporative Coolers in Texas", _ASHRAE Jour._, February 1963, pp. 65-70.

(11) Palmer, W. C., "Meteorological Drought", _Research Paper No. 45_, U. S. Weather Bureau, Washington, D. C., pp. 1-3.

(12) Portig, Wilfred H., _Atlas of the Climates of Texas_, Bureau of Engineering Research, The University of Texas at Austin, June 1962, 33 pp.

(13) Stout, G. E., and S. A. Changnon, Jr., "Climatology of Hail in the Central United States", _CHIAA Report No. 38_, Crop Hail Insurance Actuarial Association, Chicago, Illinois, February 1968, 48 pp.

(14) _Texas Almanac_, 1968-1969 edition, A. H. Belo Corporation, Dallas, Texas, pp. 12-30, 344-389, 456-460.

(15) Texas Highway Department, Travel and Information Division, _Texas Facts_, Austin, Texas.

(16) _Texas Industrial Commission News Release_, March 3, 1969.

(17) Texas Water Development Board, _The Texas Water Plan_, November 1968, 50 pp.

(18) U. S. Army Engineer District, Corps of Engineers, _Report on Hurricane Beulah, 8-21 September 1967_, Galveston, Texas, September 1968, 26pp.

(19) U. S. Department of Agriculture, Soil Conservation Service, State Conservation Needs Inventory Committee, _State Land Use Summary - Texas_, Temple, Texas, April 1968, unpublished.

(20) U. S. Department of Agriculture, Statistical Reporting Service, _Texas Commercial Vegetables, February 1969_, Austin, Texas, February 7, 1969.

(21) ____, _Texas Crop Report, February 1969_, Austin, Texas, February 11, 1969.

(22) ____, _Texas Commercial Vegetables, Annual Crop Summary, 1968_, Austin, Texas, December 30, 1968.

(23) ____, _Texas Crop Summary, Statistical Supplement_, Austin, Texas, January 2, 1969.

(24) ____, _1968 Texas Crop Summary_, Austin, Texas, December 23, 1968.

(25) ____, _Texas Annual Livestock and Poultry Inventory_, Austin, Texas, February 13, 1969.

BIBLIOGRAPHY

1. Carr, J. T., "Hurricanes Affecting the Texas Gulf Coast", Report No. 49, Texas Water Development Board, Austin, Texas, June 1967, 58 pp.

2. _____, "The Climate and Physiography of Texas", Report No. 53, Texas Water Development Board, Austin, Texas, July 1967, 27 pp.

3. _____, "Texas Droughts", Report 30, Texas Water Development Board, Austin, Texas, November 1966, 58 pp.

4. Cotton Economic Research, "Weather Data and Cotton Production", Research Report No. 75, The University of Texas at Austin, Austin, Texas, 1965.

5. Cry, G. W., "Effects of Tropical Cyclone Rainfall on the Distribution of Precipitation Over the Eastern and Southern United States.", ESSA Professional Paper No. 1, Environmental Science Services Administration, Washington, D. C., June 1967, 67 pp.

6. _____, "Tropical Cyclones of the North Atlantic Ocean", U. S. Weather Bureau, Technical Paper No. 55, Washington, D. C., 1965, 148 pp.

7. Environmental Science Services Administration, Environmental Data Service, Climatological Data, National Summary, Washington, D. C.

8. _____, Climatic Atlas of the United States, Washington, D. C., June 1968, 80 pp.

9. _____, Climatic Summaries of Resort Areas, Climatography of the United States No. 21-41, for Marlin and Mineral Wells, Texas.

10. _____, Climatological Summary, Climatography of the United States No. 20-41, Austin, Texas, for selected climatological substations in Texas.

11. _____, Climatography of Texas, a series of maps, statistical and descriptive information about the climate of the State, Weather Bureau Office for State Climatology, Austin, Texas.

12. _____, Climatological Data - Texas, Washington, D. C.

13. _____, Daily Normals of Temperature and Heating Degree Days, for period 1931-1960, Washington, D. C., 1963.

14. Environmental Science Services Administration, Environmental Data Service, Decadal Census of Weather Stations - Texas, Key to Meteorological Records Documentation No. No. 6.11, Washington, D. C., 1966.

15. _____, Heating Degree Day Normals, 1931-1960, Washington, D. C., 1963.

16. _____, Hourly Precipitation Data - Texas, Washington, D. C.

17. _____, Local Climatological Data, Washington, D. C., published monthly for Abilene, Amarillo, Austin, Brownsville, Corpus Christi, Dallas, Del Rio, El Paso, Fort Worth, Galveston, Houston, Lubbock, Midland, Port Arthur, San Angelo, San Antonio, Victoria, Waco and Wichita Falls.

18. _____, Monthly Normals of Temperature, Precipitation and Heating Degree Days, 1931-1960, Washington, D. C., 1962.

19. _____, Storm Data - Texas, Washington, D. C.

20. _____, Weekly Weather and Crop Bulletin, Washington, D. C.

21. Friedman, D. C., "The Prediction of Long Continuing Drought in Southwest Texas", Occasional Research Papers in Meteorology, No. 1, The Travelers Insurance Company, Hartford, Conn., 1957, 182 pp.

22. Griffiths, J. F., and R. Orton, "Agroclimatic Atlas of Texas", Part I, Precipitation Probabilities, MP-888, Texas Agricultural Experiment Station, Texas A & M University, College Station, Texas. August 1968

23. Hogan, W. L., Hurricane Carla, Leaman-Hogan Company, Houston, Texas, 1961, 192 pp.

24. Johnson, W. C., "Wind in the Southwestern Great Plains", Conservation Research Report No. 6, U. S. Department of Agriculture, Agricultural Research Service, Washington, D. C., December 1965, 65 pp.

25. Kane, J. W., "Monthly Reservoir Evaporation Rates for Texas", Report No. 64, Texas Water Development Board, Austin, Texas, October 1967, 111 pp.

26. Kohler, M. A., T. J. Nordenson, and Baker, D. R., "Evaporation Maps for the United States," Technical Paper No. 37, U. S. Weather Bureau, Washington, D. C., 1959, 13 pp.

27. Lowry, R. L., "A Study of Droughts in Texas", Bulletin 5914, Texas Board of Water Engineers, Austin, Texas, 1959, 76 pp.

28. _____, Surface Water Resources of Texas, Texas Electric Service Company, Austin, Texas, 1958, 49 Figures.

29. _____, "Excessive Rainfall in Texas", Bulletin No. 25, Reclamation Department, the State of Texas, Austin, Texas, 1934, 149 pp.

30. Mills, W. B., and E. E. Schroeder, Floods of April 28, 1966 in the Northern Part of Dallas, Texas, Geological Survey, Water Resources Division, Department of the Interior, Austin, Texas, December 1966, 74 pp.

31. Patterson, J. L., "Floods in Texas", Bulletin 6311, Texas Water Commission, Austin, Texas, 1963, 31 pp.

32. Price, W. A., "Hurricanes Affecting the Coast of Texas from Galveston to Rio Grande", U. S. Army Corps of Engineers, Technical Memorandum No. 78, Washington, D. C., 1956, 17 pp.

33. Sugg, A. L., and R. L. Carrodus, "Memorable Hurricanes of the United States since 1873", Technical Memorandum WBTM-SR-42, Weather Bureau Southern Region, Fort Worth, Texas, January 1969, 50 pp.

34. Texas Water Commission, "Symposium on Consideration of Droughts in Water Planning", Bulletin 6512, Austin, Texas, April 1965, 177 pp.

35. Texas Water Development Board, "Floods From Hurricane Beulah in South Texas and Northeastern Mexico September - October 1967", Report 83, Austin, Texas, 1968, 195 pp.

36. Texas Water Development Board, "Symposium on Consideration of Some Aspects of Storms and Floods in Water Planning", Report No. 33, Austin, Texas, November 1966, 163 pp.

37. Thomas, H. E., and others, "Effects of Drought in Central and South Texas", Geological Survey Professional Paper 372-C, U. S. Department of the Interior, Geological Survey, Washington, D. C., 1963, 31 pp.

38. ____, "Effects of Drought in the Rio Grande Basin", Geological Survey Professional Paper 372-D, U. S. Department of the Interior, Geological Survey, Washington, D. C., 1963, 59 pp.

39. Treadwell, M. E., Hurricane Carla, September 3-14, 1961, U. S. Department of Defense, Washington, D. C., 1961, 97 pp.

40. U. S. Army Corps of Engineers, Storm Rainfall in the United States, depth-area-duration data (complete through 1954).

41. U. S. Department of Agriculture, Soil Conservation Service, Land Resource Areas, Texas, Temple, Texas, 1962, 29 pp.

42. U. S. Department of Agriculture, Statistical Reporting Service, Texas Crop Weather, published weekly, Austin, Texas

43. U. S. Study Commission - Texas, Part I, The Commission Plan, March 1962, 199 pp.; Part II, Resources and Problems, March 1962, 363 pp.; Part III, The Eight Basins, March 1962, 217 pp.; Part IV, Summary and Recommendations, April 1962, 18 pp., Washington, D. C.

44. U. S. Weather Bureau, The Climate of Central and Coastal Texas Watersheds, National Weather Records Center, Asheville, North Carolina, 1961, 14 pp.

45. ____, Climatic Guide, Houston-Galveston, Texas, area, Climatography of the United States No. 40-41, Washington, D. C., September 1962.

46. ____, Climatic Summary of the United States (Bulletin W), 1930 edition, Sections 30, 31, 32 and 33, Washington, D. C., 1932.

47. ____, Climatic Summary of the United States, Texas - Supplement for 1931 Through 1952, (Bulletin W Supplement), Climatography of the United States No. 11-36, Washington, D.C., 1958.

48. ____, Climatic Summary of the United States, Texas - Supplement for 1951 Through 1960, (Bulletin W Supplement), Climatography of the United States No. 86-36, Washington, D. C., 1965.

49. ____, Climatological and Oceanographic Atlas for Mariners, Vol. I, North Atlantic Ocean, Washington, D. C., 1959.

50. ____, "Maximum 24-Hour Precipitation in the United States", Technical Paper No. 16, Washington, D. C., 1952, 284 pp.

51. ____, "Maximum Recorded United States Point Rainfall", Technical Paper No. 2, Washington, D. C., revised 1963, 56 pp.

52. ____, "Maximum Station Precipitation for 1, 2, 3, 6, 12, and 24 Hours; Part XXIV: Texas", Technical Paper No. 15, Washington, D. C., 1959.

53. U. S. Weather Bureau, "Normal Monthly Number of Days with Precipitation of 0.5, 1.0, 2.0, and 4.0 Inches or More in the Conterminous United States", Technical Paper No. 57, Washington, D. C., 1966, 52 pp.

54. ____, "North Atlantic Tropical Cyclones", Technical Paper No. 36, Washington, D. C., 1959, 214 pp.

55. ____, "Principal Tracks and Mean Frequencies of Cyclones and Anticyclones in the Northern Hemisphere", Research Paper No. 40, Washington, D. C., 1957, 60 pp.

56. _____, "Rainfall and Floods of April, May, and June 1957 in the South Central States", Technical Paper No. 33, Washington, D. C., 350 pp.

57. _____, "Rainfall Frequency Atlas of the United States for Durations from 30 Minutes to 24 Hours and Return Periods from 1 to 100 Years", Technical Paper No. 40, Washington, D. C., 1961, 115 pp.

58. _____, "Rainfall Intensity - Duration - Frequency Curves", for selected stations in the United States, Alaska, Hawaiian Islands, and Puerto Rico, Technical Paper No. 25, Washington, D. C., 1955, 53 pp.

59. _____, "Seasonal Variation of the Probable Maximum Precipitation East of the 105th Meridian for Areas from 10 to 1000 Square Miles and Durations of 6, 12, 24, and 48 Hours", Hydrometeorological Report No. 33, Washington, D. C., 1956, 58 pp.

60. _____, Substation History - Texas, Key to Meteorological Records Documentation No. 1.1, Washington, D. C., 1956, 223 pp.

61. _____, Summary of Hourly Observations, 1951-1960, Climatography of the United States No. 82-41, Washington, D. C., where the full 10-year period is not covered by the monthly data, summaries are based on the period 1956-1960; published for Amarillo, Austin Brownsville, Corpus Christi, Dallas, El Paso, Galveston, Houston, Laredo, Lubbock, Midland, San Antonio and Waco.

62. _____, "The Tornadoes at Dallas, Texas, April 2, 1957", Research Paper No. 41, U. S. Weather Bureau, Washington, D. C., 1960, 175 pp.

63. _____, "Tornado Deaths in the United States", Technical Paper No. 30, Washington, D. C., 1957, 48 pp.

64. _____, "Tornado Occurrences in the United States", Technical Paper No. 20, Washington, D. C., Revised 1960, 71 pp.

65. _____, "Two- to Ten-Day Precipitation for Return Periods of 2 to 100 Years in the Contiguous United States", Technical Paper No. 49, Washington, D. C., 1964, 29 pp.

66. _____, "Upper-Air Climatology of the United States", Technical Paper No. 32, Part I, "Averages for Isobaric Surfaces"; Part III, "Vector Winds and Shear", Washington, D. C., 1957.

67. _____, "Upper Wind Distribution Statistical Parameter Estimates", Technical Paper No. 34, Washington, D. C., 1958, 59 pp.

FREEZE DATA

STATION	Freeze threshold temperature	Mean date of last Spring occurrence	Mean date of first Fall occurrence	Mean No. of days between dates	Years of record Spring	No. of occurrences in Spring	Years of record Fall	No. of occurrences in Fall	STATION	Freeze threshold temperature	Mean date of last Spring occurrence	Mean date of first Fall occurrence	Mean No. of days between dates	Years of record Spring	No. of occurrences in Spring	Years of record Fall	No. of occurrences in Fall
ABILENE WB AIRPORT	32	03-26	11-12	231	10	10	10	10	BOERNE	32	03-25	11-09	229	30	30	29	29
	28	03-22	11-18	241	10	10	10	10		28	03-13	11-19	251	30	30	29	29
	24	03-03	11-29	271	10	10	10	10		24	02-22	12-11	292	30	29	29	22
	20	02-13	12-13	303	10	9	10	8		20	01-28	12-18	324	30	19	29	13
	16	01-24	12-24	334	10	8	10	4		16	01-13	12-27	348	30	14	29	6
ALBANY	32	04-03	11-05	216	30	30	30	30	BONHAM	32	03-30	11-07	222	25	25	24	24
	28	03-26	11-15	234	29	29	29	29		28	03-11	11-16	250	26	26	25	25
	24	03-06	11-26	265	29	29	29	29		24	02-24	12-04	283	26	26	27	24
	20	02-21	12-11	293	29	28	29	25		20	02-06	12-18	315	27	24	26	15
	16	02-02	12-18	319	29	24	29	17		16	01-27	12-25	332	28	20	26	9
ALICE CAA AIRPORT	32	02-20	12-04	287	23	21	20	17	BORGER 3 W	32	04-17	10-28	194	10	10	10	10
	28	01-21	12-27	340	22	12	20	6		28	04-03	11-05	216	10	10	10	10
	24	01-10	⊕	⊕	22	7	21	2		24	03-23	11-12	234	10	10	10	10
	20	01-06	⊕	⊕	22	4	22	0		20	03-14	11-19	250	10	10	10	10
	16	⊕	⊕	⊕	22	1	22	0		16	03-02	12-05	278	10	10	10	9
ALPINE	32	03-30	11-10	225	29	29	29	29	BRADY 2NNW	32	03-29	11-11	227	24	24	23	23
	28	03-21	11-22	246	28	28	29	29		28	03-14	11-16	247	24	24	23	23
	24	03-07	12-02	270	28	28	29	26		24	02-16	12-02	289	22	20	23	21
	20	02-07	12-16	312	28	21	29	15		20	01-31	12-15	318	22	18	23	14
	16	01-29	12-26	331	28	20	29	8		16	01-20	12-26	340	22	12	23	4
AMARILLO WB AIRPORT	32	04-20	10-28	191	10	10	10	10	BRENHAM	32	02-25	12-02	280	30	28	30	28
	28	04-09	11-04	209	10	10	10	10		28	02-08	12-15	310	30	27	30	17
	24	03-31	11-11	225	10	10	10	10		24	01-21	12-26	339	29	18	30	11
	20	03-16	11-19	248	10	10	10	10		20	01-13	12-29	350	29	11	30	4
	16	03-09	11-29	265	10	10	10	9		16	01-06	⊕	⊕	29	8	30	0
ANGLETON 2 W	32	03-02	11-30	273	30	29	30	28	BRIDGEPORT	32	03-30	11-05	220	26	26	26	26
	28	02-10	12-12	305	30	25	30	20		28	03-15	11-19	249	26	26	26	26
	24	01-16	12-23	341	30	13	30	10		24	02-24	12-02	281	26	26	26	25
	20	01-06	⊕	⊕	30	7	30	1		20	02-09	12-14	308	26	23	26	18
	16	01-04	⊕	⊕	30	4	30	1		16	01-26	12-21	329	25	20	26	9
AUSTIN WB AIRPORT	32	03-07	11-22	260	10	10	10	10	BRONSON	32	03-24	11-08	229	27	27	26	26
	28	02-05	12-07	305	10	9	10	9		28	03-06	11-15	254	26	26	26	26
	24	01-30	12-22	326	10	9	10	6		24	02-09	12-09	303	26	22	25	20
	20	01-04	⊕	⊕	10	2	10	1		20	01-23	12-22	333	26	18	26	11
	16	01-04	⊕	⊕	10	2	10	0		16	01-13	12-28	349	27	11	27	5
BALLINGER 1 SW	32	03-30	11-10	225	30	30	29	29	BROWNSVILLE WB AIRPORT	32	01-10	⊕	⊕	10	3	10	1
	28	03-15	11-16	246	30	30	29	29		28	⊕	⊕	⊕	10	1	10	0
	24	03-03	12-03	275	29	29	29	24		24	⊕	⊕	⊕	10	1	10	0
	20	02-09	12-15	309	29	25	29	17		20	⊕	⊕	⊕	10	0	10	0
	16	01-25	12-26	335	29	21	30	6		16	⊕	⊕	⊕	10	0	10	0
BALMORHEA	32	03-30	11-11	226	30	30	30	30	BROWNWOOD	32	03-20	11-18	243	30	30	30	30
	28	03-15	11-17	247	30	30	30	30		28	03-07	11-29	267	30	30	30	30
	24	02-26	12-01	278	30	29	29	28		24	02-16	12-14	301	30	27	30	21
	20	02-15	12-16	304	29	27	29	20		20	01-30	12-21	325	30	24	30	13
	16	01-29	12-26	331	29	23	29	6		16	01-16	⊕	⊕	30	15	30	3
BEAUMONT EXP FARM	32	02-18	11-24	279	10	9	10	10	CAMERON	32	03-15	11-21	251	30	30	29	29
	28	02-02	12-01	302	10	9	9	9		28	02-16	12-05	292	30	29	29	25
	24	01-09	12-23	348	9	3	7	3		24	01-29	12-18	323	30	23	30	16
	20	⊕	⊕	⊕	9	0	7	0		20	01-17	12-28	345	30	15	30	5
	16	⊕	⊕	⊕	9	0	7	0		16	01-10	⊕	⊕	30	11	30	1
BEEVILLE 5 NE	32	02-20	12-05	288	30	29	29	27	CANADIAN 1 ENE	32	04-09	10-30	204	26	26	28	28
	28	01-29	12-23	328	30	22	29	12		28	03-28	11-05	222	24	24	27	27
	24	01-16	12-28	346	30	13	29	4		24	03-20	11-14	239	25	25	27	27
	20	01-07	⊕	⊕	30	8	30	2		20	03-13	11-21	253	26	26	28	28
	16	01-06	⊕	⊕	30	5	30	0		16	02-28	12-03	278	25	24	27	23
BIG SPRING	32	03-29	11-11	227	10	10	10	10	CANYON	32	04-19	10-28	192	10	10	10	10
	28	03-19	11-20	246	9	9	10	10		28	04-13	11-03	204	10	10	10	10
	24	03-11	12-03	267	8	8	10	10		24	03-29	11-08	224	10	10	10	10
	20	02-10	12-16	309	8	7	10	8		20	03-17	11-16	244	10	10	9	9
	16	01-25	12-29	338	9	5	10	2		16	03-08	11-27	264	10	10	9	9
BLANCO	32	03-25	11-03	223	10	10	10	10	CARRIZO SPRINGS	32	02-22	11-26	277	10	10	10	9
	28	03-13	11-16	248	10	10	10	10		28	02-01	12-05	307	10	8	10	8
	24	02-26	12-03	280	10	10	10	10		24	01-15	12-24	343	10	4	9	3
	20	01-27	12-11	318	10	7	10	7		20	⊕	⊕	⊕	10	1	10	1
	16	01-14	⊕	⊕	10	4	10	1		16	⊕	⊕	⊕	10	1	10	0

STATION	Freeze threshold temperature	Mean date of last Spring occurrence	Mean date of first Fall occurrence	Mean No. of days between dates	Years of record Spring	No. of occurrences in Spring	Years of record Fall	No. of occurrences in Fall
CENTER	32	03-17	11-06	234	10	10	10	10
	28	02-28	11-24	269	10	10	10	10
	24	02-07	12-01	297	10	8	10	10
	20	01-24	12-17	327	10	7	10	6
	16	01-06	⊕	⊕	10	3	10	0
CENTERVILLE	32	03-14	11-09	240	20	20	22	22
	28	03-02	11-26	269	19	19	21	21
	24	02-08	12-09	304	21	18	21	18
	20	01-24	12-22	332	23	15	23	9
	16	01-12	12-25	347	23	9	23	5
CHILDRESS FAA AIRPORT	32	04-01	11-07	220	10	10	10	10
	28	03-27	11-13	231	10	10	10	10
	24	03-13	11-23	255	10	10	10	10
	20	03-02	12-04	277	10	10	10	10
	16	02-10	12-21	314	10	10	10	5
CLARKSVILLE 2 E	32	03-26	11-05	224	10	10	10	10
	28	03-11	11-12	246	10	10	9	9
	24	02-18	12-03	288	10	9	9	9
	20	02-02	12-10	311	10	9	9	7
	16	01-20	12-24	338	9	6	9	4
CLEBURNE	32	03-26	11-10	229	30	30	30	30
	28	03-09	11-22	258	30	30	30	29
	24	02-20	12-09	292	30	28	30	25
	20	02-05	12-16	314	30	27	30	18
	16	01-22	12-26	338	30	19	30	8
COLEMAN	32	03-28	11-05	222	10	10	10	10
	28	03-14	11-17	248	10	10	10	10
	24	03-03	12-01	273	10	10	10	9
	20	02-12	12-12	303	9	8	10	8
	16	01-24	12-22	332	8	6	10	4
COLLEGE STATION 7 SW	32	03-06	11-27	266	26	26	24	23
	28	02-12	12-13	304	26	24	24	15
	24	01-28	12-24	330	26	18	24	9
	20	01-15	12-28	347	26	13	24	5
	16	01-06	⊕	⊕	26	7	25	2
COLORADO CITY	32	04-05	11-03	212	10	10	10	10
	28	03-27	11-12	230	10	10	9	9
	24	03-05	11-23	263	10	10	9	9
	20	02-23	12-04	284	10	10	9	9
	16	01-31	12-27	330	10	8	10	4
CONROE	32	03-06	11-23	262	10	10	10	10
	28	02-05	12-03	301	10	9	10	9
	24	01-30	12-20	324	10	7	10	5
	20	01-06	12-28	356	10	3	10	2
	16	⊕	⊕	⊕	10	1	10	0
CORPUS CHRISTI WB AP	32	02-09	12-12	306	10	9	10	6
	28	01-15	12-29	348	10	4	10	2
	24	⊕	⊕	⊕	10	1	10	0
	20	⊕	⊕	⊕	10	1	10	0
	16	⊕	⊕	⊕	10	0	10	0
CORSICANA	32	03-13	11-20	252	30	30	29	29
	28	02-23	12-07	287	30	30	29	23
	24	02-06	12-14	311	30	28	29	19
	20	01-24	12-24	334	30	22	29	10
	16	01-15	12-29	348	30	14	29	3
COTULLA FAA AIRPORT	32	02-24	11-25	274	10	10	10	10
	28	01-30	12-12	316	10	10	10	7
	24	01-17	12-23	340	10	5	10	4
	20	01-09	⊕	⊕	10	2	10	0
	16	⊕	⊕	⊕	10	1	10	0
CROCKETT	32	03-16	11-09	238	10	10	10	10
	28	03-03	11-30	272	10	10	10	9
	24	02-04	12-04	303	10	8	10	9
	20	01-25	12-21	330	10	7	10	5
	16	01-08	12-28	354	10	3	10	2

STATION	Freeze threshold temperature	Mean date of last Spring occurrence	Mean date of first Fall occurrence	Mean No. of days between dates	Years of record Spring	No. of occurrences in Spring	Years of record Fall	No. of occurrences in Fall
CROSBYTON	32	04-11	11-02	205	29	29	29	29
	28	04-02	11-12	224	29	29	29	29
	24	03-26	11-19	238	29	29	28	28
	20	03-05	11-29	269	29	29	28	28
	16	02-22	12-16	297	29	27	28	17
CRYSTAL CITY	32	03-02	11-27	270	10	10	10	10
	28	02-11	12-10	302	10	10	10	7
	24	01-19	12-26	341	10	5	10	3
	20	01-07	⊕	⊕	10	2	10	1
	16	⊕	⊕	⊕	10	1	10	0
CUERO	32	03-08	11-27	264	22	22	22	21
	28	02-07	12-13	309	21	16	22	16
	24	01-20	12-24	338	21	11	22	7
	20	01-12	⊕	⊕	21	9	21	1
	16	01-07	⊕	⊕	21	5	22	1
DALHART EXP STA	32	04-25	10-16	174	23	23	23	23
	28	04-11	10-29	201	22	22	23	23
	24	04-03	11-06	217	23	23	23	23
	20	03-25	11-15	235	23	23	23	23
	16	03-14	11-22	253	23	23	23	23
DALHART FAA AIRPORT	32	04-23	10-25	185	10	10	10	10
	28	04-16	10-29	196	10	10	10	10
	24	04-05	11-05	214	10	10	10	10
	20	03-24	11-14	235	10	10	10	10
	16	03-14	11-19	250	10	10	10	10
DALLAS WB AIRPORT	32	03-18	11-12	239	10	10	10	10
	28	02-25	12-01	279	10	10	10	10
	24	02-10	12-11	304	10	9	10	7
	20	01-31	12-27	330	10	9	10	3
	16	01-15	12-28	347	10	6	10	2
DANEVANG 2 SE	32	02-23	12-05	285	30	28	30	26
	28	02-06	12-17	314	30	23	30	17
	24	01-13	12-28	349	30	11	30	5
	20	01-06	⊕	⊕	30	6	30	1
	16	01-04	⊕	⊕	30	4	30	1
DEL RIO 3 S	32	02-10	12-10	303	20	19	20	17
	28	01-28	12-23	329	20	15	20	8
	24	01-15	⊕	⊕	20	10	20	1
	20	01-08	⊕	⊕	20	7	20	0
	16	⊕	⊕	⊕	19	1	20	0
DENISON DAM	32	03-26	11-10	229	21	21	21	21
	28	03-07	11-23	261	21	21	21	21
	24	02-17	12-08	294	21	21	21	17
	20	02-02	12-19	320	21	17	21	11
	16	01-22	12-23	335	21	15	21	8
DILLEY	32	02-25	12-02	280	29	28	30	27
	28	01-30	12-14	318	29	20	30	19
	24	01-13	12-28	349	29	11	30	6
	20	01-06	⊕	⊕	30	6	30	1
	16	⊕	⊕	⊕	30	3	30	0
DUBLIN	32	03-26	11-17	236	30	30	30	30
	28	03-11	11-26	260	30	30	30	30
	24	02-19	12-09	293	30	28	30	27
	20	02-08	12-20	315	30	26	30	14
	16	01-21	12-28	341	30	19	30	7
EAGLE PASS	32	02-20	12-03	286	29	29	30	29
	28	01-27	12-14	321	29	23	29	19
	24	01-12	12-29	351	29	14	29	5
	20	01-04	⊕	⊕	30	5	29	1
	16	⊕	⊕	⊕	30	3	29	0
EASTLAND	32	03-29	11-09	225	27	27	29	29
	28	03-18	11-20	247	26	26	28	28
	24	03-02	12-03	276	26	25	28	26
	20	02-12	12-15	306	27	25	28	18
	16	01-27	12-24	331	27	20	28	8

FREEZE DATA

STATION	Freeze threshold temperature	Mean date of last Spring occurrence	Mean date of first Fall occurrence	Mean No. of days between dates	Years of record Spring	No. of occurrences in Spring	Years of record Fall	No. of occurrences in Fall
EDEN 1	32	04-03	11-07	218	22	22	22	22
	28	03-21	11-15	239	22	22	22	22
	24	02-28	12-03	278	22	21	21	20
	20	02-10	12-13	306	22	20	21	14
	16	01-23	12-22	333	22	13	21	7
EL CAMPO	32	02-17	12-02	288	10	9	10	10
	28	02-01	12-13	315	9	8	10	7
	24	01-15	12-28	347	9	5	10	2
	20	01-04	⊕	⊕	9	1	10	0
	16	01-04	⊕	⊕	9	1	10	0
EL PASO WB AIRPORT	32	03-14	11-12	243	10	10	10	10
	28	03-09	11-21	257	10	10	10	10
	24	02-13	12-04	294	10	10	10	9
	20	01-29	12-15	320	10	8	10	8
	16	01-07	12-29	356	10	2	10	3
ENCINAL 3 NW	32	02-14	12-08	297	29	25	30	23
	28	01-25	12-18	327	29	19	30	16
	24	01-08	12-28	354	29	10	29	5
	20	01-03	⊕	⊕	29	3	30	1
	16	⊕	⊕	⊕	30	1	30	0
FALFURRIAS	32	02-10	12-10	303	30	26	30	25
	28	01-20	12-23	337	30	17	29	11
	24	01-07	⊕	⊕	30	6	29	2
	20	01-04	⊕	⊕	30	4	30	0
	16	⊕	⊕	⊕	30	1	30	0
FLATONIA 2 W	32	03-04	12-04	275	27	26	29	27
	28	02-08	12-14	309	27	24	29	18
	24	01-21	12-25	338	30	18	30	11
	20	01-14	⊕	⊕	30	13	30	3
	16	01-05	⊕	⊕	30	7	30	1
FLINT	32	03-08	11-22	259	19	19	18	17
	28	02-19	12-07	291	19	19	18	15
	24	02-07	12-17	313	19	17	18	11
	20	01-21	12-29	342	19	11	18	5
	16	01-17	12-30	347	19	10	18	2
FOLLETT	32	04-20	10-28	191	10	10	10	10
	28	04-10	11-05	209	10	10	10	10
	24	03-28	11-08	225	10	10	10	10
	20	03-19	11-14	240	9	9	10	10
	16	03-06	11-26	265	9	9	10	10
FT STOCKTON KFST RADIO	32	04-02	11-10	222	30	30	30	30
	28	03-18	11-18	245	30	30	30	30
	24	02-28	11-27	272	30	30	30	30
	20	02-10	12-15	308	29	24	30	17
	16	01-25	12-27	336	29	18	29	6
FORT WORTH WB AIRPORT	32	03-20	11-15	240	10	10	10	10
	28	03-07	11-29	267	10	10	10	9
	24	02-07	12-11	307	10	9	10	8
	20	01-28	12-21	327	10	9	10	5
	16	01-17	12-27	344	10	6	10	3
FREDERICKSBURG	32	03-27	11-03	221	10	10	10	10
	28	03-05	11-08	248	10	10	10	10
	24	02-28	12-03	278	10	10	10	10
	20	01-30	12-15	319	10	9	10	5
	16	01-13	12-29	350	10	4	10	2
GAINESVILLE	32	03-29	11-06	222	30	30	29	29
	28	03-16	11-18	247	30	30	30	30
	24	02-26	12-05	282	30	30	30	28
	20	02-10	12-13	306	30	27	29	20
	16	01-28	12-21	327	30	24	29	12
GALVESTON WB CITY	32	01-25	12-23	332	10	7	10	4
	28	01-09	⊕	⊕	10	3	10	0
	24	⊕	⊕	⊕	10	1	10	0
	20	⊕	⊕	⊕	10	1	10	0
	16	⊕	⊕	⊕	10	0	10	0
GALVESTON WB AIRPORT	32	01-27	12-24	331	10	8	10	4
	28	01-18	⊕	⊕	10	6	10	0
	24	⊕	⊕	⊕	10	1	10	0
	20	⊕	⊕	⊕	10	1	10	0
	16	⊕	⊕	⊕	10	0	10	0
GATESVILLE	32	03-24	11-09	230	22	22	21	21
	28	03-01	11-27	271	20	20	21	20
	24	02-15	12-13	301	20	19	21	16
	20	01-31	12-24	327	19	16	21	10
	16	01-12	12-29	351	19	7	22	3
GOLIAD	32	02-22	12-01	282	10	10	10	10
	28	02-07	12-12	308	10	9	10	7
	24	01-16	12-26	344	10	4	10	4
	20	⊕	⊕	⊕	10	1	10	0
	16	⊕	⊕	⊕	10	1	10	0
GRAHAM	32	04-03	11-02	213	29	29	30	30
	28	03-20	11-16	241	29	29	30	30
	24	03-04	11-26	267	29	29	30	30
	20	02-16	12-12	299	29	28	30	24
	16	01-30	12-20	324	29	25	29	11
GREENVILLE 7 NW	32	03-20	11-12	237	30	30	30	30
	28	03-07	11-30	268	30	30	30	28
	24	02-16	12-13	300	30	29	30	24
	20	02-03	12-19	319	30	26	30	15
	16	01-20	12-26	340	30	18	30	9
HALLETTSVILLE	32	03-08	11-15	252	10	10	10	10
	28	02-19	12-06	290	9	9	10	10
	24	01-31	12-18	321	9	7	10	6
	20	01-12	12-25	347	10	4	10	2
	16	⊕	⊕	⊕	10	1	10	0
HARLINGEN	32	01-24	12-26	336	28	17	29	8
	28	01-11	12-29	352	29	10	29	4
	24	01-04	⊕	⊕	29	4	29	0
	20	⊕	⊕	⊕	29	0	30	0
	16	⊕	⊕	⊕	29	0	30	0
HASKELL	32	03-31	11-12	226	27	27	27	27
	28	03-19	11-21	247	27	27	27	27
	24	03-07	12-06	274	27	26	27	25
	20	02-21	12-14	296	29	27	27	19
	16	02-05	12-24	322	28	22	29	10
HENDERSON	32	03-11	11-13	247	10	10	10	10
	28	02-25	12-01	279	10	10	10	10
	24	01-29	12-11	316	10	7	10	8
	20	01-24	12-26	336	10	7	10	4
	16	01-08	⊕	⊕	10	4	10	0
HENRIETTA	32	03-28	11-11	228	29	29	30	30
	28	03-17	11-24	252	29	29	30	30
	24	03-04	12-07	278	29	29	30	27
	20	02-16	12-15	302	30	28	30	19
	16	01-29	12-25	330	30	23	30	8
HEREFORD	32	04-23	10-24	184	10	10	10	10
	28	04-12	11-01	203	10	10	10	10
	24	03-30	11-12	227	10	10	10	10
	20	03-16	11-18	247	10	10	10	10
	16	03-08	11-26	263	10	10	10	10
HICO	32	03-29	11-02	218	10	10	10	10
	28	03-12	11-12	245	10	10	10	10
	24	02-21	11-30	282	10	10	10	9
	20	02-05	12-13	311	10	9	10	8
	16	01-17	12-22	339	10	6	10	5
HONDO	32	03-03	11-21	263	29	29	29	29
	28	02-16	12-09	296	29	27	29	21
	24	01-25	12-18	327	28	18	29	13
	20	01-13	12-25	346	28	13	29	6
	16	01-03	⊕	⊕	28	4	29	1

FREEZE DATA

STATION	Freeze threshold temperature	Mean date of last Spring occurrence	Mean date of first Fall occurrence	Mean No. of days between dates	Years of record Spring	No. of occurrences in Spring	Years of record Fall	No. of occurrences in Fall
HOUSTON WB CITY	32	02-04	12-10	309	10	8	10	8
	28	01-25	12-23	332	10	8	10	4
	24	01-06	⊕	⊕	10	3	10	0
	20	⊕	⊕	⊕	10	1	10	0
	16	⊕	⊕	⊕	10	1	10	0
HOUSTON WB AIRPORT	32	02-12	12-04	295	10	9	10	10
	28	01-27	12-15	322	10	8	10	6
	24	01-04	⊕	⊕	10	2	10	1
	20	⊕	⊕	⊕	10	1	10	0
	16	⊕	⊕	⊕	10	1	10	0
HUNTSVILLE	32	03-09	11-27	263	30	30	30	28
	28	02-08	12-11	306	30	28	30	23
	24	01-24	12-24	334	30	19	30	13
	20	01-14	12-29	349	30	14	30	4
	16	01-08	⊕	⊕	30	8	30	1
IOWA PARK EXP STA	32	04-01	11-02	215	10	10	10	10
	28	03-20	11-07	232	10	10	10	10
	24	03-09	11-28	264	10	10	10	10
	20	02-16	12-09	296	10	9	10	10
	16	01-29	12-20	325	10	9	10	4
JACKSBORO	32	04-02	11-10	222	10	10	10	10
	28	03-14	11-20	251	10	10	10	10
	24	03-03	11-30	272	10	10	10	10
	20	02-10	12-13	306	10	9	10	8
	16	01-28	12-22	328	10	9	10	4
JUNCTION	32	04-02	11-02	214	27	27	26	26
	28	03-19	11-12	238	27	27	26	26
	24	02-27	11-24	270	27	27	26	25
	20	02-11	12-08	300	27	24	25	19
	16	01-23	12-23	334	26	16	25	8
JUNCTION FAA AIRPORT	32	03-30	10-30	214	10	10	10	10
	28	03-15	11-03	233	10	10	10	10
	24	03-04	11-17	258	10	10	10	10
	20	02-09	12-05	299	10	10	10	8
	16	01-28	12-20	326	10	7	10	5
KARNACK	32	03-18	11-04	231	10	10	10	10
	28	03-03	11-17	259	10	10	10	10
	24	02-08	12-04	299	10	8	10	10
	20	01-24	12-18	328	10	7	10	6
	16	01-10	12-28	352	10	4	10	2
KAUFMAN	32	03-21	11-15	239	30	30	30	30
	28	03-04	12-02	273	30	30	30	28
	24	02-11	12-13	305	30	28	30	22
	20	02-01	12-21	323	30	24	30	15
	16	01-20	12-26	340	30	18	30	7
KENEDY	32	03-05	11-26	266	10	10	10	10
	28	02-11	12-05	297	9	8	9	9
	24	01-13	12-20	341	10	4	9	4
	20	01-09	12-24	349	10	3	9	2
	16	⊕	⊕	⊕	10	1	9	0
KERRVILLE	32	04-04	11-06	216	30	30	30	30
	28	03-19	11-13	239	30	30	30	30
	24	03-06	11-28	267	30	30	29	28
	20	02-16	12-10	297	30	28	29	20
	16	01-22	12-21	333	30	17	29	10
KIRBYVILLE FOR SERVICE	32	03-22	11-07	230	29	29	28	28
	28	03-01	11-20	264	30	30	27	26
	24	02-02	12-10	311	29	25	27	21
	20	01-14	12-22	342	30	15	27	11
	16	01-05	⊕	⊕	30	5	27	1
KNOX CITY	32	04-01	11-09	222	10	10	10	10
	28	03-22	11-19	242	10	10	10	10
	24	03-08	11-28	265	10	10	10	10
	20	02-24	12-10	289	10	10	10	9
	16	01-25	12-22	331	10	7	10	5
LAMPASAS	32	03-30	11-08	223	30	30	30	30
	28	03-18	11-18	245	30	30	30	30
	24	02-23	12-02	282	30	30	30	28
	20	02-04	12-15	314	30	25	30	19
	16	01-15	12-24	343	30	12	30	7
LAREDO WB AIRPORT	32	01-31	12-19	322	10	8	10	5
	28	01-09	⊕	⊕	10	2	10	1
	24	⊕	⊕	⊕	10	1	10	0
	20	⊕	⊕	⊕	10	1	10	0
	16	⊕	⊕	⊕	10	0	10	0
LEVELLAND	32	04-13	10-30	200	18	18	20	20
	28	04-02	11-07	219	17	17	19	19
	24	03-16	11-16	245	17	17	18	18
	20	03-06	11-22	261	17	17	18	18
	16	02-24	12-06	285	17	17	18	15
LIBERTY	32	03-01	11-15	259	30	30	30	30
	28	02-10	12-04	297	30	26	30	28
	24	01-19	12-21	336	30	17	30	15
	20	01-08	⊕	⊕	30	8	30	3
	16	01-05	⊕	⊕	30	5	30	1
LINDALE 5 SE	32	03-12	11-11	244	10	10	10	10
	28	03-03	11-27	269	10	10	10	9
	24	02-08	12-08	303	10	9	10	8
	20	01-24	12-18	328	10	7	10	6
	16	01-19	⊕	⊕	10	6	10	0
LIVINGSTON 2 NNE	32	03-14	11-17	248	24	24	24	24
	28	02-20	12-02	285	23	23	24	21
	24	01-28	12-14	320	23	17	24	16
	20	01-16	12-26	344	23	12	24	6
	16	01-07	⊕	⊕	23	6	24	1
LLANO	32	03-27	11-10	228	30	30	30	30
	28	03-09	11-21	257	30	30	30	30
	24	02-14	12-04	293	30	28	29	27
	20	01-28	12-19	325	29	20	29	13
	16	01-15	12-27	346	30	13	29	4
LONGVIEW	32	03-15	11-12	242	29	29	30	30
	28	02-25	12-01	279	29	29	28	26
	24	02-04	12-12	311	29	24	28	19
	20	01-24	12-25	335	29	18	28	9
	16	01-10	⊕	⊕	28	11	28	2
LUBBOCK WB AIRPORT	32	04-11	11-01	204	10	10	10	10
	28	04-01	11-10	223	10	10	10	10
	24	03-16	11-16	245	10	10	10	10
	20	03-09	11-21	257	10	10	10	10
	16	02-27	12-06	282	10	10	10	8
LUFKIN FAA AIRPORT	32	03-18	11-04	231	10	10	10	10
	28	03-01	11-14	258	10	10	10	10
	24	02-04	12-06	305	10	8	10	10
	20	01-24	12-19	329	10	7	10	5
	16	01-04	12-28	358	10	2	10	2
LULING	32	03-03	11-24	266	30	30	30	29
	28	02-12	12-10	301	30	27	30	22
	24	01-25	12-22	331	30	19	30	15
	20	01-11	⊕	⊕	30	11	30	3
	16	01-05	⊕	⊕	30	6	30	1
MADISONVILLE	32	03-09	11-12	248	10	10	10	10
	28	02-20	12-01	284	10	9	10	10
	24	01-31	12-05	310	10	8	10	8
	20	01-20	12-25	339	10	7	9	4
	16	01-06	⊕	⊕	10	3	9	0
MANSFIELD DAM	32	03-15	11-15	245	10	10	10	10
	28	02-28	12-01	276	10	10	10	10
	24	02-07	12-17	313	10	8	10	7
	20	01-16	12-28	346	10	4	10	2
	16	01-04	⊕	⊕	10	2	10	1

STATION	Freeze threshold temperature	Mean date of last Spring occurrence	Mean date of first Fall occurrence	Mean No. of days between dates	Years of record Spring	No. of occurrences in Spring	Years of record Fall	No. of occurrences in Fall
MARATHON	32	04-05	11-01	210	19	19	18	18
	28	03-24	11-10	231	19	19	17	17
	24	03-11	11-21	255	19	19	17	17
	20	02-28	11-29	274	18	18	17	16
	16	02-14	12-14	303	16	14	16	12
MARSHALL	32	03-14	11-13	244	25	25	25	25
	28	02-26	12-02	279	25	25	26	24
	24	02-09	12-14	308	26	22	24	17
	20	01-25	12-24	333	26	18	27	10
	16	01-13	12-29	350	28	13	29	4
MATAGORDA NO 2	32	02-13	12-14	304	29	26	30	21
	28	01-27	12-22	329	29	19	30	13
	24	01-08	⊕	⊕	30	9	29	2
	20	01-06	⊕	⊕	30	6	29	1
	16	⊕	⊕	⊕	30	3	29	0
MAURBO	32	02-23	12-02	282	10	10	10	9
	28	01-28	12-07	313	10	8	10	9
	24	01-15	12-26	345	10	5	10	4
	20	⊕	⊕	⊕	10	1	10	0
	16	⊕	⊕	⊕	10	1	10	0
MC ALLEN	32	01-30	12-10	314	10	6	10	7
	28	01-19	12-27	342	10	4	10	2
	24	⊕	⊕	⊕	10	1	10	0
	20	⊕	⊕	⊕	10	1	10	0
	16	⊕	⊕	⊕	10	0	10	0
MC CAMEY	32	03-26	11-11	230	29	29	28	28
	28	03-12	11-22	255	29	29	29	29
	24	02-17	12-02	288	28	28	29	27
	20	02-10	12-19	312	28	23	29	17
	16	01-22	12-26	338	27	17	29	8
MC COOK	32	02-05	12-08	306	10	8	10	8
	28	01-18	12-23	339	10	4	10	4
	24	01-08	⊕	⊕	10	2	10	0
	20	01-08	⊕	⊕	10	2	10	0
	16	⊕	⊕	⊕	10	1	10	0
MC KINNEY	32	04-01	11-05	218	10	10	10	10
	28	03-12	11-14	247	10	10	10	10
	24	02-21	12-01	283	10	10	10	10
	20	02-03	12-09	309	9	8	10	9
	16	01-23	12-26	337	9	7	10	4
MEMPHIS	32	04-05	11-02	211	27	27	27	27
	28	03-26	11-12	231	27	27	29	29
	24	03-14	11-21	252	28	28	28	28
	20	02-27	12-02	278	28	27	27	25
	16	02-15	12-16	304	28	26	26	15
MEXIA	32	03-16	11-21	250	30	30	29	29
	28	02-22	12-06	287	30	29	29	25
	24	02-08	12-16	311	30	27	29	19
	20	01-21	12-26	339	30	19	30	9
	16	01-14	⊕	⊕	30	13	30	3
MIAMI	32	04-19	10-23	187	29	29	30	30
	28	04-08	10-31	206	29	29	30	30
	24	03-28	11-09	226	29	29	30	30
	20	03-22	11-16	239	29	29	29	29
	16	03-13	11-27	259	28	28	29	28
MIDLAND WB AIRPORT	32	03-28	11-15	232	10	10	10	10
	28	03-18	11-20	247	10	10	10	10
	24	03-04	11-30	271	10	10	10	10
	20	02-14	12-18	307	10	10	10	6
	16	01-23	12-29	340	10	6	10	3
MIDLAND 4 ENE	32	04-02	11-06	218	23	23	24	24
	28	03-26	11-19	238	24	24	24	24
	24	03-08	11-30	267	24	24	24	23
	20	02-22	12-17	298	23	22	24	15
	16	01-27	12-27	334	23	17	24	5
MINERAL WELLS FAA AP	32	03-25	11-06	226	10	10	10	10
	28	03-11	11-22	256	10	10	10	10
	24	02-16	12-04	291	10	10	10	9
	20	02-04	12-14	313	10	9	10	8
	16	01-22	12-27	339	10	7	10	3
MISSION	32	01-26	12-18	326	30	20	30	17
	28	01-09	⊕	⊕	30	9	29	2
	24	⊕	⊕	⊕	30	3	29	0
	20	⊕	⊕	⊕	30	2	30	0
	16	⊕	⊕	⊕	30	0	30	0
MONTAGUE	32	03-29	11-06	222	10	10	10	10
	28	03-18	11-16	243	10	10	10	10
	24	03-07	11-26	264	10	10	10	10
	20	02-20	12-09	292	9	9	10	10
	16	02-02	12-19	320	9	9	10	7
MOUNT LOCKE	32	04-23	10-26	186	10	10	10	10
	28	04-08	11-04	210	10	10	10	10
	24	03-23	11-18	240	10	10	10	10
	20	03-15	11-25	255	10	10	10	10
	16	02-20	12-08	291	10	9	10	7
MOUNT PLEASANT	32	03-25	11-09	229	28	28	30	30
	28	03-08	11-18	255	28	28	29	28
	24	02-24	12-03	282	29	29	29	23
	20	02-01	12-13	315	28	24	29	19
	16	01-19	12-22	337	28	17	29	10
MULESHOE 1	32	04-20	10-20	183	30	30	30	30
	28	04-12	11-02	204	30	30	30	30
	24	03-29	11-12	228	30	30	30	30
	20	03-20	11-17	242	30	30	30	30
	16	03-08	11-23	260	30	30	30	30
MUNDAY	32	04-06	11-07	215	10	10	10	10
	28	03-24	11-14	235	10	10	10	10
	24	03-10	11-25	260	10	10	10	10
	20	02-27	12-10	286	10	10	10	9
	16	02-02	12-16	317	10	8	10	6
NACOGDOCHES	32	03-16	11-10	239	30	30	30	30
	28	02-26	11-26	273	30	30	30	30
	24	02-06	12-11	308	30	25	30	22
	20	01-23	12-24	335	30	17	30	12
	16	01-12	12-29	351	30	12	30	4
NEW BRAUNFELS	32	03-11	11-26	260	29	29	28	28
	28	02-16	12-11	298	30	28	29	23
	24	01-24	12-23	333	30	18	28	12
	20	01-12	12-29	351	30	12	29	4
	16	01-06	⊕	⊕	30	6	29	1
PALACIOS FAA AIRPORT	32	02-12	12-04	295	10	9	10	10
	28	01-25	12-18	327	10	7	10	7
	24	01-06	⊕	⊕	10	3	10	0
	20	⊕	⊕	⊕	10	1	10	0
	16	⊕	⊕	⊕	10	1	10	0
PALESTINE	32	03-14	11-15	246	10	10	10	10
	28	02-22	12-01	282	10	9	10	10
	24	02-02	12-17	318	10	9	10	8
	20	01-24	12-25	335	10	7	10	5
	16	01-14	⊕	⊕	10	5	10	1
PARIS	32	03-27	11-10	228	30	30	30	30
	28	03-10	11-26	261	29	29	30	29
	24	02-22	12-08	289	29	29	30	24
	20	02-07	12-14	310	29	27	30	20
	16	01-25	12-25	334	29	21	30	11
PECOS	32	03-31	11-08	222	24	24	27	27
	28	03-18	11-14	241	25	25	26	26
	24	02-28	11-22	267	26	26	26	26
	20	02-14	12-10	299	25	23	26	23
	16	01-27	12-20	327	25	21	26	11

FREEZE DATA

STATION	Freeze threshold temperature	Mean date of last Spring occurrence	Mean date of first Fall occurrence	Mean No. of days between dates	Years of record Spring	No. of occurrences in Spring	Years of record Fall	No. of occurrences in Fall
PIERCE 1 E	32	03-06	11-27	266	30	30	30	28
	28	02-08	12-14	309	30	26	30	21
	24	01-21	12-22	335	30	16	30	11
	20	01-09	⊕	⊕	30	10	30	3
	16	01-05	⊕	⊕	30	5	30	1
PLAINVIEW	32	04-11	11-04	207	30	30	30	30
	28	04-02	11-11	223	30	30	30	30
	24	03-25	11-19	239	30	30	30	30
	20	03-15	11-28	258	30	30	30	30
	16	02-24	12-11	290	30	29	30	21
PORT ARTHUR	32	01-30	12-08	312	10	7	10	8
	28	01-21	12-21	334	10	6	10	5
	24	01-04	⊕	⊕	10	2	10	0
	20	⊕	⊕	⊕	10	1	10	0
	16	⊕	⊕	⊕	10	1	10	0
PORT ARTHUR WB AIRPORT	32	02-18	11-30	285	10	9	10	10
	28	01-26	12-15	323	10	8	10	7
	24	01-04	12-27	357	10	2	10	3
	20	⊕	⊕	⊕	10	1	10	0
	16	⊕	⊕	⊕	10	1	10	0
PORT ISABEL	32	01-09	⊕	⊕	10	2	10	0
	28	⊕	⊕	⊕	10	1	10	0
	24	⊕	⊕	⊕	10	1	10	0
	20	⊕	⊕	⊕	9	0	10	0
	16	⊕	⊕	⊕	9	0	10	0
PORT LAVACA NO 2	32	02-18	12-08	293	10	10	10	9
	28	01-23	12-22	333	10	6	10	5
	24	01-04	⊕	⊕	10	2	10	0
	20	⊕	⊕	⊕	10	1	10	0
	16	⊕	⊕	⊕	10	1	10	0
PORT O CONNOR	32	02-06	12-20	317	10	9	10	6
	28	01-15	⊕	⊕	10	5	10	1
	24	01-04	⊕	⊕	10	2	10	0
	20	⊕	⊕	⊕	10	1	10	0
	16	⊕	⊕	⊕	10	0	10	0
POSSUM KINGDOM DAM	32	03-22	11-10	233	10	10	10	10
	28	03-04	11-30	271	10	10	10	10
	24	02-10	12-12	305	10	9	10	7
	20	01-21	12-23	336	10	6	10	4
	16	01-13	⊕	⊕	10	4	10	1
POTEET	32	03-06	11-19	258	10	10	10	10
	28	02-18	12-02	287	10	9	10	10
	24	01-19	12-20	335	10	6	10	5
	20	⊕	12-28	⊕	10	1	10	2
	16	⊕	⊕	⊕	10	1	10	0
PRESIDIO	32	03-18	11-13	240	29	29	29	29
	28	02-26	11-24	271	30	30	29	29
	24	02-03	12-07	307	29	26	28	26
	20	01-24	12-21	331	29	21	28	14
	16	01-07	⊕	⊕	29	9	28	1
QUANAH 5 SE	32	04-03	11-04	215	30	30	30	30
	28	03-28	11-13	230	30	30	30	30
	24	03-19	11-25	251	30	30	30	30
	20	03-03	12-04	276	30	30	30	28
	16	02-22	12-17	298	30	29	30	20
RAYMONDVILLE	32	01-25	12-22	331	30	20	28	13
	28	01-15	12-27	346	30	12	29	5
	24	01-04	⊕	⊕	30	4	29	0
	20	⊕	⊕	⊕	30	2	29	0
	16	⊕	⊕	⊕	30	0	29	0
RIO GRANDE CITY 3 W	32	02-11	12-13	305	28	22	27	20
	28	01-19	12-25	340	27	13	27	7
	24	01-06	⊕	⊕	27	6	27	2
	20	01-03	⊕	⊕	28	3	27	0
	16	⊕	⊕	⊕	28	1	27	0
RISING STAR	32	03-28	11-06	223	10	10	10	10
	28	03-15	11-17	247	10	10	10	10
	24	02-20	12-04	287	10	10	10	7
	20	02-09	12-14	308	10	9	10	7
	16	01-17	12-27	344	10	6	10	3
ROCKSPRINGS	32	03-28	11-17	234	17	17	18	18
	28	03-05	12-02	272	16	16	17	17
	24	02-15	12-13	301	13	12	16	10
	20	01-31	12-26	329	13	10	19	7
	16	01-23	12-29	340	12	7	19	3
ROSCOE	32	04-06	11-06	214	23	23	24	24
	28	03-29	11-16	232	23	23	24	24
	24	03-15	11-22	252	23	23	24	24
	20	02-27	12-07	283	22	22	24	21
	16	02-10	12-17	310	22	20	24	15
RUSK	32	03-16	11-08	237	10	10	10	10
	28	03-05	11-23	263	10	10	10	10
	24	02-16	12-01	288	10	9	9	9
	20	01-26	12-24	332	10	8	9	4
	16	01-18	12-29	345	10	6	10	2
SAN ANGELO WB AIRPORT	32	03-25	11-13	233	27	27	29	29
	28	03-10	11-23	258	27	27	29	29
	24	02-22	12-10	291	27	27	29	22
	20	02-02	12-20	321	27	22	29	13
	16	01-18	12-29	345	27	15	29	4
SAN ANTONIO WB AIRPORT	32	03-03	11-26	268	10	10	10	10
	28	02-14	12-08	297	10	10	10	9
	24	01-22	12-23	335	10	6	10	4
	20	01-04	12-28	358	10	2	10	2
	16	⊕	⊕	⊕	10	1	10	0
SAN BENITO	32	01-13	12-26	347	10	4	10	3
	28	⊕	⊕	⊕	10	1	10	1
	24	⊕	⊕	⊕	10	1	10	0
	20	⊕	⊕	⊕	10	1	10	0
	16	⊕	⊕	⊕	10	0	10	0
SAN MARCOS	32	03-13	11-19	251	29	29	29	29
	28	02-22	12-02	283	27	27	30	27
	24	01-26	12-16	324	27	21	30	19
	20	01-16	12-26	344	28	14	30	7
	16	01-06	⊕	⊕	28	8	30	3
SEALY	32	02-27	12-02	278	28	27	29	28
	28	02-02	12-12	313	28	24	29	21
	24	01-23	12-24	335	28	18	29	11
	20	01-11	12-29	352	28	10	29	3
	16	01-08	⊕	⊕	28	7	29	1
SEMINOLE	32	04-10	11-02	206	27	27	28	28
	28	03-31	11-11	225	26	26	27	27
	24	03-21	11-19	243	26	26	27	27
	20	03-04	11-28	269	27	27	27	25
	16	02-21	12-15	297	27	25	27	17
SEYMOUR	32	04-04	10-31	210	29	29	27	27
	28	03-26	11-13	232	27	27	25	25
	24	03-14	11-24	255	28	28	23	23
	20	02-27	12-08	284	25	25	22	18
	16	02-03	12-14	314	24	20	21	12
SHERMAN PUMP STATION	32	03-20	11-07	232	10	10	10	10
	28	03-09	11-26	262	10	10	10	10
	24	02-14	12-07	296	10	10	10	9
	20	02-02	12-18	319	10	9	10	7
	16	01-27	12-28	335	10	8	10	3
SMITHVILLE	32	03-11	11-15	249	10	10	10	10
	28	02-19	11-30	284	10	10	10	10
	24	01-24	12-18	328	10	7	10	6
	20	01-08	12-28	354	10	3	10	2
	16	01-04	⊕	⊕	10	2	10	1

FREEZE DATA

STATION	Freeze threshold temperature	Mean date of last Spring occurrence	Mean date of first Fall occurrence	Mean No. of days between dates	Years of record Spring	No. of occurrences in Spring	Years of record Fall	No. of occurrences in Fall
SNYDER	32	04-05	11-04	213	29	29	29	29
	28	03-28	11-13	230	30	30	28	28
	24	03-13	11-24	256	30	30	28	28
	20	02-24	12-06	285	30	29	28	23
	16	02-10	12-17	310	30	25	28	16
SPEARMAN	32	04-22	10-24	185	28	28	29	29
	28	04-09	11-01	206	28	28	29	29
	24	04-01	11-09	222	28	28	28	28
	20	03-21	11-17	241	29	29	28	28
	16	03-12	11-25	258	29	29	28	28
SPUR 1 WNW	32	04-17	10-28	194	10	10	10	10
	28	04-04	11-05	215	10	10	10	10
	24	03-23	11-16	238	10	10	10	10
	20	03-19	11-24	250	9	9	10	10
	16	03-04	12-07	278	9	9	10	8
SUBSTATION 14	32	03-27	11-15	233	21	21	20	20
	28	03-07	11-29	267	21	21	20	20
	24	02-19	12-15	299	21	19	20	12
	20	02-02	12-22	323	21	16	20	8
	16	01-23	⊕	⊕	21	14	20	1
SUGAR LAND	32	02-14	11-29	288	10	9	10	10
	28	01-27	12-17	324	10	8	9	6
	24	01-10	12-29	353	10	4	9	1
	20	⊕	⊕	⊕	10	1	10	0
	16	⊕	⊕	⊕	10	1	10	0
SULPHUR SPRINGS	32	03-24	11-03	224	10	10	10	10
	28	03-10	11-17	252	10	10	10	10
	24	02-16	12-01	288	10	9	10	10
	20	01-29	12-11	316	10	9	10	7
	16	01-24	12-25	335	10	7	10	4
TAHOKA	32	04-06	11-05	213	22	22	24	24
	28	03-29	11-14	230	21	21	24	24
	24	03-12	11-25	258	22	22	23	23
	20	02-22	12-07	288	22	21	23	19
	16	02-13	12-16	306	22	20	23	16
TAYLOR	32	03-14	11-18	249	30	30	30	30
	28	02-23	12-06	286	30	29	30	28
	24	02-03	12-14	314	30	24	30	19
	20	01-17	12-26	343	30	15	30	8
	16	01-13	⊕	⊕	30	13	30	2
TEMPLE	32	03-10	11-24	259	28	28	29	29
	28	02-19	12-09	293	28	25	29	23
	24	02-06	12-15	312	28	25	28	18
	20	01-19	12-27	342	28	15	27	7
	16	01-12	⊕	⊕	29	11	27	2
THROCKMORTON	32	04-04	11-07	217	10	10	10	10
	28	03-22	11-16	239	10	10	10	10
	24	03-05	11-26	266	10	10	10	10
	20	02-11	12-10	302	10	9	10	9
	16	01-28	12-21	327	10	9	10	5
UVALDE	32	03-09	11-18	254	30	30	30	30
	28	02-20	12-02	285	30	28	30	29
	24	01-26	12-18	326	30	19	30	15
	20	01-11	12-27	350	30	11	30	5
	16	01-04	⊕	⊕	30	4	30	0
VAN HORN	32	03-31	11-05	219	18	18	17	17
	28	03-24	11-13	234	18	18	18	18
	24	03-06	11-22	261	18	18	18	18
	20	02-12	12-05	296	18	18	18	16
	16	01-31	12-16	319	18	15	18	13

STATION	Freeze threshold temperature	Mean date of last Spring occurrence	Mean date of first Fall occurrence	Mean No. of days between dates	Years of record Spring	No. of occurrences in Spring	Years of record Fall	No. of occurrences in Fall
VEGA	32	04-21	10-21	183	30	30	30	30
	28	04-08	11-01	207	30	30	30	30
	24	04-02	11-10	222	30	30	29	29
	20	03-22	11-18	241	30	30	29	29
	16	03-12	11-28	261	30	30	30	29
VICTORIA WB AIRPORT	32	02-06	12-08	305	10	9	10	8
	28	01-21	12-19	332	10	6	10	6
	24	01-14	⊕	⊕	10	4	10	1
	20	⊕	⊕	⊕	10	1	10	0
	16	⊕	⊕	⊕	10	1	10	0
WACO WB AIRPORT	32	03-16	11-18	247	10	10	10	10
	28	02-22	12-05	286	10	10	10	10
	24	01-30	12-14	318	10	9	10	7
	20	01-20	12-22	336	10	6	10	4
	16	01-04	⊕	⊕	10	2	10	0
WAXAHACHIE	32	03-25	11-05	225	10	10	10	10
	28	03-11	11-14	248	10	10	10	10
	24	02-17	12-05	291	9	8	10	9
	20	01-30	12-17	321	8	7	10	5
	16	01-18	12-26	342	9	6	10	4
WEATHERFORD	32	03-28	11-07	224	10	10	10	10
	28	03-15	11-19	249	10	10	10	10
	24	02-23	11-30	280	10	10	9	9
	20	02-08	12-10	305	10	9	9	8
	16	01-26	12-22	330	10	8	8	3
WESLACO 2 E	32	01-22	12-13	325	10	5	10	6
	28	⊕	12-26	⊕	10	1	10	3
	24	⊕	⊕	⊕	10	1	10	0
	20	⊕	⊕	⊕	10	0	10	0
	16	⊕	⊕	⊕	10	0	10	0
WHITNEY DAM	32	03-18	11-10	237	10	10	10	10
	28	03-04	11-25	266	10	10	10	10
	24	02-04	12-10	309	10	10	10	8
	20	01-24	12-16	326	10	8	10	6
	16	01-10	⊕	⊕	10	3	10	1
WICHITA FALLS WB AP	32	04-04	11-06	216	10	10	10	10
	28	03-20	11-14	239	10	10	10	10
	24	03-07	12-06	274	10	10	10	10
	20	02-13	12-12	302	10	10	10	9
	16	01-28	12-20	326	10	9	10	6
WINK FAA AIRPORT	32	03-26	11-09	228	10	10	10	10
	28	03-11	11-18	252	10	10	10	10
	24	03-08	11-24	261	10	10	10	10
	20	02-06	12-09	306	10	10	10	10
	16	01-29	12-27	332	10	8	10	4
WINTER HAVEN EXP STA	32	02-24	12-01	280	10	10	10	10
	28	01-28	12-20	326	10	7	10	6
	24	01-09	12-27	352	10	3	10	3
	20	⊕	⊕	⊕	10	1	10	1
	16	⊕	⊕	⊕	10	1	10	0
YSLETA	32	04-03	11-01	212	22	22	22	22
	28	03-26	11-09	228	22	22	22	22
	24	03-08	11-18	255	22	22	22	22
	20	02-20	12-02	285	22	22	21	19
	16	02-06	12-11	308	22	20	21	14

Data in the above table are based on the period 1931-1960, or that portion of this period for which data are available.

⊕ When the frequency of occurrence in either spring or fall is one year in ten, or less, mean dates are not given.

Means have been adjusted to take into account years of non-occurrence.

A freeze is a numerical substitute for the former term "killing frost" and is the occurrence of a minimum temperature at or below the threshold temperature of 32°, 28°, etc.

Freeze data tabulations in greater detail are available and can be reproduced at cost.

*NORMALS BY CLIMATOLOGICAL DIVISIONS

Taken from "Climatography of the United States No. 81-4, Decennial Census of U. S. Climate"

STATIONS (By Divisions)	TEMPERATURE (°F)													PRECIPITATION (In.)												
	JAN	FEB	MAR	APR	MAY	JUNE	JULY	AUG	SEPT	OCT	NOV	DEC	ANN	JAN	FEB	MAR	APR	MAY	JUNE	JULY	AUG	SEPT	OCT	NOV	DEC	ANN
HIGH PLAINS																										
AMARILLO WB AIRPORT	36.7	41.3	47.6	57.5	66.5	77.0	80.6	79.5	71.8	60.5	46.0	39.3	58.7	.65	.62	.82	1.32	3.37	2.89	2.34	2.58	1.89	1.76	.66	.77	19.67
BROWNFIELD NO 2	.	.	.	.	.	.	.	.	.	.	.	.	.	.59	.70	.59	1.01	2.50	2.24	2.11	1.65	2.37	2.03	.49	.53	16.81
CROSBYTON	39.7	42.7	49.3	59.5	67.9	77.7	80.3	79.7	72.2	61.6	48.3	41.9	60.1	.92	.76	.72	1.62	3.43	2.47	2.52	2.14	2.22	2.72	.84	.96	21.32
DALHART FAA AIRPORT	.	.	.	.	.	.	.	.	.	.	.	.	.	.44	.40	.66	1.28	2.78	2.01	2.53	2.33	1.46	1.50	.48	.46	16.33
DIMMITT 6 E	.	.	.	.	.	.	.	.	.	.	.	.	.	.53	.42	.58	1.09	2.81	2.50	2.40	1.98	1.94	1.87	.56	.68	17.36
GARDEN CITY 1 E	.	.	.	.	.	.	.	.	.	.	.	.	.	.74	.63	.56	1.27	2.37	1.76	2.14	1.65	1.87	1.67	.71	.90	16.27
LAMESA 1 SSE	41.5	45.3	51.9	61.2	69.8	79.0	80.5	79.9	73.1	62.9	50.1	43.2	61.5	.75	.61	.61	1.04	2.39	1.78	2.35	1.44	2.43	2.01	.74	.78	16.93
LUBBOCK 9 N	41.0	44.7	51.3	60.8	68.9	77.9	79.7	78.8	71.9	62.1	49.5	42.9	60.8	.69	.58	.72	1.16	3.16	2.29	2.17	1.61	2.24	1.87	.50	.68	17.67
LUBBOCK WB AIRPORT	39.2	43.1	49.6	59.5	68.2	77.5	79.5	78.8	71.4	61.2	47.9	41.0	59.7	.68	.57	.73	1.15	3.18	2.53	2.01	1.68	2.36	2.00	.54	.65	18.08
MIAMI	35.2	39.0	45.8	56.4	65.1	75.3	80.0	79.2	71.0	59.9	45.5	37.9	57.5	.70	.67	.97	1.61	3.89	3.03	2.38	2.45	2.00	1.89	.67	.89	21.15
MIDLAND WB AIRPORT	44.0	48.3	55.2	64.9	73.4	81.2	82.9	82.2	75.4	65.9	52.5	45.9	64.3	.80	.60	.36	.83	2.08	1.63	1.88	1.48	1.79	1.66	.47	.66	14.24
MULESHOE 1	36.7	40.2	46.6	56.6	65.7	75.6	78.1	76.9	69.6	58.6	45.0	38.7	57.4	.65	.48	.62	1.02	2.80	2.48	2.37	2.21	1.99	1.62	.55	.65	17.44
PAMPA NO. 2	.	.	.	.	.	.	.	.	.	.	.	.	.	.53	.69	.79	1.50	3.62	3.00	2.48	2.39	1.91	1.83	.61	.81	20.16
PERRYTON 5 NNE	.	.	.	.	.	.	.	.	.	.	.	.	.	.61	.64	1.05	1.53	3.46	2.81	3.21	2.38	1.67	1.62	.73	.69	20.40
PLAINVIEW	39.6	43.1	49.2	58.8	67.3	77.0	79.4	78.4	71.5	61.3	48.4	41.8	59.7	.80	.62	.68	1.53	3.18	2.58	2.47	2.05	2.02	1.64	.59	.82	18.98
SPEARMAN	33.9	30.3	44.6	55.1	64.4	74.8	79.6	78.8	70.7	59.1	44.4	36.9	56.7	.75	.80	1.19	1.69	3.61	3.24	2.72	2.71	1.74	1.71	.90	.80	21.86
VEGA	35.9	39.5	45.6	55.2	64.2	74.2	78.2	76.9	69.4	58.5	45.1	38.4	56.8	.67	.59	.89	1.37	2.88	2.43	2.56	2.79	1.73	1.52	.68	.85	18.96
DIVISION	38.2	41.9	48.4	58.1	66.8	76.4	79.7	78.7	71.4	60.8	47.3	40.4	59.0	.66	.60	.72	1.36	3.05	2.45	2.41	2.12	1.97	1.83	.61	.73	18.38
LOW ROLLING PLAINS																										
ABILENE WB AIRPORT	44.6	48.4	55.0	64.3	71.7	80.3	83.2	83.0	75.9	66.2	53.0	46.1	64.3	.88	1.09	1.04	2.27	4.33	2.67	2.28	1.47	2.07	2.85	1.11	1.26	23.32
ASPERMONT 2 SSW	.	.	.	.	.	.	.	.	.	.	.	.	.	.75	1.11	.83	2.06	3.72	2.26	2.02	2.19	2.24	2.46	1.10	.99	21.73
BALLINGER 1 SW	45.2	48.6	55.4	65.4	72.8	81.1	83.9	83.9	76.9	67.1	53.6	46.7	65.1	1.30	1.06	1.03	2.46	4.07	2.15	1.49	1.68	3.00	2.22	1.06	1.11	22.63
CHILDRESS FAA AIRPORT	.	.	.	.	.	.	.	.	.	.	.	.	.	.82	.88	1.01	2.14	4.11	3.10	1.90	2.08	2.06	2.04	.65	1.02	21.81
CHILLICOTHE	.	.	.	.	.	.	.	.	.	.	.	.	.	1.04	1.05	1.37	2.19	4.66	3.23	1.86	1.79	2.61	2.58	1.05	1.13	24.56
COLEMAN	47.5	51.1	57.4	66.3	72.9	80.2	83.5	83.7	77.0	67.1	55.2	48.9	65.9	1.56	1.29	1.25	2.90	4.49	2.73	2.38	1.94	3.04	2.49	1.31	1.44	26.82
CROWELL	.	.	.	.	.	.	.	.	.	.	.	.	.	.96	1.11	1.16	2.08	4.97	2.60	1.86	1.85	2.50	2.54	1.22	1.14	23.99
HASKELL	.	.	.	.	.	.	.	.	.	.	.	.	.	.91	1.34	.92	1.93	3.99	2.84	2.38	1.92	2.04	2.27	1.33	1.33	23.20
MEMPHIS	39.1	42.8	49.5	60.1	68.8	79.1	83.5	83.0	74.6	63.0	48.8	41.4	61.1	.	.	.	.	.	.	.	.	.	.	.	.	.
MUNDAY	.	.	.	.	.	.	.	.	.	.	.	.	.	.96	1.34	1.13	2.35	3.82	2.68	2.23	1.73	2.42	2.46	1.38	1.41	23.91
POST	.	.	.	.	.	.	.	.	.	.	.	.	.	.77	.88	.58	1.51	3.10	2.17	2.22	1.56	2.23	2.04	.74	.80	18.60
QUANAH 5 SE	40.3	43.9	51.2	62.2	70.9	80.9	84.8	84.5	76.1	65.1	50.6	42.4	62.7	.86	1.00	1.28	2.70	4.19	3.02	1.84	2.07	2.71	2.61	1.14	1.28	24.70
ROTAN	.	.	.	.	.	.	.	.	.	.	.	.	.	.87	.99	.87	1.80	3.77	2.32	2.02	1.69	1.92	2.15	1.08	.95	20.43
SEYMOUR	41.2	45.0	52.3	63.0	71.8	81.1	84.9	85.3	77.0	65.5	51.0	43.1	63.4	1.11	1.40	1.39	2.13	3.94	3.07	2.33	1.73	2.56	2.60	1.35	1.34	24.95
SHAMROCK	.	.	.	.	.	.	.	.	.	.	.	.	.	.72	.91	1.08	2.20	4.55	3.01	2.13	2.26	2.09	2.38	.80	.98	23.11
SNYDER	42.2	45.5	52.4	62.8	71.3	80.5	82.7	82.3	74.9	64.6	51.1	44.1	62.9	.71	.83	.76	1.39	3.57	2.16	2.46	1.70	1.98	2.14	.92	.89	19.51
SPUR 1 WNW	42.2	46.1	52.8	62.6	70.6	79.5	81.9	81.2	74.0	63.9	51.1	44.2	62.5	.69	.85	.71	1.52	3.03	2.60	2.22	2.12	2.53	2.22	.84	.91	20.24
STAMFORD 1	.	.	.	.	.	.	.	.	.	.	.	.	.	.90	1.28	.92	1.89	4.15	2.96	1.97	1.55	2.55	2.54	1.34	1.32	23.37
WICHITA FALLS WB AP	42.8	46.8	53.8	64.0	72.2	81.8	85.6	85.7	77.4	66.4	52.5	44.6	64.5	1.12	1.38	1.54	2.53	4.60	3.20	1.97	1.79	2.33	2.87	1.42	1.45	26.20
DIVISION	42.6	46.3	53.2	63.4	71.4	80.5	83.8	83.6	76.0	65.5	52.1	44.7	63.7	.96	1.05	1.05	2.12	4.06	2.72	2.14	1.85	2.40	2.38	1.07	1.19	22.80
NORTH CENTRAL																										
ALBANY	44.3	47.7	54.2	64.4	72.0	81.0	84.4	84.8	77.1	66.7	53.3	46.6	64.7	1.07	1.54	1.23	2.40	4.60	2.41	2.60	2.14	2.60	2.65	1.34	1.37	25.95
ARTHUR CITY	.	.	.	.	.	.	.	.	.	.	.	.	.	2.89	3.24	3.66	4.81	5.21	3.58	3.62	2.46	3.45	2.99	3.14	3.03	42.08
BRECKENRIDGE	.	.	.	.	.	.	.	.	.	.	.	.	.	1.20	1.43	1.35	2.45	4.11	2.65	2.21	1.79	2.58	2.44	1.43	1.52	25.16
BRIDGEPORT	.	.	.	.	.	.	.	.	.	.	.	.	.	1.71	1.91	2.10	3.22	4.51	2.88	2.04	1.61	2.33	2.99	1.87	2.15	29.32
BROWNWOOD	45.7	49.4	56.0	65.1	72.6	80.7	84.2	84.1	77.0	67.4	54.6	47.8	65.4	1.83	1.73	1.75	2.95	4.50	3.00	1.82	1.71	2.77	2.49	1.41	1.78	27.74
CAMERON	50.3	53.6	59.8	67.7	74.9	81.6	84.7	84.9	79.5	70.9	58.6	52.5	68.3	2.90	2.71	2.45	3.91	3.96	2.88	1.62	1.99	2.88	2.62	2.84	3.09	33.85
CARROLLTON	.	.	.	.	.	.	.	.	.	.	.	.	.	2.35	2.57	2.61	4.27	5.09	3.45	2.12	1.71	2.49	2.94	2.74	2.45	34.79
CLEBURNE	47.2	50.6	57.5	66.1	73.1	81.3	84.8	85.1	78.3	68.4	56.0	49.3	66.5	2.04	2.64	2.35	3.79	4.46	2.92	2.17	2.25	2.78	2.94	2.28	2.58	33.20
CLIFTON 9 E	.	.	.	.	.	.	.	.	.	.	.	.	.	2.43	2.57	2.29	4.08	4.43	3.02	2.20	1.70	2.95	2.51	2.31	2.71	33.20
COMANCHE	.	.	.	.	.	.	.	.	.	.	.	.	.	1.90	1.83	1.68	3.12	4.45	2.93	2.09	1.44	2.74	2.73	1.74	1.80	28.45
CORSICANA	47.0	49.9	56.4	65.2	73.2	81.6	85.3	85.4	78.7	69.2	56.1	49.2	66.4	2.86	3.00	2.88	4.32	4.98	3.12	1.90	2.19	2.67	2.96	2.97	3.21	37.06
DALLAS WB AIRPORT	45.9	49.5	56.1	65.0	72.9	81.3	84.9	85.0	77.9	67.8	54.9	48.1	65.8	2.32	2.55	2.85	4.00	4.83	3.24	1.94	1.93	2.82	2.70	2.70	2.67	34.55
DENTON 2 SE	45.4	49.0	55.6	64.5	72.0	80.6	84.3	84.9	77.9	67.8	54.7	47.5	65.4	1.90	2.42	2.51	3.69	5.03	3.16	2.05	1.73	2.37	2.46	2.05	2.19	31.56
DUBLIN	45.1	48.2	54.6	63.9	71.4	79.7	83.5	84.1	77.0	67.7	54.4	47.4	64.8	2.06	2.21	1.79	3.49	5.40	2.88	2.09	1.62	3.13	2.71	2.03	2.26	31.67
DUNDEE 6 NNW	.	.	.	.	.	.	.	.	.	.	.	.	.	1.24	1.48	1.56	2.12	4.19	2.87	1.94	1.58	2.39	2.93	1.47	1.49	25.26
FORT WORTH WB AIRPORT	45.5	49.2	55.9	64.8	72.7	81.5	85.4	85.4	78.4	67.9	54.8	47.7	65.8	2.04	2.24	2.51	3.60	4.59	2.98	1.75	1.68	2.54	2.59	2.46	2.35	31.33
GAINESVILLE	44.0	48.1	54.9	64.3	72.2	80.7	84.8	84.9	77.8	67.0	53.7	46.1	64.9	2.07	2.51	2.64	3.75	5.09	3.79	2.38	2.45	2.63	2.91	2.09	2.23	34.54
GRAHAM	43.5	47.3	54.3	63.9	71.6	80.7	84.5	84.9	77.3	66.5	52.9	45.8	64.4	1.31	1.62	1.79	2.84	4.22	3.23	1.87	1.87	2.88	2.61	1.66	1.51	27.41
GREENVILLE 7 NW	43.6	46.9	53.9	63.5	71.5	79.9	83.3	83.7	77.1	67.0	53.4	46.2	64.2	2.89	3.31	3.56	4.75	5.72	4.20	3.18	2.27	2.92	3.21	3.54	3.26	42.81
HENRIETTA	42.3	46.1	52.9	63.6	72.2	81.8	86.4	86.7	78.6	67.0	52.6	44.7	64.6	.	.	.	.	.	.	.	.	.	.	.	.	.
HEWITT 1 SE	.	.	.	.	.	.	.	.	.	.	.	.	.	2.07	2.52	2.21	3.90	4.36	2.43	1.94	2.00	3.07	2.45	2.34	2.58	31.87
HONEY GROVE	.	.	.	.	.	.	.	.	.	.	.	.	.	3.34	3.41	3.69	4.84	5.24	4.28	3.19	2.52	3.47	3.52	3.32	3.18	44.00
KAUFMAN	44.7	48.2	55.0	64.0	72.0	80.6	84.5	84.9	78.3	68.0	54.2	47.2	65.1	2.94	3.07	2.96	4.47	5.30	3.67	2.29	2.01	2.89	3.18	3.18	3.32	39.28
MC GREGOR	.	.	.	.	.	.	.	.	.	.	.	.	.	2.34	2.41	2.29	4.10	4.51	2.68	2.19	1.65	2.84	2.44	2.26	2.54	32.25
MEXIA	46.9	50.0	56.4	65.3	72.8	80.2	83.9	84.7	78.9	69.3	56.0	49.3	66.1	3.05	3.14	2.90	3.97	4.90	3.28	2.13	1.87	3.23	2.72	3.17	3.27	37.63
PARIS	42.7	46.1	52.8	62.6	70.7	79.4	83.1	83.3	76.4	66.2	52.7	45.4	63.5	3.08	3.45	3.47	5.19	5.40	4.26	3.65	2.73	3.52	3.22	3.78	3.36	45.11
PUTNAM	.	.	.	.	.	.	.	.	.	.	.	.	.	1.06	1.08	1.19	2.69	4.32	2.45	2.04	1.87	2.35	2.73	1.25	1.26	24.29
SHERMAN PUMP STATION	42.5	46.1	53.2	63.3	71.4	80.4	84.2	84.4	77.1	66.6	52.7	45.2	63.9	2.47	3.37	2.99	4.38	5.24	3.75	2.78	2.61	2.89	3.23	2.70	2.64	39.05
TAYLOR	49.1	52.6	58.9	67.3	74.7	81.6	84.5	84.9	79.1	70.3	57.7	51.3	67.7	2.56	2.81	2.20	3.52	3.69	3.52	1.83	2.30	3.31	2.72	2.60	2.93	33.99
TEMPLE	47.9	51.5	57.9	66.4	73.9	81.3	84.6	85.1	79.3	70.0	57.0	50.2	67.1	2.44	2.65	2.24	3.81	4.63	3.10	1.94	2.03	2.83	2.70	2.80	2.77	33.94
THROCKMORTON	44.3	48.1	54.9	64.9	72.3	81.3	84.9	85.3	77.6	67.1	53.6	46.1	65.0	1.10	1.32	1.27	2.44	4.22	3.17	1.99	1.59	2.21	2.73	1.42	1.37	24.83
WACO WB AIRPORT	48.0	51.5	57.9	66.6	74.3	82.0	85.3	85.5	79.0	69.4	56.8	50.3	67.2	2.27	2.42	2.35	3.88	4.57	2.67	1.99	1.68	2.77	2.55	2.19	2.74	32.08
WEATHERFORD	43.6	46.7	53.3	63.0	70.9	79.8	83.8	84.3	76.6	66.0	52.7	45.8	63.9	2.14	2.46	2.16	3.58	5.39	3.00	1.89	1.67	2.70	3.04	1.82	2.12	31.97
DIVISION	45.4	48.9	55.7	64.8	72.5	80.8	84.5	84.8	77.8	67.8	54.7	47.6	65.5	2.18	2.40	2.21	3.69	4.75	3.23	2.22	1.93	2.79	2.84	2.28	2.41	32.50
EAST TEXAS																										
ANDERSON	.	.	.	.	.	.	.	.	.	.	.	.	.	3.73	3.29	2.77	4.27	4.49	3.26	3.01	2.91	3.00	2.85	3.48	3.70	40.76
BON WIER 2 E	.	.	.	.	.	.	.	.	.	.	.	.	.	4.97	4.85	4.44	5.44	5.82	4.52	5.31	4.28	3.40	2.90	4.92	5.97	56.82
HENDERSON	.	.	.	.	.	.	.	.	.	.	.	.	.	4.24	4.01	4.01	4.42	5.44	3.41	3.00	2.95	3.11	3.31	4.02	4.86	46.78
HUNTSVILLE	50.8	53.5	59.4	67.2	74.5	81.2	83.5	83.6	78.3	70.2	58.2	52.5	67.7	3.90	3.97	3.28	4.40	4.71	4.26	3.67	2.81	3.25	3.27	4.34	4.29	46.15
JEFFERSON	.	.	.	.	.	.	.	.	.	.	.	.	.	4.20	4.03	3.93	4.62	5.28	3.02	3.26	2.67	3.08	2.78	4.39	4.74	46.00
KIRBYVILLE FOR SERVICE	51.9	54.9	59.7	66.9	73.4	79.5	81.6	81.5	77.2	68.3	57.5	52.7	67.1	4.73	4.57	4.05	4.66	5.55	4.31	5.87	3.78	3.24	2.86	4.58	5.96	54.16
LINDALE 5 SE	47.0	50.3	56.5	64.9	72.3	79.7	82.7	82.8	77.1	67.8	55.6	49.3	65.5	3.88	3.67	3.91	4.89	5.45	3.27	3.26	2.91	2.74	3.23	4.21	4.77	46.19
LONGVIEW	47.7	50.8	57.0	65.7	73.6	81.2	84.2	84.3	78.3	68.3	55.7	49.2	66.3	4.27	3.76	3.84	4.79	5.67	3.36	3.52	2.56	2.62	3.07	4.12	4.58	46.16
LUFKIN FAA AIRPORT	49.9	52.8	58.6	66.6	74.1	81.0	83.4	83.4	78.2	69.0	56.9	51.3	67.1	.	.	.	.	.	.	.	.	.	.	.	.	.
MARSHALL	47.1	50.5	56.7	65.5	73.3	80.8	83.8	83.8	78.1	68.3	55.3	49.0	66.0	4.75	4.01	4.40	4.43	5.25	3.18	3.28	2.53	2.70	3.01	4.41	5.01	46.96

*NORMALS BY CLIMATOLOGICAL DIVISIONS

Taken from "Climatography of the United States No. 81-4, Decennial Census of U. S. Climate"

STATIONS (By Divisions)	TEMPERATURE (°F)													PRECIPITATION (In.)												
	JAN	FEB	MAR	APR	MAY	JUNE	JULY	AUG	SEPT	OCT	NOV	DEC	ANN	JAN	FEB	MAR	APR	MAY	JUNE	JULY	AUG	SEPT	OCT	NOV	DEC	ANN
EAST TEXAS - CONTINUED																										
MOUNT PLEASANT	45.0	47.8	54.2	63.4	71.4	79.7	82.9	82.9	76.3	66.3	53.1	46.5	64.1	.	.	.	.	.	.	.	.	.	.	.	.	.
NACOGDOCHES	48.4	51.7	57.4	65.2	72.4	79.4	82.3	82.4	77.0	67.9	55.7	50.0	65.8	4.46	4.21	3.73	4.68	5.33	3.31	3.57	2.69	2.95	3.15	4.71	5.25	48.04
NAPLES 1 SW	.	.	.	.	.	.	.	.	.	.	.	.	.	4.20	3.80	4.29	5.16	5.11	3.35	3.23	2.98	2.46	3.11	4.16	4.41	46.26
PALESTINE	48.6	51.7	57.7	65.9	73.1	80.3	83.1	83.3	77.7	69.0	56.8	50.6	66.5	3.52	3.21	3.40	3.81	4.79	2.88	2.64	2.45	2.96	2.96	3.93	4.10	40.65
RIVERSIDE	.	.	.	.	.	.	.	.	.	.	.	.	.	3.98	3.84	3.38	4.48	4.78	3.94	3.02	2.41	3.08	2.66	4.51	4.18	44.26
ROCKLAND 1 N	.	.	.	.	.	.	.	.	.	.	.	.	.	5.39	4.25	3.48	4.59	5.40	3.84	4.10	2.88	2.97	3.24	4.63	5.08	49.85
TRINIDAD 1 SW	.	.	.	.	.	.	.	.	.	.	.	.	.	2.99	2.94	3.10	4.16	5.21	3.11	1.84	2.28	2.69	3.03	3.16	3.50	38.01
VALLEY JUNCTION	.	.	.	.	.	.	.	.	.	.	.	.	.	2.85	2.94	2.48	3.88	3.96	3.02	2.51	2.16	2.62	3.03	3.16	3.19	35.80
DIVISION	48.7	51.7	57.8	65.8	73.1	80.2	83.0	83.1	77.6	68.4	56.3	50.4	66.4	4.22	3.82	3.58	4.56	5.12	3.35	3.42	2.91	2.95	3.15	4.26	4.62	45.64
TRANS PECOS																										
ALPINE	46.7	50.7	55.7	63.3	70.8	77.6	77.4	77.0	71.9	65.0	53.9	48.5	63.2	.81	.45	.28	.51	1.45	2.30	2.78	2.29	2.25	1.21	.42	.67	15.42
BALMORHEA	47.3	51.5	57.6	65.9	73.2	80.9	81.1	80.1	74.9	66.3	54.4	48.5	65.1	.75	.58	.36	.71	1.41	1.47	1.55	1.35	1.95	1.42	.51	.62	12.68
EL PASO WB AIRPORT	42.9	49.1	54.9	63.4	71.9	81.0	81.9	80.4	74.5	64.4	51.2	44.1	63.3	.46	.41	.35	.29	.40	.69	1.29	1.19	1.14	.85	.33	.49	7.89
PRESIDIO	49.8	55.9	62.1	70.3	79.0	86.9	86.5	85.5	80.1	70.8	57.0	50.3	69.5	.49	.20	.16	.34	.74	.97	1.28	1.21	1.37	.81	.34	.40	8.31
DIVISION	46.0	50.4	56.3	64.4	72.2	79.8	80.3	79.5	73.9	65.4	53.0	46.9	64.0	.68	.41	.46	.56	1.26	1.34	1.80	1.65	1.73	1.18	.40	.56	11.64
EDWARDS PLATEAU																										
BLANCO	48.7	51.7	57.7	65.7	73.3	80.5	83.1	83.1	77.4	68.5	56.4	50.6	66.4	2.36	2.54	2.19	3.52	3.77	2.94	2.74	2.28	4.00	3.15	2.16	2.61	34.26
BOERNE	49.9	53.1	58.8	65.9	72.5	78.8	81.1	81.3	76.4	68.4	57.0	51.5	66.2	2.22	2.35	1.88	2.72	4.19	2.90	2.50	2.31	4.04	2.75	1.58	2.23	31.67
CARR RANCH	.	.	.	.	.	.	.	.	.	.	.	.	.	1.77	1.88	1.71	2.65	3.62	3.12	2.08	2.15	3.51	2.96	1.61	1.79	28.85
DEL RIO WB AP	51.3	56.2	63.0	71.6	78.4	84.4	86.2	85.8	79.9	71.4	58.8	52.3	70.0	.89	.88	.82	1.36	2.73	2.29	1.31	1.52	2.61	1.98	.62	.82	17.83
KERRVILLE	46.8	50.1	56.2	64.0	71.3	78.3	80.8	80.9	75.2	66.1	54.2	48.3	64.4	1.86	2.07	2.02	3.07	4.19	2.99	2.32	1.85	4.22	3.14	1.52	2.25	31.50
LAMPASAS	47.3	50.5	56.9	65.2	72.5	80.2	83.4	83.6	77.1	67.8	55.4	49.2	65.8	1.90	2.20	1.89	3.36	4.58	2.99	1.90	2.00	3.15	2.52	2.17	2.35	31.01
LLANO	47.2	50.8	57.8	66.6	74.3	82.5	85.3	85.3	78.6	68.7	55.5	48.8	66.8	1.62	1.91	1.45	2.95	3.99	2.58	2.02	1.72	3.44	2.38	1.73	1.80	27.59
MENARD	.	.	.	.	.	.	.	.	.	.	.	.	.	1.18	1.17	.95	2.28	3.29	2.20	2.22	1.92	3.30	1.99	1.01	1.13	22.64
PAINT ROCK	.	.	.	.	.	.	.	.	.	.	.	.	.	1.29	1.04	1.09	2.36	3.73	1.89	1.89	1.74	3.01	2.20	1.10	1.12	22.46
SAN ANGELO WB AIRPORT	46.9	50.5	57.0	67.1	74.3	82.5	84.8	85.1	78.3	68.3	54.8	47.7	66.4	.97	.90	.93	2.01	3.20	1.82	1.41	1.27	2.66	1.83	.78	.85	18.63
STERLING CITY	.	.	.	.	.	.	.	.	.	.	.	.	.	.87	.94	1.02	1.66	2.77	2.03	2.26	1.77	2.22	2.03	.77	1.18	19.52
DIVISION	48.8	52.5	58.5	66.7	73.9	80.7	83.1	83.1	77.4	68.6	56.4	50.3	66.6	1.60	1.59	1.46	2.42	3.54	2.67	2.26	1.98	3.11	2.33	1.28	1.67	25.30
SOUTH CENTRAL																										
AUSTIN WB AIRPORT	50.4	53.8	59.7	67.7	75.1	81.9	84.5	84.7	79.1	70.7	58.8	52.7	68.3	2.35	2.58	2.13	3.55	3.71	3.22	2.18	1.94	3.44	2.83	2.12	2.53	32.58
BEEVILLE 5 NE	55.7	58.5	64.3	71.2	76.9	82.1	84.3	84.1	80.2	73.2	63.0	57.3	70.9	2.08	1.76	1.78	2.44	3.24	2.99	2.87	2.20	3.64	2.48	1.90	2.23	29.61
BRENHAM	50.8	53.9	59.8	67.7	75.1	81.5	84.1	84.4	79.2	70.9	59.0	52.9	68.3	3.22	3.11	2.74	3.58	4.18	3.29	2.88	2.88	2.88	2.86	3.81	3.69	39.12
COLUMBUS	.	.	.	.	.	.	.	.	.	.	.	.	.	3.01	3.41	2.78	4.29	4.35	3.34	2.93	3.01	3.77	3.11	3.29	3.69	40.98
CORPUS CHRISTI WB AP	57.4	60.4	65.2	71.7	77.5	82.3	84.1	84.2	80.8	74.5	64.1	59.2	71.8	1.63	1.70	1.44	2.14	2.99	2.39	2.32	2.77	4.40	2.76	1.72	2.08	28.34
FLATONIA 2 W	53.1	56.3	62.2	69.2	76.0	81.9	84.3	84.6	79.4	72.0	60.6	55.0	69.6	2.36	2.78	2.19	4.05	4.26	3.64	2.84	2.38	3.82	2.80	2.57	2.83	36.52
GOLIAD	.	.	.	.	.	.	.	.	.	.	.	.	.	2.24	2.43	2.07	2.60	4.05	3.29	2.97	2.58	3.67	3.25	2.05	2.43	33.63
GONZALES	.	.	.	.	.	.	.	.	.	.	.	.	.	2.15	2.32	1.96	3.55	3.71	2.91	2.79	2.43	3.54	2.42	2.42	2.53	32.73
HALLETTSVILLE	53.5	56.7	62.3	69.2	75.7	81.5	84.3	84.5	79.5	72.0	61.0	55.5	69.6	.	.	.	.	.	.	.	.	.	.	.	.	.
HONDO	52.3	55.9	61.9	69.4	76.1	82.6	84.9	85.0	79.7	71.2	59.4	53.6	69.3	1.93	1.85	1.59	2.56	4.27	3.03	2.36	2.22	3.66	2.59	1.44	1.70	29.20
KARNES CITY	.	.	.	.	.	.	.	.	.	.	.	.	.	2.05	1.89	1.58	2.50	4.07	2.99	2.39	3.11	4.32	2.82	2.01	2.20	31.93
LULING	51.1	54.4	60.8	68.5	76.1	82.7	85.0	85.3	79.7	71.2	59.2	53.1	68.9	2.26	2.41	1.96	3.59	3.78	3.30	2.83	1.92	3.40	2.50	2.35	2.33	32.63
NEW BRAUNFELS	53.1	56.6	62.9	69.9	76.6	82.8	85.1	85.3	80.1	72.1	60.7	55.0	70.0	2.41	2.42	2.17	3.15	3.82	3.41	2.23	2.08	3.45	3.05	2.08	2.27	32.54
RIOMEDINA 2 N	.	.	.	.	.	.	.	.	.	.	.	.	.	1.59	1.81	1.53	2.44	3.58	3.07	2.24	2.15	3.34	2.27	1.21	1.71	26.94
RUNGE	.	.	.	.	.	.	.	.	.	.	.	.	.	1.78	1.75	1.65	2.48	3.90	3.21	2.27	2.55	3.42	2.61	2.07	2.09	29.78
SAN ANTONIO WB AIRPORT	52.0	55.4	61.0	68.2	75.3	81.9	84.0	83.8	78.6	70.6	59.5	53.7	68.7	1.74	1.65	1.67	2.82	3.45	2.95	2.09	2.36	3.49	2.50	1.37	1.75	27.84
SAN MARCOS	50.5	53.4	59.8	67.5	74.9	81.5	83.8	83.8	78.7	70.2	58.3	52.2	67.9	2.39	2.72	2.14	3.31	3.57	3.53	2.53	2.13	3.95	3.25	2.07	2.29	33.88
SEALY	53.9	56.7	62.2	69.6	76.2	82.1	84.2	84.4	79.5	72.1	61.4	55.8	69.8	3.34	3.12	2.80	3.87	4.45	3.54	3.77	3.24	3.71	3.38	3.22	4.20	42.64
SEGUIN	52.5	55.6	62.0	69.1	76.0	82.0	84.1	84.3	79.2	71.4	59.9	54.2	69.2	2.03	2.24	1.93	3.17	3.55	3.18	2.20	2.02	3.60	3.05	1.71	2.17	30.85
SMITHVILLE	51.8	55.0	61.0	68.6	75.9	82.2	85.1	85.4	79.9	71.9	59.7	53.7	69.2	2.83	2.76	2.32	3.88	3.94	3.87	2.69	2.44	3.29	2.40	2.97	2.83	36.22
WOODSBORO	.	.	.	.	.	.	.	.	.	.	.	.	.	2.03	2.06	1.79	2.97	3.77	2.75	2.95	3.50	4.83	2.91	1.83	2.37	33.76
DIVISION	53.5	56.7	62.3	69.6	76.3	82.3	84.6	84.7	79.9	72.3	61.0	55.4	70.0	2.34	2.37	2.10	3.04	3.78	3.13	2.68	2.57	3.61	2.86	2.21	2.56	32.47
UPPER COAST																										
ANGLETON 2 W	54.7	57.6	61.8	68.5	75.0	80.5	82.3	82.3	78.8	71.5	61.7	57.1	69.3	3.63	3.84	3.18	3.20	3.90	3.51	5.53	4.82	5.44	3.80	3.70	4.61	49.16
BAYTOWN	.	.	.	.	.	.	.	.	.	.	.	.	.	4.43	4.22	3.25	3.80	4.72	3.75	5.53	4.21	4.83	3.73	4.48	4.76	51.71
BEAUMONT EXP FARM	53.8	56.3	61.0	68.0	75.0	80.9	82.7	82.8	78.7	70.6	60.2	55.3	68.8	.	.	.	.	.	.	.	.	.	.	.	.	.
BEAUMONT FILTER PLANT	54.3	56.9	61.8	69.3	76.5	82.7	84.4	84.0	79.8	71.8	60.6	55.6	69.8	4.43	4.50	3.23	4.44	5.05	4.35	5.95	5.43	4.67	3.17	4.04	5.03	54.29
DANEVANG 2 SE	55.5	58.0	63.1	69.8	76.2	81.4	83.3	83.4	79.5	72.5	62.5	57.3	70.2	2.76	3.02	2.63	3.13	3.48	3.41	3.74	4.51	4.00	3.91	2.62	3.29	40.50
EDNA 3 SW	.	.	.	.	.	.	.	.	.	.	.	.	.	2.65	2.79	2.41	3.18	3.88	3.38	3.45	3.18	4.10	3.49	2.65	2.72	37.88
GALVESTON WB AIRPORT	54.4	56.6	61.4	68.6	76.0	82.1	83.6	83.6	80.2	73.4	62.7	56.8	70.0	3.76	3.35	3.03	2.97	3.01	2.49	4.87	4.52	5.76	3.19	3.84	4.42	45.21
GALVESTON WB CITY	54.9	56.8	61.4	68.5	75.8	81.7	83.1	83.3	80.1	73.5	63.0	57.2	69.9	3.46	2.88	2.86	2.59	2.79	2.65	4.79	4.39	5.09	2.86	3.56	3.89	41.81
HOUSTON WB CITY	54.6	57.1	62.4	69.3	76.2	82.2	83.9	84.1	79.8	72.4	61.6	56.5	70.0	3.72	3.21	2.40	3.42	4.43	3.83	5.15	3.55	3.81	3.60	4.04	4.10	45.26
HOUSTON WB AIRPORT	53.6	55.8	61.3	68.5	76.0	81.6	83.0	83.2	79.2	71.4	60.8	55.7	69.2	3.78	3.44	2.67	3.24	4.32	3.69	4.29	4.27	4.26	3.77	3.86	4.36	45.95
LIBERTY	53.9	56.5	61.6	68.3	75.2	81.1	83.2	83.0	78.7	70.8	60.2	55.2	69.0	4.78	4.21	3.13	4.33	4.49	4.18	5.05	4.01	4.00	3.74	4.08	5.16	51.16
MATAGORDA NO 2	56.3	58.7	63.0	69.5	76.3	81.9	83.7	83.8	80.0	73.2	63.3	58.2	70.7	3.03	3.06	2.37	3.22	3.29	2.70	3.53	4.24	5.03	3.51	2.95	3.65	40.58
PIERCE 1 E	53.2	55.9	61.1	68.4	75.5	81.1	83.0	82.9	78.9	71.5	60.7	55.2	69.0	3.17	3.16	2.61	2.84	3.92	3.51	3.80	3.66	3.87	3.93	3.34	3.38	41.19
PORT ARTHUR	53.9	56.3	61.3	68.8	76.1	82.2	83.5	83.6	79.7	72.0	61.0	55.4	69.5	4.36	4.18	3.79	4.27	4.64	4.46	6.96	5.06	5.49	3.43	3.77	4.94	55.35
PORT ARTHUR WB AIRPORT	53.6	56.0	61.1	68.2	74.7	80.6	81.9	82.3	78.2	70.3	59.7	54.8	68.5	4.23	4.45	3.44	3.94	4.94	4.29	6.00	5.49	4.88	2.88	3.46	5.09	53.09
SUGAR LAND	53.3	55.9	61.4	68.8	75.9	81.6	83.5	83.6	79.6	72.2	60.8	55.2	69.3	3.56	3.55	2.55	3.67	4.53	3.28	4.23	3.77	3.85	3.82	4.16	4.14	45.11
VICTORIA WB AIRPORT	55.4	58.0	63.2	70.0	76.2	81.3	83.2	83.4	79.1	72.6	62.4	57.2	70.2	2.34	2.34	2.32	2.62	4.12	3.04	3.61	3.13	4.23	3.48	2.36	2.61	36.20
DIVISION	54.4	57.0	61.9	68.9	75.7	81.5	83.3	83.2	79.3	71.9	61.3	56.2	69.7	3.62	3.57	2.83	3.35	4.09	3.61	4.70	4.48	4.57	3.65	3.55	4.17	44.94
SOUTHERN																										
ALICE	.	.	.	.	.	.	.	.	.	.	.	.	.	1.53	1.37	1.30	2.03	2.96	2.98	1.86	2.39	4.48	2.69	1.46	1.65	26.70
BIG WELLS	.	.	.	.	.	.	.	.	.	.	.	.	.	1.08	1.19	.92	1.91	3.11	2.43	1.56	1.82	2.81	2.19	.92	1.28	21.22
CARRIZO SPRINGS	54.3	58.2	64.5	72.3	78.8	84.7	86.6	86.5	81.5	73.4	61.7	55.5	71.5	1.02	1.03	.80	1.68	3.58	2.54	1.83	2.31	2.82	2.29	.83	1.17	21.90
COTULLA	.	.	.	.	.	.	.	.	.	.	.	.	.	1.26	1.35	1.07	1.92	3.18	2.35	1.38	1.80	2.74	2.09	.86	1.43	21.43
DILLEY	53.6	56.9	63.4	71.2	77.8	83.5	85.6	85.8	81.0	72.8	61.2	55.1	70.7	1.33	1.47	.90	2.10	3.21	2.91	2.13	1.77	2.65	2.09	1.22	1.56	23.34
EAGLE PASS	52.4	57.1	64.2	72.8	79.5	85.8	87.6	87.3	81.4	72.9	59.9	53.2	71.2	1.09	.92	.70	1.87	3.26	2.43	2.04	2.34	2.50	1.73	.66	1.00	20.54
ENCINAL 3 NW	54.8	58.4	65.0	72.9	79.1	84.6	86.5	86.3	81.3	73.4	62.2	56.0	71.7	1.23	1.27	.94	1.63	3.46	2.38	1.57	1.90	2.81	2.18	.93	1.29	21.59
FALFURRIAS	58.5	61.6	67.2	74.1	80.0	84.7	86.5	86.3	82.1	75.1	65.2	59.7	73.4	1.52	1.23	1.02	1.93	2.79	2.43	1.49	2.69	4.54	2.03	1.00	1.46	24.13
GEORGE WEST	.	.	.	.	.	.	.	.	.	.	.	.	.	1.59	1.59	1.48	2.15	3.13	2.80	2.04	2.33	3.51	2.59	1.46	2.10	26.77
LAREDO WB AIRPORT	57.7	61.7	68.1	75.2	80.8	85.7	87.5	87.5	82.7	75.5	64.8	59.2	73.9	1.13	.89	.62	1.65	2.79	1.96	1.36	1.70	2.86	1.63	.87	1.17	18.63

*NORMALS BY CLIMATOLOGICAL DIVISIONS

Taken from "Climatography of the United States No. 81-4, Decennial Census of U. S. Climate"

TEMPERATURE (°F)   PRECIPITATION (In.)

STATIONS (By Divisions) SOUTHERN – CONTINUED	JAN	FEB	MAR	APR	MAY	JUNE	JULY	AUG	SEPT	OCT	NOV	DEC	ANN	JAN	FEB	MAR	APR	MAY	JUNE	JULY	AUG	SEPT	OCT	NOV	DEC	ANN
SARITA 7 E	•	•	•	•	•	•	•	•	•	•	•	•	•	1.91	1.62	1.42	1.68	3.21	2.18	2.01	2.50	5.16	2.31	1.12	1.49	26.61
THREE RIVERS	•	•	•	•	•	•	•	•	•	•	•	•	•	1.51	1.59	1.30	2.22	3.32	2.93	1.96	2.30	3.41	2.30	1.49	2.02	26.35
WHITSETT 2 SW	•	•	•	•	•	•	•	•	•	•	•	•	•	1.52	1.69	1.25	2.36	3.42	2.57	1.94	2.16	3.94	1.75	1.53	2.03	26.16
DIVISION	55.5	59.2	65.4	72.8	79.1	84.5	86.5	86.4	81.5	73.8	62.6	56.7	72.1	1.24	1.17	.95	1.90	3.12	2.55	1.79	2.14	3.13	2.09	.98	1.27	21.98
LOWER VALLEY																										
BROWNSVILLE WB AIRPORT	61.4	64.0	67.9	73.9	79.0	82.7	84.0	84.1	81.2	75.9	67.6	62.9	73.7	1.35	1.48	1.04	1.55	2.36	2.96	1.68	2.77	4.99	3.53	1.32	1.72	26.75
HARLINGEN	61.5	64.3	68.7	75.0	79.9	83.6	85.2	85.4	81.9	76.2	67.6	62.6	74.3	1.48	1.22	1.03	1.66	3.14	2.58	1.89	3.08	4.57	2.68	1.25	1.51	26.09
MISSION	60.2	63.2	68.6	75.4	80.5	84.0	85.5	85.8	82.2	75.9	66.5	61.2	74.1	1.35	1.00	.88	1.61	2.22	1.94	1.58	1.66	3.24	2.05	.74	1.02	19.29
PORT ISABEL	62.2	64.3	67.7	73.2	78.5	82.4	83.7	83.7	81.9	77.3	69.6	64.3	74.1	1.66	1.35	1.15	1.69	1.98	2.49	1.21	2.19	4.97	3.05	1.75	2.31	25.80
RAYMONDVILLE	60.5	63.2	68.2	74.6	79.7	83.3	84.9	85.0	81.5	75.3	66.3	61.6	73.7	1.83	1.15	1.30	1.45	3.48	2.46	1.94	3.00	4.65	2.57	1.37	1.33	26.53
WESLACO 2 E	•	•	•	•	•	•	•	•	•	•	•	•	•	1.66	1.03	1.07	1.53	2.70	2.46	1.71	2.67	4.13	2.10	1.02	1.26	23.34
DIVISION	61.0	63.7	68.4	74.8	79.7	83.4	84.8	85.0	81.8	76.0	67.3	62.3	74.1	1.57	1.19	1.09	1.56	2.72	2.51	1.65	2.56	4.21	2.59	1.16	1.46	23.93

* Normals for the period 1931-1960. Divisional normals may not be the arithmetical averages of individual stations published, since additional data for shorter period stations are used to obtain better areal representation.

TEMPERATURE   PRECIPITATION

Jan.	Feb.	Mar.	Apr.	May	June	July	Aug.	Sept.	Oct.	Nov.	Dec.	Annual	Jan.	Feb.	Mar.	Apr.	May	June	July	Aug.	Sept.	Oct.	Nov.	Dec.	Annual

CONFIDENCE LIMITS

In the absence of trend or record changes, the chances are 9 out of 10 that the true mean will lie in the interval formed by adding and subtracting the values in the following table from the means for any station in the State. Because of the wider variation in mean precipitation, the corresponding monthly means and annual mean must be substituted for "p" in the precipitation table below to obtain mean precipitation confidence limits.

| 1.4 | 1.2 | 1.2 | .0 | .6 | .6 | .5 | .5 | .8 | 1.0 | 1.0 | .9 | .4 | $.31\sqrt{p}$ | $.33\sqrt{p}$ | $.30\sqrt{p}$ | $.36\sqrt{p}$ | $.44\sqrt{p}$ | $.48\sqrt{p}$ | $.44\sqrt{p}$ | $.42\sqrt{p}$ | $.49\sqrt{p}$ | $.46\sqrt{p}$ | $.35\sqrt{p}$ | $.37\sqrt{p}$ | $.40\sqrt{p}$ |

COMPARATIVE DATA

Data in the following table are the mean temperature and average precipitation for Beeville, Texas, for the period 1906-1930 and are included in this publication for comparative purposes.

| 55.2 | 58.5 | 64.7 | 70.9 | 76.6 | 82.0 | 84.0 | 84.8 | 80.7 | 72.4 | 63.2 | 55.9 | 70.7 | 1.36 | 1.61 | 2.16 | 2.13 | 4.05 | 2.76 | 2.21 | 2.41 | 3.84 | 2.64 | 2.35 | 2.32 | 29.84 |

NORMALS, MEANS AND EXTREMES

(Table Revised 1969. Base Period for Climatological Normals: 1931-1960)

ABILENE, TEXAS
MUNICIPAL AIRPORT

LATITUDE 32° 26' N
LONGITUDE 99° 41' W
ELEVATION (ground) 1762 Feet

Month	Temperature Normal Daily maximum	Daily minimum	Monthly	Extremes Record highest	Year	Record lowest	Year	Normal degree days	Precip. Normal total	Precip. Max monthly	Year	Min monthly	Year	Max in 24 hrs	Year	Snow/Sleet Mean total	Max monthly	Year	Max in 24 hrs	Year	RH 6 AM	Noon	6 PM	Mtd.
	(b)	(b)	(b)					(b)	(b)															
J	56.4	32.8	44.6	84	1966	7	1966	642	0.88	4.35	1968	T	1967	2.18	1961	1.5	6.6	1966	5.0	1947	64	53	47	5
F	60.5	36.3	48.4	82	1965	15	1966	470	1.09	2.84	1940	0.04	1962	1.74	1964	0.7	8.4	1956	4.3	1964	65	52	44	5
M	68.4	41.6	55.0	96	1967	15	1962	347	1.04	2.39	1963	0.03	1943	1.97	1943	0.7	5.7	1967	5.7	1950+	61	48	40	5
A	77.2	51.3	64.3	97	1967	35	1968	114	2.27	6.80	1966	T	1961	3.75	1957	T	T	1961	T	1950+	62	46	39	5
M	83.4	60.0	71.7	101	1967	47	1956	0	4.33	13.19	1957	0.15	1956	2.70	1956	0.0	0.0		0.0		71	51	43	5
J	91.7	68.9	80.3	101	1967+	62	1964	0	2.67	9.60	1961	0.03	1959	3.66	1959	0.0	0.0		0.0		66	47	43	5
J	94.3	72.1	83.2	104	1964	62	1964	0	2.28	7.15	1968	T	1968	3.74	1960	0.0	0.0		0.0		55	36	36	5
A	94.1	71.9	83.0	105	1964	58	1967	0	1.47	5.87	1954	T	1943	3.79	1950	0.0	0.0		0.0		57	44	42	5
S	87.4	64.4	75.9	104	1965	41	1967	99	2.07	7.86	1961	T	1956	6.70	1961	0.0	0.0		0.0		62	46	50	5
O	78.6	53.8	66.1	98	1968	36	1968+	99	2.11	4.98	1959	0.00	1952	5.10	1959	T	T	1967	0.0		62	45	51	5
N	65.3	40.7	53.0	90	1965	24	1966	366	1.11	4.60	1968	0.00	1949	1.99	1968	0.4	8.1	1968	3.5	1967	67	49	51	5
D	58.0	34.2	46.1	84	1964	9	1966	586	1.26	2.83	1960	0.02	1950	2.30	1946	0.4	4.3	1946	4.2	1946	64	51	49	5
YR	76.3	52.3	64.3	107	MAY 1967	7	JAN 1966	2624	23.32	13.19	MAY 1957	0.00	OCT 1952+	6.70	SEP 1961	4.2	8.4	FEB 1956	5.7	MAR 1962	64	50	44	5

For period August 1963 through the current year.

Means and extremes in the above table are from the existing location (or last comparable location). Annual extremes have been exceeded at other locations as follows:
Highest temperature 111 in August 1943; lowest temperature -9 in January 1947; maximum monthly precipitation 15.70 in August 1914; maximum precipitation in 24 hours 6.78 in May 1908; maximum monthly snowfall 9.5 in February 1890; lowest temperature 9.5 in January 1919 and earlier date(s).

NORMALS, MEANS AND EXTREMES

(Table Revised 1975. Base Period for Climatological Normals: 1941-1970)

Station: ABILENE, TEXAS MUNICIPAL AIRPORT
Latitude: 32° 25' N   Longitude: 99° 41' W   Elevation (ground): 1784 feet
Standard time used: CENTRAL

Month	Temp Normal Daily maximum	Daily minimum	Monthly	Extremes Record highest	Year	Record lowest	Year	Degree days Heating	Cooling	Precip Normal	Max monthly	Year	Min monthly	Year	Max 24 hrs	Year	Snow Mean monthly	Year	Max monthly	Year	Max 24 hrs	Year
	(a)																					
J	55.4	31.7	43.7	88	1969	-1	1973	660	0	1.02	4.35	1968	T	1967	2.18	1961	13.5	1973	13.5	1973	7.5	1973
F	59.9	35.9	47.9	85	1974	11	1965	479	0	0.97	2.84	1940	0.04	1962	1.74	1964	8.4	1956	8.4	1956	4.3	1964
M	67.3	41.7	54.5	97	1974	15	1965	354	29	0.98	3.28	1973	0.03	1943	1.97	1943	7.3	1970	7.3	1970	6.1	1970
A	77.2	52.7	65.2	97	1972	25	1973	104	110	2.47	6.80	1966	0.15	1961	3.75	1957	0.0	1950	0.0		0.0	
M	83.9	60.8	72.4	102	1967	37	1967	11	240	3.86	13.19	1957	0.15	1956	2.85	1969	0.0		0.0		0.0	
J	91.6	69.0	80.3	106	1972	47	1964	0	459	2.82	9.60	1961	0.00	1959	3.66	1959	0.0		0.0		0.0	
J	95.1	72.4	83.9	105	1974	62	1972	0	586	2.34	7.15	1968	T	1968	3.74	1960	0.0		0.0		0.0	
A	95.3	71.9	83.6	106	1969	58	1967	0	577	2.05	8.18	1969	0.04	1943	3.88	1969	0.0		0.0		0.0	
S	87.5	64.6	76.1	104	1965	32	1974	89	333	2.26	11.23	1974	T	1956	6.70	1961	0.0		0.0		0.0	
O	78.0	54.2	66.1	98	1968	15	1970	123	89	2.60	4.98	1959	0.00	1952	5.10	1959	0.0		0.0		0.0	
N	66.2	42.0	54.2	90	1965	15	1970	336	0	1.20	4.60	1968	0.00	1968	1.99	1968	8.1	1968	8.1	1968	3.5	1967
D	58.2	34.5	46.4	84	1964	11	1966	577	0	1.02	2.83	1960	T	1972	2.30	1946	4.3	1946	4.3	1946	4.2	1946
YR	76.4	52.6	64.5	107	MAY 1967	-1	JAN 1973	2610	2466	23.59	13.19	MAY 1957	0.00	OCT 1952	6.70	SEP 1961	13.5	JAN 1973	13.5	JAN 1973	7.5	JAN 1973

Means and extremes above are from existing and comparable exposures. Annual extremes have been exceeded at other sites in the locality as follows:
Highest temperature 111 in August 1943; lowest temperature -9 in January 1947; maximum monthly precipitation 15.70 in August 1914; maximum precipitation in 24 hours 6.78 in May 1908; maximum snowfall 8.0 in January 1919 and earlier.

REFERENCE NOTES APPLYING TO TABLES APPEAR ON THE PAGE FOLLOWING LAST TABLE.

(Caution: Letters and symbols may have different meanings in 1941-1970 tables than in earlier tables. See notes.)

NORMALS, MEANS AND EXTREMES

LATITUDE 35° 14' N
LONGITUDE 101° 42' W
ELEVATION (ground) 3604 Feet

(Table Revised 1969. Base Period for Climatological Normals: 1931-1960)

Month	Temperature Normal Daily maximum (b)	Daily minimum (b)	Monthly (b)	Extremes Record highest	Year	Record lowest	Year	Normal degree days (b)	Precip. Normal total (b)	Maximum monthly	Year	Minimum monthly	Year	Maximum in 24 hrs	Year	Snow,Sleet Mean total	Maximum monthly	Year	Maximum in 24 hrs	Year	Rel. hum. 6 AM	Noon	6 PM	Wind Mean hourly speed	Prevailing direction	Fastest mile Speed	Direction	Year	Pct. possible sunshine	Mean sky cover	Clear	Partly cloudy	Cloudy	Precip .01"+	Snow 1.0"+	Thunderstorms	Heavy fog	Temp Max 90°+	Max 32°-	Min 32°-	Min 0°-
(a)	(b)	(b)	(b)	8		8		(b)	(b)	28		28		28		27	28		28		7	7	7	27	15	28	28		27	27	27	27	27	27	27	27	27	27	7	7	7	7
J	49.8	23.5	36.7	79	1967	-9	1963	877	0.65	2.33	1968	T	1967+	1.74	1968	3.7	12.9	1960	6.7	1960	63	42	42	13.2	SW	62	NE	1953	70	5.0	13	7	11	4	4	3	3	0	0	4	7	7
F	54.9	27.6	41.3	88	1963	3	1967	664	0.62	1.83	1948	T	1943	1.05	1951	3.2	17.3	1964	11.0	1964	64	47	42	14.3	SW	70	NW	1956	68	5.2	11	7	10	4	4	*	3	0	2	23	3
M	62.7	32.7	47.6	90	1963	15	1967	546	0.82	2.82	1957	T	1950+	0.99	1961+	2.3	14.7	1961	9.8	1957	51	50	39	15.5	SW	72	SW	1950	71	5.1	12	7	10	4	1	1	2	0	1	15	0
A	72.0	43.0	57.5	98	1965	18	1973	252	1.32	3.74	1942	0.19	1964	1.57	1942	0.4	6.4	1947	5.1	1947	51	35	28	15.4	SW	74	SW	1942	72	5.0	11	9	10	5	*	3	1	1	0	2	0
M	80.1	52.9	66.5	99	1964+	33	1967	56	3.37	9.81	1951	0.01	1953	6.75	1951	0.0			0.0		64	39	41	14.8	S	84	SW	1947	71	4.9	11	10	10	8	0	8	2	1	0	0	0
J	91.0	62.9	77.0	104	1968	43	1964	0	2.89	10.73	1965	T		6.15	1960	0.0			0.0		67	40	42	14.4	S	65	S	1953	75	4.4	11	11	6	8	0	9	1	11	0	0	0
J	94.2	67.0	80.6	102	1966+	55	1964	0	2.34	7.59	1960	0.12	1946	4.09	1943	0.0			0.0		62	38	43	12.4	S	66	SW	1948	77	4.5	13	12	6	9	0	10	1	23	0	0	0
A	93.0	66.0	79.5	104	1964	48	1966	0	2.58	5.02	1950	0.39	1947	4.26	1947	0.0			0.0		62	38	46	12.3	S	65	SE	1946	77	4.6	13	10	6	9	0	9	1	17	0	0	0
S	85.4	58.2	71.8	98	1965	39	1965	18	1.89	5.02	1950	0.24	1947	3.42	1941	T	0.7	1945	0.7	1941	73	46	43	13.1	S	68	NE	1942	76	3.7	17	7	6	5	0	4	2	7	0	0	0
O	76.1	46.2	60.5	94	1968	30	1966	205	1.76	7.64	1941	0.00	1952	3.45	1948	T	0.7	1941	0.7	1952	65	37	34	12.9	SW	68	NW	1949	75	3.7	17	7	7	5	*	2	2	2	0	*	0
N	60.1	31.8	46.0	82	1966	15	1961	570	0.66	2.26	1961	T	1960+	1.20	1952	1.5	13.6	1941	12.2	1952	65	47	42	13.1	SW	56	NW	1949	73	4.2	15	8	7	3	3	1	2	0	*	11	*
D	52.2	26.3	39.3	76	1965+	-3	1961	797	0.77	4.52	1959	0.01	1956+	3.11	1943	2.2	8.5	1943	4.7	1943	70	49	48	13.0	SW	62	NE	1953	68	4.8	13	8	10	4	1	*	2	0	3	26	1
YR	72.5	44.8	58.7	104	JUN. 1968+	-9	JAN. 1963	3985	19.67	10.73	JUN. 1965	0.00	OCT. 1952	6.75	MAY 1951	13.3	17.3	FEB. 1964	12.2	NOV. 1952	62	72	44	13.7	SW	84	MAY NE	1949	73	4.6	161	104	100	66	4	48	24	69	10	104	3

Means and extremes in the above table are from the present or last comparable instrumental locations. Annual extremes have been exceeded at other locations as follows:
Highest temperature 108 in June 1953; lowest temperature -16 in February 1899; maximum monthly snowfall 28.7 in February 1903; maximum snowfall in 24 hours 20.6 in March 1934.

NORMALS, MEANS AND EXTREMES

(Table Revised 1975. Base Period for Climatological Normals: 1941-1970)

Month	Temp Normal Daily maximum	Daily minimum	Monthly	Extremes Record highest	Year	Record lowest	Year	Degree days Heating	Cooling	Precip Water equivalent Normal	Maximum monthly	Year	Minimum monthly	Year	Maximum in 24 hrs	Year	Snow, Ice pellets Maximum monthly	Year	Maximum in 24 hrs	Year	Rel. hum. Hour 00	06	12	18	Wind Mean speed	Prevailing direction	Fastest mile Speed	Direction	Year	Pct. possible sunshine	Mean sky cover	Clear	Partly cloudy	Cloudy	Precip .01"+	Snow 1.0"+	Thunderstorms	Heavy fog ¼ mi	Temp Max 90°+	Max 32°-	Min 32°-	Min 0°-
(a)				14		14				34	34		34		34		34		34		13	13	13	13	33	15	34	34		33	33	33	33	33	33	33	33	33	13	13	13	13
J	49.4	22.5	36.0	79	1970	-9	1963	899	0	0.54	2.33	1968	T	1967	1.74	1968	12.9	1960	6.7	1960	63	69	48	49	13.1	NE	62	NE	1953	68	5.1	11	7	11	4	4	*	3	0	4	26	2
F	53.0	26.4	39.7	88	1963	-9	1971	708	0	0.58	1.83	1948	T	1943	1.28	1971	17.3	1964	13.0	1971	63	70	47	44	14.5	NW	70	NW	1956	69	5.2	11	8	10	4	4	*	2	0	2	22	*
M	60.0	31.2	45.6	94	1971	6	1967	601	0	0.77	3.99	1973	T	1950	2.27	1973	14.7	1961	10.8	1957	55	66	43	37	15.5	SW	74	SW	1950	72	5.1	12	9	10	5	1	3	1	0	1	14	0
A	70.9	42.1	56.5	98	1972	18	1973	275	20	1.23	3.74	1942	0.19	1964	1.57	1942	6.4	1947	5.1	1947	52	64	35	29	15.4	SW	74	SW	1942	73	5.0	11	9	10	6	*	5	1	2	0	3	0
M	79.2	51.9	65.6	99	1974	30	1970	99	99	2.83	9.81	1951	0.01	1953	6.75	1951	0.0		0.0		57	70	37	35	14.8	S	84	SW	1947	73	4.8	11	11	9	8	0	8	1	12	0	0	0
J	88.0	61.2	74.6	104	1970	43	1964	0	298	3.45	10.73	1965	T	1953	6.15	1960	0.0		0.0		60	72	42	37	14.8	S	75	NW	1974	77	4.3	11	12	8	9	0	9	*	22	0	0	0
J	91.4	65.9	78.7	104	1970	54	1972	0	425	2.95	7.59	1960	0.12	1946	4.09	1943	0.0		0.0		60	72	42	37	12.5	S	66	SW	1948	78	4.5	13	13	6	8	0	10	1	22	0	0	0
A	90.4	64.7	77.6	104	1964	52	1972	0	391	2.93	5.02	1950	0.39	1947	4.26	1945	0.0		0.0		62	80	46	46	11.9	S	65	NE	1942	77	4.9	13	15	7	8	0	9	1	16	0	0	0
S	82.9	56.7	69.8	100	1970	32	1970	164	36	1.93	5.02	1950	0.24	1947	3.42	1970	T	1945	T	1945	55	80	51	46	13.0	S	68	NW	1970	74	3.9	16	7	7	6	0	4	2	7	0	1	0
O	72.9	46.1	59.5	94	1973	12	1972	206	0	1.83	7.64	1941	0.00	1952	3.45	1948	3.9	1970	2.9	1970	53	73	47	43	13.2	SW	68	NW	1970	74	3.8	17	7	7	5	*	2	2	2	0	1	0
N	60.0	34.2	46.3	82	1973	12	1972	561	0	0.53	2.26	1961	T	1960	1.29	1971	13.6	1941	12.2	1952	57	75	47	49	13.2	SW	56	NW	1970	73	4.3	15	7	8	3	1	1	1	0	*	12	0
D	51.5	25.5	38.5	76	1965	-3	1961	822	0	0.73	4.52	1959	0.01	1953	3.11	1943	8.5	1943	7.4	1943	55	70	49	48	13.0	SW	62	NE	1953	67	4.8	13	8	10	4	1	*	3	0	3	26	*
YR	70.8	43.9	57.4	104	JUL 1970	-9	JAN 1963	4183	1433	20.28	10.73	JUN 1965	0.00	OCT 1952	6.75	MAY 1951	17.3	FEB 1964	13.5	FEB 1971	52	72	44	40	13.7	SW	84	MAY SW	1949	73	4.6	160	104	101	67	4	48	12	66	12	106	2

Means and extremes above are from existing and comparable exposures. Annual extremes have been exceeded at other sites in the locality as follows:
Highest temperature 108 in June 1953; lowest temperature -16 in February 1899; maximum monthly snowfall 28.7 in February 1903; maximum snowfall in 24 hours 20.6 in March 1934.

Average station pressure: Elev. 3604 feet m.s.l.
J 890.5, F 891.3, M 885.9, A 888.0, M 888.1, J 889.6, J 892.7, A 892.5, S 892.5, O 893.3, N 891.0, D 889.8, YR 890.4

REFERENCE NOTES APPLYING TO TABLES APPEAR ON THE PAGE FOLLOWING LAST TABLE.
(Caution: Letters and symbols may have different meanings in 1941-1970 tables than in earlier tables. See notes.)

AUSTIN, TEXAS
MUNICIPAL AIRPORT
LATITUDE 30° 18' N
LONGITUDE 97° 42' W
ELEVATION (ground) 597 Feet

NORMALS, MEANS AND EXTREMES
(Table Revised 1969. Base Period for Climatological Normals: 1931-1960)

Month	Normal Daily maximum	Normal Daily minimum	Normal Monthly	Record highest	Year	Record lowest	Year	Normal degree days	Normal total precip	Max monthly precip	Year	Min monthly precip	Year	Max in 24 hrs	Year	Snow mean total	Snow max monthly	Year	Snow max 24 hrs	Year	RH 6AM	RH Noon	RH 6PM	Wind mean hourly speed	Prevailing direction	Fastest mile speed	Dir	Year	Mean sky cover	Pct possible sunshine	Clear	Partly cloudy	Cloudy
J	60.3	40.5	50.4	86	1963	12	1963	468	2.35	7.94	1968	0.07	1942	3.44	1965	0.5	7.0	1944	7.0	1944	72	58	56	9.9	S	47	N	1962+	6.3	48	9	6	16
F	64.0	43.5	53.8	87	1962	22	1967	325	2.58	6.39	1958	0.22	1954	3.73	1958	0.5	6.0	1966	6.0	1966	70	56	51	10.2	S	57	N	1947	6.1	51	8	6	14
M	70.6	48.7	59.7	96	1967	23	1963	223	2.13	6.13	1945	0.28	1963	2.68	1951	0.1	2.0	1965	2.0	1965	69	51	47	10.9	S	44	NW	1957	6.1	54	8	8	14
A	78.0	57.3	67.7	98	1963	39	1962	51	3.55	9.93	1957	0.10	1961	3.86	1942	0.0	0.3		0.0		78	55	55	10.9	SSE	44	NE	1957	6.3	53	7	9	14
M	85.2	64.9	75.1	99	1967	52	1968+	3	3.71	9.98	1965	T	1960	3.51	1946	0.0	0.0		0.0		81	60	57	10.9	SSE	49	NE	1966	6.3	59	7	12	12
J	92.0	71.7	81.9	100	1967+	55	1964	0	3.22	11.43	1961	0.00	1962	6.50	1964	0.0	0.0		0.0		79	55	52	9.6	S	49	SE	1956	5.2	70	8	16	6
J	95.1	73.9	84.5	103	1964+	64	1968+	0	2.18	8.40	1961	0.00	1962	5.46	1961	0.0	0.0		0.0		72	47	42	8.7	S	43	S	1953	4.7	78	11	15	5
A	95.6	73.7	84.7	105	1962	61	1967	0	1.94	8.11	1962	0.00	1952	4.58	1945	0.0	0.0		0.0		71	47	42	8.3	S	47	N	1961	4.6	77	12	13	6
S	89.7	68.5	79.1	102	1963	47	1967	0	3.44	12.31	1942	0.07	1947	4.61	1942	0.0	0.0		0.0		77	56	55	8.1	S	45	NE	1961	4.8	69	12	10	8
O	81.9	59.5	70.7	95	1963	39	1957	31	2.83	7.31	1946	0.01	1952	7.22	1960	0.0	0.0		0.0		73	51	51	8.1	SSE	47	NW	1951	4.5	68	14	10	7
N	69.6	47.9	58.8	89	1966+	31	1966+	225	2.12	7.31	1944	T	1949	3.98	1946	T	1.0	1959	1.0	1959	76	57	57	9.1	S	48	NW	1951	4.5	65	13	7	12
D	62.8	42.6	52.7	84	1966	21	1966	388	2.53	5.91	1944	0.00	1950	4.02	1953	T	T	1963+	T	1963+	74	59	59	9.2	S	49	NW	1956	5.9	51	10	6	15
YR	78.7	57.7	68.3	105	AUG. 1962	12	JAN. 1963	1711	32.58	12.31	OCT. 1960	0.00	JUL. 1962+	7.22	OCT. 1960	1.1	7.0	JAN. 1944	7.0	JAN. 1944	74	55	52	9.4	S	57	N	FEB. 1947	5.5	62	117	118	130

Mean number of days (Table 1):

Month	Precipitation .01 in or more	Snow Sleet 1.0 in or more	Thunderstorms	Heavy fog	90° and above	Max 32° and below	Min 32° and below	Min 0° and below
J	8	*	1	4	0	1	11	0
F	7	*	1	3	0	1	6	0
M	7	*	2	2	0	*	2	0
A	7	0	5	1	1	0	0	0
M	8	0	5	1	7	0	0	0
J	6	0	5	*	21	0	0	0
J	4	0	3	*	29	0	0	0
A	5	0	4	*	29	0	0	0
S	7	0	3	1	14	0	0	0
O	6	0	2	2	4	0	*	0
N	6	0	1	2	0	*	7	0
D	7	*	2	4	0	1	7	0
YR	81	*	40	23	107	1	25	0

Means and extremes in the above table are from the present instrumental locations. Annual extremes have been exceeded at other locations as follows:
Highest temperature 109 in August 1923; lowest temperature -2 in January 1949; maximum monthly precipitation 20.78 in September 1921; maximum precipitation in 24 hours 19.03 in September 1921; maximum snowfall 9.7 in November 1937; maximum snowfall in 24 hours 9.7 in November 1937.

NORMALS, MEANS AND EXTREMES
(Table Revised 1975. Base Period for Climatological Normals: 1941-1970)

Month	Normal Daily maximum	Normal Daily minimum	Normal Monthly	Record highest	Year	Record lowest	Year	Heating degree days	Cooling degree days	Normal precip	Max monthly precip	Year	Min monthly precip	Year	Max in 24 hrs	Year	Snow mean total	Snow max monthly	Year	Snow max 24 hrs	Year	RH 00	RH 06	RH 12	RH 18	Wind mean speed	Prevailing dir	Fastest mile speed	Dir	Year	Mean sky cover	Pct possible sunshine	Clear	Partly cloudy	Cloudy
J	60.0	39.3	49.7	90	1971	20	1971	483	8	1.88	7.94	1968	0.04	1971	3.44	1965	0.4	7.0	1944	7.0	1944	73	79	61	58	9.9	S	47	N	1962	6.3	48	9	6	16
F	63.8	42.8	53.3	97	1962	22	1971	344	16	3.09	6.39	1958	0.28	1954	3.73	1958	0.4	6.0	1966	6.0	1966	70	79	54	51	10.2	S	57	N	1947	6.1	51	8	6	14
M	70.7	48.2	59.5	98	1971	23	1973	223	52	1.89	4.98	1945	0.10	1972	2.86	1951	0.1	2.0	1965	2.0	1965	69	77	48	44	11.0	SSE	44	NE	1957	6.0	54	8	8	14
A	79.0	58.2	68.6	99	1972	39	1973	44	152	3.49	9.98	1957	0.81	1960	4.49	1970	0.0	0.0		0.0		76	80	53	52	10.8	SSE	44	NE	1957	6.3	54	7	8	15
M	85.1	65.1	75.1	100	1967	47	1970	0	316	3.97	9.98	1965	T	1965	6.50	1964	0.0	0.0		0.0		80	87	59	55	10.0	SSE	49	NE	1966	6.3	59	7	13	11
J	91.7	71.4	81.6	103	1967	53	1970	0	498	3.13	11.43	1961	0.00	1967	5.09	1964	0.0	0.0		0.0		78	88	55	52	9.6	S	49	SE	1956	5.2	70	8	15	7
J	95.4	73.7	84.6	103	1969	64	1970	0	608	1.88	8.90	1961	0.07	1962	5.46	1961	0.0	0.0		0.0		73	86	48	44	8.6	S	43	SE	1969	4.8	77	11	14	6
A	95.9	73.5	84.7	102	1962	61	1967	0	617	2.20	8.10	1952	T	1952	4.58	1945	0.0	0.0		0.0		73	86	47	44	8.1	S	47	N	1961	4.8	75	13	13	5
S	89.4	68.4	79.0	98	1963	39	1970	0	417	3.66	12.31	1960	0.07	1947	6.74	1973	0.0	0.0		0.0		77	85	57	57	8.1	S	45	NW	1961	4.7	65	12	10	8
O	81.3	58.9	70.1	95	1963	27	1966	29	177	3.02	7.91	1946	T	1970	7.22	1960	0.0	0.0		0.0		76	85	51	53	8.2	SSE	47	NW	1967	4.7	67	15	9	7
N	69.3	47.0	59.1	89	1966		1974	205	28	2.04	5.91	1944	T	1953	3.98	1946	T	1.0	1959	1.0	1959	76	82	57	58	9.1	S	48	NW	1951	5.3	57	11	7	12
D	63.0	41.6	52.3	86	1955		1966	399	16	2.22	5.91	1944	0.00	1953	4.02	1953	T	1.0	1971	1.0	1971	74	80	60	59	9.2	S	49	NW	1956	5.9	51	11	7	15
YR	78.8	57.4	68.1	105	AUG 1969	12	JAN 1963	1737	2908	32.49	12.31	OCT 1960	0.00	JUL 1962	7.22	OCT 1960	7.0	7.0	JAN 1944	7.0	JAN 1944	75	83	56	53	9.4	S	57	N	FEB 1947	5.5	61	115	116	134

Mean number of days and Average station pressure (Table 2):

Month	Precipitation .01 in or more	Snow, ice pellets 1.0 in or more	Thunderstorms	Heavy fog, visibility ¼ mile or less	90° and above	Max 32° and below	Min 32° and below	Min 13° and below	Avg station pressure (mb)
J	8	*	1	4	*	1	10	0	997.1
F	7	*	2	3	0	1	5	0	997.7
M	7	*	3	1	1	0	1	0	990.5
A	8	0	5	1	1	0	0	0	990.2
M	7	0	7	1	7	0	0	0	990.2
J	6	0	4	*	20	0	0	0	991.6
J	5	0	4	*	29	0	0	0	993.6
A	6	0	4	*	29	0	0	0	993.4
S	7	0	3	1	15	0	0	0	992.6
O	6	0	2	2	6	0	*	0	992.6
N	7	*	1	3	1	*	5	0	996.1
D	7	*	1	4	0	1	11	0	996.1
YR	82	*	41	24	104	1	22	0	994.0

Elev. 621 feet m.s.l.

Means and extremes above are from existing and comparable exposures. Annual extremes have been exceeded at other sites in the locality as follows:
Highest temperature 109 in August 1923; lowest temperature -2 in January 1949; maximum monthly precipitation 20.78 in September 1921; maximum precipitation in 24 hours 19.03 in September 1921; maximum snowfall 9.7 in November 1937; maximum snowfall in 24 hours 9.7 in November 1937.

REFERENCE NOTES APPLYING TO TABLES APPEAR ON THE PAGE FOLLOWING LAST TABLE.
(Caution: Letters and symbols may have different meanings in 1941-1970 tables than in earlier tables. See notes.)

NORMALS, MEANS AND EXTREMES

(Table Revised 1969. Base Period for Climatological Normals: 1931-1960)

LATITUDE 25° 54' N
LONGITUDE 97° 26' W
ELEVATION (ground) 19 Feet

Month	Temperature Normal — Daily maximum	Daily minimum	Monthly	Extremes Record highest	Year	Record lowest	Year	Normal degree days	Precipitation Normal total	Max. monthly	Year	Min. monthly	Year	Max. in 24 hrs.	Year	Snow, Sleet Mean total	Rel. hum. 6 AM	Noon	6 PM	Mean hourly speed	Prevailing dir.	Fastest mile speed	Dir.	Year	Mean sky cover	Pct. possible sunshine	Avg. daily solar radiation
(a)	(b)	(b)	(b)	2	2	2	2	(b)	(b)	29		29		29		29	2	2	2	26	14	26	26	26	26	26	15
J	70.5	52.2	61.4	80	1967	32	1968	205	1.35	5.11	1945	T	1956	2.95	1958	T	88	72	75	11.9	SSE	46	S	1953	6.6	46	284
F	73.2	54.7	64.0	85	1967	32	1968	106	1.48	10.25	1958	T	1954	4.98	1958	T	89	65	70	12.6	SSE	55	NW	1965	6.4	49	335
M	76.8	59.0	67.9	87	1967	35	1968	74	1.04	4.27	1944	0.03	1944	1.85	1940	T	86	63	71	13.7	SE	69	S	1950	6.7	55	408
A	82.3	65.5	73.9	93	1967	51	1968	0	1.55	5.85	1946	0.01	1967	3.69	1956	0.0	88	63	71	13.5	SE	52	SE	1960	6.6	55	470
M	87.1	70.8	79.0	99	1967	53	1968	0	2.36	6.05	1966	0.01	1955+	4.13	1966	0.0	88	60	67	13.9	SE	66	SE	1968	5.9	66	556
J	90.7	74.7	82.7	95	1968	66	1967+	0	2.96	13.06	1942	T		8.18	1942	0.0	86	60	67	12.8	E	52	E	1954	5.2	72	612
J	92.5	75.5	84.0	96	1967	70	1967	0	1.68	5.59	1949	T	1962	3.62	1949	0.0	86	55	65	11.9	SE	37	E	1961	4.7	81	630
A	92.8	75.3	84.1	96	1968+	63	1967	0	2.77	7.08	1944	0.07	1964	4.39	1951	0.0	85	59	65	10.9	SE	57	NW	1945	4.9	76	567
S	89.8	72.6	81.2	93	1968	56	1967	0	4.99	19.26	1967	0.07	1959	12.19	1967	0.0	88	60	70	10.9	SE	69	NE	1967	4.9	67	475
O	85.1	66.6	75.9	91	1968	52	1967	0	3.53	17.12	1958	0.34	1961	6.67	1958	0.0	86	61	70	9.8	SE	47	NE	1961	4.7	67	414
N	77.0	58.2	67.6	88	1968	41	1968	66	1.32	6.26	1957	0.01	1949	3.64	1957	0.0	81	63	72	9.8	SSE	42	NW	1961	5.3	53	304
D	72.2	53.5	62.9	87	1968	34	1967	149	1.72	9.45	1940	0.02	1948	5.69	1940	T	86	63	71	11.1	NNW	45	NW	1950	6.6	46	254
YR	82.5	64.9	73.7	99	MAY 1967	32	FEB. 1968+	600	26.75	19.26	SEP. 1967	T	APR. 1967+	12.19	SEP. 1967	T	85	62	70	12.0	SE	69	NE	SEP. 1967	5.8	61	442

Mean number of days (Table 1)

Month	Sunrise to sunset Clear	Partly cloudy	Cloudy	Precipitation .01 inch or more	Snow, Sleet 1.0 inch or more	Thunderstorms	Heavy fog	Temp. Max 90° and above	Max 32° and below	Min 32° and below	Min 0° and below
J	7	8	16	7	0	1	5	0	0	2	0
F	8	5	15	7	0	1	5	0	0	1	0
M	9	6	16	5	0	1	4	0	0	0	0
A	6	11	14	5	0	2	4	0	0	0	0
M	8	13	10	6	0	2	1	4	0	0	0
J	6	17	5	7	0	*	*	11	0	0	0
J	11	15	5	4	0	2	*	25	0	0	0
A	11	14	7	6	0	4	*	29	0	0	0
S	12	12	7	6	0	4	1	23	0	0	0
O	12	9	10	6	0	1	1	10	0	0	0
N	12	8	10	6	0	1	3	3	0	0	0
D	7	8	16	7	0	*	5	0	0	2	0
YR	99	137	129	71	0	24	25	103	0	2	0

Means and extremes in the above table are from the existing location (or last comparable location). Annual extremes have been exceeded at other locations as follows: Lowest temperature 12 in February 1899; maximum monthly precipitation 30.57 in September 1920; minimum monthly precipitation 0.00 in September 1886; fastest mile of wind 106 from Northwest in September 1933.

NORMALS, MEANS AND EXTREMES

(Table Revised 1975. Base Period for Climatological Normals: 1941-1970)

Month	Temperatures °F Normal — Daily maximum	Daily minimum	Monthly	Extremes Record highest	Year	Record lowest	Year	Normal degree days Heating	Cooling	Precipitation in inches Normal	Max. monthly	Year	Min. monthly	Year	Max. in 24 hrs.	Year	Rel. hum. pct. 06	12	18	Mean speed m.p.h.	Prevailing dir.	Fastest mile speed m.p.h.	Dir.	Year	Mean sky cover	Pct. possible sunshine	Avg. station pressure mb.
(a)	8	8		8		8				35	35		35		35		8	8	8	32	14	32	32	32	32	32	20
J	69.5	51.0	60.3	93	1971	27	1973	225	79	1.35	5.11	1956	T	1967	2.95	1958	87	69	75	11.8	SSE	46	S	1953	6.3	44	1017.6
F	72.7	54.1	63.4	90	1971	26	1973	151	106	1.48	10.25	1958	T	1973	4.98	1958	86	65	70	12.2	SSE	55	NW	1965	6.1	49	1018.5
M	76.6	58.8	67.7	95	1972	35	1968	89	173	0.69	4.27	1940	T	1943	1.85	1940	85	61	68	13.6	SE	47	S	1950	6.7	51	1011.2
A	83.1	71.4	75.7	97	1973	45	1974	0	297	2.51	5.85	1946	0.01	1969	4.56	1969	86	60	69	13.5	SE	52	SE	1960	6.6	55	1012.5
M	90.6	75.0	79.3	102	1974	60	1974	0	443	2.80	6.69	1967	0.01	1955	4.56	1969	86	60	61	12.6	SE	52	E	1954	5.3	73	1012.5
J		75.0		97				0	534	2.66	13.06	1942	T	1942	8.18	1942	89	61	61								1011.7
J	92.8	75.5	84.4	99	1969	69	1969	0	601	1.19	5.59	1949	T	1962	3.62	1949	86	56	66	11.7	SE	37	E	1961	4.9	81	1014.4
A	93.0	75.7	84.4	100	1970	63	1967	0	604	2.66	7.74	1969	0.02	1974	4.39	1959	90	57	62	10.9	SE	57	NW	1945	4.9	76	1013.9
S	89.7	75.2	82.4	91	1971	56	1967	0	498	5.23	17.26	1967	0.34	1961	6.67	1958	89	62	70	9.7	SE	69	NE	1967	4.8	67	1012.3
O	84.7	66.6	75.7	92	1973	45	1974	35	337	3.32	17.12	1958	0.01	1949	6.67	1958	88	61	71	9.7	SE	47	NE	1961	4.8	66	1012.3
N	76.6	58.1	68.1	94	1970	39	1970	128	77	1.24	6.26	1957	0.01	1969	3.64	1957	85	61	71	10.9	SE	42	NW	1961	5.3	53	1016.4
D	72.3	53.3	62.8	87	1970	27	1973	145		1.24	9.45	1940	T	1969	5.69	1940	84	63	71	11.0	NNW	45	NW	1950	6.5	46	1017.3
YR	82.5	65.0	73.8	102	MAY 1974	26	FEB. 1973	650	3874	25.09	19.26	SEP. 1967	T	MAR 1971	12.19	SEP. 1967	87	61	69	11.8	SE	69	NE	SEP. 1967	5.8	62	1014.4

Mean number of days (Table 2)

Month	Sunrise to sunset Clear	Partly cloudy	Cloudy	Precipitation .01 inch or more	Snow, Ice pellets 1.0 inch or more	Thunderstorms	Heavy fog, visibility ¼ mile or less	Temp. Max 90° and above	Max 32° and below	Min 32° and below	Min 0° and below
J	6	8	17	7	0	1	6	*	0	8	0
F	7	6	15	6	0	1	5	*	0	8	0
M	8	6	17	5	0	1	4	2	0	1	0
A	6	11	13	4	0	2	1	4	0	0	0
M	5	15	11	5	0	3	*	8	0	0	0
J	8	16	6	5	0	3	*	19	0	0	0
J	11	15	5	4	0	2	*	25	0	0	0
A	11	15	5	5	0	4	*	26	0	0	0
S	11	13	7	7	0	4	*	17	0	0	*
O	11	13	7	6	0	2	1	7	0	0	0
N	11	8	11	6	0	1	3	*	0	0	0
D	7	8	16	6	0	1	6	0	0	*	0
YR	95	139	131	72	0	24	27	104	0	2	0

Means and extremes above are from existing and comparable exposures. Annual extremes have been exceeded at other sites in the locality as follows: Lowest temperature 12 in February 1899; maximum monthly precipitation 30.57 in September 1920; minimum monthly precipitation 0.00 in September 1886; fastest mile of wind 106 from Northwest in September 1933.

REFERENCE NOTES APPLYING TO TABLES APPEAR ON THE PAGE FOLLOWING LAST TABLE.
(Caution: Letters and symbols may have different meanings in 1941-1970 tables than in earlier tables. See notes.)

NORMALS, MEANS AND EXTREMES

CORPUS CHRISTI, TEXAS
INTERNATIONAL AIRPORT

LATITUDE 27° 46' N
LONGITUDE 97° 30' W
ELEVATION (ground) 41 Feet

(Table Revised 1969. Base Period for Climatological Normals: 1931-1960)

Month	Temperature Normal Daily max	Daily min	Monthly	Record highest	Year	Record lowest	Year	Precip. Normal total	Max monthly	Year	Min monthly	Year	Max 24 hrs	Year	Snow,Sleet Mean total	RH 6AM	Noon	6PM	Mean sky cover	Pct poss. sunshine	Clear	Partly cloudy	Cloudy	Precip .01"+	Thunderstorms	Heavy fog	90°+	Min 32° below
J	67.4	47.4	57.4	87	1965	25	1967	1.63	10.78	1958	0.13	1942	6.38	1958	0.1	90	69	72	6.7	47	7	7	17	8	1	5	0	4
F	69.8	50.9	60.4	83	1967	29	1967	1.70	5.24	1954	0.01	1954	2.99	1958	T	90	66	69	6.5	51	7	6	15	8	1	5	0	2
M	74.4	55.9	65.2	88	1968+	29	1965	1.44	4.01	1945	0.07	1955	2.67	1945	0.0	87	63	66	6.7	56	7	8	16	6	2	4	0	1
A	80.4	63.0	71.7	97	1964	43	1966	2.14	8.04	1956	T	1940	7.19	1956	0.0	90	67	73	6.7	57	7	10	13	6	3	2	0	0
M	85.9	69.1	77.5	98	1967+	54	1965	2.99	9.38	1968	0.17	1961	4.65	1968	0.0	92	71	75	6.8	63	6	13	12	6	5	1	1	0
J	90.5	74.0	82.3	97	1964	59	1964	2.39	8.36	1968	0.17	1944	3.89	1968	0.0	93	66	69	5.8	73	9	15	6	6	6	*	14	0
J	93.5	74.6	84.1	101	1964	64	1967	2.32	10.99	1942	0.00	1957	3.73	1961	0.0	93	59	64	4.8	83	12	14	6	5	5	*	25	0
A	93.8	74.5	84.2	101	1966+	64	1967	2.77	12.64	1953	0.10	1952	5.52	1953	0.0	93	60	66	4.8	84	12	14	7	6	5	*	23	0
S	90.1	71.4	80.8	100	1965	54	1967	4.40	20.33	1967	0.78	1953	8.76	1967	0.0	93	63	69	5.1	69	11	11	8	9	7	1	12	0
O	84.5	64.5	74.5	93	1966	40	1964	2.76	10.66	1960	0.00	1952	7.25	1960	0.0	90	59	66	5.1	65	12	8	9	6	3	3	1	0
N	73.8	54.3	64.1	92	1965	36	1966	1.72	8.53	1947	T	1949	3.44	1947	T	89	63	71	5.8	58	14	9	12	6	1	5	0	0
D	68.8	49.5	59.2	89	1964	28	1967	2.08	7.80	1960	0.01	1950	3.86	1960	T	88	65	70	6.5	48	8	7	16	7	1	3	0	2
YR	81.1	62.4	71.8	101	JUL. 1964	25	JAN. 1967	28.34	20.33	SEP. 1967	0.00	JUL. 1957+	8.76	SEP. 1967	0.1	90	64	69	5.8	64	106	120	139	75	32	27	82	8

ø For period February 1964 through current year.

Means and extremes in the above table are from the existing or comparable location(s). Annual extremes have been exceeded at other locations as follows:
Highest temperature 105 in July 1934; lowest temperature 11 in February 1899; maximum snowfall 6.0 in January 1897; maximum snowfall in 24 hours 5.0 in January 1897; fastest mile of wind 110 from Northeast in September 1919.

NORMALS, MEANS AND EXTREMES
(Table Revised 1975. Base Period for Climatological Normals: 1941-1970)

Month	Normal Daily max	Daily min	Monthly	Record highest	Year	Record lowest	Year	Heating deg. days	Cooling deg. days	Precip Normal	Max monthly	Year	Min monthly	Year	Max 24 hrs	Year	Snow Mean monthly	Pct poss. sunshine	Clear	Partly cloudy	Cloudy	Precip .01"+	Thunderstorms	Heavy fog	90°+	Min 32° below	Avg sta. pressure mb
J	66.5	46.1	56.3	91	1971	25	1973	304	34	1.58	10.78	1958	0.03	1958	6.38	1958	1.2	47	7	7	17	8	1	6	*	3	1017.2
F	69.8	49.3	59.6	91	1974	29	1965	199	48	1.55	5.24	1954	T	1974	2.99	1958	1.1	52	6	7	15	7	1	5	*	2	1018.0
M	75.5	54.2	64.9	97	1965	31	1965	120	117	1.10	4.80	1974	T	1956	2.67	1945	T	56	6	10	15	6	2	4	1	1	1010.5
A	82.1	62.8	72.5	97	1964	39	1971	39	238	2.15	8.04	1956	T	1956	7.19	1956	0.0	56	6	13	11	5	3	2	2	0	1010.2
M	86.6	69.3	77.9	99	1973	54	1970	7	400	3.17	9.38	1968	0.13	1973	4.65	1968	0.0	63	5	13	13	6	5	*	5	0	1010.1
J	91.2	73.6	82.4	99	1972	59		0	522	2.67	8.36	1968	T	1968	4.72	1968	0.0	74	8	16	6	5	5	*	16	0	1011.4
J	94.4	75.2	84.8	101	1964	64	1967	0	614	1.88	10.99	1942	0.00	1961	3.73	1961	0.0	83	6	14	6	5	5	*	27	0	1013.8
A	94.8	75.4	85.1	100	1969	64	1967	0	623	3.20	12.64	1953	0.10	1952	6.38	1953	0.0	78	6	12	7	6	4	*	25	0	1011.9
S	90.0	71.4	80.7	100	1965	54	1967	0	480	4.20	20.33	1967	0.78	1953	8.76	1967	0.0	70	7	13	9	7	5	1	14	0	1011.9
O	84.1	63.7	73.9	93	1966	40	1969	81	283	2.77	10.66	1960	0.01	1947	7.25	1960	0.0	65	8	10	12	6	3	3	4	*	1016.3
N	75.2	54.6	64.9	93	1972	33	1973	219	78	1.53	8.53	1947	T	1947	3.44	1947	T	57	12	9	9	6	1	3	*	1	1016.1
D	69.3	48.9	59.1	90		18	1973		37	1.53	7.80	1960	0.01	1960	3.86	1960	T	48	8	7	16	7	1	5	*		1016.8
YR	81.6	62.4	71.9	101	JUL 1964	18	DEC 1973	930	3474	28.53	20.33	SEP 1967	0.00	JAN 1940	8.76	SEP 1967	1.2	64	105	119	141	76	31	29	96	7	1014.0

Means and extremes above are from existing and comparable exposures. Annual extremes have been exceeded at other sites in the locality as follows:
Highest temperature 105 in July 1934; lowest temperature 11 in February 1899; maximum monthly snowfall 6.0 in January 1897; maximum monthly snowfall in 24 hours 5.0 in January 1897.

REFERENCE NOTES APPLYING TO TABLES APPEAR ON THE PAGE FOLLOWING LAST TABLE.
(Caution: Letters and symbols may have different meanings in 1941-1970 tables than in earlier tables. See notes.)

NORMALS, MEANS AND EXTREMES

LOVE FIELD

LATITUDE 32° 51' N
LONGITUDE 96° 51' W
ELEVATION (ground) 481 Feet

(Table Revised 1969. Base Period for Climatological Normals: 1931-1960)

Month	Temp Normal Daily max	Daily min	Monthly	Extremes Rec highest	Year	Rec lowest	Year	Normal degree days	Precip Normal total	Max monthly	Year	Min monthly	Year	Max 24 hrs	Year	Snow Mean total	Max monthly	Year	Max 24 hrs	Year
J	55.8	36.0	45.9	82	1967	8	1963+	601	2.32	8.46	1949	0.30	1963	5.14	1964	1.4	7.4	1964	7.4	1964
F	59.5	39.4	49.5	87	1963+	16	1960	440	2.85	7.68	1945	0.30	1963	3.45	1965	0.5	2.9	1961	2.0	1961
M	67.0	45.2	56.1	91	1967	17	1962	319	2.85	9.53	1945	0.14	1945	5.33	1945	0.2	2.0	1962	2.0	1962
A	75.4	54.6	65.0	99	1963	42	1960	90	4.00	15.40	1966	0.72	1959	5.10	1957	0.0	0.0		0.0	
M	82.7	63.1	72.9	98	1967	53	1964	6	4.83	13.74	1957	1.31	1966	6.22	1959	0.0	0.0		0.0	
J	90.9	71.7	81.3	102	1960	65	1967	0	3.24	12.18	1941	0.28	1941	4.09	1951	0.0	0.0		0.0	
J	94.5	75.3	84.9	105	1967+	61	1966+	0	1.94	8.52	1962	T	1947+	5.39	1945	0.0	0.0		0.0	
A	95.0	75.0	85.0	105	1964	65	1962	0	1.93	10.41	1964	0.02	1963	9.18	1947	0.0	0.0		0.0	
S	88.3	67.4	77.9	103	1963	46	1967	0	2.82	10.67	1964	0.01	1963	4.81	1965	0.0	0.0		0.0	
O	78.8	56.8	67.8	94	1963	38	1959	62	2.70	11.38	1959	0.18	1949+	6.52	1959	0.0	0.0		0.0	
N	65.7	44.1	54.9	86	1965	21	1960	321	2.70	9.91	1946	0.13	1950	4.99	1946	T	T	1968+	T	1968+
D	58.0	38.1	48.1	82	1966	10	1963	524	2.67	8.50	1960	0.28	1960	3.36	1960	0.2	2.2	1963	2.2	1963
YR	76.0	55.6	65.8	107	AUG 1964	8	JAN 1963+	2363	34.55	15.40	APR 1966	T	SEP 1956+	9.18	AUG 1947	2.3	7.4	JAN 1964	7.4	JAN 1964

Month	Rel humidity Mld 6AM	Noon	6PM	Wind mean hourly	Prevailing dir	Fastest mile speed	Direction	Year	Mean sky cover	Pct possible sunshine	Clear	Partly cloudy	Cloudy	Precip .01 in+	Snow 1.0 in+	Thunderstorms	Heavy fog	Max 90°+	Max 32°-	Min 32°-	Min 0°-
J	71/77	60	56	10.5	S	50	SW	1966	6.1	50	10	5	16	7	1	1	*	0	2	15	0
F	67/76	56	56	11.2	S	61	SW	1948	5.9	56	9	6	13	7	1	1	*	0	*	10	0
M	65/75	54	48	12.8	S	59	W	1961	6.0	58	10	7	14	8	*	3	*	*	0	4	0
A	69/83	58	53	12.3	S	58	W	1962	5.9	57	8	9	13	9	0	4	*	1	0	*	0
M	72/85	58	53	11.2	S	51	S	1952	6.0	61	8	10	13	9	0	7	*	6	0	0	0
J	71/82	56	56	10.1	SSE	43	N	1948	4.8	73	12	11	7	6	0	5	*	19	0	0	0
J	55/78	51	44	10.1	S	53	S	1948	4.4	77	14	10	7	5	0	4	*	26	0	0	0
A	54/77	50	44	9.7	SSE	47	SE	1954	4.1	77	15	10	6	5	0	4	*	26	0	0	0
S	72/82	56	52	9.3	SE	67	NW	1954	4.3	67	14	7	9	6	0	3	*	12	0	0	0
O	68/81	56	56	9.3	S	61	NW	1960	4.3	67	13	8	10	5	0	2	*	3	0	1	0
N	72/81	57	56	10.2	S	49	W	1957	4.9	62	13	6	11	6	*	1	1	0	*	6	0
D	72/77	60	58	10.4	SSE	47	N	1957+	5.6	55	11	6	14	6	*	1	2	0	1	9	0
YR	69/79	55	51	10.9	S	65	N	JUN 1954	5.2	65	141	92	132	80	1	41	8	93	3	39	0

ø For period August 1959 through the current year.
Means and extremes in the above table are from the existing or comparable location(s). Annual extremes have been exceeded at other locations as follows:
Highest temperature 111 in July 1954; lowest temperature -3 in January 1930; maximum monthly snowfall 9.0 in January 1918; fastest mile of wind 77 from North in July 1936.

NORMALS, MEANS AND EXTREMES

Station: DALLAS-FORT WORTH, TEXAS REGIONAL AIRPORT
Standard time used: CENTRAL Latitude: 32° N Longitude: 97° 02' W Elevation (ground): 551 feet Year: 1974

(Table Revised 1975. Base Period for Climatological Normals: 1941-1970)

Month	Temp Normal Daily max	Daily min	Monthly	Rec highest	Year	Rec lowest	Year	Deg days Heating	Cooling	Precip Normal	Max monthly	Year	Min monthly	Year	Max 24 hrs	Year	Snow Max monthly	Year	Max 24 hrs	Year
J	55.7	33.9	44.8	88	1969	4	1964	626	0	1.80	3.60	1968	0.19	1971	2.36	1964	12.1	1964	12.1	1964
F	59.8	37.6	48.7	87	1969	12	1971	456	0	2.36	6.20	1965	0.10	1963	4.00	1966	2.5	1966	2.5	1966
M	66.6	43.3	55.0	95	1974	19	1965	335	25	2.54	6.39	1968	0.92	1959	4.55	1962	2.5	1962	2.5	1962
A	76.3	54.1	65.2	95	1972	30	1973	88	94	4.30	12.19	1957	1.06	1964	4.86	1957	0.0		0.0	
M	82.8	62.1	72.5	96	1967	41	1954	0	236	4.47	12.64	1962	0.40	1964	3.11	1965	0.0		0.0	
J	90.8	70.3	80.6	105	1972	51	1964	0	468	3.05	6.94	1966	0.09	1965	3.22	1966	0.0		0.0	
J	95.5	74.0	84.8	106	1964	59	1972	0	614	1.84	11.13	1973	0.01	1973	3.53	1970	0.0		0.0	
A	96.1	74.0	84.9	106	1964	56	1967	0	617	2.26	6.26	1970	0.00	1973	4.48	1965	0.0		0.0	
S	88.5	66.0	77.7	102	1963	46	1971	60	381	3.15	9.52	1964	0.20	1956	4.76	1965	0.0		0.0	
O	79.2	56.0	67.6	96	1963	37	1966	141	141	2.68	9.22	1959	0.20	1970	5.91	1957	0.0		0.0	
N	67.5	44.1	55.8	89	1965	24	1970	287	11	2.03	6.23	1964	0.20	1970	2.83	1974	2.9	1974	2.6	1974
D	58.7	37.0	47.9	84	1966	10	1963	530	0	1.82	6.99	1971	0.21	1955	3.10	1963	2.5	1963	2.5	1963
YR	76.5	54.4	65.5	108	AUG 1964	4	JAN 1964	2382	2587	32.30	12.64	MAY 1957	0.01	AUG 1973	5.91	OCT 1959	12.1	JAN 1964	12.1	JAN 1964

Month	Rel humidity Hour 00/06	12	18	Wind mean speed	Prevailing dir	Fastest mile m.p.h.	Direction	Year	Mean sky cover	Pct sunshine	Clear	Partly cloudy	Cloudy	Precip .01 in+	Snow 1.0 in+	Thunderstorms	Heavy fog ¼ mi	Max 90°+	Max 32°-	Min 32°-	Min 0°-	Avg sta pressure mb
J	76/82	61	59	11.5	S	46	36	1957	6.2	—	10	5	16	7	1	1	2	0	2	14	0	999.2
F	73/81	57	58	11.9	S	51	35	1962	5.8	—	9	5	14	7	*	1	2	0	*	10	0	999.0
M	75/81	51	51	13.4	S	55	29	1954	6.1	—	9	7	14	7	*	2	1	*	0	4	0	991.9
A	85/85	53	52	13.1	S	55	14	1970	6.1	—	8	11	12	8	0	4	1	1	0	*	0	994.1
M	80/88	52	57	11.4	S	52	32	1955	6.0	—	8	12	12	8	0	7	*	3	0	0	0	991.8
J	74/86	57	52	11.1	S	—	—	—	4.8	—	12	10	8	6	0	6	*	19	0	0	0	993.1
J	67/81	50	45	9.8	S	65	36	1961	4.3	—	15	9	6	5	0	5	*	27	0	0	0	994.9
A	70/89	51	60	9.8	S	73	36	1959	4.3	—	15	10	6	5	0	5	1	25	0	0	0	994.8
S	80/89	61	60	9.8	S	53	27	1957	4.5	—	14	8	8	6	0	4	*	11	0	0	0	998.2
O	77/83	56	60	10.9	S	50	34	1957	5.1	—	14	5	12	5	0	3	3	0	0	2	0	997.5
N	77/81	60	61	11.3	S	53	32	1968	5.6	—	13	6	11	7	*	2	3	0	*	6	0	997.2
D	—	—	—	—	S	—	—	—	—	—	—	—	—	—	—	—	—	—	—	—	—	—
YR	75/86	57	55	11.2	S	73	36	AUG 1959	5.2	—	141	90	134	79	1	45	11	87	2	40	0	995.6

Means and extremes above are from existing and comparable exposures. Annual extremes have been exceeded at other sites in the locality as follows:
Highest temperature 112 in August 1936; lowest temperature -8 in February 1899; maximum monthly precipitation 17.64 in April 1922; minimum monthly precipitation 0.00 in November 1903; maximum precipitation in 24 hours 9.57 in September 1932.

REFERENCE NOTES APPLYING TO TABLES APPEAR ON THE PAGE FOLLOWING LAST TABLE.
(Caution: Letters and symbols may have different meanings in 1941-1970 tables than in earlier tables. See notes.)

DEL RIO, TEXAS
INTERNATIONAL AIRPORT

LATITUDE 29° 22' N
LONGITUDE 100° 55' W
ELEVATION (ground) 1026 Feet

NORMALS, MEANS AND EXTREMES
(Table Revised 1969. Base Period for Climatological Normals: 1931-1960)

Temperature

Month	Normal Daily maximum	Normal Daily minimum	Normal Monthly	Extremes Record highest	Year	Extremes Record lowest	Year	Normal degree days (b)
(a)	(b)	(b)	(b)	6	6	6	6	(b)
J	63.7	38.8	51.3	84	1964	19	1964	438
F	69.2	43.1	56.2	82	1965	24	1967	258
M	77.2	48.8	63.0	97	1967	27	1965	147
A	85.4	57.7	71.6	102	1965	43	1968+	18
M	91.0	65.7	78.4	103	1967	51	1967	0
J	96.7	72.1	84.4	102	1967	61	1967	0
J	98.5	73.8	86.2	106	1964	68	1968+	0
A	98.0	73.5	85.8	107	1964	64	1966	0
S	91.5	68.3	79.9	103	1965+	51	1964	0
O	83.3	59.4	71.4	94	1967	38	1967	31
N	71.2	46.3	58.8	91	1968+	23	1966+	218
D	64.7	39.8	52.3	84	1964	19	1964	394
YR	82.5	57.3	70.0	107 AUG. 1964		19 JAN. 1964		1504

Precipitation

Month	Normal total (b)	Maximum monthly	Year	Minimum monthly	Year	Maximum in 24 hrs.	Year
J	0.89	0.94	1966	0.02	1967	0.65	1966
F	0.58	1.72	1965	0.29	1967	0.79	1965
M	0.82	1.15	1968	0.11	1963	0.70	1968
A	1.36	4.36	1966	1.11	1964	3.32	1966
M	2.73	2.85	1963	0.65	1963	1.86	1963
J	2.21	3.37	1966	0.03	1966	2.57	1966
J	1.31	3.07	1968	0.05	1966	2.21	1968
A	1.52	2.10	1966	T	1963	1.61	1966
S	2.61	15.79	1964	1.20	1964	4.50	1964
O	1.98	2.47	1964	0.43	1965	1.45	1964
N	0.62	2.19	1968+	0.19	1964	1.08	1968+
D	0.82	1.09	1965	T	1968	0.95	1965
YR	17.83	15.79 SEP. 1964		T DEC. 1968+		4.50 SEP. 1964	

Snow, Sleet

Month	Mean total	Maximum monthly	Year	Maximum in 24 hrs.	Year
(6)	6	6		6	
J	T	T	1966	T	1966
F	0.6	2.2	1966	2.2	1966
M	0.6	3.0	1965	3.0	1965
A	0.0	0.0		0.0	
M	0.0	0.0		0.0	
J	0.0	0.0		0.0	
J	0.0	0.0		0.0	
A	0.0	0.0		0.0	
S	0.0	0.0		0.0	
O	T	0.0	1967	0.0	1967
N	0.0	T	1967+	T	1967+
D	0.0	0.0	1968+	0.0	1968+
YR	1.2	3.0 MAR. 1965		3.0 MAR. 1965	

Relative humidity / Wind / Sunshine / Mean sky cover / Mean number of days

(Standard time used: CENTRAL)

Relative humidity (6 AM / Noon / 6 PM):
J 62 53 44 · F 62 48 42 · M 60 43 39 · A 64 40 40 · M 70 44 47 · J 73 48 49 · J 78 53 49 · A 77 52 48 · S 53 74 34 · O 58 77 38 · N 71 83 49 · D 69 78 52 · YR 64 77 54 43

Mean hourly wind speed: 8.5, 9.6, 10.9, 10.7, 10.4, 9.4, 9.0, 8.7, 8.7, 8.2, 8.3 — YR 9.6

Fastest mile — Speed / Direction / Year:
28 31 1965 · 31 33 1965 · 52 35 1964 · 26 15 1965 · 38 36 1968 · 35 04 1964 · 35 05 1964 · 29 12 1964 · 46 33 1967 · 31 33 1967 · 35 29 1968 — YR 52 33 MAR. 1964

Mean sky cover (sunrise to sunset, tenths): 5.3, 5.5, 5.2, 6.0, 6.6, 6.8, 5.7, 4.2, 3.5, 3.9, 5.5, 5.7 — YR 5.5

Mean number of days — Clear / Partly cloudy / Cloudy (sunrise to sunset):
12/6/13 · 10/6/12 · 14/7/17 · 8/7/17 · 6/8/18 · 5/16/8 · 14/12/8 · 10/9/11 · 15/7/11 — YR 117/109/139

Mean number of days: Precipitation .01 inch or more — 4,5,5,5,6,3,5,5,8,3,3,3 — 55; Thunderstorms — *,1,2,5,7,4,4,4,5,1,1 — 35; Heavy fog — 3,2,1,1,*,0,*,0,*,*,3,4 — 14

Temperatures Max. 90° and above: 0,0,2,9,26,30,29,16,2,*,0 — 125; Max. 32° and below: 0; Min. 32° and below: 8,5,2,0,0,0,0,0,0,*,6 — 21; Min. 0° and below: 0

Ø Extremes are for the period March 1963 through the current year.
Means and extremes in the above table are from the existing or comparable location(s). Annual extremes have been exceeded at other locations as follows:
Highest temperature 111° in July 1960+; lowest temperature 11° in February 1951; maximum monthly precipitation 0.00 in August 1952; minimum precipitation in 24 hours 8.88 in June 1935; maximum snowfall 4.7 in January 1926; maximum snowfall in 24 hours 4.7 in January 1926; and fastest mile of wind 62 from W in March 1935.

NORMALS, MEANS AND EXTREMES
(Table Revised 1975. Base Period for Climatological Normals: 1941-1970)

Temperatures °F

Month	Normal Daily maximum	Normal Daily minimum	Normal Monthly	Extremes Record highest	Year	Extremes Record lowest	Year
(a)				12	12	12	12
J	63.4	38.1	50.8	89	1972	17	1972
F	68.6	42.8	55.7	91	1971	21	1971
M	76.4	48.8	62.0	102	1971	22	1970
A	85.1	58.9	72.0	102	1963	33	1970
M	90.2	66.1	78.1	104	1974	55	1970
J	96.1	72.4	84.3	108	1974	67	1973
J	99.2	74.2	86.7	106	1964	71	1973
A	98.5	73.6	86.0	106	1969	73	1970
S	91.9	68.5	80.2	103	1968	48	1970
O	83.1	59.6	71.4	98	1964	26	1970
N	72.2	46.9	59.6	91	1963	21	1973
D	65.0	39.5	52.3	85	1973	17	1972
YR	82.5	57.4	70.0	109 AUG. 1969		17 JAN. 1972	

Normal Degree days Base 65° F

Month	Heating	Cooling
J	449	8
F	283	22
M	160	88
A	16	226
M	0	409
J	0	579
J	0	673
A	0	654
S	0	456
O	34	220
N	184	22
D	394	0
YR	1523	3343

Precipitation in inches — Water equivalent

Month	Normal	Maximum monthly	Year	Minimum monthly	Year	Maximum in 24 hrs.	Year
	12	12		12		12	
J	0.60	1.04	1971	0.00	1971	0.87	1969
F	1.02	1.72	1974	T	1974	1.09	1969
M	0.72	1.39	1974	T	1974	0.92	1974
A	1.57	5.46	1970	0.14	1970	2.57	1970
M	2.41	2.85	1970	0.20	1964	1.86	1964
J	2.03	4.87	1974	T	1974	2.57	1971
J	1.02	3.07	1968	0.04	1970	2.21	1968
A	1.22	1.22	1971	0.50	1971	3.81	1969
S	3.07	15.79	1964	T	1964	5.52	1964
O	2.05	3.36	1969	1.06	1969	7.60	1964
N	0.65	1.36	1970	T	1973	1.84	1969
D	0.52	1.47	1974	T	1974	1.37	1974
YR	16.88	15.79 SEP. 1964		0.00 FEB.		7.60 OCT. 1964	

Snow, Ice pellets

Month	Maximum monthly	Year	Maximum in 24 hrs.	Year
	12		12	
J	T	1973	T	1973
F	2.7	1973	2.7	1973
M	3.0	1965	3.0	1965
A	0.00		0.00	
M	0.0		0.0	
J	0.0		0.0	
J	0.0		0.0	
A	0.0		0.0	
S	0.0		0.0	
O	0.0		0.0	
N	T	1967	T	1967
D	T	1969	T	1969
YR	3.0 MAR. 1965		3.0 MAR. 1965	

Relative humidity pct. (Local time) / Wind / Sky cover / Mean number of days

Relative humidity (Hour 00 / 06 / 12 / 18):
J 64 75 55 45 · F 61 73 50 41 · M 57 76 43 36 · A 67 81 52 35 · M 64 85 55 41 · J 56 77 51 36 · J 72 84 54 41 · A 72 84 59 43 · S 70 85 58 52 · O 68 76 52 44 · N (…) · D (…) — YR 65 78 54 44

Mean wind speed (m.p.h.): 8.6, 9.6, 10.9, 10.9, 10.4, 11.5, 11.1, 10.1, 9.3, 9.1, 8.6, 8.5 — YR 9.9

Fastest mile — Speed (m.p.h.) / Direction / Year:
37 30 1972 · 30 31 1973 · 52 35 1964 · 48 36 1968 · 38 32 1969 · 60 14 1970 · 35 13 1970 · 41 33 1967 · 46 33 1972 · 43 29 1969 — YR 60 14 AUG. 1970

Mean sky cover, tenths (sunrise to sunset): 5.6, 5.5, 5.3, 6.4, 6.7, 4.8, 5.2, 5.0, 5.3, 5.5 — YR 5.5

Mean number of days — Clear / Partly cloudy / Cloudy (sunrise to sunset):
11/6/14 · 11/6/12 · 11/6/15 · 7/9/14 · 6/8/15 · 9/13/9 · 12/12/8 · 11/10/10 · 10/12/9 · 11/12/8 · 11/5/14 — YR 120/105/140

Mean number of days: Precipitation .01 inch or more — 5,5,4,5,6,3,6,6,8,4,4,4 — 59; Snow, Ice pellets 1.0 inch or more — *; Thunderstorms — *,*,2,4,8,4,4,4,2,1,* — 35; Heavy fog, visibility ¼ mile or less — 4,4,1,1,*,*,0,0,*,1,3,5 — 15

Temperatures Max. 90° and above: 0,*,2,7,14,24,29,27,16,2,*,0 — 123; Max. 32° and below: *; Min. 32° and below: 8,4,1,0,0,0,0,0,0,*,1,4 — 18; Min. 0° and below: 0

Average station pressure mb. — Elev. 1027 feet m.s.l.:
981.9, 982.6, 975.1, 976.5, 974.6, 974.6, 977.9, 978.0, 977.5, 980.9, 981.0, 981.4 — YR 978.6

Means and extremes above are from existing and comparable exposures. Annual extremes have been exceeded at other sites in the locality as follows:
Highest temperature 111° in July 1960; lowest temperature 11° in February 1951; maximum precipitation in 24 hours 8.88" in June 1935; maximum monthly snowfall 4.7" in January 1926; maximum snowfall in 24 hours 4.7" in January 1926; and fastest mile of wind 62 from W in March 1935.

REFERENCE NOTES APPLYING TO TABLES APPEAR ON THE PAGE FOLLOWING LAST TABLE.
(Caution: Letters and symbols may have different meanings in 1941-1970 tables than in earlier tables. See notes.)

LATITUDE 31° 48' N
LONGITUDE 106° 24' W
ELEVATION (ground) 3918 Feet

NORMALS, MEANS AND EXTREMES
(Table Revised 1969. Base Period for Climatological Normals: 1931-1960)

Temperature

Month	Normal Daily Max (b)	Normal Daily Min (b)	Normal Monthly (b)	Record Highest	Year	Record Lowest	Year
J	56.3	29.5	42.9	75	1963	- 8	1962
F	62.4	35.7	49.1	79	1962	11	1963
M	69.4	40.3	54.9	88	1967	15	1965
A	78.2	48.5	63.4	98	1965	33	1968
M	86.9	56.9	71.9	100	1964	51	1966
J	95.4	66.5	81.0	106	1968	64	1968
J	94.9	68.9	81.9	106	1963	58	1966
A	93.0	67.7	80.4	102	1966+	45	1968
S	87.5	61.4	74.5	97	1964	27	1967
O	78.8	50.0	64.4	92	1968+	23	1968+
N	66.3	36.1	51.2	81	1966	11	1964
D	57.5	30.7	44.1	73	1966	- 8	—
YR	77.2	49.4	63.3	106 JUN 1968+		- 8 JAN 1962	

Normal degree days & Precipitation (inches)

Month	Normal degree days (b)	Precip Normal total (b)	Max monthly	Year	Min monthly	Year	Max in 24 hrs	Year
J	685	0.46	1.84	1949	0.00	1967+	0.61	1960
F	445	0.41	1.42	1944	0.00	1943	0.87	1956
M	319	0.35	2.26	1958	T	1966+	1.72	1961
A	105	0.29	1.23	1942	0.00	1955	1.08	1966
M	0	0.29	1.92	1941	0.00	1962	1.23	1942
J	0	0.40	2.67	1966	0.00	1964+	1.45	1966
J	0	1.29	5.53	1968	0.17	1965	2.63	1968
A	0	1.19	4.11	1957	T	1962	2.00	1957
S	0	1.14	6.29	1958	0.00	1959+	2.89	1941
O	84	0.85	4.31	1945	0.00	1952	1.77	1952
N	414	0.43	1.63	1961	0.00	1964+	1.19	1943
D	648	0.49	1.73	1960	0.00	1955+	1.05	1946
YR	2700	7.89	6.29 SEP 1958		0.00 JAN 1967+		2.89 SEP 1941	

Snow, Sleet

Month	Mean total	Max monthly	Year	Max in 24 hrs	Year
J	1.3	8.3	1949	4.2	1949
F	0.8	8.9	1956	7.2	1956
M	0.4	7.3	1958	7.3	1958
A	T	T	1967+	T	1967+
M	0.0	0.0		0.0	
J	0.0	0.0		0.0	
J	0.0	0.0		0.0	
A	0.0	0.0		0.0	
S	0.0	0.0		0.0	
O	0.0	0.0		0.0	
N	0.8	7.8	1961	7.8	1961
D	1.0	10.1	1960	7.1	1951
YR	4.3	10.1 DEC 1960		7.8 NOV 1961	

Sunshine, sky cover, mean number of days, solar radiation

Month	Pct possible sunshine	Mean sky cover	Clear	Partly cloudy	Cloudy	Precip .01+	Snow 1.0+	Thunderstorms	Heavy fog	Max 90+	Max 32 & below	Min 32 & below	Min 0 & below	Solar radiation (langleys)
J	77	4.6	14	8	7	3	1	*	1	0	1	21	*	337
F	81	4.1	14	7	7	3	*	*	1	0	*	13	0	435
M	84	4.1	14	8	7	2	0	1	*	0	0	3	0	551
A	86	3.7	16	8	6	1	0	1	*	1	0	*	0	661
M	89	3.2	20	8	4	2	0	2	*	15	0	0	0	725
J	88	2.9	20	7	3	4	0	5	1	26	0	0	0	734
J	79	4.6	12	13	6	8	0	10	0	28	0	0	0	676
A	80	4.3	14	12	5	8	0	11	0	23	0	0	0	637
S	85	3.1	19	6	5	4	0	4	*	7	0	0	0	569
O	85	3.0	20	6	5	3	*	2	*	1	0	*	0	470
N	77	3.5	18	5	7	3	*	*	1	0	*	6	0	361
D	77	4.1	16	7	8	4	1	*	1	0	1	19	*	310
YR	83	3.8	196	95	74	44	1	36	2	101	1	64	*	539

Wind — Fastest mile YR: speed 70, direction NW, MAY 1950+; Mean hourly YR 10.0; Prevailing direction YR N.
Relative humidity (5AM/11AM/5PM/11PM, Mountain std time) YR: 52 / 36 / 27 / 41.

ø Beginning Sept. 1, 1960.
Means and extremes are from the existing location. Annual extremes have been exceeded at other locations as follows:
Highest temperature 109 in June and July 1960; maximum precipitation in 24 hours 6.50 in July 1881;
maximum snowfall in 24 hours 8.4 in November 1906.

NORMALS, MEANS AND EXTREMES
(Table Revised 1975. Base Period for Climatological Normals: 1941-1970)

Average station pressure 882.1 mb. Elev. 3916 feet m.s.l.

Temperature °F

Month	Normal Daily Max (a)	Normal Daily Min	Normal Monthly	Record Highest	Year	Record Lowest	Year
J	57.0	30.2	43.6	80	1970	- 8	1962
F	62.5	34.3	48.4	83	1972	14	1971
M	68.5	40.3	54.6	88	1971	17	1973
A	78.5	49.1	63.9	98	1965	24	1973
M	87.2	57.2	72.2	100	1964	34	1967
J	94.9	65.7	80.3	108	1973	51	1968
J	94.6	69.9	82.3	106	1963	59	1973
A	92.8	68.2	80.5	105	1969	56	1969
S	87.5	61.0	74.2	96	1973	32	1970
O	78.5	49.5	64.0	84	1973	25	1970
N	66.1	37.0	51.6	80	1973	17	1973
D	57.8	30.9	44.4	73	1968	10	1973
YR	77.2	49.5	63.4	108 JUN 1973		- 8 JAN 1962	

Normal Degree days (base 65°F)

Month	Heating	Cooling
J	663	0
F	465	0
M	328	6
A	89	56
M	0	223
J	0	459
J	0	536
A	0	481
S	0	276
O	92	60
N	402	0
D	639	0
YR	2678	2098

Precipitation in inches (Water equivalent)

Month	Normal	Max monthly	Year	Min monthly	Year	Max in 24 hrs	Year
J	0.39	1.84	1949	0.00	1960	0.61	1960
F	0.42	1.42	1944	0.00	1943	0.87	1956
M	0.39	2.26	1958	T	1972	1.72	1958
A	0.24	1.23	1942	0.00	1962	1.08	1966
M	0.22	1.92	1941	0.00	1962	1.23	1941
J	0.60	2.67	1966	0.00	1969	1.45	1966
J	1.53	5.53	1968	0.17	1965	2.63	1968
A	1.16	4.11	1957	T	1962	2.00	1957
S	1.16	6.68	1974	0.00	1959	2.52	1958
O	0.78	4.31	1945	0.00	1952	1.77	1945
N	0.32	1.63	1961	0.00	1943	1.19	1943
D	0.50	1.73	1960	0.00	1955	1.05	1946
YR	7.77	6.68 SEP 1974		0.00 APR 1973		2.63 JUL 1968	

Snow, Ice pellets

Month	Max monthly	Year	Max in 24 hrs	Year
J	8.3	1949	4.6	1949
F	8.9	1956	7.3	1956
M	7.3	1958	7.3	1958
A	T	1971	T	1971
M	0.0		0.0	
J	0.0		0.0	
J	0.0		0.0	
A	0.0		0.0	
S	0.0		0.0	
O	0.0		0.0	
N	7.8	1961	7.8	1961
D	10.1	1960	7.1	1951
YR	10.1 DEC 1960		7.8 NOV 1961	

Mean number of days & sunshine

Month	Pct possible sunshine	Mean sky cover (tenths)	Clear	Partly cloudy	Cloudy	Precip .01+	Snow 1.0+	Thunderstorms	Heavy fog ≤¼ mi	Max 90+	Max 32 & below	Min 32 & below	Min 0 & below
J	78	4.6	14	8	9	3	1	*	1	0	1	19	*
F	84	4.3	14	7	7	3	*	*	*	0	*	15	0
M	84	4.3	16	8	8	2	0	1	*	0	0	*	0
A	89	3.8	18	9	5	2	0	1	*	2	0	*	0
M	89	3.2	20	9	4	2	0	2	*	14	0	0	0
J	88	2.8	20	7	3	3	0	5	0	25	0	0	0
J	79	4.6	12	13	6	8	0	10	0	27	0	0	0
A	80	4.2	14	12	5	7	0	11	0	22	0	0	0
S	84	3.2	17	9	5	5	0	4	*	10	0	0	0
O	83	3.1	19	7	5	3	0	2	*	1	0	*	0
N	78	3.5	17	7	6	2	*	*	*	0	*	7	0
D	83	4.2	15	8	8	4	*	*	*	0	*	18	*
YR	83	3.8	193	101	71	45	2	36	2	102	1	63	*

Wind — Fastest mile YR: speed 70, direction NW, MAY 1950; Mean speed YR 9.7 m.p.h.
Relative humidity pct. (local time 05/11/17/23) YR: 53 / 34 / 26 / 42.

Means and extremes above are from existing and comparable exposures. Annual extremes have been exceeded at other sites in the locality as follows:
Highest temperature 109 in June and July 1960; maximum precipitation in 24 hours 6.50 in July 1881;
maximum snowfall in 24 hours 8.4 in November 1906.

REFERENCE NOTES APPLYING TO TABLES APPEAR ON THE PAGE FOLLOWING LAST TABLE.
(Caution: Letters and symbols may have different meanings in 1941-1970 tables than in earlier tables. See notes.)

NORMALS, MEANS AND EXTREMES

FORT WORTH, TEXAS — GREATER SOUTHWEST INTERNATIONAL AP
LATITUDE 32° 50' N LONGITUDE 97° 03' W ELEVATION (ground) 537 Feet

(Table Revised 1969. Base Period for Climatological Normals: 1931-1960)

Month	Normal Daily max	Normal Daily min	Normal Monthly	Record highest	Year	Record lowest	Year	Normal degree days	Precip. Normal total	Max monthly	Year	Min monthly	Year	Max 24 hrs	Year	Snow Mean total	Max monthly	Year	Max 24 hrs	Year	RH Mid.	6 AM	Noon	6 PM	Wind mean hourly	Prev. dir	Fastest speed	Dir	Year	Mean sky cover	Clear	Partly cloudy	Cloudy	Precip .10"+	Snow 1.0"+	Tstorms	Heavy fog	90°+	Max 32°-	Min 32°-	Min 0°-	Solar radiation
J	56.0	34.9	45.5	80	1967	4	1964	614	2.04	3.60	1965	0.28	1967	2.36	1961	1.9	12.1	1964	12.1	1964	74	80	61	59	12.0	S	46	36	1957	6.1	10	5	16	7	1	1	2	0	1	5	0	253
F	59.8	38.6	49.2	78	1967	18	1965	448	2.24	6.20	1965	0.15	1963	4.06	1965	2.4	12.1	1966+	12.1	1966+	74	78	57	51	12.0	S	51	36	1962+	5.6	10	5	13	7	*	2	1	0	1	13	0	324
M	67.3	44.4	55.9	90	1967	19	1965	319	2.51	6.39	1968	0.11	1956	2.99	1958	0.3	2.5	1962	2.5	1962	76	85	51	51	13.8	S	55	29	1954	5.7	10	7	14	7	1	4	1	0	*	0	0	426
A	75.6	53.9	64.8	89	1964	46	1973	99	3.60	12.19	1957	0.92	1957	4.55	1957	0.0	0.0		0.0		76	85	55	53	13.7	S	53	14	1959	6.2	8	9	12	9	*	7	2	2	0	0	0	481
M	82.8	62.6	72.7	96	1967	46	1967	0	4.59	12.64	1961	1.06	1961	4.86	1965	0.0	0.0		0.0		80	87	64	61	12.3	S	55	14	1955	6.0	8	11	12	9	0	7	1	0	0	0	0	552
J	91.6	71.3	81.5	98	1967+	51	1964	0	2.98		1964	0.40	1964	3.11	1964	0.0	0.0		0.0		76	84	56	52	11.5	S	52	32	1955	4.6	12	11	7	6	0	6	*	18	0	0	0	634
J	95.9	74.9	85.4	105	1964+	59	1967	0	1.75	6.36	1962	0.09	1965	3.22	1958	0.0	0.0		0.0		66	79	50	44	10.3	S	65	36	1961	4.3	15	10	6	5	0	6	0	27	0	0	0	623
A	96.2	74.6	85.4	108	1964	56	1967	0	1.68	3.64	1958	0.02	1961	2.42	1968	0.0	0.0		0.0		66	78	49	44	10.0	S	73	36	1959	4.0	15	9	7	5	0	4	0	25	0	0	0	581
S	89.3	67.5	78.4	102	1963	46	1971	0	2.59	9.22	1965	0.23	1956	4.76	1959	0.0	0.0		0.0		74	85	57	53	10.3	S	73	11	1961+	4.3	15	8	7	5	0	3	0	10	0	0	0	476
O	79.4	56.4	67.9	96	1963	37	1966	65	2.59	9.22	1959	0.20	1959	5.91	1959	0.0	0.0		0.0		75	82	56	59	10.4	S	44	27	1957	4.1	16	7	8	5	0	3	1	1	0	3	0	399
N	67.5	43.5	55.8	88	1963	25	1970	324	2.46	6.23	1964	0.49	1960	2.83	1964	T	T	1968+	T	1968+	76	80	56	59	11.2	S	50	34	1957	5.2	12	6	12	6	0	1	1	0	0	1	0	287
D	58.2	37.2	47.7	84	1966	10	1963	536	2.35	4.22	1960	0.21	1955	1.80	1963	0.2	2.6	1963	2.5	1963	76	80	59	60	11.7	S	53	32	1968	5.6	13	6	12	7	*	1	2	0	*	10	0	240
YR	76.5	55.0	65.8	108	AUG. 1964	4	JAN. 1964	2405	31.33	12.64	MAY 1961	0.02	AUG. 1961	5.91	OCT. 1959	3.3	12.1	JAN. 1964	12.1	JAN. 1964	73	82	56	54	11.7	S	73	36	AUG. 1959	5.1	144	94	127	79	1	45	11	85	2	47	0	440

ø For period July 1963 through the current year.
Means and extremes in the above table are from the existing location. Annual extremes have been exceeded at other locations as follows:
Highest temperature 112 in August 1936; lowest temperature −8 in February 1899; maximum monthly precipitation 17.64 in April 1922;
minimum monthly precipitation 0.00 in November 1903; maximum precipitation in 24 hours 9.57 in September 1932.

NORMALS, MEANS AND EXTREMES

Station: DALLAS-FORT WORTH, TEXAS — REGIONAL AIRPORT
Standard time used: CENTRAL Latitude: 32° 54' N Longitude: 97° 02' W Elevation (ground): 551 feet Year: 1974

(Table Revised 1975. Base Period for Climatological Normals: 1941-1970)

Month	Normal Daily max	Normal Daily min	Normal Monthly	Record highest	Year	Record lowest	Year	Heating (Base 65°F)	Cooling	Precip Normal	Max monthly	Year	Min monthly	Year	Max 24 hrs	Year	Snow Max monthly	Year	Max 24 hrs	Year	RH Hr 00	06	12	18	Wind mean m.p.h.	Prev. dir	Fastest m.p.h.	Dir	Year	Mean sky cover	Clear	Partly cloudy	Cloudy	Precip .01"+	Snow 1.0"+	Tstorms	Heavy fog	90°+	Max 32°-	Min 32°-	Min 0°-	Avg station pressure mb	Elev. feet m.s.l.
J	55.7	33.9	44.8	88	1969	4	1964	626	0	1.80	3.60	1968	0.19	1971	2.36	1961	12.1	1964	12.1	1964	76	82	61	59	11.5	S	46	36	1957	6.2	10	5	16	7	1	1	2	0	0	14	0	999.2	
F	59.8	37.6	48.7	87	1969	12	1971	456	0	2.36	6.20	1965	0.15	1963	4.06	1965	2.9	1966	2.9	1966	79	81	57	52	12.3	S	51	36	1962	5.7	10	5	14	7	*	2	1	0	0	10	0	999.0	
M	66.6	43.3	55.0	96	1974	19	1965	335	25	2.54	6.39	1968	0.10	1956	2.99	1958	2.5	1962	2.5	1962	72	81	51	51	13.4	S	55	29	1954	6.1	10	7	14	8	1	4	1	0	*	4	0	991.9	
A	76.3	54.1	65.2	95	1972	30	1973	88	94	4.30	12.19	1957	0.92	1957	4.55	1957	0.0		0.0		80	85	55	53	13.6	S	53	14	1959	6.1	8	9	12	9	*	6	2	2	0	*	0	991.4	
M	82.8	62.1	72.5	96	1967	41	1972	0	236	4.47	12.64	1957	1.06	1961	4.86	1965	0.0		0.0		80	86	59	57	12.0	S	55	14	1955	5.8	8	11	12	10	0	7	1	0	0	0	0	991.1	
J	90.8	70.3	80.6	105	1972	51	1964	0	468	3.05	6.94	1964	0.40	1962	3.11	1964	0.0		0.0		74	84	57	60	11.6	S	52	32	1955	4.8	12	10	8	6	0	6	*	19	0	0	0	993.1	
J	95.5	74.0	84.8	106	1974	59	1972	0	614	1.84	4.22	1973	0.09	1973	3.22	1958	0.0		0.0		67	83	50	45	9.8	S	65	36	1961	4.3	15	9	7	5	0	5	0	27	0	0	0	994.9	
A	96.1	73.7	84.9	108	1964	56	1967	0	617	2.26	11.13	1970	0.01	1970	3.30	1968	0.0		0.0		68	83	48	45	9.8	S	73	36	1959	4.5	15	8	8	5	0	5	0	25	0	0	0	994.8	
S	88.5	66.8	77.7	102	1963	37	1971	60	381	3.15	9.52	1964	0.23	1956	5.91	1959	0.0		0.0		78	86	56	60	9.9	S	73	11	1961	4.5	14	7	9	6	0	4	1	11	0	0	0	994.8	
O	79.2	56.0	67.6	96	1963	37	1966	141	11	2.68	9.23	1959	0.20	1959	5.91	1959	0.0		0.0		78	83	56	60	9.9	S	44	27	1957	4.5	13	8	10	5	0	3	2	0	0	0	0	995.2	
N	67.5	44.1	55.8	88	1966	24	1970	287	0	2.03	6.23	1964	0.21	1970	2.83	1964	T	1974	T	1974	76	83	56	61	10.9	S	50	34	1957	5.2	13	5	12	6	2	2	1	0	0	2	0	995.2	
D	58.7	37.0	47.9	84	1966	4	1963	530	0	1.82	6.99	1971	0.21	1955	3.10	1963	2.6	1963	2.5	1963	74	81	60	61	11.3	S	53	32	1968	5.6	13	6	13	7	*	1	3	0	*	9	0	997.2	
YR	76.5	54.4	65.5	108	AUG 1964	4	JAN 1964	2382	2587	32.30	12.64	MAY 1957	0.01	AUG 1973	5.91	OCT 1959	12.1	JAN 1964	12.1	JAN 1964	75	84	57	55	11.2	S	73	36	AUG 1959	5.2	141	90	134	79	2	45	11	87	2	40	0	995.6	2

Means and extremes above are from existing and comparable exposures. Annual extremes have been exceeded at other sites in the locality as follows:
Highest temperature 112 in August 1936; lowest temperature −8 in February 1899; maximum monthly precipitation 17.64 in April 1922; minimum monthly
precipitation 0.00 in November 1903; maximum precipitation in 24 hours 9.57 in September 1932.

REFERENCE NOTES APPLYING TO TABLES APPEAR ON THE PAGE FOLLOWING LAST TABLE.
(Caution: Letters and symbols may have different meanings in 1941-1970 tables than in earlier tables. See notes.)

NORMALS, MEANS AND EXTREMES
(Table Revised 1969. Base Period for Climatological Normals: 1931-1960)

LATITUDE 29° 18' N
LONGITUDE 94° 48' W
ELEVATION (ground) 7 Feet

Month	Temperature Normal Daily maximum	Normal Daily minimum	Normal Monthly	Extremes Record highest	Year	Record lowest	Year	Normal degree days	Precip. Normal total	Max monthly	Year	Min monthly	Year	Max in 24 hrs	Year	Snow,Sleet Mean total	Max monthly	Year	Max in 24 hrs	Year	RH 6 AM	Noon	6 PM	Wind Mean hourly speed	Prev. dir.	Fastest mile Speed	Dir.	Year	Pct. sun	Precip .01"+	Snow 1.0"+	Max 90+	Max 32-	Min 32-	Min 0-
(a)	(b)	(b)	(b)					(b)	(b)																										
J	60.5	49.3	54.9	76	1953+	11	1886	350	3.46	10.39	1899	0.02	1909	5.38	1923	T	0.9	1940	0.9	1940	84	80	80	11.8	53	S	1915	49	10	0*	0	*	2	0
F	62.4	51.2	56.8	83	1932	—	1899	258	2.88	8.29	1881	0.09	1954	6.55	1952	0.2	15.4	1895	15.4	1895	85	84	77	11.8	60	N	1927	50	9	0	0	*	1	0
M	66.6	56.2	61.4	85	1879	27	1943	189	2.86	9.39	1926	0.06	1953	4.58	1944	T	T	1932	T	1932	84	85	79	12.1	57	SE	1952	56	8	0	*	0	*	0
A	73.0	64.0	68.5	92	1938	38	1938	30	2.59	11.04	1904	0.01	1887	9.23	1904	0.0	0.0		0.0		85	84	77	12.1	66	NW	1961	61	6	0	*	0	0	0
M	80.1	71.5	75.8	93	1911	52	1954+	U	2.79	10.50	1929	T	1889	6.13	1916	0.0	0.0		0.0		86	86	77	11.5	62	SE	1959	68	6	0	*	1	0	0
J	85.6	77.5	81.7	98	1903	66	1961	U	2.65	15.49	1919	T	1907	12.56	1961	0.0	0.0		0.0		83	85	73	10.7	62	SE	1921	76	5	0	1	0	0	0
J	87.3	78.9	83.1	101	1932	66	1910	U	4.79	18.74	1900	0.00	1962+	14.35	1900	0.0	0.0		0.0		80	81	73	9.8	68	NW	1943	72	9	3	0	0	0	0
A	87.5	79.0	83.3	100	1924	67	1966+	U	4.39	19.08	1915	0.04	1902	9.05	1915	0.0	0.0		0.0		79	81	73	9.4	91	E	1915	71	9	5	0	0	0	0
S	84.6	75.5	80.1	96	1927	52	1942	U	5.09	26.01	1885	0.00	1924	11.65	1901	0.0	0.0		0.0		78	80	71	10.1	100	NE	1900	69	9	2	0	0	0	0
O	78.5	68.4	73.5	94	1952+	41	1925+	0	2.86	17.78	1871	0.03	1952+	14.10	1901	0.0	0.0		0.0		76	80	75	10.3	66	SE	1949	73	6	*	0	0	0	0
N	68.6	57.4	63.0	85	1886	26	1940	138	3.56	16.18	1940	0.03	1903	9.01	1940	T	T		T		82	84	77	11.3	54	NW	1950	61	8	0	0	0	*	0
D	62.7	51.6	57.2	80	1918	18	1880	270	3.89	10.28	1887	0.23	1889	5.43	1964	0.0	0.2	1880	0.2	1880	83	86	80		50	NW	1954	49	10	*	*	0	1	0
YR	74.8	65.1	69.9	101	JUL 1932	8	FEB 1899	1235	41.81	26.01	SEP 1885	0.00	AUG 1902	14.35	JUL 1900	0.2	15.4	FEB 1895	15.4	FEB 1895	82	83	76	11.0	100	NE	SEP 1900	64	96	*	11	*	4	0

ø 100 m.p.h. recorded at 6:15 p.m. Sept. 8 just before anemometer blew away. Maximum velocity estimated 120 from NE between 7:30 and 8:30 p.m.

% Through 1967. $ Through 1964.

NORMALS, MEANS AND EXTREMES
(Table Revised 1975. Base Period for Climatological Normals: 1941-1970)

Average station pressure mb.
Elev. 54 feet m.s.l.

Month	Temperature °F Normal Daily max	Normal Daily min	Normal Monthly	Extremes Record highest	Year	Record lowest	Year	Degree days Heating	Cooling	Precip. Water equiv. Normal	Max monthly	Year	Min monthly	Year	Max in 24 hrs	Year	Snow, Ice pellets Max monthly	Year	Max in 24 hrs	Year	RH Hr 00	06	12	18	Wind Mean speed	Prev dir	Fastest mile Speed	Dir	Year	Pct sun	Precip .01"+	Snow 1.0"+	Max 90+	Max 32-	Min 32-	Min 0-
(a)																																				
J	59.4	48.3	53.9	77	1969	11	1886	365	20	3.02	10.39	1909	0.02	1909	5.38	1923	2.5	1973	2.5	1973	84	84	77	80	11.6	53	S	1915	49	10	0*	0	*	2	0
F	61.5	50.9	56.2	83	1932	27	1899	273	27	2.60	8.29	1954	0.06	1887	6.55	1973	15.4	1895	15.4	1895	85	84	77	79	11.8	60	N	1927	50	9	0	0	*	1	0
M	66.0	56.2	61.0	85	1879	38	1943	187	63	2.63	9.49	1953	0.01	1887	8.12	1972	T	1932	T	1932	85	85	79	77	11.9	50	SE	1952	56	8	0	*	0	*	0
A	73.3	65.0	69.2	92	1953	57	1938	20	146	2.63	11.04	1887	T	1889	9.23	1904	0.0		0.0		85	86	77	73	12.1	57	NW	1961	61	6	0	0	*	0	0
M	80.2	71.8	75.9	93	1911	66	1903	0	338	3.16	10.50	1889	T	1907	12.55	1961	0.0		0.0		86	86	73	73	11.5	66	SE	1959	68	6	0	*	1	0	0
J	85.2	77.4	81.3	99	1918	67	1910	0	489	4.05	15.49	1907	0.00	1962+	14.35	1900	0.0		0.0		83	85	73	73	10.7	62	SE	1921	76	5	1	0	0	0	0
J	87.4	79.0	83.2	101	1932	66	1962	0	564	4.41	18.74	1900	0.00	1902	9.05	1915	0.0		0.0		80	81	70	73	9.8	68	NW	1943	72	9	3	0	0	0	0
A	87.6	79.1	83.4	100	1924	67	1915	0	567	4.40	19.08	1915	0.04	1924	11.65	1901	0.0		0.0		78	81	68	75	9.4	91	E	1915	71	9	5	0	0	0	0
S	84.0	75.3	79.7	94	1927	52	1942	0	450	5.60	26.01	1885	T	1952+	14.10	1901	0.0		0.0		76	80	65	72	10.3	66	NE	1949	69	8	1	0	0	0	0
O	78.0	68.1	73.1	94	1952	41	1925	12	263	3.16	16.18	1940	0.03	1903	9.41	1940	0.0		0.0		81	84	72	77	10.1	54	SE	1950	72	6	*	0	0	0	0
N	68.8	58.2	63.5	85	1886	26	1940	105	60	3.67	16.18	1961	0.23	1889	5.43	1964	0.0		0.0		82	86	72	80	11.3	50	NW	1954	61	8	0	*	0	*	0
D	62.7	51.5	57.1	80	1918	18	1880	262	17	3.67	10.28	1887	0.23	1889	5.43	1964	0.2	1880	0.2	1880	82	86	77	80		49	NW	1954	49	10	*	*	0	1	0
YR	74.5	65.0	69.8	101	JUL 1932	8	FEB 1899	1224	3004	42.20	26.01	SEP 1885	0.00	AUG 1902	14.35	JUL 1900	15.4	FEB 1895	15.4	FEB 1895	81	83	72	76	11.0	100	NE	SEP 1900	64	96	*	11	*	4	0

ø 100 m.p.h. recorded at 6:15 p.m. Sept. 8 just before anemometer blew away. Maximum velocity estimated 120 from NE between 7:30 and 8:30 p.m.

REFERENCE NOTES APPLYING TO TABLES APPEAR ON THE PAGE FOLLOWING LAST TABLE.
(Caution: Letters and symbols may have different meanings in 1941-1970 tables than in earlier tables. See notes.)

NORMALS, MEANS AND EXTREMES

(Table Revised 1969. Base Period for Climatological Normals: 1931-1960)

HOUSTON, TEXAS
WILLIAM P. HOBBY AIRPORT

LATITUDE 29° 39' N
LONGITUDE 95° 17' W
ELEVATION (ground) 50 Feet

Month	Temperature Normal Daily maximum	Daily minimum	Monthly	Extremes Record highest	Year	Record lowest	Year	Normal degree days	Precipitation Normal total	Maximum monthly	Year	Minimum monthly	Year	Maximum in 24 hrs.	Year	Snow, Sleet Mean total	Maximum monthly	Year	Maximum in 24 hrs.	Year	Relative humidity 6 AM	Noon	6 PM	Mid.	Wind Mean hourly speed	Prevailing direction	Fastest mile Speed	Direction	Year	Pct. of possible sunshine	Mean sky cover sunrise to sunset	Clear	Partly cloudy	Cloudy	.01 inch or more	1.0 inch or more	Thunderstorms	Heavy fog	Max. 90° and above	32° and below	Min. 32° and below	0° and below
(a)	(b)	(b)	(b)	9		9		(b)	(b)	36		36		36		36		36		36		8	8	8	8	20	15	6	6		7	20	20	20	20	36	36	30	23	8	8	8	8
J	63.6	43.6	53.6	83	1963+	17	1963+	384	3.78	10.51	1944	0.86	1952	3.66	1944	0.2	3.1	1949	3.1	1949	81	64	67	7.0	11.9	NNW	32	33	1964	47	7.0	7	4	20	11	*	2	7	8	*	6	8	0
F	65.5	46.0	55.8	87	1962	25	1962	288	3.44	11.33	1959	0.09	1954	3.44	1966	0.2	4.4	1960	4.4	1960	84	64	60	6.8	12.1	SSE	40	36	1963	56	6.7	7	5	16	10	*	3	6	3	1	0	8	0
M	71.8	50.8	61.3	88	1965	28	1965	192	2.67	11.42	1953	0.07	1953	5.00	1957	T	1.0	1968	1.0	1968	82	59	62	6.6	12.7	SSE	35	30	1968+	57	6.8	13	13	8	0	4	1	0	0	0	0	0
A	78.0	59.0	68.5	93	1962	38	1962	36	3.24	8.07	1966	0.36	1937	5.18	1966	0.0	0.0		0.0		86	63	68	7.0	13.0	SSE	38	16	1966+	54	6.7	13	12	5	8	0	5	4	1	1	0	0	0
M	85.7	66.2	76.0	93	1967	52	1967	0	4.32	13.24	1968	T	1937	6.35	1968	0.0	0.0		0.0		89	61	61	6.1	11.1	SSE	41	17	1966+	62	5.9	17	13	1	9	0	4	2	6	0	0	0
J	91.1	72.0	81.6	100	1963	59	1963	0	3.69	14.66	1960	0.08	1934	8.51	1934	0.0	0.0		0.0		88	61	65	5.6	10.2	SSE	44	17	1965	72	5.6	17	13	0	8	0	7	1	19	0	0	0
J	92.1	73.8	83.0	101	1964	64	1967	0	4.29	12.38	1942	0.07	1962	6.34	1943	0.0	0.0		0.0		88	59	64	5.9	8.8	S	40	36	1964	74	5.9	17	9	5	9	0	10	*	26	0	0	0
A	92.8	73.6	83.2	106	1962	64	1967	0	4.27	18.51	1945	0.38	1956+	15.65	1945	0.0	0.0		0.0		87	59	63	5.9	8.4	S	54	36	1963	76	5.9	15	15	9	7	0	9	1	26	0	0	0
S	89.1	69.3	79.2	98	1963	50	1967	0	4.26	15.40	1953	0.14	1953	5.62	1944	0.0	0.0		0.0		85	60	65	6.3	9.1	ESE	54	36	1963	66	5.7	15	12	9	9	0	7	3	15	0	0	0
O	82.3	60.5	71.4	96	1964	32	1964	6	3.77	22.31	1949	T	1952	10.25	1949	0.0	0.0		0.0		83	56	59	5.1	9.8	ESE	35	31	1952	73	4.7	12	10	9	6	0	4	2	3	0	0	0
N	71.1	50.5	60.8	88	1963	32	1963	183	3.86	14.36	1946	0.18	1949	7.89	1943	0.0	0.0		0.0		82	60	68	6.1	11.2	SSE	35	14	1963	60	5.2	12	10	8	8	0	3	5	*	1	0	0	0
D	65.4	45.9	55.7	82	1964	19	1963	307	4.36	9.80	1949	0.78	1958	4.39	1964	T	T	1961+	T	1961+	84	64	69	6.7	11.3	SSE	35	31	1968+	50	5.8	18	8	5	10	*	2	6		4	13	0
YR	79.0	59.3	69.2	106	AUG. 1962	17	JAN. 1963+	1396	45.95	22.31	OCT. 1949	T	OCT. 1952+	15.65	AUG. 1945	0.4	4.4	FEB. 1960	4.4	FEB. 1960	84	87	60	66	10.8	SSE	54	34	AUG. 1963	62	6.2	87	126	152	103	*	59	42	95	13	0

Means and extremes in the above table are from comparable exposures for periods through the current year except as noted hereafter. Annual extremes have been exceeded at other locations as follows: Highest temperature 108 in August 1909; lowest temperature 5 in January 1940 and earlier; fastest mile of wind 84 from Northwest in March 1926.

Ø For record August 1960 through the current year.

NORMALS, MEANS AND EXTREMES

(Table Revised 1975. Base Period for Climatological Normals: 1941-1970)

Station: HOUSTON, TEXAS   INTERCONTINENTAL AIRPORT   Standard time used: CENTRAL   Latitude: 29° 58' N   Longitude: 95° 21' W   Elevation (ground): 96 feet

Month	Temperatures °F Normal Daily maximum	Daily minimum	Monthly	Extremes Record highest	Year	Record lowest	Year	Normal Degree days Base 65°F Heating	Cooling	Precipitation in inches Water equivalent Normal	Maximum monthly	Year	Minimum monthly	Year	Maximum in 24 hrs.	Year	Snow, Ice pellets Maximum monthly	Year	Maximum in 24 hrs.	Year	Relative humidity pct. Hour 00	Hour 06	Hour 12	Hour 18	Wind Mean speed m.p.h.	Prevailing direction	Fastest mile Speed m.p.h.	Direction	Year	Pct. of possible sunshine	Mean sky cover, tenths, sunrise to sunset	Sunrise to sunset Clear	Partly cloudy	Cloudy	Precipitation .01 inch or more	Snow, ice pellets 1.0 inch or more	Thunderstorms	Heavy fog, visibility ¼ mile or less	Max. 90° and above	32° and below	Min. 32° and below	0° and below	Average station pressure mb.
(a)				5		5				5	5		5		5		5		5		5	5	5	5	5	15	5	5		5	5	5	5	5	5	5	5	5	5	5	5	5	2
J	62.6	41.5	52.1	83	1972	19	1973	416	16	3.57	7.68	1974	0.36	1971	2.00	1973	2.0	1973	2.0	1973	87	89	68	73	7.8	NNW	29	33	1973	40	7.2	8	8	15	11	*	2	8	0	6	9	2	1015.9
F	66.0	44.6	55.3	83	1974	25	1971	294	22	3.54	3.40	1973	0.55	1974	1.55	1973	2.8	1973	1.4	1973	83	86	56	58	8.0	SSE	35	29	1974	55	6.5	8	7	13	10	*	4	4	0	1	5	1009.9
M	71.8	49.8	60.8	90	1971	25	1971	189	59	2.68	8.52	1972	1.21	1971	7.47	1972	0.0		0.0		88	90	59	62	9.0	SSE	32	24	1972	50	6.4	8	8	15	9	0	4	3	0	0	1009.7
A	79.4	59.3	69.4	89	1973	30	1973	23	155	3.54	7.15	1973	1.68	1970	2.54	1973	0.0		0.0		89	90	59	61	9.5	SSE	30	30	1973	55	6.1	8	11	11	8	0	7	2	0	1011.7
M	85.9	65.6	75.8	93	1969	46	1970	0	335	5.10	14.39	1970	3.41	1973	4.69	1970	0.0		0.0		90	92	58	61	7.0	SSE	45	30	1973	65	5.9	7	13	11	9	0	8	2	1	1010.7
J	91.3	70.9	81.1	99	1969	52	1970	0	483	4.52	13.46	1973	0.26	1970	6.61	1973	0.0		0.0		89	92	58	60	7.0	SSE	52	30	1973	70	5.5	6	14	10	11	0	11	0	18	1010.7
J	93.8	72.8	83.3	101	1969	62	1972	0	567	4.12	6.77	1971	1.42	1971	3.99	1971	0.0		0.0		89	93	58	62	6.3	S	46	10	1969	70	6.0	6	15	10	9	0	12	0	26	1012.6
A	94.3	72.4	83.4	101	1969	61	1970	0	570	4.35	6.95	1973	2.03	1972	2.87	1972	0.0		0.0		93	95	62	69	5.1	SSE	32	18	1973	65	6.6	6	14	11	8	0	13	0	22	1012.4
S	90.1	68.2	79.2	92	1971	51	1972	0	426	4.05	9.38	1973	4.51	1972	4.06	1970	0.0		0.0		89	94	58	66	6.8	SSE	37	33	1973	55	6.0	8	11	11	8	0	9	1	11	1012.5
O	83.5	58.3	70.9	92	1971	39	1970	24	207	4.04	7.90	1970	1.30	1969	3.55	1970	0.0		0.0		87	89	57	64	7.3	SSE	35	33	1972	65	5.5	11	8	12	6	0	4	4	0	1015.1
N	73.0	47.1	60.1	90	1974	21	1973	155	38	4.04	7.90	1974	1.54	1972	3.55	1974	0.0		0.0		89	90	59	63	7.8	SSE	35	31	1972	65	5.6	10	8	12	7	0	4	4	0	5	1015.1
D	65.8	43.4	54.6	83	1974	21	1973	333	11	4.04	7.33	1973	0.64	1973	3.43	1971	0.0		0.0		86	89	63	75	7.8	SSE	35	31	1973	73	5.8	8	12	11	9	0	5	7	0	5	1015.4
YR	79.8	58.0	68.9	101	AUG. 1969	19	JAN. 1973	1434	2889	48.19	14.39	MAY 1970	0.26	JUN 1970	7.47	MAR 1972	2.8	FEB. 1973	2.0	JAN 1973	89	91	61	68	7.4	SSE	52	10	JUL 1973	57	6.2	91	112	162	107	1	72	39	82	1	22	0	1013.0

Ø Extremes for period June 1969 to date.
Means and extremes above are from existing and comparable exposures. Annual extremes have been exceeded at other sites in the locality as follows:
Highest temperature 108 in August 1909; lowest temperature 5 in January 1940 and earlier; maximum monthly precipitation 22.31 in October 1949;
minimum monthly precipitation Trace in October 1952 and May 1937; maximum precipitation in 24 hours 15.65 in August 1945; maximum monthly snowfall
4.4 in February 1960; maximum snowfall in 24 hours 4.4 in February 1960; fastest mile of wind 84 from NW in March 1926.

REFERENCE NOTES APPLYING TO TABLES APPEAR ON THE PAGE FOLLOWING LAST TABLE.

(Caption Letters and symbols may have different meanings in 1941-1970 tables than in earlier tables. See note.)

NORMALS, MEANS AND EXTREMES

(Table Revised 1969. Base Period for Climatological Normals: 1931-1960)

LATITUDE 33° 39' N
LONGITUDE 101° 49' W
ELEVATION (ground) 3254 Feet

Month	Normal Daily maximum	Normal Daily minimum	Normal Monthly	Extremes Record highest	Year	Extremes Record lowest	Year	Normal degree days	Precipitation Normal total	Precipitation Max monthly	Year	Precipitation Min monthly	Year	Precipitation Max in 24 hrs.	Year	Snow,Sleet Mean total	Snow,Sleet Max monthly	Year	Snow,Sleet Max in 24 hrs.	Year	Rel.hum. 6AM	Rel.hum. Noon	Rel.hum. 6PM	Wind Mean hourly speed	Wind Prevailing direction	Wind Fastest mile Speed	Wind Fastest mile Direction	Wind Fastest mile Year	Mean sky cover	Pct. sunshine	Clear	Partly cloudy	Cloudy	Precip .01"+	Snow 1.0"+	Thunderstorms	Heavy fog	Max 90°+	Max 32°-	Min 32°-	Min 0°-
(a)	(b)	(b)	(b)	22	22	22	22	(b)	(b)	22	22	22	22	22	22	20	20	20	20	21	21	21	19	14	20	20	20		22		22	22	22	22	22	22	22	21	21	21	21
J	53.0	25.4	39.2	80	1967+	-16	1963	800	0.68	4.05	1949	0.00	1967	0.93	1949	1.8	8.1	1958	7.8	1958	66	48	50	13.1	SW	59	28	1965	5.0		13	7	11	4	1	*	2	0	3	26	**
F	57.3	28.8	43.1	86	1962	-8	1960	613	0.57	2.51	1961	T	1955+	2.15	1961	3.2	16.8	1956	12.1	1961	64	43	43	14.7	SW	58	25	1960	4.9		11	7	10	4	1	*	2	0	3	20	**
M	64.9	34.3	49.6	90	1963	2	1948	484	0.73	3.23	1958	0.10	1959+	1.70	1968	1.9	14.3	1958	7.7	1968	55	34	34	16.1	SW	69	34	1957	5.1		12	7	10	4	1	*	1	1	6	13	0
A	74.5	44.4	59.5	96	1948	22	1968	201	1.15	3.48	1957	0.10	1962	1.92	1966	T	0.0	1957	0.1	1957	56	33	33	16.1	SW	58	25	1957	5.0		11	9	10	4	0	1	1	*	2	2	0
M	82.1	54.3	68.2	104	1967	30	1967	31	3.18	7.80	1949	0.45	1962	5.14	1963	0.0	0.0		0.0		62	44	38	15.3	S	70	36	1952	5.0		11	12	8	7	0	3	1	9	0	0	0
J	91.4	63.6	77.5	107	1957	44	1947	0	2.53	7.95	1967	T	1953	5.70	1967	0.0	0.0		0.0		64	41	38	15.1	S	63	05	1952	4.1		11	12	7	8	0	9	1	22	0	1	0
J	92.4	66.5	79.5	107	1952	51	1952	0	2.01	5.37	1954	0.19	1954	2.40	1960	0.0	0.0		0.0		63	40	40	12.1	S	64	25	1950	4.4		14	12	5	8	0	8	*	22	0	0	0
A	91.7	65.8	78.8	106	1966	52	1956	0	1.68	8.85	1966	0.05	1960	3.78	1966	0.0	0.0		0.0		63	40	40	10.7	S	46	16	1956	4.0		16	10	5	6	0	6	1	21	0	0	0
S	84.4	58.4	71.4	103	1948	38	1965	18	2.36	5.68	1965	0.00	1952	2.80	1965	0.0	0.0		0.0		67	44	44	11.4	S	45	36	1953	3.5		18	6	6	5	0	4	1	9	0	0	0
O	74.8	47.5	61.2	93	1965	25	1949	174	2.00	5.83	1960	0.00	1952	3.90	1960	0.2	0.2	1967	0.2	1967	65	44	44	12.0	S	65	25	1957	4.2		18	6	7	3	0	3	1	1	0	0	1
N	62.4	33.3	47.9	86	1965	-1	1957	513	0.54	2.67	1968	T	1960	1.57	1957	0.3	3.5	1957	3.5	1957	65	47	47	12.6	WSW	59	25	1955	4.3		16	8	6	3	*	3	2	0	3	13	*
D	54.7	27.3	41.0	81	1955	1	1950	744	0.65	1.47	1959	T	1955	1.12	1960	1.3	6.3	1960	6.3	1960	65	47	48	13.0	WSW	58	25	1957	4.6		14	7	10	3	*	2	2	0	1	25	0
YR	73.6	45.8	59.7	107	JUL 1958+	-16	JAN 1963	3578	18.08	8.85	AUG 1966	0.00	JAN 1967+	5.70	JUN 1967	8.5	16.8	FEB 1956	12.1	FEB 1961	63	74	41	13.5	S	70	36	MAY 1952	4.5		165	106	94	60	3	44	15	83	6	101	1

NORMALS, MEANS AND EXTREMES

(Table Revised 1975. Base Period for Climatological Normals: 1941-1970)

Month	Normal Daily maximum	Normal Daily minimum	Normal Monthly	Extremes Record highest	Year	Extremes Record lowest	Year	Normal degree days Heating	Normal degree days Cooling	Precipitation Normal	Precip Max monthly	Year	Precip Min monthly	Year	Precip Max in 24 hrs.	Year	Snow Max monthly	Year	Snow Max in 24 hrs.	Year	Rel.hum. Hour 00	Hour 06	Hour 12	Hour 18	Wind Mean speed	Prevailing direction	Fastest mile Speed	Direction	Year	Mean sky cover	Pct. sunshine	Clear	Partly cloudy	Cloudy	Precip .01"+	Snow 1.0"+	Thunderstorms	Heavy fog	Max 90°+	Max 32°-	Min 32°-	Min 0°-
(a)				28		28				28	28		28		28		26		26		27	27	27	18	25	14	26	26	26	26		28	28	28	28	28	28	28	28	27	27	27	27
J	53.4	24.8	39.1	83	1972	-16	1963	803	0	0.55	4.05	1949	0.00	1967	0.93	1949	9.4	1973	7.8	1958	64	72	49	46	12.6	SW	59	28	1965	5.1	64	13	7	11	3	1	*	2	0	3	26	**
F	57.0	28.3	42.7	86	1962	-8	1960	624	0	0.50	2.51	1955	T	1955+	2.15	1955	16.8	1956	12.0	1956	63	71	48	41	14.1	SW	58	25	1960	4.9	71	11	7	10	3	1	*	2	0	3	20	**
M	64.8	34.8	49.8	94	1971	2	1948	508	0	0.89	3.23	1958	0.10	1959	1.70	1968	14.3	1958	10.0	1958	56	68	34	34	15.5	SW	69	34	1957	5.0	76	11	9	11	4	1	*	1	1	2	12	0
A	74.8	45.1	60.0	96	1972	22	1968	190	40	1.08	3.48	1957	0.10	1961	1.92	1966	0.3	1973	0.3	1973	55	66	37	30	15.6	SW	58	25	1957	4.8	76	11	8	11	4	0	1	1	*	0	3	0
M	82.5	54.6	68.5	104	1947	30	1967	29	138	3.17	7.80	1949	0.32	1962	5.14	1963	0.0		0.0		62	62	44	36	14.7	S	70	36	1952	5.0	73	11	12	8	7	0	3	1	9	0	0	0
J	90.6	63.6	77.1	107	1957	44	1947	0	363	2.78	7.95	1967	T	1973	5.70	1967	0.0		0.0		62	62	41	36	14.3	S	63	05	1952	4.1	86	13	12	5	7	0	9	1	19	0	1	0
J	92.4	66.9	79.7	107	1958	51	1952	0	456	2.23	5.37	1960	0.05	1970	2.75	1960	0.0		0.0		64	64	39	34	11.6	S	64	25	1950	4.4	80	14	12	5	6	0	7	*	22	0	0	0
A	92.0	65.5	78.8	106	1966	52	1956	0	415	1.87	8.85	1966	T	1960	3.78	1966	0.0		0.0		67	75	45	39	10.2	S	46	16	1956	4.0	78	16	10	5	6	0	6	*	18	0	0	0
S	83.8	58.2	71.0	103	1948	38	1965	162	188	2.19	5.68	1974	0.00	1954	2.80	1965	0.0		0.0		67	78	46	41	11.5	S	45	36	1953	3.8	74	18	6	6	5	0	4	1	10	0	0	0
O	74.8	48.4	61.6	93	1965	25	1970	486	0	2.67	5.83	1960	0.00	1952	3.90	1960	0.2	1972	0.2	1972	66	74	47	47	12.1	S	65	25	1957	4.3	73	17	7	7	5	0	3	2	2	0	2	0
N	63.1	34.4	48.8	86	1965	-1	1957	486	0	0.49	2.67	1968	T	1973	1.57	1972	5.9	1972	5.9	1972	65	66	44	44	12.5	WSW	59	25	1955	4.1	73	15	8	7	3	*	3	2	0	2	13	*
D	55.2	27.4	41.3	81	1955	1	1950	735	0	0.61	1.47	1959	T	1960	1.12	1960	9.9	1960	6.3	1960	65	71	47	47	12.5	WSW	58	25	1957	4.7	71	14	7	10	3	*	1	2	0	1	24	1
YR	73.6	45.8	59.7	107	JUL 1958	-16	JAN 1963	3545	1647	18.41	8.85	AUG 1966	0.00	JAN 1967	5.70	JUN 1967	16.8	FEB 1956	12.1	FEB 1961	63	74	46	41	13.0	S	70	36	MAY 1952	4.5	76	163	103	99	60	3	45	16	79	6	98	1

Average station pressure
Elev. 3241 feet m.s.l. — 2
903.9, 904.7, 899.1, 901.2, 900.6, 902.2, 904.9, 905.5, 904.9, 906.2, 906.4, 903.3, YR 903.4

REFERENCE NOTES APPLYING TO TABLES APPEAR ON THE PAGE FOLLOWING LAST TABLE.

(Caution: Letters and symbols may have different meanings in 1941-1970 tables than in earlier tables. See notes.)

NORMALS, MEANS AND EXTREMES

MIDLAND, TEXAS
MIDLAND-ODESSA REGIONAL AP
LATITUDE 31° 56' N
LONGITUDE 102° 12' W
ELEVATION (ground) 2851 Feet

(Table Revised 1969. Base Period for Climatological Normals: 1931-1960)

Month	Temperature Normal Daily max	Daily min	Monthly	Extremes Record highest	Year	Record lowest	Year	Normal degree days	Precip Normal total	Max monthly	Year	Min monthly	Year	Max 24 hrs	Year
J	57.1	30.9	44.0	80	1966	5	1966	651	0.80	3.66	1949	0.00	1967	1.15	1958
F	62.9	35.4	48.3	81	1964+	14	1965	468	0.60	1.71	1968	0.03	1965	1.22	1965
M	68.5	41.5	55.2	92	1965	14	1965	322	0.36	1.62	1968	T	1963+	1.37	1968
A	78.5	52.2	64.9	98	1968	36	1967	90	0.83	2.85	1949	0.00	1964	1.25	1966
M	86.2	60.6	73.4	102	1967	51	1964	0	2.08	4.99	1959	0.08	1953	4.75	1968
J	93.1	69.3	81.2	106	1964	60	1967	0	1.63	3.93	1949	0.05	1951	2.12	1961
J	94.5	71.2	82.9	106	1967	56	1967+	0	1.88	6.73	1961	T	1954	5.99	1961
A	94.5	70.2	82.1	107	1964	58	1967+	0	1.48	3.84	1961	0.17	1967+	2.41	1965
S	87.7	63.1	75.4	100	1965+	45	1967+	0	1.79	4.06	1962	0.14	1960	2.16	1962
O	78.6	52.5	65.5	95	1966	24	1966	87	1.66	4.09	1953	0.00	1952	1.79	1959
N	66.3	38.7	52.5	88	1963	12	1958	381	0.47	2.32	1968	0.00	1950+	1.21	1968
D	59.6	32.2	45.9	81	1964	2	1966	592	0.66	1.88	1960	T	1958+	0.93	1960
YR	77.2	51.5	64.3	107 AUG 1964		2 JAN 1966		2591	14.24	6.73 JUL 1961		0.00 JAN 1967+		5.99 JUL 1961	

Month	Snow Sleet Mean total	Max monthly	Year	Max 24 hrs	Year	RH 6AM	Noon	6PM	Wind Mean hrly	Prevailing dir	Fastest mile Speed	Dir	Year	Mean sky cover	Avg daily solar rad
J	1.3	7.7	1949	5.9	1955	62	47	40	9.5	S	41	27	1965+	5.1	286
F	0.9	3.5	1965	2.5	1965+	66	46	38	11.0	SW	67	25	1960	5.1	361
M	0.3	3.5	1962	3.5	1962	56	37	27	11.8	SSE	48	28	1963+	5.1	470
A	T	T	1949	T	1949	52	28	21	12.1	SSE	44	29	1961	4.9	549
M	0.0	0.0		0.0		61	40	32	11.9	SSE	52	31	1959	4.7	601
J	0.0	0.0		0.0		60	43	32	11.6	SSE	58	24	1966+	3.9	618
J	0.0	0.0		0.0		51	39	34	10.0	SSE	35	02	1967+	4.6	654
A	0.0	0.0		0.0		58	42	34	9.4	SSE	37	31	1966	4.0	564
S	0.0	0.0		0.0		71	52	44	9.4	SSE	40	36	1959	4.0	495
O	T	T	1958	T	1958	69	46	32	9.5	SSE	35	34	1967	3.5	405
N	0.3	4.5	1968	2.0	1968	69	36	43	9.5	SSE	37	01	1967	4.3	312
D	0.4	4.8	1960	2.7	1960	65	44	43	9.6	SW	39	29	1968	4.6	289
YR	3.2	7.7 JAN 1949		5.9 JAN 1955		61	42	35	10.4	SSE	67	25 FEB 1960		4.5	467

Mean number of days — Sunrise to sunset: Clear / Partly cloudy / Cloudy; Precipitation .01 inch or more; Snow Sleet 1.0 inch or more; Thunderstorms; Heavy fog; Temperatures Max 90° and above, Max 32° and below, Min 32° and below, Min 0° and below.

Month	Clear	Partly cloudy	Cloudy	Precip .01+	Snow 1.0+	Thunder	Heavy fog	Max 90+	Max 32-	Min 32-	Min 0-
J	13	6	12	4	1	*	3	0	1	21	0
F	11	7	10	4	1	1	3	0	*	19	0
M	11	8	10	3	1	1	1	*	*	8	0
A	12	9	9	4	*	3	*	1	0	*	0
M	13	9	9	6	0	6	1	11	0	0	0
J	15	10	5	5	0	6	*	22	0	0	0
J	12	11	8	5	0	7	*	27	0	0	0
A	16	11	6	5	0	6	0	23	0	0	0
S	16	7	7	4	0	3	1	8	0	0	0
O	15	6	9	3	*	1	1	3	0	5	0
N	15	6	10	3	*	*	2	0	1	19	0
D	15	6	10			*	3		1		0
YR	166	96	103	50	2	35	15	99	2	72	0

ø For period July 1963 through the current year.
Means and extremes in the above table are from the present or last comparable instrumental locations. Annual extremes have been exceeded at other locations as follows:
Lowest temperature -11 in February 1933; maximum monthly precipitation 8.18 in September 1932.

NORMALS, MEANS AND EXTREMES

(Table Revised 1975. Base Period for Climatological Normals: 1941-1970)

Month	Temp Normal Daily max	Daily min	Monthly	Extremes Record highest	Year	Record lowest	Year	Degree days Heating	Cooling	Precip Normal	Max monthly	Year	Min monthly	Year	Max 24 hrs	Year
J	57.8	29.4	43.6	84	1974	10	1971	663	0	0.59	3.66	1958	0.00	1967	1.15	1958
F	62.1	33.5	47.8	87	1972	10	1972	482	0	0.56	1.79	1969	0.01	1971	1.22	1965
M	69.1	39.2	54.3	95	1971	20	1971	349	17	0.59	2.86	1970	T	1972	2.20	1970
A	79.4	49.4	64.4	98	1972	34	1970	98	77	0.85	2.85	1966	0.00	1964	2.85	1966
M	86.5	58.1	72.3	102	1974	48	1970	0	230	2.16	4.99	1968	0.08	1953	4.75	1968
J	92.8	66.7		106	1970	59	1971	0	447	1.49	3.93	1969	0.05	1951	2.54	1969
J	95.0	69.5	82.3	106	1964	56	1967	0	536	1.82	6.73	1961	T	1954	5.99	1961
A	94.4	68.2	81.8	107	1964	56	1967+	0	521	1.54	4.43	1974	0.17	1967	2.41	1965
S	88.0	62.8	75.4	100	1965+	42	1971	0	312	1.38	6.14	1974	0.00	1952	2.92	1974
O	79.2	52.4	65.8	95	1966	20	1973	81	105	1.49	4.09	1953	0.00	1950	1.79	1959
N	67.5	39.1	53.3	88	1963	11	1973	356	0	0.49	2.32	1968	T	1968	1.21	1968
D	60.1	31.6	45.9	81	1970	11	1973	592	0	0.52	1.88	1960	0.00	1958	0.93	1960
YR	77.7	50.1	63.9	107 AUG 1964		10 JAN 1971		2621	2250	13.51	6.73 JUL 1961		0.00 JAN 1967		5.99 JUL 1961	

Month	Snow ice pellets Max monthly	Year	Max 24 hrs	Year	RH Hr 00	06	12	18	Wind Mean mph	Prev dir	Fastest Speed	Dir	Year	Mean sky cover	Avg sta pressure mb
J	7.7	1949	6.8	1974	61	69	45	39	9.9	S	41	27	1965	5.1	917.2
F	3.5	1965	2.5	1965	61	69	43	36	9.9	SW	67	25	1960	5.1	918.2
M	5.9	1970	5.0	1970	55	63	36	28	11.1	SSE	48	28	1961	4.8	912.7
A	T	1973	T	1973	52	65	30	24	12.5	SSE	44	29	1961	4.7	913.0
M	0.0		0.0		57	65	38	30	11.9	SSE	52	31	1959	4.7	913.2
J	0.0		0.0		59	74	41	37	11.9	SSE	58	24	1966	3.9	913.0
J	0.0		0.0		53	69	40	31	10.4	SSE	40	33	1972	4.6	916.7
A	0.0		0.0		60	74	44	37	9.9	SSE	37	31	1966	4.4	916.5
S	0.0		0.0		71	83	46	42	9.6	SSE	44	36	1959	4.1	918.7
O	T	1958	T	1958	69	84	46	42	9.7	SSE	45	30	1974	3.8	917.4
N	4.5	1968	2.0	1968	68	81	44	43	9.5	S	40	35	1973	4.3	916.9
D	4.8	1960	2.7	1960	64	71	43	42	9.8	SW				4.6	
YR	7.7 JAN 1949		6.8 JAN 1974		61	73	42	36	10.8	SSE	67	25 FEB 1960		4.5	915.9

Mean number of days:

Month	Clear	Partly cloudy	Cloudy	Precip .01+	Snow 1.0+	Thunder	Heavy fog vis ¼	Max 90+	Max 32-	Min 32-	Min 0-
J	13	6	12	4	1	*	3	0	1	19	0
F	12	6	10	4	1	1	2	0	*	16	0
M	13	8	10	3	1	1	1	*	*	7	0
A	13	9	9	5	*	3	1	1	0	1	0
M	15	9	7	5	0	6	*	12	0	0	0
J	12	10	6	5	0	6	*	20	0	0	0
J	14	12	7	5	0	6	1	27	0	0	0
A	15	11	8	5	0	6	1	19	0	0	0
S	17	6	9	3	0	3	2	2	0	0	0
O	15	6	10	3	*	1	3	0	0	6	0
N	14	7		3	*	*		0	*	17	0
YR	166	97	102	51	2	35	15	92	2	65	0

Elev. 2862 feet m.s.l.

Means and extremes above are from existing and comparable exposures. Annual extremes have been exceeded at other sites in the locality as follows:
Lowest temperature -11 in February 1933; maximum monthly precipitation 8.18 in September 1932.

REFERENCE NOTES APPLYING TO TABLES APPEAR ON THE PAGE FOLLOWING LAST TABLE.
(Caution: Letters and symbols may have different meanings in 1941-1970 tables than in earlier tables. See notes.)

NORMALS, MEANS AND EXTREMES

(Table Revised 1969. Base Period for Climatological Normals: 1931-1960)

JEFFERSON COUNTY AIRPORT
LATITUDE 29° 57' N LONGITUDE 94° 01' W ELEVATION (ground) 16 Feet

Month	Temp Normal Daily max	Temp Normal Daily min	Temp Normal Monthly	Extreme Record highest	Yr	Extreme Record lowest	Yr	Normal degree days	Precip Normal total	Precip Max monthly	Yr	Precip Min monthly	Yr	Precip Max 24 hrs	Yr	Snow/Sleet Mean total	Snow Max monthly	Yr	Snow Max 24 hrs	Yr
J	63.5	43.7	53.6	80	1965	14	1962	384	4.23	9.57	1961	0.64	1957	4.92	1961	T	T	1964+	T	1964+
F	65.8	46.2	56.0	84	1962	25	1968+	274	4.45	11.76	1959	0.36	1954	5.05	1965	0.5	4.4	1960	4.4	1960
M	71.3	50.8	61.1	86	1962	27	1967	192	3.44	8.83	1957	0.06	1965	4.21	1968	0.1	1.4	1968	1.4	1968
A	78.2	58.3	68.2	90	1965	37	1961	39	3.94	7.38	1968	0.35	1965	7.23	1968	0.0	0.0		0.0	
M	83.8	65.6	74.7	92	1963+	49	1961	0	4.94	9.63	1966	0.55	1967	7.18	1954	0.0	0.0		0.0	
J	89.6	71.5	80.6	99	1963	60	1966+	0	4.29	14.05	1961	1.06	1961	10.20	1961	0.0	0.0		0.0	
J	91.0	72.7	81.9	100	1963	61	1967	0	6.00	18.71	1959	0.63	1956	4.87	1964	0.0	0.0		0.0	
A	91.2	73.3	82.3	107	1962	62	1967	0	5.49	17.26	1966	0.98	1968	8.45	1966	0.0	0.0		0.0	
S	88.4	68.0	78.2	99	1963	45	1967	0	4.88	18.15	1963	0.50	1953	13.17	1963	0.0	0.0		0.0	
O	81.7	58.8	70.3	95	1963	37	1964	22	2.88	6.59	1957	0.00	1963	5.63	1966	0.0	0.0		0.0	
N	70.7	48.4	59.7	87	1961	29	1968	207	3.46	10.42	1961	0.15	1967	7.26	1961	0.0	0.0		0.0	
D	64.9	44.6	54.8	81	1964+	17	1963	329	5.09	12.47	1956	1.32	1954	6.24	1956	T	T	1953	T	1953
YR	78.3	58.5	68.5	107	AUG 1962	14	JAN 1962	1447	53.09	18.71	JUL 1959	0.00	OCT 1963	13.17	SEP 1963	0.6	4.4	FEB 1960	4.4	FEB 1960

Relative humidity (Standard time used: CENTRAL), Wind, Sky, Sunshine and Mean number of days — Table 1:

Month	RH 6AM	RH Noon	RH 6PM	RH (eve)	Wind mean hourly	Prevailing dir	Fastest mile speed	Dir	Yr	Mean sky cover	Pct poss sunshine	Clear	Partly cloudy	Cloudy	Precip .01"+	Snow 1.0"+	Tstms	Heavy fog	Max 90+	Max 32↓	Min 32↓	Min 0↓
J	87	89	69	78	11.6	N	50	S	1957	6.9	43	7	6	18	9	0	2	5	0	*	8	0
F	86	86	62	71	11.3	S	59	SW	1963	6.3	50	8	6	15	9	*	3	8	0	0	8	0
M	87	89	62	73	12.3	S	66	SW	1964+	6.4	52	7	9	15	7	*	3	7	0	0	4	0
A	89	91	64	73	12.7	S	60	NW	1966+	6.2	52	5	9	16	7	0	4	5	*	0	2	0
M	90	92	62	70	11.7	S	51	NW	1957	6.8	58	6	13	12	8	0	6	3	8	0	0	0
J	91	92	62	69	9.1	S	54	NW	1963	6.1	68	7	16	7	11	0	8	1	19	0	0	0
J	92	92	62	69	8.0	S	66	SW	1964	6.4	66	5	17	9	12	0	13	*	26	0	0	0
A	89	90	62	67	7.8	S	73	E	1966	6.0	64	6	15	10	9	0	12	*	24	0	0	0
S	88	90	60	72	8.6	NE	56	SW	1968	6.0	63	6	15	10	9	0	6	1	13	0	0	0
O	88	89	56	70	9.7	N	65	NW	1963	5.5	68	12	8	13	6	0	3	4	2	0	0	0
N	86	88	54	71	10.7	N	56	NE	1963	5.9	60	12	5	16	8	0	2	3	0	0	*	0
D	87	89	60	78	11.0	N	69	S	1953	5.7	47	8	6	16	10	0	2	7	0	*	5	0
YR	88	90	62	72	10.3	S	73	E	AUG 1964	6.0	59	90	127	148	103	*	63	41	87	*	19	0

Annual extremes have been exceeded at other locations as follows:
Lowest temperature 11 in January 1930; maximum monthly precipitation 24.25 inches in July 1943;
maximum precipitation in 24 hours 17.76 in July 1943; fastest mile of wind 91 from Northeast in August 1940.
⊘ Record for partial year, June-December 1953, considered in extracting precipitation extremes.

NORMALS, MEANS AND EXTREMES

(Table Revised 1975. Base Period for Climatological Normals: 1941-1970)

Month	Temp Normal Daily max	Temp Normal Daily min	Temp Normal Monthly	Extreme Record highest	Yr	Extreme Record lowest	Yr	Heating deg days	Cooling deg days	Precip Normal	Water eq Max monthly	Yr	Min monthly	Yr	Max 24 hrs	Yr	Snow Max monthly	Yr	Snow Max 24 hrs	Yr
J	61.5	42.4	52.0	80	1965	14	1962	420	17	4.06	9.57	1961	0.60	1971	4.92	1961	T	1953	3.0	1973
F	65.0	45.1	55.1	84	1962	22	1973	302	25	4.24	11.15	1959	0.36	1954	5.05	1965	4.4	1960	4.4	1960
M	70.5	49.7	60.1	90	1974	27	1968	203	51	3.05	8.83	1957	0.06	1955	5.28	1973	1.4	1968	1.4	1968
A	78.3	59.4	68.9	90	1965	37	1970	35	150	4.19	9.63	1973	0.35	1973	10.09	1973	0.0		0.0	
M	84.3	65.7	75.0	93	1973	49	1971	0	310	4.94	9.63	1966	0.55	1960	7.18	1954	0.0		0.0	
J	89.7	71.7	80.8	99	1963	57	1974	0	474	4.81	14.05	1961	0.96	1961	10.20	1961	0.0		0.0	
J	92.0	74.0	83.0	103	1970	61	1967	0	558	5.89	18.71	1959	0.63	1956	4.87	1964	0.0		0.0	
A	92.6	73.6	83.1	107	1962	62	1967	0	561	5.69	17.26	1966	0.98	1968	8.45	1966	0.0		0.0	
S	88.6	69.2	78.9	99	1963	45	1967	0	417	5.34	18.15	1963	0.50	1953	13.17	1963	0.0		0.0	
O	81.3	58.5	69.9	95	1963	37	1964	35	187	3.71	6.59	1957	0.00	1963	5.63	1970	0.0		0.0	
N	70.9	49.4	60.2	87	1963	29	1969	184	40	4.26	10.42	1961	0.15	1967	7.26	1961	0.0		0.0	
D	64.2	44.1	54.2	82	1970	17	1963	342	8	4.89	12.47	1956	1.32	1954	6.24	1956	T	1953	T	1953
YR	78.3	58.6	68.5	107	AUG 1962	14	JAN 1962	1518	2798	55.07	18.71	JUL 1959	0.00	OCT 1963	13.17	SEP 1963	4.4	FEB 1960	4.4	FEB 1960

Relative humidity (Local time), Wind, Sky, Sunshine, Mean number of days, Average station pressure — Table 2:

Month	RH 00	RH 06	RH 12	RH 18	Wind mean speed m.p.h	Prevailing dir	Fastest mile speed	Dir	Yr	Mean sky cover	Pct poss sunshine	Clear	Partly cloudy	Cloudy	Precip .01"+	Snow 1.0"+	Tstms	Heavy fog vis ¼ mi or less	Max 90+	Max 32↓	Min 32↓	Min 0↓	Avg sta pressure mb
J	88	90	71	79	11.3	N	50	S	1957	7.1	42	6	7	18	10	*	3	8	0	*	7	0	1018.8
F	85	88	52	71	11.9	S	62	SE	1963	6.4	52	6	9	14	9	*	3	6	0	0	4	0	1019.9
M	88	89	54	72	12.2	S	60	SW	1964	6.4	52	7	8	15	8	*	4	6	0	0	1	0	1015.4
A	88	91	54	72	12.4	S	60	NW	1964	6.0	63	6	8	16	7	0	6	3	*	0	0	0	1015.3
M	90	92	53	70	10.7	S	72	NW	1971	5.9	66	5	15	12	8	0	7	1	2	0	0	0	1012.5
J	91	93	52	70	9.1	S	66	SW	1963	5.8	64	7	17	7	11	0	13	*	18	0	0	0	1013.9
J	93	94	54	72	8.0	S	73	SW	1964	5.6?	65	5	15	9	12	0	12	*	24	0	0	0	1015.8
A	94	95	55	75	7.5	S	56	E	1968	5.6	57	8	11	10	10	0	13	*	23	0	0	0	1015.4
S	90	91	58	73	8.8	NE	60	SW	1956	5.9	57	7	8	8	10	0	3	1	12	0	0	0	1013.4
O	87	89	50	59	10.6	N	65	NW	1963	5.6	57	10	?	12	7	0	3	1	2	0	0	0	1018.7
N	88	90	59	79	10.9	N	56	NE	1963	6.0	47	9	7	16	8	0	2	5	0	0	1	0	1018.5
D	89	89	60	59	10.2	N	69	SW	1953						10	0	2	7	0	*	4	0	1018.3
YR	89	91	54	73	10.2	S	74	SW	MAY 1971	6.0	58	93	126	146	104	*	64	41	81	*	17	0	1016.2

Average station pressure elev.: 22 feet m.s.l.

Means and extremes above are from existing and comparable exposures. Annual extremes have been exceeded at other sites in the locality as follows:
Lowest temperature 11 in January 1930; maximum monthly precipitation 24.25 inches in July 1943;
fastest mile of wind 91 from Northeast in August 1940.

REFERENCE NOTES APPLYING TO TABLES APPEAR ON THE PAGE FOLLOWING LAST TABLE.

(Caution: Letters and symbols may have different meanings in 1941-1970 tables than in earlier tables. See notes.)

SAN ANGELO, TEXAS
MATHIS FIELD

LATITUDE 31° 22' N
LONGITUDE 100° 30' W
ELEVATION (ground) 1903 Feet

NORMALS, MEANS AND EXTREMES

(Table Revised 1969. Base Period for Climatological Normals: 1931-1960)

Month	Temperature – Normal – Daily maximum	Temperature – Normal – Daily minimum	Temperature – Normal – Monthly	Temperature – Extremes – Record highest	Year	Temperature – Extremes – Record lowest	Year	Normal degree days	Precipitation – Normal total	Precipitation – Maximum monthly	Year	Precipitation – Minimum monthly	Year	Precipitation – Maximum in 24 hrs.	Year	Snow, Sleet – Mean total	Snow, Sleet – Maximum monthly	Year	Snow, Sleet – Maximum in 24 hrs.	Year	Relative humidity – Mid.	Relative humidity – 6 AM	Relative humidity – Noon	Relative humidity – 6 PM	Wind – Mean hourly speed	Wind – Prevailing direction	Wind – Fastest mile – Speed	Wind – Fastest mile – Direction	Wind – Fastest mile – Year	Pct. of possible sunshine	Mean sky cover sunrise to sunset	Mean number of days – Sunrise to sunset – Clear	Partly cloudy	Cloudy	Precipitation .01 inch or more	Snow, Sleet 1.0 inch or more	Thunderstorms	Heavy fog	Temperatures – Max. – 90° and above	Temperatures – Max. – 32° and below	Temperatures – Min. – 32° and below	Temperatures – Min. – 0° and below
																					Standard time used: CENTRAL																					
(a)	(b)	(b)	(b)	9		9		(b)	(b)	21		21		21		21	21		21		8	8	8	8	19	14	20	20			20	20	20	20	21	21	21	21	8	8	8	8
J	59.4	34.3	46.9	85	1963	6	1963	567	0.97	3.65	1961	0.00	1967	2.49	1961	0.8	7.0	1949	5.1	1949	65	73	50	45	10.3	SW	44	27	1960+		5.3	12	7	12	4	*	*	1	0	1	20	0
F	62.7	38.3	50.5	86	1962	14	1966+	412	0.90	2.86	1958	0.04	1954	1.92	1958	0.9	4.7	1961	4.1	1966	64	73	48	41	10.9	SSW	48	29	1960+		5.2	11	6	11	4	*	1	1	0	*	13	0
M	70.2	43.7	57.0	97	1967	13	1962	288	0.93	5.00	1953	T	1963	4.65	1953	0.2	3.1	1962+	3.1	1962+	57	67	42	35	12.3	SSW	58	27	1961		5.2	12	8	11	3	*	2	1	2	0	5	0
A	80.6	53.6	67.1	100	1961	31	1968	66	2.01	4.42	1949	0.10	1967	2.35	1957	0.0	0.0		0.0		59	71	42	35	12.3	S	69	29	1957		5.2	11	8	11	5	0	4	*	8	0	*	0
M	86.5	62.1	74.3	107	1960	35	1967	0	3.20	7.10	1957	0.26	1962	2.12	1958	0.0	0.0		0.0		66	78	48	40	11.7	S	60	02	1963		5.2	11	10	10	7	0	7	*	16	0	0	0
J	94.4	70.5	82.5	109	1960	48	1964	0	1.82	5.82	1961	0.07	1956+	2.86	1961	0.0	0.0		0.0		64	78	48	40	11.3	S	57	02	1955		4.2	14	11	5	5	0	4	*	24	0	0	0
J	96.9	72.7	84.8	111	1960	58	1968	0	1.41	7.21	1959	T	1963	2.95	1959	0.0	0.0		0.0		56	73	41	35	9.6	S	40	16	1952		4.2	14	11	6	4	0	5	*	29	0	0	0
A	97.5	72.6	85.1	107	1964+	55	1961	0	1.27	5.05	1966	T	1959	2.81	1966	0.0	0.0		0.0		58	74	42	37	9.0	S	43	02	1949		3.9	16	10	5	4	0	5	0	28	0	0	0
S	90.4	66.1	78.3	103	1965	45	1963	0	2.66	9.20	1959	0.24	1954	6.03	1959	0.0	0.0		0.0		73	83	55	50	8.9	S	44	09	1959+		4.2	15	8	7	6	0	4	*	12	0	0	0
O	81.3	55.3	68.3	98	1968+	35	1966+	68	1.83	5.29	1959	0.00	1952	5.11	1959	0.0	0.0		0.0		72	81	48	46	9.1	S	60	29	1960		3.8	17	6	8	4	0	2	1	4	0	0	0
N	67.6	41.9	54.8	90	1965	22	1966+	318	0.78	3.55	1968	0.00	1950+	1.54	1968	0.7	8.8	1968	5.8	1968	73	81	53	53	9.8	SW	66	29	1958		4.5	14	6	10	4	*	1	1	*	0	3	0
D	60.2	35.2	47.7	83	1966	10	1966	536	0.85	2.06	1959	T	1968+	1.08	1949	0.2	1.9	1967	1.4	1967	72	77	52	52	9.8	W	43	30	1965		4.9	13	7	11	3	*	*	1	0	*	16	0
YR	79.0	53.9	66.4	111	JUL. 1960	6	JAN. 1963	2255	18.63	9.20	SEP. 1959	0.00	JAN. 1967+	6.03	SEP. 1959	2.8	8.8	NOV. 1968	5.8	NOV. 1968	65	76	47	42	10.4	S	69	29	APR. 1957		4.7	160	98	107	55	1	36	7	123	2	56	0

Means and extremes are from existing, or last comparable instrumental location. Annual extremes have been exceeded at other or present locations as follows:
Maximum monthly precipitation 27.65 in September 1936; maximum precipitation in 24 hours 11.75 in September 1936; maximum monthly snowfall 13.0 in January 1926 and earlier date(s).

NORMALS, MEANS AND EXTREMES

(Table Revised 1975. Base Period for Climatological Normals: 1941-1970)

Month	Temperatures °F – Normal – Daily maximum	Temperatures °F – Normal – Daily minimum	Temperatures °F – Normal – Monthly	Temperatures °F – Extremes – Record highest	Year	Temperatures °F – Extremes – Record lowest	Year	Normal Degree days Base 65 °F – Heating	Normal Degree days Base 65 °F – Cooling	Precipitation in inches – Water equivalent – Normal	Maximum monthly	Year	Minimum monthly	Year	Maximum in 24 hrs.	Year	Snow, Ice pellets – Maximum monthly	Year	Snow, Ice pellets – Maximum in 24 hrs.	Year	Relative humidity pct. – Hour 00	Hour 06	Hour 12	Hour 18	Wind – Mean speed m.p.h.	Wind – Prevailing direction	Wind – Fastest mile – Speed m.p.h.	Wind – Fastest mile – Direction	Wind – Fastest mile – Year	Pct. of possible sunshine	Mean sky cover, tenths, sunrise to sunset	Mean number of days – Sunrise to sunset – Clear	Partly cloudy	Cloudy	Precipitation .01 inch or more	Snow, Ice pellets 1.0 inch or more	Thunderstorms	Heavy fog, visibility ¼ mile or less	Temperatures °F – Max. – 90° and above (b)	Temperatures °F – Max. – 32° and below	Temperatures °F – Min. – 32° and below	Temperatures °F – Min. – 0° and below	Average station pressure mb.
																					(Local time)																						Elev. 1908 feet m.s.l.
(a)				15		15					27		27		27		27		27		14	14	14	14	25	14	26	26			26	26	26	26	27	27	27	27	14	14	14	14	2
J	59.1	33.6	46.4	90	1969	6	1963	577	0	0.81	3.65	1961	0.00	1967	2.49	1961	7.7	1973	5.1	1949	66	73	50	46	10.3	SW	44	27	1960		5.3	12	7	12	4	*	*	1	*	1	18	0	950.7
F	63.2	37.5	50.4	87	1972	12	1971	413	0	0.79	2.86	1958	0.01	1974	1.92	1958	5.8	1973	4.1	1966	63	72	47	40	10.8	SSW	48	29	1960		5.2	11	6	11	4	*	1	1	0	*	11	0	950.9
M	70.7	43.5	57.1	97	1974	13	1962	287	42	0.87	5.00	1953	T	1972	4.65	1953	3.1	1962	3.1	1962	60	70	42	35	12.3	SSW	58	27	1961		5.0	12	8	11	4	*	2	1	2	0	5	0	944.3
A	80.4	54.0	67.2	103	1972	25	1973	74	140	1.66	4.42	1949	0.10	1967	3.32	1971	0.0		0.0		61	72	42	35	12.3	S	75	28	1969		5.1	11	8	11	5	0	4	*	6	0	1	0	946.2
M	86.5	62.4	74.5	107	1960	35	1967	0	298	2.70	7.10	1957	0.26	1962	2.12	1958	0.0		0.0		66	79	46	40	11.4	S	60	02	1963		5.2	11	10	10	6	0	7	*	14	0	0	0	944.8
J	93.4	69.8	81.6	110	1969	48	1964	0	498	1.88	5.82	1961	0.07	1956	2.86	1961	0.0		0.0		64	79	47	39	11.2	S	57	02	1955		4.3	14	10	6	5	0	5	*	23	0	0	0	946.2
J	96.9	72.4	84.7	111	1960	58	1968	0	611	1.23	7.21	1959	T	1970	2.95	1959	0.0		0.0		57	74	41	35	9.8	S	40	16	1952		4.4	14	10	7	4	0	4	*	29	0	0	0	948.6
A	96.9	72.0	84.5	107	1969	55	1961	0	605	1.44	8.13	1971	T	1959	3.00	1971	0.0		0.0		63	76	44	40	9.1	S	43	02	1949		4.2	15	10	6	5	0	5	*	26	0	0	0	948.7
S	88.4	65.1	76.8	103	1965	45	1963	0	354	2.74	9.20	1959	0.24	1954	6.03	1959	0.0		0.0		76	85	57	53	9.0	S	44	09	1959		4.5	14	8	8	7	0	4	*	12	0	0	0	948.6
O	79.6	54.7	67.2	98	1968	31	1970	73	141	1.86	6.59	1969	0.00	1952	5.11	1959	0.0		0.0		76	84	53	52	9.2	S	60	29	1960		4.2	16	6	9	5	0	2	1	3	0	*	0	951.4
N	68.5	42.5	55.5	90	1965	20	1970	298	13	0.85	3.55	1968	0.00	1950	1.54	1968	8.8	1968	5.8	1968	74	81	53	54	9.8	SW	66	29	1958		4.6	14	6	10	4	*	1	1	*	0	4	0	950.8
D	61.4	35.2	48.3	84	1970	10	1972	518	0	0.70	2.24	1969	T	1973	1.12	1969	1.9	1967	1.4	1967	71	78	51	51	9.8	W	43	30	1965		4.9	13	7	11	3	*	*	1	0	*	14	0	950.4
YR	78.7	53.6	66.2	111	JUL 1960	6	JAN 1963	2240	2702	17.53	9.20	SEP 1959	0.00	JAN 1967	6.03	SEP 1959	8.8	NOV 1968	5.8	NOV 1968	66	77	48	43	10.4	S	75	28	APR 1969		4.7	157	96	112	56	1	37	8	115	2	52	0	948.5

Means and extremes above are from existing and comparable exposures. Annual extremes have been exceeded at other sites in the locality as follows:
Maximum monthly precipitation 27.65 in September 1936; maximum precipitation in 24 hours 11.75 in September 1936; maximum monthly snowfall 13.0 in January 1926 and earlier.

REFERENCE NOTES APPLYING TO TABLES APPEAR ON THE PAGE FOLLOWING LAST TABLE.
(Caution: Letters and symbols may have different meanings in 1941-1970 tables than in earlier tables. See notes.)

INTERNATIONAL AIRPORT

LATITUDE 29° 32' N
LONGITUDE 98° 28' W
ELEVATION (ground) 788 Feet

NORMALS, MEANS AND EXTREMES
(Table Revised 1969. Base Period for Climatological Normals: 1931-1960)

Temperature, Degree Days, Precipitation, Snow/Sleet

Month	Normal Daily max	Normal Daily min	Normal Monthly	Record highest	Year	Record lowest	Year	Normal degree days	Precip. Normal total	Max monthly	Year	Min monthly	Year	Max in 24 hrs	Year	Snow Mean total	Snow Max monthly	Year	Snow Max 24 hrs	Year
J	62.3	41.6	52.0	87	1943	0	1949	428	1.74	8.52	1968	0.18	1967	3.18	1968	0.3	4.7	1949	4.7	1949
F	66.1	44.7	55.4	92	1959	6	1951	286	1.65	6.43	1965	0.03	1954	2.34	1965	0.3	3.5	1966	3.5	1966
M	72.4	49.6	61.0	97	1950+	21	1948+	195	1.67	4.19	1957	0.03	1961+	2.36	1945	T	T	1965+	T	1965+
A	78.8	57.6	68.2	99	1963+	34	1966+	30	2.82	9.32	1946	0.14	1955	2.77	1946	0.0	0.0		0.0	
M	85.2	65.4	75.3	101	1967+	44	1954	3	3.45	8.22	1957	0.17	1961	4.29	1957	0.0	0.0		0.0	
J	91.6	72.2	81.9	103	1960	53	1964	0	2.95	8.26	1951	0.01	1967	6.18	1951	0.0	0.0		0.0	
J	94.0	73.9	84.0	106	1954	62	1967	0	2.09	8.19	1958	T	1944	6.97	1958	0.0	0.0		0.0	
A	94.2	73.4	83.8	106	1962+	61	1966	0	2.36	6.15	1950	0.00	1952	5.57	1950	0.0	0.0		0.0	
S	88.5	68.7	78.6	102	1951	41	1942	0	3.49	15.78	1946	0.06	1947	5.29	1947	0.0	0.0		0.0	
O	81.5	59.6	70.6	95	1962+	34	1966+	31	2.50	4.97	1942	T	1952	5.29	1942	0.0	0.0		0.0	
N	70.4	48.5	59.5	91	1962+	23	1959	204	1.37	4.51	1944	T	1966	4.00	1962	T	0.3	1957	0.3	1957
D	65.4	42.0	53.7	90	1955	14	1950	363	1.75	2.95	1949	0.03	1950	2.89	1944	T	0.2	1964	0.2	1964
YR	79.2	58.1	68.7	106 (AUG 1962+)		0 (JAN 1949)		1546	27.84	15.78 (SEP 1946)		0.00 (AUG 1952)		6.97 (JUL 1958)		0.6	4.7 (JAN 1949)		4.7 (JAN 1949)	

Relative Humidity, Wind, Sky Cover, Mean Number of Days, Solar Radiation

Month	R.H. Mid.	6 AM	Noon	6 PM	Wind mean hourly speed	Prevailing direction	Fastest mile speed	Direction	Year	Pct. sunshine	Mean sky cover	Clear	Partly cloudy	Cloudy	Precip .01"+	Thunderstorms	Snow 1.0"+	Heavy fog	Max 90°+	Max 32°& below	Min 32°& below	Min 0°& below	Avg daily solar radiation (langleys)
J	76	81	60	58	9.1	N	56	NW	1953	49	6.3	9	6	16	8	1	*	5	0	*	8	*	273
F	76	81	57	54	9.8	NE	56	N	1954	50	6.3	9	6	15	8	1	*	5	0+	*	4	0	341
M	72	79	53	48	10.5	SE	57	NE	1955	57	6.2	8	7	15	7	2	0	2	1	0	2	0	410
A	77	83	56	52	10.5	SE	57	NW	1953	57	6.2	7	8	13	7	4	0	3	3	0	0	0	439
M	81	88	57	54	10.3	SE	73	NW	1946	57	5.8	7	11	13	8	6	0	1	10	0	0	0	510
J	80	88	54	50	9.1	SE	59	NW	1953	77	5.4	10	11	6	6	6	0	*	23	0	0	0	594
J	75	88	49	43	8.5	SSE	52	NE	1943	77	4.7	10	15	6	4	4	0	*	29	0	0	0	623
A	73	86	44	44	8.3	SE	74	NE	1952	76	4.7	12	14	5	5	3	0	0	28	0	0	0	572
S	78	86	54	51	8.3	SE	49	NW	1952	68	4.6	11	9	9	7	4	0	*	17	0	0	0	478
O	77	84	55	56	8.8	NE	43	NW	1945	68	4.6	13	7	8	5	2	0	2	4	0	0	0	398
N	76	80	55	58	8.8	N	43	NW	1950	53	5.5	10	6	13	6	2	*	3	*	1	1	0	290
D	76	80	57	58	8.6	N	48	NW	1947	53	5.9	10	6	15	7	1	*	5	*	1	6	0	243
YR	76	84	54	52	9.3	SE	74	NE	AUG 1942	62	5.6	113	116	136	79	35	*	23	117	*	21	*	431

(Number of years of record: 26 for most elements; 27 for Extremes; 15 for Prevailing wind direction; 16 for solar radiation.)

Means and extremes in the above table are from the existing location (or last comparable location). Annual extremes have been exceeded at other locations as follows:
Highest temperature 107 in August 1909; maximum precipitation in 24 hours 7.08 in October 1913; maximum monthly snowfall 6.4 in January 1926;
maximum snowfall in 24 hours 5.0 in January 1940.

NORMALS, MEANS AND EXTREMES
(Table Revised 1975. Base Period for Climatological Normals: 1941-1970)

Temperature, Degree Days, Precipitation, Snow/Ice Pellets

Month	Normal Daily max	Normal Daily min	Normal Monthly	Record highest	Year	Record lowest	Year	Heating	Cooling	Precip. Normal	Max monthly	Year	Min monthly	Year	Max 24 hrs	Year	Snow Max monthly	Year	Snow Max 24 hrs	Year
J	61.6	39.8	50.7	89	1971	0	1949	451	6	1.66	8.52	1968	0.04	1971	3.18	1968	4.7	1949	4.7	1949
F	66.1	43.1	54.5	92	1959	6	1951	310	16	2.06	6.43	1965	0.03	1951	2.34	1965	3.5	1966	3.5	1966
M	72.5	49.1	60.8	100	1971	21	1948	194	64	1.54	4.19	1957	0.14	1961	2.36	1945	T	1965	T	1965
A	80.3	57.8	69.1	100	1963	34	1964	31	169	2.77	9.32	1946	0.17	1955	2.77	1946	0.0		0.0	
M	86.2	65.7	76.0	103	1967	44	1954	3	341	3.57	11.24	1972	0.01	1967	6.53	1972	0.0		0.0	
J	92.4	72.0	82.2	103	1960	53	1964	0	516	2.79	10.44	1972	T	1944	6.18	1951	0.0		0.0	
J	95.6	73.8	84.7	106	1954	62	1967	0	611	1.69	8.19	1974	0.00	1952	6.97	1958	0.0		0.0	
A	95.9	73.4	84.7	106	1962	61	1966	0	611	2.41	11.41	1974	0.06	1973	5.57	1950	0.0		0.0	
S	89.8	68.8	79.3	102	1951	41	1942	0	429	3.71	15.78	1946	T	1946	7.28	1973	0.0		0.0	
O	81.8	58.2	70.0	95	1962	34	1970	32	202	2.64	5.39	1972	T	1952	4.33	1974	0.0		0.0	
N	70.1	48.2	59.7	90	1955	23	1959	179	20	1.77	4.51	1974	T	1965	4.00	1962	0.3	1957	0.3	1957
D	64.6	41.8	53.2	90	1955	14	1950	373	7	1.46	4.51	1965	0.03	1965	2.89	1944	0.2	1964	0.2	1964
YR	79.8	57.8	68.8	106 (AUG 1962)		0 (JAN 1949)		1570	2994	27.54	15.78 (SEP 1946)		0.00 (AUG 1952)		7.28 (SEP 1973)		4.7 (JAN 1949)		4.7 (JAN 1949)	

Relative Humidity, Wind, Sky Cover, Mean Number of Days, Station Pressure

Month	R.H. Hour 00	06	12	18	Wind mean speed mph	Prevailing direction	Fastest mile mph	Direction	Year	Pct. sunshine	Mean sky cover	Clear	Partly cloudy	Cloudy	Precip .01"+	Thunderstorms	Snow 1.0"+	Heavy fog vis ¼ mi or less	Max 90°+	Max 32°& below	Min 32°& below	Min 0°& below	Avg station pressure (mb)
J	76	81	60	58	9.1	N	56	NW	1953	49	6.3	9	6	16	8	1	*	6	0	*	8	*	990.7
F	75	80	57	54	9.8	NE	56	N	1954	50	6.1	8	6	14	8	1	*	3	0+	*	4	0	991.6
M	72	79	53	47	10.5	SE	57	NE	1955	58	6.4	8	7	15	7	2	0	1	1	0	2	0	984.4
A	76	83	58	51	10.6	SE	57	NW	1953	58	6.4	7	11	13	7	4	0	1	3	0	0	0	986.1
M	81	87	65	54	10.2	SE	73	NW	1946	58	5.9	7	15	8	8	6	0	*	10	0	0	0	986.3
J	80	88	54	51	9.1	SE	59	NW	1953	77	5.0	11	14	6	6	6	0	*	22	0	0	0	985.5
J	75	87	50	46	8.5	SSE	52	NE	1943	77	4.8	11	16	6	4	4	0	*	29	0	0	0	987.7
A	74	86	48	45	8.3	SE	74	NE	1952	74	5.2	12	12	8	6	4	0	*	28	0	0	0	987.7
S	78	86	55	53	8.3	SE	49	NW	1952	68	4.8	11	12	10	8	4	0	2	17	0	0	0	988.5
O	76	84	55	56	8.8	NE	43	NW	1945	68	5.2	11	12	12	6	2	0	3	4	0	0	0	990.4
N	76	81	57	58	8.8	N	43	NW	1950	56	5.5	10	7	15	6	1	*	5	*	*	1	0	990.9
D	76	80	57	58	8.6	N	48	NW	1947	52	5.9	10	6	15	7	1	*	6	*	1	6	0	990.0
YR	76	84	55	52	9.3	SE	74	NE	AUG 1942	62	5.6	110	117	138	80	36	*	23	114	*	21	*	987.9

(Elevation 794 feet m.s.l.)

Means and extremes above are from existing and comparable exposures. Annual extremes have been exceeded at other sites in the locality as follows:
Highest temperature 107 in August 1909; maximum monthly snowfall 6.4 in January 1926; maximum snowfall in 24 hours 5.0 in January 1940.

REFERENCE NOTES APPLYING TO TABLES APPEAR ON THE PAGE FOLLOWING LAST TABLE.
(Caution: Letters and symbols may have different meanings in 1941-1970 tables than in earlier tables. See notes.)

VICTORIA, TEXAS
FOSTER FIELD

LATITUDE 28° 51' N
LONGITUDE 96° 55' W
ELEVATION (ground) 104 Feet

NORMALS, MEANS AND EXTREMES

(Table Revised 1969. Base Period for Climatological Normals: 1931-1960)

ø Extremes for the period June 1961 through the current year.
Means and extremes are from the existing or last comparable instrumental locations. Annual extremes have been exceeded at other locations as follows:
Highest temperature 110 in July 1939; lowest temperature 9 in January 1930; minimum monthly precipitation 0.00 in November 1945 and earlier date(s);
maximum monthly snowfall 5.0 in January 1940.

% The station did not operate 24 hours daily during part of its history.
Fog and thunderstorm data therefore may be incomplete.

NORMALS, MEANS AND EXTREMES

(Table Revised 1975. Base Period for Climatological Normals: 1941-1970)

Means and extremes above are from existing and comparable exposures. Annual extremes have been exceeded at other sites in the locality as follows:
Highest temperature 110 in July 1939; lowest temperature 9 in January 1930; minimum monthly precipitation 0.00 in November 1945 and earlier. Maximum
monthly snowfall 5.0 in January 1940.

REFERENCE NOTES APPLYING TO TABLES APPEAR ON THE PAGE FOLLOWING LAST TABLE.
(Caution: Letters and symbols may have different meanings in 1941-1970 tables than in earlier tables. See notes.)

974

NORMALS, MEANS AND EXTREMES
(Table Revised 1969. Base Period for Climatological Normals: 1931-1960)

LATITUDE 31° 37' N
LONGITUDE 97° 13' W
ELEVATION (ground) 501 Feet

Month	Temperature Normal Daily maximum	Daily minimum	Monthly	Extremes Record highest	Year ø	Record lowest	Year ø	Normal degree days	Precipitation Normal total	Maximum monthly	Year	Minimum monthly	Year	Maximum in 24 hrs.	Year	Snow, Sleet Mean total	Maximum monthly	Year	Maximum in 24 hrs.	Year	Rel. humidity A.M.	Noon	6 PM	Mid.	Wind Mean hourly speed	Prevailing direction	Fastest mile Speed	Direction	Year	Mean sky cover sunrise to sunset	Clear	Partly cloudy	Cloudy	Precip .01"+	Snow 1.0"+	Tstorms	Heavy fog	Max 90°+	Max 32°-	Min 32°-	Min 0°-
(a)	(b)	(b)	(b)					(b)	(b)																																
J	58.2	37.8	48.0	78	1964	7	1964	536	2.27	5.83	1961	0.29	1959	2.24	1961	1.0	7.0	1949	7.0	1949	76	63	61		12.6	S	49	32	1952	6.3	9	6	16	8	*	1	3	0	1	12	0
F	61.8	41.0	51.5	76	1966	19	1966	389	2.42	4.55	1944	0.44	1954	2.14	1965	0.5	4.8	1966	4.8	1966	73	59	54		12.7	S	58	36	1954	6.0	9	7	14	8	*	3	2	0	0	9	0
M	69.1	47.7	57.9	96	1967	22	1965	270	2.35	4.55	1965	0.04	1954	3.07	1946	0.1	1.0	1962+	1.0	1962+	75	58	55		13.6	S	65	27	1952	6.4	10	7	14	7	*	4	1	1	0	4	0
A	77.4	55.7	66.6	94	1966+	35	1966+	66	3.88	13.37	1957	0.65	1956	5.09	1957	0.0	0.0		0.0		75	58	58		13.7	S	62	36	1953	6.2	8	8	15	8	0	6	1	1	0	0	0
M	84.1	63.8	74.3	95	1967	49	1967	3	4.57	15.00	1957	0.72	1953	7.18	1953	0.0	0.0		0.0		83	65	63		12.2	S	60	36	1952	6.1	11	10	13	8	0	8	1	5	0	0	0
J	92.4	71.5	82.0	103	1967	52	1964	0	2.67	12.06	1961	0.27	1953+	4.21	1947	0.0	0.0		0.0		77	52	48		12.2	S	69	09	1961	4.8	14	13	6	6	0	5	*	22	0	0	0
J	96.0	74.6	85.3	106	1968	62	1968	0	1.99	7.68	1961	T	1963	2.61	1961	0.0	0.0		0.0		60	44	39		11.4	S	60	36	1953	4.3	14	11	5	4	0	4	*	29	0	0	0
A	96.6	74.3	85.5	106	1964	60	1967	0	1.68	7.34	1952	0.00	1952	4.80	1958	0.0	0.0		0.0		62	41	46		10.6	S	60	05	1951	4.1	14	11	6	5	0	5	0	29	0	0	0
S	90.1	67.9	79.0	100	1964	48	1967	0	2.33	5.69	1957	T	1956	4.57	1957	0.0	0.0		0.0		70	53	51		10.4	S	52	34	1960	4.3	15	8	9	6	0	3	1	13	0	0	0
O	81.2	57.6	69.4	93	1965	36	1964	43	2.55	7.36	1956	0.13	1952	3.63	1959	0.0	0.0		0.0		71	51	51		11.3	S	62	29	1953	4.5	14	7	12	6	0	2	1	2	0	0	0
N	68.2	45.4	56.8	87	1965	28	1964	270	2.19	6.24	1952	0.04	1966	4.26	1952	T	T	1968+	T	1968+	75	57	59		11.3	S	52	32	1954	5.1	12	6	14	6	0	1	1	0	0	1	0
D	60.8	39.7	50.3	82	1966	14	1963	456	2.74	7.03	1960		1950	3.11	1945	0.1	2.0	1946	2.0	1946	77	61	63		11.7	S	52	32	1954	5.7	11	6	14	6	*	1	2	0	*	8	0
YR	78.1	56.3	67.2	106	AUG. 1964+	7	JAN. 1964	2030	32.08	15.00	MAY 1965	0.00	SEP. 1956+	7.18	MAY 1953	1.7	7.0	JAN. 1949	7.0	JAN. 1949	72	56	53		11.9	S	69	09	JUN. 1961	5.3	135	98	132	76	1	45	13	101	1	35	0

ø For period December 1963 through the current year.
Means and extremes in the above table are from the existing location (or last comparable location). Annual extremes have been exceeded at other locations as follows:
Highest temperature 111 in August 1936; lowest temperature -5 in January 1936; maximum monthly snowfall 13.0 in December 1929 and earlier date(s).

NORMALS, MEANS AND EXTREMES
(Table Revised 1975. Base Period for Climatological Normals: 1941-1970)

Month	Temperature Normal Daily maximum	Daily minimum	Monthly	Extremes Record highest	Year	Record lowest	Year	Normal Degree days Base 65°F Heating	Cooling	Precipitation Water equivalent Normal	Maximum monthly	Year	Minimum monthly	Year	Maximum in 24 hrs.	Year	Snow, ice pellets Maximum monthly	Year	Maximum in 24 hrs.	Year	Rel. humidity Hour 00	Hour 06	Hour 12	Hour 18	Wind Mean speed m.p.h.	Prevailing direction	Fastest mile Speed m.p.h.	Direction	Year	Mean sky cover tenths	Clear	Partly cloudy	Cloudy	Precip .01"+	Snow 1.0"+	Tstorms	Heavy fog ≤¼ mi	Max 90°+	Max 32°-	Min 32°-	Min 0°-
(a)					11		11									32					11	11	11	11	25	14	26	26		29	31	31	31	31	31	31	31	11	11	11	11
J	57.4	36.6	47.0	88	1971	11	1973	558	0	1.87	5.83	1961	0.03	1971	2.24	1961	7.0	1949	7.0	1949	76	80	64	62	12.2	S	49	32	1952	6.3	9	6	16	7	*	3	3	0	1	12	0
F	61.5	40.3	50.9	85	1972	19	1971	401	6	2.38	4.55	1944	0.17	1972	2.14	1965	4.8	1966	4.8	1966	74	79	59	54	12.4	S	58	36	1954	6.0	9	7	14	8	*	4	2	0	0	9	0
M	68.4	46.0	57.2	100	1971	22	1965	280	38	2.36	4.55	1945	0.04	1956	3.07	1946	1.0	1962	1.0	1962	72	78	58	55	13.5	S	65	27	1952	6.4	10	7	14	8	*	4	1	1	0	4	0
A	77.9	56.0	67.0	94	1972	28	1971	56	125	4.02	13.37	1957	0.65	1956	5.09	1957	0.0		0.0		73	81	60	56	13.5	S	62	36	1953	6.2	8	8	14	8	0	6	1	1	0	0	0
M	84.6	64.5	74.5	98	1974	41	1971	0	295	4.60	15.00	1957	0.72	1953	7.18	1953	0.0		0.0		82	85	61	61	12.4	S	60	36	1952	6.1	11	10	12	8	0	8	1	6	0	0	0
J	91.9	71.8	81.9	103	1972	52	1964	0	507	2.73	12.06	1961	0.27	1953	4.21	1947	0.0		0.0		71	82	52	48	12.1	S	69	09	1961	4.8	14	13	7	6	0	5	*	22	0	0	0
J	96.2	75.0	85.6	107	1974	61	1970	0	639	1.47	8.55	1971	T	1963	4.49	1973	0.0		0.0		63	77	41	41	11.2	S	60	36	1953	4.4	14	11	6	4	0	4	*	28	0	0	0
A	96.7	74.7	85.7	112	1969	60	1967	0	642	1.81	8.91	1974	T	1952	4.80	1958	0.0		0.0		66	79	49	46	10.9	S	60	05	1951	4.3	13	11	5	5	0	3	*	28	0	0	0
S	90.4	68.3	79.4	105	1969	48	1967	0	417	3.19	5.69	1957	0.72	1970	5.72	1974	0.0		0.0		70	84	48	51	9.7	S	52	34	1960	4.4	15	8	7	5	0	3	*	15	0	0	0
O	80.7	57.7	69.1	95	1969	24	1970	51	178	2.55	9.24	1973	0.26	1973	3.63	1959	0.0		0.0		75	83	56	57	10.1	S	62	29	1953	4.5	14	8	12	5	0	3	1	2	0	0	0
N	66.5	46.2	57.5	88	1969	19	1964	241	16	2.27	6.24	1952	0.13	1966	4.26	1952	T		T		77	82	61	62	11.1	S	52	32	1954	5.1	12	6	12	6	0	2	1	0	0	1	0
D	60.5	39.1	49.8	82	1966	14	1963	471	0	2.01	7.03	1960	0.04	1950	3.11	1945	2.0	1946	2.0	1946	77	82	62	63	11.5	S	52	32	1954	5.7	11	6	14	6	*	1	3	0	*	8	0
YR	77.8	56.4	67.1	112	AUG 1969	4	JAN 1973	2058	2863	31.26	15.00	MAY 1965	0.00	SEP 1956	7.18	MAY 1953	7.0	JAN 1949	7.0	JAN 1949	73	81	57	54	11.7	S	69	09	JUN 1961	5.3	134	101	130	77	1	45	13	103	1	31	*

Average station pressure mb.: 997.9
Elev. 508 feet m.s.l.

(Station pressure values by month:)
Month	Average station pressure mb.
J	1001.2
F	1001.2
M	994.3
A	996.3
M	993.9
J	995.3
J	997.1
A	997.5
S	996.8
O	1000.7
N	1000.2
D	1000.0
YR	997.9

Means and extremes above are from existing and comparable exposures. Annual extremes have been exceeded at other sites in the locality as follows:
Lowest temperature -5 in January 1949; maximum monthly snowfall 13.0 in December 1929 and earlier.

REFERENCE NOTES APPLYING TO TABLES APPEAR ON THE PAGE FOLLOWING LAST TABLE.
(Caution: Letters and symbols may have different meanings in 1941-1970 tables than in earlier tables. See notes.)

WICHITA FALLS, TEXAS
MUNICIPAL AIRPORT

LATITUDE	33° 58' N
LONGITUDE	98° 29' W
ELEVATION (ground)	994 Feet

NORMALS, MEANS AND EXTREMES

(Table Revised 1969. Base Period for Climatological Normals: 1931-1960)

Month	Temperature Normal Daily maximum	Daily minimum	Monthly	Extremes Record highest	Year	Record lowest	Year	Normal degree days	Precipitation Normal total
J	53.9	31.6	42.8	82	1967	-5	1966	698	1.12
F	58.4	35.1	46.8	88	1962	13	1962	518	1.38
M	66.4	41.1	53.8	98	1967	9	1965	378	1.54
A	76.4	52.5	64.0	97	1967	29	1962	120	2.53
M	83.7	60.6	72.2	104	1967	41	1966	6	4.60
J	93.2	70.3	81.8	105	1968	51	1964	0	3.20
J	97.9	73.3	85.6	111	1964	58	1965+	0	1.97
A	98.3	73.1	85.7	113	1964	54	1962	0	1.79
S	89.9	64.9	77.4	107	1965	43	1967	99	2.33
O	78.8	53.9	66.4	98	1963	33	1967	381	2.87
N	64.9	40.0	52.5	89	1965	22	1964	381	1.42
D	55.9	33.2	44.6	87	1966	-5	1968	632	1.45
YR	76.5	52.4	64.5	113	AUG. 1964	-5	JAN. 1966	2832	26.20

Means and extremes in the above table are from existing or comparable instrumental locations. Annual extremes have been exceeded at other locations as follows: Lowest temperature -12 in January 1947.

NORMALS, MEANS AND EXTREMES

(Table Revised 1975. Base Period for Climatological Normals: 1941-1970)

Month	Temperature °F Normal Daily maximum	Daily minimum	Monthly	Extremes Record highest	Year	Record lowest	Year	Normal Degree days Base 65°F Heating	Cooling	Precipitation Normal
J	53.5	29.4	41.5	87	1969	-5	1966	729	0	1.07
F	58.1	33.6	45.9	88	1962	-5	1971	535	0	1.16
M	65.8	39.2	52.5	102	1971	9	1965	409	22	1.62
A	76.4	51.1	63.8	102	1972	27	1970	112	91	3.16
M	83.5	59.8	71.7	104	1967	40	1973	13	239	4.58
J	93.9	68.6	81.3	105	1972	51	1964	0	489	3.39
J	99.2	72.3	85.8	111	1964	54	1970	0	645	2.16
A	99.3	71.6	85.5	113	1964	54	1962	0	636	1.77
S	90.4	63.6	77.0	107	1965	41	1970	0	360	3.00
O	79.3	52.7	66.0	102	1972	32	1970	92	123	2.68
N	66.0	39.7	52.9	89	1966	17	1968	369	0	1.35
D	56.2	32.2	44.2	87	1966	-5	1968	645	0	1.28
YR	77.0	51.2	64.1	113	AUG. 1964	-5	JAN. 1966	2904	2611	27.22

Means and extremes above are from existing and comparable exposures. Annual extremes have been exceeded at other sites in the locality as follows: Lowest temperature -12 in January 1947.

REFERENCE NOTES APPLYING TO TABLES APPEAR ON THE PAGE FOLLOWING LAST TABLE.

(Caution: Letters and symbols may have different meanings in 1941-1970 tables than in earlier tables. See notes.)

Reference notes applying to Normals, Means, and Extremes tables for 1931–1960 base period.

(a) Length of record, years.
(b) Climatological standard normals (1931-1960).
*   Less than one half.
+   Also on earlier dates, months or years.
T   Trace, an amount too small to measure.
     Below-zero temperatures are preceded by a minus sign.
     The prevailing direction for wind in the Normals, Means, and Extremes table is from records through 1963.

To 8 compass points only.

Unless otherwise indicated, dimensional units used in this bulletin are: temperature in degrees F.; precipitation, including snowfall, in inches; wind movement in miles per hour; and relative humidity in percent. Degree day totals are the sums of the negative departures of average daily temperatures from 65° F. Sleet was included in snowfall totals beginning with July 1948. Heavy fog reduces visibility to 1/4 mile or less.

Sky cover is expressed in a range of 0 for no clouds or obscuring phenomena to 10 for complete sky cover. The number of clear days is based on average cloudiness 0-3; partly cloudy days 4-7; and cloudy days 8-10 tenths.

Solar radiation data are the averages of direct and diffuse radiation on a horizontal surface. The langley denotes one gram calorie per square centimeter. Averages in the lower table for some months may be for more than the listed number of years.

&   Figures instead of letters in a direction column indicate direction in tens of degrees from true North; i. e., 09 - East, 18 - South, 27 - West, 36 - North, and 00 - Calm. Resultant wind is the vector sum of wind directions and speeds divided by the number of observations. If figures appear in the direction column under "Fastest mile" the corresponding speeds are fastest observed 1 - minute values.

Reference notes applying to Normals, Means, and Extremes tables for 1941–1970 base period.

(a) Length of record, years, through the current year unless otherwise noted, based on January data.
(b) 70° and above at Alaskan stations.
*   Less than one half.
T   Trace.

NORMALS - Based on record for the 1941-1970 period.
DATE OF AN EXTREME - The most recent in cases of multiple occurrence.

PREVAILING WIND DIRECTION - Record through 1963.
WIND DIRECTION - Numerals indicate tens of degrees clockwise from true north. 00 indicates calm.

FASTEST MILE WIND - Speed is fastest observed 1-minute value when the direction is in tens of degrees.

%   The station did not operate 24 hours daily during a part of its history. Fog and thunderstorm data may be incomplete.

Mean Maximum Temperature (°F.), January

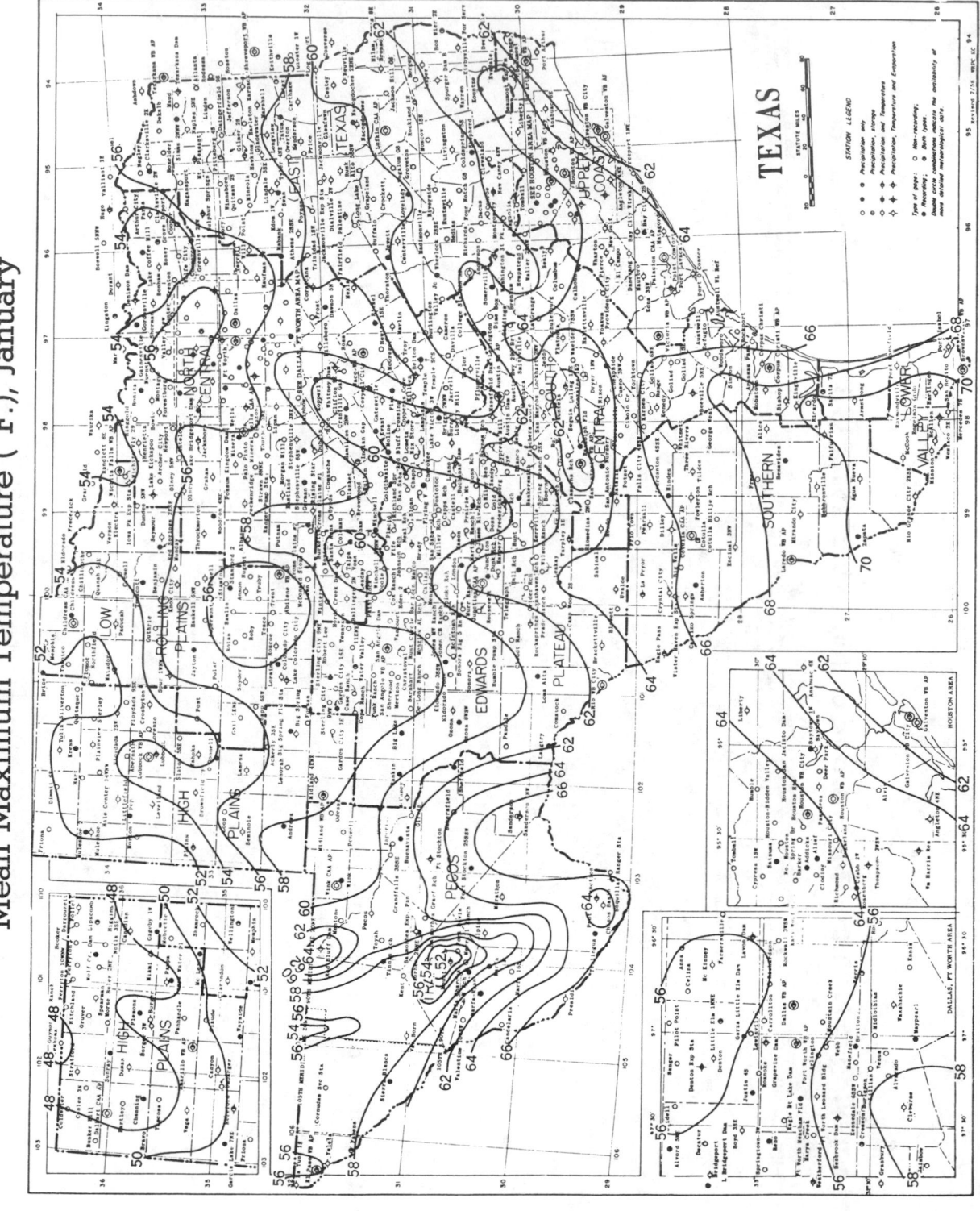

TEXAS

STATUTE MILES

STATION LEGEND

Based on period 1931-52

Isolines are drawn through points of approximately equal value. Caution should be used in interpolating on these maps, particularly in mountainous areas.

Mean Minimum Temperature (°F.), January

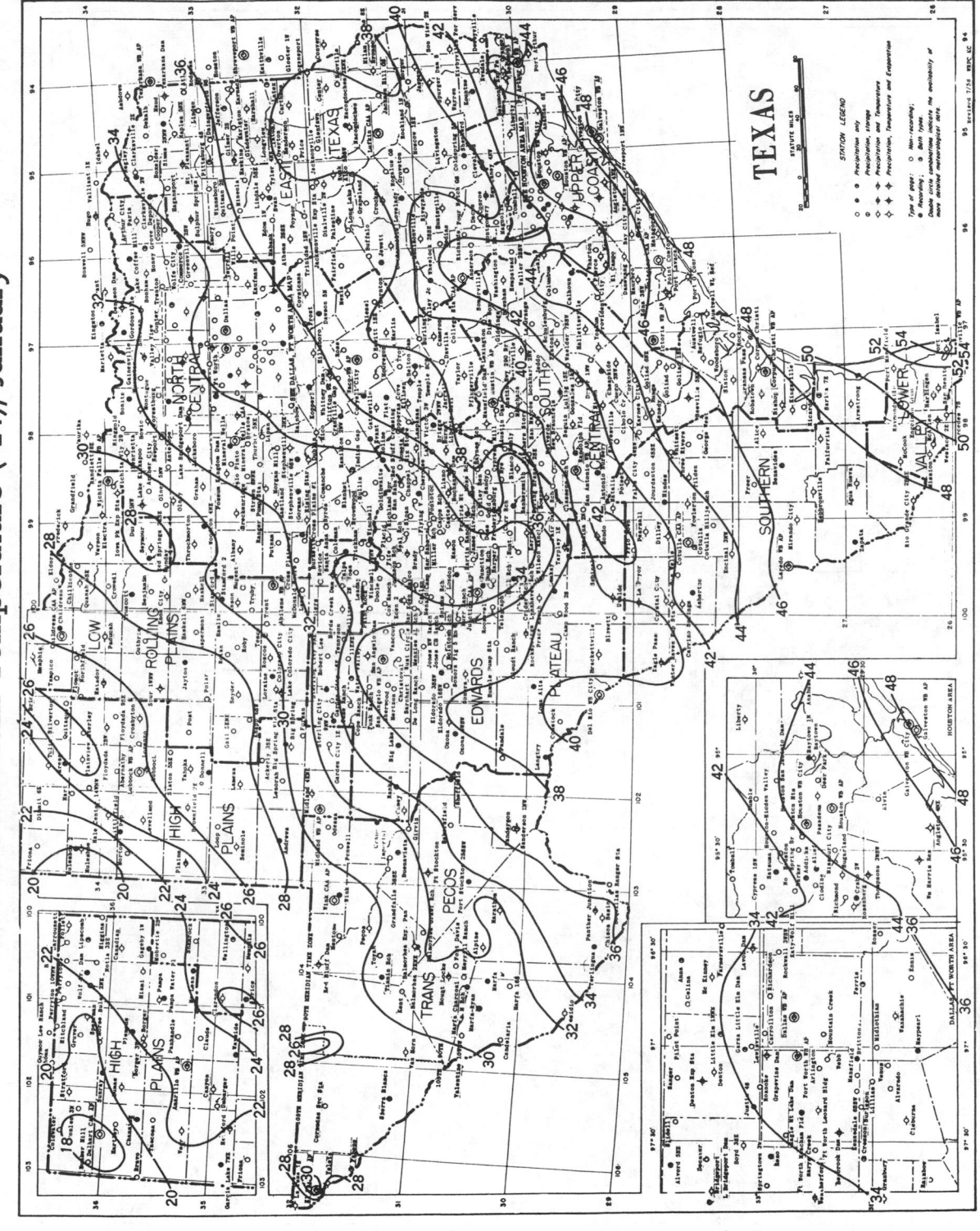

TEXAS

STATUTE MILES

STATION LEGEND

Based on period 1931-52

Isolines are drawn through points of approximately equal value. Caution should be used in interpolating on these maps, particularly in mountainous areas.

979

Mean Maximum Temperature (°F.), July

TEXAS

STATION LEGEND

Based on period 1931-52

Isolines are drawn through points of approximately equal value. Caution should be used in interpolating on these maps, particularly in mountainous areas.

Mean Minimum Temperature (°F.), July

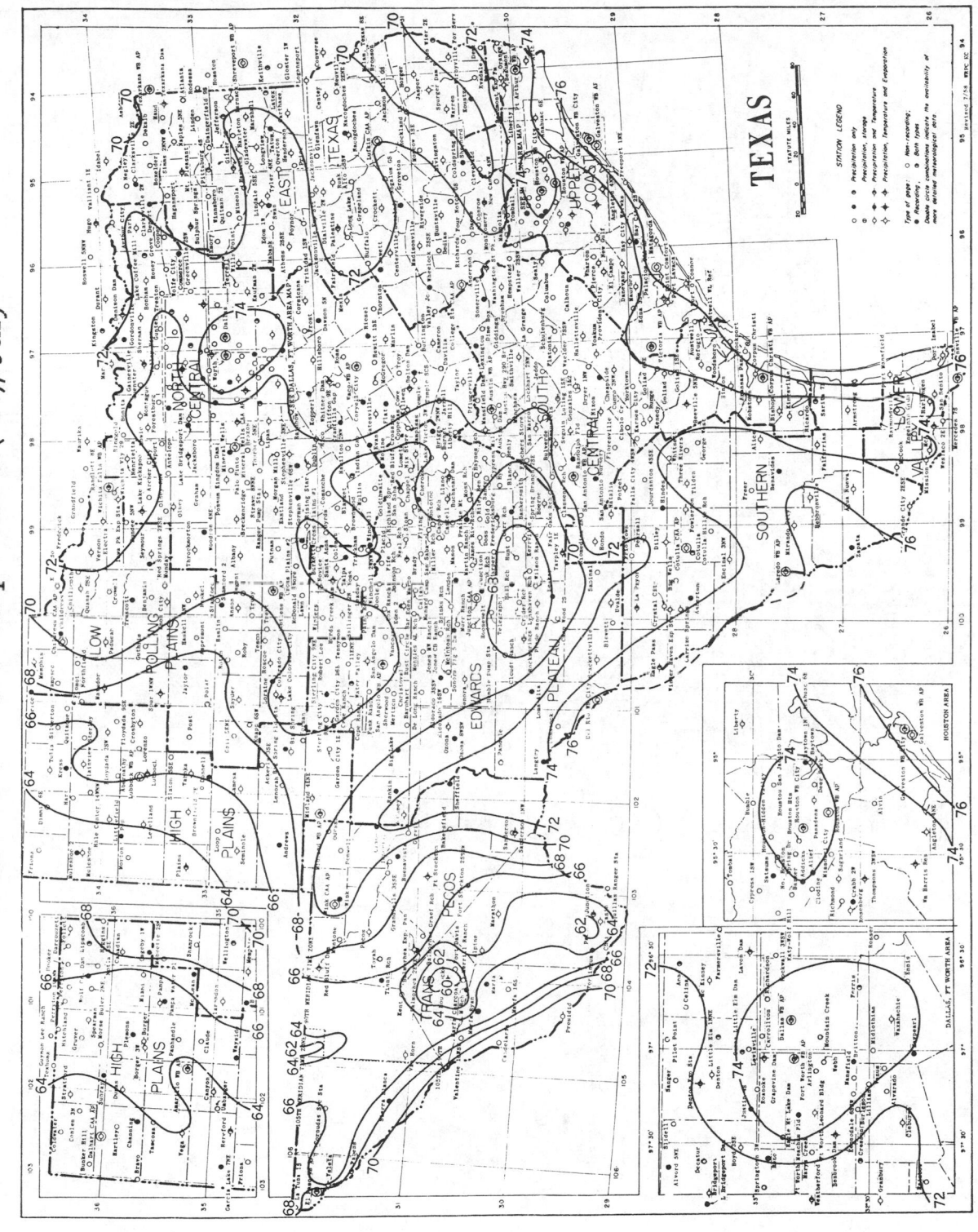

TEXAS

Based on period 1931-52

Isolines are drawn through points of approximately equal value. Caution should be used in interpolating on these maps, particularly in mountainous areas.

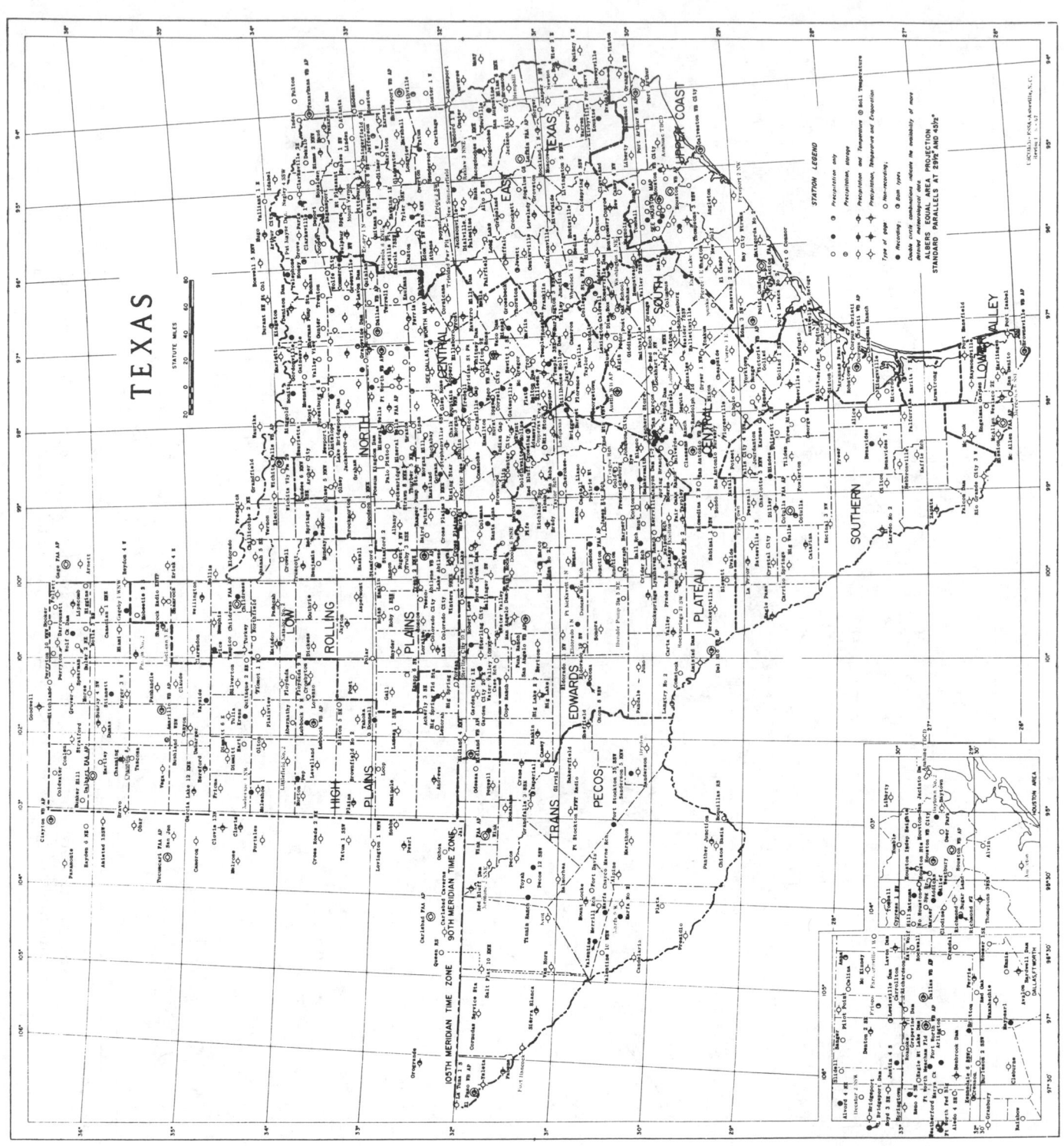

TEXAS

STATUTE MILES

CLIMATES OF THE STATES

UTAH

(Normals, Means and Extremes tables revised 1972 and 1975. Basic report revised February 1960.)

Climate of Utah

Merle Brown, Weather Bureau State Climatologist

The topography of Utah is extremely varied, with most of the State being mountainous. A series of mountains (including the Wasatch Range), which run generally north and south through the middle of Utah, and the Uinta Mountains, which extend east and west through the northeast portion, are the principal ranges. Crest lines of these mountains are mostly above 10,000 feet above sea level. Less extensive ranges are scattered over the remainder of the State. The lowest area is the Virgin River Valley in the extreme southwestern part, its elevation being between 2,500 and 3,500 feet above sea level, and the highest point is Kings Peak in the Uinta Mountains, which rises to 13,498 feet above sea level.

Practically all of eastern Utah is drained by the Colorado River and its principal tributary within the State, the Green River, although neither rises within its borders. Western Utah is almost entirely within the Great Basin, without outlet to the sea. The largest rivers in this area are the Bear, Weber, Jordan, Provo, and Sevier, the first three of which empty into Great Salt Lake. The Sevier River drains the west-central area and empties into Sevier Lake, a brackish saline basin of southwestern Utah.

The main streams in the eastern portion of the State flow through canyons or very narrow, confined mountain valleys and finally into desert canyons. Some narrow meadows, usually in native grass, and only a few small local areas of farmland are subject to overflow. Nearly all the main highways and railroads, as well as residential areas are above flood levels. Highest flow occurs in the streams in this region in May and June during spring runoff from melting snow.

The most serious floods in Utah have occurred in the Great Lake Basin, particularly in the Weber River drainage on the western slopes of the Wasatch Mountains. During the past 100 years approximately 300 flash-floods, resulting from high intensity rainfall accompanying thunderstorms, and 135 snowmelt floods have been recorded. Some have been very limited in area and extent of damage, while others have been highly destructive in cities and towns and agricultural areas. However, severe floods are not likely to occur in any given locality more than once in several years, or even several decades.

Great Salt Lake, in northwestern Utah, occupies the bottom of the Great Basin, the largest closed basin in North America. The flatter part of this drainage area is below 4,500 feet in elevation, the Lake itself being about 4,200 feet above sea level. Great Salt Lake, which has an average maximum length of about 75 miles and an average maximum width of 50 miles, is the largest lake at its elevation (or higher) in the World[1]. In glacial times it was a fresh water lake occupying an area 346 miles long and 145 miles wide; but due to increased evaporation and/or reduced precipitation, it gradually shrunk in size and the salinity increased until it finally became the Great Salt Lake of our time[1]. Since this large body of water now has no drainage outlet, the salt content is high, averaging about 25 percent. Thus, the Lake, which never freezes over, provides a moderating effect throughout the year on temperatures in its immediate vicinity.

Essentially, Utah's climate is determined by its distance from the equator; its elevation above sea level; the location of the State with respect to the average storm paths over the Intermountain Region; and its distance from the principal moisture sources of the area, namely, the Pacific Ocean and the Gulf of Mexico. Also, the mountain ranges over the western United States, particularly the Sierra Nevada and Cascade Ranges and the Rocky Mountains, have a marked influence on the climate of the State. Pacific storms, before reaching Utah, must first cross the Sierras or Cascades. As the moist air is forced to rise over these high mountains, a large portion of the original moisture falls as precipitation. Thus, the prevailing westerly air currents reaching Utah are comparatively dry, resulting in light precipitation over most of the State.

Precipitation varies greatly, from an average of less than 5 inches annually over the Great Salt Lake Desert (west of Great Salt Lake), to more than 40 inches in some parts of the Wasatch Mountains. The average annual precipitation in the leading agricultural areas is between 10 and 15 inches, necessitating the practice of irrigation for the economic production of most crops. However, the mountains, where winter snows form the chief reservoirs of moisture, are conveniently adjacent to practically all farming areas, and there is usually sufficient water for most lands under irrigation. The areas of the State below an elevation of 4,000 feet, all in the southern part, receive generally less than 10 inches of moisture annually.

Northwestern Utah, over and along the mountains, receives appreciably more precipitation in a year than is received at similar elevations over the rest of the State, primarily due to terrain and the direction of normal storm tracks. The bulk of the moisture occurring over that area can be attributed to the movement of Pacific storms through the region during the winter and spring months. In summer northwestern Utah is comparatively dry. The eastern portion receives appreciable rain from summer thundershowers, which are usually associated with moisture-laden air masses moving in from the Gulf of Mexico.

Snowfall is moderately heavy in the mountains, especially over the northern part. This is conducive to a large amount of winter sports activity, including skiing and hunting. While the principal population centers along the base of the mountains receive more snow, as a rule, than many middle and northeastern sections of the United States, a deep snow cover seldom remains long on the ground.

Runoff from melting mountain snow usually reaches its peak in April, May, or early June, and sometimes causes flooding along the lower streams. However, damaging floods of this kind are infrequent. Flash floods from summer thunderstorms are more frequent, but they affect only small, local areas.

There are definite variations in temperature with altitude and with latitude. Naturally, the mountains and the elevated valleys have the cooler climates, with the lower areas of the State having the higher temperatures. There is about a 3°F. decrease in mean annual temperature for each 1,000 feet increase in altitude, and approximately 1.5° to 2.0°F. decrease in average yearly temperature for each 1° increase in latitude. Thus, weather stations in the southern tier of counties generally have average annual temperatures 6° to 8° higher than those at similar altitudes over the extreme northern counties.

Temperatures above 100°F. occur occasionally in summer in nearly all parts of the State. However, low humidity makes these high temperatures more bearable in Utah than in more humid regions. During the warmer season of the year, air conditioning is used in a large percent of the commercial establishments over the State, but only a small portion of the family dwellings are air conditioned. Due to the dryness of the air, evaporative coolers operate very efficiently in Utah's climate. The highest temperature of record in the State is 116°F. observed June 28, 1892, at Saint George.

Temperatures below zero during winter and early spring are uncommon in most areas of the State, and prolonged periods of extremely cold weather are rare. This is primarily due to the mountains east and north of the State, which act as a barrier to intensely cold continental arctic air masses. The lowest temperature of record is -50°F.

Utah experiences relatively strong insolation during the day and rapid nocturnal cooling, resulting in wide daily ranges in temperature. At Richfield, for example, the average annual daily variation from high to low is 35°F.; and on individual days a range of 50° to 60°F. is not uncommon. Even after the hottest days, nights are usually cool over the State.

On clear nights the colder air usually accumulates, by drainage, on the valley bottoms, while the foothills and bench areas remain relatively warm. For this reason, the higher lands at the edges of the valleys are devoted ordinarily to the more valuable and delicate fruits, berries, and vegetables, while the hardier grains and vegetables are planted in the bottom lands.

Owing to the varied topography of the State, there are no orderly nor extensive zones of equal length of growing season between the last freeze in spring and the first in fall. There are, however, from 4 1/2 to 5 months of freeze-free growing weather in the State's principal agricultural areas. A difference of 2 weeks in the growing season is often noted in the same valley between the bottom lands and the adjacent farming lands at the foot of the mountains.

Sunny skies prevail most of the year in Utah. For example, there is an average of about 65 to 75 percent of the possible amount of sunshine at Salt Lake City during spring, summer, and fall. In winter Salt Lake City has about 50 percent of the possible sunshine.

During the late fall and winter months, anticyclones tend to settle over the Great Basin area for as long as several weeks at a time. Under these conditions, smoke and haze accumulate in the lower levels of the stagnant air over the valleys of northwestern Utah, frequently becoming an obstruction to visibility. This is also true of fog which may persist for 2 to 3 weeks at a time.

Wind speeds are usually light to moderate, ranging normally below 20 miles per hour. There were only eight known tornadoes in Utah during the period 1916-1958, and those that were reported caused only slight damage. However, strong winds occur occasionally, sometimes attaining damaging proportions in local areas, particularly in the vicinity of the canyon mouths along the western slopes of the Wasatch Mountains. Duststorms occur occasionally, principally over western Utah. These storms are usually associated with the movement of low pressure disturbances through the area during the spring months.

Hailstorms may damage fruit and vegetables in limited areas during spring and summer, although the hail is usually small.

Utah is not a large agricultural state, even though appreciable crops, livestock, and dairy products are produced within its boundaries. Only 4 percent of the land is under cultivation, but approximately 85 percent of the land area is utilized for livestock grazing purposes. Livestock represent the largest portion of cash farm income in the State. The largest crop is wheat, most of it being "winter" or "dryland" wheat. Other principal crops are barley, oats, hay, potatoes, corn, and sugarbeets. Lesser crops include other grains, fruits, vegetables, berries, melons, dry beans, and alfalfa and sugarbeets for seed. Range feeds and dryland crops in nonirrigable areas, particularly in the southern portion, often suffer from lack of moisture.

Mining and manufacturing are the two other basic industries in Utah. The State ranks high in the quantity and value of minerals it produces each year, mainly copper, lead, zinc, gold, and silver. Because of the dry climate, several companies have found it economically feasible to produce salt from the brine of the Great Salt Lake by the evaporation process.

Salt Lake City is the commercial, industrial, and financial center of Utah. Three-fourths of the State's population is concentrated within a 100-mile radius of that city, and well over one-half the people reside within 50 miles of Salt Lake City.

Tourists come to Utah primarily to visit historic Salt Lake City; to see the Great Salt Lake; to tour the park areas, including Zion National Park, Cedar Breaks National Monument, and Bryce Canyon National Park; and to fish in the cool mountain streams. Persons traveling in the State during the winter and early spring months should be prepared for cold weather and snow. When crossing the less-frequently traveled desert areas of the western portion, motorists should carry a supply of fresh water as a safeguard, particularly during the summer months.

REFERENCES

(1) R. A. Hart, "The Great Salt Lake", (unpublished)

(2) Elroy Nelson, "Agriculture in Utah's Economy", Chap. 3 of Utah's Economic Patterns, University of Utah Press, Salt Lake City, Utah, 1956, pp. 23-32 (p. 32).

(3) Weather Bureau Technical Paper No. 15 - Maximum Station Precipitation for 1,2,3,6,12 and 24 Hours.

(4) Weather Bureau Technical Paper No. 16 - Maximum 24-Hour Precipitation in the United States. Washington, D. C. 1952.

(5) Weather Bureau Technical Paper No. 25 - Rainfall Intensity-Duration-Frequency Curves. For selected stations in the United States, Alaska, Hawaiian Islands, and Puerto Rico.

BIBLIOGRAPHY

(A) Climatic Summary of the United States (Bulletin W) 1930 edition, Sections 20 and 21. U. S. Weather Bureau

(B) Climatic Summary of the United States, Utah - Supplement for 1931 through 1952 (Bulletin W Supplement). U. S. Weather Bureau

(C) Climatological Data - Utah. U. S. Weather Bureau

(D) Climatological Data National Summary. U. S. Weather Bureau

(E) Hourly Precipitation Data - Utah. U. S. Weather Bureau

(F) Local Climatological Data, U. S. Weather Bureau, for Milford and Salt Lake City, Utah.

STATION	Freeze threshold temperature	Mean date of last Spring occurrence	Mean date of first Fall occurrence	Mean No. of days between dates	Years of record Spring	No. of occurrences in Spring	Years of record Fall	No. of occurrences in Fall
ALTON	32	06-05	09-25	112	30	30	30	30
	28	05-20	10-07	140	30	30	30	30
	24	04-30	10-21	174	30	30	30	30
	20	04-19	11-08	203	30	30	30	30
	16	04-05	11-15	224	30	30	30	30
ALUNITE	32	05-28	09-27	122	26	26	27	27
	28	05-11	10-08	152	26	26	27	27
	24	04-24	10-22	182	26	26	27	27
	20	04-07	11-01	208	26	26	27	27
	16	03-23	11-11	232	26	26	27	27
BEAVER	32	06-04	09-18	106	28	28	26	26
	28	05-11	09-30	142	27	27	26	26
	24	04-30	10-12	166	27	27	25	25
	20	04-10	10-27	200	27	27	25	25
	16	03-28	11-08	225	27	27	25	25
BLANDING	32	05-18	10-14	148	28	28	28	28
	28	04-30	10-24	178	28	28	28	28
	24	04-09	11-04	208	28	28	28	28
	20	03-29	11-12	228	28	28	28	28
	16	03-16	11-19	248	27	27	27	27
BLUFF	32	04-17	10-25	190	26	26	23	23
	28	04-01	11-05	218	26	26	23	23
	24	03-20	11-12	237	26	26	23	23
	20	02-23	11-24	274	26	25	23	23
	16	02-03	12-04	304	26	23	22	22
CEDAR CITY PH	32	05-15	10-05	143	28	28	29	29
	28	04-28	10-14	169	28	28	29	29
	24	04-17	10-29	195	28	28	28	28
	20	03-30	11-06	221	28	28	27	27
	16	03-14	11-18	248	28	28	27	27
CORINNE	32	05-13	10-02	142	25	25	24	24
	28	04-24	10-18	177	25	25	24	24
	24	04-03	10-30	211	25	25	24	24
	20	03-24	11-09	230	24	24	24	24
	16	03-11	11-21	255	24	24	23	22
DESERET	32	05-21	09-19	121	30	30	26	26
	28	05-04	09-29	148	30	30	26	26
	24	04-24	10-14	173	30	30	26	26
	20	04-07	10-26	202	30	30	26	26
	16	03-25	11-07	226	30	30	26	26
DUCHESNE	32	05-28	09-20	115	29	29	29	29
	28	05-15	09-29	137	29	29	29	29
	24	04-27	10-10	167	29	29	28	28
	20	04-15	10-25	193	29	29	28	28
	16	04-01	11-03	215	29	29	28	28
ELBERTA	32	05-19	09-23	128	26	26	26	26
	28	04-30	10-10	164	26	26	26	26
	24	04-13	10-23	193	26	26	26	26
	20	04-02	11-06	219	26	26	26	26
	16	03-17	11-17	245	25	25	26	26
EMERY	32	05-21	09-30	132	30	30	29	29
	28	04-30	10-13	166	30	30	29	29
	24	04-21	10-25	187	30	30	29	29
	20	04-06	11-04	212	30	30	29	29
	16	03-22	11-13	236	30	30	29	29

STATION	Freeze threshold temperature	Mean date of last Spring occurrence	Mean date of first Fall occurrence	Mean No. of days between dates	Years of record Spring	No. of occurrences in Spring	Years of record Fall	No. of occurrences in Fall
ESCALANTE	32	05-15	09-30	138	27	27	25	25
	28	04-27	10-17	172	27	27	25	25
	24	04-14	10-26	195	27	27	25	25
	20	03-28	11-06	222	27	27	24	24
	16	03-04	11-14	255	26	26	24	24
FARMINGTON	32	04-29	10-14	168	30	30	26	26
	28	04-11	10-24	196	29	29	26	26
	24	03-30	11-07	222	29	29	26	26
	20	03-16	11-19	248	29	29	26	26
	16	03-01	11-28	272	29	29	26	25
FILLMORE	32	05-08	10-10	156	29	29	25	25
	28	04-21	10-20	182	29	29	26	26
	24	04-05	10-31	209	29	29	26	26
	20	03-24	11-12	233	29	29	26	26
	16	03-11	11-22	256	29	29	26	26
FT DUCHESNE	32	05-23	09-24	124	21	21	23	23
	28	05-07	10-07	153	20	20	23	23
	24	04-24	10-17	176	20	20	23	23
	20	04-12	10-29	200	20	20	23	23
	16	03-30	11-08	223	20	20	23	23
GOVERNMENT CREEK	32	05-24	09-24	124	26	26	21	21
	28	05-05	10-12	160	26	26	21	21
	24	04-20	10-26	190	26	26	21	21
	20	04-08	11-08	214	26	26	20	20
	16	03-25	11-17	237	25	25	20	20
GREENRIVER	32	05-03	10-08	158	30	30	30	30
	28	04-22	10-17	178	30	30	30	30
	24	04-06	10-28	205	30	30	29	29
	20	03-29	11-04	220	29	29	29	29
	16	03-17	11-14	243	29	29	29	29
HANKSVILLE CAA AP	32	05-01	10-04	156	27	27	24	24
	28	04-17	10-16	182	27	27	24	24
	24	04-09	10-27	201	25	25	24	24
	20	04-01	11-05	218	25	25	24	24
	16	03-20	11-14	239	25	25	24	24
HEBER	32	06-13	08-29	77	29	29	24	24
	28	05-20	09-16	119	29	29	24	24
	24	05-02	10-03	154	28	28	24	24
	20	04-19	10-19	183	28	28	24	24
	16	04-05	10-30	208	28	28	24	24
IBAPAH	32	06-15	08-28	74	23	23	23	23
	28	06-02	09-15	105	23	23	23	23
	24	05-14	09-27	136	22	22	22	22
	20	04-21	10-07	169	22	22	22	22
	16	04-08	10-23	198	22	22	22	22
KANAB PH	32	05-09	10-19	163	30	30	26	26
	28	04-18	10-30	195	29	29	26	26
	24	04-02	11-08	220	29	29	26	26
	20	03-13	11-22	254	29	29	26	26
	16	02-23	12-02	282	29	28	25	23
LAKETOWN	32	06-13	09-07	87	30	30	27	27
	28	05-25	09-24	122	30	30	27	27
	24	05-04	10-06	156	30	30	27	27
	20	04-19	10-21	185	30	30	27	27
	16	04-03	11-04	215	30	30	27	27

FREEZE DATA

STATION	Freeze threshold temperature	Mean date of last Spring occurrence	Mean date of first Fall occurrence	Mean No. of days between dates	Years of record Spring	No. of occurrences in Spring	Years of record Fall	No. of occurrences in Fall
LEVAN	32	05-17	10-01	137	30	30	27	27
	28	04-30	10-14	168	30	30	27	27
	24	04-12	10-30	201	30	30	27	27
	20	03-31	11-09	223	30	30	27	27
	16	03-13	11-22	253	30	30	27	26
LOA	32	06-13	09-08	87	26	26	27	27
	28	06-02	09-18	108	25	25	27	27
	24	05-12	10-01	142	24	24	27	27
	20	04-28	10-08	163	24	24	27	27
	16	04-15	10-22	190	24	24	26	26
LOGAN USAC	32	05-03	10-15	165	30	30	25	25
	28	04-12	10-31	202	30	30	25	25
	24	04-01	11-07	220	30	30	25	25
	20	03-22	11-19	242	30	30	25	25
	16	03-11	11-28	262	30	30	25	24
LUCIN AP	32	05-14	09-23	133	18	18	13	13
	28	05-01	10-05	157	18	18	12	12
	24	04-21	10-18	180	18	18	12	12
	20	04-08	10-24	199	17	17	11	11
	16	03-21	11-18	243	17	17	11	11
MANTI	32	05-19	09-25	129	28	28	26	26
	28	05-06	10-11	158	28	28	26	26
	24	04-16	10-27	195	28	28	26	26
	20	04-04	11-06	217	28	28	26	26
	16	03-20	11-15	240	28	28	26	26
MIDVALE	32	05-11	10-01	144	30	30	27	27
	28	04-19	10-19	183	30	30	27	27
	24	04-02	10-31	211	30	30	27	27
	20	03-21	11-09	233	30	30	27	27
	16	03-10	11-23	258	30	30	27	27
MILFORD WB AP	32	05-21	09-26	128	29	29	28	28
	28	05-01	10-06	158	29	29	28	28
	24	04-18	10-18	182	29	29	28	28
	20	04-04	10-29	208	29	29	28	28
	16	03-22	11-06	229	29	29	28	28
MOAB	32	04-18	10-18	183	29	29	30	30
	28	04-03	10-26	206	29	29	30	30
	24	03-27	11-08	226	29	29	30	30
	20	03-16	11-15	245	29	29	30	30
	16	02-19	11-30	285	28	27	29	28
MODENA	32	05-25	09-30	128	30	30	30	30
	28	05-06	10-10	157	30	30	30	30
	24	04-21	10-23	186	30	30	30	30
	20	04-07	11-01	208	30	30	30	30
	16	03-23	11-13	235	30	30	30	30
MONTICELLO	32	05-23	10-08	138	22	22	22	22
	28	05-04	10-20	168	21	21	22	22
	24	04-21	10-27	189	21	21	22	22
	20	04-10	11-06	210	21	21	22	22
	16	03-29	11-14	229	20	20	22	22
MORGAN	32	06-06	09-06	91	29	29	27	27
	28	05-17	09-19	125	29	29	27	27
	24	04-26	10-05	162	29	29	27	27
	20	04-11	10-17	189	29	29	27	27
	16	03-31	11-03	218	29	29	27	27
MORONI	32	05-26	09-18	115	26	26	25	25
	28	05-11	10-05	147	25	25	24	24
	24	04-22	10-17	178	25	25	24	24
	20	04-06	10-27	204	25	25	24	24
	16	03-23	11-10	232	25	25	24	24
MYTON	32	05-20	09-25	128	19	19	19	19
	28	05-07	10-12	158	19	19	18	18
	24	04-16	10-22	189	18	18	17	17
	20	04-08	11-02	207	18	18	16	16
	16	03-29	11-09	225	18	18	16	16
OAK CITY	32	05-14	10-07	146	25	25	25	25
	28	04-25	10-19	178	25	25	25	25
	24	04-09	10-31	205	25	25	25	25
	20	03-28	11-11	228	25	25	25	25
	16	03-14	11-23	254	25	25	25	25
OGDEN SUGAR FACTORY	32	05-07	10-14	159	29	29	29	29
	28	04-18	10-25	189	29	29	29	29
	24	04-02	11-05	216	29	29	29	29
	20	03-18	11-15	242	29	29	29	29
	16	03-06	11-25	264	29	29	28	28
PANGUITCH	32	06-17	08-23	67	27	27	29	29
	28	06-08	09-14	98	27	27	29	29
	24	05-23	09-26	127	26	26	29	29
	20	05-06	10-09	157	26	26	28	28
	16	04-16	10-24	190	25	25	28	28
PARK CITY	32	06-06	09-14	100	30	30	30	30
	28	05-20	10-01	133	30	30	30	30
	24	05-02	10-11	162	30	30	30	30
	20	04-24	10-23	182	29	29	30	30
	16	04-07	11-08	215	27	27	30	30
PARK VALLEY	32	05-21	10-02	135	20	20	20	20
	28	05-06	10-14	162	20	20	19	19
	24	04-26	10-22	179	20	20	19	19
	20	04-09	11-03	208	20	20	19	19
	16	03-20	11-10	236	20	20	19	19
PAROWAN	32	05-19	10-03	137	25	25	24	24
	28	05-03	10-13	163	25	25	24	24
	24	04-19	10-26	190	25	25	24	24
	20	04-02	11-05	217	25	25	24	24
	16	03-18	11-15	242	25	25	24	24
PIUTE DAM	32	05-24	09-27	126	27	27	27	27
	28	05-08	10-05	150	27	27	27	27
	24	04-23	10-19	179	27	27	23	23
	20	04-07	10-27	202	27	27	22	22
	16	03-21	11-12	236	27	27	19	19
PRICE	32	05-08	10-03	148	29	29	29	29
	28	04-27	10-16	172	29	29	29	29
	24	04-11	10-25	197	29	29	28	28
	20	03-29	11-07	223	29	29	28	28
	16	03-18	11-16	244	29	29	27	27
PROVO AIRPORT	32	05-20	09-23	126	30	30	29	29
	28	04-28	10-10	166	30	30	29	29
	24	04-06	10-26	204	30	30	28	28
	20	03-21	11-08	232	29	29	28	28
	16	03-08	11-19	256	29	29	27	27

FREEZE DATA

STATION	Freeze threshold temperature	Mean date of last Spring occurrence	Mean date of first Fall occurrence	Mean No. of days between dates	Years of record Spring	No. of occurrences in Spring	Years of record Fall	No. of occurrences in Fall
RICHFIELD RDO KSVC	32	05-20	09-26	129	25	25	27	27
	28	05-06	10-05	152	25	25	27	27
	24	04-19	10-21	185	25	25	27	27
	20	04-02	10-30	211	25	25	27	27
	16	03-23	11-09	231	25	25	27	27
RIVERDALE PH	32	05-07	10-09	155	19	19	18	18
	28	04-23	10-22	183	19	19	18	18
	24	03-31	11-09	223	19	19	18	18
	20	03-21	11-20	245	19	19	18	18
	16	03-03	11-25	267	19	19	18	17
ST GEORGE CAA AP	32	04-03	10-30	210	28	28	28	28
	28	03-20	11-11	236	28	28	28	28
	24	02-20	11-20	272	28	28	28	28
	20	02-03	12-06	306	28	26	28	26
	16	01-15	12-23	343	28	16	28	12
SALT LAKE CITY WB AP	32	04-12	11-01	202	30	30	30	30
	28	03-26	11-13	232	30	30	30	30
	24	03-14	11-23	254	30	30	30	30
	20	02-25	11-29	276	30	30	30	29
	16	02-08	12-16	311	30	29	30	21
SCIPIO	32	06-03	09-17	106	27	27	27	27
	28	05-18	09-25	131	27	27	27	27
	24	04-30	10-06	159	27	27	27	27
	20	04-19	10-18	183	27	27	27	27
	16	03-30	11-05	220	27	27	27	27
SNAKE CREEK P H	32	06-10	09-05	86	27	27	27	27
	28	05-25	09-17	115	27	27	27	27
	24	05-06	10-08	156	27	27	27	27
	20	04-18	10-19	184	27	27	27	27
	16	04-06	11-03	211	27	27	27	27
SOLDIER SUMMIT	32	06-20	08-14	55	27	27	28	28
	28	06-12	09-08	89	27	27	28	28
	24	05-22	09-27	128	27	27	28	28
	20	05-05	10-14	162	27	27	27	27
	16	04-16	10-26	193	27	27	27	27
SPANISH FORK 1 S	32	04-28	10-13	168	27	27	27	27
	28	04-09	10-31	206	27	27	27	27
	24	03-31	11-13	227	27	27	27	27
	20	03-21	11-19	243	27	27	27	27
	16	03-05	11-27	267	27	27	27	26
THOMPSONS	32	04-23	10-22	181	28	28	26	26
	28	04-06	10-30	207	27	27	26	26
	24	04-01	11-09	222	27	27	26	26
	20	03-15	11-20	250	27	27	26	25
	16	02-28	12-01	276	27	27	26	25
TOOELE	32	04-28	10-15	169	29	29	29	29
	28	04-08	11-03	209	29	29	29	29
	24	03-28	11-13	230	29	29	29	29
	20	03-17	11-20	248	29	29	29	29
	16	03-03	11-27	269	29	29	28	28
UTAH LAKE LEHI	32	05-13	09-27	137	21	21	21	21
	28	04-24	10-12	171	21	21	21	21
	24	04-09	10-24	198	20	20	21	21
	20	03-26	11-06	226	20	20	21	21
	16	03-02	11-21	265	20	20	21	21
VERNAL RS	32	05-29	09-25	118	22	22	22	22
	28	05-11	10-04	146	22	22	22	22
	24	04-25	10-19	177	22	22	22	22
	20	04-12	10-27	198	22	22	22	22
	16	04-03	11-06	218	22	22	22	21
WENDOVER CAA AP	32	04-17	10-24	190	26	26	25	25
	28	04-04	11-07	217	26	26	25	25
	24	03-26	11-13	232	26	26	25	25
	20	03-06	11-21	260	26	26	25	25
	16	02-10	12-04	297	26	26	25	25

Data in the above table are based on the period 1921-1950, or that portion of this period for which data are available.

Means have been adjusted to take into account years of non-occurrence.

A freeze is a numerical substitute for the former term "killing frost" and is the occurrence of a minimum temperature at or below the threshold temperature of 32°, 28°, etc.

Freeze data tabulations in greater detail are available and can be reproduced at cost.

*MEAN TEMPERATURE AND PRECIPITATION

STATION	JANUARY		FEBRUARY		MARCH		APRIL		MAY		JUNE		JULY		AUGUST		SEPTEMBER		OCTOBER		NOVEMBER		DECEMBER		ANNUAL	
	Temperature	Precipitation	Temperature	Precipitation	Temperature	Precipitation	Temperature	Precipitation	Temperature	Precipitation	Temperature	Precipitation	Temperature	Precipitation	Temperature	Precipitation	Temperature	Precipitation	Temperature	Precipitation	Temperature	Precipitation	Temperature	Precipitation	Temperature	Precipitation
WESTERN																										
DESERET	25.4	.42	31.6	.40	39.5	.63	48.7	.79	56.9	.75	65.0	.47	74.1	.47	72.3	.54	63.2	.34	51.4	.64	37.1	.45	29.3	.59	49.5	6.49
MILFORD WB AP	23.8	.57	30.9	.71	39.2	1.06	47.8	.76	56.8	.74	65.8	.45	74.0	.77	71.7	.81	62.6	.40	50.1	.88	37.4	.54	28.0	.75	49.0	8.44
MODENA	25.8	.81	31.1	.74	38.3	.98	48.8	.73	55.2	.62	63.4	.44	71.9	.98	69.9	1.46	62.2	.64	50.3	1.03	37.5	.55	29.4	.87	48.7	9.85
PARK VALLEY	23.9	1.05	27.4	.86	34.9	.75	44.8	1.00	53.5	.98	61.2	.90	72.1	.93	70.5	.78	61.2	.65	49.6	.64	35.2	.84	27.5	1.05	46.8	10.43
WENDOVER CAA AP	27.1	.31	32.7	.30	41.7	.38	52.2	.58	61.7	.58	70.1	.49	80.0	.34	77.8	.40	66.8	.35	53.5	.51	38.1	.27	30.3	.31	52.7	4.82
DIVISION	25.7	.72	30.2	.53	37.9	.80	48.1	.85	56.0	.82	64.4	.58	74.2	.67	72.3	.75	63.2	.46	51.1	.71	37.2	.62	28.9	.71	49.1	8.22
DIXIE																										
GUNLOCK PH		1.34		1.26		1.53		.82		.50		.27		1.06		1.16		.78		.84		.69		1.48		11.73
ST GEORGE PH	38.7	.99	44.2	1.01	51.5	.92	60.4	.51	68.1	.33	75.9	.22	83.5	.62	81.8	.67	74.5	.64	61.9	.61	48.0	.56	40.5	1.14	60.8	8.22
ZION NP	39.4	1.59	43.7	1.70	50.0	1.72	59.1	1.21	67.8	.73	77.0	.58	84.3	.86	82.2	1.31	76.3	1.06	64.0	1.06	50.5	.98	41.9	1.80	61.4	14.60
DIVISION	38.9	1.15	43.1	.99	49.5	1.31	59.5	.86	67.1	.47	75.6	.34	83.4	.81	81.3	.92	74.7	.75	62.9	.63	48.9	.61	40.5	1.23	60.5	10.07
NORTH CENTRAL																										
ALPINE		1.42		1.29		1.23		1.56		1.29		1.02		.61		.95		.51		1.36		1.16		1.61		14.01
BRIGHAM CITY	25.9	1.98	31.0	1.60	39.0	1.81	50.2	2.13	59.1	1.74	67.0	1.41	76.3	.46	73.8	.58	64.2	.97	53.0	1.59	38.6	1.86	30.8	1.80	50.7	17.93
CORINNE	23.8	1.52	29.1	1.34	38.1	1.55	48.8	1.75	57.1	1.72	65.0	1.12	74.2	.50	72.2	.46	62.8	.81	51.6	1.19	37.3	1.55	29.3	1.72	49.1	15.23
COTTONWOOD WEIR		2.10		1.92		2.56		2.49		2.10		1.40		.76		1.23		.79		1.85		2.03		1.97		21.20
ELBERTA	26.3	.72	32.0	.77	39.6	.92	49.4	.89	57.6	1.02	65.7	.78	74.7	.84	73.3	.80	63.9	.48	51.9	1.03	38.1	.78	30.4	.89	50.2	9.92
EUREKA		1.47		1.36		1.55		1.26		1.19		.92		1.07		1.15		.58		1.31		.97		1.65		14.48
FARMINGTON	28.1	2.28	33.5	1.79	41.0	2.05	50.9	2.32	59.0	1.76	66.4	1.41	75.6	.45	73.9	1.02	64.7	.69	53.7	1.69	39.6	1.89	32.4	2.04	51.6	19.39
LEWISTON	20.6	1.85	26.1	1.49	34.8	1.83	46.3	2.16	54.3	2.12	61.3	1.51	69.9	.54	68.1	.89	59.0	.97	48.1	1.54	34.6	1.53	26.4	1.82	45.8	18.25
LOGAN UTAH STATE UNIV	23.6	1.65	28.4	1.37	36.6	1.86	51.4	2.16	56.5	1.89	63.8	1.35	73.5	.43	72.1	.71	63.1	.92	51.5	1.54	36.4	1.55	28.5	1.60	48.8	17.03
LOWER AMERICAN FORK PH	28.9	1.59	33.6	1.47	40.9	1.61	50.9	1.75	59.4	1.48	67.4	1.23	76.8	.79	75.4	1.11	67.3	.64	55.4	1.61	40.5	1.43	32.6	1.78	52.4	16.49
MIDVALE	27.8	1.23	33.1	1.26	40.5	1.64	50.3	1.61	58.4	1.33	66.5	1.00	76.2	.74	74.4	.96	64.7	.51	53.3	1.26	39.3	1.41	32.0	1.28	51.4	14.23
OGDEN SUGAR FACTORY	26.2	1.70	31.5	1.45	39.5	1.59	50.0	2.16	58.5	1.55	66.0	1.31	75.4	.56	73.4	.79	64.2	.83	53.0	1.67	38.7	1.64	31.3	1.82	50.6	17.07
PAYSON		1.47		1.23		1.52		1.42		1.39		.93		.82		.91		.77		1.38		1.37		1.63		14.84
PROVO RADIO KOVO		1.42		1.22		1.27		1.21		1.12		.84		.65		.86		.58		1.43		1.20		1.40		13.20
RICHMOND		1.96		1.62		1.90		2.26		2.06		1.51		.60		.69		.95		1.64		1.75		1.91		18.85
RIVERDALE PH	26.2	1.78	31.2	1.51	39.3	1.65	49.6	2.04	57.7	1.58	65.6	1.31	74.5	.54	72.7	.78	63.4	.79	52.5	1.64	38.4	1.57	30.8	1.76	50.2	16.95
SALT LAKE CITY WB AP	26.5	1.20	33.4	1.23	41.1	1.66	50.1	1.76	58.9	1.56	67.1	.91	76.6	.61	74.4	.97	64.2	.74	52.9	1.34	39.3	1.42	31.5	1.34	51.3	14.74
SANTAQUIN PH	27.3	1.67	31.6	1.77	38.8	2.22	49.0	2.10	57.1	1.80	65.4	1.13	74.9	.87	73.4	1.08	64.7	.77	52.4	1.87	38.3	1.79	31.1	1.82	50.3	18.89
SPANISH FORK PH	28.3	1.74	33.1	1.49	40.9	1.96	51.2	1.85	59.3	1.41	68.0	1.07	76.6	.78	74.8	.85	66.2	.72	54.4	1.57	40.3	1.62	32.4	1.81	52.1	16.87
TOOELE	28.2	1.28	32.4	1.54	40.1	1.80	50.3	1.81	58.8	1.47	67.1	1.07	76.7	.81	74.9	.94	65.5	.62	53.2	1.37	39.1	1.56	32.0	1.45	51.5	15.72
UTAH LAKE LEHI	25.0	.85	30.7	.84	38.3	.93	48.1	.91	56.3	.88	64.0	.75	72.3	.71	70.7	.96	61.3	.43	50.0	98	36.7	.83	29.4	1.05	48.6	10.12
DIVISION	26.6	1.57	30.8	1.23	38.1	1.59	49.1	1.70	57.3	1.49	65.1	1.15	74.6	.70	73.0	.85	64.3	.68	52.4	1.29	38.4	1.44	30.2	1.55	50.0	15.24
SOUTH CENTRAL																										
ALTON	25.6	1.67	28.1	1.64	33.4	1.53	42.9	1.13	50.6	.70	58.1	.66	65.5	1.43	64.1	1.62	58.1	1.39	47.9	1.27	36.7	.85	29.2	1.93	45.0	15.82
BEAVER	26.2	.84	30.5	.98	37.3	1.11	46.0	.99	53.8	.94	61.5	.71	69.2	1.39	67.6	1.56	60.3	.76	48.7	1.00	36.3	.60	29.5	.89	47.2	11.77
CEDAR CITY PH	28.5	.88	32.6	.84	39.6	1.18	49.0	.94	56.9	.82	65.8	.50	73.3	1.03	71.7	1.04	64.4	.63	52.7	1.14	39.8	.88	32.2	.93	50.5	10.81
ESCALANTE	26.0	1.20	31.1	.86	38.9	.91	47.3	.58	55.2	.67	63.1	.58	70.3	1.35	67.9	1.95	61.4	1.11	50.5	1.11	37.4	.64	29.0	1.23	48.2	12.19
FILLMORE	28.6	1.38	33.4	1.48	41.3	1.83	51.0	1.39	59.3	1.32	68.2	.88	77.3	.79	75.6	.91	67.0	.52	54.9	1.15	40.4	1.26	32.3	1.35	52.4	14.26
KANAB PH	34.2	1.50	38.0	1.53	44.6	1.24	53.2	.84	60.9	.48	69.1	.40	76.6	.89	74.7	1.50	68.7	.95	57.5	.92	45.0	.72	36.8	1.65	54.9	12.62
KANOSH		1.25		1.51		1.73		1.23		1.46		.78		.77		.99		.47		1.02		1.29		1.29		13.79
LEVAN	25.4	1.16	30.6	1.20	39.0	1.58	48.7	1.35	56.0	1.26	64.0	.88	72.9	.74	71.6	.92	63.7	.61	52.0	1.29	38.4	1.09	29.9	1.38	49.4	13.46
LOA	22.4	.36	26.1	.26	33.0	.48	41.7	.33	50.0	.52	57.7	.61	64.5	1.15	62.6	1.33	55.7	.69	45.3	.77	32.6	.30	24.3	.40	43.0	7.20
MANTI	24.9	1.02	29.2	1.11	37.6	1.34	47.1	1.15	54.7	1.07	62.5	.91	69.8	.80	68.4	.83	60.9	.57	50.2	1.14	36.8	.93	28.8	1.14	47.6	12.01
MORONI		.88		.87		.88		.77		.76		.74		.69		.94		.61		.91		.74		1.09		9.88
OAK CITY	28.2	.98	33.4	1.06	41.1	1.31	50.6	1.24	59.0	1.20	68.2	.81	78.0	.54	76.2	.88	67.1	.56	55.2	1.17	39.5	1.02	31.7	1.10	52.4	11.87
ORDERVILLE		1.77		1.85		1.49		1.07		.66		.47		.80		1.24		1.16		1.14		.98		1.96		14.59
PANGUITCH	22.1	.56	26.4	.51	33.8	.75	43.0	.60	50.1	.56	57.1	.57	63.9	1.50	62.4	1.57	55.6	.81	45.4	.85	33.5	.39	25.2	.66	43.2	9.33
PAROWAN		.88		.96		1.39		.99		.96		.54		1.17		1.17		.68		1.00		.85		1.00		11.59
PIUTE DAM	26.6	.57	31.2	.49	38.0	.76	47.3	.58	55.1	.65	63.5	.62	71.4	.91	69.9	.97	62.2	.75	51.3	.76	38.1	.54	30.2	.59	48.7	8.19
RICHFIELD RADIO KSVC	28.0	.61	32.6	.62	40.2	.87	48.5	.68	56.7	.80	64.5	.62	71.8	.84	70.0	.83	61.9	.49	51.0	.61	38.0	.58	30.8	.64	49.5	8.19
SALINA		.90		.86		1.08		.89		.97		.77		.72		.74		.47		.71		.78		.77		9.66
SCIPIO	24.3	1.10	29.6	1.18	38.0	1.28	47.4	1.00	55.3	.97	63.3	.84	71.4	.75	69.5	.90	60.7	.53	49.7	1.14	36.1	.97	28.5	1.21	47.8	11.87
DIVISION	26.6	1.18	29.9	1.09	37.3	1.37	46.5	1.00	54.2	.87	62.5	.70	70.0	1.12	68.0	1.27	60.9	.75	50.3	.98	37.6	.90	29.6	1.18	47.8	12.41
NORTHERN MOUNTAINS																										
CITY CREEK WATER PLT		2.83		2.64		3.22		3.13		2.35		1.69		.84		1.06		.71		2.21		2.48		2.78		25.95
COALVILLE	22.3	1.27	27.1	1.23	33.9	1.57	43.6	1.19	51.3	1.28	58.0	1.07	66.0	.92	63.9	1.17	56.1	.64	46.7	1.17	33.5	1.27	26.1	1.20	44.0	13.98
HEBER	20.3	1.82	25.3	1.64	34.0	1.40	44.3	1.16	52.1	1.07	59.0	.93	66.9	.83	65.5	1.08	57.6	.72	47.9	1.31	34.7	1.44	25.7	1.98	44.4	15.38
HIAWATHA	23.1	1.00	26.6	.85	33.5	1.03	45.7	.83	52.6	1.04	61.6	1.02	69.4	1.27	67.2	1.88	60.3	.99	48.6	1.27	34.1	.73	26.1	1.07	45.7	12.98
LAKETOWN	20.4	.87	21.9	.83	29.2	.91	42.9	1.03	50.1	1.23	57.0	1.05	65.2	.55	63.4	.86	55.4	.77	45.5	1.05	32.6	.85	25.6	.78	42.4	10.78
MORGAN	22.5	1.61	27.7	1.54	35.3	1.69	45.9	1.40	53.8	1.28	61.0	1.04	68.9	.53	66.8	.88	58.0	.58	48.0	1.28	34.2	1.42	26.4	1.70	45.7	14.95
MOUNTAIN DELL DAM		2.16		2.18		2.46		2.40		2.17		1.43		.84		.97		.89		2.05		2.22		2.29		22.06
SILVER LAKE BRIGHTON		5.17		5.19		5.47		3.75		2.66		2.13		1.52		1.76		1.36		2.98		4.38		5.49		41.86
SNAKE CREEK PH	20.9	3.24	24.8	2.76	32.0	2.57	42.6	1.70	50.5	1.47	57.9	1.11	65.6	.92	63.8	1.16	56.0	.85	46.1	1.62	32.5	2.17	25.5	3.23	43.2	22.80
SOLDIER SUMMIT	17.4	1.69	20.3	1.54	27.6	1.47	37.6	1.11	46.1	1.04	53.3	.97	61.4	1.36	59.9	1.65	52.3	.98	42.0	1.24	28.6	1.20	21.0	1.67	39.0	15.92
WOODRUFF	14.2	.52	17.8	.61	26.4	.71	38.6	.76	47.2	1.09	54.2	.96	62.1	.86	60.1	1.01	51.9	.70	41.9	1.02	27.8	.56	19.7	.55	38.5	9.35
DIVISION	20.3	2.11	23.6	1.79	30.2	2.21	41.3	1.63	49.5	1.42	56.5	1.19	65.0	1.00	63.3	1.38	55.5	1.04	45.2	1.58	32.1	1.75	24.6	2.22	42.3	19.32
UINTA BASIN																										
DUCHESNE	16.7	.58	22.9	.54	34.8	.73	46.1	.61	54.5	.87	61.8	.81	69.2	.97	67.1	1.27	59.0	.84	47.6	.97	32.5	.42	22.5	.67	44.6	9.28
FORT DUCHESNE	14.7	.41	21.0	.36	34.5	.40	47.1	.60	56.0	.70	63.9	.71	70.9	.52	68.7	.73	60.6	.71	48.6	.86	32.7	.42	21.5	.59	45.0	7.01
JENSEN		.53		.44		.64		.81		.61		.73		.50		.76		.69		.96		.45		.76		7.88
MYTON	15.3	.30	22.3	.32	35.5	.42	47.9	.54	56.8	.54	64.4	.61	72.0	.64	70.1	.88	62.1	.60	49.8	.70	33.3	.34	21.8	.50	45.9	6.39
VERNAL AP	15.4	.55	22.1	.49	34.2	.65	45.9	.93	54.6	.73	61.9	.72	69.5	.54	67.2	.80	59.4	.67	47.9	.88	32.6	.50	21.4	.76	44.3	8.22
DIVISION	16.7	.53	22.9	.46	34.4	.58	46.7	.71	55.6	.66	63.4	.70	71.1	.65	68.6	.99	60.7	.74	48.8	.95	33.5	.47	22.7	.68	45.4	8.12
SOUTHEAST																										
BLANDING	26.6	1.05	32.4	1.20	39.2	1.08	48.0	.94	56.7	.76	65.7	.58	72.3	.99	70.5	1.24	62.7	1.47	51.8	1.29	38.6	.82	30.2	1.35	49.6	12.77
BLUFF	30.9	.65	37.8	.72	46.2	.66	56.0	.52	64.5	.37	73.6	.31	80.7	.57	78.5	.84	70.0	1.00	57.3	.72	42.1	.51	33.3	.82	55.9	7.69
EMERY	24.0	.51	28.2	.38	36.3	.47	45.6	.36	53.7	.50	61.3	.55	68.4	.80	66.2	1.26	59.3	.73	48.7	.76	35.5	.32	27.0	.58	46.2	7.22
GREENRIVER AWY	22.7	.37	32.0	.35	42.7	.46	53.7	.44	63.0	.36	71.5	.52	79.8	.54	77.1	.81	67.7	.53	54.0	.68	37.5	.35	28.2	.47	52.5	5.91
HANKSVILLE CAA AP	25.6	.32	33.1	.26	43.3	.30	53.7	.28	62.5	.36	71.8	.35	79.0	.73	76.7	.87	67.8	.47	54.8	.62	39.0	.30	29.7	.35	53.1	5.21
LA SAL		1.07		.89		.98		1.09		.79		.74		1.41		1.67		1.24		1.32		.77		.93		12.90
MOAB	29.7	.57	36.5	.62	46.1	.76	56.7	.78	65.4	.57	73.8	.47	80.7	.55	77.6	.87	69.4	.78	56.8	1.05	41.6	.67	33.0	.71	55.6	8.40
MONTICELLO	25.4	1.10	29.2	.92	35.4	1.04	45.2	.91	53.5	.78	61.9	.63	69.2	1.38	67.0	2.00	60.4	1.47	49.8	1.63	36.5	.80	28.5	1.15	46.8	13.81
PRICE	23.3	.74	29.6	.59	39.2	.69	49.5	.61	58.6	.66	66.9	.73	74.3	.95	72.2	1.10	64.4	.83	52.1	.91	37.0	.48	27.8	.96	49.6	9.25
THOMPSONS	26.1	.73	32.5	.61	41.6	.72	52.3	.75	61.5	.55	70.9	.52	78.6	.60	76.1	.98	68.0	.79	55.4	1.01	39.5	.47	30.0	.81	52.7	8.54
DIVISION	25.9	.68	31.1	.58	40.5	.64	51.4	.59	60.0	.54	68.7	.50	76.4	.75	73.8	1.10	64.8	.78	53.6	.91	38.7	.53	29.2	.72	51.2	8.32

* Averages for period 1931 - 1955, except for stations marked WB and Blanding which are "normals" based on period 1921 - 1950. Divisional means may not be the arithmetical average of individual stations published, since additional data from shorter period stations are used to obtain better areal representation.

NORMALS, MEANS AND EXTREMES
(Table Revised 1972. Base Period for Climatological Normals: 1931-1960)

Station: MILFORD, UTAH   MILFORD AIRPORT   Standard time used: MOUNTAIN   Latitude: 38° 26' N   Longitude: 113° 01' W   Elevation (ground): 5028 feet

Month	Temperature Normal Daily max	Daily min	Monthly	Extremes Record highest	Year	Record lowest	Year	Normal heating degree days (Base 65°)	Precipitation Normal total	Max monthly	Year	Min monthly	Year	Max in 24 hrs	Year	Snow, Ice pellets Mean total	Max monthly	Year	Max in 24 hrs	Year
J	37.1	12.1	24.6	65	1949	-28	1949	1252	0.57	1.63	1969	0.17	1968	0.87	1957	8.7	29.8	1949	11.8	1949
F	42.2	17.2	29.7	73	1963	-27	1949	988	0.74	1.50	1962	0.13	1950	0.74	1971	7.8	18.8	1971	7.8	1971
M	52.4	24.5	38.5	78	1966	-14	1966	822	1.03	1.83	1952	T	1955	1.16	1969	4.2	19.3	1969	10.4	1969
A	63.6	31.8	47.7	87	1959	12	1970+	519	0.86	1.80	1964	T	1955	0.86	1964	2.1	16.0	1953	8.2	1953
M	73.3	39.6	56.5	94	1967	18	1967	279	0.69	1.58	1951	0.01	1961+	0.91	1954	0.8	3.9	1953	3.8	1953
J	83.5	46.9	65.4	105	1961+	30	1966	87	0.43	2.43	1967	T	1958	1.04	1949	0.0	0.0		0.0	
J	92.1	55.5	73.8	104	1960+	38	1964	0	0.70	1.42	1955	0.01	1958	1.24	1967	0.0	0.0		0.0	
A	90.1	54.3	72.2	102	1958	34	1968	0	0.73	1.69	1955	0.03	1955	0.78	1955	0.0	0.0		0.0	
S	81.3	44.1	63.0	98	1950	24	1959	99	0.43	2.60	1970	0.00	1957+	1.33	1970	0.4	8.4	1965	6.7	1965
O	67.1	34.2	50.7	90	1963	-2	1971	443	0.77	2.47	1971	0.00	1952	1.20	1951	1.9	17.4	1971	8.3	1971
N	51.3	22.5	36.1	76	1958	-13	1958	867	0.52	2.10	1963	0.06	1956	1.01	1963	4.4	14.0	1963	6.4	1963
D	41.4	15.0	28.2	65	1958	-23	1968	1141	0.71	2.45	1966	0.05	1955	1.00	1949	6.8	27.4	1949	16.3	1949
YR	64.7	33.1	48.9	105	JUN 1970	-28	JAN 1949	6497	8.00	2.60	SEP 1967	0.00	OCT 1952	1.33	SEP 1970	42.7	29.8	JAN 1949	16.3	DEC 1949

Means and extremes above are from existing and comparable exposures. Annual extremes have been exceeded at other sites in the locality as follows: Lowest temperature -34 in January 1937; maximum precipitation 3.75 in October 1946 and earlier; maximum precipitation in 24 hours 1.92 in October 1916.

NORMALS, MEANS AND EXTREMES
(Table Revised 1975. Base Period for Climatological Normals: 1941-1970)

Average station pressure 846.4 mb.   Elev. 5033 feet m.s.l.

Month	Temperature Normal Daily max	Daily min	Monthly	Extremes Record highest	Year	Record lowest	Year	Normal Degree days Base 65°F Heating	Cooling	Precipitation Normal	Max monthly	Year	Min monthly	Year	Max in 24 hrs	Year	Snow, Ice pellets Max monthly	Year	Max in 24 hrs	Year
J	38.4	12.9	25.7	65	1953	-28	1949	1218	0	0.61	1.63	1969	0.05	1972	0.87	1957	29.8	1949	11.8	1949
F	44.2	18.6	31.4	73	1963	-27	1949	941	0	0.70	1.50	1962	0.01	1972	0.74	1971	18.8	1971	10.2	1971
M	52.7	23.7	38.2	78	1972	-14	1966	834	0	1.04	1.83	1952	0.00	1972	1.16	1969	19.3	1973	10.4	1969
A	62.9	31.4	47.2	87	1959	18	1967	534	10	0.90	2.48	1973	0.02	1972	1.06	1973	24.4	1973	11.2	1973
M	73.5	39.3	56.5	94	1967	18	1951	274	88	0.61	1.58	1951	T	1963	0.91	1954	3.9	1953	3.8	1953
J	83.5	46.8	65.4	105	1970	30	1974	82	288	0.56	2.43	1967	T	1974	1.04	1949	0.0		0.0	
J	92.8	55.8	74.3	104	1960+	38	1964	0	242	0.51	1.42	1967	0.01	1958	1.24	1967	0.0		0.0	
A	90.3	54.8	72.6	102	1958	34	1968	0	60	0.68	2.52	1972	0.03	1957	1.05	1972	0.0		0.0	
S	81.6	44.1	63.0	98	1950	24	1959	120	0	0.61	2.60	1970	T	1967	1.33	1970	8.4	1965	6.7	1965
O	68.0	34.5	51.3	90	1963	-2	1971	443	0	0.78	2.61	1972	0.00	1972	1.20	1951	17.4	1971	6.4	1971
N	52.0	22.5	37.3	76	1965	-13	1958	831	0	0.52	2.10	1963	0.06	1956	1.01	1963	14.0	1963	8.3	1963
D	41.3	15.8	28.6	65	1972	-32	1972	1128	0	0.73	2.45	1966	0.05	1955	1.03	1949	30.6	1972	16.3	1949
YR	65.1	33.3	49.2	105	JUN 1970	-32	DEC 1972	6412	688	8.40	2.61	OCT 1972	0.00	MAR 1972	1.33	SEP 1970	30.6	DEC 1972	16.3	DEC 1949

Means and extremes above are from existing and comparable exposures. Annual extremes have been exceeded at other sites in the locality as follows: Lowest temperature -34 in January 1937; maximum precipitation 3.75 in October 1946 and earlier; maximum precipitation in 24 hours 1.92 in October 1916.

REFERENCE NOTES APPLYING TO TABLES APPEAR ON THE PAGE FOLLOWING LAST TABLE.
(Caution: Letters and symbols may have different meanings in 1941-1970 tables than in earlier tables. See notes.)

NORMALS, MEANS AND EXTREMES

Station: SALT LAKE CITY, UTAH — INTERNATIONAL AIRPORT   Standard time used: MOUNTAIN   Latitude: 40° 46' N   Longitude: 111° 58' W   Elevation (ground): 4220 feet

(Table Revised 1972. Base Period for Climatological Normals: 1931-1960)

Month	Temperature Normal Daily maximum	Daily minimum	Monthly	Extremes Record highest	Year	Record lowest	Year	Normal heating degree days (Base 65°)	Precipitation Normal total	Maximum monthly	Year	Minimum monthly	Year	Maximum in 24 hrs.	Year	Snow, ice pellets Mean total	Maximum monthly	Year	Maximum in 24 hrs.	Year	Relative humidity Hour 05	Hour 11	Hour 17	Hour 23	Wind Mean speed	Prevailing direction	Fastest mile Speed	Direction	Year	Pct. of possible sunshine	Mean sky cover sunrise to sunset	Sunrise to sunset Clear	Partly cloudy	Cloudy	Precipitation .01 inch or more	Snow, ice pellets 1.0 inch or more	Thunderstorms	Heavy fog	Temperatures Max. 90° and above	32° and below	Min. 32° and below	0° and below	Average daily solar radiation - langleys
(a)	(b)	(b)	(b)	12		12		(b)	(b)	43		43		43		43	43		43		12	12	12	12	42	32	36	36	36	34	36	43	43	43	43	43	43	43	12	12	12	12	4
J	36.8	17.5	27.2	61	1971	-18	1963	1172	1.35	3.14	1940	0.09	1961	1.36	1940	13.2	32.3	1937	9.7	1962	77	69	68	77	7.6	SSE	52	SE	1950+	47	7.1	6	6	18	10	9	*	4	0	10	27	2	175
F	42.0	22.9	32.5	65	1961	-4	1960	910	1.19	3.22	1936	0.12	1946	1.05	1969	27.9	27.9	1969	8.7	1944	77	64	58	67	8.2	SSE	56	SE	1954	55	6.5	6	8	15	11	9	1	2	0	4	24	*	291
M	52.0	28.8	40.4	78	1966	2	1960	763	1.56	3.67	1944	0.10	1956	1.83	1952	35.6	35.6	1952	15.4	1965+	70	50	44	61	9.4	SSE	71	NW	195-	64	6.4	8	10	13	10	7	1	*	0	1	22	0	417
A	61.8	36.6	49.2	85	1962	22	1971	459	1.76	3.67	1944	0.45	1934	2.41	1957	6.4	26.4	1957	16.1	1974	65	39	37	58	9.5	SSE	57	NW	195-	67	6.3	7	11	12	9	4	1	*	0	0	10	0	497
M	71.0	43.8	58.9	92	1961+	25	1962	233	1.40	3.37	1957	T	1934	2.03	1947	0.5	5.3	1965+	5.3	1965	64	31	28	53	9.3	SSE	63	NW	1963	73	5.6	8	11	12	6	4	*	*	0	0	2	0	627
J	83.7	51.0	67.4	104	1961	35	1961	84	0.98	2.93	1947	0.01	1946+	1.88	1968+	T	T	1968+	T	1968	63	28	33	53	9.2	SSE	63	SW	1968	78	4.4	14	10	6	6	0	1	0	8	0	0	0	650
J	94.1	59.6	76.9	107	1960	40	1968	0	0.58	2.52	1962	T	1963	2.35	1962	0.0	0.0		0.0		51	26	26	41	9.5	SSE	49	W	1936	83	3.5	17	10	4	4	0	6	0	26	0	0	0	702
A	90.8	58.2	74.5	103	1960	37	1965+	5	0.87	3.66	1968	T	1944	1.96	1932	0.0	0.0		0.0		50	23	35	54	9.0	SE	58	SW	1946	82	3.5	16	11	4	5	0	8	0	21	0	0	0	578
S	80.3	48.5	64.4	100	1967	27	1965	81	0.53	2.80	1970	T	1951+	2.19	1973	0.1	0.1	1971	T	1970	52	25	28	46	8.5	SE	61	NW	1950	83	3.5	17	10	3	4	0	4	0	14	0	0	0	485
O	67.2	38.2	51.7	89	1963	16	1971	419	1.15	3.51	1946	0.00	1952	1.47	1946	1.0	16.0	1931	8.5	1971	68	35	40	65	7.4	SE	67	NW	1937	74	4.5	15	8	8	6	0	2	0	4	0	1	0	335
N	50.0	28.1	39.1	75	1967	-5	1959	777	1.10	3.82	1964	0.01	1939	1.13	1948	8.5	18.5	1948	13.4	1948	68	42	58	72	7.7	SSE	63	NW	1950	54	6.1	8	8	14	7	4	2	*	*	0	19	*	219
D	39.0	21.2	30.1	67	1969	-10	1961	1082	1.24	3.22	1962	0.28	1962	1.41	1964	11.8	34.3	1948	13.4	1948	79	58	59	79	7.4	SSE	54	S	1955	44	7.1	6	6	18	8	7	*	3	0	1	28	1	156
YR	64.1	37.7	50.9	107	JUL. 1960	-18	JAN. 1963	6052	13.90	4.90	APR. 1944	0.00	OCT. 1952	2.41	MAR. 1944	55.7	35.6	MAR. 1952	15.4	MAR. 1944	67	46	42	62	8.7	SSE	71	NW	MAR. 1954	69	5.4	131	103	131	88	17	35	11	59	26	139	4	3428

Means and extremes above are from existing and comparable exposures. Annual extremes have been exceeded at other sites in the locality as follows: Lowest temperature -30 in February 1933; maximum monthly precipitation 5.81 in November 1875; maximum precipitation in 24 hours 2.72 in May 1901; maximum monthly snowfall 39.1 in December 1948.

NORMALS, MEANS AND EXTREMES

(Table Revised 1975. Base Period for Climatological Normals: 1941-1970)

Month	Temperature °F Normal Daily maximum	Daily minimum	Monthly	Extremes Record highest	Year	Record lowest	Year	Normal Degree days Base 65° F Heating	Cooling	Precipitation in inches Normal	Maximum monthly	Year	Minimum monthly	Year	Maximum in 24 hrs.	Year	Water equivalent monthly maximum	Year	monthly minimum	Year	Snow, Ice pellets year maximum	Year	Maximum monthly	Year	Maximum in 24 hrs.	Year	Relative humidity pct. Hour 05	Hour 11	Hour 17	Hour 23	Wind Mean speed m.p.h.	Prevailing direction	Fastest mile Speed m.p.h.	Direction	Year	Pct. of possible sunshine	Mean sky cover, tenths, sunrise to sunset	Sunrise to sunset Clear	Partly cloudy	Cloudy	Precipitation .01 inch or more	Snow, Ice pellets 1.0 inch or more	Thunderstorms	Heavy fog, visibility 1/4 mile or less	Temperatures °F Max. 90° and above	32° and below	Min. 32° and below	0° and below	Average station pressure mb. Elev. 4227 feet m.s.l.
(a)				15		15					46		46		46		46		46		46		46		46		15	15	15	15	45	32	39	39	39	37	39	46	46	46	46	46	46	46	15	15	15	3	2
J	37.4	18.5	28.0	61	1971	-18	1963	1147	0	1.27	3.14	1940	0.09	1961	1.36	1963	3.14	1940	0.09	1961	32.3	1937	32.3	1937	9.7	1962	76	67	69	76	7.7	SSE	52	SE	1950	47	7.1	6	7	18	10	9	*	4	0	11	27	3	873.0
F	43.4	23.3	33.4	69	1972	-4	1960	885	0	1.19	3.22	1936	0.12	1946	1.05	1958	3.22	1936	0.12	1946	27.9	1969	27.9	1969	8.7	1944	75	63	58	63	8.2	SSE	56	SE	1954	55	6.6	6	9	13	9	9	1	2	0	4	24	*	873.4
M	50.8	28.3	39.6	78	1966	2	1960	787	0	1.63	3.67	1944	0.10	1956	1.83	1952	3.67	1944	0.10	1956	26.4	1952	26.4	1952	15.4	1952	64	44	39	57	9.5	SSE	71	NW	1964	63	6.5	7	10	14	9	7	1	*	0	1	21	0	867.9
A	61.8	36.4	49.2	85	1962	22	1967	474	0	2.12	3.67	1944	0.45	1934	2.41	1957	3.67	1944	0.45	1934	16.2	1957	16.2	1957	16.1	1974	61	39	35	51	9.5	SSE	57	NW	1957	66	6.3	7	10	13	7	4	1	*	0	0	10	0	869.5
M	71.3	44.2	57.8	92	1961	25	1962	237	11	1.49	3.37	1957	T	1934	2.03	1965	3.37	1957	T	1934	5.3	1965	5.3	1965	5.3	1965	53	30	27	46	9.3	SSE	63	NW	1963	73	5.5	8	10	13	6	4	*	*	0	0	1	0	870.3
J	81.3	51.1	66.2	104	1961	35	1961	88	124	1.30	2.93	1947	0.01	1934	1.88	1962	2.93	1947	0.01	1934	T	1974	T	1974	T	1974	40	31	27	51	9.0	SE	63	S	1963	78	4.4	14	14	6	6	0	1	0	9	0	0	0	870.3
J	92.8	60.5	76.7	107	1960	40	1968	0	363	0.70	2.52	1962	0.00	1963	2.35	1962	2.52	1962	0.00	1963	0.0		0.0		0.0		51	26	19	41	9.4	SSE	49	W	1936	84	3.4	17	11	4	4	0	6	0	26	0	0	0	871.0
A	90.3	59.7	74.5	103	1960	37	1965	5	300	0.93	3.66	1968	0.05	1932	1.96	1932	3.66	1968	0.05	1932	0.0		0.0		0.0		61	29	28	54	9.5	SE	58	SW	1946	83	3.5	17	11	3	5	0	8	0	21	0	0	0	872.9
S	80.3	49.7	64.6	100	1967	27	1965	105	19	0.68	2.80	1970	0.00	1951+	2.19	1973	2.80	1970	0.00	1951	0.0		0.0		0.0		51	25	28	43	8.5	SE	61	NW	1950	83	3.5	17	11	3	4	0	4	0	14	0	0	0	872.9
O	66.4	38.4	52.4	89	1963	16	1971	402	11	1.16	3.51	1946	0.00	1952	1.47	1946	3.51	1946	0.00	1952	19.5	1971	19.5	1971	8.5	1971	68	40	42	65	7.8	SE	67	NW	1950	74	4.5	16	9	8	6	0	2	0	4	0	1	0	872.5
N	50.0	28.1	39.1	75	1967	-5	1959	777	0	1.31	3.82	1964	0.01	1939	1.13	1952	3.82	1964	0.01	1939	18.1	1973	18.1	1973	13.4	1972	71	58	58	71	7.5	SSE	63	NW	1955	54	6.1	8	8	14	7	4	2	*	*	0	18	*	874.5
D	39.0	21.5	30.3	67	1969	-15	1972	1076	0	1.39	3.82	1962	0.28	1962	1.82	1962	3.82	1962	0.28	1962	35.2	1972	35.2	1972	18.1	1972	79	72	73	79	7.5	SSE	54	S	1955	44	7.1	6	6	18	8	7	*	3	0	1	28	1	874.7
YR	63.8	38.2	51.0	107	JUL. 1960	-18	JAN. 1963	5983	927	15.17	4.90	APR. 1944	0.00	OCT. 1952	2.41	APR. 1957	4.90	APR. 1944	0.00	OCT. 1952	35.6	MAR. 1952	35.6	MAR. 1952	18.1	DEC. 1972	67	46	42	62	8.7	SSE	71	NW	MAR. 1954	70	5.4	129	104	132	88	18	35	11	60	26	134	4	871.9

Means and extremes above are from existing and comparable exposures. Annual extremes have been exceeded at other sites in the locality as follows: Lowest temperature -30 in February 1933; maximum monthly precipitation 5.81 in November 1875; maximum precipitation in 24 hours 2.72 in May 1901; maximum monthly snowfall 39.1 in December 1948.

REFERENCE NOTES APPLYING TO TABLES APPEAR ON THE PAGE FOLLOWING LAST TABLE. (Caution: Letters and symbols may have different meanings in 1941-1970 tables than in earlier tables. See notes.)

Reference notes applying to Normals, Means, and Extremes tables for 1931–1960 base period.

(a) Length of record, years, based on January data, Other months may be for more or fewer years if there have been breaks in the record.

(b) Climatological standard normals (1931-1960).

* Less than one half.

+ Also on earlier dates, months, or years.

T Trace, an amount too small to measure.

Below zero temperatures are preceded by a minus sign. The prevailing direction for wind in the Normals, Means, and Extremes table is from records through 1963.

‡ ≥ 70° at Alaskan stations.

Unless otherwise indicated, dimensional units used in this bulletin are: temperature in degrees F.; precipitation, including snowfall, in inches; wind movement in miles per hour; and relative humidity in percent. Heating degree day totals are the sums of negative departures of average daily temperatures from 65° F. Cooling degree day totals are the sums of positive departures of average daily temperatures from 65° F. Sleet was included in snowfall totals beginning with July 1948. The term "Ice pellets" includes solid grains of ice (sleet) and particles consisting of snow pellets encased in a thin layer of ice. Heavy fog reduces visibility to 1/4 mile or less.

Sky cover is expressed in a range of 0 for no clouds or obscuring phenomena to 10 for complete sky cover. The number of clear days is based on average cloudiness 0-3, partly cloudy days 4-7, and cloudy days 8-10 tenths.

Solar radiation data are the averages of direct and diffuse radiation on a horizontal surface. The langley denotes one gram calorie per square centimeter.

& Figures instead of letters in a direction column indicate direction in tens of degrees from true North; i.e., 09 - East, 18 - South, 27 - West, 36 - North, and 00 - Calm. Resultant wind is the vector sum of wind directions and speeds divided by the number of observations. If figures appear in the direction column under "Fastest mile" the corresponding speeds are fastest observed 1-minute values.

\# To 8 compass points only.

% Through 1964. The station did not operate 24 hours daily. Fog and thunderstorm data may be incomplete.

$ Through 1966.

** The National Weather Service considers the accuracy of solar radiation data questionable; therefore, publication is suspended pending determination of corrected values.

Reference notes applying to Normals, Means, and Extremes tables for 1941–1970 base period.

(a) Length of record, years, through the current year unless otherwise noted, based on January data.

(b) 70° and above at Alaskan stations.

* Less than one half.

T Trace.

NORMALS - Based on record for the 1941-1970 period.

DATE OF AN EXTREME - The most recent in cases of multiple occurrence.

PREVAILING WIND DIRECTION - Record through 1963.

WIND DIRECTION - Numerals indicate tens of degrees clockwise from true north. 00 indicates calm.

FASTEST MILE WIND - Speed is fastest observed 1-minute value when the direction is in tens of degrees.

% Through 1964. The station did not operate 24 hours daily. Fog and thunderstorm data may be incomplete.

$ Through 1966.

c Through 1971 plus 1974.

Mean Annual Precipitation, Inches

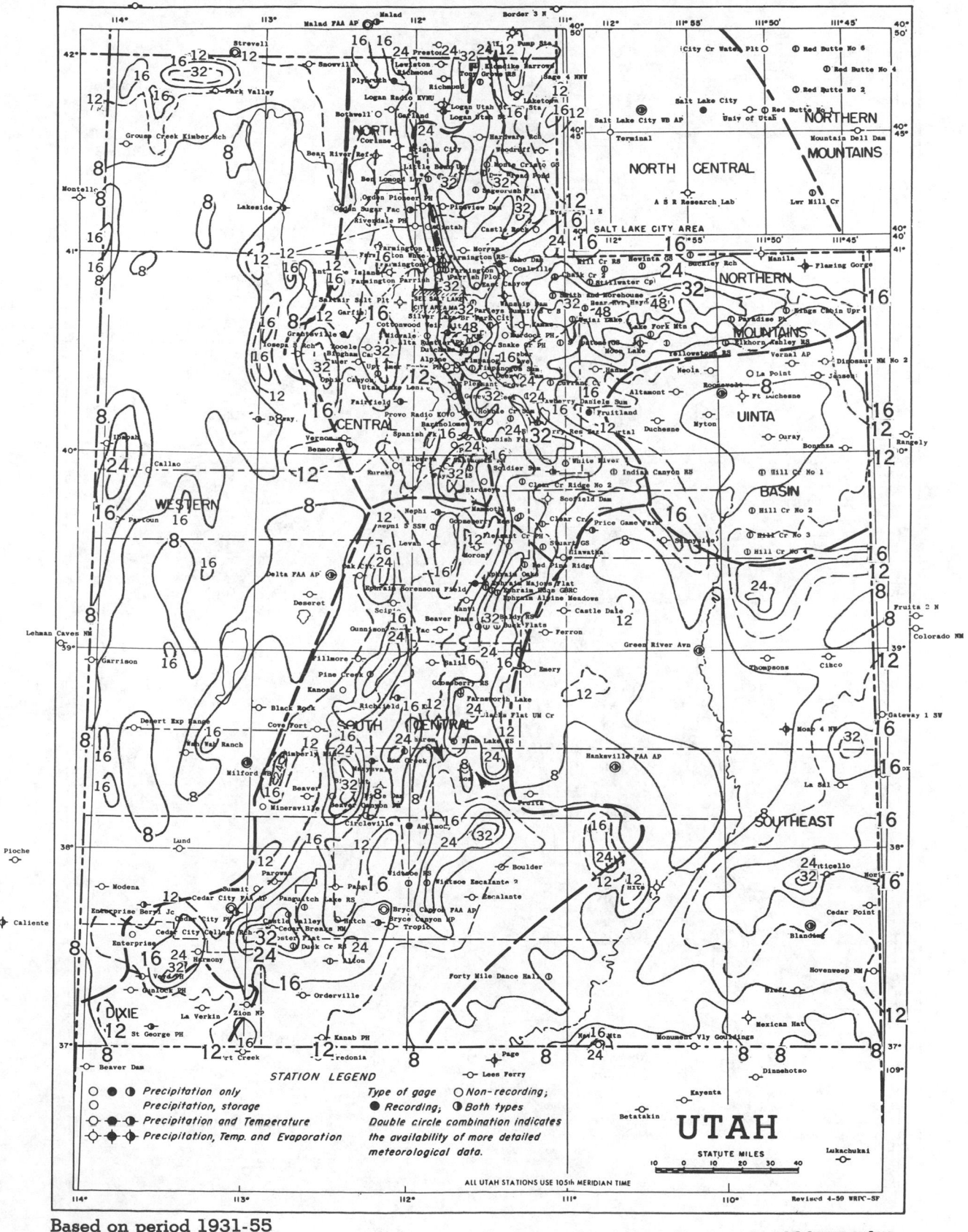

Based on period 1931-55

Isolines are drawn through points of approximately equal value. Caution should be used in interpolating on these maps, particularly in mountainous areas.

STATION LEGEND

○ ● ◐ Precipitation only   Type of gage   ○ Non-recording;
○ Precipitation, storage   ● Recording;
◒ Precipitation and Temperature   ◑ Both types
◓ Precipitation, Temp. and Evaporation   Double circle combination indicates
                                          the availability of more detailed
                                          meteorological data.

ALL UTAH STATIONS USE 105th MERIDIAN TIME

Revised 4-59 WBPC-SF

UTAH

STATUTE MILES
10   0   10   20   30   40

Mean Maximum Temperature (°F.), January

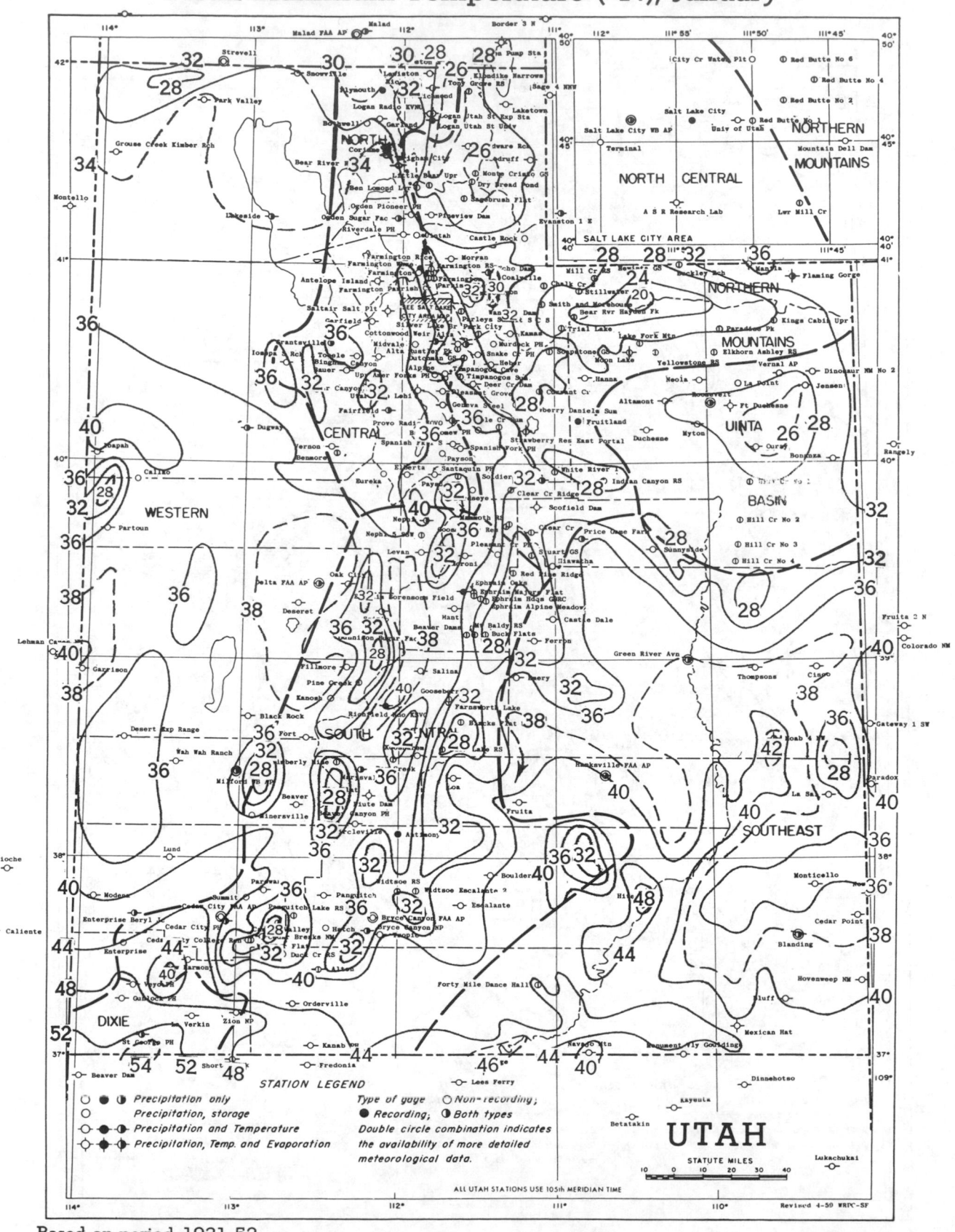

Based on period 1931-52

Isolines are drawn through points of approximately equal value. Caution should be used in interpolating on these maps, particularly in mountainous areas.

Mean Minimum Temperature (°F.), January

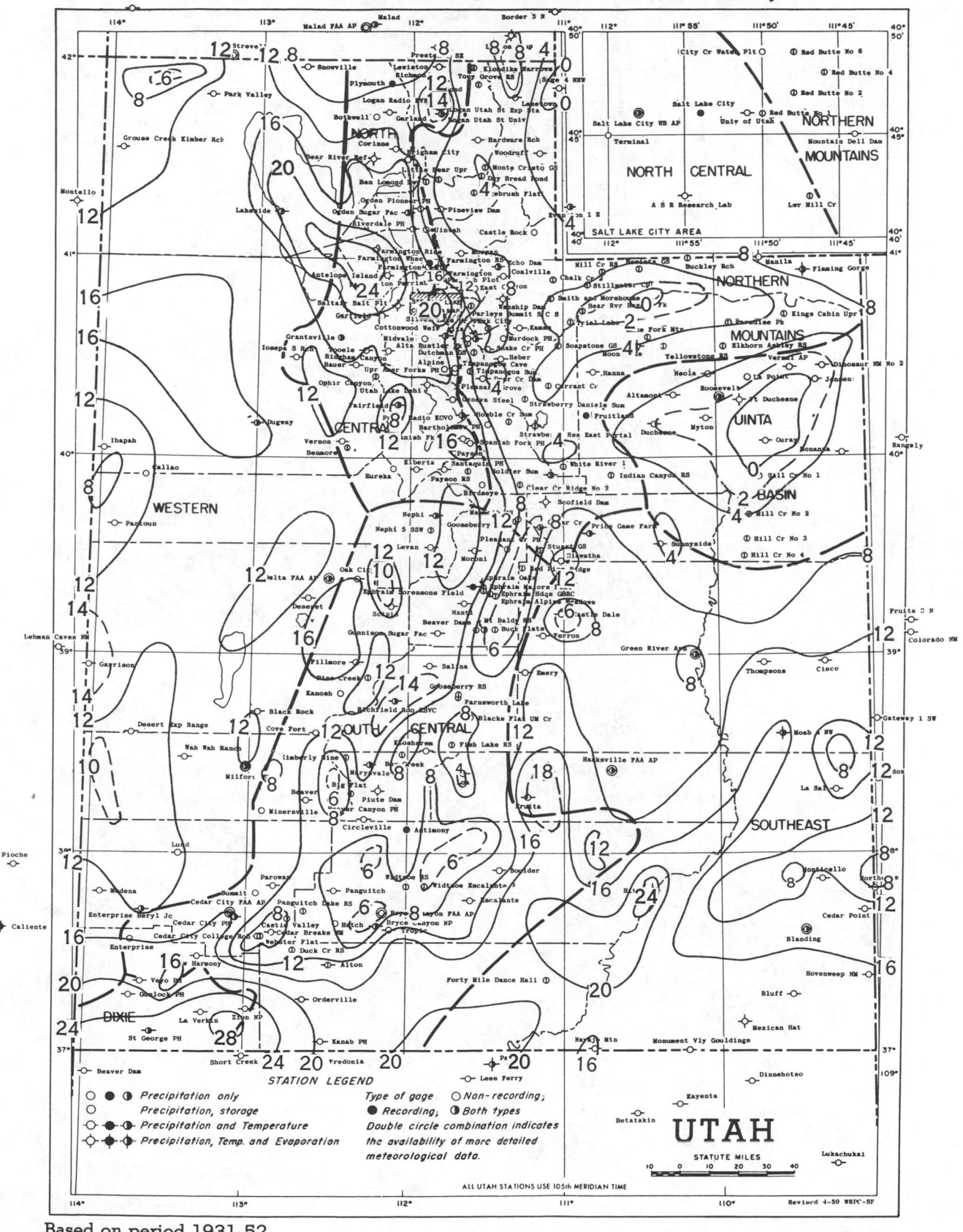

Based on period 1931-52

Isolines are drawn through points of approximately equal value. Caution should be used in interpolating on these maps, particularly in mountainous areas.

Mean Maximum Temperature (°F.), July

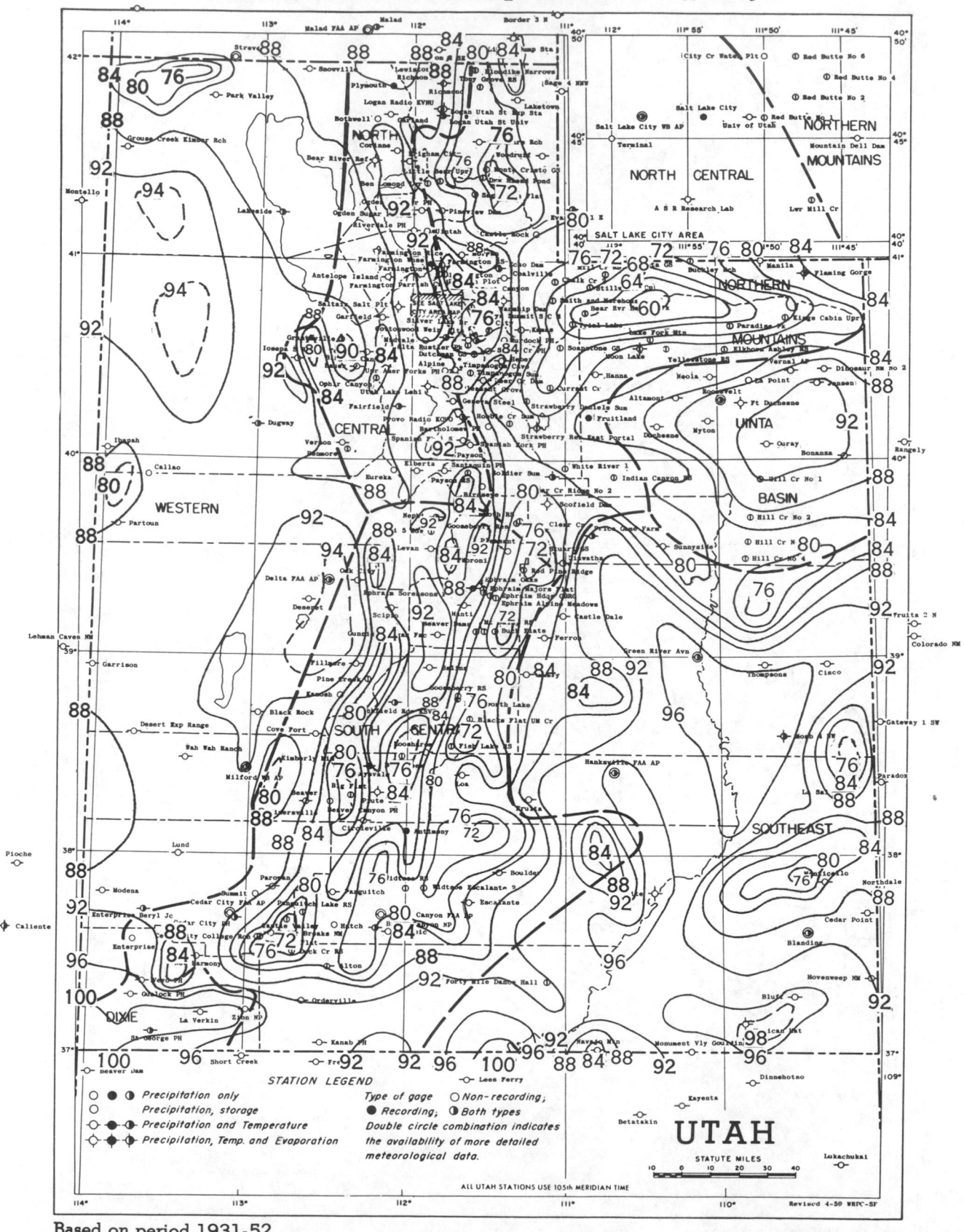

UTAH

STATION LEGEND

○ ● ◖	Precipitation only
○	Precipitation, storage
○ ● ◖	Precipitation and Temperature
◇ ◆ ◗	Precipitation, Temp. and Evaporation

Type of gage ○ Non-recording;
● Recording; ◖ Both types
Double circle combination indicates
the availability of more detailed
meteorological data.

STATUTE MILES
10 0   10   20   30   40

ALL UTAH STATIONS USE 105th MERIDIAN TIME

Revised 4-59 WBPC-SF

Based on period 1931-52
Isolines are drawn through points of approximately equal value.   Caution should be used
in interpolating on these maps, particularly in mountainous areas.

Mean Minimum Temperature (°F.), July

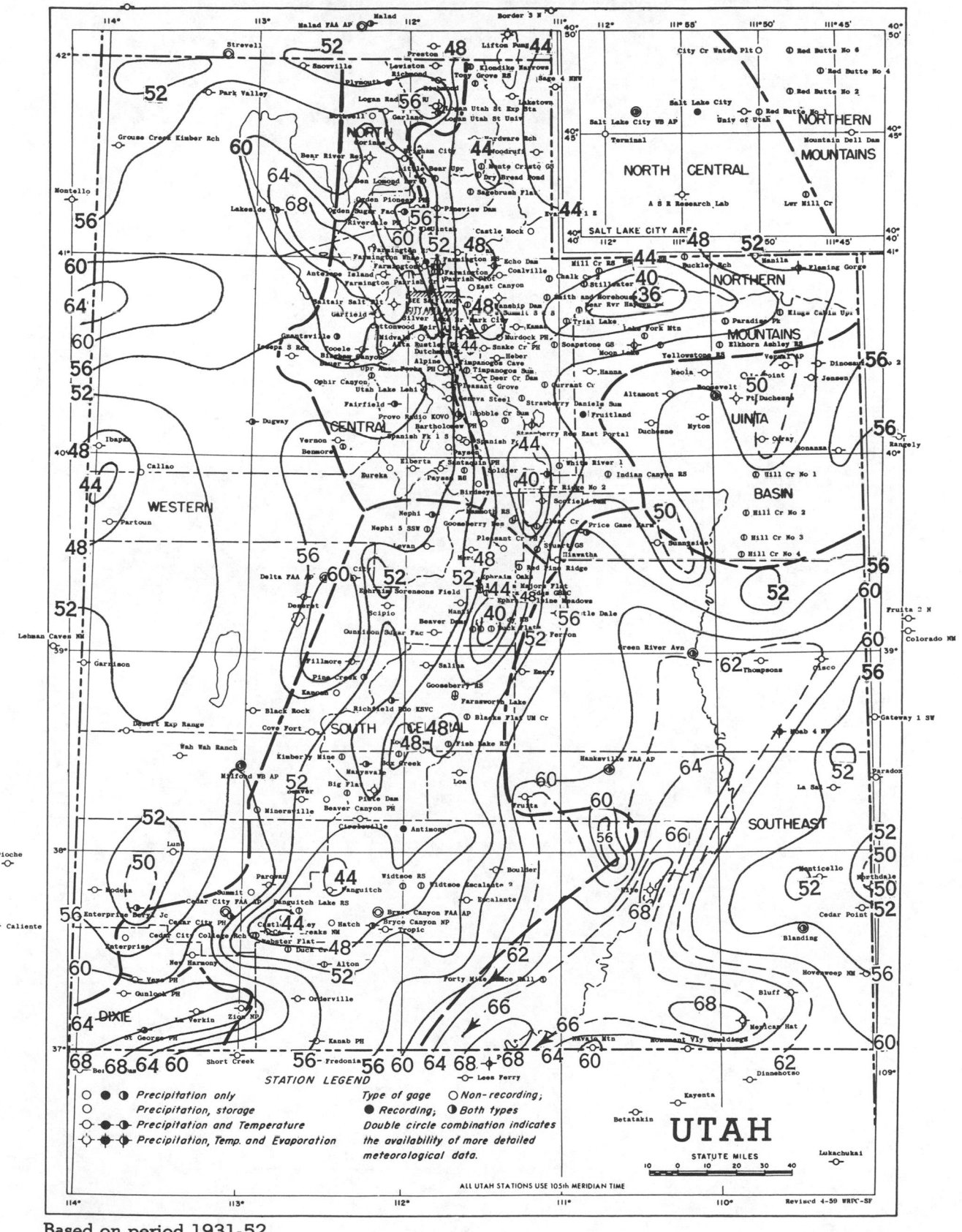

UTAH

STATION LEGEND

○ ◐ ● Precipitation only
○ ◐ ● Precipitation, storage
○ ◐ ● Precipitation and Temperature
○ ◐ ● Precipitation, Temp. and Evaporation

Type of gage ○ Non-recording;
● Recording; ◑ Both types
Double circle combination indicates
the availability of more detailed
meteorological data.

ALL UTAH STATIONS USE 105th MERIDIAN TIME

STATUTE MILES
10 0 10 20 30 40

Revised 4-59 WBPC-SF

Based on period 1931-52
Isolines are drawn through points of approximately equal value. Caution should be used
in interpolating on these maps, particularly in mountainous areas.

CONFIDENCE LIMITS

In the absence of trend or record changes, the chances are 9 out of 10 that the true mean will lie in the interval formed by adding and subtracting the values in the following table from the means for any station in the State. Because of the wider variation in mean precipitation, the corresponding monthly means and annual mean must be substituted for "p" in the precipitation table below to obtain mean precipitation confidence limits.

2.0	.29√p	2.1	.24√p	1.2	.27√p	1.2	.27√p	1.1	.34√p	1.1	.36√p	.7	.26√p	.5	.31√p	.7	.34√p	1.0	.30√p	1.3	.29√p	1.6	.28√p	.4	.30√p

COMPARATIVE DATA

Data in the following table are the mean temperature and average precipitation for Logan (Utah State Agricultural College), Utah for the period 1906 - 1930 and are included in this publication for comparative purposes :

23.7	1.80	28.6	1.69	37.5	1.94	47.1	2.01	55.1	2.00	64.0	1.02	72.8	.71	71.0	.78	61.2	1.44	49.4	1.90	37.2	1.24	25.4	1.32	47.8	17.85

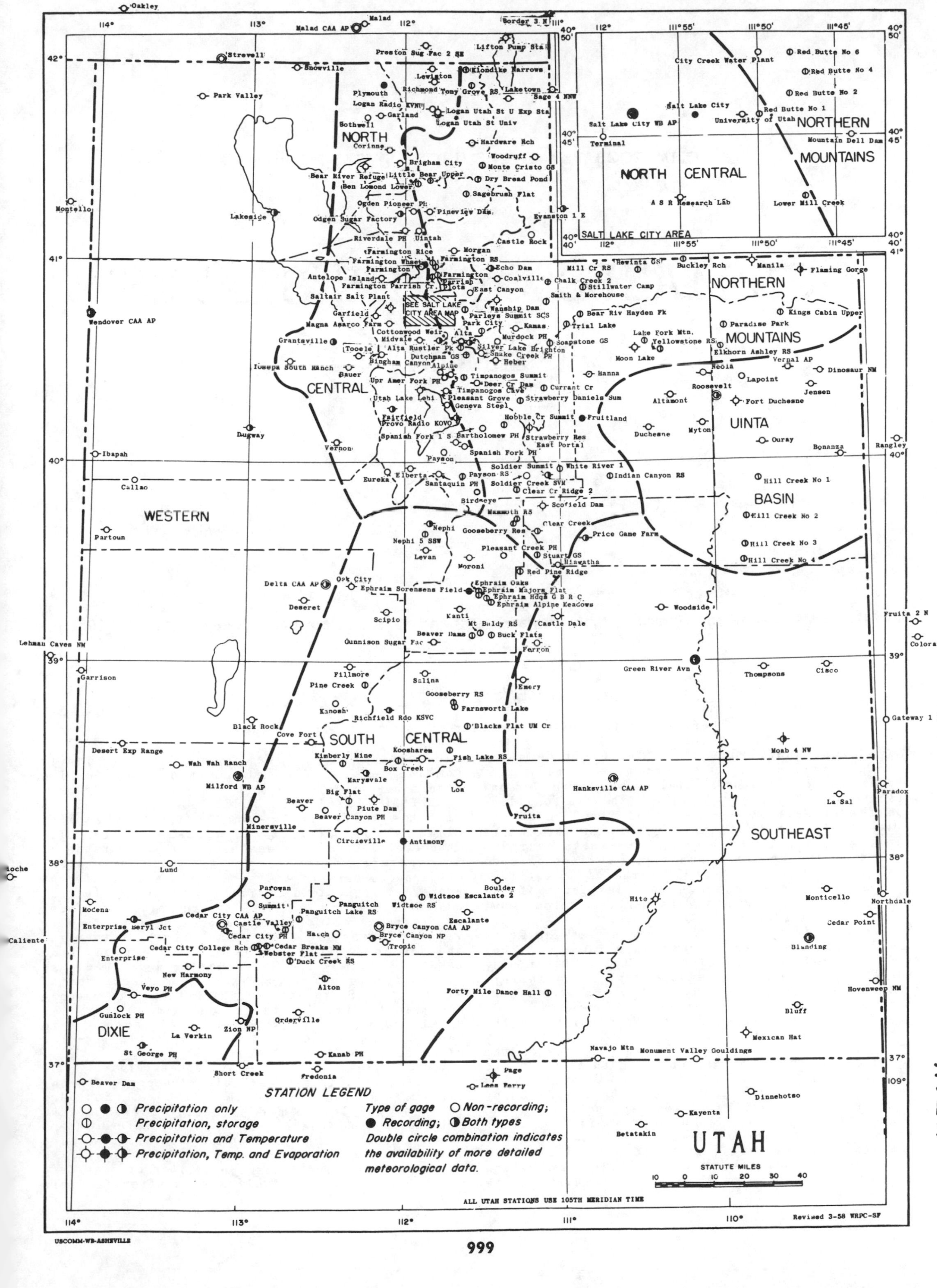
SALT LAKE CITY AREA
NORTH CENTRAL
NORTHERN MOUNTAINS
City Creek Water Plant
Red Butte No 6
Red Butte No 4
Red Butte No 2
Red Butte No 1
Salt Lake City
Salt Lake City WB AP
University of Utah
Terminal
Mountain Dell Dam
A S R Research Lab
Lower Mill Creek
NORTH
CENTRAL
WESTERN
NORTHERN
MOUNTAINS
UINTA
BASIN
SOUTH
CENTRAL
SOUTHEAST
DIXIE
Oakley
Malad
Malad CAA AP
Strevell
Snowville
Park Valley
Border 3 N
Lifton Pump Sta
Preston Sug Fac 2 SE
Klondike Narrows
Lewiston
Plymouth
Richmond
Tony Grove RS
Laketown
Sage 4 NNW
Logan Radio KVNU
Logan Utah St U Exp Sta
Logan Utah St Univ
Garland
Bothwell
Corinne
Hardware Rch
Woodruff
Brigham City
Monte Cristo GS
Bear River Refuge
Little Bear Upper
Ben Lomond Lower
Dry Bread Pond
Sagebrush Flat
Montello
Lakeside
Ogden Pioneer PH
Ogden Sugar Factory
Pineview Dam
Evanston 1 E
Riverdale PH
Uintah
Castle Rock
Farmington Rice
Morgan
Farmington Wheat
Farmington RS
Farmington
Echo Dam
Antelope Island
Coalville
Farmington Parrish Cr. Plots
East Canyon
Saltair Salt Plant
Wendover CAA AP
Garfield
SEE SALT LAKE CITY AREA MAP
Wanship Dam
Parleys Summit SCS
Magna Asarco Farm
Park City
Kamas
Grantsville
Cottonwood Weir
Midvale
Alta
Murdock PH
Tooele
Alta Rustler Pk
Silver Lake Brighton
Dutchman GS
Snake Creek PH
Bingham Canyon
Alpine
Heber
Iosepa South Ranch
Bauer
Timpanogos Summit
Upr Amer Fork PH
Deer Cr Dam
Timpanogos Cave
Utah Lake Lehi
Pleasant Grove
Geneva Steel
Fairfield
Provo Radio KOVO
Dugway
Hobble Cr Summit
Spanish Fork 1 S
Bartholomew PH
Vernon
Ibapah
Payson
Spanish Fork PH
Eureka
Elberta
Payson RS
Santaquin PH
Callao
Mill Cr RS
Hewinta GS
Buckley Rch
Manila
Flaming Gorge
Chalk Creek 2
Stillwater Camp
Smith & Morehouse
Bear Riv Hayden Fk
Kings Cabin Upper
Trial Lake
Paradise Park
Lake Fork Mtn.
Yellowstone RS
Soapstone GS
Moon Lake
Elkhorn Ashley RS
Vernal AP
Dinosaur NM
Hanna
Neola
Lapoint
Jensen
Currant Cr
Roosevelt
Fort Duchesne
Strawberry Daniels Sum
Altamont
Fruitland
Duchesne
Myton
Ouray
Strawberry Res East Portal
Bonanza
Rangley
Soldier Summit
White River 1
Indian Canyon RS
Soldier Creek SVM
Clear Cr Ridge 2
Hill Creek No 1
Birdseye
Scofield Dam
Mammoth RS
Partoun
Nephi
Gooseberry Res
Clear Creek
Nephi 5 SSW
Price Game Farm
Hill Creek No 2
Pleasant Creek PH
Levan
Moroni
Stuart GS
Hiawatha
Hill Creek No 3
Hill Creek No 4
Red Pine Ridge
Delta CAA AP
Oak City
Ephraim Oaks
Ephraim Sorensens Field
Ephraim Majors Flat
Ephraim Hdqs G B R C
Ephraim Alpine Meadows
Deseret
Woodside
Scipio
Manti
Fruita 2 N
Mt Baldy RS
Castle Dale
Beaver Dams
Buck Flats
Gunnison Sugar Fac
Colorado NM
Lehman Caves NM
Ferron
Green River Avn
Garrison
Fillmore
Salina
Thompsons
Cisco
Pine Creek
Emery
Gooseberry RS
Farnsworth Lake
Kanosh
Richfield Rdo KSVC
Gateway 1 SW
Black Rock
Blacks Flat UM Cr
Cove Fort
Desert Exp Range
Koosharem
Moab 4 NW
Kimberly Mine
Fish Lake RS
Wah Wah Ranch
Box Creek
Milford WB AP
Marysvale
Loa
Hanksville CAA AP
Paradox
Big Flat
Beaver
Piute Dam
La Sal
Beaver Canyon PH
Fruita
Minersville
Circleville
Antimony
Loche
Lund
Boulder
Parowan
Monticello
Panguitch
Widtsoe Escalante 2
Widtsoe RS
Hite
Northdale
Mocena
Summit
Cedar City CAA AP
Panguitch Lake RS
Escalante
Cedar Point
Enterprise Beryl Jct
Castle Valley
Bryce Canyon CAA AP
Bryce Canyon NP
Cedar City PH
Hatch
Tropic
Blanding
Caliente
Cedar City College Rch
Cedar Breaks NM
Webster Flat
Enterprise
Duck Creek RS
New Harmony
Alton
Forty Mile Dance Hall
Veyo PH
Hovenweep NM
Gunlock PH
Bluff
Orderville
Zion NP
La Verkin
Mexican Hat
St George PH
Navajo Mtn
Monument Valley Gouldings
Kanab PH
Short Creek
Fredonia
Page
Beaver Dam
Lees Ferry
Dinnehotso
Kayenta
Betatakin
STATION LEGEND
Precipitation only
Precipitation, storage
Precipitation and Temperature
Precipitation, Temp. and Evaporation
Type of gage ○ Non-recording;
● Recording; ◑ Both types
Double circle combination indicates the availability of more detailed meteorological data.
UTAH
STATUTE MILES
10 0 10 20 30 40
ALL UTAH STATIONS USE 105TH MERIDIAN TIME
Revised 3-58 WRPC-SF
114°
113°
112°
111°
110°
109°
42°
41°
40°
39°
38°
37°
112°
111°55'
111°50'
111°45'
40° 50'
40° 45'
40° 40'
USCOMM-WB-ASHEVILLE

UTAH

CLIMATES OF THE STATES

VERMONT

(Normals, Means and Extremes tables revised 1973 and 1975. Basic report revised December 1959.)

Climate of Vermont

Robert E. Lautzenheiser, Weather Bureau State Climatologist

PHYSICAL DESCRIPTION: -- "The Green Mountain State" occupies 9,609 square miles, fully one-seventh of New England's total area. Though Vermont is the only New England state without a coastline on the Atlantic Ocean, most of its boundary is water. The Connecticut River forms the entire eastern border. Lake Champlain marks over 100 miles of the western boundary. Vermont extends southward from near the 45° parallel of latitude almost 160 miles to about 20 miles south of the 43d parallel. Vermont widens northward from about 40 to 90 miles across.

The terrain is hilly to mountainous. The Green Mountains extend the length of the State. They rise to their highest elevation at Mt. Mansfield, 4,393 feet above sea level. Many peaks in this range rise to over 3,000 feet, as do several others in eastern Vermont. Elevations of less than 500 feet above sea level are mostly confined to the lowlands paralleling Lake Champlain in the west and to the central and southern portions of the Connecticut Valley in the east. Much of the State ranges from 500 to 2,000 feet in elevation. The glacier of the great Ice Age accounts for many topographical features, lakes, and soils. Inland waters cover more than 300 square miles.

Two-thirds of Vermont is forest, contained in National, State, municipal, and private reserves and in farm woodlands. A considerable area, especially in the north, is sparsely settled. The mountains, hills, lakes, streams, and forests combine to make Vermont a state noted for its scenic beauty.

GENERAL CLIMATIC FEATURES: -- Vermont shares with the other New England states in the chief climatic characteristics. These include: (1) Changeableness of the weather, (2) large range of temperature, both daily and annual, (3) great differences between the same seasons in different years, (4) equable distribution of precipitation, and (5) considerable diversity from place to place. The regional climatic influences are modified in Vermont by varying elevations, types of terrain, and distances from the Atlantic Ocean and from Lake Champlain. The State has been divided into three climatological divisions (Western, Northeastern, and Southeastern) which take into account the main features of these modifying factors, in a general way. To take all local factors into consideration would require an impractical number of areal divisions.

Vermont lies in the "prevailing westerlies", the belt of generally eastward air movement which encircles the globe in middle latitudes. Embedded in this circulation are extensive masses of air originating in higher or lower latitudes and interacting to produce low-pressure storm systems. Relative to most other sections of the country, a large number of such storms pass over or near Vermont. The majority of air masses affecting this

State belong to three types: (1) Cold, dry air pouring down from subarctic North America, (2) warm, moist air streaming up on a long overland journey from the Gulf of Mexico and other subtropical waters, and (3) cool, damp air moving in from the North Atlantic. Because the atmospheric flow is usually from a westerly direction, Vermont is more influenced by the first two types than it is by the third. In other words, the Atlantic Ocean sometimes affects Vermont, but does not dominate its climate.

The procession of contrasting air masses and the relatively frequent passage of "Lows" bring about on the average a twice-weekly alternation from fair to cloudy or stormy conditions, attended by often abrupt changes in temperature, moisture, sunshine, wind direction and speed. There is no regular or persistent rhythm to this sequence, and it is interrupted by periods during which the weather patterns continue the same for several days, infrequently for several weeks. Vermont weather, however, is cited for variety rather than monotony. Changeability is also one of its features on a longer time-scale. That is, the same month or season will exhibit varying characteristics over the years, sometimes in close alternation, and sometimes arranged in similar groups for successive years. A "normal" month, season, or year is indeed the exception rather than the rule.

The basic climate, as outlined above, obviously does not result from the predominance of any single controlling weather regime, but is rather the integrated effect of a variety of weather patterns. Hence, "weather averages" in Vermont usually are not sufficient for important planning purposes without further climatological analysis.

The Western Division is a relatively narrow band running the full length of the State west of the Green Mountains. This Division is least affected by Atlantic Ocean influences. Because its northern portion is moderated by Lake Champlain it can be included with southwestern Vermont even though its north-south extension is so long. The Northeastern Division is the largest of the three and includes the northeastern, north-central, and east-central portions of Vermont, excepting a narrow strip in the Connecticut River Valley in the east-central portion. This strip is included as a part of the Southeastern Division because of its lower elevation.

TEMPERATURE: -- The annual mean temperature is near 43°F. in the Northeastern Division, 44°F. in the Southeastern, and 46°F. in the Western. Averages vary also within the divisions. Elevation, slope, and other local environmental aspects, including urbanization, all have an effect. As an extreme example of the effect of altitude, a comparison between the summit station on Mt. Mansfield with Enosburg Falls is interesting. Though these stations are about the same distance from Lake Champlain, the average temperature for the year 1958 on Mt. Mansfield was only slightly above freezing, 32.8°F.; Enosburg Falls, at 3,500 feet lower elevation, was nearly 10° warmer, with 42.0°F.; and Enosburg Falls is about 25 miles north of Mt. Mansfield. The highest temperature of record in the State is 105°F. observed July 4, 1911 at Vernon; the lowest, -50°F., December 30, 1933, at Bloomfield.

Summer temperatures are delightfully comfortable as a rule. They are also reasonably uniform over the State, excepting topographical extremes. Long-period means for July average near 70°F. in the Western Division and near 68°F. in the other Divisions. Average daily minima in July are in the 50's over nearly the entire State. The average daily maxima reach only near 80°F. Hot days with maxima of 90°F. or higher average less than 10 per year at most stations. The frequency varies from place to place and from year to year. In the coolest summers, they range, in frequency of occurrence, from none at many stations to only a few at the warmest stations. In the warmest years many stations still have less than 10, but the frequency ranges up to as high as 30 at the warmer sites. Even after one of these hot days the temperature is likely to fall to 60°F. or lower during the night. The average daily range is 20° to 30° in summer, with the variation averaging a little more in the south than in the north. The diurnal range may reach 40°F. or more during cool, dry weather in valleys and lowlands. Late spring or early fall freeze may be a threat at a few of the more susceptible areas.

Temperatures from place to place vary more in winter than in summer. The Northeastern Division average in January is near 17°F. The Southeastern Division average is near 19°F. and the Western Division, 21°F. The daily temperature range is less in winter than in summer, averaging near 20°F. Days with subzero readings are common at most stations in winter. They number from 10 to 40 per year in the southern portion and from 20 to 50 in the north. The number exceeds 60 at some stations in the coldest winters and may be less than 10 at other stations in the mildest winters.

The growing season for vegetation subject to injury from freezing temperature averages 130 to 150 days in much of the Western Division and along the Connecticut River in the Southeastern Division. Elsewhere, and including the extreme southern portion of the Western Division the season varies from 100 to 130 days. Local topography causes exceptions and some localities have growing seasons as short as 80 to 90 days. The growing season begins in May and ends in September for most of the State.

PRECIPITATION: -- Vermont's precipitation, fortunately, is well distributed through the year. The summer months ordinarily receive adequate amounts for growing crops over the entire State. Winter precipitation is noticeably less than summer rainfall in the northern and western portions of the State. This difference is greater in those areas than in any other part of New England. New England as a whole is noted for the even distribution of its precipitation throughout the year, an effect due to the influence of the Atlantic Ocean. "Wet" or "dry" seasons, climatic characteristics of most parts of the World, are not, normally, conditions with which this section has to contend. This ocean influence is still strongly felt in southeastern Vermont, but it becomes weaker with increasing distance from the ocean. Low-pressure, or frontal, storm systems are the principal year-round moisture producers. When this activity ebbs somewhat in summer, bands or patches of thunderstorms increase in activity, more than making up the difference. Though brief and often of small extent, the thunderstorms produce the heaviest local rainfall intensities. They sometimes cause minor washouts of roads and soils. Rains of 1 to 2 inches in 1 hour can be expected at least once in a 10-year period.

Variations in monthly totals are extreme, ranging from none to over 10 inches. Such large fluctuations are rare. A large majority of monthly totals falls in the range of from 50 to 200 percent of normal. As prolonged droughts are infrequent, irrigation water is available during the fairly common shorter dry spells of summer. Similarly widespread floods are infrequent. However, torrential rains on November 2-3, 1927, caused flood damage estimated at $26 million. Other floods of note occurred in 1801, 1826, 1830, 1886, 1895, 1897, 1909, 1913, 1936, and 1947. Floods occur most often in the spring when they are caused by rainfall and melting snow. Stages of spring over-bank flooding are frequently increased by ice jams. Local flash floods result on occasions from short period summer storms between May and November.

The mean annual runoff in the streams ranges from about 10 inches in portions of the Lake Champlain drainage to 40 inches in southern Vermont. The Connecticut River forms the eastern border and its tributaries drain the major portion of Vermont. In the northwest portion, rivers drain into Lake Champlain or directly to the St. Lawrence. A small area in southwest Vermont drains to the Hudson River.

Total annual precipitation averages nearly 45 inches in the Southeastern Division and nearly 38 inches in the other divisions. Individual means vary considerably from station to station, especially within the Southeastern Division. Bellows Falls, with less than 41 inches per year, and Searsburg Station, with 55 inches, are less than 30 miles apart. The mountainous character of much of the State largely accounts for the variability from place to place.

Occasionally freezing rain occurs, coating exposed surfaces with troublesome ice. Most areas can expect at least one such occurrence in a winter. Frequency of days with measurable precipitation is between 120 and 160 days per year. As much as 6 inches of rain in 24 hours is rare in Vermont. Most stations have never recorded that much in a single day. However, Somerset received 8.77 inches in 24 hours during the flooding rains of November 1927.

SNOWFALL: -- Average annual total snowfall is from 55 to 65 inches in much of the Western Division and also in parts of the Connecticut River Valley. Elsewhere the annual averages vary greatly. They range upward to as much as 100 inches and, at a few stations, 100 to 125 inches. Topographical differences cause large variations in a short distance. As an example, Bennington has only about 55 inches per year, while Somerset, with over 120 inches, is only about 15 miles away but at a much higher elevation.

Snowfall is highly variable from season to season. It also varies for the same month in different years as well as from place to place. Variations in seasonal totals are mostly from about 50 to 150 percent of the long-period average. Totals for the least snowy seasons range from 25 to 50 percent of the greatest seasonal amounts. Month to month variations are much greater. Burlington's maximum monthly total is 34.3 inches in February 1958, but only 1.3 inches fell in that month in 1957.

The average number of days with 1 inch or more of snowfall in a season varies from near 20 to 40. The frequency increases with elevation. Most winters have several snowstorms of 5 inches or more per year. Storms of this magnitude may temporarily disrupt transportation.

One of the heaviest single snowstorms of record was that of March 11-14, 1888, known as the "Great Blizzard". Amounts in the southwestern part of the State ranged from 40 to 50 inches and in the southeastern part, from 30 to 40 inches. Drifts of 15 to 40 feet high were reported. Most of northern Vermont received from 20 to 30 inches in this storm. However, snowfalls of 20 inches or more are unusual in any part of the State. The heaviest 24-hour falls of record at many stations do not exceed 25 inches.

Snow cover is continuous throughout the winter season as a rule. Depth of snow on the ground reaches its maximum for much of the State in the latter part of February. At the highest elevations, however, the date falls in the middle of March. Water stored in the snow is an important

contribution to the water supply. Spring melting is usually too gradual to produce serious flooding.

OTHER CLIMATIC FEATURES: -- Sunshine averages near 50 percent of possible on a year-round basis, but varies with topography. Data is not sufficient to describe this in detail. Higher elevations and peaks are much more cloudy, especially in winter, probably reducing the percentage to as low as 40 in local areas. Sunshine is most abundant during the summer season.

Heavy fog occurrence varies remarkably with location and topography but, again, not enough data are available to describe this in detail. Persistent fogs are sometimes experienced on the higher elevations. The duration of fog diminishes over flat and valley locations. But the shorter duration heavy ground fogs of early morning occur frequently at susceptible places in these areas. The number of days with fog probably varies from 10 to 60 per year over the State, except possibly even more on the highest mountain peaks.

WINDS AND STORMS: -- Vermont lies in the region of prevailing westerlies -- wind from the northwest in winter, and from the southwest in the warmer part of the year. But because the rugged topography has a strong influence on the direction of the wind, many areas have prevailing winds paralleling a valley. The major valleys tend to lie in a north-south direction. Thus prevailing winds may be from the north in winter and from the south in the warmer seasons in those areas.

Coastal storms, or "northeasters", are well known to New England. Their influence on Vermont is minimized by its inland location. They remain a factor, however, especially in the Southeastern Division. They generate very strong winds and heavy rain or snow. Some of the heavier snows are produced by these storms.

Storms of tropical origin may occasionally affect Vermont in summer or fall, but only rarely contain destructive winds. The very severe, rapidly moving hurricane of September 1938 is best remembered. Its path crossed the entire State, from near Wilder to Burlington. However, Vermont is far enough inland so that, usually, winds are considerably weakened by the time tropical storms reach the State, and are generally only light to moderate. Rainfall associated with these storms may, however, remain heavy.

Tornadoes are not common phenomena. Yet, on a per unit area basis, Vermont ranks with many other states in frequency of tornado occurrence. One or more of these most violent storms may occur in a year. Historical accounts suggest that the most notable Vermont tornado occurred on June 23, 1782. Entering the State at the southwest corner, it traveled northward and eastward and crossed into New Hampshire near Weathersfield. Fortunately, most tornadoes are very small, affecting a very localized area. Due to the extent of forested or sparsely settled areas, a large percentage that do occur are probably neither seen, recorded, nor do appreciable damage. They may occur even in the northern portion of the State. About 73 percent occur between May 15 and September 15. About 78 percent strike between 2 and 7 p.m. The peak months are June and July and the peak hour is 5 to 6 p.m. The chance of a tornado striking any given spot is extremely small.

Thunder and hailstorms also have a frequency maximum from midspring to early fall. Thunderstorms occur on 20 to 30 days per year. The most severe are attended by hail. Hail can damage or even ruin field crops, break glass, dent automobiles, and damage other vulnerable exposed objects. The size of an area struck by a hailstorm, however, is usually small. Glaze and icestorms of winter can make travel hazardous. These are usually of brief duration. At least one ice storm may be expected each year. A few widespread and prolonged ice storms have occurred. Besides affecting travel and transport, they also break trees or limbs, utility lines and poles. In such structural design as steel towers, possible ice load should be considered. The ice load also magnifies the wind stress by increasing the area exposed to the wind.

CLIMATE AND ECONOMY: -- Activities in Vermont are profoundly influenced by climate. Tree growth is especially favored. Covering two-thirds of the area, forests are a major scenic attraction. The spectacular coloration of foliage in the autumn is of special interest, drawing countless visitors. Lumbering and related wood products are leading industries. The ample supply of rainfall provides not only for timber growth but also the huge amount of water required in making of paper and other manufactures. Favored industries also include the manufacturing of machinery, textiles, and leather, and stone, clay, and glass products. A great diversity of other interests takes advantage of the abundant water supply. A large portion of the State's electrical power comes from a well developed hydroelectric system.

Climate is a significant factor in Vermont agriculture. Principal farm specialties include dairying, poultry raising, tree fruit, and truck gardening. Fresh milk and milk products are the leading farm outputs. These amount to one-third of New England's total dairy production. Apples are the most prolific of the tree fruits, with quality production an important commercial pursuit. Vermont is the leading state in top quality maple syrup and sugar production. Strawberries are an important truck product. A large acreage is devoted to pasture and hay, and to oats and corn.

Climate is particularly important to a major industry, the tourist and vacation trade, amounting to over $100 million annually. Summer camps abound on the shores of many of the State's 400 lakes and ponds. Abundant game and teeming lakes and streams draw sportsmen from far and near. Skiing, with related winter sports, is a very important seasonal attraction, made possible by the abundant snowfall. The winter sports industry has grown rapidly in recent years making Vermont a four-season vacation area.

SELECTED REFERENCES

General:

1. National Planning Association: The Economic State of New England (1954).

2. U. S. Dept. Agriculture: Atlas of American Agriculture (1936)

3. --- : Climate and Man Yearbook of Agriculture for 1941, Part 5, Climatic data, with special reference to agriculture in the United States.*

Specialized:

1. Brooks, C. F.: "New England Snowfall", Monthly Weather Review, Vol. 45 (1917).

2. --- : "The Rainfall of New England - General Statement", Journ. N. Eng. Water Works Assoc., Vol. 44 (1930).

3. Brown, Rodger A.: "Twisters in New England", unpublished manuscript, Antioch College, 1957.*

4. Church, P. E.: "A Geographical Study of New England Temperatures", Geogr. Review, Vol. 26 (1936).

5. Eustis, R. S.: "Winds over New England in relation to topography", Bull. Amer. Met. Soc., Vol. 23 (1942).

6. Galway, Joseph G.: "A Statistical Study of New England Snowfall", unpublished manuscript of U. S. Weather Bureau (1954).*

7. Goodnough, X. H.: "Rainfall in New England", Journ. N. Eng. Water Works Assoc., Vols. 29 (1915), 35 (1921) and 40 (1926).*

8. Palmer, Robert S.: "Agricultural Drought in New England". Technical Bulletin 97, Agricultural Experiment Station, U. of New Hampshire, Durham, N. H. (1958).

9. Perley, S.: Historic Storms of New England (1891).*

* References marked with an asterisk are useful sources of data; the others are principally studies of the important climatic elements.

4. --- : Soil (Yearbook of Agriculture for 1957).

5. --- : Climatological Data, New England (issued monthly and annually, 1888 ---; pub. under various other titles previous to Jan. 1921).*

10. Stone, R. G.: "Distribution of snow depths over New York and New England", Trans. Amer. Geophy. Union (1940).

11. --- : "The average length of the season with snow cover of various depths in New England", Trans. Amer. Geophy. Union (1944).

12. Upton, W.: "Characteristics of the New England Climate", Annals Harvard Astron. Obser. (1890).

13. U. S. Weather Bureau: Tabulations of frequencies of various climatic elements for various selected stations. Available on microfilm at library of Weather Bureau State Climatologist, 1900 Post Office Bldg., Boston 9, Mass.

14. Weber, J. H.: "The Rainfall in New England. Historical Statement. Annual Rainfall. Seasonal Rainfall. Mean Monthly Rainfall of Southern New England. Maximum and Minimum Rainfall of Southern New England". Journ. N. Eng. Water Works Assn., Vol. 44 (1930).

15. White, C. V.: "Rainfall in New England", Journ. N. Eng. Water Works Assn., Vols. 56 (1942) and 57 (1943).*

16. Weather Bureau Technical Paper No. 15 - Maximum Station Precipitation for 1, 2, 3, 6, 12, and 24 Hours.

17. Weather Bureau Technical Paper No. 16 - Maximum 24-Hour Precipitation in the United States. Washington, D. C. 1952.

18. Weather Bureau Technical Paper No. 25 - Rainfall Intensity-Duration-Frequency Curves. For selected stations in the United States, Alaska, Hawaiian Islands, and Puerto Rico.

BIBLIOGRAPHY

(A) Climatic Summary of the United States (Bulle-tim W) 1930 edition, Section 84 (New Hampshire and Vermont). U. S. Weather Bureau

(B) Climatic Summary of the United States, New England - Supplement for 1931 through 1952 (Bulletin W Supplement). U. S. Weather Bureau

(C) Climatological Data - New England, U. S. Weather Bureau

(D) Climatological Data National Summary. U. S. Weather Bureau

(E) Hourly Precipitation Data - New England. U. S. Weather Bureau

(F) Local Climatological Data. U. S. Weather Bureau, for Burlington, Vermont.

FREEZE DATA

STATION	Freeze threshold temperature	Mean date of last Spring occurrence	Mean date of first Fall occurrence	Mean No. of days between dates	Years of record Spring	No. of occurrences in Spring	Years of record Fall	No. of occurrences in Fall	STATION	Freeze threshold temperature	Mean date of last Spring occurrence	Mean date of first Fall occurrence	Mean No. of days between dates	Years of record Spring	No. of occurrences in Spring	Years of record Fall	No. of occurrences in Fall
BELLOWS FALLS	32	05-16	10-02	139	20	20	20	20	DORSET 1 SSW	32	05-29	09-14	108	10	10	10	10
	28	04-27	10-12	168	20	20	20	20		28	05-12	09-24	135	10	10	10	10
	24	04-15	10-28	195	20	20	20	20		24	05-03	10-04	154	10	10	10	10
	20	03-28	11-12	229	20	20	20	20		20	04-19	10-16	180	10	10	10	10
	16	03-22	11-23	246	20	20	20	20		16	04-01	11-03	216	10	10	10	10
BLOOMFIELD	32	06-02	09-16	107	30	30	30	30	ENOSBURG FALLS	32	05-27	09-20	117	25	25	25	25
	28	05-19	09-27	131	30	30	30	30		28	05-12	09-29	140	25	25	25	25
	24	05-06	10-11	158	30	30	30	30		24	04-29	10-11	165	25	25	25	25
	20	04-22	10-20	181	30	30	30	30		20	04-16	10-25	192	25	25	25	25
	16	04-13	11-02	203	30	30	30	30		16	04-06	11-08	216	25	25	25	25
BURLINGTON WB	32	05-08	10-03	148	30	30	30	30	NORTHFIELD NORWICH U	32	05-27	09-17	113	30	30	30	30
	28	04-22	10-18	179	30	30	30	30		28	05-15	09-29	137	30	30	30	30
	24	04-13	11-03	204	30	30	30	30		24	04-28	10-11	165	30	30	30	30
	20	04-03	11-15	227	30	30	30	30		20	04-16	10-25	193	30	30	30	30
	16	03-24	11-25	246	30	30	30	30		16	04-03	11-09	219	30	30	30	30
CAVENDISH	32	05-30	09-13	106	30	30	30	30	RUTLAND	32	05-15	09-24	131	30	30	30	30
	28	05-14	09-29	138	30	30	30	30		28	05-01	10-06	159	30	30	30	30
	24	05-02	10-08	159	29	29	30	30		24	04-19	10-21	185	30	30	30	30
	20	04-18	10-21	186	29	29	30	30		20	04-06	11-04	212	30	30	30	30
	16	04-05	11-04	213	29	29	30	30		16	03-28	11-17	234	30	30	30	30
CHELSEA	32	06-02	09-10	100	30	30	30	30	ST JOHNSBURY	32	05-21	09-23	125	30	30	30	30
	28	05-19	09-25	129	30	30	30	30		28	05-09	10-03	148	30	30	30	30
	24	05-07	10-06	152	30	30	30	30		24	04-24	10-18	177	30	30	30	30
	20	04-21	10-19	181	30	30	30	30		20	04-10	11-01	204	30	30	30	30
	16	04-10	11-02	206	30	30	30	30		16	04-02	11-12	225	30	30	30	30
CORNWALL	32	05-06	10-07	154	30	30	30	30	SOMERSET	32	06-08	08-30	83	30	30	30	30
	28	04-25	10-18	176	30	30	30	30		28	05-24	09-23	122	30	30	30	30
	24	04-13	11-05	206	30	30	30	30		24	05-11	10-06	149	30	30	30	30
	20	04-02	11-12	225	30	30	30	30		20	04-24	10-19	179	30	30	30	30
	16	03-24	11-22	242	30	30	30	30		16	04-17	10-30	196	30	30	30	30
									WOODSTOCK	32	05-28	09-18	113	30	30	30	30
										28	05-14	09-29	138	30	30	30	30
										24	04-29	10-11	165	30	30	30	30
										20	04-15	10-25	193	30	30	30	30
										16	04-04	11-08	218	30	30	30	30

Data in the above table are based on the period 1921-1950, or that portion of this period for which data are available.

Means have been adjusted to take into account years of non-occurrence.

A freeze is a numerical substitute for the former term "killing frost" and is the occurrence of a minimum temperature at or below the threshold temperature of 32°, 28°, etc.

Freeze data tabulations in greater detail are available and can be reproduced at cost.

TROPICAL CYCLONE DATA HAVING IMPORTANCE FOR THIS STATE IS INCLUDED IN STATISTICS AND CHARTS ON PAGES 1161 THROUGH 1164 .

*MEAN TEMPERATURE AND PRECIPITATION

STATION	JAN Temp	JAN Precip	FEB Temp	FEB Precip	MAR Temp	MAR Precip	APR Temp	APR Precip	MAY Temp	MAY Precip	JUN Temp	JUN Precip	JUL Temp	JUL Precip	AUG Temp	AUG Precip	SEP Temp	SEP Precip	OCT Temp	OCT Precip	NOV Temp	NOV Precip	DEC Temp	DEC Precip	ANN Temp	ANN Precip
VERMONT																										
NORTHEASTERN																										
BLOOMFIELD	16.3	2.40	17.4	2.16	27.3	2.47	40.7	3.18	53.1	3.60	62.4	4.17	67.0	4.23	65.0	3.75	57.3	3.99	46.7	3.29	34.4	3.44	20.1	2.58	42.3	39.26
CHELSEA	16.8	2.48	17.5	2.17	28.0	2.66	40.9	3.14	53.1	3.57	62.4	3.67	67.2	4.09	64.8	3.28	56.9	3.56	46.5	3.02	34.4	3.21	20.8	2.46	42.4	37.31
NEWPORT	15.6	2.35	17.0	2.24	26.6	2.51	40.3	3.00	53.5	3.07	63.1	3.73	67.5	4.21	65.5	3.42	57.5	3.71	46.9	3.07	34.2	3.02	19.7	2.49	42.3	36.82
NORTHFIELD NORWICH UNI	17.8	2.20	19.0	2.06	28.9	2.46	41.4	2.60	53.6	3.14	62.8	3.33	67.5	3.53	65.0	3.20	57.3	3.23	47.3	2.67	35.3	3.04	21.5	2.38	43.1	33.84
ROCHESTER		3.02		2.98		3.61		3.63		3.98		3.81		4.37		3.71		3.92		3.34		3.81		3.26	44.4	43.44
SAINT JOHNSBURY	17.5	2.53	19.5	2.17	29.6	2.51	42.8	2.87	55.7	3.30	65.1	3.86	69.6	3.52	67.3	3.35	59.1	3.53	48.5	2.87	35.8	3.14	21.7	2.59	44.4	36.24
WEST BURKE		2.76		2.50		2.76		3.16		3.56		3.84		3.83		3.40		4.00		3.28		3.42		3.08		39.59
DIVISION	16.5	2.41	17.7	2.20	27.8	2.60	41.0	3.06	53.7	3.43	62.9	3.87	67.6	3.95	65.3	3.48	57.4	3.73	47.0	3.13	34.7	3.24	20.4	2.60	42.7	37.70
WESTERN																										
BURLINGTON WB AIRPORT	17.9	1.89	18.1	1.53	29.3	2.19	42.3	2.63	55.4	2.89	65.5	3.57	70.4	3.75	68.1	3.01	59.9	3.14	48.2	2.89	36.4	2.85	22.8	1.88	44.5	32.22
CORNWALL	20.7		21.7		31.6		44.8		57.3		66.3		71.3		69.2		61.2		50.4		38.1		24.7		46.4	
RUTLAND	21.5	2.55	22.4	2.11	32.0	2.70	44.6	3.00	56.3	3.71	65.1	4.25	69.5	4.70	67.2	3.45	59.6	3.76	49.5	2.96	38.0	3.14	25.3	2.34	45.9	38.67
DIVISION	20.6	2.46	21.5	2.11	31.2	2.74	44.0	3.11	56.2	3.49	65.2	3.78	70.0	4.17	67.7	3.59	60.0	3.65	49.5	3.04	37.6	3.10	24.5	2.55	45.7	37.79
SOUTHEASTERN																										
BELLOWS FALLS		3.22		2.70		3.33		3.78		3.04		3.57		3.80		3.22		3.57		2.85		4.04		3.03		40.95
CAVENDISH	19.0	3.35	20.0	3.10	29.7	3.72	42.6	3.80	55.2	3.85	64.0	3.96	68.5	4.37	65.9	3.29	58.0	3.54	47.2	3.27	35.3	3.85	22.5	3.22	44.0	43.32
MAYS MILL		4.25		3.60		4.78		4.40		4.93		4.16		4.41		4.14		4.05		3.89		4.84		4.20		52.04
READSBORO 1 SSE		4.04		3.24		4.34		4.39		4.62		4.20		4.29		4.05		4.36		3.59		4.27		3.90		49.79
SEARSBURG MOUNTAIN		3.97		3.29		4.40		4.76		5.02		4.78		4.68		4.79		5.18		4.42		4.61		4.18		54.08
SEARSBURG STATION		4.58		3.74		4.79		4.87		5.02		4.47		4.61		4.08		4.75		4.14		4.88		4.67		55.01
SOMERSET	17.6	4.56	17.2	3.71	25.4	4.86	37.8	4.69	50.0	4.74	58.8	4.71	62.8	4.29	60.7	4.11	53.7	4.93	44.6	4.11	32.8	4.78	20.2	4.54	40.1	54.00
VERNON		3.38		2.71		3.70		3.82		4.01		3.82		4.06		3.08		3.32		2.80		3.28		2.70		43.43
WHITE RIVER JUNCTION 1		2.86		2.43		2.86		3.27		3.41		3.47		4.47		4.15		4.57		3.89		4.57		4.44		37.54
WHITINGHAM 3 W		4.34		3.46		4.57		4.58		4.77		4.30		4.77												52.11
WILDER	19.6	2.59	20.7	2.12	30.4	2.53	43.2	3.12	55.2	3.31	64.1	3.34	68.7	4.16	66.5	3.03	58.7	3.29	48.0	2.75	36.3	3.21	23.1	2.51	44.5	35.96
WOODSTOCK 3 ENE		3.30		2.78		3.49		3.57		3.72		3.73		4.22		3.31		3.49		3.12		3.76		3.19		41.68
DIVISION	19.2	3.60	19.8	3.03	29.3	3.87	41.9	3.90	54.2	3.97	63.2	4.00	67.8	4.14	65.5	3.51	57.7	3.95	47.4	3.38	35.5	4.09	22.6	3.50	43.7	44.94

* Averages for period 1931-1955, except for stations marked WB which are "normals" based on period 1921-1950. Divisional means may not be the arithmetical average of individual stations published, since additional data from shorter period stations are used to obtain better areal representation.

CONFIDENCE LIMITS

In the absence of trend or record changes, the chances are 9 out of 10 that the true mean will lie in the interval formed by adding and subtracting the values in the following table from the means for any station in the State. Because of the wider variation in mean precipitation, the corresponding monthly means and annual mean must be substituted for "p" in the precipitation table below to obtain mean precipitation confidence limits.

JAN		FEB		MAR		APR		MAY		JUN		JUL		AUG		SEP		OCT		NOV		DEC		ANN	
1.8	$.22\sqrt{p}$	1.5	$.18\sqrt{p}$	1.7	$.23\sqrt{p}$	1.1	$.23\sqrt{p}$	.9	$.26\sqrt{p}$	.7	$.30\sqrt{p}$	.7	$.31\sqrt{p}$	.9	$.29\sqrt{p}$	.8	$.33\sqrt{p}$	1.0	$.32\sqrt{p}$	1.1	$.29\sqrt{p}$	1.3	$.28\sqrt{p}$	.4	$.30\sqrt{p}$

COMPARATIVE DATA

Data in the following table are the mean temperature and average precipitation for Northfield, Vermont for the period 1906-1930 and are included in this publication for comparative purposes:

JAN		FEB		MAR		APR		MAY		JUN		JUL		AUG		SEP		OCT		NOV		DEC		ANN	
15.9	2.00	15.7	2.37	26.8	3.64	39.4	2.54	51.2	2.76	60.1	3.38	65.5	3.22	62.4	3.12	55.8	3.01	45.5	3.04	33.3	2.88	20.4	2.11	41.0	34.07

NORMALS, MEANS AND EXTREMES

(Table Revised 1973. Base Period for Climatological Normals: 1931-1960)

Station: BURLINGTON, VERMONT   INTERNATIONAL AIRPORT   Standard time used: EASTERN   Latitude: 44° 28′ N   Longitude: 73° 09′ W   Elevation (ground): 332 feet

Ø For period December 1964 through the current year.
Means and extremes above are from existing and comparable exposures. Annual extremes have been exceeded at other sites in the locality as follows:
Minimum precipitation 0.15 in October 1924; maximum precipitation in 24 hours 4.49 in November 1927; maximum snowfall in 24 hours 24.2 in
January 1934; highest temperature 101 in August 1944; lowest temperature -30 in January 1957.

NORMALS, MEANS AND EXTREMES

(Table Revised 1975. Base Period for Climatological Normals: 1941-1970)

Average station pressure mb.   Elev. 340 feet m.s.l.

Means and extremes above are from existing and comparable exposures. Annual extremes have been exceeded at other sites in the locality as follows:
Minimum monthly precipitation 0.15 in October 1924; maximum precipitation in 24 hours 4.49 in November 1927; maximum snowfall in 24 hours 24.2 in
January 1934; highest temperature 101 in August 1944; lowest temperature -30 in January 1957.

REFERENCE NOTES APPLYING TO TABLES APPEAR ON THE PAGE FOLLOWING LAST TABLE.
(Caution: Letters and symbols may have different meanings in 1941-1970 tables than in earlier tables. See notes.)

Reference notes applying to Normals, Means, and Extremes tables for 1931–1960 base period.

(a) Length of record, years, based on January data. Other months may be for more or fewer years if there have been breaks in the record.

(b) Climatological standard normals (1931-1960).

• Less than one half.

+ Also on earlier dates, months, or years.

T Trace, an amount too small to measure.

Below zero temperatures are preceded by a minus sign. The prevailing direction for wind in the Normals, Means, and Extremes table is from records through 1963.

‡ ≥ 70° at Alaskan stations.

Unless otherwise indicated, dimensional units used in this bulletin are: temperature in degrees F.; precipitation, including snowfall, in inches; wind movement in miles per hour; and relative humidity in percent. Heating degree day totals are the sums of negative departures of average daily temperatures from 65° F. Cooling degree day totals are the sums of positive departures of average daily temperatures from 65° F. Sleet was included in snowfall totals beginning with July 1948. The term "ice pellets" includes solid grains of ice (sleet) and particles consisting of snow pellets encased in a thin layer of ice. Heavy fog reduces visibility to 1/4 mile or less.

Sky cover is expressed in a range of 0 for no clouds or obscuring phenomena to 10 for complete sky cover. The number of clear days is based on average cloudiness 0–3, partly cloudy days 4–7, and cloudy days 8–10 tenths.

Solar radiation data are the averages of direct and diffuse radiation on a horizontal surface. The langley denotes one gram calorie per square centimeter.

& Figures instead of letters in a direction column indicate direction in tens of degrees from true North; i.e., 09 - East, 18 - South, 27 - West, 36 - North, and 00 - Calm. Resultant wind is the vector sum of wind directions and speeds divided by the number of observations. If figures appear in the direction column under "Fastest mile" the corresponding speeds are fastest observed 1-minute values.

To 8 compass points only.

** The National Weather Service considers the accuracy of solar radiation data questionable; therefore, publication is suspended pending determination of corrected values.

Reference notes applying to Normals, Means, and Extremes tables for 1941–1970 base period.

(a) Length of record, years, through the current year unless otherwise noted, based on January data.

(b) 70° and above at Alaskan stations.

* Less than one half.

T Trace.

NORMALS - Based on record for the 1941-1970 period.

DATE OF AN EXTREME - The most recent in cases of multiple occurrence.

PREVAILING WIND DIRECTION - Record through 1963.

WIND DIRECTION - Numerals indicate tens of degrees clockwise from true north. 00 indicates calm.

FASTEST MILE WIND - Speed is fastest observed 1-minute value when the direction is in tens of degrees.

Mean Maximum Temperature (°F.), January

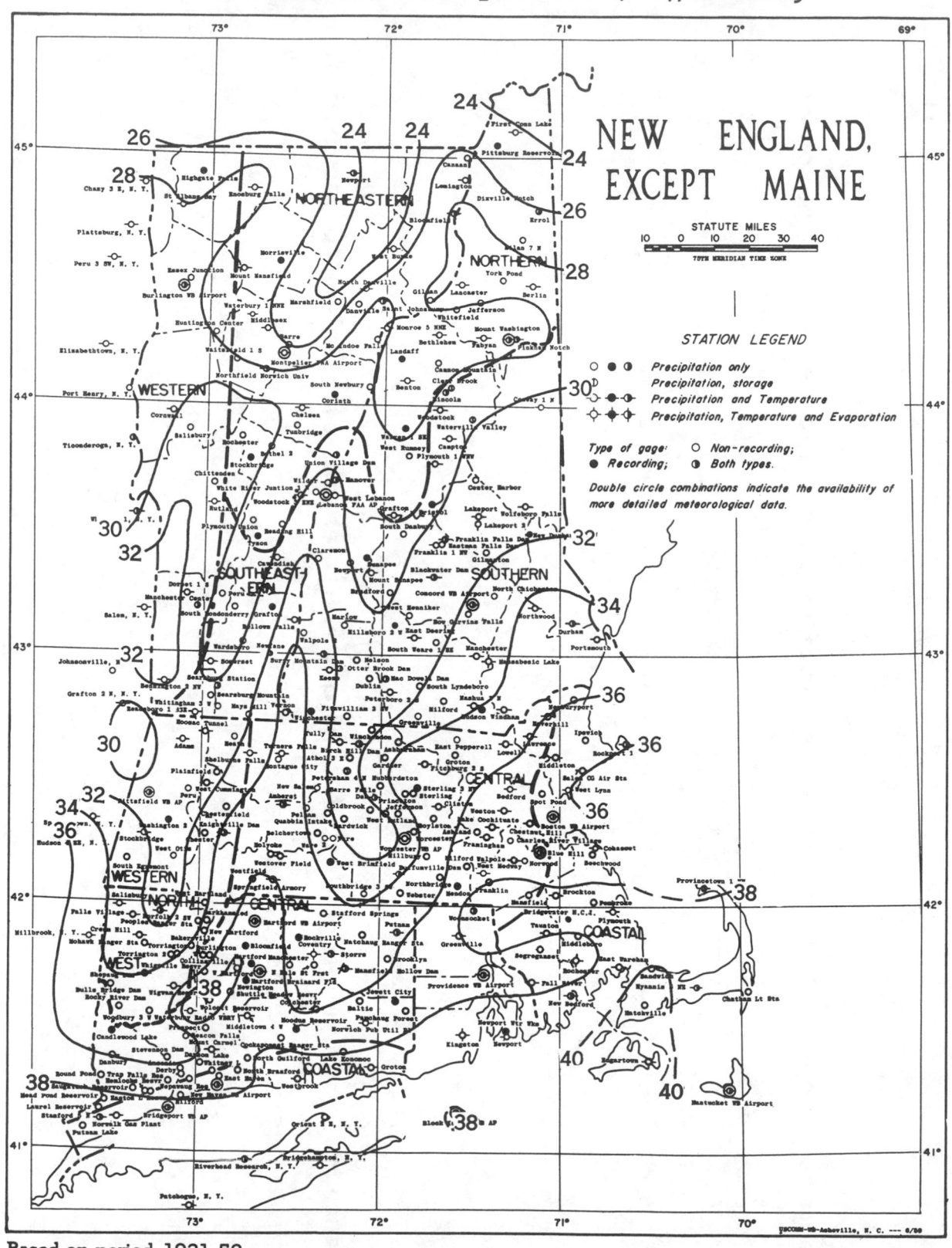

Based on period 1931-52

Isolines are drawn through points of approximately equal value. Caution should be used in interpolating on these maps, particularly in mountainous areas.

Mean Minimum Temperature (°F.), January

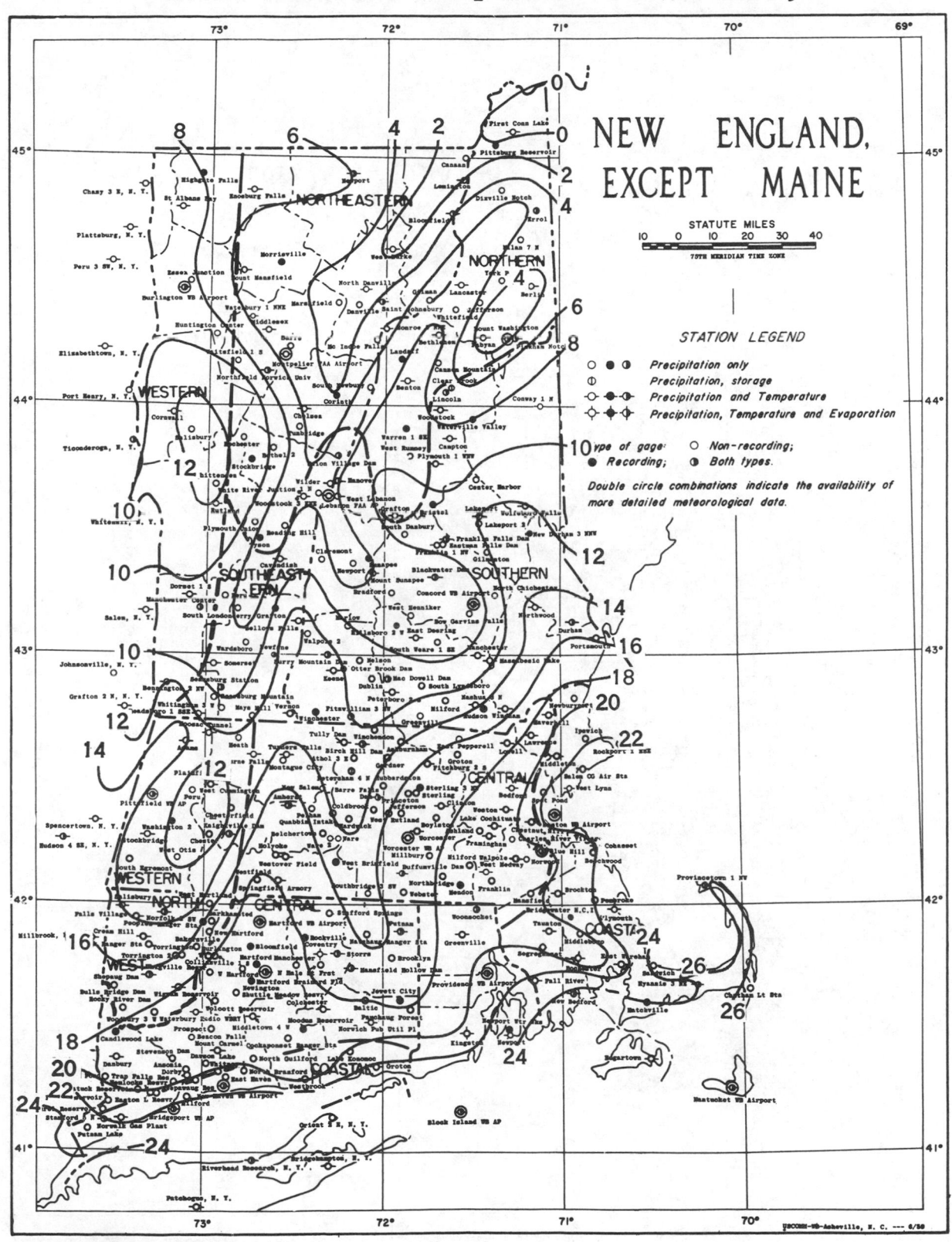

NEW ENGLAND, EXCEPT MAINE

STATUTE MILES
10   0   10   20   30   40

75TH MERIDIAN TIME ZONE

STATION LEGEND

○ ● ◐   *Precipitation only*
◍   *Precipitation, storage*
◒ ◓ ◑   *Precipitation and Temperature*
◇ ◈   *Precipitation, Temperature and Evaporation*

Type of gage: ○ Non-recording;
● Recording;   ◑ Both types.

Double circle combinations indicate the availability of
more detailed meteorological data.

Based on period 1931-52

Isolines are drawn through points of approximately equal value.   Caution should be used
in interpolating on these maps, particularly in mountainous areas.

USCOMM-WB-Asheville, N. C. --- 6/59

Mean Maximum Temperature (°F.), July

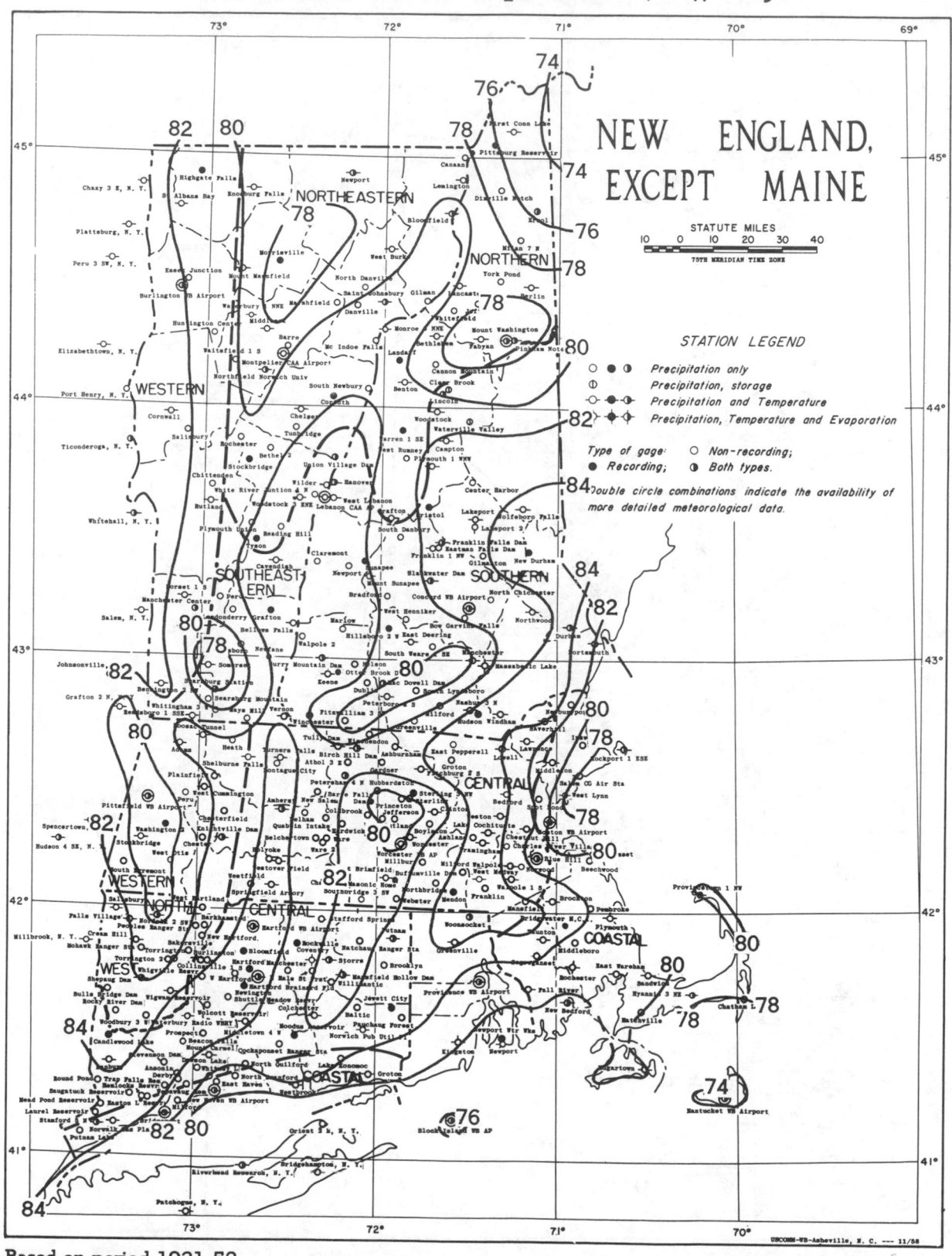

NEW ENGLAND, EXCEPT MAINE

STATUTE MILES
10 0 10 20 30 40
75TH MERIDIAN TIME ZONE

STATION LEGEND

○ ● ◑ Precipitation only
◐ Precipitation, storage
◒ ◓ Precipitation and Temperature
✦ Precipitation, Temperature and Evaporation

Type of gage: ○ Non-recording;
● Recording; ◐ Both types.

Double circle combinations indicate the availability of more detailed meteorological data.

Based on period 1931-52

Isolines are drawn through points of approximately equal value. Caution should be used in interpolating on these maps, particularly in mountainous areas.

Mean Minimum Temperature (°F.), July

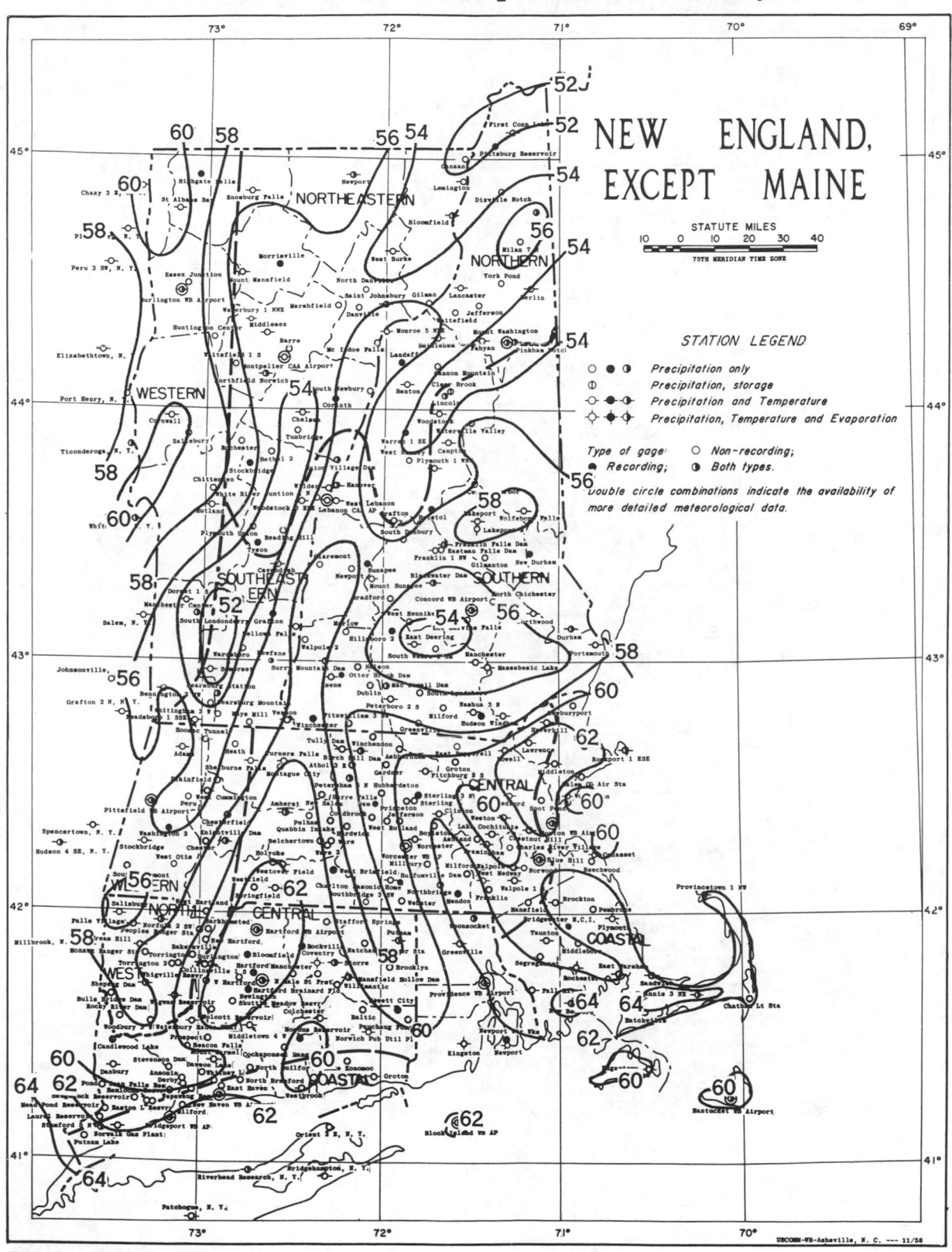

NEW ENGLAND, EXCEPT MAINE

STATUTE MILES

75TH MERIDIAN TIME ZONE

STATION LEGEND

○ ● ◐	Precipitation only
◑	Precipitation, storage
◌─ ●─ ◐─	Precipitation and Temperature
◇ ◆ ◈	Precipitation, Temperature and Evaporation

Type of gage: ○ Non-recording;
● Recording; ◐ Both types.

Double circle combinations indicate the availability of
more detailed meteorological data.

Based on period 1931-52

Isolines are drawn through points of approximately equal value. Caution should be used
in interpolating on these maps, particularly in mountainous areas.

1011

Mean Annual Precipitation, Inches

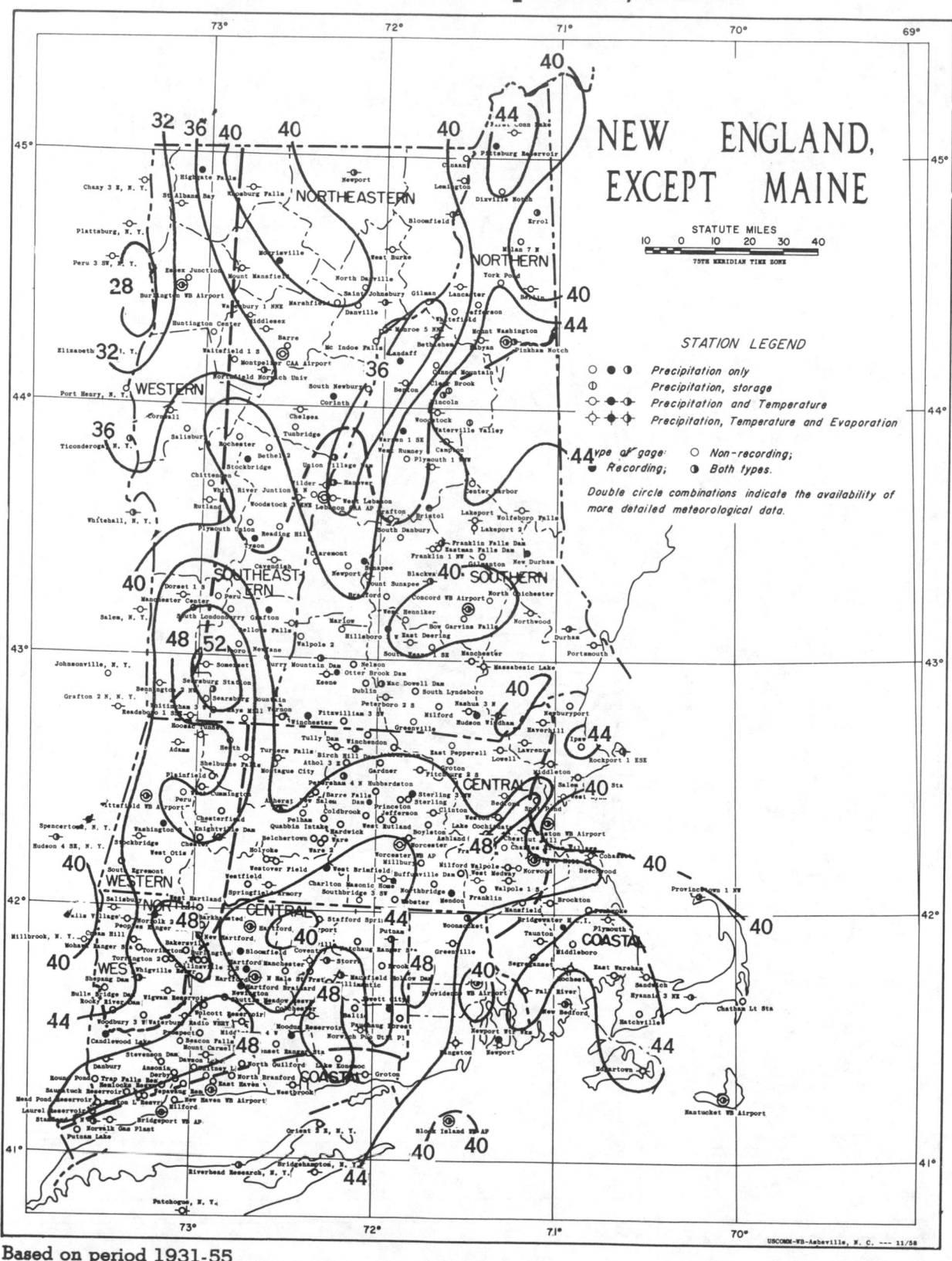

NEW ENGLAND, EXCEPT MAINE

STATUTE MILES

75TH MERIDIAN TIME ZONE

STATION LEGEND

○ ◑ ● Precipitation only
◍ Precipitation, storage
◌ ◐ ◉ Precipitation and Temperature
◇ ◆ ◈ Precipitation, Temperature and Evaporation

Type of gage: ○ Non-recording; ● Recording; ◑ Both types.

Double circle combinations indicate the availability of more detailed meteorological data.

Based on period 1931-55

Isolines are drawn through points of approximately equal value. Caution should be used in interpolating on these maps, particularly in mountainous areas.

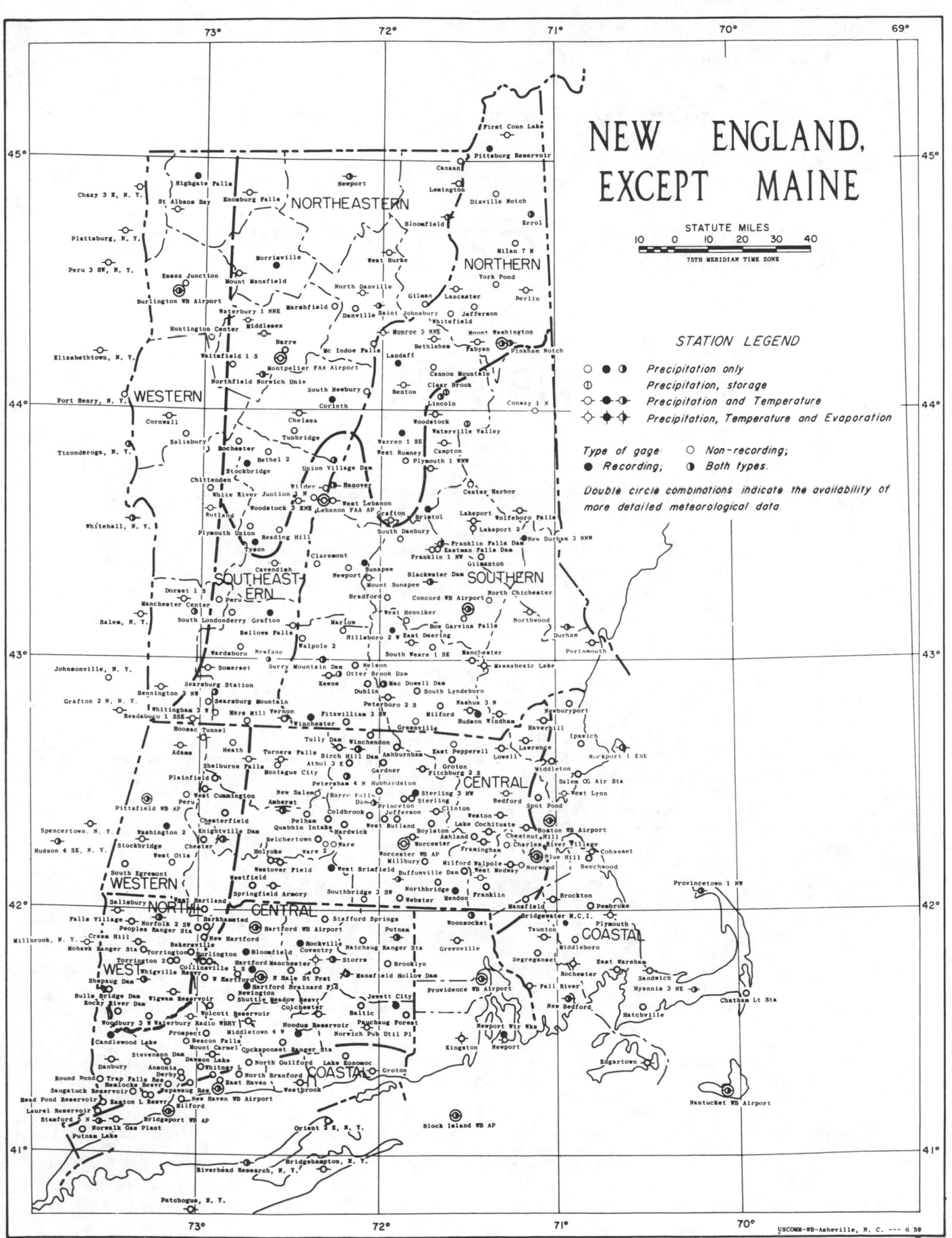

NEW ENGLAND, EXCEPT MAINE

STATUTE MILES

10   0   10   20   30   40

75TH MERIDIAN TIME ZONE

STATION LEGEND

○ ● ◑	*Precipitation only*	
◍	*Precipitation, storage*	
○– ●– ◑–	*Precipitation and Temperature*	
◇ ◆ ◈	*Precipitation, Temperature and Evaporation*	

Type of gage: ○ *Non-recording;*
Recording;   ◑ *Both types.*

Double circle combinations indicate the availability of
more detailed meteorological data.

CLIMATES OF THE STATES

VIRGINIA

(Normals, Means and Extremes tables revised 1971 and 1975. Basic report revised March 1971.)

Curtis W. Crockett, NOAA Climatologist - Virginia

Virginia is located on the east coast of the North American continent between latitudes 36-1/2° and 39-1/2° north. The State is triangular in shape with the longest north-south distance of about 200 miles and the longest east-west distance more than 400 miles. There are 40,815 square miles of area within the State of which 1,200 square miles are inland waters.

The State is composed of 3 natural topographic regions, namely: the Tidewater or coastal plains area, the Piedmont plateau or middle Virginia, and the western mountain region. Natural regions of lesser extent include the "Fall Line", located between Tidewater Virginia and the Piedmont region; the Blue Ridge Mountains that serve as the eastern boundary of the great Shenandoah Valley; the Shenandoah Valley; and the Appalachian plateau, in southwestern Virginia.

Tidewater Virginia extends westward from the Atlantic Coast and west shore of the Chesapeake Bay to the "Fall Line." The "Fall Line" extends from Great Falls in the north, southward through Richmond to Emporia. It is divided into necks or peninsulas by 4 principal rivers and by numerous estuaries that open into the Chesapeake Bay. There are numerous peninsulas, wide estuaries, and many swamp areas. The principal rivers include the Potomac, Rappahannock, York, and the James. Tidewater extends up these rivers to near the "Fall Line." The James and Potomac Rivers are navigable by medium sized ships across Tidewater Virginia.

The Piedmont region is more than 200 miles wide in southern Virginia, but the Virginia section becomes quite narrow in the north. This region from east to west becomes more rolling and hilly with a few isolated mountains and ridges appearing a few miles east of the Blue Ridge. Elevations in general range from about 300 feet above sea level in the east to about 1,000 feet in the west. The James, the largest river crossing this region, divides it into two parts.

West of the Piedmont, the Blue Ridge Mountains traverse the State from southwest to northeast. They range from narrow ridges in the north to a high, wide plateau southwest from Roanoke. Elevations range generally from 1,500 to 3,500 feet. Mt. Rogers, in western Grayson County, towers to 5,719 feet, the highest point in the State.

A great valley west of the Blue Ridge extends from Tennessee through Scott and Washington Counties in the south, northeastward to the northern-most point of the State. It embraces 6 separate valleys of which the largest is the Shenandoah. Elevations range mostly from 1,000 to 2,000 feet. This great valley of Virginia is well drained. The north is drained by the north and south forks of the Shenandoah River, thence into the Potomac; the central portion, by the Cow Pasture and Jackson Rivers flowing southeastward into the James; and the southwestern half of the valley is drained by the Roanoke River, the New River, and three forks of the Holston River. The New River drains northwestward into West Virginia and the Ohio River Basin. The Holston drains southwestward into the Tennessee River.

The Appalachian Plateau in southwestern Virginia is divided into many sharp ridges and deep valleys. Large coal beds underlie the area.

The climate of Virginia is determined by its proximity to the Atlantic Ocean, latitude, and topography. The State is in the zone of prevailing westerly movement of the earth's atmosphere, in or near the mean path of winter storm tracks, and in the mean path of tropical, moist air from the southwest Atlantic and Gulf of Mexico much of the summer and early fall seasons. The mountains provide the usual elevation effects on temperatures, which are distinctly lower in this section, and there are wide variations over short distances as elevations change. Summers in the mountains are comparatively cool, and winters are more severe. In addition, these mountains produce various steering, blocking, and modifying effects on storms and general air movements in their vicinity. Temperature variations within the State due to latitude alone are very small, yearly averages are only 2° to 3° higher in the south than in the north. The longitudinal variations, however, show a sharper contrast, from the mountain extremes in the west toward an ocean influence in the east. The prevalence of winds with a westerly component prevents the extension of ocean influences very far westward from the coast.

Annual temperature averages, by divisions, range from 54° in the Southwestern Mountain Division to nearly 59° in Tidewater Virginia. As might be expected, the highest temperatures of record in the State have been recorded in the Piedmont Plateau, and the lowest in the higher mountain sections of the central and southwest.

The growing season, based on average dates of the last freeze in spring and the first in fall, range from around 140 days in parts of Tazewell County to a little over 250 days in the Norfolk area. Cold air drainage is an important factor in determining the growing season. The first and last freezes of the season usually occur with large surface high-pressure systems where clear skies and light winds are conducive to large radiational losses of heat. The cold air layer next to the ground becomes more dense and flows from the ridges and higher elevations into the valleys and lower elevations. The temperature at nearby locations under such conditions may differ by several degrees.

Virginia lies in the zone of prevailing westerlies where the general motion is from west to east. Southerly and northerly winds are about equally frequent, reflecting the progression of weather systems over the State. The Appalachian mountains, however, act to deflect these winds to some extent with northeasterly and south-

westerly directions occurring frequently. Local winds are also created by such other factors as differential heating, air drainage, local terrain, and proximity to bodies of water. During the cold season a more intense circulation is present with frequent storms and outbreaks of cold polar air. Northerly winds are most common during this season. The storm track is well north of the State during the warm season and southerly wind with light speeds prevail.

Summers in Virginia are usually warm and humid, and several hot and humid periods usually occur each year. Principal sources of moisture are the Gulf of Mexico and the Atlantic Ocean. Relative humidity, the usual measure of moisture, varies inversely with temperature -- high in the morning and low in the afternoon. Average values are not appreciably different over the State, but Tidewater locations have a much higher frequency of humidity and temperature values in the range where human discomfort occurs. For example, consider the frequency of temperature greater than 80° and relative humidity greater than 70 percent. Norfolk has an average of 228 hours each year in this category; Roanoke has only 23.

The annual precipitation based on the period 1931-60 ranges from about 35 to 50 inches. The heaviest amounts occur in the extreme southwest, the southeast, and the south-central areas. Minimum amounts are found in the sheltered valleys west of the Blue Ridge Mountains. Precipitation is well distributed throughout the year without distinct wet and dry periods. Maximum rainfall occurs in the summer months and minimum in the fall months. Precipitation during the cold season is associated with migratory low-pressure storms. The amounts are quite evenly distributed during this season in comparison to the warm season when showers and thundershowers account for most of the rainfall. Excessive rainfall usually occurs in the fall season with the passage of hurricanes. Hurricane rainfall in excess of 8 inches at some location in the State can be expected about once every 6 years.

Snow is common in winter, but normally without damaging consequences. Average seasonal amounts range from less than 10 inches in Tidewater Virginia to around 20 inches west of the Blue Ridge, and up to 30 inches on the mountains. Snow for individual seasons may range from nearly none up to the record amount of 98 inches observed at Mountain Lake in 1913-14. A month with snow of 10 inches or more has occurred about once every 4 years in the Tidewater area, about once every 2 years in Piedmont areas, and almost yearly in mountain locations. Occasionally, a major snowstorm will occur with snow depths up to, but usually much less than, the record of 42 inches which fell at Big Meadows in March 1962. Such a storm with snow depths greater than 10 inches usually causes considerable damage to trees, interrupts electric and telephone service, blocks highways, and generally hinders the normal way of life.

The greater portion of the State lies in the Atlantic drainage. The extreme southwestern portion drains to the Ohio Basin. Floods occur in all months of the year. The greatest frequency occurs in late winter and early spring; snowmelt occasionally is a factor. July is the month of least flooding. A second period of high water shows up in late summer and fall in the Piedmont and Tidewater sections associated mainly with tropical storms. Intense convectional storms in summer occasionally cause local flash floods.

Virginia is also subject to drought periods. Drought may be defined broadly as a prolonged and abnormal moisture deficiency. Some portion of Virginia sustains real damage from drought on the average of 1 year out of 3, but rarely are all crops affected for the entire season. Equitable distribution of ample precipitation has seldom, if ever, occurred over all of Virginia for an entire season.

Almost every year some sections undergo periods of insufficient rainfall during which time crops make little growth or actually sustain damage. Normal precipitation during the months of May through August just about equals the moisture lost through evaporation from the soil and vegetation.

The drought of 1930 is generally considered to have been the most serious in the past century. In more recent years some parts of the State suffered disastrous or near disastrous drought conditions. The variety of Virginia's crops and their widely variable moisture requirements are factors which have sustained Virginia's agriculture during droughts.

Although Virginia is normally favored with abundant rainfall, the frequency of dry spells has prompted the use of supplemental irrigation. Although the percentage of crops under irrigation only totals around 1 percent, the acreage is expected to approximately triple during the decade of the 1970's. The use of irrigation for crops has been found economical mainly for tobacco, truck crops, fruit, and slightly for corn. Virginia's water use by industry, agriculture, electric utilities, and municipalities is enormous, in addition to the extensive use for recreational purposes. Several impoundments have been made on Virginia streams and rivers and many more are in the planning stage. The lakes created by these dams provide a more uniform supply and lead to increased use of water.

There are 11 major river basins or portions thereof in Virginia. Two of the largest, the James and Potomac Rivers, traverse the three main natural topographic divisions of the State. Nine of these rivers originate within the State, the exceptions being the New River whose headwaters lie within a small portion of northwestern North Carolina, and the Potomac which originates in West Virginia.

Of Virginia's 26 million plus acres, approximately 87 percent are classed as agricultural or potentially suited for agriculture, 10 percent are submarginal for agriculture, and the remaining approximate 3 percent consists of inland water areas. To some extent, all significant staple agricultural products of the nation are grown within the State.

Milder winters and longer growing seasons have made possible the extensive truck farming industry in the southeastern counties and on the eastern shore.

Marine life is abundant, and commercial fisheries constitute one of Virginia's important industries. Extensive forests and mineral resources have created diversified manufacturing and allied chemical industries which are scattered throughout Virginia's river valleys, favorably situated to adequate water supply.

Thunderstorms occur on the average of 32 to 50 days each year, the greater number occurring in the mountains of extreme southwestern Virginia, decreasing in number toward the northeastern part of the State. About 85 percent of the annual total occur during the period May to September. Only a small percentage of these can be classed as severe, however. Thunderstorms exact a sizable annual toll of damage, when accompanied by severe lightning, wind, or hail. One of the most destructive single hailstorms was reported near Winchester in Frederick County on June 19, 1944, when losses amounting to more than $500 thousand were sustained mainly to the fruit crops. In general, damage from hail over the years has amounted to more than 5 times the damage from tornadoes.

Tornadoes are local storms of short duration and usually small dimensions, formed of winds rotating at high speeds. Sometimes the tornado is visible as a funnel extending from a thunderstorm cloud. Wind speeds have been estimated up to 300 m.p.h. in tornadoes, and extreme damage occurs wherever the funnel touches the ground. Good sight observations of tornadoes are rare in Virginia as most funnels are hidden by the usually heavy precipitation occurring with the associated thunderstorm. Approximately 4 tornadoes are reported in Virginia each year. These tornadoes occur mainly east of the Blue Ridge, but a few have been observed in the mountains. On May 2, 1929, a series of tornadoes which unquestionably caused the greatest loss in lives and property of any Virginia has ever had, first struck at Rye Cove in Scott County, where 13 fatalities occurred when a rural high school was demolished. At least 4 or 5 other tornadoes occurred later the same day in Rappahannock, Bath, Culpeper, Fauquier, and Loudoun Counties. A total of 22 lives were lost, and property damage approximated $500 thousand.

A hurricane is a tropical storm with winds of at least 74 m.p.h., which blow in a large spiral around a relatively calm center. This center is called the "eye" and is unique to hurricanes. The "eye" is bordered by hurricane force winds and torrential rains. Virginia has been affected by hurricanes since the early settlement days, but most have decreased in intensity before entering the State. Even though a hurricane may not enter the State, it can be much more destructive by passing closely offshore and maintaining its intense circulation. High winds are not the prime cause of destruction. High tides along with waves and currents, and flooding from the torrential rains cause immense damage.

About 80 percent of the hurricanes occur during August, September, and October. An average of about 2 hurricanes (2.3) each year come close enough to affect Virginia, but less than one (0.6) enters the State. A hurricane has entered the State in less than half of the years in the past century. Two have been observed in the State about 1 year in 10.

The three most destructive hurricanes affecting Virginia were Camille in August 1969, and August 22-23, 1933, hurricane, and Hazel in October 1954. The destruction associated with Camille was mostly the results of excessive rainfall (up to 27 inches) which caused flash floods and earth slides on the eastern slopes of the Blue Ridge. Total damage exceeded $100 million, and fatalities numbered 151. The August 1933 hurricane moved from the southeast inland south of Norfolk. Norfolk reported winds of 70 m.p.h. and tides 9.7 feet above mean low water. There were 18 fatalities and monetary damage, adjusted to 1969, amounted to $79 million. Hazel in October 1954 moved almost due north through central Virginia. This hurricane maintained its intense circulation. Richmond reported winds of 68 m.p.h., and Washington National Airport reported winds of 78 m.p.h. Widespread damage occurred, with total damage, adjusted to 1969, of about $25 million, with 13 fatalities.

Middle latitude storms sometimes develop south of Virginia and move northward along the Virginia coast. These storms, although usually weaker than hurricanes, produce similar type of damage. This type of storm, often referred to as a "Northeaster," generally occurs from late fall through the spring months. They often account for considerable damage from high tides, strong east or northeast winds, and heavy rain, mainly in Tidewater Virginia.

TROPICAL CYCLONE DATA HAVING IMPORTANCE FOR THIS STATE IS INCLUDED IN STATISTICS AND CHARTS ON PAGES 1161 THROUGH 1164.

BIBLIOGRAPHY

Bailey, M. H., "Monthly Precipitation--Amount Probabilities for Selected Stations in Virginia," ESSA Technical Memorandum WBTM-ER-30, Washington, D. C., 1968, 13 pp.

_____, and Tinga, J. H., "Extreme Temperatures in Virginia," Research Division Report 128, Virginia Polytechnic Institute, Blacksburg, Va., 1968.

Chapman, Dorthy J., "Storms of Tropical Origin That Have Affected the Norfolk Area," Unpublished, Norfolk, Va., 1970.

Environmental Science Services Administration, "Tropical Cyclone Rainfall," ESSA Professional Paper I, Silver Spring, Md., 1967, 67 pp.

Flora, Snowden D., Hailstorms of the U.S., University of Oklahoma Press, Norman, Okla., 1956.

_____, Tornadoes of the United States, University of Oklahoma Press, Norman, Okla., Revised Edition, 1954.

Hoyt, William G., and Langbein, Walter B., Floods, Princeton University Press, Princeton, N. J., 1955.

NOAA, Environmental Data Service, "Climatic Summaries of Resort Areas," Climatography of the United States, No. 21-44, Silver Spring, Md., irregular.

_____, Climatological Data--National Summary, Silver Spring, Md., monthly plus annual summary.

_____, "Climatological Data--Virginia," Silver Spring, Md., Monthly plus annual summary.

_____, "Climatological Substation Summaries," Climatography of the United States, No. 20-44, Silver Spring, Md., irregular.

_____, "Hourly Precipitation Data--Virginia," Silver Spring, Md., monthly plus annual summary.

_____, "Local Climatological Data," Washington National Airport, Lynchburg, Dulles Airport, Norfolk, Roanoke, and Richmond; Silver Spring, Md., monthly plus annual summary.

_____, "Monthly Normals of Temperature, Precipitation, and Heating Degree Days, Virginia," Climatography of the United States, No. 81-38, Silver Spring, Md.

NOAA, Environmental Data Service, Storm Data, Silver Spring, Md., monthly plus annual summary.

NOAA, National Ocean Survey, "Tides in Hampton Roads," Unpublished, Norfolk, Va., 1970.

Rice, K. A., "Climate of Virginia, Climates of the States, Virginia," Climatography of the United States, No. 60-44, U.S. Weather Bureau, Washington, D. C., 1959.

Tannehill, Ivan R., Drought, Its Causes and Effects, Princeton University Press, Princeton, N. J., 1947.

_____, Hurricanes, Princeton University Press, Princeton, N. J., 9th Revised Edition, 1956.

Tinga, J. H., and Bailey, M. H., "Freeze Probabilities in Virginia and Protection Practices," Research Report 119, Virginia Polytechnic Institute, Blacksburg, Va., 1967, 20 pp.

U.S. Department of Agriculture, Soils--Yearbook of Agriculture, 1957, Washington, D. C., 1957, 784 pp.

_____, Water--Yearbook of Agriculture, 1955, Washington, D. C., 1955, 751 pp.

U.S. Weather Bureau, Climatic Summary of the United States (Bulletin W), Sections 91, 93, 94, Washington, D. C., 1930, 64 pp.

_____, Climatic Summary of the United States, Virginia, Supplement for 1931 through 1952 (Bulletin W Supplement), Washington, D. C., 43 pp.

BIBLIOGRAPHY

————, Climatic Summary of the United States, Virginia, Supplement for 1951 through 1960 (Bulletin W Supplement), Washington, D. C., 67 pp.

————, The Climatic Handbook for Washington, D. C., its Technical Paper No. 8, Washington, D. C., 1949, 235 pp.

————, "Summary of Hourly Observations, for Roanoke, Washington National Airport, Norfolk, and Richmond, "Climatography of the United States, No. 82-44, Washington, D. C.

————, "Tropical Cyclones of the North Atlantic Ocean," its Technical Paper No. 55, Silver Spring, Md., 1965, 148 pp.

Van Bavel, C. H. M., and Lillard, J. H., "Agricultural Drought in Virginia," Technical Bulletin 128, Virginia Polytechnic Institute, Blacksburg, Va., 1957, 38 pp.

Virginia Polytechnic Institute, "A Handbook of Agronomy," its Bulletin No. 97, Revised, Blacksburg, Va., 1966, 167 pp.

Virginia Polytechnic Institute, "Soils of Virginia," its Bulletin No. 203, Revised, Blacksburg, Va., Dec. 1966.

————, "Soil, Virginia's Basic Natural Resources," its Bulletin No. 253, Blacksburg, Va., Jan. 1962.

FREEZE DATA

PROBABILITY OF SELECTED TEMPERATURES ON OR AFTER GIVEN DATES IN SPRING AND ON OR BEFORE GIVEN DATES IN FALL

STATION	TEMP	PROBABILITY – SPRING					NO. OF DAYS BETWEEN 50% P'S	PROBABILITY – FALL					SPRING Y O	FALL Y O
		90%	70%	50%	30%	10%		10%	30%	50%	70%	90%		
ASHLAND HANOVER COUNTY	36	APR 20	APR 28	MAY 3	MAY 8	MAY 15	160	SEP 25	OCT 4	OCT 10	OCT 16	OCT 25	22 22	23 23
	32	APR 6	APR 15	APR 21	APR 27	MAY 5	178	OCT 3	OCT 11	OCT 16	OCT 22	OCT 30	22 22	22 22
	28	MAR 22	MAR 30	APR 5	APR 11	APR 19	208	OCT 18	OCT 25	OCT 30	NOV 4	NOV 12	22 22	22 22
	24	MAR 8	MAR 18	MAR 24	MAR 31	APR 10	232	OCT 27	NOV 11	NOV 11	NOV 17	NOV 26	22 22	22 22
	20	FEB 26	MAR 8	MAR 15	MAR 22	APR 2	252	NOV 6	NOV 15	NOV 22	NOV 28	DEC 7	22 22	22 22
	16	JAN 17	FEB 13	FEB 24	MAR 6	MAR 19	284	NOV 19	NOV 29	DEC 5	DEC 13	*	22 20	22 20
BACK BAY WILDLIFE REFUGE	36	MAR 21	MAR 27	MAR 31	APR 4	APR 10	226	OCT 30	NOV 7	NOV 12	NOV 18	NOV 26	17 17	17 17
	32	MAR 4	MAR 12	MAR 17	MAR 23	MAR 31	251	NOV 10	NOV 18	NOV 23	NOV 29	DEC 6	17 17	17 17
	28	FEB 9	FEB 22	MAR 4	MAR 13	MAR 26	277	NOV 20	NOV 29	DEC 6	DEC 13	DEC 26	17 17	17 16
	24	JAN 24	FEB 6	FEB 16	FEB 25	MAR 11	299	NOV 26	DEC 5	DEC 12	DEC 23	*	17 17	17 13
	20	*	JAN 28	FEB 8	FEB 18	MAR 3	*	DEC 7	DEC 18	*	*	*	17 15	17 7
	16	*	*	JAN 11	FEB 2	FEB 18	*	DEC 10	*	*	*	*	17 9	17 3
BALCONY FALLS ROCKBRIDGE COUNTY	36	APR 14	APR 22	APR 28	MAY 4	MAY 12	166	SEP 27	OCT 5	OCT 11	OCT 16	OCT 24	19 19	19 19
	32	APR 7	APR 15	APR 21	APR 26	MAY 4	186	OCT 10	OCT 18	OCT 24	OCT 30	NOV 7	20 20	19 19
	28	MAR 17	MAR 25	MAR 31	APR 6	APR 15	216	OCT 24	OCT 29	NOV 2	NOV 6	NOV 11	20 20	19 19
	24	MAR 3	MAR 14	MAR 21	MAR 28	APR 8	237	OCT 28	NOV 6	NOV 13	NOV 20	NOV 29	20 20	19 19
	20	FEB 17	FEB 28	MAR 8	MAR 16	MAR 27	262	NOV 8	NOV 18	NOV 25	DEC 2	DEC 11	20 20	19 19
	16	JAN 28	FEB 13	FEB 22	MAR 2	MAR 14	289	NOV 23	DEC 2	DEC 8	DEC 14	DEC 22	18 17	19 19
BEDFORD BEDFORD COUNTY	36	APR 14	APR 24	MAY 1	MAY 8	MAY 19	157	SEP 24	OCT 1	OCT 5	OCT 10	OCT 17	30 30	31 31
	32	APR 5	APR 14	APR 20	APR 25	MAY 4	179	SEP 30	OCT 10	OCT 16	OCT 23	NOV 4	30 30	31 31
	28	MAR 20	MAR 30	APR 5	APR 12	APR 21	209	OCT 16	OCT 25	OCT 31	NOV 6	NOV 14	30 30	31 31
	24	FEB 28	MAR 12	MAR 21	MAR 29	APR 11	238	OCT 28	NOV 7	NOV 14	NOV 21	DEC 1	30 30	31 31
	20	FEB 14	FEB 27	MAR 8	MAR 17	MAR 30	264	NOV 15	NOV 22	NOV 27	DEC 2	DEC 9	30 30	31 31
	16	JAN 24	FEB 10	FEB 21	MAR 3	MAR 18	291	NOV 25	DEC 3	DEC 9	DEC 15	DEC 25	31 30	31 30
BERRYVILLE CLARKE COUNTY	36	MAY 5	MAY 12	MAY 17	MAY 22	MAY 29	134	SEP 14	SEP 22	SEP 28	OCT 4	OCT 13	18 18	19 19
	32	APR 15	APR 26	MAY 3	MAY 11	MAY 21	157	SEP 23	OCT 1	OCT 7	OCT 13	OCT 21	18 18	19 19
	28	APR 3	APR 14	APR 22	APR 30	MAY 11	177	OCT 4	OCT 11	OCT 17	OCT 22	OCT 29	18 18	19 19
	24	MAR 24	MAR 31	APR 5	APR 11	APR 18	212	OCT 24	OCT 30	NOV 3	NOV 7	NOV 12	18 18	19 19
	20	MAR 10	MAR 19	MAR 26	APR 2	APR 11	231	OCT 28	NOV 6	NOV 12	NOV 19	NOV 27	18 18	19 19
	16	FEB 22	MAR 4	MAR 11	MAR 18	MAR 28	261	NOV 15	NOV 22	NOV 27	DEC 2	DEC 9	18 18	19 19
BIG MEADOWS PAGE COUNTY	36	MAY 9	MAY 16	MAY 20	MAY 25	JUN 1	127	SEP 9	SEP 18	SEP 24	SEP 30	OCT 9	30 30	31 31
	32	APR 17	APR 30	MAY 9	MAY 17	MAY 30	146	SEP 20	SEP 27	OCT 2	OCT 6	OCT 13	30 30	31 31
	28	APR 10	APR 19	APR 25	MAY 2	MAY 11	176	OCT 3	OCT 11	OCT 18	OCT 24	NOV 2	30 30	31 31
	24	MAR 30	APR 9	APR 15	APR 22	MAY 1	196	OCT 12	OCT 21	OCT 28	NOV 3	NOV 13	30 30	31 31
	20	MAR 19	MAR 29	APR 6	APR 13	APR 23	214	OCT 23	NOV 1	NOV 6	NOV 12	NOV 20	30 30	31 31
	16	MAR 6	MAR 16	MAR 24	MAR 31	APR 11	237	NOV 1	NOV 10	NOV 16	NOV 23	DEC 2	30 30	31 31
BLACKSBURG MONTGOMERY COUNTY	36	APR 20	MAY 2	MAY 10	MAY 18	MAY 30	141	SEP 11	SEP 21	SEP 28	OCT 5	OCT 15	31 31	31 31
	32	APR 14	APR 23	APR 30	MAY 6	MAY 16	161	SEP 22	OCT 1	OCT 8	OCT 15	OCT 24	31 31	31 31
	28	APR 2	APR 12	APR 19	APR 26	MAY 6	183	OCT 3	OCT 13	OCT 19	OCT 26	NOV 4	31 31	31 31
	24	MAR 17	MAR 27	APR 3	APR 10	APR 19	214	OCT 20	OCT 28	NOV 9	NOV 9	NOV 17	31 31	31 31
	20	FEB 22	MAR 9	MAR 19	MAR 29	APR 12	242	OCT 30	NOV 9	NOV 16	NOV 22	DEC 2	31 31	31 31
	16	FEB 7	FEB 22	MAR 3	MAR 12	MAR 25	268	NOV 7	NOV 19	NOV 26	DEC 4	DEC 16	31 30	31 31
BLACKSTONE NOTTOWAY COUNTY	36	APR 10	APR 18	APR 23	APR 29	MAY 7	181	OCT 6	OCT 15	OCT 21	OCT 27	NOV 5	25 25	25 25
	32	MAR 27	APR 3	APR 8	APR 13	APR 20	206	OCT 17	OCT 25	OCT 31	NOV 6	NOV 14	25 25	25 25
	28	MAR 12	MAR 20	MAR 25	MAR 30	APR 6	230	OCT 23	NOV 3	NOV 10	NOV 18	NOV 28	25 25	25 25
	24	FEB 26	MAR 7	MAR 14	MAR 20	MAR 29	255	NOV 12	NOV 19	NOV 24	NOV 29	DEC 6	25 25	25 25
	20	JAN 29	FEB 16	FEB 26	MAR 4	MAR 24	278	NOV 15	NOV 25	DEC 1	DEC 7	DEC 17	25 24	25 25
	16	*	FEB 6	FEB 15	FEB 23	MAR 5	304	DEC 1	DEC 9	DEC 16	DEC 26	*	25 22	25 19
BOHANNAN MATHEWS COUNTY	36	APR 5	APR 15	APR 22	APR 30	MAY 10	184	OCT 8	OCT 17	OCT 23	OCT 29	NOV 6	21 21	21 21
	32	MAR 28	APR 5	APR 11	APR 17	APR 25	207	OCT 25	OCT 31	NOV 4	NOV 8	NOV 14	21 21	21 21
	28	MAR 14	MAR 23	MAR 29	APR 4	APR 13	231	NOV 2	NOV 9	NOV 15	NOV 20	NOV 28	21 21	21 21
	24	MAR 2	MAR 10	MAR 16	MAR 22	MAR 30	253	NOV 10	NOV 18	NOV 24	NOV 30	DEC 8	20 20	21 21
	20	FEB 13	FEB 23	MAR 2	MAR 9	MAR 19	283	NOV 23	DEC 3	DEC 10	DEC 18	*	21 21	22 20
	16	JAN 19	FEB 5	FEB 14	FEB 24	MAR 9	313	DEC 6	DEC 15	DEC 24	*	*	21 20	22 13
BOYKINS SOUTHAMPTON COUNTY	36	APR 11	APR 21	APR 28	MAY 5	MAY 14	170	OCT 3	OCT 11	OCT 15	OCT 20	OCT 27	17 17	16 16
	32	MAR 28	APR 8	APR 16	APR 24	MAY 5	190	OCT 5	OCT 16	OCT 23	OCT 31	NOV 11	17 17	17 17
	28	MAR 9	MAR 18	MAR 24	MAR 30	APR 8	226	OCT 18	OCT 28	NOV 5	NOV 12	NOV 22	17 17	17 17
	24	MAR 2	MAR 11	MAR 17	MAR 23	APR 1	242	OCT 30	NOV 7	NOV 14	NOV 20	DEC 2	16 16	17 16
	20	FEB 5	FEB 16	FEB 23	MAR 2	MAR 12	274	NOV 6	NOV 16	NOV 24	DEC 3	*	16 16	17 15
	16	JAN 8	JAN 28	FEB 8	FEB 18	MAR 4	310	NOV 28	DEC 8	DEC 15	DEC 25	*	16 15	17 13
BRISTOL WASHINGTON COUNTY	36	APR 12	APR 22	APR 30	MAY 7	MAY 18	173	OCT 7	OCT 14	OCT 20	OCT 25	NOV 2	19 19	18 18
	32	APR 9	APR 16	APR 21	APR 26	MAY 3	188	OCT 15	OCT 22	OCT 26	OCT 30	NOV 6	19 19	18 18
	28	MAR 22	MAR 31	APR 6	APR 12	APR 21	208	OCT 22	OCT 27	OCT 31	NOV 3	NOV 8	19 19	18 18
	24	FEB 24	MAR 10	MAR 20	MAR 29	APR 12	231	OCT 24	NOV 1	NOV 6	NOV 11	NOV 19	19 19	17 17
	20	FEB 17	FEB 28	MAR 8	MAR 16	MAR 27	255	NOV 10	NOV 18	NOV 24	DEC 8	*	19 19	18 18
	16	JAN 21	FEB 10	FEB 21	MAR 4	MAR 20	292	NOV 20	DEC 2	DEC 10	DEC 18	*	19 18	18 17
BUCHANAN BOTETOURT COUNTY	36	APR 19	APR 29	MAY 6	MAY 14	MAY 24	152	SEP 23	SEP 30	OCT 5	OCT 10	OCT 17	31 31	31 31
	32	APR 15	APR 22	APR 27	MAY 2	MAY 10	170	SEP 30	OCT 9	OCT 14	OCT 20	OCT 28	31 31	31 31
	28	MAR 30	APR 8	APR 14	APR 19	APR 28	196	OCT 12	OCT 21	OCT 27	NOV 2	NOV 11	31 31	31 31
	24	MAR 11	MAR 20	MAR 27	APR 2	APR 12	227	OCT 28	NOV 4	NOV 9	NOV 14	NOV 21	31 31	31 31
	20	FEB 20	MAR 3	MAR 11	MAR 19	MAR 31	254	NOV 3	NOV 13	NOV 20	NOV 27	DEC 7	31 31	31 31
	16	FEB 2	FEB 17	FEB 26	MAR 8	MAR 21	279	NOV 15	NOV 25	DEC 2	DEC 9	DEC 19	31 30	31 31
BURKES GARDEN TAZEWELL COUNTY	36	MAY 10	MAY 20	MAY 27	JUN 3	JUN 14	103	AUG 11	AUG 27	SEP 7	SEP 18	OCT 4	31 31	31 31
	32	MAY 1	MAY 9	MAY 15	MAY 21	MAY 29	135	SEP 14	SEP 22	SEP 27	OCT 3	OCT 10	31 31	31 31
	28	APR 15	APR 26	MAY 3	MAY 11	MAY 21	155	SEP 21	SEP 29	OCT 5	OCT 11	OCT 19	30 30	31 31
	24	APR 5	APR 13	APR 19	APR 25	MAY 3	181	OCT 2	OCT 11	OCT 17	OCT 24	NOV 2	30 30	31 31
	20	MAR 20	MAR 30	APR 6	APR 13	APR 23	206	OCT 12	OCT 22	OCT 29	NOV 5	NOV 15	30 30	31 31
	16	FEB 23	MAR 8	MAR 17	MAR 27	APR 9	240	OCT 23	NOV 4	NOV 12	NOV 20	DEC 1	30 30	31 31
CAPE HENRY	36	MAR 14	MAR 23	MAR 29	APR 3	APR 12	232	NOV 2	NOV 10	NOV 16	NOV 22	NOV 30	29 28	29 29
	32	FEB 26	MAR 9	MAR 17	MAR 24	APR 4	259	NOV 18	NOV 26	DEC 1	DEC 7	DEC 15	30 30	28 28
	28	FEB 2	FEB 17	FEB 28	MAR 10	MAR 26	289	NOV 29	DEC 7	DEC 14	DEC 23	*	30 30	28 21
	24	*	FEB 2	FEB 13	FEB 22	MAR 8	*	DEC 5	DEC 14	*	*	*	30 27	28 14
	20	*	*	JAN 31	FEB 9	FEB 20	*	DEC 12	*	*	*	*	30 21	28 9
	16	*	*	*	JAN 15	FEB 9	*	*	*	*	*	*	30 10	28 0

1020

FREEZE DATA

PROBABILITY OF SELECTED TEMPERATURES ON OR AFTER GIVEN DATES IN SPRING AND ON OR BEFORE GIVEN DATES IN FALL

STATION	TEMP	PROBABILITY - SPRING					NO. OF DAYS BETWEEN 50% P'S	PROBABILITY - FALL					SPRING		FALL	
		90%	70%	50%	30%	10%		10%	30%	50%	70%	90%	Y	O	Y	O
CATAWBA ROANOKE COUNTY	36	APR 17	APR 27	MAY 4	MAY 11	MAY 20	154	SEP 23	SEP 30	OCT 5	OCT 10	OCT 17	31	31	31	31
	32	APR 6	APR 16	APR 24	MAY 1	MAY 12	178	OCT 3	OCT 12	OCT 19	OCT 25	NOV 4	31	31	31	31
	28	MAR 19	MAR 28	APR 4	APR 11	APR 21	213	OCT 20	OCT 28	NOV 3	NOV 9	NOV 17	31	31	31	31
	24	MAR 6	MAR 16	MAR 23	MAR 30	APR 8	237	OCT 31	NOV 9	NOV 15	NOV 21	NOV 29	31	31	31	31
	20	FEB 16	MAR 1	MAR 10	MAR 19	APR 1	260	NOV 11	NOV 19	NOV 25	DEC 1	DEC 10	31	31	31	31
	16	FEB 4	FEB 20	MAR 1	MAR 11	MAR 24	277	NOV 20	NOV 28	DEC 3	DEC 9	DEC 16	31	30	31	31
CHARLOTTE COURT HOUSE CHARLOTTE COUNTY	36	APR 11	APR 21	APR 28	MAY 4	MAY 14	167	OCT 2	OCT 8	OCT 12	OCT 16	OCT 23	22	22	23	23
	32	APR 1	APR 11	APR 17	APR 24	MAY 4	188	OCT 7	OCT 16	OCT 22	OCT 29	NOV 6	22	22	23	23
	28	MAR 15	MAR 25	APR 1	APR 8	APR 18	213	OCT 17	OCT 25	OCT 31	NOV 5	NOV 13	22	22	23	23
	24	MAR 3	MAR 13	MAR 19	MAR 26	APR 5	240	OCT 29	NOV 8	NOV 14	NOV 21	NOV 30	22	22	23	23
	20	FEB 13	FEB 25	MAR 6	MAR 14	MAR 26	262	NOV 8	NOV 17	NOV 23	NOV 29	DEC 8	22	22	23	23
	16	*	FEB 7	FEB 19	MAR 1	MAR 14	290	NOV 19	NOV 29	DEC 6	DEC 13	DEC 25	22	19	23	22
CHARLOTTESVILLE 2W ALBEMARLE COUNTY	36	APR 4	APR 11	APR 16	APR 21	APR 27	190	OCT 8	OCT 17	OCT 23	OCT 30	NOV 8	31	31	31	31
	32	MAR 27	APR 4	APR 9	APR 14	APR 22	211	OCT 24	OCT 31	NOV 6	NOV 11	NOV 19	31	31	31	31
	28	MAR 6	MAR 17	MAR 24	MAR 31	APR 10	235	OCT 29	NOV 8	NOV 14	NOV 21	DEC 1	31	31	31	31
	24	FEB 20	MAR 4	MAR 12	MAR 21	APR 1	262	NOV 19	NOV 25	NOV 29	DEC 4	DEC 10	31	31	31	31
	20	FEB 9	FEB 21	MAR 1	MAR 9	MAR 21	279	NOV 20	NOV 29	DEC 5	DEC 12	DEC 22	31	31	31	30
	16	JAN 5	FEB 5	FEB 15	FEB 24	MAR 9	306	DEC 2	DEC 11	DEC 18	*	*	31	28	31	20
CHARLOTTESVILLE 1 W ALBEMARLE COUNTY	36	APR 10	APR 19	APR 25	MAY 1	MAY 10	172	SEP 28	OCT 7	OCT 14	OCT 21	OCT 31	19	19	20	20
	32	MAR 31	APR 8	APR 14	APR 19	APR 27	194	OCT 10	OCT 19	OCT 25	OCT 31	NOV 9	19	19	20	20
	28	MAR 18	MAR 26	MAR 31	APR 5	APR 13	218	OCT 23	OCT 30	NOV 4	NOV 8	NOV 15	19	19	20	20
	24	MAR 6	MAR 13	MAR 18	MAR 23	MAR 30	244	NOV 5	NOV 12	NOV 17	NOV 23	NOV 30	19	19	20	20
	20	FEB 6	FEB 21	MAR 3	MAR 13	MAR 27	271	NOV 13	NOV 23	NOV 29	DEC 5	DEC 14	19	19	20	20
	16	JAN 23	FEB 7	FEB 16	FEB 24	MAR 8	299	NOV 27	DEC 6	DEC 12	DEC 19	*	19	18	20	18
CHASE CITY MECKLENBURG COUNTY	36	APR 8	APR 16	APR 23	APR 29	MAY 8	180	OCT 9	OCT 16	OCT 20	OCT 25	NOV 1	18	18	19	19
	32	MAR 27	APR 4	APR 10	APR 15	APR 23	202	OCT 19	OCT 25	OCT 29	NOV 3	NOV 9	18	18	19	19
	28	MAR 6	MAR 17	MAR 24	MAR 31	APR 10	230	OCT 26	NOV 3	NOV 9	NOV 14	NOV 22	18	18	19	19
	24	FEB 21	MAR 4	MAR 13	MAR 21	APR 1	252	NOV 2	NOV 12	NOV 20	NOV 27	DEC 8	18	18	19	19
	20	JAN 28	FEB 13	FEB 25	MAR 8	MAR 25	281	NOV 18	NOV 27	DEC 3	DEC 10	DEC 21	18	18	19	19
	16	*	JAN 28	FEB 10	FEB 21	MAR 9	309	DEC 2	DEC 10	DEC 16	DEC 27	*	18	16	19	14
CHATHAM PITTSYLVANIA COUNTY	36	APR 14	APR 23	APR 30	MAY 6	MAY 15	161	SEP 27	OCT 3	OCT 8	OCT 13	OCT 19	31	31	31	31
	32	APR 4	APR 13	APR 19	APR 25	MAY 3	183	OCT 3	OCT 12	OCT 19	OCT 25	NOV 3	31	31	31	31
	28	MAR 24	MAR 31	APR 5	APR 10	APR 17	209	OCT 19	OCT 26	OCT 31	NOV 5	NOV 12	31	31	31	31
	24	FEB 25	MAR 10	MAR 18	MAR 27	APR 8	241	OCT 30	NOV 8	NOV 14	NOV 21	NOV 30	31	31	31	31
	20	FEB 11	FEB 23	MAR 3	MAR 11	MAR 23	273	NOV 14	NOV 24	DEC 1	DEC 7	DEC 17	31	31	31	31
	16	JAN 16	FEB 8	FEB 17	FEB 27	MAR 14	295	NOV 26	DEC 4	DEC 9	DEC 15	DEC 24	31	29	31	30
CHERITON NORTHAMPTON COUNTY	36	MAR 27	APR 6	APR 14	APR 21	MAY 1	206	OCT 25	NOV 1	NOV 6	NOV 11	NOV 18	31	31	31	31
	32	MAR 16	MAR 25	APR 1	APR 7	APR 17	229	NOV 2	NOV 11	NOV 16	NOV 22	NOV 30	31	31	31	31
	28	FEB 24	MAR 7	MAR 15	MAR 22	APR 2	260	NOV 17	NOV 25	NOV 30	DEC 6	DEC 15	31	31	31	30
	24	FEB 1	FEB 17	FEB 25	MAR 5	MAR 16	290	NOV 25	DEC 5	DEC 12	DEC 21	*	31	29	31	25
	20	*	JAN 28	FEB 8	FEB 18	MAR 5	324	DEC 3	DEC 15	DEC 29	*	*	31	28	31	17
	16	*	*	JAN 27	FEB 9	FEB 23	0	DEC 16	*	*	*	*	31	20	31	8
CHILHOWIE SMYTH COUNTY	36	MAY 2	MAY 12	MAY 20	MAY 27	JUN 6	124	SEP 9	SEP 16	SEP 21	SEP 26	OCT 3	19	19	19	19
	32	APR 23	MAY 2	MAY 9	MAY 16	MAY 26	146	SEP 17	SEP 26	OCT 2	OCT 8	OCT 16	19	19	19	19
	28	APR 16	APR 22	APR 27	MAY 2	MAY 9	171	OCT 5	OCT 11	OCT 15	OCT 19	OCT 25	19	19	19	19
	24	MAR 27	APR 5	APR 12	APR 18	APR 27	195	OCT 12	OCT 19	OCT 24	OCT 29	NOV 4	18	18	19	19
	20	MAR 6	MAR 18	MAR 27	APR 5	APR 17	218	OCT 21	OCT 27	OCT 31	NOV 4	NOV 10	18	18	19	19
	16	FEB 17	MAR 3	MAR 13	MAR 23	APR 6	248	OCT 29	NOV 9	NOV 16	NOV 24	DEC 4	18	18	19	19
COLUMBIA GOOCHLAND COUNTY	36	APR 19	APR 29	MAY 6	MAY 13	MAY 23	152	SEP 25	OCT 1	OCT 5	OCT 9	OCT 15	30	30	31	31
	32	APR 12	APR 20	APR 26	MAY 2	MAY 10	172	OCT 1	OCT 9	OCT 15	OCT 20	OCT 28	30	30	31	31
	28	MAR 25	APR 4	APR 11	APR 18	APR 28	198	OCT 12	OCT 20	OCT 26	NOV 1	NOV 10	30	30	31	31
	24	MAR 10	MAR 19	MAR 25	APR 1	APR 10	227	OCT 27	NOV 3	NOV 7	NOV 12	NOV 19	30	30	31	31
	20	FEB 25	MAR 6	MAR 13	MAR 19	MAR 28	254	NOV 9	NOV 17	NOV 22	NOV 28	DEC 5	30	30	31	31
	16	FEB 7	FEB 19	FEB 28	MAR 9	MAR 21	279	NOV 20	NOV 28	DEC 4	DEC 9	DEC 17	30	30	31	31
CULPEPER CULPEPER COUNTY	36	APR 19	APR 27	MAY 3	MAY 8	MAY 16	157	SEP 25	OCT 2	OCT 7	OCT 12	OCT 20	31	31	31	31
	32	APR 5	APR 14	APR 20	APR 26	MAY 4	181	OCT 3	OCT 12	OCT 18	OCT 24	NOV 2	30	30	31	31
	28	MAR 24	MAR 31	APR 6	APR 11	APR 18	207	OCT 14	OCT 23	OCT 30	NOV 5	NOV 14	31	31	31	31
	24	MAR 8	MAR 18	MAR 25	APR 1	APR 11	232	OCT 28	NOV 6	NOV 12	NOV 17	NOV 26	31	31	31	31
	20	FEB 25	MAR 6	MAR 12	MAR 18	MAR 27	254	NOV 3	NOV 14	NOV 21	NOV 28	DEC 8	31	31	31	31
	16	FEB 7	FEB 19	FEB 27	MAR 7	MAR 19	280	NOV 20	NOV 28	DEC 4	DEC 9	DEC 17	31	31	31	31
DALE ENTERPRISE ROCKINGHAM COUNTY	36	APR 23	MAY 3	MAY 10	MAY 17	MAY 28	145	SEP 20	SEP 27	OCT 2	OCT 7	OCT 15	31	31	31	31
	32	APR 12	APR 21	APR 28	MAY 5	MAY 14	166	SEP 27	OCT 5	OCT 11	OCT 16	OCT 24	31	31	31	31
	28	MAR 30	APR 9	APR 17	APR 24	MAY 4	193	OCT 12	OCT 21	OCT 27	NOV 2	NOV 11	31	31	31	31
	24	MAR 14	MAR 25	APR 1	APR 9	APR 20	220	OCT 26	NOV 2	NOV 7	NOV 12	NOV 19	31	31	31	31
	20	MAR 2	MAR 12	MAR 19	MAR 26	APR 5	243	NOV 1	NOV 10	NOV 17	NOV 23	DEC 2	31	31	31	31
	16	FEB 11	FEB 24	MAR 5	MAR 14	MAR 27	270	NOV 12	NOV 23	NOV 30	DEC 8	DEC 19	31	31	31	31
DANVILLE PITTSYLVANIA COUNTY	36	APR 8	APR 18	APR 24	MAY 1	MAY 10	173	OCT 2	OCT 9	OCT 14	OCT 19	OCT 27	31	31	31	31
	32	MAR 24	APR 5	APR 13	APR 21	MAY 3	196	OCT 12	OCT 21	OCT 26	NOV 1	NOV 9	31	31	31	31
	28	MAR 12	MAR 22	MAR 29	APR 5	APR 16	223	OCT 27	NOV 2	NOV 7	NOV 11	NOV 17	31	31	31	31
	24	FEB 21	MAR 6	MAR 14	MAR 23	APR 4	248	NOV 3	NOV 12	NOV 17	NOV 23	DEC 2	31	31	31	31
	20	FEB 5	FEB 18	FEB 27	MAR 7	MAR 19	279	NOV 19	NOV 27	DEC 3	DEC 9	DEC 16	31	31	31	31
	16	*	JAN 28	FEB 9	FEB 20	MAR 6	307	NOV 28	DEC 7	DEC 13	DEC 20	*	31	27	31	27
DIAMOND SPRINGS	36	MAR 23	APR 2	APR 9	APR 17	APR 27	209	OCT 24	OCT 31	NOV 4	NOV 8	NOV 15	29	29	28	28
	32	MAR 9	MAR 18	MAR 24	MAR 30	APR 8	237	NOV 2	NOV 10	NOV 16	NOV 22	NOV 30	31	31	31	31
	28	FEB 25	MAR 7	MAR 13	MAR 20	MAR 29	262	NOV 20	NOV 26	NOV 30	DEC 5	DEC 11	31	31	31	31
	24	JAN 24	FEB 12	FEB 22	MAR 4	MAR 16	295	NOV 30	DEC 8	DEC 14	DEC 23	*	31	29	31	24
	20	*	JAN 31	FEB 10	FEB 20	MAR 5	323	DEC 8	DEC 16	DEC 30	*	*	31	28	31	16
	16	*	*	JAN 24	FEB 7	FEB 21	0	DEC 16	*	*	*	*	31	18	31	7
ELKWOOD CULPEPER COUNTY	36	APR 22	MAY 1	MAY 8	MAY 15	MAY 24	147	SEP 20	SEP 27	OCT 2	OCT 7	OCT 14	22	22	23	23
	32	APR 9	APR 19	APR 26	MAY 3	MAY 13	169	SEP 30	OCT 7	OCT 12	OCT 17	OCT 24	23	23	23	23
	28	APR 2	APR 10	APR 15	APR 20	APR 28	193	OCT 10	OCT 19	OCT 25	OCT 31	NOV 9	23	23	23	23
	24	MAR 15	MAR 24	MAR 30	APR 5	APR 15	218	OCT 22	OCT 29	NOV 3	NOV 8	NOV 15	23	23	23	23
	20	MAR 4	MAR 13	MAR 20	MAR 26	APR 5	241	NOV 2	NOV 10	NOV 16	NOV 23	DEC 1	22	22	23	23
	16	FEB 14	FEB 25	MAR 5	MAR 13	MAR 25	271	NOV 16	NOV 25	DEC 1	DEC 7	DEC 15	22	22	23	23

FREEZE DATA

PROBABILITY OF SELECTED TEMPERATURES ON OR AFTER GIVEN DATES IN SPRING AND ON OR BEFORE GIVEN DATES IN FALL

STATION	TEMP	PROBABILITY – SPRING					NO. OF DAYS BETWEEN 50% P'S	PROBABILITY – FALL					SPRING Y O	FALL Y O
		90%	70%	50%	30%	10%		10%	30%	50%	70%	90%		
FALLS CHURCH FAIRFAX COUNTY	36	APR 14	APR 26	MAY 4	MAY 13	MAY 25	157	SEP 25	OCT 3	OCT 8	OCT 13	OCT 21	21 21	21 21
	32	APR 8	APR 16	APR 21	APR 27	MAY 4	182	OCT 1	OCT 13	OCT 20	OCT 26	NOV 5	21 21	20 20
	28	MAR 25	APR 1	APR 6	APR 11	APR 18	209	OCT 21	OCT 27	NOV 1	NOV 5	NOV 11	21 21	20 20
	24	MAR 8	MAR 18	MAR 24	MAR 31	APR 9	233	OCT 29	NOV 6	NOV 12	NOV 17	NOV 25	21 21	20 20
	20	FEB 21	MAR 2	MAR 9	MAR 16	MAR 25	262	NOV 12	NOV 20	NOV 26	DEC 2	DEC 10	21 21	20 20
	16	JAN 25	FEB 11	FEB 20	MAR 2	MAR 15	293	NOV 27	DEC 5	DEC 10	DEC 16	DEC 26	21 20	20 19
FARMVILLE CUMBERLAND COUNTY	36	APR 21	MAY 1	MAY 8	MAY 15	MAY 24	147	SEP 19	SEP 27	OCT 2	OCT 7	OCT 15	31 31	31 31
	32	APR 16	APR 24	APR 29	MAY 4	MAY 12	164	SEP 27	OCT 4	OCT 10	OCT 15	OCT 22	31 31	31 31
	28	APR 2	APR 11	APR 17	APR 23	MAY 1	188	OCT 8	OCT 17	OCT 22	OCT 28	NOV 5	31 31	31 31
	24	MAR 17	MAR 26	APR 1	APR 7	APR 15	218	OCT 22	OCT 31	NOV 5	NOV 11	NOV 19	31 31	31 31
	20	MAR 6	MAR 15	MAR 22	MAR 28	APR 7	239	OCT 30	NOV 9	NOV 16	NOV 22	DEC 2	31 31	31 31
	16	FEB 17	FEB 28	MAR 8	MAR 16	MAR 27	263	NOV 9	NOV 19	NOV 26	DEC 3	DEC 13	31 31	31 31
FLOYD FLOYD COUNTY	36	APR 22	MAY 4	MAY 13	MAY 21	JUN 2	129	AUG 23	SEP 8	SEP 19	OCT 1	OCT 17	31 31	31 31
	32	APR 15	APR 26	MAY 4	MAY 12	MAY 23	155	SEP 18	SEP 29	OCT 6	OCT 14	OCT 24	30 30	31 31
	28	APR 1	APR 11	APR 19	APR 26	MAY 6	182	SEP 27	OCT 10	OCT 18	OCT 27	NOV 9	30 30	31 31
	24	MAR 14	MAR 26	APR 3	APR 12	APR 23	213	OCT 15	OCT 26	NOV 2	NOV 9	NOV 20	31 31	31 31
	20	MAR 2	MAR 15	MAR 24	APR 2	APR 16	231	OCT 22	NOV 2	NOV 10	NOV 17	NOV 29	31 31	31 31
	16	FEB 16	MAR 2	MAR 13	MAR 23	APR 6	254	NOV 5	NOV 15	NOV 22	NOV 30	DEC 10	31 31	31 31
FORT LEE PRINCE GEORGE COUNTY	36	APR 16	APR 24	APR 29	MAY 4	MAY 12	165	OCT 2	OCT 7	OCT 11	OCT 15	OCT 21	21 21	20 20
	32	APR 3	APR 12	APR 18	APR 24	MAY 2	191	OCT 14	OCT 21	OCT 26	OCT 31	NOV 7	21 21	20 20
	28	MAR 11	MAR 24	APR 3	APR 12	APR 25	217	OCT 25	NOV 1	NOV 6	NOV 10	NOV 18	21 21	20 20
	24	MAR 4	MAR 13	MAR 20	MAR 26	APR 5	242	NOV 2	NOV 11	NOV 17	NOV 23	DEC 1	20 20	20 20
	20	FEB 6	FEB 20	MAR 1	MAR 10	MAR 24	274	NOV 16	NOV 24	NOV 30	DEC 5	DEC 14	20 20	20 20
	16	JAN 24	FEB 7	FEB 15	FEB 23	MAR 6	304	DEC 1	DEC 10	DEC 16	DEC 23	DEC 7	20 19	20 18
FREDERICKSBURG SPOTSYLVANIA COUNTY	36	APR 20	APR 27	MAY 2	MAY 7	MAY 14	157	SEP 24	OCT 1	OCT 6	OCT 12	OCT 19	30 30	29 29
	32	APR 8	APR 17	APR 22	APR 28	MAY 6	178	OCT 3	OCT 11	OCT 17	OCT 23	OCT 31	30 30	29 29
	28	MAR 23	APR 2	APR 9	APR 16	APR 26	202	OCT 14	OCT 22	OCT 28	NOV 3	NOV 12	30 30	29 29
	24	MAR 9	MAR 19	MAR 26	APR 2	APR 12	230	OCT 28	NOV 5	NOV 11	NOV 16	NOV 24	30 30	30 30
	20	FEB 20	MAR 4	MAR 13	MAR 22	APR 3	258	NOV 13	NOV 21	NOV 26	DEC 1	DEC 9	30 30	30 30
	16	FEB 5	FEB 17	FEB 25	MAR 5	MAR 17	285	NOV 23	DEC 1	DEC 7	DEC 12	DEC 20	30 30	30 30
GALAX CARROLL COUNTY	36	MAY 1	MAY 10	MAY 16	MAY 22	MAY 31	131	SEP 10	SEP 19	SEP 24	SEP 30	OCT 9	15 15	16 16
	32	APR 19	MAY 1	MAY 9	MAY 17	MAY 29	149	SEP 22	SEP 29	OCT 5	OCT 10	OCT 17	15 15	16 16
	28	APR 5	APR 14	APR 21	APR 27	MAY 7	177	OCT 3	OCT 10	OCT 15	OCT 20	OCT 28	15 15	16 16
	24	MAR 27	APR 3	APR 8	APR 13	APR 21	202	OCT 14	OCT 21	OCT 27	NOV 1	NOV 9	15 15	16 16
	20	MAR 4	MAR 16	MAR 25	APR 2	APR 14	229	OCT 30	NOV 5	NOV 9	NOV 14	NOV 20	15 15	16 16
	16	FEB 21	MAR 4	MAR 12	MAR 20	MAR 31	254	NOV 4	NOV 14	NOV 21	NOV 29	DEC 9	15 15	15 15
GLEN LYN GILES COUNTY	36	APR 18	APR 29	MAY 7	MAY 15	MAY 26	150	SEP 20	SEP 28	OCT 4	OCT 10	OCT 18	31 31	31 31
	32	APR 12	APR 20	APR 26	MAY 2	MAY 11	175	OCT 4	OCT 12	OCT 18	OCT 24	NOV 1	31 31	31 31
	28	MAR 30	APR 5	APR 9	APR 14	APR 20	204	OCT 17	OCT 24	OCT 30	NOV 4	NOV 12	31 31	31 31
	24	MAR 6	MAR 17	MAR 24	MAR 31	APR 11	230	OCT 26	NOV 3	NOV 9	NOV 14	NOV 23	31 31	30 30
	20	FEB 17	MAR 3	MAR 13	MAR 23	APR 6	252	NOV 3	NOV 13	NOV 20	NOV 27	DEC 7	31 31	30 30
	16	FEB 10	FEB 23	MAR 5	MAR 14	MAR 27	272	NOV 15	NOV 25	DEC 2	DEC 9	DEC 21	31 31	30 29
HALIFAX HALIFAX COUNTY	36	APR 9	APR 17	APR 23	APR 28	MAY 6	171	OCT 1	OCT 7	OCT 11	OCT 15	OCT 21	23 23	24 24
	32	MAR 28	APR 8	APR 15	APR 23	MAY 3	192	OCT 8	OCT 18	OCT 24	OCT 30	NOV 9	23 23	24 24
	28	MAR 14	MAR 23	MAR 29	APR 5	APR 14	220	OCT 22	OCT 30	NOV 4	NOV 10	NOV 18	23 23	24 24
	24	FEB 26	MAR 9	MAR 17	MAR 25	APR 5	242	OCT 30	NOV 8	NOV 14	NOV 20	NOV 29	23 23	24 24
	20	FEB 7	FEB 20	MAR 1	MAR 10	MAR 24	270	NOV 9	NOV 19	NOV 26	DEC 3	DEC 14	23 23	24 24
	16	JAN 9	FEB 4	FEB 14	FEB 24	MAR 10	298	NOV 26	DEC 3	DEC 9	DEC 16	*	23 21	24 20
HOLLAND NANSEMOND COUNTY	36	APR 16	APR 24	APR 29	MAY 4	MAY 12	168	OCT 3	OCT 9	OCT 14	OCT 19	OCT 26	31 31	31 31
	32	APR 4	APR 13	APR 19	APR 25	MAY 4	189	OCT 12	OCT 20	OCT 25	OCT 31	NOV 8	31 31	31 31
	28	MAR 14	MAR 25	APR 1	APR 8	APR 18	221	OCT 25	NOV 2	NOV 8	NOV 13	NOV 22	31 31	31 31
	24	FEB 19	MAR 4	MAR 13	MAR 22	APR 4	253	NOV 7	NOV 15	NOV 21	NOV 26	DEC 5	31 31	31 31
	20	FEB 2	FEB 17	FEB 26	MAR 7	MAR 20	282	NOV 17	NOV 27	DEC 5	DEC 14	*	31 30	31 28
	16	JAN 16	FEB 5	FEB 15	FEB 25	MAR 11	307	DEC 6	DEC 13	DEC 19	DEC 29	*	31 29	31 23
HOPEWELL PRINCE GEORGE COUNTY	36	APR 8	APR 16	APR 21	APR 26	MAY 4	181	OCT 7	OCT 14	OCT 19	OCT 24	OCT 31	31 31	31 31
	32	MAR 26	APR 3	APR 9	APR 15	APR 23	204	OCT 16	OCT 24	OCT 30	NOV 4	NOV 12	31 31	31 31
	28	MAR 8	MAR 18	MAR 24	MAR 31	APR 10	229	OCT 26	NOV 3	NOV 8	NOV 14	NOV 21	31 31	31 31
	24	FEB 20	MAR 3	MAR 12	MAR 20	APR 1	254	NOV 6	NOV 15	NOV 21	NOV 27	DEC 6	31 31	31 31
	20	FEB 5	FEB 17	FEB 25	MAR 6	MAR 18	285	NOV 21	NOV 30	DEC 7	DEC 14	DEC 24	31 31	31 30
	16	JAN 13	JAN 31	FEB 10	FEB 20	MAR 5	312	DEC 1	DEC 10	DEC 19	*	*	31 29	31 19
HOT SPRINGS BATH COUNTY	36	APR 27	MAY 7	MAY 15	MAY 22	JUN 1	135	SEP 14	SEP 22	SEP 27	OCT 2	OCT 10	31 31	31 31
	32	APR 17	APR 26	MAY 3	MAY 9	MAY 19	157	SEP 25	OCT 2	OCT 7	OCT 13	OCT 20	31 31	31 31
	28	APR 8	APR 17	APR 22	APR 28	MAY 7	181	OCT 3	OCT 13	OCT 20	OCT 27	NOV 6	31 31	31 31
	24	MAR 24	APR 2	APR 8	APR 15	APR 24	210	OCT 21	OCT 29	NOV 4	NOV 10	NOV 19	31 31	31 31
	20	MAR 7	MAR 17	MAR 24	MAR 31	APR 11	236	OCT 29	NOV 8	NOV 15	NOV 21	DEC 1	31 31	31 31
	16	FEB 23	MAR 4	MAR 11	MAR 18	MAR 28	260	NOV 10	NOV 20	NOV 26	DEC 2	DEC 12	31 31	31 31
JOHN H KERR DAM MECKLENBURG COUNTY	36	APR 12	APR 20	APR 25	MAY 1	MAY 9	173	OCT 1	OCT 9	OCT 15	OCT 21	OCT 29	22 22	23 23
	32	MAR 30	APR 10	APR 18	APR 25	MAY 6	188	OCT 7	OCT 16	OCT 23	OCT 29	NOV 7	22 22	23 23
	28	MAR 13	MAR 24	APR 1	APR 9	APR 20	213	OCT 16	OCT 24	OCT 31	NOV 6	NOV 15	22 22	23 23
	24	MAR 4	MAR 11	MAR 17	MAR 22	MAR 30	242	OCT 27	NOV 6	NOV 14	NOV 21	DEC 2	22 22	23 23
	20	FEB 2	FEB 17	FEB 27	MAR 9	MAR 23	274	NOV 10	NOV 21	NOV 28	DEC 6	DEC 18	22 22	23 22
	16	*	JAN 29	FEB 13	FEB 24	MAR 11	298	NOV 20	DEC 1	DEC 8	DEC 17	*	22 18	23 21
LANGLEY FIELD	36	MAR 24	APR 2	APR 9	APR 15	APR 25	208	OCT 17	OCT 27	NOV 3	NOV 10	NOV 19	31 31	31 31
	32	MAR 6	MAR 17	MAR 25	APR 2	APR 13	237	NOV 2	NOV 10	NOV 17	NOV 23	DEC 3	31 31	31 30
	28	FEB 16	MAR 1	MAR 10	MAR 20	APR 2	264	NOV 13	NOV 22	NOV 29	DEC 5	DEC 16	31 31	31 30
	24	JAN 27	FEB 11	FEB 20	MAR 1	MAR 14	293	NOV 24	DEC 3	DEC 10	DEC 17	*	31 30	31 27
	20	*	JAN 26	FEB 8	FEB 20	MAR 7	325	DEC 7	DEC 16	DEC 30	*	*	31 26	31 16
	16	*	*	JAN 19	FEB 2	FEB 17	361	DEC 13	*	*	*	*	31 19	31 9
LAWRENCEVILLE BRUNSWICK COUNTY	36	APR 17	APR 26	MAY 2	MAY 8	MAY 16	161	SEP 30	OCT 6	OCT 10	OCT 13	OCT 19	28 28	29 29
	32	APR 9	APR 18	APR 23	APR 29	MAY 8	181	OCT 7	OCT 15	OCT 21	OCT 26	NOV 3	28 28	29 29
	28	MAR 22	MAR 31	APR 6	APR 12	APR 21	209	OCT 21	OCT 27	NOV 1	NOV 5	NOV 11	28 28	28 78
	24	MAR 3	MAR 14	MAR 21	MAR 29	APR 9	238	OCT 30	NOV 8	NOV 14	NOV 20	NOV 26	29 29	28 28
	20	FEB 16	FEB 28	MAR 8	MAR 17	MAR 29	262	NOV 10	NOV 18	NOV 25	DEC 1	DEC 11	29 29	28 27
	16	JAN 29	FEB 14	FEB 23	MAR 4	MAR 18	286	NOV 17	NOV 28	DEC 6	DEC 14	DEC 31	29 28	28 26

FREEZE DATA

PROBABILITY OF SELECTED TEMPERATURES ON OR AFTER GIVEN DATES IN SPRING AND ON OR BEFORE GIVEN DATES IN FALL

STATION	TEMP	PROBABILITY – SPRING					NO. OF DAYS BETWEEN 50% P'S	PROBABILITY – FALL					SPRING		FALL	
		90%	70%	50%	30%	10%		10%	30%	50%	70%	90%	Y	O	Y	O
LEXINGTON ROCKBRIDGE COUNTY	36	APR 18	APR 29	MAY 6	MAY 13	MAY 24	151	SEP 23	SEP 30	OCT 4	OCT 9	OCT 16	31	31	31	31
	32	APR 11	APR 20	APR 26	MAY 2	MAY 11	168	SEP 27	OCT 5	OCT 11	OCT 17	OCT 25	31	31	31	31
	28	MAR 24	APR 3	APR 10	APR 17	APR 26	200	OCT 12	OCT 21	OCT 27	NOV 2	NOV 10	31	31	31	31
	24	MAR 7	MAR 18	MAR 25	APR 2	APR 13	228	OCT 25	NOV 3	NOV 8	NOV 14	NOV 22	31	31	31	31
	20	FEB 22	MAR 4	MAR 12	MAR 19	MAR 29	253	NOV 5	NOV 14	NOV 20	NOV 26	DEC 4	31	31	31	31
	16	FEB 1	FEB 17	FEB 27	MAR 9	MAR 22	279	NOV 16	NOV 26	DEC 3	DEC 10	DEC 20	31	30	31	31
LINCOLN LOUDOUN COUNTY	36	APR 15	APR 23	APR 29	MAY 5	MAY 13	165	SEP 25	OCT 5	OCT 11	OCT 18	OCT 27	31	31	31	31
	32	APR 3	APR 11	APR 17	APR 23	MAY 2	187	OCT 3	OCT 13	OCT 21	OCT 28	NOV 8	31	31	31	31
	28	MAR 16	MAR 25	APR 1	APR 8	APR 17	218	OCT 23	OCT 31	NOV 5	NOV 10	NOV 17	31	31	31	31
	24	MAR 3	MAR 13	MAR 21	MAR 28	APR 7	244	NOV 6	NOV 14	NOV 20	NOV 25	DEC 3	31	31	31	31
	20	FEB 20	MAR 3	MAR 11	MAR 19	MAR 30	265	NOV 16	NOV 24	DEC 1	DEC 7	DEC 15	31	31	31	31
	16	FEB 5	FEB 18	FEB 27	MAR 8	MAR 22	286	NOV 26	DEC 4	DEC 10	DEC 16	DEC 26	31	31	31	30
LOUISA LOUISA COUNTY	36	APR 19	APR 28	MAY 5	MAY 11	MAY 20	155	SEP 25	OCT 2	OCT 7	OCT 11	OCT 18	31	31	30	30
	32	APR 8	APR 18	APR 24	MAY 1	MAY 10	173	SEP 29	OCT 8	OCT 14	OCT 21	OCT 29	31	31	30	30
	28	MAR 25	APR 3	APR 10	APR 17	APR 26	200	OCT 11	OCT 20	OCT 27	NOV 2	NOV 11	31	31	30	30
	24	MAR 13	MAR 22	MAR 28	APR 3	APR 11	227	OCT 28	NOV 4	NOV 10	NOV 15	NOV 22	31	31	30	30
	20	FEB 26	MAR 8	MAR 15	MAR 22	APR 1	253	NOV 8	NOV 17	NOV 23	NOV 29	DEC 8	31	31	30	30
	16	FEB 6	FEB 20	MAR 1	MAR 10	MAR 22	279	NOV 22	NOV 29	DEC 5	DEC 10	DEC 18	31	30	30	30
LURAY PAGE COUNTY	36	APR 30	MAY 10	MAY 17	MAY 24	JUN 2	136	SEP 13	SEP 23	SEP 30	OCT 7	OCT 17	29	29	29	29
	32	APR 19	APR 30	MAY 8	MAY 16	MAY 27	154	SEP 23	OCT 2	OCT 9	OCT 15	OCT 24	29	29	29	29
	28	APR 5	APR 14	APR 20	APR 26	MAY 5	184	OCT 6	OCT 15	OCT 21	OCT 28	NOV 6	28	28	29	29
	24	MAR 22	MAR 31	APR 7	APR 13	APR 22	212	OCT 24	NOV 5	NOV 10	NOV 17		28	28	29	29
	20	MAR 4	MAR 14	MAR 21	MAR 28	APR 6	239	OCT 28	NOV 8	NOV 15	NOV 22	DEC 2	28	28	29	29
	16	FEB 23	MAR 5	MAR 11	MAR 18	MAR 27	263	NOV 17	NOV 24	NOV 29	DEC 4	DEC 11	28	28	29	29
LYNCHBURG MUNICIPAL AIRPORT	36	APR 9	APR 17	APR 23	APR 29	MAY 8	176	OCT 2	OCT 10	OCT 16	OCT 21	OCT 29	31	31	31	31
	32	MAR 25	APR 4	APR 11	APR 18	APR 28	200	OCT 15	OCT 22	OCT 28	NOV 2	NOV 10	31	31	31	31
	28	MAR 5	MAR 16	MAR 24	MAR 31	APR 11	229	OCT 23	NOV 2	NOV 8	NOV 15	NOV 25	31	31	31	31
	24	FEB 22	MAR 5	MAR 13	MAR 21	APR 1	253	NOV 2	NOV 13	NOV 21	NOV 29	DEC 10	31	31	31	31
	20	FEB 1	FEB 16	FEB 27	MAR 10	MAR 26	274	NOV 19	NOV 27	DEC 3	DEC 8	DEC 17	31	31	31	30
	16	JAN 1	FEB 5	FEB 16	FEB 26	MAR 12	299	NOV 25	DEC 5	DEC 12	DEC 20	*	31	28	31	27
MANASSAS PRINCE WILLIAM COUNTY	36	APR 15	APR 25	MAY 3	MAY 10	MAY 20	156	SEP 22	SEP 30	OCT 6	OCT 12	OCT 20	22	22	23	23
	32	APR 2	APR 12	APR 19	APR 25	MAY 5	183	OCT 4	OCT 13	OCT 19	OCT 25	NOV 3	22	22	23	23
	28	MAR 20	MAR 29	APR 5	APR 12	APR 22	210	OCT 20	OCT 27	NOV 1	NOV 7	NOV 14	22	22	23	23
	24	MAR 9	MAR 17	MAR 22	MAR 28	APR 5	235	OCT 28	NOV 6	NOV 12	NOV 18	NOV 27	22	22	23	23
	20	FEB 21	MAR 4	MAR 11	MAR 19	MAR 30	259	NOV 8	NOV 18	NOV 25	DEC 2	DEC 12	22	22	23	23
	16	FEB 2	FEB 16	FEB 24	MAR 4	MAR 15	286	NOV 22	DEC 1	DEC 7	DEC 14	DEC 23	22	21	23	23
MARTINSVILLE HENRY COUNTY	36	APR 15	APR 27	MAY 6	MAY 15	MAY 27	149	SEP 16	SEP 25	OCT 2	OCT 9	OCT 18	31	31	31	31
	32	APR 8	APR 19	APR 26	MAY 4	MAY 15	173	SEP 28	OCT 9	OCT 16	OCT 23	NOV 3	31	31	31	31
	28	MAR 27	APR 8	APR 16	APR 24	MAY 5	191	OCT 5	OCT 16	OCT 24	OCT 31	NOV 11	31	31	31	31
	24	MAR 5	MAR 19	MAR 29	APR 8	APR 22	225	OCT 22	NOV 1	NOV 9	NOV 16	NOV 26	31	31	31	31
	20	FEB 21	MAR 8	MAR 18	MAR 28	APR 12	246	OCT 31	NOV 11	NOV 19	NOV 27	DEC 9	31	31	31	31
	16	FEB 8	FEB 21	MAR 2	MAR 11	MAR 25	271	NOV 8	NOV 20	NOV 28	DEC 6	DEC 22	31	31	31	29
MONTEREY HIGHLAND COUNTY	36	MAY 2	MAY 14	MAY 23	JUN 1	JUN 13	121	SEP 5	SEP 14	SEP 21	SEP 27	OCT 7	30	30	29	29
	32	APR 22	MAY 4	MAY 12	MAY 20	MAY 31	143	SEP 17	SEP 26	OCT 2	OCT 9	OCT 18	30	30	29	29
	28	APR 6	APR 19	APR 28	MAY 7	MAY 20	167	SEP 26	OCT 5	OCT 12	OCT 18	OCT 28	30	30	29	29
	24	MAR 23	APR 5	APR 15	APR 24	MAY 7	189	OCT 2	OCT 13	OCT 21	OCT 29	NOV 10	29	29	29	29
	20	MAR 12	MAR 23	MAR 31	APR 8	APR 19	219	OCT 18	OCT 28	NOV 5	NOV 12	NOV 23	29	29	29	29
	16	MAR 4	MAR 14	MAR 21	MAR 28	APR 7	241	OCT 30	NOV 10	NOV 17	NOV 24	DEC 4	29	29	29	29
MOUNT WEATHER LOUDOUN COUNTY	36	APR 13	APR 22	APR 28	MAY 5	MAY 14	167	SEP 30	OCT 7	OCT 12	OCT 17	OCT 24	26	26	26	26
	32	APR 6	APR 16	APR 24	MAY 1	MAY 11	178	OCT 2	OCT 12	OCT 19	OCT 26	NOV 5	30	30	30	30
	28	MAR 25	APR 4	APR 11	APR 18	APR 29	203	OCT 14	OCT 24	OCT 31	NOV 7	NOV 16	30	30	31	31
	24	MAR 6	MAR 19	MAR 27	APR 5	APR 18	231	OCT 28	NOV 7	NOV 13	NOV 20	NOV 29	30	30	31	31
	20	FEB 27	MAR 12	MAR 20	MAR 29	APR 10	250	NOV 7	NOV 17	NOV 25	DEC 2	DEC 12	31	31	31	31
	16	FEB 17	MAR 1	MAR 10	MAR 18	MAR 31	269	NOV 22	NOV 29	DEC 4	DEC 9	DEC 16	31	31	31	31
NASSAWADOX NORTHAMPTON COUNTY	36	APR 5	APR 17	APR 23	APR 29	MAY 8	185	OCT 10	OCT 19	OCT 25	OCT 31	NOV 9	15	14	15	15
	32	MAR 24	APR 3	APR 8	APR 13	APR 20	209	OCT 23	OCT 29	NOV 3	NOV 8	NOV 14	15	14	15	15
	28	MAR 4	MAR 17	MAR 23	MAR 30	APR 7	237	NOV 3	NOV 10	NOV 15	NOV 20	NOV 28	15	14	15	15
	24	FEB 17	MAR 3	MAR 10	MAR 17	MAR 26	262	NOV 14	NOV 22	NOV 27	DEC 1	DEC 11	15	14	15	15
	20	JAN 25	FEB 11	FEB 21	MAR 1	MAR 13	293	NOV 26	DEC 5	DEC 11	DEC 18	DEC 30	15	14	15	14
	16	*	JAN 30	FEB 12	FEB 23	MAR 11	—	DEC 9	DEC 16	*	*	*	15	13	15	5
NEW CANTON BUCKINGHAM COUNTY	36	APR 25	MAY 3	MAY 9	MAY 15	MAY 23	148	SEP 22	SEP 29	OCT 4	OCT 9	OCT 17	31	31	31	31
	32	APR 16	APR 23	APR 29	MAY 4	MAY 12	168	SEP 29	OCT 8	OCT 14	OCT 21	OCT 29	31	31	31	31
	28	MAR 30	APR 8	APR 15	APR 21	MAY 1	194	OCT 11	OCT 20	OCT 26	OCT 31	NOV 9	31	31	31	31
	24	MAR 15	MAR 25	APR 1	APR 8	APR 19	219	OCT 25	NOV 1	NOV 6	NOV 10	NOV 17	31	31	31	31
	20	MAR 3	MAR 13	MAR 20	MAR 27	APR 6	240	OCT 30	NOV 9	NOV 15	NOV 21	DEC 1	31	31	31	31
	16	FEB 11	FEB 24	MAR 4	MAR 13	MAR 26	270	NOV 14	NOV 23	NOV 29	DEC 6	DEC 15	31	31	31	31
NEWPORT NEWS	36	MAR 15	MAR 26	APR 3	APR 11	APR 22	219	OCT 24	NOV 2	NOV 8	NOV 14	NOV 23	22	22	22	22
	32	MAR 7	MAR 17	MAR 23	MAR 30	APR 8	243	NOV 7	NOV 15	NOV 21	NOV 26	DEC 4	22	22	22	22
	28	FEB 20	MAR 4	MAR 12	MAR 21	APR 2	266	NOV 14	NOV 25	DEC 3	DEC 10	DEC 24	22	22	23	22
	24	JAN 23	FEB 12	FEB 23	MAR 6	MAR 22	292	NOV 25	DEC 4	DEC 12	DEC 20	*	22	21	23	20
	20	*	JAN 25	FEB 7	FEB 17	MAR 3	320	DEC 5	DEC 14	DEC 24	*	*	22	18	23	13
	16	*	*	JAN 15	FEB 6	FEB 25	—	DEC 13	*	*	*	*	22	12	23	3
NORFOLK REGIONAL AIRPORT	36	MAR 15	MAR 27	APR 4	APR 12	APR 24	219	OCT 24	NOV 2	NOV 9	NOV 15	NOV 24	31	31	31	31
	32	MAR 5	MAR 15	MAR 22	MAR 29	APR 8	244	NOV 2	NOV 13	NOV 21	NOV 28	DEC 9	31	31	31	31
	28	FEB 16	FEB 27	MAR 7	MAR 15	MAR 26	271	NOV 19	NOV 27	DEC 3	DEC 9	DEC 19	31	31	31	30
	24	JAN 20	FEB 8	FEB 18	FEB 27	MAR 13	298	NOV 27	DEC 6	DEC 13	DEC 23	*	31	29	31	24
	20	*	JAN 18	FEB 3	FEB 14	MAR 1	—	DEC 11	DEC 18	*	*	*	31	24	31	15
	16	*	*	*	FEB 2	FEB 20	—	DEC 17	*	*	*	*	31	15	31	6
PAINTER ACCOMAC COUNTY	36	APR 8	APR 18	APR 25	MAY 2	MAY 13	182	OCT 6	OCT 16	OCT 24	OCT 31	NOV 11	31	31	31	31
	32	MAR 25	APR 4	APR 11	APR 18	APR 29	207	OCT 19	OCT 28	NOV 4	NOV 10	NOV 19	31	31	31	31
	28	MAR 14	MAR 23	MAR 30	APR 6	APR 16	228	OCT 29	NOV 7	NOV 13	NOV 20	NOV 29	31	31	31	31
	24	FEB 21	MAR 4	MAR 11	MAR 18	MAR 30	264	NOV 12	NOV 23	NOV 30	DEC 7	DEC 18	31	31	31	31
	20	FEB 1	FEB 17	FEB 25	MAR 4	MAR 15	291	NOV 27	DEC 6	DEC 13	DEC 22	*	31	29	31	24
	16	*	JAN 26	FEB 7	FEB 17	MAR 2	322	DEC 10	DEC 17	DEC 26	*	*	31	27	31	17

FREEZE DATA

PROBABILITY OF SELECTED TEMPERATURES ON OR AFTER GIVEN DATES IN SPRING AND ON OR BEFORE GIVEN DATES IN FALL

STATION	TEMP	PROBABILITY – SPRING 90%	70%	50%	30%	10%	NO. OF DAYS BETWEEN 50% P'S	PROBABILITY – FALL 10%	30%	50%	70%	90%	SPRING Y O	FALL Y O
PARTLOW SPOTSYLVANIA COUNTY	36	APR 30	MAY 9	MAY 16	MAY 22	JUN 1	133	SEP 15	SEP 21	SEP 26	SEP 30	OCT 7	18 18	19 19
	32	APR 21	APR 29	MAY 4	MAY 10	MAY 18	155	SEP 24	OCT 1	OCT 6	OCT 10	OCT 17	18 18	19 19
	28	APR 7	APR 15	APR 21	APR 27	MAY 5	177	OCT 1	OCT 9	OCT 15	OCT 20	OCT 28	17 17	19 19
	24	MAR 24	APR 2	APR 9	APR 16	APR 26	201	OCT 13	OCT 22	OCT 27	NOV 2	NOV 11	18 18	19 19
	20	MAR 6	MAR 17	MAR 25	APR 2	APR 13	226	OCT 22	OCT 31	NOV 6	NOV 12	NOV 20	18 18	19 19
	16	FEB 21	MAR 5	MAR 13	MAR 21	APR 1	252	NOV 2	NOV 13	NOV 20	NOV 28	DEC 8	18 18	19 19
PENNINGTON GAP LEE COUNTY	36	APR 27	MAY 6	MAY 11	MAY 17	MAY 25	143	SEP 19	SEP 26	OCT 1	OCT 5	OCT 12	31 31	29 29
	32	APR 12	APR 22	APR 29	MAY 6	MAY 16	164	SEP 28	OCT 5	OCT 10	OCT 15	OCT 22	31 31	29 29
	28	APR 1	APR 9	APR 15	APR 20	APR 28	194	OCT 14	OCT 21	OCT 26	OCT 31	NOV 8	31 31	29 29
	24	MAR 19	MAR 27	APR 2	APR 8	APR 17	214	OCT 21	OCT 28	NOV 2	NOV 7	NOV 15	31 31	29 29
	20	FEB 22	MAR 6	MAR 15	MAR 23	APR 4	244	OCT 29	NOV 7	NOV 14	NOV 21	NOV 30	31 31	29 29
	16	FEB 12	FEB 25	MAR 6	MAR 15	MAR 28	263	NOV 4	NOV 16	NOV 24	DEC 2	DEC 14	31 31	29 29
PHILPOT DAM HENRY COUNTY	36	APR 14	APR 25	MAY 2	MAY 10	MAY 21	164	OCT 1	OCT 9	OCT 13	OCT 18	OCT 26	17 17	17 17
	32	APR 2	APR 11	APR 17	APR 23	MAY 2	189	OCT 9	OCT 17	OCT 23	OCT 29	NOV 6	17 17	17 17
	28	MAR 17	MAR 25	APR 1	APR 7	APR 16	215	OCT 22	OCT 29	NOV 2	NOV 7	NOV 14	17 17	17 17
	24	MAR 4	MAR 13	MAR 20	MAR 26	APR 4	241	OCT 30	NOV 9	NOV 16	NOV 23	DEC 4	17 17	18 18
	20	FEB 8	FEB 23	MAR 6	MAR 17	APR 2	265	NOV 10	NOV 20	NOV 26	DEC 3	DEC 13	17 17	18 18
	16	JAN 26	FEB 9	FEB 19	MAR 1	MAR 16	295	NOV 28	DEC 6	DEC 11	DEC 17	DEC 27	17 17	18 17
PIEDMONT FIELD STATION ORANGE COUNTY	36	APR 12	APR 23	APR 30	MAY 8	MAY 19	165	SEP 29	OCT 6	OCT 12	OCT 17	OCT 25	24 24	24 24
	32	MAR 31	APR 10	APR 17	APR 24	MAY 4	189	OCT 7	OCT 16	OCT 23	OCT 29	NOV 7	24 24	24 24
	28	MAR 22	APR 1	APR 7	APR 14	APR 23	208	OCT 21	OCT 28	NOV 1	NOV 6	NOV 13	24 24	24 24
	24	MAR 7	MAR 16	MAR 23	MAR 29	APR 8	237	OCT 31	NOV 9	NOV 15	NOV 21	NOV 30	24 24	24 24
	20	FEB 25	MAR 7	MAR 14	MAR 21	APR 1	255	NOV 9	NOV 18	NOV 24	NOV 30	DEC 9	24 24	24 24
	16	FEB 5	FEB 18	FEB 26	MAR 5	MAR 16	285	NOV 24	DEC 2	DEC 8	DEC 14	DEC 23	24 23	24 24
PULASKI PULASKI COUNTY	36	APR 15	APR 28	MAY 8	MAY 17	MAY 30	144	SEP 16	SEP 23	SEP 29	OCT 4	OCT 11	21 21	22 22
	32	APR 9	APR 18	APR 25	MAY 1	MAY 10	167	SEP 24	OCT 3	OCT 9	OCT 15	OCT 23	21 21	22 22
	28	APR 2	APR 11	APR 17	APR 23	MAY 1	185	OCT 2	OCT 12	OCT 19	OCT 26	NOV 5	21 21	22 22
	24	MAR 9	MAR 21	MAR 30	APR 7	APR 19	217	OCT 18	OCT 27	NOV 2	NOV 9	NOV 18	21 21	22 22
	20	FEB 25	MAR 7	MAR 14	MAR 21	APR 1	245	OCT 26	NOV 6	NOV 14	NOV 22	DEC 4	21 21	22 22
	16	FEB 13	FEB 26	MAR 6	MAR 15	MAR 28	267	NOV 10	NOV 20	NOV 28	DEC 5	DEC 15	21 21	22 22
QUANTICO PRINCE WILLIAM COUNTY	36	APR 8	APR 17	APR 23	APR 30	MAY 9	179	OCT 4	OCT 13	OCT 19	OCT 25	NOV 3	31 31	31 31
	32	MAR 24	APR 4	APR 12	APR 20	MAY 1	202	OCT 19	OCT 26	OCT 31	NOV 5	NOV 13	31 31	31 31
	28	MAR 9	MAR 20	MAR 27	APR 3	APR 14	231	NOV 1	NOV 8	NOV 13	NOV 18	NOV 26	31 31	31 31
	24	FEB 23	MAR 5	MAR 12	MAR 19	MAR 29	259	NOV 12	NOV 20	NOV 26	DEC 1	DEC 9	31 31	31 31
	20	FEB 3	FEB 16	FEB 25	MAR 5	MAR 18	284	NOV 21	NOV 30	DEC 6	DEC 12	DEC 21	31 31	31 31
	16	JAN 18	FEB 7	FEB 18	FEB 28	MAR 14	258	NOV 26	DEC 6	DEC 13	DEC 22	*	31 29	31 26
RICHMOND BYRD FIELD	36	APR 4	APR 14	APR 20	APR 26	MAY 6	181	OCT 5	OCT 13	OCT 18	OCT 24	NOV 1	31 31	31 31
	32	MAR 24	APR 2	APR 9	APR 16	APR 26	206	OCT 17	OCT 26	NOV 1	NOV 7	NOV 16	31 31	31 31
	28	MAR 5	MAR 17	MAR 24	APR 1	APR 13	233	OCT 27	NOV 6	NOV 12	NOV 19	NOV 28	31 31	31 31
	24	FEB 22	MAR 4	MAR 11	MAR 18	MAR 28	261	NOV 12	NOV 21	NOV 27	DEC 3	DEC 13	31 31	31 31
	20	FEB 5	FEB 18	FEB 26	MAR 7	MAR 19	283	NOV 21	NOV 29	DEC 6	DEC 12	DEC 24	31 31	31 29
	16	JAN 1	JAN 30	FEB 9	FEB 17	MAR 1	312	DEC 5	DEC 12	DEC 18	DEC 26	*	31 28	31 24
ROANOKE WOODRUM AIRPORT	36	APR 14	APR 23	APR 29	MAY 5	MAY 13	165	SEP 27	OCT 6	OCT 11	OCT 17	OCT 26	31 31	31 31
	32	MAR 28	APR 6	APR 12	APR 19	APR 28	194	OCT 7	OCT 17	OCT 23	OCT 30	NOV 8	31 31	31 31
	28	MAR 13	MAR 22	MAR 28	APR 4	APR 13	223	OCT 24	NOV 1	NOV 6	NOV 12	NOV 20	31 31	31 31
	24	FEB 20	MAR 4	MAR 13	MAR 21	APR 2	255	NOV 10	NOV 18	NOV 23	NOV 29	DEC 6	31 31	31 31
	20	FEB 4	FEB 20	MAR 2	MAR 12	MAR 26	274	NOV 16	NOV 25	DEC 1	DEC 7	DEC 15	31 30	31 31
	16	JAN 23	FEB 9	FEB 19	FEB 27	MAR 12	296	NOV 27	DEC 5	DEC 12	DEC 19	*	31 29	31 26
ROCKY KNOB FLOYD COUNTY	36	APR 21	MAY 1	MAY 9	MAY 16	MAY 27	155	SEP 25	OCT 4	OCT 11	OCT 17	OCT 26	21 21	21 21
	32	APR 6	APR 17	APR 24	MAY 2	MAY 13	182	OCT 8	OCT 17	OCT 23	OCT 29	NOV 7	21 21	20 20
	28	MAR 28	APR 7	APR 14	APR 21	MAY 2	200	OCT 18	OCT 26	OCT 31	NOV 6	NOV 13	20 20	20 20
	24	MAR 16	MAR 26	APR 2	APR 9	APR 19	219	OCT 22	OCT 31	NOV 7	NOV 13	NOV 23	20 20	21 21
	20	FEB 27	MAR 10	MAR 18	MAR 26	APR 5	245	OCT 30	NOV 10	NOV 18	NOV 25	DEC 6	20 20	22 22
	16	FEB 11	FEB 26	MAR 8	MAR 18	APR 2	265	NOV 12	NOV 21	NOV 28	DEC 4	DEC 13	20 20	22 22
ROCKY MOUNT FRANKLIN COUNTY	36	APR 15	APR 26	MAY 3	MAY 11	MAY 22	157	SEP 26	OCT 3	OCT 7	OCT 12	OCT 18	29 29	25 25
	32	APR 3	APR 13	APR 21	APR 29	MAY 10	178	OCT 1	OCT 10	OCT 16	OCT 22	OCT 31	29 29	25 25
	28	MAR 14	MAR 26	APR 4	APR 13	APR 25	212	OCT 15	OCT 25	NOV 2	NOV 9	NOV 20	29 29	25 25
	24	MAR 2	MAR 13	MAR 21	MAR 29	APR 9	239	OCT 30	NOV 8	NOV 15	NOV 21	DEC 1	30 30	25 25
	20	FEB 15	FEB 28	MAR 9	MAR 18	APR 1	263	NOV 15	NOV 22	NOV 27	DEC 2	DEC 9	30 30	25 25
	16	JAN 27	FEB 12	FEB 22	MAR 4	MAR 18	287	NOV 22	NOV 30	DEC 6	DEC 11	DEC 20	30 29	25 24
SALTVILLE SMYTH COUNTY	36	APR 17	APR 26	MAY 2	MAY 8	MAY 17	158	SEP 26	OCT 3	OCT 7	OCT 12	OCT 19	31 31	31 31
	32	APR 3	APR 12	APR 18	APR 24	MAY 2	189	OCT 9	OCT 18	OCT 24	OCT 30	NOV 7	31 31	31 31
	28	MAR 17	MAR 28	APR 4	APR 11	APR 22	210	OCT 18	OCT 25	OCT 31	NOV 5	NOV 12	31 31	31 31
	24	FEB 27	MAR 12	MAR 20	MAR 29	APR 10	238	OCT 29	NOV 7	NOV 13	NOV 19	NOV 28	31 31	31 31
	20	FEB 17	MAR 1	MAR 9	MAR 18	MAR 29	259	NOV 5	NOV 16	NOV 23	NOV 30	DEC 11	31 31	31 31
	16	FEB 2	FEB 18	FEB 27	MAR 9	MAR 22	284	NOV 23	DEC 1	DEC 8	DEC 14	DEC 26	31 30	31 29
STAUNTON AUGUSTA COUNTY	36	APR 13	APR 25	MAY 3	MAY 12	MAY 23	155	SEP 20	SEP 29	OCT 5	OCT 10	OCT 19	31 31	30 30
	32	APR 8	APR 17	APR 24	MAY 1	MAY 10	175	OCT 3	OCT 11	OCT 16	OCT 22	OCT 30	31 31	30 30
	28	MAR 24	APR 4	APR 11	APR 19	APR 30	203	OCT 14	OCT 24	OCT 31	NOV 7	NOV 17	31 31	31 31
	24	MAR 7	MAR 18	MAR 25	APR 2	APR 13	230	OCT 26	NOV 4	NOV 10	NOV 16	NOV 25	31 31	31 31
	20	FEB 26	MAR 9	MAR 16	MAR 24	APR 4	250	NOV 3	NOV 13	NOV 21	NOV 28	DEC 8	31 31	31 31
	16	FEB 9	FEB 22	MAR 2	MAR 10	MAR 21	276	NOV 19	NOV 27	DEC 3	DEC 8	DEC 16	31 30	31 31
STUART PATRICK COUNTY	36	APR 12	APR 22	APR 28	MAY 4	MAY 14	169	SEP 28	OCT 7	OCT 14	OCT 20	OCT 29	29 29	29 29
	32	MAR 20	APR 3	APR 12	APR 21	MAY 5	196	OCT 11	OCT 19	OCT 25	OCT 30	NOV 8	29 29	29 29
	28	MAR 12	MAR 24	APR 2	APR 11	APR 23	218	OCT 21	OCT 30	NOV 6	NOV 13	NOV 23	29 29	27 27
	24	FEB 27	MAR 13	MAR 22	APR 1	APR 15	242	NOV 3	NOV 12	NOV 19	NOV 25	DEC 5	29 29	28 28
	20	FEB 14	FEB 27	MAR 8	MAR 17	MAR 30	267	NOV 15	NOV 24	NOV 30	DEC 5	DEC 14	29 29	28 28
	16	JAN 15	FEB 8	FEB 21	MAR 5	MAR 21	291	NOV 23	DEC 3	DEC 9	DEC 17	*	29 27	28 26
SUFFOLK NANSEMOND COUNTY	36	APR 4	APR 13	APR 19	APR 26	MAY 5	189	OCT 11	OCT 19	OCT 25	OCT 31	NOV 9	22 22	23 23
	32	MAR 24	APR 2	APR 9	APR 15	APR 25	208	OCT 21	OCT 30	NOV 3	NOV 8	NOV 16	23 23	23 23
	28	MAR 10	MAR 17	MAR 22	MAR 27	APR 3	237	OCT 29	NOV 8	NOV 14	NOV 21	NOV 30	23 23	23 23
	24	FEB 18	MAR 1	MAR 9	MAR 17	MAR 29	264	NOV 13	NOV 22	NOV 28	DEC 4	DEC 13	23 23	23 23
	20	FEB 1	FEB 13	FEB 22	MAR 2	MAR 15	297	NOV 28	DEC 8	DEC 16	DEC 27	*	22 22	23 18
	16	*	JAN 26	FEB 6	FEB 15	FEB 28	325	DEC 11	DEC 18	DEC 28	*	*	22 19	23 12

FREEZE DATA

PROBABILITY OF SELECTED TEMPERATURES ON OR AFTER GIVEN DATES IN SPRING AND ON OR BEFORE GIVEN DATES IN FALL

STATION	TEMP	PROBABILITY – SPRING					NO. OF DAYS BETWEEN 50% P'S	PROBABILITY – FALL					SPRING	FALL
		90%	70%	50%	30%	10%		10%	30%	50%	70%	90%	Y O	Y O
SUNNYBANK NORTHUMBERLAND COUNTY	36	MAR 28	APR 9	APR 16	APR 22	APR 30	192	OCT 12	OCT 20	OCT 25	OCT 30	NOV 7	17 16	16 16
	32	MAR 22	MAR 31	APR 5	APR 10	APR 16	208	OCT 17	OCT 25	OCT 30	NOV 5	NOV 12	17 16	16 16
	28	MAR 7	MAR 17	MAR 23	MAR 28	APR 4	240	NOV 2	NOV 11	NOV 18	NOV 24	DEC 4	17 16	16 16
	24	FEB 12	FEB 27	MAR 8	MAR 16	MAR 27	268	NOV 16	NOV 25	DEC 1	DEC 6	DEC 15	17 16	16 16
	20	FEB 1	FEB 14	FEB 22	MAR 1	MAR 11	294	DEC 2	DEC 8	DEC 13	DEC 19	*	17 16	16 14
	16	*	JAN 28	FEB 9	FEB 20	MAR 6	314	DEC 6	DEC 13	DEC 20	*	*	17 15	16 10
TIMBERVILLE ROCKINGHAM COUNTY	36	APR 25	MAY 6	MAY 13	MAY 20	MAY 30	140	SEP 16	SEP 24	SEP 30	OCT 5	OCT 13	22 22	23 23
	32	APR 13	APR 23	APR 30	MAY 6	MAY 16	163	SEP 26	OCT 5	OCT 10	OCT 16	OCT 24	23 23	23 23
	28	APR 4	APR 12	APR 18	APR 23	MAY 1	185	OCT 7	OCT 15	OCT 20	OCT 26	NOV 2	23 23	23 23
	24	MAR 21	MAR 30	APR 4	APR 10	APR 18	214	OCT 24	OCT 31	NOV 4	NOV 9	NOV 16	23 23	23 23
	20	MAR 6	MAR 17	MAR 24	MAR 31	APR 10	234	OCT 27	NOV 6	NOV 13	NOV 20	DEC 1	22 22	23 23
	16	FEB 19	MAR 1	MAR 9	MAR 16	MAR 26	264	NOV 15	NOV 23	NOV 28	DEC 4	DEC 12	22 22	23 23
URBANNA MIDDLESEX COUNTY	36	APR 5	APR 12	APR 17	APR 22	APR 29	190	OCT 9	OCT 18	OCT 24	OCT 30	NOV 8	23 23	24 24
	32	MAR 25	APR 1	APR 5	APR 10	APR 16	213	OCT 23	OCT 30	NOV 4	NOV 9	NOV 16	23 23	24 24
	28	MAR 9	MAR 17	MAR 22	MAR 27	APR 4	237	OCT 30	NOV 8	NOV 14	NOV 20	NOV 29	23 23	24 24
	24	FEB 21	MAR 3	MAR 10	MAR 17	MAR 27	266	NOV 16	NOV 25	DEC 1	DEC 7	DEC 16	23 23	24 24
	20	JAN 23	FEB 13	FEB 22	MAR 2	MAR 13	293	NOV 26	DEC 6	DEC 12	DEC 19	DEC 30	23 21	24 23
	16	JAN 5	JAN 29	FEB 8	FEB 18	MAR 3	318	DEC 6	DEC 15	DEC 23	*	*	23 21	24 15
WALKERTON KING & QUEEN COUNTY	36	APR 18	APR 26	MAY 1	MAY 6	MAY 14	162	SEP 27	OCT 4	OCT 10	OCT 15	OCT 23	31 31	31 31
	32	APR 8	APR 16	APR 21	APR 27	MAY 4	181	OCT 6	OCT 13	OCT 19	OCT 24	OCT 31	31 31	31 31
	28	MAR 21	MAR 30	APR 5	APR 12	APR 21	208	OCT 19	OCT 26	OCT 30	NOV 4	NOV 11	31 31	31 31
	24	MAR 8	MAR 17	MAR 24	MAR 30	APR 9	235	OCT 30	NOV 8	NOV 14	NOV 21	NOV 29	31 31	31 31
	20	FEB 20	MAR 2	MAR 10	MAR 17	MAR 28	262	NOV 14	NOV 21	NOV 27	DEC 2	DEC 10	31 31	31 31
	16	FEB 5	FEB 18	FEB 26	MAR 7	MAR 19	285	NOV 25	DEC 3	DEC 8	DEC 14	DEC 22	31 31	31 30
WARRENTON FAUQUIER COUNTY	36	APR 9	APR 19	APR 26	MAY 3	MAY 13	173	OCT 1	OCT 10	OCT 16	OCT 21	OCT 30	20 20	20 20
	32	APR 2	APR 9	APR 14	APR 19	APR 26	199	OCT 19	OCT 26	OCT 30	NOV 4	NOV 11	20 20	20 20
	28	MAR 22	MAR 30	APR 4	APR 10	APR 17	218	OCT 25	NOV 2	NOV 8	NOV 13	NOV 21	20 20	20 20
	24	MAR 7	MAR 15	MAR 21	MAR 27	APR 5	246	NOV 4	NOV 15	NOV 22	NOV 29	DEC 10	20 20	20 20
	20	FEB 22	MAR 4	MAR 12	MAR 19	MAR 30	263	NOV 18	NOV 25	NOV 30	DEC 4	DEC 11	19 19	20 20
	16	FEB 1	FEB 13	FEB 22	MAR 2	MAR 15	291	NOV 26	DEC 4	DEC 10	DEC 16	DEC 24	19 19	20 20
WARSAW RICHMOND COUNTY	36	APR 12	APR 21	APR 27	MAY 4	MAY 13	169	SEP 27	OCT 6	OCT 13	OCT 19	OCT 29	27 27	26 26
	32	APR 2	APR 10	APR 15	APR 20	APR 27	194	OCT 11	OCT 20	OCT 26	NOV 1	NOV 10	27 27	25 25
	28	MAR 13	MAR 24	MAR 31	APR 7	APR 17	219	OCT 26	NOV 1	NOV 5	NOV 9	NOV 15	27 27	25 25
	24	FEB 24	MAR 8	MAR 16	MAR 23	APR 4	251	NOV 8	NOV 16	NOV 22	NOV 28	DEC 6	27 27	25 25
	20	FEB 7	FEB 21	MAR 2	MAR 11	MAR 24	276	NOV 20	NOV 27	DEC 3	DEC 8	DEC 16	27 27	25 25
	16	JAN 25	FEB 11	FEB 22	MAR 4	MAR 19	296	DEC 2	DEC 10	DEC 15	DEC 22	*	27 26	25 21
WASHINGTON RAPPAHANNOCK COUNTY	36	MAY 3	MAY 10	MAY 15	MAY 19	MAY 26	138	SEP 17	SEP 25	SEP 30	OCT 5	OCT 13	16 16	15 15
	32	APR 17	APR 27	MAY 4	MAY 12	MAY 22	159	SEP 22	OCT 3	OCT 10	OCT 18	OCT 29	16 16	15 15
	28	APR 4	APR 12	APR 17	APR 23	MAY 1	187	SEP 29	OCT 12	OCT 21	OCT 30	NOV 12	16 16	15 15
	24	MAR 19	MAR 28	APR 4	APR 10	APR 19	215	OCT 20	OCT 29	NOV 5	NOV 12	NOV 21	15 15	15 15
	20	MAR 8	MAR 17	MAR 23	MAR 29	APR 7	233	OCT 26	NOV 4	NOV 11	NOV 18	NOV 27	15 15	15 15
	16	FEB 25	MAR 7	MAR 13	MAR 20	MAR 30	253	NOV 2	NOV 14	NOV 21	NOV 30	DEC 16	15 15	15 14
WASHINGTON NATIONAL AIRPORT	36	MAR 31	APR 9	APR 14	APR 20	APR 29	200	OCT 19	OCT 26	OCT 31	NOV 5	NOV 11	31 31	31 31
	32	MAR 15	MAR 24	MAR 30	APR 4	APR 13	223	OCT 25	NOV 2	NOV 8	NOV 13	NOV 21	31 31	31 31
	28	MAR 1	MAR 12	MAR 19	MAR 26	APR 5	250	NOV 13	NOV 19	NOV 24	NOV 29	DEC 6	31 31	31 31
	24	FEB 17	MAR 1	MAR 9	MAR 17	MAR 29	272	NOV 23	DEC 1	DEC 6	DEC 11	DEC 20	31 31	31 30
	20	JAN 30	FEB 16	FEB 25	MAR 6	MAR 18	291	NOV 29	DEC 7	DEC 13	DEC 19	*	31 29	31 28
	16	*	JAN 30	FEB 11	FEB 21	MAR 7	314	DEC 6	DEC 14	DEC 22	*	*	31 27	31 18

FREEZE DATA

PROBABILITY OF SELECTED TEMPERATURES ON OR AFTER GIVEN DATES IN SPRING AND ON OR BEFORE GIVEN DATES IN FALL

STATION	TEMP	PROBABILITY – SPRING					NO. OF DAYS BETWEEN 50% P'S	PROBABILITY – FALL					SPRING		FALL	
		90%	70%	50%	30%	10%		10%	30%	50%	70%	90%	Y	O	Y	O
WEST POINT NEW KENT COUNTY	36	APR 11	APR 20	APR 26	MAY 2	MAY 11	170	SEP 30	OCT 8	OCT 13	OCT 18	OCT 25	17	17	17	17
	32	APR 2	APR 9	APR 15	APR 20	APR 28	191	OCT 8	OCT 17	OCT 23	OCT 29	NOV 7	17	17	17	17
	28	MAR 21	MAR 29	APR 2	APR 7	APR 15	216	OCT 21	OCT 29	NOV 4	NOV 10	NOV 18	17	17	17	17
	24	MAR 3	MAR 13	MAR 20	MAR 27	APR 5	243	NOV 3	NOV 12	NOV 18	NOV 23	DEC 2	17	17	17	17
	20	FEB 21	MAR 3	MAR 9	MAR 16	MAR 26	265	NOV 14	NOV 23	NOV 29	DEC 6	DEC 15	16	16	17	17
	16	JAN 31	FEB 12	FEB 20	FEB 28	MAR 12	292	NOV 28	DEC 4	DEC 9	DEC 14	*	16	16	17	15
WILLIAMSBURG YORK COUNTY	36	APR 11	APR 20	APR 26	MAY 2	MAY 10	175	OCT 4	OCT 12	OCT 18	OCT 23	NOV 1	30	30	31	31
	32	APR 1	APR 11	APR 18	APR 26	MAY 6	192	OCT 12	OCT 21	OCT 27	NOV 2	NOV 11	30	30	31	31
	28	MAR 15	MAR 25	APR 1	APR 8	APR 17	224	OCT 28	NOV 5	NOV 11	NOV 16	NOV 24	30	30	31	31
	24	MAR 1	MAR 11	MAR 18	MAR 24	APR 3	249	NOV 8	NOV 16	NOV 22	NOV 28	DEC 6	30	30	31	31
	20	FEB 14	FEB 26	MAR 5	MAR 13	MAR 25	273	NOV 19	NOV 27	DEC 3	DEC 9	DEC 15	30	30	31	31
	16	JAN 19	FEB 7	FEB 16	FEB 25	MAR 10	307	DEC 4	DEC 12	DEC 20	*	*	30	28	31	22
WINCHESTER FREDERICK COUNTY	36	APR 13	APR 23	APR 29	MAY 6	MAY 15	162	SEP 24	OCT 2	OCT 8	OCT 14	OCT 22	31	31	31	31
	32	APR 3	APR 12	APR 18	APR 24	MAY 3	185	OCT 3	OCT 13	OCT 20	OCT 26	NOV 5	31	31	31	31
	28	MAR 22	APR 1	APR 8	APR 15	APR 25	208	OCT 19	OCT 27	NOV 2	NOV 8	NOV 16	31	31	31	31
	24	MAR 8	MAR 17	MAR 24	MAR 31	APR 10	237	NOV 2	NOV 10	NOV 16	NOV 22	NOV 30	31	31	31	31
	20	FEB 23	MAR 6	MAR 13	MAR 20	MAR 31	260	NOV 15	NOV 22	NOV 28	DEC 3	DEC 11	31	31	31	31
	16	FEB 5	FEB 17	FEB 25	MAR 5	MAR 17	286	NOV 24	DEC 2	DEC 8	DEC 14	DEC 24	31	31	31	30
WISE WISE COUNTY	36	APR 29	MAY 12	MAY 18	MAY 25	JUN 3	131	SEP 12	SEP 20	SEP 26	OCT 2	OCT 10	16	15	16	16
	32	APR 12	APR 27	MAY 5	MAY 13	MAY 24	156	SEP 23	OCT 8	OCT 14	OCT 22	OCT 30	16	15	16	16
	28	APR 8	APR 19	APR 25	MAY 1	MAY 9	176	OCT 7	OCT 14	OCT 18	OCT 23	OCT 30	16	15	16	16
	24	MAR 18	MAR 31	APR 6	APR 13	APR 22	201	OCT 12	OCT 19	OCT 24	OCT 29	NOV 5	16	15	16	16
	20	FEB 23	MAR 12	MAR 22	MAR 31	APR 13	230	OCT 25	NOV 2	NOV 7	NOV 12	NOV 19	16	15	16	16
	16	FEB 14	MAR 2	MAR 11	MAR 20	APR 1	254	NOV 5	NOV 14	NOV 20	NOV 25	DEC 4	16	15	16	16
WOODSTOCK SHENANDOAH COUNTY	36	APR 20	APR 30	MAY 6	MAY 13	MAY 22	147	SEP 17	SEP 25	SEP 30	OCT 6	OCT 14	31	31	31	31
	32	APR 8	APR 18	APR 25	MAY 2	MAY 12	169	SEP 27	OCT 5	OCT 11	OCT 17	OCT 25	31	31	31	31
	28	MAR 26	APR 4	APR 10	APR 16	APR 25	199	OCT 13	OCT 19	OCT 26	NOV 2	NOV 13	31	31	31	31
	24	MAR 10	MAR 20	MAR 26	APR 2	APR 12	229	OCT 29	NOV 5	NOV 10	NOV 15	NOV 22	31	31	31	31
	20	FEB 22	MAR 5	MAR 13	MAR 21	APR 1	255	NOV 7	NOV 17	NOV 23	NOV 30	DEC 10	31	31	31	31
	16	FEB 6	FEB 21	MAR 2	MAR 11	MAR 23	276	NOV 19	NOV 27	DEC 3	DEC 9	DEC 17	31	30	31	31
WYTHEVILLE WYTHE COUNTY	36	APR 26	MAY 6	MAY 13	MAY 20	MAY 30	136	SEP 10	SEP 20	SEP 26	OCT 3	OCT 12	31	31	31	31
	32	APR 8	APR 19	APR 27	MAY 5	MAY 17	163	SEP 20	SEP 30	OCT 7	OCT 14	OCT 23	31	31	31	31
	28	MAR 31	APR 10	APR 17	APR 24	MAY 4	186	SEP 30	OCT 12	OCT 20	OCT 28	NOV 9	31	31	31	31
	24	MAR 15	MAR 27	APR 5	APR 14	APR 26	210	OCT 13	OCT 24	NOV 1	NOV 9	NOV 20	31	31	31	31
	20	FEB 24	MAR 9	MAR 18	MAR 28	APR 10	239	OCT 21	NOV 3	NOV 12	NOV 21	DEC 5	31	31	31	31
	16	FEB 8	FEB 24	MAR 6	MAR 15	MAR 29	263	NOV 2	NOV 15	NOV 24	DEC 3	DEC 16	31	30	31	31

Data in the above table are based on the period 1940-70, or that portion of this period for which data are available.

* Preselected probability value greater than the probability of the temperature threshold being reached or the computed date fell before January 1 for Spring or after December 31 for Fall.

Y Number of years with available observations.

O Number of years with observed threshold temperature.

—— Insufficient data

A freeze is a numerical substitute for the former term "killing frost" and is the occurrence of a minimum temperature at or below the threshold occurrence of 32°, 28°, 24°, 20°, and 16°. The 36° temperature has been included in the table since temperatures at ground level may be at freezing or lower. Data are adjusted to take into account the years of non-occurrence.

*NORMALS BY CLIMATOLOGICAL DIVISIONS

Taken from "Climatography of the United States No. 81-4,   Decennial Census of U. S. Climate"

TEMPERATURE (°F)   PRECIPITATION (In.)

STATIONS (By Divisions)	JAN	FEB	MAR	APR	MAY	JUNE	JULY	AUG	SEPT	OCT	NOV	DEC	ANN	JAN	FEB	MAR	APR	MAY	JUNE	JULY	AUG	SEPT	OCT	NOV	DEC	ANN
TIDEWATER																										
CAPE HENRY WSO	42.6	42.4	47.9	56.9	66.1	74.5	78.1	77.5	73.1	63.2	53.0	44.2	60.0	2.98	3.25	3.36	2.93	3.04	3.54	5.00	5.36	3.95	2.87	2.80	2.76	41.84
COLONIAL BEACH	37.5	38.2	45.0	55.6	65.5	73.8	77.7	76.3	70.1	59.4	48.2	38.9	57.2	3.24	2.43	3.20	3.14	3.13	3.23	4.55	4.49	3.77	2.98	2.44	2.95	39.55
DIAMOND SPRINGS	43.7	44.0	50.0	59.1	67.9	75.7	79.0	77.9	73.2	63.2	53.5	44.6	61.0	3.63	3.45	3.93	3.37	3.66	3.79	6.19	6.58	4.48	3.17	3.33	2.96	48.54
EMPORIA 1 WNW	•	•	•	•	•	•	•	•	•	•	•	•	•	3.16	3.34	3.57	3.43	4.02	4.11	6.18	5.03	4.01	2.64	2.79	2.99	45.09
FREDERICKSBURG NATL PK	37.1	38.0	45.0	55.6	65.3	73.3	77.2	75.5	69.1	58.3	47.3	37.9	56.6	3.20	2.38	3.28	3.10	3.56	3.34	4.89	5.10	3.45	3.35	2.84	2.80	41.29
HOPEWELL	41.5	42.7	49.3	59.7	68.5	76.1	79.2	77.8	72.2	61.3	51.0	41.9	60.1	3.07	2.76	3.16	3.34	3.97	4.23	5.86	5.10	3.73	2.88	2.84	2.78	43.68
LANGLEY AIR FORCE BASE	41.5	42.0	48.3	57.7	67.2	75.4	79.0	77.9	72.7	62.1	51.6	42.6	59.8	3.23	3.02	3.42	3.10	3.48	3.61	5.33	5.13	3.98	2.87	2.79	2.57	41.95
NORFOLK WSO	41.2	41.6	48.0	58.0	67.5	75.6	78.8	77.5	72.6	62.0	51.4	42.5	59.7	3.33	3.21	3.45	3.16	3.36	3.61	5.92	5.97	4.22	2.92	3.05	2.74	44.94
QUANTICO IS	36.5	37.5	44.5	55.5	65.2	73.5	77.3	75.7	72.6	58.3	47.0	37.4	56.8	3.16	2.36	3.22	3.08	3.47	3.26	4.58	4.91	3.29	2.95	2.79	2.80	39.87
WALLACETON LK DRUMMOND	•	•	•	•	•	•	•	•	•	•	•	•	•	3.64	3.65	3.95	3.76	3.98	4.49	6.73	5.92	4.37	3.20	3.45	3.28	50.42
DIVISION	40.3	41.1	47.4	57.2	66.6	74.6	78.1	76.8	71.3	60.8	50.5	41.3	58.8	3.37	3.02	3.53	3.23	3.61	3.57	5.41	5.37	3.88	3.06	2.98	2.84	43.87
EASTERN PIEDMONT																										
CLARKSVILLE	•	•	•	•	•	•	•	•	•	•	•	•	•	3.66	3.32	3.90	3.81	3.61	4.13	4.94	5.31	3.85	2.73	3.26	3.10	45.62
COLUMBIA	37.7	38.8	45.8	56.3	65.9	73.9	77.5	75.9	69.4	58.1	47.2	38.3	57.1	3.24	2.64	3.47	3.49	3.79	3.21	4.63	4.29	3.78	2.94	2.73	2.94	41.15
FARMVILLE 2 N	39.4	40.7	47.0	57.2	65.9	73.8	77.5	75.9	69.6	58.7	48.1	39.3	57.8	3.80	3.25	3.98	3.64	4.02	4.31	4.02	4.69	3.80	2.96	3.09	3.18	45.54
NEW CANTON	38.7	39.7	46.2	56.4	65.1	72.4	76.0	74.4	68.4	57.8	47.7	39.0	56.8	3.60	2.87	3.75	3.68	3.76	3.26	4.57	4.52	3.56	2.88	2.93	3.31	42.69
RICHMOND WSO	38.7	39.9	47.7	58.1	67.0	75.1	78.1	76.0	70.2	58.7	48.5	39.7	58.1	3.46	2.90	3.42	3.15	3.72	3.75	5.61	5.54	3.65	3.00	3.04	2.97	44.21
DIVISION	38.8	39.9	46.6	56.9	66.1	73.9	77.4	75.9	69.7	58.7	48.1	39.2	57.6	3.51	2.97	3.71	3.53	3.69	3.67	5.09	4.78	3.64	2.92	3.00	3.09	43.60
WESTERN PIEDMONT																										
BEDFORD	•	•	•	•	•	•	•	•	•	•	•	•	•	3.46	2.92	4.11	3.48	4.08	4.45	4.48	5.13	3.44	3.00	2.90	3.42	44.87
CHARLOTTESVILLE 2 W	37.2	38.4	45.3	56.6	66.3	73.7	77.3	76.0	69.9	59.5	48.4	38.7	57.3	3.30	2.78	3.90	3.69	3.89	3.98	5.58	4.70	4.24	3.37	3.01	3.43	45.87
CHATHAM 2 NE	39.2	40.4	46.8	57.1	66.2	73.9	76.9	75.5	69.6	58.9	47.8	39.4	57.6	3.44	3.08	3.80	3.56	3.67	3.79	4.38	4.50	4.10	2.88	2.96	3.30	43.46
DANVILLE-BRIDGE ST	41.1	42.2	48.7	59.2	68.0	75.7	78.5	77.4	71.4	60.5	49.4	41.0	59.4	3.51	3.18	3.87	3.65	4.18	3.64	4.55	4.25	4.05	2.87	2.91	3.27	43.93
HALIFAX 1 N	•	•	•	•	•	•	•	•	•	•	•	•	•	3.35	3.13	3.87	3.75	3.84	4.11	4.90	4.35	3.74	2.85	3.03	3.18	44.10
LYNCHBURG WSO	37.6	38.9	45.5	56.2	65.3	73.0	76.3	74.8	68.7	58.2	47.0	38.5	56.7	3.29	2.65	3.61	3.14	3.21	4.06	4.21	4.41	3.36	2.64	2.58	3.14	40.30
RANDOLPH	•	•	•	•	•	•	•	•	•	•	•	•	•	3.63	3.14	3.88	3.91	4.10	3.60	4.64	4.81	4.47	2.87	3.29	3.18	45.52
ROCKY MOUNT	38.7	39.6	45.8	56.0	64.9	72.1	75.7	74.4	67.8	58.2	47.3	39.2	56.6	3.21	2.98	3.84	3.43	4.13	3.83	4.49	4.14	4.06	3.32	2.68	3.10	43.21
DIVISION	39.0	40.0	46.3	56.9	66.0	73.5	76.6	75.3	69.2	58.9	47.9	39.4	57.4	3.42	3.04	3.94	3.68	3.96	4.08	4.82	4.63	4.07	3.11	2.91	3.32	44.98
NORTHERN																										
CULPEPER	36.8	37.9	45.5	56.1	65.5	73.3	76.9	75.1	69.0	58.2	47.0	37.4	56.6	3.03	2.55	3.38	3.64	4.10	3.68	4.80	4.38	3.62	3.14	2.98	2.87	42.17
LINCOLN	35.7	36.7	44.0	55.0	65.3	73.5	77.4	76.0	69.0	58.3	47.0	36.9	56.2	3.03	2.74	3.78	3.53	4.16	3.59	4.26	4.74	3.33	3.47	2.88	3.12	42.63
MOUNT WEATHER	30.4	31.1	38.4	49.3	59.8	68.5	72.5	71.2	64.6	54.0	42.6	32.3	51.2	2.67	2.33	3.26	3.31	4.29	3.56	3.83	4.49	3.77	3.70	2.89	2.63	40.73
RIVERTON	•	•	•	•	•	•	•	•	•	•	•	•	•	2.40	1.99	2.86	2.88	3.87	3.21	3.38	3.76	2.86	3.41	2.37	2.38	35.37
WASHINGTON NAT AP WSO	36.9	37.8	44.8	55.7	65.8	74.2	78.2	76.5	69.7	59.0	47.7	38.1	57.0	3.03	2.47	3.21	3.15	4.14	3.21	4.15	4.90	3.83	3.07	2.84	2.78	40.78
WINCHESTER 1 N	34.9	36.1	43.0	54.2	64.3	72.2	76.1	74.5	68.0	57.3	46.0	36.2	55.2	2.39	2.12	3.16	3.11	4.10	3.70	4.24	4.16	2.97	3.45	2.59	2.48	38.47
WOODSTOCK	35.9	36.9	43.6	54.4	63.8	71.4	75.2	73.5	67.0	56.6	45.9	36.5	55.1	2.26	1.74	2.86	2.83	3.81	3.83	3.80	4.26	2.83	2.90	2.13	2.22	35.47
DIVISION	34.8	35.7	42.7	53.7	63.6	71.5	75.3	73.7	67.2	56.6	45.5	35.8	54.7	2.81	2.35	3.37	3.33	4.13	3.67	4.31	4.56	3.43	3.45	2.83	2.80	41.04
CENTRAL MOUNTAIN																										
BALCONY FALLS	•	•	•	•	•	•	•	•	•	•	•	•	•	3.42	2.98	3.99	3.61	3.71	3.51	4.08	4.62	3.80	3.33	2.99	3.21	43.25
BUCHANAN	38.4	39.9	46.2	56.5	65.2	72.7	76.0	74.8	68.8	58.1	46.8	38.5	56.8	3.25	2.93	3.81	3.35	4.04	4.32	4.41	4.67	3.71	3.42	3.09	3.20	44.20
CATAWBA SANATORIUM	36.4	37.2	43.3	54.1	62.8	69.8	73.0	72.0	66.3	56.5	45.5	37.3	54.5	3.13	2.90	3.61	3.38	4.02	4.56	4.51	3.68	3.17	2.59	2.93		42.50
CLIFTON FORGE	•	•	•	•	•	•	•	•	•	•	•	•	•	2.99	2.85	3.92	3.08	3.98	3.71	3.90	4.37	3.31	2.94	2.74	2.95	40.74
DALE ENTERPRISE	35.0	36.0	42.2	52.8	62.3	69.9	73.4	71.8	66.0	55.7	44.5	35.8	53.8	2.16	1.95	2.87	2.59	4.00	3.79	4.57	4.31	3.14	2.63	2.10	2.23	36.34
HOT SPRINGS	32.6	33.6	40.1	50.9	60.1	67.2	70.3	68.9	62.7	52.6	41.6	33.4	51.2	3.01	2.84	3.93	3.12	3.67	4.17	4.18	4.65	3.22	2.94	2.77	2.88	41.38
LEXINGTON	37.4	38.6	45.1	55.5	64.2	71.6	75.1	73.8	67.7	56.9	46.0	37.7	55.8	3.08	2.75	3.61	3.03	3.55	3.58	4.18	4.03	3.39	2.87	2.66	2.91	39.64
NORTH RIVER DAM	•	•	•	•	•	•	•	•	•	•	•	•	•	2.68	2.43	3.34	3.21	3.93	4.21	5.02	4.89	3.69	3.04	2.71	2.61	41.76
ROANOKE	38.9	40.0	46.2	56.6	65.7	73.2	76.4	75.1	68.6	58.3	47.2	39.2	57.1	3.11	2.87	3.64	3.10	4.10	4.06	4.44	4.86	3.68	3.44	2.72	3.10	43.12
ROANOKE WSO	38.1	39.2	45.5	56.4	65.7	73.4	76.6	75.4	69.1	58.2	46.7	38.4	56.9	3.12	2.86	3.53	3.10	3.79	3.80	4.25	4.63	3.26	3.21	2.70	2.98	41.23
STAUNTON SEWAGE PLANT	35.7	36.4	42.6	53.5	63.3	71.1	74.5	72.7	66.8	56.3	45.4	36.6	54.6	2.67	2.11	3.17	2.98	3.74	3.69	4.01	3.82	3.29	2.83	2.41	2.70	37.42
TIMBERVILLE 3 E	•	•	•	•	•	•	•	•	•	•	•	•	•	2.18	2.03	2.96	2.72	3.76	3.52	4.25	4.04	3.40	2.61	2.13	2.17	35.77
DIVISION	36.0	37.0	43.3	53.9	63.1	70.5	73.9	72.5	66.4	56.1	45.0	36.5	54.5	2.90	2.61	3.49	3.07	3.86	3.87	4.35	4.44	3.37	3.02	2.61	2.83	40.43

Taken from "Climatography of the United States No. 81-4, Decennial Census of U. S. Climate"

TEMPERATURE (°F) PRECIPITATION (In.)

STATIONS (By Divisions)	JAN	FEB	MAR	APR	MAY	JUNE	JULY	AUG	SEPT	OCT	NOV	DEC	ANN	JAN	FEB	MAR	APR	MAY	JUNE	JULY	AUG	SEPT	OCT	NOV	DEC	ANN
SOUTHWESTERN MOUNTAIN																										
BURKES GARDEN	33.2	34.3	39.6	49.7	57.8	64.9	67.9	67.0	61.3	51.4	40.6	33.6	50.1	3.90	3.62	4.41	3.48	3.94	4.07	4.81	4.32	3.23	2.39	2.99	3.37	44.53
DAMASCUS	.	.	.	.	.	.	.	.	.	.	.	.	.	3.93	3.91	4.30	3.59	4.23	4.12	5.96	4.43	2.90	2.34	2.81	3.28	45.80
GLEN LYN	.	.	.	.	.	.	.	.	.	.	.	.	.	2.88	2.85	3.53	3.03	3.39	3.61	4.38	3.91	2.56	2.34	2.27	2.48	37.23
MENDOTA	.	.	.	.	.	.	.	.	.	.	.	.	.	4.27	3.90	4.59	3.70	3.91	3.67	5.36	4.38	2.79	2.45	2.93	3.54	45.49
PENNINGTON GAP	36.1	37.6	43.6	53.9	63.1	70.5	73.7	72.8	67.3	56.5	44.0	36.8	54.7	5.28	4.88	5.48	3.92	3.85	4.21	5.60	4.24	2.76	2.44	3.42	4.36	50.44
RADFORD 5 SW	.	.	.	.	.	.	.	.	.	.	.	.	.	2.90	2.62	3.33	2.85	3.45	3.23	4.37	3.37	2.75	2.43	2.20	2.61	36.11
SALTVILLE 1 N	36.6	37.7	43.7	54.7	64.2	71.8	74.7	73.6	67.9	57.0	44.8	36.8	55.3	3.89	3.68	4.27	3.19	4.18	3.80	4.81	4.29	2.92	2.33	2.64	3.24	43.24
WYTHEVILLE 1 S	36.2	37.1	42.9	53.0	61.6	68.9	71.9	70.8	65.0	55.0	44.0	36.3	53.6	2.79	2.71	3.38	2.91	3.61	3.16	4.36	3.88	2.90	2.17	2.24	2.55	36.66
DIVISION	35.6	36.8	42.7	53.2	62.0	69.3	72.3	71.3	65.5	55.2	43.7	36.0	53.6	3.65	3.50	4.15	3.45	3.88	3.99	4.92	4.30	3.11	2.52	2.69	3.27	43.40

* Normals for the period 1931-1960. Divisional normals may not be the arithmetical average of individual stations published, since additional data for shorter period stations are used to obtain better areal representation.

CONFIDENCE - LIMITS

In the absence of trend or record changes, the chances are 9 out of 10 that the true mean will lie in the interval formed by adding and subtracting the values in the following table from the means for any station in the State. Because of the wider variation in mean precipitation, the corresponding monthly means and annual mean must be substituted for "p" in the precipitation table below to obtain mean precipitation confidence limits.

| 1.4 | 1.4 | 1.4 | .7 | .8 | .7 | .6 | .6 | 1.0 | .8 | .8 | 1.2 | .3 | $.27\sqrt{p}$ | $.24\sqrt{p}$ | $.25\sqrt{p}$ | $.25\sqrt{p}$ | $.33\sqrt{p}$ | $.29\sqrt{p}$ | $.34\sqrt{p}$ | $.42\sqrt{p}$ | $.38\sqrt{p}$ | $.37\sqrt{p}$ | $.34\sqrt{p}$ | $.26\sqrt{p}$ | $.32\sqrt{p}$ |

COMPARATIVE DATA

Data in the following table are the mean temperature and average precipitation for Staunton D. and B. Institute, Virginia, for the period 1906-1930 and are included in this publication for comparative purposes.

| 36.3 | 36.9 | 45.4 | 53.8 | 63.5 | 70.4 | 74.3 | 72.7 | 67.9 | 56.1 | 45.5 | 36.9 | 55.0 | 2.83 | 2.15 | 2.93 | 3.01 | 2.79 | 4.21 | 4.48 | 4.20 | 2.79 | 2.89 | 2.19 | 2.61 | 37.08 |

NORMALS, MEANS AND EXTREMES
(Table Revised 1971. Base Period for Climatological Normals: 1931-1960)

Station: LYNCHBURG, VIRGINIA MUNICIPAL AIRPORT Standard time used: EASTERN Latitude: 37° 20' N Longitude: 79° 12' W Elevation (ground): 916 feet

Month	Temperature Normal Daily maximum	Normal Daily minimum	Normal Monthly	Extremes Record highest	Year	Record lowest	Year	Precipitation Normal total	Max monthly	Year	Min monthly	Year	Max in 24 hrs	Year	Snow, ice pellets Mean total	Max monthly	Year	Max 24 hrs	Year	Relative humidity 01	07	13	19	Wind Mean speed	Prevailing direction	Fastest mile Speed	Direction	Year	Pct. possible sunshine	Mean sky cover	Clear	Partly cloudy	Cloudy	Precip .01	.01 snow	Thunderstorms	Heavy fog	Temp Max ≥90	Max ≤32	Min ≤32	Min ≤0	Avg solar radiation
(a)	(b)	(b)	(b)	7		7			26		26		26		26	26		26		7	7	7	7	20	9	26	26	26	26	26	26	26	26				22	22	7	7	7	7	
J	46.2	29.0	37.6	74 1967	-4 1970			3.29	7.92 1950	0.76 1956	2.13 1948	5.8	31.8 1966	10.9 1966		70 51 59 58	8.8	SW	38 S	1959+	51 6.2										11 15 13 11 7			5	24	7	
F	48.3	29.5	38.1	74 1965	0 1965			2.65	5.83 1964	0.04 1968	1.90 1965	4.8	14.7 1967	12.8 1967		51 46 46 48	9.0	SW	43 S	1961	55 6.0										11 13 11 7 7			2	23	*	
M	55.7	35.3	45.5	85 1971	7 1960			3.61	4.66 1960	1.15 1960	2.43 1960	4.1	24.9 1960	13.4 1960		71 46 52 50	9.4	SW	43 SW	1950	58 6.0										11 13 11 8 4			*	13	1	
A	67.8	45.2	56.5	93 1960	24 1970+			3.23	6.06 1951	1.13 1963	2.45 1951	0.4	2.80 1957	4.8 1957		58 45 46 50	9.2	SW	56 SW	1958	62 6.6										10 11 9 8 8			0	1	0	
M	76.6	53.7	65.2	93 1969	31 1966			4.06	6.02 1949	0.67 1957	3.47 1949	0.0	0.00	0.0		62 50 50 53	7.1	SW	56 SW	1951	65 5.7										10 11 10 8 8			0	0	0	
J	83.8	62.1	73.0	99 1964	42			4.06	8.50 1949	0.67	3.23	0.0	0.00	0.0			7.1	SW																			
J	86.6	65.9	76.3	99 1966	50 1963			4.21				0.0					6.8	SW	43 NW	1954+	61 5.9													7	0	0	0	
A	84.5	64.7	74.6	98 1968	45 1965			4.41				0.0					6.6	NW	46 N	1952	61 5.5													6	0	0	0	
S	79.1	58.2	68.7	93 1973	36 1967			3.36				0.0					6.9	NE	45 NE	1956	63 5.2													3	0	2	0	
O	69.0	47.4	58.2	85 1970+	21 1969			2.64				T	T 1951				7.5	SW	43 NW	1954	62 4.7													*	0	4	0	
N	57.0	36.9	47.0	81 1968	7 1970			2.58				0.9	11.6 1968	6.7 1968			8.1	SW	41 NW	1949	56 5.5													0	1	11	1	
D	47.0	29.9	38.5	71 1966	7 1968			3.14	6.74 1948			3.7	31.8 1966	12.7 1969			8.0	SW	45 SE	1950	53 5.8													0	2	21	1	
YR	66.8	46.5	56.7	99 JUL 1966	-4 JAN 1970			40.30	11.36 AUG 1952	0.33 DEC 1965	4.98 OCT 1954	19.4	31.8 JAN 1966	13.4 MAR 1969		78 51 62	8.0	SW	56 N	MAY 1959+	59 5.8										112 107 146 119	40	22	23	100	1	

Ø For period July 1963 through the current year.

% For period July 1963 through the current year. Means and extremes above are from existing and comparable exposures. Annual extremes have been exceeded at other sites in the locality as follows: Highest temperature 106 in July 1936; lowest temperature -7 in January 1912; maximum monthly precipitation 14.87 in August 1928; minimum monthly precipitation .03 in November 1890; maximum precipitation in 24 hours 7.59 in August 1928; maximum snowfall in 24 hours 16.4 in January 1922; fastest mile wind 62 W in April 1917+.

(b) Data through 1966. Station operated less than 24 hours per day prior to August 1962. Fog and thunderstorm data may be incomplete.

\# To 8 compass points only.

NORMALS, MEANS AND EXTREMES
(Table Revised 1975. Base Period for Climatological Normals: 1941-1970)

Month	Temperature °F Normal Daily maximum	Normal Daily minimum	Normal Monthly	Extremes Record highest	Year	Record lowest	Year	Normal Degree days Base 65°F Heating	Cooling	Precipitation in inches Normal	Max monthly	Year	Min monthly	Year	Water equiv Max	Year	Water equiv Min	Year	Max in 24 hrs	Year	Snow, ice pellets Max monthly	Year	Max 24 hrs	Year	Relative humidity 01	07	13	19	Wind Mean speed	Prevailing dir	Fastest mile m.p.h.	Direction	Year	Pct. possible sunshine	Mean sky cover	Clear	Partly cloudy	Cloudy	Precip .01	snow .01	Thunderstorms	Heavy fog	Max ≥90	Max ≤32	Min ≤32	Min ≤0	Avg station pressure mb	Elev 937 ft m.s.l.
(a)				11		11					30		30		30		30		30		30		30		11	11	11	11	20	9	30	30	30	30	30	30	30	30	30	30	22	22	11	11	11	11		2
J	45.8	27.3	36.6	74 1967	-4 1972			880	0	2.77	4.82 1964	0.76 1956	2.13 1948						5.8	31.8 1966	10.9 1966		72 53 59 59	8.8	SW	45 W	1971	50 5.2	9 7 15	12	*	5	24	22	11		985.8 983.3
F	47.8	28.3	38.1	77 1972	0 1965			753	0	2.79	5.70 1972	0.64 1968	1.96 1972						4.8	14.7 1967	12.8 1967		53 48 53	9.0	SW	43 S	1961	55 5.0	9 6 9 13	11	*	2	22	13		982.4 982.2
M	56.2	34.8	45.5	85 1973	7 1960			605	0	3.46	4.66 1960	0.74 1969	2.43 1960						4.1	24.9 1960	13.4 1960		72 48 53 52	9.4	SW	43 SW	1950	58 5.0	9 8 13 10	11	1	*	13	1		982.6
A	68.1	45.0	56.6	88 1974	24 1972			260	8	3.22	9.07 1951	1.15 1963	2.45 1951						0.0	2.80 1957	4.8 1957		46 62	9.0	SW	56 SW	1958	62 4.8	8 8 13 11	10	3	0	1	0		983.9
M	76.6	53.7	65.2	99 1969	31 1966			85	91	3.43	8.58 1972	1.36 1971	3.47 1949						0.0	0.0	0.0		80 50 61 65	7.1	SW	56 SW	1951	65 5.8	8 8 12 11	11	5	0	0	0		
J	83.5	61.6	72.6	99 1964	41 1972			85	232	3.21	8.58 1972	0.67 1972	3.03 1972						0.0	0.0	0.0		83 55 66	7.1	SW	56 SW	1951	66 5.8	8 8 12	7	0	0	0		
J	86.1	65.5	75.8	99 1966	50 1963			0	335	4.05	8.45 1945	1.75 1963	3.18 1973						0.0	0.0	0.0		87 58 71 75	6.8	SW	43 NW	1954	61 5.9	8 8 11 12	11	7	0	0	0		983.9
A	84.5	64.3	74.4	98 1968	50 1965			0	291	4.05	11.36 1952	1.36 1952	3.47 1960						0.0	0.0	0.0		85 57 75 75	6.6	SW	46 N	1952	61 5.3	8 7 11 11	11	5	0	0	0		985.1
S	78.7	57.5	68.1	93 1973	36 1967			33	126	3.30	7.48 1957	0.38 1957	4.98 1954						0.0	0.0	0.0		85 53 70	7.5	SW	41 NE	1954	62 4.8	8 6 8 14	8	1	0	*	3		987.5
O	69.0	47.0	58.0	85 1970	17 1969			234	22	2.60	7.40 1971	0.78 1971	3.10 1971						T	T	0.0		85 50 64	8.1	SW	43 NW	1949	56 4.8	11 6 14	6	*	0	1	7		984.8
N	57.3	36.7	47.0	83 1974	7 1970			540	0	2.66	5.78 1972	0.90 1960	2.65 1951						0.9	11.6 1968	6.7 1968		77 50 64	8.1	SW	43 NW	1949	56 5.5	7 7 14 9	9	1	*	0	11	1		983.4
D	46.9	28.7	37.8	71 1971	-4 1968			843	0	3.21	7.15 1973	0.33 1965	3.03 1948						3.7	31.8 1966	12.7 1969		75 55 64	8.0	SW	45 SE	1950	51 5.0	10 7 14	7	*	0	2	19	1		
YR	66.7	45.9	56.3	99 JUL 1966	-4 JAN 1972			4233	1100	38.27	11.36 AUG 1952	0.33 DEC 1965	6.27 JUN 1972						19.4	31.8 JAN 1966	13.4 MAR 1969		79 53 64	8.0	SW	56 N	MAY 1958	59 5.8	105 113 147 120		41	40	8	18	94	1		983.9

Means and extremes above are from existing and comparable exposures. Annual extremes have been exceeded at other sites in the locality as follows: Highest temperature 106 in July 1936; lowest temperature -7 in January 1912; maximum monthly precipitation 14.87 in August 1928; minimum monthly precipitation .03 in November 1890; maximum precipitation in 24 hours 7.59 in August 1928; maximum snowfall in 24 hours 16.4 in January 1922; fastest mile wind 62 W in April 1917+.

REFERENCE NOTES APPLYING TO TABLES APPEAR ON THE PAGE FOLLOWING LAST TABLE.
(Caution: Letters and symbols may have different meanings in 1941-1970 tables than in earlier tables. See notes.)

NORMALS, MEANS AND EXTREMES
(Table Revised 1971. Base Period for Climatological Normals: 1931-1960)

Station: NORFOLK, VIRGINIA   NORFOLK REGIONAL AIRPORT   Standard time used: EASTERN   Latitude: 36°54'N   Longitude: 76°12'W   Elevation (ground): 22 feet   22 feet

Month	Normal Daily maximum	Normal Daily minimum	Normal Monthly	Extremes Record highest	Year	Record lowest	Year	Normal heating degree days (Base 65°)	Precipitation Normal total
(a)	(b)	(b)	(b)					(b)	(b)
J	50.2	32.2	41.2	78	1966	10	1965	738	3.33
F	51.0	32.7	41.6	78	1965	10	1965	655	3.21
M	57.2	38.7	48.0	85	1968+	20	1950	533	3.45
A	68.0	47.9	58.0	97	1960	36	1964	216	3.16
M	76.3	57.5	67.5	97	1956	45	1967	37	3.36
J	84.9	66.3	75.6	101	1964			0	3.61
J	87.6	69.6	78.8	103	1952	57	1962	0	5.92
A	86.2	68.8	77.5	99	1968	52	1965	0	5.97
S	80.9	64.3	72.6	98	1954	45	1967	9	4.22
O	70.1	53.1	62.0	95	1954	29	1965	136	2.92
N	61.0	41.8	51.4	85	1950	20	1965	408	3.05
D	51.8	33.1	42.5	77	1956+	8	1962	698	2.74
YR	68.9	50.5	59.7	103	JUL 1952	8	FEB 1965	3421	44.94

Means and extremes above are from existing and comparable exposures. Annual extremes have been exceeded at other sites in the locality as follows:
Highest temperature 105 in August 1918; lowest temperature 2 in February 1895; maximum monthly precipitation 15.61 in August 1942; minimum monthly precipitation 0.04 in October 1874; maximum snowfall 18.6 in December 1892; maximum snowfall in 24 hours 17.7 in December 1892; fastest mile wind 80 W in June 1925.

\# To 8 compass points only.

NORMALS, MEANS AND EXTREMES
(Table Revised 1975. Base Period for Climatological Normals: 1941-1970)

Station: NORFOLK, VIRGINIA   NORFOLK REGIONAL AIRPORT   Standard time used: EASTERN   Latitude: 36°54'N   Longitude: 76°12'W   Elevation (ground): 24 feet   30 feet

Month	Normal Daily maximum	Normal Daily minimum	Normal Monthly	Extremes Record highest	Year	Record lowest	Year	Heating degree days Base 65°F	Cooling degree days	Water equivalent Normal
(a)										
J	48.8	32.2	40.5	78	1970	8	1972	760	0	3.35
F	50.2	32.7	41.4	78	1972	8	1965	661	0	3.31
M	57.7	38.9	48.1	85	1968	20	1950	532	0	3.72
A	67.6	47.9	57.8	89	1960	20	1964	226	0	2.71
M	76.2	57.2	66.7	97	1956	36	1966	53	106	3.34
J	83.5	65.5	74.5	101	1964	45	1967	0	285	3.62
J	86.9	69.9	78.3	103	1952	56	1972	0	412	5.70
A	84.9	68.9	76.9	99	1968	52	1965	0	369	5.92
S	79.6	63.9	71.8	98	1954	43	1965	9	213	4.20
O	70.1	53.3	61.7	95	1954	29	1965	141	38	3.06
N	60.5	42.6	51.6	86	1974	20	1955	402	0	2.94
D	50.6	34.0	42.3	79	1971	14	1962	704	0	3.11
YR	68.0	50.6	59.3	103	JUL 1952	8	JAN 1972	3488	1441	44.68

Means and extremes above are from existing and comparable exposures. Annual extremes have been exceeded at other sites in the locality as follows:
Highest temperature 105 in August 1918; lowest temperature 2 in February 1895; maximum monthly precipitation 15.61 in August 1942; minimum monthly precipitation 0.04 in October 1874; maximum snowfall 18.6 in December 1892; maximum snowfall in 24 hours 17.7 in December 1892; fastest mile wind 80 W in June 1925.

REFERENCE NOTES APPLYING TO TABLES APPEAR ON THE PAGE FOLLOWING LAST TABLE.
(Caution: Letters and symbols may have different meanings in 1941-1970 tables than in earlier tables. See notes.)

NORMALS, MEANS AND EXTREMES
(Table Revised 1971. Base Period for Climatological Normals: 1931-1960)

Station: RICHMOND, VIRGINIA   BYRD FIELD   Standard time used: EASTERN   Latitude: 37° 30' N   Longitude: 77° 20' W   Elevation (ground): 164 feet

Month	Temperature — Normal Daily maximum (b)	Daily minimum (b)	Monthly (b)	Extremes — Record highest	Year	Record lowest	Year	Normal heating degree days (Base 65°) (b)	Precipitation — Normal total (b)	Max monthly	Year	Min monthly	Year	Max in 24 hrs	Year	Snow, Ice pellets — Mean total	Max monthly	Year	Max in 24 hrs	Year
J	48.3	29.0	38.7	80	1950	-12	1940	815	3.46	5.95	1962	1.08	1951	3.31	1962	5.6	28.5	1940	21.6	1940
F	50.6	29.2	39.9	83	1932+	-10	1936	703	2.90	5.61	1944	0.98	1969	1.87	1969	3.3	17.1	1967	9.2	1947
M	59.1	36.3	47.7	93	1938	10	1960	546	3.42	5.85	1944	0.94	1966	2.04	1942	3.0	19.7	1960	12.1	1962
A	70.4	45.8	58.1	96	1960+	26	1964+	219	3.15	5.32	1952	0.64	1963	2.30	1952	0.1	2.0	1940	0.0	1940
M	79.3	54.6	67.0	100	1941	31	1956	53	3.72	7.73	1946	0.87	1965	2.30	1958	0.0	0.0		0.0	
J	86.7	63.4	75.1	104	1952	40	1952	0	3.75	9.24	1938	0.91	1963	4.61	1963	0.0	0.0		0.0	
J	89.6	66.7	78.1	104	1936+	51	1965+	0	5.61	18.87	1955	0.52	1963	5.73	1969	0.0	0.0		0.0	
A	86.5	65.4	76.0	102	1953	46	1934	0	5.54	14.10	1955	0.52	1943	8.79	1955	0.0	0.0		0.0	
S	80.6	58.6	70.2	103	1954	37	1963	36	3.65	8.49	1945	0.69	1954	3.82	1955	0.0	0.0		0.0	
O	70.6	46.7	58.7	99	1941	21	1962	214	3.00	8.78	1961	0.30	1963	6.50	1961	T	T	1954	T	1954
N	59.8	37.1	48.5	86	1950	10	1933	495	3.04	7.64	1959	0.36	1963	4.07	1956	0.5	7.3	1953	7.3	1953
D	49.1	29.5	39.7	78	1956	-1	1942	784	2.97	6.88	1957	0.72	1965	3.16	1958	2.2	12.5	1958	7.5	1966
YR	59.4	46.9	58.1	104	JUN 1952+	-12	JAN 1940	3865	44.21	18.87	JUL 1945	0.30	OCT 1963	8.79	AUG 1955	14.7	28.5	JAN 1940	21.6	JAN 1940

Means and extremes above are from existing and comparable exposures. Annual extremes have been exceeded at other sites in the locality as follows:
Highest temperature 107 in August 1918; minimum monthly precipitation 0.11 in November 1890 and earlier.

\# To 8 compass points only.

NORMALS, MEANS AND EXTREMES
(Table Revised 1975. Base Period for Climatological Normals: 1941-1970)

Month	Temperature — Normal Daily maximum	Daily minimum	Monthly	Extremes — Record highest	Year	Record lowest	Year	Normal Degree days Base 65°F — Heating	Cooling	Precipitation — Water equivalent — Normal	Max monthly	Year	Min monthly	Year	Max in 24 hrs	Year	Snow, Ice pellets — Max monthly	Year	Max in 24 hrs	Year
J	47.4	27.6	37.5	80	1950	-12	1940	853	0	2.86	5.95	1962	1.08	1951	3.31	1962	28.5	1940	21.6	1940
F	49.9	29.4	39.4	83	1936	-10	1936	717	0	3.01	5.61	1944	0.98	1968	1.91	1973	17.1	1967	9.2	1947
M	58.2	35.5	46.9	93	1938	10	1960	569	8	3.38	5.85	1944	0.94	1966	2.04	1942	19.7	1960	12.1	1962
A	70.3	45.2	57.8	96	1964	26	1964	226	10	2.77	5.32	1952	0.64	1963	2.07	1952	2.0	1940	0.0	1940
M	78.4	54.5	66.5	104	1941	40	1956	64	111	3.42	8.87	1972	0.87	1965	2.53	1972	0.0		0.0	
J	85.4	63.0	74.2	104	1952	40	1967	0	276	3.52	9.24	1938	0.91	1960	4.61	1963	0.0		0.0	
J	88.2	67.5	77.9	104	1936	51	1965	0	400	5.63	18.87	1955	0.52	1963	5.73	1969	0.0		0.0	
A	86.6	66.3	76.3	103	1953	46	1934	0	350	5.06	14.10	1955	0.52	1943	8.79	1955	0.0		0.0	
S	80.2	59.3	70.0	103	1954	35	1974	21	171	3.58	8.49	1945	0.69	1954	3.82	1955	0.0		0.0	
O	71.2	47.4	59.3	101	1962	10	1962	203	27	2.94	9.39	1971	0.30	1965	6.50	1961	T	1972	T	1972
N	60.6	37.3	49.0	86	1974	10	1933	480	0	3.20	7.64	1959	0.36	1956	4.07	1959	7.3	1953	7.3	1953
D	49.1	28.8	39.0	80	1971	-1	1942	806	0	3.22	7.07	1973	0.72	1965	3.16	1966	12.5	1958	7.5	1966
YR	68.8	46.7	57.8	104	JUN 1952+	-12	JAN 1940	3939	1353	42.59	18.87	JUL 1945	0.30	OCT 1963	8.79	AUG 1955	28.5	JAN 1940	21.6	JAN 1940

Means and extremes above are from existing and comparable exposures. Annual extremes have been exceeded at other sites in the locality as follows:
Highest temperature 107 in August 1918; minimum monthly precipitation 0.11 in November 1890 and earlier.

REFERENCE NOTES APPLYING TO TABLES APPEAR ON THE PAGE FOLLOWING LAST TABLE.
(Caution: Letters and symbols may have different meanings in 1941-1970 tables than in earlier tables. See notes.)

NORMALS, MEANS AND EXTREMES

(Table Revised 1971. Base Period for Climatological Normals: 1931-1960)

Station: ROANOKE, VIRGINIA   WOODRUM AIRPORT   Standard time used: EASTERN   Latitude: 37 19' N   Longitude: 79 58' W   Elevation (ground): 1149 feet

Month	Normal Daily max	Normal Daily min	Normal Monthly	Record highest	Year	Record lowest	Year	Normal heating degree days (Base 65°)	Precipitation Normal total	Maximum monthly	Year	Minimum monthly	Year	Maximum in 24 hrs	Year	Snow, ice pellets Mean total	Maximum monthly	Year	Maximum in 24 hrs	Year
(a)		(b)	(b)					(b)	(b)											
J	47.0	29.1	38.1	73	1967	-1	1970	834	3.12	5.20	1966	0.50	1965	2.71	1966	7.7	41.6	1966	13.7	1960
F	49.1	29.3	39.2	69	1967	-5	1970	722	2.86	7.17	1965	0.56	1960	2.54	1960	7.1	27.0	1960	15.7	1960
M	56.3	34.5	45.4	86	1968	14	1965	614	3.53	7.25	1963	0.43	1966	2.57	1967	4.1	30.7	1960	10.7	1960
A	67.2	44.6	56.4	87	1970	24	1969	261	3.19	5.15	1952	0.87	1963	1.83	1971	0.1	T	1971	7.3	1971
M	76.0	53.4	64.7	93	1949	31	1966	55	3.79	5.40	1960	0.97	1950	3.23	1963	0.0	0.0		0.0	
J	83.4	61.3	72.4	97	1968	40	1966	0	3.80	6.67	1949	1.46	1960	2.72	1966	0.0	0.0		0.0	
J	86.0	65.2	75.6	100	1966	50	1965	0	4.25	7.85	1949	1.13	1957	3.35	1944	0.0	0.0		0.0	
A	84.9	63.3	74.1	95	1965	43	1965	0	4.63	9.12	1949	1.16	1944	3.45	1955	0.0	0.0		0.0	
S	79.5	57.5	68.5	95	1954	36	1974	51	3.21	7.25	1966	0.44	1959	3.41	1966	T	T	1953	1.0	1953
O	70.0	46.7	58.4	84	1969	23	1969	229	3.21	6.36	1963	0.27	1963	3.00	1968	T	1.0	1957	1.0	1957
N	57.3	36.0	46.7	72	1968	17	1970	549	2.98	7.10	1948	0.44	1948	3.40	1948	1.8	13.8	1968	10.0	1968
D	47.5	29.2	38.4	72	1967	4	1968	825	3.80	6.67		0.18	1965	3.23	1965	5.2	22.6	1966	16.4	1966
YR	57.8	45.9	56.9	100	JUL 1966	-1	JAN 1970	4150	41.23	9.12	AUG 1949	0.18	DEC 1965	6.41	OCT 1968	26.2	41.2	JAN 1966	17.4	MAR 1960

Ø For period September 1964 through the current year.
Means and extremes above are from existing and comparable exposures. Annual extremes have been exceeded at other sites in the locality as follows:
Highest temperature 105 in July 1936; lowest temperature -12 in December 1917; maximum monthly precipitation 12.91 in August 1940; minimum monthly precipitation .16 in November 1931.

NORMALS, MEANS AND EXTREMES

(Table Revised 1975. Base Period for Climatological Normals: 1941-1970)

Month	Normal Daily max	Normal Daily min	Normal Monthly	Record highest	Year	Record lowest	Year	Normal Degree days Base 65°F Heating	Cooling	Precipitation Normal	Water equivalent Maximum monthly	Year	Minimum monthly	Year	Maximum in 24 hrs	Year	Snow, ice pellets Maximum monthly	Year	Maximum in 24 hrs	Year
(a)					10		10				27		27		27				27	
J	45.6	27.2	36.4	75	1974	-4	1972	887	0	2.74	5.20	1949	0.60	1957	2.71	1968	41.2	1966	13.7	1966
F	47.9	28.3	38.1	78	1972	1	1970	753	0	3.09	7.17	1960	0.56	1959	2.54	1954	27.6	1960	15.7	1960
M	56.3	34.3	45.3	88	1968	4	1965	611	0	3.30	5.92	1973	0.43	1966	2.57	1967	30.3	1960	7.3	1971
A	67.9	43.9	55.9	88	1974	22	1972	283	10	2.80	5.39	1971	0.87	1963	1.83	1971	T	1971	T	
M	76.1	52.7	64.4	93	1969	31	1966	101	83	3.47	8.42	1950	1.27	1971	3.98	1973	0.0		0.0	
J	83.0	60.4	71.7	97	1968	40	1966	0	205	3.51	7.55	1972	1.46	1972	3.98	1972	0.0		0.0	
J	85.9	64.4	75.2	100	1966	48	1972	0	316	3.74	7.85	1949	1.13	1957	2.72	1966	0.0		0.0	
A	84.9	63.3	74.1	96	1973	43	1965	0	282	4.15	9.12	1949	1.16	1959	3.35	1970	0.0		0.0	
S	79.5	56.5	68.0	95	1973	36	1974	32	122	3.42	7.25	1966	0.44	1959	3.45	1966	T	1953	T	1953
O	69.9	45.6	57.8	84	1973	11	1972	229	12	2.48	6.36	1968	0.27	1963	3.00	1962	1.0	1957	1.0	1957
N	57.6	35.8	46.7	82	1974	11	1968	549	0	3.11	6.36	1948	0.44	1948	3.40	1968	13.8	1968	10.0	1968
D	46.6	28.1	37.4	75	1968	-4	1968	856	0	...	7.10	1948	0.18	1965	...	1948	22.6	1966	16.4	1966
YR	66.8	45.0	55.9	100	JUL 1966	-4	JAN 1972	4307	1030	39.03	9.12	AUG 1949	0.18	DEC 1965	6.41	OCT 1968	41.2	JAN 1966	17.4	MAR 1960

Means and extremes above are from existing and comparable exposures. Annual extremes have been exceeded at other sites in the locality as follows:
Highest temperature 105 in July 1936; lowest temperature -12 in December 1917; maximum monthly precipitation 12.91 in December 1917; maximum monthly precipitation 12.91 in August 1940; minimum monthly precipitation .16 in November 1931.

REFERENCE NOTES APPLYING TO TABLES APPEAR ON THE PAGE FOLLOWING LAST TABLE.

(Caution: Letters and symbols may have different meanings in 1941-1970 tables than in earlier tables. See notes.)

NORMALS, MEANS AND EXTREMES
(Table Revised 1971. Base Period for Climatological Normals: 1931-1960)

Station: WASHINGTON, D.C.   DULLES INTERNATIONAL AIRPORT
Latitude: 38° 57' N   Longitude: 77° 27' W   Standard time used: EASTERN   Elevation (ground): 290 feet

Month	Temperature Normal (Daily max / Daily min / Monthly)	Extremes Record highest	Year	Record lowest	Year
J	Ø	73	1967	-8	1970
F	Ø	73	1967	-6	1967
M	Ø	83	1969	10	1969
A	Ø	89	1969	17	1969
M	Ø	97	1969	28	1966
J	Ø	100	1964	41	1964
J	Ø	103	1966	47	1970
A	Ø	98	1968	44	1968
S	Ø	97	1964	35	1967
O	Ø	89	1963	15	1969
N	Ø	78	1968	14	1967+
D	Ø	75	1970	0	1967
YR		103	JUL. 1966	-8	JAN. 1970

Precipitation (Mean total / Maximum monthly / Year / Minimum monthly / Year / Maximum in 24 hrs / Year):

Month	Max monthly	Year	Min monthly	Year	Max 24 hrs	Year
J	5.56	1964	1.12	1967	1.46	1964
F	4.50	1967	0.68	1968	1.60	1970
M	4.38	1966	1.03	1966	2.18	1966
A	4.38	1970	0.93	1967	1.87	1968
M	8.59	1963	0.94	1966	2.98	1963
J	6.00	1970	1.25	1963	2.36	1968
J	9.28	1967	1.79	1966	3.89	1966
A	9.31	1966	0.72	1966	5.54	1967
S	9.01	1966	0.46	1963	3.97	1963
O	7.83	1963	0.46	1963	3.33	1963
N	6.74	1964	0.42	1965	1.84	1964
YR	9.39	SEP. 1966	T	OCT. 1963	5.54	SEP. 1966

Snow, Ice pellets (Mean total / Maximum monthly / Year / Maximum in 24 hrs / Year):

Month	Mean total	Max monthly	Year	Max 24 hrs	Year
J	7.0	19.0	1966	9.6	1966
F	8.2	18.0	1967	11.7	1967
M	4.0	10.4	1969	7.2	1969
A	0.1	T	1970	0.7	1970
M	0.0	0.0	1963	0.0	1963
J	0.0	0.0		0.0	
J	0.0	0.0		0.0	
A	0.0	0.0		0.0	
S	0.0	0.0		0.0	
O	0.0	0.0	1967	0.0	1967
N	2.2	11.4	1966	11.4	1966
D	8.1	24.2	1969	12.1	1969
YR	29.6	24.2	DEC. 1969	12.1	DEC. 1969

Ø Normals have not been established for this station.

NORMALS, MEANS AND EXTREMES
(Table Revised 1975. Base Period for Climatological Normals: 1941-1970)

Month	Daily maximum	Daily minimum	Monthly	Record highest	Year	Record lowest	Year	Heating	Cooling
J	41.2	23.0	32.1	73	1970	-8	1970	1020	0
F	43.4	24.1	33.8	75	1967	-6	1967	874	0
M	52.1	30.9	41.8	83	1964	10	1960	719	0
A	65.0	41.1	53.1	89	1969	17	1969	357	0
M	74.5	50.6	62.6	97	1969	28	1966	131	57
J	82.7	59.4	71.1	100	1964	40	1964	5	188
J	86.6	64.1	75.3	103	1966	47	1970	0	319
A	85.0	62.1	73.6	98	1967	44	1968	0	267
S	78.2	55.0	66.9	97	1964	30	1967	43	109
O	67.8	43.5	55.9	89	1970	15	1969	291	0
N	55.6	34.7	44.7	84	1974	11	1974	609	0
D	43.3	26.4	34.0	75	1966	-5	1975	961	0
YR	64.7	42.7	53.7	103	JUL. 1966	-8	JAN. 1970	5010	940

Precipitation in inches — Water equivalent (Normal / Maximum monthly / Year / Minimum monthly / Year / Maximum in 24 hrs / Year):

Month	Normal	Max monthly	Year	Min monthly	Year	Max 24 hrs	Year
J	2.84	5.56	1964	1.12	1967	1.70	1971
F	2.48	5.44	1967	0.68	1968	1.66	1972
M	3.48	5.24	1967	1.04	1966	2.18	1964
A	2.96	4.35	1970	0.93	1967	2.58	1970
M	3.68	8.47	1971	0.90	1964	2.08	1973
J	3.61	18.19	1972	0.94	1966	11.88	1963
J	4.12	6.00	1970	1.25	1963	2.36	1963
A	4.25	9.28	1967	1.79	1966	3.89	1967
S	3.29	9.39	1966	0.62	1966	5.54	1966
O	2.74	9.12	1971	0.72	1963	3.39	1971
N	3.06	7.83	1963	0.46	1965	3.33	1963
D	3.47	6.74	1974	0.42	1965	2.02	1974
YR	40.11	18.19	JUN. 1972	T	OCT. 1963	11.88	JUN. 1972

Snow, Ice pellets (Maximum monthly / Year / Maximum in 24 hrs / Year):

Month	Max monthly	Year	Max 24 hrs	Year
J	19.0	1966	15.4	1966
F	18.0	1967	11.7	1967
M	10.4	1969	7.2	1969
A	T	1970	T	1970
M	0.0	1963	0.0	1963
J	0.0		0.0	
J	0.0		0.0	
A	0.0		0.0	
S	0.0	1972	0.4	1972
O	0.4	1967	0.4	1967
N	11.4	1966	11.4	1966
D	24.2	1966	12.1	1966
YR	24.2	DEC. 1966	15.4	JAN. 1971

REFERENCE NOTES APPLYING TO TABLES APPEAR ON THE PAGE FOLLOWING LAST TABLE.
(Caution: Letters and symbols may have different meanings in 1941-1970 tables than in earlier tables. See notes.)

NORMALS, MEANS AND EXTREMES
(Table Revised 1971. Base Period for Climatological Normals: 1931-1960)

Station: WASHINGTON, D. C.   NATIONAL AIRPORT   Standard time used: EASTERN   Latitude: 38° 51' N   Longitude: 77° 02' W   Elevation (ground): 10 feet

Means and extremes above are from existing and comparable exposures. Annual extremes have been exceeded at other sites in the locality as follows:
Highest temperature 106 in July 1930+; lowest temperature -15 in February 1899; maximum monthly precipitation 17.45 in September 1934; maximum
precipitation in 24 hours 7.31 in August 1928; maximum monthly snowfall 35.2 in February 1899; maximum snowfall in 24 hours 25.0 in January 1922.

Solar radiation data have been recorded at several locations in the vicinity of
Washington. Instruments have been at the Observational Test and Development
Center Sterling Virginia since October 1960, elevations (m.s.l.) 276 ft. to
7-13-64 and 281 ft. thereafter.

NORMALS, MEANS AND EXTREMES
(Table Revised 1975. Base Period for Climatological Normals: 1941-1970)

Means and extremes above are from existing and comparable exposures. Annual extremes have been exceeded at other sites in the locality as follows:
Highest temperature 106 in July 1930+; lowest temperature -15 in February 1899; maximum monthly precipitation 17.45 in September 1934; maximum
precipitation in 24 hours 7.31 in August 1928; maximum monthly snowfall 35.2 in February 1899; maximum snowfall in 24 hours 25.0 in January 1922.

REFERENCE NOTES APPLYING TO TABLES APPEAR ON THE PAGE FOLLOWING LAST TABLE.
(Caution: Letters and symbols may have different meanings in 1941-1970 tables than in earlier tables. See notes.)

Reference notes applying to Normals, Means, and Extremes tables for 1931–1960 base period.

(a) Length of record, years, based on January data. Other months may be for more or fewer years if there have been breaks in the record.

(b) Climatological standard normals (1931-1960).

* Less than one half.

+ Also on earlier dates, months, or years.

T Trace, an amount too small to measure.

- Below zero temperatures are preceded by a minus sign. The prevailing direction for wind in the Normals, Means, and Extremes table is from records through 1963.

‡ >70° at Alaskan stations.

Unless otherwise indicated, dimensional units used in this bulletin are: temperature in degrees F.; precipitation, including snowfall, in inches; wind movement in miles per hour; and relative humidity in percent. Heating degree day totals are the sums of negative departures of average daily temperatures from 65° F. Cooling degree day totals are the sums of positive departures of average daily temperatures from 65° F. Sleet was included in snowfall totals beginning with July 1948. The term "Ice pellets" includes solid grains of ice (sleet) and particles consisting of snow pellets encased in a thin layer of ice. Heavy fog reduces visibility to 1/4 mile or less.

Sky cover is expressed in a range of 0 for no clouds or obscuring phenomena to 10 for complete sky cover. The number of clear days is based on average cloudiness 0-3, partly cloudy days 4-7, and cloudy days 8-10 tenths.

Solar radiation data are the averages of direct and diffuse radiation on a horizontal surface. The langley denotes one gram calorie per square centimeter.

& Figures instead of letters in a direction column indicate direction. 1: tens of degrees from true North; i.e., 09 - East, 18 - South, 27 - West, 36 - North, and 00 - Calm. Resultant wind is the vector sum of wind directions and speeds divided by the number of observations. If figures appear in the direction column under "Fastest mile" the corresponding speeds are fastest observed 1-minute values.

Reference notes applying to Normals, Means, and Extremes tables for 1941–1970 base period.

(a) Length of record, years, through the current year unless otherwise noted, based on January data.

(b) 70° and above at Alaskan stations.

T Trace.

NORMALS - Based on record for the 1941-1970 period.

DATE OF AN EXTREME - The most recent in cases of multiple occurrence.

PREVAILING WIND DIRECTION - Record through 1963.

WIND DIRECTION - Numerals indicate tens of degrees clockwise from true north. 00 indicates calm.

FASTEST MILE WIND - Speed is fastest observed 1-minute value when the direction is in tens of degrees.

% Data through 1966. Station operated less than 24 hours per day prior to August 1962. Fog and thunderstorm data may be incomplete.

Mean Maximum Temperature (°F), January

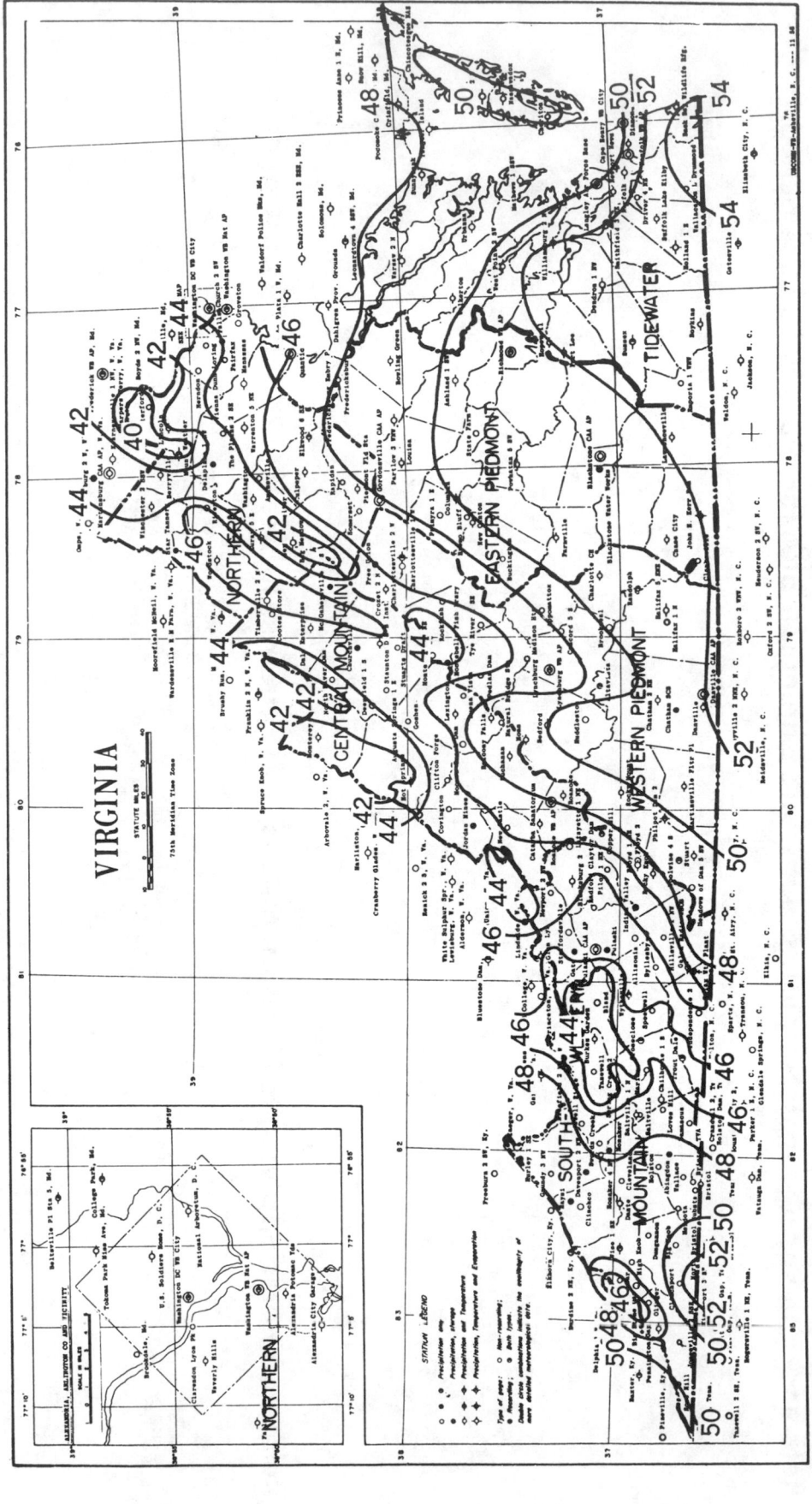

Based on period 1931-52

Isolines are drawn through points of approximately equal value. Caution should be used in interpolating on these maps, particularly in mountainous areas.

Mean Minimum Temperature (°F), January

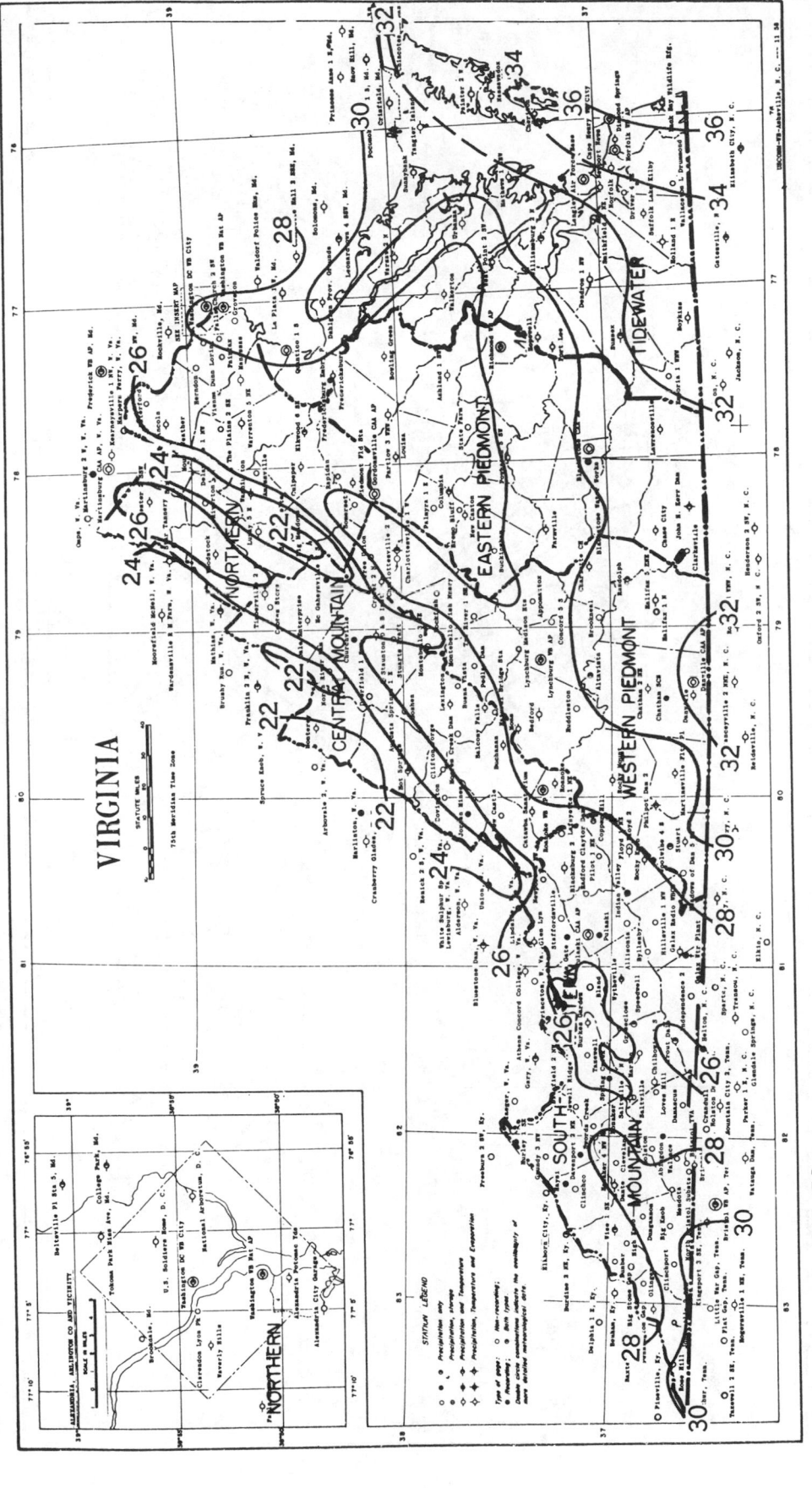

VIRGINIA

Based on period 1931-52

Isolines are drawn through points of approximately equal value. Caution should be used in interpolating on these maps, particularly in mountainous areas.

1037

Mean Maximum Temperature (°F), July

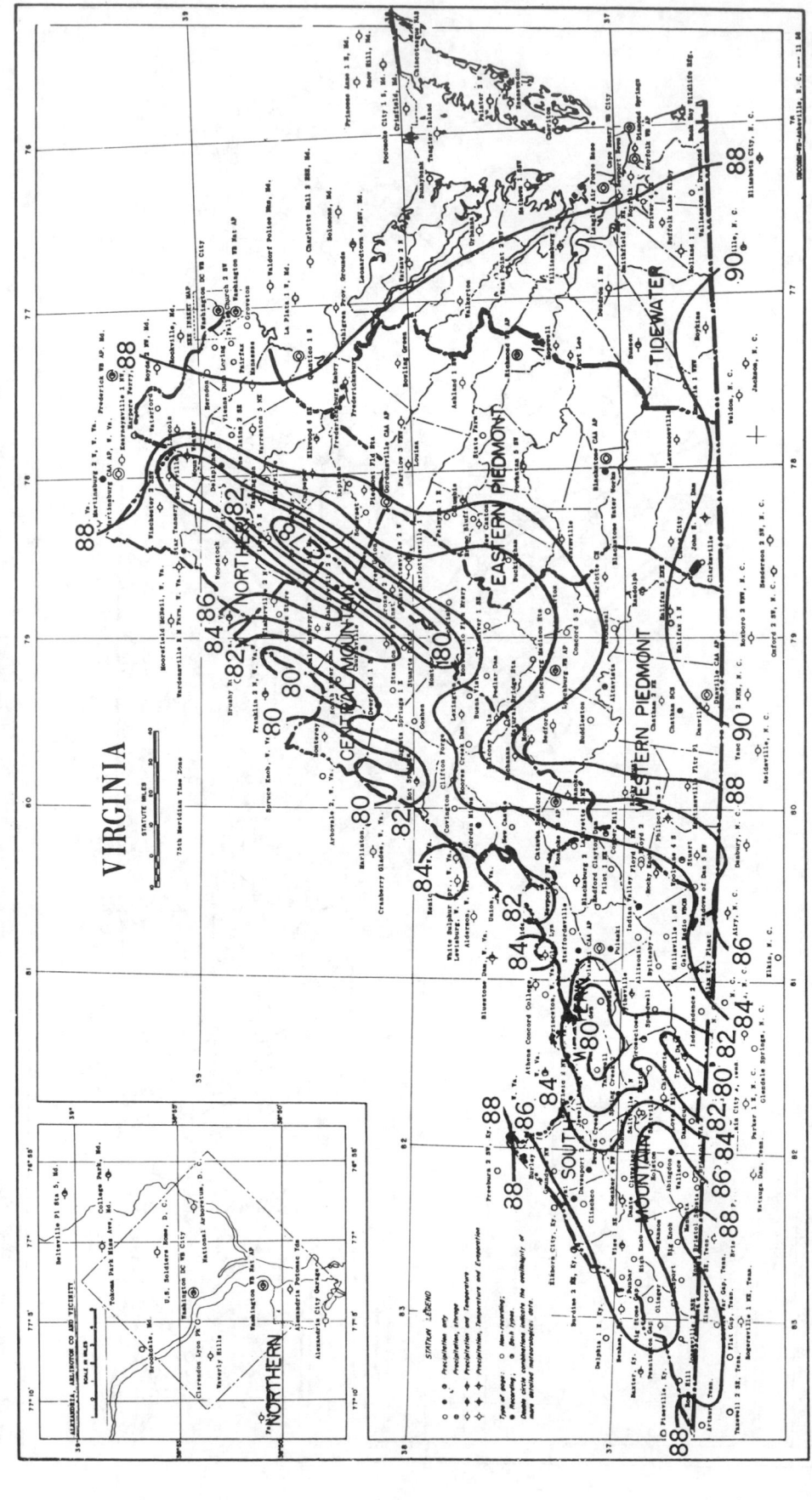

Based on period 1931-52

Isolines are drawn through points of approximately equal value. Caution should be used in interpolating on these maps, particularly in mountainous areas.

Mean Minimum Temperature (°F), July

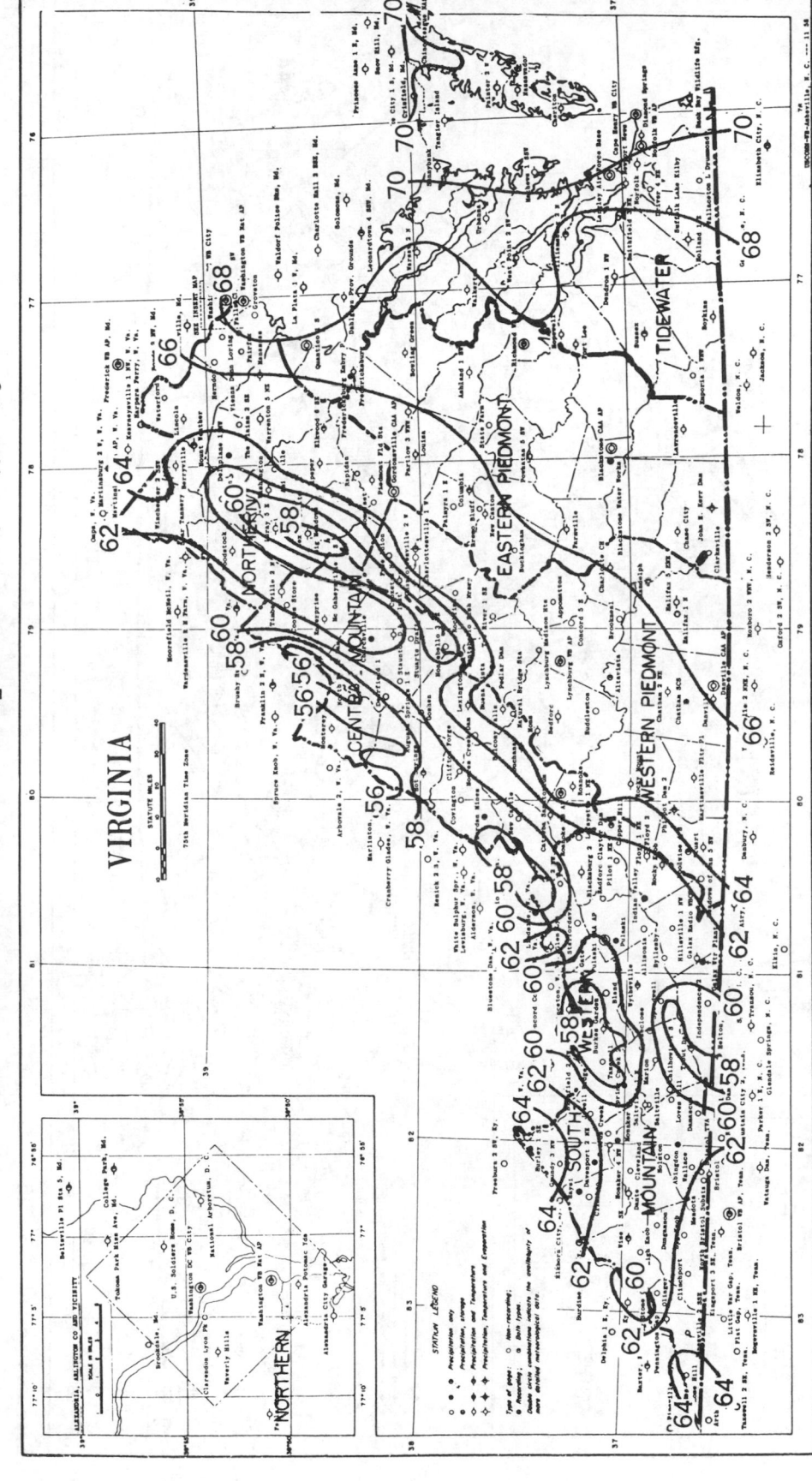

Based on period 1931-52

Isolines are drawn through points of approximately equal value. Caution should be used in interpolating on these maps, particularly in mountainous areas.

Mean Annual Precipitation, Inches

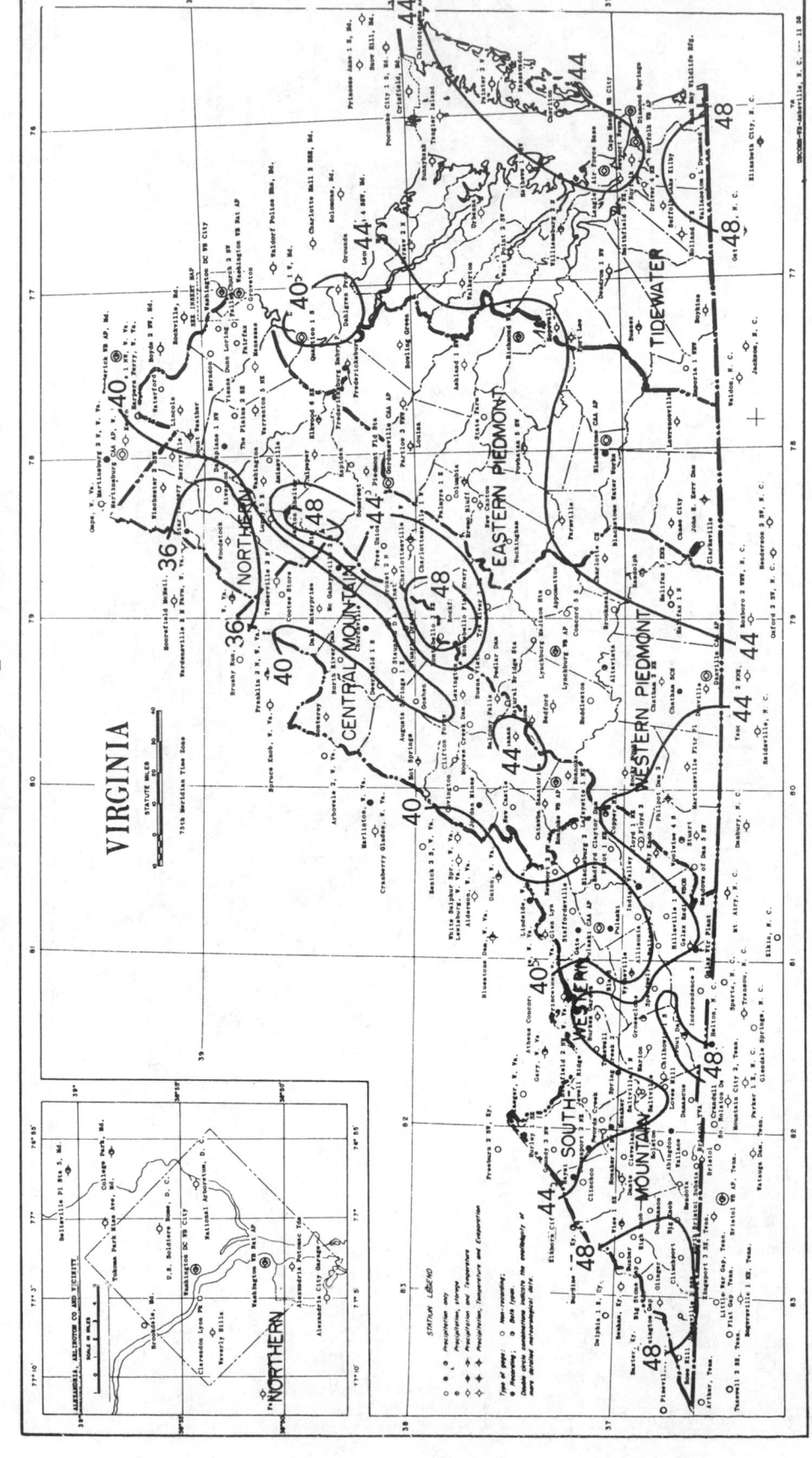

Based on period 1931-55

Isolines are drawn through points of approximately equal value. Caution should be used in interpolating on these maps, particularly in mountainous areas.

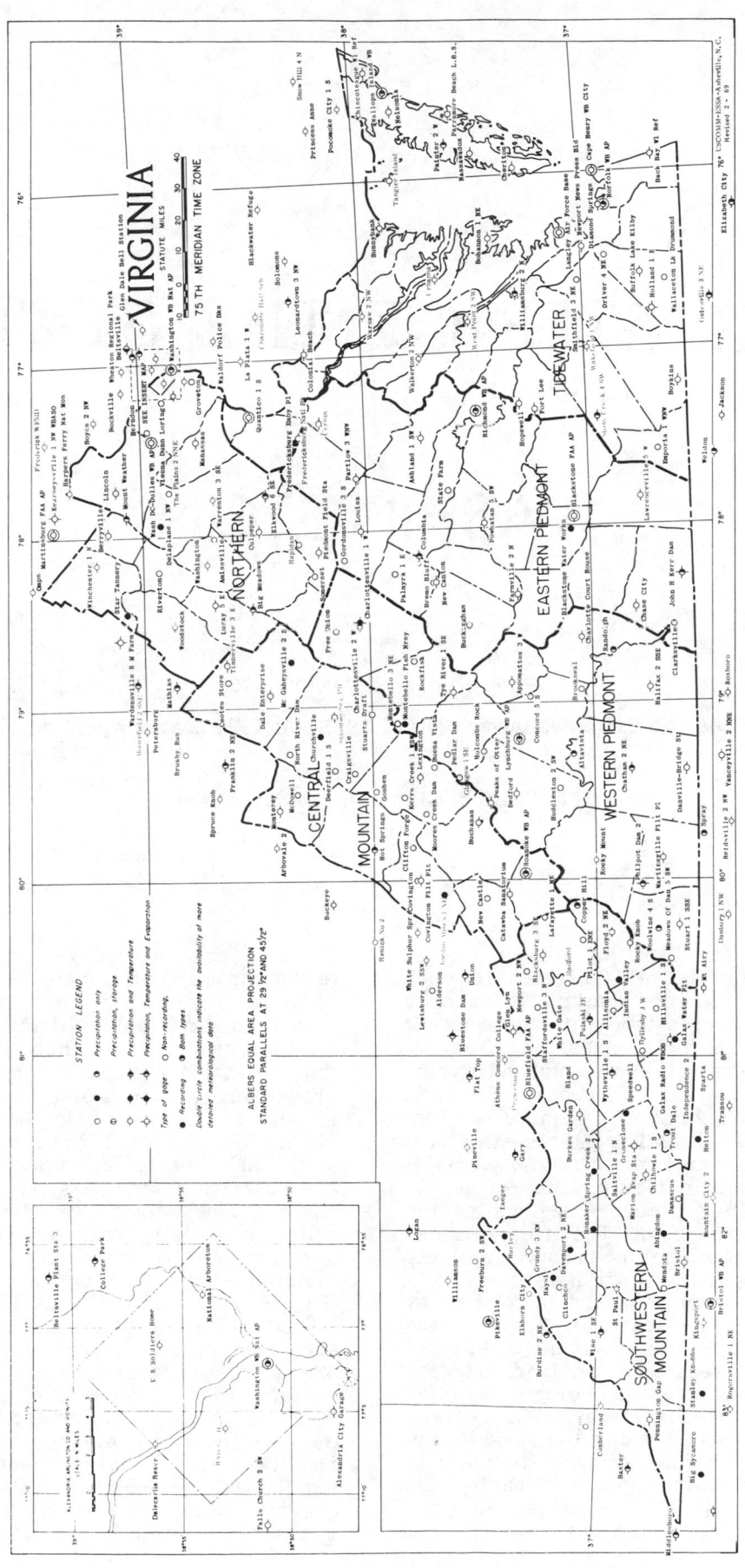

VIRGINIA

STATUTE MILES

10 20 30 40

75 TH MERIDIAN TIME ZONE

ALBERS EQUAL AREA PROJECTION
STANDARD PARALLELS AT 29 1/2°AND 45 1/2°

STATION LEGEND

Precipitation only

Precipitation, storage

Precipitation and Temperature

Precipitation, Temperature and Evaporation

Type of gage   O Non-recording   ● Recording   ◐ Both types

Double circle combinations indicate the availability of more
detailed meteorological data

NORTHERN

CENTRAL MOUNTAIN

WESTERN PIEDMONT

EASTERN PIEDMONT

TIDEWATER

SOUTHWESTERN MOUNTAIN

1041

CLIMATES OF THE STATES

WASHINGTON

First Printed February 1960

(Normals, Means and Extremes tables revised 1965 and 1975. Basic report revised April 1965.)

Earl L. Phillips, Climatologist for Washington

The location of the State of Washington on the windward coast in mid-latitudes is such that the climatic elements combine to produce a predominantly marine-type climate west of the Cascade Mountains, while east of the Cascades, the climate possesses both continental and marine characteristics. Considering its northerly latitude, 46° to 49°, Washington's climate is mild.

There are several climatic controls which have a definite influence on the climate, namely: (a) terrain, (b) Pacific Ocean, and (c) semi-permanent high and low pressure regions located over the north Pacific Ocean. The effect of these various controls combine to produce entirely different conditions within short distances.

Washington's western boundary is formed by the Pacific Ocean. The seasonal change in the temperature of the ocean is less than the seasonal change in the temperature of the land, thus the ocean is warmer in winter and cooler in summer than the adjoining land surfaces. The average temperature of the water along the coast and in the Strait of Juan de Fuca ranges from 45° in January to 53° in July; however, during the summer, some of the shallow bays and protected coves are five to ten degrees warmer.

There are two ranges of mountains parallel to the coast and athwart to the prevailing direction of moist air moving inland from over the ocean. The first orographic lifting and major release of moisture occurs along the western slope of the Coastal Range. The second area of heavy orographic precipitation is along the windward slopes of the Cascade Range. The Cascade Mountains, 90 to 125 miles inland and 4,000 to 10,000 feet in elevation, are a topographic and climatic barrier separating the State into eastern and western Washington. The higher, wider and more rugged sections are in the northern part of the State. Some of the highest isolated volcanic peaks are Mt. Rainier (14,408 ft.), Mt. Adams (12,307 ft.) and Mt. Baker (10,730 ft.). These and other high peaks are snowcapped throughout the year. The only break in the Cascade Range is the narrow and scenic Columbia River gorge.

Warming and drying of air as it descends along the lee (eastern) slopes of the Cascade Range results in near desert conditions in the lowest section of the Columbia Basin. Another orographic lifting of the air occurs as it flows eastward from the lowest elevations of the

Inland Basin toward the Rocky Mountains. This lifting of air results in a gradual increase in precipitation from the lowest section of the basin to the higher elevations along the eastern border of the State.

The location and intensity of the semi-permanent high and low pressure areas over the north Pacific Ocean have a definite influence on the climate. Air circulates in a clockwise direction around the semi-permanent high pressure cell and in a counter-clockwise direction around the semi-permanent low pressure cell. During the spring and summer, the low pressure cell becomes weak and moves north of the Aleutian Islands. At the same time, the high pressure area spreads over most of the north Pacific Ocean. A circulation of air around the high pressure center brings a prevailing westerly and northwesterly flow of comparatively dry, cool and stable air into the Pacific Northwest. As the air moves inland, it becomes warmer and drier which results in a dry season beginning in the late spring and reaching a peak in mid-summer.

In the fall and winter, the Aleutian low pressure center intensifies and moves southward reaching a maximum intensity in midwinter. At the same time, the high pressure area becomes weaker and moves southward. A circulation of air around these two pressure centers over the ocean brings a prevailing southwesterly and westerly flow of air into the Pacific Northwest. This air from over the ocean is moist and near the temperature of the water. Condensation occurs as the air moves inland over the cooler land and rises along the windward slopes of the mountains. This results in a wet season beginning in October, reaching a peak in winter, then gradually decreasing in the spring.

Although the Cascade Range divides the State into two major climatic regions, there are several distinct climatic areas within each of these regions.

WESTERN WASHINGTON.--West of the Cascade Mountains, summers are cool and comparatively dry and winters are mild, wet and cloudy. The average number of clear or only partly cloudly days each month varies from four to eight in winter, eight to fifteen in spring and fall, and fifteen to twenty in summer. The percent of possible sunshine received each month ranges from approximately 25 percent in winter to 60 percent in summer. In the interior valleys, measurable rainfall is recorded on 150 days each year and on 190 days in the mountains and along the coast. Thunderstorms over the lower elevations occur on 4 to 8 days each year and over the mountains on 7 to 15 days. Damaging hail storms rarely, if ever, occur in most localities of western Washington. During July and August, the driest months, it is not unusual for two to four weeks to pass with only a few showers; however, in December and January, the wettest months, precipitation is frequently recorded on 20 to 25 days or more each month. The range in annual precipitation is from approximately 20 inches in an area northeast of the Olympic Mountains to 150 inches along the southwestern slopes of these mountains. Snowfall is light in the lower elevations and heavy in the mountains.

During the wet season, rainfall is usually of light to moderate intensity and continuous over a period of time rather than heavy downpours for brief periods. Maximum rainfall intensities to expect in one out of ten years are: .6 to 1.0 inch in one hour; 1.0 to 2.5 inches in three hours; 1.5 to 5.0 inches in six hours; and 2.0 to 7.0 inches in twelve hours. The heavier intensities occur along the windward slopes of the mountains.

During the latter half of the summer and early fall, the lower valleys are sometimes filled with fog or low clouds until noon, while at the same time, the higher elevations are sunny. The strongest winds are generally from the south or southwest and occur during the late fall and winter. In the interior valleys, wind velocities can be expected to reach 40 to 50 m.p.h. each winter and 75 to 90 m.p.h. once in 50 years. The daily variation in relative humidity in January is from approximately 87 percent at 4 a.m. to 78 percent at 4 p.m., and in July, from 85 percent at 4 a.m. to 47 percent at 4 p.m. During periods of easterly winds, the relative humidity occasionally drops to 25 percent or lower. The highest summer and lowest winter temperatures are usually recorded during periods of easterly winds. The total evaporation for the warm season, May through September, as measured by a Weather Bureau evaporation pan at Seattle, is 25 inches with an average of 7 inches in July.

In order to describe the climate of western Washington in more detail, the area has been divided into 5 regions.

WEST OLYMPIC - COASTAL.--This area includes the coastal plains and the western slope of the Coastal Range from the Columbia River to the Strait of Juan de Fuca. The Olympic Mountains, located on the northern section of the Olympic Peninsula, tower to nearly 8,000 feet in dome-like structures, deeply carved by rivers. The Willapa Hills, elevation 1,000 to 3,000 feet, form a continuous ridge from the Chehalis River valley to the Columbia River. This area receives the full force of storms moving inland from over the ocean, thus heavy precipitation and winds of gale force occur frequently during the winter season. Wind velocities in the lower elevations can be expected to reach 90 to 100 m.p.h. once in one-hundred years. Wind data from a well-exposed site on a ridge near the ocean, elevation 2,000 feet, indicates that wind velocities in excess of 100 m.p.h. occur in the higher elevations almost every winter.

The "rainforest" area along the southwestern and western slopes of the Olympic Mountains receives the heaviest precipitation in the continental United States (48). Annual precipitation ranges from 70 to 100 inches over the Coastal Plains to 150 inches or more along the windward slopes of the mountains. The greatest annual precipitation recorded in the "rainforest" area is 184 inches at Wynoochee Oxbow, elevation 600 feet. The heaviest rainfall during a single storm was 12 inches in 24 hours; 23.5 inches in 48 hours; 28.6 inches in 72 hours; and 35 inches in four days recorded at Quinault Ranger Station, January 21-24, 1935. On Blue Glacier, elevation 6,900 feet and near the summit of Mt. Olympus, 149 inches of precipitation were recorded between August 1957 and July 1958. The total snowfall for this period was 542 inches.

During this same period, precipitation at lower elevation stations was approximately 15 percent below normal.

Winter season snowfall ranges from 10 to 30 inches in the lower elevations and between 250 and 500 inches in the higher mountains. In the lower elevations, snow melts rather quickly and depths seldom exceed 6 to 15 inches. In midwinter, the snowline in the Olympic Mountains and the Willapa Hills is between 1,500 and 3,000 feet above sea level. The higher ridges are covered with snow from November until June. The average maximum temperature in July is near 70° F. along the coast and 75° F. in the foothills and minimum temperatures are near 50° F. In winter, the warmer areas are near the coast. In January, maximum temperatures range from 43° to 48° F. and minimum temperatures from 32° to 38° F.

NORTHEAST OLYMPIC - SAN JUAN.--This area includes the lower elevations along the northeastern slope of the Olympic Mountains extending eastward along the Strait of Juan de Fuca from near Port Angeles to Whidbey Island and then northward into the San Juan Islands. The Olympic Mountains and the extension of the Coastal Range on Vancouver Island shields this area from winter storms moving inland from over the ocean. This belt in the "rain shadow" of the Olympic Mountains is the driest area in western Washington. The average annual precipitation ranges from about 18 inches near Sequim, Port Townsend and Coupeville to between 25 and 30 inches in the vicinity of Everett on the east, Port Angeles on the west and Olga in the San Juan Islands on the north. Measurable precipitation is recorded on three to five days each month in summer and on 17 to 22 days in winter.

Another factor which distinguishes this belt from other localities in the Puget Sound region is the rate of rainfall. This area frequently receives drizzle or light rain while other localities are experiencing light to moderate rainfall. Snowfall is light in the lower elevations adjacent to the water increasing with distance from the water and rise in terrain.

This area is considered to receive slightly more sunshine and have less cloudiness than other localities in Puget Sound, however, the difference is not in proportion to the decrease in precipitation. During the latter half of the summer and early fall, fog banks from over the ocean and Strait of Juan de Fuca result in considerable fog and morning cloudiness in the lower elevations.

The average July maximum temperature ranges from 65° F. near the water to 70° F. or 75° F. inland, and the minimum temperature is near 50° F. Maximum temperatures seldom exceed 90° F. In Janaury, maximum temperatures are in the 40's and minimums in the lower 30's. Minimum temperatures between -5° and -8° F. have been recorded; however, the minimum temperature seldom drops below 15° to 20° F. The coldest weather is usually associated with an outbreak of cold air from the interior of Canada. The average date of the last freezing temperature in the spring ranges from the latter half of March near the water to the last of April in agricultural areas 100 to 300 feet above sea level and a few miles inland. The first freezing temperature in the fall is about the first of November.

PUGET SOUND - LOWLANDS.--This area includes a narrow strip of land along the west side of Puget Sound southward from the Strait of Juan de Fuca to the vicinity of Centralia and Chehalis and a somewhat wider strip along the east side of the Sound extending northward to the Canadian Border. Variations in the temperature, length of the growing season, fog, rainfall and snowfall are due to such factors as distance from the Sound, the rolling terrain and air from over the ocean reaching this area through the Strait of Juan de Fuca and the Chehalis River valley. Occasionally, in the winter season, cold air from the interior of Canada flows southward through the Fraser River canyon and over the northern Puget Sound lowlands.

The prevailing direction of the wind is south or southwest during the wet season and northwest in summer. The average wind velocity is less than 10 m.p.h. Although this is the most densely populated and industrialized area in the State, there is sufficient wind most of the year to disperse air pollutants released into the atmosphere. Air pollution is usually most noticeable in the late fall and winter season, under conditions of clear skies, light wind and a sharp temperature inversion. These conditions only prevail a few days before a weather system moves through removing the pollution by wind and rain.

Annual precipitation ranges from 32 to 35 inches from the Canadian Border to Seattle then gradually increasing to 45 inches in the vicinity of Centralia. The winter season snowfall ranges from 10 to 20 inches. Both rainfall and snowfall increase with a slight increase in elevation and distance from the Sound. Snow generally melts rather quickly and depths seldom exceed 6 to 15 inches. The greatest snow depth recorded in Seattle is 29 inches. Most of this area is near the eastern edge of the "rain shadow" of the Olympic Mountains.

The average January maximum temperature ranges from 41° to 45° F. and minimum temperatures from 28° to 32° F. With an increase in distance from the Sound, winter temperatures decrease and summer temperatures increase. Minimum temperatures ranging from 0° to -10° F. have been recorded; however, temperatures seldom drop lower than 10° to 15° F. During July, the average maximum temperature ranges from 73° F. near the Canadian border to 78° F. in the vicinity of Olympia and the minimum temperature is near 50° F. Maximum temperatures have reached 100° F.; however, in an average summer, 90° or higher is only recorded on 3 to 5 days. The growing season is from the latter half of April until the middle of October.

EAST OLYMPIC - CASCADE FOOTHILLS.--This area includes foothills along the eastern slope of the Coastal Range, foothills along the western slope of the Cascade Mountains and the valley separating these ridges from the vicinity of Chehalis to the Columbia River. The easterly movement of moist air from over the ocean produces down-slope winds in foothills along the eastern slope of the Coastal Range and upslope winds in the foothills along the western slope of the Cascade Mountains. Precipitation is heavier along the windward slopes than in the valley or along the lee slopes. The average annual precipitation ranges from 40 inches in the lower valleys near the Columbia River to 90 inches at stations 800 to 1,000 feet above sea level and along the western slope of the Cascade range. Annual snowfall increases from less than 10 inches in the lower valleys to 50 inches in elevations 500 to 800 feet.

The Columbia River gorge permits an exchange of air between eastern and western Washington. The direction and speed of air movement through the gorge is determined primarily by the pressure gradient between the eastern and western slopes of the mountains. In summer, the flow of air is usually from west to east, and in winter from east to west. During the winter season, easterly winds in the gorge sometimes reach gale force. Rather severe ice storms or "silver thaws", as they are frequently called, occur in a narrow area westward from the gorge to the vicinity of Vancouver. The "silver thaws" are the result of rain falling through a layer of cold air flowing westward through the gorge.

In January, the average maximum temperature ranges from 38° to 45° F., and the minimum from 25° to 32° F. Minimum temperatures have dropped to between 0° and -15° F, however, minimum temperatures lower than 5° to 10° F. occur infrequently. In July, the average maximum temperature ranges from 75° to 80° F. and the minimum is near 50° F. Maximum temperatures have reached 100° to 105° F.; however, it is unusual for afternoon temperatures to exceed 90° on more than 8 to 15 days in the summer season. The hottest weather occurs during periods of dry easterly winds. The average date of the last freezing temperature in the spring ranges from the middle of April in the warmer valleys to the middle of May in the colder localities. In the fall, freezing temperatures can be expected after the middle of October.

CASCADE MOUNTAINS - WEST.--This area includes the western slope of the Cascade range from an elevation of approximately 1,000 feet to the summit and extending from the Columbia River to the Canadian border. Daily temperature and precipitation reporting stations have been limited to elevations below 5,500 feet. Snow course measurements consisting of snow depth and water content of the snow pack are available for some of the higher elevations. Orographic lifting of the moisture-ladened southwesterly and westerly winds results in heavy precipitation in this area. The annual precipitation ranges from 60 to 100 inches or more. Indications are that the heaviest precipitation probably occurs along the slopes of east-west mountain valleys which become more narrow as the elevation increases along the windward slopes of the Cascades. Annual precipitation in some of the wetter areas has reached 140 inches in one out of ten years.

The average winter season snowfall ranges from 50 to 75 inches in the lower elevations gradually increasing with elevation to between 400 and 600 inches at 4,000 to 5,500 feet. Some

of the greatest seasonal snowfalls and snow depths in the United States have been recorded on the slopes of Mt. Rainier and Mt. Baker. The greatest seasonal snowfall recorded at Mt. Rainier-Paradise Ranger Station (elevation 5,500 ft.) was 1,000 inches in 1955-56. These and other high peaks above 7,000 or 8,000 feet remain snow-capped throughout the summer. Snowfall usually begins in the higher elevations in September, gradually working down to 3,000 feet by the last of October. The snowline in midwinter varies from 1,500 to 2,000 feet above sea level. Although snowfall continues until late spring, the maximum depth is usually reached during the first half of March. At this season of the year, snow depths above 3,000 feet range from 10 to 25 feet. The density of the snow pack increases from approximately 30 percent water the first of December to 45 percent water in March. In elevations above 5,000 feet, snow remains on the ground until the last of June or first of July.

The average January maximum temperature ranges from 40° F. in the lower elevations to 30° F. at the 5,500 foot elevation. Minimum temperatures range from 30° F. in the lower elevations to 20° F. in the higher elevations. Minimum temperatures from 0° to -17° F. have been recorded in the higher elevations. The lowest temperatures occur during periods of easterly winds. The average July maximum temperature ranges from 75° in the lower elevations to the lower 60's in the higher elevations. The minimum temperature is in the 40's. Above 4,000 feet, minimum temperatures occasionally drop below freezing in midsummer. In general, the temperature decreases approximately 3° F. with each 1,000 feet increase in elevation.

EASTERN WASHINGTON.--This section of the State is part of the large inland basin between the Cascade and Rocky Mountains. In an easterly and northerly direction, the Rocky Mountains shield the inland basin from the winter season's cold air masses traveling southward across Canada. In a westerly direction, the Cascade range forms a barrier to the easterly movement of moist air in winter and comparatively mild air in winter and cool air in summer. Some of the air from each of these source regions reaches this section of the State and produces a climate which has some of the characteristics of both continental and marine types. Most of the air masses and weather systems crossing eastern Washington are traveling under the influence of the prevailing westerly winds. Infrequently, dry continental air masses enter the inland basin from the north or east. In the summer season, this air from over the continent results in low relative humidity and high temperatures while in winter, clear cold weather prevails. Extremes in both summer and winter temperatures generally occur when the inland basin is under the influence of air from over the continent.

East of the Cascades, summers are warmer, winters are colder and precipitation is less than in western Washington.

The average number of clear or only partly cloudy days each month varies from 5 to 10 in winter, 12 to 18 in spring and fall, and 20 to 28 in summer. The percent of possible sunshine received each month is from 20 to 30 percent in winter, 50 to 60 percent in spring and fall and 80 to 85 percent in summer. The number of hours of sunshine possible on a clear day ranges from approximately 8 in December to 16 in June. In the driest areas, rainfall is recorded on 70 days each year and on 120 days or more in the higher elevations near the eastern border and along the eastern slope of the Cascades.

Annual precipitation ranges from 7 to 9 inches near the confluence of the Snake and Columbia Rivers, 15 to 30 inches along the eastern border and 75 to 90 inches near the summit of the Cascade Mountains. During July and August, it is not unusual for 4 to 8 weeks to pass with only a few scattered showers. Thunderstorms can be expected on 1 to 3 days each month from April through September. Most thunderstorms in the warmest months occur as isolated cells covering only a few square miles. A few damaging hail storms are reported each summer. Maximum rainfall intensities to expect in one out of ten years are: .6 of an inch in one hour; 1.0 inch in three hours; 1.0 to 1.5 inches in six hours; and 1.2 to 2.0 inches in twelve hours.

During the coldest months, a loss of heat by radiation at night and moist air crossing the Cascades and mixing with the colder air in the inland basin results in cloudiness, fog and occasional freezing drizzle. A "chinook" wind which produces a rapid rise in temperature occurs a few times each winter. Frost penetration in the soil depends to some extent on the vegetative cover, snow cover and the duration of low temperatures. In an average winter, frost in the soil can be expected to reach a depth of 10 to 20 inches. During a few of the colder winters with little or no snow cover, frost has reached a depth of 25 to 35 inches.

During most of the year, the prevailing direction of the wind is from the southwest or west. The frequency of northeasterly winds is greatest in the fall and winter. Wind velocities ranging from 4 to 12 m.p.h. can be expected 60 to 70 percent of the time, 13 to 24 m.p.h., 15 to 24 percent of the time and 25 m.p.h. or higher, 1 to 2 percent of the time. The highest wind velocities are from the southwest or west and are frequently associated with rapidly moving weather systems. Extreme wind velocities at 30 feet above the ground can be expected to reach 50 m.p.h. at least once in 2 years; 60 to 70 m.p.h. once in 50 years and 80 m.p.h. once in 100 years.

During the growing season, April through September, the average evaporation from a Class A evaporation pan is from 35 to 52 inches. Monthly evaporation in midsummer ranges from 9 to 12 inches. Annual evaporation from lakes and reservoirs is estimated at 26 inches in the mountains and 34 to 42 inches in other localities. The average relative humidity in January is approximately 85 percent at 4 a.m. and 75 percent at 4 p.m. and in July, 65 percent at 4 a.m. and 27 percent at 4 p.m.

In order to describe the climate in more detail, eastern Washington has been divided into 5 sections:

EAST SLOPE - CASCADES.--This area extends from the summit of the Cascades eastward for distances varying from 25 to 75 miles and from the Canadian border to the Columbia River.

In an easterly direction, the elevation decreases from the summit of the Cascade range to approximately 2,000 feet above sea level. One of the outstanding features of the climate is the decrease in precipitation along the eastern slope of the mountains as the distance from the summit increases and the elevation decreases. For example, within a distance of 20 miles, the average annual precipitation decreases from 92 inches at Stampede Pass (elevation 3,958 ft.) to 22 inches at Cle Elum (elevation 1,920 ft.).

The average winter season snowfall decreases from approximately 400 inches near the summit of the mountains to about 75 inches at 2,000 feet above sea level. In elevations above 3,000 feet snow can be expected in October; however, it generally does not accumulate on the ground until after the first of November. In the lower elevations, snow reaches a depth of two to five feet in January or February and in the higher elevations, 10 to 20 feet by the first of March. The density of the snow pack increases from approximately 30 percent water at the beginning of the winter season to 45 percent water by mid-March. In the higher elevations, snow remains on the ground until June or July. Several large irrigation reservoirs are located in valleys along the eastern slope of the Cascades. Melting of the snow provides irrigation water for orchards and other agricultural areas in the Okanogan, Wenatchee, Methow, Yakima and Columbia River valleys.

The average January maximum temperature varies from 25° to 35° F. and the minimum temperature from 15° to 25° F. Minimum temperatures ranging from 0° to -15 ° F. are recorded almost every winter and minimum temperatures have dropped to -30° F. in the colder valleys. In July, the average maximum temperature ranges from 70° to 85° F. and the minimum temperature from 45° to 50° F. In the lower elevations, maximum temperatures exceed 90° F. on 15 to 20 days each summer and 80° F. or higher is usually recorded in the higher elevations. In elevations below 3,000 feet, maximum temperatures have reached 100° to 105° F. A cool mountain breeze in the late afternoon results in rapid cooling after sunset.

OKANOGAN - BIG BEND.--This area includes fruit producing valleys along the Okanogan, Methow and Columbia Rivers, grazing land along the southern Okanogan highlands, the Waterville Plateau and part of the channeled scablands. The elevation varies from approximately 1,000 feet in the lower river valleys to 3,000 feet over the Waterville Plateau and Okanogan highlands. North-south ranges of mountains extending into southern British Columbia reach elevations of 4,000 to 5,000 feet within a few miles of the Okanogan River. The annual precipitation increases from 11 inches in the valley to 16 inches over some of the Plateau. Winter season snowfall varies from 30 to 70 inches. Both rainfall and snowfall increase in the higher elevations. Snow can be expected after the first of November and to remain on the ground from the first of December until March or April. Snow accumulates to a depth of 10 to 20 inches in the valleys and over the Waterville Plateau increasing to 40 inches in the higher grazing areas.

The average January maximum temperature is between 28° to 32° F. and the minimum temperature varies from 15° to 20° F. Minimum temperatures from 0° to -15° F. are recorded on a few nights each winter and -30° has been recorded in the colder localities. Occasional outbreaks of cold air from Canada moving southward through the valleys result in a late spring or early fall freeze. In July, the average maximum temperature is between 85° to 90°, and the minimum is in the lower 50's. Maximum temperatures reach 100° F. or higher on a few afternoons each summer and 105° to 113° F. have been recorded. Thunderstorms occur on 10 to 15 days each summer and a few damaging hail storms are reported in the fruit-producing valleys.

The average date of the last freezing temperature in the spring is the latter half of April in the warmer fruit-producing valleys along the Columbia and Okanogan Rivers, the middle of May in the colder valleys along the Wenatchee and Methow Rivers and the last of May over the Waterville Plateau and the higher rangelands. The first freezing temperature in the fall usually occurs in the latter half of September on the Waterville Plateau and by the middle of October in the warmer fruit-producing valleys.

CENTRAL BASIN.--The Central Basin includes the Ellensburg valley, the central plains area in the Columbia basin south from the Waterville Plateau to the Oregon border and east to near the Palouse River. The elevation increases from approximately 400 feet at the confluence of the Snake and Columbia Rivers to 1,300 feet near the Waterville Plateau and to 1,800 feet along the eastern edge of this area. This is the lowest and driest section in eastern Washington. Annual precipitation ranges from 7 inches in the drier localities along the southern slopes of the Saddle Mountains, Frenchman Hills and east of Rattlesnake Mountains to 15 inches in the vicinity of the Blue Mountains. Summer precipitation is usually associated with thunderstorms. During July and August, it is not unusual for four to six weeks to pass without measurable rainfall.

The winter season snowfall is from 10 to 35 inches. Snow can be expected after the first of December and to remain on the ground for periods varying from a few days to two months between mid-December and the last of February. Other than in the Ellensburg valley, snow depths seldom exceed 8 to 15 inches. The Central Basin is subject to "chinook" winds which produce a rapid rise in temperature. A few damaging hail storms are reported in the agricultural areas each summer.

The average January maximum temperature is near 30° F. in the colder localities in the Columbia Basin and 40° F. in the lower Yakima valley and minimum temperatures are between 15° to 25° F. Minimum temperatures between 0° to -10° F. are recorded almost every winter and temperatures from -15° F. to -30° F. have been recorded.

In July, the average maximum temperature is in the lower 90's and the minimum temperature is in the upper 50's. The record high temperature for the State, 118° F., was recorded on July 24, 1928, at Wahluke, located along the southern slope of the Saddle Mountains and again on August 5, 1961, at Ice Harbor Dam on the Snake River. Maximum temperatures reach 100° to 105° F. on a few afternoons each summer. The last freezing temperature in the spring occurs during the latter half of April in the Yakima valley and the latter half of May in the colder localities of the Columbia Basin. The first freezing temperature in the fall is usually recorded between mid-September and mid-October.

NORTHEASTERN.--The northeastern and higher elevations of the Okanogan highlands, the Selkirk Mountains, and the lower elevations southward to the vicinity of the Spokane River are included in the northeastern area. Ranges of mountains in this section of the State are separated by narrow north-south valleys. The elevation increases from 2,000 feet in the valleys to 6,000 feet along the higher ridges. Most of the temperature and precipitation records are from stations located in the valleys. The average annual precipitation increases in a northeasterly direction from 17 inches in the Spokane area to 28 inches in the northeastern corner of the State.

Winter season snowfall in the valleys varies from 40 to 80 inches. Both rainfall and snowfall increases along the slopes of the mountains. Snow can be expected in the higher elevations in October and in the lower valleys by the last of November. In the lower elevations, snow reaches a depth of 15 to 30 inches and remains on the ground most of the time from the first of December until March. The few snow survey reports available for elevations above 5,000 feet indicate 6 to 8 feet of snow on the ground the first of April and 4 to 5 feet the first of May.

Cold continental air moving southward through Canada will occasionally cross the higher mountains and follow the north-south valleys into the Columbia Basin. On clear, calm winter nights, the loss of heat by radiation from over a snow cover produces ideal conditions for low temperatures. The lowest temperature in the State, -42° F., was recorded January 20, 1937, at Deer Park. In January, the average maximum temperature is near 30° F. and the minimum temperature is 15° F. Minimum temperatures from -10° F. to -20° F. are recorded almost every winter and temperatures ranging from -25° to -42° F. have been recorded in the colder valleys. In July, the average maximum temperature is 85° to 90° and the minimum temperature 45° to 50° F. Maximum temperatures reach 100° F. on a few afternoons each summer and temperatures between 105° to 110° F. have been recorded. Temperatures in the mountains decrease three to five degrees Fahrenheit with each 1,000 feet increase in elevation. The average date of the last freezing temperature in the spring is the last of April in the warmer valleys and the last of June in the colder localities. In the fall, the first freezing temperatures can be expected in the colder valleys by the first of September and before mid-October in the warmer areas.

PALOUSE - BLUE MOUNTAINS.--This area includes counties along the eastern border of the

State south from Spokane to the Oregon border and west to near Walla Walla. The elevation increases from 1,000 feet in the vicinity of Walla Walla to 3,500 feet in the Palouse Hills and to 6,000 feet in the Blue Mountains. Precipitation increases as the elevation increases in an easterly direction across this area. Annual precipitation is between 10 to 20 inches over most of the agricultural section increasing to 40 inches or more in the higher elevations of the Blue Mountains. The average winter season snowfall varies from 20 to 40 inches. Snow can be expected in November and to remain on the ground for periods ranging from a few days to two months between the first of December and March. Snowfall and the depth on the ground increase along the slopes of the mountains.

The average January maximum temperature is near 34° F. in the Palouse Hills and 38° in the Snake and Walla Walla River valleys. The average minimum temperature varies from 20° to 25°. Minimum temperatures between 0° to -15° F. are recorded on a few nights each winter and temperatures ranging from -25° to -35° F. have been recorded. In July, the average maximum temperature is in the upper 80's and the minimum is in the mid-50's. Maximum temperatures usually reach 100° F. on a few afternoons and temperatures from 105° to 112° F. have been recorded.

The last freezing temperature in the spring is the last of April in the Walla Walla and Snake River valleys and the last of May in the Palouse Hills. The first freezing temperatures usually occur the last of September or first of October.

RIVERS.--The Columbia River, draining approximately 259,000 square miles in the Pacific Northwest and second only to the Mississippi River in volume flow, enters near the northeastern corner of the State and flows in a semi-circular pattern through eastern Washington. Before reaching the Pacific Ocean, it forms most of the boundary between Washington and Oregon, draining all of eastern Washington and the western slope of the Cascade Mountains between Mt. Rainier and the southern border. In addition to providing water for vast irrigation and hydroelectric projects, the Columbia River is a navigable stream for ocean vessels to ports at Vancouver and Portland and for river barges into eastern Washington. Principal tributaries of the Columbia in Washington include the Pend Oreille, Spokane, Snake and Cowlitz Rivers.

Although some overflow may be expected in Washington in most years, severe flooding occurs infrequently. In recent years, the most severe flooding in the Columbia River basin occurred in 1948 and 1950, while some of the other notable flood years have been 1894, 1897, 1913, 1916, 1928, 1933.

In the Columbia River basin in eastern Washington, winter floods are rare. They may occur at times, however, especially in local areas as a result of a combination of moderate snow cover, warm southerly winds and heavy rains. Annual peak flows occur in the spring and early summer as the winter snow pack melts.

In western Washington, the Snoqualmie, Skagit, Stillaguamish, Chehalis and other streams drain into Puget Sound, the Strait of Juan de Fuca and the Pacific Ocean. There are two periods of high flow in the streams of western Washington especially in the Puget Sound region and in the Cowlitz River basin. One occurs during the winter months, coinciding with the period of maximum precipitation and the other in the spring or early summer caused by the seasonal rise in temperature with the resultant melting of snow accumulations in the higher elevations augmented at times by rainfall. In western Washington, some of the most significant overflows occurred in 1909, 1917, 1921, 1932, 1933, 1934, 1951 and 1959.

FORESTRY AND AGRICULTURE.--Land utilization is determined to a large extent by the terrain, soil and the climate. The mountainous areas over the entire State and a major portion of the lowlands west of the Cascades are in timber. Forest vegetation varies from the large Douglas fir, spruce, hemlock and cedar with a dense undergrowth of fern and moss in the rainforest on the Olympic Peninsula to the open stands of Ponderosa pine in eastern Washington. Lumbering and forestry management are major activities in many areas.

West of the Cascades, agriculture is confined to the river valleys and well drained areas in the Puget Sound lowlands. The climate is favorable for growing berry crops, cool season vegetable crops, flower bulbs, certified seed potatoes and grass. Dairying and poultry production are important sources of income to the Puget Sound area. Reservoirs on the windward slopes of the mountains provide an abundance of water for metropolitan areas and hydroelectric projects have been developed along several rivers.

The major agricultural areas are in eastern Washington. Agriculture is highly specialized in some localities and diversified in others. The fruit producing areas are in irrigated valleys along the Okanogan, Columbia, Wenatchee and Yakima Rivers. The Okanogan highlands, northeastern valleys and channeled scablands are devoted to grazing. The major wheat producing areas include the Big Bend, Waterville Plateau, Palouse Hills and Horse Heaven Hills. Dryland farming practices are generally followed in the small grain section. In addition to the older irrigated sections of the Yakima and Walla Walla valleys, a major irrigation project has been developed in the Central Basin. The more important crops grown in the irrigated sections include sugar beets, potatoes, alfalfa, corn, onions, beans, peppermint, spearmint, hops and a variety of vegetable crops.

Ordinarily, drouth is not a problem in Washington agriculture. The dry season begins at approximately the same time each summer and agricultural activities are planned accordingly.

RECREATION.--Tourist business and recreational activities are rapidly becoming an important source of income. The climate, mountains, ocean beaches, lakes, rivers, national parks and forest areas permit a vast range of recreational activities. In the mountains the ski season begins in

November and continues until late spring. The season for camping, hiking and fishing in the higher mountain lakes and streams begins as the snow melts and continues until early fall.

In the fall, hunters flock to the mountains seeking their limit of deer, elk and other game, while those looking for birds scatter over the lowlands. In summer, the numerous lakes and warm sunny days east of the Cascades are inviting to many, while to others, the cooler marine air and ocean beaches in western Washington are a welcome relief from summer heat in other sections of the country.

REFERENCES

(1)   Atlas of Washington Agriculture 1963, U. S. Department of Agriculture and Washington State Department of Agriculture

(2)   Surface Water Temperatures at Tide Station, Pacific Coast (1952), U. S. Coast & Geodetic Survey

(3)   Weather Bureau Technical Paper No. 13, Mean monthly and annual Evaporation from Free Water Surfaces for the United States and the West Indies. Washington, D. C. 1950

(4)   Weather Bureau Technical Paper No. 16, Maximum 24-hour Precipitation in the United States (1952)

(5)   Weather Bureau Technical Paper No. 28, Rainfall intensities for Local Drainage Design in Western United States (1956)

(6)   Weather Bureau Technical Paper No. 40, Rainfall Frequency Atlas of the United States (1961)

(7)   Weather Bureau Technical Paper No. 25, Rainfall - Intensity - Duration - Frequency Curves for Selected Stations in the United States, Alaska, Hawaiian Islands, and Puerto Rico (1955)

(8)   Weather Bureau Technical Paper No. 37, Evaporation Maps for the United States (1959)

(9)   Washington State Freeze Circular No. 400, U. S. Weather Bureau and Washington State University, Washington Agricultural Experiment Station (1962)

(10)   Distribution of Extreme Winds in the United States, H.C.S. Thom, published Journal of the Structural Division Proceedings of the American Society of Civil Engineers, April 1960

(11)   Summary of Snow Survey Measurements in the State of Washington 1915-1960, U. S. Soil Conservation Service and State of Washington, Department of Conservation, Division of Water Resources.

BIBLIOGRAPHY

(A)   Climatic Summary of the United States (Bulletin W) 1930 edition, Sections 1 and 2, U. S. Weather Bureau

(B)   Climatic Summary of the United States, Washington--Supplement for 1931-through 1952 (Bulletin W Supplement), U. S. Weather Bureau

(C)   Climatological Data - Washington, U. S. Weather Bureau

(D)   Climatological Data National Summary, U. S. Weather Bureau

(E)   Hourly Precipitation Data - Washington, U. S. Weather Bureau

(F)   Precipitation Data from Storage-Gage Stations, U. S. Weather Bureau

(G)   Local Climatological Data, U. S. Weather Bureau for Olympia, Seattle, Spokane, Stampede Pass, Tatoosh Island, Walla Walla and Yakima

(H)   Summary of Hourly Observations Spokane, Seattle-Tacoma Airport and Yakima

(I)   Climatic Guide for Seattle, Washington and Adjacent Puget Sound Area, U. S. Weather Bureau (1961).

STATION	Freeze threshold temperature	Mean date of last Spring occurrence	Mean date of first Fall occurrence	Mean No. of days between dates	Years of record Spring	No. of occurrences in Spring	Years of record Fall	No. of occurrences in Fall	STATION	Freeze threshold temperature	Mean date of last Spring occurrence	Mean date of first Fall occurrence	Mean No. of days between dates	Years of record Spring	No. of occurrences in Spring	Years of record Fall	No. of occurrences in Fall
ABERDEEN	32	04-16	10-30	197	30	30	30	30	CENTRALIA	32	04-24	10-23	182	30	30	30	30
	28	03-09	11-29	265	30	30	30	24		28	03-24	11-14	235	30	30	30	28
	24	01-28	12-18	324	30	21	30	15		24	02-10	12-05	298	30	25	30	23
	20	01-14	12-26	346	29	15	30	6		20	01-22	12-18	330	30	20	30	15
	16	01-08	●	●	29	10	30	2		16	01-16	12-25	343	30	17	30	7
ANACORTES	32	03-12	11-19	252	28	28	30	29	CHEWELAH 2 S	32	06-02	08-25	84	29	29	30	30
	28	02-08	12-09	304	28	21	30	22		28	05-03	09-19	139	29	29	30	30
	24	01-23	12-20	331	28	18	30	12		24	04-04	10-02	181	29	29	30	30
	20	01-18	12-28	344	28	17	30	4		20	03-16	10-27	225	29	29	30	30
	16	01-11	●	●	28	10	30	3		16	03-01	11-13	257	29	28	30	30
ANATONE	32	05-23	09-27	127	25	25	26	26	CHIEF JOSEPH DAM	32	04-29	10-13	167	10	10	10	10
	28	05-03	10-11	161	25	25	25	25		28	04-07	10-30	206	10	10	10	10
	24	03-31	10-29	212	24	24	25	25		24	03-24	11-07	228	10	10	10	10
	20	03-14	11-13	244	24	22	24	24		20	03-06	11-19	258	10	10	10	10
	16	03-04	11-25	266	24	22	23	22		16	02-16	12-05	292	10	8	10	8
BATTLE GROUND	32	05-05	10-11	159	19	19	20	20	CLALLAM BAY 1 NNE	32	04-25	10-30	188	10	10	10	10
	28	03-29	11-04	220	19	19	20	20		28	03-10	12-12	277	10	9	10	5
	24	02-19	11-26	280	19	18	20	17		24	02-10	12-20	313	10	9	10	4
	20	01-30	12-19	323	18	15	20	11		20	01-14	●	●	10	4	10	1
	16	01-15	12-27	346	18	11	20	3		16	●	●	●	10	1	10	1
BELLINGHAM 2 N	32	05-10	10-05	148	10	10	10	10	CLARKSTON HEIGHTS	32	05-01	10-06	158	21	21	23	23
	28	04-13	10-28	198	10	10	10	10		28	04-04	10-24	203	21	21	23	23
	24	03-20	11-16	241	10	10	10	9		24	03-15	11-06	236	21	21	23	23
	20	02-20	12-05	288	10	8	10	8		20	02-26	11-27	274	21	21	23	20
	16	01-27	12-15	322	10	8	10	5		16	02-11	12-12	304	21	18	23	15
BELLINGHAM FAA AP	32	05-08	10-05	150	30	30	30	30	CLEARBROOK	32	05-06	10-05	152	30	30	30	30
	28	04-06	10-28	205	30	30	30	30		28	04-05	10-31	209	30	30	30	30
	24	03-07	11-19	257	30	30	30	28		24	02-21	11-21	273	30	25	30	27
	20	01-30	12-09	313	30	25	30	21		20	01-27	12-08	315	30	23	30	21
	16	01-22	12-15	327	30	19	30	16		16	01-20	12-22	336	30	19	30	10
BENTON CITY 2 NW	32	04-22	10-17	178	19	19	21	21	CLEARWATER	32	05-02	10-11	162	14	14	15	15
	28	04-06	10-29	206	19	19	20	20		28	03-23	11-08	230	13	13	15	14
	24	03-20	11-14	239	19	19	20	20		24	02-13	12-01	291	13	10	14	11
	20	03-02	11-24	267	19	19	20	19		20	01-30	12-26	330	13	9	14	4
	16	02-04	12-10	309	17	14	18	13		16	01-14	12-26	346	13	5	13	2
BICKLETON	32	05-23	10-10	140	27	27	25	25	CLE ELUM	32	05-24	09-07	106	27	27	27	27
	28	04-27	10-28	184	27	27	25	25		28	05-05	10-02	150	27	27	27	27
	24	04-04	11-12	222	27	27	25	25		24	04-11	10-25	197	27	27	27	27
	20	03-13	11-21	253	26	25	25	25		20	03-25	11-11	231	26	26	26	25
	16	02-19	12-07	291	24	22	25	19		16	03-06	11-25	264	26	26	25	23
BLAINE 1 E	32	04-21	10-21	183	29	29	30	30	COLFAX 1 NW	32	05-17	09-17	123	30	30	30	30
	28	03-23	11-10	232	29	29	30	29		28	04-21	10-06	168	30	30	30	30
	24	02-17	11-30	286	29	24	30	27		24	03-29	10-25	210	30	30	30	30
	20	01-23	12-14	325	29	20	30	19		20	03-11	11-07	241	30	29	30	29
	16	01-21	12-27	340	29	18	30	8		16	02-19	12-02	286	30	26	30	25
BOTHELL	32	05-12	10-05	146	27	27	26	26	COLVILLE AP	32	05-20	09-23	126	30	30	30	30
	28	04-10	10-31	204	27	27	26	26		28	04-28	10-05	160	30	30	30	30
	24	03-02	11-17	260	27	25	26	23		24	04-04	10-22	201	30	30	30	30
	20	02-02	12-10	311	27	20	26	16		20	03-15	11-08	238	30	30	30	30
	16	01-20	12-21	335	27	17	26	10		16	02-27	11-26	272	30	29	30	29
BUCKLEY 1 NE	32	04-15	10-27	195	29	29	30	30	CONCONULLY	32	05-20	09-27	130	29	29	30	30
	28	03-14	11-14	245	29	29	30	29		28	04-22	10-14	175	29	29	30	30
	24	02-08	12-07	302	29	23	30	22		24	04-07	10-30	206	29	29	30	30
	20	01-20	12-21	335	29	18	30	15		20	03-22	11-12	235	29	29	30	30
	16	01-16	12-27	345	29	17	30	5		16	03-14	11-23	254	29	29	30	29
BUMPING LAKE	32	06-16	08-14	59	30	30	30	30	CONCRETE	32	04-05	11-10	219	29	29	30	30
	28	06-01	09-09	100	30	30	30	30		28	03-03	11-29	271	29	27	30	27
	24	05-03	10-13	163	30	30	30	30		24	02-06	12-11	308	29	22	30	19
	20	04-12	11-06	208	30	30	30	30		20	01-19	12-21	336	27	17	28	10
	16	04-01	11-21	234	30	30	29	28		16	01-16	12-26	344	27	14	28	4
CEDAR LAKE	32	04-20	10-30	193	29	29	30	30	COULEE DAM 1 SW	32	04-23	10-23	183	10	10	10	10
	28	03-19	11-19	245	29	27	30	29		28	04-06	11-03	211	10	10	10	10
	24	02-18	12-08	293	29	25	30	22		24	03-15	11-17	247	10	10	10	10
	20	02-01	12-18	320	29	20	30	14		20	02-25	12-01	279	10	10	10	9
	16	01-20	12-27	341	29	18	30	5		16	02-16	12-07	294	10	8	10	7

STATION	Freeze threshold temperature	Mean date of last Spring occurrence	Mean date of first Fall occurrence	Mean No. of days between dates	Years of record Spring	No. of occurrences in Spring	Years of record Fall	No. of occurrences in Fall	STATION	Freeze threshold temperature	Mean date of last Spring occurrence	Mean date of first Fall occurrence	Mean No. of days between dates	Years of record Spring	No. of occurrences in Spring	Years of record Fall	No. of occurrences in Fall
COUPEVILLE 1 S	32	04-09	11-02	207	10	10	10	10	EPHRATA FAA AP	32	04-22	10-25	186	10	10	10	10
	28	03-14	11-16	247	10	10	10	9		28	04-12	11-02	204	10	10	10	10
	24	02-01	12-06	308	10	8	10	8		24	03-13	11-13	245	10	10	10	10
	20	02-01	12-17	319	10	8	10	5		20	02-25	11-28	276	10	10	10	9
	16	01-22	12-24	336	10	6	10	3		16	02-15	12-03	291	10	9	10	8
CUSHMAN DAM	32	04-18	11-02	198	10	10	10	10	EVERETT	32	04-05	10-31	209	30	30	30	30
	28	03-13	11-23	255	10	10	10	9		28	03-10	11-19	254	30	30	30	27
	24	02-17	12-10	296	10	9	10	7		24	01-30	12-10	314	29	21	30	20
	20	01-29	12-24	329	10	7	10	3		20	01-18	12-20	336	29	17	30	12
	16	01-23	⊕	⊕	10	7	10	1		16	01-16	12-26	344	29	16	30	5
DALLESPORT 9 N	32	04-15	10-22	190	10	10	10	10	FORKS 1 E	32	04-24	10-22	181	30	30	30	30
	28	03-31	11-04	218	10	10	10	10		28	03-20	11-20	245	30	30	29	27
	24	02-28	11-12	257	10	9	10	10		24	02-11	12-13	305	30	23	29	19
	20	02-10	12-01	294	10	8	10	8		20	01-23	12-28	339	30	18	29	6
	16	02-01	12-21	323	10	8	10	3		16	01-13	⊕	⊕	30	14	29	2
DARRINGTON RANGER STA	32	05-11	10-12	154	28	28	29	29	FRIDAY HARBOR	32	04-09	10-30	204	13	13	12	12
	28	04-04	11-03	213	27	27	28	28		28	03-07	11-20	258	13	11	12	11
	24	03-10	11-21	256	25	25	28	27		24	02-04	12-09	308	11	8	11	6
	20	02-13	12-05	295	23	20	25	19		20	01-21	12-24	337	11	7	11	2
	16	01-31	12-11	314	23	18	25	16		16	01-16	⊕	⊕	11	5	11	1
DAVENPORT	32	05-24	09-24	123	29	29	30	30	GOLDENDALE	32	05-25	09-23	121	30	30	29	29
	28	04-30	10-12	165	29	29	30	30		28	04-29	10-17	171	30	30	29	29
	24	03-31	10-29	212	29	29	30	30		24	03-30	10-31	215	30	30	29	29
	20	03-10	11-09	244	29	29	30	30		20	03-09	11-19	255	29	28	29	27
	16	02-24	11-24	273	29	25	30	30		16	02-11	12-08	300	27	23	29	20
DAYTON 1 WSW	32	05-03	10-07	157	30	30	30	30	GRAPEVIEW	32	03-31	11-15	229	29	29	29	29
	28	04-06	10-27	204	30	30	30	30		28	02-18	12-12	297	29	26	29	22
	24	03-13	11-11	243	30	30	30	29		24	01-18	12-26	342	29	17	29	7
	20	02-24	11-28	277	30	28	30	26		20	01-13	⊕	⊕	29	15	29	2
	16	02-10	12-07	300	30	27	30	21		16	01-06	⊕	⊕	29	7	29	2
DEER PARK 2 E	32	06-03	08-22	80	30	30	29	29	HARTLINE	32	05-11	10-06	148	26	26	30	30
	28	05-17	09-22	128	30	30	29	29		28	04-14	10-19	188	26	26	30	30
	24	04-20	10-12	175	30	30	29	29		24	03-25	11-04	224	25	25	30	30
	20	03-30	10-27	211	29	29	28	28		20	03-01	11-20	264	24	24	29	29
	16	03-13	11-15	247	27	27	27	27		16	02-17	12-01	287	24	23	29	27
DIABLO DAM	32	04-13	11-03	204	26	26	27	27	HATTON 8 E	32	05-20	09-29	132	29	29	30	30
	28	03-18	11-20	247	26	26	27	26		28	04-28	10-14	169	29	29	30	30
	24	02-22	12-06	287	26	25	27	21		24	04-03	10-28	208	29	29	30	30
	20	02-07	12-15	311	26	19	27	15		20	03-04	11-16	257	29	28	30	29
	16	01-29	12-22	327	26	18	27	10		16	02-13	11-29	289	28	25	29	24
ELECTRON HEADWORKS	32	05-24	09-12	111	16	16	17	17	KENNEWICK	32	04-13	10-17	187	30	30	30	30
	28	04-28	10-22	177	16	16	17	17		28	03-31	10-31	214	30	30	30	30
	24	03-20	11-14	239	16	16	17	15		24	03-15	11-17	247	30	30	30	28
	20	02-22	12-07	288	16	15	17	12		20	02-24	11-28	277	30	28	30	24
	16	02-03	12-23	323	16	12	17	7		16	02-05	12-10	308	29	23	30	18
ELLENSBURG	32	05-04	09-28	147	26	26	27	27	KENNEWICK 10 SW	32	04-20	10-20	183	10	10	10	10
	28	04-19	10-15	179	26	26	27	27		28	04-08	11-06	212	10	10	10	10
	24	04-01	10-28	210	26	26	27	27		24	03-13	11-21	253	10	10	10	10
	20	03-19	11-14	240	26	26	27	27		20	02-18	11-28	283	10	9	10	8
	16	02-20	11-29	282	26	24	27	24		16	02-01	12-06	308	10	8	10	7
ELMA	32	05-08	10-15	160	10	10	10	10	KENT	32	04-25	10-21	179	10	10	10	10
	28	04-02	11-07	219	10	10	10	10		28	03-26	11-08	227	10	10	10	10
	24	02-11	12-07	299	10	8	10	6		24	02-19	11-26	280	10	9	10	9
	20	01-29	12-19	324	10	8	10	4		20	01-31	12-14	317	10	8	10	7
	16	01-18	⊕	⊕	10	6	10	1		16	01-20	12-21	335	10	7	10	3
ELWHA R S	32	04-18	11-06	202	10	10	10	10	KID VALLEY	32	05-05	10-22	170	19	19	20	20
	28	03-06	11-28	267	10	9	10	9		28	03-30	11-10	225	19	19	20	19
	24	02-09	12-15	309	10	8	10	5		24	02-28	12-02	277	18	18	20	16
	20	01-23	12-19	330	10	7	10	4		20	02-05	12-20	318	17	14	20	12
	16	01-17	12-24	341	10	5	10	3		16	01-21	12-26	339	17	12	20	4
EPHRATA	32	04-13	10-16	186	29	29	30	30	KOSMOS	32	05-11	09-28	140	10	10	10	10
	28	03-29	11-02	218	29	29	30	30		28	04-19	10-25	189	10	10	10	10
	24	03-05	11-17	257	29	29	30	30		24	03-19	11-16	242	10	10	10	9
	20	02-22	11-29	280	29	28	30	28		20	02-03	12-06	306	10	8	10	7
	16	02-09	12-06	300	29	24	30	24		16	02-03	12-20	320	10	8	10	4

STATION	Freeze threshold temperature	Mean date of last Spring occurrence	Mean date of first Fall occurrence	Mean No. of days between dates	Years of record Spring	No. of occurrences in Spring	Years of record Fall	No. of occurrences in Fall
LACROSSE 3 ESE	32	05-24	09-10	109	30	30	30	30
	28	05-09	10-01	145	30	30	30	30
	24	04-03	10-18	198	30	30	30	30
	20	03-10	11-02	237	30	30	30	29
	16	02-17	11-17	273	30	25	30	27
LAKE CLE ELUM	32	05-16	09-14	121	10	10	10	10
	28	05-10	10-08	151	10	10	10	10
	24	04-13	10-27	197	10	10	10	10
	20	03-31	11-10	224	10	10	10	10
	16	03-09	11-21	257	10	10	10	9
LAKE KACHESS	32	05-27	09-09	105	10	10	10	10
	28	05-05	10-20	168	10	10	10	10
	24	04-10	11-01	205	10	10	10	10
	20	03-26	11-11	230	10	10	10	10
	16	03-11	11-27	261	10	10	10	9
LAKE KEECHELUS	32	05-30	09-14	107	29	29	30	30
	28	05-08	10-11	156	29	29	30	30
	24	04-13	11-06	207	29	29	30	30
	20	03-31	11-23	237	29	29	30	28
	16	03-23	12-06	258	29	29	30	23
LAKESIDE (Chelan)	32	04-09	10-29	203	27	27	27	27
	28	03-24	11-12	233	27	27	27	27
	24	03-03	11-28	270	27	27	27	25
	20	02-17	12-06	292	27	26	27	24
	16	02-05	12-15	313	27	22	27	17
LANDSBURG	32	05-12	10-14	155	29	29	30	30
	28	04-03	11-08	219	29	29	30	30
	24	02-22	11-30	281	29	26	30	25
	20	01-28	12-16	322	29	20	30	18
	16	01-19	12-23	338	29	18	30	8
LAURIER	32	05-21	09-20	122	28	28	28	28
	28	04-30	10-03	156	28	28	28	28
	24	04-09	10-26	200	27	27	28	28
	20	03-26	11-07	226	27	27	28	28
	16	03-12	11-24	257	27	27	28	28
LEAVENWORTH 3 S	32	05-14	09-30	139	29	29	30	30
	28	04-21	10-15	177	29	29	29	29
	24	03-27	11-04	222	29	29	29	28
	20	03-14	11-20	251	27	27	26	26
	16	02-27	12-02	278	26	26	25	24
LIND 3 NE	32	05-14	10-02	141	30	30	29	29
	28	04-18	10-17	182	30	30	29	29
	24	03-27	11-02	220	30	30	29	29
	20	03-05	11-17	257	30	29	29	28
	16	02-14	11-30	289	30	26	29	25
LONGVIEW	32	04-26	10-25	182	30	30	30	30
	28	03-25	11-11	231	30	30	30	29
	24	02-16	12-09	296	30	27	30	21
	20	01-26	12-20	328	30	22	30	11
	16	01-14	12-22	342	30	16	30	9
MARIETTA 3 NNW	32	04-25	10-13	171	26	26	26	26
	28	03-24	11-03	224	26	26	26	26
	24	02-23	11-22	272	26	24	26	25
	20	01-25	12-13	322	26	17	26	17
	16	01-24	12-25	335	26	17	26	8
MC MILLIN RESERVOIR	32	05-06	10-19	166	10	10	10	10
	28	04-07	11-07	214	10	10	10	10
	24	02-23	11-28	278	10	10	10	9
	20	02-01	12-16	318	10	8	10	5
	16	01-29	12-24	329	10	8	10	2
METALINE FALLS	32	05-14	09-21	130	19	19	20	20
	28	04-28	10-17	172	19	19	20	20
	24	04-05	10-27	205	19	19	20	20
	20	03-24	11-14	235	19	19	20	20
	16	03-09	11-24	260	19	19	20	19
MONROE	32	04-18	10-31	196	10	10	10	10
	28	03-14	11-19	250	10	10	10	9
	24	02-16	11-28	285	10	9	10	9
	20	01-28	12-12	318	10	8	10	6
	16	01-20	12-24	338	10	7	10	3
MOSES LAKE 3 E	32	05-15	10-01	139	12	12	13	13
	28	04-19	10-09	173	12	12	13	13
	24	04-08	10-24	199	12	12	13	13
	20	03-18	11-15	242	12	12	13	13
	16	02-21	11-22	274	12	12	13	12
MOUNT ADAMS RANGER STA	32	06-05	09-07	94	21	21	22	22
	28	05-15	09-29	137	21	21	22	22
	24	04-12	10-29	200	21	21	22	22
	20	03-17	11-12	240	20	20	22	22
	16	02-25	12-07	285	19	18	22	20
MOUNT BAKER LODGE	32	05-21	10-06	138	17	17	13	13
	28	05-01	10-29	181	16	16	13	13
	24	04-15	11-06	205	15	15	13	13
	20	04-01	11-16	229	15	15	13	13
	16	03-05	12-01	271	15	14	13	13
MOXEE CITY 10 E	32	05-19	09-15	119	10	10	10	10
	28	05-02	10-08	159	10	10	10	10
	24	04-14	10-30	199	10	10	10	10
	20	03-17	11-10	238	10	10	10	10
	16	02-18	11-23	278	10	10	10	9
MUD MOUNTAIN DAM	32	04-26	10-29	186	10	10	10	10
	28	03-24	11-24	245	10	10	10	9
	24	03-02	12-06	279	10	9	10	7
	20	01-30	12-24	328	10	8	10	3
	16	01-21	12-24	337	10	7	10	2
NESPELEM 2 S	32	05-25	09-21	119	27	27	26	26
	28	05-06	10-06	153	27	27	26	26
	24	04-09	10-27	201	27	27	26	26
	20	03-16	11-07	236	26	26	26	26
	16	02-28	11-22	267	26	26	26	24
NEWPORT	32	06-04	08-22	79	10	10	10	10
	28	05-16	09-22	129	10	10	10	10
	24	04-29	10-17	171	10	10	10	10
	20	04-07	11-03	210	10	10	10	10
	16	03-22	11-14	237	10	10	10	10
NORTH HEAD WB	32	02-03	12-20	320	23	16	22	13
	28	01-22	12-27	339	23	14	22	4
	24	01-10	⊕	⊕	23	10	22	1
	20	01-05	⊕	⊕	23	4	22	1
	16	⊕	⊕	⊕	23	2	22	0
OAKVILLE	32	05-04	10-14	163	28	28	29	29
	28	03-26	11-08	227	28	28	28	27
	24	02-22	11-29	280	28	24	28	22
	20	01-29	12-18	323	28	19	28	12
	16	01-15	12-24	343	28	15	28	6
ODESSA	32	05-16	09-25	132	30	30	30	30
	28	04-26	10-10	167	30	30	30	30
	24	04-02	10-28	209	30	30	30	30
	20	03-15	11-10	240	30	30	30	30
	16	02-19	11-23	277	30	28	30	30
OLGA 2 SE	32	03-24	11-15	236	30	30	30	30
	28	02-10	12-06	299	30	24	30	24
	24	01-26	12-20	328	30	22	30	12
	20	01-22	12-24	336	30	20	30	7
	16	01-14	⊕	⊕	30	15	30	3
OLYMPIA PRIEST PT PK	32	04-18	11-05	201	25	25	25	25
	28	03-08	11-23	260	25	24	24	22
	24	01-31	12-15	318	25	18	24	16
	20	01-16	12-21	339	25	14	24	9
	16	01-13	12-28	349	25	13	24	3

STATION	Freeze threshold temperature	Mean date of last Spring occurrence	Mean date of first Fall occurrence	Mean No. of days between dates	Years of record Spring	No. of occurrences in Spring	Years of record Fall	No. of occurrences in Fall
OLYMPIA WB AIRPORT	32	05-08	10-15	160	10	10	10	10
	28	04-09	11-05	210	10	10	10	10
	24	03-02	11-24	267	10	9	10	8
	20	02-12	12-13	304	10	9	10	6
	16	01-26	12-23	331	10	8	10	3
OROVILLE 1 S	32	04-22	10-14	175	30	30	29	29
	28	04-02	10-27	208	30	30	29	29
	24	03-15	11-14	244	30	30	29	29
	20	03-02	11-30	273	30	30	29	26
	16	02-15	12-07	295	29	26	29	25
OTHELLO	32	05-06	10-06	153	12	12	14	14
	28	04-15	10-22	190	12	12	14	14
	24	04-06	10-31	208	12	12	14	14
	20	03-09	11-13	249	12	12	13	13
	16	02-16	11-21	278	12	11	13	13
PALMER 3 SE	32	04-14	11-11	211	10	10	10	10
	28	03-06	11-29	268	10	9	10	8
	24	02-05	12-19	317	10	8	10	4
	20	01-30	12-24	328	10	8	10	3
	16	01-20	⊕	⊕	10	7	10	1
PARKWAY 6 S CRYSTAL	32	06-04	09-10	98	10	10	10	10
	28	05-11	10-16	158	10	10	10	10
	24	04-23	11-01	192	10	10	10	10
	20	03-29	11-14	230	10	10	10	10
	16	03-11	12-06	270	10	10	10	8
POMEROY	32	05-10	09-30	143	29	29	30	30
	28	04-15	10-17	185	29	29	30	30
	24	03-21	11-03	227	29	29	30	30
	20	03-08	11-23	260	29	28	30	27
	16	02-13	12-04	294	29	24	30	23
PORT ANGELES	32	03-26	11-19	238	29	29	29	28
	28	02-16	12-11	298	29	24	29	21
	24	01-23	12-22	333	29	19	29	8
	20	01-18	12-28	344	29	16	29	4
	16	01-11	⊕	⊕	29	11	29	1
PORT TOWNSEND	32	03-11	11-20	254	29	28	28	28
	28	02-06	12-10	307	29	21	28	18
	24	01-18	12-25	341	28	16	28	9
	20	01-15	12-27	346	28	14	28	5
	16	01-10	⊕	⊕	28	10	28	2
PROSSER 4 NE	32	05-08	10-16	161	10	10	10	10
	28	04-13	10-28	198	10	10	10	10
	24	03-25	11-10	230	10	10	10	10
	20	02-18	11-22	277	10	9	10	9
	16	02-13	12-06	296	10	9	10	8
PULLMAN 2 NW	32	04-26	10-10	167	26	26	27	27
	28	03-30	10-26	210	26	26	27	27
	24	03-09	11-09	245	26	25	27	27
	20	02-26	11-29	276	26	25	27	25
	16	02-20	12-06	289	26	23	27	21
PUYALLUP 2 W EXP STA	32	05-01	10-13	165	29	29	29	29
	28	03-30	11-04	219	29	29	29	29
	24	02-22	11-23	274	29	28	29	25
	20	01-30	12-13	317	28	21	29	18
	16	01-19	12-23	338	28	17	29	8
QUILCENE 2 SW	32	04-29	10-14	168	30	30	30	30
	28	03-26	11-08	227	30	30	30	30
	24	02-13	12-03	293	30	23	30	27
	20	01-22	12-16	328	30	19	30	17
	16	01-16	12-26	344	30	17	30	5
QUINAULT RANGER STATION	32	04-08	11-15	221	22	22	23	23
	28	02-23	12-10	290	22	17	21	14
	24	02-01	12-23	325	21	14	20	6
	20	01-20	12-27	341	20	10	19	2
	16	01-08	⊕	⊕	20	5	19	1
QUINCY 3 S	32	05-04	10-04	153	10	10	10	10
	28	04-14	10-24	193	10	10	10	10
	24	04-01	11-01	214	10	10	10	10
	20	03-13	11-14	246	10	10	10	10
	16	02-23	12-04	284	10	10	10	9
RAINIER LONGMIRE	32	05-23	10-06	136	10	10	10	10
	28	05-02	10-25	176	10	10	10	10
	24	04-08	11-17	223	10	10	10	10
	20	03-16	11-30	259	10	10	10	9
	16	02-23	12-09	289	10	9	10	7
REPUBLIC	32	06-02	09-06	96	27	27	27	27
	28	05-18	09-25	130	27	27	27	27
	24	04-26	10-04	161	24	24	27	27
	20	04-05	10-27	205	22	22	26	26
	16	03-23	11-05	227	22	22	25	25
RICHLAND	32	04-13	10-22	192	10	10	10	10
	28	03-25	10-31	220	10	10	10	10
	24	03-01	11-17	261	10	10	10	9
	20	02-10	11-27	290	10	9	10	8
	16	01-31	12-10	313	10	8	10	6
RIMROCK TIETON DAM	32	06-06	09-10	96	29	29	29	29
	28	05-15	10-07	145	29	29	29	29
	24	04-10	10-30	203	29	29	29	29
	20	03-25	11-13	233	29	29	29	29
	16	03-13	11-27	259	29	29	29	27
RITZVILLE	32	05-17	10-01	137	29	29	29	29
	28	04-29	10-19	173	29	29	29	29
	24	04-01	11-02	215	29	29	29	29
	20	03-05	11-19	259	29	28	28	27
	16	02-17	12-01	287	28	24	28	25
ROSALIA	32	05-09	10-01	145	30	30	30	30
	28	04-13	10-15	185	29	29	30	30
	24	03-22	11-03	226	29	29	30	30
	20	03-03	11-18	260	29	28	30	29
	16	02-20	11-29	282	29	26	29	26
RUFF 3 SW	32	05-27	09-27	123	21	21	22	22
	28	05-07	10-13	159	21	21	22	22
	24	04-06	10-30	207	21	21	22	22
	20	03-19	11-13	239	21	21	22	22
	16	02-13	12-03	293	21	19	21	18
SEATTLE TACOMA WB AP	32	04-06	11-04	212	10	10	10	10
	28	03-03	11-25	267	10	9	10	9
	24	02-08	12-20	315	10	8	10	4
	20	01-22	12-20	332	10	7	10	4
	16	01-19	12-26	341	10	7	10	2
SEATTLE U OF W	32	03-23	11-18	240	30	30	30	30
	28	02-04	12-13	312	30	23	30	21
	24	01-22	12-24	336	30	19	30	10
	20	01-15	12-28	347	30	15	30	5
	16	01-09	⊕	⊕	30	11	30	2
SEATTLE WB AIRPORT	32	03-23	11-11	233	10	10	10	10
	28	02-19	12-02	286	10	9	10	8
	24	01-26	12-16	324	10	8	10	5
	20	01-20	12-21	335	10	7	10	4
	16	01-17	⊕	⊕	10	6	10	1
SEATTLE WB CITY	32	02-23	11-30	280	30	29	30	25
	28	01-23	12-18	329	30	20	30	15
	24	01-16	12-27	345	30	17	30	6
	20	01-11	⊕	⊕	30	13	30	2
	16	01-05	⊕	⊕	30	6	30	1
SEDRO WOOLLEY 1 E	32	04-16	10-26	193	30	30	30	30
	28	03-17	11-14	242	30	30	30	29
	24	02-08	11-29	294	30	24	30	26
	20	01-23	12-14	325	30	20	30	18
	16	01-18	12-23	339	30	18	30	9

FREEZE DATA

STATION	Freeze threshold temperature	Mean date of last Spring occurrence	Mean date of first Fall occurrence	Mean No. of days between dates	Years of record Spring	No. of occurrences in Spring	Years of record Fall	No. of occurrences in Fall
SEQUIM	32	04-16	10-30	197	29	29	30	30
	28	03-20	11-27	252	29	29	30	29
	24	02-10	12-13	306	27	22	30	19
	20	01-23	12-23	334	27	17	30	9
	16	01-15	12-27	346	27	13	29	5
SHELTON	32	05-06	10-21	168	10	10	10	10
	28	04-02	11-19	231	10	10	10	9
	24	02-15	12-07	295	10	9	10	8
	20	01-30	12-20	324	10	8	10	4
	16	01-12	12-24	346	10	6	10	2
SNOQUALMIE FALLS	32	05-11	10-12	154	30	30	30	30
	28	04-08	11-06	212	30	30	30	30
	24	03-01	12-02	276	30	28	30	23
	20	02-05	12-18	316	30	23	29	16
	16	01-24	12-24	334	30	18	29	8
SPIRIT LAKE RS	32	06-07	10-04	119	18	18	19	19
	28	05-09	11-03	178	18	18	19	19
	24	04-03	11-27	238	18	17	19	19
	20	03-24	12-06	257	18	17	17	15
	16	03-12	12-17	280	18	17	16	8
SPOKANE WB AIRPORT	32	04-25	10-11	169	30	30	30	30
	28	04-10	10-24	197	30	30	30	30
	24	03-16	11-10	239	30	30	30	30
	20	02-23	11-22	272	30	27	30	28
	16	02-17	12-06	292	30	26	30	24
STAMPEDE PASS WB	32	06-06	09-27	113	10	10	10	10
	28	05-11	10-25	167	10	10	10	10
	24	04-14	11-13	213	10	10	10	10
	20	03-29	11-20	236	10	10	10	10
	16	03-04	12-01	272	10	9	10	8
STARTUP 1 E	32	04-24	10-25	184	29	29	29	29
	28	03-18	11-11	238	29	28	28	26
	24	02-08	12-05	300	29	24	28	20
	20	01-20	12-15	329	29	19	28	16
	16	01-18	12-27	343	28	16	28	6
STEHEKIN 3 NW	32	05-08	09-30	145	29	29	30	30
	28	04-17	10-18	184	29	29	30	30
	24	03-24	11-07	228	29	29	30	30
	20	03-10	11-27	262	27	27	28	25
	16	02-21	12-07	289	27	26	27	23
STOCKDILL RANCH	32	06-21	08-10	50	29	29	30	30
	28	05-30	09-12	105	28	28	30	30
	24	05-12	09-29	140	28	28	30	30
	20	04-12	10-22	193	28	28	29	29
	16	03-27	11-06	224	28	28	29	28
SUNNYSIDE	32	04-30	10-05	158	28	28	29	29
	28	04-10	10-25	198	28	28	29	29
	24	03-24	11-03	224	28	28	29	29
	20	02-27	11-22	268	28	26	29	27
	16	02-08	12-09	304	28	21	29	20
TATOOSH ISLAND WB	32	02-01	12-20	322	30	20	30	11
	28	01-20	12-28	342	30	16	30	4
	24	01-10	⊕	⊕	30	12	30	2
	20	01-04	⊕	⊕	30	4	30	1
	16	⊕	⊕	⊕	30	2	30	0
TIETON INTAKE	32	05-22	09-27	128	29	29	30	30
	28	05-01	10-18	170	29	29	28	28
	24	04-04	10-30	209	29	29	27	27
	20	03-21	11-13	237	28	28	26	26
	16	03-02	11-28	271	28	28	25	21
TRINIDAD	32	04-18	10-22	187	16	16	17	17
	28	03-26	11-05	224	16	16	17	17
	24	03-07	11-20	258	16	16	17	17
	20	02-21	12-04	286	16	15	17	16
	16	01-30	12-09	313	16	14	17	12

STATION	Freeze threshold temperature	Mean date of last Spring occurrence	Mean date of first Fall occurrence	Mean No. of days between dates	Years of record Spring	No. of occurrences in Spring	Years of record Fall	No. of occurrences in Fall
TRINIDAD 2 SSE	32	04-19	10-18	182	10	10	10	10
	28	04-11	11-02	205	10	10	10	10
	24	03-20	11-12	237	10	10	10	10
	20	02-21	12-02	284	10	10	10	8
	16	02-12	12-11	302	10	8	10	7
VANCOUVER	32	03-25	11-13	233	30	30	30	30
	28	02-13	12-04	294	30	28	30	24
	24	01-24	12-17	327	30	22	30	16
	20	01-18	12-26	342	30	18	30	5
	16	01-11	⊕	⊕	30	13	30	2
VASHON ISLAND	32	04-07	11-11	218	21	21	22	22
	28	02-19	12-12	296	21	19	22	15
	24	01-19	12-25	340	21	13	22	9
	20	01-16	12-28	346	20	10	20	3
	16	01-09	⊕	⊕	20	6	20	1
WALLA WALLA 3 W ENT LA	32	04-24	10-21	180	10	10	10	10
	28	04-02	10-27	208	10	10	10	10
	24	03-26	11-10	229	10	10	10	10
	20	02-12	11-26	287	10	9	10	8
	16	01-31	12-09	312	10	8	10	6
WALLA WALLA FAA AP	32	04-09	10-28	202	10	10	10	10
	28	03-21	11-07	231	10	10	10	10
	24	03-03	11-21	263	10	10	10	9
	20	02-12	11-29	290	10	8	10	8
	16	01-29	12-10	315	10	8	10	6
WALLA WALLA WB CITY	32	03-27	11-02	220	30	30	30	30
	28	03-12	11-15	248	30	30	30	30
	24	02-18	12-02	287	30	28	30	24
	20	02-06	12-10	307	30	26	30	19
	16	01-26	12-20	328	30	21	30	12
WAPATO	32	04-15	10-16	184	29	29	30	30
	28	04-03	10-29	209	29	29	30	30
	24	03-14	11-07	238	29	29	30	30
	20	02-23	11-27	277	29	28	30	26
	16	02-04	12-12	311	29	23	30	20
WATERVILLE	32	05-10	10-03	146	30	30	30	30
	28	04-18	10-16	181	30	30	30	30
	24	03-31	11-03	217	30	30	30	30
	20	03-19	11-17	243	29	29	30	30
	16	03-05	11-26	266	29	29	30	29
WENATCHEE	32	04-14	10-19	188	30	30	30	30
	28	03-28	11-04	221	30	30	30	30
	24	03-09	11-14	250	30	30	30	30
	20	02-19	12-01	285	30	30	30	28
	16	02-10	12-11	304	30	27	30	21
WHITE SALMON 4 NE	32	04-23	11-02	193	22	22	22	21
	28	03-29	11-16	232	22	22	22	21
	24	03-01	12-02	276	22	21	22	20
	20	02-11	12-12	304	21	18	20	14
	16	01-28	12-21	327	20	15	20	11
WHITE SWAN	32	05-14	10-03	142	26	26	26	26
	28	04-20	10-23	186	26	26	25	25
	24	03-11	11-02	216	26	26	25	25
	20	03-14	11-20	251	26	26	24	24
	16	02-13	11-30	290	25	21	23	19
WILBUR	32	05-28	09-16	111	30	30	30	30
	28	05-02	10-06	157	30	30	30	30
	24	04-10	10-25	198	29	29	30	30
	20	03-10	11-09	244	29	29	30	29
	16	02-24	11-21	270	29	26	29	28
WILLAPA HARBOR	32	04-17	11-04	201	19	19	19	19
	28	03-14	11-26	257	19	19	19	16
	24	02-05	12-21	319	19	16	19	9
	20	01-15	⊕	⊕	19	10	19	1
	16	01-07	⊕	⊕	19	6	19	1

FREEZE DATA

STATION	Freeze threshold temperature	Mean date of last Spring occurrence	Mean date of first Fall occurrence	Mean No. of days between dates	Years of record Spring	No. of occurrences in Spring	Years of record Fall	No. of occurrences in Fall	STATION	Freeze threshold temperature	Mean date of last Spring occurrence	Mean date of first Fall occurrence	Mean No. of days between dates	Years of record Spring	No. of occurrences in Spring	Years of record Fall	No. of occurrences in Fall
WILSON CREEK	32	05-20	09-27	130	20	20	22	22	WINTHROP 1 WSW	32	05-18	09-22	127	30	30	30	30
	28	04-28	10-09	164	20	20	22	22		28	05-01	10-02	154	30	30	30	30
	24	04-05	10-27	205	20	20	22	22		24	04-08	10-20	195	30	30	30	30
	20	03-13	11-06	238	20	20	22	22		20	03-27	11-01	219	30	30	30	30
	16	02-21	11-21	273	20	19	22	21		16	03-15	11-17	247	30	30	30	30
WIND RIVER	32	05-23	09-29	129	30	30	30	30	YAKIMA WB AIRPORT	32	04-21	10-15	177	30	30	30	30
	28	04-30	10-19	172	30	30	30	30		28	04-05	10-25	203	30	30	30	30
	24	03-22	11-13	236	30	29	30	29		24	03-15	11-09	239	30	29	30	30
	20	02-23	12-07	287	30	27	30	22		20	02-21	11-25	277	30	28	30	27
	16	02-01	12-18	320	30	23	29	13		16	02-06	12-07	304	30	25	30	25

Data in the above table are based on the period 1931-1960, or that portion of this period for which data are available.

⊕ When the frequency of occurrence in either spring or fall is one year in ten or less, mean dates are not given.

Means have been adjusted to take into account years of non-occurrence.

A freeze is a numerical substitute for the former term "killing frost" and is the occurrence of a minimum temperature at or below the threshold temperature of 32°, 28°, etc.

Freeze data tabulations in greater detail are available and can be reproduced at cost.

* NORMALS BY CLIMATOLOGICAL DIVISIONS

Taken from "Climatography of the United States No. 81-6, Decennial Census of U. S. Climate"

Station	TEMPERATURE (°F)													PRECIPITATION (In.)												
	Jan.	Feb.	Mar.	Apr.	May	June	July	Aug.	Sept.	Oct.	Nov.	Dec.	Annual	Jan.	Feb.	Mar.	Apr.	May	June	July	Aug.	Sept.	Oct.	Nov.	Dec.	Annual
WEST OLYMPIC-COASTAL 01																										
ABERDEEN	39.7	41.8	44.1	48.7	53.3	57.1	60.1	60.6	58.7	52.7	45.2	41.9	50.3	12.70	10.23	9.19	5.56	3.43	2.70	1.51	1.79	3.71	8.13	11.09	14.50	84.54
ABERDEEN 20 NNE	•	•	•	•	•	•	•	•	•	•	•	•	•	19.14	15.17	14.41	8.90	5.38	3.98	2.37	2.49	5.88	12.93	16.59	21.84	129.08
CLALLAM BAY 1 NNE	38.3	39.7	42.0	46.4	50.5	54.0	56.3	56.5	54.5	49.7	43.4	40.7	47.7	13.55	9.81	8.43	5.25	3.15	2.44	1.65	1.52	3.15	7.85	11.16	14.29	82.25
CUSHMAN DAM	37.1	39.6	42.8	49.0	55.6	59.8	64.7	64.4	60.4	52.6	43.6	39.7	50.8	16.65	12.31	10.41	6.07	3.33	2.44	1.28	1.38	3.74	9.73	14.73	18.18	100.25
FORKS 1 E	38.7	40.3	42.5	46.9	52.1	56.0	59.7	60.0	57.6	51.5	44.1	40.6	49.2	17.49	14.12	12.69	8.33	4.89	3.69	2.50	2.25	5.11	11.70	15.08	19.25	117.10
NASELLE	•	•	•	•	•	•	•	•	•	•	•	•	•	17.18	14.28	13.60	7.01	4.32	3.59	1.62	2.03	4.55	11.10	15.03	20.14	114.45
QUINAULT RIVER RS	•	•	•	•	•	•	•	•	•	•	•	•	•	19.95	15.94	14.22	9.18	5.90	4.31	2.60	2.79	6.00	13.27	17.52	22.75	134.43
SPRUCE	•	•	•	•	•	•	•	•	•	•	•	•	•	18.70	14.90	13.08	8.76	5.29	4.17	2.46	2.53	5.32	12.86	16.77	21.74	126.58
TATOOSH ISLAND WB	42.0	43.1	44.2	47.5	51.1	53.9	55.5	56.0	54.8	51.9	47.2	44.4	49.3	10.82	8.70	8.34	5.23	3.00	2.84	2.34	1.98	3.55	8.22	10.51	12.16	77.69
WILLAPA HARBOR	40.5	42.9	45.1	49.6	54.2	57.8	61.4	61.8	59.6	53.7	46.3	42.8	51.3	12.37	10.43	9.89	5.94	3.68	3.17	1.46	1.73	3.55	8.50	11.25	14.59	86.56
DIVISION	39.5	41.3	43.6	48.1	52.8	56.4	59.7	59.9	57.7	52.6	45.1	41.8	49.9	14.21	11.11	10.08	6.26	3.83	3.02	1.84	1.91	4.06	9.30	12.53	15.77	93.92
NORTHEAST OLYMPIC-SAN JUAN 02																										
ANACORTES	39.7	42.1	45.3	50.6	55.5	59.4	62.4	62.1	58.7	52.7	45.5	42.5	51.4	3.40	2.53	2.39	1.52	1.27	1.51	.78	.87	1.45	2.68	3.48	3.82	25.70
CHIMACUM 4 S	•	•	•	•	•	•	•	•	•	•	•	•	•	3.63	3.26	2.40	2.03	1.83	1.92	.76	.93	1.17	2.32	3.41	4.19	27.85
COUPEVILLE 1 S	38.3	40.4	43.7	48.7	53.5	57.5	60.9	60.9	57.0	50.5	43.7	40.8	49.7	2.01	1.59	1.62	1.12	1.25	1.31	.67	.69	1.19	1.73	2.25	2.30	17.73
OLGA 2 SE	38.6	40.7	43.7	48.6	53.3	57.0	59.6	59.7	56.7	50.9	44.2	41.2	49.5	4.04	2.94	2.51	1.65	1.35	1.51	.88	.95	1.61	3.04	3.91	4.39	28.78
PORT ANGELES	38.6	40.3	42.6	47.2	52.0	55.7	58.8	58.7	56.1	50.1	43.7	41.0	48.7	3.87	3.06	1.99	1.08	.89	.96	.48	.58	1.10	2.48	3.77	4.35	24.61
PORT TOWNSEND	39.4	41.5	44.5	49.4	54.4	58.1	61.3	61.3	57.9	51.6	45.1	42.0	50.5	2.20	1.66	1.57	1.12	1.36	1.42	.68	.73	1.11	1.68	2.41	2.40	18.34
SEQUIM	37.9	40.1	42.8	47.8	53.1	57.2	60.4	60.8	57.4	50.6	43.6	40.2	49.3	2.18	1.73	1.30	.93	.97	1.15	.47	.59	.95	1.57	2.31	2.66	16.81
DIVISION	38.8	40.9	43.7	48.6	53.5	57.3	60.3	60.4	57.1	51.4	44.3	41.3	49.8	3.09	2.35	1.92	1.23	1.16	1.28	.65	.73	1.25	2.30	3.14	3.47	22.57
PUGET SOUND LOWLANDS 03																										
BELLINGHAM 2 N	36.8	39.5	43.0	48.0	53.2	57.8	61.0	60.6	56.7	50.1	43.1	39.6	49.1	4.14	3.22	3.11	2.26	1.82	1.93	.99	1.10	1.98	3.64	4.51	4.89	33.59
BLAINE 1 ENE	36.7	39.2	43.1	48.4	54.3	59.0	62.4	61.5	56.9	50.1	42.7	39.4	49.5	5.72	4.30	3.68	2.52	2.03	1.97	1.16	1.22	2.24	4.54	5.33	6.17	40.88
BREMERTON	•	•	•	•	•	•	•	•	•	•	•	•	•	6.36	4.79	3.65	2.27	1.38	1.35	.58	.68	1.45	3.60	5.85	6.70	38.66
CENTRALIA	39.2	42.1	45.4	51.0	56.5	60.5	65.1	64.4	60.6	53.0	44.8	41.6	52.0	6.36	5.45	4.66	2.68	1.91	1.92	.73	1.02	1.86	4.50	6.69	7.75	45.53
CLEARBROOK	35.1	38.5	43.1	48.8	54.3	58.8	62.1	61.4	57.6	50.3	42.3	38.4	49.2	5.75	4.52	4.47	3.18	2.63	2.58	1.46	1.58	2.92	5.27	5.72	6.74	46.82
EVERETT	38.6	41.0	44.6	49.6	54.7	58.9	62.4	61.8	57.9	51.8	44.4	41.2	50.6	4.45	3.58	3.33	2.39	2.26	2.25	.93	1.12	1.98	3.54	4.55	4.86	35.24
MONROE	38.0	40.9	44.7	50.3	55.9	60.2	64.1	63.6	59.2	52.2	44.2	40.7	51.2	6.03	5.01	4.59	3.21	3.01	2.53	1.04	1.34	2.49	4.65	6.32	6.54	46.76
OLYMPIA WB AIRPORT	38.1	40.9	44.2	50.0	55.1	59.1	63.9	63.4	58.5	51.4	43.8	40.7	50.8	7.85	6.62	5.40	2.96	2.01	1.79	.76	.89	2.09	5.28	7.67	9.05	52.37
PUYALLUP 2 W EXP STA	38.6	41.3	44.5	50.0	55.5	59.8	63.7	63.1	58.8	51.9	43.8	40.8	51.0	5.63	4.66	4.14	2.64	2.02	1.81	.81	.96	2.03	3.95	5.45	6.40	40.50
SEATTLE-TACOMA WB AP	38.3	40.8	43.8	49.2	55.5	59.8	64.9	64.1	59.9	52.4	43.9	40.8	51.1	5.73	4.24	3.79	2.40	1.73	1.58	.81	.95	2.05	4.02	5.35	6.29	38.94
SEATTLE U OF W	40.1	42.6	45.9	51.3	57.0	61.4	65.7	65.2	61.1	54.1	46.0	42.7	52.8	5.02	3.93	3.28	2.16	1.84	1.62	.74	.75	1.72	3.42	5.01	5.47	34.96
SEATTLE WB AIRPORT	38.2	41.6	45.4	51.3	57.4	62.2	66.7	65.6	60.5	52.6	44.2	40.4	52.2	5.46	4.21	3.53	2.15	1.58	1.43	.66	.81	1.83	3.50	5.22	5.73	36.11
SEATTLE WB CITY	41.2	43.6	46.4	51.8	57.4	61.4	65.6	65.0	61.2	54.4	46.9	43.8	53.2	5.19	3.90	3.32	1.97	1.59	1.41	.63	.74	1.65	3.28	5.00	5.42	34.10
SEDRO WOOLLEY	38.0	41.0	44.7	50.1	55.4	59.3	62.5	62.3	58.4	51.8	44.0	40.4	50.7	5.57	4.33	4.65	3.30	2.56	2.78	1.33	1.38	3.01	4.91	5.87	6.38	46.07
DIVISION	38.3	40.9	44.4	49.8	55.4	59.7	63.6	63.0	58.9	52.0	44.2	41.0	50.9	5.59	4.44	3.96	2.54	2.00	1.89	.89	1.01	2.06	4.11	5.58	6.32	40.39
EAST OLYMPIC-CASCADE FOOTHILLS 04																										
BROOKLYN	•	•	•	•	•	•	•	•	•	•	•	•	•	11.14	9.47	8.59	5.11	3.18	2.53	.93	1.43	3.12	7.86	10.21	13.25	76.82
BUCKLEY 1 NE	37.8	40.4	43.8	49.0	54.6	58.7	63.1	62.6	58.3	50.8	43.2	40.3	50.2	5.59	4.71	4.94	3.86	3.13	3.36	1.25	1.41	2.77	5.15	6.23	6.91	49.31
CONCRETE 1 E	36.7	40.1	44.5	51.3	57.3	61.1	65.4	65.2	61.2	53.3	43.7	39.2	51.6	8.80	7.03	6.76	4.12	2.87	2.75	1.30	1.50	3.57	7.03	9.10	10.38	65.21
DARRINGTON RANGER STA	•	•	•	•	•	•	•	•	•	•	•	•	•	11.79	9.37	8.13	5.30	3.43	3.20	1.36	1.50	3.92	8.23	11.14	13.14	80.51
GRAPEVIEW	39.5	41.7	45.1	50.7	56.6	60.7	64.8	64.8	60.4	52.9	45.2	42.0	52.0	8.33	6.56	5.51	3.24	1.96	1.62	.75	.94	2.04	5.16	7.61	9.29	53.01
KALAMA 5 ENE	•	•	•	•	•	•	•	•	•	•	•	•	•	9.08	6.99	7.58	4.35	3.25	2.73	.93	1.51	2.54	5.94	8.59	10.48	63.97
LANDSBURG	36.9	39.8	43.3	48.4	53.8	57.9	61.9	61.3	57.4	50.3	42.6	39.3	49.4	6.97	5.63	5.76	4.02	3.23	3.31	1.34	1.64	3.39	5.80	7.26	8.13	56.48
LONGVIEW	38.1	41.2	44.6	50.0	55.3	59.7	64.1	64.0	60.7	53.1	44.5	40.7	51.3	5.81	4.87	4.85	2.72	2.30	2.14	.75	1.26	2.01	4.40	6.43	7.56	45.10
NEWHALEM	33.8	37.5	42.0	49.5	56.4	60.4	65.5	65.1	60.7	51.6	41.9	36.9	50.1	10.88	8.77	7.45	4.75	2.94	2.70	1.50	1.76	4.14	8.96	11.04	13.33	78.22
OAKVILLE	38.4	41.0	44.1	49.5	54.7	59.1	63.7	63.5	59.6	52.2	44.0	40.8	50.9	8.49	6.56	5.70	3.34	2.28	1.86	.65	1.10	2.25	5.46	7.42	9.44	54.55
PALMER 3 SE	35.9	38.6	41.9	47.8	53.9	57.9	63.1	62.6	58.6	51.2	42.9	38.7	49.4	11.23	9.38	10.33	7.54	5.81	5.35	2.20	2.49	5.21	9.30	11.95	13.75	94.54
QUILCENE 2 SW	36.8	39.8	43.5	49.3	55.0	59.2	63.5	62.8	58.5	50.9	42.6	39.0	50.1	8.09	6.51	4.40	3.07	2.52	2.40	.98	1.01	1.49	3.84	7.26	9.41	50.98
RANDLE 1 E	•	•	•	•	•	•	•	•	•	•	•	•	•	8.38	6.98	6.36	3.98	2.97	2.75	.72	1.33	2.67	6.32	8.73	10.60	61.79
SHELTON	38.4	40.8	44.0	49.7	55.7	59.9	64.3	63.9	59.6	51.9	43.9	40.6	51.1	10.37	8.06	6.83	3.89	2.21	1.73	.80	1.06	2.32	6.09	9.26	11.67	64.29
SNOQUALMIE FALLS	37.7	40.2	43.5	49.2	54.5	58.6	63.0	62.4	57.7	50.7	43.3	40.3	50.1	7.85	6.35	6.14	4.00	3.20	3.21	1.29	1.43	3.18	6.15	8.38	9.12	60.30
STARTUP 1 E	•	•	•	•	•	•	•	•	•	•	•	•	•	7.59	6.31	6.26	4.74	4.32	3.89	1.70	1.77	3.72	6.72	8.40	8.73	64.15
VANCOUVER	38.8	42.4	46.7	52.6	58.2	62.6	67.5	67.2	63.5	55.2	45.7	41.7	53.5	5.63	4.43	4.00	2.31	2.02	1.88	.46	.74	1.64	3.58	5.64	6.67	39.00
DIVISION	36.4	39.3	43.0	49.0	54.8	59.0	63.6	63.2	59.0	51.3	42.8	39.1	50.0	8.24	6.74	6.25	4.15	3.10	2.91	1.13	1.40	2.96	6.22	8.46	9.96	61.52
CASCADE MOUNTAINS-WEST 05																										
CEDAR LAKE	34.4	37.0	39.9	45.8	51.7	55.6	60.7	60.2	56.7	49.9	41.7	37.6	47.6	13.11	10.52	11.17	7.69	6.00	5.53	2.17	2.57	5.54	10.52	13.80	15.61	104.23
LAKE KEECHELUS	25.8	28.6	33.0	39.8	47.4	53.6	60.4	59.9	54.4	45.4	34.9	30.0	42.8	9.86	8.04	6.96	3.78	2.70	2.35	.77	.93	2.83	6.94	10.54	12.46	68.16
RAINIER LONGMIRE	30.3	32.8	35.8	42.1	49.5	54.6	61.2	60.3	56.0	47.6	37.8	33.4	45.1	10.92	8.98	8.32	5.11	4.12	3.63	1.35	1.75	3.92	8.63	11.91	13.79	82.43
SCENIC	•	•	•	•	•	•	•	•	•	•	•	•	•	11.19	9.89	8.84	5.42	4.00	3.27	1.21	1.21	3.82	8.35	11.92	13.84	82.96
STAMPEDE PASS WB	23.5	26.6	30.0	36.5	43.9	48.9	56.2	55.6	51.9	42.4	31.4	27.0	39.5	12.03	10.15	10.60	5.60	4.25	4.09	1.46	2.04	4.39	8.81	12.58	16.19	92.19
WIND RIVER	32.0	35.5	40.2	46.6	53.2	58.1	63.5	62.5	58.0	49.4	39.7	35.0	47.8	16.05	12.53	11.49	6.26	3.74	2.54	1.01	1.12	2.96	8.74	14.68	18.39	99.51
DIVISION	28.5	31.2	34.7	40.9	47.6	52.6	59.2	58.5	54.2	46.1	36.2	31.7	43.5	12.61	10.16	9.67	5.80	4.26	3.92	1.47	1.81	4.22	9.09	12.99	15.00	91.00
EAST SLOPE CASCADES 06																										
BUMPING LAKE	23.1	26.5	30.9	38.1	45.6	50.9	58.2	57.7	51.9	43.0	32.3	26.6	40.4	7.85	6.03	4.82	2.29	1.77	1.59	.53	.64	1.52	4.28	7.25	9.25	47.82
LAKE CLE ELUM	26.0	29.9	35.7	43.6	51.2	57.1	64.3	63.4	56.9	46.8	35.5	30.3	45.1	5.87	4.31	3.67	1.57	1.31	1.09	.36	.42	1.32	3.61	6.00	6.96	36.49
LAKE KACHESS	25.7	29.6	35.1	42.7	50.3	56.1	63.4	62.7	56.3	46.2	34.9	29.8	44.4	8.36	6.67	5.77	2.86	2.12	1.88	.61	.71	2.10	5.28	8.21	10.37	54.94
LAKE WENATCHEE	•	•	•	•	•	•	•	•	•	•	•	•	•	6.90	5.32	3.61	1.50	1.20	1.31	.38	.51	1.30	3.88	6.77	7.99	40.67
LEAVENWORTH 3 S	•	•	•	•	•	•	•	•	•	•	•	•	•	3.97	2.81	2.10	.87	.91	1.01	.20	.44	.75	2.21	4.05	4.59	23.91
MOUNT ADAMS RANGER STA	28.4	32.5	38.0	45.4	52.9	58.7	65.2	63.4	57.4	47.5	36.6	32.1	46.5	8.49	6.05	5.07	2.45	1.67	1.29	.19	.42	1.26	4.14	7.25	9.14	47.42
RIMROCK TIETON DAM	25.6	29.8	35.3	43.4	50.5	56.1	63.0	61.7	56.1	46.3	35.1	29.8	44.4	4.21	2.89	2.41	1.16	.96	1.02	.34	.50	.76	2.36	4.21	5.30	26.12
STEHEKIN 3 NW	•	•	•	•	•	•	•	•	•	•	•	•	•	5.28	4.34	3.01	1.19	1.00	.90	.39	.46	1.22	3.38	5.88	6.75	33.80
TIETON INTAKE	•	•	•	•	•	•	•	•	•	•	•	•	•	3.02	2.11	1.77	1.04	.89	1.10	.33	.45	.53	1.53	2.56	3.84	19.17
WINTHROP 1 WSW	18.4	24.4	35.3	47.2	55.2	61.3	68.2	66.3	58.6	47.2	32.6	22.9	44.8	2.04	1.59	.89	.67	1.01	1.23	.52	.48	.66	1.02	1.94	2.50	14.55
DIVISION	24.6	29.1	35.8	44.3	52.1	57.7	64.7	63.3	56.9	46.6	34.6	28.5	44.9	5.08	3.85	3.06	1.42	1.24	1.23	.39	.49	1.10	2.91	4.92	6.14	31.83
OKANOGAN-BIG BEND 07																										
CHELAN	25.5	30.4	40.6	50.6	59.4	65.6	72.9	71.2	62.8	50.9	37.3	30.3	49.8	1.40	1.10	.89	.72	.86	1.07	.24	.31	.56	.98	1.50	1.60	11.23
CONCONULLY	20.7	26.2	35.4	46.0	54.3	60.0	66.8	65.2	58.2	46.6	32.5	25.2	44.8	1.49	1.31	1.18	1.07	1.37	1.87	.57	.75	.75	1.16	1.75	1.75	15.02
DAVENPORT	24.3	29.0	37.8	46.7	54.5	59.9	67.4	65.6	58.8	47.6	34.6	28.5	46.2	2.11	1.58	1.50	1.07	1.41	1.42	.53	.54	.92	1.42	2.02	2.20	16.72
HARRINGTON 5 S	•	•	•	•	•	•	•	•	•	•	•	•	•	1.51	1.03	1.06	.81	1.12	1.29	.40	.36	.82	1.37	1.52	1.65	12.94
MANSFIELD	•	•	•	•	•	•	•	•	•	•	•	•	•	1.36	1.09	.79	.77	.97	1.36	.30	.41	.60	.85	1.33	1.43	11.26
NESPELEM 2 S	23.9	28.8	38.5	48.0	56.0	61.9	69.1	67.2	60.1	48.7	34.7	28.8	47.1	1.58	1.11	1.05	1.05	1.23	1.54	.51	.48	.73	1.19	1.55	1.55	13.57
ODESSA	27.2	32.4	41.1	49.3	57.3	63.5	70.9	68.6	61.1	50.2	37.1	31.1	49.2	1.35	.96	.97	.66	.96	1.11	.42	.26	.60	.97	1.26	1.29	10.81
SPRAGUE	•	•	•	•	•	•	•	•	•	•	•	•	•	1.81	1.32	1.32	.94	1.16	1.14	.33	.42	.83	1.55	1.84	2.04	14.70
WATERVILLE	22.2	27.3	35.9	46.0	53.7	59.0	66.6	65.2	58.4	47.4	33.6	26.3	45.1	1.33	1.15	.87	.68	1.04	1.37	.22	.44	.63	.88	1.42	1.54	11.57
WILBUR	24.5	29.2	38.2	47.0	55.0	60.6	67.6	65.6	58.6	47.9	35.3	29.1	46.6	1.66	1.13	1.03	.83	1.21	1.20	.41	.35	.70	1.25	1.57	1.59	12.93
DIVISION	24.4	29.4	39.0	48.6	56.8	62.6	70.0	68.0	60.6	49.2	35.4	28.8	47.7	1.51	1.16	1.01	.87	1.11	1.35	.41	.46	.67	1.13	1.54	1.64	12.86

*NORMALS BY CLIMATOLOGICAL DIVISIONS

Taken from "Climatography of the United States No. 81-6, Decennial Census of U. S. Climate"

TEMPERATURE (°F) PRECIPITATION (In.)

Station	T Jan	T Feb	T Mar	T Apr	T May	T June	T July	T Aug	T Sept	T Oct	T Nov	T Dec	T Annual	P Jan	P Feb	P Mar	P Apr	P May	P June	P July	P Aug	P Sept	P Oct	P Nov	P Dec	P Annual
CENTRAL BASIN 08																										
BICKLETON	·	·	·	·	·	·	·	·	·	·	·	·	·	1.83	1.57	1.15	.72	.75	.88	.18	.17	.36	1.03	1.81	2.18	12.63
EPHRATA	·	·												1.05	.74	.67	.56	.72	.92	.20	.25	.46	.77	.98	1.10	8.42
GOLDENDALE	29.4	34.6	41.1	48.0	54.7	60.2	66.9	65.1	59.4	49.7	38.5	33.2	48.4	2.93	2.03	1.70	.85	.79	.91	.15	.21	.62	1.64	2.53	3.05	17.41
HATTON 8 E	28.1	33.6	41.9	49.8	57.7	64.0	71.9	69.6	62.1	50.7	37.8	32.1	49.9	1.22	.92	.87	.70	.79	.86	.21	.22	.49	1.13	1.22	1.31	9.94
KAHLOTUS 4 SW	·	·												1.27	1.06	.89	.72	.83	.93	.16	.22	.46	1.08	1.23	1.52	10.37
KENNEWICK	31.8	37.5	46.1	54.5	62.3	68.3	75.1	72.4	64.6	53.7	41.2	36.0	53.6	1.05	.82	.56	.48	.54	.63	.17	.14	.33	.73	.95	1.09	7.49
LIND 3 NE	27.8	33.2	41.5	49.6	57.7	63.8	71.8	69.8	62.3	51.0	37.5	31.7	49.8	1.13	.97	.81	.67	.85	1.04	.26	.30	.57	1.11	1.12	1.28	10.11
MILL CREEK	·	·												4.68	3.84	4.52	3.56	3.16	2.78	.68	.62	1.66	3.96	4.85	5.25	39.56
PROSSER 4 NE	29.8	35.7	43.6	51.5	58.8	64.6	70.8	68.9	62.6	52.2	39.4	33.8	51.0	.94	.75	.61	.55	.57	.76	.17	.19	.37	.92	.94	1.00	7.77
RITZVILLE	27.0	32.1	40.6	48.8	56.8	63.1	71.1	69.3	62.1	50.6	36.9	31.2	49.1	1.44	1.04	1.01	.74	.94	1.15	.29	.33	.65	1.19	1.38	1.51	11.67
SUNNYSIDE	30.2	36.0	44.2	52.6	60.5	66.2	72.1	69.7	63.2	52.6	39.7	33.8	51.7	.89	.66	.48	.42	.51	.84	.18	.21	.35	.72	.76	.88	6.90
WALLA WALLA WB CITY	33.2	38.4	46.0	53.8	61.0	67.2	76.0	73.8	66.0	55.1	42.3	37.8	54.2	1.89	1.52	1.59	1.40	1.49	1.22	.21	.30	.78	1.53	1.72	1.85	15.50
WAPATO	30.2	36.6	44.9	53.8	62.0	67.8	74.8	72.5	64.7	53.3	39.7	34.0	52.9	1.04	.73	.53	.45	.45	.76	.15	.21	.33	.57	.87	1.02	7.11
WENATCHEE	26.2	31.5	42.3	52.2	60.5	66.5	73.4	71.5	63.5	51.1	37.7	31.1	50.6	1.16	.98	.67	.47	.65	.91	.11	.37	.47	.73	1.14	1.34	9.00
WILSON CREEK	·	·												1.09	.78	.69	.54	.82	1.08	.28	.20	.52	.88	1.12	1.17	9.17
YAKIMA WB AIRPORT	27.5	34.0	42.0	50.5	58.5	64.4	71.0	68.6	61.3	50.5	37.4	31.5	49.8	1.19	.87	.62	.47	.54	.81	.13	.20	.35	.60	.96	1.12	7.86
DIVISION	29.1	34.8	43.3	51.7	59.6	65.7	72.9	70.9	63.6	52.3	39.0	33.1	51.3	1.29	.99	.83	.63	.71	.88	.19	.23	.45	.92	1.18	1.33	9.63
NORTHEASTERN 09																										
CHEWELAH 2 S	23.3	28.5	38.2	47.2	55.2	60.7	66.6	64.4	57.3	46.7	34.6	28.4	45.9	2.52	1.83	1.70	1.29	1.67	1.58	.64	.59	1.08	1.85	2.34	2.73	19.82
DEER PARK 2 E	23.8	28.4	36.3	45.9	53.8	59.4	65.9	63.5	56.5	46.1	33.8	28.4	45.2	2.90	2.11	1.86	1.53	1.67	1.52	.47	.58	1.18	2.22	2.75	3.24	22.03
IRENE MT WAUCONDA	·	·												1.10	.83	.80	1.05	1.75	2.47	.93	.97	1.41	1.20	1.15		14.63
LAURIER	22.7	27.6	36.9	47.9	56.8	62.4	69.4	67.4	59.2	46.6	33.6	26.9	46.5	3.50	2.65	2.48	1.71	1.99	1.89	.61	.80	1.47	2.76	3.51	3.79	27.16
NEWPORT	24.0	28.5	36.0	45.2	53.1	59.0	65.4	63.1	56.0	45.8	33.8	28.5	44.9	2.11	1.52	1.35	1.32	1.68	2.28	.98	.95	1.11	1.79	1.95	2.22	19.26
NORTHPORT	25.1	29.5	39.0	49.4	58.0	63.6	70.2	67.9	59.9	48.7	35.6	29.4	48.0	2.16	1.55	1.43	1.27	1.69	2.19	.80	.84	1.23	2.04	1.99	2.27	19.46
SPOKANE WB AIRPORT	25.3	30.0	38.1	47.3	56.2	61.9	70.5	68.0	60.9	49.1	35.7	30.1	47.8	2.44	1.86	1.50	.91	1.21	1.49	.38	.41	.75	1.57	2.24	2.43	17.19
WELLPINIT	·	·												2.57	1.90	1.80	1.45	1.68	1.67	.49	.63	1.10	1.79	2.34	2.75	20.17
DIVISION	23.4	28.2	36.7	46.5	54.6	60.2	66.9	64.7	57.3	46.5	33.7	27.7	45.5	2.40	1.77	1.65	1.34	1.66	1.86	.71	.79	1.16	1.99	2.33	2.58	20.24
PALOUSE-BLUE MOUNTAINS 10																										
COLFAX 1 NW	29.2	33.4	40.2	48.2	55.3	60.8	67.8	65.9	59.0	49.4	37.9	33.1	48.4	2.55	1.93	2.14	1.54	1.41	1.64	.46	.47	1.12	2.01	2.66	3.04	20.97
DAYTON 1 WSW	31.6	36.2	43.0	50.3	57.3	63.1	70.7	68.8	62.0	52.0	40.2	35.3	50.9	2.43	1.90	2.12	1.56	1.46	1.51	.40	.36	.82	1.89	2.38	2.70	19.53
LACROSSE 3 ESE	29.7	34.7	42.2	49.4	56.4	62.7	70.4	68.0	60.7	50.3	38.4	33.5	49.7	1.84	1.37	1.33	.96	.96	1.13	.28	.36	.68	1.35	1.74	2.05	14.05
POMEROY	31.4	35.8	42.4	49.5	56.4	62.3	69.7	67.8	61.2	51.4	39.8	35.0	50.2	2.02	1.63	1.66	1.27	1.29	1.49	.32	.38	.85	1.50	1.91	2.26	16.58
PULLMAN 2 NW	·	·												2.67	2.10	2.12	1.49	1.46	1.54	.39	.52	1.08	1.91	2.47	2.74	20.49
ROSALIA	27.2	31.5	38.8	47.1	56.0	60.1	68.0	66.3	58.8	48.5	36.8	31.3	47.4	2.20	1.64	1.66	1.26	1.49	1.53	.40	.52	1.13	1.82	2.20	2.46	18.31
DIVISION	29.8	34.3	41.2	49.1	56.3	61.9	69.7	67.8	60.7	50.6	38.7	33.6	49.5	2.18	1.73	1.80	1.34	1.43	1.57	.39	.46	.95	1.69	2.13	2.40	18.07

* Normals for the period 1931-1960. Divisional normals may not be the arithmetical averages of individual stations published, since additional data for shorter period stations are used to obtain better areal representation.

TEMPERATURE / PRECIPITATION

CONFIDENCE LIMITS

In the absence of trend or record changes, the chances are 9 out of 10 that the true mean will lie in the interval formed by adding and subtracting the values in the following table from the means for any station in the State. Because of the wider variation in mean precipitation, the corresponding monthly means and annual mean must be substituted for "p" in the precipitation table below to obtain mean precipitation confidence limits.

	Jan	Feb	Mar	Apr	May	June	July	Aug	Sept	Oct	Nov	Dec	Annual
TEMPERATURE	1.7	1.4	.9	.8	.8	.5	.5	.5	.7	.7	.8	1.1	.5
PRECIPITATION	.34√p	.35√p	.30√p	.25√p	.28√p	.34√p	.26√p	.25√p	.35√p	.35√p	.40√p	.33√p	.32√p

COMPARATIVE DATA

Data in the following table are the mean temperature and average precipitation for Tatoosh Island (W.B.O.), Washington, for the period 1906 - 1930 and are included in this publication for comparative purposes:

	Jan	Feb	Mar	Apr	May	June	July	Aug	Sept	Oct	Nov	Dec	Annual
TEMPERATURE	40.9	42.6	44.0	46.6	50.0	53.0	54.9	55.1	54.0	51.1	47.0	43.2	48.5
PRECIPITATION	10.56	8.21	6.89	4.86	3.27	2.61	1.07	1.84	4.14	8.45	10.28	11.40	73.58

NORMALS, MEANS AND EXTREMES
(Table Revised 1965. Base Period for Climatological Normals: 1931-1960)

OLYMPIA, WASHINGTON
MUNICIPAL AIRPORT
LATITUDE 46° 58' N
LONGITUDE 122° 54' W
ELEVATION (ground) 190 FEET

Month	Temperature Normal Daily maximum	Daily minimum	Monthly	Extremes Record highest	Year	Record lowest	Year	Precipitation Normal total	Normal degree days	Max monthly	Year	Min monthly	Year	Max in 24 hrs	Year	Snow, Sleet Mean total	Max monthly	Year	Max in 24 hrs	Year	Relative humidity 4:00 A.M. PST	10:00 A.M. PST	4:00 P.M. PST	10:00 P.M. PST	Wind Mean hourly speed	Prevailing direction	Fastest mile Speed	Direction	Year	Mean sky cover sunrise to sunset	Pct of possible sunshine	Days Clear	Partly cloudy	Cloudy	Precip .01"+	Snow/Sleet 1.0"+	Thunderstorms	Heavy fog	Max 90+	Max 32-	Min 32-	Min 0-
(a)	(b)	(b)	(b)	5		5		(b)	(b)	23		23		23		23	23		23		5	5	5	5	16	15				20	23	23	23	23	23	23	23	23	5	5	5	5
J	45.1	31.4	38.1	60	1962	5	1962	7.85	834	19.84	1953	0.84	1949	2.94	1964	6.9	45.0	1950	13.7	1950	89	87	78	89	7.2	SSW				8.5		2	5	24	20	2	*	9	0	1	18	5
F	49.6	32.2	40.9	64	1962	12	1962	6.62	675	13.18	1961	2.54	1964	4.93	1951	4.5	17.7	1949	13.7	1950	92	86	71	87	7.1	SSW				8.2		2	5	21	18	1	*	8	0	0	11	1
M	54.4	34.0	44.2	72	1964	14	1960	5.40	645	10.13	1950	2.39	1958	2.41	1948	2.2	20.6	1951	6.6	1951	91	84	65	87	7.5	SSW				8.2		3	7	22	18	1	*	7	0	0	7	0
A	62.3	37.6	50.0	81	1962	26	1964	3.16	450	4.78	1948	2.37	1956	1.58	1962	T	7.0	1955+	C		87	75	60	84	7.2	SW				7.5		4	9	17	15	0	*	5	0	0	1	0
M	68.6	43.6	56.1	89	1963	34	1961	2.01	307	5.83	1948	2.15	1947	1.54	1948	T	C		C		91	69	58	82	6.4	SW				7.1		5	10	16	10	0	1	4	*	0	0	0
J	72.6	45.5	59.1	94	1961	35	1961	1.79	177	6.48	1946	2.04	1945	1.20	1954	C	C		C		90	67	54	80	6.4	SSW				7.0		5	11	14	10	0	1	3	3	1	5	0
J	79.7	48.0	63.9	100	1961	38	1963	0.76	68	2.68	1955	0.00	1960+	0.96	1955	0.0	C.0		C.0		89	64	49	78	6.4	SSW				5.2		11	10	10	5	0	1	4	3	0	0	0
A	78.9	47.8	63.4	100	1960	33	1960	0.89	71	3.17	1962	0.06	1946	1.22	1962	0.0	C.0		C.0		90	70	53	83	5.7	SW				5.5		8	12	11	6	0	1	4	2	0	0	0
S	72.1	44.4	58.6	94	1960	27	1961	2.09	198	4.26	1959	1.06	1959	1.79	1965	0.0	C.0		C.0		92	77	57	87	5.4	SW				6.1		9	8	13	10	0	1	5	*	0	1	0
O	62.3	40.5	51.4	82	1961	18	1961	5.28	426	9.93	1947	2.55	1952	2.69	1942	T	12.4	1946	4.33	1962	92	88	70	92	5.4	SSW				6.7		6	8	17	15	*	1	15	0	0	1	0
N	52.3	36.5	43.8	69	1961	18	1961	7.67	636	15.51	1962	1.39	1962	4.33	1962	1.5	12.4	1946	10.2	1955	91	85	75	90	5.7	SSW				8.5		3	6	20	19	1	1	11	0	1	11	0
D	47.5	33.9	40.7	58	1962	-3	1964	9.05	753	12.59	1955	2.28	1944	3.83	1956	1.9	16.5	1964	10.2	1955	91	90	84	90	7.4	SSW				8.9		1	4	26	20	1	*	10	0	1	15	0
YR	62.2	39.3	50.8	100	JUL 1961+	3	DEC 1964	52.37	5236	19.84	JAN 1953	0.00	AUG 1946	4.93	FEB 1951	15.5	45.0	JAN 1950	13.7	JAN 1950	91	79	65	86	6.6	SSW				7.4		50	90	225	163	5	4	93	6	2	87	0

Means and extremes in the above table are from the existing location. Annual extremes have been exceeded at other locations as follows: Highest temperature 103 in July 1941; lowest temperature -1 in November 1955 and earlier date; maximum monthly precipitation 27.12 in December 1933.

NORMALS, MEANS AND EXTREMES
(Table Revised 1975. Base Period for Climatological Normals: 1941-1970)

Station: OLYMPIA AIRPORT   OLYMPIA, WASHINGTON   Standard time used: PACIFIC
Latitude: 46 58 N   Longitude: 122 54 W   Elevation (ground): 195 feet

Month	Temperatures °F Normal Daily maximum	Daily minimum	Monthly	Extremes Record highest	Year	Record lowest	Year	Normal Degree days Base 65°F Heating	Cooling	Precipitation Normal	Max monthly	Year	Min monthly	Year	Max in 24 hrs	Year	Snow, ice pellets Max monthly	Year	Max in 24 hrs	Year	Relative humidity pct. Hour 04	Hour 10	Hour 16	Hour 22	Wind Mean speed m.p.h.	Prevailing direction	Fastest mile Speed m.p.h.	Direction	Year	Mean sky cover	Pct possible sunshine	Days Clear	Partly cloudy	Cloudy	Precip .01"+	Snow/ice pellets 1.0"+	Thunderstorms %	Heavy fog %	Max 90+	Max 32-	Min 32-	Min 0-	Avg station pressure mb.	Elev. 200 feet m.s.l.
(a)				15		15				33	33		33		33		33		33		15	15	15	12	23	15	26			30		33	33	33	33	33	31	31	15	15	15	15		2
J	44.0	30.4	37.2	60	1962	-7	1972	862	0	7.93	19.84	1953	0.84	1949	3.32	1949	58.7	1969	20.9	1969	91	88	81	89	7.7	SSW	55	18	1957	8.6		2	4	25	20	3	*	8	0	0	16	*	1008.7	
F	49.0	32.0	40.5	60	1962	-1	1972	672	0	5.57	13.18	1961	1.71	1973	4.93	1951	17.7	1949	9.9		91	80	71	81	7.7	SSW	18	26	1964	8.3		2	5	21	18	1	*	7	0	0	14	*	1008.8	
M	53.0	32.8	43.2	76	1960	14	1960	676	0	4.81	10.13	1950	0.48	1965	3.92	1972	20.6	1951	9.1	1962	90	81	62	80	7.3	SSW	23	14	1962	8.1		3	7	21	18	1	*	6	0	0	9	0	1006.1	
A	59.5	36.5	48.0	81	1969	26	1973	504	0	3.14	5.87	1948	0.37	1956	2.31	1965	2.2	1972	1.8		88	67	58	74	6.7	SW	49	9	1962	7.8		4	9	17	15	0	*	5	*	0	1	0	1012.8	
M	67.2	40.8	54.0	94	1964	31	1966	341	14	1.88	5.83	1948	0.15	1947	1.54	1948	T	1974	T		90	67	54	79	6.6	SSW	39	25	1969	7.1		5	10	16	11	0	1	3	*	0	0	0	1010.7	
J	71.9	45.8	—	94	1970	35	—	197	46	1.57	6.48	1945	0.04	1945	1.38	1968	0.0		0.0		90	67	54	77	6.6	SSW	32		1949	7.0		11	9	10	10	0	1	3	*	1	0	0	1010.8	
J	78.4	48.7	63.6	100	1961	35	1961	89	46	0.70	2.68	1955	T	1960	1.22	1974	0.0		0.0		90	63	48	72	6.1	SSW	29	18	1957	5.2		11	10	10	5	0	1	4	0	3	0	0	1010.7	
A	77.2	48.4	62.8	100	1960	35	1973	103	35	1.17	5.45	1958	0.00	1962	1.22	1962	0.0		0.0		93	66	49	80	5.5	SW	26	27	1964	5.8		9	12	10	5	0	1	5	0	3	0	0	1010.4	
S	71.3	44.5	57.9	94	1961	25	1972	198	12	2.12	5.23	1959	0.00	1972	2.69	1972	T	1972	T	1972	94	74	52	86	5.0	SSW	58	23	1962	6.7		8	13	14	10	0	1	6	0	1	0	0	1009.4	
O	61.2	40.2	50.6	82	1971	18	1971	446	0	5.28	10.08	1947	0.85	1952	2.69	1962	T	1971	9.0	1946	93	82	62	89	5.6	SSW	23	23	1968	7.7		6	9	16	14	0	1	14	*	0	1	0	1011.4	
N	51.3	35.3	43.3	69	1970	18	1961	651	0	7.58	15.51	1962	1.39	1962	4.33	1962	12.4	1946	9.0	1946	92	87	75	90	6.0	SSW	55	18	1962	8.9		4	8	18	21	1	1	11	0	1	6	0	1005.9	
D	45.8	33.1	39.5	58	1965	-3	1972	791	0	8.19	14.32	1970	2.28	1944	3.83	1956	21.4	1968	11.9		92	90	85	92	6.7	SSW	41	18	1969	8.9		3	6	23	21	2	*	10	0	0	14	*	1009.2	
YR	61.0	39.1	50.1	100	JUL 1961	-7	JAN 1972	5530	101	50.74	19.84	JAN 1953	0.00	AUG 1946	4.93	FEB 1951	58.7	JAN 1969	20.5	JAN 1969	91	78	64	85	6.7	SSW	60	18	NOV 1958	7.4		51	88	226	163	6	5	90	3	7	87	*	1009.6	

Means and extremes above are from existing and comparable exposures. Annual extremes have been exceeded at other sites in the locality as follows: Highest temperature 103 in July 1941; maximum monthly precipitation 27.12 in December 1933.

REFERENCE NOTES APPLYING TO TABLES APPEAR ON THE PAGE FOLLOWING LAST TABLE.
(Caution: Letters and symbols may have different meanings in 1941-1970 tables than in earlier tables. See notes.)

NORMALS, MEANS AND EXTREMES
(Table Revised 1965. Base Period for Climatological Normals: 1931-1960)

SEATTLE, WASHINGTON
SEATTLE-TACOMA AIRPORT
LATITUDE 47° 27' N
LONGITUDE 122° 18' W
ELEVATION (ground) 386 FEET

Month	Temperature — Normal Daily maximum	Daily minimum	Monthly	Extremes Record highest	Year	Record lowest	Year	Normal degree days	Precipitation Normal total	Max. monthly	Year	Min. monthly	Year	Max. in 24 hrs.	Year	Snow,Sleet Mean total	Max. monthly	Year	Max. in 24 hrs.	Year	Rel. hum. 4:00 A.M.	10:00 A.M.	4:00 P.M.	10:00 P.M.	Wind Mean hourly speed	Prevailing direction	Mean sky cover	Clear	Partly	Cloudy	Precip .01"+	Snow 1.0"+	Tstorms	Heavy fog	Max 90+	Max 32−	Min 32−	Min 0−	Solar radiation
(a)	(b)	(b)	(b)	5		5		(b)	(b)	20		20		20		20	20		20		5	5	5	5	16	15	20	20	20	20	20	20	20	20	5	5	5	5	12
J	43.6	33.0	38.3	61	1960	12	1963	828	5.73	12.92	1953	0.86	1949	2.22	1950	6.1	57.2	1950	21.4	1950	80	79	74	77	10.8	SSW	8.3	3	4	24	19	2	*	5	0	1	13	0	79
F	47.0	36.2	40.8	68	1963	18	1962	678	4.24	9.11	1961	1.66	1964	3.41	1961	2.5	13.1	1949	7.2	1962	81	77	67	74	10.9	SW	8.1	3	5	22	17	1	*	4	0	0	6	0	143
M	51.3	36.4	43.9	71	1964	18	1951	657	3.79	8.40	1950	2.07	1954	3.14	1951	2.3	18.2	1951	5.6	1951	81	72	63	75	11.4	SW	8.1	4	6	21	17	1	1	1	0	0	5	0	261
A	58.2	40.1	49.2	77	1963	30	1961	474	2.43	3.75	1948	0.33	1956	1.15	1959	T	T		T	1964+	86	71	57	72	11.0	SW	7.8	5	9	20	14	0	1	1	0	0	1	0	389
M	65.6	45.8	55.8	93	1963	30	1962	295	1.66	3.76	1948	0.35	1941	1.53	1954	0.0	0.0		0.0		84	67	55	72	9.9	SW	7.1	5	9	17	11	0	1	3	*	0	*	0	503
J	69.9	49.7	59.8	90	1961	41	1962	159	1.58	3.90	1946	0.13	1951			0.0	0.0		0.0		84	67	55	72	9.9	SW	7.1	5	9	17	11	0	1	1	0	0	0	0	514
J	75.6	54.1	64.9	97	1960	46	1960	56	0.81	2.10	1955	T	1960+	0.74	1957	0.0	0.0		0.0		86	66	50	69	9.3	SW	5.2	10	11	10	5	0	1	3	1	0	0	0	558
A	74.6	53.6	64.1	99	1960	45	1960	62	0.95	2.17	1950	0.17	1952	1.36	1950	0.0	0.0		0.0		86	68	52	71	8.9	SW	5.5	11	10	10	7	0	1	4	1	0	0	0	455
S	69.3	50.5	59.9	97	1963	35	1961+	162	2.05	4.60	1959	0.32	1952	1.77	1953	0.0	0.0		0.0		89	76	58	78	9.1	N	4.9	9	8	13	10	0	1	8	1	0	0	0	331
O	60.3	44.4	52.4	80	1961	33	1961	391	4.02	8.95	1947	1.00	1947	2.27	1947	T	T	1956+	T	1956+	89	82	76	83	9.5	S	4.7	7	7	20	15	0	1	9	0	0	3	0	192
N	49.6	38.1	43.9	74	1960	23	1963	633	5.35	9.69	1963	1.11	1952	3.41	1959	1.5	13.7	1946	4.7	1955	85	82	76	83	10.2	S	8.3	3	4	23	19	2	1	7	0	1	5	0	103
D	45.9	35.7	40.8	60	1964	10	1964	750	6.29	9.50	1955	3.75	1955	2.53	1959	1.6	7.6	1964		1964	84	80	83		10.8	SSW	8.7	3	4	26	20	1	*	6	0	1	10	0	66
YR	59.2	42.9	51.1	99	AUG. 1960	10	DEC. 1964	5145	38.94	12.92	JAN. 1953	T	JUL. 1960+	3.41	NOV. 1959+	14.0	57.2	JAN. 1950	21.4	JAN. 1950	85	76	64	77	10.2	SSW	7.4	54	82	229	164	5	8	54	3	2	40	0	300

Means and extremes in the above table are from the existing and comparable locations. Extremes have been exceeded at other locations as follows: Lowest temperature 0 in January 1950.

NORMALS, MEANS AND EXTREMES
(Table Revised 1975. Base Period for Climatological Normals: 1941-1970)

Station: SEATTLE-TACOMA AIRPORT, SEATTLE, WASHINGTON
Standard time used: PACIFIC
Latitude: 47° 27' N
Longitude: 122° 13' W
Elevation (ground): 400 feet

Month	Temp Daily maximum	Daily minimum	Monthly	Record highest	Year	Record lowest	Year	Heating	Cooling	Precip Normal	Max. monthly	Year	Min. monthly	Year	Max. 24 hrs.	Year	Snow Max. monthly	Year	Max. 24 hrs.	Year	RH 04	10	16	22	Mean speed	Prev. dir.	Fastest mile m.p.h.	Dir.	Year	Pct. sunshine	Sky cover	Clear	Partly	Cloudy	Precip .01"+	Snow 1.0"+	Tstorms	Heavy fog	Max 90+	Max 32−	Min 32−	Min 0−	Station pressure mb
(a)				−1	−5	15					30		30			30	30		30		15	15	15	15	26	15	7	7		9	30	30	30	30	30	30	30	30	15	15	15	15	2
J	43.4	33.0	38.2	61	1960	12	1972	831	0	5.79	12.92	1953	0.71	1949	2.41	1967	57.2	1950	21.4	1950	80	78	75	78	10.4	SSW	45	SW	1971	21.8	8.5	3	3	25	20	2	*	5	0	0	11	0	999.2
F	48.5	36.6	42.3	70	1968	18	1962	636	0	4.19	9.11	1961	1.08	1964	3.41	1951	13.1	1949	7.2	1962	78	66	65	76	10.2	SW	36	SW	1971	43	8.0	3	4	21	16	1	*	4	0	0	4	0	1000.7
M	51.5	37.2	44.3	71	1964	18	1951	648	0	3.61	8.40	1950	1.55	1965	1.85	1956	18.2	1951	5.6	1951	81	62	57	75	10.5	SW	41	SW	1972	50	7.8	4	6	21	17	1	1	3	0	*	4	0	999.8
A	57.0	40.5	48.7	77	1972	30	1963	489	0	2.46	4.12	1972	0.35	1956	1.77	1972	2.3	1972	2.3	1972	82	59	54	72	9.9	SW	38	SW	1965	57	7.4	5	7	20	14	1	1	1	*	0	1	0	1000.4
M	64.1	46.0	54.9	94	1963	30	1962	313	11	1.70	4.76	1948	0.35	1947	1.83	1968	0.0		0.0		83	59	53	68	9.2	SW	32	SW	1968	54	7.1	6	10	15	10	0	1	3	*	0	*	0	1001.4
J	69.0	50.6	59.8	94	1970	41	1962	167	11	1.53	3.90	1946	0.13	1951	1.75	1951	0.0		0.0		83	57	54	67	9.0	SW	29	SW	1974	54	7.1	5	11	14	9	0	1	6	1	0	0	0	1000.7
J	73.8	53.8	63.8	97	1961	45	1973	80	65	0.71	2.10	1955	T	1960	0.84	1960	0.0		0.0		83	49	44	66	8.5	SW	24	SW	1973	67	5.7	11	12	8	5	0	1	7	1	0	0	0	1001.7
A	73.1	53.8	63.4	99	1960	45	1968	82	45	1.08	2.17	1974	0.01	1974	1.75	1974	0.0		0.0		83	49	45	67	8.1	SW	25	SW	1972	67	5.9	10	12	9	6	0	1	8	1	0	0	0	1001.2
S	68.7	50.4	59.6	89	1967	35	1972	170	8	1.99	4.60	1959	0.10	1953	1.77	1947	2.0	1972	2.0	1972	86	58	50	72	8.3	N	30	N	1970	59	5.2	9	8	13	10	0	1	7	*	0	0	0	1000.9
O	59.4	44.6	52.2	81	1971	30	1971	397	0	3.91	8.95	1947	0.72	1972	2.27	1972	T	1971	T	1971	84	68	60	80	8.9	S	31	S	1973	52	6.2	7	6	18	14	0	1	6	0	0	3	0	1002.2
N	50.4	38.8	44.6	72	1970	23	1961	612	0	5.88	9.69	1963	1.11	1952	3.41	1959	13.7	1946	9.4	1955	84	75	70	84	9.3	S	26	S	1970	38	8.4	2	5	23	18	1	*	6	0	0	3	0	996.6
D	45.4	35.5	40.5	60	1970	23	1965	760	0	5.94	9.50	1955	3.75	1955	2.53	1959	13.1	1968	13.0	1968	82	81	78	81	10.1	SSW	42	S	1969	16	8.8	2	3	26	21	1	*	6	0	0	8	0	1000.0
YR	58.8	43.3	51.1	99	AUG. 1960	12	DEC. 1972	5185	129	38.79	12.92	JAN. 1953	T	JUL. 1960	3.41	NOV. 1959	57.2	JAN. 1950	21.4	JAN. 1950	83	74	62	75	9.3	SW	45	SW	JAN. 1971	49	7.4	57	80	228	161	5	7	48	3	0	31	0	1000.4

Means and extremes above are from existing and comparable exposures. Annual extremes have been exceeded at other sites in the locality as follows: Highest temperature 100 in June 1955 and earlier; lowest temperature zero in January 1950; maximum monthly precipitation 15.33 in December 1933; minimum monthly precipitation 0.00 in July 1922 and earlier. Maximum precipitation in 24 hours 3.52 in December 1921; maximum snowfall in 24 hours 21.5 in February 1916; highest wind (fastest observed 1-minute speed) 55 from 20 degrees in February 1958.

REFERENCE NOTES APPLYING TO TABLES APPEAR ON THE PAGE FOLLOWING LAST TABLE.
(Caution: Letters and symbols may have different meanings in 1941-1970 tables than in earlier tables. See notes.)

NORMALS, MEANS AND EXTREMES

(Table Revised 1965. Base Period for Climatological Normals: 1931-1960)

SEATTLE, WASHINGTON
FEDERAL OFFICE BUILDING
LATITUDE 47° 36' N
LONGITUDE 122° 20' W
ELEVATION (ground) 14 FEET

Month	Temperature Normal — Daily maximum	Daily minimum	Monthly	Extremes Record highest	Year	Record lowest	Year	Normal degree days	Precipitation Normal total	Max monthly	Year	Min monthly	Year	Max in 24 hrs	Year	Snow,Sleet Mean total	Max monthly	Year	Max in 24 hrs	Year	Mean sky cover sunrise to sunset	Pct of possible sunshine	Clear	Partly cloudy	Cloudy	Precip .01 or more	Snow/Sleet 1.0 or more	Thunderstorms	Heavy fog	Max 90 and above	Max 32 and below	Min 32 and below	Min 0 and below	Avg daily solar radiation-langleys
(a)	(b)	(b)	(b)	31	31	31		(b)	(b)	31		31		31		31	31		31		24	31	24	24	24	31	31	24		31	31	31	31	11
J	45.6	36.8	41.2	66	1935	11	1950	738	5.19	10.93	1953	1.43	1949	2.46	1935	4.3	31.0	1950	11.5	1943	8.0	28.8	3	5	23	18	1	*		0	0	18	0	71
F	48.8	38.3	43.6	68	1938	12	1950	599	3.90	7.75	1961	1.29	1934	2.69	1961	1.4	10.4	1949	7.6	1937	7.7	34.7	3	6	19	16	*	*		0	*	7	0	130
M	52.7	40.1	46.4	75	1941	22	1955	577	3.32	4.55	1950	1.22	1944	2.53	1950	0.8	7.5	1951	5.5	1960	7.4	42	4	8	19	16	*	*		0	0	3	0	242
A	59.4	44.1	51.8	87	1947	31	1936	396	1.97	4.56	1937	0.41	1956	1.97	1953	T	1.0	1936	1.0	1972	6.9	47	5	9	16	13	*	*		0	0	*	0	356
M	65.7	49.0	57.4	92	1963+	35	1954	242	1.59	3.68	1948	0.35	1952	1.35	1937	T	0.0		0.0		6.9	52	6	10	15	11	0	*		*	0	0	0	440
J	69.6	53.1	61.4	100	1955	43	1954	117	1.41	3.68	1964	0.23	1964	1.22	1964	0.0	0.0		0.0		6.4	49	7	8	15	9	0	1		*	0	0	0	470
J	75.1	56.1	65.6	100	1941	48	1948	50	0.63	1.81	1954	T	1960+			0.0	0.0		0.0		4.9	63	12	10	9	5	0	1		1	0	0	0	501
A	73.9	56.1	65.0	97	1960	48	1955	47	0.74	2.03	1941	0.12	1955		1955	0.0	0.0		0.0		5.3	56	10	10	11	6	0	1		1	0	0	0	440
S	69.0	51.2	60.6	92	1937	42	1957	129	1.65	3.43	1945	0.80	1952	1.97	1947	0.0	0.0		0.0		5.6	53	8	8	13	8	0	1		*	0	0	0	313
O	60.4	45.4	52.9	78	1952	30	1935	329	3.28	7.40	1947			3.20	1941	T	T	1971	4.5	1960	7.2	37	6	6	21	14	0	*		*	0	1	0	174
N	51.8	41.9	46.9	70	1949	13	1955	543	5.00	9.40	1942	0.48	1947	3.20	1937	0.8	5.5	1955	5.5	1955	8.0	28	3	6	21	17	*	*		0	*	2	1	95
D	48.0	39.5	43.8	65	1939	11	1964	657	5.42	10.41	1939	1.00	1944	3.31	1937	0.8	6.9	1968	10.0	1968	8.1	23	3	5	23	19	*	*		0	*	3	3	60
YR	60.0	46.4	53.2	100 JUN 1955+		11 DEC 1964+		4424	34.10	10.93 JAN 1953		T JUL 1960+		3.31 DEC 1937		8.1	31.0 JAN 1950		11.5 JAN 1943		6.8	45	71	93	201	152	6	6	24	2	15		0	274

Means and extremes in the above table are from the existing location. Annual extremes have been exceeded at other locations as follows: Lowest temperature 3 in January 1893; maximum monthly precipitation 15.33 in December 1933; minimum monthly precipitation 0.00 in July 1922 and earlier dates; maximum precipitation in 24 hours 3.52 in December 1921; maximum snowfall 35.4 in February 1916; maximum snowfall in 24 hours 21.5 in February 1916.
Ø Based on the period of record 1934 – 1957.

NORMALS, MEANS AND EXTREMES

(Table Revised 1975. Base Period for Climatological Normals: 1941-1970)

Station: SEATTLE, WASHINGTON — 2725 MONTLAKE BLVD EAST
Standard time used: PACIFIC — Latitude: 47 39 N — Longitude: 122 18 W — Elevation (ground): 19 feet — Average station pressure Elev. 28 feet m.s.l.

Month	Temperature °F Normal — Daily maximum	Daily minimum	Monthly	Extremes Record highest	Year	Record lowest	Year	Degree days Base 65°F Heating	Cooling	Precipitation Water equivalent Normal	Max monthly	Year	Min monthly	Year	Max in 24 hrs	Year	Snow,Ice pellets Max monthly	Year	Max in 24 hrs	Year	Mean sky cover (tenths)	Pct possible sunshine	Fastest mile Speed m.p.h.	Direction	Year	Clear	Partly	Cloudy	Precip .01 or more	Snow 1.0 or more	Thunderstorms	Heavy fog	Max 90 and above	Max 32 and below	Min 32 and below	Min 0 and below
(a)				41		41					41		41		41		41		41		24		31	31		24	24	24	24	24	24	24	24	24	24	24
J	44.7	34.7	39.7	65	1935	11	1950	784	0	5.17	10.93	1953	1.43	1949	2.46	1935	31.0	1950	11.5	1943	8.0	28	63	SW	1951	3	5	23	20	2	*	2	0	1	6	0
F	50.0	36.9	43.5	74	1968	12	1950	602	0	3.93	7.75	1961	1.29	1934	2.69	1961	10.4	1949	7.6	1937	7.7	34	64	SW	1934	3	6	19	17	*	*	*	0	*	2	0
M	53.0	38.0	45.5	75	1961	22	1955	605	0	3.24	4.55	1950	1.22	1944	2.32	1950	7.5	1951	5.5	1960	7.4	42	60	SW	1956	4	8	19	17	*	*	*	0	0	1	0
A	59.0	41.8	50.4	87	1947	31	1936	438	0	2.41	4.56	1937	0.16	1956	2.23	1972	1.0	1936	1.0	1972	6.9	47	65	S	1943	5	9	16	14	*	*	1	0	0	*	0
M	65.8	47.1	56.5	92	1963	35	1954	269	6	1.57	3.68	1948	0.34	1937	1.35	1937	0.0		0.0		6.4	52	60	SW	1963	7	10	14	10	0	1	1	*	0	0	0
J	70.4	52.1	61.3	100	1941	43	1954	133	22	1.57	3.68	1964	0.23	1964	1.22	1949	0.0		0.0		6.4	49	54	SW	1936	7	8	15	9	0	1	*	0	0	0
J	75.8	55.8	65.7	100	1960	48	1971	62	83	0.87	1.95	1960	T	1974	1.22	1954	0.0		0.0		4.9	63	38	SW	1939	12	10	9	6	0	1	*	1	0	0	0
A	74.5	55.3	64.9	97	1967	48	1973	58	55	0.88	4.28	1968	0.12	1968	1.91	1968	0.0		0.0		5.3	56	44	SW	1964	10	10	11	6	0	1	*	1	0	0	0
S	69.3	51.8	60.6	92	1952	40	1972	149	10	1.75	3.43	1959	0.15	1969	1.94	1963	0.0		0.0		5.6	53	55	SW	1958	9	8	13	9	0	1	*	*	0	0	0
O	61.8	46.5	54.2	78	1970	30	1935	335	0	3.43	7.43	1947	0.48	1947	1.97	1947	T	1971	4.5	1960	7.2	37	65	SW	1962	5	6	21	18	0	*	1	0	1	0
N	51.0	40.4	45.7	70	1949	10	1955	579	0	5.34	9.40	1942	0.48	1952	3.20	1937	5.5	1955	5.5	1955	8.0	28	60	SW	1959	3	5	23	18	1	*	*	0	*	3	0
D	46.6	37.4	42.0	65	1939	10	1968	713	0	5.35	10.41	1939	1.00	1944	3.31	1937	13.5	1968	10.0	1968	8.1	23	60	SW	1959	3	5	23	20	1	*	*	0	1	3	0
YR	60.2	44.8	52.5	100 JUN 1941		10 DEC 1968		4727	183	35.65	10.93 JAN 1953		T AUG 1967		3.31 DEC 1937		31.0 JAN 1950		11.5 JAN 1943		6.8	45	65 SW OCT 1962			71	93	201	151	3	6	24	2	2	14	0

Means and extremes above are from existing and comparable exposures. Annual extremes have been exceeded at other sites in the locality as follows: Lowest temperature 3 in January 1893; maximum monthly precipitation 15.33 in December 1933; minimum monthly precipitation 0.00 in July 1922 and earlier dates; maximum precipitation in 24 hours 3.52 in December 1921; maximum snowfall 35.4 in February 1916; maximum snowfall in 24 hours 21.5 in February 1916.

REFERENCE NOTES APPLYING TO TABLES APPEAR ON THE PAGE FOLLOWING LAST TABLE.
(Caution: Letters and symbols may have different meanings in 1941-1970 tables than in earlier tables. See notes.)

NORMALS, MEANS AND EXTREMES
(Table Revised 1965. Base Period for Climatological Normals: 1931-1960)

SPOKANE, WASHINGTON
GEIGER FIELD
LATITUDE 47° 37' N
LONGITUDE 117° 31' W
ELEVATION (ground) 2357 FEET

Temperature, Degree Days, Precipitation and Snow

Month	Normal Daily maximum	Normal Daily minimum	Normal Monthly	Extremes Record highest	year	Extremes Record lowest	year	Normal degree days	Precip. Normal total	Precip. Max monthly	year	Min monthly	year	Max in 24 hrs	year	Snow Mean total	Max monthly	year	Max in 24 hrs	year
(yrs)	(b)	(b)	(b)	5		5		(b)	(b)	17		17		17		17	17		17	
J	31.4	19.2	25.3	49	1962	-13	1962	1231	2.44	4.96	1954	0.50	1949	1.48	1954	19.4	56.9	1950	13.0	1950
F	37.4	22.5	30.0	56	1962	-11	1963	980	1.86	3.94	1963	0.98	1964	1.11	1961	9.1	20.6	1949	8.0	1951
M	47.0	29.1	38.1	71	1960	2	1960	932	1.50	3.75	1950	0.45	1965	0.91	1950	5.5	15.3	1951	5.3	1960
A	58.6	35.9	47.3	80	1960	24	1964	531	0.91	3.08	1948	0.08	1956	0.89	1962	0.6	6.2	1964	4.9	1964
M	69.3	43.1	56.2	87	1961	30	1960	285	1.21	3.06	1948	0.45	1964	1.67	1948	T	6.1	1954	6.1	1954
J	74.5	49.3	61.9	97	1961	34	1962	138	1.49	3.06	1964	0.16	1960	2.07	1964	T	T	1954		
J	85.6	55.4	70.5	102	1960	39	1962	9	0.38	1.29	1948	T	1960	0.79	1955	0.0	0.0			
A	74.7	52.9	68.0	108	1961	35	1961	25	0.75	1.46	1964	T	1950	1.09	1959	0.0	0.0			
S	60.1	38.0	49.1	93	1963	32	1963	168	0.75	2.05	1959	0.06	1950	0.92	1948	0.0	0.1	1957	0.1	1957
O	42.9	28.5	35.7	85	1963+	24	1960+	493	1.57	4.05	1960	0.34	1965	0.98	1950	0.6	6.1	1955	6.1	1955
N	35.9	24.2	30.1	58	1959	-2	1959	879	2.05	4.64	1960	0.34	1956	1.41	1960	6.1	24.7	1955	14.7	1955
D	30.1	24.2	30.1	51	1959	-20	1964	1082	2.43	5.13	1964	1.21	1951	1.60	1951	16.5	42.0	1951	12.0	1951
YR	58.4	37.1	47.8	108	AUG 1961	-20	DEC 1964	6655	17.19	5.71	MAY 1948	T	JUL 1960+	2.07	JUN 1964+	57.0	56.9	JAN 1950	13.0	JAN 1950

∅ Data for August - December 1959 considered in extracting temperature extremes above.
Means and extremes in the above table are from the existing location. Annual extremes have been exceeded at other locations as follows:
Lowest temperature -30 in January 1888; maximum monthly precipitation 5.85 in November 1897; minimum monthly precipitation 0.00 in July 1883; maximum precipitation in 24 hours 2.22 in June 1888.

NORMALS, MEANS AND EXTREMES
(Table Revised 1975. Base Period for Climatological Normals: 1941-1970)

Station: SPOKANE, WASHINGTON INTERNATIONAL AIRPORT
Latitude: 47° 38' N
Longitude: 117° 32' W
Elevation (ground): 2356 feet
Standard time used: PACIFIC

Temperature, Degree Days, Precipitation and Snow

Month	Normal Daily maximum	Normal Daily minimum	Normal Monthly	Extremes Record highest	year	Extremes Record lowest	year	Normal Degree days Base 65°F Heating	Cooling	Water equiv. Normal	Max monthly	year	Min monthly	year	Max in 24 hrs	year	Snow Max monthly	year	Max in 24 hrs	year
(yrs)	(a)			15		15					27		27		27		27		27	
J	31.1	19.6	25.4	59	1971	-19	1969	1228	0	2.47	4.96	1949	0.50	1959	1.48	1954	56.9	1950	13.0	1950
F	39.0	25.3	32.2	60	1972	-12	1972	918	0	1.63	3.75	1961	0.58	1965	1.11	1963	20.6	1949	8.0	1951
M	46.2	28.8	37.5	72	1960	17	1960	853	0	1.53	3.08	1950	0.31	1965	0.89	1950	15.3	1951	5.3	1962
A	57.0	35.2	46.1	80	1968	24	1966	567	8	1.12	2.20	1962	0.45	1956	1.67	1962	6.2	1964	4.9	1964
M	66.5	42.8	54.7	93	1966	30	1954	327	39	1.46	5.71	1948	0.16	1964	2.07	1964	3.5	1967	3.5	1967
J	73.6	49.4	61.5	100	1973	34	1962	144		1.36	3.06	1964			2.07	1964				
J	84.3	54.1	69.7	103	1967	38	1971	21	167	0.40	1.29	1948	T	1973	0.79	1955	0.0			
A	82.5	54.0	68.0	103	1961	34	1968	47	140	0.58	1.73	1965	0.06	1950	1.12	1973	T	1965	T	1965
S	72.5	46.7	59.6	93	1973	25	1973	196	34	0.83	2.05	1950	0.05	1965	0.98	1965	0.1	1957	0.1	1957
O	58.1	37.5	47.8	80	1965	17	1935	533	0	1.42	4.05	1960	0.34	1956	1.41	1960	6.1	1949	6.1	1949
N	41.8	29.2	35.5	65	1965	-2	1955	825	0	2.20	5.10	1973	0.34	1956	1.41	1960	24.7	1955	12.1	1951
D	33.9	24.0	29.0	58	1968	-25	1968	1116	0	2.37	5.13	1964	1.21	1951	1.60	1951	42.0	1951	12.0	1951
YR	57.2	37.3	47.3	108	AUG 1961	-25	DEC 1968	6835	388	17.42	5.71	MAY 1948	T	JUL 1973	2.07	JUN 1964	56.9	JAN 1950	13.0	JAN 1950

Means and extremes above are from existing and comparable exposures. Annual extremes have been exceeded at other sites in the locality as follows:
Lowest temperature -30 in January 1888; maximum monthly precipitation 5.85 in November 1897; minimum monthly precipitation 0.00 in July 1883;
maximum precipitation in 24 hours 2.22 in June 1888.

REFERENCE NOTES APPLYING TO TABLES APPEAR ON THE PAGE FOLLOWING LAST TABLE.
(Caution: Letters and symbols may have different meanings in 1941-1970 tables than in earlier tables. See notes.)

NORMALS, MEANS AND EXTREMES
(Table Revised 1965. Base Period for Climatological Normals: 1931-1960)

STAMPEDE PASS, WASHINGTON
LATITUDE 47° 17' N
LONGITUDE 121° 20' W
Elevation (ground) 3958 FEET

Month	Normal Daily maximum	Normal Daily minimum	Normal Monthly	Extremes Record highest	Year	Extremes Record lowest	Year	Normal degree days	Precip Normal total	Precip Max monthly	Year	Precip Min monthly	Year	Precip Max 24 hrs	Year	Snow Mean total	Snow Max monthly	Year	Snow Max 24 hrs	Year	RH 4:00 A.M.	RH 10:00 A.M.	RH 4:00 P.M.	Mean sky cover	Clear	Partly cloudy	Cloudy	Precip .01"+	Snow 1.0"+	Thunderstorms	Heavy fog	Max 90°+	Max 32°-	Min 32°-	Min 0°-
(a)	(b)	(b)	(b)	21		21		(b)	(b)	21		21		21		21	21		21		20	21	21	21	21	21	21	21	21	21	21	21	21	21	21
J	27.5	19.5	23.5	56	1962	-11	1950	1287	12.03	29.02	1953	4.37	1960	5.74	1945	78.5	192.9	1946	34.4	1946	93	92	91	8.6	3	2	26	22	21	0	24	0	23	30	2
F	31.1	22.5	26.6	57	1962	-9	1950	1075	10.15	20.78	1961	2.99	1962	3.86	1951	79.3	181.7	1949	36.2	1949	93	92	89	8.5	3	3	23	22	15	0	23	0	16	27	*
M	34.8	25.1	30.0	61	1960	-6	1960	1085	10.60	19.54	1955	3.45	1958	3.34	1954	74.0	154.8	1955	25.3	1954	90	89	85	8.1	3	3	25	22	15	*	21	0	15	29	0
A	42.6	30.6	36.5	66	1960	16	1960	855	6.43	19.57	1959	1.12	1948	5.41	1962	34.7	107.2	1970	18.2	1948	89	82	78	8.0	3	5	22	16	10	*	21	0	4	23	0
M	51.0	36.7	43.9	81	1958	20	1954	654	4.25	9.12	1948	0.91	1949	2.10	1948	11.2	42.5	1951	8.8	1951	89	76	71	8.0	4	6	20	14	2	*	21	0	1	11	0
J	56.7	42.9	48.0	86	1961	28	1950	483	4.49	7.98	1947	1.03	1951	2.04	1952	1.8	9.4	1954	6.5	1950	91	76	71	7.3	6	5	19	12	1	*	21	0	0	1	0
J	65.4	47.0	56.2	87	1958	30	1955	273	1.46	4.04	1955	0.36	1960	1.17	1961	T	2.1	1962	2.1	1964	87	67	50	4.8	14	7	10	8	0	1	17	0	0	0	0
A	64.5	46.7	55.6	90	1958	34	1964	291	2.04	6.19	1964	0.26	1955	1.95	1964	T	2.1	1964	1.7	1961	87	71	65	5.6	11	5	14	9	0	1	19	0	0	0	0
S	59.8	43.2	51.3	87	1944	28	1944	393	4.39	15.48	1947	0.92	1957	4.47	1946	0.7	6.0	1946	6.0	1946	87	71	66	5.4	11	5	14	12	*	1	20	*	1	11	*
O	48.1	36.6	42.4	73	1952	22	1951	701	9.81	25.43	1947	1.32	1952	5.32	1947	17.0	47.0	1961	13.7	1961	88	81	79	7.2	6	3	22	17	4	*	23	0	11	26	*
N	35.0	27.2	31.1	58	1949	-9	1955	1008	12.58	29.06	1962	0.93	1945	7.94	1962	64.0	138.5	1945	32.8	1945	93	92	91	8.4	3	3	24	23	11	*	23	0	11	26	*
D	30.9	23.0	27.0	53	1957	-16	1964	1178	16.19	29.06	1953	6.00	1960	7.01	1951	81.1	163.3	1949	32.6	1951	94	94	94	8.9	2	2	27	24	14	0	26	0	21	30	3
YR	45.7	33.3	39.5	90 AUG 1958		-16 DEC 1964		9283	92.19	29.06 DEC. 1953		0.19 JUL. 1960		7.94 NOV. 1962		450.5	192.9 JAN. 1946		36.2 FEB. 1949		90	82	78	7.4	70	54	241	206	85	7	252	*	94	190	3

NORMALS, MEANS AND EXTREMES
(Table Revised 1975. Base Period for Climatological Normals: 1941-1970)

Month	Normal Daily maximum	Normal Daily minimum	Normal Monthly	Extremes Record highest	Year	Extremes Record lowest	Year	Degree days Heating	Degree days Cooling	Precip Normal	Precip Max monthly	Year	Precip Min monthly	Year	Precip Max 24 hrs	Year	Snow Max monthly	Year	Snow Max 24 hrs	Year	RH 04	RH 10	RH 16	RH 22	Mean sky cover	Clear	Partly cloudy	Cloudy	Precip .01"+	Snow 1.0"+	Thunderstorms	Heavy fog	Max 90°+	Max 32°-	Min 32°-	Min 0°-	Avg station pressure mb	Elev feet m.s.l.
(a)				31		31				31	31		31		31		31		31		30	31	31	29	29	29	29	29	31	31	21	31	31	31	31	31		3967
J	27.3	18.9	23.1	59	1968	-11	1950	1299	0	12.89	30.42	1955	4.37	1969	5.74	1945	192.9	1946	37.7	1946	94	92	91	89	8.7	2	3	26	23	16	0	24	0	23	30	2		875.4
F	32.1	23.5	27.8	57	1962	-9	1950	1042	0	10.22	20.78	1961	7.14	1958	3.86	1951	181.7	1949	36.2	1949	92	91	88	87	8.5	3	3	24	21	14	*	23	0	16	29	*		877.4
M	35.1	24.9	30.0	61	1968	-6	1960	1085	0	8.93	19.54	1955	2.56	1960	3.34	1954	154.8	1955	25.3	1954	93	91	87	84	8.4	3	4	24	21	14	0	21	0	14	29	0		873.7
A	41.5	29.5	35.5	67	1968	16	1960	885	0	6.63	19.50	1959	1.12	1957	5.41	1962	107.2	1970	18.8	1970	89	82	78	75	8.4	3	5	20	16	10	*	21	0	5	23	0		880.2
M	50.3	35.9	43.1	81	1958	20	1954	679	0	4.17	9.12	1948	0.98	1962	2.76	1948	42.5	1951	8.8	1951	90	75	70	70	7.8	4	6	20	15	3	*	21	0	1	11	0		880.0
J	57.0	41.3	49.2	86	1961	28	1950	474	0	3.96	7.98	1947	0.37	1947	2.76	1974	9.4	1954	6.5	1974	89	75	70	70	7.3	6	5	19	15	3	1	21	0	1	1	0		880.0
J	65.7	46.9	56.2	88	1971	30	1955	281	8	1.54	4.04	1955	0.36	1960	1.96	1972	6.4	1971	6.4	1971	86	66	59	59	4.8	13	8	10	8	*	1	17	0	0	0	0		882.0
A	64.4	47.0	55.7	90	1958	34	1964	296	8	2.35	7.14	1968	0.36	1955	1.95	1964	T	1964	T	1964	81	70	63	63	5.4	12	6	12	11	*	2	19	*	0	1	0		881.2
S	59.4	43.2	51.3	87	1944	28	1944	411	0	4.89	15.48	1947	0.92	1957	4.47	1961	26.7	1961	11.9	1961	87	70	66	66	5.7	10	6	14	11	*	1	20	*	1	11	*		881.3
O	47.6	35.0	41.1	73	1970	22	1951	722	0	8.69	23.55	1947	1.32	1952	5.32	1947	47.0	1961	13.7	1961	91	81	80	80	7.4	6	5	20	16	5	*	23	0	12	25	1		881.3
N	35.0	27.2	31.1	58	1949	-15	1955	1017	0	12.31	29.06	1958	0.93	1958	7.94	1962	138.5	1945	32.6	1951	94	92	91	91	8.6	3	5	23	20	11	*	23	0	11	30	*		877.1
D	29.8	22.2	26.0	53	1957	-21	1968	1209	0	14.31	29.06	1953	6.20	1953	7.01	1951	163.3	1949	32.6	1951	95	95	94	94	9.1	2	2	27	24	15	0	26	0	21	30	3		877.3
YR	45.4	33.0	39.2	90 AUG 1958		-21 DEC 1968		9400	16	91.06	30.42 JAN 1969		0.19 JUL 1960		7.94 NOV 1962		192.9 JAN 1946		37.7 JAN 1946		90	81	78	78	7.5	66	58	241	206	88	7	252	*	93	188	3		878.8

REFERENCE NOTES APPLYING TO TABLES APPEAR ON THE PAGE FOLLOWING LAST TABLE.
(Caution: Letters and symbols may have different meanings in 1941-1970 tables than in earlier tables. See notes.)

NORMALS, MEANS AND EXTREMES

(Table Revised 1965. Base Period for Climatological Normals: 1931-1960)

TATOOSH ISLAND, WASHINGTON
LATITUDE 48° 23' N
LONGITUDE 124° 44' W
ELEVATION (ground) 101 FEET

Month	Temp Normal Daily max	Temp Normal Daily min	Temp Normal Monthly	Rec highest	Yr	Rec lowest	Yr	Normal degree days	Precip Normal total	Precip Max monthly	Yr	Precip Min monthly	Yr	Precip Max 24 hr	Yr	Snow Mean total	Snow Max monthly	Yr	Snow Max 24 hr	Yr	RH 4 AM	RH 10 AM	RH 4 PM	RH 10 PM	Wind Mean hourly	Prevail dir	Fastest mile speed	Dir	Yr	Pct poss. sun	Mean sky cover	Days Clear	Days Ptly cloudy	Days Cloudy	Precip .01+	Snow 1.0+	Tstorms	Heavy fog	Max 90+	Max 32−	Min 32−	Min 0−
Yrs.	(b)	(b)	(b)	62		62		(b)	(b)	62		62		62		62	62		62		62	47	62	26	33	24	62	62		54	58	62	62	62	62	62	62	62	62	62	62	62
J	45.2	38.8	42.0	64	1935	14	1950+	713	10.82	22.57	1935	1.84	1949	3.67	1935	3.4	32.3	1950	9.6	1950	84	82	82	82	20.0	E	87	S	1921	26	8.0	4	4	23	22	1	*	1	0	1	4	0
F	46.6	39.5	43.1	64	1941	16	1923	613	8.70	21.16	1961	1.43	1920	4.57	1923	1.5	14.0	1923	12.0	1923	85	81	80	82	17.7	E	84	NE	1916	36	7.5	5	5	19	18	1	1	1	0	*	4	0
M	47.9	40.4	44.2	69	1915	25	1955+	645	8.34	18.16	1916	2.08	1922	4.76	1916	1.2	24.7	1951	10.8	1951	85	81	80	82	15.7	E	91	E	1928	39	7.5	5	6	20	20	1	*	1	0	0	1	0
A	51.6	43.3	47.5	75	1931	33	1961+	525	5.23	10.79	1937	0.68	1956	3.70	1959	T	0.6	1929	0.6	1929	87	82	79	83	13.6	W	73	E	1962+	44	7.3	5	7	18	17	*	*	2	0	0	0	0
M	55.2	47.0	51.1	81	1926	36	1922	431	3.00	8.05	1948	0.54	1924	2.22	1922	0.0	0.0		0.0		89	81	81	84	11.6	W	66	S	1951	47	7.2	5	8	18	14	0	*	3	0	0	0	0
J	57.8	50.0	53.9	84	1903	43	1903	333	2.84	7.81	1922	0.11	1922	2.75	1946	0.0	0.0		0.0		93	86	84	89	10.0	SW	73	S	1962	45	7.2	4	8	18	12	0	*	5	0	0	0	0
J	59.5	51.5	55.5	88	1924	44	1911	295	2.34	7.73	1932	0.07	1922	3.72	1945	0.0	0.0		0.0		94	90	87	97	10.1	S	53	SW	1913	47	6.8	6	8	17	10	0	*	11	0	0	0	0
A	60.1	51.8	56.0	78	1944	45	1956+	279	1.98	5.06	1948	0.07	1916	2.30	1920	0.0	0.0		0.0		96	90	86	94	9.9	S	53	S	1961	44	7.0	5	7	18	11	0	*	16	0	0	0	0
S	59.2	50.4	54.8	80	1949	40	1908	306	3.55	12.31	1920	0.17	1929	3.79	1903	0.0	T	1956	T	1956	90	86	85	90	11.4	S	68	NE	1959+	48	6.5	7	7	16	11	0	*	11	0	0	0	0
O	55.8	48.0	51.9	77	1932	33	1935	406	8.22	17.36	1921	2.50	1952	5.91	1930	T	T		T		86	85	84	87	15.2	E	73	S	1956	39	7.1	6	6	19	17	0	1	6	0	0	0	0
N	50.5	43.9	47.2	68	1936	19	1955	534	10.51	22.17	1954	2.85	1936	4.38	1955	0.4	11.8	1911	6.8	1911	86	83	83	83	18.3	E	94	S	1942	27	8.0	5	5	22	21	*	1	2	0	*	*	0
D	47.6	41.1	44.4	61	1962+	14	1964	639	12.16	21.00	1923	3.61	1914	4.03	1914	1.1	20.2	1964	7.0	1964	83	83	83	83	19.7	E	85	S	1940+	23	8.1	3	5	23	23	*	1	1	0	*	1	0
YR	53.1	45.5	49.3	88 JUL. 1924		14 DEC. 1964+		5719	77.69	22.57 JAN. 1935		0.02 JUL. 1922		5.91 OCT. 1930		7.6	32.3 JAN. 1950		12.0 FEB. 1923		89	84	83	86	14.4	E	94	S	NOV. 1942	40	7.3	58	76	231	197	3	5	59	0	2	8	0

Means and extremes in the above table are from the existing location. Annual extremes have been exceeded at other locations as follows: Lowest temperature 7 in January 1893; maximum monthly precipitation 25.84 in December 1886; minimum monthly precipitation 0.01 in July 1889; maximum snowfall in 24 hours 13.8 in February 1893.

Station moved or discontinued. See other stations for the   this state.

NORMALS, MEANS AND EXTREMES

(Table Revised 1965. Base Period for Climatological Normals: 1931-1960)

WALLA WALLA, WASHINGTON
POST OFFICE BUILDING
LATITUDE 46° 02' N
LONGITUDE 118° 20' W
ELEVATION (ground) 949 FEET

Month	Normal Daily maximum	Normal Daily minimum	Normal Monthly	Extremes Record highest	Year	Extremes Record lowest	Year	Normal degree days	Precipitation Normal total
J	39.0	27.4	33.2	69	1935	-16	1957+	986	1.89
F	44.9	31.8	38.4	71	1963	-14	1950	745	1.52
M	54.4	37.6	46.0	79	1960+	13	1955	589	1.59
A	63.8	43.7	53.8	93	1926	19	1936	342	1.40
M	72.0	49.6	61.0	99	1936+	28	1954	177	1.49
J	78.7	55.6	67.2	106	1961+	41	1916	45	1.22
J	89.2	62.7	76.0	112	1928	46	1921		0.21
A	86.6	61.0	73.8	113	1961	45	1918		0.30
S	77.4	54.0	66.0	103	1955+	26	1926	87	0.78
O	64.6	45.6	55.1	88	1934	15	1935	310	1.53
N	48.8	37.0	42.3	77	1934	1	1955	681	1.72
D	43.8	31.7	37.8	73	1921	-14	1919	843	1.85
YR	63.6	44.7	54.2	113	AUG. 1961	-16	JAN. 1957+	4805	15.50

January-May and September-December 1956 considered in compiling wind data in above table.
Means and extremes in the above table are from the existing location. Annual extremes have been exceeded at other locations as follows:
Lowest temperature -29 in January 1875; maximum monthly precipitation 5.15 in November 1897; maximum precipitation in 24 hours 2.74 in
May 1906; fastest mile of wind 70 from Southwest in December 1912.

NORMALS, MEANS AND EXTREMES

(Table Revised 1975. Base Period for Climatological Normals: 1941-1970)

Average station pressure mb. 991 — Elev. 949 feet m.s.l.

Month	Normal Daily maximum	Normal Daily minimum	Normal Monthly	Extremes Record highest	Year	Extremes Record lowest	Year	Degree days Heating	Degree days Cooling	Precipitation Normal
J	39.3	27.5	33.4	71	1968	-16	1957	990	0	2.07
F	46.9	33.4	40.2	71	1963	-14	1950	694	0	1.40
M	54.1	37.1	45.6	78	1960	13	1955	601	0	1.37
A	62.6	43.0	52.8	92	1926	19	1936	375	0	1.43
M	70.9	49.6	60.3	99	1936	28	1954	175	29	1.58
J	78.7	55.6	67.2	106	1961	46	1966	66	115	1.18
J	88.9	62.3	75.6	112	1928	46	1971	5	334	0.33
A	86.0	61.2	73.6	113	1961	45	1918	12	279	0.45
S	77.4	53.9	65.7	103	1955	26	1926	89	80	0.85
O	63.9	45.4	54.7	88	1934	15	1935	334	0	1.49
N	49.2	36.2	42.7	77	1934	1	1955	669	0	1.50
D	42.5	31.4	37.0	73	1921	-14	1919	868	0	1.97
YR	63.4	44.7	54.1	113	AUG. 1961	-16	JAN. 1957	4835	862	16.01

January-May and September-December 1956 considered in compiling wind and sunshine data in above table.
Means and extremes above are from existing and comparable exposures. Annual extremes have been exceeded at other sites in the locality as follows:
Lowest temperature -29 in January 1875; maximum precipitation in 24 hours 2.74 in May 1906; fastest mile of wind 70 from Southwest in December 1912.

REFERENCE NOTES APPLYING TO TABLES APPEAR ON THE PAGE FOLLOWING LAST TABLE.

(Caution: Letters and symbols may have different meanings in 1941-1970 tables than in earlier tables. See notes.)

NORMALS, MEANS AND EXTREMES

(Table Revised 1965. Base Period for Climatological Normals: 1931-1960)

YAKIMA, WASHINGTON
MUNICIPAL AIRPORT
LATITUDE 46° 34' N
LONGITUDE 120° 32' W
ELEVATION (ground) 1031 FEET

Month	Normal Daily maximum	Normal Daily minimum	Normal Monthly	Extremes Record highest	Year	Extremes Record lowest	Year	Normal degree days	Precip. Normal total
(a)	(b)		(b)		18		18	(b)	(b)
J	36.5	18.5	27.5	64	1953	-21	1950	1163	1.19
F	44.7	24.0	34.0	69	1947	-25	1950	868	0.87
M	55.5	28.7	42.0	80	1960	-1	1960	713	0.62
A	65.8	35.2	50.5	88	1947	21	1964+	435	0.47
M	74.1	42.8	58.5	98	1958+	25	1954	220	0.54
J	80.0	48.8	64.4	103	1961+	33	1962	56	0.81
J	88.9	53.1	71.0	105	1960	35	1954	0	0.13
A	86.5	50.6	68.6	108	1961	35	1960	12	0.20
S	78.8	43.8	61.3	100	1949	29	1965+	144	0.35
O	65.6	35.4	50.5	85	1958	17	1949	450	0.60
N	48.1	26.7	37.4	70	1955	-3	1955	828	0.96
D	39.7	23.3	31.5	62	1959	-17	1964	1039	1.12
YR	63.7	35.8	49.8	108	1961	-25	1950	5941	7.86
				AUG.		FEB.			

Means and extremes in the above table are from the existing and comparable locations. Annual extremes have been exceeded at other locations as follows: highest temperature 111 in July 1928.

NORMALS, MEANS AND EXTREMES

(Table Revised 1975. Base Period for Climatological Normals: 1941-1970)

Station: **YAKIMA, WASHINGTON**
MUNICIPAL AIRPORT

Standard time used: PACIFIC   Latitude: 46° 34' N   Longitude: 120° 32' W   Elevation (ground): 1052 feet

Month	Normal Daily maximum	Normal Daily minimum	Normal Monthly	Extremes Record highest	Year	Extremes Record lowest	Year	Normal Degree days Base 65°F Heating	Cooling	Precip. Normal
(a)				28		28				
J	36.4	18.6	27.5	68	1971	-21	1950	1163	0	1.33
F	46.1	25.2	35.7	69	1947	-25	1950	820	0	0.78
M	54.8	28.8	41.8	80	1960	-1	1960	719	0	0.58
A	64.1	34.8	49.5	88	1947	21	1972	465	0	0.51
M	73.1	42.6	57.9	98	1958	25	1954	239	19	0.55
J	79.7	49.3	64.5	103	1961	33	1962	94	79	0.73
J	88.1	53.3	70.7	108	1971	34	1971	20	197	0.16
A	85.9	50.6	68.1	110	1971	35	1960	37	148	0.25
S	78.3	44.3	61.3	100	1949	26	1971	147	36	0.31
O	64.7	35.4	50.1	85	1958	17	1971	462	0	0.62
N	48.5	28.3	38.4	70	1955	-3	1955	798	0	1.07
D	39.1	23.5	31.3	67	1972	-17	1964	1045	0	1.15
YR	63.2	36.3	49.8	110	1971	-25	1950	6009	479	8.00
				AUG.		FEB.				

Means and extremes above are from existing and comparable exposures. Annual extremes have been exceeded at other sites in the locality as follows: Highest temperature 111 in July 1928.

REFERENCE NOTES APPLYING TO TABLES APPEAR ON THE PAGE FOLLOWING LAST TABLE.

(Caution: Letters and symbols may have different meanings in 1941-1970 tables than in earlier tables. See notes.)

Reference notes applying to Normals, Means, and Extremes tables for 1931–1960 base period.

(a) Length of record, years.
(b) Climatological standard normals (1931-1960).
* Less than one half.
+ Also on earlier dates, months or years.
T Trace, an amount too small to measure. Below-zero temperatures are preceded by a minus sign.

To 8 compass points only.

Unless otherwise indicated, dimensional units used in these tables are: temperature in degrees F.; precipitation, including snowfall, in inches; wind movement in miles per hour; and relative humidity in percent. Monthly heating degree day totals are the sums of the negative departures of average daily temperatures from 65° F. Sleet was included in snowfall totals beginning with July 1948. Heavy fog reduces visibility to 1/4 mile or less.

Sky cover is expressed in a range of 0 for no clouds or obscuring phenomena to 10 for complete sky cover. The number of clear days is based on average cloudiness 0-3; partly cloudy days 4-7; and cloudy days 8-10 tenths.

& Figures instead of letters in a direction column indicate direction in tens of degrees from true North; i. e., 09=East, 18=South, 27=West, and 36=North. If figures appear in the direction column under "Fastest mile", the corresponding speeds are fastest observed 1-minute values.

Solar radiation data are the averages of direct and diffuse radiation on a horizontal surface. The langley denotes one gram calorie per square centimeter. Averages in the lower table for some months may be for more than the listed number of years.

% The station does not operate 24 hours daily. Fog and thunderstorm data therefore may be incomplete.

Reference notes applying to Normals, Means, and Extremes tables for 1941–1970 base period.

(a) Length of record, years, through the current year unless otherwise noted, based on January data.
(b) 70° and above at Alaskan stations.
T Trace.

NORMALS - Based on record for the 1941-1970 period.
DATE OF AN EXTREME - The most recent in cases of multiple occurrence.
PREVAILING WIND DIRECTION - Record through 1962.
WIND DIRECTION - Numerals indicate tens of degrees clockwise from true north. 00 indicates calm.
FASTEST MILE WIND - Speed is fastest observed 1-minute value when the direction is in tens of degrees.

% Through 1964. The station did not operate 24 hours daily. Fog and thunderstorm data may therefore be incomplete.
c Through 1962.
% Record for 1934-64.
∅ For the period 1934-1957.

Mean Annual Precipitation, Inches

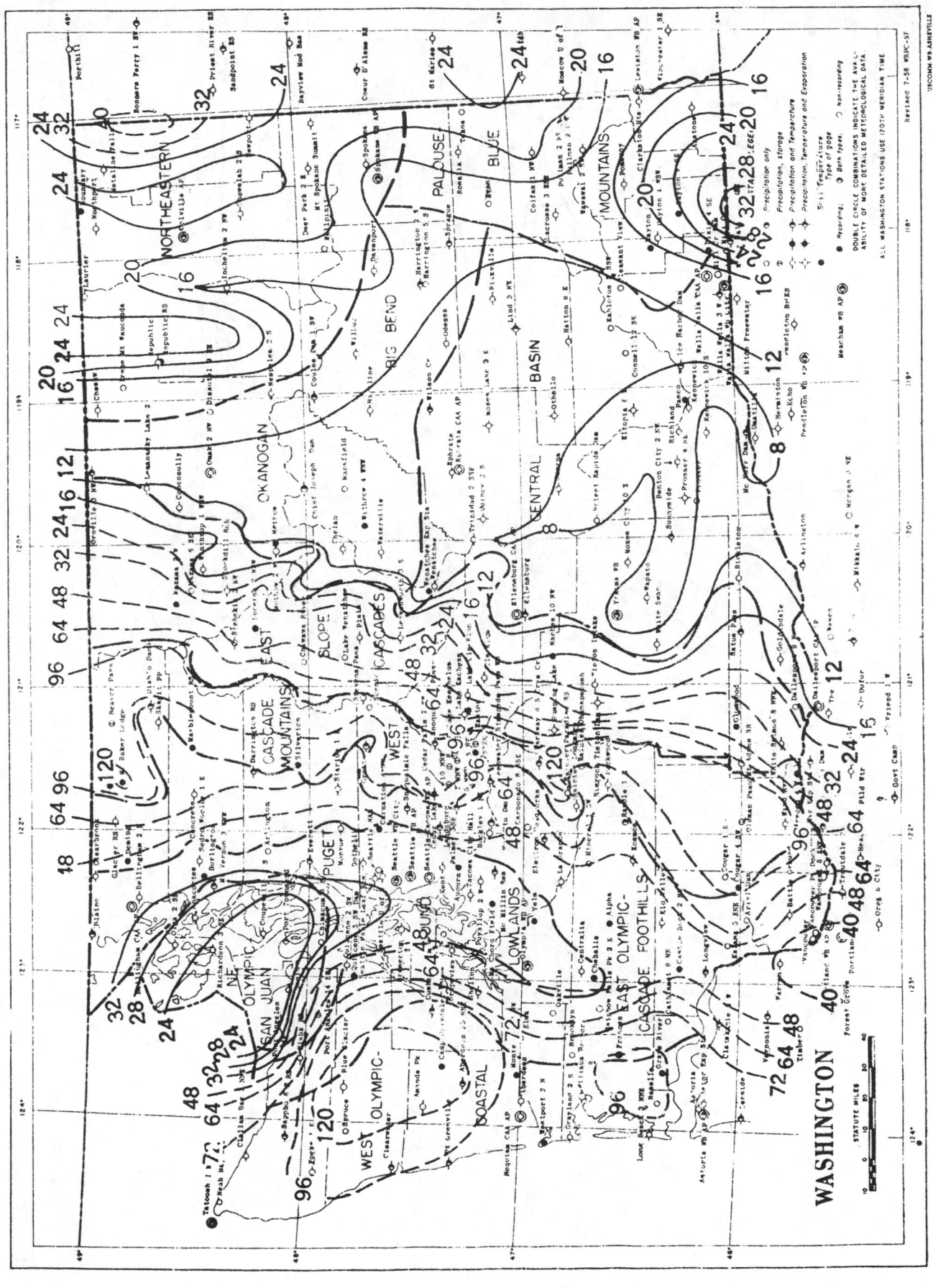

WASHINGTON

Based on period 1931-55

Isolines are drawn through points of approximately equal value. Caution should be used in interpolating on these maps, particularly in mountainous areas.

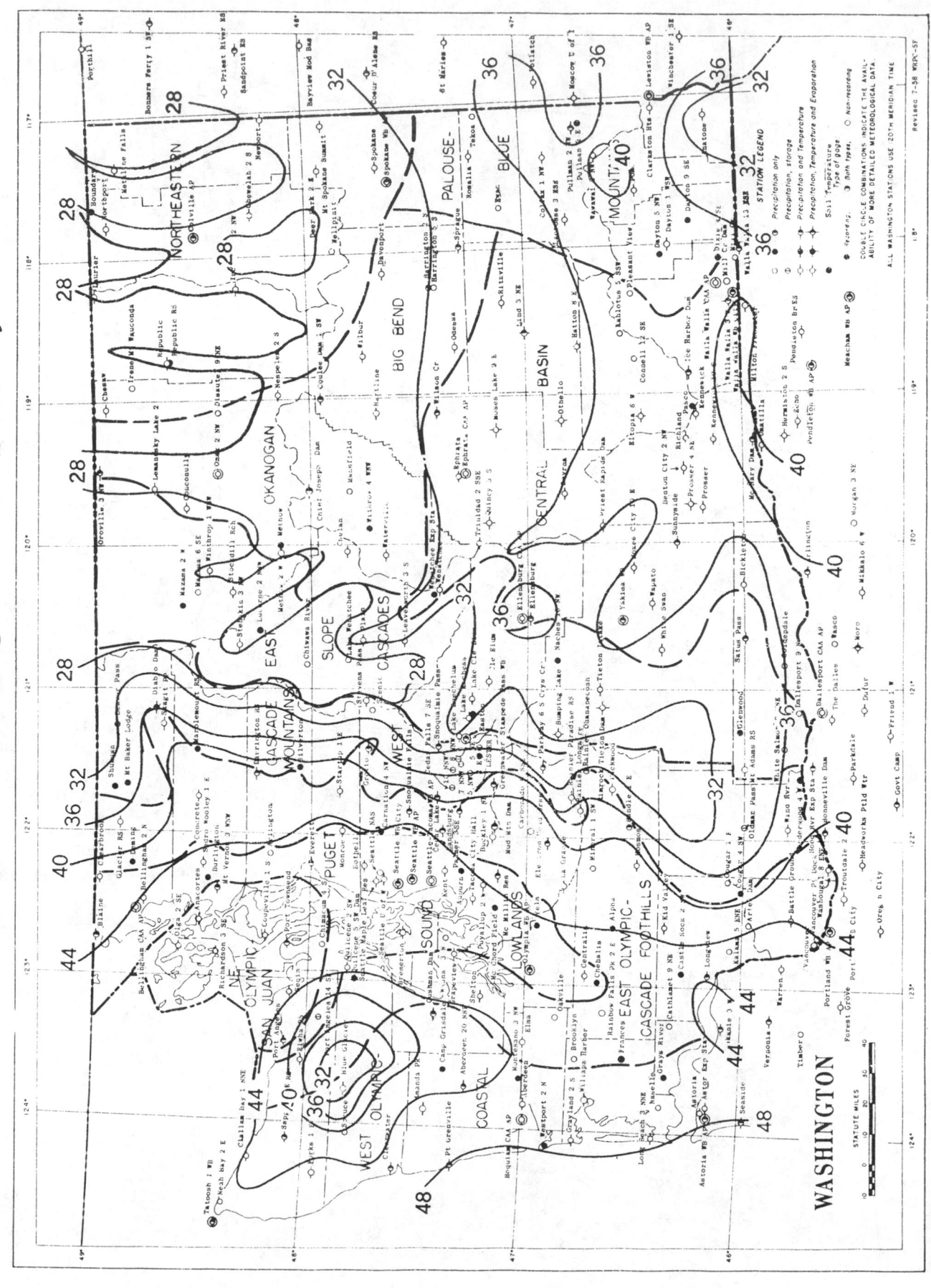

Mean Maximum Temperature (°F.), January

WASHINGTON

Based on period 1931-52

Isolines are drawn through points of approximately equal value. Caution should be used in interpolating on these maps, particularly in mountainous areas.

STATION LEGEND

Mean Minimum Temperature (°F.), January

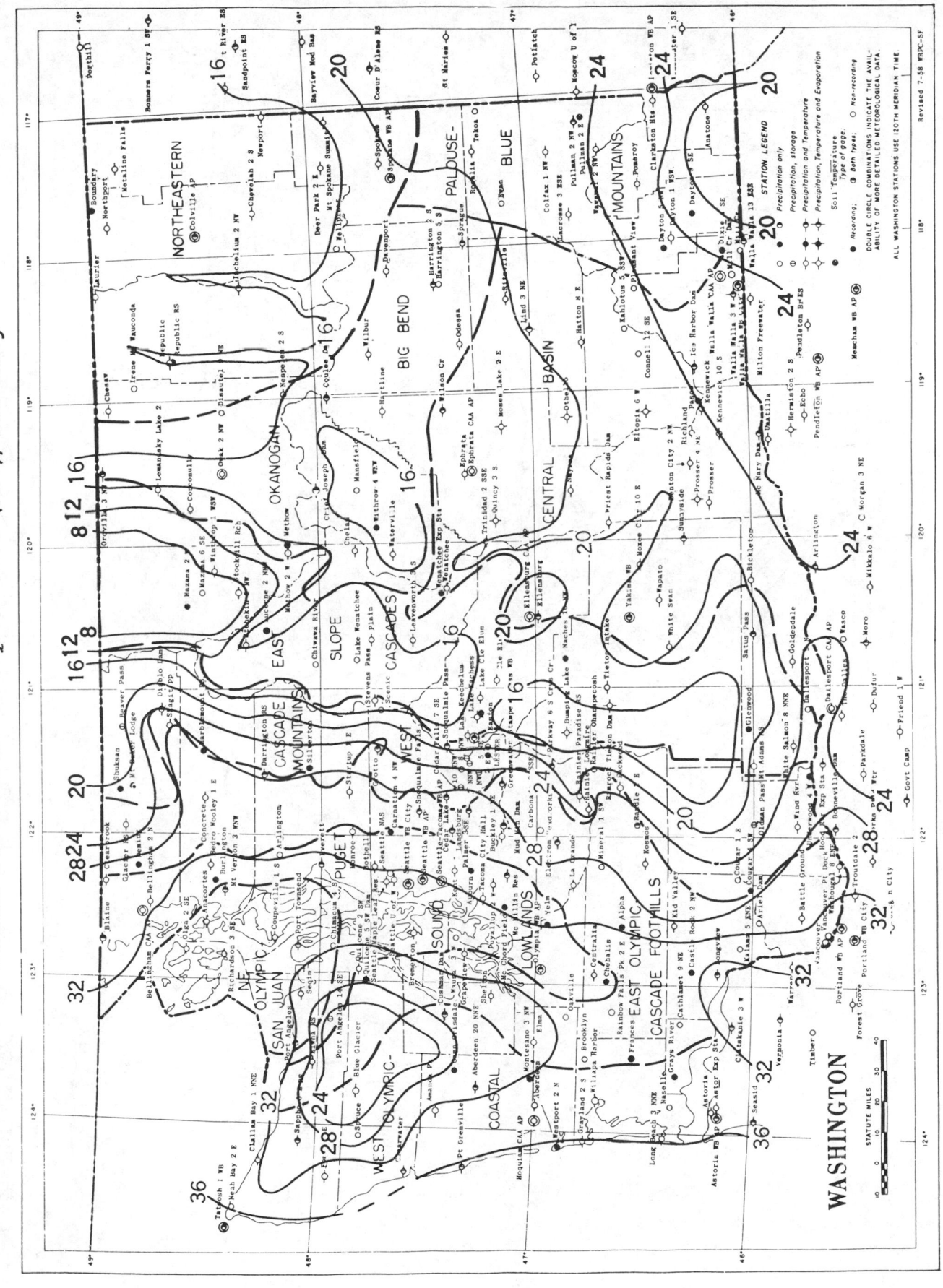

Based on period 1931-52

Isolines are drawn through points of approximately equal value. Caution should be used in interpolating on these maps, particularly in mountainous areas.

Mean Maximum Temperature (°F.), July

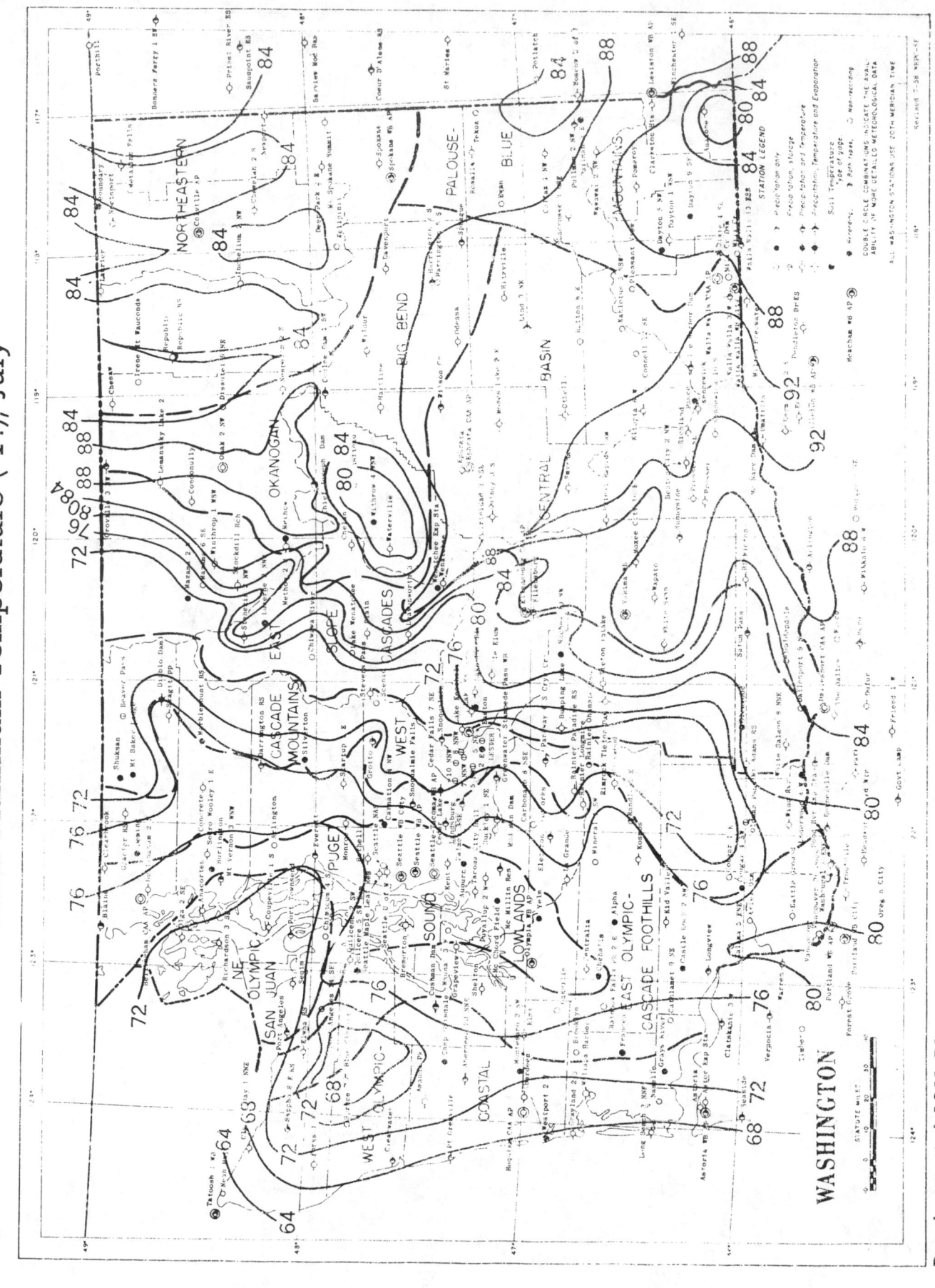

WASHINGTON

Based on period 1931-52

Isolines are drawn through points of approximately equal value. Caution should be used in interpolating on these maps, particularly in mountainous areas.

Mean Minimum Temperature (°F.), July

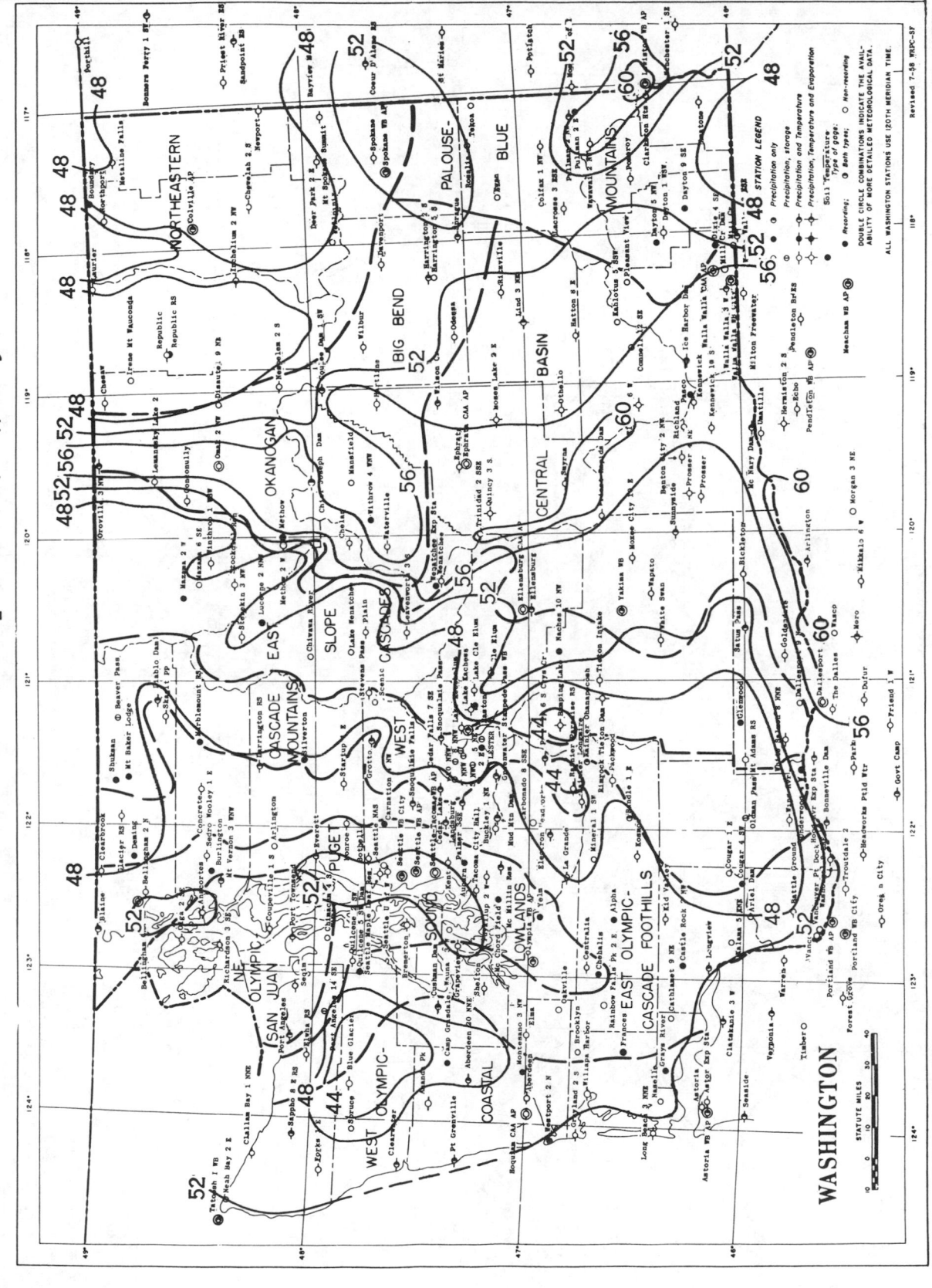

WASHINGTON

Based on period 1931-52

Isolines are drawn through points of approximately equal value. Caution should be used in interpolating on these maps, particularly in mountainous areas.

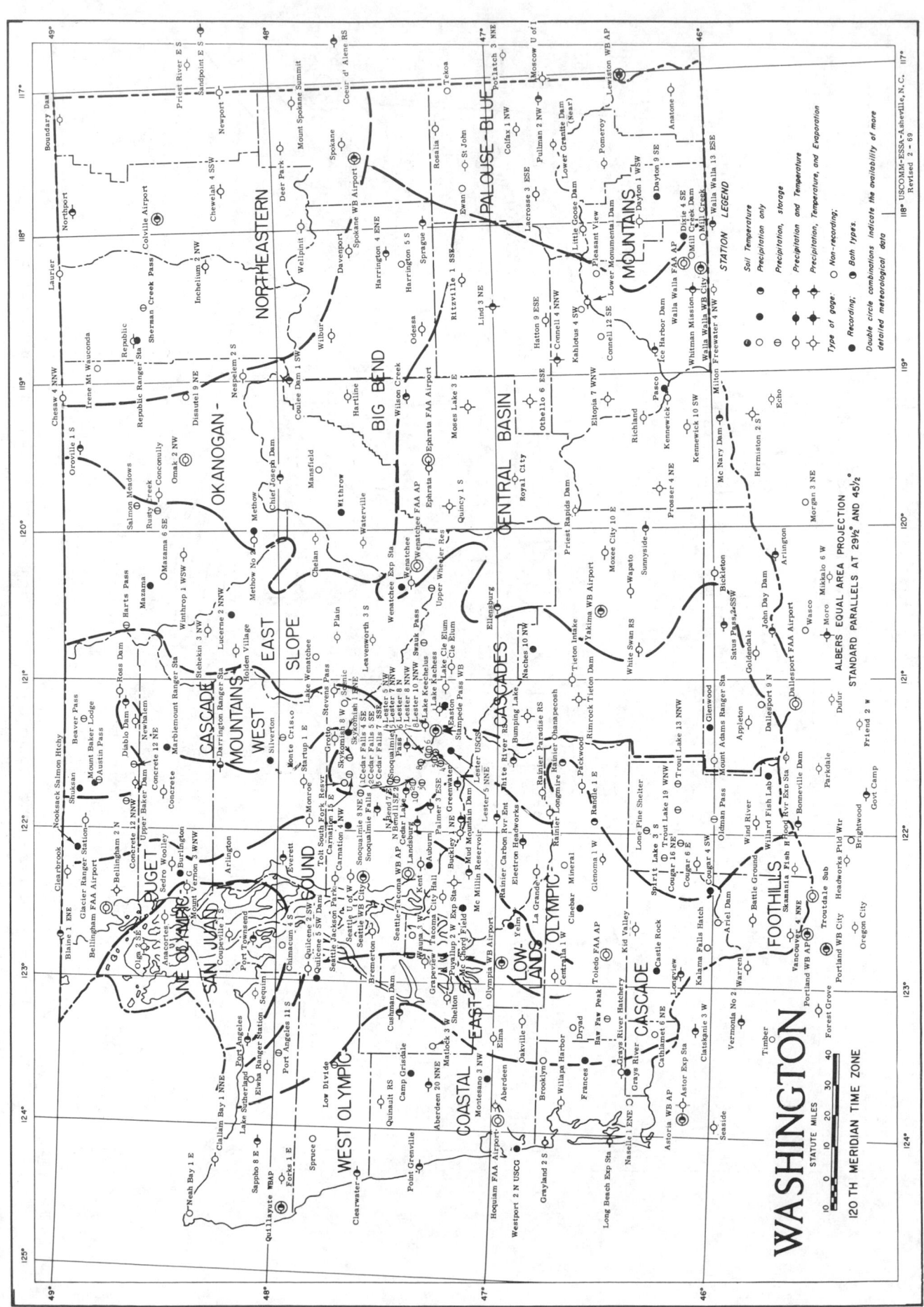

WASHINGTON

STATUTE MILES

10 0 10 20 30 40

120 TH MERIDIAN TIME ZONE

ALBERS EQUAL AREA PROJECTION
STANDARD PARALLELS AT 29½° AND 45½°

STATION LEGEND

Soil Temperature
Precipitation only
Precipitation, storage
Precipitation, Temperature, and Evaporation
Precipitation, Temperature

Type of gage: ● Recording; ○ Non-recording; ⊖ Both types.

Double circle combinations indicate the availability of more detailed meteorological data

USCOMM-ESSA-Asheville, N.C.
Revised 2 - 69

1072

CLIMATES OF THE STATES

WEST VIRGINIA

(Normals, Means and Extremes tables revised 1972 and 1975. Basic report revised February 1960.)

Climate of West Virginia

Victor T. Horn, Weather Bureau State Climatologist

James K. McGuire, Northeastern Area Climatologist

TOPOGRAPHIC FEATURES--The diversity of climatic conditions in West Virginia can be understood best only with some background knowledge of the topography.

West Virginia has an area of over 24,000 square miles, and its main portion is roughly oblong in shape. From southwest to northeast, the oblong is about 200 miles in length; width averages a little over one-half the length. There are two projections: one, the Northeastern Panhandle, which juts eastward between Maryland and Virginia; the other, the Northern Panhandle, is a narrow strip stretching northward along the Ohio River between Ohio and Pennsylvania. The easternmost extremity of the State is about 150 miles from the Atlantic Ocean and the southwestern corner adjacent to Kentucky is nearly 400 miles away from the ocean. As a result, West Virginia lies beyond the immediate climatic effect of the Atlantic, and its climate is much more of the continental than it is of the maritime type. The most important aspect of this type of climate is the marked temperature contrast between summer and winter.

Furthermore, the physical configuration of the State accentuates its interior location. Excluding the Northeastern Panhandle, the State lies in the Allegheny Plateau; but becuase the Appalachian Mountains are the most pronounced feature of the eastern part of the plateau, it is more appropriate to treat the main part of the State in two parts.

The eastern third of the plateau is part of the Appalachian Mountain chain and contains the highest land in the State. Peak elevations in this area range from about 2,500 feet to 4,860 feet (above sea level) at Spruce Knob, the highest point in West Virginia. The central and western thirds of the plateau slope generally westward to the Ohio River which lies at about 550 to 650 feet above sea level. In the north and west, the Allegheny Plateau has been well cut by weather and stream erosion into rounded hills and many fertile and winding valleys. In the south, the plateau has not been eroded as much, and is characterized by flat-topped hills with precipitous slopes. The nature of the terrain and the general topography -- the eastern border of the plateau containing the highest land -- have important climatological effects that will be indicated below.

The foregoing has excluded the Northeastern Panhandle. This is marked by long ridges and valleys, oriented southwest-northeast, intersected by the winding courses of the Potomac River and its tributaries. The main stream of the Potomac with its North Branch forms the northern border of this part of the State. Summit elevations exceed 4,000 feet (above sea level), but the land in general slopes eastward away from the main ridgeline to the west and finally reaches the lowest elevation in the State of 274 feet at Harpers

Ferry. This section lies in the Atlantic Ocean drainage and is drained by the Potomac River. The remainder of the State drains into the Ohio River, whose principal subbasins from north to south are the Monongahela (which flows northward to join the Allegheny River at Pittsburgh, Pa., to form the Ohio River), Little Kanawha, Kanawha, Guyandot, and the Big Sandy. These flow in a general north to west direction from the mountain belt, across the plateau to the main stream which forms most of the State's western border.

CLIMATIC FEATURES--It has been necessary to describe West Virginia's topography in some detail because its physical features considerably modify the effects of the major climatic controls. Briefly, the State's latitudinal position (from about 37° 15' N. Lat. in the south to 40° in the north) places it in the zone of prevailing westerly winds, which are frequently interrupted by northward and southward surges of relatively warm and cold air, respectively. These atmospheric movements are accompanied by the passage of high and low-pressure areas; the latter are the large-dimension storms, known as extratropical cyclones, which are most common in the United States in the colder half-year. West Virginia lies near the average path of the extratropical cyclones that move in a general easterly direction across the United States. In the warmer half-year, the State is affected by the showers and thunderstorms that occur in the broad current of air that tends to sweep north-eastward from the Gulf of Mexico.

The State has a moderately severe winter climate, accentuated and prolonged in the mountains, with frequent alternations of fair and stormy weather. Summer is marked by hot and showery weather: the heat is less pronounced in the mountains, but they are more subject to thunderstorms and have fewer clear days the year-round. Little more can be said in the way of general climatic characteristics because there are marked variations in temperature, precipitation, and the other weather elements, due to the rugged topography, occurring not only between the mountains and plateau areas but even between different parts of the same county. For example, appreciable differences exist between the bottoms and upper slopes of the numerous valleys that entrench the Allegheny Plateau.

For climatological purposes, the State has been divided into six divisions. They are: (1) Northeastern, comprising the projection into the Potomac drainage basin; (2) North Central, embracing most of the northern part of the plateau; (3) Northwestern, made up of the adjacent strip along the Ohio and the panhandle extending thence northward; (4) Southwestern, covering the remainder of the Ohio Valley and stretching back over the major portion of the southern plateau; (5) Central, which includes the main mountainous area; and (6) Southern, occupying the small remainder of the plateau and the mountain country at the lower end of the State. The exact position and area of each of these divisions are shown on the maps accompanying this article. They delineate the more important climatic zones, but cannot be taken to represent all the local differences mentioned above.

TEMPERATURE--The maps of January and July mean monthly maximum and minimum temperature illustrate the winter and summer thermal patterns. Despite several considerable differences, the maps share a common feature: there is about as much temperature contrast across the State from east to west as there is in twice the distance from north to south. This condition prevails throughout the year, though it varies in magnitude with the seasons and cannot be expected to hold every day. Here the general effect of the topography is clear: locations in

the mountainous belt, regardless of their latitude, tend to have lower temperatures than those in the rest of the State. Average winter minimum temperatures range from the low 20's in the mountains of the Central and Northeastern Divisions, and in the Northern Panhandle, to near 30°F. in the extreme southern and southwestern corners of the State, while average winter maximum readings are in the middle and upper 40's, except in the mountains and in the Northern Panhandle where they are close to 40°F. In summer, maximum temperatures average over 85°F. everywhere except in the mountains, where they are 5° to 10° cooler; average minimum temperatures during this season range from the middle 50's in the mountains to the middle 60's elsewhere.

Spring and autumn mean temperatures average in the 50's, with similar geographical variations. The average date of the last freezing temperature in spring ranges from mid-April in the southwest to mid-May in the mountains; the average first occurrence of 32°F. in the fall similarly varies from late October to late September. A table accompanying this article gives more information for specific places on the occurrence of 32° and other low temperatures.

Despite what has been said about the coolness of the mountains, they can on occasion be as hot as any other part of West Virginia. Temperatures near or over 100°F. have been recorded at all observing stations in the State, up to 112° at Martinsburg in the Northeastern Division. On the other hand, very low temperatures (below -30°) have been observed only in the mountains and in the North Central Division, down to -37°F. at Lewisburg. Of course, these are extremes, and do not represent usual winter conditions. Cold waves, with near or subzero temperatures, come on an average of three times a winter, but as a rule do not last more than 2 or 3 days.

HUMIDITY AND FOG CONDITIONS--Because of the varied topography and associated differences in local climates, it is difficult to generalize about the humidity conditions over the State. Relative humidity averages from the Weather Bureau Office at Parkersburg may be taken as representative of conditions in the Ohio Valley and the western part of the plateau. At this location, nighttime and early morning relative humidity averages about 80 percent, being somewhat less in spring (near 74 percent) and higher in late summer and autumn (about 84 percent). The maximum in late summer and autumn is associated with the occasional occurrence of nocturnal and morning fogs in the river bottoms at this time of year. Midday values are moderate, about 50 to 60 percent for all months, so that there is usually a sharp decrease in the relative humidity from sunrise to noon. Only infrequently will there occur a spell of oppressively hot, muggy weather in the summer, lasting as long as 2 weeks or more, when a steady flow of warm, humid air from the Gulf of Mexico is pumped northward, induced by a more-or-less stationary high pressure center off the Southeastern Coast.

At Charleston, in the southern part of the State, the midday humidities average practically the same as those at Parkersburg. The morning values are about the same in winter and spring, but average higher in summer and fall when there is a higher frequency of fog conditions.

At Elkins, which is representative of the high valleys of the central mountain area, 1 a.m. and 7 a.m. relative humidity averages are quite high (80 to 95 percent), reflecting valley fog conditions in the early morning hours. Since these values are accompanied by moderate air temperatures, the high humidities cause comparatively little discomfort. During the rest of the day, humidities are generally at comfortable levels.

At Petersburg, on the eastern side of the main Appalachian ridge, relative humidities in the morning and midday average somewhat lower than at Elkins for all seasons of the year.

Fog conditions over the State are complicated as to their causes and distribution. The valley fogs, just mentioned, are usually of the radiation type, and occur characteristically when a high-pressure area is centered over or near the State. This situation is most common in late summer and fall. Low cloudiness and fog in the mountains are generally orographic in nature, that is, the result of moist winds moving upslope, so that there is usually a great difference in cloud and fog conditions on opposite sides of a ridge.

PRECIPITATION (INCLUDING SNOWFALL)--The map of mean annual precipitation exhibits some interesting features. It will be noticed that yearly amounts average the greatest in the Central Division -- in excess of 50 (and even 60) inches. West of this belt of heavy precipitation, amounts decrease to about 40 inches along the Ohio River. East of it, there is a much more abrupt decrease to close to 30 inches in the western part of the Northeastern Division, with an increase to about 40 inches in the extreme eastern tip of the State.

This pattern can be directly related to the fact that the rain and snow-producing atmospheric currents generally move across West Virginia on an eastward course. As they approach the mountains, these air currents are subject to orographic lifting, which acts to "trigger" potential precipitation or to intensify the rain or snow that may already be falling. As a result, average annual precipitation increases from the Ohio eastward to the Appalachians. On the other side of the mountains, there is the well-marked "rain shadow" where the air currents descend the leeward slopes and precipitation is correspondingly reduced, to increase only when more favorable topographic influences are encountered farther eastward and where the influence of the ocean and coastal storms is more pronounced.

Mean annual snowfall exhibits the same features, but to a more remarkable degree. The mountain belt receives over 60 inches of snow a year, on the average. Pickens, at an elevation of 2,700 feet (above sea level), located near the middle of the western boundary of the Central Division, had an average annual snowfall of 115 inches for a recent period of 14 years. Amounts over 20 inches have been experienced everywhere else, except in that part of the State west of longitude 81° 30 'W. which usually receives about 15 inches. The Northeastern Division averages about 20 to 30 inches yearly; much of this occurs with the coastal storms. These are very heavy producers of snow and occasionally strike this portion, but only infrequently affect the area farther inland.

It is very unusual for a relatively small and compact area the size of West Virginia to exhibit such great differences in snowfall. From Charleston to Pickens there is a sevenfold increase in average annual snowfall over an airline distance of only 75 miles. Furthermore, the heavy snowfall at elevations under 5,000 feet (above sea level) is unusual here in the East, for an area located south of 40° north latitude.

In winter, roads may be blockaded by heavy falls of snow, particularly in the mountain country. The snow, as a general rule, does not remain on the ground for extended periods over most of the State. Except in the higher portions of the plateau and in the mountains themselves, the snow cover does not persist for anything like the duration of the winter. In other words, the snowstorms are usually followed by thawing periods and there is no large-scale melting in the spring of a seasonally accumulated snowpack.

SUNSHINE AND CLOUDINESS--West Virginia lies in a cloudy belt. Percentage of possible sunshine is only about 40 in winter, increasing to somewhat over 60 percent in early autumn. Cloudiness is most pronounced over the mountains. The average annual number of clear days ranges from about 80 in the mountains to about 120 in the western portion. Conversely, cloudy days average fewest (about 140) in the west and increase by 10 to 20 percent in the mountain belt. In addition to cloudiness, the hours of sunshine are reduced by fog, particularly in the river valleys.

WINDS AND STORMS--As stated previously, the prevailing winds blow from westerly directions. There is a tendency outside of the mountain belt for southerly or southwesterly winds during summer and fall. Thunderstorms occur on an average of 40 to 50 days per year, being more frequent in the mountains. June and July are the months of most frequent occurrence. Violent local winds accompanying thunderstorms are experienced every year in some part of the State, but tornadoes are rare. In the 43 years ending with 1958, a total of 13 tornadoes struck the State; almost all the deaths and destruction recorded from such storms during this period were due to one very severe tornado that struck Shinnston and nearby towns on June 23, 1944; all the other tornadoes were comparatively minor. The most outstanding hailstorm reported in the State caused losses of $200,000 to building and crops in the northern part of Preston County on July 18, 1926. The climatological records show that destructive hailstorms occur on an average of about three per year in West Virginia. Hailstorms are most serious in their economic effects on the fruit growing areas of the Northeastern Panhandle and, to a lesser extent, on the burley tobacco growing areas of the southwestern part of the State.

Though hurricanes have damaged the State, principally as a result of heavy rains, it is uncommon for this type of storm to strike West Virginia with full force. The remnants of the hurricanes which have affected the State have been more noted for their accompanying heavy rainfalls than for any high winds produced. In the Northeastern Panhandle, there have been sizeable losses from fruit drop caused by winds accompanying the passage of a hurricane, but such losses were due more to the circumstance that the fruits were at the stage of development when droppage is likely to occur rather than to any unusually high intensity of the wind.

Much more frequent and costly is the damage from intense large-area storms -- that is to say, from exceptionally strong specimens of the ordinary LOWS that affect the State quite frequently during the colder half of the year. The great storm of November 1950 is an example of this sort. Such storms produce high winds and heavy rain or snow; they paralyze commercial and agricultural activities and cause widespread major damage with deaths and injuries. Under proper conditions, they lead to flooding and damage to the river towns.

Warm-season thunderstorms, mostly those of June and July, often yield intense local rainfall and cause flash flooding in the narrow valleys that cut through the plateau and mountain districts. Greatest precipitation amounts recorded in 24 hours or less at officially recognized precipitation-measuring stations have exceeded 5.00 inches in all six climatological divisions and have exceeded 6.00 inches in divisions 1, 2, and 3 (Northeastern,

North Central, and Northwestern); amounts in excess of 10.0 inches (in 24 hours or less) have been accepted for locations in those same three divisions. Perhaps the outstanding example of intense local rainfall due to thunderstorm activity was the occurrence of a deluge of 19.0 inches in 2 hours and 10 minutes at Rockport in Wood County on July 18, 1889. More recently, 31 fatalities and damage exceeding $10 million resulted from flash flooding caused by heavy thunderstorm rainfall (amounts up to 14 inches) on the night of June 24-25, 1950, in parts of Doddridge, Gilmer, Harrison, Lewis, Pleasants, Ritchie, Tyler, and Upshur Counties. The Petersburg-Moorefield area was hard hit by flash floods on the night of June 17-18, 1949, when up to 12 inches of rain fell in 24 hours. The climatological records for the past quarter century show that this kind of severe local flood, caused by heavy thunderstorm rainfall, is likely to occur in some part of the State every year.

In contrast to flash flooding on the smaller streams due principally to heavy local thundershowers in the warm season, flooding in the larger streams is almost exclusively a cold season phenomenon. Of the 58 floods recorded on the Ohio River in the Parkersburg area since 1832, 54 have occurred during the months from December to April, inclusive. The ideal setup for the cold season floods requires the soil to be well saturated from previous rains, a good snow cover, and a more-or-less stationary front lying northeast-southwest across the State. Along this front separating two contrasting air masses, a succession of "waves" may move northeastward, resulting in copious warm rains for a period of at least several days and a rapid melting of the snow cover. Hoyt and Langbein point out that the Ohio River basin is unique in relation to storm tracks across the United States in that it lies directly in the path of many of the large-scale cyclonic storms which, in the cold half-year and under the conditions just outlined, may bring about the interaction of polar and tropical air masses and consequent excessive and prolonged rainfall simultaneously with the melting of any snow cover present. The Potomac Basin is also subject to winter floods, but they are generally of lesser magnitude than those on the Ohio.

The Ohio River exceeds flood stage more frequently than any of its tributary streams, but severe overflow is infrequent. Since the turn of the century, severe and extensive overflow along the Ohio occurred in March 1913 and 1936 and January 1937. Disastrous floods occurred in the Big Sandy and Guyandot River basins in January 1957. Some other notable flood years in tributary basins have been 1901, 1912, 1916, 1917, 1918, 1926, 1929, 1932, 1935, and 1940.

ECONOMIC ASPECTS--There are several ways in which the State's climate may be related to the activities of its citizens. The farm population, 460,000, according to the 1950 census, represents about 23 percent of the total population. There are about 68,000 farms with an average of 107 acres. In 1957, poultry raising accounted for 28 percent of the total cash receipts from farm sales. The two other most important types of agriculture were the raising of cattle, sheep, and hogs (29 percent of the cash receipts) and dairy farming (23 percent). Other activities are fruit growing, the cultivation of field crops, lumbering, and raising greenhouse and nursery products.

All these agricultural activities are dependent, to a greater or lesser degree, on the weather and climate. For example, broiler and turkey production is a major activity in portions of the Northeastern Panhandle where the yearly extremes of heat and cold are not severe. The important commercial fruit-growing business in the Northeastern Panhandle is favored by the combination of relatively cool winters and frost-free conditions on the higher slopes in spring.

There are 10 million acres of forest land in West Virginia, or 65 percent of the total land area, of which approximately 1 million acres are owned by the Federal and State governments. About one-third of the remainder consists of farm woodlots. The rest is held for nonagricultural and industrial purposes. The forests are predominantly hardwood, with coniferous or softwood spruce occupying only about 3 percent of the total wooded area. The moderate climate and abundant rainfall help to account for the rapid growth and healthy development of the hardwood trees.

In recent years, many varied kinds of manufacturing activities have been attracted to West Virginia. In numerous phases of their operations, they rely upon an ample water supply in the State's principal streams which maintain an adequate flow owing to the abundant and generally dependable precipitation. Furthermore, the Ohio River is a major commercial artery, not only for West Virginia, but also for the neighboring States, and its status as such owes much to the rain and snow that fall in the West Virginia headwaters. Worker efficiency is promoted by the climate in that it is characterized by weather changes that stimulate bodily well-being, without being so severe as to strain the physique. Also in recent years, more and more summer vacationists and weekend visitors from nearby States have been attracted by the temperatures that prevail in the West Virginia mountains, especially at night. The post-war years have witnessed a general upsurge in winter sports, and ski-slope developers have taken advantage of favorable snow conditions in the Beckley, Davis, Morgantown, and Terra Alta sections which have some of the few such installations south of the Mason and Dixon line.

All in all, the climate of West Virginia may be summarized as favorable to human activity, with occasional periods in summer and winter that are extreme but rarely prolonged. The State is usually favored by ample precipitation, though by the same token subject to considerable cloudiness; is strongly influenced by its geographical position and topographic features; and is marked by a diversity of local climates the most striking of which is that of the mountain belt.

REFERENCES

(1) Flora, S. D., *Tornadoes of the United States*, 2nd rev. ed., University of Oklahoma Press, 1954.

(2) *Hailstorms of the United States*, University of Oklahoma Press, 1956.

(3) Hoyt, W. G. and Langbein, W. B., *Floods*, Princeton University Press, 1955.

(4) U. S. Department of Agriculture, *Climate and Man*, *The Yearbook of Agriculture for 1941*, Government Printing Office, 1942 (esp. pp. 1182-1190 on the Climate of West Virginia).

(5) West Virginia, Dept. of Agriculture, *West Virginia Agricultural Statistics*, 1956, Charleston, W. Va., 1955.

(6) West Virginia University, Agricultural Extension Service, *Cash Receipts from Farm Sales in West Virginia 1956 and 1957*, Morgantown, W. Va., 1958.

(7) West Virginia, Geological and Economic Survey, *Natural Resources of West Virginia*, Morgantown, W. Va., 1952.

(8) West Virginia, Industrial and Publicity Commission, *West Virginia*, Charleston, W. Va., n.d.

(9) West Virginia, State Planning Board, *Water Resources of West Virginia*, Morgantown, W. Va., December 1937.

(10) Weather Bureau Technical Paper No. 15 - Maximum Station Precipitation for 1, 2, 3, 6, 12, and 24 Hours.

(11) Weather Bureau Technical Paper No. 16 - Maximum 24-Hour Precipitation in the United States. Washington, D. C. 1952.

(12) Weather Bureau Technical Paper No. 25 - Rainfall Intensity-Duration-Frequency Curves. For selected stations in the United States, Alaska, Hawaiian Islands, and Puerto Rico.

(13) Weather Bureau Technical Paper No. 29 - Rainfall Intensity-Frequency Regime. Washington, D. C.

BIBLIOGRAPHY

(A) Climatic Summary of the United States (Bulletin W) 1930 edition, Sections 72 and 73. U. S. Weather Bureau

(B) Climatic Summary of the United States, West Virginia-Supplement for 1931 through 1952 (Bulletin W Supplement). U. S. Weather Bureau

(C) Climatological Data - West Virginia. U. S. Weather Bureau

(D) Climatological Data National Summary. U. S. Weather Bureau

(E) Hourly Precipitation Data - West Virginia. U. S. Weather Bureau

(F) Local Climatological Data, U. S. Weather Bureau for Charleston, Elkins, Huntington and Parkersburg, West Virginia.

FREEZE DATA

STATION	Freeze threshold temperature	Mean date of last Spring occurrence	Mean date of first Fall occurrence	Mean No. of days between dates	Years of record Spring	No. of occurrences in Spring	Years of record Fall	No. of occurrences in Fall
BAYARD	32	05-30	09-14	107	29	29	30	30
	28	05-18	09-28	133	28	28	30	30
	24	04-29	10-09	163	28	28	30	30
	20	04-17	10-21	188	28	28	30	30
	16	03-29	11-08	223	28	28	30	30
BENSON	32	05-15	10-05	143	26	26	27	27
	28	05-01	10-14	166	26	26	27	27
	24	04-19	10-27	190	26	26	27	27
	20	04-05	11-06	214	26	26	27	27
	16	03-20	11-22	248	26	26	27	27
BENS RUN	32	04-21	10-27	189	30	30	30	30
	28	04-05	11-07	217	30	30	30	30
	24	03-22	11-17	240	29	29	30	30
	20	03-08	11-30	267	29	29	30	30
	16	03-05	12-09	279	29	29	30	30
BLUEFIELD 1	32	04-24	10-15	174	29	29	30	30
	28	04-10	10-28	201	29	29	30	30
	24	03-29	11-11	227	29	29	30	30
	20	03-15	11-24	253	29	29	30	30
	16	03-03	12-05	277	29	29	30	28
BUCKHANNON	32	05-06	10-10	157	30	30	30	30
	28	04-17	10-21	187	30	30	30	30
	24	04-03	11-03	214	30	30	30	30
	20	03-16	11-20	249	30	30	30	30
	16	03-09	12-02	268	30	30	30	29
CAIRO 3 S	32	05-02	10-09	160	26	26	26	26
	28	04-22	10-17	177	26	26	26	26
	24	04-10	10-31	204	25	25	26	26
	20	03-22	11-13	236	25	25	26	26
	16	03-11	11-25	259	24	24	26	26
CHARLESTON 1	32	04-18	10-28	193	28	28	27	27
	28	04-01	11-09	222	28	28	27	27
	24	03-19	11-22	248	27	27	27	26
	20	03-09	12-03	270	27	27	27	25
	16	03-02	12-16	289	27	27	27	21
CLARKSBURG 1	32	05-02	10-12	163	28	28	28	28
	28	04-20	10-24	188	28	28	28	28
	24	04-08	11-03	210	28	28	28	28
	20	03-22	11-15	238	28	28	28	28
	16	03-11	11-28	262	28	28	28	28
ELKINS	32	05-10	10-07	150	30	30	30	30
	28	04-26	10-18	175	30	30	30	30
	24	04-11	10-30	202	30	30	30	30
	20	03-27	11-13	231	30	30	30	30
	16	03-16	11-27	256	29	29	30	30
FAIRMONT	32	04-28	10-15	169	30	30	30	30
	28	04-15	10-29	196	30	30	30	30
	24	03-31	11-09	223	30	30	30	30
	20	03-15	11-24	254	30	30	30	30
	16	03-07	12-05	273	30	30	29	29
FLAT TOP	32	05-09	10-05	149	12	12	13	13
	28	04-26	10-21	178	12	12	13	13
	24	04-09	11-02	206	12	12	13	13
	20	04-01	11-14	227	12	12	13	13
	16	03-24	11-25	246	12	12	13	13
GARY	32	04-26	10-17	174	30	30	30	30
	28	04-16	10-29	196	30	30	30	30
	24	03-30	11-08	223	30	30	30	30
	20	03-13	11-23	255	30	30	30	29
	16	03-02	12-07	280	30	30	30	27
HASTINGS	32	04-30	10-16	169	16	16	16	16
	28	04-16	10-31	198	16	16	16	16
	24	04-04	11-11	221	16	16	16	16
	20	03-16	11-18	247	16	16	16	16
	16	03-12	12-02	265	16	16	16	15
HUNTINGTON 1	32	04-14	10-27	196	29	29	28	28
	28	03-28	11-09	226	29	29	28	28
	24	03-13	11-23	254	29	29	28	28
	20	03-06	12-03	272	29	29	28	26
	16	02-20	12-11	294	29	27	28	23
LOGAN	32	04-19	11-01	195	12	12	13	13
	28	04-04	11-09	219	12	12	13	13
	24	03-17	11-28	256	12	12	13	13
	20	03-08	12-08	275	12	12	13	13
	16	02-24	12-15	294	12	11	13	11
LONDON LOCKS	32	04-18	11-01	197	14	14	13	13
	28	04-05	11-11	219	14	14	12	12
	24	03-20	11-24	249	14	14	12	12
	20	03-12	12-06	268	14	14	12	11
	16	03-04	12-07	277	14	14	12	11
MADISON	32	04-24	10-18	176	17	17	18	18
	28	04-10	10-30	203	17	17	18	18
	24	03-25	11-09	230	17	17	18	18
	20	03-12	11-24	258	17	17	17	17
	16	03-03	12-01	273	17	17	17	17
MANNINGTON 1 N	32	05-10	10-07	150	30	30	30	30
	28	04-26	10-16	173	30	30	30	30
	24	04-14	10-28	197	30	30	30	30
	20	03-27	11-12	230	30	30	30	30
	16	03-17	11-26	253	30	30	30	30
MARTINSBURG CAA AP	32	04-23	10-17	177	29	29	30	30
	28	04-08	11-03	209	29	29	29	29
	24	03-22	11-15	238	29	29	29	29
	20	03-10	11-29	264	29	29	29	29
	16	03-01	12-10	284	29	29	29	25
MOOREFIELD MCNEILL	32	05-14	10-01	140	28	28	28	28
	28	04-30	10-11	164	28	28	27	27
	24	04-17	10-20	186	28	28	26	26
	20	04-02	10-31	212	28	28	26	26
	16	03-17	11-17	245	28	28	26	26
NEW CUMBERLAND DAM 9	32	05-04	10-17	166	30	30	30	30
	28	04-22	10-31	192	30	30	30	30
	24	04-10	11-13	217	30	30	30	30
	20	03-23	11-26	248	30	30	30	30
	16	03-11	12-04	268	30	30	30	28
NEW MARTINSVILLE	32	05-02	10-15	166	29	29	29	29
	28	04-19	10-30	194	29	29	29	29
	24	04-01	11-09	223	29	29	29	29
	20	03-18	11-21	248	28	28	29	29
	16	03-07	11-30	268	28	28	28	28
PARKERSBURG	32	04-16	10-21	189	30	30	30	30
	28	04-02	11-08	220	30	30	30	30
	24	03-17	11-22	250	30	30	30	30
	20	03-08	12-01	268	30	30	30	29
	16	03-01	12-12	286	30	30	30	26
PARSONS	32	05-08	10-11	156	27	27	26	26
	28	04-26	10-21	178	27	27	26	26
	24	04-14	11-02	202	27	27	26	26
	20	03-27	11-13	231	25	25	25	25
	16	03-15	11-24	255	25	25	25	25
PETERSBURG WB CITY	32	04-30	10-05	158	12	12	12	12
	28	04-19	10-19	184	12	12	12	12
	24	04-03	11-03	213	12	12	12	12
	20	03-18	11-16	243	12	12	12	12
	16	03-06	12-03	272	12	12	12	12
PIEDMONT	32	05-01	10-10	162	29	29	30	30
	28	04-15	10-25	192	29	29	30	30
	24	03-28	11-08	226	29	29	30	30
	20	03-13	11-25	257	29	29	30	30
	16	03-05	12-08	277	29	29	30	30
PT PLEASANT	32	04-23	10-22	182	29	29	29	29
	28	04-07	11-05	212	29	29	28	28
	24	03-23	11-15	238	29	29	28	28
	20	03-11	11-24	258	29	29	27	27
	16	03-01	12-07	281	29	28	27	24
RAINELLE	32	05-20	10-01	134	20	20	21	21
	28	04-30	10-11	164	20	20	21	21
	24	04-22	10-19	180	20	20	21	21
	20	04-07	11-01	208	20	20	21	21
	16	03-19	11-14	241	20	20	21	21

FREEZE DATA

STATION	Freeze threshold temperature	Mean date of last Spring occurrence	Mean date of first Fall occurrence	Mean No. of days between dates	Years of record Spring	No. of occurrences in Spring	Years of record Fall	No. of occurrences in Fall
RAVENSWOOD DAM 22	32	04-28	10-12	167	27	27	28	28
	28	04-15	10-25	194	27	27	28	28
	24	04-01	11-07	220	27	27	28	28
	20	03-16	11-18	247	27	27	28	28
	16	03-08	12-02	269	26	26	27	26
RICHWOOD	32	05-08	10-05	150	19	19	17	17
	28	04-26	10-18	175	19	19	16	16
	24	04-14	10-28	197	19	19	16	16
	20	04-03	11-16	227	19	19	15	15
	16	03-19	11-24	250	18	18	15	15
ROMNEY	32	05-06	10-05	152	28	28	29	29
	28	04-23	10-16	177	28	28	28	28
	24	04-04	10-30	209	28	28	28	28
	20	03-17	11-16	244	28	28	28	28
	16	03-08	11-30	268	28	28	28	28
SPENCER	32	04-30	10-12	165	29	29	29	29
	28	04-14	10-24	193	28	28	29	29
	24	04-01	11-06	219	28	28	29	29
	20	03-17	11-18	246	28	28	29	29
	16	03-07	11-30	268	28	28	29	28
SUTTON 3 SE	32	05-02	10-12	163	29	29	30	30
	28	04-18	10-27	192	28	28	29	29
	24	04-04	11-04	214	28	28	29	29
	20	03-18	11-16	242	28	28	29	29
	16	03-11	12-02	266	28	28	29	27
TERRA ALTA 1	32	05-27	09-24	120	24	24	23	23
	28	05-15	10-06	144	23	23	22	22
	24	04-24	10-21	180	23	23	21	21
	20	04-10	11-03	207	22	22	21	21
	16	03-27	11-16	235	22	22	20	20
WARDENSVILLE R M FRM	32	05-12	10-02	143	25	25	27	27
	28	04-27	10-10	167	24	24	27	27
	24	04-15	10-24	192	24	24	25	25
	20	03-27	11-11	229	24	24	24	24
	16	03-13	11-23	255	23	23	23	23
WELLSBURG 3 NE	32	05-13	10-04	143	30	30	30	30
	28	04-29	10-17	172	30	30	30	30
	24	04-19	10-31	196	30	30	30	30
	20	04-02	11-13	225	30	30	30	30
	16	03-15	11-26	256	30	30	30	30
WHEELING WARWD D 12	32	05-05	10-16	164	30	30	30	30
	28	04-16	11-04	202	29	29	30	30
	24	04-03	11-15	226	29	29	30	30
	20	03-20	11-25	250	29	29	30	30
	16	03-08	12-07	274	29	29	30	28
WHITE SULPHUR SPRGS	32	05-11	10-05	147	30	30	29	29
	28	04-27	10-14	170	28	28	29	29
	24	04-14	10-22	192	27	27	29	29
	20	03-30	11-04	219	26	26	29	29
	16	03-15	11-19	249	26	26	29	29
WILLIAMSON	32	04-11	10-29	201	29	29	28	28
	28	03-26	11-10	228	29	29	28	28
	24	03-11	11-27	262	29	29	28	26
	20	03-03	12-06	278	29	29	28	24
	16	02-20	12-13	297	29	28	27	20
WINFIELD LOCKS	32	04-19	10-24	187	11	11	12	12
	28	04-05	11-12	221	11	11	12	12
	24	03-17	11-24	251	11	11	12	12
	20	03-12	12-04	267	11	11	12	12
	16	03-10	12-10	275	11	11	11	10

Data in the above table are based on the period 1921-1950, or that portion of this period for which data are available.

Means have been adjusted to take into account years of non-occurrence.

A freeze is a numerical substitute for the former term "killing frost" and is the occurrence of a minimum temperature at or below the threshold temperature of 32°, 28°, etc.

Freeze data tabulations in greater detail are available and can be reproduced at cost.

MEAN TEMPERATURE AND PRECIPITATION

STATION	JANUARY Temperature	JANUARY Precipitation	FEBRUARY Temperature	FEBRUARY Precipitation	MARCH Temperature	MARCH Precipitation	APRIL Temperature	APRIL Precipitation	MAY Temperature	MAY Precipitation	JUNE Temperature	JUNE Precipitation	JULY Temperature	JULY Precipitation	AUGUST Temperature	AUGUST Precipitation	SEPTEMBER Temperature	SEPTEMBER Precipitation	OCTOBER Temperature	OCTOBER Precipitation	NOVEMBER Temperature	NOVEMBER Precipitation	DECEMBER Temperature	DECEMBER Precipitation	ANNUAL Temperature	ANNUAL Precipitation
NORTHWESTERN																										
BELLEVILLE DAM 20		3.60		2.83		4.07		3.63		3.96		4.25		4.24		4.02		2.91		2.14		2.67		2.95		41.27
BENS RUN	35.0	3.65	35.5	2.67	43.1	3.94	53.6	3.59	63.9	4.38	73.3	4.93	76.7	4.25	75.0	4.65	68.6	2.90	57.0	2.35	44.2	2.84	35.2	3.25	55.1	43.40
CAIRO 3 S	35.3	3.91	35.8	2.92	43.4	4.27	53.6	3.67	63.0	4.21	71.5	4.90	74.4	4.78	73.2	4.12	66.9	3.42	56.6	2.58	44.0	3.03	35.0	3.08	54.4	44.89
CRESTON	35.1	4.05	35.4	3.11	42.8	4.43	52.9	3.52	62.8	4.08	71.0	4.91	74.6	4.32	73.3	4.43	67.2	2.96	56.0	2.57	43.5	2.98	35.0	3.30	54.1	44.66
MCMECHEN DAM 13		3.36		2.57		3.88		3.39		3.99		4.12		4.15		3.93		2.90		2.37		2.66		2.66		39.08
NEW CUMBERLAND DAM 9	32.6	3.06	32.5	2.21	40.3	3.60	51.0	3.42	61.7	3.56	70.9	4.10	74.4	3.71	72.8	3.63	66.6	2.98	55.5	2.61	43.1	2.59	33.6	2.44	52.9	37.91
NEW MARTINSVILLE	34.7	3.73	35.0	2.92	42.8	4.05	53.2	3.60	63.8	4.21	72.9	4.54	76.1	4.47	74.6	4.57	68.0	3.10	56.8	2.52	44.1	2.91	34.8	3.40	54.7	44.02
PARKERSBURG WB CITY	34.4	3.17	35.5	2.65	43.6	3.54	53.8	3.08	63.5	3.50	72.4	4.18	75.7	4.16	74.0	4.15	68.4	2.99	56.9	2.12	45.0	2.67	36.1	2.90	54.9	39.11
WASHINGTON DAM 19		3.40		2.65		3.77		3.34		3.62		3.97		3.93		3.98		2.82		1.91		2.45		2.65		38.49
WELLSBURG 3 NE	32.6	3.10	32.6	2.32	40.6	3.91	50.8	3.72	61.0	3.85	70.2	4.25	73.4	4.42	71.9	3.94	65.4	3.38	53.9	2.67	42.0	2.75	33.1	2.52	52.3	40.83
WHEELING WARWOOD DAM 1	32.5	3.10	32.1	2.45	39.7	3.73	50.9	3.40	61.8	3.95	71.3	4.55	75.0	3.71	73.4	3.51	66.8	3.07	55.4	2.30	42.8	2.64	33.3	2.54	52.9	38.95
DIVISION	34.1	3.50	34.3	2.66	41.9	3.93	52.5	3.51	62.7	3.97	71.7	4.54	75.0	4.19	73.5	4.11	67.2	3.07	56.0	2.44	43.6	2.73	34.4	2.90	53.9	41.55
NORTH CENTRAL																										
ABERDEEN		3.97		3.22		4.30		3.96		4.42		4.99		5.30		5.17		3.56		3.03		3.08		3.48		48.48
BENSON	35.1	4.18	35.2	3.40	42.2	4.43	52.5	3.83	62.2	4.27	70.6	5.05	73.8	4.90	72.2	4.35	65.9	3.52	55.3	2.85	43.1	3.12	34.9	3.62	53.6	47.52
BUCKHANNON 2 W	35.1	4.16	34.8	3.46	41.9	4.80	52.1	3.87	61.1	4.51	69.2	5.64	72.0	5.57	70.5	4.93	64.4	3.46	54.3	3.15	43.0	3.25	35.0	3.67	52.8	50.47
CLARKSBURG 1	33.0	3.47	32.7	2.78	40.2	3.58	51.1	3.36	61.6	3.97	70.5	4.40	73.6	4.43	72.0	4.49	65.3	3.28	53.8	2.52	41.5	2.65	33.1	2.89	52.4	41.82
CRAWFORD		3.82		3.10		4.43		3.62		4.41		4.98		5.66		4.88		3.33		2.98		2.99		3.34		47.54
FAIRMONT	34.3	3.59	34.3	2.77	41.9	3.84	53.2	3.52	63.3	4.12	71.8	4.63	75.0	4.39	73.4	4.26	67.1	3.04	55.9	2.61	43.6	2.79	34.6	3.07	54.0	42.63
GLENVILLE	36.8	4.25	37.4	3.49	44.7	4.43	55.0	3.76	64.4	4.37	72.9	5.21	76.0	5.24	74.7	4.77	68.6	3.54	57.4	2.88	45.1	3.29	36.5	3.49	55.8	48.72
GRAFTON 1 NE	34.9	3.89	34.9	2.89	42.1	4.36	51.8	3.79	61.4	4.51	69.6	4.80	72.8	5.01	71.4	4.77	65.6	3.26	55.2	2.99	43.4	3.10	34.9	3.42	53.7	46.79
HORNER		4.06		2.99		4.08		3.65		4.19		5.50		5.11		4.52		3.54		3.04		3.10		3.41		47.19
HOULT LOCK 15		3.78		2.90		3.97		3.60		4.18		4.86		4.29		4.35		3.04		2.61		2.84		3.01		43.43
JANE LEW		3.83		2.94		4.13		3.74		4.33		4.97		5.03		4.74		3.52		2.85		3.08		3.20		46.36
LAKE LYNN		3.28		2.40		3.78		3.51		4.07		4.43		4.08		4.46		3.02		2.65		2.56		2.60		40.84
MANNINGTON 1 N	33.7	4.03	34.0	3.07	41.4	4.45	51.5	4.10	61.3	4.59	69.6	5.24	72.5	4.97	71.0	4.54	64.4	3.29	54.1	2.74	42.3	3.11	33.6	3.43	52.4	47.56
MIDDLEBOURNE 2 ESE		3.41		2.54		3.78		3.62		4.27		4.75		4.73		4.59		3.21		2.51		2.79		2.91		43.11
MORGANTOWN LOCK AND DA		3.58		2.68		4.00		3.64		4.09		4.50		4.08		4.32		3.04		2.67		2.83		2.87		42.30
PHILIPPI		4.02		3.22		4.26		3.66		4.45		4.47		5.14		4.90		3.46		3.18		3.17		3.35		47.28
ROANOKE		3.90		3.11		4.23		3.50		4.23		5.05		4.87		4.72		3.39		3.00		3.08		3.38		46.46
VANDALIA		4.08		3.31		4.37		3.69		4.37		5.08		5.23		4.85		3.46		2.95		3.10		3.47		47.96
WESTON	36.6	4.43	36.5	3.54	43.7	4.76	54.0	3.85	63.8	4.54	72.6	5.45	75.4	5.45	74.0	4.73	67.9	3.70	57.2	3.03	45.0	3.35	36.2	3.78	55.2	50.61
DIVISION	35.1	3.95	35.2	3.11	42.5	4.27	52.8	3.72	62.6	4.30	71.1	4.92	74.1	4.82	72.7	4.58	66.4	3.33	55.8	2.79	43.6	3.04	35.1	3.34	53.9	46.17
SOUTHWESTERN																										
CHARLESTON WB AP	36.4	3.99	38.2	3.50	44.9	4.16	55.0	3.74	63.7	3.78	72.0	3.93	75.4	5.45	73.6	4.55	68.6	2.94	57.4	2.81	45.8	3.17	38.1	2.98	55.8	45.00
CHARLESTON 1		3.80		3.13		4.31		3.29		3.74		3.75		5.39		3.95		2.88		2.37		2.90		2.85		42.36
CLAY 1		4.03		3.25		4.59		3.61		3.92		4.79		5.94		4.67		3.21		2.74		2.96		3.55		47.26
HOGSETT GALLIPOLIS DAM		3.78		2.82		4.43		3.49		3.76		4.06		4.01		3.43		2.56		1.99		2.71		2.82		39.86
HUNTINGTON WB CITY	38.0	3.61	39.0	3.06	47.2	4.08	57.3	3.42	65.7	3.82	74.1	4.34	76.9	4.82	75.4	3.37	70.5	2.88	59.2	2.43	46.7	2.79	39.1	3.17	57.4	41.79
LOGAN		3.86		3.39		4.80		3.39		4.02		4.54		4.84		4.12		2.84		2.31		2.76		3.20		44.07
RAVENSWOOD DAM 22	36.6	3.58	36.9	2.78	44.1	3.82	54.7	3.30	63.9	3.58	72.6	3.90	75.6	4.40	74.1	3.53	68.1	2.56	57.6	2.06	45.0	2.58	36.4	2.85	55.5	38.94
SPENCER	36.1	4.10	36.2	3.40	43.6	4.33	53.9	3.62	63.2	3.92	71.5	4.61	74.7	4.46	73.2	4.17	67.2	2.90	56.2	2.69	44.3	2.85	36.0	3.18	54.7	44.23
WILLIAMSON	38.6	3.67	39.2	3.29	46.1	4.66	56.5	3.42	66.1	3.92	74.6	4.41	77.7	5.19	76.4	4.53	70.2	2.61	58.9	2.12	46.2	2.76	38.7	3.21	57.4	43.79
WINFIELD LOCKS	36.4	4.00	36.6	2.94	44.0	4.17	54.4	3.39	64.4	3.46	73.3	3.95	76.6	4.40	75.0	3.97	69.1	3.02	58.0	2.02	45.6	2.70	36.6	2.77	55.8	40.79
DIVISION	36.9	3.88	37.3	3.07	44.6	4.35	55.1	3.41	64.5	3.83	73.0	4.30	76.2	4.87	74.9	3.90	68.8	2.87	57.7	2.25	45.2	2.76	36.9	3.02	55.9	42.51
CENTRAL																										
ARBOVALE 2		3.38		3.14		4.05		3.09		3.90		4.41		4.68		3.92		2.77		2.42		2.66		2.81		41.23
BAYARD	29.8	4.11	29.4	3.26	36.0	4.52	46.1	3.96	56.0	4.75	64.0	4.89	67.0	4.82	65.3	4.76	58.9	3.29	49.1	3.23	38.2	2.98	29.7	3.35	47.5	47.92
BECKLEY V A HOSPITAL	34.9	3.77	35.1	3.37	41.7	4.58	51.5	3.59	60.2	4.09	67.5	4.67	70.7	5.20	69.4	4.69	63.8	3.08	53.8	2.68	42.4	2.60	34.7	3.25	52.1	45.57
ELKINS AIRPORT	32.2	3.22	32.5	3.05	39.3	3.79	49.1	3.36	58.2	4.25	66.5	5.26	70.0	5.14	68.2	3.83	63.0	3.28	51.7	2.86	40.8	2.87	32.9	3.13	50.4	44.04
FLAT TOP	31.5	3.77	31.7	3.58	37.9	4.67	48.1	3.46	57.4	3.96	64.8	4.55	67.7	4.82	66.7	4.51	61.4	2.63	51.6	2.52	39.2	2.65	31.6	3.32	49.1	44.44
STONY RIVER DAM		3.36		2.95		3.17		3.82		4.59		4.63		4.04		4.59		2.96		3.43		2.84		3.08		43.46
THOMAS		4.45		3.99		5.17		4.37		5.44		5.70		5.78		5.25		3.81		3.52		3.42		4.10		55.00
DIVISION	32.3	4.21	32.2	3.67	38.9	4.94	49.0	3.93	58.3	4.82	66.1	5.11	69.3	5.46	67.8	4.99	62.0	3.31	52.0	3.11	40.3	3.14	32.2	3.66	50.0	50.35
SOUTHERN																										
BLUEFIELD 1	37.1	3.40	37.3	3.32	43.8	4.23	53.5	3.08	62.3	3.91	69.6	3.96	72.2	4.72	71.0	4.27	65.6	2.63	55.8	2.38	44.5	2.59	36.8	2.93	54.1	41.42
GARY	37.1	3.44	37.2	3.25	43.7	4.31	53.8	3.24	63.0	3.96	70.9	4.05	74.0	5.25	73.0	4.08	66.9	2.72	55.9	2.29	44.1	2.26	36.7	3.07	54.7	41.92
UNION	34.2	3.03	34.7	2.67	41.5	3.71	51.6	2.55	60.8	3.45	68.5	3.86	71.7	4.30	70.4	3.46	64.5	2.44	54.0	2.09	41.8	2.24	33.4	2.55	52.3	36.35
WHITE SULPHUR SPRINGS	34.3	3.19	35.0	2.68	42.1	4.11	52.3	2.98	62.1	3.48	69.8	3.70	72.7	4.25	71.2	3.91	64.7	2.65	54.3	2.27	42.1	2.62	33.7	2.69	52.9	38.53
DIVISION	35.8	3.25	36.4	2.96	43.0	4.01	53.2	2.98	62.5	3.71	70.3	3.85	73.2	4.49	72.0	3.93	66.0	2.54	55.5	2.19	43.6	2.41	35.2	2.85	53.9	39.17
NORTHEASTERN																										
HARPERS FERRY		2.69		2.23		3.43		3.44		3.73		3.45		3.88		4.48		3.44		3.65		2.93		2.96		40.31
KEARNEYSVILLE 1 NW	34.2	2.68	35.0	2.09	42.9	3.24	53.0	3.24	63.4	3.77	71.7	3.47	75.7	3.51	73.7	4.28	66.7	3.41	56.0	3.40	44.8	2.84	34.9	2.78	54.3	38.71
MARTINSBURG CAA AP	33.5	2.71	34.0	2.02	41.9	3.38	52.5	3.29	63.5	3.87	72.2	3.17	76.7	3.46	74.4	4.46	66.8	3.18	55.7	3.36	43.9	2.75	34.4	2.82	54.1	38.47
PIEDMONT	32.8	2.98	33.2	2.02	40.9	3.84	51.5	3.22	61.8	3.98	69.7	4.14	73.6	3.42	72.1	4.11	65.3	3.00	55.1	2.82	42.8	2.27	33.1	2.45	52.7	38.25
WARDENSVILLE R M FARM	32.9	2.11	33.3	1.71	40.8	3.01	50.6	2.89	61.1	3.72	69.0	3.61	72.9	3.78	71.1	4.84	64.2	2.77	53.9	2.94	42.7	2.36	32.9	2.17	52.1	35.91
DIVISION	33.8	2.42	34.4	1.86	41.9	3.11	51.9	2.89	62.2	3.69	70.3	3.75	74.2	3.66	72.3	4.19	65.5	2.85	55.0	2.98	43.6	2.42	34.1	2.31	53.3	36.13

* **Averages for period 1931-1955, except for stations marked Wb which are "normals" based on period 1921-1950. Divisional means may not be the arithmetical average of individual stations published, since additional data from shorter period stations are used to obtain better areal representation.**

CONFIDENCE LIMITS

In the absence of trend or record changes, the chances are 9 out of 10 that the true mean will lie in the interval formed by adding and subtracting the values in the following table from the means for any station in the State.

1.9	.48	1.6	.38	1.9	.44	1.0	.44	.9	.71	.8	.59	.6	.63	.8	.77	1.0	.46	1.0	.67	1.0	.43	1.4	.48	.3	2.32

COMPARATIVE DATA

Data in the following table are the mean temperature and average precipitation for Spencer, West Virginia for the period 1906 - 1930 and are included in this publication for comparative purposes :

33.2	4.22	34.5	3.24	43.5	3.97	52.5	3.45	61.7	3.67	69.0	4.34	72.8	4.31	71.8	4.18	66.1	3.00	54.9	3.35	43.8	2.75	35.4	3.44	53.3	43.92

NORMALS, MEANS AND EXTREMES
(Table Revised 1972. Base Period for Climatological Normals: 1931-1960)

Station: CHARLESTON, WEST VIRGINIA — KANAWHA AIRPORT Standard time used: EASTERN Latitude: 38° 22′ N Longitude: 81° 36′ W Elevation (ground): 939 feet

Month	Temp Normal Daily max	Daily min	Monthly	Extremes Record highest	Year	Record lowest	Year	Normal heating degree days (Base 65°)	Precip Normal total	Max monthly	Year	Min monthly	Year	Max in 24 hrs	Year	Snow Mean total	Max monthly	Year	Max in 24 hrs	Year	RH 01	RH 07	RH 13	RH 19	Wind Mean speed	Prevailing dir	Fastest mile Speed	Dir	Year	Mean sky cover	Clear	Partly cloudy	Cloudy	Precip .01+	Snow 1.0+	Thunderstorms	Heavy fog	Max 90°+	Max 32°-	Min 32°-	Min 0°-	Solar
J	45.2	27.9	36.6	79	1950	-12	1963	880	4.32	9.11	1950	1.15	1961	1.91	1961	8.5	19.8	1966	13.0	1966	74	77	62	65	7.7	WSW	45	25	1951	7.7	4	7	20	15	3	1	5	0	7	23	1	
F	46.2	28.0	37.5	77	1950	-6	1968	770	3.53	6.89	1956	0.64	1951	2.45	1951	7.8	21.8	1964	8.3	1960	77	76	59	61	8.1	WSW	40	29	1952	7.5	4	7	17	14	2	1	4	0	4	19	*	
M	56.5	33.3	44.5	87	1954	4	1950	648	4.34	6.80	1967	0.64	1967	2.86	1967	4.4	18.3	1960	9.9	1954	69	74	53	55	8.0	SW	45	32	1955	7.4	4	8	19	15	1	2	3	0	1	16	0	
A	67.9	43.3	55.6	91	1970	19	1972	300	3.68	6.46	1948	1.19	1948	2.72	1948	0.3	0.5	1964	5.5	1959	68	75	50	53	8.0	SW	45	27	1953	7.0	5	8	17	14	*	4	3	*	0	5	0	
M	76.6	52.8	64.7	93	1969+	26	1966	96	3.71	6.59	1968	0.95	1964	2.17	1967	0.0	T	1963	0.2	1963	80	86	56	61	6.4	SW	50	32	1951	7.0	6	10	16	13	0	7	4	1	0	*	0	
J	83.2	60.8	72.0	98	1953	35	1966	9	3.69	6.43	1950	0.70	1962	2.24	1962	0.0	0.0		0.0		86		54		5.2	SW	46	29	1957	6.5	6	14	12	11	0	7	9	6	0	0	0	
J	85.6	64.2	74.9	102	1954	46	1963	0	5.67	13.54	1961	2.46	1965	5.60	1961	0.0	0.0		0.0		90	90	61	66	5.5	SW	50	20	1952	6.8	4	15	14	13	0	10	10	8	0	0	0	
A	84.3	63.3	73.8	100	1953	41	1965	0	3.95	10.45	1958	4.17	1958	4.17	1958	0.0	0.0		0.0		92	92	61	69	4.9	SW	50	29	1956	6.4	4	15	14	10	0	6	16	5	0	0	0	
S	79.5	56.8	68.2	102	1953	34	1963+	63	2.92	7.61	1971	0.09	1963	2.40	1961	0.0	0.0		0.0		91	91	60	70	4.5	SW	35	29	1950	6.1	7	11	12	9	0	3	20	3	0	0	0	
O	69.3	45.2	57.3	92	1951	17	1962	254	2.58	6.11	1961	0.05	1963	2.48	1961	0.2	2.8	1961	2.8	1961	84	88	56	63	4.5	SW	45	25	1954+	6.4	9	7	12	9	*	1	17	*	0	3	0	
N	55.3	35.2	45.3	85	1948	-6	1950	591	2.79	6.27	1962	0.64	1962	2.27	1962	2.9	25.8	1950	15.1	1950	75	80	61	66	5.9	SW	40	29	1953	7.2	5	7	18	11	1	1	11	0	1	13	0	
D	45.6	37.1	37.1	80	1971	-2	1963	865	3.25	5.73	1956+	0.45	1965	2.10	1948	5.1	18.6	1962	11.2	1967	75	78	62	66	7.3	SW	55	25	1953	7.5	4	6	20	13	2	*	5	0	5	21	*	
YR	66.0	45.1	55.6	102	JUL. 1954+	-12	JAN. 1963	4476	44.43	13.54	JUL. 1961	0.09	OCT. 1963	5.60	JUL. 1961	29.2	25.8	NOV. 1950	15.1	NOV. 1950	80	82	55	62	6.6	SW	55	25	DEC. 1953+	6.9	60	117	188	147	9	42	111	23	18	101	1	

Means and extremes above are from existing and comparable exposures. Annual extremes have been exceeded at other sites in the locality as follows: Highest temperature 108 in July 1931; lowest temperature -17 in December 1917 (early records which were maintained irregularly show -24 in January 1857); minimum monthly precipitation 0.00 in October 1897.

NORMALS, MEANS AND EXTREMES
(Table Revised 1975. Base Period for Climatological Normals: 1941-1970)

Month	Temp Normal Daily max	Daily min	Monthly	Extremes Record highest	Year	Record lowest	Year	Normal Degree Days Base 65°F Heating	Cooling	Precip Normal	Max monthly	Year	Min monthly	Year	Max in 24 hrs	Year	Water equiv Max monthly	Year	Min monthly	Year	Snow Mean monthly	Max monthly	Year	Max in 24 hrs	Year	RH 01	RH 07	RH 13	RH 19	Wind Mean speed	Prevailing dir	Fastest mile Speed	Dir	Year	Mean sky cover	Clear	Partly cloudy	Cloudy	Precip .01+	Snow 1.0+	Thunderstorms	Heavy fog	Max 90°+	Max 32°-	Min 32°-	Min 0°-	Avg station pressure mb
J	43.6	25.3	34.5	79	1950	-12	1963	946	0	3.39	9.11	1950	1.15	1961	1.91	1970	9.11	1950	1.15	1961	8.5	19.8	1966	13.0	1966	74	77	62	65	7.6	WSW	45	29	1951	7.7	4	6	21	15	2	1	5	0	7	22	1	985.6
F	46.2	26.8	36.5	77	1950	-6	1968	798	0	3.11	6.89	1956	0.64	1951	2.45	1968	6.89	1956	0.64	1951	7.8	21.8	1964	8.3	1960	76	76	59	61	8.0	WSW	40	29	1952	7.5	4	8	18	14	1	1	4	0	4	19	*	983.8
M	55.7	33.8	44.5	87	1954	4	1950	642	0	4.03	6.80	1967	1.43	1967	2.86	1967	6.80	1967	1.43	1967	4.4	18.3	1960	9.9	1954	69	74	53	55	8.0	WSW	45	32	1955	7.4	4	8	19	14	1	2	3	0	1	15	0	981.4
A	67.9	43.3	55.6	91	1974	19	1972	287	14	3.33	6.46	1948	1.19	1948	2.72	1948	6.46	1948	1.19	1948	0.3	0.5	1964	5.5	1959	68	75	48	53	7.9	SW	45	27	1953	7.1	5	8	17	13	*	4	3	*	0	5	0	980.4
M	76.6	52.8	64.7	93	1969	26	1966	113	97	3.48	6.59	1968	0.95	1974	2.48	1967	6.59	1968	0.95	1974	0.0	T	1963	0.2	1963	80	86	56	56	6.4	SW	50	32	1951	6.7	5	11	15	13	0	7	4	1	0	*	0	980.7
J	83.4	60.6	72.0	98	1953	33	1972	14	220	3.31	6.43	1950	0.70	1966	2.24	1962	6.43	1950	0.70	1966	0.0	0.0		0.0		86	86	54	61	5.6	SW	50	32	1957	6.5	6	14	12	11	0	7	9	5	0	0	0	980.7
J	85.6	64.3	75.0	102	1954	46	1963	0	310	5.04	13.54	1961	2.16	1974	5.60	1961	13.54	1961	2.16	1974	0.0	0.0		0.0		90	90	61	67	5.1	SW	50	20	1952	6.8	4	14	13	13	0	10	7	7	0	0	0	983.4
A	84.4	62.8	73.6	100	1953	41	1965	0	267	3.68	10.45	1958	0.66	1957	4.17	1958	10.45	1958	0.66	1957	0.0	0.0		0.0		92	92	58	69	4.8	SW	50	29	1956	6.4	4	15	12	10	0	6	12	5	0	0	0	984.8
S	79.0	55.9	67.5	102	1953	34	1974	46	121	2.94	7.61	1971	0.65	1971	2.48	1961	7.61	1971	0.65	1971	0.0	0.0		0.0		91	91	53	70	4.5	SW	45	29	1950	5.9	7	11	12	9	0	3	11	3	0	0	0	987.2
O	67.9	44.8	56.4	92	1951	17	1962	267	19	2.45	6.11	1961	0.11	1963	2.35	1973	6.11	1961	0.09	1963	0.2	2.8	1961	2.8	1961	84	88	55	63	5.3	SW	45	25	1954	6.2	8	7	12	8	*	1	12	1	0	4	0	984.8
N	55.8	35.0	45.4	85	1948	-6	1950	588	0	2.81	6.27	1962	0.64	1962	2.27	1962	6.27	1962	0.64	1962	2.9	25.8	1950	15.1	1950	75	80	56	66	5.7	SW	45	29	1953	7.2	5	7	18	11	1	1	13	0	1	13	0	984.8
D	45.2	27.2	36.2	80	1971	-2	1963	893	0	3.18	6.35	1972	0.45	1965	2.10	1972	6.35	1972	0.45	1965	5.1	18.6	1962	11.2	1967	75	78	62	66	7.3	SW	55	25	1953	7.6	4	7	20	9	2	*	5	0	5	21	1	983.2
YR	66.0	44.4	55.2	102	JUL 1954	-12	JAN 1963	4590	1055	40.75	13.54	JUL 1961	0.09	OCT 1963	5.60	JUL 1961	13.54	JUL 1961	0.09	OCT 1963	29.2	25.8	NOV 1950	15.1	NOV 1950	80	83	56	62	6.5	SW	55	25	DEC 1953	6.9	57	118	190	149	9	43	100	22	17	100	1	983.5

Elev. 951 feet m.s.l.

Means and extremes above are from existing and comparable exposures. Annual extremes have been exceeded at other sites in the locality as follows: Highest temperature 108 in July 1931; lowest temperature -17 in December 1917 (early records which were maintained irregularly show -24 in January 1857); minimum monthly precipitation 0.00 in October 1897.

REFERENCE NOTES APPLYING TO TABLES APPEAR ON THE PAGE FOLLOWING LAST TABLE.
(Caution: Letters and symbols may have different meanings in 1941-1970 tables than in earlier tables. See notes.)

NORMALS, MEANS AND EXTREMES
(Table Revised 1972. Base Period for Climatological Normals: 1931-1960)

Station: ELKINS-RANDOLPH COUNTY AIRPORT — ELKINS, WEST VIRGINIA
Standard time used: EASTERN Latitude: 38° 53' N Longitude: 79° 51' W Elevation (ground): 1948 feet

Month	Temp Normal Daily max	Daily min	Monthly	Extremes Record highest	Year	Record lowest	Year	Normal heating degree days (Base 65°)	Precip Normal total	Max monthly	Year	Min monthly	Year	Max in 24 hrs	Year	Snow/Ice pellets Mean total	Max monthly	Year	Max in 24 hrs	Year	RH 01	07	13	19	Wind Mean speed	Prevailing dir	Fastest mile Speed	Dir	Year	Mean sky cover	Pct poss sunshine
J	43.0	22.0	32.5	76	1950	-20	1963	1008	3.62	6.09	1949	1.05	1967	1.85	1967	15.3	25.2	1971	18.7	1971	81	81	64	72	7.3	W	46	27	1965	7.8	
F	44.0	21.9	33.0	70	1954+	-21	1963	896	3.27	6.41	1957	1.12	1952	1.12	1971	15.3	29.5	1952	8.6	1971	78	78	60	67	8.0	W	55	32	1955	7.7	
M	51.1	27.8	39.5	84	1954	-4	1960	791	4.14	8.85	1963	1.39	1951	2.02	1966	12.1	33.5	1960	8.3	1971+	82	81	52	63	8.0	WNW	46	30	1958	7.5	
A	63.1	37.0	50.2	86	1960	13	1954	444	3.48	6.90	1965	1.02	1971	2.04	1968	1.4	5.0	1966	4.8	1966	81	83	50	56	7.9	NW	50	27	1963	7.4	
M	72.4	46.1	59.3	87	1966+	20	1963	198	4.57	7.67	1966+	1.45	1970+	2.86	1968	0.0	0.0		0.7	1963	85	89	53	58	6.7	SSE	46	30	1967	7.0	
J	79.6	54.5	67.1	93	1952	29	1966	48	5.19	7.74	1951	1.66	1960	1.81	1967	0.0	0.0		0.0		95	93	55	68	4.9	NW	35	36	1964	7.1	
J	82.3	57.9	70.1	95	1954	39	1963+	9	5.49	9.30	1958	2.27	1965	1.92	1970	0.0	0.0		0.0		97	96	55	72	4.3	NW	37	30	1963+	6.9	
A	80.9	56.6	68.8	95	1948	34	1945	25	5.06	6.41	1954	3.21	1969	3.21	1969	0.0	0.0		0.0		97	98	61	77	4.1	NW	40	32	1969+	6.6	
S	75.6	49.6	62.6	97	1953	27	1963	135	3.07	6.29	1971	0.89	1971	2.68	1966	0.0	0.0		0.0		96	97	60	82	5.0	NNW	29	26	1967	6.7	
O	65.0	38.4	51.7	86	1951+	11	1952	400	3.03	8.43	1954	0.31	1963	2.02	1967	0.1	0.5	1968	0.5	1968	91	92	56	76	5.0	NW	35	29	1955	7.3	
N	52.7	28.7	40.7	80	1958+	0		729	2.75	6.21	1945	1.25	1953	1.38	1963	8.1	13.2	1970	6.7	1970	84	86	60	76	6.9	WNW	33	28	1968+	7.8	
D	44.0	22.4	33.2	76	1951	-11	1962	992	3.08	5.84	1969	0.90	1951	2.22	1969	15.0	34.9	1969	17.8	1967	85	84	67	76	6.9	W				7.2	
YR	62.9	38.6	50.7	97	SEP 1953	-21	FEB 1963	5675	45.92	9.30	JUL 1958	0.31	OCT 1963	3.21	AUG 1969	71.9	34.9	DEC 1969	18.7	JAN 1971	87	89	59	70	6.3	NW	55	25	FEB 1958	7.2	

Mean number of days — Sunrise to sunset (Clear / Partly cloudy / Cloudy), Precipitation .01" or more, Snow/Ice pellets 1.0" or more, Thunderstorms, Heavy fog, Max 90°+ , Max 32° below, Min 32° below, Min 0° below:

Month	Clear	Partly	Cloudy	Precip .01+	Snow 1.0+	Tstorms	Heavy fog	Max 90°+	Max 32°↓	Min 32°↓	Min 0°↓
J	3	7	21	18	5	*	2	0	10	26	3
F	5	7	18	17	4	1	2	0	6	24	2
M	4	7	20	17	4	2	2	0	3	23	*
A	4	8	18	15	1	4	5	0	*	11	0
M	4	10	16	15	0	6	6	0	0	2	0
J	3	12	15	12	0	8	11	*	0	*	0
J	3	12	15	13	0	8	13	1	0	0	0
A	5	11	13	13	0	7	13	1	0	0	0
S	5	9	12	10	0	3	15	*	0	1	0
O	4	6	14	10	0	1	10	0	*	10	0
N	3	6	17	13	2	*	3	0	2	20	*
D	4	6	21	16	5	*	2	0	7	26	1
YR	51	108	206	165	23	44	81	2	29	143	6

Means and extremes above are from existing and comparable exposures. Annual extremes have been exceeded at other sites in the locality as follows: Highest temperature 99 in August 1918; lowest temperature -28 in December 1917; maximum monthly precipitation 11.10 in July 1907; minimum monthly precipitation 0.26 in October 1924; maximum precipitation in 24 hours 5.45 in July 1935; maximum monthly snowfall 37.6 in November 1950; maximum snowfall in 24 hours 18.8 in November 1913; fastest mile of wind 72 from the northwest in July 1951.

NORMALS, MEANS AND EXTREMES
(Table Revised 1975. Base Period for Climatological Normals: 1941-1970)

Month	Temp Normal Daily max	Daily min	Monthly	Extremes Record highest	Year	Record lowest	Year	Deg days Heating	Cooling	Precip Normal	Max monthly	Year	Min monthly	Year	Max 24 hrs	Year	Snow Mean total	Max monthly	Year	Max 24 hrs	Year	RH 01	07	13	19	Wind Mean speed	Prev dir	Fastest mile mph	Dir	Year	Mean sky cover	Avg sta press mb
J	40.7	19.2	30.0	76	1950	-20	1963	1085	0	3.29	6.09	1949	1.05	1967	1.85	1967	15.3	25.2	1971	18.7	1971	81	80	63	71	7.3	W	46	27	1965	7.8	947.6
F	42.0	20.4	31.4	70	1972	-21	1963	941	0	2.92	5.68	1957	0.92	1952	1.38	1968	15.3	29.5	1952	8.6	1971	78	79	60	67	8.0	W	55	32	1955	7.7	945.5
M	50.5	27.0	38.8	86	1954	-4	1960	812	0	3.93	8.95	1963	1.39	1951	2.94	1966	12.1	33.5	1971	8.3	1971	82	82	52	61	8.9	WNW	46	30	1958	7.6	944.3
A	62.5	36.7	49.7	86	1960	13	1954	459	0	3.62	6.90	1965	1.02	1971	2.02	1972	1.4	5.0	1966	7.4	1973	81	82	50	56	8.2	NW	50	27	1963	7.4	944.8
M	71.8	45.3	58.2	88	1974	20	1970	236	25	4.27	7.67	1966	1.45	1970	2.86	1968	0.0	0.0		0.7	1963	85	86	53	58	6.7	SSE	46	30	1967	7.0	943.7
J	78.0	53.3	65.7	93	1960	29	1966	63	84	4.78	7.74	1974	1.66	1960	2.44	1963	0.0	0.0		0.0		95	93	55	68	4.9	NW	35	36	1964	7.1	946.2
J	80.3	57.1	68.7	95	1954	39	1963	20	135	4.94	9.30	1958	1.93	1958	1.92	1974	0.0	0.0		0.0		97	95	60	72	4.3	NW	37	30	1963	6.9	948.2
A	79.1	55.5	67.4	95	1948	34	1945	36	111	4.02	6.29	1971	1.67	1957	3.21	1969	0.0	0.0		0.0		97	97	62	78	4.1	NW	40	32	1969	6.7	948.2
S	74.0	48.8	61.4	97	1953	27	1963	139	34	3.18	6.29	1971	0.31	1959	2.68	1959	0.0	0.0		0.0		96	96	60	83	5.0	NNW	29	26	1972	6.7	949.0
O	64.2	37.2	50.7	86	1954	11	1955	420	0	2.72	8.43	1954	0.31	1963	2.08	1954	0.1	1.2	1972	2.4	1973	91	91	57	73	5.0	NW	35	29	1955	7.3	950.4
N	52.3	29.2	40.8	80	1958	0	1953	724	0	3.25	6.02	1953	1.45	1953	1.38	1963	8.1	13.2	1970	6.7	1970	84	84	60	76	6.9	WNW	33	28	1968	7.8	947.4
D	42.2	21.0	31.6	76	1951	-11	1962	1035	0	3.25	6.02	1972	0.90	1965	2.22	1970	15.0	34.9	1969	17.8	1967	85	85	68	76	6.9	W				7.2	945.3
YR	61.5	37.7	49.6	97	SEP 1953	-21	FEB 1963	5975	389	43.22	9.30	JUL 1958	0.31	OCT 1963	3.21	AUG 1969	34.9	DEC 1969		18.7	JAN 1971	87	87	59	70	6.3	NW	55	25	FEB 1958	7.2	946.8

Elev 1997 feet m.s.l.

Mean number of days — Sunrise to sunset (Clear / Partly cloudy / Cloudy), Precipitation .01" or more, Snow/Ice pellets 1.0" or more, Thunderstorms, Heavy fog visibility ¼ mile or less, Max 90°+, Max 32° below, Min 32° below, Min 0° below:

Month	Clear	Partly	Cloudy	Precip .01+	Snow 1.0+	Tstorms	Heavy fog	Max 90°+	Max 32°↓	Min 32°↓	Min 0°↓
J	3	7	21	18	4	*	2	0	9	26	3
F	4	7	18	17	4	1	2	0	6	24	2
M	4	7	20	17	4	2	2	0	3	22	*
A	4	8	18	15	1	4	5	0	*	11	0
M	4	10	16	14	0	6	6	*	0	2	0
J	3	12	15	13	0	8	11	1	0	*	0
J	3	12	15	13	0	8	13	1	0	0	0
A	5	11	13	11	0	7	17	1	0	0	0
S	5	9	12	10	0	3	15	*	0	1	0
O	4	6	14	10	0	1	10	0	*	10	*
N	3	6	17	13	2	*	3	0	2	20	1
D	4	6	21	16	4	*	2	0	7	26	1
YR	51	108	206	167	22	44	81	2	28	143	6

Means and extremes above are from existing and comparable exposures. Annual extremes have been exceeded at other sites in the locality as follows: Highest temperature 99 in August 1918; lowest temperature -28 in December 1917; maximum monthly precipitation 11.10 in July 1907; minimum monthly precipitation 0.26 in October 1924; maximum precipitation in 24 hours 5.45 in July 1935; maximum monthly snowfall 37.6 in November 1950; maximum snowfall in 24 hours 18.8 in November 1913; fastest mile of wind 72 from the northwest in July 1951.

REFERENCE NOTES APPLYING TO TABLES APPEAR ON THE PAGE FOLLOWING LAST TABLE.
(Caution: Letters and symbols may have different meanings in 1941-1970 tables than in earlier tables. See notes.)

NORMALS, MEANS AND EXTREMES

(Table Revised 1972. Base Period for Climatological Normals: 1931-1960)

Station: HUNTINGTON, WEST VIRGINIA — TRI-STATE AIRPORT — Standard time used: EASTERN — Latitude: 38° 22' N — Longitude: 82° 33' W — Elevation (ground): 827 feet

Month	Temperature Normal Daily max	Daily min	Monthly	Extremes Record highest	Year	Record lowest	Year	Normal heating degree days (Base 65°)	Precip Normal total	Max monthly	Year	Min monthly	Year	Max in 24 hrs	Year	Snow Mean total	Max monthly	Year	Max in 24 hrs	Year
J	45.3	27.9	36.6	74	1967+	-15	1963	880	3.65	3.36	1965	1.12	1970	1.11	1966	7.4	12.9	1966	8.2	1966
F	47.1	28.2	37.7	74	1971	-6	1970	764	3.04	5.66	1962	0.53	1968	2.43	1966	6.6	14.3	1964	6.5	1971
M	54.7	34.9	44.8	85	1973	10	1964	636	4.20	7.54	1967	1.26	1966	3.52	1963	4.1	11.0	1966	7.9	1967
A	67.5	43.9	55.7	90	1963	22	1964	294	3.67	6.56	1965	1.06	1963	1.91	1965	T	T	1970+	T	1970+
M	76.6	52.5	64.6	92	1963+	27	1966	99	3.89	6.81	1966	0.93	1965	2.20	1967	0.0	0.0		0.0	
J	83.3	60.6	72.0	95	1966	40	1966	12	4.10	4.80	1971	0.41	1966	1.57	1971	0.0	0.0		0.0	
J	86.0	64.3	75.2	96	1966	46	1968	0	4.50	8.57	1962	2.18	1964	4.27	1962	0.0	0.0		0.0	
A	85.1	62.9	74.0	100	1964	43	1965	0	2.82	6.21	1964	0.68	1962	2.90	1964	0.0	0.0		0.0	
S	80.2	56.1	68.2	94	1964	36	1964+	63	2.54	5.64	1968	1.14	1968	2.74	1968	0.0	0.0		0.0	
O	69.7	44.9	57.3	86	1962	16	1962	257	1.85	5.23	1970	0.96	1963	1.82	1970	T	T	1969	T	1969
N	56.0	34.6	44.9	79	1968	8	1964	585	2.46	4.71	1967	0.96	1967	1.81	1964	1.9	4.6	1969	4.4	1969
D	46.1	28.6	37.4	76	1971	-5	1963	856	2.81	4.80	1965	0.31	1965	1.66	1970	5.3	13.2	1970	5.7	1967
YR	66.5	44.9	55.8	100 AUG 1964		-15 JAN 1963		4446	39.53	8.57 JUL 1962		0.31 OCT 1965	T	4.27 JUL 1962		25.3	14.3 FEB 1964		8.2 JAN 1966	

Ø Extremes for December 1961 through the current year.
Means and extremes above are from existing and comparable exposures. Annual extremes have been exceeded at other sites in the locality as follows:
Highest temperature 108 in July 1930; maximum monthly precipitation 9.90 in July 1961; maximum monthly snowfall 19.6 in November 1950; maximum snowfall in 24 hours 12.0 in February 1960.

NORMALS, MEANS AND EXTREMES

(Table Revised 1975. Base Period for Climatological Normals: 1941-1970)

Month	Temperature Normal Daily max	Daily min	Monthly	Extremes Record highest	Year	Record lowest	Year	Normal Degree days Base 65°F Heating	Cooling	Precip Normal	Max monthly	Year	Min monthly	Year	Max in 24 hrs	Year	Snow Max monthly	Year	Max in 24 hrs	Year
J	42.9	25.6	34.3	74	1967	-15	1963	952	0	3.15	5.57	1974	1.12	1970	2.03	1966	12.9	1966	8.2	1966
F	45.1	26.8	36.1	76	1972	-6	1970	809	0	2.90	5.66	1962	0.53	1968	2.43	1966	14.3	1964	6.5	1971
M	54.7	33.8	44.3	85	1973	10	1964	649	7	4.07	7.54	1967	0.86	1966	3.43	1966	11.0	1966	7.9	1967
A	67.5	43.8	55.7	90	1963	22	1972	293	49	3.26	6.56	1965	0.93	1963	1.91	1965	0.8	1974	0.6	1974
M	76.2	52.7	64.5	92	1963	27	1966	115	99	3.82	4.86	1974	0.93	1965	2.60	1974	T	1963	T	1963
J	83.4	61.3	72.4	95	1966	40	1966	11	233	3.37	4.86	1974	0.41	1966	1.80	1974	0.0		0.0	
J	85.7	64.8	75.3	96	1968	46	1968	0	319	4.19	8.57	1962	1.37	1962	4.27	1962	0.0		0.0	
A	84.6	63.1	73.9	100	1964	43	1965	0	279	3.34	5.66	1962	0.68	1962	2.90	1964	0.0		0.0	
S	79.0	56.3	67.7	96	1973	35	1964	46	127	2.86	5.64	1968	1.14	1970	2.74	1968	0.0		0.0	
O	68.8	45.4	57.1	86	1962	18	1964	265	20	2.86	5.23	1970	0.82	1973	1.82	1970	0.4	1974	0.4	1974
N	55.4	35.6	45.5	79	1968	8	1964	585	0	2.85	5.17	1973	0.96	1972	2.28	1973	4.6	1969	4.4	1969
D	44.6	27.4	36.0	76	1971	-5	1963	899	0	2.97	5.52	1972	0.31	1965	2.40	1973	13.2	1970	5.7	1967
YR	65.7	44.7	55.2	100 AUG 1964		-15 JAN 1963		4624	1098	38.88	9.26 MAY 1974		T OCT 1963		4.27 JUL 1962		14.3 FEB 1964		8.2 JAN 1966

Average station pressure: Elev. 838 feet m.s.l. = 987.2 mb (annual)

Means and extremes above are from existing and comparable exposures. Annual extremes have been exceeded at other sites in the locality as follows:
Highest temperature 108 in July 1930; maximum monthly precipitation 9.90 in July 1961; maximum monthly snowfall 19.6 in November 1950; maximum snowfall in 24 hours 12.0 in February 1960.

REFERENCE NOTES APPLYING TO TABLES APPEAR ON THE PAGE FOLLOWING LAST TABLE.
(Caution: Letters and symbols may have different meanings in 1941-1970 tables than in earlier tables. See notes.)

NORMALS, MEANS AND EXTREMES
(Table Revised 1972. Base Period for Climatological Normals: 1931-1960)

Station: PARKERSBURG, WEST VIRGINIA   FEDERAL BUILDING   Standard time used: EASTERN   Latitude: 39° 16' N   Longitude: 81° 34' W   Elevation (ground): 615 feet

Month	Normal Daily max	Normal Daily min	Normal Monthly	Record highest	Year	Record lowest	Year	Normal heating degree days (Base 65°)	Precip Normal total	Max monthly	Year	Min monthly	Year	Max in 24 hrs	Year
J	42.9	26.3	34.6	78	1950	-16	1936	942	3.34	8.99	1937	0.51	1931	2.97	1913
F	44.5	26.5	35.5	77	1932	-27	1899	826	2.83	7.04	1897	0.58	1968	2.89	1945
M	52.6	32.8	42.7	89	1929	-3	1943	691	3.53	9.75	1963	0.48	1910	3.46	1963
A	65.9	42.9	54.1	91	1925	15	1923	339	3.25	6.75	1948	0.55	1900	3.40	1920
M	75.4	52.5	64.0	96	1914	29	1914	115	3.70	10.00	1968	0.85	1939	3.40	1905
J	83.4	61.7	72.6	99	1895	38	1945	6	4.27	8.63	1928	0.92	1966	3.58	1932
J	86.3	65.1	75.8	104	1930	47	1947	0	4.11	12.05	1958	0.73	1901	4.81	1947
A	85.0	63.8	74.4	106	1918	42	1965	0	3.78	3.60	1957	0.73	1957	3.60	1909
S	79.3	56.8	68.1	102	1953	32	1942	60	2.71	8.41	1890	0.46	1897	3.00	1935
O	68.4	45.7	57.0	91	1927	20	1895	264	2.05	6.48	1905	0.11	1924	3.40	1954
N	54.3	35.3	45.8	86	1961	4	1929	606	2.36	5.59	1927	0.46	1904	3.22	1900
D	44.0	27.6	35.8	77	1971	-10	1917	905	2.84	5.51	1944	0.46	1965	2.69	1948
YR	65.1	44.8	55.0	106	AUG. 1918	-27	FEB. 1899	4754	38.77	12.05	JUL. 1958	0.07	OCT. 1897	4.81	JUL. 1947

Month	Snow Mean total	Snow Max monthly	Year	Snow Max in 24 hrs	Year	RH 01	RH 07	RH 13	RH 19	Wind Mean speed	Fastest mile Speed	Direction	Year	Pct of possible sunshine	Mean sky cover
J	6.9	26.4	1948	14.0	1966	82	66	59	74	7.2	49	W	1971	31	7.3
F	6.2	25.7	1894	13.0	1914	81	63	55	71	7.6	45	W	1967	36	6.9
M	4.7	22.0	1902	11.5	1902	79	58	49	64	7.8	47	SE	1973	42	6.8
A	0.7	8.7	1918	8.7	1918	74	54	42	58	7.3	47	SW	1920	49	6.0
M	T	0.1	1895	0.1	1895	78	56	49	60	6.0	43	NW	1914	56	5.3
J	0.0	0.0		0.0		80	52	46	66	5.4	49	NW	1934	60	5.1
J	0.0	0.0		0.0		83	57	52	67	5.1	62	NW	1926	62	5.0
A	0.0	0.0		0.0		84	71	51	70	4.9	37	NW	1955	62	4.8
S	0.0	0.0		0.0		84	72	57	71	5.1	61	NW	1954	54	5.6
O	0.1	3.7	1925	3.5	1925	81	57	51	71	5.5	38	NW	1932	54	5.5
N	1.8	16.8	1950	16.8	1950	81	64	57	76	6.5	66	SW	1954	37	6.7
D	4.5	29.1	1890	18.3	1890	80	64		76	6.7	41	NW	1968	29	7.3
YR	24.3	34.6	NOV 1950	18.3	DEC 1890	80	56	68		6.3	66	NW	NOV 1954	48	6.0

Month	Clear	Partly cloudy	Cloudy	Precip .01"+	Snow 1.0"+	Thunderstorms	Heavy fog	Max 90°+	Max 32°-	Min 32°-	Min 0°-
J	6	6	19	15	2	*	0	0	7	23	1
F	6	7	15	15	1	1	*	0	6	21	1
M	8	8	15	14	1	2	*	0	2	15	*
A	9	8	13	13	*	4	*	*	*	4	0
M	10	10	11	12	0	6	1	1	0	*	0
J	12	11	8	11	0	9	1	5	0	0	0
J	12	11	8	10	0	7	2	8	0	0	0
A	14	10	10	9	0	4	2	6	0	0	0
S	12	8	11	9	0	1	3	3	0	*	0
O	11	7	16	11	*	*	4	*	0	2	0
N	7	6	19	13	*	*	2	0	1	12	0
D	6	6	19		1	*	1	0	6	21	*
YR	109	103	153	142	8	44	12	22	22	98	2

Ø Through 1948.   ¢ Through 1964. $ Through 1960.

NORMALS, MEANS AND EXTREMES
(Table Revised 1975. Base Period for Climatological Normals: 1941-1970)

Month	Normal Daily max	Normal Daily min	Normal Monthly	Record highest	Year	Record lowest	Year	Heating degree days Base 65°	Cooling degree days Base 65°	Water equiv. Normal	Max monthly	Year	Min monthly	Year	Max in 24 hrs	Year
J	41.3	24.4	32.9	78	1950	-16	1936	995	0	3.08	8.99	1937	0.91	1931	2.97	1913
F	43.7	25.6	34.7	77	1932	-27	1899	848	0	2.77	7.04	1897	0.58	1968	2.89	1945
M	53.1	32.7	42.9	89	1929	-3	1943	685	0	3.75	9.77	1963	0.10	1910	3.46	1963
A	65.9	43.2	54.6	91	1925	15	1923	320	8	3.45	6.75	1948	0.77	1900	3.45	1968
M	75.5	52.4	63.9	96	1914	29	1914	120	86	3.56	10.00	1968	0.53	1939	3.00	1905
J	83.1	61.0	72.1	99	1895	36	1972	0	221	4.01	8.63	1928	0.92	1966	3.58	1932
J	86.6	64.7	75.2	104	1930	47	1947	0	316	4.28	12.05	1958	0.69	1901	4.81	1947
A	85.0	63.0	74.0	106	1918	42	1965	0	276	3.34	10.42	1955	0.73	1957	3.60	1909
S	78.6	56.5	67.6	102	1953	32	1942	46	118	2.80	6.48	1890	0.46	1897	3.40	1954
O	68.4	45.0	57.0	91	1927	24	1895	268	20	2.11	6.48	1905	0.17	1924	3.40	1954
N	54.3	35.7	45.0	86	1971	4	1929	600	0	2.52	5.59	1927	0.46	1905	3.22	1972
D	43.2	26.9	35.1	77	1971	-10	1917	927	0	2.77	5.51	1944	0.46	1965	2.69	1948
YR	64.8	44.3	54.6	106	AUG 1918	-27	FEB 1899	4817	1045	38.44	12.05	JUL 1958	0.07	OCT 1897	4.81	JUL 1947

Month	Snow Mean total	Snow Max monthly	Year	Snow Max in 24 hrs	Year	RH 01	RH 07	RH 13	RH 19	Wind Mean speed	Fastest mile Speed	Direction	Year	Pct possible sunshine	Mean sky cover
J	6.9	26.4	1948	14.0	1966	82	66	59	74	7.2	49	W	1971	32	7.3
F	6.2	25.7	1894	13.0	1914	81	63	55	71	7.6	45	W	1967	36	6.9
M	4.7	22.0	1902	11.5	1902	79	58	49	64	7.8	47	SE	1973	43	6.8
A	0.7	8.7	1918	8.7	1918	74	54	42	58	7.3	47	SW	1920	49	6.0
M	T	0.1	1895	0.1	1895	78	56	49	60	6.0	43	NW	1914	56	5.4
J	0.0	0.0		0.0		80	52	46	66	5.4	49	NW	1934	60	5.1
J	0.0	0.0		0.0		83	57	52	67	5.1	62	NW	1926	62	5.0
A	0.0	0.0		0.0		84	71	51	70	4.9	37	NW	1955	59	4.8
S	0.0	0.0		0.0		84	72	57	71	5.1	61	NW	1954	54	5.6
O	0.1	3.7	1925	3.5	1925	81	57	51	71	5.5	38	NW	1932	54	5.5
N	1.8	16.8	1950	16.8	1950	81	64	57	76	6.5	66	SW	1954	37	6.7
D	4.5	29.1	1890	18.3	1890	80	64		76	6.7	42	W	1972	29	7.3
YR	34.6	34.6	NOV 1950	18.3	DEC 1890	80	56	68		6.3	66	NW	NOV 1954	48	6.0

Month	Clear	Partly cloudy	Cloudy	Precip .01"+	Snow 1.0"+	Thunderstorms	Heavy fog, visibility ¼ mile or less	Max 90°+	Max 32°-	Min 32°-	Min 0°-
J	6	6	19	15	2	*	0	0	7	23	1
F	6	7	15	15	1	1	*	0	6	21	1
M	8	8	15	14	1	2	*	0	2	15	*
A	9	8	13	13	*	4	1	*	*	4	0
M	10	10	11	12	0	6	1	1	0	*	0
J	12	11	8	11	0	9	2	5	0	0	0
J	12	11	8	10	0	7	2	8	0	0	0
A	14	10	10	9	0	4	3	6	0	0	0
S	12	8	11	9	0	1	4	3	0	*	0
O	11	7	16	11	*	*	1	*	0	2	0
N	7	6	19	13	*	*	2	0	1	12	0
D	6	6	19		1	*	1	0	6	21	*
YR	109	103	153	142	7	44	12	22	22	97	2

Average station pressure: 637 mb. Elev. 537 feet m.s.l.

REFERENCE NOTES APPLYING TO TABLES APPEAR ON THE PAGE FOLLOWING LAST TABLE.
(Caution: Letters and symbols may have different meanings in 1941-1970 tables than in earlier tables. See notes.)

Reference notes applying to Normals, Means, and Extremes tables for 1931–1960 base period.

(a) Length of record, years, based on January data. Other months may be for more or fewer years if there have been breaks in the record.

(b) Climatological standard normals (1931-1960).

* Less than one half.

+ Also on earlier dates, months, or years.

T Trace, an amount too small to measure.

Below zero temperatures are preceded by a minus sign. The prevailing direction for wind in the Normals, Means, and Extremes table is from records through 1963.

‡ $\geq 70°$ at Alaskan stations.

Unless otherwise indicated, dimensional units used in this bulletin are: temperature in degrees F.; precipitation, including snowfall, in inches; wind movement in miles per hour; and relative humidity in percent. Heating degree day totals are the sums of negative departures of average daily temperatures from 65° F. Cooling degree day totals are the sums of positive departures of average daily temperatures from 65° F. Sleet was included in snowfall totals beginning with July 1948. The term "Ice pellets" includes solid grains of ice (sleet) and particles consisting of snow pellets encased in a thin layer of ice. Heavy fog reduces visibility to 1/4 mile or less.

Sky cover is expressed in a range of 0 for no clouds or obscuring phenomena to 10 for complete sky cover. The number of clear days is based on average cloudiness 0-3, partly cloudy days 4-7, and cloudy days 8-10 tenths.

Solar radiation data are the averages of direct and diffuse radiation on a horizontal surface. The langley denotes one gram calorie per square centimeter.

& Figures instead of letters in a direction column indicate direction in tens of degrees from true North; i.e., 09 - East, 18 - South, 27 - West, 36 - North, and 00 - Calm. Resultant wind is the vector sum of wind directions and speeds divided by the number of observations. If figures appear in the direction column under "Fastest mile" the corresponding speeds are fastest observed 1-minute values.

% Through 1964. The station did not operate 24 hours daily. Fog and thunderstorm data therefore may be incomplete.

To 8 compass points only.

Reference notes applying to Normals, Means, and Extremes tables for 1941–1970 base period.

(a) Length of record, years, through the current year unless otherwise noted, based on January data.

(b) 70° and above at Alaskan stations.

* Less than one half.

T Trace.

NORMALS - Based on record for the 1941-1970 period.

DATE OF AN EXTREME - The most recent in cases of multiple occurrence.

PREVAILING WIND DIRECTION - Record through 1963.

WIND DIRECTION - Numerals indicate tens of degrees clockwise from true north; 00 indicates calm.

FASTEST MILE WIND - Speed is fastest observed 1-minute value when the direction is in tens of degrees.

% Through 1964. The station did not operate 24 hours daily. Fog and thunderstorm data therefore may be incomplete.

∅ 1955 through 1958 and 1963 through 1968.

¢ Through 1964.

$ Through 1960.

Mean Annual Precipitation, Inches

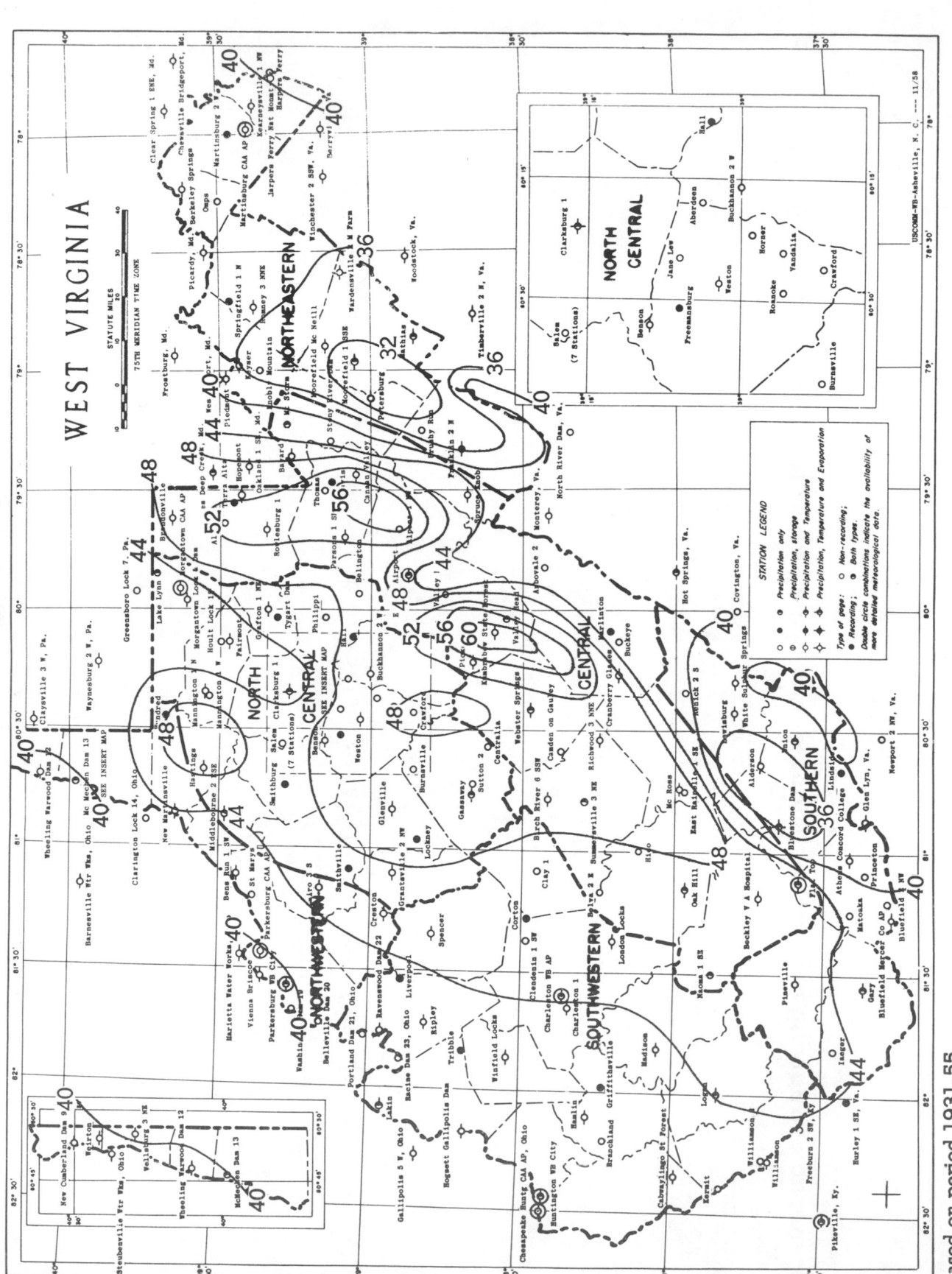

WEST VIRGINIA

Based on period 1931-55

Isolines are drawn through points of approximately equal value. Caution should be used in interpolating on these maps, particularly in mountainous areas.

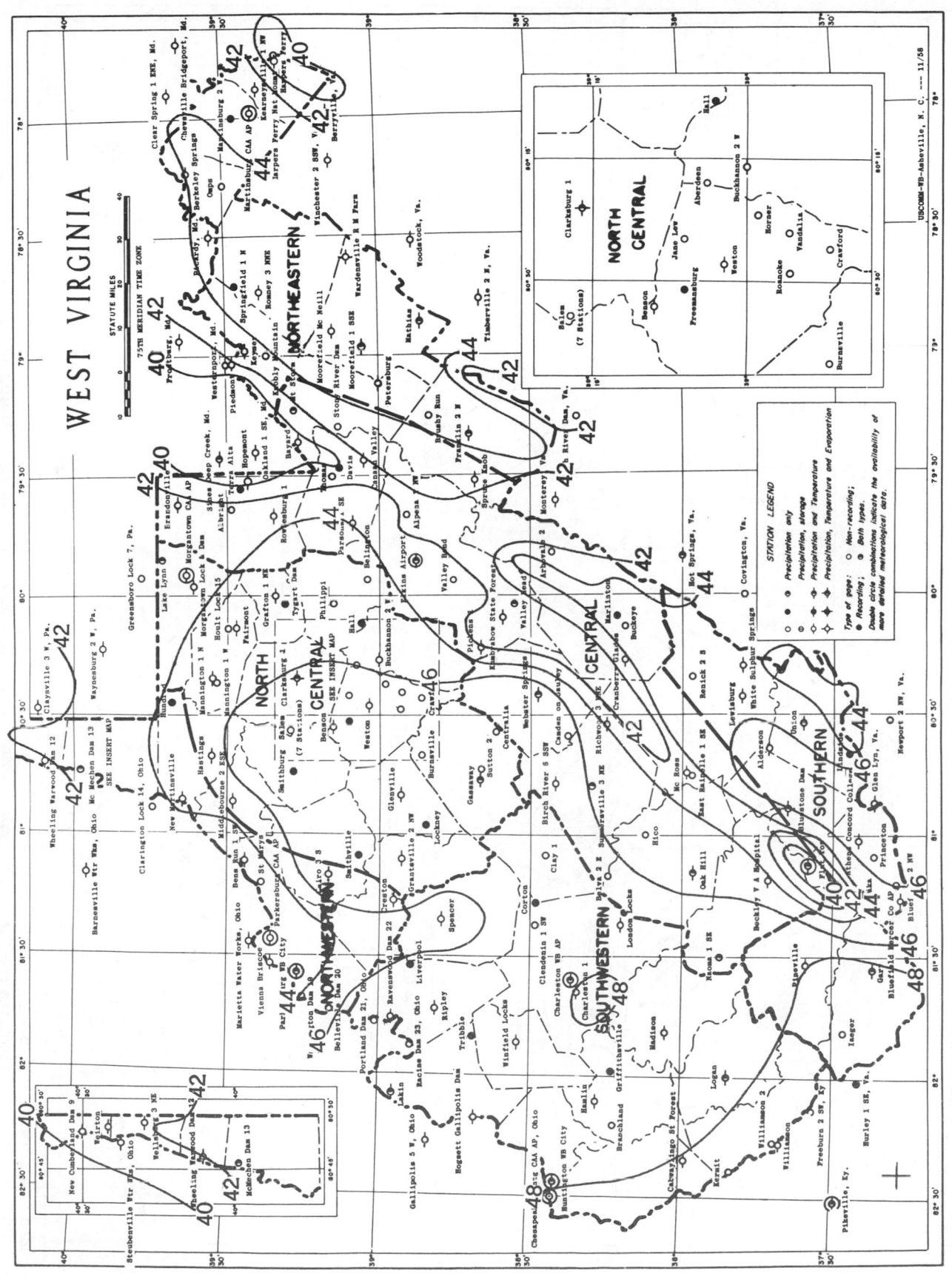

Mean Maximum Temperature (°F.), January

WEST VIRGINIA

Based on period 1931-52

Isolines are drawn through points of approximately equal value. Caution should be used in interpolating on these maps, particularly in mountainous areas.

STATION LEGEND

Mean Minimum Temperature (°F.), January

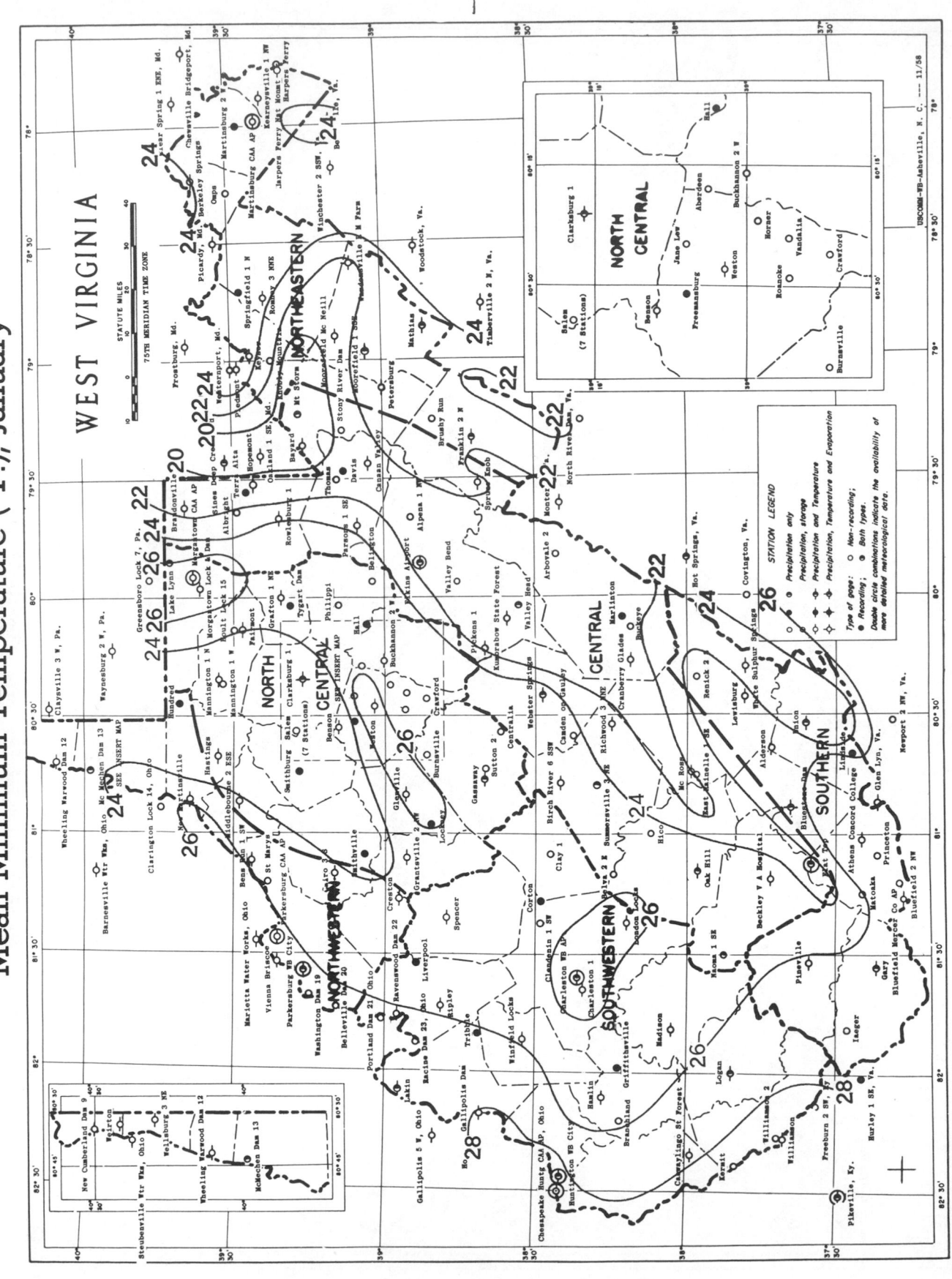

Based on period 1931-52

Isolines are drawn through points of approximately equal value. Caution should be used in interpolating on these maps, particularly in mountainous areas.

1088

Mean Maximum Temperature (°F.), July

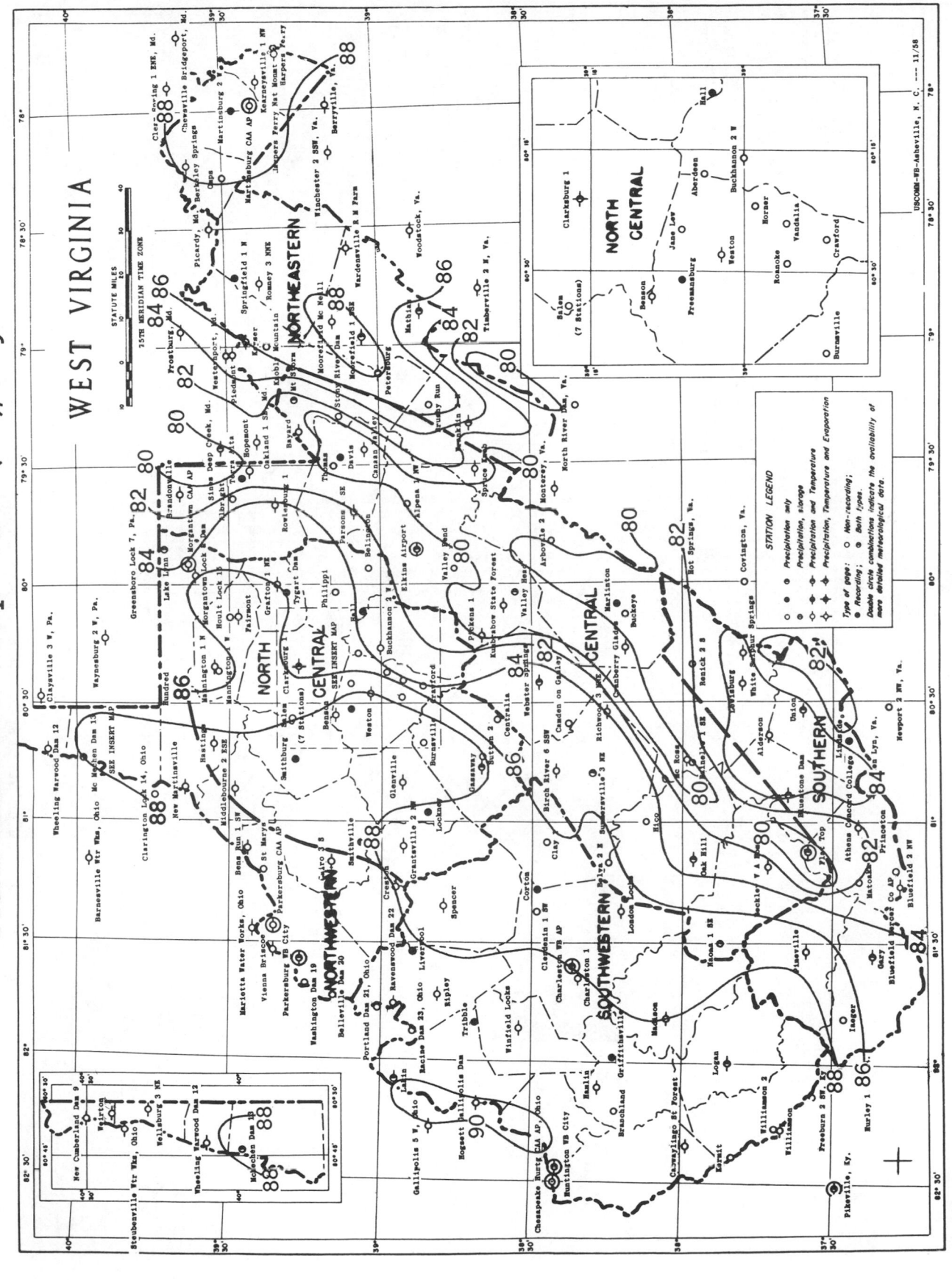

WEST VIRGINIA

Based on period 1931-52

Isolines are drawn through points of approximately equal value. Caution should be used in interpolating on these maps, particularly in mountainous areas.

Mean Minimum Temperature (°F.), July

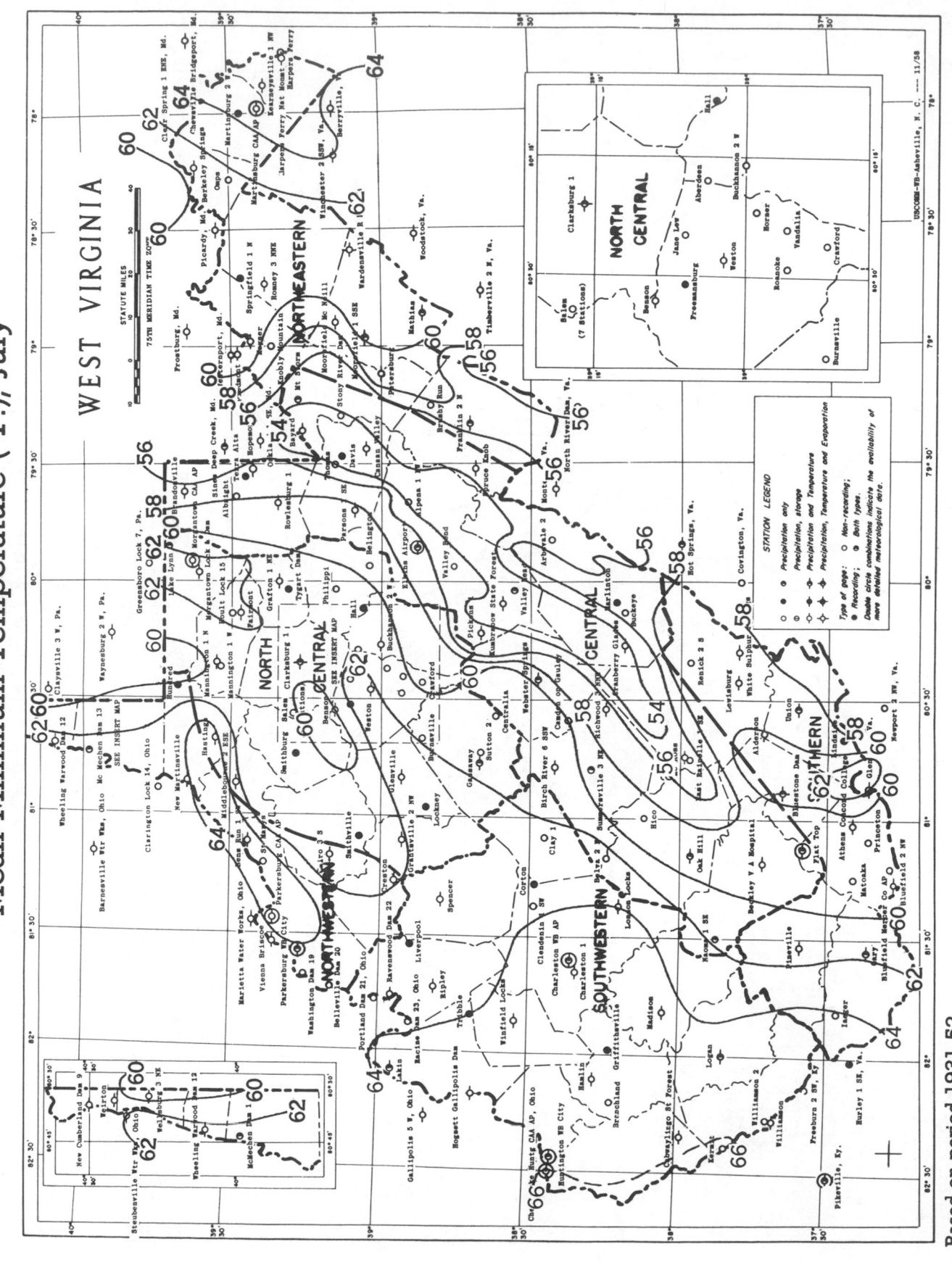

WEST VIRGINIA

STATION LEGEND

Based on period 1931-52

Isolines are drawn through points of approximately equal value. Caution should be used in interpolating on these maps, particularly in mountainous areas.

CLIMATES OF THE STATES

WISCONSIN

(Normals, Means and Extremes tables revised 1972 and 1975. Basic report revised February 1960.)

Climate of Wisconsin

Paul J. Waite, Weather Bureau State Climatologist

Wisconsin lies in the upper Midwest between Lake Superior, Upper Michigan, Lake Michigan, and the Mississippi and Saint Croix Rivers. Its greatest length is 320 miles, greatest width 295 miles, and total area 56,066 square miles.

Glaciation has largely determined the topography and soils of the State, excepting the 13,360 square miles of driftless area in southwestern Wisconsin. The various glaciations created a rolling terrain with nearly 9,000 lakes and several areas of marshes and swamps.

Elevations range from about 600 feet above sea level along the Lake Superior and Lake Michigan shores and in the Mississippi flood plain in southwestern Wisconsin to nearly 1,950 feet above sea level at Rib and Strawberry Hills.

The Northern Highlands, a plateau extending across northern Wisconsin, is an area of about 15,000 square miles with elevations from 1,000 to 1,800 feet above sea level. This area is the location of many lakes and the origin of most of the major streams in the State. The slope down to the narrow Lake Superior plain is quite steep.

A comparatively flat, crescent shaped lowland lies immediately south of the Northern Highlands embodying nearly one-fourth of Wisconsin.

The eastern ridges and lowlands to the southeast of the central plains is the most densely populated with the highest concentration of industry and most available land in farms.

The western uplands of southwestern Wisconsin west of the ridges and lowlands and south of the central plains contains approximately one-fourth of the State. This area is the roughest section of the State rising 200 to 350 feet above the central plains and 100 to 200 feet above the Eastern Ridges and Lowlands. The Mississippi River bluffs rise 230 to 650 feet.

The Wisconsin climate is typically continental with some modification by Lakes Michigan and Superior. The cold, snowy winters favor a variety of winter sports, and the warm summers appeal to thousands of vacationers each year. About two-thirds of the annual precipitation falls during the growing season (freeze-free period). It is normally adequate for vegetation, although drought is occasionally reported. This climate is most favorable for dairy farming; the primary crops being corn, small grains, hay, and vegetables. The rapid succession of storms moving from west to east and southwest to northeast account for the stimulating climate.

The average annual temperature varies from 39.0°F. at Winter to 49.5°F. at Beloit. The highest temperature ever recorded in Wisconsin was 114°F. at Wisconsin Dells on July 13, 1936, and the lowest temperature on record is -54°F. reported from Danbury on January 24, 1922.

During more than one-half of the winters temperatures fall to -40°F. or lower, and almost every winter -30°F. or colder is reported from northern Wisconsin. Summer temperatures above 90°F. or higher average 2 to 4 days in northern counties to about 14 days in southern districts. During marked cool outbreaks in the summer months, the central lowlands occasionally report freezing temperatures. Wisconsin temperatures have averaged about 1°F. warmer since 1931 as compared to the 1891-1930 period.

The freeze-free season averages around 80 days per year in the upper northeast and north-central lowlands to about 180 days in the Milwaukee area. The pronounced moderating effect of Lake Michigan is well illustrated by the fact that the growing season of 140 to 150 days along the east-central coastal area is of the same duration as in the southwestern Wisconsin valleys. The short growing season in the central portion of the State is attributed to a number of factors, among them being an inward cold air drainage and the low heat capacities of the peat and sandy soils. The average date of last spring freeze ranges from early May along the Lake Michigan coastal area and southern counties to early June in the northernmost counties. The first autumn freezes occur in late August and early September in northern, and central lowlands to mid-October along the Lake Michigan coast line. However, July freeze is not unusual in the north and central Wisconsin lowlands.

The long-term mean (1931-55) annual precipitation totals 30 to 34 inches over most of the Western Uplands and Northern Highlands, diminishing to about 28 inches along most of the Wisconsin coastal area bordering Lake Michigan and 28 to 30 inches over most of the Wisconsin Central Plain and Lake Superior Coastal area. The higher average annual precipitation coincides generally with the highest elevations, particularly to the windward slopes of the Western Uplands and Northern Highlands. Thunderstorms average about 30 per year in northern Wisconsin to about 40 per year in southern counties, occurring mostly in the summer. Occasional hail, wind, and lightning damage are reported.

The average seasonal snowfall varies from about 30 inches at Beloit to well over 100 inches in northern Iron County along the steep western slope of the Gogebic Range. The heavy snowfall along the Gogebic Range is a result of the prevailing cold northerly winter winds blowing across the relatively warm Lake Superior. Relatively greater average snowfall is recorded over the Western Uplands and Eastern Ridges than in adjacent lowland areas. The mean dates of the first snowfall of consequence, an inch or more, varies from early November in northern localities to around December 1 in southern Wisconsin counties. Average annual duration of snowcover ranges from 85 days in southern-most Wisconsin to more than 140 days along Lake Superior. The snow cover acts as a protective insulation for grasses, autumn seeded grains, and other wintering vegetation.

The drainage of Wisconsin is into Lake Superior, Lake Michigan, and the Mississippi River. The Mississippi and St. Croix Rivers form most of the western boundary. About one-half of the northwestern portion of the State is drained through the Chippewa River, while the ramainder of this region drains directly into the Mississippi or the St. Croix and into Lake Superior. The Wisconsin River has its source at a small lake nearly 1,600 feet above mean sea level on the Upper Michigan boundary and drains most of central Wisconsin.

Most of the Wisconsin River tributaries also spring from the many lakes in the north. Except for the Rock River, a Mississippi River tributary which flows through northern Illinois, eastern Wisconsin drains into Lake Michigan, a large part through Green Bay.

Most of the streams and lakes in Wisconsin are ice-covered from late November to late March. Snow covers the ground in practically all the winter months, except in the extreme southern areas. Flooding is most frequent, and most serious during April, due to the melting of snow associated with spring rains. During this period, flood conditions are often aggravated by ice jams which back up the flood waters. Excessive rains of the thunderstorm type sometimes produce tributary flooding or flash flooding along the smaller streams and creeks. Major flooding occurs on the Mississippi River, on the average, about 3 years in 10.

The most notable floods along the Mississippi occurred in 1880, 1951, and 1952. Important overflow in tributary basins occurred in 1905, 1912, 1916, 1920, 1922, 1934, 1935, 1938, 1941, 1944, 1945, 1946, 1948, 1950, and 1951.

Tornado occurrences have averaged four per year for the period 1916-58 although better observations and public awareness have increased in recent years resulting in more tornadoes being reported. Most of the very destructive Wisconsin tornadoes occur in the northwestern quarter of the State. Wisconsin tornado frequency is highest in June and July, followed in order by April, May, and September.

The fertile soils, gently rolling topography, and climate favor intensified dairy farming in southern and eastern Wisconsin. Most southeastern counties have over 90 percent of the land area in farms as contrasted to the northern third of the State where much less than 50 percent of the land area is in farms.

Farmland constitutes about 43 percent of the total land area of Wisconsin. The primary crops, of hay, oats, and corn utilize 93 percent of the land farmed. Milk is the largest single source of farm income, now surpassing the combined farm income of all other products. Wisconsin farm income for the past several years has ranked among the first 10 states of the Nation. Wisconsin is among the country's leading states in the following commercial vegetables harvested: green peas, sweet corn, cucumbers for pickles, snap beans, beets, cabbage for sauerkraut, carrots, and tomatoes. Marshy areas have been utilized for cranberry growing in central and scattered northern localities, with state production now second in the Nation. Since freeze is a hazard through all summer months, water sources for flooding are a necessity.

About 50 percent of the State is forested and supplies about one-third of the pulpwood used in the paper industry. Since much of the northern areas are of low fertility and with a short growing season, reforestation continues with improved lumber prospect. Naturally, wood and food processing industries have become important as well as machine manufacture.

Wisconsin's stimulating climate has created an energetic population concentrated mostly in the southern and eastern districts of the State, with northern districts sparsely settled. With adequate transportation the recreational facilities of the northern counties are coming into greater usage.

REFERENCES

(1) Finley, Robert W., 1957. Geography of Wisconsin. Regents of the University of Wisconsin.

(2) Martin, Lawrence. 1932. The Physical Geography of Wisconsin. Wisconsin Survey Bulletin No. XXXVI.

(3) Thom, H.C.S. 1957. Probabilities of One-Inch Snowfall Thresholds for the United States. Monthly Weather Review. Vol. 85, No. 8.

(4) U. S. Department of Agriculture. 1941. Climate of the States, Wisconsin. pp. 1199 and 1200.

(5) U. S. Department of Agriculture and Wisconsin Department of Agriculture. 1957. Cranberries of Wisconsin. Bull. No. 70.

(6) U. S. Department of Agriculture and Wisconsin Department of Agriculture. 1954. Wisconsin Agriculture in Mid-Century. Bulletin No. 325.

(7) Weather Bureau Technical Paper No. 16 - Maximum 24-Hour Precipitation in the United States. Washington, D. C. 1952.

(8) Weather Bureau Technical Paper No. 25 - Rainfall Intensity-Duration-Frequency Curves. For selected stations in the United States, Alaska, Hawaiian Islands and Puerto Rico.

BIBLIOGRAPHY

(A) Climatic Summary of the United States (Bulletin W) 1930 edition, Sections 47, 48 and 49. U. S. Weather Bureau

(B) Climatic Summary of the United States, Wisconsin - Supplement for 1931 through 1952 (Bulletin W Supplement). U. S. Weather Bureau

(C) Climatological Data - Wisconsin. U. S. Weather Bureau

(D) Climatological Data National Summary. U. S. Weather Bureau

(E) Hourly Precipitation Data - Wisconsin. U. S. Weather Bureau

(F) Local Climatological Data, U. S. Weather Bureau for Green Bay, La Crosse, Madison and Milwaukee, Wisconsin.

FREEZE DATA

STATION	Freeze threshold temperature	Mean date of last Spring occurrence	Mean date of first Fall occurrence	Mean No. of days between dates	Years of record Spring	No. of occurrences in Spring	Years of record Fall	No. of occurrences in Fall	STATION	Freeze threshold temperature	Mean date of last Spring occurrence	Mean date of first Fall occurrence	Mean No. of days between dates	Years of record Spring	No. of occurrences in Spring	Years of record Fall	No. of occurrences in Fall
AMERY HYDRO PLT	32	05-15	09-19	127	28	28	28	28	DODGEVILLE	32	05-03	10-08	158	10	10	10	10
	28	05-04	10-04	153	28	28	28	28		28	04-23	10-26	186	10	10	10	10
	24	04-23	10-17	177	28	28	28	28		24	04-03	11-05	216	10	10	10	10
	20	04-08	10-27	202	28	28	27	27		20	03-29	11-11	228	10	10	10	10
	16	03-28	11-09	225	28	28	28	28		16	03-19	11-19	245	10	10	9	9
ANTIGO	32	05-17	09-29	135	30	30	30	30	EAU CLAIRE CAA AP	32	05-05	10-04	151	30	30	30	30
	28	05-02	10-10	161	30	30	30	30		28	04-21	10-18	180	30	30	30	30
	24	04-20	10-22	185	30	30	30	30		24	04-08	10-31	205	30	30	30	30
	20	04-10	11-04	209	30	30	30	30		20	03-31	11-07	222	30	30	30	30
	16	03-30	11-13	228	30	30	30	30		16	03-23	11-14	237	30	30	30	30
ASHLAND EXP FARM	32	05-30	09-16	109	30	30	30	30	FOND DU LAC	32	05-11	10-09	151	30	30	30	30
	28	05-19	09-28	131	30	30	30	30		28	04-25	10-18	176	29	29	30	30
	24	05-11	10-11	153	30	30	30	30		24	04-10	10-31	204	29	29	30	30
	20	04-22	10-25	186	30	30	30	30		20	03-31	11-09	223	29	29	30	30
	16	04-10	11-06	211	30	30	30	30		16	03-23	11-20	242	29	29	30	30
BELOIT COLLEGE	32	05-01	10-13	165	30	30	30	30	GRAND RIVER LOCK	32	05-13	09-26	135	30	30	30	30
	28	04-16	10-25	193	30	30	30	30		28	04-30	10-02	155	30	30	30	30
	24	04-01	11-05	218	30	30	30	30		24	04-16	10-17	184	30	30	29	29
	20	03-24	11-16	236	30	30	30	30		20	04-03	10-27	207	30	30	29	29
	16	03-17	11-25	253	30	30	30	30		16	03-22	11-10	233	30	30	29	29
BIG ST GERMAIN DAM	32	06-06	09-05	91	30	30	29	29	GRANTSBURG CAA AP	32	05-24	09-21	121	30	30	30	30
	28	05-25	09-23	121	30	30	29	29		28	05-13	10-01	141	30	30	30	30
	24	05-12	10-08	149	30	30	28	28		24	04-30	10-12	165	30	30	30	30
	20	05-02	10-18	170	30	30	28	28		20	04-19	10-27	190	30	30	30	30
	16	04-21	10-31	194	30	30	28	28		16	04-09	11-04	209	29	29	30	30
BLAIR	32	05-19	09-24	128	25	25	25	25	GREEN BAY WB AP	32	05-06	10-13	161	30	30	30	30
	28	05-07	10-04	149	25	25	25	25		28	04-20	10-26	188	30	30	30	30
	24	04-24	10-14	173	25	25	25	25		24	04-06	11-06	214	30	30	30	30
	20	04-08	10-27	202	25	25	25	25		20	03-28	11-15	232	30	30	30	30
	16	03-27	11-06	224	24	24	25	25		16	03-22	11-23	246	30	30	30	30
BRODHEAD	32	05-09	10-03	147	30	30	30	30	HANCOCK EXP FARM	32	05-17	09-30	135	29	29	30	30
	28	04-26	10-14	171	30	30	30	30		28	05-06	10-07	154	29	29	30	30
	24	04-13	10-29	199	30	30	30	30		24	04-26	10-20	177	28	28	30	30
	20	03-30	11-09	224	30	30	30	30		20	04-15	10-31	199	28	28	30	30
	16	03-22	11-19	242	30	30	30	30		16	04-03	11-06	217	28	28	30	30
BRULE ISL	32	06-10	09-04	87	15	15	14	14	HATFIELD DAM	32	05-23	09-17	116	29	29	29	29
	28	05-20	09-17	120	14	14	14	14		28	05-11	09-26	138	30	30	29	29
	24	05-12	10-02	143	13	13	14	14		24	04-30	10-08	161	30	30	29	29
	20	04-27	10-18	174	13	13	14	14		20	04-19	10-19	183	30	30	29	29
	16	04-14	11-02	202	13	13	14	14		16	04-05	10-30	208	30	30	29	29
BURNETT 2 NW	32	05-12	10-01	142	29	29	28	28	HILLSBORO	32	05-16	09-28	135	30	30	29	29
	28	04-29	10-13	167	29	29	28	28		28	05-03	10-06	156	30	30	29	29
	24	04-14	10-28	197	28	28	28	28		24	04-18	10-18	183	30	30	29	29
	20	03-30	11-06	221	28	28	28	28		20	04-05	10-31	208	30	30	29	29
	16	03-27	11-19	237	28	28	28	28		16	03-27	11-12	230	30	30	29	29
CODDINGTON EXP FARM	32	06-02	08-30	89	29	29	29	29	KEWAUNEE	32	05-06	10-14	161	21	21	22	22
	28	05-18	09-14	119	29	29	29	29		28	04-16	10-24	191	21	21	22	22
	24	05-05	09-28	146	29	29	28	28		24	04-04	11-05	215	21	21	22	22
	20	04-24	10-14	174	29	29	28	28		20	03-29	11-18	235	21	21	20	20
	16	04-11	10-26	198	29	29	28	28		16	03-22	11-26	249	21	21	19	19
CRIVITZ HIGH FALLS	32	05-26	09-21	119	25	25	25	25	LA CROSSE WB CITY	32	05-01	10-08	161	30	30	30	30
	28	05-13	09-29	139	25	25	25	25		28	04-19	10-19	183	30	30	30	30
	24	05-02	10-13	163	25	25	25	25		24	04-05	11-01	210	30	30	30	30
	20	04-20	10-25	188	25	25	24	24		20	03-28	11-11	228	30	30	30	30
	16	04-10	11-04	207	25	25	24	24		16	03-17	11-19	247	30	30	30	30
CUMBERLAND	32	05-15	09-24	133	19	19	19	19	LAKE MILLS	32	05-10	10-08	151	30	30	30	30
	28	05-06	10-08	155	19	19	19	19		28	04-26	10-17	174	30	30	30	30
	24	04-22	10-18	179	19	19	18	18		24	04-11	11-01	205	30	30	30	30
	20	04-10	10-27	200	19	19	18	18		20	04-01	11-11	224	30	30	30	30
	16	04-01	11-06	219	19	19	18	18		16	03-24	11-21	242	30	30	29	29
DANBURY	32	05-31	09-12	105	28	28	29	29	LANCASTER	32	05-07	10-10	156	28	28	29	29
	28	05-18	09-25	130	27	27	29	29		28	04-21	10-22	184	28	28	29	29
	24	05-05	10-09	156	28	28	28	28		24	04-06	11-02	210	28	28	29	29
	20	04-25	10-16	174	27	27	28	28		20	03-29	11-10	226	28	28	29	29
	16	04-13	10-29	199	27	27	27	27		16	03-23	11-16	238	28	28	29	29
DARLINGTON	32	05-12	09-29	140	30	30	29	29	LAONA RS	32	06-03	09-02	92	17	17	19	19
	28	05-03	10-09	159	28	28	28	28		28	05-24	09-27	127	17	17	17	17
	24	04-18	10-20	185	28	28	28	28		24	05-09	10-07	151	19	19	16	16
	20	04-04	11-01	211	28	28	28	28		20	04-23	10-23	183	17	17	16	16
	16	03-24	11-11	233	28	28	28	28		16	04-10	10-30	203	16	16	14	14

FREEZE DATA

STATION	Freeze threshold temperature	Mean date of last Spring occurrence	Mean date of first Fall occurrence	Mean No. of days between dates	Years of record Spring	No. of occurrences in Spring	Years of record Fall	No. of occurrences in Fall
LONG LAKE DAM	32	06-11	08-24	73	30	30	30	30
	28	05-31	09-14	106	30	30	29	29
	24	05-17	09-28	135	30	30	29	29
	20	05-08	10-10	155	30	30	29	29
	16	04-26	10-26	183	30	30	29	29
MADISON WB	32	04-26	10-19	177	30	30	30	30
	28	04-10	10-31	204	30	30	30	30
	24	03-31	11-09	223	30	30	30	30
	20	03-23	11-16	237	30	30	30	30
	16	03-18	11-24	251	30	30	30	30
MANITOWOC	32	04-29	10-19	173	30	30	30	30
	28	04-14	10-31	200	30	30	30	30
	24	04-04	11-10	220	30	30	30	30
	20	03-27	11-19	237	30	30	30	30
	16	03-19	11-26	252	30	30	30	30
MARINETTE	32	05-12	10-03	143	30	30	30	30
	28	04-25	10-16	173	30	30	30	30
	24	04-16	10-30	197	30	30	30	30
	20	04-06	11-10	218	30	30	29	29
	16	03-26	11-17	237	30	30	29	29
MARSHFIELD EXP STA	32	05-17	09-27	133	30	30	30	30
	28	05-03	10-06	155	30	30	30	30
	24	04-20	10-19	182	30	30	30	30
	20	04-08	10-31	205	30	30	30	30
	16	03-29	11-10	227	30	30	30	30
MATHER 3 NW	32	05-18	09-21	126	30	30	30	30
	28	05-07	10-01	147	30	30	30	30
	24	04-23	10-10	170	29	29	30	30
	20	04-09	10-27	201	30	30	29	29
	16	03-30	11-04	219	30	30	28	28
MEDFORD	32	05-19	09-23	126	29	29	27	27
	28	05-10	10-02	145	28	28	27	27
	24	04-27	10-12	168	30	30	27	27
	20	04-15	10-29	197	29	29	27	27
	16	04-01	11-06	219	29	29	27	27
MELLEN 2 NE	32	06-03	09-10	99	22	22	21	21
	28	05-19	09-26	130	22	22	21	21
	24	05-12	10-10	151	21	21	21	21
	20	04-27	10-25	181	20	20	20	20
	16	04-14	11-01	201	20	20	20	20
MENASHA LOCKS	32	05-05	10-08	155	30	30	30	30
	28	04-18	10-25	190	30	30	30	30
	24	04-07	11-07	214	30	30	30	30
	20	03-30	11-14	229	30	30	30	30
	16	03-24	11-22	244	30	30	30	30
MERRILL	32	05-22	09-23	124	29	29	29	29
	28	05-08	10-02	146	28	28	28	28
	24	04-24	10-14	173	29	29	28	28
	20	04-14	10-29	199	28	28	27	27
	16	04-04	11-09	219	28	28	27	27
MILWAUKEE WB CITY	32	04-20	10-25	188	24	24	25	25
	28	04-05	11-02	210	24	24	25	25
	24	03-28	11-11	228	24	24	25	25
	20	03-24	11-21	243	24	24	25	25
	16	03-12	11-28	261	24	24	25	25
MINOCQUA DAM	32	05-28	09-22	117	28	28	28	28
	28	05-15	10-03	140	28	28	28	28
	24	05-04	10-16	166	28	28	28	28
	20	04-23	10-27	187	28	28	28	28
	16	04-14	11-05	205	28	28	27	27
MONDOVI	32	05-17	09-28	134	24	24	23	23
	28	05-03	10-04	154	22	22	23	23
	24	04-20	10-18	181	25	25	21	21
	20	04-12	10-30	202	24	24	21	21
	16	03-30	11-08	223	24	24	20	20
NEILLSVILLE	32	05-14	09-28	138	29	29	27	27
	28	05-04	10-05	154	29	29	28	28
	24	04-20	10-19	182	29	29	28	28
	20	04-05	11-03	212	29	29	27	27
	16	03-29	11-11	227	29	29	27	27
NEW LONDON	32	05-10	09-30	143	29	29	30	30
	28	04-26	10-13	169	29	29	30	30
	24	04-11	10-24	196	28	28	29	29
	20	04-02	11-06	219	28	28	29	29
	16	03-26	11-16	235	29	29	29	29
OCONTO	32	05-17	09-27	133	29	29	30	30
	28	05-02	10-11	162	29	29	29	29
	24	04-18	10-29	194	29	29	29	29
	20	04-03	11-06	217	28	28	29	29
	16	03-29	11-14	229	29	29	28	28
OSHKOSH BUCKSTAFF OB	32	05-09	10-04	148	30	30	30	30
	28	04-26	10-17	174	30	30	30	30
	24	04-12	10-28	199	30	30	30	30
	20	03-30	11-07	222	30	30	30	30
	16	03-22	11-21	244	30	30	30	30
PARK FALLS	32	05-27	09-16	112	30	30	30	30
	28	05-10	10-01	144	30	30	30	30
	24	04-29	10-12	167	30	30	30	30
	20	04-17	10-27	192	30	30	30	30
	16	04-05	11-05	213	30	30	30	30
PINE RIVER	32	05-11	10-03	145	27	27	26	26
	28	05-01	10-11	164	27	27	27	27
	24	04-16	10-26	194	27	27	27	27
	20	04-04	11-05	215	27	27	26	26
	16	03-29	11-17	233	27	27	26	26
PLYMOUTH	32	05-09	10-11	155	30	30	30	30
	28	04-25	10-21	179	30	30	30	30
	24	04-11	11-02	205	29	29	30	30
	20	04-01	11-11	224	27	27	30	30
	16	03-27	11-20	238	26	26	30	30
PORTAGE LOCK	32	04-29	10-11	165	29	29	30	30
	28	04-17	10-23	189	30	30	30	30
	24	04-05	11-04	213	30	30	30	30
	20	03-30	11-12	227	30	30	30	30
	16	03-21	11-19	244	30	30	30	30
PRAIRIE DU CHIEN	32	04-27	10-09	165	30	30	30	30
	28	04-16	10-21	187	30	30	30	30
	24	04-05	11-03	212	30	30	30	30
	20	03-26	11-09	228	30	30	30	30
	16	03-17	11-17	245	30	30	30	30
PRAIRIE DU SAC 2 N	32	04-26	10-16	173	30	30	30	30
	28	04-12	10-27	198	30	30	30	30
	24	03-31	11-08	222	30	30	30	30
	20	03-25	11-15	236	30	30	30	30
	16	03-18	11-24	251	30	30	30	30
PRENTICE 5 W	32	06-05	08-29	85	30	30	30	30
	28	05-22	09-18	119	30	30	29	29
	24	05-07	09-28	144	30	30	29	29
	20	04-24	10-12	171	30	30	29	29
	16	04-11	10-27	199	30	30	29	29
RACINE	32	04-28	10-20	174	30	30	30	30
	28	04-18	10-31	197	30	30	30	30
	24	03-31	11-10	224	30	30	30	30
	20	03-23	11-20	242	30	30	30	30
	16	03-14	11-28	259	30	30	30	30
REST LAKE	32	05-28	09-16	111	24	24	22	22
	28	05-18	09-28	133	23	23	21	21
	24	05-08	10-13	158	23	23	21	21
	20	04-26	10-28	185	24	24	22	22
	16	04-14	11-08	208	24	24	19	19
RICHLAND CENTER	32	05-10	09-28	142	29	29	29	29
	28	05-01	10-09	161	29	29	29	29
	24	04-18	10-24	189	29	29	29	29
	20	04-03	11-03	214	29	29	29	29
	16	03-24	11-15	237	29	29	29	29
RIVER FALLS	32	05-14	09-26	135	30	30	30	30
	28	05-04	10-08	157	30	30	30	30
	24	04-20	10-18	181	30	30	30	30
	20	04-11	10-30	203	30	30	30	30
	16	03-26	11-08	227	30	30	30	30

FREEZE DATA

STATION	Freeze threshold temperature	Mean date of last Spring occurrence	Mean date of first Fall occurrence	Mean No. of days between dates	Years of record Spring	No. of occurrences in Spring	Years of record Fall	No. of occurrences in Fall
SHAWANO	32	05-18	09-26	131	26	26	25	25
	28	05-06	10-07	154	26	26	26	26
	24	04-19	10-21	185	26	26	26	26
	20	04-08	11-02	208	26	26	26	26
	16	03-30	11-16	231	26	26	26	26
SHEBOYGAN	32	04-28	10-19	174	30	30	29	29
	28	04-17	10-31	197	30	30	29	29
	24	04-03	11-09	220	30	30	29	29
	20	03-29	11-19	236	30	30	29	29
	16	03-21	11-27	252	30	30	29	28
SOLON SPRINGS	32	06-01	09-13	104	28	28	26	26
	28	05-19	09-23	127	29	29	27	27
	24	05-09	10-08	152	29	29	26	26
	20	04-27	10-14	170	29	29	26	26
	16	04-15	10-29	197	29	29	26	26
SPARTA	32	05-10	09-29	142	14	14	15	15
	28	04-24	10-08	167	14	14	15	15
	24	04-11	10-23	195	14	14	15	15
	20	03-30	11-06	220	14	14	15	15
	16	03-21	11-15	239	14	14	15	15
SPOONER EXP FARM	32	05-24	09-20	120	29	29	28	28
	28	05-10	09-30	143	29	29	28	28
	24	04-30	10-12	165	30	30	28	28
	20	04-17	10-23	189	30	30	28	28
	16	04-07	11-05	212	30	30	28	28
STANLEY 1 E	32	05-16	09-23	130	28	28	26	26
	28	05-06	10-04	151	28	28	26	26
	24	04-16	10-18	185	28	28	26	26
	20	04-05	11-03	212	28	28	24	24
	16	03-29	11-11	227	27	27	24	24
STEVENS PT	32	05-11	10-01	142	30	30	30	30
	28	04-28	10-12	167	30	30	30	30
	24	04-15	10-26	195	30	30	30	30
	20	04-03	11-07	217	30	30	30	30
	16	03-27	11-16	234	30	30	30	30
STURGEON BAY EXP FRM	32	05-17	10-02	137	27	27	29	29
	28	04-30	10-18	171	27	27	29	29
	24	04-17	11-02	199	27	27	29	29
	20	04-04	11-14	224	27	27	28	28
	16	03-26	11-25	244	27	27	28	28
SUPERIOR BONG AP	32	05-12	10-04	145	30	30	30	30
	28	04-26	10-16	173	30	30	30	30
	24	04-14	10-29	197	30	30	30	30
	20	04-04	11-09	218	30	30	30	30
	16	03-30	11-13	229	30	30	30	30
VIROQUA	32	05-06	10-05	152	24	24	26	26
	28	04-25	10-19	177	25	25	25	25
	24	04-10	10-30	203	26	26	25	25
	20	04-01	11-09	222	27	27	25	25
	16	03-23	11-16	239	27	27	25	25
WASHINGTON ISL	32	05-17	10-17	153	19	19	19	19
	28	05-02	10-31	182	18	18	19	19
	24	04-15	11-09	208	18	18	19	19
	20	04-08	11-18	225	18	18	19	19
	16	03-28	11-28	245	18	18	19	19
WATERTOWN	32	05-04	10-10	159	30	30	30	30
	28	04-21	10-23	186	30	30	30	30
	24	04-02	11-05	217	30	30	30	30
	20	03-27	11-12	230	30	30	30	30
	16	03-18	11-23	249	30	30	30	30
WAUKESHA WTR WRKS	32	05-08	10-08	153	30	30	28	28
	28	04-23	10-20	180	30	30	28	28
	24	04-09	11-02	208	30	30	28	28
	20	03-31	11-10	224	30	30	28	28
	16	03-20	11-21	247	30	30	29	29
WAUPACA	32	05-11	10-02	144	29	29	28	28
	28	04-29	10-13	167	29	29	28	28
	24	04-13	10-26	196	29	29	28	28
	20	04-04	11-08	218	29	29	28	28
	16	03-24	11-20	240	29	29	28	28
WAUSAU	32	05-14	09-30	139	30	30	30	30
	28	05-03	10-10	160	30	30	30	30
	24	04-17	10-24	190	30	30	30	30
	20	04-08	11-09	215	30	30	30	30
	16	03-28	11-14	231	30	30	30	30
WEST BEND	32	05-10	10-08	151	25	25	25	25
	28	04-24	10-20	179	25	25	25	25
	24	04-10	10-31	204	25	25	25	25
	20	03-31	11-10	224	25	25	25	25
	16	03-23	11-20	242	25	25	25	25
WEYERHAUSER	32	05-20	09-22	125	30	30	29	29
	28	05-10	10-01	145	30	30	30	30
	24	04-26	10-15	172	30	30	30	30
	20	04-13	10-28	198	30	30	30	30
	16	04-02	11-04	216	30	30	30	30
WILLIAMS BAY YERKES OBS	32	05-02	10-10	160	30	30	30	30
	28	04-24	10-26	185	29	29	30	30
	24	04-06	11-01	209	29	29	30	30
	20	03-27	11-12	230	29	29	30	30
	16	03-21	11-22	246	29	29	30	30
WINTER PK RES	32	06-03	09-10	99	15	15	15	15
	28	05-24	09-25	124	14	14	15	15
	24	05-16	10-01	139	14	14	15	15
	20	04-28	10-12	168	14	14	15	15
	16	04-12	10-30	200	14	14	15	15
WISCONSIN DELLS	32	05-08	10-03	148	19	19	20	20
	28	04-27	10-15	171	19	19	20	20
	24	04-17	10-25	191	19	19	20	20
	20	04-07	11-04	211	19	19	20	20
	16	03-25	11-15	235	19	19	20	20
WISCONSIN RAPIDS	32	05-19	09-20	124	30	30	30	30
	28	05-08	09-29	144	30	30	30	30
	24	04-26	10-12	169	30	30	30	29
	20	04-15	10-23	191	30	30	29	29
	16	04-01	11-06	219	30	30	29	29

Data in the above table are based on the period 1921-1950, or that portion of this period for which data are available.

Means have been adjusted to take into account years of non-occurrence.

A freeze is a numerical substitute for the former term "killing frost" and is the occurrence of a minimum temperature at or below the threshold temperature of 32°, 28°, etc.

Freeze data tabulations in greater detail are available and can be reproduced at cost.

*MEAN TEMPERATURE AND PRECIPITATION

STATION	JAN Temp	JAN Precip	FEB Temp	FEB Precip	MAR Temp	MAR Precip	APR Temp	APR Precip	MAY Temp	MAY Precip	JUN Temp	JUN Precip	JUL Temp	JUL Precip	AUG Temp	AUG Precip	SEP Temp	SEP Precip	OCT Temp	OCT Precip	NOV Temp	NOV Precip	DEC Temp	DEC Precip	ANN Temp	ANN Precip
NORTHWEST DIVISION																										
AMERY BLACK BROOK HYDRO	12.3	.81	14.8	.88	26.2	1.46	43.4	2.24	56.1	3.42	65.6	4.99	71.1	3.24	68.6	3.69	59.6	3.00	48.2	1.80	30.9	1.58	17.4	.87	42.9	27.98
ASHLAND EXP FARM	13.7	1.01	15.5	.76	24.8	1.27	38.9	2.26	50.2	3.44	60.7	4.55	69.6	3.85	65.3	3.95	56.8	2.64	46.3	2.18	30.9	1.93	18.7	.91	41.0	28.75
DANBURY 1 SE	10.5	1.07	13.6	.97	25.2	1.65	41.7	2.36	54.3	3.65	64.3	5.13	69.9	3.52	67.0	4.44	57.8	3.18	46.6	2.13	29.7	1.78	15.7	1.04	41.4	30.92
HOLCOMBE		1.04		.86		1.64		2.89		3.76		4.61		3.73		3.68		3.72		2.05		1.64		1.04		30.66
LADYSMITH		.90		.87		1.66		2.53		3.50		4.65		3.63		3.76		3.19		2.18		1.74		1.03		29.64
SOLON SPRINGS	12.0		14.5		25.3		41.3		53.6		63.3		69.0		66.3		57.1		46.8		30.0		16.7		41.3	
SPOONER EXP FARM	12.4	.81	14.5	.70	24.8	1.41	42.7	2.23	55.5	3.28	65.0	4.39	70.5	3.79	67.8	3.91	58.5	3.16	47.5	1.88	30.4	1.63	17.2	.90	42.4	28.09
SUPERIOR 7 SE	12.9	-	15.5		25.4		39.4		49.6		59.3		67.0		66.2		56.8		46.6		30.7		18.2		40.6	
WEYERHAUSER 1 N	12.9		15.5		26.1		42.5		54.8		63.8		69.1		66.5		57.7		46.9		30.4		17.3		42.0	
WINTER 6 NNW	10.3	.99	11.3	.92	21.9	1.51	39.0	2.40	52.1	3.73	61.6	4.88	66.5	4.06	63.6	4.29	54.7	3.19	43.9	2.04	28.5	1.90	14.7	.93	39.0	30.84
DIVISION	12.4	1.00	15.0	.90	25.4	1.59	41.6	2.52	53.8	3.61	63.4	4.77	69.1	3.76	66.8	4.01	57.7	3.12	47.1	2.06	30.5	1.78	17.4	1.02	41.7	30.14
NORTH CENTRAL DIVISION																										
BIG SAINT GERMAIN DAM	12.4	1.26	14.2	1.04	23.8	1.53	39.0	2.11	52.2	3.56	61.5	4.73	66.1	4.11	63.6	3.91	55.4	3.86	44.9	2.27	28.9	2.17	16.6	1.17	39.9	31.72
FLAMBEAU RESERVOIR		1.10		1.09		1.59		2.48		3.81		5.33		4.31		4.06		3.31		2.22		2.08		1.15		32.53
LONG LAKE DAM	12.6	1.43	13.6	1.19	23.1	1.65	38.6	2.44	51.8	3.38	61.4	4.47	66.0	4.14	63.6	4.02	55.4	3.63	44.7	2.44	29.1	2.36	16.8	1.32	39.7	32.47
MEDFORD	13.5	1.36	15.3	1.20	25.6	1.85	41.9	2.45	54.2	3.96	63.6	5.21	68.4	3.46	66.9	4.15	57.7	3.79	46.6	2.15	30.3	2.17	17.6	1.46	41.8	33.21
MELLEN 2 N	13.5		14.7		24.3		40.0		52.1		62.0		67.7		64.9		56.8		46.3		30.8		18.2		40.9	
MERRILL		1.17		.96		1.44		2.26		3.71		4.64		3.21		3.80		3.66		2.15		2.11		1.06		30.17
PARK FALLS	12.7	1.19	14.3	1.04	24.7	1.61	40.5	2.63	53.4	3.56	62.9	5.68	68.1	4.27	65.4	4.40	56.6	3.33	45.7	2.29	29.2	2.00	16.5	1.19	40.8	33.19
PHELPS DEERSKIN DAM		1.39		1.12		1.44		2.38		3.32		4.70		3.85		3.96		3.78		2.43		2.23		1.28		31.88
PRENTICE 5 W	12.9	1.45	14.6	1.27	25.0	2.04	41.3	2.70	53.6	4.01	62.7	5.40	67.3	3.92	64.7	4.21	56.4	3.99	46.2	2.39	30.0	2.10	17.3	1.30	41.0	34.78
RHINELANDER	13.1	1.33	14.6	1.26	24.8	1.64	40.6	2.18	53.5	3.40	63.3	4.81	68.3	3.80	65.6	3.80	57.0	3.50	46.4	2.34	30.3	2.00	17.7	1.20	41.3	31.26
WAUSAU OLD POST OFFICE	16.9	1.43	18.3	1.35	28.8	1.91	44.5	2.66	57.3	3.75	67.1	4.76	72.1	3.55	69.5	4.04	60.9	3.54	49.2	2.38	33.3	2.22	21.0	1.31	44.9	32.90
DIVISION	13.4	1.28	15.2	1.11	25.0	1.68	41.0	2.47	53.6	3.67	63.2	4.95	68.1	3.99	65.8	4.00	57.2	3.59	46.7	2.28	30.5	2.07	17.9	1.20	41.5	32.29
NORTHEAST DIVISION																										
ANTIGO	16.1	1.30	17.4	1.03	27.0	1.51	42.5	2.47	55.2	3.46	64.5	4.58	69.4	3.58	67.0	3.79	58.8	3.60	47.9	2.28	32.0	1.97	19.8	1.08	43.1	30.65
BREAKWATER		1.30		1.15		1.72		2.26		3.19		3.82		3.44		2.99		1.96		2.31		1.22		28.62	
BRULE ISLAND		1.40		1.27		1.70		2.16		3.33		4.46		3.58		3.27		3.37		2.08		2.31		1.23		30.16
CRIVITZ HIGH FALLS	16.6	1.30	17.7	1.09	27.2	1.64	42.2	2.60	54.8	2.82	65.0	3.83	70.0	3.05	67.5	3.18	58.9	3.13	48.0	1.96	33.1	2.29	20.3	1.25	43.4	28.14
MARINETTE	20.4	1.59	21.5	1.27	30.0	1.65	43.2	2.37	55.2	2.78	66.0	3.75	71.9	2.71	69.5	3.04	61.5	3.14	50.6	2.17	35.8	2.43	24.4	1.29	45.8	28.19
OCONTO	18.6	1.56	19.8	1.37	29.4	1.74	43.4	2.55	55.1	2.48	65.7	3.41	70.9	2.42	68.7	2.90	60.3	3.07	49.4	2.02	35.0	2.24	23.0	1.29	44.9	27.05
SHAWANO	17.5	1.65	19.2	1.43	29.3	1.77	44.6	2.66	57.0	3.19	66.9	4.19	71.7	2.85	69.0	3.66	60.4	3.13	48.8	2.15	34.1	2.33	21.7	1.58	45.0	30.59
DIVISION	17.0	1.47	18.3	1.25	27.7	1.66	42.6	2.45	54.7	3.07	64.7	4.08	69.6	3.14	67.3	3.29	58.9	3.23	48.2	2.11	33.3	2.25	21.2	1.28	43.6	29.28
WEST CENTRAL DIVISION																										
BLAIR	15.0	1.31	17.3	1.12	29.2	1.91	45.0	2.68	57.4	3.58	67.3	4.54	72.1	3.94	69.5	3.66	60.7	3.70	49.2	2.06	32.8	1.89	19.6	1.05	44.6	31.44
EAU CLAIRE	15.7	1.05	18.4	1.06	29.5	1.90	45.4	2.88	58.8	3.52	68.8	4.61	74.3	3.33	71.6	3.70	62.2	3.43	50.3	2.06	33.3	1.82	20.5	1.06	45.7	30.42
HATFIELD DAM	15.2	.99	17.7	.85	28.3	1.67	44.2	2.73	56.9	3.95	66.3	5.11	71.0	3.36	68.6	3.46	60.0	3.46	48.9	2.26	32.8	1.83	20.2	1.00	44.2	30.67
LA CROSSE WB AP	15.7	1.22	19.3	1.11	31.6	1.86	46.6	2.31	59.0	3.27	68.6	3.87	74.0	3.21	71.4	3.29	62.3	3.82	50.8	1.93	34.3	1.81	20.5	1.22	46.2	28.92
MATHER 3 NW	15.2	1.31	17.2	1.12	28.1	1.91	43.7	2.68	56.2	4.00	65.8	4.67	70.2	3.32	67.7	3.10	58.9	3.61	48.2	2.29	32.2	2.16	19.5	1.23	43.6	31.40
RIVER FALLS	13.3	1.00	16.3	.92	27.8	1.80	44.5	2.54	57.2	3.85	66.9	4.69	72.2	3.71	69.7	3.20	60.6	3.30	49.2	1.90	32.0	1.63	18.6	1.17	44.0	29.71
DIVISION	15.2	1.05	18.2	.97	29.0	1.85	45.1	2.70	57.6	3.73	67.3	4.78	72.3	3.62	69.9	3.50	58.9	3.31	49.8	1.98	33.0	1.75	19.9	1.08	44.9	30.32
CENTRAL DIVISION																										
CODDINGTON 1 E	15.2	1.05	16.7	.95	27.3	1.55	42.8	2.87	54.7	3.63	64.1	5.31	68.9	3.26	66.3	3.36	58.1	3.59	47.3	2.41	32.0	2.25	19.1	1.00	42.7	31.23
HANCOCK EXP FARM	16.5	1.06	18.3	.98	28.7	1.51	44.5	2.61	57.0	3.59	67.2	4.64	72.3	3.12	69.5	3.03	60.8	3.61	49.7	2.29	33.2	2.17	20.4	1.06	44.8	29.67
MARSHFIELD EXP FARM	14.8	1.31	16.7	1.10	27.1	1.71	43.1	2.79	55.3	3.69	64.9	4.85	69.8	3.22	67.5	3.90	59.0	3.47	47.9	2.44	31.8	2.02	19.1	1.14	43.1	31.64
MAUSTON		1.15		1.18		1.79		2.72		3.37		4.24		3.52		3.31		3.36		2.10		2.00		1.03		29.77
MONTELLO	19.0	1.21	20.6	1.12	31.1	1.64	45.7	2.81	57.5	3.13	67.3	4.43	71.8	3.10	69.5	3.18	61.3	3.35	50.6	1.95	35.2	2.02	22.6	1.21	46.0	29.15
NEW LONDON	18.2	1.55	19.7	1.32	30.1	1.95	45.1	3.00	57.3	3.25	67.4	4.47	72.3	2.86	69.6	2.95	61.3	3.44	50.2	2.15	34.7	2.29	22.1	1.39	45.7	30.62
STEVENS POINT	16.7	1.54	18.2	1.36	29.0	1.77	44.8	2.76	57.6	3.79	67.7	4.88	72.8	3.03	70.2	3.61	61.5	3.57	49.7	2.17	33.5	2.15	20.7	1.36	45.2	31.99
WAUPACA	18.0	1.21	19.3	1.02	29.7	1.57	45.0	2.79	57.6	3.36	67.4	4.64	72.4	2.88	69.8	3.34	61.7	3.37	50.3	2.13	34.8	2.10	22.1	1.12	45.7	29.53
WISCONSIN RAPIDS	15.4	1.14	17.0	1.07	29.7	1.69	43.4	2.68	56.1	3.66	66.4	4.88	71.2	3.10	68.6	3.39	59.5	3.67	48.2	2.30	32.1	2.17	19.5	1.21	43.8	30.99
DIVISION	16.8	1.23	18.7	1.15	29.0	1.72	44.5	2.84	56.4	3.51	66.4	4.61	71.4	3.19	68.9	3.35	60.4	3.37	49.5	2.19	33.5	2.10	20.8	1.18	44.7	30.44

* Averages for period 1931-1955, except for stations marked WB which are "normals" based on period 1921-1950. Divisional means may not be the arithmetical average of individual stations published, since additional data from shorter period stations are used to obtain better areal representation.

*MEAN TEMPERATURE AND PRECIPITATION

STATION	Jan Temp	Jan Precip	Feb Temp	Feb Precip	Mar Temp	Mar Precip	Apr Temp	Apr Precip	May Temp	May Precip	Jun Temp	Jun Precip	Jul Temp	Jul Precip	Aug Temp	Aug Precip	Sep Temp	Sep Precip	Oct Temp	Oct Precip	Nov Temp	Nov Precip	Dec Temp	Dec Precip	Ann Temp	Ann Precip
EAST CENTRAL DIVISION																										
APPLETON	18.5	1.36	19.7	1.32	29.7	1.70	44.0	2.59	56.5	2.86	67.1	4.09	72.5	2.80	70.2	2.82	61.7	3.23	50.3	1.92	34.9	2.13	22.7	1.41	45.7	28.23
FOND DU LAC	20.4	1.43	22.0	1.35	32.0	1.84	46.3	2.45	57.9	2.82	68.0	4.08	73.0	3.25	70.8	3.34	62.6	3.23	51.6	1.99	36.2	2.06	23.7	1.38	47.0	29.22
GREEN BAY WB AP	16.1	1.29	17.3	1.36	28.5	1.76	41.8	2.51	54.4	2.53	64.7	3.57	69.9	2.59	67.8	3.03	60.2	2.87	48.4	1.80	33.5	1.94	20.1	1.26	43.6	26.51
KEWAUNEE	21.2	1.51	21.9	1.40	30.8	1.65	42.1	2.50	51.9	2.42	61.7	3.35	69.2	2.82	68.5	2.78	60.7	2.94	50.4	1.75	36.3	2.26	25.0	1.51	45.0	26.89
MANITOWOC	22.3	1.53	23.2	1.44	31.4	1.90	43.4	2.64	54.1	2.63	64.5	3.82	71.4	2.38	69.9	3.02	61.7	3.20	51.1	2.05	37.1	2.19	25.9	1.45	46.3	28.25
OSHKOSH	19.0	1.42	20.3	1.23	30.2	1.63	44.6	2.59	56.9	2.64	67.5	4.06	72.8	2.78	70.7	3.18	62.3	3.25	50.9	1.85	35.2	2.14	22.7	1.35	46.1	28.12
PLYMOUTH	20.7		21.8		30.7		44.1		55.2		65.4		71.2		69.7		61.5		50.8		36.0				45.9	
SHEBOYGAN	21.7	1.77	22.6	1.57	31.8	2.01	43.5	2.41	53.7	2.99	64.5	4.01	72.0	2.75	70.8	3.00	63.0	3.11	51.8	2.22	37.1	2.18	25.4	1.74	46.5	29.76
STURGEON BAY EXP FARM	19.1	1.34	19.2	1.33	28.2	1.76	41.2	2.43	52.1	2.46	62.6	3.20	69.0	2.75	67.4	2.86	59.3	3.25	48.7	2.16	35.0	2.36	23.8	1.32	43.8	27.22
DIVISION	20.0	1.46	21.3	1.33	30.6	1.78	43.9	2.58	54.9	2.71	65.3	3.83	71.3	2.90	69.7	3.01	61.5	3.10	50.8	1.98	35.9	2.16	23.9	1.42	45.8	28.26
SOUTHWEST DIVISION																										
DARLINGTON	20.6	1.39	23.5	1.08	33.6	2.07	47.1	2.80	57.9	3.59	67.9	4.94	72.5	3.82	70.0	4.28	62.0	3.63	51.3	2.32	36.1	2.18	23.9	1.42	47.2	33.52
HILLSBORO	18.2	1.23	20.6	1.15	30.8	1.97	45.6	2.85	57.3	3.47	67.2	4.56	72.1	3.67	69.4	3.46	61.1	3.93	50.1	2.24	34.6	2.29	22.0	1.20	45.8	32.02
LANCASTER	19.9	1.32	22.6	1.13	32.7	2.33	47.2	2.73	59.0	3.73	68.7	5.20	73.9	3.86	71.6	3.60	63.4	3.78	52.5	2.32	36.0	2.16	23.6	1.42	47.6	33.58
PRAIRIE DU CHIEN	20.9	1.22	23.5	1.11	34.1	2.19	49.1	2.71	60.9	3.82	70.6	5.11	75.4	3.80	73.1	4.09	64.7	3.76	53.1	2.07	37.1	2.09	24.8	1.32	48.9	33.29
PRAIRIE DU SAC 2 N	19.2	1.18	21.2	1.04	31.1	1.59	45.7	2.41	58.0	3.05	68.4	4.16	73.8	3.45	71.3	3.37	62.5	3.55	51.2	1.91	35.4	1.92	23.1	1.21	46.7	28.79
RICHLAND CENTER	20.1	1.24	22.8	1.17	33.0	2.17	47.3	2.65	58.9	3.48	68.7	5.20	73.4	4.03	71.0	3.73	62.7	4.02	51.5	2.31	36.0	2.13	23.7	1.30	47.4	33.43
VIROQUA	17.2	1.23	19.8	1.13	30.4	1.97	45.6	2.56	57.6	3.69	68.2	4.79	73.4	4.16	71.0	3.32	62.0	3.84	50.7	2.09	33.9	1.90	21.2	1.17	45.9	31.85
DIVISION	19.3	1.22	22.4	1.11	32.2	2.03	47.2	2.81	58.4	3.56	68.4	5.10	73.3	3.94	71.0	3.66	62.5	3.56	51.7	2.14	35.5	1.96	23.0	1.27	47.1	32.36
SOUTH CENTRAL DIV																										
ARLINGTON		1.48		1.20		1.72		2.63		3.35		3.94		3.61		3.59		4.06		2.03		2.05		1.34		31.00
BELOIT	23.3	1.64	25.5	1.29	35.4	2.03	49.0	2.60	60.1	3.46	70.1	4.55	74.9	3.75	72.5	3.80	64.7	3.82	53.9	2.34	38.5	2.33	26.5	1.61	49.5	33.22
BRODHEAD 1 SW	21.4	1.77	23.6	1.36	33.7	2.21	47.6	2.90	59.2	3.42	69.5	4.43	74.5	3.46	72.0	4.25	63.6	3.77	52.3	2.28	36.7	2.36	24.5	1.65	48.2	33.86
LAKE MILLS	21.2	1.66	22.8	1.21	32.4	1.92	46.5	2.67	57.9	3.21	67.8	4.33	73.1	3.53	71.2	2.95	62.9	3.51	52.1	2.25	36.9	2.34	24.5	1.60	47.4	31.18
MADISON WB AP	19.1	1.31	21.9	1.13	32.5	1.83	45.7	2.49	57.5	3.27	67.4	4.02	73.0	3.30	70.7	2.89	62.1	3.99	50.4	2.08	35.3	2.29	23.0	1.40	46.6	30.00
MADISON WB CITY	19.3	1.47	21.9	1.27	32.4	2.03	45.9	2.49	57.7	3.21	67.7	4.02	73.1	3.40	71.8	3.33	62.9	4.11	51.7	2.00	36.2	2.20	23.5	1.44	46.9	30.71
PORTAGE	20.6	1.48	22.7	1.25	32.7	1.95	45.7	2.82	59.6	3.02	69.4	4.21	74.4	3.41	71.8	3.33	63.7	3.90	52.5	1.93	36.9	2.11	24.2	1.36	48.0	30.77
STOUGHTON		1.58		1.24		1.90		2.55		3.21		4.05		3.24		3.32		3.28		2.14		2.10				30.07
WATERTOWN	22.2	1.71	23.9	1.32	33.3	2.09	46.9	2.73	58.4	3.10	68.5	4.10	73.7	3.40	71.5	3.13	63.1	3.59	52.2	2.26	37.1	2.13	25.2	1.60	48.0	31.26
WISCONSIN DELLS	18.6	1.31	20.8	1.21	30.9	2.02	45.3	2.80	57.6	3.03	67.6	4.24	73.0	3.66	70.6	3.48	61.7	3.87	50.7	2.16	34.8	2.13	25.2	1.24	46.2	31.15
DIVISION	20.8	1.53	22.9	1.23	32.8	1.93	46.9	2.67	58.4	3.23	68.4	4.26	73.5	3.51	71.1	3.45	63.0	3.68	52.1	2.13	36.6	2.14	24.2	1.47	47.6	31.23
SOUTHEAST DIVISION																										
MILWAUKEE WB AP	21.9	1.58	24.2	1.27	33.3	2.19	44.3	2.39	54.3	2.98	64.9	3.22	71.3	2.43	69.9	2.62	62.6	3.33	51.4	1.97	37.3	2.11	25.7	1.48	46.8	27.57
PORT WASHINGTON		1.58		1.35		1.84		2.50						2.74		2.80		2.94		2.03		1.95		1.50		27.71
RACINE	24.9	1.99	26.2	1.55	34.4	2.63	45.8	2.58	55.8	3.84	66.9	3.65	73.2	2.96	72.1	3.26	64.6	3.26	53.6	1.93	39.0	2.34	27.6	1.96	48.7	31.95
WAUKESHA	21.3	1.84	22.9	1.35	32.1	2.14	45.2	2.43	56.3	3.53	66.8	3.90	72.3	3.30	70.6	3.08	62.3	2.99	51.3	2.02	38.4	2.29	24.6	1.61	48.8	30.48
WEST BEND	20.9	1.82	22.3	1.43	31.4	2.07	44.9	2.51	56.3	2.91	66.8	4.20	72.2	3.21	70.0	2.94	62.0	3.29	51.1	2.15	36.3	2.15	24.2	1.55	46.5	30.23
WILLIAMS BAY	21.8	1.96	23.4	1.32	32.7	2.42	46.3	2.68	57.7	3.59	68.0	4.08	73.3	3.80	71.5	3.53	63.6	3.36	52.6	2.17	36.9	2.45	24.9	1.75	47.7	33.11
DIVISION	22.1	1.79	24.2	1.32	32.8	2.03	46.0	2.81	56.5	3.36	67.1	4.22	72.6	3.60	71.0	3.14	63.0	2.95	52.5	1.97	37.1	2.07	25.2	1.75	47.5	31.67

* Averages for period 1931-1955, except for stations marked WB which are "normals" based on period 1921-1950. Divisional means may not be the arithmetical average of individual stations published, since additional data from shorter period stations were used to obtain better areal representation.

CONFIDENCE LIMITS

In the absence of trend or record changes, the chances are 9 out of 10 that the true mean will lie in the interval formed by adding and subtracting the values in the following table from the means for any station in the State.

2.0	.21	1.7	.18	1.7	.26	1.3	.47	1.2	.71	1.2	.86	.9	.55	.9	.66	1.1	.57	1.3	.47	1.3	.55	1.7	.22	.6	2.51

COMPARATIVE DATA

Data in the following table are the mean temperature and average precipitation for Hancock, Wisconsin for the period 1906 – 1930 and are included in this publication for comparative purposes :

13.1	1.13	17.1	1.21	29.8	1.63	44.4	3.15	55.9	4.16	65.6	4.53	71.1	3.27	68.6	3.38	61.1	3.95	48.0	2.28	34.0	1.74	19.6	1.19	44.0	31.62

NORMALS, MEANS AND EXTREMES
(Table Revised 1972. Base Period for Climatological Normals: 1931-1960)

Station: GREEN BAY, WISCONSIN   AUSTIN STRAUBEL FIELD   Standard time used: CENTRAL   Latitude: 44° 29' N   Longitude: 88° 08' W   Elevation (ground): 682 feet

Month	Normal Daily max	Normal Daily min	Normal Monthly	Record highest	Year	Record lowest	Year	Normal heating degree days (Base 65°)	Precip Normal total	Max monthly	Year	Min monthly	Year	Max in 24 hrs	Year	Snow Mean total	Max monthly	Year	Max in 24 hrs	Year
J	25.1	8.5	16.8	48	1964	-27	1963	1494	1.15	2.64	1950	0.31	1961	1.05	1950	9.7	20.8	1971	8.6	1971
F	26.9	9.3	18.1	54	1964	-26	1971	1313	1.08	3.56	1953	0.04	1969	1.78	1966	7.7	20.6	1962	9.2	1962
M	36.0	20.4	28.2	73	1967	-27	1962	1141	1.34	2.66	1951	0.46	1967	0.75	1960	8.6	22.2	1972	10.1	1957
A	52.4	33.9	43.2	87	1962	14	1964	654	2.46	5.52	1953	0.98	1954	1.75	1957	1.7	8.6	1957	3.9	1957
M	65.3	44.2	54.8	88	1966	25	1966	335	3.06	7.75	1960	0.89	1960	2.73	1960	0.2	2.6	1960	2.2	1960
J	75.4	55.5	65.5	97	1971	35	1966	99	3.36	8.47	1967	1.05	1964	2.65	1964	0.0	0.0		0.0	
J	80.2	59.0	70.5	97	1966	40	1965	28	2.71	6.50	1950	1.87	1960	2.95	1955	0.0	0.0		0.0	
A	80.0	57.8	68.5	96	1965	38	1967	50	2.75	5.50	1951	0.90	1955	2.22	1951	0.0	0.0		0.0	
S	71.3	49.6	60.5	88	1970	27	1966	174	2.92	7.80	1965	0.70	1967	2.99	1964	T	T	1965	T	1965
O	59.6	39.1	49.4	88	1963	15	1966	484	1.91	5.00	1954	T	1952	3.68	1954	0.1	1.7	1959	1.6	1959
N	41.2	27.1	34.2	70	1964	5	1964	924	1.91	3.52	1957	0.39	1953	1.25	1971	3.9	12.6	1971	6.5	1971
D	29.1	14.8	22.0	62	1970	-19	1967	1333	1.18	3.15	1971	0.10	1960	1.55	1968	9.0	26.7	1968	7.2	1968
YR	53.7	34.9	44.3	97 JUN. 1971+		-29 MAR. 1962		8029	25.83	8.47 JUN. 1967		T OCT. 1952		3.68 OCT. 1954		40.9	26.7 DEC. 1968		10.1 MAR. 1964	

Means and extremes above are from existing and comparable exposures. Annual extremes have been exceeded at other sites in the locality as follows:
Highest temperature 104 in July 1936; lowest temperature -36 in January 1888; maximum monthly precipitation 9.70 in May 1918; maximum precipitation in 24 hours 4.41 in June 1914; maximum snowfall 32.1 in March 1923; maximum snowfall in 24 hours 22.0 in January 1889.

NORMALS, MEANS AND EXTREMES
(Table Revised 1975. Base Period for Climatological Normals: 1941-1970)

Elev. 702 feet m.s.l.   Average station pressure 989.9 mb.

Month	Normal Daily max	Normal Daily min	Normal Monthly	Record highest	Year	Record lowest	Year	Heating	Cooling	Precip Normal	Max monthly	Year	Min monthly	Year	Max in 24 hrs	Year
J	23.9	6.9	15.4	48	1973	-28	1972	1538	0	1.09	2.64	1950	0.31	1961	1.05	1950
F	27.2	9.3	18.0	54	1964	-26	1963	1316	0	1.01	3.56	1953	0.31	1951	1.78	1966
M	37.1	20.0	28.6	73	1967	-29	1962	1128	0	1.68	2.66	1951	0.46	1957	0.97	1960
A	54.0	33.5	43.8	90	1962	11	1972	636	12	2.59	5.52	1953	0.98	1953	1.75	1957
M	65.8	43.1	54.5	90	1964	21	1966	338	76	3.10	8.21	1973	0.82	1972	3.28	1960
J	75.5	53.2	64.5	97	1971	35	1966	91	152	3.41	8.47	1967	1.05	1964	2.65	1964
J	80.7	57.7	69.2	96	1966	40	1974	22	152	3.09	6.50	1972	1.85	1959	2.95	1959
A	79.1	56.3	67.7	96	1973	38	1967	54	138	2.62	5.50	1951	0.90	1951	2.62	1951
S	69.8	48.0	58.9	94	1973	27	1974	191	18	3.24	7.80	1965	0.46	1952	2.99	1964
O	59.6	38.7	49.2	88	1963	15	1966	490	0	1.93	3.68	1954	T	1952	3.68	1954
N	41.8	26.4	34.1	70	1964	5	1964	927	0	1.88	3.52	1957	0.39	1953	1.41	1971
D	28.6	13.2	20.9	62	1970	-21	1972	1367	0	1.27	3.15	1971	0.10	1959	1.55	1968
YR	53.6	33.8	43.7	97 JUL. 1974		-29 MAR. 1962		8098	386	27.01	8.47 JUN. 1967		T OCT. 1952		3.68 OCT. 1954	

Means and extremes above are from existing and comparable exposures. Annual extremes have been exceeded at other sites in the locality as follows:
Highest temperature 104 in July 1936; lowest temperature -36 in January 1888; maximum monthly precipitation 9.70 in May 1918; maximum precipitation in 24 hours 4.41 in June 1914; maximum snowfall 32.1 in March 1923; maximum snowfall in 24 hours 22.0 in January 1889.

REFERENCE NOTES APPLYING TO TABLES APPEAR ON THE PAGE FOLLOWING LAST TABLE.
(Caution: Letters and symbols may have different meanings in 1941-1970 tables than in earlier tables. See notes.)

NORMALS, MEANS AND EXTREMES

(Table Revised 1972. Base Period for Climatological Normals: 1931-1960)

Station: LA CROSSE, WISCONSIN   MUNICIPAL AIRPORT   Standard time used: CENTRAL   Latitude: 43° 52' N   Longitude: 91° 15' W   Elevation (ground): 651 feet

Month	Temperature Normal Daily maximum (b)	Daily minimum (b)	Monthly (b)	Extremes Record highest	Year'	Record lowest	Year	Normal heating degree days (Base 65°) (b)	Precipitation total Normal total	Maximum monthly	Year	Minimum monthly	Year	Maximum in 24 hrs. c	Year	Snow, Ice pellets Mean total	Maximum monthly	Year	Maximum in 24 hrs. c	Year	Relative humidity Hour 00	Hour 06	Hour 12	Hour 18	Wind Mean speed	Prevailing direction	Fastest mile Speed	Direction	Year	Mean sky cover Sunrise to sunset	Sunrise to sunset Clear	Partly cloudy	Cloudy	Precipitation .01 inch or more	Snow, Ice pellets 1.0 inch or more	Thunderstorms	Heavy fog	Temperatures Max. 90° and above	32° and below	Temperatures Min. 32° and below	0° and below	Avg daily solar radiation - langleys
(a)	(b)	(b)	(b)	7		7		(b)	21	21		21		18		21	21		18		7	7	7	7	21	13	18	18		5	17	17	17	21	21	21	21	7	7	7	7	7
J	25.6	7.4	16.5	50	1968	-32	1970	1504	1.19	2.86	1967	0.19	1962	1.31	1967	9.5	26.5	1971	8.7	1963	75	76	66	74	9.1	S	45	32	1962	6.7	8	7	17	8	9	*	1	0	24	31	14	7
F	28.4	9.8	19.4	51	1966	-36	1971	1277	1.05	2.58	1959	0.05	1969	1.06	1966	7.8	31.0	1959	10.9	1959	74	75	61	65	9.2	NW	37	34	1963	6.6	7	7	13	8	6	*	1	0	17	28	9	4
M	39.4	21.5	30.5	80	1967	-13	1965	1070	2.07	3.82	1951	0.30	1958	1.64	1959	7.8	33.5	1959	15.7	1959	77	82	61	62	9.8	NW	40	34	1960	6.6	7	7	17	10	4	1	1	0	6	24	1	1
A	57.2	36.9	47.0	90	1970	18	1971	540	2.75	6.79	1954	1.07	1966	3.84	1954	1.4	10.2	1973	7.3	1973	75	81	54	53	11.4	NW	53	25	1964	6.5	7	7	16	12	2	3	1	*	0	8	0	0
M	69.3	48.5	59.2	94	1967	26	1974	245	3.76	8.83	1960	1.07	1958	2.72	1960	T	0.8	1960	0.8	1960	77	82	53	53	10.6	S	58	09	1952	6.2	7	7	16	11	1	6	1	1	0	2	0	0
J	79.1	58.5	68.8	97	1970+	42	1969+	69	4.20	9.53	1968	1.66	1955	2.94	1967	0.0	0.0		0.0		85	87	54	56	8.8	S	58	34	1961	6.2	7	7	16	11	0	8	1	3	0	0	0	0
J	84.6	63.0	73.8	98	1965	44	1953	12	3.75	9.16	1953	0.16	1967	3.72	1952	0.0	0.0		0.0		88	89	57	57	7.7	S	52	27	1961	5.6	9	11	11	10	0	8	2	5	0	0	0	0
A	81.7	60.9	71.6	98	1968	40	1965	19	3.63	7.85	1959	0.79	1955	3.92	1962	0.0	0.0		0.0		89	94	55	59	7.5	S	40	32	1963	5.5	9	11	11	10	0	6	4	4	0	0	0	0
S	72.4	51.0	62.2	95	1971	28	1967	153	3.48	10.52	1965	0.42	1968	2.25	1968	T	T	1965+	T	1965+	88	92	58	67	8.6	S	52	27	1961	5.8	9	11	10	8	0	5	4	2	0	0	0	0
O	61.4	40.6	51.1	84	1963	17	1969	437	2.19	4.47	1954	0.02	1952	2.17	1966	0.1	1.4	1959	1.2	1959	80	86	59	67	9.9	S	39	34	1964	5.7	11	12	13	7	*	2	2	1	0	4	0	0
N	42.5	26.0	34.2	68	1965	-9	1958	924	1.94	2.98	1965	0.09	1962	2.38	1958	3.7	13.0	1957	11.0	1957	82	86	62	73	10.5	S	46	18	1958	7.1	6	6	18	9	7	1	2	1	0	22	6	0
D	29.7	13.8	21.8	59	1965	-22	1968	1339	1.15	2.55	1971	0.30	1962+	0.93	1968	8.6	26.6	1968	9.1	1966	83	83	73	76	9.2	S	43	34	1957	7.1	7	6	19	7	9	1	1	*	0	28	28	4
YR	56.1	36.5	46.3	99	JUL. 1965	-36	FEB. 1971	7589	31.16	10.52	SEP. 1965	0.02	OCT. 1952	3.94	JUN. 1967	42.2	33.5	MAR. 1959	15.7	MAR. 1959	81	84	61	64	9.5	S	63	32	AUG. 1963	6.3	95	97	173	109	12	41	22	16	70	151	29

Means and extremes above are from existing and comparable exposures. Annual extremes have been exceeded at other sites in the locality as follows:
Highest temperature 108 in July 1936; lowest temperature -43 in January 1873; maximum monthly precipitation 12.09 in October 1900; minimum monthly precipitation 0.01 in December 1943; maximum precipitation in 24 hours 7.23 in October 1900; maximum monthly snowfall 39.6 in January 1929; fastest mile of wind 69 from Southwest in October 1949.

NORMALS, MEANS AND EXTREMES

(Table Revised 1975. Base Period for Climatological Normals: 1941-1970)

Month	Temperature °F Normal Daily maximum	Daily minimum	Monthly	Extremes Record highest	Year	Record lowest	Year	Normal Degree days Base 65°F Heating	Cooling	Precipitation in inches Water equivalent Normal	Maximum monthly	Year	Minimum monthly	Year	Maximum in 24 hrs. c	Year	Snow, Ice pellets Maximum monthly	Year	Maximum in 24 hrs. c	Year	Relative humidity pct. Hour 00	Hour 06	Hour 12	Hour 18	Wind Mean speed m.p.h.	Prevailing direction	Fastest mile Speed m.p.h.	Direction	Year	Mean sky cover, tenths, sunrise to sunset	Pct. of possible sunshine	Sunrise to sunset Clear	Partly cloudy	Cloudy	Precipitation .01 inch or more	Snow, Ice pellets 1.0 inch or more	Heavy fog, visibility ¼ mile or less	Thunderstorms	Temperatures Max. 90° and above	32° and below	Temperatures Min. 32° and below	0° and below	Average station pressure mb.
(a)				10		10					24		24		18		24		18		10	10	10	10	24	13	18	18		5		17	17	17	24	24	24	24	10	10	10	10	2
J	25.0	7.1	16.1	52	1973	-32	1970	1516	0	0.96	2.86	1967	0.19	1962	1.31	1967	26.5	1971	8.7	1963	77	71	67	71	8.8	S	45	32	1962	6.7		8	7	17	8	9	2	*	0	24	31	13	994.3
F	29.7	10.3	20.0	51	1966	-36	1971	1260	0	0.87	2.58	1959	0.05	1969	1.06	1966	31.0	1959	10.9	1959	76	64	64	66	8.9	NW	37	34	1963	6.6		7	7	13	7	6	2	*	0	17	28	8	995.1
M	40.3	22.1	31.2	80	1967	-13	1965	1051	0	2.02	3.82	1951	0.30	1958	1.64	1959	33.5	1959	15.7	1959	78	63	60	63	9.4	NW	40	34	1960	6.6		7	7	17	10	4	1	1	0	6	25	1	989.0
A	57.8	37.4	47.6	90	1970	11	1972	522	0	2.63	7.31	1973	0.40	1966	3.84	1954	17.0	1973	7.3	1973	77	56	57	58	11.0	NW	53	25	1964	6.5		7	7	16	11	1	3	1	*	0	7	0	989.7
M	69.2	48.7	59.0	94	1967	26	1974	224	38	3.70	9.53	1960	1.07	1958	2.72	1960	0.0	1960	0.0	1960	77	52	56	59	10.2	S	58	08	1952	6.2		7	7	16	12	*	6	1	1	0	2	0	987.5
J	78.4	58.5	68.5	97	1967	40	1970	39	144	4.44	9.53	1968	1.66	1955	2.94	1967	0.0		0.0		85	59	59		8.5	S	63	04	1963	6.2		7	11	12	11	0	8	1	3	0	0	0	987.9
J	83.0	62.5	72.8	101	1974	44	1965	10	252	3.52	9.16	1953	0.16	1967	3.72	1952	0.0		0.0		87	58	58	58	7.8	S	52	27	1961	5.6		10	9	12	10	0	7	2	5	0	0	0	991.4
A	81.7	61.0	71.4	98	1968	40	1967	17	215	3.02	8.63	1973	0.79	1955	3.92	1962	0.0		0.0		89	57	64		7.5	S	40	32	1963	5.5		9	11	12	10	0	7	4	4	0	0	0	992.1
S	71.8	51.8	61.8	95	1971	28	1971	121	34	3.38	10.52	1965	0.42	1968	2.25	1968	T	1974	T		90	64	69	67	8.3	S	52	27	1961	5.8		9	11	11	8	0	5	4	2	0	0	0	992.3
O	61.6	41.7	51.7	84	1969	15	1972	420	12	2.05	4.47	1954	0.02	1952	2.17	1966	1.4	1957	1.2	1957	86	57	67	75	9.5	S	39	34	1964	5.7		11	12	8	7	*	2	2	1	0	6	0	992.9
N	43.0	27.8	35.4	68	1965	-5	1958	888	0	1.45	2.98	1965	0.09	1962	2.38	1958	13.0	1957	11.0	1957	87	62	67	74	10.0	S	46	18	1958	7.1		6	7	18	9	7	2	2	1	0	21	6	993.0
D	29.6	14.0	21.8	59	1965	-22	1968	1339	0	1.04	2.55	1971	0.30	1962	0.93	1968	26.6	1968	9.1	1966	84	77	77		8.8	S	43	34	1957	7.1		7	6	19	9	9	3	1	*	0	29	4	993.1
YR	55.9	36.9	46.4	101	JUL. 1974	-36	FEB. 1971	7417	695	29.08	10.52	SEP. 1965	0.02	OCT. 1952	3.94	JUN. 1967	33.5	MAR. 1959	15.7	MAR. 1959	81	85	62	65	9.0	S	63	32	AUG. 1963	6.3		95	97	173	110	12	41	22	16	67	151	27	991.8

Means and extremes above are from existing and comparable exposures. Annual extremes have been exceeded at other sites in the locality as follows:
Highest temperature 108 in July 1936; lowest temperature -43 in January 1873; maximum monthly precipitation 12.09 in October 1900; minimum monthly precipitation 0.01 in December 1943; maximum precipitation in 24 hours 7.23 in October 1900; maximum monthly snowfall 39.6 in January 1929; fastest mile of wind 69 from Southwest in October 1949.

Elev. 672 feet m.s.l.

REFERENCE NOTES APPLYING TO TABLES APPEAR ON THE PAGE FOLLOWING LAST TABLE.
(Caution: Letters and symbols may have different meanings in 1941-1970 tables than in earlier tables. See notes.)

NORMALS, MEANS AND EXTREMES
(Table Revised 1972. Base Period for Climatological Normals: 1931-1960)

Station: MADISON, WISCONSIN   TRUAX FIELD   Standard time used: CENTRAL   Latitude: 43° 08' N   Longitude: 89° 20' W   Elevation (ground): 858 feet

Month	Normal Daily maximum	Normal Daily minimum	Normal Monthly	Extremes Record highest	Year	Extremes Record lowest	Year	Normal heating degree days (Base 65°)	Precip. Normal total	Precip. Max monthly	Year	Precip. Min monthly	Year	Precip. Max in 24 hrs.	Year	Snow Mean total	Snow Max monthly	Year	Snow Max in 24 hrs.	Year	RH Hour 00	RH 06	RH 12	RH 18	Wind Mean speed	Prevailing direction	Fastest mile Speed	Direction	Year	Mean sky cover	Pct. possible sunshine	Clear	Partly cloudy	Cloudy	Precip .01"+	Snow 1.0"+	Thunderstorms	Heavy fog	Max 90°+	Max 32°-	Min 32°-	Min 0°-	Avg daily solar radiation (langleys)
(yrs)	(b)	(b)	(b)	12	12	12		(b)	(b)	32		32		23		23	23		23		12	12	12	12	25	14	25	25	25	25	25	25	25	25	25	12	23	25	12	12	12	12	53
J	25.9	9.1	17.5	54	1967	-30	1963	1473	1.40	2.43	1950	0.19	1961	1.27	1961	9.3	21.9	1971	11.6	1971	76	77	69	72	10.5	NNW	68	E	1947	6.7	49	8	6	17	10	3	*	3	0	22	30	12	152
F	28.2	10.8	19.5	56	1962	-30	1965	1274	1.13	2.77	1953	0.08	1958	1.55	1962	6.2	16.1	1962	10.3	1962	75	79	65	68	10.7	NNW	57	SW	1948	6.5	53	7	6	15	8	2	*	2	0	16	28	6	227
M	38.0	20.2	29.1	56	1967	-29	1967	1113	1.84	3.42	1961	0.42	1958	1.37	1961	10.0	25.4	1959	13.0	1959	77	79	63	61	11.4	NW	73	SW	1954	6.8	55	7	7	17	11	3	2	2	0	8	27	2	320
A	54.0	34.8	44.4	87	1962	16	1962	618	2.57	4.86	1955	0.96	1947	2.15	1955	1.0	3.2	1961+	2.9	1961+	75	79	56	56	10.6	NW	73	SW	1947	6.6	53	7	7	16	11	*	4	1	*	*	15	0	396
M	66.4	45.8	56.1	91	1962	20	1966	310	3.34	6.26	1960	0.98	1971	3.64	1966	T	0.7	1966	0.7	1966	79	83	55	53	10.5	S	59	SW	1950	6.4	65	8	10	13	11	0	6	1	*	0	4	0	473
J	77.0	55.2	66.1	95	1962	33	1964	102	3.95	8.15	1963	1.84	1961	3.67	1961	0.0	0.0		0.0		81	83	58	58		S	72	NW	1951	6.1	70	9	12	10	10	0	7	1	4	0	0	0	518
J	82.2	60.0	71.1	98	1965	36	1965	25	3.58	10.93	1950	1.38	1946	5.25	1950	0.0	0.0		0.0		85	88	57	59	8.2	S	72	NW	1951	5.0	70	9	11	11	9	0	7	1	3	0	0	0	535
A	80.2	58.7	69.5	95	1965	35	1968	40	3.37	6.77	1965	0.49	1948	2.90	1965	0.0	0.0		0.0		88	92	61	61	8.1	S	47	W	1955	5.3	63	10	11	10	9	0	6	2	2	0	0	0	459
S	71.6	49.4	60.0	90	1971+	25	1968+	174	3.32	9.51	1941	0.40	1952	5.55	1959	T	T	1965	T	1965	88	92	60	67	8.8	S	52	W	1948	5.6	63	11	8	11	9	0	4	2	1	0	2	0	351
O	60.1	39.7	49.9	90	1963	15	1969	474	2.21	5.55	1959	0.34	1952	2.01	1967	0.1	0.9	1967	0.9	1967	80	86	60	67	9.5	S	73	SE	1951	5.7	58	11	7	13	9	0	2	2	*	0	3	0	244
N	42.3	26.4	34.7	76	1964	-11	1964	930	2.14	3.94	1961	0.36	1962	2.32	1954	2.9	8.9	1954	6.8	1954	80	86	67	76	9.7	S	56	SE	1967	6.6	41	6	7	18	9	3	1	2	0	3	18	0	146
D	29.8	14.4	22.1	62	1962	-22	1962	1330	1.31	3.64	1971	0.25	1960	1.66	1970	9.3	20.8	1970	16.0	1970	81	82	72	76	10.3	W	65	SW	1949	7.1	41	6	6	18	11	3	*	3	0	18	29	6	116
YR	54.6	35.4	45.0	98 JUL	1965	-30 JAN	1963	7863	30.16	10.93 JUL	1950	0.06 OCT	1952	5.25 JUL	1950	38.8	25.4 MAR	1959	16.0 DEC	1970	80	83	62	65	10.0	S	77	SW MAY	1950	6.3	58	97	97	171	115	12	40	22	10	67	166	26	328

Means and extremes above are from existing and comparable exposures. Annual extremes have been exceeded at other sites in the locality as follows: Highest temperature 107 in July 1936; lowest temperature -37 in January 1951; minimum monthly precipitation T in October 1889 and earlier; maximum precipitation in 24 hours 5.31 in September 1941; maximum monthly snowfall 31.8 in January 1929.

NORMALS, MEANS AND EXTREMES
(Table Revised 1975. Base Period for Climatological Normals: 1941-1970)

Average station pressure 2 mb.   Elev. 866 feet m.s.l.   (station pressure 984.1)

Month	Normal Daily maximum	Normal Daily minimum	Normal Monthly	Extremes Record highest	Year	Extremes Record lowest	Year	Degree days Heating	Cooling	Precip Normal	Max monthly	Year	Min monthly	Year	Max in 24 hrs.	Year	Snow Max monthly	Year	Snow Max in 24 hrs.	Year	RH Hour 00	RH 06	RH 12	RH 18	Wind Mean speed	Prevailing direction	Fastest mile m.p.h.	Direction	Year	Mean sky cover	Pct. sunshine	Clear	Partly	Cloudy	Precip .01"+	Snow 1.0"+	Thunderstorms	Heavy fog ¼ mi or less	Max 90°+	Max 32°-	Min 32°-	Min 0°-
(yrs)	15			15		15				35	35		35		26		26		26		15	15	15	15	28	14	28	28	28	28	28	28	28	28	26	26	26	28	15	15	15	15
J	25.4	8.2	16.8	54	1967	-30	1963	1494	0	1.25	2.45	1961	0.19	1961	1.27	1961	21.9	1971	11.6	1971	76	77	69	73	10.5	NNW	68	E	1947	6.7	49	8	6	17	10	3	*	3	0	21	30	12
F	29.5	11.1	20.3	56	1962	-28	1962	1252	0	0.95	2.77	1958	0.08	1958	1.55	1962	16.1	1962	10.3	1962	76	79	66	69	10.7	NNW	57	W	1948	6.5	54	7	6	15	8	2	*	2	0	17	27	7
M	39.2	20.3	29.8	56	1967	-25	1962	1079	0	2.09	5.04	1958	0.38	1958	2.52	1971	25.4	1959	13.0	1959	77	80	63	63	11.7	NW	70	SW	1954	6.7	54	6	7	17	11	3	2	2	*	8	27	1
A	54.0	34.6	44.3	87	1962	12	1972	591	0	2.66	7.11	1973	0.98	1947	2.15	1955	17.4	1973	12.9	1973	75	80	56	57	11.3	NW	73	SW	1947	6.5	52	8	9	13	12	1	4	1	0	*	14	0
M	67.3	45.3	56.3	91	1962	20	1966	297	18	3.18	8.15	1946	0.81	1973	3.64	1966	3.4	1966	2.9	1966	77	80	57	57	10.8	S	59	W	1947	6.4	58	7	10	14	11	0	5	1	*	0	4	0
J	76.9	54.6	65.8	95	1962	31	1972	72	96	4.33	10.93	1950	1.38	1946	3.67	1961	0.0		0.0		81	83	57	57	9.2	S	72	NW	1951	6.1	64	6	11	13	10	0	7	1	3	0	0	0
J	81.4	59.8	70.1	98	1955	36	1965	14	172	3.81	10.93	1950	0.70	1948	5.25	1950	0.0		0.0		85	87	57	59	8.1	S	72	NW	1951	5.5	69	6	11	11	9	0	8	1	4	0	0	0
A	78.9	58.3	68.7	95	1965	35	1968	39	154	3.05	9.51	1941	0.49	1952	2.90	1965	0.0		0.0		87	90	62	62	8.0	S	47	W	1955	5.5	62	7	11	10	9	0	6	2	3	0	0	0
S	70.9	48.5	59.7	90	1973	25	1974	173	14	3.36	9.51	1941	0.49	1952	3.57	1961	0.9	1967	0.9	1967	88	90	62	65	8.9	S	52	W	1948	5.8	57	8	13	9	9	1	5	2	1	0	1	0
O	59.8	39.0	48.9	90	1972	15	1972	499	0	2.16	5.94	1971	0.06	1952	2.01	1967	0.9	1967	6.8	1954	80	86	68	73	9.6	S	73	SE	1951	5.7	53	7	8	16	10	1	2	2	*	0	9	0
N	43.0	26.4	34.7	76	1964	-11	1964	909	0	1.87	3.94	1961	0.34	1962	2.32	1954	8.9	1954	6.8	1954	82	83	68	77	10.8	S	56	SE	1967	6.4	40	7	8	15	10	3	1	3	0	3	21	1
D	29.8	14.0	21.9	62	1970	-22	1962	1336	0	1.47	3.64	1971	0.25	1960	1.66	1970	20.8	1970	16.0	1970	81	82	73	77	10.2	W	65	SW	1949	7.2	40	6	6	19	12	3	*	3	0	18	29	5
YR	55.0	34.8	44.9	98 JUL	1965	-30 JAN	1963	7730	460	30.25	10.93 JUL	1950	0.06 OCT	1952	5.25 JUL	1950	25.4 MAR	1959	16.0 DEC	1970	81	84	62	65	10.0	S	77	SW MAY	1950	6.4	57	93	96	176	117	12	41	22	10	65	163	25

Means and extremes above are from existing and comparable exposures. Annual extremes have been exceeded at other sites in the locality as follows: Highest temperature 107 in July 1936; lowest temperature -37 in January 1951; minimum monthly precipitation T in October 1889 and earlier; maximum precipitation in 24 hours 5.31 in September 1941; maximum monthly snowfall 31.8 in January 1929.

REFERENCE NOTES APPLYING TO TABLES APPEAR ON THE PAGE FOLLOWING LAST TABLE.
(Caution: Letters and symbols may have different meanings in 1941-1970 tables than in earlier tables. See notes.)

NORMALS, MEANS AND EXTREMES

(Table Revised 1972. Base Period for Climatological Normals: 1931-1960)

Station: MILWAUKEE, WISCONSIN
GENERAL MITCHELL FIELD
Standard time used: CENTRAL Latitude: 42° 57' N Longitude: 87° 54' W Elevation (ground): 672 feet

Month	Temperature Normal Daily max	Normal Daily min	Normal Monthly	Record highest	Year	Record lowest	Year	Normal heating degree days (Base 65°)	Precip. Normal total
J	28.3	12.8	20.6	57	1967+	-24	1963	1376	1.83
F	30.2	14.6	22.4	51	1961	-15	1971	1193	1.40
M	38.8	23.2	31.0	77	1967	-10	1962	1054	2.31
A	53.1	34.1	43.6	85	1962+	13	1971	642	2.53
M	65.0	42.9	53.4	89	1962	21	1966	372	3.16
J	73.9	52.6	63.3	95	1965+	36	1969+	135	3.64
J	78.9	58.4	68.7	98	1963	40	1965	43	2.95
A	77.7	57.8	67.8	95	1964	44	1965+	47	3.06
S	70.7	49.9	60.3	93	1961+	31	1961	174	2.72
O	60.1	39.9	50.0	89	1963	21	1960	471	2.10
N	44.1	27.5	35.8	74	1964+	-6	1964	876	2.18
D	32.0	17.1	24.6	63	1970	-15	1963	1252	1.63
YR	54.3	35.9	45.1	98 JUL 1963		-24 JAN 1963		7635	29.51

Means and extremes above are from existing and comparable exposures. Annual extremes have been exceeded at other sites in the locality as follows:
Highest temperature 105 in July 1934; lowest temperature -25 in January 1875; maximum monthly precipitation 10.03 in June 1917; maximum precipitation in 24 hours 5.76 in June 1917; maximum monthly snowfall 52.6 in January 1918; maximum snowfall in 24 hours 20.3 in February 1924.

NORMALS, MEANS AND EXTREMES

(Table Revised 1975. Base Period for Climatological Normals: 1941-1970)

Month	Temperature Normal Daily max	Normal Daily min	Normal Monthly	Record highest	Year	Record lowest	Year	Precip. Normal
J	27.3	11.4	19.4	57	1967+	-24	1963	1.63
F	30.3	14.6	22.5	51	1961	-15	1971	1.13
M	39.4	24.0	31.4	77	1967	-10	1962	2.24
A	54.6	34.7	44.7	85	1962	13	1971	2.76
M	65.4	44.7	54.2	89	1962	21	1966	2.88
J	75.3	53.6	64.5	95	1965	36	1969	3.58
J	80.4	59.3	69.9	98	1963	40	1965	3.41
A	79.7	58.7	69.2	99	1973	44	1965	2.68
S	71.8	51.0	61.0	93	1963	31	1948	3.02
O	61.4	40.6	51.0	89	1963	21	1960	1.98
N	44.4	28.5	36.5	74	1964	-6	1959	2.01
D	31.5	16.8	24.2	63	1965	-15	1958	1.75
YR	55.1	36.3	45.7	99 AUG 1973		-24 JAN 1963		29.07

Average station pressure: 693 feet m.s.l.

Means and extremes above are from existing and comparable exposures. Annual extremes have been exceeded at other sites in the locality as follows:
Highest temperature 105 in July 1934; lowest temperature -25 in January 1875; maximum monthly precipitation 10.03 in June 1917; maximum precipitation in 24 hours 5.76 in June 1917; maximum monthly snowfall 52.6 in January 1918; maximum snowfall in 24 hours 20.3 in February 1924.

REFERENCE NOTES APPLYING TO TABLES APPEAR ON THE PAGE FOLLOWING LAST TABLE.

(Caution: Letters and symbols may have different meanings in 1941-1970 tables than in earlier tables. See notes.)

Reference notes applying to Normals, Means, and Extremes tables for 1931–1960 base period.

(a) Length of record, years, based on January data.
Other months may be for more or fewer years if
there have been breaks in the record.

(b) Climatological standard normals (1931-1960).

* Less than one half.

+ Also on earlier dates, months, or years.

T Trace, an amount too small to measure.

Below zero temperatures are preceded by a minus sign.
The prevailing direction for wind in the Normals,
Means, and Extremes table is from records through
1963.

‡ ≥ 70° at Alaskan stations.

Unless otherwise indicated, dimensional units used in this bulletin are: temperature in degrees F.;
precipitation, including snowfall, in inches; wind movement in miles per hour; and relative humidity
in percent. Heating degree day totals are the sums of negative departures of average daily temperatures from 65° F. Cooling degree day totals are the sums of positive departures of average daily
temperatures from 65° F. Sleet was included in snowfall totals beginning with July 1948. The term
"Ice pellets" includes solid grains of ice (sleet) and particles consisting of snow pellets encased
in a thin layer of ice. Heavy fog reduces visibility to 1/4 mile or less.

Sky cover is expressed in a range of 0 for no clouds or obscuring phenomena to 10 for complete sky
cover. The number of clear days is based on average cloudiness 0-3, partly cloudy days 4-7, and
cloudy days 8-10 tenths.

Solar radiation data are the averages of direct and diffuse radiation on a horizontal surface. The langley
denotes one gram calorie per square centimeter.

& Figures instead of letters in a direction column indicate direction in tens of degrees from true North;
i.e., 09 - East, 18 - South, 27 - West, 36 - North, and 00 - Calm. Resultant wind is the vector sum of
wind directions and speeds divided by the number of observations. If figures appear in the direction
column under "Fastest mile" the corresponding speeds are fastest observed 1-minute values.

To 8 compass points only.

** The National Weather Service considers the accuracy of solar
radiation data questionable; therefore, publication is suspended
pending determination of corrected values.

Reference notes applying to Normals, Means, and Extremes tables for 1941–1970 base period.

(a) Length of record, years, through the
current year unless otherwise noted,
based on January data.

(b) 70° and above at Alaskan stations.

* Less than one half.

T Trace.

NORMALS - Based on record for the 1941-1970 period.
DATE OF AN EXTREME - The most recent in cases of multiple
occurrence.
PREVAILING WIND DIRECTION - Record through 1963.
WIND DIRECTION - Numerals indicate tens of degrees clockwise
from true north. 00 indicates calm.
FASTEST MILE WIND - Speed is fastest observed 1-minute value
when the direction is in tens of degrees.

$ Through 1967.

c Through September 1968.

Mean Annual Precipitation, Inches

Based on period 1931-55

Isolines are drawn through points of approximately equal value. Caution should be used in interpolating on these maps.

Mean Maximum Temperature (°F.), January

WISCONSIN

Based on period 1931-52

Isolines are drawn through points of approximately equal value. Caution should be used in interpolating on these maps.

Mean Minimum Temperature (°F.), January

Based on period 1931-52

Isolines are drawn through points of approximately equal value. Caution should be used in interpolating on these maps.

Mean Maximum Temperature (°F.), July

Based on period 1931-52

Isolines are drawn through points of approximately equal value. Caution should be used in interpolating on these maps.

Mean Minimum Temperature (°F.), July

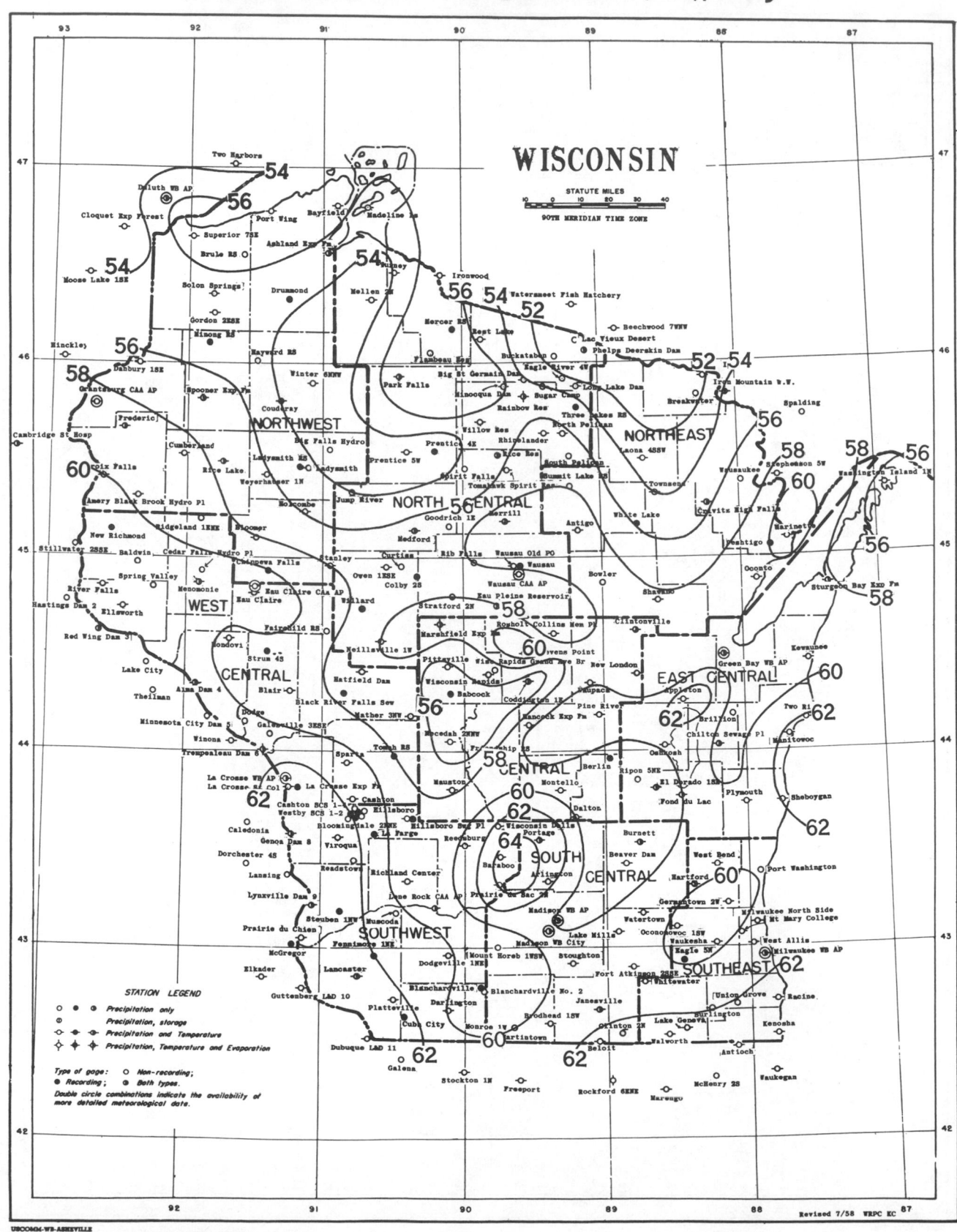

Based on period 1931-52

Isolines are drawn through points of approximately equal value. Caution should be used
in interpolating on these maps.

WISCONSIN

STATUTE MILES

90TH MERIDIAN TIME ZONE

STATION LEGEND

○ ◑ ●   Precipitation only

◎   Precipitation, storage

◔ ◕ ◒   Precipitation and Temperature

✧ ✦   Precipitation, Temperature and Evaporation

Type of gage: ○ Non-recording;

● Recording;   ◑ Both types.

Double circle combinations indicate the availability of
more detailed meteorological data.

Revised 7/59 WRPC KC

USCOMM-WB-ASHEVILLE

1109

CLIMATES OF THE STATES

WYOMING

(Normals, Means and Extremes tables revised 1972 and 1975. Basic report revised February 1960.)

Climate of Wyoming

A. R. Lowers, Weather Bureau State Climatologist

TERRAIN.--Wyoming is a name of Delaware Indian origin and is variously interpreted as "large plains" or "end of the Plains". Thus, the name describes the State's outstanding topographic feature. There are, of course, several mountain ranges, but the mountains themselves cover less area than the high plains. The topography and variations in elevation make it difficult to divide the State into homogeneous, climatological areas.

The mean elevation is given as about 6,700 feet above sea level. Even excluding the mountain ranges, the average elevation over the southern portion is well over 6,000 feet, while much of the northern portion is some 2,500 feet lower. The lowest point, 3,125 feet, is near the northeast corner where the Belle Fourche River crosses the State line to South Dakota. The highest point is Gannett Peak at 13,785 feet, which is part of the Wind River Range in the west-central portion. Since the mountain ranges lie in a general north to south direction, they are perpendicular to the prevailing westerlies. Therefore, the mountain ranges provide effective barriers which force the air currents moving in from the Pacific Ocean to drop much of their moisture. It naturally follows that the mountain ranges and the western slopes receive the greatest amount of precipitation. Outside of the mountains, the State is considered semiarid.

RIVERS AND FLOODS.--The Continental Divide splits the State from near the northwest corner to a point along the southern border about midway. This leaves most of the drainage areas to the east. Precipitation drains into three great river systems: the Columbia, the Colorado, and the Missouri. The Snake with its tributaries in the northwest portion flows to the Columbia; the Green River draining most of the Southwest portion joins the Colorado; the Yellowstone, Wind River, Big Horn, Tongue, and Powder drainage areas cover most of the north portion and flow northward into Montana, entering the Missouri there; the Belle Fourche, Cheyenne, and Niobrara covering the east-central portion, flow eastward; and the Platte (mostly North Platte), draining all of the southeast, flows eastward over Nebraska. There is a relatively small area along the southwest border that is drained by the Bear going to the Great Salt Lake. In the south-central portion west of Rawlins, there is an area called the Great Divide Basin which extends from near Rawlins to nearly 100 miles westward and about 50 miles in a north to south direction. Part of this is often referred to as the Red Desert. There is no drainage from the Great Divide Basin. Precipitation here, which averages only 7 to 10 inches annually, follows usually dry creekbeds to ponds or small lakes, also often dry.

Snow accumulates to considerable depths in the high mountains of Wyoming and many of the streams fed by the melting snow furnish ample quantities of water for irrigation of thousands of acres of land both within and without the State, as well as furnish the necessary water for the production of electric power, and for domestic use.

Rapid runoff from heavy thunderstorm rains causes flash flooding on the headwater streams of the State, and when the time of these storms coincides with the melting of the snowpack, the flooding is intensified. When such overflow occurs in the vicinity of urban communities situated near or on the streams considerable damage results.

LAND USES.--Most of the State has been subjected to erosion for tens of thousands of years and less than 10 percent is covered with a mantle of recent (geologically speaking) water-transported soil. The lack of such soil as well as moisture is why much of the natural vegetation is composed of hardy plants, such as sagebrush, greasewood, and short grass. Low relative humidity and high rate of evaporation add their influence to the types of vegetation. Quite a number of abandoned homesteads of onetime enthusiastic settlers bear silent testimony to the lack of moisture. Even so, dryland farming in addition to irrigation is carried on successfully in some areas. Approximately 42 percent of the State's total area is privately-owned land, the majority of which is used for grazing although some is timberland. Most of the remainder is under Federal or State ownership. The fact that most of the State is still Government-owned attests to the semiarid climate making the land less attractive to homesteading. Nearly 4 percent of the State is considered as cultivated cropland including both irrigated and nonirrigated. Another 13 percent is covered with forests, while parks and recreational areas take up about 4 percent.

However, the majority of the State is used for grazing and has a general appearance of dryness most of the time. The more abundant spring moisture brings a greener landscape often with myriad, varicolored, wild flowers. As the season merges into summer, grasses and flowers turn brown, but continue to furnish food for livestock. Native grasses are nutritious, although mostly scant. There are some very fine grazing areas with luxuriant grasses, especially in or near the mountains. In general though, the grass is scant and large ranches are required for herds of cattle or sheep in numbers sufficient for profitable operation. In some sections stockmen claim that it requires at least 50 acres for 1 cow. The average for most cattle grazing areas is in the range of 35 to 40 acres per cow. Much of the southwest portion, a part of which is commonly called the Red Desert, has no water consistently available to livestock during the summer months. That section is a valuable winter grazing area for sheep when snow can supply the necessary water. The mountain areas provide timber and a storage place for the winter snows which in the spring and summer feed lakes and reservoirs, and supply water used in the irrigation districts. Most of the irrigated land is located in the valleys of the following river systems and their tributaries: North Platte, Wind River, Big Horn, Tongue, and Green. The main crops of a region are a good index to the climate. Principal crops in the irrigation districts of Wyoming are sugar beets, beans, potatoes, and hay. On the nonirrigated land the principal crops are hay and small grains, such as wheat, barley, and oats.

PRECIPITATION.--Like the other states in the western part of the country, precipitation varies considerably from one location to another. The period of maximum precipitation occurs in the spring and early summer for most of the State. It is greater over the mountain ranges and usually at the higher elevations, although elevation alone is not the only influence. For example, over most of the southwest portion, where the elevation ranges from about 6,500 to 8,500 feet above sea level, annual precipitation values vary from 7 to 10 inches, while at much lower elevations over the northeast portion and along the east border, where elevations are mostly in the range from 4,000 to 5,500 feet, annual averages are generally from 12 to 16 inches. The relatively dry southwest portion is a high plateau nearly surrounded by mountain ranges.

The Big Horn Basin provides a striking example of the effect of mountain ranges in blocking the flow of moisture laden air from the east as well as from the west. The lower portion of the Basin has an annual precipitation in the range of 5 to 8 inches, and it is the driest part of the State. The station showing the least amount is Deaver at 4,105 feet with an annual mean of only 5.31 inches. In the southern part of the Basin, Worland at 4,061 feet has an annual mean of 8.15 inches as compared with Termopolis at 4,313 feet and 11.15 inches. There is another good example in the southeastern part of the State where Laramie at 7,236 feet has an annual mean of 11.16 inches, while 30 miles to the west, Centennial at 8,074 feet received 16.47 inches.

The station receiving the most precipitation and having a longterm mean is Snake River at the south entrance of Yellowstone Park where the annual amount is 31.53 inches. There are a number of other places with shorter records which show much higher amounts. Beckler River Ranger Station located near the southwest corner of Yellowstone Park and now inoperative as a weather station, but with a record covering 22 years, has an annual average of 38.22 inches. Standpipe gages have been erected in 16 locations at higher elevations where it is impossible to secure observers. While these have not been in existence long enough to determine reliable means, several of the standpipe gages show very high averages. The one at North French Creek at 10,100 feet in the Medicine Bow Mountains has a 10-year average of 46.05 inches. At Valley 6 W in the Absaroka Mountains of the northwest portion, the 6-year average at the 9,000-foot level is 44.32 inches.

SHOWERS AND THUNDERSTORMS.--During the summer months showers are quite frequent but light and often amount to only a few hundredths of an inch. Occasionally there will be some very heavy rain associated with thunderstorms covering a few square miles. There are usually several local storms each year with from 1 to 2 inches of rain in a 24-hour period. On rarer occasions 24-hour amounts range from 3 to 5 inches. The greatest 24-hour total recorded for any place in Wyoming is 5.50 inches and that occurred at Dull Center located nearly 50 miles southwest of Newcastle on May 31, 1927.

SNOW AND BLIZZARDS.--Snow falls frequently from November through May. Generally snowfall at lower elevations is light to moderate. About five times a year on the average, stations at the lower elevations will have snowfall exceeding 5 inches. Falls of 10 to 15 inches or more for any one storm are occasional but infrequent outside of the mountains. Of course, wind will frequently accompany or follow a snowstorm and pile the snow into drifts several feet deep. Frequently the snow drifts so much that it is difficult to obtain an accurate measurement of snowfall. An unusually heavy snow occurred at Sheridan on the 3d and 4th of April 1955 during which the snow amounted to 39.0 inches with a water equivalent of 4.30 inches and blizzard conditions extended over a 43-hour period. Wind with the snow will quite often cause blizzard or near blizzard conditions in parts of the State for a few hours; however, it is uncommon for a severe blizzard to last nearly 3 days such as the one early in January of 1949 which was mainly over the southeast portion.

The total annual amount of snow varies considerably over the State as does the rainfall. At the lower elevations of the east portion, the range is mostly from 60 to 70 inches annually. Over the drier southwest portion, amounts vary from 45 to 55 inches at most places. Snow is very light in the Big Horn Basin with annual averages from 15 to 20 inches over the lower portion to 30 to 40 inches on the sides of the Basin where elevations range from 5,000 to 6,000 feet. Of course the mountains receive a great deal more and over the higher ranges annual amounts are well over 200 inches. At Beckler River Ranger Station in the southwest corner of Yellowstone Park, the snowfall averaged 262 inches for a 20-year period.

AIR MASSES.--The weather map pattern most favorable for precipitation is one showing a storm center over or a little to the south of the State. This would normally provide a condition where relatively cool air at the surface is overrun by warmer air aloft having fairly high humidity. The numerous low pressure systems that first show up on the continent in the Alberta area of Canada and move to the southeast over the northern plains frequently cause very strong and gusty winds over Wyoming but not much precipitation, especially at the lower elevations. Studies of wind flow patterns indicate that Wyoming is covered most of the time by air from the Pacific. A smaller percentage of time the State is covered by cold air masses that move down from Canada, but these usually modify rapidly after reaching this area. It is seldom that air reaches Wyoming from the Gulf of Mexico, the source credited with most of the precipitation over the Mississippi Valley.

TEMPERATURES.--Because of the elevation, Wyoming has a relatively cool climate. Above the 6,000-foot level the temperature rarely exceeds 100°F. The warmest parts of the State are the lower portion of the Big Horn Basin, the lower elevations of the central and northeast portions, and along the east border. The highest recorded temperature was 114°F. on July 12, 1900, at Basin in the Big Horn Basin. The average high temperature at Basin in July is 92°F. For most of the State, mean high temperatures in July range between 85° and 90°F. With increasing elevation, average values drop rapidly. A few places in the mountains at about the 9,000-foot level show an average high in July close to 70°F. Summer nights are almost invariably cool, even though daytime readings may be quite high at times. For most places outside of the mountains, the mean low temperature in July is in the range from 50° to 60°F. Of course, the mountains and high valleys are much cooler with average lows in the middle of the summer in the 30's and low 40's with occasional drops below freezing.

In the wintertime it is characteristic to have rapid and frequent changes between mild and cold spells. Usually there are less than 10 cold waves during a winter, and frequently less than half that number for most of the State. The majority of cold waves move southward on the east side of the Divide, with only an occasional cold wave for the west side. Sometimes only the northeast portion will be affected by the cold air as it slides on to the east over the plains. Many of the cold waves are not accompanied by enough snow to cause severe conditions. In January--the coldest month generally--mean minimum temperatures range mostly from 5° to 15°F. In the western valleys mean values go down to about 5° below zero. The record low for the State is -63°F. observed February 9, 1933, at Moran in Teton County. During warm spells in the winter, nighttime temperatures frequently remain above freezing. Chinooks are common along the eastern slopes.

Numerous valleys provide ideal pockets for the collection of cold air drainage at night. Protecting mountain ranges prevent the wind from stirring the air, and the colder heavier air continues to deepen in valleys often sending readings well below zero. It is common to have temperatures in the valleys considerably lower than on the nearby mountain sides. Big Piney in the Green River Valley is such a location. Mean January temperatures in the Big Horn Basin show the variation between readings in the lower part of a valley and those higher up. At Worland and Basin in the lower portion of the Big Horn Basin, not far from the 4,000-foot level, the mean minimum temperature for January is zero, while Cody close to 5,000 feet on the west side of the valley has a mean January minimum of 11°F. Except for the occasional cold waves and an infrequent blizzard, the winters are not severe. Even January, the coldest month, has occasional mild periods when maximum readings will reach the 50's and 60's.

GROWING SEASON.--Early freezes in the fall and late in the spring are characteristic of the Wyoming climate. This has the effect of seemingly long winters and short growing seasons. However, it is a country of rapid changes through the fall, winter, and spring seasons, with frequent variations from cold to mild periods. The average growing season (freeze-free period) for the principal agricultural areas is approximately 125 days. A computation for the last 10 years shows a range in the growing season from 115 days around Riverton to 132 at Worland. Records at the first-order Weather Bureau stations located at Casper, Cheyenne, Lander, and Sheridan show an average close to 130 days. For the hardier plants which can stand a temperature down to 28°F. or slightly lower, the growing season in the agricultural areas east of the Divide is approximately 145 days. In the mountains and high valleys freezing temperatures may occur any time during the summer. For tender plants there is practically no growing season in such areas as the upper Green River Valley, the Star Valley, and Jackson Hole. At Farson near Sandy Creek, a tributary of the Green River, the last 10-year average shows only 42 days between the last report of a temperature of 29° to 32°F. in early summer to the first freeze in late summer. For the places like the Star Valley and Jackson Hole, the time is even shorter.

SUNSHINE.--For most of the State, sunshine ranges from approximately 60 percent of the possible amount during the winter to about 75 percent during the summer. Mountain areas receive less, and in the wintertime the estimated amount over the northwestern mountains is about 45 percent. Although the average amount of sunshine is less in winter, the low point on the annual variations is not during the coldest month (January or February). One low period of sunshine comes in November or December, and another in April or May. These periods of low sunshine correspond fairly close to the periods of greatest temperature changes, i.e., in the late fall when average temperatures are dropping rapidly and in the spring when the average is climbing rapidly. To be sure, sunshine will not be much higher during the coldest months, but cold air masses are apt to be more stable at that time, and frontal activity is followed by a slightly longer period of sunshine. In the summertime when sunshine is greatest--not only in time but also intensity--it is characteristic for the forenoons to be mostly clear. Cumulus clouds develop nearly every day and frequently blot out the sun for a portion of the afternoons. Because of the altitude--providing less atmosphere for the sun's rays to penetrate--and because of the very small amount of fog, haze, and smoke, the intensity of sunshine is unusually high.

SEVERE STORMS.--Hailstorms are the most destructive type of local storm for this State, and every year damage to crops and property from hail amounts to many thousands of dollars. Occasionally one of these hailstorms will pass over a city with considerably more damage. The one that hit Cheyenne on August 2, 1957, caused property damage amounting to approximately $1 1/4 million, but that was exceptional. Most of the hailstorms pass over the open rangeland and destruction is much less. In small areas of crop producing land, some farmers occasionally lose an entire crop by hail.

Tornadoes occur over Wyoming, but records show they are much less frequent and destructive as compared with the region to the east. The relatively small amount of destruction is partly due to the fact that most of Wyoming is open range country and sparsely populated. However, records point to the fact that the tornadoes which occur here are somewhat smaller and have a shorter duration. Many of them touch the ground for only a few minutes before receding into the clouds. Of course, any tornado can be very destructive. A tabulation covering a 43-year period ending with 1958 shows a total of 78 tornadoes in the State with 3 deaths and damage totaling a little over $1/2 million. Most of these occurred over the east portion. The season for these local storms extends from April through September. June has the greatest number on the average with May next.

WIND.--Wind is an important factor of the Wyoming climate. This is largely due to the high elevation and the enormous stretches of rolling plains. Over the higher south portion average annual wind speeds range from 12 to 14 miles per hour. Much of the north portion, being considerably lower in elevations has a lower average speed. There are some favorable locations which are protected from the wind by mountain ranges. Lander is one such place and has an annual average of only 7.0 m.p.h., at the airport station. The records for the city of Lander show even lighter winds. The lower portion of the Big Horn Basin at such places as Worland and Basin also has very light winds. However, most of Wyoming is quite windy, and during the colder months from November through March there are frequent periods when the wind reaches 30 to 40 miles per hour with occasional gusts much higher. Prevailing directions in the different localities vary from west-southwest through west to northwest. In many localities winds are so strong and constant from those directions that trees show a definite lean toward the east or southeast.

HUMIDITY AND EVAPORATION.--The average relative humidity is quite low and, while this has a distinct advantage in providing delightful summer weather, it is related to the rather low amount of moisture. During the warmer part of the summer days, the average drops to about 25 to 30 percent, and on a few occasions it will be as low as 5 to 10 percent. Late at night when the temperature is lowest the humidity will generally be up to 65 to 75 percent. This results in an average diurnal variation of about 40 to 45 percent during the summer, but in the winter the variation is much less. Low relative humidity, high percentage of sunshine, and rather high average winds add their influence in causing a large amount of evaporation. Because of frequent spells of freezing weather before May 1 and after September 30, it is difficult to obtain consistent records of evaporation for more than the 5-month period from May through September. For this period, the average amount of evaporation is approximately 41 inches, as determined from evaporation pans at a few selected locations. The overall range is from a little more than 31 inches at Archer and Sheridan Field Station to near 47 inches at Farson and Boysen Dam.

REFERENCES

(1) Climate of the States - Wyoming, 1941, Agricultural Yearbook Separate No 1866.

(2) Wyoming. Contribution of several writers under WPA. Oxford University Press - New York, 1941.

(3) The Dark Missouri. Henry C. Hart. University of Wisconsin Press, 1957.

(4) The Missouri Valley. Rufus Terral. Yale University Press, 1947.

(5) Tornado Occurrences in the United States. Technical Paper No. 20.

(6) Weather Bureau Technical Paper No. 16 - Maximum 24-Hour Precipitation in the United States. Washington, D. C. 1952.

(7) Weather Bureau Technical Paper No. 25 - Rainfall Intensity-Duration-Frequency Curves. For selected stations in the United States, Alaska, Hawaiian Islands, and Puerto Rico.

BIBLIOGRAPHY

(A) Climatic Summary of the United States (Bulletin W) 1930 edition, Sections 12, 13 and 14. U. S. Weather Bureau

(B) Climatic Summary of the United States, Wyoming Supplement for 1931 through 1952 (Bulletin W Supplement). U. S. Weather Bureau

(C) Climatological Data - Wyoming. U. S. Weather Bureau

(D) Climatological Data National Summary. U. S. Weather Bureau

(E) Hourly Precipitation Data - Wyoming. U. S. Weather Bureau

(F) Local Climatological Data, U. S. Weather Bureau for Casper, Cheyenne, Lander and Sheridan, Wyoming.

FREEZE DATA

STATION	Freeze threshold temperature	Mean date of last Spring occurrence	Mean date of first Fall occurrence	Mean No. of days between dates	Years of record Spring	No. of occurrences in Spring	Years of record Fall	No. of occurrences in Fall
BASIN	32	05-12	09-25	136	30	30	30	30
	28	04-27	10-05	161	30	30	30	30
	24	04-15	10-16	185	30	30	30	30
	20	04-05	10-27	204	30	30	30	30
	16	03-27	11-02	220	30	30	30	30
BUFFALO BILL DAM	32	05-08	10-06	151	30	30	30	30
	28	04-25	10-17	176	30	30	30	30
	24	04-12	10-27	198	30	30	30	30
	20	04-06	11-05	214	30	30	30	30
	16	03-27	11-13	231	30	30	30	30
CASPER WB AP	32	05-18	09-25	130	30	30	29	29
	28	05-05	10-02	150	29	29	29	29
	24	04-24	10-15	174	29	29	29	29
	20	04-13	10-22	192	29	29	29	29
	16	04-02	10-28	209	29	29	29	29
CHEYENNE WB AP	32	05-20	09-27	130	30	30	30	30
	28	05-09	10-04	149	30	30	30	30
	24	04-30	10-15	169	30	30	30	30
	20	04-19	10-25	189	30	30	30	30
	16	04-06	11-05	213	30	30	30	30
CHUGWATER	32	06-03	09-13	102	30	30	30	30
	28	05-17	09-22	127	30	30	30	30
	24	05-06	10-03	150	30	30	30	30
	20	04-24	10-12	172	30	30	30	30
	16	04-15	10-19	187	30	30	30	30
DIVERSION DAM	32	05-24	09-18	117	30	30	30	30
	28	05-07	09-27	143	30	30	30	30
	24	04-26	10-09	165	30	30	30	30
	20	04-17	10-17	183	30	30	30	30
	16	04-06	10-28	205	30	30	30	30
DIXON	32	06-10	09-01	83	29	29	28	28
	28	05-28	09-19	114	29	29	29	28
	24	05-13	09-30	141	29	29	29	29
	20	04-29	10-13	167	29	29	29	29
	16	04-09	10-25	199	29	29	29	28
DULL CENTER	32	05-19	09-20	125	23	23	24	24
	28	05-06	09-30	147	23	23	24	24
	24	04-27	10-10	167	22	22	24	24
	20	04-13	10-22	192	22	22	24	24
	16	04-04	10-30	209	22	22	24	23
EVANSTON 1 E	32	06-18	08-06	49	30	30	30	30
	28	06-07	09-10	95	30	30	30	30
	24	05-15	09-24	131	30	30	30	30
	20	05-01	10-06	158	30	30	30	30
	16	04-18	10-23	188	30	30	30	30
FARSON	32	06-17	08-14	59	27	27	27	27
	28	06-05	09-01	89	27	27	25	25
	24	05-23	09-23	123	26	26	25	25
	20	05-06	09-27	144	26	26	25	25
	16	04-26	10-10	168	25	25	24	24
GILLETTE	32	05-21	09-27	129	29	29	28	28
	28	05-07	10-04	150	28	28	29	29
	24	04-24	10-15	174	27	27	29	29
	20	04-13	10-23	193	28	28	29	29
	16	04-08	11-01	207	27	27	29	29
GROVER 2 S (Known as Afton)	32	06-27	07-15	18	29	29	30	30
	28	06-16	08-09	54	29	29	30	30
	24	05-23	09-11	110	30	30	30	30
	20	05-08	09-24	138	30	30	30	30
	16	04-17	10-10	176	27	27	29	29
KAYCEE	32	05-30	09-17	110	10	10	11	11
	28	05-15	09-26	134	10	10	11	11
	24	04-30	10-05	158	10	10	11	11
	20	04-18	10-10	175	10	10	11	11
	16	04-05	10-20	198	10	10	11	11
KEMMERER	32	06-17	08-16	60	17	17	18	18
	28	05-29	09-17	111	17	17	17	17
	24	05-18	09-26	131	17	17	17	17
	20	04-24	10-06	165	17	17	17	17
	16	04-11	10-25	196	17	17	16	16

STATION	Freeze threshold temperature	Mean date of last Spring occurrence	Mean date of first Fall occurrence	Mean No. of days between dates	Years of record Spring	No. of occurrences in Spring	Years of record Fall	No. of occurrences in Fall
KIRTLEY	32	05-29	09-21	115	30	30	30	30
	28	05-11	09-28	140	29	29	30	30
	24	05-02	10-10	161	29	29	30	30
	20	04-24	10-19	178	28	28	30	30
	16	04-15	10-27	195	28	28	30	30
LAGRANGE	32	05-24	09-21	120	23	23	25	25
	28	05-12	10-01	142	22	22	25	25
	24	04-27	10-11	167	22	22	24	24
	20	04-17	10-20	185	22	22	24	24
	16	04-10	10-28	201	22	22	23	23
LARAMIE	32	05-29	09-19	113	30	30	30	30
	28	05-16	09-28	135	30	30	29	29
	24	05-04	10-10	159	30	30	29	29
	20	04-27	10-20	176	30	30	29	29
	16	04-16	10-26	193	30	30	29	29
LEO 6 SW	32	06-07	08-30	84	19	19	19	19
	28	05-20	09-26	129	20	20	20	20
	24	05-03	10-03	153	21	21	20	20
	20	04-25	10-14	172	21	21	20	20
	16	04-13	10-24	195	21	21	20	20
LOOKOUT 14 NE	32	06-03	09-07	96	15	15	16	16
	28	05-21	09-23	125	15	15	16	16
	24	05-05	10-02	150	15	15	16	16
	20	04-30	10-10	163	15	15	16	16
	16	04-22	10-20	181	15	15	16	16
LOVELL	32	05-14	09-21	129	30	30	29	29
	28	05-03	10-03	154	30	30	29	29
	24	04-22	10-14	176	30	30	29	29
	20	04-11	10-24	196	30	30	29	29
	16	04-02	11-01	213	30	30	29	29
MARSHALL 7 SW	32	06-20	08-18	60	27	27	27	27
	28	06-10	09-01	83	26	26	25	25
	24	05-22	09-22	122	26	26	25	25
	20	05-11	10-01	144	25	25	25	25
	16	04-27	10-10	166	25	25	25	25
METZ RANCH	32	05-20	09-19	121	27	27	25	25
	28	05-11	09-30	143	26	26	25	25
	24	04-22	10-08	169	25	25	25	25
	20	04-17	10-21	187	24	24	25	25
	16	04-05	10-30	208	23	23	25	25
MOOSE 3 NW	32	06-25	07-24	29	15	15	15	15
	28	06-16	08-20	65	15	15	15	15
	24	05-21	09-21	123	15	15	15	15
	20	05-04	10-06	155	15	15	15	15
	16	04-21	10-25	187	15	15	15	15
NEWCASTLE	32	05-15	09-29	137	30	30	30	30
	28	05-05	10-08	156	30	30	30	30
	24	04-20	10-20	183	30	30	30	30
	20	04-11	10-27	198	30	30	30	30
	16	04-03	11-03	214	30	30	30	30
PINE BLUFFS	32	05-20	09-21	124	30	30	30	30
	28	05-07	09-29	146	30	30	30	30
	24	04-27	10-14	171	30	30	30	30
	20	04-15	10-23	191	30	30	30	30
	16	04-07	10-29	204	30	30	30	30
PINEDALE	32	06-23	07-27	34	20	20	21	21
	28	06-14	08-21	69	20	20	21	21
	24	05-25	09-10	108	19	19	21	21
	20	05-12	09-20	131	19	19	21	21
	16	04-30	10-02	155	19	19	21	21
RIVERTON	32	05-22	09-20	121	29	29	29	29
	28	05-09	09-28	142	29	29	29	29
	24	04-26	10-07	164	28	28	29	29
	20	04-17	10-19	185	28	28	29	29
	16	04-02	10-29	210	28	28	29	29
ROCKYPOINT 2 SW	32	05-22	09-22	123	28	28	28	28
	28	05-13	09-30	141	28	28	28	28
	24	04-29	10-11	165	28	28	28	28
	20	04-18	10-21	186	28	28	28	28
	16	04-08	10-29	204	28	28	28	28

STATION	Freeze threshold temperature	Mean date of last Spring occurrence	Mean date of first Fall occurrence	Mean No. of days between dates	Years of record Spring	No. of occurrences in Spring	Years of record Fall	No. of occurrences in Fall	STATION	Freeze threshold temperature	Mean date of last Spring occurrence	Mean date of first Fall occurrence	Mean No. of days between dates	Years of record Spring	No. of occurrences in Spring	Years of record Fall	No. of occurrences in Fall
ROSS	32	05-27	09-15	110	26	26	26	26	SOUTH PASS CITY	32	06-27	07-08	11	30	30	30	30
	28	05-07	09-30	146	26	26	26	26		28	06-21	07-29	38	30	30	30	30
	24	04-26	10-07	164	26	26	26	26		24	06-09	08-23	76	30	30	30	30
	20	04-18	10-15	180	26	26	25	25		20	05-23	09-13	113	30	30	30	30
	16	04-09	10-22	196	26	26	25	25		16	05-06	09-26	143	30	30	30	30
SARATOGA	32	06-10	08-29	80	30	30	30	30	THERMOPOLIS	32	05-22	09-17	118	28	28	29	29
	28	05-26	09-16	112	30	30	30	30		28	05-05	09-27	145	28	28	28	28
	24	05-14	09-27	137	30	30	30	30		24	04-22	10-11	173	28	28	27	27
	20	04-29	10-09	162	30	30	30	30		20	04-14	10-23	193	28	28	27	27
	16	04-21	10-21	183	30	30	30	30		16	04-02	10-31	213	28	28	26	26
SHERIDAN FIELD STA	32	05-21	09-21	123	30	30	30	30	WORLAND	32	05-13	09-23	133	30	30	30	30
	28	05-04	10-01	150	30	30	30	30		28	04-30	10-04	157	30	30	30	30
	24	04-20	10-18	181	30	30	30	30		24	04-17	10-16	182	30	30	30	30
	20	04-11	10-26	198	30	30	30	30		20	04-09	10-26	201	30	30	30	30
	16	04-04	11-01	211	30	30	30	30		16	03-28	11-03	220	30	30	30	30
									YELLOWSTONE PARK	32	06-08	09-07	91	30	30	30	30
										28	05-21	09-19	120	30	30	30	30
										24	05-05	10-03	151	30	30	30	30
										20	04-23	10-15	176	30	30	30	30
										16	04-14	10-26	196	30	30	30	30

Data in the above table are based on the period 1921-1950, or that portion of this period for which data are available.

Means have been adjusted to take into account years of non-occurrence.

A freeze is a numerical substitute for the former term "killing frost" and is the occurrence of a minimum temperature at or below the threshold temperature of 32°, 28°, etc.

Freeze data tabulations in greater detail are available and can be reproduced at cost.

*MEAN TEMPERATURE AND PRECIPITATION

STATION	JANUARY		FEBRUARY		MARCH		APRIL		MAY		JUNE		JULY		AUGUST		SEPTEMBER		OCTOBER		NOVEMBER		DECEMBER		ANNUAL	
	Temperature	Precipitation	Temperature	Precipitation	Temperature	Precipitation	Temperature	Precipitation	Temperature	Precipitation	Temperature	Precipitation	Temperature	Precipitation	Temperature	Precipitation	Temperature	Precipitation	Temperature	Precipitation	Temperature	Precipitation	Temperature	Precipitation	Temperature	Precipitation
YELLOWSTONE DRAINAGE																										
LAKE YELLOWSTONE		1.87		1.54		1.58		1.46		1.64		2.18		1.48		1.38		1.39		1.21		1.48		1.79		19.00
LAMAR RS	12.3	.95	17.0	.76	23.6	.89	35.6	1.00	45.1	1.52	51.5	2.13	58.6	1.30	56.6	1.29	49.0	1.21	39.6	.99	25.3	.82	16.6	.87	35.9	13.73
TOWER FALLS		1.29		.95		1.30		1.19		1.77		2.29		1.43		1.30		1.29		1.25		1.11		1.17		16.34
YELLOWSTONE PARK	18.3	1.12	21.6	.91	26.9	1.31	38.1	1.20	46.9	1.71	53.8	2.26	62.9	1.18	61.1	1.19	52.4	1.25	42.6	1.17	29.1	1.02	22.2	1.18	39.7	15.50
DIVISION	16.7	1.13	20.0	.92	25.9	1.26	37.1	1.25	45.9	1.54	52.8	2.03	60.9	1.38	58.9	1.20	50.8	1.24	41.2	1.09	27.9	.95	20.8	1.04	38.2	15.03
SNAKE DRAINAGE																										
ALTA 1 NNW	18.1	1.51	21.5	1.43	27.0	1.51	37.9	1.37	47.1	2.07	53.9	2.34	63.0	1.04	61.4	1.19	53.2	1.29	43.7	1.53	29.5	1.37	22.3	1.56	39.9	18.21
BEDFORD 2 SE	16.7	2.14	21.2	1.87	27.1	2.00	38.1	1.56	47.3	2.00	53.5	2.17	61.6	.93	60.4	1.05	52.8	1.25	43.2	1.58	28.9	1.73	20.7	2.24	39.3	20.52
GROVER 2 S	13.1	1.52	17.6	1.44	24.6	1.61	37.0	1.49	47.3	1.85	53.0	2.14	60.4	1.10	58.7	1.10	51.3	1.21	42.1	1.61	27.5	1.50	17.0	1.62	37.5	18.19
JACKSON	14.1	1.47	18.2	1.41	26.0	1.28	38.4	1.16	46.5	1.51	53.6	1.59	61.0	.78	58.9	1.19	51.4	1.08	41.8	1.09	27.1	1.10	17.9	1.67	37.9	15.33
MORAN	10.3	2.24	14.0	2.18	20.2	2.05	32.8	1.69	43.2	1.89	50.4	1.86	57.6	1.02	55.5	1.33	48.0	1.32	38.7	1.43	24.8	1.80	15.2	2.40	34.2	21.21
SNAKE RIVER	13.0	3.76	16.8	3.55	22.4	3.63	33.5	2.20	42.7	2.44	50.5	2.56	58.6	1.43	56.5	1.39	48.2	1.65	38.6	1.91	24.1	2.91	16.2	4.10	35.1	31.53
DIVISION	13.4	2.78	17.5	2.56	24.0	2.45	35.8	1.68	45.2	2.15	51.9	2.44	59.8	1.06	58.0	1.29	50.3	1.44	40.8	1.72	26.6	2.07	17.6	2.74	36.7	24.38
GREEN AND BEAR DRAINAGE																										
BORDER 3 N	11.4	1.23	15.4	1.05	24.1	1.09	38.5	1.11	48.4	1.25	55.1	1.33	62.9	.80	61.0	.80	52.9	.82	42.7	1.24	28.0	1.07	17.4	1.25	38.2	13.04
DIXON	17.3	.93	22.1	.69	30.3	1.20	41.8	1.36	50.4	1.34	58.2	.98	65.4	1.23	63.2	1.19	54.9	.82	44.5	1.34	30.8	.82	22.5	1.05	41.8	12.95
EVANSTON 1 E	17.1	.75	20.5	.85	26.7	.94	38.5	1.02	47.3	1.09	54.4	1.07	62.5	.74	60.9	.98	52.9	.64	42.7	.98	28.6	.85	21.3	.74	39.5	10.65
GREEN RIVER	19.3	.48	24.3	.54	32.6	.70	43.9	1.04	53.3	1.19	61.6	.92	69.7	.60	67.4	.87	58.1	.64	46.6	1.00	32.0	.49	23.4	.40	44.4	8.87
KENDALL		1.77		1.44		1.54		1.21		1.96		1.94		1.21		1.23		1.12		1.13		1.32		1.69		17.56
ROCK SPRINGS CAA AP	18.0	.46	22.8	.60	29.2	.66	40.5	1.10	50.9	1.07	59.8	.79	68.8	.68	66.9	.76	56.4	.75	44.1	1.00	30.4	.54	21.5	.41	42.5	8.82
SAGE 4 NNW		.68		.61		.67		.75		1.17		1.04		.69		.83		.63		.90		.65		.58		9.20
DIVISION	15.4	.74	19.1	.68	26.7	.81	38.9	.93	48.2	1.21	56.5	1.08	64.1	.85	61.8	.89	53.3	.70	42.6	1.00	28.2	.68	19.7	.72	39.5	10.29
BIG HORN AND WIND RIVER DRAINAGE																										
BASIN	16.4	.26	23.2	.18	34.9	.36	48.1	.75	58.0	1.07	66.2	1.13	74.5	.61	71.7	.29	61.1	.68	49.3	.47	32.6	.34	21.5	.22	46.5	6.36
BUFFALO BILL DAM	27.9	.44	29.8	.49	34.4	.82	44.9	1.36	52.9	1.96	60.2	1.97	68.9	1.08	68.2	.79	60.4	1.03	51.2	.72	38.9	.45	32.1	.33	47.5	11.44
CODY AP	24.8	.36	27.8	.38	33.8	.66	45.0	1.11	53.7	1.54	60.9	1.71	69.9	1.01	67.5	.71	58.1	.88	48.8	.62	35.3	.43	28.5	.28	46.2	9.69
DEAVER	16.7	.20	22.7	.13	31.9	.22	45.3	.49	55.2	.89	63.3	1.18	72.0	.55	68.8	.34	57.9	.54	46.3	.36	31.2	.18	21.5	.23	44.4	5.31
DIVERSION DAM	20.5	.23	24.2	.23	32.1	.49	43.1	1.30	52.0	1.85	60.1	1.49	68.6	.78	66.1	.66	57.1	1.00	46.5	.91	31.9	.38	24.0	.19	43.9	9.51
DUBOIS	21.9	.39	24.0	.35	29.1	.58	38.4	1.10	46.9	1.42	54.0	1.37	61.5	.96	59.4	.70	51.8	1.01	43.1	.83	31.3	.41	25.0	.33	40.5	9.45
FORT WASHAKIE 2 S		.34		.47		.90		1.92		2.30		1.53		.68		.49		1.08		1.11		.64		.30		11.76
LANDER WB AP	16.8	.50	22.9	.61	31.3	1.14	42.1	2.41	52.4	2.57	61.5	1.35	70.4	.88	68.4	.60	57.5	1.23	44.6	1.53	30.0	.87	20.4	.49	43.2	14.18
LOVELL	17.8	.32	23.3	.27	32.6	.47	45.8	.59	55.8	1.10	63.6	1.53	72.1	.63	68.8	.42	57.5	.66	46.8	.60	32.6	.32	22.8	.28	45.0	7.19
PAVILLION	19.6	.21	25.1	.19	33.9	.47	45.3	1.25	54.6	1.72	62.8	1.31	71.1	.75	68.5	.56	59.4	.84	47.6	.90	31.6	.36	23.0	.18	45.2	8.74
POWELL	19.8	.15	24.9	.12	33.0	.22	45.4	.41	54.9	1.02	62.6	1.31	71.0	.59	68.7	.47	58.8	.65	47.9	.37	32.5	.19	24.2	.10	45.3	5.60
RIVERTON	15.4	.22	22.0	.27	32.9	.52	44.5	1.37	53.9	1.73	62.4	1.31	70.3	.67	67.7	.47	58.0	.78	46.4	.88	29.6	.47	19.4	.21	43.5	8.90
SUNSHINE 3 SW	21.7	.56	22.8	.53	26.8	1.17	37.6	2.35	46.0	3.04	52.8	2.78	60.8	1.90	59.2	1.25	51.5	1.44	43.1	1.13	31.0	.81	24.5	.47	39.8	17.43
THERMOPOLIS	20.2	.45	25.4	.36	34.3	.67	46.3	1.72	55.4	1.98	63.7	1.68	72.1	.71	70.0	.57	59.8	1.03	48.0	1.06	32.9	.62	24.5	.30	46.1	11.15
WORLAND	15.5	.29	21.8	.23	33.1	.41	46.0	1.05	56.0	1.44	64.3	1.47	72.4	.81	69.5	.42	58.8	.74	47.1	.65	31.4	.39	20.0	.25	44.7	8.15
DIVISION	20.0	.35	24.6	.36	32.2	.64	43.8	1.39	53.1	1.79	61.1	1.55	69.8	.93	67.5	.66	58.0	.91	47.2	.82	32.6	.47	24.2	.29	44.5	10.16
POWDER, LITTLE MISSOURI, & TONGUE DRAINAGE																										
MIDWEST	24.0	.59	27.3	.58	33.9	1.07	45.2	1.65	54.6	2.23	64.0	1.67	73.3	1.18	71.2	.70	61.1	1.00	49.9	.97	35.3	.65	27.8	.55	47.3	12.84
NINE MILE CREEK	21.8	.52	25.2	.43	31.4	.93	43.2	1.63	52.2	2.07	60.9	1.77	70.6	1.06	69.1	.69	58.8	.99	47.7	.78	33.0	.74	25.8	.56	45.0	12.17
ROCKYPOINT 2 NE	19.2	.89	22.1	.75	28.7	1.33	42.5	1.94	52.7	2.59	61.4	3.39	71.4	1.38	69.7	1.09	58.9	1.34	47.3	.95	31.7	.84	23.3	.72	44.1	17.21
SHERIDAN WB AP	20.1	.75	23.2	.61	31.6	1.37	43.5	2.26	52.9	2.64	61.4	2.60	70.6	1.38	68.6	.82	57.9	1.45	46.6	1.27	33.1	.93	24.0	.67	44.5	16.75
SHERIDAN FLD STA	19.6	.55	22.4	.62	30.2	1.22	44.1	2.00	54.1	2.81	62.0	2.86	71.6	1.41	69.8	.89	58.9	1.43	47.7	1.18	32.7	.70	24.1	.51	44.8	16.18
DIVISION	20.3	.62	23.5	.55	29.5	1.18	40.4	1.89	51.1	2.41	59.5	2.56	68.9	1.34	67.0	.92	56.8	1.29	46.3	1.01	32.4	.77	24.7	.57	43.4	15.11
BELLE FOURCHE DRAINAGE																										
COLONY	21.7	.55	24.3	.45	30.7	.93	45.0	1.66	55.3	2.23	64.2	3.29	74.2	1.41	72.0	1.15	61.4	1.26	49.6	.83	34.6	.67	26.1	.47	46.6	14.90
GILLETTE 2 E	21.9	.70	25.0	.50	30.9	1.15	43.3	1.71	52.9	2.31	61.3	2.51	72.2	1.27	70.2	.86	59.9	1.17	48.7	.72	33.8	.73	26.1	.64	45.5	14.27
SUNDANCE		.74		.62		1.14		1.70		2.55		3.38		1.66		1.09		1.38		.99		.78		.69		16.72
DIVISION	20.6	.64	24.1	.56	29.4	1.03	42.7	1.80	52.5	2.56	61.2	2.96	71.0	1.58	69.2	1.15	58.6	1.33	47.7	.89	32.7	.73	24.5	.59	44.5	15.82
CHEYENNE AND NIOBRARA DRAINAGE																										
DULL CENTER 1 SE	23.6	.27	26.5	.27	32.8	.55	44.7	1.55	54.3	2.37	63.7	2.24	73.4	1.45	70.7	1.47	60.4	.98	48.6	.88	34.5	.42	26.3	.29	46.6	12.74
KIRTLEY	22.7	.58	25.1	.63	30.5	.99	42.3	2.41	51.9	3.02	60.8	3.12	70.5	1.55	68.9	1.37	59.2	1.33	47.6	1.16	33.7	.59	26.0	.54	44.9	17.29
LUSK	24.1	.54	26.9	.52	32.8	.80	43.6	1.95	52.8	2.71	62.3	2.86	70.9	1.62	68.9	1.19	59.1	1.18	48.0	.96	34.4	.63	27.0	.50	45.9	15.46
NEWCASTLE	23.2	.45	26.4	.36	32.2	.81	44.8	1.44	55.0	2.32	64.5	2.57	74.4	1.56	71.9	1.70	61.2	.90	49.4	.83	34.8	.47	27.2	.41	47.1	13.82
ROCHELLE 3 E		.29		.28		.78		1.72		2.06		2.25		1.30		1.38		.91		.82		.38		.35		12.52
SPENCER 10 NE		.45		.49		.94		1.66		2.27		2.46		1.57		1.40		.91		.88		.43		.40		13.86
DIVISION	22.2	.41	25.8	.41	31.5	.79	44.1	1.58	53.4	2.37	62.5	2.45	72.3	1.32	70.4	1.24	58.9	1.00	48.2	.81	33.8	.47	25.7	.37	45.7	13.22
PLATTE DRAINAGE																										
ARCHER		.39		.33		.74		1.63		2.60		2.68		1.76		1.71		1.26		.87		.44		.28		14.69
CASPER WB AP	22.3	.69	25.9	.72	32.4	1.13	43.0	1.92	53.1	2.32	62.0	1.48	71.1	1.17	68.9	.86	58.1	1.22	46.4	1.11	33.3	.82	25.5	.65	45.1	14.09
CHEYENNE WB AP	25.5	.56	27.7	.65	31.8	1.22	41.1	2.14	50.1	2.46	60.3	2.10	68.1	1.96	66.8	1.61	57.4	1.20	46.4	1.13	35.1	.70	28.7	.52	44.9	16.25
CHUGWATER	27.5	.55	29.3	.57	33.9	1.05	43.8	2.14	52.5	2.42	61.8	2.47	69.7	1.85	68.0	1.11	58.8	1.14	48.4	.90	36.1	.62	30.1	.66	46.7	15.48
ELK MOUNTAIN		.75		.90		1.38		1.71		1.31		.93		.73		.92		.94		.81		.70		.64		11.72
ENCAMPMENT 10 ESE	21.7	.86	23.4	.81	28.9	1.57	39.8	1.80	48.8	1.65	57.0	1.35	64.8	1.19	63.3	1.34	55.5	1.13	45.2	1.25	32.3	.91	25.4	.87	42.2	14.73
FORT LARAMIE 11 NNW		.44		.50		.82		2.04		2.50		2.48		1.49		1.29		1.29		.97		.48		.54		14.84
FOXPARK	15.1	1.38	16.9	1.39	21.5	1.84	30.6	1.81	39.5	1.59	48.3	1.63	55.1	1.61	54.0	1.51	46.5	1.22	37.4	1.00	24.3	1.20	18.6	1.06	34.0	17.24
HECLA		.47		.56		1.18		2.13		2.50		1.91		1.88		1.35		1.11		.89		.62		.47		15.07
LARAMIE	22.6	.38	25.2	.40	29.8	.75	39.5	1.36	48.5	1.46	58.0	1.50	65.6	1.75	63.8	1.25	55.4	.81	45.2	.65	32.3	.44	26.2	.41	42.7	11.16
LEO 6 SW		.66		.64		.98		1.37		1.40		1.08		.91		.96		.83		1.02		.72		.57		11.14
MARSHALL 7 SW	14.6	.63	18.4	.56	25.2	.87	37.1	1.17	46.3	1.49	55.7	1.25	63.9	1.27	61.4	1.11	52.1	.78	41.2	.71	26.3	.47	19.1	.50	38.4	10.81
PATHFINDER DAM	22.5	.30	25.2	.38	32.1	.58	43.6	1.20	53.2	1.66	62.8	1.18	72.2	1.07	70.2	.86	60.4	.85	49.2	.90	34.9	.48	27.3	.31	46.1	9.77
PINE BLUFFS	26.4	.29	29.5	.31	34.8	.70	45.6	1.63	54.6	2.45	64.4	2.55	72.4	1.78	70.6	1.70	61.3	1.25	49.6	.59	36.0	.41	29.1	.36	47.9	14.02
SARATOGA	20.4	.47	23.5	.41	29.6	.73	40.1	1.04	49.1	1.24	57.8	1.06	65.8	.87	63.9	.96	55.2	.80	44.5	.91	30.9	.54	24.1	.50	42.1	9.53
SOUTH PASS CITY		.98		.98		1.22		1.58		1.26		1.44		.83		.92		.79		1.00		.91		.84		12.75
TORRINGTON EXP FARM	26.5	.29	29.6	.42	35.6	.76	46.4	1.86	55.8	2.44	65.3	2.85	72.9	1.43	70.5	.98	60.7	1.04	49.7	.72	36.3	.47	29.1	.42	48.2	13.68
WHEATLAND 1 N	29.1	.28	21.1	.39	36.3	.67	46.9	1.89	55.9	2.18	65.3	2.08	73.8	1.17	71.9	1.04	62.2	1.19	51.3	.79	38.5	.42	32.2	.36	49.5	12.46
YODER		.32		.39		.74		1.84		2.23		2.63		1.54		1.07		1.09		.77		.41		.37		13.40
DIVISION	23.2	.57	25.9	.57	30.8	.95	42.1	1.63	50.9	2.04	60.1	2.00	68.4	1.37	66.7	1.10	57.5	1.04	46.7	.90	33.3	.63	26.6	.53	44.4	13.33

* Averages for period 1931 - 1955, except for stations marked WB and Rock Springs CAA AP which are "normals" based on period 1921 - 1950. Divisional means may not be the arithmetical average of individual stations published, since additional data from shorter period stations are used to obtain better areal representation.

NORMALS, MEANS AND EXTREMES
(Table Revised 1972. Base Period for Climatological Normals: 1931-1960)

Station: CASPER, WYOMING — CASPER AIR TERMINAL
Standard time used: MOUNTAIN Latitude: 42° 55' N Longitude: 106° 28' W Elevation (ground): 5338 feet

Month	Temperature Normal — Daily maximum (b)	Daily minimum (b)	Monthly (b)	Extremes — Record highest	Year	Record lowest	Year Ø	Normal heating degree days (Base 65°) (b)	Precipitation Normal total (b)	Maximum monthly	Year	Minimum monthly	Year	Maximum in 24 hrs.	Year	Snow, Ice pellets Mean total	Maximum monthly	Year	Maximum in 24 hrs.	Year	Relative humidity Hour 05	11	17	23	Wind Mean speed	Prevailing direction	Fastest mile Speed	Direction	Year
J	32.8	13.9	23.4	60	1971	-23	1970	1290	0.56	0.78	1952	T	1971	0.42	1964	6.4	16.5	1965	8.0	1964	66	56	59	67	16.8	SW	58	SW	1954
F	36.1	16.0	26.3	73	1957	-16	1971	1084	0.59	1.01	1957	T	1950	0.40	1956	6.7	23.4	1952	10.4	1952	69	56	55	69	16.0	SW	58	SW	1957
M	41.5	21.1	32.1	73	1966	-21	1965	1020	1.03	2.43	1953	0.15	1956	1.00	1958	14.3	32.9	1954	14.6	1954	70	53	48	68	14.1	SW	81	SW	1956
A	55.5	30.7	43.1	80	1969	-4	1966	657	1.69	3.40	1954	0.25	1954	1.57	1964	11.7	24.7	1971	13.2	1964	76	48	42	68	14.4	WSW	54	SW	1959
M	66.1	40.1	53.1	91	1969	17	1968	381	2.03	5.59	1971	0.30	1957	2.07	1952	1.8	23.9	1950	14.1	1950	76	44	39	67	11.8	WSW	58	SW	1967
J	77.2	48.9	63.1	101	1970	37	1969+	129	1.25	3.75	1956	0.03	1959	2.07	1956	0.4	3.0	1956	3.0	1969+	78	43	36	66	11.1	SW	52	SW	1959
J	87.1	56.3	71.7	99	1966	—	—	6	1.00	3.05	1951	0.11	1971	1.33	1951	0.0	0.0	—	0.0	—	69	31	25	54	9.9	WSW	46	SW	1962+
A	85.1	55.0	70.1	99	1969	33	1965	16	0.72	1.13	1957	0.07	1950	0.75	1950	T	T	1964	T	1964	62	29	24	55	10.4	SW	50	SW	1954
S	74.1	45.3	59.7	91	1969+	19	1965	192	0.90	3.11	1950	0.07	1956	1.80	1965	0.9	8.8	1971	4.5	1971	62	35	30	58	11.2	WSW	55	SW	1965
O	61.0	35.5	48.3	82	1970	-3	1971	524	0.84	2.49	1962	0.54	1962	2.49	1962	4.4	13.1	1971	8.2	1971	67	45	43	54	12.7	SW	55	SW	1954
N	43.8	24.0	33.9	71	1965	-8	1964	942	0.71	1.30	1965	0.03	1956	0.54	1956	9.0	19.7	1956	9.2	1967	54	41	41	67	14.7	SW	50	SW	1970
D	36.6	18.0	27.3	59	1965	-31	1964	1169	0.48	1.04	1952	0.03	1955	0.47	1955	8.7	17.8	1967	8.2	1953	63	59	62	69	16.3	SW	63	SW	1955
YR	58.3	33.7	46.0	101	JUN. 1970	-31	DEC. 1964	7410	11.80	5.59	MAY 1971	T	OCT. 1965+	2.49	OCT. 1962	69.3	32.9	MAR. 1954	14.6	MAR. 1954	70	46	43	64	13.0	SW	81	MAR. 1956

Mean number of days / Mean sky cover / solar radiation (right portion):

Month	Mean sky cover sunrise to sunset	Sunrise to sunset — Clear	Partly cloudy	Cloudy	Precipitation .01 inch or more	Snow, Ice pellets 1.0 inch or more	Thunderstorms	Heavy fog	Temperatures Max. 90° and above	32° and below	Min. 32° and below	0° and below	Avg. daily solar radiation langleys
(number of years) 21	21	21	21	21	21	21	21	21	7	7	7	7
J	6.5	7	7	17	7	3	0*	1	0	11	29	5	5
F	6.5	7	7	16	8	4	*	1	0	8	27	2	1
M	6.7	6	9	16	9	4	*	1	0	7	27	*	3
A	6.8	6	9	15	9	3	1	2	0	1	21	*	*
M	—	6	10	15	10	1	5	1	0	*	9	0	0
J	5.3	10	11	9	9	*	8	1	2	0	1	0	0
J	3.9	16	10	5	5	0	8	*	12	0	0	0	0
A	4.2	15	9	7	6	0	6	*	12	0	0	0	*
S	4.3	15	8	7	6	3	3	1	1	0	5	*	2
O	5.1	12	8	11	6	2	1	1	0	*	18	0	5
N	6.2	7	8	14	7	3	*	1	0	5	30	1	—
D	6.3	8	8	15	7	3	0	1	0	14	30	3	—
YR	5.7	114	105	146	91	22	35	9	27	47	190	18	18

Ø For period October 1964 through the current year.
Means and extremes above are from existing and comparable exposures. Annual extremes have been exceeded at other sites in the locality as follows:
Highest temperature 104 in July 1954; lowest temperature -40 in January 1949; maximum monthly precipitation 5.75 in April 1941; maximum precipitation in 24 hours 3.09 in April 1941; maximum snowfall 39.3 in January 1949; maximum snowfall in 24 hours 20.6 in May 1946.

NORMALS, MEANS AND EXTREMES
(Table Revised 1975. Base Period for Climatological Normals: 1941-1970)

Month	Temperatures °F Normal — Daily maximum (a)	Daily minimum	Monthly	Extremes — Record highest	Year	Record lowest	Year	Normal Degree days Base 65°F Heating	Cooling	Precipitation in inches Water equivalent Normal	Maximum monthly	Year	Minimum monthly	Year	Maximum in 24 hrs.	Year	Snow, Ice pellets Maximum monthly	Year	Maximum in 24 hrs.	Year	Relative humidity pct. Hour 05	11	17	23	Wind Mean speed m.p.h.	Prevailing direction	Fastest mile Speed m.p.h.	Direction	Year	Mean sky cover sunrise to sunset	Sunrise to sunset — Clear	Partly cloudy	Cloudy	Precipitation .01 inch or more	Snow, Ice pellets 1.0 inch or more	Thunderstorms	Heavy fog, visibility ¼ mile or less	Temperatures Max. 90° and above	32° and below	Min. 32° and below	0° and below	Average station pressure mb.
(years)				10		10					24		24		24		24		24		10	10	10	10	13		21			24	24	24	24	24	24	24	24	10			10	2
J	33.6	12.7	23.2	50	1971	-40	1972	1296	0	0.50	0.99	1952	T	1972	0.53	1972	19.2	1972	9.7	1972	68	59	61	68	17.0	SW	58	SW	1954	6.7	7	7	17	7	3	0*	1	0	13	28	2	834.6
F	37.7	15.9	26.8	51	1972	-17	1972	1070	0	0.50	1.01	1955	0.15	1955	0.40	1952	23.8	1952	10.4	1952	59	57	56	70	15.3	SW	58	SW	1957	6.7	7	7	14	8	4	*	1	0	9	27	2	830.8
M	41.6	19.4	30.8	73	1966	-4	1966	1054	0	0.71	2.43	1954	0.25	1954	1.00	1958	32.9	1954	16.5	1964	57	53	48	69	14.4	WSW	81	SW	1956	6.8	6	9	16	9	4	1	1	0	6	21	*	831.7
A	52.7	29.3	42.7	80	1969	-4	1970	669	0	1.45	3.92	1974	0.20	1973	3.00	1973	56.3	1973	14.1	1973	76	50	43	71	13.6	WSW	54	SW	1959	6.6	6	9	14	9	4	2	2	*	1	21	*	834.6
M	61.9	39.3	51.9	101	1970	17	1968	388	6	1.44	5.59	1971	0.30	1967	1.67	1969	3.0	1969	3.0	1969	77	41	35	65	11.1	SW	52	SW	1959	5.2	15	15	14	10	1	6	1	1	*	9	0	837.0
J	76.3	47.4	61.9	101	1974	28	1969	147	54	—	3.75	1956	0.03	1956	1.67	1969	0.0	—	0.0	—	71	32	26	56	10.1	WSW	52	SW	1974	4.0	15	11	5	5	0	9	*	1	0	1	0	840.0
J	85.6	54.5	71.0	99	1973	30	1972	13	199	0.95	1.57	1951	0.11	1951	1.33	1951	0.0	—	0.0	—	71	32	26	55	10.4	SW	50	SW	1965	4.5	14	11	6	5	0	11	*	11	0	0	0	840.2
A	85.6	53.3	69.4	99	1969+	31	1971	16	159	0.57	1.57	1973	0.07	1973	1.01	1973	T	1964	T	1964	69	31	26	55	10.3	SW	53	SW	1954	4.2	14	9	8	5	0	8	*	10	0	0	0	839.4
S	74.1	33.3	53.7	91	1971	19	1971	229	40	0.87	3.28	1973	0.07	1956	2.28	1973	8.8	1965	8.2	1965	68	38	32	64	12.3	WSW	55	SW	1965	5.2	12	9	9	6	2	3	1	1	0	5	*	839.0
O	61.4	22.9	44.8	82	1973	-8	1964	536	0	0.92	2.49	1955	0.54	1955	2.49	1955	19.9	1956	9.7	1956	70	55	57	68	14.6	SW	49	SW	1970	5.1	14	9	8	6	2	*	1	0	4	16	*	835.9
N	44.8	22.2	33.9	71	1965	-31	1964	933	0	0.68	1.30	1965	0.03	1956	0.47	1967	17.8	1967	8.2	1967	58	59	62	68	14.8	SW	63	SW	1955	5.6	11	8	11	7	3	0	1	0	16	25	1	835.3
D	36.2	16.2	26.2	51	1973	-31	1964	1203	0	0.49	1.04	1952	0.03	—	0.47	1974	17.8	—	3.00	—	77	41	35	62	16.2	SW	—	SW	—	6.3	8	8	15	7	3	0	1	0	15	29	5	—
YR	58.4	32.4	45.4	101	JUN. 1970+	-40	JAN. 1972	7555	458	11.22	5.59	MAY 1971	T	OCT. 1965	3.00	APR. 1974	56.3	APR. 1973	16.5	APR. 1973	71	47	44	65	13.1	SW	81	MAR. 1956	5.7	113	107	145	91	23	34	9	25	48	187	18	836.7

Means and extremes above are from existing and comparable exposures. Annual extremes have been exceeded at other sites in the locality as follows:
Highest temperature 104 in July 1954; maximum monthly precipitation 5.75 in April 1941; maximum precipitation in 24 hours 3.09 in April 1941;
maximum monthly snowfall 39.3 in January 1949; maximum snowfall in 24 hours 20.6 in May 1946.

REFERENCE NOTES APPLYING TO TABLES APPEAR ON THE PAGE FOLLOWING LAST TABLE.
(Caution: Letters and symbols may have different meanings in 1941-1970 tables than in earlier tables. See notes.)

NORMALS, MEANS AND EXTREMES

(Table Revised 1972. Base Period for Climatological Normals: 1931-1960)

Station: CHEYENNE, WYOMING   MUNICIPAL AIRPORT   Standard time used: MOUNTAIN   Latitude: 41° 09' N   Longitude: 104° 49' W   Elevation (ground): 6126 feet

Means and extremes above are from existing and comparable exposures. Annual extremes have been exceeded at other sites in the locality as follows:
Highest temperature 100 in June 1954 and earlier; lowest temperature -38 in January 1875; maximum monthly precipitation 7.66 in April 1900;minimum
monthly precipitation .00 in September 1879; maximum precipitation in 24 hours 4.70 in July 1896; maximum monthly snowfall 46.5 in April 1905;
fastest mile of wind 71 from West in December 1923. Data for September - December 1959 considered in extracting temperature extremes above.

NORMALS, MEANS AND EXTREMES

(Table Revised 1975. Base Period for Climatological Normals: 1941-1970)

Means and extremes above are from existing and comparable exposures. Annual extremes have been exceeded at other sites in the locality as follows:
Highest temperature 100 in June 1954 and earlier; lowest temperature -38 in January 1875; maximum monthly precipitation 7.66 in April 1900;minimum
monthly precipitation .00 in September 1879; maximum precipitation in 24 hours 4.70 in July 1896; maximum monthly snowfall 46.5 in April 1905.

REFERENCE NOTES APPLYING TO TABLES APPEAR ON THE PAGE FOLLOWING LAST TABLE.
(Caution: Letters and symbols may have different meanings in 1941-1970 tables than in earlier tables. See notes.)

NORMALS, MEANS AND EXTREMES
(Table Revised 1972. Base Period for Climatological Normals: 1931-1960)

Station: LANDER, WYOMING   HUNT FIELD   Standard time used: MOUNTAIN   Latitude: 42° 49' N   Longitude: 108° 44' W   Elevation (ground): 5563 feet

Month	Normal Daily max	Normal Daily min	Normal Monthly	Record highest	Year	Record lowest	Year	Normal heating degree days (Base 65°)	Precip Normal total	Max monthly	Year	Min monthly	Year	Max in 24 hrs	Year	Snow Mean total	Max monthly	Year	Max 24 hrs	Year	RH 05	RH 11	RH 17	RH 23	Wind mean speed	Prevailing dir	Fastest speed	Dir	Year	% sunshine	Mean sky cover	Clear	Partly cloudy	Cloudy	Precip .01+	Snow 1.0+	Thunderstorms	Heavy fog	Max 90+	Max 32-	Min 32-	Min 0-	Solar radiation
J	30.8	7.7	19.3	63	1963	-37	1963	1417	0.46	1.65	1949	T	1952	0.81	1962	7.7	26.5	1962	13.8	1962	66	58	58	66	6.3	SW	73	SW	1967	67	6.1	8	10	13	6	3	0	*	0	14	30	8	223
F	36.3	11.8	24.1	68	1951	-28	1949	1145	0.69	2.18	1955	T	1970	0.88	1955	12.2	43.8	1955	13.8	1955	67	56	53	64	6.3	SW	77	SW	1957	70	6.0	7	10	11	6	4	0	*	0	9	28	4	319
M	44.5	19.9	32.2	76	1966	-16	1960	1017	1.15	2.20	1949	0.35	1970	1.26	1949	14.2	40.3	1970	15.4	1963	66	53	47	61	7.1	SW	74	SW	1970	74	6.1	8	10	14	8	4	1	*	0	5	29	2	451
A	55.8	30.5	43.2	82	1962	10	1966	654	2.45	5.46	1957	0.78	1948	2.16	1948	21.2	45.2	1971	21.9	1971	65	47	41	57	8.0	SW	61	SW	1955	68	6.4	6	11	14	9	4	1	0	0	1	20	0	525
M	66.0	39.8	52.9	91	1954	18	1954	381	2.65	2.65	1954	0.49	1967	2.65	1957	1.9	18.4	1947	18.4	1947	67	45	39	57	7.5	SW	56	SW	1956+	66	6.4	6	11	14	9	1	5	*	*	*	5	0	574
J	76.0	48.0	62.0	100	1954	25	1951	153	1.36	6.88	1947	T	1971+	3.56	1947	0.0	0.0		0.0		65	42	35	54	7.7	SW	61	SW	1960	72	5.2	10	11	9	7		8	0	2	0	*	0	644
J	86.8	55.4	70.6	101	1971	39	1968	6	0.77	1.72	1962	0.05	1963	0.71	1962	0.0	0.0		0.0		56	34	28	44	7.6	SW	57	W	1959	77	4.1	14	12	5	6		9	*	10	0	0	0	633
A	85.0	55.0	68.8	101	1951	35	1962	19	0.89	4.62	1950	0.10	1970+	0.89	1966	0.0	0.0		0.0		54	32	27	43	7.1	SW	56	W	1962+	77	4.3	13	12	6	6		6	*	7	0	0	0	568
S	73.3	44.6	59.0	93	1966	10	1965	204	1.03	3.58	1971	0.01	1958	1.97	1950	2.3	23.6	1965	14.4	1965	60	39	34	50	7.1	SW	56	W	1966	73	4.8	12	10	8	5	1	3	*	1	1	3	0	462
O	60.4	33.8	47.1	85	1963	-15	1971	555	1.21	2.08	1957	0.01	1949	1.71	1958	4.0	32.5	1971	19.4	1966	63	45	42	57	6.2	SW	70	SW	1950	72	4.8	14	9	8	5	2	*	1	0	3	14	1	346
N	42.9	19.0	31.0	68	1953	-22	1968+	1020	0.92	1.51	1955	0.01	1958	1.16	1958	11.9	32.5	1971	23.1	1958	68	59	58	66	5.7	N	75	N	1958	62	5.7	8	10	12	5	3	0	1	0	6	28	4	235
D	34.5	11.7	23.1	61	1969	-22	1968+	1299	0.42		1954	0.03	1955	0.98	1955	8.7	26.3	1967	15.7	1967					5.8	SW	73	SW	1964	65	5.7	9	11	12	5	3	0	1	0	13	31	4	192
YR	57.5	31.4	44.4	101	JUL. 1954	-37	JAN. 1963	7870	13.58	6.88	JUN. 1947	T	JUN. 1971+	3.56	JUN. 1947	99.6	45.2	APR. 1971	23.1	NOV. 1958	64	47	44	57	7.0	SW	77	SW	FEB. 1957	71	5.4	115	126	124	70	24	32	3	20	48	187	19	431

Means and extremes above are from existing and comparable exposures. Annual extremes have been exceeded at other sites in the locality as follows: Highest temperature 102 in July 1935; lowest temperature -40 in February 1936 and earlier; maximum monthly precipitation 7.19 in April 1900; minimum monthly precipitation 0.00 in November 1939 and earlier; maximum precipitation in 24 hours 3.66 in May 1924; maximum monthly snowfall 58.9 in April 1945.

NORMALS, MEANS AND EXTREMES
(Table Revised 1975. Base Period for Climatological Normals: 1941-1970)

Elevation: 5558 feet m.s.l.   Average station pressure: 828.5 mb

Month	Normal Daily max	Normal Daily min	Normal Monthly	Record highest	Year	Record lowest	Year	Heating	Cooling	Precip Normal	Max monthly	Year	Min monthly	Year	Max 24 hrs	Year	Water equiv Max monthly	Year	Min monthly	Year	Max 24 hrs	Year	Snow Max monthly	Year	Max 24 hrs	Year	RH 05	RH 11	RH 17	RH 23	Wind mean speed	Prevailing dir	Fastest speed	Dir	Year	% sunshine	Sky cover	Clear	Partly cloudy	Cloudy	Precip .01+	Snow 1.0+	Thunderstorms	Heavy fog	Max 90+	Max 32-	Min 32-	Min 0-
J	31.1	8.0	19.6	63	1963	-37	1963	1407	0	0.48	1.65	1949	T	1952	0.81	1962	1.65	1949	T	1952	0.98	1973	26.5	1962	13.8	1962	66	58	58	65	6.4	SW	73	SW	1967	67	6.1	8	10	13	5	4	0	1	0	15	30	8
F	37.6	11.0	19.3	68	1951	-28	1949	1106	0	0.66	2.18	1955	0.02	1970	0.88	1955	2.10	1955	0.02	1970	0.89	1968	43.8	1955	13.8	1955	67	56	53	64	6.2	SW	77	SW	1957	70	6.0	7	10	11	5	5	0	*	0	8	28	4
M	43.5	19.3	31.4	76	1966	-16	1960	1042	0	1.18	3.02	1973	0.35	1973	1.26	1949	3.46	1960	0.10	1957	2.21	1966	40.3	1973	20.3	1966	66	53	47	61	7.2	SW	72	SW	1972	74	6.1	7	11	13	7	5	1	*	0	4	29	2
A	55.4	30.4	43.2	82	1962	-16	1966	663	0	2.16	5.46	1957	0.35	1957	2.16	1948	4.68	1973	0.35	1957	1.71	1958	45.2	1971	21.9	1971	67	46	40	58	8.0	SW	61	SW	1955	68	6.2	6	12	13	9	5	2	*	0	2	20	*
M	65.9	39.7	52.8	91	1954	18	1954	382	0	2.59	6.03	1974	0.31	1971	2.75	1957	6.03	1974	0.31	1971	1.16	1949	18.4	1947	18.4	1947	67	44	39	57	8.0	SW	56	SW	1960	66	6.4	6	12	13	9	1	5	*	*	0	5	0
J	74.9	47.4	61.2	100	1954	25	1954	150	36	1.93	6.88	1947	T	1971	3.56	1947	6.88	1947	T	1971	1.51	1955	0.0		0.0		64	42	34	53	7.8	SW	61	SW	1960	72	5.2	9	12	9	7		8	0	2	0	*	0
J	86.4	55.7	70.0	101	1971	39	1968	9	182	0.61	2.10	1973	0.05	1963	0.98	1973	2.10	1973	0.05	1963			0.0		0.0		56	34	28	44	7.7	SW	57	W	1959	77	4.3	14	13	5	6		9	*	10	0	0	0
A	84.3	53.7	69.0	101	1951	35	1962	14	138	0.42	1.77	1966	0.01	1968	1.77	1966	1.77	1966	0.01	1968			0.0		0.0		56	33	27	43	7.1	SW	56	W	1966	77	4.3	13	13	6	6		6	*	7	0	0	0
S	73.0	43.7	58.3	93	1966	10	1965	225	27	1.05	4.68	1973	0.01	1973	1.71	1966	4.68	1973	0.01	1973			23.6	1965	19.4	1966	64	39	34	50	6.2	SW	56	W	1966	73	4.9	12	10	7	5	2	3	*	1	0	3	0
O	60.2	33.0	46.8	85	1963	-15	1971	564	0	0.87	2.08	1957	0.01	1957	1.71	1958	2.08	1957	0.01	1957			32.5	1971	19.4	1966	65	45	41	58	5.7	SW	75	N	1958	62	4.7	12	10	9	5	2	*	1	0	3	14	1
N	43.3		31.4	68	1953			1005	0	0.45	1.51	1955	0.03	1949	0.98	1955	1.51	1955	0.03	1955			32.5	1971	15.7	1967	61	58	57	66	5.8	SW	73	SW	1964	64	5.7	10	10	12	5	3	0	1	0	13	31	5
D	34.3	11.6	23.0	61	1969	-25	1972	1302	0	0.45	1.51	1955	0.03	1955	0.98	1955	1.51	1955	0.03	1955			26.3	1967	15.7	1967					5.8	SW	73	SW	1964	64	5.7	10	10	12	5	3	0	1	0	13	31	5
YR	57.5	31.3	44.4	101	JUL. 1954	-37	JAN. 1963	7869	383	13.84	6.88	JUN. 1947	T	JUN. 1971	3.56	JUN. 1947	6.88	JUN. 1947	T	JUN. 1971	23.1	NOV 1958	45.2	APR 1973	23.1	NOV 1958	64	47	44	57	7.0	SW	80	SW	MAR 1972	71	5.4	115	126	124	71	25	32	3	20	49	187	20

Means and extremes above are from existing and comparable exposures. Annual extremes have been exceeded at other sites in the locality as follows: Highest temperature 102 in July 1935; lowest temperature -40 in February 1936 and earlier; maximum monthly precipitation 7.19 in April 1900; minimum monthly precipitation 0.00 in November 1939 and earlier; maximum precipitation in 24 hours 3.66 in May 1924.

REFERENCE NOTES APPLYING TO TABLES APPEAR ON THE PAGE FOLLOWING LAST TABLE.
(Caution: Letters and symbols may have different meanings in 1941-1970 tables than in earlier tables. See notes.)

NORMALS, MEANS AND EXTREMES
(Table Revised 1972. Base Period for Climatological Normals: 1931-1960)

Station: SHERIDAN, WYOMING   SHERIDAN COUNTY AIRPORT   Standard time used: MOUNTAIN   Latitude: 44° 46' N   Longitude: 106° 58' W   Elevation (ground): 3964 feet

Temperature (°F)

Month	Normal Daily maximum	Normal Daily minimum	Normal Monthly	Record highest	Year	Record lowest	Year	Normal heating degree days (Base 65°)
(a)	(b)	(b)				7		(b)
J	34.0	8.6	21.3	67	1966	-25	1968	1355
F	36.4	11.1	23.8	65	1966	-16	1971	1154
M	42.9	19.1	31.0	75	1966	-23	1965	1054
A	56.3	30.8	43.6	81	1968	3	1966	642
M	66.6	40.4	53.6	92	1969	17	1967	366
J	75.4	48.4	61.9	100	1970	31	1969	150
J	87.1	55.5	71.3	103	1966	35	1971	25
A	85.6	53.4	69.5	100	1971+	34	1966	31
S	74.1	42.9	58.5	96	1971	15	1965	219
O	62.3	33.2	47.8	87	1971	-1	1971	539
N	46.0	20.8	33.4	77	1965	-9	1966	948
D	38.5	14.0	26.3	70	1965	-29	1964	1200
YR	58.8	31.5	45.2	103	JUL. 1966	-29	DEC. 1964	7683

Precipitation (inches) / Snow, Ice pellets

Month	Normal total	Max monthly	Year	Min monthly	Year	Max in 24 hrs.	Year	Snow Mean total	Snow Max monthly	Year	Snow Max in 24 hrs.	Year
J	0.64	1.70	1948	0.12	1961	0.88	1948	9.9	23.9	1963	11.2	1963
F	0.74	2.68	1955	T	1955	1.10	1971	11.3	35.0	1955	11.1	1955
M	1.42	3.26	1946	0.40	1960	2.85	1946	12.9	36.8	1954	13.3	1954
A	2.16	4.80	1963	0.60	1958	3.84	1948	9.2	39.6	1955	26.7	1955
M	2.57	5.20	1970	0.30	1957	3.44	1943	1.7	11.0	1943	7.3	1954
J	2.57	9.54	1944	0.28	1971	3.44	1969	0.2	4.0	1969	4.0	1969
J	1.19	3.78	1958	0.08	1959	2.28	1959	0.0	0.0		0.0	
A	0.90	3.02	1968	T	1970	1.71	1970	0.0	0.0		0.0	
S	1.17	3.08	1951	0.06	1949	1.33	1947	0.8	9.5	1970	5.5	1970
O	1.13	3.16	1971	0.02	1953	1.88	1953	3.6	14.8	1971	7.8	1949
N	0.80	2.03	1942	0.15	1942	0.86	1942	8.6	25.8	1964	12.0	1964
D	0.62	2.03	1955	0.23	1957+	0.54	1946	9.6	27.6	1955	8.6	1955
YR	15.91	9.54	JUN. 1944	T	AUG. 1970	3.84	APR. 1948	67.8	39.6	APR. 1955	26.7	APR. 1955

∅ For period October 1964 through the current year.

Means and extremes above are from existing and comparable exposures. Annual extremes have been exceeded at other sites in the locality as follows:
Highest temperature 106 in July 1954; lowest temperature -41 in December 1919; maximum precipitation in 24 hours 4.41 in July 1923; maximum monthly snowfall 42.6 in April 1927.

NORMALS, MEANS AND EXTREMES
(Table Revised 1975. Base Period for Climatological Normals: 1941-1970)

Average station pressure mb.: Elev. 3968 feet m.s.l. — YR 877.8

Temperatures °F

Month	Normal Daily maximum	Normal Daily minimum	Normal Monthly	Record highest	Year	Record lowest	Year
(a)				10		10	
J	33.5	8.5	21.0	70	1974	-30	1972
F	38.0	13.8	25.9	67	1957	-16	1972
M	43.1	18.9	31.0	75	1972	-23	1965
A	56.6	30.9	43.8	81	1968	3	1966
M	66.0	40.6	53.3	92	1969	17	1967
J	74.3	47.8	61.1	100	1970	31	1969
J	86.1	54.6	70.4	103	1966	35	1971
A	85.3	53.0	69.2	100	1971	34	1966
S	72.9	42.9	57.9	96	1971	15	1965
O	62.5	33.1	47.8	87	1971	-1	1971
N	46.0	20.8	33.4	77	1965	-9	1966
D	37.6	13.4	25.5	70	1973	-30	1972
YR	58.5	31.5	45.0	103	JUL 1966	-30	DEC 1972

Normal Degree days Base 65°F / Precipitation water equivalent

Month	Heating	Cooling	Normal (precip.)
J	1364	0	0.69
F	1095	0	0.77
M	1054	0	1.21
A	642	0	2.12
M	375	0	2.45
J	168	51	2.99
J	28	195	1.07
A	31	161	0.95
S	245	30	1.28
O	533	0	1.17
N	948	0	0.92
D	1225	0	0.69
YR	7708	446	16.16

Means and extremes above are from existing and comparable exposures. Annual extremes have been exceeded at other sites in the locality as follows:
Highest temperature 106 in July 1954; lowest temperature -41 in December 1919; maximum precipitation in 24 hours 4.41 in July 1923; maximum monthly snowfall 42.6 in April 1927.

REFERENCE NOTES APPLYING TO TABLES APPEAR ON THE PAGE FOLLOWING LAST TABLE.
(Caution: Letters and symbols may have different meanings in 1941-1970 tables than in earlier tables. See notes.)

Reference notes applying to Normals, Means, and Extremes tables for 1931–1960 base period.

(a) Length of record, years, based on January data. Other months may be for more or fewer years if there have been breaks in the record.

(b) Climatological standard normals (1931-1960).

* Less than one half.

+ Also on earlier dates, months, or years.

T Trace, an amount too small to measure.

Below zero temperatures are preceded by a minus sign. The prevailing direction for wind in the Normals, Means, and Extremes table is from records through 1963.

‡ ≥70° at Alaskan stations.

Unless otherwise indicated, dimensional units used in this bulletin are: temperature in degrees F.; precipitation, including snowfall, in inches; wind movement in miles per hour; and relative humidity in percent. Heating degree day totals are the sums of negative departures of average daily temperatures from 65° F. Cooling degree day totals are the sums of positive departures of average daily temperatures from 65° F. Sleet was included in snowfall totals beginning with July 1948. The term "ice pellets" includes solid grains of ice (sleet) and particles consisting of snow pellets encased in a thin layer of ice. Heavy fog reduces visibility to 1/4 mile or less.

Sky cover is expressed in a range of 0 for no clouds or obscuring phenomena to 10 for complete sky cover. The number of clear days is based on average cloudiness 0-3, partly cloudy days 4-7, and cloudy days 8-10 tenths.

Solar radiation data are the averages of direct and diffuse radiation on a horizontal surface. The langley denotes one gram calorie per square centimeter.

& Figures instead of letters in a direction column indicate direction in tens of degrees from true North; i.e., 09 - East, 18 - South, 27 - West, 36 - North, and 00 - Calm. Resultant wind is the vector sum of wind directions and speeds divided by the number of observations. If figures appear in the direction column under "Fastest mile" the corresponding speeds are fastest observed 1-minute values.

To 8 compass points only.

** The National Weather Service considers the accuracy of solar radiation data questionable; therefore, publication is suspended pending determination of corrected values.

Reference notes applying to Normals, Means, and Extremes tables for 1941–1970 base period.

(a) Length of record, years, through the current year unless otherwise noted, based on January data.

(b) 70° and above at Alaskan stations.

* Less than one half.

T Trace.

NORMALS - Based on record for the 1941-1970 period.

DATE OF AN EXTREME - The most recent in cases of multiple occurrence.

PREVAILING WIND DIRECTION - Record through 1963.

WIND DIRECTION - Numerals indicate tens of degrees clockwise from true north. 00 indicates calm.

FASTEST MILE WIND - Speed is fastest observed 1-minute value when the direction is in tens of degrees.

Mean Maximum Temperature (°F.), January

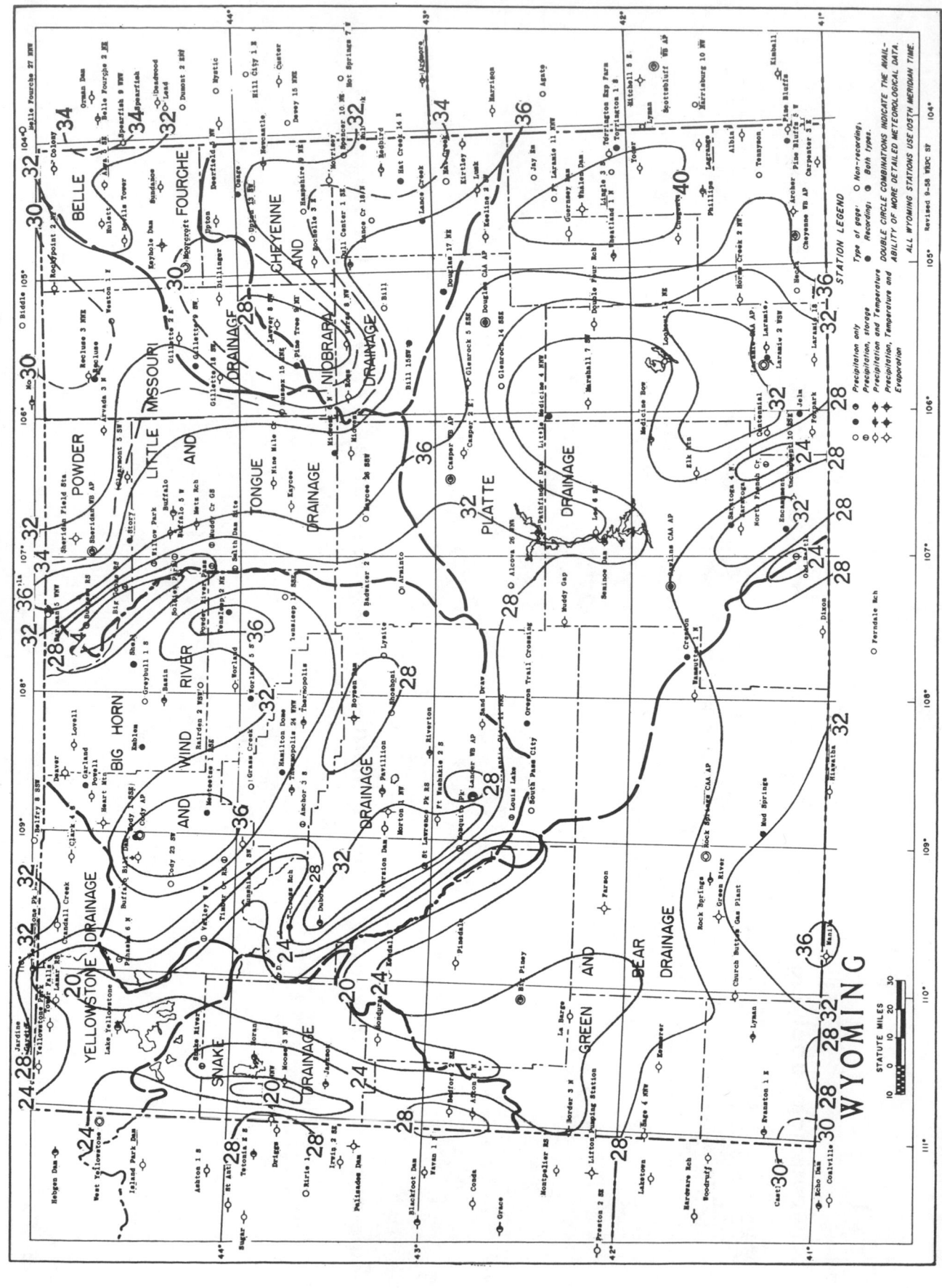

WYOMING

Based on period 1931-52

Isolines are drawn through points of approximately equal value. Caution should be used in interpolating on these maps, particularly in mountainous areas.

Mean Minimum Temperature (°F.), January

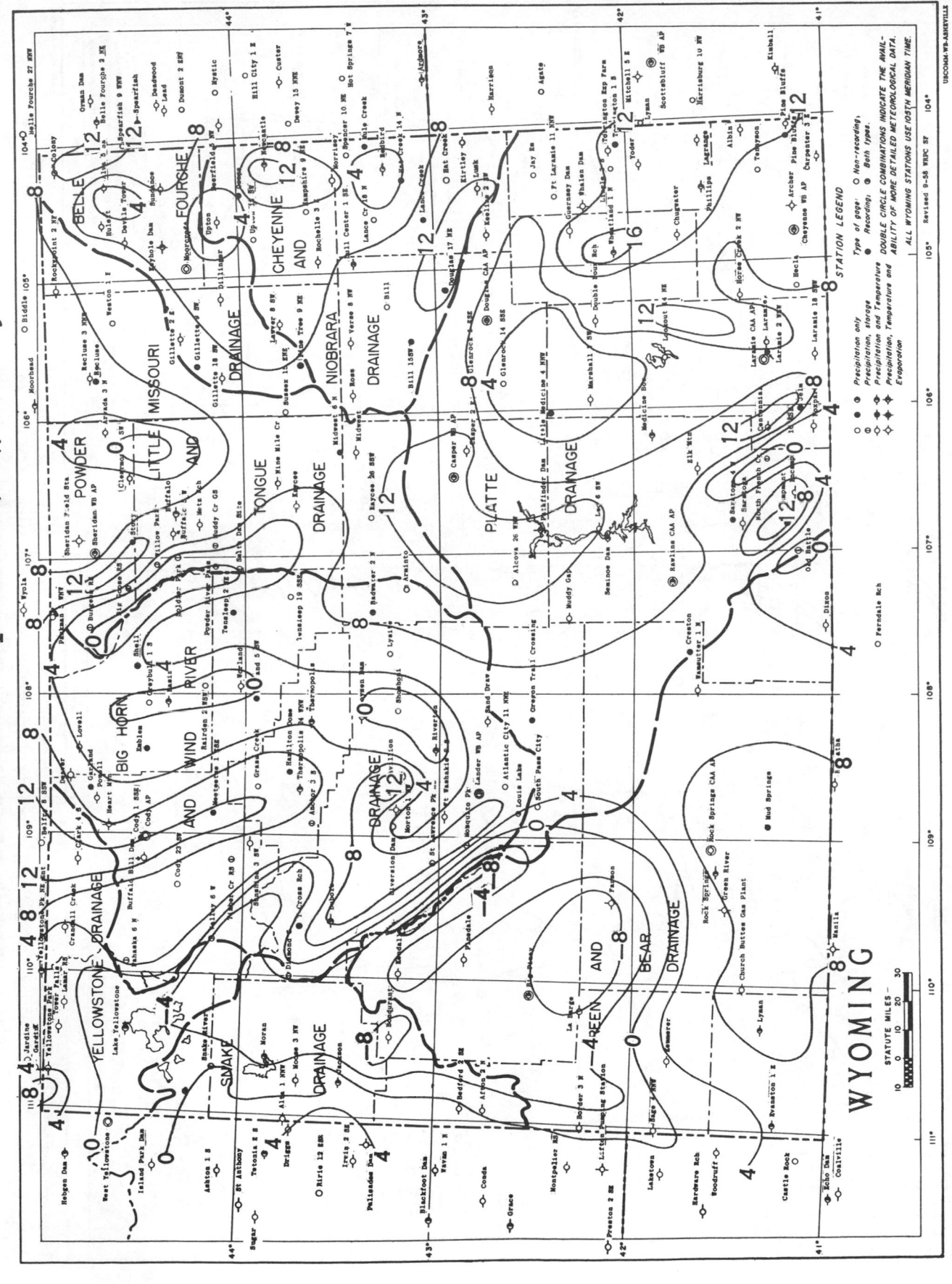

WYOMING

Based on period 1931-52

Isolines are drawn through points of approximately equal value. Caution should be used in interpolating on these maps, particularly in mountainous areas.

STATION LEGEND

Type of gage:
○ Precipitation only
◐ Precipitation, storage
● Precipitation and Temperature
◑ Precipitation, Temperature and Evaporation

○ Non-recording
● Recording
◐ Both types.

DOUBLE CIRCLE COMBINATIONS INDICATE THE AVAIL-ABILITY OF MORE DETAILED METEOROLOGICAL DATA.

ALL WYOMING STATIONS USE 105TH MERIDIAN TIME.

USCOMM-WB-ASHEVILLE   Revised 9-58 WBrc SV

Mean Maximum Temperature (°F.), July

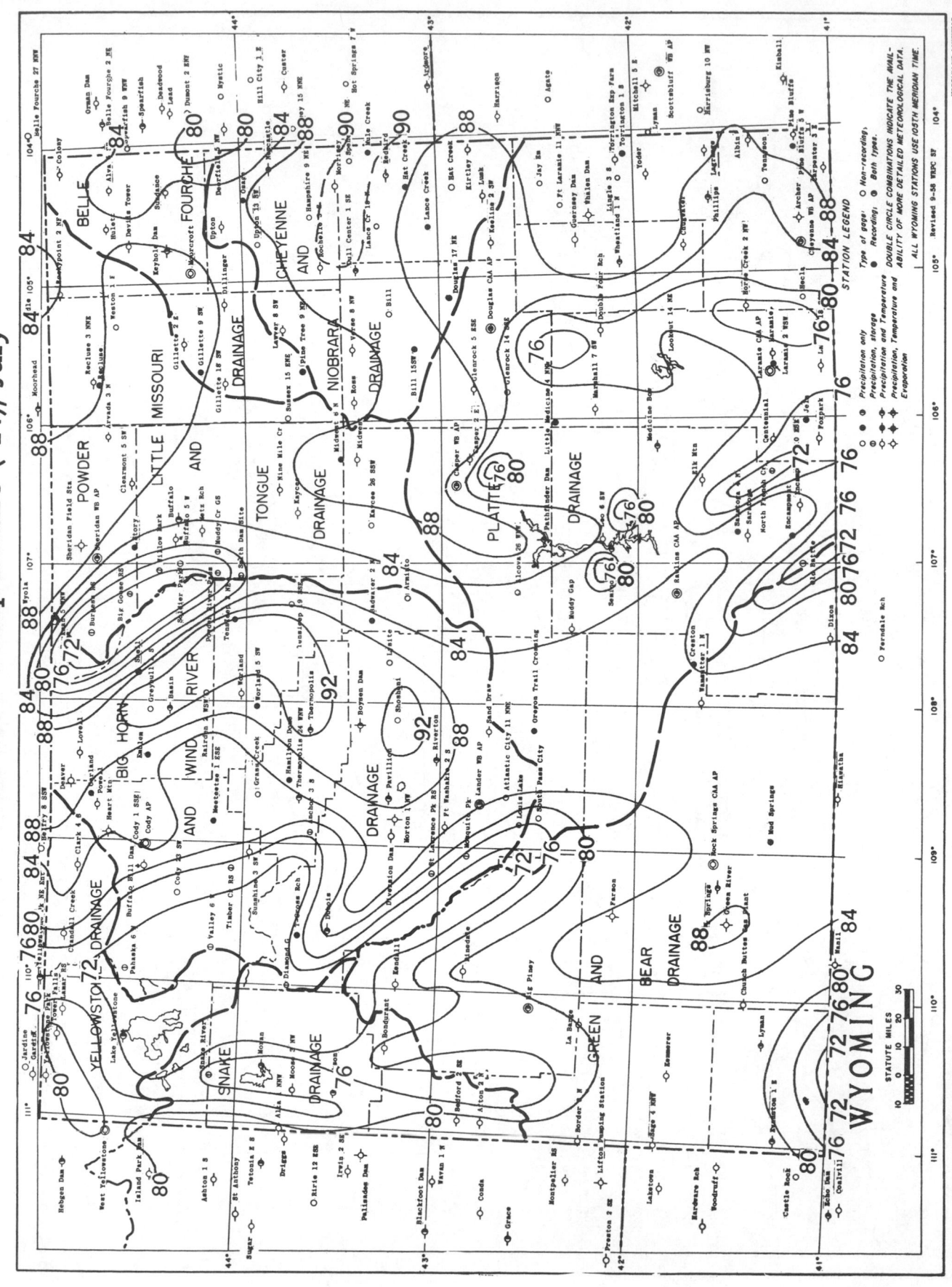

Based on period 1931-52

Isolines are drawn through points of approximately equal value. Caution should be used in interpolating on these maps, particularly in mountainous areas.

Mean Minimum Temperature (°F.), July

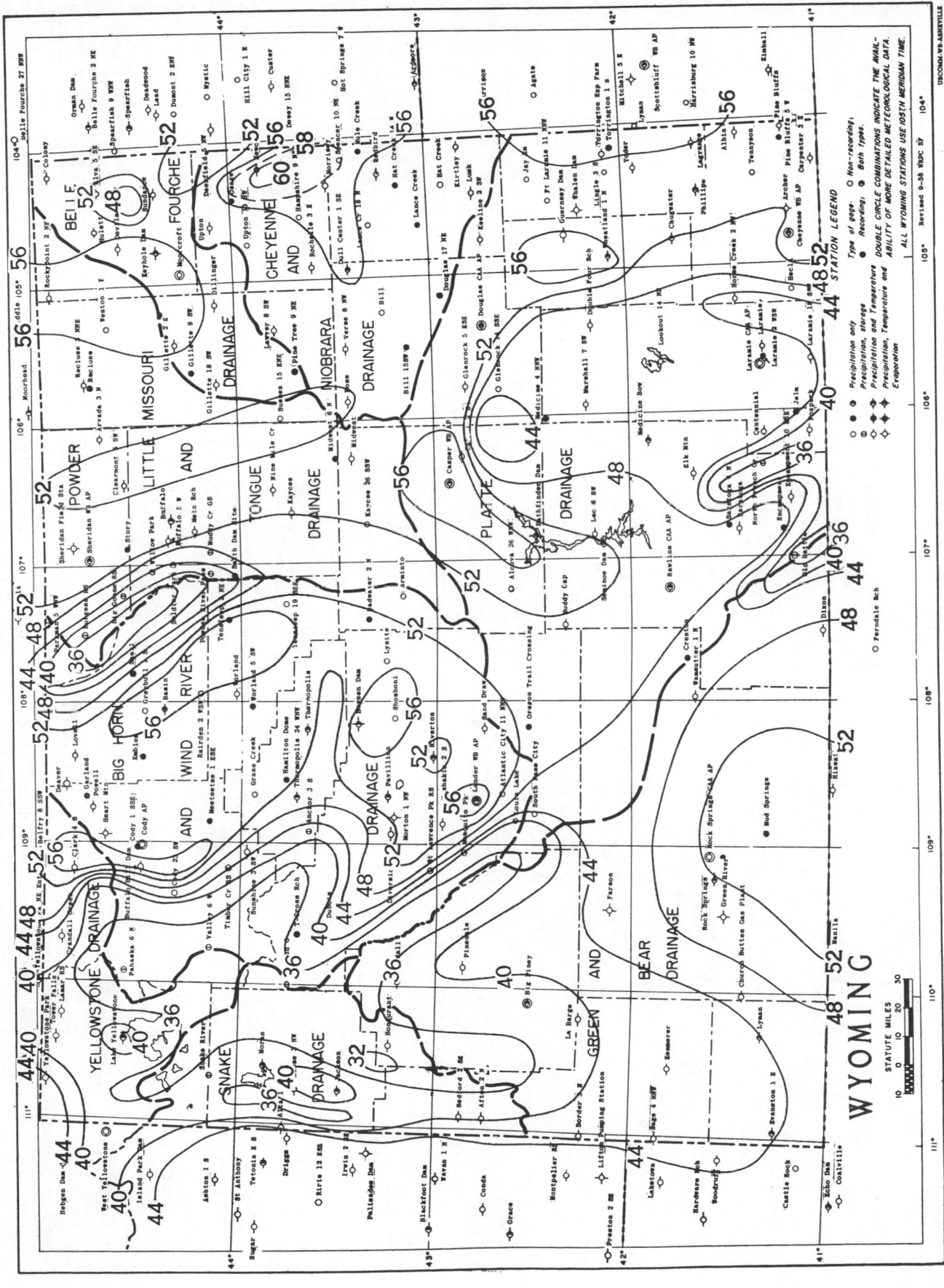

WYOMING

Based on period 1931-52

Isolines are drawn through points of approximately equal value. Caution should be used in interpolating on these maps, particularly in mountainous areas.

Mean Annual Precipitation, Inches

WYOMING

Based on period 1931-55

Isolines are drawn through points of approximately equal value. Caution should be used in interpolating on these maps, particularly in mountainous areas.

CONFIDENCE LIMITS

In the absence of trend or record changes, the chances are 9 out of 10 that the true mean will lie in the interval formed by adding and subtracting the values in the following table from the means for any station in the State. Because of the wider variation in mean precipitation, the corresponding monthly means and annual mean must be substituted for "p" in the precipitation table below to obtain mean precipitation confidence limits.

2.6	.19√p	2.3	.17√p	1.4	.19√p	1.4	.31√p	1.1	.31√p	1.1	.37√p	.7	.28√p	.6	.30√p	1.0	.37√p	1.4	.33√p	1.7	.25√p	1.9	.19√p	.6	.28√p

COMPARATIVE DATA

Data in the following table are the mean temperature and average precipitation for Yellowstone Park, Wyoming for the period 1906 - 1930 and are included in this publication for comparative purposes :

17.6	1.36	21.0	1.01	27.1	1.38	36.5	1.40	44.7	1.98	53.5	1.63	61.0	1.34	59.0	1.28	49.9	1.38	39.8	1.61	28.8	1.22	19.2	1.22	38.2	16.81

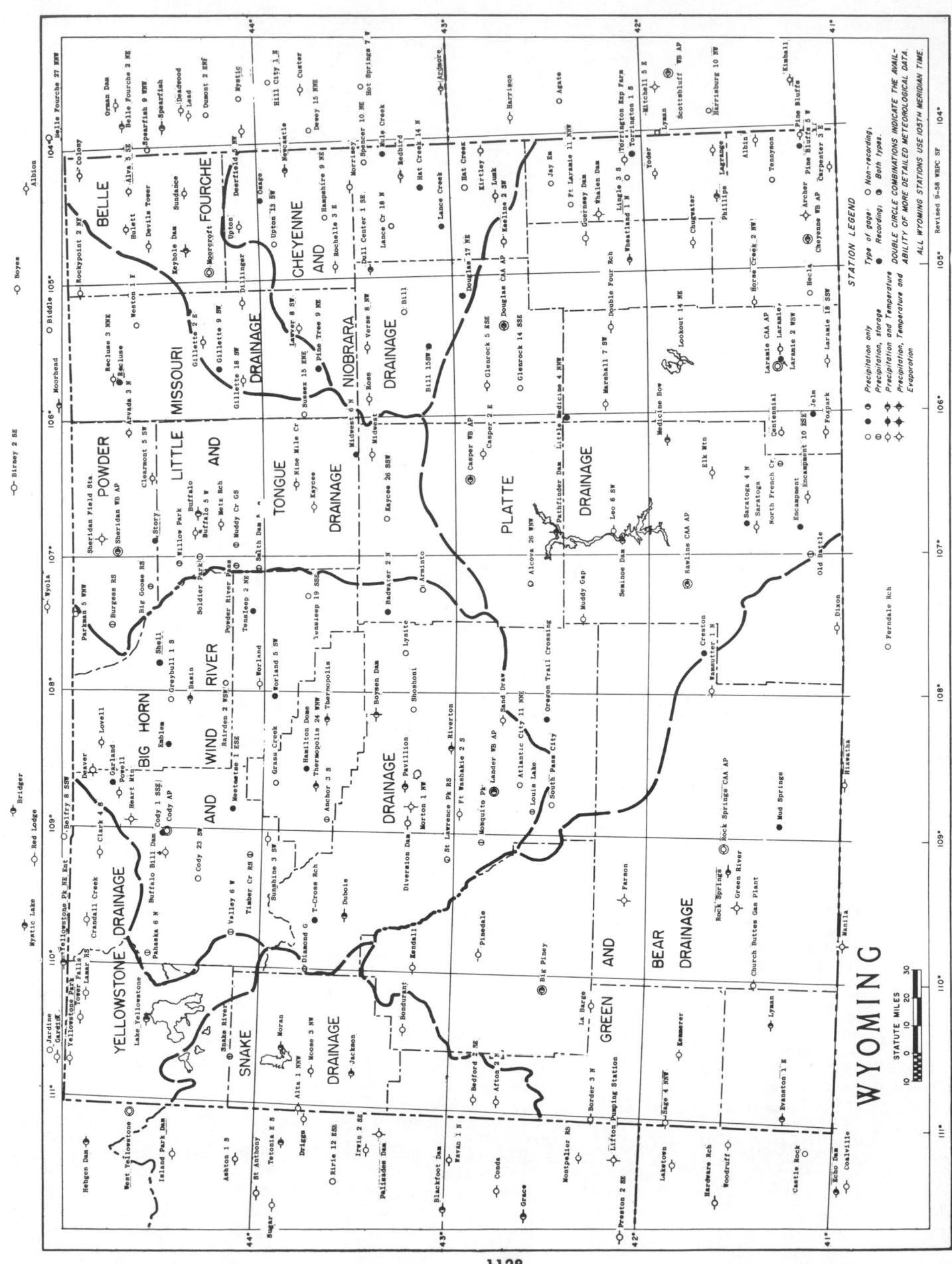

WYOMING

STATUTE MILES

10   0   10   20   30

STATION LEGEND

	Type of gage	○ Non-recording
●	Precipitation only	
◐	Precipitation, storage	
⊙	Precipitation and Temperature	⊕ Recording;
⊕	Precipitation, Temperature and Evaporation	◑ Both types.

DOUBLE CIRCLE COMBINATIONS INDICATE THE AVAIL-
ABILITY OF MORE DETAILED METEOROLOGICAL DATA.

ALL WYOMING STATIONS USE 105TH MERIDIAN TIME.

Revised 9-58 WBPC SF   USCOMM-WB-ASHEVILLE

1128

CLIMATES OF THE STATES

Puerto Rico and U. S. Virgin Islands

(Normals, Means and Extremes tables revised 1953, 1970 and 1975. Basic report revised June 1970.)

Robert J. Calvesbert, ESSA Commonwealth Climatologist

INTRODUCTION

Puerto Rico and the U. S. Virgin Islands are tropical, hilly islands which lie directly in the path of the easterly trade winds throughout the year.

This single introductory sentence opens the door to understanding a great deal about the climate of these islands. Since all are islands, daily temperature ranges are relatively small, at least close to the coasts due to the tempering effect of the nearby waters. The larger islands are influenced by land and sea breezes. Since they are hilly islands, there are rather sizable variations in rainfall and temperature over relatively short distances. The rugged aspect of the terrain also causes wide local variations in wind speed and direction due to sheltering and channeling effects. Location in the tropics induces warm temperatures and raises some question as to their location with respect to tropical storms and hurricanes. Naturally, we would expect the tropical warmth to be moderated considerably by the trade winds.

PUERTO RICO

Location: Puerto Rico is the easternmost and smallest of the Greater Antilles and is shaped somewhat like a brick with its southeast corner broken off. It lies between 18° 31'N and 17° 55'N latitude and 65° 37'W and 67° 17'W longitude. From its easternmost to its westernmost tip it is about 109 miles, while from its northernmost to its southernmost tip it is about 40 miles. The small area is a little more than 3,400 square miles which makes it slightly larger than the States of Delaware and Rhode Island combined. In addition to the main island, Puerto Rico also embraces three secondary and a number of smaller islands.

Mona Island lies nearly 40 miles to the west of Puerto Rico proper, covers an area of about 21 square miles and is, for the most part, uninhabited. Vieques Island, the largest of the three secondary islands, lies a short distance to the east of the main island and covers an area of slightly more than 50 square miles. This is the most densely populated of the three with about 15,000 inhabitants. It is primarily an agricultural community, although efforts are being directed toward development as a tourist resort in coming years. A large portion of Vieques Island at the east and west ends is used as a U. S. military reservation for training exercises. Culebra Island lies to the north of the eastern tip of Vieques Island and covers an area of about 10 square miles. It, too, is partially a military reservation and has a total population of less than 1,000. Vieques and Culebra Islands are the westernmost of the Lesser Antilles which extend from Puerto Rico in a sweeping southeasterly curve to the coast of South America.

Topography: Viewed from the horizon in any direction, Puerto Rico presents a rather rugged profile with peaks to more than 3,000 feet in the east and to more than 4,000 feet in the west. El Toro, rising to 3,535 feet, is the highest in the east and Cerro de Punta reaches an elevation of 4,389 feet in the west. There is a main divide running roughly east-west across the Island. In the eastern third it is closer to the northern coast, and in the western two-thirds it lies closer to the southern coast. The divide is over 3,000 feet above sea level for the most part, with passes where the north-south roads cross the Island at an elevation of about 2,000 feet. With this configuration, there are coastal plains varying from 8 to 12 miles along the northern coast and 2 to 8 miles along the southern coast. There is a gradual rise to the peaks in the north, while in the south the descent to the sea is rather abrupt. In the western end the mountains spread out fanwise filling the width of the Island. In the east the divide curves to the northeast corner where there is a detached group of peaks known as the Luquillos.

Mona Island has a plateau-like surface which descends abruptly to the sea in cliffs 150 to 200 feet high. A small area of about 1,000 acres in the southwestern corner is sandy beach.

Vieques Island is somewhat spindle-shaped, 21 by 6 miles, with its longer axis running east-northeast to west-southwest. It is mostly level, although somewhat hilly in the central sections with elevations rising to 981 feet at Mt. Pirata in the southwest.

Culebra Island with a very irregular shoreline is slightly higher than Mona Island. Mount Resaca at 646 feet in the central part of the Island is the highest point. The entire island is quite hilly, and very little level land is to be found.

U. S. VIRGIN ISLANDS

Location: The U. S. Virgin Islands are composed of three major islands together with a number of smaller islands and cays totaling about 50. The three of primary importance are St. Thomas, where the capital is located; St. Croix, the largest; and St. John, the smallest.

These islands follow Vieques Island and Culebra Island in the path of the Lesser Antilles toward South America. St. Thomas lies some 38 miles east of Puerto Rico and about 1,500 miles southeast of New York. St. John lies a few miles east of St. Thomas and St. Croix is located about 40 miles south of St. Thomas and St. John.

With an area of about 28 square miles, St. Thomas is the second largest of the U. S. Virgin Islands. This island lies between latitudes 18° 23'N and 18° 18'N and longitudes 65° 03'W and 64° 50'W. It is about 5 miles from its northernmost to its southernmost points and a little more than 12 miles from its eastern to western extremities.

The smallest of the three principal islands is St. John, with an area of only about 20 square miles. It is also the least populated. St. John lies between latitudes 18° 23' N and 18° 18'N, and longitudes 64° 48'W and 64° 40'W. This island extends about 5 miles from its northern to southern tips and about 8 miles from its easternmost to westernmost points.

Somewhat apart from the others, the largest of the three islands is St. Croix which has an area of 84 square miles. It lies between latitudes 17° 47'N and 17° 41'N, and longitudes 64° 54'W and 64° 34'W. The Island extends some 19 miles from east to west and 6 miles from north to south.

Topography: St. Thomas has an extremely irregular coastline and is very hilly with practically no flatland. The highest hills are generally found near the center of the Island, with Crown Mountain at 1,550 feet the highest point. The Island is relatively small and many of the peaks rise above 1,000 feet. This results in rather steep slopes over all the island, so that rainfall runoff is quite rapid and there are no permanent streams or rivers.

Like St. Thomas, St. John has an extremely irregular shoreline and a very hilly topography. It has a number of peaks over 1,000 feet, topped by Bordeaux Mountain at 1,297 feet in the eastern portion of the island. Slopes are quite steep over all of the island, and there are very few areas of flatland. There are no permanent rivers or creeks.

St. Croix is the largest of the three U. S. Virgin Islands. The topography is somewhat different from the other two with a broad expanse of low, relatively flatland running along the southern two-thirds of the Island. A range of hills, ranging in elevation from about 500 feet to more than 1,000 feet, topped by Mount Eagle at 1,165 feet, runs along the northern coast. In the eastern end of St. Croix is found another group of slightly lower hills with a maximum elevation of about 860 feet. The relatively small area covered by hills on St. Croix results in rather steep slopes down to the Caribbean in the north and to the level areas to the south.

PUERTO RICO

With nearly 2.8 million inhabitants Puerto Rico has a population density slightly less than that of Rhode Island. With so many people to feed and situated so far from the mainland, it is quite obvious that agriculture is of major importance to the overall economy.

Agricultural products grown include sugarcane (the principal crop), coffee, tobacco, pineapple, bananas, plantains, and many subsistence crops. Chickens are raised for food and eggs, and cattle for milk and meat.

Rainfall is of prime importance to all growing crops and a favorable wet-dry season relationship is vital for sugarcane. In areas where rainfall is deficient, irrigation is necessary.

The Island is earnestly striving to increase the standard of living, provide more jobs and, in general, to make Puerto Rico a more desirable place to live by stimulating the growth of the industrial community of the Island. As a result, industry has grown by leaps and bounds over the past years.

Climate is important to industry with its need for water for cooling and other purposes; its need to exhaust smoke and gasses, sometimes noxious, into the air; its sensitivity to moisture and salt content of the air both in its manufacturing processes and in the deterioration of plant and machinery.

Over the years tourism is playing an increasingly greater role in the overall economic picture with great strides forward being taken each year. The number of tourists has increased markedly year after year, and many cruise ships are making San Juan a port of call during the season.

As in any tourist resort, a favorable climate is one of the greatest assets. Abundant sunshine, comfortable swimming water, refreshing breezes, and a reliable small temperature range are desirable. Careful consideration of anticipated climatic conditions may well be the determining factor in whether a vacation is pleasant or not.

U. S. VIRGIN ISLANDS

Agriculture is not as important in the U. S. Virgin Islands as it is in Puerto Rico. St. Croix is the only one of the U. S. Virgin Islands with any sizable expanse of flatland suitable for farming. Here sugarcane, which was the principal crop, has been abandoned. Subsistence crops are now a minor effort. Some cattle are raised for milk and meat.

In St. Croix, industrial growth has become a significant factor in the island's economy. With the downgrading of agriculture, industrial complexes are being expanded to include the petrochemical industry and refinement of aluminum. Light industrial plants and the manufacture of rum are the other industrial activities in St. Croix and St. Thomas. St. John has no industrial development and remains primarily a National Park.

Tourism is the biggest factor in the Virgin Islands' economy. It has, over the past years, undergone a vast increase in the numbers of cruise ships, especially at St. Thomas and St. Croix. Hotel facilities have been increased on both islands.

One of the principal causes of concern in the U. S. Virgin Islands is the short supply of water.

Rainfall, while above 40 inches annually over most of the area, is insufficient. This is due partially to a high evaporation rate and the rapid runoff from the steep slopes on St. Thomas and St. John and, to a certain extent, on St. Croix.

In an effort to utilize available water efficiently, most homes and business establishments catch rainwater on the roofs and pipe it to cisterns. The runway at the airport at St. Thomas is also used as a catchment area. On St. Thomas and St. John it is common to see the entire side of a hill cemented to act as a catchment area. Generally, during the drier portion of the year, it is necessary to carry water by barge from Puerto Rico. Installation of a sea-water distillation unit on St. Thomas and St. Croix has helped alleviate the water shortage but it remains a significant factor in the development of the island's economy.

PRECIPITATION

PUERTO RICO

Rainfall in Puerto Rico varies markedly from place to place over relatively short distances. Measurements of rainfall are made daily at approximately 100 Weather Bureau cooperative climatological stations.

The majority of all Puerto Rico's rainfall is orographic in nature. Moisture laden air from the ocean is carried by the trade winds inland and forced to ascend over the mountains and is cooled, thus causing condensation in the form of rainfall. The majority of these orographic showers are rather brief. Abundant sunshine prevails throughout the year--even during the so called rainy seasons.

There are two rainfall-producing mechanisms in Puerto Rico: easterly waves and cold fronts. During the period from May through November, Puerto Rico experiences easterly waves. These are migratory wave-like disturbances moving from east to west within the basic easterly air current, generally more slowly than this current in which they are imbedded. An easterly wave of slight intensity may produce little more than greater-than-normal cloudiness. However, an intense easterly wave, especially if it is moving slowly, may bring one or several cloudy, rainy days with it which can produce rains sufficient to cause flooding.

Tropical storms and hurricanes occasionally develop in the easterly waves and may cause torrential rains on Puerto Rico. Fortunately visitations by these storms are infrequent, although each year there are generally one or more scares where "every eye" is focused on the sky and "every ear" is glued to the radio for news of the storm.

The other major rainfall-producing situation occurs during the winter months, generally from about November to April. Occasionally during this period, the trailing edge of a cold front which has swept across the continental United States penetrates far enough south to have a definite effect upon Puerto Rico's rainfall. The degree of effect depends upon the intensity and rate of progression of the cold front. A weak or dissipating cold front may cause only cloudier-than-normal skies while a strong, active slowly moving front is capable of bringing heavy and continuing rainfall which may last for several days. Occasionally the front moves over the entire Island, bringing rainfall to most areas. In this case the topographic effects combine with the frontal mechanism. At other times the front may only penetrate a short distance into Puerto Rico, generally over the northwestern corner. In those cases that section of the Island is subjected to heavy rains, while the remainder of the Island enjoys pleasant weather.

The marked difference in rainfall over relatively short distances is illustrated by a glimpse at the annual rainfall map. In the area of El Yunque an annual average for a 10-year period of 183.51 inches is noted. Along the western portion of the southern coast there is an area of less than 40 inches, with the lowest being 34.62 inches at Santa Rita. Thus, over a distance of only about 60 miles there is a difference of nearly 150 inches annually. At the La Mina El Yunque station the total for one year has reached 253.79 inches. In contrast, stations along the south coast have experienced occasional years when less than 15 inches were received.

The geographical distribution of the rainfall over the Island shows four areas of heavy rainfall. The heaviest of these is centered over El Yunque in the Luquillo Mountains in the northeastern section. While a little more than 180 inches, on the average, has been recorded part way down the slope, it may be well be that at the peak an estimate of a 200-inch mean annual total would not be unrealistic. The second area of heavy rainfall lies to the southwest of El Yunque in the San Lorenzo area. Here the mean annual rainfall exceeds 120 inches. Farther west, over the highest peaks is another area of heavy rainfall. Guineo Reservoir and Toro Negro Hydroelectric Plant, with mean annual totals of 113.41 and 105.95 inches, respectively, are in the center of this region. Farther west around Maricao, another area of copious rainfall above 100 inches exists. The lowest annual averages are along the southern coast from Aguirre westward.

Rainfall patterns on the outlying islands of Culebra, Vieques, and Mona are not well defined due to a lack of sufficient observations, but existing data indicate about 40 to 50 inches of rain annually.

The effect of topography and winds upon the rainfall is well illustrated in the isohyets (lines of equal rainfall). In general, rainfall along the northern, or windward coast, is greater than along the southern, or leeward coast. As indicated earlier, in the north the slope of the hills to the peaks is a gradual one, while the drop to the ocean on the south is quite precipitous. This is reflected in the isohyets where there is a rather gradual increase in the amount of rainfall on the northern slopes up to the divide. In the south, where the topographic slope is much greater, rainfall shows a sharp decrease from the ridge to the coast.

The distribution of rainfall over the year does not show an absolute wet season - dry season relationship, but only a relatively dry season and a relatively wet season. The length of the dry season varies somewhat with location of the Island. In the northern portion of Puerto Rico the dry season is generally a little shorter than it is in the southern section of the Island, with a narrow intermediate belt in between. The difference in the length of this relatively dry season is produced by variations in the beginning of the dry season. This is because the onset, in May, of the relatively rainy season is the same over the entire Island. In the north the dry season normally begins in February and ends in April, while in the south the dry season sets in during December.

In most areas there is a transitional period of about 1 month between these seasons. Naturally, there are isolated exceptions to this general rule--the dry season is not <u>always</u> from December to April nor is the wet season <u>always</u> from May to November. Some of the heaviest rainfalls have occurred during the so-called relatively dry season. In practically every instance, however, on the average the driest month is either February or March. For the most part, May may be considered to be the rainiest month in the northern half of the Island, while September or October is the rainiest month in the south.

The number of days with measurable rain follows the isohyetal (rainfall) pattern, with the largest number in areas of greatest rainfall. Once again there is the pattern of an increase from north to south to the area of heaviest rainfall, and then a rather sharp decrease to the south coast where the lowest average rainfall is noted. The average number of rainy days varies from just under 300 in the El Yunque area to less than 100 along the southern coast, and below 50 at the driest locations.

The matter of anticipated rainfall intensity is of the utmost importance to a great many people, including the engineer, farmer, and of course, all those with an interest in flood situations. About once in 100 years each section of Puerto Rico may expect a rainfall of at least 8.50 inches during a 24-hour period. The area of the highest peaks in the southwest may receive as much as 18 inches in a 24-hour period. Twelve-hour rainfall in this area may reach as high as 15 inches, while the 1-hour rainfall may be as much as 6.50 inches.

U. S. VIRGIN ISLANDS

Rainfall in the U. S. Virgin Islands is of the same nature as that in Puerto Rico, falling most frequently in the form of brief showers. The rainfall-producing mechanisms are essentially the same as in Puerto Rico except in the matter of degree.

Orographic lifting of the moisture laden air

over the hilly terrain of these islands is the most frequent cause of rainfall. However, due to the smaller elevations and smaller size of the islands, there is a less marked variation in annual amounts. The higher mean annual totals are between 50 and 60 inches at the higher elevations, and the variation between the greatest and least average value is not as marked as it is in Puerto Rico. Clouds formed by forced ascent of the wind over small and narrow islands, as is the case for St. Thomas and St. Croix, lean to the leeward, so that most of the rain from them falls in the ocean to the lee of the island. Easterly wave passages are important contributors to the rainfall of the Virgin Islands during the months from May through November. Like Puerto Rico, the U. S. Virgin Islands lie in the path of the tropical storms and hurricanes which form over the ocean to the east of the Lesser Antilles. As in Puerto Rico, they are relatively infrequent. While cold frontal passages affect the rainfall regime of the Virgin Islands, the frequency of fronts is less and their intensity is more likely to be diminished and less effective than in Puerto Rico.

Annual rainfall values indicate differences in rainfall from location to location with higher elevations generally receiving greater amounts. On St. Thomas and St. John, on the basis of the limited data available, annual averages of between 40 and 60 inches appear reasonable. On St. Croix there is a more noticeable variation from place to place. This Island has the greatest annual rainfall, in excess of 50 inches in the northwestern corner. There are some indications that stations in a small area along the central portion of the southern coast of St. Croix receive about 40 to 45 inches. A narrow finger of between 25 and 35 inches extends northeast to southwest over the flatlands south of the hills in the western portion of the Island. Annual rainfall averages less than 30 inches in the eastern end of St. Croix, possibly as low as 20 inches.

As in Puerto Rico, there is no sharply defined wet-dry season relationship. Records available for the three islands indicate a relatively wet-relatively dry season distribution similar to that found in the southern portion of Puerto Rico. The relatively dry period extends from about December through June. Occasionally, quite heavy rainfall occurs during the so-called drier months. The driest month on St. Thomas and St. John usually is February or March and the wettest month September or October, as in the southern sections of Puerto Rico. On St. Croix, the month with the heaviest rainfall, on the average, ranges from September through November.

The number of days with measurable rainfall over the Virgin Islands, based on a few known-to-be reliable stations, range from a little less than 200 days annually at the higher rainfall stations to less than 100 days annually at the stations with lowest rainfall.

TEMPERATURE

PUERTO RICO

Mean temperatures in Puerto Rico have a very small range between the warmest and coldest months. The smallest range is generally found in areas near the coast, with only 4.5° at Patillas. In the interior a slightly larger range is observed, with 7.3° at Humacao the largest. The normal range in downtown San Juan is only 5.7° between the warmest month, August - 80.8°, and the coolest months, January and February - 75.1°.

These small annual temperature ranges are due in part to the fact that Puerto Rico is an island surrounded by waters whose temperature changes but little from the warmest to coolest season (82.5° in September to 77.9° in February and March). It is also due to its location, only about 1,100 miles north of the equator, and the resultant small differences in the energy received from the sun from season to season. The difference between the length of the longest day (13 hours 13 minutes) and the shortest days (11 hours 2 minutes) is only a little over 2 hours. Also there is not much change during the year in the height of the sun above the southern horizon. The coolest average temperatures are found in the area of the higher peaks, while the warmer average temperatures are observed along the coastal regions. The lowest mean annual temperature is 67.0° at Guineo Reservoir, while the highest is 81.0° at Guayama on the southern coast. Mean annual maximum temperatures range from 89.2° at Dos Bocas to 74.6° at Guineo Reservoir. There are two areas of high average maximum temperature. One is along the southern or leeward coast and the other is in the Dos Bocas area. Mean annual minimum temperatures range from 74.4° at the Roosevelt Roads Naval Station at the eastern end of the Island to 59.3° at Guineo Reservoir. Once again the higher values are found along the coastal areas.

The daily range of temperature or the difference between the daytime maximum and nighttime minimum, varies with location on the island. In the areas along the northern and eastern coasts, the mean daily range is usually between 10° and 15°, with increasing values inland away from the tempering effect of the oceans. The mean daily range is between 15° and 20° in the southeast and in excess of 20° in the west and southwest. Maximum values of a little over 25° are found in the Caguas Valley and at Utuado, where the highest annual mean daily range, 26.3° is found. There is no significant variation in this pattern from month to month. San Juan has the lowest annual mean daily range, 9.6°.

Afternoon temperatures in the 90's are not unusual in some sections of Puerto Rico. The number of days with temperatures of 90° or higher during a year may reach 200 or more annually along the southern coast and more than 100 at some stations in lower portions of the interior and along the western coast. Along the northern and eastern coasts, however, extremely warm temperatures are rather infrequent due to the tempering effect of the waters of the ocean. At San Juan City the annual average number of days with 90° or more is only 9 and other stations close to the water record less than 100 annually. Due to the effect of altitude there are several stations where the annual number of 90° days is usually less than 1 every 2 years, such as Garzas Dam, Barranquitas, Cidra, and Aibonito. At Guineo Reservoir the maximum temperature has never gone above 85°. There are some locations where the record maximum temperature is 100° or more. The highest temperature ever recorded in Puerto Rico was 103° at San Lorenzo in August 1906. At San Juan City the record high of 96° occurred in March 1958. At this time the Island was experiencing one of the worst droughts in history, and the soil over the island was practically without moisture. An interruption in the normal easterly windflow brought the extremely warm, dry air from the south over San Juan. Practically all of San Juan's higher temperatures occur during periods of southerly windflow, when air is warmed during passage overland.

Over practically the entire Island the lowest mean maximum temperatures occur in January, although there are a few areas where they are found in February. The warmest daytime temperatures are recorded in the northern section of the island and the Cayey-Patillas-Guayama area of the southeast in September, in the southwest in July, and elsewhere in August. At San Juan it is interesting to note that July daily maximum temperatures do not generally reach the extremes that they do during June and August.

Freezing temperatures are unknown in Puerto Rico, even at the higher elevations. In fact, the coolest temperature ever recorded is 40° at Aibonito in March 1911. Most of the hilly areas have experienced temperatures in the 40's, while in the coastal areas and foothill sections the coolest temperatures on record are generally between 50° and 60°. In downtown San Juan, which is literally surrounded by water, the coolest temperature recorded is 62°. Almost without exception the lowest mean minimum temperatures occur during February. Over most of Puerto Rico the highest mean minimum temperatures have been recorded during August with a few scattered areas reporting them in June, July, or even in September.

U. S. VIRGIN ISLANDS

As in Puerto Rico, one of the most striking features of the temperature regime in the U. S. Virgin Islands is the relatively small variation in temperature from the coolest to the warmest months, ranging from about 5° to 7°.

Due to the small size of the islands and the location of all stations within a few miles of the water, the mean daily range is quite small. It varies from 9.1° at Charlotte Amalie to 15.1° at Wintberg. For these same reasons extremes of temperature are not as great as they are in Puerto Rico, and relatively few days have temperatures of 90° or above. Since the extent of land areas is so small, the overland air passage is quite short and there is not sufficient time for extreme heating to take place, regardless of the wind direction. On St. Croix, Annas Hope has had a temperature as high as 99°.

During the warmest months, maximum temperatures average about 87° to 89°, with nighttime temperatures falling to about 74° to 78°, and a little lower at the higher elevations. In the winter, daily maximum temperatures are generally in the low 80's and nighttime minima in the high 60's or low 70's.

The highest mean maximum temperatures are found in August, while the lowest mean maxima fall either in January or February. The lowest mean minimum temperatures are observed in January and February, and the highest mean minimum temperatures are generally in July or August.

WIND

One of the outstanding features of the wind in Puerto Rico and the U. S. Virgin Islands is the steadiness of the trade winds. They blow almost without exception from an easterly direction, i.e., between north-northeast and south-southeast. These islands are under the influence of three wind regimes, with the trade winds primary and the others superimposed. Being surrounded by water, the land and sea breeze effect is important in most coastal areas, but it is not as noticeable at places farther in the interior. The trade winds, modified somewhat by the land and sea breeze, pass inland to a formidable barrier of hills where they are lifted over the top or pushed aside through narrow passes and valleys until their basic characteristics have become quite confused. Night winds are lighter than the daytime winds. About daybreak the wind speed begins to pick up, reaching a maximum late in the morning or early afternoon. The return to the lighter nighttime winds begins later in the afternoon, usually about 4 p.m. The afternoon decrease appears to be more leisurely than the morning increase.

The highest mean maximum wind speeds occur during July, with average peak speeds reaching more than 18 m.p.h. at downtown San Juan and at Ramey Air Force Base in northwestern Puerto Rico. Alexander Hamilton Field, St. Croix, V. I., follows with maximum values slightly above 16 m.p.h. while the other stations have mean speeds several miles per hour slower. At several of the stations a secondary maximum appears in March or April, while the lightest winds are during the autumn in either October or November.

A measure of the degree of variability of daily mean maximum wind speed from strongest to weakest months is illustrated in the following table, followed by a similar table for the daily mean minimum wind speed.

MEAN MAXIMUM WIND SPEED (m.p.h.) AND LOCAL TIME

Station and Period of Record	Strongest Month		Weakest Month	
WBO San Juan, P. R. (1931-42)	18.4	2 PM	13.8	2 PM
WBFO San Juan, P. R. (1957-60)	14.8	2-3 PM	11.5	2-3 PM
Ramey AFB, P. R. (1940-55)	18.6	1 PM	11.7	1-2 PM
Santa Isabel AP, P. R. (1940-45 and 1951-54)	13.9	1 PM	10.0	2 PM
Alex. Hamilton Fld., St. Croix, V. I. (1954-58)	16.1	11 AM	11.0	Noon-1 PM
Truman Field, St. Thomas, V. I. (1953-58)	14.9	Noon	9.9	11 AM

MEAN MINIMUM WIND SPEED (m.p.h.) AND LOCAL TIME

Station and Period of Record	Strongest Month		Weakest Month	
WBO San Juan, P. R. (1931-42)	9.3	5 AM	5.5	8 AM
WBFO San Juan, P. R. (1957-60)	5.3	7 AM	3.5	6 AM
Ramey AFB, P. R. (1940-55)	9.1	3 AM	5.1	3 AM
Santa Isabel AP, P. R. (1940-45 and 1951-54)	3.7	1 AM & 3 AM	2.9	1 AM
Alex. Hamilton Fld., St. Croix, V. I. (1954-58)	7.5	4 AM	4.8	4 AM & 7 AM
Truman Field, St. Thomas, V. I. (1953-58)	7.4	3-4 AM	3.7	4 AM

From the preceding tables it is apparent that even with the modifications of the trade winds by the land and sea breezes and local topography, the overall pattern shows much similarity. The six stations for which these summaries were prepared are variously located, with three exposures on the north coast of Puerto Rico, one on the south coast of Puerto Rico, one on the south coast of St. Croix, V. I., and the other in a rather sheltered location on St. Thomas.

ANNUAL PREVAILING WIND DIRECTION

Station	2 AM	8 AM	2 PM	8 PM	Year
			(Local Time)		
WBFO San Juan, P. R.	SE	ESE	ENE	ENE	ENE
Santa Isabel AP, P. R.	NE	NE	SE	NE	NE
Ramey AFB, P. R.	E	E	ENE	E	E
Roosevelt Roads, P. R.	E	E	E	E	E
Alex. Hamilton Field, St. Croix, V. I.	ENE	ENE	ESE	ENE	ENE
Truman Field, St. Thomas, V. I.	E	E	ESE	E	E

This table shows that even though the most frequent direction through the year may be in the northeastern quadrant, the distribution through the day shows a regular variation which might have been expected from the location of the stations. The WBFO San Juan, P. R., and Ramey AFB, P. R., on the northern coast show evidence, with that at San Juan the more pronounced, of a more northerly component of the wind during the afternoon as a result of the sea breeze effect blowing onshore. At San Juan the southerly component during the early morning hours reflects the strength of the land breeze. At Santa Isabel AP, P. R., and on St. Croix, V. I., the situation is the reverse of that at San Juan and Ramey AFB. Northeasterly winds predominate, except in the afternoon when the sea breeze becomes strong enough to bring about an alteration in the direction of the trade winds to a more southerly direction. The station in St. Thomas lies in a very sheltered position on the lee side of the hills to the south. Here again, the sea breeze effect is illustrated by a slight variation in direction at the 2 p. m. observation. However, it is not as marked as at the two other less sheltered south coast exposures.

At Mayaguez on the western end of Puerto Rico, the effect of the land and sea breezes is quite different than it is in other locations on the island. The sea breeze acts in an almost opposite direction and lessens the strength of the trade winds to such an extent that it frequently becomes dominant and a westerly wind is observed. This is quite infrequent in other sections of the islands.

Another indication of the land and sea breeze effects may be found in the following table of "Mean Annual Percentage Frequencies of Wind Direction" at several stations.

MEAN ANNUAL PERCENTAGE FREQUENCIES OF WIND DIRECTION

	WBFO San Juan, Puerto Rico	Ramey AFB, Puerto Rico	Roosevelt Roads, Puerto Rico	Alex. Hamilton Fld., St. Croix, Virgin Islands	Truman Field St. Thomas, Virgin Islands
N	0.9	1.0	3.7	1.7	0.8
NNE	1.7	1.7	3.5	2.4	1.1
NE	6.4	7.5	15.4	14.8	6.0
ENE	28.5	21.8	22.3	31.8	23.3
E	9.7	30.5	24.7	11.7	33.6
ESE	13.4	13.6	8.6	19.5	13.2
SE	9.8	9.9	4.6	7.5	5.6
SSE	6.0	1.9	1.1	2.2	1.0
S	4.2	2.0	1.7	0.7	1.0
SSW	1.8	0.8	0.9	0.3	0.4
SW	1.5	1.1	1.0	0.2	0.1
WSW	0.8	0.3	0.9	0.1	0.1
W	0.3	0.6	0.5	0.1	0.2
WNW	0.3	0.3	0.3	0.1	0.2
NW	0.5	0.7	1.4	0.2	0.3
NNW	0.7	0.4	2.3	0.4	0.3
CALM	13.5	5.9	7.1	6.3	12.7

At two stations in the preceding table, where the degree of shift in wind direction as a result of the sea breeze is the greatest, WBFO San Juan, P. R., and Alexander Hamilton Field, St. Croix, the frequency distribution is bimodal. That is, there are two peaks, with the greatest from east-northeast and the secondary one from east-southeast, while the frequency of east winds is somewhat less. The wind passes through east twice--during the transitional period from east-northeast to east-southeast in the case of Alexander Hamilton Field, and from east-southeast to east-northeast at San Juan, and vice-versa. At Ramey AFB and Truman Field, where the degree of shift in wind direction due to land and sea breeze is smaller, a distribution with a single mode (peak) from the east is observed. At Roosevelt Roads no directional shift occurs.

At Santa Isabel Airport, on the southern coast of Puerto Rico, the annual frequency distribution indicates that the shift from a northeasterly direction to a southeasterly direction begins to take place about 8 a.m. It reaches the peak about 1 to 2 p.m. when nearly half of the observations have a southeasterly wind. The return to northeasterly then takes place slowly, reaching the peak during the early morning hours shortly after midnight.

At Ramey AFB, in the northwestern corner of Puerto Rico, east is the most frequent wind during the night, with a gradual shift to east-northeast beginning at about 9 a.m. The shift back to east begins about 4 to 5 p.m., reaching a maximum in the early morning hours when a larger number of southeasterly winds are observed.

At Mayaguez, during the rainy season, from May through December, the predominant westerly wind does not appear until 11 a.m., which is 1 hour later than in the dry season (January through April) and it disappears at 5 p.m., 1 hour earlier than in the dry season.

Factors which interrupt the trade wind flow are the same as those discussed above as rainfall-producing mechanisms; the frontal passages and easterly wave passages. As the cold front approaches, a shift to a more southerly direction is noted, and then as the front passes a gradual shift through the southwest and northwest quadrants back to northeast. The easterly wave passage normally does not bring westerly winds but is usually characterized by an east-northeast wind ahead of the wave and a change to east-southeast following its passage.

During recent years hourly observations taken at the WBFO San Juan, Alexander Hamilton Field, St. Croix, V. I., and Truman Field, St. Thomas, V. I., indicate that less than 1 percent of all wind observations were above 24 m.p.h., and less than 5 percent were above 18 m.p.h.

Of course it must be realized that since these islands lie in the path of tropical storms and hurricanes, occasional winds of extreme speed are experienced. At San Juan, during the passage of the hurricane known locally as "San Felipe" in September 1928 the Weather Bureau's ane-mometer blew away after reaching an extreme velocity of 160 m.p.h. This is the highest value recorded in Puerto Rico to date. Winds of 110 m.p.h. are expected to occur once every century on the average.

EVAPORATION

Evaporation in Puerto Rico and the U. S. Virgin Islands, due to warm temperatures and rather constant wind flow, is fairly high as indicated by measurements at several locations in Puerto Rico and one at St. Croix, V. I., over a period of years. The San Juan annual average evaporation is 81.59 inches compared to annual values of 81.40 at Lajas and 79.85 inches at Aguirre. Higher elevation interior stations have somewhat less: Gurabo with 62.50 inches, Corozal 53.13 inches, and Adjuntas 50.93 inches. On St. Croix measurements made at the Annas Hope Agricultural Experiment Station averaged 72.69 inches per year.

It is interesting to note that at the coastal stations the evaporation is more than the average annual rainfall. At all locations the distribution throughout the year is similar, with maximum monthly values in the spring and early summer and the lowest monthly averages during November and December.

MOISTURE IN THE AIR

The relative humidity in the islands is rather high, averaging somewhere near 80 percent over the course of the year.

The relatively small differences in the dewpoint temperature over the area indicate that the air mass overlying the area is quite similar with respect to moisture content throughout the year.

Another interesting aspect of the distribution of the dewpoint is the small diurnal change. At San Juan, for example, the annual average dewpoint at 2 a.m., 8 a.m., 2 p.m., and 8 p.m. is 70°, thus illustrating the fact that the changes throughout the day are small. This also appears to be true at the other locations, with a small increase to the 2 p.m. value which is most marked at Santa Isabel along the southern coast. Here the winds most of the time are from overland, but the sea breeze during the afternoon brings slightly more moisture from the ocean.

The highest relative humidity values are generally found during the night when temperatures are the lowest. As the temperatures begin to rise, the relative humidity begins to fall, reaching its lowest point at about the time of the maximum temperature. High relative humidities are recorded during rainfall and when occasional cold fronts or easterly waves pass over.

The combination of high temperatures and fairly high relative humidity would usually result in physical discomfort. However, in Puerto Rico and the U. S. Virgin Islands there are several factors which greatly influence personal comfort and make the area normally very pleasant. The most noteworthy factor which lowers the sensible temperatures is the consistency

(Local Time)	San Juan, Puerto Rico		Ramey AFB, Puerto Rico		Santa Isabel Airport, Puerto Rico		St. Thomas, Virgin Islands		St. Croix, Virgin Islands	
2 AM	RH	DP	RH	DP	RH	DP	RH	DP	RH	DP
March	82	66	83	67	87	63	79	67	84	67
Sept.	89	73	87	73	92	72	84	74	86	74
Annual Avg.	86	70	86	70	90	68	81	70	84	70
8 AM										
March	80	66	80	66	72	64	73	67	76	68
Sept.	83	73	81	72	79	72	76	74	79	75
Annual Avg.	81	70	82	70	77	70	74	70	78	71
2 PM										
March	63	67	65	69	61	67	64	67	64	68
Sept.	70	73	70	75	71	75	68	74	71	75
Annual Avg.	67	70	69	72	66	72	66	71	67	72
8 PM										
March	74	67	78	70	76	65	74	67	80	68
Sept.	81	73	82	74	86	73	80	74	85	74
Annual Avg.	78	70	81	72	82	70	77	70	82	71

of the trade winds. The greatest beneficial effect normally is felt during the afternoon hours when temperatures are highest. Even at night when temperatures fall to their lowest level, a mild breeze is still experienced. This factor is an important one to consider when a home, hospital, industrial plant, or any building for that matter, is being designed. Proper orientation and proper location of parking areas, and care in landscaping may well spell the difference between comfort and discomfort for all concerned.

Also of importance in the matter of comfort is the normal diurnal march of temperature and relative humidity, which progress in opposite directions, with the lowest relative humidities at the time of maximum temperature. Indeed, the frequency of occasions with high temperature at the same time as high relative humidity is quite small. A study of a 5-year period of record shows that only one-tenth of 1 percent of the hourly observations at San Juan had temperatures higher than 84° with simultaneous relative humidities of 80 percent or higher.

Generally speaking, relative humidity values of 90 percent or above during the nighttime are not infrequent. During the day, values from 60 percent to the low 70's are predominant. Extremely low relative humidity is almost never experienced, with the lowest readings generally in the 50's. During the 5-year period noted above, only about three-tenths of 1 percent of the hourly observations at San Juan had relative humidities below 50 percent.

VISIBILITY

Visibility normally presents no problem in Puerto Rico and the U. S. Virgin Islands, with very few occurrences of observations with critical values. During the 10-year period from 1951 through 1960 at San Juan, less than one-half of 1 percent of the observations had values of less than 3 miles. During 1960 at Alexander Hamilton Field, St. Croix, V. I., and Truman Field, St. Thomas, V. I., only about two-tenths of 1 percent of the hourly observations recorded visibilities of less than 3 miles. At Ramey AFB only three-tenths of 1 percent of all observations had visibility less than 3 miles in a 14-year period.

The comparatively rare occurrences of low visibilities are associated with periods of heavy rainfall. Fog does not enter as a factor. However, industrialization is now beginning to cause a decrease in visibility due to air pollution under certain atmospheric conditions.

FOG

Fog, in the sense that it is known elsewhere with visibility reduced to almost zero, is unknown in Puerto Rico and the Virgin Islands. The only type of fog found in the area is radiation fog, which may form in the early morning hours and is rapidly dissipated. This usually occurs in the interior valleys. No serious problems to either air or surface travel are encountered.

HAIL

PUERTO RICO

Hail is a relatively rare event and evokes much curiosity at each occurrence since many persons on the Island have never experienced it. It is believed to fall somewhere in Puerto Rico at some time during each year. Past, incomplete records indicate that at one time or another hail has fallen in just about all areas, but a greater frequency of occurrence is noted in the Las Marias, Coloso, Lares, and Maricao areas.

The hailstones are generally small, usually about the size of peas and in most cases cause no serious damage. However, there have been a few reports of hailstones up to 1 inch in diameter and of hail damage amounting to thousands of dollars to crops and buildings.

Hail has not fallen at the Weather Bureau Office in San Juan, but on May 6, 1926 a severe local hailstorm accompanied by sharp lightning and heavy thunder occurred about 2 miles away, at Santurce. Rough solid ice hailstones, some as large as hens eggs, fell during the storm which lasted about 15 minutes. Damage to paper-covered roofs resulted and trees were uprooted by the high winds which accompanied the storm.

U. S. VIRGIN ISLANDS

In the U. S. Virgin Islands, hail is even less frequent than in Puerto Rico. In January of 1969 a severe local hailstorm with hailstones up to 1 1/2 inches in diameter occurred. This was the first hailstorm on record in the U. S. Virgin Islands.

DROUGHT

PUERTO RICO

Periods of deficient rainfall are not uncommon in Puerto Rico and almost every year some section of the Island suffers to some extent for varying periods of time. Due to the importance of agriculture to Puerto Rico's economy deficient rainfall, even over a restricted area, can have a serious impact. The most susceptible periods for a serious deficiency of rainfall are late fall, winter, and early spring, when rainfall is normally the lowest.

Rainfall distribution in Puerto Rico is most uneven. During the 1957 drought, the 6-month rainfall, December-May, at Rio Blanco Upper totaled 56.31 inches. In contrast, rainfall for those 6 months at Yauco was only 4.36 inches, an amount totally inadequate to satisfy requirements. On the south coast when the farmers speak of the "drought season" they are referring to the months from December through April. Irrigation for growing crops is a must in this season. On the other hand, the sugar plantations take advantage of the dry weather to mature their cane and carry out harvesting and grinding operations. A true drought, resulting from a prolonged rainfall deficiency, will adversely affect the economy of the island once or twice during every decade. One of the most extreme droughts on recent record occurred during 1967 through May 1968. The period 1964 through May 1965 was also one of extreme drought, particularly in the south coastal section. Prior to this a short but severe drought started in late 1956 and ended with June rains of 1957. In all these cases, agriculture and the dairy industry suffered the worst effects when pastures burned up, irrigation supplies became exhausted, and entire agricultural zones in non-irrigated areas lacked sufficient soil moisture to support crops. Reservoirs remained empty. Near the end of the 1967-68 drought the San Juan Metropolitan area went on strict water rationing until the May rains replenished water storage supplies. Water tables over most of the island were lowered to record levels and many wells were lost through intrusion of salt water. Some of the south coast sections received only about 35 percent of their normal rainfall during the year 1964. Individual stations received as little as 13 inches of rain. The island rainfall average for the 1967 drought year was only 43 inches compared to a long term average of 69 inches. This was the lowest rainfall on record since the beginning of observations in 1899.

U. S. VIRGIN ISLANDS

Drought in the Virgin Islands occurs about as often and is just as damaging as it is in Puerto Rico. None of the three islands has any significant running rivers or streams and only St. Croix has an underground water source in a few sections. Water for irrigation is not available in quantity at any time. Large storage reservoirs do not exist so the Virgin Islands are, to some extent, more at the mercy of "Mother Nature" than is Puerto Rico where there are adequate facilities for water storage.

In the Virgin Islands, as noted above, a variety of rainfall catchment areas, such as housetops, sides of hills, and even airport runways, are used the year round in an effort to catch and store as much of the precious rainwater as possible. During times of drought the water supply reaches dangerously low levels and the hauling of potable water in barges from Puerto Rico becomes a necessity. On St. John there is at present, a small private sea water distillation unit, and additional municipal units are in operation at St. Thomas and St. Croix.

FLOODS

The time of greatest likelihood of flooding naturally is at the time of maximum rainfall expectancy, roughly from May through November or December. The hilly nature of Puerto Rico and the Virgin Islands and the steep slope of the waterways from their basins in the mountainous areas to their outlets into the ocean dictates that many of the flood situations will be of the flash flood type. Thus with a relatively short warning period possible, eternal vigilance for possible flood-producing situations must be maintained.

The orographic effect, when carried to maximum development, is capable of depositing large quantities of water over the area with resulting occasional flooding. This situation existed during the May 1960 flood in the Virgin Islands, when thunderstorms formed over the northern Virgin Islands and remained practically stationary. They depositied up to 14.56 inches of rainfall during a 3-day period, with the greatest amounts falling on the last day.

Hurricanes and tropical storms, when they pass over the area or close by, are a serious flood-producing threat. Two of the three major floods in Puerto Rico occurred during hurricanes passages over the Island. The first occurred on August 8, 1899, as hurricane "San Ciriaco" buffeted Puerto Rico, and the other on September 13-14, 1928 in hurricane "San Felipe". In September 1960, hurricane "Donna" passed to the northeast of Puerto Rico without bringing dangerous winds, but the heavy rainfall which followed deposited torrential amounts, which caused widespread floods and accounted for 107 deaths, 136 injuries, and considerable agricultural and property losses.

Cold frontal action may also be responsible for floods. During late November and early December 1960 such activity was sufficient to produce as much as 17.25 inches of rainfall at Yaurel in the southeastern corner of Puerto Rico. This occurred during a 6-day period, while 11 other stations in Puerto Rico received more than 14 inches. In the Virgin Islands the rainfall during this same 6-day period totaled more than 12 inches at several stations on St. Thomas and St. Croix. Agricultural and property damage resulted together with the loss of 4 lives in Puerto Rico. Damages were slight with no loss of life in the Virgin Islands.

Areas of Puerto Rico most susceptible to flooding include the flatlands in the vicinity of the rivers. The annual average is about one

serious local flood together with one or more lesser floods either in Puerto Rico or the U. S. Virgin Islands. During 1960, however, there were three major floods: One occurred on St. Thomas and St. John in May; the next in September 1960, associated with hurricane Donna, affected a large portion of Puerto Rico, principally the northern coast from Manati to the east and in the southeastern sections of the Island; and the third flood, during December 1960, affected both Puerto Rico and the Virgin Islands.

CLOUD COVER

Cloud observations taken at WBFO San Juan and Ramey AFB in Puerto Rico, Alexander Hamilton Field on St. Croix, V. I., and at Truman Field, St. Thomas, V. I., show a high degree of similarity. As with wind speed, the minimum cloudiness occurs during the hours of darkness with increasing amounts after sunrise. Maximum cloudiness, averaged over the year, occurs between 10 a.m. and 5 p.m. at Alexander Hamilton Field, and from 11 a.m. to 3 p.m. at Truman Field. At San Juan the maximum occurs a little later between 2 p.m. and 3 p.m., although there is only a small variation from 10 a.m. to 7 p.m. Ramey AFB has a later maximum, with the highest values from 5 to 7 p.m.

The seasonal variation of cloudiness shows a double maximum at all stations, in May or June and again in September or October, with the June maximum somewhat more pronounced. At all four stations the lowest daily average cloudiness is in March, with the nighttime minima reaching their lowest point at that time.

Daytime cloud cover is greatest at Ramey AFB with a peak value of 8.6-tenths at 5 p.m., in June. The San Juan maximum value is 7.5-tenths at 4 p.m., in June. At Alexander Hamilton Field a double maximum is noted, with a value of 6.8-tenths at 8 a.m., and again at 5 p.m., in June with a slight decrease between those hours. The Truman Field record indicates a maximum of 6.4-tenths at 11 a.m., and at noon during May. During the night the minimum value noted is 1.9-tenths at Truman Field at 4 a.m., during March.

Actual observational data are not available for the interior. However, cloudiness in the interior is somewhat higher during the day than at the coastal stations due to the convective buildups, and somewhat lower during the night.

THUNDERSTORMS

While thunderstorms are not completely unknown during the winter months, the period of greatest likelihood falls between May and November. Due to the small size of the Virgin Islands, thunderstorm activity is less frequent there than in Puerto Rico, where the larger land mass and higher hills are more favorable for thunderstorm development especially in the western Cordilleras.

The average number of thunderstorms per year reported over various locations in Puerto

Rico: Ponce-12, San Juan WFBO-10, Roosevelt Roads-11, Mayaguez-35, and Ramey AFB-22. At St. Thomas thunder is heard on an average of about 13 hours per year; at St. Croix, 45 hours.

TROPICAL CYCLONES

Hurricanes and tropical storms are an important feature of Puerto Rico and U.S. Virgin Islands climate during summer and early autumn. The tropical cyclone season in the North Atlantic region extends from June through November. Because of seasonal shifts in favored locations of tropical cyclone development, the Puerto Rico-Virgin Islands area is outside the main paths of these most severe tropical atmospheric disturbances, except from August through the first half of October. A few "off-season" tropical cyclones have, however, slightly "brushed" the area at infrequent intervals.

Those hurricanes and tropical storms which do severely affect Puerto Rico and the Virgin Islands develop over the waters of the southern North Atlantic to the east of the Lesser Antilles. The movements of the storms are usually towards the west and northwest. They may pass either to the south or to the north of the islands, and occasionally directly over them.

A map is included which shows the paths of the more severe hurricanes which have passed directly over Puerto Rico since 1893. These are:

SOME HURRICANE PATHS ACROSS PUERTO RICO

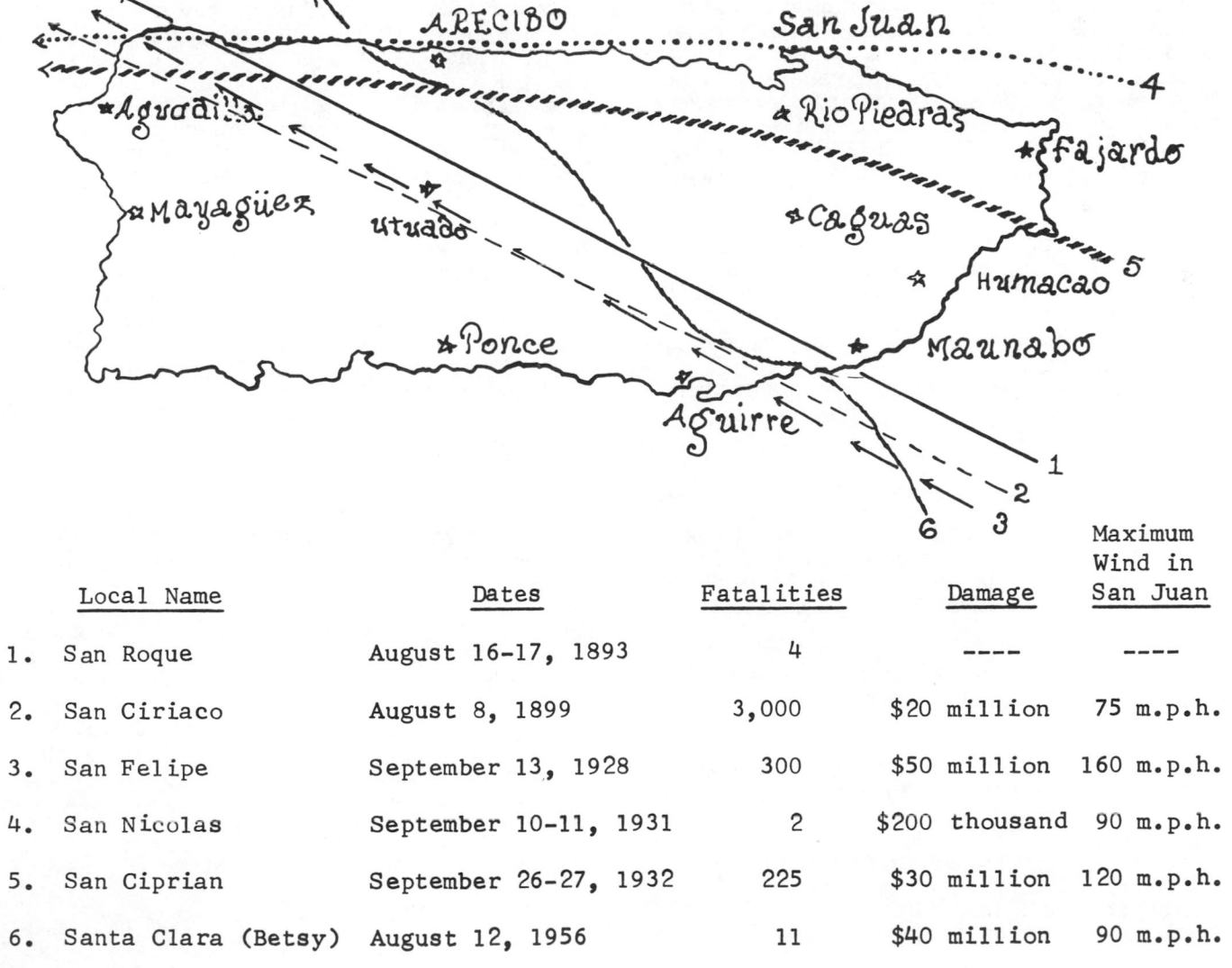

	Local Name	Dates	Fatalities	Damage	Maximum Wind in San Juan
1.	San Roque	August 16-17, 1893	4	----	----
2.	San Ciriaco	August 8, 1899	3,000	$20 million	75 m.p.h.
3.	San Felipe	September 13, 1928	300	$50 million	160 m.p.h.
4.	San Nicolas	September 10-11, 1931	2	$200 thousand	90 m.p.h.
5.	San Ciprian	September 26-27, 1932	225	$30 million	120 m.p.h.
6.	Santa Clara (Betsy)	August 12, 1956	11	$40 million	90 m.p.h.

Additional "near misses" of intense hurricanes or tropical storms, which produce little wind damage, may cause extensive rain (flooding) and/or tide damage.

An outstanding example of the fringe effects of hurricanes on the area occurred during the passage of Donna in September 1960. The hurricane center passed west-northwestward to the north of Puerto Rico and the Virgin Islands on September 5. Winds reached only 62 m.p.h. at St. Thomas and 42 m.p.h. at San Juan, in gusts, and little wind damage was sustained. Torrential rains developed during the evening of the 5th after the center had passed. Rains of up to 10 to 15 inches fell over much of the southeastern half of Puerto Rico during the night and produced the highest flash floods ever known on many streams. Several million dollars of crop and property damage was reported and 107 persons perished-- mostly by drowning.

The extent of destruction and loss of life caused by a tropical storm or hurricane is dependent upon several factors, both meteorological and in the field of public awareness.

The intensity of the storm, exemplified by the strength of the maximum winds (which have reached speeds of near 160 m.p.h. at San Juan), is of major importance, together with the forward speed and size of the hurricane. Longer durations of extreme intensity wind will generally cause more destruction. Slowly moving storms also increase the possibility of excessive rainfall. Fortunately the area covered by extremely high winds in hurricanes in tropical latitudes is usually very small, and very slow forward speeds are rare, so that high winds are usually sustained for only a very few hours at any one place.

Elements affecting the extent of damage and loss of life which may be controlled include: 1) the accuracy of warnings and the time of advance notice, and 2) the public response to issued warnings.

The accuracy of hurricane forecasts has been continually improved and the time of advance notice has increased during recent years, as a result of advanced communication facilities, aircraft reconnaissance observations, satellite pictures, and land-based radar. The structure and behavior of tropical cyclones is also now much better known, and this too has improved accuracy of forecasts and lengthened advance notice.

After warnings have been issued, the utilization of this information in holding damage, injury, and loss of life to a minimum, devolves upon the public and the local authorities. Complete cooperation with the authorities in taking adequate protective and precautionary measures and in keeping informed concerning latest advices are necessary for individual safety and the reduction of destruction.

CONCLUSION

Climate is closely linked with man's everyday life, his degree of comfort and happiness, his ability to work efficiently, and indeed his very survival. The tropical marine climate of Puerto Rico and the Virgin Islands can be one of the most compatible and enjoyable in the world. However, the key to this is man's willingness to plan his activities in consonance with the climate and not in spite of it. Whether it is the choice of an industrial or home site, structural design or orientation, vacation date or trip itinerary - a well founded decision must include complete consideration of the climatic factor. This is true more so in the tropics than perhaps any other areas of the world. Where the weather factor is involved in a tropical environment the balance between human comfort or discomfort, between economic success or failure, or between safe and compatible building design can be a delicate one. Through effective planning and intelligent application of climatic considerations to life in the Caribbean, man can truly say he has found his tropical paradise.

BIBLIOGRAPHY

BOGART, D. B., T. ARNOW, and J. W. Crooks, Recursos Hidrologicos Boletin Num. 1, Problemas de Abasto de Aguas en Puerto Rico y Programa de Investigaciones de Recursos Hidrologicos.

------------------------, Water Resources of Puerto Rico, A Progress Report, 1964.

BRISCOE, C. B., Weather in the Luquillo Mountains of Puerto Rico, Institute of Tropical Forestry, Rio Piedras, Puerto Rico.

CALVESBERT, R. J., 1966, The Climatology of Meteorological Drought in Puerto Rico, Proceedings of Conference on Climatology and Related Fields in the Caribbean, University of the West Indies, Kingston, Jamaica.

COLON, J. A., 1958, A study of surface winds in the vicinity of Mayaguez, Puerto Rico: U. S. Weather Bureau.

CRY, G. W., 1961, North Atlantic Tropical Cyclones 1960: U. S. Weather Bureau, Mariner's Weather Log, pp. 1-7.

DUNN, G. E. & B. I. Miller, (Director and Research Meteorologist, respectively, at the National Hurricane Center, Miami, Fla.) 1960, Atlantic Hurricanes, Louisiana State University Press, Baton Rouge.

BIBLIOGRAPHY

FASSIG, O. L., The Trade Winds of the Eastern Caribbean. (In American Geophysical Union. Transactions, fourteenth annual meeting, 1933).

------------- The Climate of Porto Rico. (In P. R. Rev. Pub. Health, v. 4, 1928, p. 199).

GRAY, RICHARD W., U. S. Weather Bureau, La Distribucion Anual y Geografica de Lluvia en Puerto Rico, pp. 47, Almanaque Agricola de Puerto Rico 1944.

HARRIS, F. MILES, U. S. Weather Bureau, Almanaque Agricola de Puerto Rico 1947, La Temperature de Puerto Rico, pp. 75.

JONES, Clarence F. and PICO, RAFAEL, Symposium on the Geography of Puerto Rico, University of Puerto Rico Press 1955.

PICO, RAFAEL, The Geographic Regions of Puerto Rico, University of Puerto Rico 1950.

PUIG-THOMAS, JUAN, MANUEL LEDESMA and JOSE L. GARCIA DE QUEVEDO, 1960, Puerto Rico Nuclear Center-Meteorology Report.

QUINONES, MIGUEL A., High Intensity Rainfall and Major Floods in Puerto Rico.

RIEHL, HERBERT, Tropical Meteorology. New York, McGraw-Hill, 1954. p. 392.

SALIVIA, LUIS A., Historia de Los Temporales de Puerto Rico (1508-1949) San Juan, P. R. 1950.

SMEDLEY, DAVID, Climate of Puerto Rico and U. S. Virgin Islands, Climates of the States, Climatography of the United States No. 60-52, 1961. ESSA, Environmental Data Service.

STONE, ROBERT G., The New York Academy of Sciences, Scientific Survey of Porto Rico and the Virgin Islands, Volume XIX - Part 1 - Meteorology of the Virgin Islands.

TANNEHILL, IVAN RAY, Hurricanes, Princeton University Press.

WARD, ROBERT DeC. and CHARLES F. BROOKS, Handbuch der Klimatologie, Band II, Teil I, Westindien, Climatology of the West Indies, 1934.

Agricultural Yearbook Separate No. 1869 - Climate of the States, The West Indies Islands (Including Puerto Rico). ESSA, Weather Bureau

Climatography of the United States No. 30-52, Summary of Hourly Observations, San Juan, P. R., ESSA, Weather Bureau

Department of Tourism and Trade - Government of the Virgin Islands, Annual Report - July 1, 1959--June 30, 1960.

Economic Development Administration, Office of Economic Research, General Economics Division, Selected Statistics on the Visitors and Hotel Industry in Puerto Rico - 1959-60.

Local Climatological Data, Monthly and Annual, San Juan, P. R., Alexander Hamilton Field, St. Croix, V. I. and Santa Isabel Airport, P. R. ESSA, Environmental Data Service.

Miscellaneous Mimeographed Summaries of Climatological Observations for the Puerto Rico-Virgin Islands Area.

Monthly Weather Review. ESSA, Weather Bureau.

Puerto Rico Planning Board, Bureau of Economics and Statistics, Statistical Yearbook, Historical Statistics.

The West Indies and Caribbean Year Book, Thomas Skinner & Co. (Publishers), Limited.

U. S. Air Force, Uniform Summary of Surface Weather Observations-Station #11603 Aguadilla, P. R., Ramey AFB - Period of Record March 1940-March 1955, less May 1946.

U. S. Navy, Summary of Monthly Aerological Records, San Juan, P. R. - Period of Record March 1945-February 1959.

U. S. Navy, Summary of Monthly Aerological Records, Roosevelt Roads, P. R.

Unpublished Series of Temperature and Rainfall Maps. ESSA, Weather Bureau.

Unpublished special summaries for Ramey AFB, Santa Isabel, AP, Puerto Rico, Alexander Hamilton Field, St. Croix, V. I., Harry S. Truman Field, St. Thomas, V. I. and San Juan, Puerto Rico. ESSA, Weather Bureau.

Weather Bureau Technical Paper No. 13 - Mean Monthly and Annual Evaporation from Free Water Surfaces for the United States, Alaska, Hawaii and the West Indies. Washington, D. C. 1950. ESSA, Weather Bureau.

Weather Bureau Technical Paper No. 32, Upper-Air Climatology of the United States. ESSA, Weather Bureau.

Weather Bureau Technical Paper No. 36, North Atlantic Tropical Cyclones. ESSA, Weather Bureau.

BIBLIOGRAPHY

Weather Bureau Technical Paper No. 42, Generalized Estimates of Probable Maximum Precipitation and Rainfall, Frequency Data for Puerto Rico and Virgin Islands for Areas to 400 Square Miles, Durations to 24 Hours and Return Periods from 1 to 100 years. ESSA, Weather Bureau.

Weekly Weather and Crop Bulletin, Puerto Rico. ESSA, Weather Bureau.

Additional Publications:

Climatic Summary of the United States (Bulletin W) 1930 Edition, Section 106. ESSA, Weather Bureau.

Climatic Summary of the United States - Puerto Rico and the Virgin Islands - Supplement for 1931 through 1952 (Bulletin W Supplement). ESSA, Weather Bureau.

Climatic Summary of the United States - Puerto Rico and the United States. Virgin Islands - Supplement for 1951 through 1960 (Bulletin W Supplement). ESSA, Weather Bureau.

Climatological Data - Puerto Rico and the Virgin Islands. ESSA, Environmental Data Service.

Climatological Data National Summary. ESSA, Environmental Data Service.

Taken from "Climatography of the United States No. 81-4, Decennial Census of U. S. Climate"

TEMPERATURE (°F)   PRECIPITATION (In.)

STATIONS (By Divisions)	JAN	FEB	MAR	APR	MAY	JUNE	JULY	AUG	SEPT	OCT	NOV	DEC	ANN	JAN	FEB	MAR	APR	MAY	JUNE	JULY	AUG	SEPT	OCT	NOV	DEC	ANN
NORTH COASTAL																										
ARECIBO 2 ESE	73.9	73.6	74.6	76.0	78.1	79.4	79.9	80.2	80.2	79.5	77.7	75.6	77.4	4.25	3.27	2.57	3.77	7.40	4.24	4.14	4.96	4.98	5.58	5.58	4.79	55.53
DORADO 4 W	74.5	74.5	75.4	76.8	78.6	79.9	79.8	80.3	80.2	79.4	77.7	75.8	77.7	5.83	3.25	2.48	4.41	6.78	5.26	7.02	6.65	5.69	5.31	6.28	6.18	65.14
GARROCHALES	.	.	.	.	.	.	.	.	.	.	.	.	.	4.93	3.27	2.72	4.17	6.46	4.58	4.71	5.82	5.08	5.29	5.78	5.03	57.84
QUEBRADILLAS	.	.	.	.	.	.	.	.	.	.	.	.	.	4.07	3.29	3.18	4.68	8.04	5.00	3.98	5.47	5.46	5.34	5.72	4.69	58.92
SAN JUAN CITY	75.1	75.1	76.0	77.0	78.8	79.9	80.1	80.8	80.7	80.2	78.6	76.7	78.3	4.13	2.70	2.07	3.89	7.16	5.83	6.02	6.34	6.04	5.24	6.05	4.89	60.36
SAN JUAN WB AIRPORT 2	74.4	74.4	75.3	76.6	78.7	80.0	80.4	80.9	80.5	80.0	78.2	76.2	78.0	4.70	2.90	2.20	3.72	7.12	5.66	6.25	7.13	6.76	5.83	6.49	5.45	64.21
DIVISION	73.9	73.7	74.8	76.2	78.2	79.4	79.7	80.1	79.9	79.2	77.5	75.5	77.7	4.75	3.31	2.72	4.32	7.36	5.66	6.06	6.72	6.08	5.81	5.88	5.28	64.09
SOUTH COASTAL																										
AGUIRRE	77.0	76.8	77.6	78.9	80.4	81.6	82.2	82.5	82.2	81.7	80.4	78.5	80.0	1.07	1.29	.76	2.37	4.94	4.73	4.17	5.28	5.93	5.77	4.36	2.10	42.77
CENTRAL SAN FRANCISCO	.	.	.	.	.	.	.	.	.	.	.	.	.	.74	1.24	.72	1.83	4.20	3.08	2.90	4.55	6.20	5.19	3.40	1.24	35.29
ENSENADA	.	.	.	.	.	.	.	.	.	.	.	.	.	.70	1.28	1.02	1.95	4.13	2.69	2.60	3.72	5.71	4.55	3.88	1.66	33.89
PONCE 4 E	76.1	75.8	76.6	78.2	79.7	80.9	81.5	81.6	81.4	80.7	79.4	77.6	79.1	.86	1.07	.61	2.28	4.49	3.51	2.74	4.69	5.93	5.31	3.70	1.34	36.53
POTOLA	.	.	.	.	.	.	.	.	.	.	.	.	.	.79	1.05	.67	2.07	4.71	3.34	3.00	4.34	6.42	5.70	3.46	1.45	37.00
SABATER	.	.	.	.	.	.	.	.	.	.	.	.	.	.82	1.12	.75	2.22	4.55	4.43	3.70	4.74	5.93	5.99	3.71	1.96	39.92
SANTA ISABEL	.	.	.	.	.	.	.	.	.	.	.	.	.	.86	.95	.50	1.70	4.42	3.92	2.74	4.44	5.91	5.14	3.11	1.38	35.07
SANTA RITA	.	.	.	.	.	.	.	.	.	.	.	.	.	.59	1.40	.93	2.28	4.15	2.77	2.46	4.19	5.72	5.12	3.75	1.75	35.11
YAUCO 1 S	.	.	.	.	.	.	.	.	.	.	.	.	.	.72	1.46	.87	2.02	4.82	3.22	2.75	4.46	6.50	5.68	3.74	1.41	37.65
DIVISION	76.1	75.9	76.7	78.2	79.7	80.9	81.6	81.6	81.4	80.8	79.5	77.7	79.0	.87	1.21	.79	2.11	4.51	3.50	3.02	4.71	6.12	5.51	3.74	1.71	37.76
NORTHERN SLOPES																										
CALERO CAMP	.	.	.	.	.	.	.	.	.	.	.	.	.	3.18	2.68	2.92	4.37	7.68	7.07	4.80	6.14	5.78	6.30	4.45	3.69	59.06
CANOVANAS 2 N	74.0	73.7	74.9	76.6	78.5	79.5	79.6	80.6	79.9	79.2	77.5	75.5	77.5	3.40	3.57	2.82	4.50	8.25	7.08	7.80	7.65	7.23	6.18	6.79	6.01	72.84
FAJARDO	76.0	75.9	77.0	78.4	80.0	81.2	81.7	81.9	81.3	80.5	79.2	77.5	79.2	3.40	2.78	2.32	4.32	7.97	6.12	6.06	6.80	7.75	8.01	6.08	4.40	66.01
ISABELA 4 SW	73.5	73.4	74.5	75.5	77.1	78.3	79.1	79.1	78.9	78.3	77.0	75.1	76.7	3.80	3.34	3.25	4.84	8.44	7.54	5.00	6.68	6.90	6.52	5.42	4.32	66.05
DIVISION	74.3	74.1	75.3	76.8	78.4	79.6	80.0	80.3	79.9	79.3	77.7	75.7	77.5	4.17	3.28	2.80	4.50	8.22	6.56	5.85	6.77	7.11	6.61	5.87	4.85	66.50
SOUTHERN SLOPES																										
CABO ROJO	.	.	.	.	.	.	.	.	.	.	.	.	.	2.33	2.19	3.23	4.92	7.33	4.00	6.47	8.29	8.48	7.20	5.84	3.74	64.02
GUAYABAL RESERVOIR	.	.	.	.	.	.	.	.	.	.	.	.	.	1.36	1.91	1.49	3.76	6.26	4.37	3.98	6.59	9.45	8.76	5.58	2.18	55.69
GUAYAMA	78.5	78.4	79.1	80.3	81.3	82.3	82.7	83.3	82.9	82.5	81.4	79.8	81.0	2.09	1.81	1.27	3.26	6.85	6.80	5.75	6.55	8.69	7.70	5.57	3.50	59.84
HUMACAO	73.0	73.4	75.1	77.2	78.8	80.2	80.2	80.3	79.8	78.8	76.6	74.1	77.3	4.23	3.30	2.87	5.22	10.26	9.53	8.65	9.41	10.87	9.96	8.16	5.59	88.05
JOSEFA	.	.	.	.	.	.	.	.	.	.	.	.	.	1.62	1.57	1.01	2.84	6.02	5.78	4.63	6.18	7.46	6.89	5.89	3.15	53.04
JUANA DIAZ CAMP	76.5	76.3	77.0	78.2	79.5	80.8	81.8	81.5	80.7	80.2	79.1	77.7	79.1	1.22	1.52	.95	2.64	5.00	3.89	3.51	5.56	8.00	6.85	4.50	2.00	45.64
MAYAGUEZ NUCLEAR CTR	74.4	74.8	75.7	76.9	78.3	79.3	79.3	79.6	79.5	78.8	77.2	75.7	77.5	1.91	1.64	3.50	4.68	8.45	8.52	9.60	10.08	11.01	8.50	5.63	2.60	76.12
PATILLAS DAM	.	.	.	.	.	.	.	.	.	.	.	.	.	3.01	2.28	1.69	3.88	8.65	8.57	6.95	7.78	9.53	9.06	6.56	4.36	72.32
SAN GERMAN	74.6	74.6	75.4	76.9	78.6	80.0	79.9	80.0	79.9	79.2	77.6	75.8	77.7	2.52	2.64	3.09	5.25	7.19	3.89	5.91	7.73	9.96	9.21	6.74	3.90	68.03
YABUCOA 1 NE	.	.	.	.	.	.	.	.	.	.	.	.	.	4.38	3.45	2.70	5.13	10.46	8.76	7.94	9.44	10.51	10.47	8.22	6.80	88.26
DIVISION	75.4	75.4	76.3	77.7	79.1	80.3	80.5	80.8	80.4	79.7	78.3	76.6	78.3	2.59	2.27	2.13	4.09	7.79	6.40	6.34	7.79	9.49	8.61	6.33	3.93	67.71
EASTERN INTERIOR																										
CARITE DAM	69.6	69.4	70.4	71.1	72.2	73.3	73.8	74.7	74.6	74.1	73.0	71.2	72.3	4.20	3.40	2.57	5.26	9.66	8.79	8.92	8.65	10.86	9.71	7.18	5.39	84.59
CAYEY 1 E	69.1	69.3	71.0	73.0	74.6	75.5	76.0	76.2	75.8	74.8	73.0	70.9	73.3	3.28	2.66	1.81	3.85	6.66	5.36	6.15	6.35	7.89	7.03	4.99	3.89	59.92
CIDRA 1 E	.	.	.	.	.	.	.	.	.	.	.	.	.	4.17	3.10	2.18	4.61	7.05	6.11	7.25	7.32	8.14	7.07	5.61	4.37	66.98
JAJOME ALTO	.	.	.	.	.	.	.	.	.	.	.	.	.	3.69	3.25	2.49	4.50	8.06	7.54	7.76	7.46	9.79	8.71	6.01	4.90	74.16
JUNCOS 1 NNE	.	.	.	.	.	.	.	.	.	.	.	.	.	2.74	2.51	1.52	3.34	7.41	7.32	6.36	7.85	9.28	7.58	5.43	4.38	65.72
RIO BLANCO UPPER	.	.	.	.	.	.	.	.	.	.	.	.	.	10.67	9.68	6.65	10.50	16.41	14.67	14.00	15.63	17.06	16.05	14.19	13.67	159.18
DIVISION	70.7	70.5	71.8	73.4	75.1	76.1	76.5	76.9	76.7	75.9	74.3	72.2	74.1	4.81	3.94	2.81	5.22	9.57	8.48	8.12	9.11	10.47	9.46	7.63	6.24	85.80
WESTERN INTERIOR																										
COLOSO	74.1	73.9	75.0	76.5	78.4	79.6	80.2	80.3	79.9	79.2	77.8	75.8	77.6	2.38	2.10	2.99	5.95	11.64	11.75	9.77	11.44	10.20	7.98	5.02	2.55	83.77
GARZAS DAM	.	.	.	.	.	.	.	.	.	.	.	.	.	2.90	2.97	3.39	6.15	10.97	7.34	8.18	11.98	12.89	13.63	7.66	3.96	92.02
GUAJATACA DAM	.	.	.	.	.	.	.	.	.	.	.	.	.	3.82	3.22	3.48	6.22	11.21	7.87	5.82	7.87	8.99	8.46	5.64	4.24	76.84
GUINEO RESERVOIR	.	.	.	.	.	.	.	.	.	.	.	.	.	4.09	4.28	4.37	8.35	16.45	9.08	7.70	12.63	15.76	16.17	8.76	5.00	112.63
LARES 3 SE	71.9	71.7	72.8	73.8	75.3	76.4	76.9	77.0	76.5	75.3	73.3		74.9	3.22	3.06	4.03	8.28	15.96	8.59	7.54	10.62	12.12	11.80	6.81	3.85	95.88
MARICAO	.	.	.	.	.	.	.	.	.	.	.	.	.	2.67	2.94	4.62	7.56	12.85	8.36	9.95	14.06	14.16	13.70	7.20	3.84	101.91
TORO NEGRO PLANT 2	.	.	.	.	.	.	.	.	.	.	.	.	.	4.34	4.47	3.85	7.47	14.95	7.95	6.46	11.92	15.56	15.65	8.22	4.88	105.72
DIVISION	71.0	70.9	71.6	73.2	75.0	76.4	76.7	77.0	76.6	75.9	74.4	72.3	74.1	3.55	3.37	3.46	6.57	11.65	7.23	6.86	10.12	11.43	10.77	7.02	4.33	86.32
OUTLYING ISLANDS																										
DIVISION	76.7	76.6	77.4	78.8	80.4	81.7	82.3	82.6	82.1	81.5	80.0	77.9	79.8	2.17	1.65	1.60	2.62	4.69	3.83	3.46	4.28	5.27	5.22	4.78	3.36	42.74

*NORMALS BY CLIMATOLOGICAL DIVISIONS

Taken from "Climatography of the United States No. 81-4, Decennial Census of U. S. Climate"

TEMPERATURE (°F)   PRECIPITATION (In.)

STATIONS (By Divisions)	JAN	FEB	MAR	APR	MAY	JUNE	JULY	AUG	SEPT	OCT	NOV	DEC	ANN	JAN	FEB	MAR	APR	MAY	JUNE	JULY	AUG	SEPT	OCT	NOV	DEC	ANN
VIRGIN ISLANDS																										
CHARLOTTE AMALIE 2	77.1	76.8	77.7	78.9	80.3	81.9	82.5	82.9	82.6	81.9	80.5	78.7	80.2	2.87	1.91	1.59	2.53	4.95	3.30	3.47	4.88	6.08	5.51	4.39	3.27	44.75
DIVISION	76.5	76.3	77.2	78.4	79.7	81.3	81.8	82.1	81.4	80.8	79.6	77.7	79.3	2.71	1.78	1.47	2.45	4.54	3.30	3.73	4.49	6.18	5.56	4.84	3.33	44.32

* Normals for the period 1931-1960. Divisional normals may not be the arithmetical averages of individual stations published, since additional data for shorter period stations are used to obtain better areal representation.

ADDITIONAL MEAN DATA FOR VIRGIN ISLANDS NOT INCLUDED IN "NORMALS" TABULATION*

STATIONS		JAN	FEB	MAR	APR	MAY	JUNE	JULY	AUG	SEPT	OCT	NOV	DEC	ANN	JAN	FEB	MAR	APR	MAY	JUNE	JULY	AUG	SEPT	OCT	NOV	DEC	ANN
ALEX HAMILTON AP FAA	MEAN	76.6	76.7	77.5	79.0	80.4	81.8	81.9	82.2	81.5	81.0	79.6	77.9	79.7	2.23	2.19	1.73	2.83	4.31	3.10	3.51	4.58	6.65	5.45	4.65	3.34	44.57
	YEARS	13	13	13	13	13	13	14	14	14	14	14	14		13	13	13	13	13	13	14	14	14	14	14	14	
ANNAS HOPE EXP STA	MEAN	76.0	75.6	76.3	77.5	79.2	80.5	81.0	81.2	80.3	79.7	78.4	76.8	78.5	2.69	1.96	1.75	2.24	3.72	2.95	3.31	4.49	6.39	5.84	5.06	3.28	43.68
	YEARS	30	29	30	31	31	31	31	31	31	30	30	30		41	41	41	41	41	41	41	41	41	41	41	41	
CRUZ BAY	MEAN	76.7	77.0	77.9	79.3	80.2	81.3	81.8	82.3	81.8	80.4	79.5	77.8	79.7	2.38	1.87	1.55	2.15	4.62	3.11	3.80	3.93	5.94	5.09	4.51	2.93	41.88
	YEARS	14	13	13	12	13	14	14	14	13	13	13	14		39	40	40	39	39	39	40	40	40	39	39	39	

*OBTAINED FROM "CLIMATIC SUMMARY OF THE UNITED STATES-SUPPLEMENT FOR 1951 THROUGH 1960."

TEMPERATURE   PRECIPITATION

JAN	FEB	MAR	APR	MAY	JUNE	JULY	AUG	SEPT	OCT	NOV	DEC	ANN	JAN	FEB	MAR	APR	MAY	JUNE	JULY	AUG	SEPT	OCT	NOV	DEC	ANN

CONFIDENCE - LIMITS

In the absence of trend or record changes, the chances are 9 out of 10 that the true mean will lie in the interval formed by adding and subtracting the values in the following table from the means for any station in the State. Because of the wider variation in mean precipitation, the corresponding monthly means and annual mean must be substituted for "p" in the precipitation table below to obtain mean precipitation confidence limits.

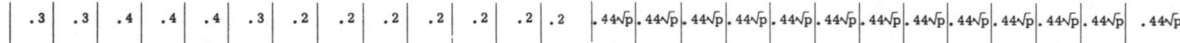

| .3 | .3 | .4 | .4 | .4 | .3 | .2 | .2 | .2 | .2 | .2 | .2 | .2 | .44√p | .44√p | .44√p | .44√p | .44√p | .44√p | .44√p | .44√p | .44√p | .44√p | .44√p | .44√p | .44√p |

COMPARATIVE DATA

Data in the following table are the mean temperature and average precipitation for Canovanas 2 N, Puerto Rico, for the period 1906-1930 and are included in this publication for comparative purposes.

| 74.3 | 74.7 | 75.5 | 77.1 | 79.6 | 80.6 | 80.6 | 81.2 | 81.0 | 80.2 | 78.0 | 75.8 | 78.2 | 6.19 | 4.03 | 4.42 | 4.74 | 7.20 | 6.95 | 8.46 | 7.84 | 7.53 | 7.44 | 8.23 | 7.43 | 80.46 |

NORMALS, MEANS AND EXTREMES
(Years of Record: 1940-1966)

Station: AGUADILLA/RAMEY AFB   Standard time used: ATLANTIC   Latitude: 18° 30'   Longitude: 67° 08'   Elevation (ground): 247 feet

Month	(a)	Means Daily maximum	Means Daily minimum	Means Monthly	Extremes Record highest	Year	Extremes Record lowest	Year	Normal heating degree days (Base 65°)	Precip. Normal total	Precip. Max monthly	Year	Precip. Min monthly	Year	Precip. Max in 24 hrs	Year	Snow Mean total	RH 00-02	RH 06-08	RH 12-14	RH 18-20	Wind Mean speed	Prevailing direction	Fastest mile Speed	Direction	Year	Precip ≥.01 days	Snow ≥1.0 days	Thunderstorms	Heavy fog	Max 90+	Max 32↓	Min 32↓	Min 0↓
(a)		27	27	27	27		27			27	27		27		27		27	27	27	27	27			18			27	27	21	21	27	27	27	27
J		80.2	68.6	74.5	87	1963	60	1948	---	2.82	10.44	---	1.12	---	3.48	1959	0.0	84	84	68	76	9.8	E	18	ENE	1950	12	0	*	*	0	0	0	0
F		80.4	68.2	74.4	89	1946	61	1965	---	2.40	6.96	---	.03	---	2.83	1956	0.0	83	83	68	76	9.8	E	53	ENE	1949	9	0	*	*	0	0	0	0
M		81.5	69.0	75.3	90	1952	62	1951+	---	2.64	9.48	---	.23	---	4.43	1963	0.0	82	81	66	76	9.8	E	38	ENE	1950	9	0	*	*	*	0	0	0
A		82.2	70.6	76.5	90	1958+	64	1946	---	4.46	14.14	---	.47	---	5.53	1958	0.0	84	81	61	78	9.4	E	36	E	1966	11	0	*	*	*	0	0	0
M		83.6	72.5	78.2	90	1964+	67	1962+	---	5.40	10.24	---	.70	---	4.40	1945	0.0	85	82	72	80	8.0	E	36	ENE	1963	14	0	2	*	*	0	0	0
J		84.9	73.4	79.3	93	1963	68	1943	---	6.22	12.34	---	1.67	---	3.61	1956	0.0	86	82	72	81	8.7	E	57	ENE	1963	14	0	8	*	*	0	0	0
J		85.3	74.2	79.8	95	1962	68	1945	---	4.23	8.03	---	.96	---	2.60	1966	0.0	87	84	72	81	10.6	E	43	E	1950	15	0	13	*	*	0	0	0
A		85.8	74.2	80.1	92	1940	68	1940	---	5.16	10.64	---	1.70	---	4.60	1966	0.0	87	84	71	82	8.8	E	54	ENE	1949	14	0	12	*	*	0	0	0
S		86.0	73.8	80.0	92	1963+	66	1942	---	4.74	8.44	---	1.35	---	3.67	1954	0.0	87	84	71	82	7.2	E	35	SSE	1965	14	0	14	*	1	0	0	0
O		85.5	73.2	79.5	90	1963+	66	1946	---	5.43	9.46	---	1.73	---	3.74	1958	0.0	87	84	69	79	6.0	E	40	SSE	1965	14	0	12	*	*	0	0	0
N		81.6	73.9	77.9	90	1963+	64	1966	---	3.80	9.62	---	.61	---	4.07	1961	0.0	85	84	66	78	7.7	E	38	ENE	1950	13	0	4	*	*	0	0	0
D		81.6	70.0	75.9	87	1963+	62	1949	---	3.86	10.04	---	.80	---	3.27	1959	0.0	85	83	70	79	9.0	E	38	E	1965	13	0	1	*	0	0	0	0
YR		83.4	71.7	77.7	95 JUL 1962		60 JAN 1948		---	51.16	14.14 APR 1958		.03 FEB		5.53 APR 1958		0.0	85	83	70	79	8.7	E	57 ENE JUN 1963			150	0	80	*	1	0	0	0

THIS TABLE CONSISTS OF DATA EXTRACTED FROM THE USAF PUBLICATION, "REVISED UNIFORM SUMMARY OF SURFACE WEATHER OBSERVATIONS." YEARS OF RECORD, 1940-1966

Station moved or discontinued. See other stations for this state.

REFERENCE NOTES APPLYING TO TABLES APPEAR ON THE PAGE FOLLOWING LAST TABLE.
(Caution: Letters and symbols may have different meanings in 1941-1970 tables than in earlier tables. See notes.)

NORMALS, MEANS AND EXTREMES

(Table Revised 1970. Base Period for Climatological Normals: 1931-1960)

NWSED, ROOSEVELT ROADS, P. R.
JAN. 49—DEC. 68   N 18°15' W 65°38'
ZONE: +4 HOURS

Temperature

Month	(n)	Daily maximum	Daily minimum	Monthly	Record highest	Year	Record lowest	Year
J	13	82.7	71.2	77.2	88	1961	62	1965
F	13	83.0	71.2	77.3	90	1964	62	1966
M	13	84.1	72.1	78.4	89	1965+	64	1949
A	13	84.5	73.3	79.2	90	1966	64	1968
M	13	85.7	75.1	80.7	91	1964+	68	1963+
J	13	86.9	76.7	82.1	92	1960	70	1967+
J	13	87.3	77.3	82.6	92	1962	71	1967+
A	13	86.0	77.5	83.0	93	1968	70	1967
S	13	87.7	76.4	82.5	92	1960+	68	1960
O	13	86.3	75.2	81.7	93	1963+	69	1965+
N	13	84.1	73.7	78.6	93	1957	66	1963
D	13	82.1	72.5	72.5	91	1966	63	1949
YR	13	85.7	74.4	80.3	93	AUG 1968+	62	FEB 1966+

Mean degree days: 0 for all months; annual 0.

Precipitation

Month	(n)	Mean	Max monthly	Year	Min monthly	Year	Max in 24 hrs	Year
J	13	3.46	10.82	1958	1.69	1965	4.26	1958
F	13	2.14	7.48	1950+	1.30	1960	1.49	1950+
M	13	2.33	5.96	1960	0.23	1958	5.96	1949
A	13	4.06	9.14	1962	0.87	1967	4.11	1960
M	13	6.21	13.28	1959	1.60	1964	4.55	1959
J	13	4.99	9.61	1962	2.30	1961	2.74	1967
J	13	5.13	9.71	1958	2.77	1967	2.97	1958
A	13	5.44	10.04	1960	2.43	1967	2.59	1960
S	13	6.56	16.27	1960	2.53	1967	7.39	1957
O	13	5.26	10.65	1949	2.01	1957	4.00	1961
N	13	5.23	9.82	1957	2.05	1964	5.48	1966
D	13	4.89	11.28	1966	1.92	1967	2.76	1965
YR	13	55.72	16.27	SEP 1960	0.23	MAR 1958	8.25	MAR 1949

Snow, Sleet: Mean 0.0, Maximum monthly 0.0, Maximum in 24 hrs 0.0 for all months; annual 0.0.

Relative humidity (mean %)

Month	(n)	0200 LST	0800 LST	1400 LST	2000 LST
J	13	79	70	66	76
F	13	80	78	65	76
M	13	80	76	63	76
A	13	81	75	66	78
M	13	84	78	70	81
J	13	82	76	69	79
J	13	81	76	69	80
A	13	82	77	68	80
S	13	84	79	69	81
O	13	84	81	68	80
N	13	83	81	67	80
D	13	81	81	68	79
YR	13	82	78	67	79

Wind / Cloud

Month	(n)	Mean speed (Kts)	Prevailing direction	Peak gust Speed (Kts)	Direction	Year	Mean cloud amount
J	13	8.8	ENE	47	ENE	1950	.5
F	13	8.5	E	43	ESE	1950+	.5
M	13	8.2	E	41	ESE	1949	.5
A	13	8.6	E	35	S	1960	.5
M	13	8.0	E	37	S	1960	.6
J	13	8.3	E	40	E	1960	.6
J	13	9.0	E	46	ENE	1960+	.6
A	13	8.3	E	44	SW	1949	.5
S	13	5.4	E	76	SW	1960	.6
O	13	5.4	E	38	S	1958	.6
N	13	6.0	E	41	NE	1949	.5
D	13	7.3	E	40	SW	1949	.5
YR	13	7.7	E	76	SW	SEP 1960	.5

Mean number of days

Month	(n)	Precip .01 in or more	Snow,Sleet 1.0 in or more	Thunderstorms	Fog	Max 90° and above	Max 32° and below	Min 32° and below	Min 0° and below
J	13	17	0	1	*	*	0	0	0
F	13	16	0	*	*	0	0	0	0
M	13	13	0	1	0	*	0	0	0
A	13	15	0	*	0	*	0	0	0
M	13	19	0	1	*	0	0	0	0
J	13	18	0	2	0	2	0	0	0
J	13	21	0	5	0	3	0	0	0
A	13	19	0	4	0	5	0	0	0
S	13	20	0	5	*	5	0	0	0
O	13	17	0	5	*	4	0	0	0
N	13	17	0	2	0	2	0	0	0
D	13	19	0	1	0	*	0	0	0
YR	13	211	0	33	*	23	0	0	0

Mean number of 3 hrly obs., Temperature

Month	(n)	73° wb and above	93° db and above	67° wb and above	80° db and above
J	13	60	0	227	59
F	13	51	0	205	58
M	13	76	0	234	82
A	13	123	0	231	92
M	13	205	0	247	126
J	13	228	0	239	164
J	13	243	0	248	201
A	13	245	0	247	207
S	13	230	0	239	181
O	13	218	0	247	160
N	13	169	0	237	115
D	13	121	0	232	80
YR	13	1969	0	2833	1525

Station moved or discontinued. See other stations for this state.

REFERENCE NOTES APPLYING TO TABLES APPEAR ON THE PAGE FOLLOWING LAST TABLE.
(Caution: Letters and symbols may have different meanings in 1941-1970 tables than in earlier tables. See notes.)

NORMALS, MEANS AND EXTREMES

(Table Revised 1953. Base Period for Climatological Normals: 1921-1950)

1953
LATITUDE 17° 42' N
LONGITUDE 64° 48' W
ELEVATION (ground) 53 feet

Month	Temperature Normal Daily maximum	Daily minimum	Monthly	Extremes Record highest	Year	Record lowest	Year	Normal degree days	Precipitation Normal total	Maximum monthly	Year	Minimum monthly	Year	Maximum in 24 hrs.	Year	Snow, Sleet, Hail Mean total	Maximum monthly	Year	Maximum in 24 hrs.	Year
	(b)	(b)	(b)	6				(b)	(b)	6				6		6				
J	82.5	69.1	75.8	88	1948	64	1953+		2.97	4.77	1952	1.17	1953	2.88	1952	0.0	0.0		0.0	
F	82.7	69.0	75.9	88	1948	62	1951		2.80	8.33	1950	0.93	1953	3.95	1950	0.0	0.0		0.0	
M	83.7	69.5	76.6	88	1949	62	1951		2.33	6.63	1949	0.17	1949	3.42	1949	0.0	0.0		0.0	
A	84.7	71.5	78.1	89	1951	65	1953+		2.21	5.38	1952	0.22	1953	1.24	1948	0.0	0.0		0.0	
M	85.8	73.7	79.8	90	1948	68	1950		4.90	6.21	1950	2.72	1950	2.84	1951	0.0	0.0		0.0	
J	87.0	75.0	81.0	91	1953	68	1953		2.93	4.76	1953	1.70	1952	3.47	1953	0.0	0.0		0.0	
J	88.2	75.5	81.9	95	1948	69	1949		2.77	6.14	1952	1.51	1947	2.00	1952	0.0	0.0		0.0	
A	88.7	75.2	82.0	93	1953	70	1947		4.04	6.67	1952	2.22	1952	2.19	1947	0.0	0.0		0.0	
S	87.8	74.1	81.0	92	1953+	68	1949		5.84	15.23	1947	2.61	1948	7.76	1948	0.0	0.0		0.0	
O	86.7	73.7	80.2	90	1953+	67	1951		5.34	8.66	1950	2.58	1950	2.35	1951	0.0	0.0		0.0	
N	85.1	72.7	78.9	89	1953+	64	1952		4.17	10.35	1953	2.43	1953	4.49	1953	0.0	0.0		0.0	
D	83.7	71.0	77.4	87	1953+	63	1949		2.05	4.56	1953	1.08	1953	1.89	1952	0.0	0.0		0.0	
Year	85.6	72.5	79.1	95	July 1948	62	March 1951+		42.35	15.23	Sept 1947	0.17	Mar. 1951	7.76	Sept 1953	0.0	0.0		0.0	

Month	Relative humidity 2:30 a.m. AST	8:30 a.m. AST	2:30 p.m. AST	8:30 p.m. AST	Wind Mean hourly speed	Prevailing direction	Mean sky cover sunrise to sunset	Mean number of days Clear	Partly cloudy	Cloudy	Precipitation .01 inch or more	Snow, Sleet, Hail 1.0 inch or more	Thunderstorms	Heavy fog	Temp. Max 90° and above	Max 32° and below	Min 32° and below	Min 0° and below
	6	6	6	6	6	6	6	6	6	6	6	6	6	6	6	6	6	6
J	81	76	66	80	12.7	ENE	5.6	5	20	6	19	0	0	0	0	0	0	0
F	84	76	63	80	12.2	ENE	5.7	5	17	6	13	0	0	0	0	0	0	0
M	83	73	63	79	12.6	ENE	5.7	5	21	5	13	0	1	0	0	0	0	0
A	85	74	66	81	12.3	ESE	5.9	4	19	7	13	0	1	0	0	0	0	0
M	87	78	70	83	12.0	ESE	7.2	2	14	15	16	0	4	0	0	0	0	0
J	84	75	69	82	13.0	ESE	6.9	1	18	11	15	0	5	0	4	0	0	0
J	86	77	70	84	13.0	ENE	7.0	1	17	13	19	0	4	0	6	0	0	0
A	87	77	69	85	12.1	ENE	6.1	2	21	8	18	0	6	0	10	0	0	0
S	88	78	73	86	11.0	ESE	7.0	1	15	13	17	0	9	0	5	0	0	0
O	90	80	73	87	9.4	ESE	6.5	3	17	11	17	0	7	0	2	0	0	0
N	90	80	72	87	9.4	ENE	6.0	4	19	7	18	0	4	0	0	0	0	0
D	86	79	69	83	11.2	ENE	5.8	5	18	8	18	0	2	0	0	0	0	0
Year	86	77	69	83	11.7	ENE	6.3	39	216	110	196	0	43	0	27	0	0	0

(a) Length of record, years.
Normal values are based on the period 1921-1950
(b)

REFERENCE NOTES APPLYING TO TABLES APPEAR ON THE PAGE FOLLOWING LAST TABLE.
(Caution: Letters and symbols may have different meanings in 1941-1970 tables than in earlier tables. See notes.)

SANTA ISABEL, PUERTO RICO
SANTA ISABEL AIRPORT
1953

LATITUDE 17° 58' N
LONGITUDE 66° 24' W
ELEVATION (ground) 28 feet

NORMALS, MEANS AND EXTREMES
(Table Revised 1953. Base Period for Climatological Normals: 1921-1950)

Temperature

Month	Normal Daily maximum (b)	Normal Daily minimum	Normal Monthly (b)	Record highest	Year	Record lowest	Year
J	83.0	64.9	74.0	89	1947	59	1951
F	82.0	64.3	73.2	88	1949+	57	1951
M	82.2	65.1	73.7	89	1948	58	1951
A	83.0	67.1	75.1	90	1947	62	1953+
M	84.0	70.4	77.2	89	1952+	60	1950
J	85.6	71.7	78.7	94	1948	67	1950
J	86.9	72.0	79.5	96	1948	67	1950
A	87.1	72.0	79.6	95	1950	68	1950
S	87.0	71.6	79.3	93	1953+	68	1952+
O	85.0	70.5	78.2	91	1947	67	1950
N	84.9	68.6	76.8	91	1948	63	1952
D	83.8	66.0	74.9	88	1953+	59	1950+
Year	84.6	68.7	76.7	96	July 1948	57	Feb. 1951

Precipitation (inches)

Month	Normal total (b)	Maximum monthly	Year	Minimum monthly	Year	Maximum in 24 hrs.	Year
J	0.74	1.19	1952	0.05	1948	0.91	1952
F	0.82	1.30	1950	0.07	1948	0.54	1950
M	0.48	1.07	1949	0.09	1951	0.44	1949
A	1.41	3.73	1952	0.22	1949	1.66	1952
M	4.29	5.28	1946	0.31	1947	1.70	1947
J	3.38	4.02	1952	0.77	1946	1.94	1946
J	3.41	6.11	1949	0.19	1946	3.97	1949
A	4.36	7.02	1952	2.46	1951	3.33	1953
S	4.54	12.37	1952	2.17	1950	6.62	1949
O	4.24	12.10	1948	2.49	1948	4.02	1950
N	3.96	7.66	1950	0.45	1952	5.17	1950
D	1.02	2.62	1948	0.10	1952	1.98	1948
Year	32.65	12.37	Sept 1952	0.05	Jan 1948	6.62	Sept 1949

Snow, Sleet, Hail

Month	Mean total	Maximum monthly	Year	Maximum in 24 hrs.	Year
J	0.0	0.0		0.0	
F	0.0	0.0		0.0	
M	0.0	0.0		0.0	
A	0.0	0.0		0.0	
M	0.0	0.0		0.0	
J	0.0	0.0		0.0	
J	0.0	0.0		0.0	
A	0.0	0.0		0.0	
S	0.0	0.0		0.0	
O	0.0	0.0		0.0	
N	0.0	0.0		0.0	
D	0.0	0.0		0.0	
Year	0.0	0.0		0.0	

Relative humidity / Wind / Sky cover

Month	2:30 a.m. AST	8:30 a.m. AST	2:30 p.m. AST	8:30 p.m. AST	Mean hourly wind speed	Prevailing direction	Mean sky cover sunrise to sunset
J	83	78	59	76	6.7	ENE	4.3
F	84	76	61	75	6.2	NE	4.9
M	83	71	61	75	7.7	ESE	4.5
A	73	62	57	66	7.1	SE	5.0
M	76	63	58	69	6.5	SE	5.5
J	75	63	58	68	7.2	SE	6.4
J	74	61	58	67	7.4	SE	6.1
A	86	71	67	79	6.9	E	5.2
S	87	76	70	85	5.8	E	6.5
O	90	78	71	86	5.4	N	6.1
N	90	80	66	85	5.1	NE	5.0
D	88	80	63	82	5.2	NE	4.6
Year	82	72	62	76	6.4	SE	5.4

Mean number of days

Month	Clear	Partly cloudy	Cloudy	Precipitation .01 inch or more	Thunderstorms	Heavy fog	Temp. 90° and above
J	13	15	3	4	0	0	0
F	11	14	3	5	0	0	0
M	14	14	3	3	0	0	0
A	11	12	7	7	0	0	0
M	4	16	11	11	5	0	0
J	4	16	10	11	6	0	6
J	7	14	10	9	6	0	10
A	9	16	6	9	8	0	10
S	5	13	12	13	12	0	10
O	6	16	9	11	15	0	2
N	9	15	6	10	18	0	1
D	12	14	5	8	1	0	0
Year	105	175	85	99	70	0	35

(Snow/Sleet 1.0 inch or more, Max. 32° and below, Min. 32° and below, Min. 0° and below: all 0)

(a) Length of record, years.
(b) Normal values are based on the period 1931-1960.

Station moved or discontinued. See other stations for this state.

REFERENCE NOTES APPLYING TO TABLES APPEAR ON THE PAGE FOLLOWING LAST TABLE.
(Caution: Letters and symbols may have different meanings in 1941-1970 tables than in earlier tables. See notes.)

NORMALS, MEANS AND EXTREMES

(Table Revised 1970. Base Period for Climatological Normals: 1931-1960)

1964

LATITUDE 18° 28' N
LONGITUDE 66° 06' W
ELEVATION (ground) 47 feet

Month	Normal Daily max.	Normal Daily min.	Normal Monthly	Extremes Record highest	Year	Extremes Record lowest	Year	Normal degree days	Precip. Normal total	Max. monthly	Year	Min. monthly	Year	Max. in 24 hrs	Year	Snow,Sleet Mean total	Snow Max. monthly	Snow Max. 24 hrs	Precip. .01"+	Snow 1.0"+	Thunderstorms	Heavy fog	Max 90°+	Max 32°-	Min 32°-	Min 0°-
(years)	(b)	(b)	(b)	66		66		(b)	(b)	66		66		66		66	66	66	66	66	66	66	66	66	66	66
J	79.9	70.3	75.1	90	1958	63	1917+		4.13	15.47	1937	0.74	1960	4.52	1934	0.0	0.0	0.0	20	0	*	0	*	0	0	0
F	80.0	70.1	75.1	92	1964	62	1917		2.70	8.13	1954	0.05	1941	4.84	1918	0.0	0.0	0.0	14	0	*	0	*	0	0	0
M	81.0	71.0	76.0	96	1958	63	1911		2.07	9.38	1927	0.29	1957	5.26	1956	0.0	0.0	0.0	14	0	*	0	*	0	0	0
A	81.8	72.2	77.0	93	1900	65	1922+		3.89	13.17	1915	0.42	1935	6.72	1915	0.0	0.0	0.0	14	0	2	0	1	0	0	0
M	84.4	74.1	78.8	94	1964+	66	1903		5.83	16.88	1936	0.71	1925	5.37	1936	0.0	0.0	0.0	16	0	5	0	2	0	0	0
J	84.2	75.4	79.9	94	1964	66	1909		7.16	12.22	1902	1.43	1945	6.47	1938	0.0	0.0	0.0	17	0	6	0	1	0	0	0
J	85.3	76.0	80.1	92	1942	70	1960+		6.02	11.19	1961	1.10	1929	4.05	1901	0.0	0.0	0.0	19	0	7	0	*	0	0	0
A	85.8	76.2	80.8	94	1955	68	1904		6.34	14.10	1944	1.53	1919	9.83	1944	0.0	0.0	0.0	20	0	10	0	1	0	0	0
S	85.5	75.6	80.7	94	1961+	69	1929+		6.04	13.68	1928	1.49	1940	8.50	1928	0.0	0.0	0.0	18	0	8	0	2	0	0	0
O	85.5	74.9	80.2	96	1963	68	1903+		5.24	12.15	1927	1.61	1915	3.87	1927	0.0	0.0	0.0	18	0	6	0	2	0	0	0
N	83.5	73.7	78.6	93	1915	65	1898		6.05	15.64	1931	1.63	1910	8.06	1931	0.0	0.0	0.0	19	0	3	0	*	0	0	0
D	81.4	72.0	76.7	90	1962+	62	1917+		4.89	15.40	1910	0.58	1963	10.55	1910	0.0	0.0	0.0	20	0	1	0	*	0	0	0
YR	83.0	73.5	78.3	96	OCT. 1963+	62	DEC. 1917+	(b)	60.36	16.88	MAY 1936	0.05	FEB. 1941	10.55	DEC. 1910	0.0	0.0	0.0	208	0	49	0	10	0	0	0

Station moved or discontinued. See other stations for this state.

REFERENCE NOTES APPLYING TO TABLES APPEAR ON THE PAGE FOLLOWING LAST TABLE.

(Caution: Letters and symbols may have different meanings in 1941-1970 tables than in earlier tables. See notes.)

NORMALS, MEANS AND EXTREMES

(Table Revised 1970. Base Period for Climatological Normals: 1931-1960)

Station: SAN JUAN, PUERTO RICO   ISLA VERDE AIRPORT   Standard time used: ATLANTIC   Latitude: 18° 26' N   Longitude: 66° 00' W   Elevation (ground): 13 feet

Month	Temperature Normal Daily maximum	Normal Daily minimum	Monthly	Extremes Record highest	Year	Record lowest	Year	Normal heating degree days (Base 65°)	Precipitation Normal total	Maximum monthly	Year	Minimum monthly	Year	Maximum in 24 hrs.	Year	Snow, Ice pellets Mean total	Maximum monthly	Maximum in 24 hrs.	Relative humidity Hour 02	Hour 08	Hour 14	Hour 20	Wind Mean speed	Prevailing direction	Fastest mile Speed	Direction	Year	Pct. of possible sunshine	Mean sky cover sunrise to sunset	Sunrise to sunset Clear	Partly cloudy	Cloudy	Precipitation .01 inch or more	Snow, Ice pellets 1.0 inch or more	Thunderstorms	Heavy fog	Temperatures Max. 90° and above	32° and below	Min. 32° and below	0° and below	Average daily solar radiation - langleys
(a)	(b)	(b)	(b)	15		15		(b)	(b)	15		15		15		14		15	14	14	14	14	14	8	14	14	14	14	14	14	14	14	14	14	14	14	14	14	14	14	14
J	81.1	67.4	74.4	90	1962	61	1962		4.70	7.49	1969	0.94	1960	5.08	1969	0.0	0.0	0.0	84	85	66	77	8.7	ENE	40	NE	1967+	66	5.2	8	18	5	17	0.0	*	0	0	*	0	0	0	
F	81.3	67.0	74.4	92	1968	62	1968		2.90	6.44	1956	0.79	1965	2.73	1969	0.0	0.0	0.0	83	83	63	75	9.1	ENE	36	NE	1965	68	5.1	7	16	5	13	0.0	*	0	1	0	0	0	0	
M	83.1	67.5	75.3	93	1958	60	1957		2.20	5.41	1958	0.86	1968	3.91	1969	0.0	0.0	0.0	83	83	61	75	9.4	ENE	37	E	1967	74	5.1	9	16	4	12	0.0	*	0	2	0	0	0	0	
A	84.2	69.2	76.7	93	1956	64	1968		3.12	6.38	1964	0.50	1968	3.15	1968	0.0	0.0	0.0	86	84	63	79	9.0	ENE	35	S	1961	61	5.7	6	15	4	14	0.0	1	0	3	0	0	0	0	
M	85.0	71.5	78.7	93	1956	66	1962		7.12	14.99	1965	1.77	1961	3.08	1965	0.0	0.0	0.0	86	81	68	79	8.6	ENE	35	NE	1969	61	6.8	5	16	10	20	0.0	4	0	4	0	0	0	0	
J	85.7	72.9	80.0	94	1968+	69	1957+		5.06	10.96	1955	2.19	1963	3.55	1955	0.0	0.0	0.0	87	82	68	79	8.6	ENE	32	NE	1956	58	6.8	2	15	13	18	0.0	6	0	3	0	0	0	0	
J	85.1	73.7	80.4	93	1955	69	1959		6.25	9.35	1961	2.80	1959	2.28	1969	0.0	0.0	0.0	87	82	67	79	9.0	ENE	34	SE	1958	65	6.3	4	18	9	21	0.0	5	0	3	0	0	0	0	
A	87.1	74.0	80.5	92	1957+	70	1956		7.13	11.76	1955	3.41	1966	5.08	1955	0.0	0.0	0.0	87	82	67	80	8.9	ENE	36	NE	1956	65	6.5	3	18	10	19	0.0	6	0	5	0	0	0	0	
S	87.0	73.2	80.5	94	1957	69	1960		6.76	10.85	1963	1.93	1961	3.08	1963	0.0	0.0	0.0	89	83	68	81	6.5	ENE	36	NE	1966	59	6.5	3	18	10	19	0.0	8	0	4	0	0	0	0	
O	87.1	72.8	82.0	94	1957+	67	1959		5.53	8.99	1960	1.63	1963	3.42	1963	0.0	0.0	0.0	97	83	67	80	6.5	ENE	34	NE	1958	58	6.3	4	18	8	18	0.0	7	0	5	0	0	0	0	
N	85.0	71.4	78.2	92	1950	66	1969		6.49	11.11	1968	3.00	1969	3.72	1968	0.0	0.0	0.0	85	82	67	80	7.4	ENE	35	NE	1961	60	6.1	4	18	8	18	0.0	3	0	1	0	0	0	0	
D	82.7	69.6	76.2	92	1955+	63	1964		5.45	10.00	1961	0.68	1963	3.54	1961	0.0	0.0	0.0	85	84	67	79	8.3	ENE	35	NE	1965+	61	5.8	4	18	8	18	0.0	1	0	*	0	0	0	0	
YR	85.0	70.9	78.0	94	SEP. 1957+	60	MAR. 1957		64.21	14.99	MAY 1965	0.50	APR. 1968	5.08	JAN. 1969+	0.0	0.0	0.0	86	82	66	78	8.5	ENE	80	NE	AUG. 1956	64	6.0	54	208	103	205	0.0	40	0	31	0	0	0	0	

Means and extremes above are from existing and comparable exposures. Annual extremes have been exceeded at other sites in the locality as follows:
Annual extremes have been exceeded at City Office locations as follows (1890 - 1964 record): Highest temperature 96 in October 1963 and earlier;
maximum monthly precipitation 16.88 in May 1936; minimum monthly precipitation 0.05 in February 1941; maximum precipitation in 24 hours 10.55 in
December 1910.

NORMALS, MEANS AND EXTREMES

(Table Revised 1975. Base Period for Climatological Normals: 1941-1970)

Month	Temperatures °F Normal Daily maximum	Normal Daily minimum	Monthly	Extremes Record highest	Year	Record lowest	Year	Normal Degree days Base 65°F Heating	Cooling	Precipitation in inches Normal	Water equivalent Maximum monthly	Year	Minimum monthly	Year	Maximum in 24 hrs.	Year	Snow, Ice pellets Maximum monthly	Year	Maximum in 24 hrs.	Relative humidity pct. Hour 02	Hour 08	Hour 14	Hour 20	Wind Mean speed m.p.h.	Prevailing direction	Fastest mile Speed m.p.h.	Direction	Year	Pct. of possible sunshine	Mean sky cover, tenths, sunrise to sunset	Sunrise to sunset Clear	Partly cloudy	Cloudy	Precipitation .01 inch or more	Snow, Ice pellets 1.0 inch or more	Thunderstorms	Heavy fog, visibility ¼ mile or less	Temperatures °F Max. 90° and above	32° and below	Min. 32° and below	0° and below	Average station pressure mb.
																															Elev. 62 feet m.s.l.											2
(a)	20	20	20	20		20					20		20		20		20		20	19	19	19	19	19	8	19	19	19	19	19	19	19	19	19	19	19	19	19	19	19	19	
J	81.9	68.8	75.4	90	1958	61	1962	0	322	3.73	7.49	1969	0.94	1960	5.08	1969	0.0		0.0	82	80	65	75	9.1	ENE	40	SE	1974	64	5.2	8	18	4	17	0.0	*	0	*	0	0	0	1015.8
F	82.1	68.4	75.3	92	1958	62	1968	0	288	2.50	6.44	1956	0.79	1965	2.73	1969	0.0		0.0	81	80	63	73	9.1	ENE	36	NE	1965	68	5.1	8	17	4	13	0.0	*	0	1	0	0	0	1014.7
M	83.4	68.6	76.3	93	1958	60	1957	0	350	2.04	5.41	1958	0.72	1968	3.91	1973	0.0		0.0	81	78	60	73	9.6	ENE	37	NE	1969	74	5.1	9	16	4	12	0.0	1	0	2	0	0	0	1014.1
A	84.4	70.6	77.5	94	1973	64	1968	0	375	3.40	8.48	1973	0.50	1968	6.37	1973	0.0		0.0	81	76	62	73	9.4	ENE	69	NE	1970	69	5.5	7	16	7	14	0.0	2	0	4	0	0	0	1014.1
M	85.6	72.5	79.2	96	1972	66	1962	0	440	6.54	14.99	1965	0.44	1971	3.08	1965	0.0		0.0	83	78	65	76	8.8	ENE	62	NE	1969	57	6.5	5	17	9	17	0.0	4	0	6	0	0	0	1013.9
J	87.0	74.0	80.5	96	1972	69	1957	0	465	5.44	10.96	1955	1.24	1971	3.55	1955	0.0		0.0	84	79	65	77	9.0	ENE	63	NE	1970	57	6.6	3	15	12	17	0.0	6	0	6	0	0	0	1013.7
J	87.0	74.8	80.9	93	1974	69	1959	0	493	6.41	9.35	1961	1.12	1974	2.28	1969	0.0		0.0	83	79	66	77	9.9	ENE	34	SE	1958	64	6.2	4	18	9	20	0.0	5	0	4	0	0	0	1014.6
A	87.5	75.1	81.3	96	1974	70	1956	0	505	6.98	11.76	1955	3.06	1972	5.08	1955	0.0		0.0	84	80	67	78	9.7	ENE	47	NE	1973	66	6.4	4	18	10	19	0.0	7	0	5	0	0	0	1013.5
S	87.6	74.6	81.1	94	1974	69	1960	0	483	6.07	10.85	1963	1.93	1961	3.08	1963	0.0		0.0	85	81	67	78	7.6	ENE	44	NE	1972	61	6.3	4	18	11	18	0.0	7	0	6	0	0	0	1011.4
O	87.4	73.7	80.6	95	1970	67	1959	0	484	5.64	15.06	1970	1.63	1971	5.00	1970	0.0		0.0	85	81	66	77	6.8	ENE	44	NE	1961	57	6.3	4	17	11	18	0.0	6	0	1	0	0	0	1011.2
N	85.0	72.3	78.7	92	1973	66	1969	0	411	5.49	11.11	1968	2.31	1971	3.72	1968	0.0		0.0	84	82	66	77	7.7	ENE	46	NE	1961	57	5.9	5	18	8	18	0.0	3	0	1	0	0	0	1012.2
D	83.1	70.5	76.8	90	1965	63	1964	0	366	4.71	10.00	1961	0.68	1963	3.54	1961	0.0		0.0	82	81	65	76	8.7	ENE	46	NE	1970	57	5.8	5	17	7	19	0.0	*	0	*	0	0	0	1014.2
YR	85.2	72.0	78.6	96	AUG. 1974	60	MAR. 1957	0	4982	59.15	15.06	OCT. 1970	0.44	MAY 1972	6.37	APR. 1973	0.0		0.0	83	80	65	76	8.8	ENE	80	NE	AUG. 1956	63	5.9	58	210	97	200	0.0	40	0	38	0	0	0	1013.8

Means and extremes above are from existing and comparable exposures.
Annual extremes at City locations (1890-1964) were: Highest temperature 96 in October 1963 and earlier; lowest temperature 62 in December 1917 and
earlier; maximum monthly precipitation 16.88 in May 1936; minimum monthly precipitation 0.05 in February 1941; maximum precipitation in 24 hours
10.55 in December 1910; fastest mile of wind 149 from East in September 1928.

REFERENCE NOTES APPLYING TO TABLES APPEAR ON THE PAGE FOLLOWING LAST TABLE.

(Caution: Letters and symbols may have different meanings in 1941-1970 tables than in earlier tables. See notes.)

Reference notes applying to Normals, Means, and Extremes tables for 1921–1950 base period.

(a) Length of record, years.
'(b) Normal values are based on the period 1921-1950, and are means adjusted to represent observations taken at the present standard location.
• Less than one-half.

- No record.
† Airport data.
‡ City Office data.
# Also on earlier dates, months, or years.
T Trace, an amount too small to measure.

Sky cover is expressed in a range of 0 for no clouds or obscuring phenomena to 10 for complete sky cover. The number of clear days is based on average cloudiness 0-3 tenths; partly cloudy days on 4-7 tenths; and cloudy days on 8-10 tenths. Monthly degree day totals are the sum of the negative departures of average daily temperatures from 65°F. Sleet was included in snowfall totals beginning with July 1948. Heavy fog also includes data referred to at various times in the past as "Dense" or "Thick". The upper visibility limit for heavy fog is 1/4 mile. Data in these tables are based on records through 1957.

Reference notes applying to Normals, Means, and Extremes tables for 1931–1960 base period.

(a) Length of record, years, based on January data. Other months may be for more or fewer years if there have been breaks in the record. Climatological standard normals (1931-1960).
- Less than one half.
+ Also on earlier dates, months, or years.
T Trace, an amount too small to measure.
Below zero temperatures are preceded by a minus sign. The prevailing direction for wind in the Normals, Means, and Extremes table is from records through 1963.
‡ ≥ 70° at Alaskan stations.

Unless otherwise indicated, dimensional units used in this bulletin are: temperature in degrees F.; precipitation, including snowfall, in inches; wind movement in miles per hour; and relative humidity in percent. Heating degree day totals are the sums of negative departures of average daily temperatures from 65°F. Cooling degree day totals are the sums of positive departures of average daily temperatures from 65°F. Sleet was included in snowfall totals beginning with July 1948. The term "Ice pellets" includes solid grains of ice (sleet) and particles consisting of snow pellets encased in a thin layer of ice. Heavy fog reduces visibility to 1/4 mile or less.

Sky cover is expressed in a range of 0 for no clouds or obscuring phenomena to 10 for complete sky cover. The number of clear days is based on average cloudiness 0-3, partly cloudy days 4-7, and cloudy days 8-10 tenths.

Solar radiation data are the averages of direct and diffuse radiation on a horizontal surface. The langley denotes one gram calorie per square centimeter.

& Figures instead of letters in a direction column indicate direction in tens of degrees from true North; i.e., 09 - East, 18 - South, 27 - West, 36 - North, and 00 - Calm. Resultant wind is the vector sum of wind directions and speeds divided by the number of observations. If figures appear in the direction column under "Fastest mile" the corresponding speeds are fastest observed 1-minute values.

To 8 compass points only.

% The station does not operate 24 hours daily. Fog and thunderstorm data therefore may be incomplete.

Reference notes applying to Normals, Means, and Extremes tables for 1941–1970 base period.

(a) Length of record, years, through the current year unless otherwise noted, based on January data.
(b) 70° and above at Alaskan stations.
- Less than one half.
T Trace.

NORMALS - Based on record for the 1941-1970 period.
DATE OF AN EXTREME - The most recent in cases of multiple occurrence.
PREVAILING WIND DIRECTION - Record through 1963.
WIND DIRECTION - Numerals indicate tens of degrees clockwise from true north. 00 indicates calm.
FASTEST MILE WIND - Speed is fastest observed 1-minute value when the direction is in tens of degrees.

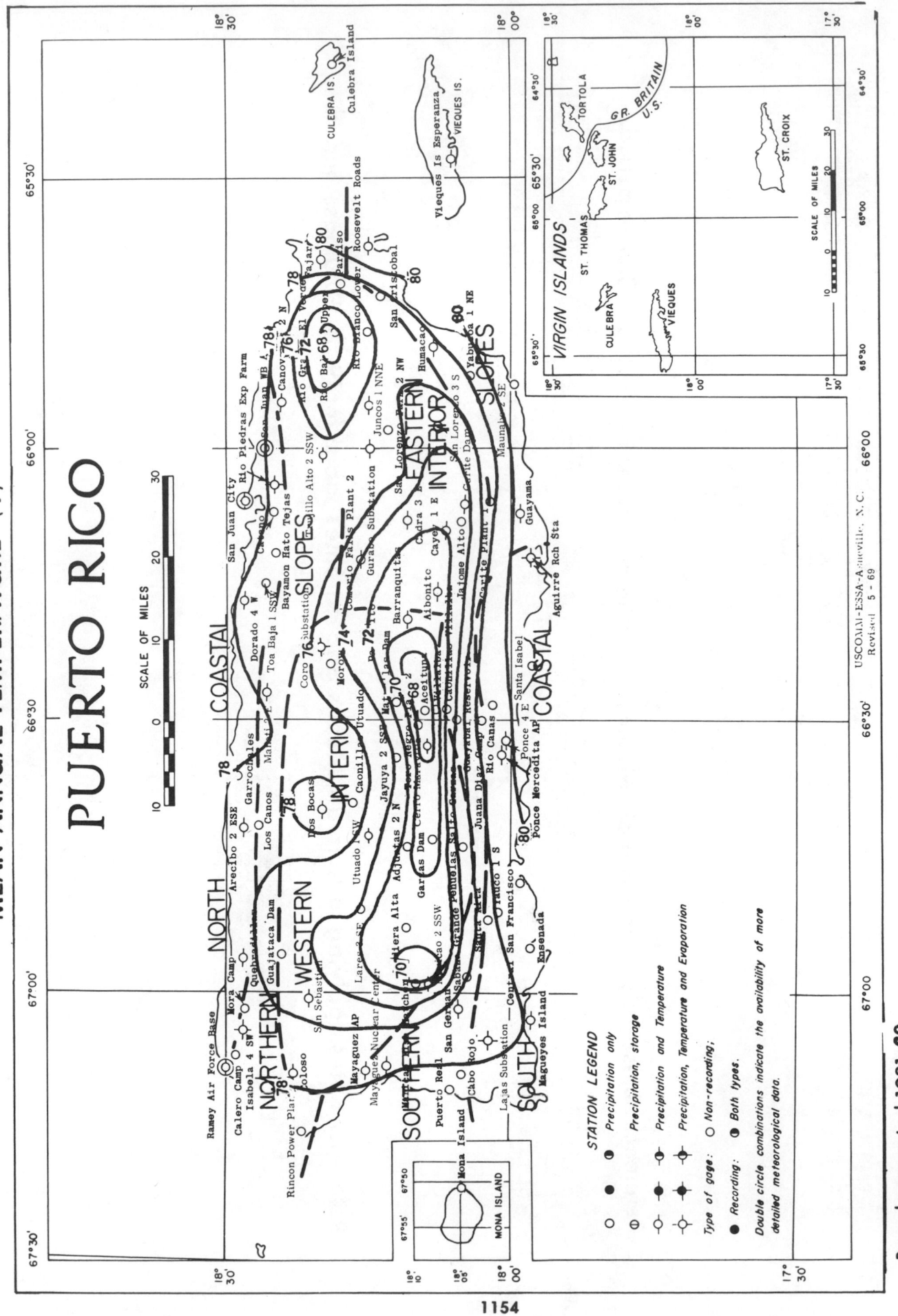

MEAN ANNUAL TEMPERATURE (°F)

PUERTO RICO

SCALE OF MILES

Based on period 1931-60

Isolines are drawn through points of approximately equal value. Caution should be used in interpolating on these maps, particularly in mountainous areas.

USCOMM-ESSA-Asheville, N.C.
Revised 5 - 69

VIRGIN ISLANDS

SCALE OF MILES

MONA ISLAND

STATION LEGEND

● Precipitation only

◐ Precipitation, storage

◑ Precipitation and Temperature

⊕ Precipitation, Temperature and Evaporation

Type of gage: ○ Non-recording;

● Recording: ● Both types.

Double circle combinations indicate the availability of more detailed meteorological data.

1154

MEAN ANNUAL PRECIPITATION, INCHES

PUERTO RICO

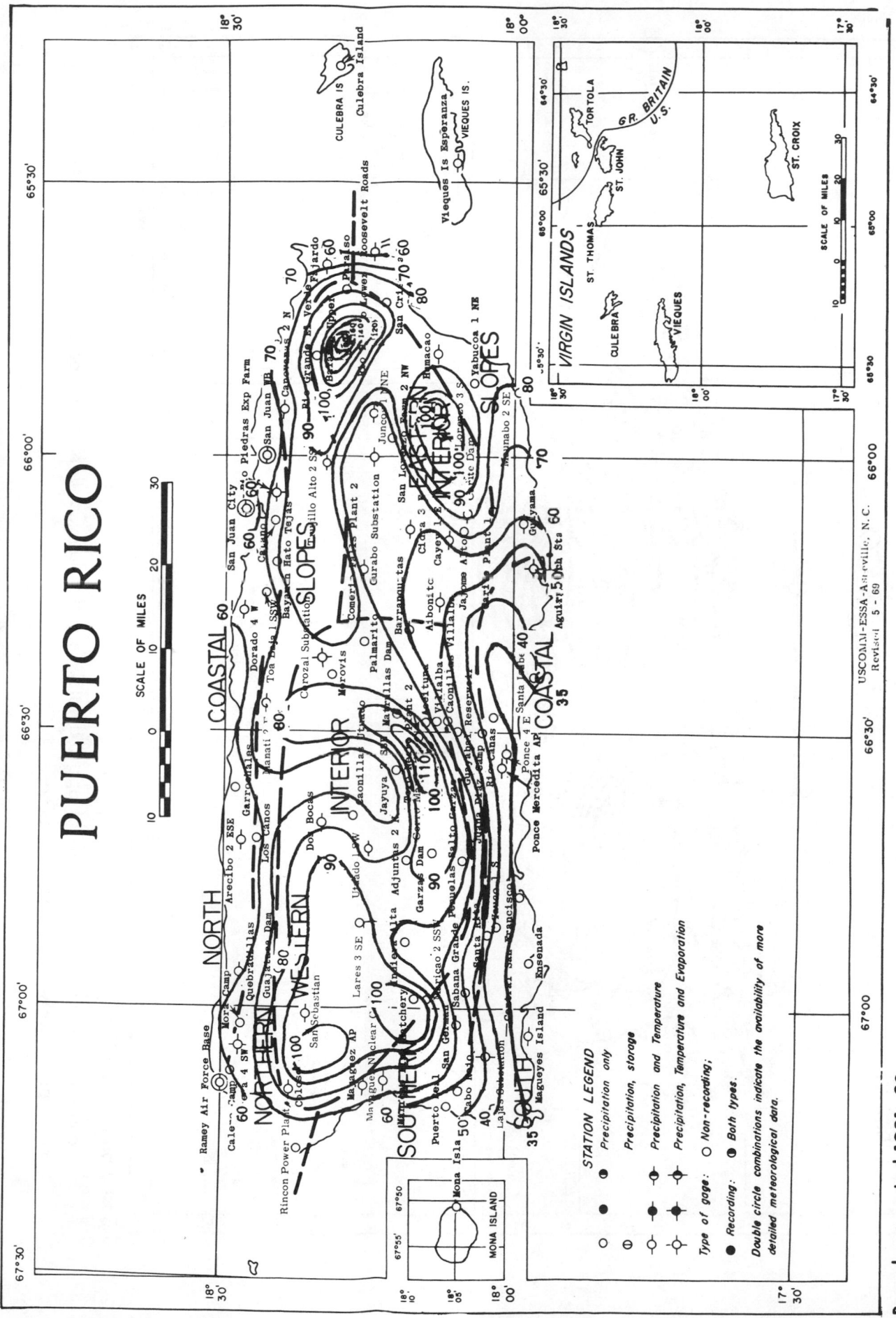

SCALE OF MILES

STATION LEGEND

	Precipitation only
	Precipitation, storage
	Precipitation and Temperature
	Precipitation, Temperature and Evaporation

Type of gage: ○ Non-recording;
Recording: ● Both types.

Double circle combinations indicate the availability of more detailed meteorological data.

MONA ISLAND

VIRGIN ISLANDS

SCALE OF MILES

USCOMM-ESSA-Asheville, N.C.
Revised 5 - 69

Based on period 1931-60

Isolines are drawn through points of approximately equal value. Caution should be used in interpolating on these maps, particularly in mountainous areas.

MEAN MAXIMUM TEMPERATURE (°F), JANUARY

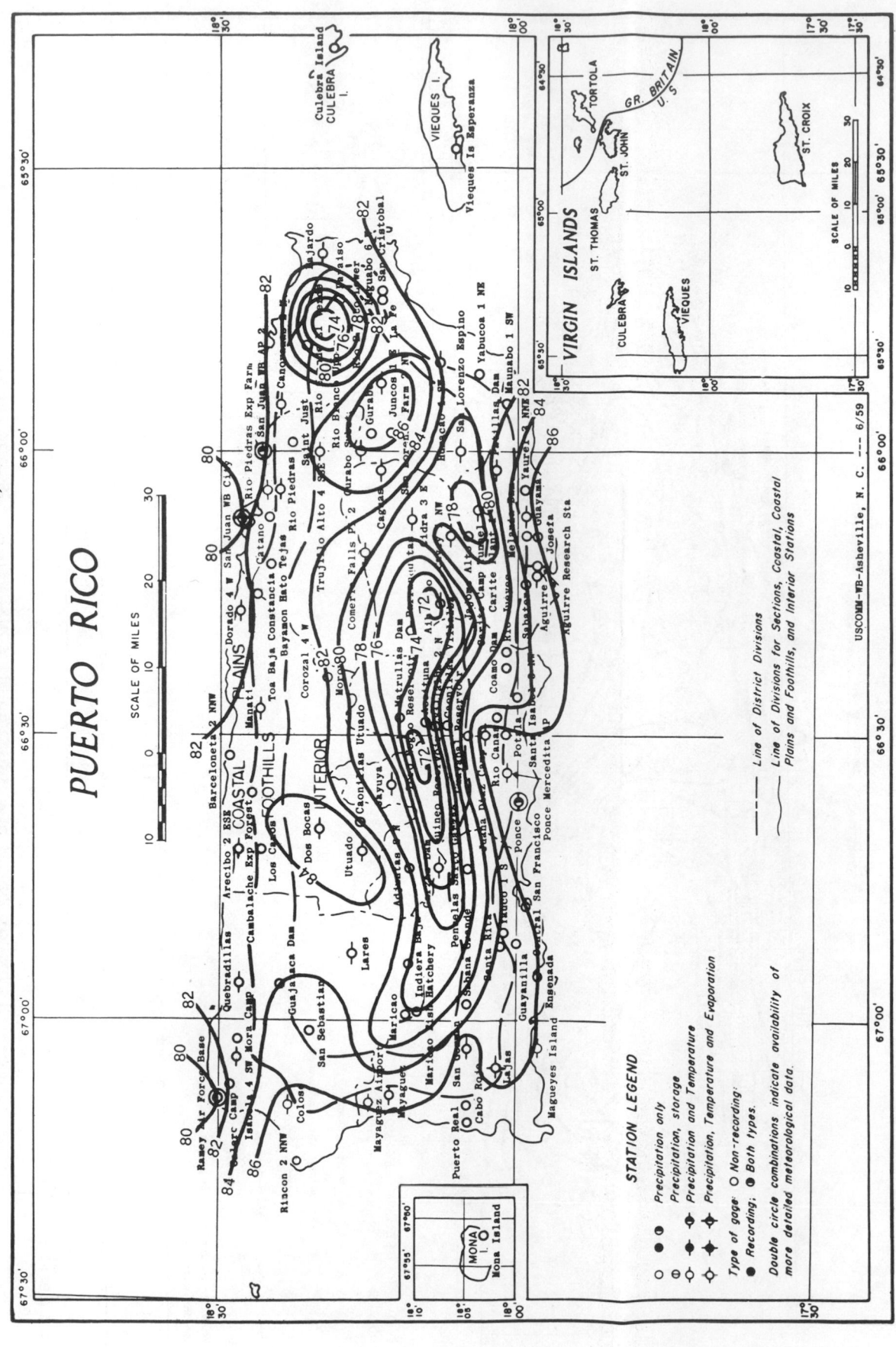

Based on period 1931-52

Isolines are drawn through points of approximately equal value. Caution should be used in interpolating on these maps, particularly in mountainous areas.

MEAN MINIMUM TEMPERATURE (°F), JANUARY

PUERTO RICO

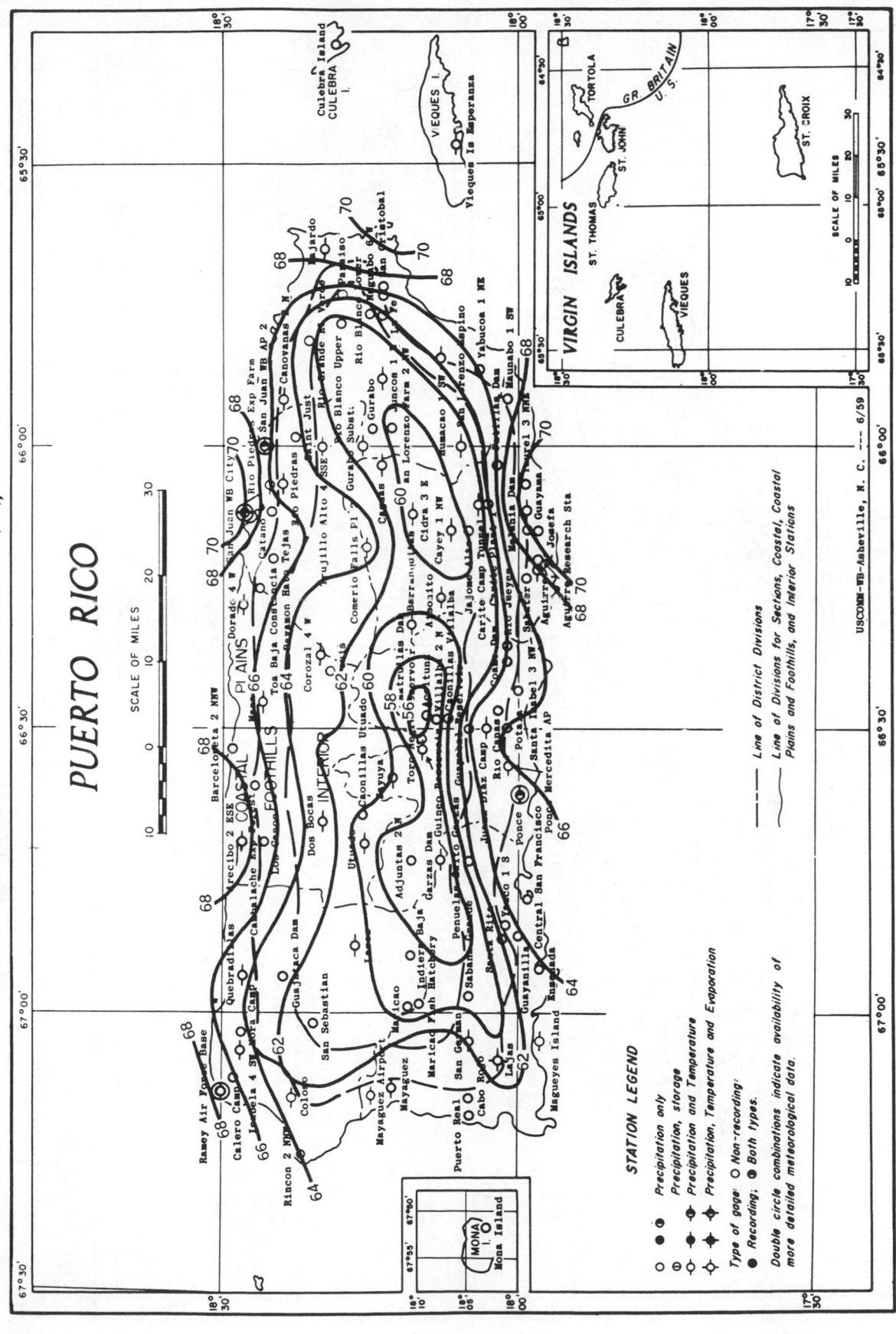

Based on period 1931-52

Isolines are drawn through points of approximately equal value. Caution should be used in interpolating on these maps, particularly in mountainous areas.

USCOMM-WB-Asheville, N. C. --- 6/59

STATION LEGEND

Precipitation only

Precipitation, storage

Precipitation and Temperature

Precipitation, Temperature and Evaporation

Type of gage: ○ Non-recording;

● Recording; ◉ Both types.

Double circle combinations indicate availability of more detailed meteorological data.

—— Line of District Divisions

—— Line of Divisions for Sections, Coastal, Coastal Plains and Foothills, and Interior Stations

VIRGIN ISLANDS

MEAN MAXIMUM TEMPERATURE (°F), JULY

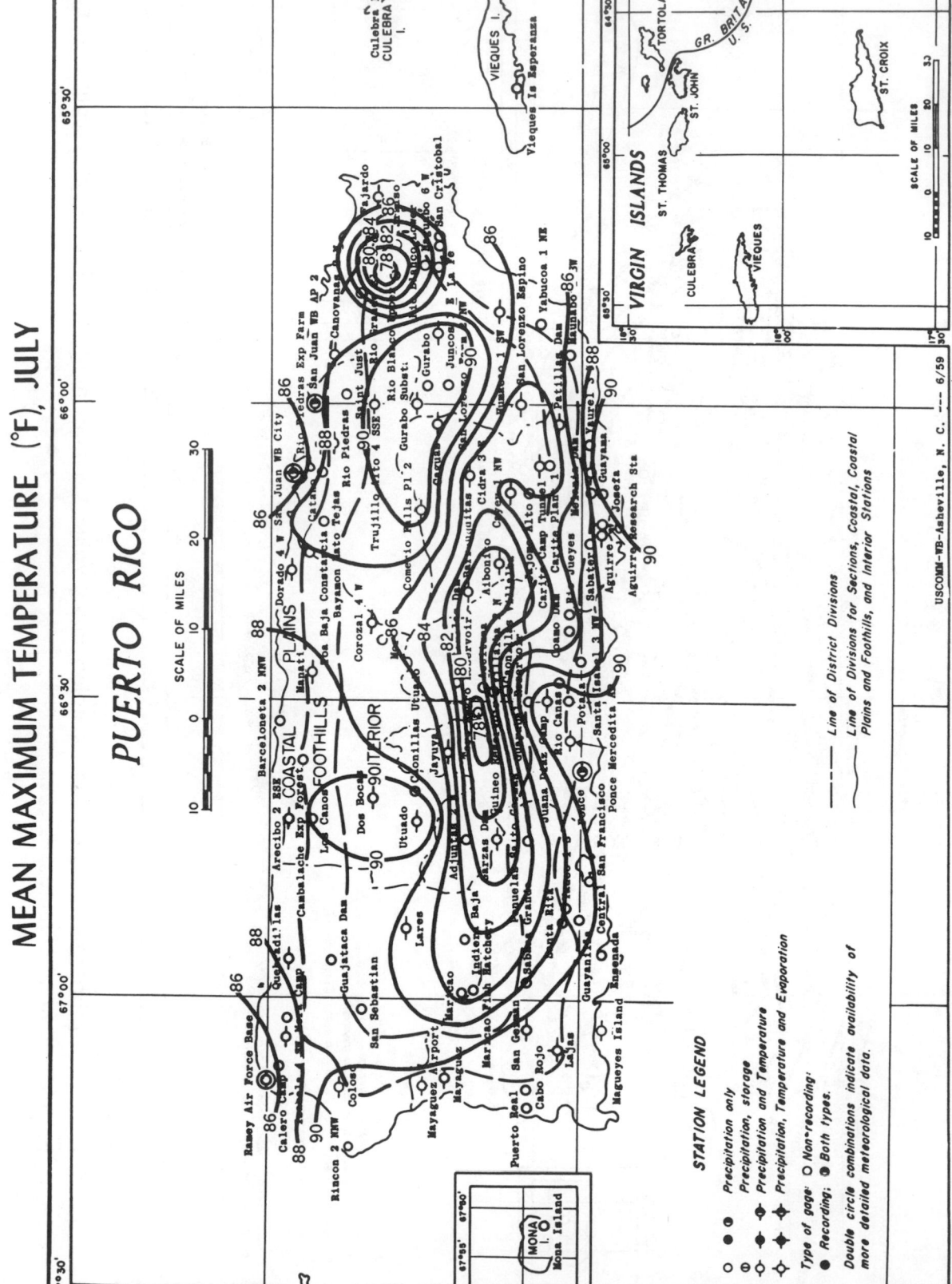

PUERTO RICO

SCALE OF MILES

STATION LEGEND

Precipitation only
Precipitation, storage
Precipitation and Temperature
Precipitation, Temperature and Evaporation

Type of gage: O Non-recording;
Recording: Both types.

Double circle combinations indicate availability of
more detailed meteorological data.

----- Line of District Divisions
—— Line of Divisions for Sections, Coastal, Coastal
Plains and Foothills, and Interior Stations

VIRGIN ISLANDS

SCALE OF MILES

USCOMM-WB-Asheville, N. C. --- 6/59

Based on period 1931-52

Isolines are drawn through points of approximately equal value. Caution should be used in interpolating on these
maps, particularly in mountainous areas.

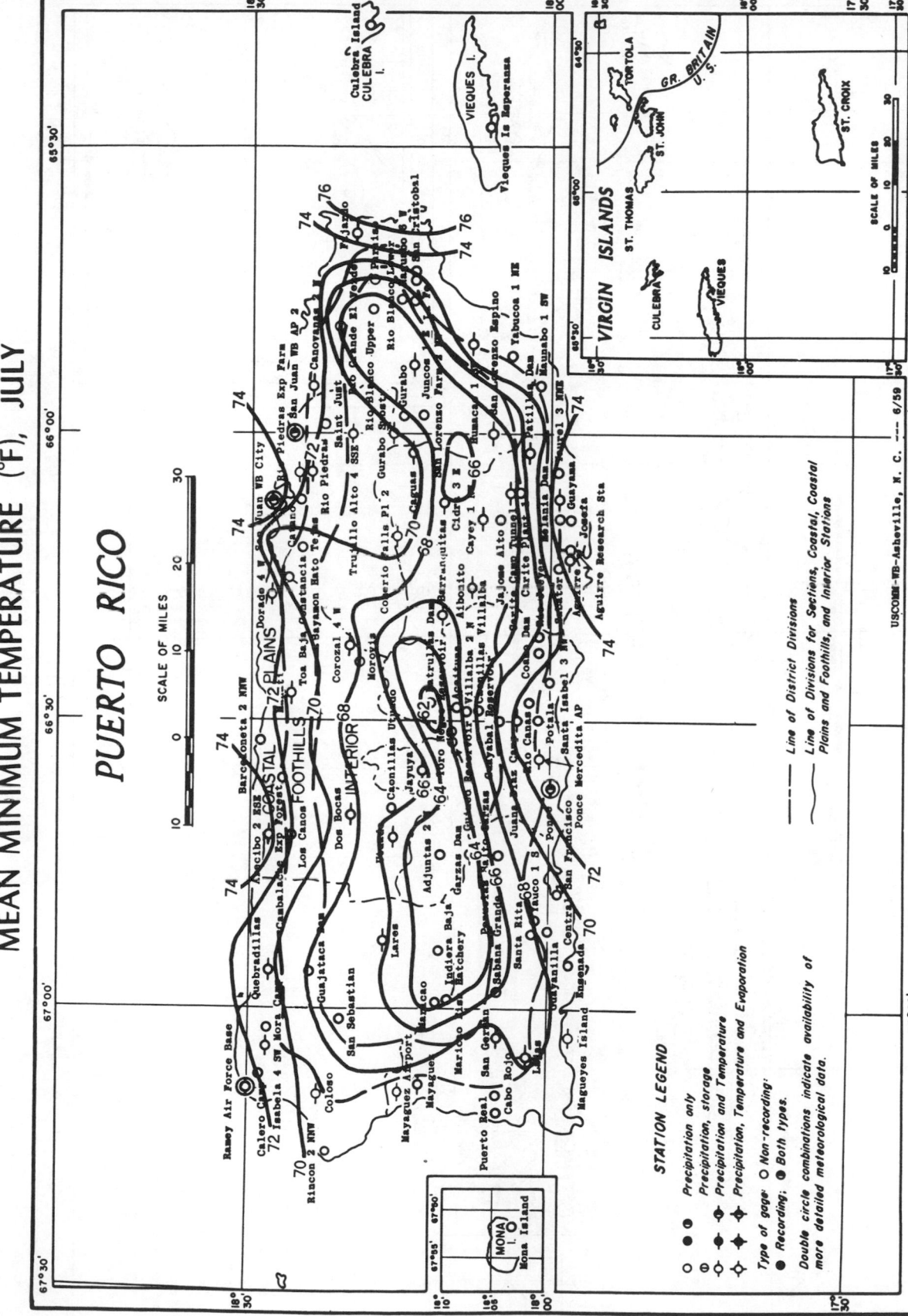

MEAN MINIMUM TEMPERATURE (°F), JULY

Based on period 1931-52

Isolines are drawn through points of approximately equal value. Caution should be used in interpolating on these maps, particularly in mountainous areas.

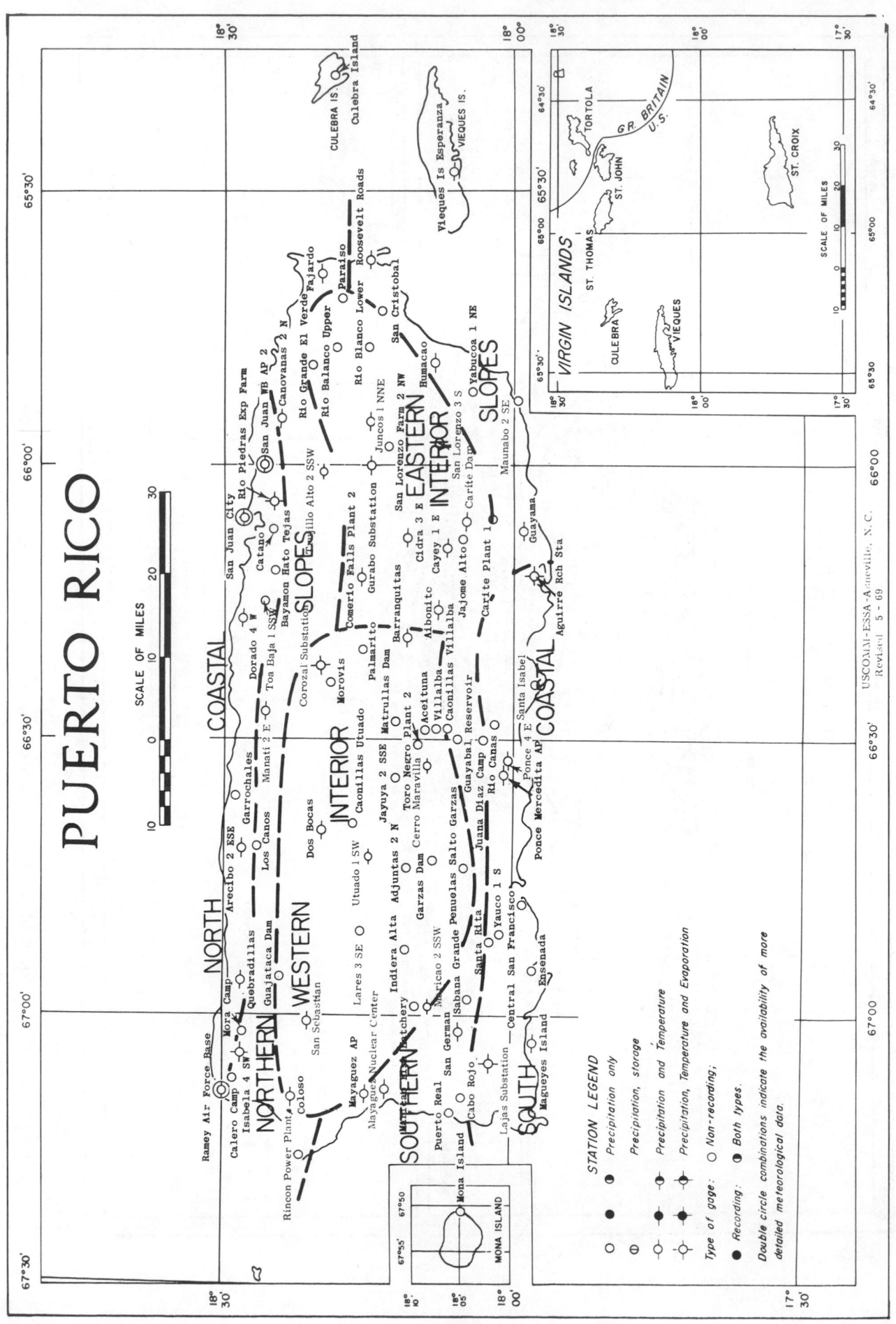

PUERTO RICO

SCALE OF MILES

10 0 10 20 30

VIRGIN ISLANDS

TORTOLA

GR. BRITAIN
U.S.

ST. JOHN

ST. THOMAS

ST. CROIX

CULEBRA

VIEQUES

SCALE OF MILES

10 0 10 20 30

CULEBRA IS.

Culebra Island

Vieques Is Esperanza

VIEQUES IS.

Roosevelt Roads

Paraiso

Fajardo

El Verde

Rio Grande

Rio Blanco Lower

Rio Blanco Upper

Rio Balanco Upper

San Cristobal

Humacao

Rio Piedras Exp Farm

Canovanas 2 N

San Juan WB AP 2

Juncos 1 NNE

Yabucoa 1 NE

San Juan City

Catano

Bayamon Hato Tejas

COASTAL

Trujillo Alto 2 SSW

San Lorenzo Farm 2 NW

San Lorenzo 3 S

EASTERN

Maunabo 2 SE

SLOPES

Corozal Substation

Comerio Falls Plant 2

Gurabo Substation

INTERIOR

Cidra 3 E

Cayey 1 E

SLOPES

Dorado 4 W

Toa Baja 1 SSW

Manati 2 E

Morovis

Palmarito

Barranquitas

Aibonito

JaJome Alto

Carite Dam

Guayama

Carite Plant 1

Aguirre Rch Sta

Garrochales

Los Canos

Dos Bocas

Caonillas Utuado

Matrullas Dam

Aceituna

Villalba

Villalba

Caonillas Villalba

Guayabal Reservoir

Santa Isabel

Arecibo 2 ESE

INTERIOR

Jayuya 2 N

Toro Negro Plant 2

Juana Diaz Camp

Ponce 4 E

COASTAL

Quebradillas

Lares 3 SE

Utuado 1 SW

Adjuntas 2 N

Cerro Maravilla

Rio Canas

Ponce Mercedita AP

Mora Camp

Guajataca Dam

Indiera Alta

Garzas Dam

Penuelas Salto Garzas

NORTH

WESTERN

Sabana Grande

Santa Rita

Isabela 4 SW

Calero Camp

Maricao 2 SSW

Yauco 1 S

Ramey Air Force Base

San Sebastian

Coloso

Rincon Power Plant

Mayaguez AP

Mayaguez Nuclear Center

Hatchery

San German

Puerto Real

Cabo Rojo

SOUTHERN

Lajas Substation — Central San Francisco

SOUTH

San Francisco

Ensenada

Magueyes Island

67°50'   67°55'

18°10'

18°05'

18°00'

MONA ISLAND

Mona Island

STATION LEGEND

Precipitation only

Precipitation, storage

Precipitation and Temperature

Precipitation, Temperature and Evaporation

Type of gage: ○ Non-recording;

Recording: ● Both types.

Double circle combinations indicate the availability of more detailed meteorological data.

USCOMM-ESSA-Asheville, N. C.
Revised 5 - 69

Number of Times Destruction was Caused
by Tropical Storms, 1901-1955

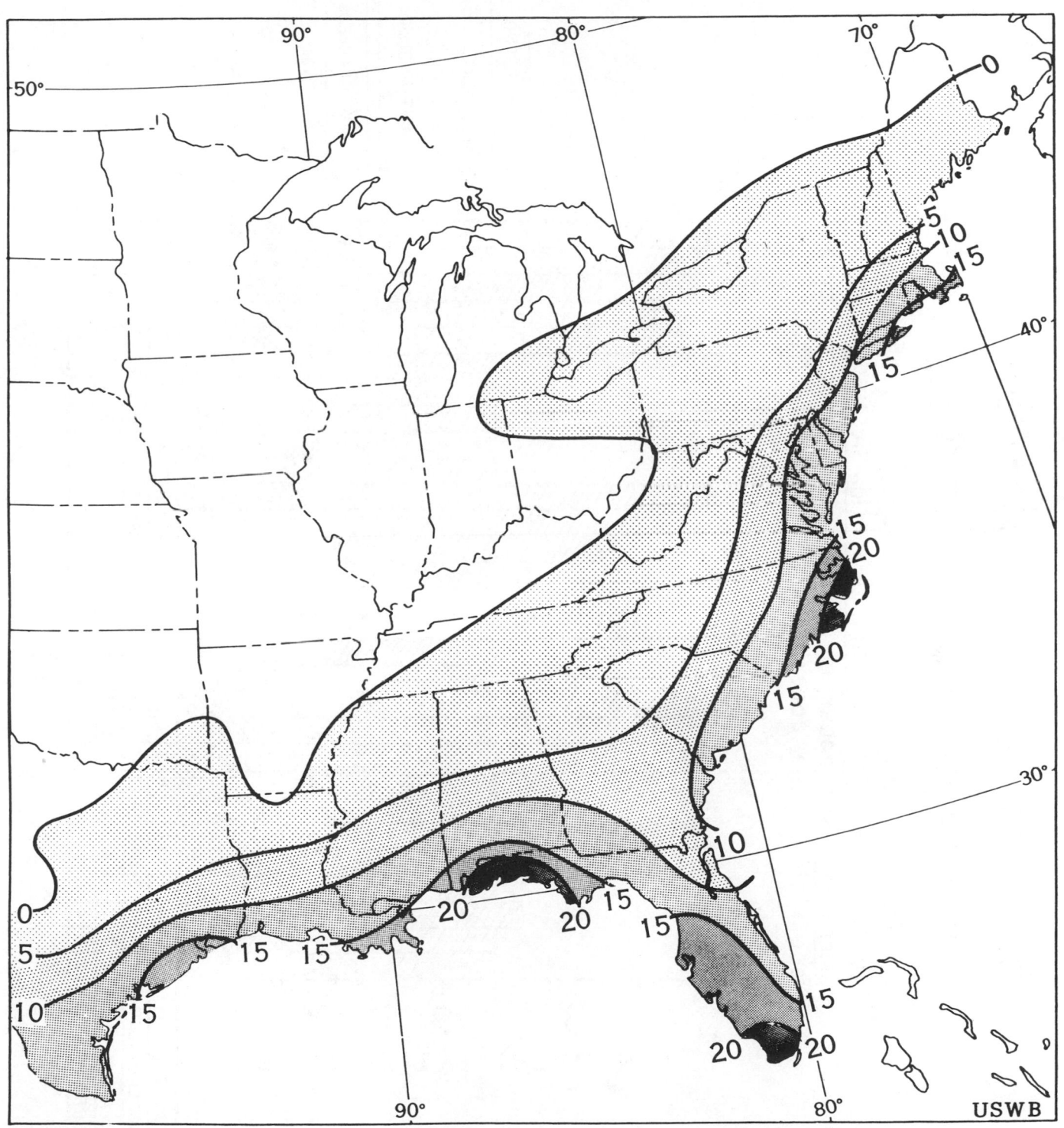

USWB

RISK OF TROPICAL CYCLONES
U.S. Gulf of Mexico Coastline

This histogram and table show the probability (percentage) that a tropical storm, hurricane, or great hurricane will occur in any one year in a 50 mile segment of the coastline.

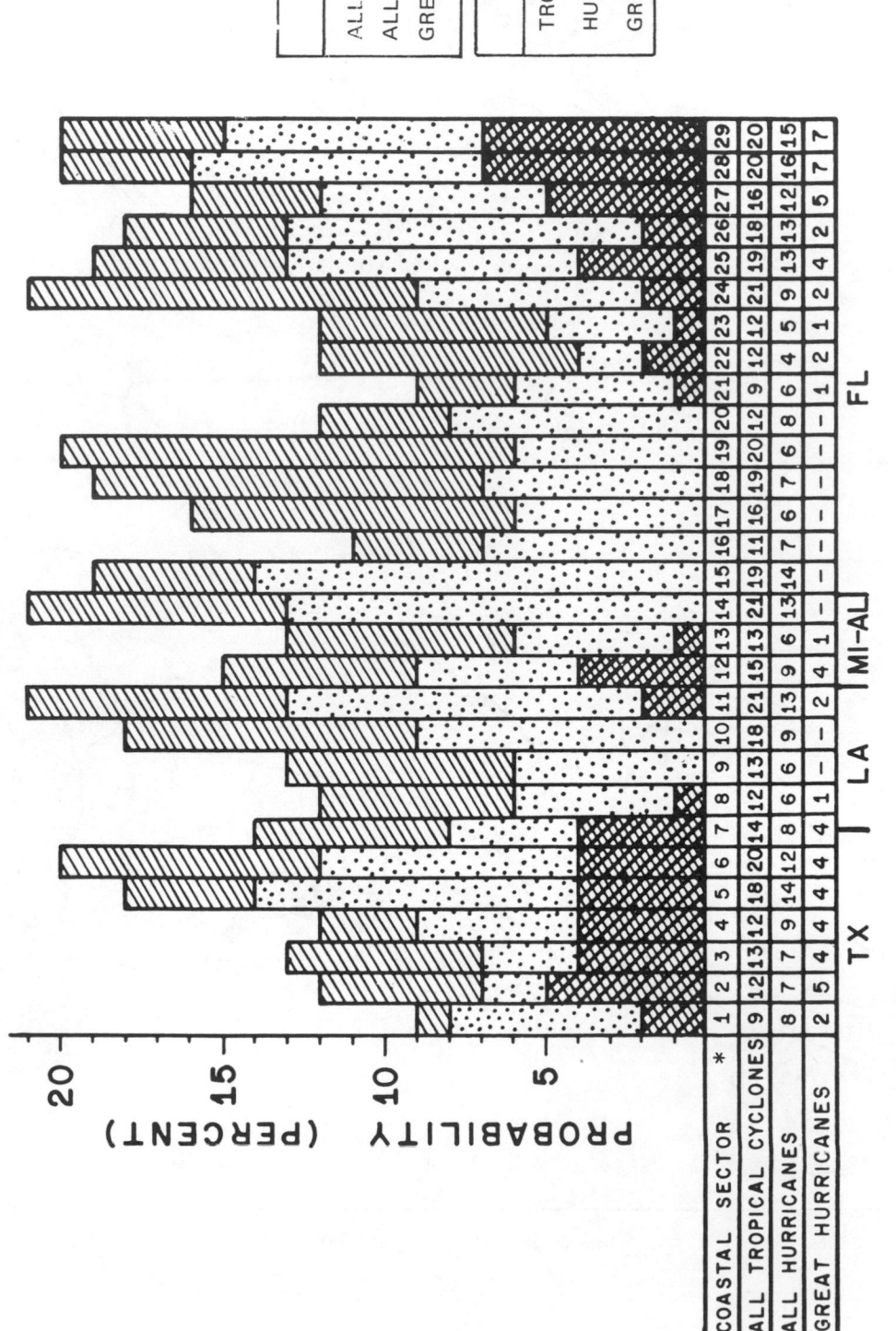

LEGEND

| | ALL TROP. CYCL. | ALL HURR. | GREAT HURR. |

WIND SPEEDS (m.p.h.)	
TROP. CYCL.	39- 73
HURR.	74-124
GREAT HURR.	≥ 125

COASTAL SECTOR	1	2	3	4	5	6	7	8	9	10	11	12	13	14	15	16	17	18	19	20	21	22	23	24	25	26	27	28	29
ALL TROPICAL CYCLONES	9	12	13	12	18	20	14	12	13	18	21	15	21	19	11	16	20	18	21	9	12	12	23	21	19	18	16	20	20
ALL HURRICANES	8	7	7	9	14	12	8	6	6	9	9	6	13	14	7	6	6	7	6	8	6	4	5	9	13	13	12	16	15
GREAT HURRICANES	2	5	4	4	4	4	4	1	-	2	4	1	1	-	-	-	-	1	-	1	2	1	2	4	2	5	7	7	

TX   LA   MI-AL   FL

(Simpson and Lawrence 1971)

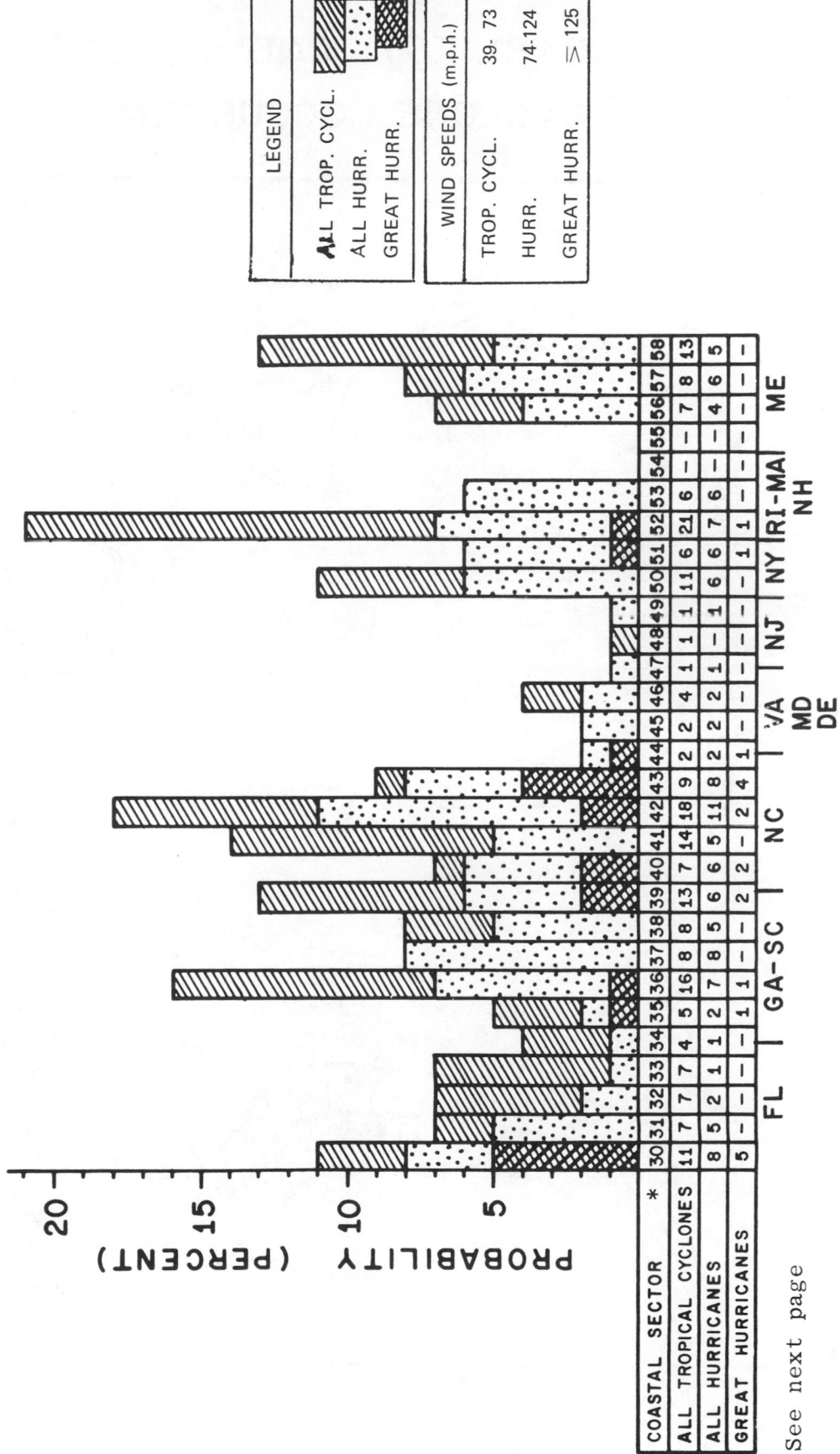

RISK OF TROPICAL CYCLONES
U.S. Atlantic Coastline

This histogram and table show the probability (percentage) that a tropical storm, hurricane, or great hurricane will occur in any one year in a 50 mile segment of the coastline.

* See next page

(Simpson and Lawrence 1971)

EARLIEST AND LATEST TROPICAL CYCLONE OCCURRENCES 1886-1970

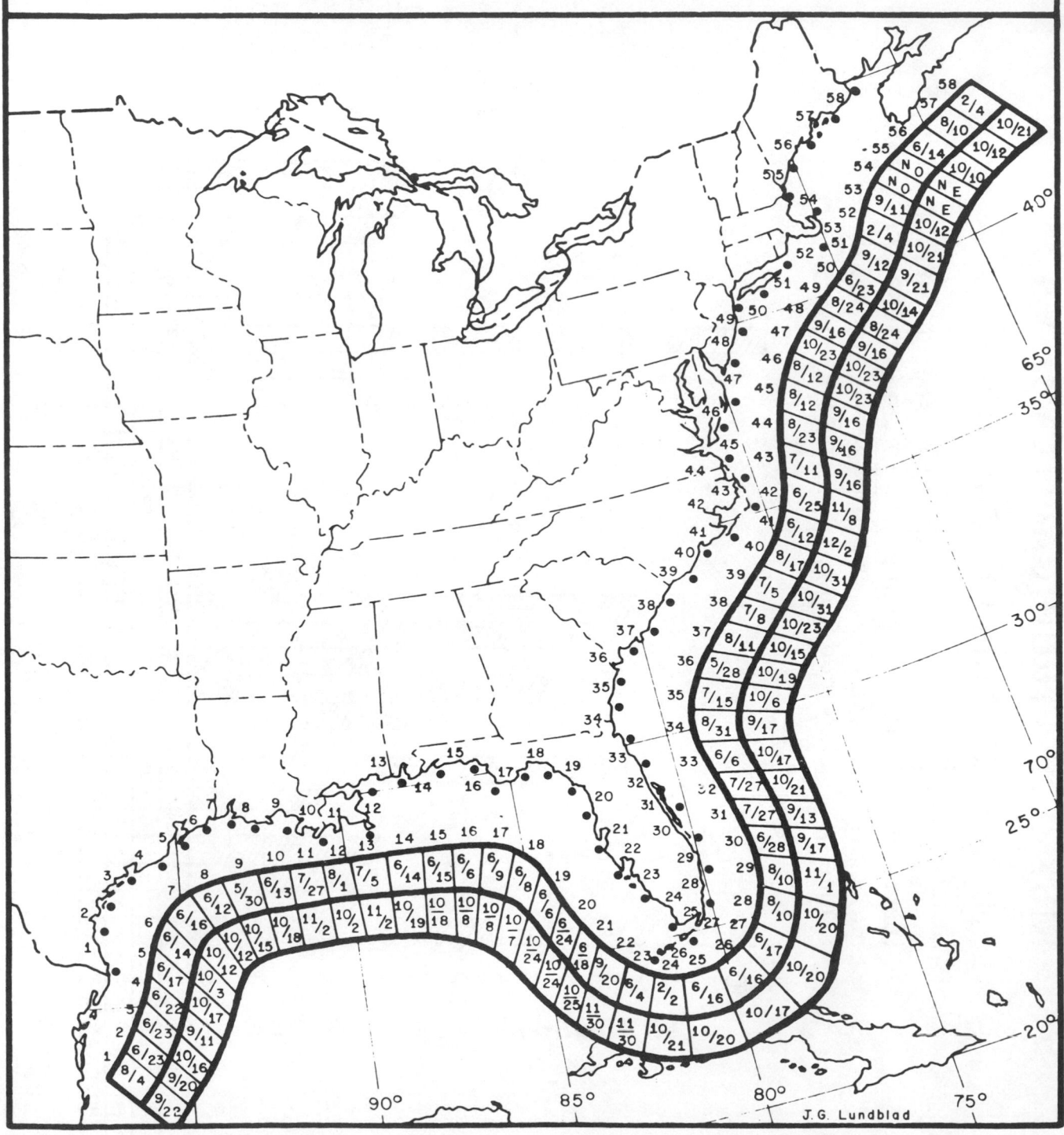

Numerals indicate coastal strips approximately 50 nautical miles in length.
(Simpson and Lawrence 1971)

THE STATE CLIMATOLOGIST PROGRAM 1954 — 1973

The State Climatologist Program of the National Weather Service was born paradoxically during an economy move in 1954 and died during an economy move in 1973. It provided one of those happy instances where more service was provided for less money. Many of the routine activities coordinated by the program were neither wholly new nor totally disbanded at the end. The main element introduced or given fresh focus, and then lost, was personalized service at the state level.

The State Climatologists cultivated contacts in agriculture, business and industry, state and local agencies, and the general public through chambers of commerce, service clubs, and student activities. They learned firsthand the wide variety of human needs for meteorological and climatological services beyond ordinary and specialized weather forecasts routinely provided by the National Weather Service. Conversely, they provided information on the potential and limitation of climatology applied to life's ordinary concerns. They conducted needed research and special data processing or recommended such projects to other agencies with resources more adequate for the job. The state climatologists were, to use common metaphors, the spearhead or sparkplug of specialized climatological services in their state or area -- the interface of the manifold suppliers and consumers of climatological information and analysis.

For day to day activities, timely weather forecasts can affect the margin of success or failure of plans. For middle to long range planning and construction processes, weather "forecasts" must be replaced by the climatological "odds" for favorable or unfavorable weather conditions and sequences of conditions. For example, engineers need historically based information on snow loading, extreme temperatures, extreme winds, extreme rainfall, extreme humidities, pollution potential, and the like. Farmers and nurserymen forced to relocate by urbanization need to consider climatological factors which affect their businesses. Planners and contractors need climatological data for Environmental Impact Statements (EIS) now required by the National Environmental Protection Act (NEPA) and comparable state laws and policies. These groups and others who know approximately what information they need can, for a fee, make a request to the National Climatic Center at Asheville, North Carolina. In its day, the State Climatologist Office served as a library and work area with consultation and advice frequently directly at hand.

The State Climatologist Program was a sign to the profession, to weatherwise public or private officials, and to prospective students that climatological services must be based on specialized education and that they have as great a significance for the social and economic life of the people as weather forecasting services. The importance has not diminished, but the capability has been diffused and the problem of timely informed delivery has been impaired. The information may be available, if one knows how to go about looking for it, but there are no fulltime professionals at the state or regional level funded by the federal government to supply and interpret the climatological data required in every state for a wide variety of public, commercial, professional, and educational uses.

Fortunately, as urged in 1973, a number of state governments and universities have taken over the full support and revitalization of the program in their state. Other states offer only partial services, and, unfortunately, some states offer no public services at all. Appendix I provides the status of service as of April 1977.

Early Climatological Activities

The routine accumulation and interpretation of climatological data was a sporadic, isolated, and uncoordinated activity from the scientific revolution of the seventeenth century to the establishment of national weather services around the world starting a little over a century ago. Prominent among early American climatological activists was Thomas Jefferson, the inspirational father of much scientific activity in the early years of the Republic. Jefferson's scientific spirit was present during the first half of the nineteenth century in the establishment of a network of military weather observation stations and several state networks to serve a variety of scientific and econimic interests. During the War of 1812, the Surgeon General specifically related the purposes of the Army network to "medical topography . . prevalent regional [health] complaints . . . change of climate . . . cultivation of soil . . [and] density of population."[12] Similarly, in 1817, the General Land Office required the local land office agents to accumulate daily weather observations in the interests of agriculture. During the second quarter century, several states, New York most successfully, established statewide climatological networks for similar purposes. At mid century, however, the scientific community still had to look forward to "exhaustive studies of American climatology."[12] Joseph Henry, America's leading physical scientist, took pioneering action from his post at the Smithsonian Institution in Washington.

Professor Henry recognized the potential for timely weather warnings over the telegraphic circuits that increased rapidly after their commercial introduction in 1845. By 1850, Henry "had circularized Congress, solicited volunteer observers, established telegraphic reporting procedures, and initiated pictorial weather maps."[8] The volunteer observer program gradually absorbed the existing state networks until, just before the Civil War, the Smithsonian telegraphic reporting network had over 600 weather observers.

The first federal weather service was organized in 1870 by the Army around its Signal Service telegraphic network. Colonel Albert J. Myer (eponymized in Fort Myer), who headed the service, was a former Army surgeon familiar with the earlier efforts in agricultural and medical climatology. Myer named the fledgling organization "the Division of Telegrams and Reports for the Benefit of Commerce." It became more commonly known as "the Weather Bureau."[8] The service utilized both military and civilian personnel, including important American scientists such as Professor Cleveland Abbe. The Weather Bureau absorbed most of the other governmental and quasi-governmental observing networks, but Myer allowed the volunteer components to wither. New leadership under General William Hazen after 1880 recognized the need for cooperative observers in defining the climate of the vast new areas opening up to agriculture in the last quarter century and the volunteer program began to be rebuilt. Monthly, later weekly or biweekly, reports of weather and crop conditions across the nation had been prepared interruptedly since 1863, officially since 1872, and continuously since 1884. In 1881, authorization was gained to (re)establish supplementary state weather services. In 1891, the Weather Bureau was transferred to civilian control under the newly established Department of Agriculture. By 1892, the Bureau and ten supplementary state programs concerned mainly with the effects of weather on agricultural production had the combined services of more than 2,000 volunteer observers.

The spirit of Jefferson remains today in the Thomas Jefferson Award, the highest federal honor for cooperative (volunteer) weather observers, who now number over 12,000. The growing pains of the system were recounted briefly on the centennial celebration of the Weather and Crop Service by Howard Pollock, Deputy Administrator of the National Oceanic and Atmospheric Administration (NOAA):

> *In order to build a store of data the Smithsonian Institution and others in the 19th century called on private citizens to document the climate of a country with many sparsely settled lands. By seeking out pioneering spirits who knew that climate ranked with soil in building a then agricul-*

tural nation, those early government officials were able to build a weather reporting network which still today is the backbone of climatological planning and design, particularly in agriculture.

The Congressional Act of October 1, 1890, establishing civilian control over the Weather Bureau, included among the specific duties of its Chief "the taking of such meteorological observations as may be necessary to establish and record the climatic conditions in the United States." The Weather Bureau subsequently tried to expand the role of climatology in relation to agriculture, commerce, and health. Almost immediately, however, depression generated budget cuts backed up by changing medical philosophy, based upon the newly developing germ theory of disease, contributed to the virtual destruction of the century long effort to develop the climatological basis of disease. (Today, many medical persons still have to learn the importance of multi-causal analysis of disease, including weather and other environmental factors.) In keeping with the Congressional mandate and its new bureaucratic home, the crop services expanded. Yet, data piled up in hard to retrieve form and the interpretation and dissemination, even in agriculture, of climatic expectations languished during the 20's and early 30's.

Machine processing of the mountains of data finally became a reality in the 1930s -- a decade after the lead of European weather services. Support from both the WPA and the Army permitted the necessary card punching, machine processing, analysis and summary of decades of records. One product of the revitalization of basic climatological work was the masterful Agricultural Yearbook of 1941, *Climate and Man*, now available as a Gale reprint. This work provided, in part, the forerunners of the present state climatologies. In 1940, the Weather Bureau was made part of the Department of Commerce.

Design of the State Climatologist Program

World War II proved the value of meteorology and climatology applied to the planning of human affairs other than agriculture and going beyond data collection and weather forecasting. For example, meteorologists and climatologists were useful in planning airports, scheduling large construction projects, timing military operations, and designing clothing and shelter appropriate to new arenas of military operation. About the same time, applied meteorology began to be appreciated as an important component of air pollution control programs.

Also, there was a renaissance of American interest in biometeorology and medical climatology. By mid century, the Weather Bureau officially recognised the need for a Climatological Services Division that went beyond the interests of the Climate and Crop Weather Division:

> *Industrial and agricultural development have placed further emphasis in the last half-century on the demands for climatological information: a need for data and interpretations useful for planning crops, housing, marketing, shipping, aviation, air conditioning, flood control, manufacturing, insuring against weather risk and many other agricultural and commercial aspects.*

The common image of climatologists in the early post war period remained that of persons who could add a column of figures and divide by 30. Atmospheric scientists, such as Dr. Helmut Landsberg, who had demonstrated the value of expertly interpreted climatological information in the war effort, wanted to deal positively with the role of professionally trained climatologists. Landsberg, in particular, wanted to attract bright young students and create professional interest in climatology by establishing a path of career advancement that mixed a variety of practical experiences with advanced education.

A chance to enhance the status of climatology came during the budget cutting of 1954 when many established ways of doing things in the climatic service of the Weather Bureau were due for drastic consolidation or elimination. Frequently, such "reassignment and personnel elimination" actions are known by the appropriate acronym, but Landsberg was named Chief of the new Climatological Services Division and he proceeded with vigorous sensitivity. He noted, "it is easy to do a job with unlimited resources, but it challenges the ingenuity to do it equally well or better, with less."

Landsberg contended that a climatologist is a professional meteorologist who specializes in climatology, not one who escapes from forecasting shift work that rotates endlessly around the clock in order to settle down to a regular five day work week. He believed that there should be distinct career opportunities to make it attractive for people to specialize in the subject. Landsberg indicated that climatologist need to learn the basic weather observation techniques, the principles of theoretical meteorology, statistics, and then proceed to become "well acquainted with many special areas to which climatology can be applied as for example, problems in agriculture, heating engineering, air conditioning, architecture, water supply, etc., so that he can serve the general public intelligently." In career development, Landsberg visualized that

> *the incipient climatologist, after finishing his formal training, will start as a field inspector*

> *so that he might become thoroughly familiar with the raw material of much of our climatological work, the station and substation observations. After that indoctrination, work as a state climatologist in one of the smaller states would get him acquainted with broad general service to the public. Further advancement to a bigger state or to one of the Weather Record Processing Centers is the next step.*

Landsberg detailed several further career steps at area and national levels, but he noted, "the real key man in all our plans is the State Climatologist." The duties were described briefly as follows:

> *Close cooperation with State agencies, land-grant colleges, agricultural experiment stations is envisaged. The duties of the State climatologists, in addition to the routine work on weather and crop bulletins, severe storm reports, and descriptive climatological summaries for the State, include analytical and developmental work. Particularly, attention will be devoted to use of climatological data for general agricultural purposes, irrigation, water supply problems, recreation, industrial and urban development planning in the State.*

Landsberg emphasized that he did not want State Climatologists to become bogged down with administration of the volunteer or full time observation networks and the checking of forms or similar sub-professional activities. The duties were visualized somewhat more fully as follows:

> *1. Maintain adequate files of records for state, and amplify such records as necessary for the purposes of state climatology.*

> *2. Foster progress for publishing timely current reports of monthly and other climatological data, preferably along state lines.*

> *3. In cooperation with the Area Climatologist and the Central Office suggest and assist with the development of new climatic studies, etc. applicable to the general economy of the state.*

> *4. Establish and maintain contacts with State agencies interested in the development or application of climatic data and records.*

5. *Maintain approved program for issuance of weekly reports on weather and crop progress.*

6. *Seek to develop interest in and use of climatological and hydrologic information by State agencies and state-wide channels of information, especially those concerned with agriculture.*

7. *Seek ways to develop State cooperation for special climatological studies and publications to be produced cooperatively at State expense. These should be planned to conform to guide lines issued by the Central Office in order to obtain a reasonably comparable content and format for all states.*

8. *Cooperate with Meteorologists in Charge of other Weather Bureau Offices, the area climatologist, the area hydrologic engineer and the Central Office, and give all possible assistance to make the climatological network and derived data of maximum use to all public interests within the state.*

9. *"Act as the personal representative of the Weather Bureau with cooperative or automatic rain gage observers in the State including travel to substations when special problems or circumstances are evident or it is deemed by the State Climatologist and the Regional Director that visits by the State Climatologist will best foster high morale and satisfactory personal relationships within the network."*

Noting that "the precise formulation of these duties is still in the making," he continued:

> *We want to encourage climatological investigations which lead to general advancement of the science and better applications of the data. We are exploring the opportunities for advanced training in climatology and hope to get authority to send some people to school.*
>
> *All of this will take time. We will spare no effort to get approval and funds for this work. In the meantime we will review all our procedures in order to assure that they are efficient and up-to-date. For a while Economy will remain an important by-word while SERVICE is our primary aim.*

Endless debate might rage whether the old Service Centers were cut too far in the effort to fund a new level of professional public services through the State Climatologist Program. In any case, the Service Centers had been slated for drastic reduction and consolidation before the State Climatologists were invented. A careful study of the various plans and options under the budgetary realities of 1954 could help quiet the debate. As Landsberg stated:

> *Most of us dislike change. Usually it is outward circumstances that force change upon us. Often these circumstances cannot be argued with—notably when they are represented by the firm figures of an appropriations act. If confronted with a reduced budget we are free to grumble, but it is better to face the facts. This is the situation we find ourselves in with regard to the climatic field service.*
> *In view of the critical fiscal problems there has not been time for much consultation, nor is there much point in bewailing difficulties in advance of experience. Instead let's make the new system work.*

Service to agriculture continued to be the major activity in most states because, fortunately, that sector of the economy was best organized to call for climatic studies in its areas of concern. Fortunately so. Louis Thompson, Associate Dean of Agriculture, Iowa State University, suggests that "the United States had so little variability in weather and grain production in the...two decades [before] 1974 that an attitude of complacency had developed. There was frequent reference in the early 1970's to the fact that technology had advanced to such a level that weather was no longer a significant factor in grain production." In early 1974, a NOAA report warned of probable weather related crop production problems, but a committee of the Department of Agriculture retorted:

> *A comprehensive study published by this Department in 1965 evaluating the effect of weather and technology on corn yield in the Corn Belt for the years 1929 through 1962 concluded that through the use of better varieties and improved cultivation and fertilization practices, man has reduced variation in yields in both good and bad weather. It seems logical to assume that continued progress had been made since that date, particularly in the use of fertilizer, improved cultural practices and the increased benefits of mechanization.*

Dean Thompson noted that "the highly variable weather of 1974 (resulting in reduced crop yields) was shocking to so many people that now there is more than usual concern over the future of our food supply."

Fortunately, there are a few experienced agricultural climatologists ready to supply advice to anyone who will listen at USDA. The State Climatologist Program contributed to the development of this vitally important experience.

One conspicuous failure occurred in the Program. Dr. Gerald Barger, operating at a level just above the State Climatologist, laid the ground work in 1956 for cooperation with the American Academy of Allergy of the American Medical Association to study the distribution and occurrence of diseases associated with weather and climate. Medical clinicians were identified in each state and it was intended that "each state climatologist should, as occasion permits, contact the listed clinician for his state, discuss the plan of the 'weather project' and offer whatever assistance possible." For reasons not yet fully understood, little came from the arrangements.

However, successes of the Program were many. Much needed research was initiated and gaps in the observation network were identified in light of the developing needs within a state. For example, the climatologist in Alaska was much concerned about the development of its potential power resources both with respect to locally important hydropower and to nationally important oil and gas development in "remote areas." The service aspect to cities and the business community was frequently impressive. To cite only one example, the State Climatologist in Indiana provided valuable two page summaries to more than 50 cities in aid of city planning and the attraction of new businesses. He also made himself highly visible through dozens of articles written for publications of local and regional interest.

New Priorities in Public Service

Whether it is the case that all good things must come to an end, it does seem the case that federal/state weather service efforts over the past 150 years operate with an organizational style that has a typical life cycle of about 20 years. In the early 1970's, the National Weather Service faced another crisis of allocating scarce resources to fulfill its diverse missions in the most efficient way. Satellites were demonstrating a vast potential in the global weather observing program and in studies of climatic change, urban climatology, air pollution trends, and, weather-crop relationships. The weather and environmental satellite programs needed significantly increased support. Also, concern for the increasing devastation of tornadoes, hurricanes, and flood suffered by a continually sprawling population pointed to the need for improved natural disaster warnings. In June 1972, tropical storm Agnes created record floods that killed 118 people and caused $3.5 billion damages making it the worst natural disaster in U.S. history in terms of property damage. An advisory panel rated the performance of the storm warning system good to outstanding, but noted deficiencies in "delivery of warnings to the public, public response to warnings, rainfall forecasts and observations, and adequate flashflood warning systems." It was desirable for the National Weather Service to fund community preparedness specialists to work directly with school boards and other community officials and service organizations to increase the density of the warning network and improve community response to warnings. Countless other pressing needs were identified, and yet as NOAA Administrator Dr. Robert White pointed out in a message on January 29, 1973:

> For some time the programs of all Federal agencies, including NOAA's, have been undergoing the closest scrutiny to determine how each can be streamlined for greater efficiency...

> The months ahead, particularly in the remainder of Fiscal Year 1973, will be difficult. In human terms, they are distressing. They are, however, unavoidable. Recognizing this, let us not equate austerity with lack of progress. Let us resolve even more firmly to continue to serve our Nation with the same effort, dedication and excellence which have always marked the performance of the NOAA family.[15]

Then, among the pertinent details that affected the National Weather Service, Dr. White announced;

> the Weather Service's programs will be curtailed principally in the areas of (surface based) observations and certain specialized services. The state climatologist program will be terminated and limited climatic services will be provided through the weather stations.[15]

One of the trade-offs was between service to the states and communities by way of the State Climatologist Program in exchange for Community Preparedness Specialists who could help improve the natural disaster warning system during ordinary weather periods and provide additional station strength or community support during the hectic and vastly overworked periods of actual weather disasters. A year later, when the planned improvements in the warning system had just begun, the system's effectiveness underwent another severe test:

> On two days—April 3 and 4, 1974—148 tornadoes struck 13 states, killing some 300 persons and injuring 6142 others. The most devastating outbreak of tornadoes ever

recorded anywhere in the world, its toll of lives was high, but timely Weather Service warnings were credited with keeping fatalities in the hundreds rather than the thousands. However, the disaster gave renewed emphasis to the need for completing the planned warning and preparedness system as swiftly as possible. [4]

One can predict with some assurance that after the value of the Community Preparedness Specialists has been demonstrated, they will be terminated at some future budget crisis with the hope that they will be funded by state or county government, as was hoped for the State Climatology Program.

In March 1973, Dr. White wrote to the state and territorial governors outlining the budget dilemma and program reassessment, and calling for continuation of the State Climatologist under state funds. One such letter to Honorable Rafael Hernandez Colon, Governor of Puerto Rico, dated March 20, 1973, said in part:

As a part of that reassessment of existing programs, it has been determined that NOAA and the National Weather Service must terminate the State Climatologist Program, under which you have had the services of a federally supported professional climatologist to work on long-term weather problems of particular interest to your own area. The National Weather Service will continue the climatological services it usually provides from all of its weather offices, but the kind of consultative assistance that the State Climatologists have provided in the past can no longer be supported by the Federal Government.

I am writing you at this time to urge you to consider the establishment of a similar position either within your government or as a part of your system of higher education. A more modest step, possibly preferable to you, may be the designation of responsibility for climatological studies within one of the existing elements of your government and the contracting of special studies, as required, to consulting meteorologists who can be found in most sections of the United States. Our climatological data files at the National Climatic Center, Asheville, North Carolina, continue to be available through NOAA's Environmental Data Service, and specially requested compilations can usually be prepared at very reasonable cost.

Climatological studies may contribute to your industrial or economic development plans, to your environmental protection programs, or to improved operations in such functions as planning, construction, or maintenance of governmental facilities. [16]

It is unfortunate that the Territory of Puerto Rico has not found the means to offer continuing public service in state climatology. However, 14 states were able to assume the entire program and another 18 states offer at least limited public service. The status of each program as of April 1977 is provided in Appendix I.

Dr. White reported the decision to terminate the State Climatologist Program at the Annual Meeting of the American **Meteorological** Society in January 1973. The announcement was greeted with some dismay, "a reaction that extended well beyond the small group whose jobs were threatened. It was felt by many that agriculture and business interests would be affected by the discontinuance of this service, and that the federal government should continue to run it as it had for many years." [7] In early February, the AMS Committee on Agricultural and Forest Meteorology protested the move in a "detailed and hard hitting report" prepared by Professor Robert Dale, himself a former State Climatologist. The report concluded:

Meteorological services have a great impact on the agriculture, business and industry of every state. While the effects of **forecasting** *services may be more obvious on a day to day basis, the effects of* **climatology** *services, as our examples have indicated, can be considerable. It is our view that the National Weather Service should maintain and provide* **both forecasting** *and* **climatology** *services for all of our states.* [10]

The Executive Committee of the AMS decided that it could not issue a public statement or otherwise try to influence the Federal Government to save the program, presumably because of tax exemption considerations, but, more importantly from the Society's point of view, also because it was in no position "to demand a reinstatement of one program without recommending which **other** program should be sacrificed." [7] The President of the Society was directed, however, to consult personally and in writing with Dr. White to ask that the decision be reconsidered and simultaneously, under existing realities, to write each Governor a letter supporting the forthcoming letter from Dr. White recommending that each state assume responsibility for its own program. By and large, the Society has been gratified by the responses from the states and by the reply of the Administrator explaining NOAA's position and expectations more fully. Dr. White responded:

The important role performed over the years by the State Climatologists is certainly recognized and fully appreciated. However, the resources now available to the National Oceanic and Atmospheric Administration are extremely limited. The need to hold Federal spending within strict limits has been emphasized by the President.

Other important programs, including part of our air pollution meteorological support program, are being curtailed to enable us to live within our budgetary and personnel allowances and to permit further progress on high priority efforts such as those directed at improving disaster warning and community preparedness. These efforts require strengthening of our forecast organization; maintaining our community preparedness work by helping communities plan for emergency actions when faced with warnings of impending severe weather and floods; maintaining, and in some cases, replacing our weather radar equipment; procuring and installing additional flash flood alarm systems; and improving our capability to obtain and disseminate satellite weather information and apply it particularly to severe weather situations.

I appreciate the arguments offered in the statement of the AMS Committee on Agricultural and Forest Meteorology and in the letter from Dr. George G. Olson of Colorado State University. The questions posed are, of course, quite valid, and we are attempting to find solutions to these problems.

Some of the more important State Climatoogist functions, but not the consultative assistance formerly provided, will be continued by reassignment to other National Weather Service offices. For example, basic meteorological data and summaries for weekly weather and crop reports are expected to become the responsibility of Weather Service Forecast Offices. Climatological data summaries will be provided by the National Climatic Center. Climatic summaries for the County Soil Survey Reports, which are used for land value appraisals, must now be prepared by State professional staffs or by private consultants.

Many States have previously recognized the problems of applying meteorological data to numerous local endeavors and have hired professional staffs to work with our State Climatologists. In these states, I have no doubt that important work will continue without Federal support. Where a local institution such as a State University has agreed to keep local records intact as a file,

we will not withdraw them. A number of universities are planning to keep such files. Many of the needs of individual or low budget researchers can be met from these data sources.

Degree-day normals and probabilities based on past data will be developed by the National Climatic Center at Asheville. Popular articles including meteorological information can be provided by many non-Federal sources. In the collection, verification and publication of severe storm data, Weather Service Offices will send out questionnaires to their respective areas of county responsibility and Weather Service Forecast Offices will prepare monthly summaries.

We expect that States, where sufficiently concerned, will provide continuing support for climatology for their respective areas. For more than a decade the Federal Government has made a substantial contribution to the States in the applications of climatology. Now that the course has been charted, others may follow as the demand requires. Of course, the extensive climatological data at the National Climatic Center at Asheville will continue to be accessible as a data source.

The move to curtail the State Climatologist service was made with great reluctance. Obviously, no one should expect that the same level of services previously provided by State Climatologists will continue to be available. As we have described, however, many of these service activities will be assumed by Weather Service Forecast Offices, Weather Service Offices, and the National Climatic Center.

Sincerely,
Robert M. White
Administrator [7]

Guide to Current Information Sources in Weather and Climate

Basic climatological services are still available to the public. As noted, about half of the states offer at least limited public service through their own State Climatologist and local files have been retained in many universities. NOAA has several specialized data and information centers under its **Environmental Data Service**. The one most useful in lieu of a State Climatologist — and the one to which they relate ultimately — is the **National Climatic Center (NCC)** in Asheville, N. C. The center was established in 1952 as the collection center and custodian of all United States weather records. It is the largest center of its kind in the world. NCC's weather records range from the contents of eighteenth century journals and diaries to the meteorological data currently generated by **NOAA's National Weather Service**, the weather services of the **Air Force, Navy**, and **Federal Aviation Administration**, the **Coast Guard**, cooperative observers, and foreign countries. Included are cloud photographs and other data obtained from environmental satellites. The center currently receives more than 100 million observations annually from around the world. For researchers in government, private institutions, industry, and the general public, NCC is a unique central source of historical weather information. The data may be stored variously in automated systems, microforms, or hard copy. Data for which there is a general demand are summarized and published for distribution to subscribers or for answering specific requests. Unfortunately, the publication schedule for some summaries has been long delayed, as in the case of the *Climates of the States*.

Climatic information available from NCC include:
1. Hourly surface observations from land stations (ceiling height, total sky cover, visibility, precipitation or other weather phenomena occuring, obstructions to vision, pressure, temperature, dew point, wind direction, and gustiness).
2. Three-hourly and six-hourly surface observations from land stations, ocean weather stations, and ships at sea (variable data content).
3. Upper air observations (balloon and rocket soundings, and pilot reports).
4. Radar observations (radar log sheets and radar scope photographs).
5. Selected weather maps and charts from the **National Meteorological Center.**
6. Special collections (large scale interdisciplinary experiments, such as **Global Atmospheric Research Program (GARP)** basic data set; **GARP Atlantic Tropical Experiment (GATE)** meteorological data; **International Field Year for the Great Lakes** data collections, solar radiation data; and many others).

Satellite data are available from the **Satellite Data Services Branch, NOAA**, World Weather Building, Room 606, Washington, D.C. 20233. The building is located at 5200 Auth Road, Camp Spring, Md., telephone (301) 763-8111.

NCC has developed special programs to respond to increased user requests since the 1973 organization. An **Information Services Division has been** established and equipped with a telephone and intercom system to enable simultaneous responses to six requests. An automated data retrieval and microform display system is being installed to speed access to much of the vast store of data. The regular telephone number is **(704) 254-0961**. The special toll free number from Washington, D.C. is 427-7919. The **Federal Telecommunications System** number is **(704) 254-0683**. Mail should be addressed to: The **National Climatic Center, NOAA**, Federal Building, Asheville, N. C. 28801.

NCC services include:
1. **Supply of publications, including reference manuals, catalogs of holdings, data reports, and atlases.**
2. **Data and map reproduction, including hard copy manuscripts, microforms, digital media, and other forms.**
3. **Analysis and preparation of statistical summaries based upon archive holdings.**
4. **Evaluation of various data records for specific analytical requirements.**
5. **Library search for bibliographic references, abstracts, and documents** -- also see Environmental Sciences Information Center (ESIC) below.
6. **Referral to organizations holding requested information.**
7. **Provision of general atmospheric sciences information.**[18]

A review conducted in 1971, when approximately 11,000 requests annually were made, indicated:

> About 12 percent of the requests received at the National Climatic Center come from government agencies, about 2 percent from state and local government, and the rest -- by far the largest group -- from businessmen, lawyers, scientists, and private citizens. Although these requests range through a broad spectrum of applications, most fall into one of the categories of health, business, litigation, or research.[5]

All work at NCC is performed on a reimbursable basis and the requester is provided a cost estimate before the work begins. The requests are expected to specify the data required, stations or geographical limits involved, whether output is to be carried on magnetic tape, microforms, punched cards, or hard copy, and other pertinent information such as a description of the problem involved. Private consulting meteorologists may be required to assist nongovernment users in specifying their needs. NCC may help with the referral and the **American Meteorological Society**, 45 Beacon Street, Boston, Mass. 02108, telephone **(617) 227-2425**, maintains a roster of certified consultants.

Visitors are welcome at NCC. Many data users find a visit advantageous. Advance notice is desirable, especially if the visitor wishes to confer with specific staff members or types of specialists. Working space and assistance are provided as needed. The various specialists may work most effectively as a team in exploring the problem with visitors and in determining how NCC may best serve their needs.

NCC supports the National Weather Service forecasting offices by assembling climatic reference handbooks to be used to respond to local inquiries. One handbook is located in each of some 300 first order offices of the weather service. The handbooks contain worldwide, national, regional, state, district, and local climatic data, together with climatic summaries of certain resort areas, statistics on weather extremes, monthly ocean and lake temperature averages, and information on the use of weather records in litigation. *Weather Almanac* published by Gale contains much of the information of these handbooks and is readily available in thousands of libraries around the nation and world.

Other major facilities in the **Environmental Data Service** are the **National Geophysical and Solar-Terrestrial Data Center**, Boulder, Colo.; the **National Oceanographic Data Center**, the **Environmental Science Information Center**, and the **Center for Experimental Design and Data Analysis**, Washington, D.C.; and the **Center for Climatic and Environmental Assessment**, Columbia, Mo.

The **Environmental Science Information Center** (ESIC) is NOAA's scientific and technical publisher, information banker, and librarian. ESIC complements the family of data centers by providing a comprehensive, single source for literature of interest within NOAA.

As NOAA's information banker, ESIC serves as a national focus for literature based information in the marine, atmospheric, and earth sciences and their related technologies. The **Technical Information Division (TID)** collects, indexes, abstracts, and announces scientific and technical publications relating to NOAA's mission and fields of service. TID exploits both internal and extra-NOAA bibliographic resources to provide environmental science information service for NOAA clients, and deposits all scientific papers produced or sponsored by NOAA in the Department of Commerce's **National Technical Information Service (NTIS)**, which uses associated bibliographic data sheets as a basis for announcements. TID also deposits resumes of NOAA's internal and sponsored research and development projects in the **Smithsonian Science Information Exchange (SIC)**, and conducts a scientific and engineering news service that brings timely word of NOAA's newsworthy contributions to more than 700 scientific and technical publications.

OASIS (Oceanic and Atmospheric Information System) is TID's information retrieval service that furnishes ready reference to the technical literature and research efforts concerning the environmental sciences and marine and coastal resources. It provides one-stop computerized searches of both NOAA and non-NOAA multi-discipline publication reference files and gives the user access to major atmospheric, marine, and solid earth bibliographic information files not available anywhere else in computer-searchable form. OASIS provides the user on a reimbursable basis with bibliographic references, abstracts, and indexing terms. Assistance is available from an Information Specialist, including one in the area of Meteorology and Climatology. The OASIS Information Specialists may be contacted at The **Environmental Science Information Center, NOAA,** 3300 Whitehaven St., Washington, D.C. 20235, telephone (202) 343-6454.

As NOAA's librarian, ESIC coordinates the agencies library services and its participation in the national network of scientific libraries. The **Atmospheric Sciences Library**, formerly the U.S. Weather Bureau Library, dates from 1870 and, with approximately 175,000 volumes and pamphlets, maintains a comprehensive collection of meteorological, climatological, and hydrological literature. It also has extensive holdings of published daily weather maps and climatic data. The Library maintains for loan microfilm of National Weather Service manuscript maps. It is open to the public for reference use only. The address is: **Atmospheric Sciences Library (D821) NOAA,** 8060 13th Street, Silver Spring, Md. 20910. A branch library in Miami has a collection dealing mainly with the specialties of the **Atlantic Oceanographic and Meteorological Laboratories** including tropical meteorology, hurricanes, experimental meteorology, and weather modification. The address is: **NOAA Miami Library**, 15 Rickenbacker Causeway, Miami, Fla. 33130. A branch library in Boulder specializes in areas that include weather modification and upper air and solar physics. The address is: **NOAA Environmental Research Laboratory Library (R51),** Boulder, Colorado 80302.

Gale Research Company has responded to the need for greater accessibility of weather and climatic information in several publications. The present volume reprints the existing *Climates of the States* with the addition of the updated 1941-1970 climatic data base for the 300-odd first order stations. *This new data base is not included in any other edition of the state climatologies including those available from the federal government*, a series now partially out of print. Many users will find the combined publication of the 1931-60 data base and the 1941-1970 data base a useful and unique feature. Gale also published and revises periodically, *Weather Almanac*, its own compilation of useful weather, climate, and air pollution information from government and other sources. *Weather Atlas of the United States*, formerly titled *Climatic Atlas of the United States*, and *Climate and Man*, the still useful 1941 Agricultural Yearbook, have been republished by Gale. These Gale publications are standard library reference works and are readily available in thousands of public, private, and university libraries across the nation and around the world. To be sure, many specialized users will need to work through state and federal agencies noted in this report, but the referencewise user can satisfy a significant number of queries through these widely available Gale publications.

J.A.R.

Bibliography

Books

1. Hughes, Patrick A., *A Century of Weather Service*, New York, Gordon and Breach, 1970.
2. Popkin, Roy, *The Environmental Science Services*, New York, Praeger, 1967.
3. Whitnah, Donald R., *A History of the United States Weather Bureau*, Urbana, U. of Illinois, 1961.

Articles

4. Cook, Ann K., "NOAA: a five-year look," *NOAA Magazine*, Vol. 5, No. 4 (October 1975), p. 7.
5. Hughes, Patrick A., "Ask Asheville," reprinted from *NOAA Magazine*, April 1971.
6. Hull, Arnold R., "Status of State Climatological Services, *"Bulletin of the American Meteorological Society* (BAMS), V. 55, No. 1, Jan. 1974, pp. 20-21.
7. Kellogg, William W., "A Report on the Discontinuation by NOAA of the State Climatologist Program," *BAMS*, V. 54, No. 8, Aug. 1973, pp. 852-53.
8. News Feature, "Weather and Crop Services Centennial," *BAMS*, V. 53, No. 11, Nov. 1972, pp. 1111-13.
9. Thompson, Louis M., "Weather Variability, Climatic Change, and Grain Production," *Science*, V. 188, May 9, 1975, p. 535.

Documents and pamphlets (located at National Climatic Center library except as noted)

10. Committee on Agricultural and Forest Meteorology, "The Need for Continuing State Climatology Services as a Function of the National Weather Service," report in American Meteorological Society archives.
11. Climatological Service Memorandum (CSM) No. 45, Aug. 31, 1954.
12. CSM No. 50, Nov. 23, 1955.
13. CSM No. 51, Jan. 6, 1956.
14. CSM No. 52, Feb. 17, 1956.
15. NOAA, "Message from the Administrator," January 29, 1973.
16. Robert M. White, Administrator, to Honorable Rafael Hernandez Colon, March 20, 1973.
17. NOAA, "Status of State Climatologist Program in each State as of 1/9/75," n.d.
18. Pamphlets on NCC (NOAA/PA 72026) 1975 (Rev); ESIC (NOAA/PA 72029) 1972; Users Guide to OASIS, May 1974.
19. Records and reminiscences of William T. Hodge of NCC and a former State Climatologist.

Appendix I
STATUS OF STATE CLIMATOLOGIST PROGRAM
IN EACH STATE AS OF APRIL, 1977

First address indicates recipient of material normally sent to State Climatologist offices on a monthly basis.

# Indicates the agency that received the material not transferred to NCC at the time State Climatologist office closed.

We have given each state a code based on their participation in the State Climatologist program. The codes are as follows:

A. States that have assumed entire program.

B. Agencies within a state receiving materials such as carbon copies of records, the monthly arrays, and general data normally sent to State Climatologist offices. In most cases these agencies are offering limited public service.

C. Agencies within a state receiving material such as carbon copies of records, monthly arrays, and general data normally sent to SC offices. No public service offered although materials are generally available for research.

D. Substation Network Specialist receives all material normally sent to SC offices. No public service offered.

Code	State	Address
D	Alabama	#Substation Network Specialist National Weather Service Office Dannelly Field Route 8, Box 600 Montgomery, AL 36108
		#Mr. Jim Wise Associate in Climatology Arctic Environmental Information & Data Ctr. 707 A Street Anchorage, AK 99501 Telephone: 907 279-4523
A	Arizona	#Dr. Robert W. Durrenberger Director, The Laboratory of Climatology Arizona State University Tempe, AZ 85281 Telephone: 602 965-6265
D	Arkansas	#Substation Network Specialist National Weather Service Forecast Office Adams Field Little Rock, AR 72202

A	California	# Mr. James Goodridge California Department of Water Resources Resources Building, Room 235 P. O. Box 388 Sacramento, CA 95802

Telephone: 916 445-1993

The original manuscript records for the substations in California will be returned to the California Department of Water Resources at the above address as soon as they have been placed on microfilm.

Northern California	WSO (SNS) 1641 Resources Building 1416 Ninth Street Sacramento, CA 95814 Attn: Folsom
Southern California	NWSFO (SWS) 11102 Federal Building 11000 Wilshire Boulevard Los Angeles, CA 90024 Attn: Sissions

A	Colorado	#Dr. Thomas McKee Department of Atmospheric Science Colorado State University Fort Collins, CO 80521

Telephone: 303 491-8545

B	Connecticut	Mr. R.E. Lautzenheiser 35 Arcadia Avenue Reading, MA 01867

Telephone: 617 944-2137

#University of Connecticut

D	Delaware	Mr. Arthur Snider Substation Network Specialist National Weather Service Office Friendship International Airport Baltimore, MD 21240

Retained at NWS offices

A	Florida	#Dr. Clark I. Cross Department of Geography University of Florida Gainesville, FL 32611

D	**Georgia**	#Substation Network Specialist National Weather Service Office Lewis B. Wilson Airport Macon, GA 31201
D	**Hawaii**	National Weather Service Hawaii Region P. O. Box 3650 Honolulu, HI 96811 # University of Hawaii Honolulu, HI
D	**Idaho**	#NWSO (SNS) Federal Office Building, Room 481 550 West Fort Street Boise, ID 83702
B	**Illinois**	#Mr. Stanley A. Changnon Illinois State Water Survey Box 232 Urbana, IL 61801 Telephone: 217 333-2210
B	**Indiana**	#Professor Lawrence A. Schaal Poultry Science Building, Room 201 Purdue University Lafayette, IN 47907 Telephone: 317 749-2891
A	**Iowa**	#Mr. Paul Waite Iowa Weather Service Municipal Airport Des Moines, IA 50321 Telephone: 515 247-4062
B	**Kansas**	#Professor L. Dean Bark Department of Physics Cardwell Hall Kansas State University Manhatten, KS 66506 Telephone: 919 532-6786
D	**Kentucky**	#National Weather Service Room 205 Agriculture Experiment Station Lexington, KY 40506 # Retained at NWS offices
B	**Louisiana**	#Professor Robert A. Muller Department of Geography & Anthropology Louisiana State University Baton Rouge, LA 70803 Telephone: 504 388-5942
B	**Maine**	Mr. E. Lautzenheiser 35 Arcadia Avenue Reading, MA 01867 Telphone 617 944-2137 # Retained at NWS offices

D	**Maryland**	Mr. Arthur Snider Substation Network Specialist National Weather Service Office Friendship International Airport Baltimore, MD 21240

The University of Maryland is receiving publications, the Palmer Drought Index and Form 1066, etc., at the following address:

\# Ms. Kathy Clark
Jull Hall, Room 101
University of Maryland
College Park, MD 20740

A	**Massachusetts**	Mr. R. E. Lautzenheiser 35 Arcadia Avenue Reading, MA 01867 Telephone: 617 944-2137
A	**Michigan**	#Dr. Fred V. Nurnberger Michigan Department of Agriculture Weather Service Steven S. Nisbet Building 1407 S. Harrison Road East Lansing, MI 48823 Telephone: 517 373-8338
A	**Minnesota**	#Mr. Earl L. Kuehnast Crops Research Building, Room 127 University of Minnesota St. Paul, MN 55101 Telephone: 612 296-6157
D	**Mississippi**	Substation Network Specialist National Weather Service Forecast Office Allen C. Thompson Field P. O. Box 5779 Jackson, MS 39208 \# Retained at NWS offices
B	**Missouri**	Professor Wayne L. Decker University of Missouri 701 Hitt Street Columbia, MO 65201 Telephone: 314 882-6591 #Atmospheric Sciences University of Missouri 701 Hitt Street Columbia, MO 65201

B	**Montana**	# Professor Joseph M. Caprio Plant and Soil Science Department Montana State University Agricultural Experiment Station Bozeman, MT 59715 Telephone: 406 994-4601
B	**Nebraska**	Mr. R. E. Myers Climatology Office 113 Nebraska Hall, 901 N. 17th Street University of Nebraska Lincoln, NE 68508 Telephone: 402 472-3471 # Climatologist Library Conservation & Survey Division University of Nebraska - Lincoln 113 Nebraska Hall, 901 N. 17th Street Lincoln, NE 68508
B	**Nevada**	#Dr. Richard O. Gifford Plant, Soil & Water Science Division College of Agriculture University of Nevada, Reno Reno, NV 89507 Telephone: 702 784-6947
A	**New Hampshire**	Mr. Gerard Pregent Department of Geography University of New Hampshire Durham, NH 03824 Telephone: 603 862-1719 #Retained at NWS Offices
B	**New Jersey**	#Professor A.V. Havens Department of Meteorology & Physical Oceanography Cook College, Rugers University P.O. Box 231 New Brunswick, NJ 08903 Telephone: 201 932-08903

B	New Mexico	Professor Iven Bennett Department of Geography The University of New Mexico Albuquerque, NM 87131

Retained at NWS offices

A	New York	#Dr. A.B. Pack New York State College of Agriculture Cornell University Ithaca, NY 14850

Telephone: 607 256-3034

D	North Carolina	Mr. Robert Daniels Substation Network Specialist National Weather Service Forecast Office P. O. Box 25879 Raleigh, NC 27611

Retained at NWS offices

B	North Dakota	#Professor J.M. Ramirez Department of Soils Walster Hall North Dakota State University Fargo, ND 58102

Telephone: 701 237-8901

D	Ohio	Mr. Lloyd Seidel Substation Network Specialist National Weather Service Office Port Columbus International Airport Columbus, OH 43219

Professor John N. Rayner offered to maintain a file of cooperative weather records in Ohio. Mr. Seidel sends the carbon copies on to Professor Rayner at the following address:

#Professor John N. Rayner
Department of Geography
The Ohio State University
1775 South College Road
Columbus, OH 43210

Telephone: 614 422-2514

B	Oklahoma	Dr. Amos Eddy
		Professor of Meteorology & Environmental Design
		The University of Oklahoma
		200 Felgar Street, Room 219
		Norman, OK 73069

Telephone: 405 325-4069

D	Oregon	National Weather Service Forecast Office
		5420 N.E. Marine Drive
		Portland, OR 97218

Oregon State University accepted the carbon copies of the Oregon sub-station records for the period January 1954-December 1972.

\# Mr. Earl Bates
National Weather Service Office
Agricultural Hall
Oregon State University
Corvallis, OR 97331

Telephone: 513 754-1557

D	Pennsylvania	Mr. John Karlock
		Substation Network Specialist
		National Weather Service Office
		Allentown-Bethleham-Easton Airport
		Allentown, PA 18103

\# National Weather Service Office
228 Walnut Street, Federal Building
P. O. Box 1185
Harrisburg, PA 17108

B	Rhode Island	Mr. R. E. Lautzenheiser
		35 Arcadia Avenue
		Reading, MA 01867

Telephone: 617 944-2137
\# Retained at NWS offices

B	South Carolina	Dr. Paul E. Lovingood
		Department of Geography
		University of South Carolina
		Columbia, SC 29208

\# Clemson University

B	South Dakota	#Professor William Lytle Agricultural Engineering Building South Dakota State University Brookings, SD Telephone: 605 688-5141
D	Tennessee	Substation Network Specialist National Weather Service Meteorological Observatory Route 1 Old Hickory, TN 37138 At the time the State Climatologist office closed, the University of Tennessee accepted part of the material. It was sent to the following address: # Dr. Henry A. Fribourg Department of Agronomy University of Tennessee P. O. Box 1071 Knoxville, TN 37901
A	Texas	#Professor John F. Griffiths Center for Applied Geosciences O&M Building, Room 1017 Texas A&M University College Station, TX 77843 Telephone: 713 845-5044
A	Utah	Mr. Arlo Richardson Department of Soils & Biometeorology Utah State University, UMC-48 Logan, UT 84322 Telephone: 801 753-7181
B	Vermont	Mr. R.E. Lautzenheiser 35 Arcadia Avenue Reading, MA 01867 Telephone: 617 944-2137 # Retained at NWS offices

D	**Virginia**	Mr. Arthur Snider Substation Network Specialist National Weather Service Office Friendship International Airport Baltimore, MD 21240 #Virginia Polytechnic Institute & State Univ., Blacksburg, VA
B	**Washington**	Dr. Howard J. Critchfield Department of Geography Western Washington State College Bellingham, WA 98225 Telephone: 206 676-3277
A	**West Virginia**	#National Weather Service Office Kanawha Airport Charleston, WV 25311
A	**Wisconsin**	#Dr. Va. L. Mitchell Wisconsin State Climatologist 1225 West Dayton Street Madison, WI 53706 Telephone: 608 263-2374
D	**Wyoming**	Mr. Dean C. Hirschi, SNS Weather Bureau Airport Station FAA Weather Bureau Building, Room 118 175 N. 24th West Salt Lake City, UT 84116
D	**Puerto Rico**	#Substation Network Specialist National Weather Service G.P.O. Box 4407 San Juan, PR 00936

Appendix II

NWS Personnel with service as State (or Territorial) Climatologist*

Alabama
   A. R. Long
   Robert M. Ferry

Alaska
   C. E. Watson
   Harold W. Searby

Arizona
   Paul C. Kangieser

Arkansas
   Robert O. Reinhold

California
   Robert F. Dale
   C. R. Elford
   Joseph W. Berry
   Clyde A. O'Dell

Colorado
   Joseph W. Berry
   Dennis S. Watts

Connecticut and Rhode Island
   Dr. A. Boyd Pack
   Joseph J. Brumbach, Jr.

Delaware and Maryland
   Howard H. Engelbrecht
   W. Joseph Moyer

Florida
   Keith Butson
   James T. Bradley

Georgia
   Horace S. Carter

Hawaii
   W. F. Feldwisch

Idaho
   David J. Stevlingson
   Kenneth A. Rice

Illinois
   Paul F. Sutton
   Lothar A. Joos
   R. W. Harms
   W. L. Denmark
   Donald E. Wverch

Indiana
   Lawrence A. Schaal

Iowa
   C. R. Elford
   Paul J. Waite

Kansas
   A. D. Robb
   Merle J. Brown

Kentucky
   A Benjamin Elam, Jr.

Louisiana
   E. J. Saltsman (also Mississippi)
   George W. Cry

Maine, Massachusetts, New Hampshire, and Vermont
   Paul C. Kangieser (also Connecticut and Rhode Isla
   Robert E. Lautzenheiser

Maryland (see Delaware)

Massachusetts (see Maine)

Michigan
   A. H. Eichmeier
   Norton B. Strommen

Minnesota
   Joseph H. Strub, Jr.
   Earl Kuehnast

Mississippi
   E. J. Saltsman (also Louisiana)

Missouri
   James D. McQuigg
   Warren M. Wisner

Montana
   Richard A. Dightman
   John W. Fassler

Nebraska
   Richard E. Myers
   Morris S. Webb

Nevada
   Merle J. Brown (also Utah)
   Dr. Clarence M. Sakamoto

New Hampshire (see Maine)

New Jersey
   Donald V. Dunlap

New Mexico
   George Von Eschen
   Frank E. Houghton

New York
   Dr. A. Boyd Pack

North Carolina
   Charles B. Charney
   Albert V. Hardy

North Dakota
A. A. Skrede
H. G. Strommel
Morton Bailey

Ohio
L. T. Pierce
Dr. Jerry M. Davis

Oklahoma
Hugo V. Lehrer
Stanley G. Holbrook
Billy R. Curry

Oregon
Gilbert L. Sternes
Stanley G. Holbrook

Pacific Territories
David I. Bluemenstock
Saul Price

Pennsylvania
Nelson M. Kauffman
Dr. James J. Rahn

Puerto Rico and Virgin Islands
David Smedley
J. V. Vaiksnoras
Robert J. Calvesbert

South Carolina
Nathan Kronberg
Norton B. Strommen
Holbrook Landers

South Dakota
William T. Hodge
Dr. A. Boyd Pack
Walter Spuhler

Tennessee
Robert R. Dickson
Morton Bailey
J. V. Vaiksnoras

Texas
Richard D. W. Blood
Robert B. Orton

Utah
Merle J. Brown (also Nevada)
E. Arlo Richardson

Vermont (see Maine)

Virginia
Daniel L. Sala
Curtis W. Crockett

Washington
Earl L. Phillips
Charles P. Ruscha, Jr.

West Virginia
Robert O. Weedfall

Wisconsin
Paul J. Waite
Marvin W. Burley
Hans E. Rosendal

Wyoming
John D. Alyea

* *Roster based on records at NCC. Personnel records at NWS contain fuller information and dates of service. The person with the earliest service in each state is listed first.*